ADVANCES IN ENZYMOLOGY

AND RELATED AREAS OF MOLECULAR BIOLOGY

Cumulative Index, Volumes 1–74

ADVANCES IN ENZYMOLOGY

AND RELATED AREAS OF MOLECULAR BIOLOGY

Cumulative Index, Volumes 1–74

AN INTERSCIENCE PUBLICATION
JOHN WILEY & SONS, INC.

Published by John Wiley & Sons, Inc., Hoboken, New Jersey.
Published simultaneously in Canada.

Library of Congress Card Number: 41-9213

Advances in Enzymology and Related Areas of Molecular Biology:
Cumulative Index, Volumes 1–74

ISBN 0-471-66087-6

Printed in the United States of America

10 9 8 7 6 5 4 3 2 1

Cumulative Index, Volumes 1–74

Page references followed by italic *t* indicate material in tables. Some subheadings also include the word "table" explicitly.

A

p-AB, **35:**38
Abbreviations, **23:**323; 407, **24:**163; 294, **26:**174–175. *See also* at beginning of each review
  alcohol dehydrogenase systems, **27:**49
  folic acid systems, **27:**418; 425
  hydrocarbon oxidations, **27:**470
p-AB-dihydropteridine condensing enzyme, **27:**426
Abequose, *See* 3,6-Dideoxy-D-galactose
p-ABG, **35:**38
Abietic acid, **17:**154
Ab initio calculations, glutathione S-transferases, **69:**19–20; 22
Abiogenic syntheses, **27:**356–375; 382–386; 388–392; 399–405
A blood group
  N-acetylgalactosaminyltransferase, **52:**68
  antigen, **52:**68
ABO blood group, **52:**116
  antigens, **52:**46; 54
ABO(H) blood types, inheritance of, **36:**134
Abortive complex, **46:**239
Abscisic acid, **56:**179
Absidia glauca, enzymic transformations of steroids by, **20:**245; 247; 271
Absolute configuration of
  4′-d_1-pyridoxamine, **35:**100
  2-t_1-glycine (table), **35:**96
  2-2d_1-glycine (table), **35:**96
  4′-t_1-pyridoxamine, **35:**101
  3-t_1-serine, **35:**97
Absolute temperature, **34:**461
Absorption
  in depancreatized dogs, **8:**180–181
  of fat and phosphatide turnover, **7:**137
  and phosphate bond energy, **1:**150–152
  ultraviolet, **39:**357
  of water by plants, **18:**48–52
Absorption maximum
  α-glycerophosphate dehydrogenase, **22:**370
Absorption spectra, **4:**270–309. *See also* Chromophoric changes; Cotton effects; EPR spectroscopy; Fluorescence, Luminescenece, Phosphorescence, Spectrophotometric determinations, and Spectroscopic studies; Relaxation Spectrometry; Spectra; specific compounds
  ADH-DPNH complex, **22:**348–350
  of D-amino acid oxidase
    after anaerobic mixing with D-alanine with rapid scan spectrophotometer, **34:**46
    upon addition of D-lactate, **34:**50; 52
    upon addition of pyruvate, **34:**51
  biotin, **3:**299
  butyryl dehydrogenase, **22:**390; 391
  changes during enzyme action, **27:**29–32
  of chlorophyll, **19:**58
  chromophores, **4:**272–274
  combination of resonators, **4:**273–276
  of cryptoporphyrin *a,* **23:**272
    of cytochromes, **23:**306–310
    of haem *a* compounds, **23:**290–292; 306–310
    of haematin *a* compounds, **23:**291
    of porphyrin *a,* **23:**272
  cytochrome reductase with DPNH, DPN, **22:**372
  of cytochromes *a* and a_3, **20:**164; 165; 192
    of cytochrome *b,* **20:**162; 163; 191
    of cytochrome *c,* **20:**162; 163; 173–175
    of cytochrome c_1, **20:**162; 163
    of imidazole acyl derivatives, **20:**73

Advances in Enzymology and Related Areas of Molecular Biology: Cumulative Index, Volumes 1–74
ISBN 0-471-66087-6

Absorption spectra (*Continued*)
of ninhydrids of asparagine, glutamine and proline, **20:**337; 338
of succinic dehydrogenase, **20:**190
cytoplasm, **3:**25; 26
of dehydropeptides, **8:**130–133; 136; 137; 138; 139; 152
of diacylaminopropionic acids, **8:**153
of dihydroporphins, **7:**345–348
of dipyrroles, **7:**358–362
DPN-ADH-hydroxylamine ternary complex, **22:**351
DPNH-MDH complex, **22:**370
DPN-hydroxylamine complex, **22:**351
of enzyme-substrate compounds, **12:**154–157
FAD, **22:**376
of FAD, effects of intermolecular bindings, **27:**3–16; 29–32
of fatty acids, **7:**172
flavines from succinic dehydrogenases, **22:**384–386
frequency shifts, **4:**273–276
fundamentals, **4:**271–278
glutamic-aspartic transaminase, effect of pH on, **22:**395
of 3-hydroxypyridine, **50:**4
intensity shifts, **4:**272–276
of LADH-Cd, **27:**79–80
of LADH-Zn-OP complex, **27:**53–54
of lipoyl dehydrogenase, **34:**53
of luciferin derivatives, **25:**138–140
of luminescent substances, **7:**256
of metalloporphins, **7:**355–358
methods of recording, **4:**270–271
of Michaelis complex of D-amino acid oxidase, **27:**19; 20
of mono-, di-, tri-, and tetrapyrroles, **7:** 358–363
nucleic acids, **3:**6–13; 21–23
in nucleic acid study, **9:**480–490
of adenosine triphosphate, **9:**487–488
of amines, **9:**77
of amino acids, **9:**77–81
of fatty acids, **9:**77
of nucleosides, **9:**482–484; 484–485; 487–490
of nucleotides, **9:**482–484; 487–490
of pepsin, **9:**143–144
of peptides, **9:**89–91
of proteins, **9:**96–97; 122
effect of irradiation, **9:**110–116
of pyrimidines, **9:**480–484
nucleus, **3:**6–8
optical (*See* Optical absorption spectra)
in photosynthesis, **25:**45–46; 54–55; 62–64; 72–73
changes, chemically induced, **25:**79–81
photoinduced, **25:**73–79
of porphins, **7:**340–350
of proteins
effect of denaturation, **24:**195–197; 228; 232
ionizing radiations, **24:**360; 369–370
of purple intermediate, using α-amino-butyric acid as substrate, **34:**43
resonance, **4:**272–278
of riboflavin tetrabutyrate, **34:**54
and structure of molecule, **4:**270–277
TPD complexes
with acetylpyridine analogue of DPN, **22:**341; 342
with DPN, **22:**341; 342
with DPNH, **22:**343
ultraviolet, **39:**347
Verdoperoxidase, **3:**141–143; 146
Absorption Spectra of Vitamins, Hormones, Enzymes (Brode), **4:**269
Acacia, fibrinoplastic action of, **16:**33; 39
Accelerator factor, *See* Factor 5
Accelerator globulin, *See* Ac-globulin; Factor 5
Accelerator-inhibitor complexes of blood, separation by centrifuging, **16:**88–89
Accelerators, of blood coagulation, **16:**85; 88; 91–92, **30:**303–305
Accelerin
and blood clotting, **10:**28
use of term for blood clotting factor, **16:**59; 60
Acceptor complex, formulation of, **30:**189
Acceptor domains
glucosamine 6-phosphate synthase, **72:** 104–106
glutamine PRPP amidotransferase active sites, **72:**99–100
Acceptors. *See also* Hill reaction
artificial electron, acceptance of electrons from photosystem 1 or 2 by, **30:**22
in polymer synthesis, **30:**187
specificity, **39:**362
Acceptor substrates, **39:**341; 347

Accessibility barrier, for β-lactamase, **28:** 266–269
Accetylcholine
tyrosine hydroxylase induction, **70:**168
tyrosine hydroxylase regulation, **70:**145
Acclimatization of organisms, **7:**583–584
Accostrep, composition, **20:**384
Acectbacter xylinum, pyruvate, phosphate dikinase in, **45:**108
Acetabularia
enucleation experiments, **16:**470–471; 475; 477; 492, **24:**11–12
enzyme synthesis in, **24:**16–20
Acetaldehyde. *See also* Aldehydges
carbonic anhydrase catalyzed hydration, **47:**190–191; 194; 206–207
in citric acid cycle, **15:**183; 207; 222–223; 245; 254; 256; 262
dehydrogenation by yeast, **10:**190
in deoxyribose synthesis, **28:**396; 421–422; 424; 431–433; 450–451
and fat formation, **8:**315–316; 326; 327–330
as hydrogen acceptor, **15:**14; 88–90; 97; 176
oxidation by chain mechanism, **15:**7
and pentose metabolism, **15:**154; 158; 167; 171
in plants, **18:**296–298
role in fat synthesis, **9:**664; 693–697
and wood molds, **8:**273–279
as xanthine oxidase substrate, **17:**299–301; 309; 323
Acetaldehyde carboligase, classification, **14:**256
Acetaldehyde-coenzyme A, **15:**222
Acetaldehyde dehydrogenase, **15:**198, **23:**388
yeast, activity of -SH group, **11:**241
Acetaldehyde hydrate, **43:**372
1-^{14}C-Acetaldehyde, in pentose synthesis, **28:**429–431
Acetaldehyde metabolism, **45:**440
blood centration, **45:**440
enzymes for, **45:**441
Acetaldehyde r DPN transhydrogenases, classification, **14:**253
Acetaldehyde reductase, classification, **14:**253
Acetals, **37:**14; 27
A-1, **37:**27; 44
electrostatic facilitation, **37:**44
general acid catalysis, **37:**24–25; 28–36; 50
intramolecular, **37:**36–51
hydrolysis, **37:**14; 15; 16; 44
A-1, **37:**15; 16
A-2, **37:**15; 16
entropy of activation, **37:**16
intermolecular, **37:**28–30; 50
intramolecular, **37:**50
metal ion catalysis, **37:**23; 34
nucleophilic participation, **37:**19; 22; 42
secondary, deuterium kinetic isotope effect, **37:**19
spontaneous hydrolysis, **37:**21–24
Acetamide
and Cotton effect of LADH-DPNH, **27:**71
effect on papain action, **15:**292
and fluoro compounds, **18:**141–142
3-L-Acetamido-4-carbethoxylbutanoic acid, synthesis of, **31:**196
2-Acetamino-2-deoxy-β-glucopyranosyl isothiocyanate, **48:**49
Acetanilide
configuration of, **23:**14–16
liver nonspecific hydroxylase action on, **19:**175; 176
Acetate, **43:**44, **51:**10. *See also* Acetic Acid; Acetic acid, Acetylation, Acetyl phosphate, and C_2 units; Acetyl coenzyme A, Acetylations, Acetyl phosphate, and Acetyl transfer
1-^{14}C-Acetate, labeling of pentoses, **28:**407; 418–419; 427–428; 433
2-^{14}C-Acetate, labeling of pentoses, **28:** 412–413; 433
^{14}C-Acetate, labeling of pentoses, **28:**434–435
activation, in mitochondria, **21:**20
radioactive, in study of carotenogenesis, **21:**330–335; 339; 340; 343; 346
in study of mitochondrial fatty acid synthesis, **21:**47–49
amination of, **18:**230
labeled, and citric acid cycle, **18:**243; 247; 249; 268–269
effect of fluoro compounds, **18:**146; 151
in lignin study, **18:**364
in plant respiration, **18:**303
bacterial fixation of carbon dioxide in, **74:**34; 36
in biosynthesis, **12:**43; 44; 49–50
in cellulose synthesis, **15:**315 (C^{14})
in cholesterol biosyntheses, **22:**282–288
labeled, distribution of carbons from, in cholesterol, **22:**282–285

^{14}C-Acetate, labeling of pentoses (*Continued*)
in squalene syntheses, **22:**287
in mevalonate, biosyntheses, **22:**290–295
as precursor, of eburicoic acid, **22:**329
of 3-methylcrotonic acid, mechanism of syntheses, **22:**286
of soyasapogenol A, **22:**329
in cholesterol synthesis, **15:**127
in citric acid cycle, **15:**183; 186–190; 194; 196–197; 201; 210; 212; 214; 218; 223–225; 231; 245; 248 ; 252 (C^{14}); 257; 258; 262
complex with carbonic anhydrase, **47:** 239–240
condensation with CO_2 to form pyruvate, **16:**268
labeled, distribution of isotope in amino acids, **16:**269; 276; 282; 283; 285; 286; 290; 294; 296–299; 301
decomposition in yeast, **10:**169–176
effect on oxalacetic decarboxylase, **12:**80
effect on sulfanilamide acetylation, **6:**258
and elastase action, **23:**246–247
and glycine, **23:**457–462
and glycolate metabolism, **23:**437–440
labeled, **23:**246–247; 404–409; 423; 426–430; 432–434; 437–440
metabolism by microorganisms, **23:** 401–434; 464
and mevalonate, **23:**482
and oxalate metabolism, **23:**447–449
photometabolism of, **23:**420–431
ethanol-derived, **45:**432
leaving liver, **45:**432
fixation in pyruvate, **6:**245
heavy acetate, **6:**245
labeled
in acetyl CoA formation, **19:**257
oxidation in plants, **19:**13
incorporation in ribose, **15:**159; 175–176
in studies of aspartic acid biosynthesis in plants, **20:**328
in studies of fatty acid synthesis in plastids, **20:**224
in studies of lactose biosynthesis, **20:**123
in studies of pectin biosynthesis, **20:**371
oxygen exchange with water, **12:**304
phosphorylation, **6:**256–257
reaction with catalase-peroxide complex, **12:**167; 169
use in biological acetylation, **12:**311–331
utilization in rubber biosynthesis, **29:**226–229
Acetate^{+}, **50:**147
Acetate, SH[acyl-carrier protein], enzyme ligase (AMP), **43:**67
Acetate activating enzyme, **15:**249–253
Acetate-activating enzyme, and fluoroacetate, **18:**130–131
and lipoic acid, **18:**342–343
Acetate activation, **17:**163; 181–183
Acetate (C^{14})
in coenzyme Q biosynthesis, **25:**296–300; 302
exchange with acetylCoA, **17:**181–182
and fluoro compounds, **18:**128
Acetate-1-C^{14}, as carbon source for yeast, **16:**286–287
Acetate condensation enzyme, activity of —SH group in, **11:**243
Acetate-dependent *Neurospora,* **17:**422–425
Acetate kinase, **43:**67, **55:**359–360
kinetic mechanism of, **45:**302
Acetate (labeled) in glucose synthesis, **14:**122
in tyrosine synthesis, **14:**120–121
Acetate metabolism, **5:**149; 150; 156
acetoacetate formation, **5:**149–150
succinate formation, **5:**150
Acetate metabolism, kinetic analysis of, **23:** 407–409
Acetate (methyl-C^{14}), synthetic use by sperm, **24:**89
Acetate (O^{18}), in acetylation of CoA, **18:** 203–204; 208
Acetate oxidation, factors affecting, **18:**245
Acetate replacing factor, **15:**214, **18:**319–320. *See also* Lipoic acid
Acetate-replacing factor, **23:**471–472
Acetazolamide, binding to carbonic anhydrase, **47:**216–218
inhibition of carbonic anhydrase, **47:**212–215; 224–225; 227–229
Acetic acid. *See also* Acetate, Acetylation, Acetyl phosphate, and C_2 units; Acetate, Citric acid cycle, and 2C-particles; Fatty acids
activated, **10:**221–222
activation in citric acid cycle, **10:**177–180; 180–182; 184–186; 188–190
and biological methylation, **12:**447–448
degradation in the animal body, **10:**176–177
effect on carotenogenesis, **21:**318

metal ion complexes of, model catalyses with, **21:**188; 189
from even-numbered fatty acids, **8:**349
and fat formation, **8:**315; 325–326; 327–330
and fatty acid catabolism, **8:**392–401
in fatty acid synthesis in plastids, **20:**224
in pectin, **20:**324; 344; 345; 349
thermodynamic data on ionization, **20:**64
fixation in pyruvic acid, **6:**245
in mold metabolism, **9:**542–543; 545–546; 547–550; 550–554; 554–560; 564–566; 570–571
utilization by yeasts, **9:**677–678; 699
and sterol synthesis, **8:**331–333
synthesis from CO_2, **11:**107–12:129
Acetic acid and pyrrole formation, **7:**329
Acetic acid bacteria, in cacao fermentation, **25:**470–475; 486; 487
Acetic anhydride, reaction with imidazole, **20:**73
Acetic thiokinase, in mitochondria, **21:**74
Acetimidate, phenyl *N*-methyl, carbonic anhydrase catalyzed hydrolysis, **47:**189; 190
N-Acetimidoylsarcosine, **50:**204
Acetl enzymes, kinetic studies of, **20:**302; 304; 309; 310; 312
N-Acetlygalactosamine, **52:**25
Acetoacetate, **5:**156; 157, **53:**225, **59:**125
and biotin enzymes, **26:**296–297; 303–304; 333
and fluoroacetate, **18:**129–130; 133; 140; 146
in lignin study, **18:**363
formation from β-hydroxybutyric acid, **15:**122–123
metabolism, **15:**183–186; 196–197; 201; 232 ; 244 (C^{14}); 262
from phenylalanine and tyrosine, **14:**74; 96–112
in phenylalanine metabolism, **19:**87; 88; 183
in photosynthesis, **19:**42–45
production, in cholesterol biosyntheses, **22:**292
and sulfanilamide acetylation, **6:**258
Acetoacetate condensing enzyme, **15:**241
Acetoacetate decarboxylase, **37:**490, **43:**254; 374, **46:**477
Acetoacetate-β-hydroxybutyrate shuttle, **23:**345–346
Acetoacetic acid, **58:**182. *See also* Fatty acids, Keto acids, and Ketone bodies
biosynthesis, **10:**221–222
formation by oxidation of fatty acids, **10:**187–188
transformation into citric acid by muscles, **10:**182–186
from ethylbenzene, **27:**517–518
and fat synthesis, **8:**336
and fatty acid catabolism, **8:**392–401
fixation of CO_2 in, **12:**46; 53–54; 91
formation in animal tissues, **6:**248
free energy changes, **6:**262–264
metabolism, and hydroxylamine reactivity, **6:**257
Acetoacetic decarboxylases, classification, **14:**256
Acetoacetic-succinyl thiophorase, in mitochondria, **21:**74
S-Acetoacetyl-*N*-acetyl thioethanol-amine, **15:**237; 238
Acetoacetyl-CoA
in carotenogenesis, **21:**338
in mevalonic acid biosynthesis, **21:**336
S-Acetoacetyl CoA, synthesis, **19:**240
Acetoacetyl-CoA reductase, location in subcellular fraction, **24:**296
Acetoacetyl-CoA thiolase, **55:**50; 57
S-Acetoacetyl coenzyme A, **15:**184; 185; 199; 233–244
Acetoacetyl-coenzyme A, in cholesterol biosynthesis, **22:**283; 284; 291
Acetoacetyl coenzyme A deacylase, **15:**233
Acetobacter
aldehyde oxidase in, **19:**202
synthesis, **19:**258
Acetobacter aceti
in cacao fermentation, **25:**472
use of acetate, **23:**411
Acetobacter melanogenum, in cacao fermentation, **25:**472
Acetobacter pasteurianum, cytochrome in, **23:**267; 307
Acetobacter peroxydans, cytochromes of, **23:**266–268; 308
Acetobacter rancens, in cacao fermentation, **25:**472
Acetobacter species, carbohydrate metabolism, **15:**161; 177
synthesis of cellulose, **15:**315
Acetobacter stimulatory factor (ASF), relation to CoA, **16:**8–12

Acetobacter suboxydans
folate metabolism, **27:**457
PN-dependent enzymes, **26:**120; 121
Acetobacter vinelandii, phosphorylating system from, **23:**353; 358
Acetobacter xylinum, **59:**75, **70:**231
synthesis of cellulose, **26:**58
γ-Aceto-γ-chloropropyl acetate, action on thiaminase, **15:**401
Acetocoenzyme A kinase, **18:**205; 342–343
Acetohydroxamic, **35:**170
acid, **35:**138; 156
Acetohydroxamic acid formation, **6:**257
α-Aceto-α-hydroxybutyrate, in isoleucine synthesis, **16:**283
Acetoin, **15:**115; 209, **17:**424
condensation, **15:**158
and fat metabolism, **8:**389–391; 400
by fermentation, **7:**578
from pyruvate, **18:**333
and sulfanilamide acetylation, **6:**258
Acetokinase, **15:**248–250, **17:**183
and ATP synthesis, **23:**326; 333; 335; 355
α-Acetolactate, **15:**209
Acetolactate, in dismutation reactions, **12:** 102–103
α-Acetolactate synthesis, allosteric regulation, **28:**64
Acetolactic acid, in valine synthesis, **16:**285
Acetolactic decarboxylase, classification, **14:**256
Acetolysis
of cellulose, **8:**256–259
of yeast mannans, **40:**243; 245; 247; 249; 251; 264
γ-Aceto-γ-mercaptopropyl acetate, action on thiaminase, **15:**401
γ-Aceto-γ-mercaptopropyl alcohol, action on thiaminase, **15:**401
Acetomonas oxydans, in cacao fermentation, **25:**472
Acetone
effect on β-glucuronidase, **16:**381
effect on pectic substances, **20:**356
methyl groups and formate, **12:**48
Acetone extraction, **34:**473
Acetone fermentation, **7:**560–573
S-(Acetonyl)glutathione, **59:**116
Acetothetin, *See* Dimethylacetothetin
Acetothiokinase, in R. *rubrum,* **23:**424
N-Acetoxy-*N*-2-acetylaminofluorence, **60:**10
N-Acetoxy-*N*-2-aminofluorine-guanine, **60:**23
3-Acetoxypyridine, carbonic anhydrase catalyzed hydrolysis
2,6–6dinitro, **47:**191; 198–200; 206–207; 211–212
2-nitro, **47:**198–200; 206–209; 211–212; 223
Acetycholine, **55:**138
as a neurotransmitter, **29:**482
Acetycholine hydrolase, *See* Acetycholine esterase
Acetylacetone, acid-base catalysis, **25:**24; 26–27
Acetylacetone indicator system, relaxation spectra, **25:**9
Acetyl adenylate, synthesis, **22:**174
Acetylalanine, absorption spectrum, **8:**131
β-Acetylaminodeoxyglucosidase, localization of, **24:**323
Acetylaminofluorenes, **60:**15
N-Acetylarginine, **40:**77
Acetylasalicylic acid, **58:**221
Acetylating agent, **58:**3
Acetylation, **6:**257–262, **34:**470. *See also* Acetate, Acetic acid, and C_2 units; Acylation
of amines, **16:**1–2; 261
of amino acids, **6:**262
biological, **12:**311–331, **15:**122; 183; 186–190; 196–199; 211–213; 220; 222; 248; 257 (*See also* Acetylcholine, Acetyl coenzyme A, Acetyl phosphate, and Acetyl transfer)
of choline, **6:**260–262
acetyl phosphatase in, **6:**260–261
adenyl pyrophosphate in, **6:**260
coenzyme properties, **6:**259–260; 261
coupling with respiration, **6:**257–258
of cysteine, **19:**242
effect on radiosensitivity, **24:**391
exopolysaccharide alginate synthesis, **70:**234–235
of galactose moiety of O side chains, **31:**99
inhibitors, **6:**259; 261
reactivators, **6:**261
role of ATP in, **12:**96; 98
of sulfanilamide, **6:**257–260; 261; 262
with adenyl pyrophosphate, **6:**258; 259
reversible inactivation, **6:**260
of trypsin, **13:**328
role of coenzyme A in, **13:**280; 311–312
N-Acetylation, **35:**161; 166

Acetylation, N-terminus, **66:**157
Acetylation tests for CoA, **16:**16
Acetylcarnitin, **60:**270
Acetyl Ch, *See* Acetylcholine
Acetyl chloride, reaction with imidiazole, **20:**72
Acetylcholine, **5:**143; 144, **17:**436. *See also* Choline
effect on action polymerization, **13:**155
effect on cholinesterase, **20:**295–298
effect on sperm, **24:**84
enzymic hydrolysis and synthesis of, **12:** 259–339
esterase (*See* Cholinesterase)
and muscle weight loss, **6:**199
and myasthenia gravis, **6:**209; 210; 211
and myotonia, **6:**213
and fluoroacetylcholine, **18:**139; 150
hydrolysis, **9:**20
hydrolysis by enzymes, **8:**461–467; 467–471; 490–491
production, **15:**184; 200–201; 205–206; 210
rate of hydrolysis, **34:**12
Acetylcholine, Enzymic Hydrolysis and Synthesis (Nachmansohn and Wilson), **12:**259
Acetylcholineesterase
adsorbed, activity of, **28:**3; 21–22
in schizophrenics, acetylthiocholine hydrolysis test for, **29:**505
Acetylcholineesterases
classification, **14:**244
Acetylcholine receptors, **43:**195; 200; 201; 202; 205; 206; 207, **55:**139
Acetylcholinesterase, **12:**262–317, **15:**29, **17:**45, **34:**449; 525, **36:**66, **37:**414; 460; 568
electron microscopy of, **33:**105
enzymic activity of, **34:**12
imidazole in, **20:**76; 77
inhibitors of, **55:**275–276
localization of, **55:**168–169
pK value, **22:**58
reaction catalyzed by
with acetylcholine, **74:**52–53; 53
with paraoxon, **74:**5–53; 53
Acetylcholinesterase (Rosenberry), **43:**103
Acetylcholinesterases
binding sites of, **24:**116
effect of radiation on, **24:**392; 400
localization of, **24:**317
multiple forms, **25:**349
steps in reaction of, **25:**14; 32
Acetylcholine synthesis, requirement of –SH groups, **11:**243
Acetyl citrate lyase, **43:**61
Acetyl-CoA, **17:**181–182, **35:**161, **50:**123, **57:**143, **60:**270
in acetylation of glutamate, **16:**261
and ATP activation, **18:**208; 230
and citric acid cycle, **18:**261
and fluoroacetyl-CoA, **18:**128–132
biosynthesis, **19:**257
formation, **19:**82
in photosynthesis, **19:**13; 43; 45
carboxylase, **35:**364–378
ATP-^{1} ^{4}C-ADP exchange, **35:**366
ATP^{3} ^{2}-P_i exchange, **35:**366
avian liver, **35:**368
bovine adipose tissue, **35:**368
E. coli carboxylase, **35:**369
enz-biotin-CO_2, **35:**366
regulation, cetyl trimethylammonium ion, **35:**375
fatty acyl-CoA derivatives, **35:**372
hypolipidemic drugs, **35:**373
palmitoyl carnitine, **35:**375
tricarboxylic acid activators, citrate and isocitrate, **35:**373
wheat germ, **35:**371
in carotenogenesis, **21:**338; 339
in mevalonic acid biosynthesis, **21:**336
synthesis and reactions, in mitochondria, **21:**20; 117
in glucogenesis, **28:**89–92
imidazole in formation, **20:**91
in oxidation of hydrocarbons, **27:**490–491; 502; 513
as rubber precursor, **29:**229
S-Acetyl CoA, **19:**239
S-Acetyl-CoA, identification of active acetate as, **16:**1–2; 18
Acetyl-CoA acetyltransferase, localization of, **24:**310
Acetyl-CoA-ACP transacylase, **36:**292–294
Acetyl-CoA acyltransferase, localization of, **24:**311
Acetyl-CoA apocarboxylase, activation of, **26:**333
Acetyl-CoA (C^{14}), in carboxylation experiment, **26:**306; 326

Acetyl-CoA carboxylase, **26:**105–107, **37:**140; 153–155; 424, **43:**93, **55:**27
allosteric regulation, **28:**54–56; 90; 92; 138
biotin enzyme, **26:**297; 333
properties, **26:**297–303; 306; 307
reaction mechanism, **26:**313; 326
Acetyl-CoA carboxylases, **45:**8–11
chemical mecahnism of, **45:**381
Acetyl-CoA deacylase, **21:**74; 409
Acetyl-CoA deacylase activity, localization of, **24:**318
Acetyl-CoA holocarboxylase, synthesis of, **26:**333; 334
Acetyl-CoA hydrolase, localization of, **24:**318
Acetyl CoA r arylamine transacetylase, classification, **14:**251
Acetyl CoA r D-glucosamine transacetylase, classification, **14:**251
Acetyl-CoA synthetase, **50:**375
location in subcellular fraction, **24:**310; 334
Acetyl-coenzyme, and choline acetylation, **12:**328
Acetyl-coenzyme A, **15:**115; 121–123; 129–131; 183–186; 188–191; 195–201; 203–207; 210–213; 219–224; 227; 229; 231–237; 239–240; 242–243; 245–259; 262; 342
and ATP, **23:**325; 335; 341; 348
in dicarboxylic acid cycle, **23:**437–440; 446
enzyme for, **14:**222; 249
and glycine metabolism, **23:**453; 455; 463
inhibition of enzyme by, **25:**224; 242
in microorganisms, **23:**402–404; 408–409; 411–414
and oxalate metabolism, **23:**447–449; 463
S-Acetyl coenzyme A, *See* Acetyl coenzyme A
Acetylcysteamine, disulfide exchange, **27:** 207
N-Acetylcysteine, thioethers, **46:**384
N-Acetylcystine, in synthesis of *N*-benzoylcystine, **19:**243
Acetyldehydroalanine, absorption spectrum, **8:**131; 136; 137; 139
and dehydropeptidases, **8:**136; 137; 139; 140; 142–143; 145; 151
preparation, **8:**124; 128
Acetyldehydroaminobutyric acid, absorption spectrum, **8:**133
Acetyldehydroleucine, absorption, spectrum, **8:**133
Acetyldehydrophenylalanine, absorption spectrum, **8:**132
and dehydropeptidases, **8:**139; 140; 144; 151
preparation, **8:**120; 125
Acetyldehydrotyrosine, absorption spectrum, **8:**133; 137
and dehydropeptidases, **8:**137; 139
Acetyldehydrovaline, absorption spectrum, **8:**133
N-Acetyl derivatives, as intermediates in amino acid metabolism, **16:**260–262
Acetyl-1,4-diaminobutane, in urine, **36:**237
S-Acetyl dihydri-α-lipoic aic, **18:**336
Acetyldjenkolic acid, **50:**122
N-Acetyldjenkolic acid, **50:**124
Acetyl enzyme, and ATP synthesis, **23:**329
Acetyl esterase, **17:**166
Acetyl esterases
classification, **14:**244
effect of phosphate inhibitors, on, **13:**332; 336
effect on cephalosporin, **28:**250
localization of, **24:**316; 317
N-Acetyl-galactosamine, **60:**102
N-acetylgalactosamine-[14], **38:**301
N-Acetyl-D-galactosamine, in polysaccharide synthesis, **26:**76–78
Acetylgalactosamine, UDP ester of, **18:**201
N-Acetylgalactosamine-6-phosphate phosphatase, **60:**262
biological role, **60:**263
β1 3*N*-Acetylgalactosamine α1r3*N*-acetylgalactosaminyltransferase, **52:**72
N-Acetylgalactosamine-4-sulfate sulfatase, **60:**256
biological role, **60:**260
biosynthesis, **60:**258
carbethoxylation, **60:**260
catalyzed reaction, **60:**258
deficiency, **60:**261
diazotization, **60:**260
genetics, **60:**260
inactivation, **60:**258
inhibition by silver ions, **60:**259
isolectric point, **60:**257
molecular weight, **60:**257
natural substrate, **60:**259
oligosaccharide substrate, **60:**259
precursor, **60:**258
properties, **60:**257
purification, **60:**257

N-Acetylgalactosamine-6-sulfate sulfatase, **60:**261
catalyzed reaction, **60:**262
inhibition, **60:**263
properties, **60:**262
purification, **60:**262
substrate specificity, **60:**262
α-*N*-Acetylgalactosaminidase, **48:**38
α-*N*-Acetyl-D-galactosaminidase, **60:**93
kinetics, **60:**100
purification, **60:**98
β-*N*-Acetylgalactosaminidase, **48:**38, **60:**246
N-Acetylgalactosaminide β1r3 galactosyltransferase, **52:**27; 61
α-*N*-Acetylgalactosaminide β1r3 galactosyltransferase, **52:**110
N-Acetylgalactosaminide α2r6 sialyltransferase, **52:**27; 110; 129; 156
α-*N*-Acetylgalactosaminide α2r6 sialyltransferase, **52:**33; 39; 42–44; 60; 98; 105; 127; 130; 137; 149
N-Acetylgalactosaminyltransferase, **52:**68–75; 109
α-*N*-Acetylgalactosaminyltransferase, **52:**110
N-Acetylgalactosamyl residues in teichoic acids, **30:**225
N-Acetyl-γ-glutamyl phosphate, **40:**69; 70
N-Acetylglucosamine, **43:**412, **52:**25; 36, **59:**62, **60:**239
linkage to asparagine in yeast mannans, **40:**243; 265
variability factor in yeast mannan, **40:**246
N-acetylglucosamine, RNA polymerase II transcription, C-terminal domain characteristics, posttranslational modifications, **71:**49–50
N-Acetyl-D-glucosamine, sugar nucleotides from, **26:**40; 49; 50; 59; 73–75
N-Acetyl-D-glucosamine-2-epimerase, **44:**88
from hog kidney, **44:**88
N-Acetyl-D-glucosamine (GNAc) in lipopolysaccharide from *Salmonella,* **31:**79
N-Acetyl-D-glucosamine-6-P-2-epimerase, from *aerobacter cloacae,* **44:**88
N-Acetylglucosamine 1-phosphate in teichoic acid, **30:**229
N-Acetylglucosamine residues in chitin, **16:**162
N-Acetylglucosamine-6-sulfate sulfatase, **60:**264
biological role, **60:**266
catalyzed reaction, **60:**265
properties, **60:**265
purification, **60:**265
substrate specificity, **60:**265
endo-β-*N*-Acetylglucosaminidase, **60:**165
assay, **60:**166
biological role, **60:**167
kinetic data, **60:**167
molecular weight, **60:**166
properties, **60:**166
substrate specificity, **60:**166
endo-γ-*N*-Acetylglucosaminidase, **48:**42
N-Acetyl-β-Glucosaminidase, **48:**62; 81
α-*N*-Acetylglucosaminidase
binding studies, **60:**246
biological role, **60:**246
biosynthesis, **60:**245
carboxyl group, **60:**245
catalyzed reactions, **60:**245
deficiency, **60:**244
genetic inactivity, **60:**246
heparan sulfate degradation, **60:**246
homogeneity, **60:**244
inhibition, **60:**246
intracellular processing, **60:**245
isoelectric point, **60:**244
maturation, **60:**245
natural substrate, **60:**245
optimum pH, **60:**246
polymerization, **60:**245
polymorphism, **60:**246
properties, **60:**244
purification, **60:**244
size heterogeneity, **60:**244
sulfate group, **60:**246
synthetic substrate, **60:**245
urinary, **60:**245
β-*N*-Acetylglucosaminidase, **60:**246
action on murein, **26:**205; 210; 212; 219
inhibited by 2-acetamido-2-deoxygglactonolactone, **36:**156
inhibited by 2-acetamido-2-deoxyglu conolactone, **36:**156
β-*N*-Acetylglucosaminidase, in semen, **24:**88
endo-β-Acetylglucosaminidases, **48:**40; 41; 42; 45; 53; 55
N-Acetylglucosaminidases
assay of, **56:**145–146
occurrence and localization of, **56:**145
properties of, **56:**147–149

N-Acetylglucosaminidases (*Continued*)
purification of, **56:**146–147
regulation of, **56:**150
role of, **56:**149
N-Acetylglucosaminide α1r3 fucosyltransferase, **52:**27; 54–56; 105; 116
N-Acetylglucosaminide α1r4 fucosyltransferase, **52:**27; 48; 49; 54
N-Acetylglucosaminide α1r6 fucosyltransferase, **52:**27; 49
β-*N*-Acetylglucosaminide α1r3 fucosyltransferase, **52:**50; 98
N-Acetylglucosaminide β1r4 galactosyltransferase, **52:**27; 56; 58; 59; 60; 61; 64–67; 97; 116; 146
N-Acetylglucosaminyl-hexanolamine-agarose, **52:**65; 146
N-Acetylglucosaminyl-1-phosphotransferase, **60:**219
α-*N*-Acetylglucosaminyl residues in teichoic acids, **30:**231
β-*N*-Acetylglucosaminyl residues in teichoic aids, **30:**231
N-Acetyl-D-glucosaminyl ribitol linkages in teichoic acids, **26:**75
N-Acetylglucosaminyltransferases, **52:**75–81
N-Acetylglucosaminyltransferases I and II, **52:**80
N-Acetylglucosamyl residues in teichoic acids, **30:**225
Acetylglutamate, **15:**371
N-Acetylglutamate, **40:**69, **50:**145, **53:**222
carbamyl phosphate synthetase, **72:**125
in pyrimidine biosynthesis, **21:**268 269; 271; 272
N-Acetylglutamate, in arginine synthesis, **16:**260; 261; 264; 265
in ornithine synthesis, **16:**231
Acetylglutamate kinase, **40:**59
Acetylglutamate phosphokinase, **50:**145
N-Acetylglutamate γ-semialdehyde, **40:**70; 71; 72
Acetylglutamate synthetase, inhibition of, **29:**133
Acetyl-L-glutamic acid, **34:**501
Acetylglutamic acid, effect of disulfide, **27:**216
Acetyl-L-glutamic acid diamide, **34:**501
N-Acetylglutamic-γ-semialdehyde, in arginine synthesis, **16:**260; 261
α-*N*-Acetylglutamic-γ-semialdehyde, in transamination reactions, **16:**194; 205; 231; 232
N-Acetyl-D-β-glutamine, synthesis of, **31:**197
N-Acetyl-L-β-glutamine, synthesis of, **31:**197
N-Acetylglutamyl phosphate, **50:**145
Acetylglutamylphosphate reductase, **40:**70
N-Acetylglutamyl semialdehyde, **50:**145
S-Acetylglutathione, action of glyoxalase II on, **19:**248
synthesis, **19:**240
Acetylglycine, configuration of, **23:**14–16
Acetylglycosamine (Ac-GN), in mureins, **26:**198; 200–204; 210; 223
linkage in oligomuropeptides, **26:**204–207; 212
Acetyl groups, effect on gelation of pectic substances, **20:**357
Acetylgucosamine, UDP ester of, **18:**201
β-*N*-Acetylhexosaminidase, **60:**219; 246
activity toward hyaluronate oligosaccharide, **60:**248
biological role, **60:**250
chondroitin-4-sulfate degradation, **60:**248
deficient activity, **60:**247
dermatan sulfate degradation, **60:**248
genetic inactivity, **60:**248
glycosaminoglycan degradation, **60:**247
glycosyl transferase activity, **60:**250
isoenzymes, **60:**247
keratan sulfate degradation, **60:**249
kinetic properties, **60:**247
optimum pH, **60:**248
purification, **60:**247
substrate, **60:**248
subunits, **60:**247
trisaccharide hydrolysis, **60:**247
N-Acetylhexosaminidases, **48:**38; 39
N-Acetyl-β-D-hexosaminidases, **48:**48; 60
β-*N*-Acetylhexosaminidases, **48:**48; 60
N-Acetyl-hydroxyornithine, **35:**159
Acetylimidazole, **54:**139
N-Acetylimidazole, **57:**18
N-Acetylimidazole, preparation and properties, **20:**84; 85
Acetyl kynurenine, **17:**418–419; 421
α-*N*-Acetylkynurenine, **15:**99
N-Acetyllactosamine, **52:**41
linkage in, **25:**447
uridine diphosphate derivative of, **25:** 449–450
Acetyl-lipase, deacylation of, **48:**339
isolation of, **48:**336

Acetyllipic acid-enzyme complex, **18:**335
Acetyl lipoic acid, **15:**220
α-*N*-Acetyl-L-Lysine Deacylase, **38:**416–417
N-Acetylmannosamine, **38:**300
N-Acetyl-D-mannosamine, carbohydrate synthesis, **26:**39; 40; 46
N-Acetylmannoseamine, formation of, **28:** 114–115
Acetyl mercaptide, **15:**189–190; 247
Acetylmethylcarbinol, **15:**209
Acetyl methyl carbinol, formation in plant tissues, **9:**623
Acetylmethylcarbinol synthesis, **6:**300
Acetylmorphine esterase, classification, **14:**244
N-Acetylmuramic acid, in mureins, **26:**200–204; 208; 212; 216–219; 223
  linkage in oligomuropeptides, **26:**204–207; 212
N-Acetylneuraminic acid, **20:**135; 137–141; 144
  in carbohydrate metabolism, **26:**37; 40; 41; 51; 56; 58
N-Acetylorinithine, **50:**145
δ-Acetylorinithine, **50:**135
Acetylorinthine δ-transaminase, **17:**397
Acetylornithinase, **17:**397
Acetylornithine, **35:**161
N-Acetylornithine, **40:**70; 71
  and *E. coli* growth rates, **40:**73; 74
  growth medium, **40:**87
N^{α}-Acetylornithine, in arginine synthesis, **16:**260–263
α-*N*-Acetylornithine, in transaminations, **16:**194; 205; 231
Acetylornithine aminotransferase, **40:**70; 71, **74:**133
  effect of streptomycin and tetracycline, **40:**81
  phylogenetic distribution of, **74:**157
  residual repression, **40:**82
Acetylornithine deacetylase, **40:**71; 72
  arginine antagonism, **40:**74
  transcriptional repression, **40:**79
  translation of message accumulation, **40:**84
Acetylornithine permeation system, repression by arginine, **40:**75
N-Acetyl-*S*-oxalylcysteamine, **57:**132
Acetylpepsin, activity, **14:**189
Acetylphenylalanine, absorption spectrum, **8:**132
N-Acetyl-L-phenylalanine ethyl ester, kinetics of hydrolysis, **20:**294; 305
Acetyl-L-phenylalanyl-diiodo-L-tyrosine, **33:**407
Acetyl-DL-phenylalanylglycine, as cosubstrate for papain, **19:**311; 312
Acetyl phosphatase
  of animal tissues, **6:**251–254
    heat stability, **6:**253
    trichloroacetic acid affecting, **6:**253
  and ATP synthesis, **23:**330–331; 379
  in choline acetylation, **6:**260–261
Acetylphosphatase, classification, **14:**245
Acetyl phosphatase, site of cleavage by, **17:**163; 166
Acetyl phosphate, **1:**109; 111; 112; 121; 128; 152–154, **6:**231–265, **15:**129–130, **17:**167; 182–183, **45:**145, **58:**12
  anabolism, **6:**233; 242–250
    and chemosynthesis, **6:**248–250
    and photosynthesis, **6:**248–250
  animal metabolism, **6:**250–262
    acetate phosphorylation, **6:**256–257
    acetylation mechanism, **6:**257–262
      choline acetylation, **6:**260–262
      coenzyme properties, **6:**259–260
    and acetyl phosphatase, **6:**251
    free energy of formation, **6:**262–265
    hydroxylamine reactivity, **6:**256
    pyruvate oxidation, **6:**254–256
  and ATP, **23:**324–327; 331; 335–337
    labeled, **23:**326
  bacterial metabolism, **6:**232–250
    and butyryl phosphate, **6:**238
    carbon dioxide fixation in lactic acid, **6:**248–250
      in pyruvate caboxyl, **6:**246
    Clostridium butylicum extracts, **6:**237–238
    condensation process, **6:**244–246
    coupling reaction, **6:**239–242
    in *Escherichia coli*, **6:**238; 246
    heavy acetate tests, **6:**245
    heavy formate tests, **6:**245
    Lactobacillus delbrueckii, **6:**233
    molybdate affecting, **6:**235; 236
    phosphoroclastic reaction, **6:**236–239
      reversibility, **6:**244
    *p*H values in, **6:**235
    pyruvate cleavage to, **6:**237–239
      reversibility in *Cl. butylicum*, **6:**246–248
    radioactive phosphate tests, **6:**246
    thermodynamics, **6:**241; 246; 249
  catabolism, **6:**233–241

Acetyl phosphate (*Continued*)
chemistry, **6:**234–236
and choline acetylase, **12:**327–331
and CO_2 fixation, **12:**93–101
in citric acid cycle, **15:**184; 188–192; 194–195; 200–207; 209–211; 220; 236; 245; 247 (C^{14}); 254; 256 (C^{14}); 259; 262
effect on sulfanilamide acetylation, **6:**258
hydrogenation to lactic acid, **6:**248
and lipoic acid, **18:**336; 337
as quasi-substrate for triosephosphate dehydrogenase, **22:**52
reductive carboxylation to pyruvate, **6:**248
Acetyl phosphate and fat metabolism, **8:** 400–401
Acetylphosphate in citric acid cycle, **10:** 189–190; 221–222
Acetyl Phosphate (Lipmann), **6:**231
Acetylphosphate r CoA transacetylase, classification, **14:**251
Acetylproline N-methylamide, configuration of, **23:**24
3-Acetylpyridine
as inhibitory analogue, **6:**132
rates of reaction with reduced diphosphopyridine nucleotide dehydrogenases, **20:**169
Acetylpyridine, inhibitory analogues of, **6:**132; 141
3-Acetylpyridine-ADH, complex with glutamate dehydrogenase, **39:**373; 381; 404
3-Acetylpyridine deamino-DPN, in isoenzyme study, **27:**271
3-Acetylpyridine-DPN, in isoenzyme study, **27:**271
3-Acetylpyridine-DPNH, enzyme-bound, Cotton effect, **27:**57; 58
Acetylpyridine DPNH, substrate in EPR study, **27:**180–181
N-Acetylserinamide, **43:**228
O-Acetylserine, **50:**123; 125; 145; 146; 147; 155
α, β-elimination of acetic acid elements from, transition state structure of, **74:**221; 222
rearrangement to *N*-acetylserine, mechanism for, **74:**216
O-Acetyl-L-serine, **50:**119
O-Acetylserine sulfhydrylase
α-aminoacrylate intermediate, **74:**196–197; 203; 213; 221–225
apoenzyme, preparation of, **74:**199
chemical mechanism, **74:**213–225; 223
acid-base chemistry of, **74:**219–220; 220
isotope effects on, **74:**220–221
p*K* for *O*-acetyl-L-serine, **74:**216
spectral studies of, **74:**213–215; 215
transition state structure of, **74:**220–221; 222
gene expression, in enteric bacteria, **74:** 189–190
isozyme A
acid-base chemistry of, **74:**219–220; 220
active site of, **74:**200–201; 201; 203
apoenzyme, preparation of, **74:**199
chemical mechanism of, **74:**213–225; 223
dynamics, **74:**225–229; 229
in enteric bacteria, **74:**189–190; 197
isotope effect studies of, **74:**211; 220–221
kinetic mechanism of, **74:**204–213
K41 mutant, **74:**202; 222
OAS: acetate lyase reaction, **74:**216–219; 217–218; 220
overexpression, of, **74:**198–199
phosphorescence studies of, **74:**226–227; 229
^{31}P-NMR spectral studies of, **74:**226; 228
purification of, **74:**199
reaction pathway for, structural changes along, **74:**229
spectral studies of, **74:**213–214; 215
static and time-resolved fluorescence studies, of, **74:**227–228
steady state kinetics of, **74:**204–207; 205; 206
stereochemistry of, **74:**225
stopped-flow kinetic studies of, **74:**209–211; 210; 212
structure of, **74:**199–204; 200; 201; 207
substrate specificity of, **74:**204
ultraviolet-visible spectral studies of, **74:**207–209; 208
isozyme B, in enteric bacteria, **74:**189–190; 197
isozymes, **74:**197–198
assays, **74:**198
from *Salmonella typhimurium,* sequence alignment between, **74:**197; 198
kinetic mechanism of, **74:**204–213; 212
absorbance spectrum of, **74:**207–209
isotope effect studies of, **74:**211

steady state kinetics of, **74:**204–207; 205; 206; 213
stopped-flow kinetic studies of, **74:**201; 209–211; 213
ultraviolet-visible spectral studies of, **74:**207–209; 208
O-acetyl-L-serine:acetate lyase reaction, **74:**216–219; 217–218
overexpression of, **74:**198–199
purification of, **74:**199
reaction catalyzed by, **74:**188
β-replacement reactions, **74:**197
from *Salmonella typhimurium,* **74:**189–191
spectral studies of, **74:**213–215; 215
^{31}P-NMR, **74:**226; 228
ultraviolet-visible, **74:**207–209; 208
stereochemistry of, **74:**225
structure of, **74:**199–204; 200; 201; 202
substrate specificity of, **74:**204
O-Acetyl-L-serine sulfydrylase, **50:**146
Acetylserotonin O-methyltransferase, **25:**416
Acetylspermidine, **36:**222
in urine, **36:**237
Acetylthiamine, **58:**3
Acetyl thiamine intermediate, in ATP synthesis, **23:**337
Acetylthiocholine, hydrolysis as a test for red cell acetylcholine esterase in schizophrenia, **29:**505
S-Acetylthioethanolamine, **15:**190
Acetyl thioglycolate, **15:**248
Acetyl transfer. *See also* Acetate, Acetyl coenzyme A, Acetylations, Acetyl phosphate, and Lipoic acid
from citrate, **15:**199–201
by CoA, **16:**18
form β-keto acids, **15:**232–244
from pyruvate, **15:**202–213
Acetyl trypsin, column chromatography, **22:**452; 453
Acetyltrypsin, properties of, **24:**136–140
Acetyl-L-trypsine ethyl ester, **34:**253
Acetyltrypsinogen, activation by pepsin, **24:**126–127
Acetyl trypsinogen, column chromatography, **22:**451–453
Acetyltyrosine, absorption spectrum, **8:**133
N-Acetyltyrosine, competitive inhibitor in tyrosine oxidation, **14:**81; 84; 93–95
Acetyl-L-tyrosine ethyl ester, **34:**460; 522
Acetyltyrosine ethyl ester, binding by thiolsubtilisin, **33:**396
hydrolysis by thiol-subtilisin, **33:**396–397
Ac-globulin
distribution in centrifugal field, **16:**88–89
effect of vitamin K on, **16:**64
equilibrium conditions among trace proteins, **16:**76
plasma type to serum type, **16:**42; 53; 61–63
purification, **16:**27; 29; 30
quantitative determination, **16:**63
relation to blood clotting, **16:**57–63; 91–92
relation to platelet activity, **16:**67; 69–71
role in prothrombin activation, **16:**25; 49–55
properties of, **30:**262 (*See also* Factor V)
Ac-globulin deficiency disease, **16:**65
Achras sapota, gutta obtained from, **29:**222
Achromobacter
glycine metabolism in, **23:**456–457
propionyl-CoA carboxylase from, **26:**305
Achromobacter fischeri, **51:**16
luminescence, **25:**144–154
Achromobacter fischeri, 118, **15:**128
Achromobacter fischeri, PN-dependent enzymes, **26:**130
Achromobacter protease, **53:**268
Achromobacter proteinase, **53:**260; 261
Acicularia, enzyme synthesis in, **24:**20
α-acid, as primary product, in D-amino acid oxidase reaction, **34:**69–71
Acid, mutagenic effect, **27:**305; 319–320; 335
Acid amides, of tobacco mosaic virus, **15:** 462–463
of virus mutant, **15:**479
Acid-amino acid ligases (peptide synthetases), localization of, **24:**335
Acid-ammonia ligases (amide synthetases), localization of, **24:**334
Acid-*n*-amylester, hydrolysis of, **34:**11
Acid anhydride bonds, enzymes acting on, **24:**328–330
Acid anhydrides, and pyrophosphorylases, **18:**201–212
Acid autolytic activity, localization of, **24:**326
Acid-base catalysis, enzymic mechanisms, **25:**24–27–25:30
relaxation methods, **25:**21–30; 32–33

Acid-base catalytic action of hydrolytic enzymes, **15:**284–285; 292–295; 297
Acid-base groups, change in virus mutation, **15:**476–479; 480–481
Acid-base homeostasis, acinar glutamine metabolism, **72:**71–75
Acid CE hydrolase, cholesterol ester hydrolases, **69:**45
Acid deoxyribonuclease, localization of, **24:**321
Acid Deoxyribonuclease, Mechanism of Action and Structure of (Bernadi), **31:**1
Acid DNase, acid-soluble oligonucleotide liberation with, **31:**20
  assay for, **31:**24–25
  bis (*p*-nitrophenyl) phosphate as substrate for, **31:**35
  catalytic properties of, **31:**30–36
  from chicken erythrocytes, **31:**4
  chromatographic properties of, **31:**35
  chromatography of, **31:**22; 28; 29
  chromatography on Amberlite IRC-50, CM-Sephadex, and Sephadex G-50 and G-100 of, **31:**33–34
  definition of, **31:**2
  degradation of native DNA by, **31:**9; 12; 13
  digestion of DNA sample with, **31:**10; 11
  dimeric structure and allosteric properties of, **31:**40–41
  DNase activity of, **31:**30–33
  early work on, **31:**3–4
  hog spleen, amino acid analysis of, **31:**37
    chromatographic purification of, **31:**26
    competitive inhibition of, by yeast tRNA, **31:**39
    electrophoresis chromatography of trypsin peptides obtained from, **31:**42
    molecular weight of, **31:**36
    physical properties of, **31:**36
    preparation of, **31:**25
    sedimentation coefficient of, **31:**36
    sedimentation patterns of, **31:**41
  inhibition of, by polyribonucleotides, **31:**36; 38–39
  inhibitors of, **31:**35
  mechanism of action of, **31:**1–49
  model of, **31:**44
  p-nitrophenyl esters as substrates for, **31:**35
  nomenclature of, **31:**2
  pancreatic DNase digestion method of, **31:**23
  pH optimum of, **31:**35
  3′-phosphates from, **31:**2
  3′-phosphate terminal nucleotides in digests of, of calf thymus DNA, **31:**23
  phosphodiesterase activity of, **31:**33–36
    as assayed on Ca (bis(*p*-nitrophenyl)phosphate)$_2$, **31:**34
  phosphomonoesterase and spleen exonuclease digestion method of, **31:**23
  physical and chemical properties of, **31:**36–38
  purification of, **31:**24–29
  rechromatography of, on CM-Sephadex C-50, **31:**30
  sedimentation coefficient for, **31:**35; 40
  specificity of, **31:**2; 18–23
  structure of, **31:**1–49
  substrates for, **31:**35
  sucrose-gradient centrifugation of, **31:**34–35
  thermal inactivation of, **31:**34; 35
  venom exonuclease digestion method of, **31:**23
Acides tricarboxyliques, cycle, role biosynth,tique (Wiame), **18:**241
Acid formation in Aspergillus niger *and in Related Molds* (Walder), **9:**537
Acid formation in molds, **9:**537–584
Acid-β-glucosidase
  activation by phospholipids, **60:**117
  assay, **60:**117
  binding site, **60:**120
  biological role, **60:**124
  biosynthesis, **60:**119
  catalytic mechanism, **60:**120
  genetics, **60:**124
  inhibition, **60:**117
  isoelectric point, **60:**120
  kinetic data, **60:**120
  Michaelis constant, **60:**120
  molecular weight, **60:**119
  properties, **60:**119
  purification, **60:**116
  self inactivation, **60:**122
  substrate, **60:**118; 125
Acid glycosidases, localization of, **24:**322
Acid hydrolases, localization of, **24:**315; 322
Acidic chromatin proteins, **38:**49
Acidic phospholipids, **60:**123
Acid invertase, **56:**155
  in maize endosperm, **56:**163
  occurrence and localization of, **56:**156
  purification of, **56:**158

in sucrose metabolism, **56:**162
in sugar cane, **56:**163
Acidity, of pectic substances, **20:**354
Acid labile sulfide, **48:**17
Acidosis, metabolic, **53:**229–230
Acidosis and kidney phosphatides, **7:**142
Acid phosphatase, **18:**118; 198, **37:**488, **40:**242, **55:**365–366, **60:**221. *See also* Phosphatase; Phosphatases
adsorbed on kaolin, **33:**271
cytochemical method for, **8:**30
in chromidia, **8:**31
in nuclei, **8:**33
effect of estrogen, on, **9:**352
histochemical methods for, **9:**590–592
effect of polycations, **40:**42
kidney bean, purple, **74:**71
localization of, **24:**305; 316; 319–320
nuclear, **27:**251–253
in particles of liver cells, **16:**367
pK value, **22:**58
prostatic, test on phosphorylases, **24:**282
in semen, **24:**65
site of cleavage by, **17:**163; 169
Acid phosphatases, low molecular weight, **68:**7
Acid phosphatases multiple forms, **25:**348
product inhibition of, **25:**178; 222; 226; 229
Acid phosphatase synthesis, effect of nucleus, **24:**20
Acid proteases, multiple forms, **25:**351
Acid proteinases, **44:**1, **53:**256
active-site-directed inhibitors, **44:**17
Acid ribonuclease, localization of, **24:**315
Acids, **34:**446; 510. *See also* Amino acids
dicarboxylic, in transaminations, **16:**199; 203; 225–226
organic, as inhibitors of β-glucuronidase, **16:**373–375
effect on pectic substances, **20:**349; 357
fatty, **34:**475; 478; 486–488
Acids from wood molds, **8:**271–272; 273–279
Acid-thiol ligases, localization of, **24:**334
Acinar glytamine metabolism, **72:**67–75
acid-base homeostasis, **72:**71–75
glutaminase ammonia amplification, **72:** 69–70
glutamine synthase as ammonia scavenger, **72:**67–69
intercellular signaling, **72:**70–71
Acinetobacter calcoaceticus, **49:**5
Acinetobacter Iwoffi, **54:**169
Acivicin, γ-glutamyl transpeptidase catalytic sites, **72:**255–262
Acivisin, **54:**101
Aconitase, **9:**412; 550; 574, **12:**90, **15:**186–191, **43:**496, **45:**354, **55:**219–220, **74:**303
and acetate metabolism, **23:**414; 434
in ATP synthesis, **23:**340–341
action on *cis* acid, **8:**363
and fat metabolism, **8:**393
chemical mechanism of, **45:**354
and citric acid cycle, **18:**244; 272–274
and fluoro compounds, **18:**118; 122–128; 132; 134; 136; 137; 141; 148; 153
in plants, **18:**298–299
in citric acid cycle, **10:**169; 190–199
kinetics of action, **10:**192–196
classification, **14:**237; 256
effect of fluorocitric acid on, **19:**359
in photosynthesis, **19:**14; 53
2-fluorocitrate and, **55:**222; 223
3-fluorocitrate and, **55:**226
kinetic mechanism of, **45:**354
in mitochondria, **21:**13; 18; 21; 74
pH optimum of in mitochondria, **33:**289
in sperm, **24:**86
localization of, **24:**332
cis-Aconitate, **51:**236
trans-Aconitate, **15:**261
cis-Aconitate, and acetate metabolism, **23:**410
Aconitate, inactivity in *E. coli* mutant, **16:**255
cis-Aconitate, in citric acid cycle, **15:**185; 192; 261
Aconitate hydratase, localization of, **24:**332; 337; 343
Aconitates, **17:**149
cis-Aconitic acid. *See also* Citric acid cycle
in citric acid metabolism, **10:**169; 194–196; 196–199
and fatty acid catabolism, **8:**392–401
cis-Aconitic acid, deuterium-labeled, **18:** 126–127
Aconitic acid in mold metabolism, **9:**549–550
cis-Aconitic acid in mold metabolism, **9:** 540–541; 573–576
Aconitic hydratases, classification, **14:**237; 256
ACP, **43:**71
intracellular localization of, **36:**305–307
ACP hydrolase, **36:**283

Acquired drug resistance, glutathione S-transferases, **69:**6–7
Acridine mutagenesis mechanism, **27:**306–309
Acridine orange, effect on production of β-lactamase, **28:**269
Acridines, mutagenic agents, **27:**296; 303–305; 325; 330; 331; 338–340
Acriflavin
chemiluminescence, **25:**128
photooxidation of Mn^{2+} inhibited by, **30:**25
Acriflavine, effect on production of β-lactamase, **28:**269
Acrolein, **36:**236, **55:**278
Acroleinaminofumaric acid, in tryptophan metabolism, **19:**95–97
Acrylamide, **34:**452
-acrylic acid, **34:**520
gel, **34:**452
Acrylates, **17:**149
Acrylic acids, substituted, from amino acids, **9:**408–409
Acrylonitrile, **35:**475, **58:**16
AcrylylCoA, **55:**210
α-A-Crystallin, *See* entries under Crystallin
N-Acteylmuramic acid (C^{14}), biosynthesis of, **26:**198
ACTH
activity and structure, **24:**107
and phosphorylase, **24:**286
adrenal-stimulating activity of, **29:**459
basic core of, **29:**460–464
biological properties of, **29:**443–468
caboxymethylcellulose chromatography of, **29:**432
depletion of adrenal ascorbic acid by, **29:**456
digestion with porcine leucine aminopeptidase, **29:**408
effect of acetylation on activity in, **29:**464
effect of racemization of amino acids on, **29:**452–453
effect on adrenal protein synthesis by, **29:**469–470
essential and non-essential regions of, **29:**467–468
general strategy of the synthesis of, **29:**410–412
glycollyl derivative of, **29:**464
importance of side chain functions for, **29:**453
importance of the amino terminal region in, **29:**464–468
isolation of, **29:**406–410
melanophore-stimulating activities of peptides of, **29:**450
properties of, **29:**446
methods of peptide synthesis of, **29:** 393–406
peptide containing D-amino acids, **29:** 452–454
porcine, synthesis of, **29:**433–435
potency ratios of, **29:**456
potentiation and prolongation of activity in, **29:**452–456
presumed binding region in, **29:**462
role of C-terminal region in, **29:**462–463
role of methionine in activity of, **29:**466
similarities in structure of α-MSH, **29:**449
stereochemical configuration of, **29:**455
stimulation of cyclic adenosine 3′,5′-monophosphate by, **29:**465–469
stimulation of release of corticosterone by, **29:**457
structure-activity relationships of, **29:** 391–477
structure of, **29:**406–410; 407; 459
synthesis of peptides related to, **29:**392
effect on β-glucuronidase, **16:**388; 397
effect on hyaluronidase inhibition, **13:**218
effect on rate of phenolsulfophthalein excretion, **13:**229
effect on ribose synthesis, **28:**417
and melanin formation, **14:**91
separation by chromatography, **14:**332; 342; 346; 347; 348; 351
Actidione, effect on microbiological formation of hydrocortisone, **20:**270
in treatment of plant diseases, **20:**384; 393–396; 400–403
Actin, **6:**151; 156–157, **42:**230, **43:**7; 17; 36; 40; 356. *See also* Actomyosin and Myosin
amino acid composition, **16:**319; 330
effect on myosin ATPase activity, **16:**388
globular, preparation, **16:**333–334
size and shape of molecule, **16:**316
polymerization, **16:**333–335; 342
reaction with myosin, **16:**320–322; 327; 343–346
relation to myosin and tropomyosin, **16:**330–331

biosynthesis of, **50:**223
effect of creatine on biosynthesis of, **50:**223–234
F-actin, **6:**157
flow birefringence, **6:**157
G-actin, **6:**157
preparation, **6:**156
properties, **6:**157
solubility, **6:**157
viscosity, **6:**157
Actinomyces albus, enzymic transformations of steroids by, **20:**259
Actinomyces antibioticus, teichoic acid from walls of, **30:**227
Actinomyces elastases, **23:**232
Actinomyces sp., β-lactamase of, **28:**242
Actinomyces streptomycini, wall ribitol teichoic acid of, **30:**231
Actinomyces violaceus, ribitol teichoic acid of, **30:**231
Actinomycete (Nocardio hydrocarbonoxydans), oxidation of hydrocarbons, **27:**488
Actinomycetes
antibiotics from, uncouple oxidative phosphorylation, **29:**289
stimulation of ATPase activity by, **29:**289
β-lactamase of, **28:**242
Actinomycetes, methylated lysine and arginine derivatives in, **42:**234
Actinomycin, **7:**466; 467; 487; 489; 491; 493, **33:**367
effect of, on uridine-^{3}H and on L-leucine-^{14}C incorporation by sensory ganglia in medium with nerve growth factor, **31:**70
effect on acid DNase by, **31:**32
incorporated of proline analogs in, **29:**104
synthesis of peptides in, **29:**105
Actinomycin D
effects on enzyme synthesis by, **29:**376
inhibition of dehydrogenase induction by, **29:**375
synthesis of gramicidin S, inhibited by, **33:**342–343
Actinomycin IV, structure of, **29:**104
Actinomyosin-like complex, in respiratory chain, **28:**161
Actin-troponin-tropomyosin complex, **57:** 207–208
Action. *See also* Actomyosin, Muscle and Myosin
and ATP in muscle contraction, **13:**152–156; 161–163
interaction with myosin and ATP, **13:**169–180
in special types of muscle, **13:**180–184
Action potential, **43:**208
Activated complex, **9:**27; 28; 29; 67
Activated Fru-P_2ase, **53:**49
Activated Hageman factor, **48:**284
Activating agents in fermentations, **7:**563
Activating enzyme, luciferase as, **44:**64
Activating groups in enzymes, **15:**29; 36–37
Activation, **34:**486
of amino acids
activation of fructose diphosphatase by reduced ferredoxins, **33:**136–138
reaction requirements (table), **33:**137
role of B_{12}, **26:**254
of apoenzymes, **26:**331–334
conduction of, in enzyme systems, **14:**184–190
of apoenzymes, **14:**183–190
or organic catalyst by substitution of groups, **14:**176–179; 183–190
of proteolytic enzymes of pancreas, relation to structure, **14:**285
energies of, **36:**112
heat of, in tryptophan r indole reaction, **10:**66–68
of acetic acid, **10:**177–182; 184–186; 188–190; 221–222
of alkaline phosphatase, by amino acids, **10:**101; 106–107
of blood coagulation, **10:**3–28
by divalent cations, **10:**89–93; 98–100; 102–104; 110–111
of enzymes by amino acids, **10:**100; 103–104; 106–107; 278
of trypsinogen, **10:**12
interfacial, **48:**324; 340; 364; 365
of monoamine oxidase by pyrrole, **7:**537
polar, **7:**76–79
volume increase, **7:**248; 249; 250
Activation effects of enzymes, and steric structure, **26:**105–107
Activation energies
on enzymes, **24:**371–373
for hydrolysis by esterases, **20:**303; 304; 313
for phophorylation of esterases, **20:**304
for reactivation of phosphorylated esterases, **20:**307; 308
slow reaction study, **27:**29–32

Activation energy, **33:**268
Arrhenius, **33:**288
for conversion of purple intermediate to fully reduced enzyme, **34:**61
for $E_{ox} \cdot S \rightarrow$ purple intermediate, **34:**60
for water permeation, **34:**467
of DEAE-cellulose-aminocylase system, **33:**271
of esterolytic reactions, **23:**150–151
Activation energy diagram
catalytic power definition, **73:**35–36
catalytic power origins, **73:**43
Activation energy of enzymic oxidoreductions, **15:**10–11; 14; 24–25; 36; 41–42
Activation energy of luminescence, **7:**237; 242; 246
Activation enzymes, in polypeptide synthesis, **24:**2–4
Activation of amino acids, **19:**235–236
Activation of enzymes
allosteric, **28:**51–52; 58; 104–114
energy of, **9:**29–39; 128; 146
effect of dipoles on, **9:**70
of oxidation-reduction potential on, **9:**41–42
of pressure on, **9:**66–69
of specific reagents, on, **9:**46–47
of substrate on, **9:**47–48
of temperature on, **9:**40–49
in simultaneous reactions, **9:**54–56
in successive reactions, **9:**49–53
entropy of, **9:**43–48; 58; 60–67
equilibrium constants, **9:**28–29
mechanism, **9:**69
thermodynamic constants, **9:**28–31
Activation of β-glucuronidase, mechanism of, **16:**376–377
of prothrombin, **16:**25; 30–31; 44–57; 58–75; 82; 91–92
Activation of intracellular proteolytic enzymes, **1:**82; 83; 85–88
Activation of prorennin, **15:**425
Activation of RNase, entropy of, **24:**455–458
Activation of zymogens, **24:**24; 126–127
Activation parameters, **39:**356
Activation reactions involving ATP, **17:** 176–186
Activator
classification, **14:**242
of maltose system, **30:**123
Activator molecules, **43:**348
Activator of lipoxidase, **8:**448–449
Activator protein, **60:**171
Activators. *See also* Kinetics, enzyme; Primers
effect on RNase, **24:**235
effect on succinic dehydrogenase, **18:**89
enzyme, **7:**39
of β-glucuronidase, **16:**370–371; 375–377; 379
interaction with inhibitors, **46:**346
mechanisms for effects, **17:**11–12; 17; 19–26; 55
in polysaccharide synthesis, **11:**300–301, **26:**59
sulfhydryl, **9:**428–430
Activators of pyruvate carboxylase, **42:**13
Activators of thiaminase, **15:**399–404
Activator system in blood coagulation, **5:**34–41
inhibitor of action, **5:**48
Active acetate, *See* Acetyl coenzyme A; S-Acetyl CoA
identification as *S*-acetyl-CoA, **16:**1–2; 18
Active aldehyde, in isoleucine synthesis, **16:**283
Active carbamyl, *See* Carbamyl phosphate
Active carbon dioxide
in photosynthesis, **19:**42; 59
in pyrimidine biosynthesis, **21:**269
Active center
of cholinesterase and related enzymes, **20:**283–318
of enzymes, imidazole in, **20:**61–92; 100
of enzyme, **27:**38–41; 48–49; 53; 55–56; 62; 226–228; 432; 445–450
primary sequence at, **27:**80–82
and sulfhydryl groups, **27:**78–82; 214–215
of YADH, **27:**85
of papain molecule, **19:**299; 301
Active centers. *See also* Active sites and Binding sites
of biologically active proteins, **24:**115–125
catalytic efficiency of, **33:**403
effect of partial loss of, **24:**120–125
of enzymes, effect of ionizing radiations, **24:**382–383
of globular proteins, **24:**128–131
of myosin molecule, **16:**336
number on molecule, **24:**118–120
of some esterases, **24:**434
Active centers in catalysts, **14:**179–190
Active core, **42:**150; 151; 154; 159; 160

Active core of enzyme molecules, **24:**106–107
Active formaldehyde, *See* N^5,N^{10}-Methylenetetrahydrofolic acid
Active formate, *See* N^{10}-Formyltetrahydrofolic acid
from glycine, **16:**276
Active fraction, expression for, **34:**493
Active glycoaldehyde, **22:**399
Active glycolaldehyde, **19:**25; 26
Active groups, in antibacterial substances, **7:**503–504
in catalases, **7:**295
of enzyme, **7:**392
Active methionine, *See* S-Adenosylmethionine
Active site, **36:**114, **39:**204; 206; 343; 346; 349
aliphatic amino acids in, **22:**56
amino acid composition, **22:**63–72
covalent labeling of, **22:**50–57
with limited amino acid reagent, **22:**53; 54; 56; 90
by prosthetic group, **22:**55; 56
by quasi-substrate, **22:**52; 56; 66; 90
with radioactive reagent, **22:**54–56; 90
by substrate moiety, **22:**50; 51; 56; 90
definition, **22:**50
and enzyme action, **22:**45–97
flexibility, **22:**76–78
of GAD, **22:**369
identification of amino acids in, without covalent labeling, **22:**57–63
location in amino acid chain, **22:**78; 79; 89
nomenclature, **22:**47–50
role, **22:**79–89
summary description, **22:**89–91
three-dimensional structure, **22:**72–79
apolar(hydrophobic), **39:**346
of biologically active proteins, simulated mutation at, **33:**381–400
of enzymes, serine residue as part of, **33:**385
synthetically formed, activity of thiolenzymes with, **33:**391–399
cationic, **39:**345
enzyme, **39:**358
hydrophobic, **39:**357; 358
of rhodanese, **39:**345
Active Site and Enzyme Action (Koshland), **22:**54
Active-site catalytic residues, cholesterol ester hydrolases regulation, **69:**68
Active site chemistry, γ-glutamyl transpeptidase, **72:**250–262
catalytic sites, **72:**254–262
light subunit catalysis, **72:**250–251
substrate binding sites, **72:**251–254
Active site-directed irreversible inhibitor, **50:**335
Active-site ionizing group, **34:**462
Active-site peptides, **51:**228
Active-site region, **48:**284
Active site residue roles, NMP kinase catalytic mechanism, **73:**111–114
Active sites. *See also* Active centers and Binding sites; Allosteric sites; Binding sites; Regulatory sites
of enzyme, and conformational change, **28:**299–301; 304
lysine as, **28:**451–452
of esterases and proteases, **24:**459; 465
of β-lactamase, **28:**260; 262
of laminarinases, **28:**347
of RNase, **24:**448–452
Active sites of enzymes, and electronic properties, **15:**40; 296–297
inhibitor studies, **15:**285–290
number per enzyme molecule, **15:**275–277; 289–290
and pH dependence of enzymes, **15:**282–284
Active site topography, **46:**256
Active site tyrosine, glutathione S-transferases, **69:**16; 17–18; 23
Active succinate, in synthesis of lysine, **16:** 296–297
Active sulfate, *See* Adenosine-3′-phosphate-5′-phosphosulfate
Active transport
into nucleus, **27:**259–260
reversal, **49:**44
Active transport in permeability, **13:**29–33; 37–39; 45–46; 47; 48–49; 51–57; 63
Activity
catalytic, **34:**472–474; 480
coefficient, **34:**463
of pancreatic lipase, **34:**488
reconstituted, **34:**480
enzymic, effect of, chemical modifications, **24:**105–160
ionizing radiations, **24:**359–413
of RNase, effect of urea on, **24:**215; 221; 237; 247; 249–250
of heme proteins and changes of side chains, **7:**322–324
of invertase, action of urethan, **7:**259

Activity (*Continued*)
of tetrapyrroles and side chain composition, **7:**314–315
Activity (biological) and protein structure, **14:**262
effect of adsorption on, **14:**332–333; 337
of enzymes, assay during isolation, **14:** 380–383
Activity of drugs, **1:**193
Activity of enzymes, and —SH groups, **11:** 229–232; 232–244
tests for, **11:**235–239
Activity of protein as enzyme, relation to amino acid composition, **15:**36–37; 296–297; 430
Activity-structure relationships of RNase, **24:**243–252; 448–455
Activity-titration curves, for reconstitution study, **28:**215–217
Actomyosin, **6:**151; 157–158, **43:**27; 36. *See also* Actin, Muscle and Myosin; Actin and Myosin
and ATP in muscle contraction, **13:**159–161
and contraction in muscle, **16:**346–347
ATPase of, **16:**336; 338–339; 345
contractility, **16:**331; 339–343
effect of ATP on, **16:**337–339; 344–345
effect of polyphosphates on, **16:**333
and relaxation of muscle, **16:**347–351
gels, interaction with ATP, **13:**173–178
heat sensitivity, **6:**163
importance of —SH groups, **11:**230–232
and product inhibition, **25:**231
sols, interaction with ATP, **13:**171–173
threads, **6:**178
viscosity, **6:**180–181
Actomyosin, SH-group, and contraction, **27:**215
Actomyosin and sperm motility, **24:**42
Actomyosin ATPase, **43:**28
Actotropomyosin, as contractile system of uterus, **16:**331
N-Actyl-D-galactosamine, linkage to sialic acid, **20:**135; 142; 143
N-Actyl-D-glucosamine, in formation of N-acetylneuraminic acid, **20:**140
Acute intermittent porphyria, *See* Prophyria
Acute lymphoblastic leukemia (ALL), asparagine synthetase metabolism, **72:**146; 151–153
Acyclovir, **54:**63
Acylamidases, classification, **14:**224; 247
Acyl anion, **58:**2
Acylase, and ATP synthesis, **23:**330–331
Acylase (hog kidney), imidazole in, **20:**76; 90
Acylases, penicillin, **28:**245
Acylated amino acids, action of peptidase on, **12:**235–237
Acylated dipeptides, action of carboxypeptidase on, **12:**233–235
Acylating agents
relation to pantetheine, **14:**58; 61; 63
thiol esters as, **16:**18–19
Acylation, *See* Acetylation, Coenzyme A, and various Acyl derivatives
effect on
pepsin, **24:**136
phosphatases, **24:**137–138
RNase, **24:**184; 222
trypsin, **24:**116–124
in identification of imidazole, **20:**68; 70; 72–75; 81; 82; 99
inactivation by, **24:**459–460
Acylation of amino groups of virus (TMV) mutants, **15:**474
Acyl azide, **34:**451
hydrophobic chains, **34:**489
Acyl carrier protein, **34:**18; 19
exopolysaccharide alginate synthesis, **70:**234
Acyl Carrier Protein (Prescott and Vagelos), **36:**269
Acylcholine acylhydrolase, *See* Cholinesterases, plasma
AcylCoA
binding site, **67:**396–398
catalytic efficiency, **67:**395–396
functional segregation in *S. cerevisiae,* **67:**411–412
Nmt binding domain, **67:**404
Nmt substrate specificity, **67:**394–399
Acyl-CoA, and oxidation of hydrocarbons, **27:**485; 486; 490–491; 495; 502; 512–513
Acyl CoA: cholesterol O-acyltransferase (ACAT), cholesterol ester hydrolases, **69:**46
Acyl-CoA dehydrogenase, β-oxidation, **26:**367
Acyl CoA dehydrogenases, as metalloflavoproteins, **17:**266; 285
Acyl CoA dehydrogenases, EPR, study, **27:** 131

Acyl CoA fatty acids, inhibitors of glucose-6-P dehydrogenase, **28:**117
AcylCoA:Nmtlp catalytic intermediate, **67:** 388–391
Acyl-CoA regulatory site of pyruvate carboxylase, **42:**37
Acyl-CoA synthetase, localization of, **24:**334; 338
Acyl coenzyme A, and glycine metabolism, **23:**455–456
Acyl coenzyme A derivatives, **15:**123; 129–131; 184; 185; 197–201; 218; 219; 233; 234; 236; 239; 249; 251–254; 256, **18:**208
S-Acyl coenzyme A derivatives, *See* Acyl coenzyme A derivatives
Acyldehydrogenase, EPR study, **27:**183
Acyl dehydrogenase, in mitochondria, **21:**74
Acyl dehydrogenases, substrate binding in, **22:**390–392
Acyl-enzyme, **52:**181, **53:**256; 257; 285
Acyl-enzyme intermediate, **48:**336
Acyl enzyme mechanism, and ATP synthesis, **23:**329–333; 376; 381
Acyl enzymes, **43:**308
  product inhibition, **25:**190
Acyl esterases, penicillins as substrates, **28:**238; 250
N-Acylglutamate derivatives, in pyrimidine biosynthesis, **21:**272
Acyl glutathione, **15:**259
N-Acyl imidazoles, preparation and properties, **20:**82–85; 87–89; 91; 92; 310
Acyl lipoic acid, in citric acid cycle, **15:**184; 219–220
ε-*N*-Acyllysine, in transaminations, **16:**232
Acyl mercaptan group, role as energy-rich group, **12:**329–330
Acyl mercaptides, **15:**189; 196; 198; 199; 253; 256
  hydroxylamine test for, **15:**190; 220; 226–227; 230; 231; 236–237; 239–240; 246–247; 250–251
Acyl-5-nucleotidase, column chromatography, **22:**483; 484
Acyloin condensation, **8:**389–391
Acyloin synthesis, **15:**209; 218–219
  mechanism of catalysis, **14:**170
Acyl phosphate, **43:**356, **51:**93; 95; 220
  in ATP synthesis, **23:**327; 336–339
Acyl phosphate, and oxidation of hydrocarbons, **27:**485
Acyl phosphate anhydrides, imidazole in formation of, **20:**91
Acyl phosphates, **15:**114. *See also* Phosphate bond, energy-rich
  in glutamine synthesis, **13:**276–278
  in peptidic linkage formation, **13:**311–315
Acyl-P intermediate of glutamine synthetase
  borohydride trapping, **72:**26–28
  positional isotope exchange, **72:**28–29
Acyl pyruvase, **16:**235
Acyl-pyruvase, **14:**107
Acyl-S-lipSH, *See* Acyl lipoic acid
Acyl transfer, **15:**209; 216; 219; 230; 248
  by esterases, **20:**301–304; 311; 315
    with imidazole and imidazole derivatives, **20:**59; 72–75; 85; 88; 91; 315
Acyl-transfer, **53:**258
Acyltransferases, location in subcellular fraction, **24:**310–311
N-Acyltriamines, toxicity of, **36:**238
Acyltrypsine, **24:**134–141
Adamantanecarboxylate, **58:**19
Adamantane group, **58:**21
Adaptation
  of enzymes attacking hydrocarbons, **10:**449; 464
    of sucrose phosphorylase, **10:**124–125
    of tryptophanase, **10:**56; 75–81
  enzymic, **18:**135; 247–248; 250
  genetic studies, **17:**416–417; 455–457
    respiratory, **17:**283–284
  by *Streptococcus hemolyticus,* 254, enzymic, **13:**67–119
Adaptation, and Enzyme Problems (Sevag), **6:**33
Adaptation experiments, in hydrocarbon oxidation, **27:**478; 481; 485; 487
Adaptins, adenosine receptors, **69:**113
Adaptive enyzme, xanthine oxidase as, **17:**322
Adaptive enzyme formation, **16:**252; 297; 377–379; 460; 492
Adaptive enzymes, **6:**34–35; 67–86
  bacterial amino acid decarboxylases, **6:**7–8
  in decomposition of creatine derivatives, **6:**80–81
  dihydroxyacetone zymase, **6:**78–79
  formation, **6:**67–68; 91
  formic hydrogenlyase, **6:**79–80
  galactozymase, **6:**69–78
  inactivation reversal, **6:**82–84
  melibiose zymase, **6:**78

Adaptive enzymes formation, **14:**377
Adaptive nature of hydrocarbon oxidizing enzymes, **27:**475–477; 494–495; 499; 519
Adaptive response to substrate, of β-glucuronidase, **16:**402
Adaptochrome, **46:**93
Adding enzymes, classification, **14:**221; 222; 223
Addison's disease, **2:**344; 345; 347
Addison's disease, pigmentation of, **14:**91; 94; 96
Additives, effect on enzyme inactivation by radiation, **24:**373–375; 378
Adducts, *See* Solubilization of matter
Adducts, NAD, **46:**239
  reduced, structures of, **46:**251
Adenase, **9:**442; 498, **17:**320
  classification, **14:**246; 250
Adenine, **18:**164; 222, **35:**469; 470, **38:**215; 216
  in air-cured tobacco, **6:**348
    inhibitory analogue of, **6:**134; 139
  in cellular transport, **54:**250
  deamination, **60:**14
  effect on ribose synthesis, **28:**408–409
  eukaryotic dihydrofolate reductase, NADPH binding, structural changes induced by, **70:**45
  function in coenzyme I, **15:**31
    of tobacco mosaic virus, **15:**460
  genetic studies, **17:**409; 426–428; 430; 454
  from hydrolysis of CoA, **16:**4
    of DNA, **16:**431–433; 457
    incorporation of C^{14}-formate, **16:**457; 458
    of nuclear, RNA, **16:**437; 439
    in transamination, **16:**237
  hypochromicity in glutamate dehydrogenase complexes, **39:**372
  in pterin synthesis, **21:**393
  synthesis of, **54:**247
Adenine, Cotton effect, **27:**55; 56; 58; 94
N^{15}-Adenine, incorporation into DNA, **16:**458
Adenine aminase, classification, **14:**246; 250
Adenine amino group, of FAD, in coenzyme binding, **22:**389–393
Adenine (C^{14})
  and ATP, **18:**231
  pteridine formation, **27:**423
  in transglycosylations, **17:**221
Adenine-8-C^{14}, incorporation into RNA and DNA, **16:**455
Adenine catabolism, **56:**92
Adeninecobamide coenzyme, electrophoresis, **26:**250
  role in isomerizations, **26:**264; 265
Adenine deaminase, **74:**72
  localization of, **24:**328; 341
Adenine in myosin, **13:**166
Adenine nucleotide binding, energy-dependent, **49:**264
Adenine nucleotide of DPNH, and coenzyme binding, **27:**61–64
Adenine nucleotides, **49:**256–258, **55:**298–306
  effect on sperm, **24:**85
  as general ligands, **46:**208–221
  inhibition of ATPase activity of F_1, **49:**245–249
  in mitochondria, **21:**10; 100
    in mitochondrial fragments, **21:**100
  in phosphate transfers, **17:**163; 164; 167; 176–186
Adenine ribonucleotide of DPNH, enzyme binding, **27:**62
Adenocarcinoma, **54:**65, **58:**68
  phosphatide content, **7:**168
Adenohypophyseal hormones, synthesis of peptides related to, **29:**392
Adenosine, **35:**478, **54:**64, **57:**162, *See* Nucelosides
  in *S*-adenosylmethionine biosynthesis, **22:**245
  cardiovascular effects of, **69:**84
  catabolism of, **54:**232
  containing nucleotides, **57:**143
  deaminase action on derivatives, **15:**80; 106–108
    metabolism in red cells, **15:**163
    in pyruvate oxidation, **15:**214
  3′5′-diphosphate, in CoA structure, **16:**14
  and DR synthesis, **28:**422
  effect on LADH-DPNH complex, **27:**61–64
    effect on Zn exchange of LADH, **27:** 64–65
    zinc complexing, **27:**69–70
  effect on sperm, **24:**85
  fluorinated analogs of, **55:**271
  relation to CoA, **16:**4; 7
Adenosine 3′,5,′-cyclic phosphate, **55:**312
Adenosine aminase, classification, **14:**246
Adenosine 3,5′-cyclic monophosphate, **50:**42; 43; 304; 306

Adenosine deaminase, **9:**488; 491; 495–499
adsorption studies of, **14:**333–334; 346; 363–364
structure, studies, **14:**308
in DR synthesis study, **28:**425
hydrogen transfers, **26:**116
intracellular distribution, **16:**463; 464; 466–468
localization of, **24:**328; 341
Adenosine deaminase (ADA) deficiency, **51:**167–210
adenine and deoxyadenine nucleotide content of RBC, **51:**175
biochemical accompaniments, **51:** 170–174
cell culture models, **51:**182–184
clinical manifestations, **51:**169–170
consequences of, **51:**169–174
and immunocompetency, **51:**198–199
immunodeficiency and, **51:**184–196
introduction to, **51:**168–169
physical-chemical properties, **51:**174–179
in purine salvage pathway, **51:**179–182
significance to biomedical sciences, **51:** 199–202
therapy, **51:**196–198
Adenosine Deaminase Deficiency and Severe Combined Immunodeficiency Disease (Thompson and Seegmiller), **51:**167
Adenosine deaminases, effect on 5,-methylthioadenosine, **25:**386
multiple forms, **25:**360
Adenosine diphosphatase, *See* ADPase
Adenosine diphosphate, *See* ADP
Adenosine diphosphate choline, synthesis by carbodiimide method, **22:**188
Adenosine diphosphate glucose, *See* ADP-Glucose
Adenosine Diphosphate Glucose Pyrophosphorylase, Regulation of (Preiss), **46:**315
Adenosine diphosphate glucose pyrophosphorylases, *See* ADP-Glucose pyrophosporylases
Adenosine-5′-diphosphate sulfurylase, **22:**209
in arsenolysis of APS, **22:**209
assay for, **22:**209
Adenosine diphosphoribose, in allosteric regulation, **28:**138
Adenosine kinase (AK), **51:**180–181
Adenosine-mediated increases in cAMP, **51:**188–190
Adenosine-mediated pyrimidine starvation, **51:**184–188
nucleotide biosynthesis, **51:**185
Adenosine monophosphate, *See* AMP
3′,5′-Adenosine monophosphate, *See* Cyclic adenosine monophosphate
Adenosine 3′,5′-monophosphate, cyclic, stimulation by ACTH of, **29:** 468–469
Adenosine 5′-monosulfate (AMS), in ternary complex formation, **27:**5–7
Adenosine nucleosidase, **54:**235
Adenosine nucleotide, in regulation of fructose-bisphoshate, **54:**135–142
AMP inhibition, **54:**136–140
ATP and ADP inhibition, **54:**140–141
by energy charge, **54:**141–142
Adenosine-5′-phosphate. *See also* Adenosine monophosphate and Adenylic acid
in citric acid cycle, **15:**184; 187; 227–229; 231; 234; 249; 251–254
2′-Adenosine phosphate, **15:**83
Adenosine-3′,5′-phosphate, action in phosphorylase *a* and *b* interconversion, **24:**279
hormonal regulation, **24:**286
Adenosine 5′-phosphate, from CoA, **16:**13; 14
Adenosine 5′-phosphate-deaminase, **16:**13
Adenosine-3′-phosphate-5′-phosphosulfate (PAPS), **22:**210
enzymic sulfate transfer, **25:**440
in estrone sulfokinase reaction, **22:**212
in phenol sulfokinase reaction, **22:**211; 212
as sulfate donor, **22:**214
sulfate transfer from, **22:**210–214
synthesis, **22:**207–210
Adenosine phosphates
from CoA, **16:**12–16
effect on sperm, **24:**85
and luciferin, **25:**129–137; 140–144
Adenosine-5′-phosphates, synthesis by phosphoramidate method, **22:**193; 194
Adenosine-3′-phosphate-5′-sulfatophosphate, structure, **22:**192
synthesis, **22:**192
Adenosine-5′-phosphosulfate, phosphorylation with APS phosphokinase, **22:**209; 210

Adenosine-5′-phosphosulfate phosphokinase, **22:**209; 210
reaction characteristics, **22:**209
reactions, **22:**207
specificity, **22:**209
Adenosine polyphosphates, **1:**109; 111; 112; 114–116. *See also* ATP; Phosphate bond, energy-rich
in glutathione synthesis, **13:**280–282
in synthesis of glutamine, **13:**276–280; 290–291
and transferase action, **13:**290–291; 297; 299; 303; 305; 311–315
Adenosine-5′-polyphosphates, synthesis by phosphoramidate method, **22:**199
Adenosine-5′-pyrophosphate. *See also* Adenosine diphosphate
structure, **22:**167
synthesis, by mixed anhydride method, **22:**166; 167
by phosphoramidate method, **22:**198
Adenosine receptors, **69:**83–120
alignment of subtypes, **69:**85–87
future research directions, **69:**105–115
function regulation, **69:**111–115
gene expression regulation, **69:**110–111
ligand binding and G-protein activation, identification of domains, [illegible]
novel forms: isolation and delineation of functions, **69:**105–108
generally, **69:**83–84
pharmacology of, **69:**84–85
primary structure of, **69:**85–91
post-translational modifications and, **69:**89–91
sequences common to other GPCRs, **69:**85–89
structure to function relationship, **69:**91–105
G-protein activation, **69:**94–99
ligand binding, **69:**91–94
regulation of expression and function, **69:**99–105
Adenosine-5′-sulfatophosphates, synthesis by carbodiimide method, **22:**192
Adenosine tetraphosphate, **43:**23
Adenosine triphosphatase, **5:**134, **10:**300, **15:**143; 250; 330; 334; 336; 349; 382, **58:**161
and cell structure, **8:**31–32
calcium-activated, **8:**45
as coupling factor 1, **23:**370–371; 378–382
mitochondrial, **23:**360–363
multiple, **23:**362–363
and oxidative phosphorylation, **23:**331; 332; 341; 348; 355–360; 364–369; 371; 372; 374–375; 378–383; 385
in photophosphorylation, **23:**386
effect of chloroacetophenone, **11:**239
in myosin preparations, **11:**230–232
thiol and enzyme activity, **11:**234–235; 243–244
effect of radiation on, **24:**388
localization of, **24:**329
in sperm, **24:**42; 87; 88; 99
ion pump, **27:**259–260
in mitochondria, **21:**30–32; 36; 37; 39; 40; 58; 74; 117; 118
effect of Versene on, **21:**5
and myosin, **4:**216; 217; 218; 220; 238
in normal and cancer tissues, **4:**243–244
Adenosine triphosphatase, B, solubilization of, **27:**265–266
Adenosine triphosphatase, in plastids, **20:**214
Adenosine triphosphatase activity and fertilization, **9:**310; 318
Adenosine triphosphatase (ATPase), **7:**17; 115
Caulobacter cell cycle, transcription regulation, **71:**18
Adenosinetriphosphatase in myosin, **6:** 147–189.
Adenosinetriphosphatase Properties of Myosin (Engelhardt), **6:**147
Adenosine triphosphatases, in plants, **18:**311
Adenosine triphosphate, **5:**122; 123; 131; 132; 133; 142; 144; 145, **68:**70. *See also* ATP; ATP; Growth control
absorption spectrum, **9:**487–488
and acetate metabolism, **23:**407; 432
biosynthesis of, **23:**323–399
chemistry of, **23:**325–327
free energy of hydrolysis, **23:**326
and glycine metabolism, **23:**454; 461
and glycolate metabolism, **23:**439; 444; 446
labeled, preparation of, **23:**326–327
and oxalate metabolism, **23:**447–449; 452
of photosynthetic bacteria, **23:**420; 424
in *S*-adenosylmethionine synthesis, **22:**244
in APS phosphokinase reaction, **22:**209
in cholesterol biosynthesis, **22:**296; 297; 301; 302; 304

in farnesyl pyrophosphate synthesis, **22:**319
fission with ATP sulfurylase, **22:**208; 209
in squalene biosynthesis, **22:**298
structure, **22:**168
synthesis, by carbodiimide method, **22:**186
by mixed anhydride method, **22:**167; 168
by oxime sulfonate method, **22:**187
by phosphoramidate method, **22:**198; 199
in amino acid and amide biosynthesis in plants, **20:**325; 331
in phospholipid synthesis in plastids, **20:**224
relation of synthesis to respiratory chain, **20:**194
in uridine diphosphogalactose metabolism, **20:**112; 122
aspartate transcarbamylase binding
asymmetry, **68:**94; 96
domain separation changes, **68:**105
from aspartyl phosphate, **18:**226
and lipoic acid, **18:**340; 342
as phosphate donor, **18:**195–197
and phrophosphorolysis, **18:**202–205; 208–214; 219–221
in plant respiration, **18:**287; 289; 302; 311–313
reaction with nicotinamide mononucleotide, **18:**193
synthesis by chloroplasts and mitochondria, **18:**58–59
binding in T form, **68:**124–125
breakdown and glycolysis, **4:**219; 225
and growth control, **4:**218–223; 225; 226
and pharmacological control, **4:**219; 224–225
chelates, in metal ion activation of enzymes, **21:**147; 148; 233; 239
in arginine synthesis, **21:**255
in carotenogenesis, **21:**346
exchange reaction with ADP, in amide bond synthesis, **21:**235; 236; 239; 240
exchange reaction with inorganic phosphate, in mitochondria, **21:**36; 74
in mitochondrial fragments, **21:**39
in purine biosynthesis, **21:**213; 234; 235; 239; 240; 247
in folic acid biosynthesis, **21:**394
hydrolysis, by metal ions, **21:**176; 193
in mitochondrial reactions, **21:**25; 27; 30; 42; 47–54
in purine biosynthesis, **21:**203–205; 207–209; 211; 213–216; 220; 229; 232–241; 247; 248; 250–252; 401
in pyrimidine biosynthesis, **21:**264; 267; 268; 271; 272; 278; 279; 281; 282; 285; 289; 421; 422
in reactions involving folic acid coenzymes, **21:**225; 226; 404; 405; 407–409; 412; 413; 424; 425
synthesis, with phosphatase models, **21:**192
conformational control, **68:**98
deamination, **9:**495–499
and dietary purines, **9:**463
effects on mold metabolism, **9:**570
as energy reservoir, **4:**218
energy supply for glycoside bond formation, **10:**128
in citric acid cycle, **10:**189–190; 214; 221–222
and fatty acid oxidation, **15:**123
in carbohydrate metabolism, **15:**141; 143; 155; 170; 172–173
in citric acid cycle, **15:**184–189; 201; 225–236; 248; 252–254 (P^{32}); 258; 262
phosphorylation, of hexoses, **15:**145–148
of hexose monophosphate, **15:**151–152
production by oxidative phosphorylation, **15:**129–131; 177–178
in synthesis of active methionine, **15:**111
diphosphopyridine nucleotide, **15:**81–82; 104–105 ($P^{32}P^{32}$)
nicotinamide mononucleotide, **15:**103
triphosphopyridine nucleotide, **15:**105
urea, **15:**320; 327–337; 339; 344–352; 363–369; 372–373; 379–383
and fertilization, **9:**251–252; 254; 274; 278; 285; 310–311; 318
and glutamine synthesis, **9:**413
in glycogen synthesis, **4:**216; 218; 222
heterotropic effects, aspartate transcarbamylase, **68:**130–135
interaction with aspartate transcarbamylase, **68:**101–103
and ketone body formation, **8:**401
methods of assay, **4:**238
and nucleic acids, **9:**505–507
priming reactions
phosphorylation of glucose, **4:**213–214
phosphorylation of glycerol, **4:**213

Adenosine triphosphate (*Continued*)
and semen metabolism, **9:**340–342; 362–364; 372
and staphylococci, **4:**186
synthesis, **4:**214–216
in synthesis of glutathione, **11:**249
utilization, **4:**216–218
Adenosine Triphosphate, Mechanisms of Synthesis of (Racher), **23:**323
Adenosine triphosphate-activated reactions, **18:**229–231
Adenosine Triphosphate and Structural Proteins in Relation to Muscle Contraction (Needham), **13:**151
Adenosine triphosphate enzymes, role of sulfhydryl group, **15:**11
Adenosine triphosphate (P^{32}), **18:**199–200
Adenosine triphosphate (P^{32}) in phosphorylase *b* to *a* reaction, **24:**269–270; 277–279; 286
in polypeptide synthesis, **24:**2
in preparation of phosphorylase *a*, **24:**265; 269; 275; 276
in sperm, **24:**40–42; 88
pathways for regeneration, **24:**45–89
Adenosine triphosphate ($P^{32}P^{32}$), **18:**202–203
Adenosine triphosphate sulfurylase, **22:**208; 209
assay for, **22:**208
in liver, **22:**209
reaction, **22:**207; 208
Adenosinetriphosphoric acid, **5:**179
Adenosyl cofactor, **66:**22
Adenosylethionine, **54:**209
S-Adenosyl-L-ethionine, **54:**223
effect on enzymes, **25:**381
enzymic cleavage of, **25:**385
enzymic synthesis of, **25:**370; 382–388
S-Adenosylethionine, as ethyl group donor, **22:**272
natural occurrence, **22:**273
structure, **22:**272
Adenosylethionine, transethylation of mercaptans by, **25:**421; 422
Adenosylhomocysteine, **54:**197; 201; 223
analogs of, **54:**211
in methylthioadenosine comparison, **54:**209
and spermidine synthase, **54:**219
S-Adenosylhomocysteine
in *S*-adenosylmethionine synthesis, **22:**245
metabolism, **22:**267
in methyl synthesis, **25:**402–404; 408
S-Adenosylhomocysteine, **51:**181; 184; 187
adenosine-mediated accumulation of, **51:** 194–196
in methionine synthesis, **21:**425; 426
S-Adenosyl-L-homocysteine, **42:**243; 252
S-Adenosyl-L-methionine, **46:**69–70
Adenosylhomocysteine enzyme, methyl synthesis, **25:**404; 413
Adenosylmethionine, **54:**196
Adenosyl methionine, **58:**112
(S)-Adenosyl methionine, **35:**470
Adenosylmethionine
analogs of, **54:**209–210
azetidine-2-carboxylic acid from, **54:**226
in cells infected by T phage, **54:**224
chemical structure of, **54:**199
concentration of, in tissues, **54:**216
decarboxylated, **54:**200; 217; 220, **56:**268
stability of, **54:**202
enzymatic conversion of, to methythioadenosine, **54:**231–232
enzymatic decomposition of, **54:**222–227
historical perspectives on, **56:**252
hydrolysis of, to methylthioadenosine, **54:**200–201; 202
mechanism of action of, **56:**253
as parent compound of methylthioadenosine, **54:**199
selenium analog of, **54:**[illegible]
S-Adenosylmethionine, **18:**202; 211, **36:**7, **43:**225
decarboxylation of, **36:**214
in methionine synthesis, **21:**426
specificity of, **36:**6
in spermidine biosynthesis, **36:**214
S-Adenosylmethionine, **15:**112
as adenosyl donor, **66:**23
as alkylating agent, **66:**23
anaerobic *E. coli* activation, **66:**34–35
in anserine synthesis, **22:**256
biosynthesis, **22:**241
from *S*-adenosylhomocysteine, **22:**245
from methionine and ATP, **22:**243–245
chemical properties, **22:**250
in choline synthesis, **22:**256
in creatine synthesis, **22:**255
decomposition products, **22:**249
discovery, **22:**241; 242
enzymic preparation, **22:**247
enzymic splitting, **22:**259

mechanism, products, **22:**259; 260
in ergosterol synthesis, **22:**257
isolation from yeast, **22:**247; 248
lability, **22:**249
in 2-methylamino-6-aminopurine synthesis, **22:**257
in methylation, of epinephrine, **22:**257
of histamine, **22:**255
of norepinephrine, **22:**256
in *N*-methylnicotinamide formation, **22:**255
in microorganisms, **22:**269; 270
modes of splitting, **22:**249; 250
reaction sites, **22:**249; 250
in spermine, spermidine biosynthesis, **22:**261; 262
stereoisomers, **22:**246
structure, **22:**241
in transmethylation of homocysteine, **22:**257
utilization by methionine mutants, **22:**269
cleavage
5′-deoxyadenosine and methionine, **66:**22
lysine 2,3-aminomutase, **66:**99
formation, **16:**275
formation of, **29:**137
hydrogen transfer, **66:**6–9
iron-sulfur cluster reactions, **66:**24–26
as methyl donor, **25:**396; 408; 410–427
methyltransferase reactions, **25:**413–423
radical enzyme activation, **66:**33–36
S-Adenosylmethionine: protein methyltransferase, distribution of various, **42:**249
S-Adenosylmethionine decarboxylase, **36:**231
in rapidly growing tissues, **36:**249
in *S. cerevisiae,* **36:**220
S-Adenosylmethionine:guanidinoacetate N-methyl-transferase, **50:**180; 190–191
biological distribution, **50:**180
S-Adenosylmethionine methyl transferase, **48:**7
S-Adenosylmethionine synthetase, **43:**259, **48:**7
destabilization mechanism, **43:**261
S-Adenosyl-L-methionine, **42:**265
effects on enzymes, **25:**381
enzymic cleavage of, **25:**385
K_m value, **42:**270
A-Adenosyl methionine, **46:**58
S-Adenosyl-L-methionine: 7-keto-8-aminopelargonic acid amino transferase, **54:**227–228
S-Adenosyl-L-methionine analogs, enzymic cleavage of, **25:**385–386
Adenosylmethionine (C^{14}), as methionine source, **25:**404
methylation by, **25:**418–419; 422
S-Adenosylmethionine-cleaving enzyme, steric requirements for, **25:**385
Adenosylmethionine decarboxylase, **54:** 218–219
in animal tissues, **56:**270–275
bovine lymphocytes, **56:**272
distribution of, **56:**272–273
half life of, **56:**273
hormonal control of, **56:**275
inhibitors of, **56:**271
in mitogen-activated lymphocytes, **56:**275
mouse and calf liver, **56:**272
mouse mammary gland, **56:**272
ornithine decarboxylase inhibitors and, **56:**274
prostate and sea urchin eggs, **56:**272
purification and characterization of, **56:**270–273
rat muscle, **56:**272
turnover and control of, **56:**273–275
bacterial, **56:**266–269
cation requirement of, **56:**267
covalently-bound pyruvate of, **56:**267
enzyme mechanisms of, **56:**268
genetic studies of, **56:**268–269
inhibition of, **56:**268
kinetics of, **56:**268
molecular weight of, **56:**267
purification and characterization of, **56:**266–268
substrate specificity of, **56:**268
historical perspectives on, **56:**266
pyruvate in, **36:**214
in spermidine biosynthesis, **36:**214
in spermidine and spermine biosynthesis, **56:**265–266
structural formula for, **56:**265
of yeast, **56:**269–270
genetics of, **56:**269–270
purification and characterization of, **56:**269
Adenosylmethionine-homocysteine transmethylase, **21:**426

S-Adenosylmethionine-requiring mutants, **22:**269
Adenosylmethionine synthetase, **56:**252; 253
S-Adenosylmethionine synthetase, reactions catalyzed by, **29:**137
Adenosylobalamin
  cofactor for rearrangements, **66:**7
  5′-deoxyadenosyl radical, **66:**27–28
  evolutionary predecessors, **66:**32–33
  hydrogen transfer, **66:**27–28
  radical intermediates, **66:**9–13
  as rearrangement cofactor, **66:**2
S-Adenosyl-*S*-*n*-propyl-L-homocysteine, **54:**209
Adenosymethionine nucleoside hydrolase, **54:**234
Adenoviruses, **52:**141
Adenyl acetate, **17:**181–182
  in ATP-acetatecoenzyme A system, **18:** 203–205; 208
Adenyl acylates, in fatty acid activation, **18:**208
Adenylase, **6:**149
2′-Adenylate, **15:**117
Adenylate cyclase, **43:**7; 17, **50:**304; 312, **53:** 8–11, **58:**119
  activation of, **58:**126
  activity, **50:**306
  GTP-binding protein, **50:**312
  muscarinic inhibition of, **55:**147–148
  NaF stimulation, **53:**10
  nucleotidyl transfer reactions with, **55:**340
  reconstitution (*See* Reconstitution of adenylate cyclase)
  solubilization, **53:**10
  thermodynamic studies, **53:**9
Adenylate deaminase, allosteric regulation, **28:**122–123
Adenylate kinase, **18:**196–197; 220, **37:**460; 505; 568; 600, **43:**13; 22; 23; 24; 32; 35; 353, **46:**246, **47:**58; 88–92; 94; 96; 124; 135
  adenosine monophosphate-binding site of, **59:**261–262
  and amino acid
    near MgATP, **59:**260
    residues, **59:**257
  and ATP, **23:**326; 327; 358; 359
  and biotin enzymes, **26:**312
  and bound mgATP, **59:**257
  classification, **14:**251
  isolated metal-ATP binding site of, **59:**255
  localization of, **24:**313; 328
  mechanism of deduced from relaxivity data, **33:**43
  and MgATP-binding site of enzyme and peptide fragment, **59:**253–257
  and MgATP conformation, **59:**256–257
  in mitochondria, **21:**31; 53; 58; 59
  NMP kinase primary structure, **73:**104
  and peptide, **59:**1–45; 255–256
  and positioning of MgATP and X-ray structure, **59:**257; 259–261
  and reaction coordinate distance, **59:**262
  reactions of thiophosphate analogs with, **43:**14
  sequence homologies of
    with F_1 adenosine triphosphate, **59:** 262–263; 265–267
    with other nucleotide-binding proteins, **59:**262–263; 265–267
    with Ras p21, **59:**262–263; 265–267
  with stereochemical inversion, **59:**253
Adenylation-deadenylylation cascade, regulation of bacterial form glutamine synthetase, **72:**32
Adenyl carbonate, **18:**202; 210
Adenylcitrulline, **18:**211–212
Adenyl cobamide (AC) coenzyme, absorption spectrum, **26:**348; 360
  effect on cobamide enzymes, **26:**346–347; 359–360; 366
  structure, **26:**362–363
Adenylic acid *a,* **15:**83
Adenylic acid deaminase, **9:**495–499
5-Adenylic acid deaminase, column chromatography, **22:**461
Adenylic acid deaminase in chromidia, **8:**31
Adenylic acids, **5:**131; 145, **15:**80–82; 83; 91; 106–108; 117; 130; 305, **19:**257. *See also* Adenosine monophosphate and Adenosine-5′-phosphate; AMP
  and acetyl phosphate anabolism, **6:**242
  biosynthesis, **21:**204; 228; 229; 232; 243; 253
    allosteric regulation, **28:**66–69
      from glucose-2-^{14}C, **28:**409
    in carotenogenesis, **21:**321; 344
    in histidine biosynthesis, **21:**403–405
    hydrolysis, by metal ions, **21:**184; 193
    in mitochondrial oxidations, **21:**31
  and depolarization, **6:**186

desaminase activity in myosin, **6:**170
formation from ATP, **6:**161–162
muscle, reaction with borate, **12:**514–515
in normal and cancer cells, **4:**208; 210; 214; 216; 222; 223; 226; 227; 238
phosphorylase reaction with, **24:**274–275
relation to amide ammonia, **13:**267
RNA, uptake of P^{32}, **16:**461–462
and semen metabolism, **9:**353
Adenylic acids, and phosphorylase *b,* **11:**304
Adenylic acids *a* and *b,* from ribonucleic acid, **16:**14
Adenylic aminases, classification, **14:**246
Adenylic deaminase in muscle, **13:**168–169
Adenylic kinase, in photosynthesis, **19:**10
Adenylic-3-phosphatase, intracellular distribution, **16:**465
Adenylic-5-phosphatase, intracellular distribution, **16:**465
Adenylleucine, **18:**202
Adenyl-L-methionine, **18:**209
Adenyl methylenediphosphonate, **43:**3
allosteric effectors, **43:**7
Adenylopyrophosphatase, *See* ATPase
Adenylosuccinase, **21:**217; 228; 254
in *E. coli,* **22:**33
in *Neurospora,* **22:**12; 29; 33; 34; 37
Adenylosuccinate lyase, **37:**446; 524; 568, **74:**314; 315
antagonistic effect, **37:**567
sequential response, **37:**567
Adenylosuccinate synthetase, **73:**57–102
allosteric regulation, **28:**122
gene structure and expression, **73:**86–94
bacteria, **73:**86–90
Dictyostelium, **73:**91
human, **73:**93–94
mouse, **73:**91–93
yeast, **73:**90–91
heavy metal tolerance, **73:**94
kinetics and mechanism, **73:**72–77
ligand binding sites, **73:**64–72
metal ions, **73:**77–79
properties, **73:**60–61
regulation, **73:**84–86
review, **73:**57–60
structure and conformational changes, **73:** 61–64
substrates and inhibitors, **73:**79–84
Adenylosuccinic acid, in purine biosynthesis, **21:**204; 228; 229; 253; 254
Adenyloxyluciferin, **18:**212
Adenyl pantoate, **18:**209
Adenyl polyphosphates, classification in enzymes system, **14:**241
Adenyl pyrophosphatase, **18:**118
classification, **14:**248
gasometric methods, **9:**601
role of metal ions, **10:**104
specificity, **10:**86
Adenyl pyrophosphate
and acetic acid fixation in pyruvic acid, **6:**245
in choline acetylation, **6:**260
in sulfanilamide acetylation, **6:**258; 259
Adenyl selenate, **18:**210
Adenyl sulfate, **18:**209–210
Adenyl-D-thiomethylribose from yeast, **10:**243
Adenylylcyclase, localization of, **24:**315
Adenylypyrophosphate, turnover studies, **7:**114–126
Aderfeldt test, *N,N′*-demthyl-*p*-phenylene-diamine oxidation in patients with mental diseases, **29:**506
ADH, *See* Alcohol dehydrogenase
Adhesion, **59:**2
Adipate, **51:**236
Adipocytes, isolated, adenosine receptors, **69:**100
Adipose tissue, **59:**372
phosphoenolpyruvate carboxykinase (GTP), **69:**218; 226–227
Adnamine
biosynthesis of, **29:**491
obtained from epinephrine, **29:**491
S-Adnosyl-3-thiopropylamine, **54:**217
AdoMet, *See* Adenosylmethionine
AdoMet 8-amino-7-oxononanoate aminotransferase, **74:**133
phylogenetic distribution of, **74:**157
ADP, **4:**208; 222; 223; 227, **5:**131; 132; 133, **15:**141; 143; 155; 172–173, **23:**323–399, **34:**520, **49:**228; 229; 241; 242; 243; 245; 247; 253; 255; 256; 257; 261, **55:**318, **57:**147, **58:**101
addition to coacervate, **27:**369
effect on LADH-DPNH complex, **27:** 61–64
in allosteric regulation, **28:**137; 138; 140; 141; 143; 144

ADP (*Continued*)
in DR metabolism, **28:**444; 445; 449
calorimetric measurements of binding to glutamate dehydrogenase, **39:**383
chelates, in metal ion activation of enzymes, **21:**147; 148
in citric acid cycle, **15:**184; 185; 188; 201; 205; 225; 227; 229–232; 248–251; 253; 254 (P^{32}); 262
and CO_2 fixation, **12:**51; 56–57
complexed with metal ion, **54:**373
in creatine kinase transition state complex, **54:**414; 415–416
dismutation of, **49:**241–242
effect on mitochondrial coenzyme Q, **21:**112
effect on thread extensibility, **6:**176
in energy transport, **14:**16
as enzyme inhibitor, **25:**223; 229–233; 242–249; 265
enzymic breakdown, **6:**161–162
exchange reaction with ATP in amide bond synthesis, **21:**235; 236; 239; 240
exclusion of guanosine triphosphate from glutamate dehydrogenase surface by, **39:**387; 411
fission with ADP sulfurylase, **22:**209
in formate transfer reactions, **21:**227; 228; 402
formation of ternary complex with glutamate dehydrogenase-L-leucine, **39:**387
in fructose-bisphosphatase inhibition, **54:**140–141
glutamine synthetase activity, **72:**13
imidazole in formation of, **20:**91
inhibition of ATPase activity of F_1, **49:** 245–249
isomerized, **6:**162
labeled
in exchange reactions, **23:**355–359
in glutathione biosynthesis, **19:**254
labeled, in contractile system, **16:**342
in mitochondrial oxidations, **21:**30; 31; 38; 39
in myosin, effect on ATP, **6:**160
in oxidative phosphorylation, **17:**88–112
as phosphate donor, **18:**196–197
in photosynthesis, **19:**23; 40; 63
polynucleotide enzyme, **17:**185
P radioactive
ATP-ADP exchange, **26:**311–312
preparation, **6:**158–159
and pyridine nucleotides, **15:**81; 82; 91; 105
red shift of 260nm band in glutamate dehydrogenase complexes, **39:**382; 384
regulation of bacterial form glutamine synthetase, **72:**33–35
relation to contraction of myosin, **16:**341–342
relation to muscle relaxation, **16:**350–351
reversal of spectral perturbation of glutamate dehydrogenase-NADPH, **39:**377; 381; 405
in schizophrenia, level of red cells, **29:**500
in succinyl-CoA metabolism, **21:**226; 227
synthesis, with phosphatase models, **21:**192
in urea synthesis, **15:**327; 330; 332; 344; 346–349; 365–366; 370; 372; 380–382
[^{3}H] ADP, **49:**242
N^6-(6-aminohexyl)-2′5′-ADP, **46:**208
N^6-(6-aminohexyl)-3′-ADP, **46:**208
ADP*, exchange of P* with ATP, **17:**183
2′,5′-ADP analogs, **46:**208–211
3′,5′-ADP analogs, **46:**208–211
ADPase
in myosin, **6:**170
specificity in adenylic acid formation, **6:**161
ADP^{32}-ATP exchange, **23:**333; 355; 356; 358–359
ADP-ATP exchange enzyme, **23:**359; 360; 371
ADP/ATP ratio, in P-enolpyruvate synthesis, **28:**92
ADP_cP, **35:**483
ADPG, and glucosyl donor in starch synthesis, **34:**314
as a group donor for glycogen synthetase, **34:**316
ADP-glucose, chemical and physical properties, **46:**365–373
ADP-glucose pathway, **46:**318
ADP-D-glucose pyrophosphorylase, action of, **26:**52
in starch synthesis, **26:**61–62
ADP-glucose pyrophosphorylases. *See also* Anti-ADP-glucose pyrophosphorylase
activation specificity, **46:**320–322
activator-inhibitor interaction, **46:**355–365
activators and inhibitors, **46:**320–323
activity in *Aphanocapsa*-6308, **46:**338
of *Aeromonads,* **46:**353
of algae, green, **46:**335
algal, **46:**339
3P-gly cerate activation, **46:**339
phosphate inhibition, **46:**339

of bacteria blue-green, **46:**335
in barley, **46:**336
in butter, **46:**336
chemical and physical properties, **46:**365
of *Chlamydomonas reinhardii,* **46:**337
of *Chlorella pyrenoidosa,* **46:**337
of *Chlorella vulgaris,* **46:**337
of *Enterobacteriaceae,* **46:**314
evolutionary relationships of, **46:**373–377
fructose-P_2 as activator, **46:**332–335
fructose-6-P as activator, **46:**328–332
of higher plants, **46:**335
activation by 3-*P*-glycerate, **46:**335
inhibition by P_i, **46:**335
of kidney been, **46:**336
of leaf, **46:**335; 336; 339
of lettuce, **46:**336
of maize, **46:**336
of maize endosperm mutants, **46:**319
modulation by energy charge state of cell, **46:**319
molecular weight, **46:**366
of peanut, **46:**336
pyruvate, as activator, **46:**328–332
regulation of, **46:**317–381
regulatory properties, **46:**319
of *Rhodospirillum,* **46:**323–328
of rice, **46:**336
of *Scenedesmus obliquus,* **46:**337
of *Serratia,* **46:**352
of sorghum, **46:**335
subunit studies, **46:**366
of sugar beet, **46:**336
of *Synechocossus*-6301, **46:**338
of tobacco, **46:**336
of tomato, **46:**336
ADP-D-glucose pyrophosphorylases, allosteric regulation, **28:**111–112
ADP-hexanolamine-agarose, **52:**148
ADP level, metabolic control by, **17:**66; 124–130
ADP-Mg, enzyme complex, Cotton effect, **27:**59
ADPNP, **35:**483
ADP PS, **35:**483
ADP-pyrophosphorylases, allosteric regulation, **28:**112
ADP r ADP transphosphatases, classification, **14:**251
ADP reductase, in DR metabolism, **28:**444
ADP-ribose, **50:**311, **57:**147
effect on LADH-DPHN complex, **27:**61–64
effect on Zn exchange of LADH, **27:** 64–65
ADP-ribosylation, **50:**304; 310–313
ADP-Ribosylation of Guanyl Nucleotide-Binding Regulatory Proteins by Bacterial Toxins (Moss and Vaughn), **61:**303
ADP-ribosyltransferase, **50:**313
ADP-ribosyl transferase, **58:**101
ADPαS, **55:**300; 303–306
inorganic thiophosphate synthesis with, **55:**319
ADPβS, **55:**299–303
ADP-sugars, **26:**61–62; 66
ADP sulfate, **43:**22
β-AdR, rhodopsin compared, adenosine receptors, **69:**104
Adrenal cortical hormones, **2:**326–352
activity tests, **2:**326; 327; 338–343; 345–347
amorphous fraction, **2:**327; 328; 331; 339
and carbohydrate metabolism, **2:**344
concentrates, **2:**327–329
of cortin activity, **2:**329; 330
isolation, **2:**329–331
nomenclature, **2:**326; 327
physiologically inactive, **2:**331
related steroids, **2:**347–352
of sex hormone activity, **2:**330
and sodium retention, **2:**345–347
steriod constituents, isolation, **2:**329–331
structure, **2:**331–336
synthesis, partial, **2:**336–338
Adrenal Cortical Hormones (Pfiffner), **2:**325
Adrenal corticosteroid biosynthesis, **47:**5
Adrenalectomy, *See* Adrenal cortical hormones, activity test
effect on renin substrate, **7:**524
Adrenal gland
dopa in, **14:**109
and phosphatide formation, **7:**145
Adrenal β-glucuronidase, **16:**376
Adrenaline, **5:**67; 72; 80–81, **34:**331
absorption spectrum, **4:**285; 286
activation of phosphorylase by, in heart, **34:**394
antagonism by ephedrine, **6:**40–41
effect in familial periodic paralysis, **6:**215; 217
effect on actin polymerization, **13:**155

Adrenaline (*Continued*)
effect on myosin ATPase, **13:**167
effect on sodium flux, **13:**52; 55
effect on fructose diphosphatase, **52:**276
effect on gluconeogenesis, **52:**314
effect on glycolytic flux, **52:**315
effect on mammalian liver glycogen synthetase, **34:**342
effect on phosphoenolpyruvate carboxykinase, **52:**298
effect on pyruvate kinase, **52:**304; 315
effect on pyruvate metabolism in mitochondria, **52:**315
effect on pyruvate transport and metabolism, **52:**292
effect on tryptophan peroxidase, **25:**377
effect on tyrosinase, **14:**81; 145; 149
peroxidase action on, **14:**151; 152–153; 155
from phenylalanine and tyrosine, **14:**74; 90; 116–119
and elimination of ketone bodies, **8:**414
epinephrine and glycogen metabolism, **4:**262
formation from tyrosine, **5:**80
hyper-reactivity to, **7:**522
inactivation, **7:**538
identification of its glucuronide, **16:**383
reduction of cytochrome *c*, **17:**350–351
in regulation of glycogen synthetase, **34:**333
and transmethylation, **12:**479
Adrenal medulla, tyrosine hydroxylase, **70:**106
Adrenal-stimulating activity, effect of ACTH chain length on, **29:**457
modification of NH_2-terminal amino acids of ACTH on, **29:**465; 468–469
of ovine, human, and porcine ACTH, **29:**459
of synthetic ACTH peptides with various chain lengths, **29:**460–461
α-Adrenergic agents, **59:**366–367
Adrenochrome, **14:**119, **58:**62
character of, **16:**113
excretion patterns, **29:**517
formation from catecholamines, **29:** 516–517
hypothesis, **29:**517
lack of evidence for formation *in vivo*, **29:**516–517
and schizophrenia, **29:**507
inhibition of ATPase by, **13:**167
from peroxidase action, **14:**152; 153
in tyrosinase action, **14:**149
Adrenocortical hormones, effect on β-glucuronidase levels, **16:**389
Adrenocorticopin, adrenal steroidogenesis of, effect of chain length on, **29:**457–460. *See also* ACTH
potency of, **29:**459
stimulation of, **29:**468
Adrenocorticotropic activity, of ACTH, **29:** 456–468
Adrenocorticotropins and Melanotropins, Structure-Activity Relationships of, Synthetic Approach (Ramachandran and Li), **29:**391
Adrenodoxin, **48:**10, **60:**46
Adrenoglomerulotropin, in pineal gland, **29:**520
structure of, **29:**521
Adrenolutin, formation of, **29:**516
in urine, **29:**518
Adrenolytic agents, **7:**547–548
Adrenosterone, **2:**329; 330; 333
Adsorbed enzymes
and DEAE-cellulose columns, **33:**271
effect of radiations on, **24:**399–401
kinetics of action, **28:**28
Adsorbed films, characteristics, **28:**12–16
Adsorbed substrates, enzyme action on, **33:** 266–269
Adsorbents. *See also* Adsorption
for column chromatography of enzymes, **22:**444–447
table of, **14:**342; 348–351
Adsorption, **34:**452; 487. *See also* Chromatography; Surfaces
of coenzyme I on alcohol dehydrogenase, **15:**41
of substrate by model enzyme, **15:**22
in enzyme isolation, **14:**387; 400–401
of enzymes, **1:**331; 332; 344; 345
conformational changes, **28:**4–5; 7; 8; 11–12; 17–21; 23–25; 31–32
at interfaces, **28:**1–40
physical properties, **28:**3
on soil particles, **28:**367; 368
on glass, β-lactamase, **28:**298; 299
in isolation of LBF, **14:**51–52; 56
of phage to bacteria, **9:**178–179; 183
process, **34:**454
in protein chromatography, **22:**433–436

studies of enzymes and other proteins, **14:**319–374
techniques, **34:**452
Adsorption and permeability, **7:**20–21
Adsorption compounds, *See* Solubilization of matter
Adsorption exchange, experiments in D_2O, **15:**13
Adsorption isotherms
relation to agglutination, **15:**53; 54
relation to chromatography, **14:**320–325
Adsorption methods, effect on activity of enzymes, **28:**3; 28–30
Adsorption of Enzymes at Interfaces: Film Formation and the Effect on Activity (James and Augenstein), **28:**1
Adsorption of ions, on a monolayer, **11:**67–69
Adsorption of organic molecules on proteins, **16:**125
Adsorption phenomena, in clotting of fibrinogen, **16:**78–79
Adsorption rates, **28:**14–15; 19–20; 28
Adsorption Studies of Enzymes and Other Proteins (Zittle), **14:**319
Adsorption theory, Langmuir's, application to hydrolytic enzymes, **15:**277–280
Aedes aegypti, **45:**401
Aequorea aequorea, light emission, **25:**148
Aerobacter, **17:**399; 413
Aerobacter aceti, **71:**189
Aerobacter aerogenes, **17:**409; 423; 430–431
allosteric enzyme of, **28:**64; 111–113; 121
β-lactamase of, **28:**268; 290; 296
aromatic poly-auxotrophs of, **16:**287–289
aryl sulfatase of, **22:**216; 218
methionine mutants, **22:**268; 269
as source of transmethylase, **22:**253
biosynthesis of glycogen in, **34:**323
pullulanase, **34:**408
cell walls, **26:**202
and cobamide coenzymes, **26:**265
diol dehydrase from, **26:**365–366
distribution of sugars, **26:**47
PN-dependent enzymes, **26:**118; 161
and citric acid cycle, **18:**247; 262–264; 269
effect of CO on, **18:**32
effect of fluoro compounds, **18:**135; 151
mutants, **18:**269; 270
cytochrome of, **23:**267; 287; 288; 315
use of acetate, **23:**412
folate metabolism, **27:**430; 456
oxidation of hydrocarbons, **27:**520
β-galactosidase, **25:**336
nucleosidase, **25:**386
guanylic acid biosynthesis in, **21:**229
histidine degradation in, **21:**407
xanthosine-5′-phosphate aminase in, **21:**251
Aerobacter aerogenes, 201, **15:**208–209
Aerobacter aerogenes, enzymic adaptation in, **13:**86; 111–112
hyaluronidase action on, **13:**230
transferases in, **13:**295
Aerobacter aerogenes, in industrial fermentation, **7:**578; 602–607
Aerobacter aerogenes mutant, amino acid metabolism in, **16:**252; 255; 257–258; 263
Aerobacter cloacae, cell walls, **26:**217–218
Aerobacter indologenes, metabolism in, **74:**21; 34–36
Aerobacter-Klebsiella group, β-lactamase of, **28:**242; 254
Aerobacter levanicum, properties of levansucrase preparations from, **30:**202–203
Aerobic, transhydrogenases, classification, **14:**227–228; 236; 253; 255
Aerobic bacteria in cellulose decomposition, **2:**245; 249–255; 260–262
Aerobic metabolism, **1:**139–148
Aerobic metabolism of spermatozoa, **24:**52–68
factors affecting, **24:**69–86
Aerobic organisms and oxygen concentration, **2:**222
and substrate concentration, **2:**222
Aerobic oxidases, **7:**295
Aerobic pathway, quinolinate synthesis, **73:**140–144
Aerobic transelectronases, classification, **14:**255
Aeroglucosedehydrogenase, **5:**181
Aeromonads, ADP-Glucose pyrophosphorylase of, **46:**353
Aeromonas, formicans, **46:**353
Aeromonas hydrophilia, molecular weight of enzymes from, **46:**367
Aeromonas proteolytica, **68:**162
Aeropentosedehydrogenase, **5:**181
Aerosol OT (AOT), **51:**32; 34; 38; 40; 49; 55–56; 58

Aesculetin, chemiluminescence, **25:**121
Aetobacter suboxydans, growth stimulation by pantothenic acid derivatives, **16:**8; 17
Affective disorder, biochemistry of, **29:**489–498
mineral metabolism in, **29:**494–497
Affinity chromatographic methods, **52:**145–154
Affinity chromatography, **43:**46, **46:**365; 385, **57:**8–10; 373, **59:**123; 137
acrylamide, **36:**50–51
advantages, **36:**33
capacity, **36:**36; 37
cells, **36:**83
cellulose, **36:**38
effect of pH, **36:**45; 51–53
elution, **36:**36; 54; 55
glass, **36:**41
insoluble carriers, **36:**37
linking reactions, **36:**43
lymphocytes, **36:**40; 41; 84
mutant enzymes, **36:**61; 69–71
nucleic acids, **36:**81
polyacrylamide, **36:**38; 50; 58; 62
polystyrene, **36:**38
porosity of adsorbent, **36:**40
protein adsorbents, **36:**51
receptors, **36:**83
ribosomes, **36:**83
steric factors, **36:**42–43; 66–67
substrates, use of, **36:**41
subunit interactions, **36:**80–81
theory, **36:**34; 55
dynamic, **46:**259
general ligand, **46:**208
gradient, **46:**262
nature of chemical reactions involved, **53:**310–311
phosphate-linked derivatives, **43:**46
purine-linked derivatives, **43:**47
ribose-linked derivatives, **43:**47
tRNA synthetases, **40:**153; 154; 155
Affinity constant, enzyme action, steady-state hypothesis, **73:**27–28. *See also* V/K constant
Affinity constants, of enzyme-substrate complexes, **9:**2; 4; 9–15; 16–22; 56–60
of hemoglobin-oxygen complex, **9:**13
Affinity labeling, **43:**46, **50:**353, **67:**27–30
Affinity-labeling, **36:**75
of aminoacyl-tRNA synthetases, **40:**143; 144
Aflatoxins, **60:**15
Agaamaglobulinemia, X-linked, **51:**201–202
Agar-Agar, *See* Polysaccharide sulfuric acid esters
Agar gels, irradiation of enzymes in, **24:**401
Agaricus campestris, coenzyme Q in, **25:**289
Agarose-heparin columns, **40:**34
Age, and variation in enzyme structure, **15:**274
Agglutination, **15:**51–64; 75
activity of influenza virus, **9:**191–195; 227–228
of bacilli by borate, **12:**514
effect of borate on, **12:**520–521
of platelets, **43:**28
Agglutinin reactions, quantitative studies, **8:**72; 77; 84
Agglutinins, **55:**3
structure, **15:**61–62
Aggragation, **34:**479
Aggregates, **39:**193
Aggregation, **43:**116; 121; 133; 204, **47:**81
of protein split-products, **9:**127–128
of tobacco mosaic virus, **9:**212–215; 217–218
of viruses, **9:**212–220
state of, in allosteric enzymes, **28:**53–56; 68; 73; 83; 90; 105; 117; 129; 132; 138–142; 144
of viruses, **1:**57
Aggregation reactions, **48:**203
Aging, **48:**130
of amylose and amylopectin, **12:**350; 351–353; 370–371; 389
of amylose, effect on enzyme action, **12:**381; 387–388; 393
and collagenase action, **23:**203–204
and elastase action, **23:**224; 247
dependence upon optical activity, **20:**1–29
effect on fumaric reductase, **18:**102
effect on succinic dehydrogenase, **18:** 75–76; 80–81; 83–84; 98
effect on mitochondrial oxidative phosphorylation, **21:**30; 32; 37; 38; 58
of erythrocytes, **48:**129
of films, **28:**5; 8; 15; 16; 23
of succinate dehydrogenase, effect of air, **28:**182–184
Aging, Possible Relation between Optical Activity and (Kuhn), **20:**1
Aging and protoplasmic colloids, **8:**26–27
Aging of sperm, effects, **24:**90

Aging of tobacco. *See also* Tobacco leaf
chemical changes in, **10:**401–409
technique, **10:**400
types of processing, **10:**327–333
Aglycon, **6:**162, **48:**31; 39; 42; 45; 46
Aglycones, **56:**144
from glucuronides, identification, **16:**382; 383
Aglycon-gentiobioside linkage, hydrolysis of, **28:**345
Aglycon group, **36:**102
alkyl group, **4:**53–59
aryl group, **4:**59–61
in glucosides, **4:**53–61; 62–64
in glycosides, **4:**42; 43; 45
structure, chain length and adsorption on substrate, **4:**54–58
Aglycoprothrombin, **46:**13
Agmatine, **15:**379, **36:**210–213; 227; 228. *See also* 1,4-Diaminobutane, arginine decarboxylase
growth requirements for, **36:**218
growth requirements of mutant strains of *E. coli,* **36:**218
mutants lacking, **36:**218–219
ureohydrolase, **36:**213
Agmatine production, **6:**4
Agmatine ureohydrolase, **36:**213
mutants lacking, **36:**219
bacteriophage growth in, **36:**219
polyamines in, **36:**219
Agonist-stimulated phosphorylation, adenosine receptors, **69:**103; 104; 112
Agonist-stimulated sequestration, adenosine receptors, **69:**103
Agrimycin 100, composition, **20:**384
in treatment of plant diseases, **20:**390
Agrimycin 500, composition, **20:**384
in treatment of plant diseases, **20:**390; 392
Agrimycin, in treatment of plant diseases, **20:**387; 388; 391; 392; 395–402
Agrisrep, composition, **20:**384
in treatment of plant, diseases, **20:**390
Agrobacterium, natural resistance to proline analogs, **29:**107
Agrobacterium lumefaciens, allosteric enzymes, **28:**111; 112
Agrobacterium tumefaciens, **34:**466, **46:**328; 329; 330–331, **70:**231
plant disease caused by, **20:**400; 402
AHA, *See* Aspartohydroxamic acid
AHH-1 cells, **68:**50
AICAR, *See* 4-Amino-5-imidazole-carboxamide ribotide
AICAR transformylase, *See* Carboxamide ribotide transformylase
AI protein, **42:**264
basic, **42:**266
AIR, *See* 5-Aminoimidazole ribotide
Air-borne plant infection, **11:**431
Air/water interface, **34:**484
Air-water interface, activity of adsorbed molecules, **28:**21–25
ALA, *See* δ-Aminolevulinic acid
Alamethicin, ion channels, **71:**214
Alanine, **15:**102; 184, **17:**409; 449, **34:**21, **35:**481, **49:**31, **51:**217, **53:**209; 210; 211, **57:**35. *See also* Amino acids, Peptides, Polypeptides, Proteins
and acetate, **23:**419; 426–427; 432–433
and glycine, **23:**457–459
and glycolate, **23:**437
activation of alkaline phosphatase by, **10:** 106–107
effect on indole production from tryptophan, **10:**58–60; 62; 77
reactivation of alkaline phosphatase by, **10:**111
activity in *Neurospora* mutant, **16:**256
biosynthesis of, **16:**275
deamination of, **16:**117
in enzymic transaminations, **16:**186; 189; 190; 192–194; 201; 203–209; 211–213; 218; 222–223; 226–229; 233–235
as N-terminal residues of human fibrinogen, **16:**37
reaction with quinones, **16:**114
recemase for, **16:**250
relation to valine metabolism, **16:**281
from tryptophane, **16:**294
in chloroplasts, **20:**209
metal binding by, **20:**56
in phosphorylated peptides of esterases, **20:**288
in plants, biosynthesis, **20:**329; 330
thermodynamic data on ionizations, **20:**64
and citric acid cycle, **18:**271; 272
CO_2 fixation in, **12:**132
incorporation into proteins, in mitochondria, **21:**44
as inhibitory analogue, **6:**133

Alanine (*Continued*)
isomers in mureins, **26:**199–207; 210–212; 216–219; 223
labeled, in aldolase and triosephosphate biosynthesis, **19:**260
transamination reactions of, **19:**338–340
metabolism in thyroxine formation, **14:**115
in pantothenic acid synthesis, **15:**253–254
in plant metabolism, **9:**400–401; 403; 407; 409
production, **6:**2; 5
racemase, **35:**106
racemization of, **14:**195
of tobacco mosaic virus, **15:**462–464
D-Alanine, **35:**93
crystallization of Michaelis complex, **27:** 16–18
reduction of enzyme, in model complex, **27:**9–10
in slow reaction, **27:**29–32
elimination in living organisms, **20:**18
synthesis, **20:**14
in transaminations, **16:**206
requirement of *Streptococcus faecalis,* **16:**250
L-Alanine
formation, **20:**14
transamination reactions of, **19:**346–351; 367; 368; 379; 380
in UDPG analog, **14:**204
β-Alanine
in biosynthesis of pantothenic acid, **19:**259
transamination reactions of, **19:**343; 367
as cleavage product of pantothenic acid, **3:**256–257
enzymatic formation, **53:**373–374
genetic studies, **17:**431; 442–445
from hydrolysis of CoA, **16:**3
in transaminations, **16:**194; 206; 237
metal binding by, **20:**56
and microbial synthesis of pantothenic acid, **7:**435–440
origin, **7:**436–437
physiological activity, **3:**259–261
in pyrimidine biosynthesis, **21:**283; 284
relating to coenzyme A, **14:**63
relation to LBF, **14:**53; 54; 57
in synthesis of pantothenic acid, **3:**258–259
DL-Alanine, resolution by papain, **19:**309
DL-Alanine, oxidation in plants, **20:**320; 322
Alanine adenylate, synthesis by carbodiimide method, **22:**191
Alanine aminotransferase, **37:**140; 490; 505; 538, **53:**221, **74:**133; 148; 149
cytosolic, species distribution of, **74:**158
evolution of, rate of, **74:**158
localization of, **24:**312
D-Alanine aminotransferase, **74:**134; 141
Alanine amino transferase, cobamide coenzyme study, **26:**364
D-Alanine carboxypeptidase, action on mureins, **26:**204; 206; 212
Alanine dehydrogenase, allosteric regulation, **28:**140
L(+)-Alanine dehydrogenase, and citric acid cycle, **18:**252–253
L-Alanine dehydrogenase, in *Bacillus subtilis* and *Escherichia coli,* **19:**340
D-Alanine dipeptide, in murein nucleotide synthesis, **26:**217–219
D-Alanine-D-glutamate transaminase, **35:**106
L-Alanine-L-glutamate transaminase, **35:**106
Alanine-glutamate transamination, **16:**204; 222–223; 228
Alanine-α-ketoglutarate reaction, free energy change in, **16:**204
Alanine metabolism, role of glutamine, **28:** 66–69
a-Alanine migration, structural specificity, **49:**47
β-Alanine production, **6:**2; 5
Alanine racemase, **55:**231–234
Bacillus subtilis, **44:**104
cycloserine inhibition, **44:**107
hydrogen exchange, **44:**111
kinetic constants, **44:**105
Lactobacillus fermenti, **44:**100
of *Pseudomonas,* **44:**100
staphylococcus aureus, **44:**100
of *streplococcus faecalis,* **44:**100
substrate specificity, **44:**102; 104
coenzymes for, **14:**182
structural data on, **74:**134; 142
β-Alanine ribotide, in pyrimidine biosynthesis, **21:**284
Alanine Transfer RNA Synthetase: Structure-Function Relationships and Molecular Recognition of Transfer RNA (Schimmel), **63:**233

Alanylalanine, from pyruvoylalanine, **16:**239
β-Alanyl-CoA, from propionic acid, **26:**368–369
S-Alanyl CoA, synthesis, **19:**240; 249
thioesterase action on, **19:**249
transacylation by, **19:**249
Alanyl-CoA hydrolase, localization of, **24:**318
Alanyldehydroalanine, absorption spectrum, **8:**131
and dehydropeptidases, **8:**136
DL-Alanyldehydroalanine, and dehydropeptidases, **8:**139; 140; 142; 144; 145
preparation, **8:**128
N-Alanylglutathione, **19:**248
S-Alanylglutathione
intramolecular aminolysis of, **19:**248
synthesis and hydrolysis, **19:**247; 248
Alanylglycine
hydrolysis, by metal ions, **21:**178
inactivity in transamination, **16:**239
L-Alanylglycine peptidase, classification, **14:**247
2-Alanyl-3-isoxalozolin-5-one, **50:**147
N-β-Alanyl-2-mercaptoethylamine, in synthesis of pantetheine, **16:**6
Alanyl peptides, rate of deamination, **16:**117
β-Alanyl peptides steric factor in enzymic hydrolysis, **15:**292
D-Alanyl residues in teichoic acids, **30:**225
Alanylthiophenol, labeled, in nonenzymic reactions, **19:**262
Alanyl-tRNA phosphatidylglycerol transferase, requirements, **40:**114
substrate, donor specificity, **40:**115; 116
alanyl-tRNA, **40:**113; 114
arginyl-tRNA, **40:**113; 114
lysyl-tRNA, **40:**114
Alanyl-tRNA UDP-MurNAc-pentapeptide transferase, **40:**127; 128; 130
substrate specificity, acceptor, **40:**134; 135
donor, **40:**132
Ala-Pro
and carboxyalkyldipeptides, **57:**34; 36
and enalaprilat structure, **57:**57–58
and peptide inhibitor analogs, **57:**27
and snake-venom peptides, **57:**29
Ala-scan analysis, **68:**25–29
Albinism. *See also* Depigmentation, Leukoderma, Melanin, Pigmentation
relation to tyrosine metabolism, **14:**92; 123
Albizziine, **50:**121; 124; 125
Albomycin, **35:**146–148; 173; 177
Albumin binding, **49:**92
Albumin-bound amino acid, **49:**92
Albumin-Ci(II)-histidine complexes, **44:**424
Albumin films, enzymic activity on, **28:**29
Albumin nucleates, solubilization of, **17:**137; 138; 143
Albumins, **17:**154, **46:**143; 145; 150. *See also* Egg white protein; Egg white proteins, Lactalbumin, Serum proteins
action of tyrosinase on, **14:**150
adsorption studies, **14:**332; 338
arginylated, **40:**100; 101
site of arginylation, **40:**96
of leucylation, **40:**99
of phenylalanylation, **40:**99
in *vivo* acceptor in arginine-transfer reaction, **40:**108
bovine serum
electrophoresis for complex formation between pepsin and, **30:**146
zone electrophoresis, of, on cellulose acetate, **30:**166
effect on fertilizability, **9:**250–251; 272–274; 278; 281
effect on fertilizing life of sperm, **9:**254
serum, virus crystallizing agent, **9:**196
egg, **19:**302 (*See* Egg albumin)
adsorption studies, **14:**322–326; 328; 329; 331; 336; 351; 360
structure studies, **14:**279–280; 283; 287; 299; 300–301; 302; 305; 309
binding of methyl orange, **16:**332
plasma, comparison with prothrombin, **16:**28
serum, **16:**44; 376–377
effect of pH on structure, **22:**426
serum, effect of pH on viscosity, **22:**426; 427
serum, **19:**293 (*See* Serum albumin)
adsorption studies, **14:**323; 325; 326; 328; 335; 336; 343; 345; 348; 350; 357–359; 360; 361; 363–364
structure studies, **14:**277; 284; 285; 305; 309
complexing between pepsin and, **30:**143
rate of synthesis, **26:**108–110
and transfer across cell, **8:**51

Albumose, whey, **15:**432
Alcalase, produced by *B. subtilis,* **33:**385
Alcaligenes eutrophus, phosphoribulokinase from, **74:**96; 97; 98; 99; 100; 110; 117; 120; 121
Alcaligenes faecalis
oxidative phosphorylation, **23:**353
use of acetate, **23:**412
pentose metabolism, **28:**411–413
Alcaligenes metalcaligenes, aryl sulfatase of, **22:**216–218
Alcaligenes species, enzymic transformations of steroids by, **20:**254; 255; 262
Alcaptonuria, **59:**137. *See also* Homogentisic acid
alcohol dehydrogenases, **59:**187–189
hereditary defect, **16:**396
and tyrosine metabolism, **14:**89; 98; 99; 102–104; 110–111; 122–123
Alcian-blue, binding by yeast cells, **40:**261
Alcohol
conversion to fat by yeasts, **9:**674–675; 699
ethyl (*See* Ethanol)
oxidation by plant tissues, **18:**309
secondary, PN-dependent reduction, **26:** 128–129
Alcohol absorption, **45:**429
Alcohol dehydrogenase
yeast, activity of –SH group, **11:**237; 239; 241
Alcohol dehydrogenases, **7:**413, **15:**11; 13; 41; 85–86; 88–90; 97–98; 114; 116; 120; 128; 200; 254–255; 257; 296, **17:**70, **26:**117; 118–119; 140; 156; 173; 174, **34:**448, **37:**410; 414; 446; 459; 470; 490; 505; 506; 538; 539; 557; 567; 569; 588; 589; 593, **45:**433–435; 441, **46:**240; 270; 271, **56:**292
anodic band, **45:**455
asymmetric binding of coenzyme, **26:** 164–165
atypical form, **45:**452
chemical mechanism of, **45:**372
chromophores, Cotton effects, **27:**58
liver, horse, Cotton effects, **27:**49–82
human, **27:**58; 82–83
methane oxidation, **27:**479
NAD-dependent, alkane oxidations, **27:**485
yeast, Cotton effect, **27:**58; 83–85
classification, **14:**253
isolation, **14:**397–398; 405
complex formation, **26:**156–165
with NADH, **26:**164–165
ternary, **26:**164; 165
constraint, **37:**593; 594
denaturation of, **23:**33; 46–50; 62
in phosphorylating system, **23:**353
and DPN, mechanism, **24:**442; 469–471
effect of radiation on, **24:**388; 391; 392; 396; 398; 401
location in subcellular fraction, **24:**295
molecular association of, **24:**119–120
role of metal, **24:**112–113
EDTA and, **56:**301
enzymic adaptation, **13:**76
inhibition of, **13:**332
genetic model for molecular forms, **45:**451
isoenzymes, **45:**451
health and disease, effect of, **45:**454
heptyl, **34:**489
inhibition by hydroxylamine, **22:**351
liver, **22:**348–354
complexes with DPN, DPNH, dissociation constants, **22:**358
complex with acetylpyridine analogue of DPNH, **22:**350; 351
complex with DPN, DPNH, effect of pH on, **22:**353
complex with DPNH, absorption and fluorescence spectra, **22:**348–351
in ternary complexes, **22:**351
yeast, **22:**354–358
complexes with DPN, DPNH, dissociation constants, **22:**358
complex with DPNH, absorption, fluorescence maxima, **22:**354
functional groups in, **22:**354–358
inactivation by irradiation, **22:**355
zinc in, **22:**355–357
inhibitors of, **45:**433; 446; 450
isozymes, **46:**255
kinetic mechanism of, **45:**316; 318
kinetic parameters, **45:**448
kinetic properties, **45:**457; 458; 459
kinetics of reaction of, **56:**325
kinetic studies, **17:**36; 42–45
liver, steps in reaction, **25:**14
and luminescence, **25:**124
multiple forms, **25:**333; 334; 362
product inhibition of, **25:**210; 224; 250–253

rate constants for enzyme-substrate reaction, **25:**13–15
yeast, product inhibition of, **25:**224; 253; 254; 260
metal centers of, properties of, **56:**320–325
metal content of, **56:**318
metalloenzyme substitutions of, **56:**304
metal substitution in, **56:**306; 321; 323
molecular properties, **45:**444–447
molecular properties of, **56:**319–320
multiple molecular forms, **45:**450
NAD-sepharose complex, **46:**267
noncatalytic role of zinc in, **56:**317
phenanthroline and, **56:**301
in photosynthesis, **19:**11
in plants, **9:**624, **18:**296–297; 303–304
polyisoprenoid, **34:**478
purification of, **45:**443; 444
reaction mechanism, **45:**449
role of zinc in, **56:**409
sequential response, **37:**588
spectral features of, **56:**322
substrate specificity, **45:**448
substrate stereospecificity, **26:**166–169
tissue distribution, **45:**442
zinc content, **45:**446
Alcohol elimination rates, **45:**429
chronic ethanol ingestion, effects of, **45:**471
fructose administration, **45:**468
glycogenosis patients, type I, **45:**469
individual variation, **45:**431
kinetics of ADH activity, **45:**434
measurement of, **45:**429; 433
Michaelis-Menten kinetics, **45:**429
rate-limiting factors, **45:**468
Alcohol-glyoxylate adducts, **57:**122
Alcoholic fermentation, **3:**349–360. *See also* Fermentation, alcoholic; Wine making; Yeasts
alcohol content obtained, **3:**359–360
role of SO_2 in, **3:**357; 367–374
sequence of microorganisms in, **3:**351–352
syruped fermentation, **3:**359–360
and temperature, **3:**357–358
Alcoholic Fermentation of the Oligosaccharides (Leibowitz and Hestrin), **5:**87
Alcohol intake, and lipoic acid, **18:**343
Alcoholism, **45:**428, **74:**236
effects, **45:**428
zinc metabolism, **45:**476
Alcohol metabolism, **45:**428; 429
acetate, ethanol-derived, **45:**432
effects, **45:**428
enzyme systems, **45:**433
enzymology of, **45:**427–476
extrahepatic metabolism, **45:**432
metabolic capacity, **45:**430
NAD+/NADH ratio, reduction in, **45:**434
organs responsible for, **45:**431
Alcohol Metabolism, Human, Enzymology of (Li, Ting-Kai), **45:**427
Alcohol-oxidizing system, effect on hydrocarbons, **27:**476
Alcohol peroxidase, methane oxidation, **27:**479
Alcohols, **51:**134. *See also* Ethanol
and autoxidation of unsaturated fatty acids, **8:**428
from cellulose, **8:**267–268
from fatty acids, **8:**334–337
course of dehydrogenation of, **5:**182–187
effect on β-glucuronidase, **16:**376; 381
effect on β-lactamases, **28:**295
effect on microbiological formation of hydrocortisone, **20:**246
effect on pectic substances, **20:**356
effect on pectinesterases, **20:**360
polyhydroxy, effect on foliar absorption of streptomycin, **20:**386
ethyl, dehydrogenation by yeast, **10:**190
from microbial oxidation of hydrocarbons, **10:**467–469
mechanism of oxidation, **15:**7; 9–11; 13–15; 17
and phenylalanine hydroxylase, **19:**182
catalase action on, **19:**115
reactions with activated amino acids, **19:**238; 250
as phosphate acceptors, **17:**165–167; 170–173
in transglycosylations, **17:**196; 197; 220; 226
polyhydric, in carbohydrate metabolism, **15:**150; 161; 167–168
n-Alcohols, primary, **34:**488
Alcohol sensitivity, **45:**453; 475
Alcoholysis, and phosphate transfer, **18:** 197–198
Alcomeal, **7:**590
Aldehyde, entropy of hydration, **43:**301
Aldehyde condensations, mechanism, **14:**168; 170; 171; 172

Aldehyde content and quality of frozen vegetable, **9:**621
Aldehyde dehydrogenase, **59:**193–194
in ATP synthesis, **23:**335; 388
classification, **14:**253
disulfiram, mode of action of, **45:**466
enzyme mechanism, **45:**465
induction of, **45:**461
inhibitors, **45:**460; 464
kinetic properties, **45:**463–465
molecular forms, affinity for acetaldehyde, **45:**459
molecular properties, **45:**462
subcellular distribution, **45:**459; 461
substract specificity, **45:**464
oxidation of hydrocarbons, **27:**485; 486
Aldehyde dehydrogenases, **15:**116; 129; 246; 254–259
Aldehyde formation in frozen vegetable tissue, **9:**617–624
Aldehyde-lyases, localization of, **24:**330
Aldehyde mutase, **15:**257
in tobacco, **6:**370
Aldehyde oxidases, **51:**109; 131
activity with cytochromes, **17:**78; 355; 364; 381–382
as metalloflavoprotein, **17:**242–247; 262; 268–270; 275; 276; 284–285; 296; 317–318
and xanthine oxidase, **17:**322–323
classification, **14:**253
in electron transfer, **19:**201
Inhibited signal on, **51:**135
liver, nonheme iron, **27:**89
Mo(V) in, EPR study, **27:**161; 167–172; 178; 183
PN-dependent, **26:**121
of potato, **9:**393
properties, **19:**202
summary and inferences, **51:**149–151
Aldehyde reductase, **46:**240
Aldehyde r O_2 transhydrogenases, classification, **14:**253
Aldehydes, **57:**141–142
acceptor, **15:**157–160
active, **15:**156–160; 166; 177
donor, **15:**157–160
formation of acyl coenzyme A, **15:**129
mechanism of autoxidation, **15:**10
mechanism of oxidation, **15:**7; 10; 12; 14
binding to carbonic anhydrase, **47:**174
carbonic anhydrase catalyzed hydration of, **47:**187; 190–191; 194; 206–207
and chemiluminescence, **25:**125; 128; 145–152; 154; 155–156; 161
from 1,2-diols, role of cobamide coenzyme, **26:**265–266; 359; 365–366
PN-dependent oxidation, **26:**119; 121; 122
from microbial oxidation of hydrocarbons, **10:**467
and phenylalanine hydroxylase, **19:**182
role in fat synthesis, **9:**664; 693–697
as substrates of xanthine oxidase, **17:** 238–241; 296–301; 309–318; 322–323
tautomerism in, **50:**7–11
in transaminations, **16:**204–206
Aldimine, **58:**43
Aldimine enzyme, Cotton effect, **27:**86–87
Aldimines, **53:**163; 170–172
pyridoxal-P-dependent, **56:**4; 7
Alditols, hydrogen bonding, **65:**231; 233
Aldohexose configuration, **44:**245
Aldo-keto isomerization, **59:**107
Aldoketomutase, in tobacco, **6:**370
Aldol and fat formation, **8:**315
Aldolase, **5:**132, **9:**471; 473, **15:**162–165; 170; 171, **23:**480, **34:**448; 452, **37:**428; 447; 459; 473; 482; 490; 506; 507; 524; 526; 540; 596, **43:**501; 502, **46:**249; 462–463, **47:**85; 137, **53:**49; 50, **54:**149–150; 181, **56:**293. *See also* Deoxyribose aldolase
action of tetrose phosphate plus triose phosphate, **16:**291
intracellular distribution, **16:**468
bacterial reaction by, **31:**140
catalytic role of zinc in, **56:**316
classification, **14:**228; 237; 256
isolation, **14:**402; 403
mode of action, **14:**183
sources, **14:**377
structure studies, **14:**267; 275; 283; 284
in sugar interconversions, **14:**208
cleavage of FDP, **28:**396; 397
correlation of hepatoma growth rate with, **29:**331
in latex, **29:**225
and DR synthesis, **28:**427
effect of radiation on, **24:**388; 398; 399
localization of, **24:**330; 335; 336
partial degradation of, **24:**125

effects of polyanions, **40:**41; 42
formation from polupeptide, **26:**91; 104–105
fructose-1,6-bisphosphate, **56:**293; 385
interaction with fructose 1,6-bisphosphatase, **53:**49–50
localization in cytoplasm, **13:**7
mechanism of action of, **56:**388
mechanism of reaction, **28:**450–452
metal center of, properties of, **56:**387–388
metal content of, **56:**385
metalloenzyme substitutions of, **56:**304
molecular properties of, **56:**385–387
nuclear-cytoplasmic ratio, **27:**262
in photosynthesis, **19:**7; 9; 24; 26; 53
in plants, **18:**292–293
rabbit liver, comparison of, with muscle aldolase, **31:**160–163; 165
control of activity of, **31:**165
evolutionary relationship of active sites of, **31:**173–175
K_m of, **31:**163
metabolic function of, **31:**165
molecular weight of, **31:**163
mutational changes of, **31:**164
properties of, **31:**163
structure of active site peptide of, **31:**163–165
rabbit muscle
action of urea on, **31:**158
active sites of
amino acid sequence at, **31:**160
evolutionary relationship of, **31:**173–175
amino-terminal amino acids of, **31:**142
carboxy-terminal amino acids of, **31:**142
conformational transitions of, **31:**157–159
cysteine residues of, **31:**154–157
dealdolization reaction catalyzed by, **31:**133
equilibrium-binding studies of, **31:**147
functional groups in, **31:**148–157
histidine residues of, **31:**152–153
inhibition of
with fluorodinitro-benzene, **31:**147
by *p*-mercuribenzoates, **31:**154
K_m of, **31:**163
lysine residues in, **31:**148
molecular weight and subunits of, **31:**143–144; 163
mutational changes of, **31:**164
non-identical subunits of, **31:**144–145
phosphate-binding sites of, **31:**145–147
photooxidation of, **31:**152
properties of, **31:**163
proposed mechanism for cleavage of fructose diphosphate by, **31:**171
structure of, **31:**142–145
structure of active site peptide of, **31:**159
substrate activation of, **31:**157–158
substrate-and analog-binding sites of, **31:**147
tyrosine residues in, **31:**148–152
relation of binding energy to catalytic efficiency, **33:**437
rhamnosidase-1-phosphate, **56:**293
in ribose synthesis, **28:**411; 415–416
spectral features of, **56:**386
Aldolase in myosin, **6:**170
Aldolases, **50:**272; 280–281. *See also* Metalloaldolases
action of carboxypeptidase on, **31:**130
classes of, **31:**128
effect of chelators on, **31:**129
general mechanism of, **31:**126–129
history of, **31:**126
isozymes of, **31:**141
mechanism of action of, **31:**168–175
organ-specific variants of, **31:**141–142
subunits of, **31:**142
mammalian
active site structure of, **31:**168–175
apolar nature of, **31:**169–170
differential synthesis during ontogeny and its reversal, **31:**167
evolutionary divergence of, **31:**174
genetic regulation of synthesis of, **31:** 166–168
nonhelicity of peptide of, **31:**168
role of functional groups of, **31:**170–173
mechanism of action of, **31:**125–181
molecular weight of class II enzymes of, **31:**129–130
multiple forms, **25:**334; 344
in photosynthesis, **25:**95
turnover number, **25:**32
reaction catalyzed by, **31:**126
in regenerating rat liver, **31:**167
Schiff base-forming, inactivation with borohydride of, from *Clostridium*, **31:**140
lack of metal ions in, **31:**131

Aldolases (*Continued*)
mechanism of action of, **31:**132–138
molecular weight of, **31:**132
from plant sources, **31:**132
postulated reaction mechanism for, **31:** 136
rate of ^{3}H exchange for, **31:**140
rate of ^{18}O exchange for, **31:**140
related enzymes of, **31:**138–141
sequence of reactions involved in, **31:**135
subunit structure of, **31:**132
Schiff base-forming (class I), distribution and properties of, **31:**131–132
stereospecific exchange of tritiated water by, **31:**127
Aldolases, Mechanism of Action (Morse and Horecker), **31:**125
Aldol cleavage enzymes, **55:**227
Aldol condensation, in isoleucine synthesis, **16:**283
in valine synthesis, **16:**286
Aldol condensations, non-enzymic mechanism of, **31:**128
Aldonolactonase, localization of, **24:**317; 318
Aldono-1:5-lactone, half-chair conformation, **36:**175
Aldonolactones
binding of, by lysozyme, **36:**176
interconversion in aqueous solution, **36:**169
mechanism of inhibition, **36:**174
specificity of action, **36:**171
as inhibitors, **48:**49
Aldose dehydrogenases, **59:**189–191
Aldose-Ketose Isomerase Reactions, Mechanism of the (Rose), **43:**491
Aldose-ketose isomerases, **43:**491–515
Aldose-ketose transformation, alkaline, study by tracers, **14:**205
enzymic, **14:**194; 195; 205–206
Aldoses, enzymes for conversion to ketoses, **24:**333
Aldosterone
microbiological formation, **20:**249
steroidogenic pathway, **68:**44–46
d-Aldosterone, microbiological formation, **20:**275; 276
Aldrin (hexachlorohexahydrodimethano-napthalene), **17:**314
Algae, **46:**99
bleaching effect, **46:**99
blue-green, **46:**38; 42; 54; 84; 92; 93
absence of chloroplasts in, **30:**4–5
chloroplasts in, **30:**4–5
blue-green, DAP in, **16:**299
brown, **46:**79
carbonic anhydrase of, **47:**151
carotenogenesis in, **21:**308; 312; 313; 323; 324
inhibition, **21:**319–320
carotenoid functions in, **21:**349–357; 359–361
carotenoid in, **21:**298; 300–303
cell walls, **26:**196; 221–222
cryptomonad, **46:**38
fertilization of, **9:**256–257
β-1,3 glucans of, **28:**328–329
intermediary metabolism, tracer experiments, **28:**338
laminarinases of, **28:**341–342; 346–347
laminarin of, **28:**327; 328; 330–331
green, **46:**54; 76; 94; 98; 335
ADP-glucose pyrophosphorylase of, **46:**335
red, **46:**38; 92
sulfuric acid esters in, **10:**227; 234–235
Algebraic analogies, graphic analogies *versus,* enzyme action, **73:**49–52
Alginase, sources, **28:**354
Alginate, *See* Exopolysaccharide alginate synthesis
Alginates, **17:**147
Alginetin, as degradation product of pectins, **20:**350
Alginic acid
as inhibitor of β-glucuronidase, **16:**374
precursors of, **28:**338
AlgR2 gene, role in TCA cycle operation, energy metabolism, and alginate synthesis, **70:**238–245
Ali-esterase
activity of, **20:**288
phosphorylated, amino acids in, **20:**288
amino acid composition near active side, **22:**65
pK value, **22:**58
Aliesterase, active center of, **24:**434
Ali-esterase in sera, **8:**462–463; 471
Alimentary canal in low vertebrata, **1:**374–378
Aliphatic acylamidases, classification, **14:**247
Aliphatic disulfides, exchanges, **27:**203
Aliphatic hydrocarbons, anaerobic oxidation, **27:**529–534
dissimilative metabolism, **27:**477–505

Alkalide, pH, **34:**461
Alkali disease, **48:**4
Alkali ions, in nuclei, **27:**256–261
Alkali metal ions and fermentation, **7:**27–30
Alkaline earth salts, solubilization of, **17:**138; 142–151; 155
Alkaline invertase, **56:**155
occurrence and locationzation of, **56:**156
purification of, **56:**158
in sucrose metabolism, **56:**162
in sugar cane, **56:**163
Alkaline phenolphthalein phosphatase, intracellular distribution, **16:**465
Alkaline phosphatase, **17:**163, **34:**308; 454–456; 480–481; 504–511; 524, **37:**485; 557, **43:**43; 332; 349; 355, **45:**357; 358, **46:**388, **52:**179, **56:**292, **57:**159; 161, **58:**4, **60:**231. *See also* Phosphatase, alkaline; Phosphatases
accessibility measurements of, **55:**470; 471
action on CoA, **16:**4; 7
intracellular distribution, **16:**417–418; 463; 466–468
relation to estrogen-stimulated growth, **16:**387; 399
activation by magnesium, **9:**590
histochemical methods, for, **9:**589–590
active centers of, **55:**457
active site regions of, **55:**465–469
conformational flexibility of, **55:**471–474
activity of metal substituted, **55:**393
amino acid composition of, **55:**463–465
amino acid sequence of, **55:**384; 465–469
ammonium sulfate precipitation of, **55:**397
anomalous difference map of, **55:**472; 473; 474
anomalous Fourier map of, **55:**471; 472; 474
apoenzyme of
cadmium binding to, **55:**398–400
magnesium binding to, **55:**405
photooxidation of, **55:**394
zinc binding to, **55:**400–404
apophosphoryl, **55:**392
architecture of molecule, **55:**455–463
arginine residues of, **55:**476
in active site, **55:**465–468
modification of, **55:**394
biosynthetic modification of, **55:**395
bone, **55:**386
burst kinetics of, **55:**389; 390
2,3-butanedione modification of, **55:**394
cadmium anomalous dispersion map of, **55:**458; 459
cadmium binding to, **55:**396
binding affinity for, **55:**398–400
B site, **55:**404; 423–425
concentrations as factor in, **55:**398; 399–400
enzymatic activity of, **55:**391–394
magnesium in, **55:**400; 401
manganese in, **55:**400; 401
migration with, **55:**422
NMR spectroscopy of, **55:**396–397; 426
pH effect on, **55:**389; 398–399
phosphates as factor in, **55:**400
phosphorylation with, **55:**405–406; 417–423; 426
A site, **55:**404; 406; 423–425
stabilization with, **55:**411
thermally induced transitions of, **55:**411
zinc in, **55:**400; 401
catalytic properties of, **55:**387–395
chemical mechanism of, **45:**358
chemical modification of, **55:**394–395
circular dichroism spectra of, **56:**348
cobalt binding to, **55:**392; 407; 408
enzymatic activity of, **55:**391–394
conformational flexibility of, **55:**454
contact residues of, **55:**471
contamination of micrococcal nuclease preparations with, **29:**192
effect of heating of, **29:**99
incorporation of norleucine in, **29:**136
incorporation of proline analogs in, **29:**98
replacement of phenylalanine by analogs in, **29:**110
copper binding to, **55:**404
enzymatic activity of, **55:**391–394
ESR spectroscopy of, **55:**409
crystal structure of, **55:**384
cytochemical method for, **8:**30
in chromidia, **8:**31
in nuclei, **8:**33
de-repression of, **55:**383
differential scanning caloriertry of, **55:**411
as dimmer, **55:**455
distribution of, **55:**322
disulfide bonds of, **55:**384
from *E. coli,* **55:**383; 384–386; 453
effect of borate on, **12:**518–520

Alkaline phosphatase (*Continued*)
use of borate in purification, **12:**509
entatic state of, **56:**314
enzymatic activity of, **55:**392; 393
equilibrium dialysis of, **55:**398
from *Escherichia coli,* **74:**71
ESR spectroscopy of, **55:**404
from eukaryotes and prokaryotes, **55:**382; 383
fluorinated amino acid incorporation into, **55:**395
function of, **55:**386
genes for, **55:**385; 386
genetic control of, **24:**9
α-helix sequences of, **55:**464
histidine ligands of, **55:**456; 468
in hydrolysis reactions, rate-limiting step in, **55:**394
hydrolytic activities of, **56:**346; 348
inactivation of, **55:**394
inhibition of, **55:**387; 474
intermediates of, **55:**388–391
interpretation of pH effects, **53:**334
intestinal, **55:**386–387
test on phosphorylases, **24:**282
isozymes of, **55:**384; 385
kinetic mechanism of, **45:**357
labeling of
^{32}P, **55:**388
^{35}S, **55:**390; 391
ligands to metals of, **55:**465; 468
location of, **55:**460
localization of, **24:**318–319; 343
lysyl residues of, **55:**395
in matrix vessels of osteoblasts, **55:**386
mechanism of action of, **55:**445–447
metal binding sites of
A, **55:**396; 404; 413; 414–415
B, **55:**396; 404–405; 416–417
C, **55:**396; 404; 405; 413; 423
^{13}C, NMR of, **55:**426–430
^{113}Cd NMR of, **55:**413–426
inorganic phosphate and, **55:**474–476
metal ion substitution at, **55:**398–407
structure and function of, **55:**395–430
metal centers of, properties of, **56:**344–350
metal content of, **56:**341–342
metalloenzyme substitutions of, **56:**304
Mg(II) binding to
activation with, **55:**397–398
competitive, **55:**405; 406
effects of, **55:**405–407
ESR spectroscopy of, **55:**408
phosphate and, **55:**405
Mg(II) content of, **55:**397; 405
enzymatic activity of, **55:**391–394
Mn(II) binding of, **55:**408–409
molecular properties of, **56:**343–344
molecular weight of, **55:**383; 384
negative cooperativity of substrate binding, **55:**442
nitration effect on, **55:**395
NMR spectroscopy of, **55:**396; 397
^{13}C, **55:**415; 426–430
^{113}Cd, **55:**396
chemical exchange effects on, **55:**416
chemical shifts in, **55:**413; 414; 415
with ordered addition of isotopes, **55:**424; 425
pH effect on, **55:**416
phosphorylation effect on, **55:**417
^{35}Cl, **55:**415
^{19}F, **55:**395
^{18}O exchange by, **55:**394
^{31}P, **55:**383; 433; 436–437; 440; 441
with ^{113}Cd-O-^{31}P coupling, **55:**424; 426
phosphoenzyme intermediates in, **55:**430–445
with phosphorylation, **55:**414
saturation transfer, **55:**444–445
noncatalytic role of zinc in, **56:**317
nuclear magnetic resonance of, **56:**349
periplasmic space location of, **55:**385
pH dependence of, **55:**387; 388–390
phenylglyoxal modification of, **55:**394
phosphatase-collodion enzymes, **34:**495–496; 507–509
phosphate and, **56:**312; 347
phosphate binding to, **55:**475; 476
stabilization with, **55:**412
phosphate contamination of, **55:**390
phosphate starvation affecting, **55:**383
phosphoseryl intermediate of, **55:**382; 386; 454
dephosphorylation of, **55:**387; 389; 392
pH dependence of, **55:**431–432; 444–445
phosphotransferase activity of, **55:**390; 391
physicochemical properties of, **55:**384–387
pK value, **22:**58
rate of reaction of, **55:**387
ribbon drawing of, **55:**459; 460

serine residues of, **55:**465
β sheet of
composition of, **55:**464
connectivity diagram of, **55:**459
interfacial residues of, **55:**471
signal peptide sequence of, **55:**384; 385
of soil, **28:**367
specificity of, **55:**382; 387
spectral features of, **56:**309; 345; 347
structural stabilization of, **55:**410; 412
subunits of, **55:**453
allosteric effects between, **55:**451
interactions between, **55:**442–443
interface between, **55:**469–471
rotational symmetry of, **55:**453
tetranitromethane modification of, **55:**395
three-dimensional structure of, **55:**453–479
transphosphorylation reactions by, **55:** 365–367; 391
tyrosine residues of, **55:**476
from various species, **55:**386–387
x-ray diffraction studies of, **55:**454
zinc binding to, **55:**395–396
affinity in, **55:**398; 400–404
cadmium effect on, **55:**403
concentration levels as factors in, **55:**403
enzymatic profile of, **55:**392
hysteresis effect with, **55:**403
magnesium effect on, **55:**402
phosphate effect on, **55:**403
pH rate profile of, **55:**388; 389; 400
at A site, **55:**405
stabilization with, **55:**410; 411
zinc content of, **55:**383; 395; 397
Alkaline phosphatase, RNA polymerase II transcription, reversible phosphorylation of CTD during transcription cycle, **71:**51
Alkaline phosphatase, SH-group, **27:**221
Alkaline Phosphatase, Solution Structure, and Mechanism (Coleman and Gettins), **55:**381
Alkaline Phosphatase From E. Coli, The Three Dimensional Structure of (Wyckoff, Handschumacher, Murthy, and Sowadski), **55:**453
Alkaline phosphatases, multiple forms, **25:**348
Alkaline proteases, multiple forms, **25:**351
Alkaline ribonuclease, localization of, **24:**314
Alkalis, effect on pectic substances, **20:**348; 349; 370
effect on succinic dehydrogenase, **20:**161
Alkaloid Biogenesis (Dawson), **8:**203
Alkaloid Biosynthesis (Leete), **32:**373
Alkaloid compounds, solubilization of, **17:**138; 143; 147; 154
Alkaloids, **9:**443–444. *See also* Nornicotine; individual alkaloids, eg, Nicotine
accumulation patterns, **8:**230–231
of *Atropa belladonna,* **8:**224–226; 236–240; 241; 242; 243
precursors of, **8:**237–240
basis for classification, **8:**208–209
biogenesis, **8:**203–251
and protein metabolism, **8:**206; 207; 212; 217; 228; 234; 238; 239; 241–245
biosynthesis, **8:**204–206
and putrescine, **8:**237–240; 242
as catalytic agents, **8:**246–247
of *Cinchona,* **8:**231–233; 235; 242
combined with organic acids, **8:**245
distribution patterns in plant body, **8:** 210–213
experimentation problem, **8:**207–209
and indole ring, **8:**235–236
inheritance of, **8:**235
isoquinoline group in, **8:**233; 235; 236
loci of synthesis, **8:**207; 208; 209–233
mechanisms of synthesis, **8:**233–241
as metabolic waste products, **8:**206–242
physiological effects in animals, **8:**207
physiological functions, **8:**206–207
and plant physiology, **8:**245–247
produced in leaves, **8:**227–231
produced in roots, **8:**213–214; 217–227
and pyridine ring, **8:**235–236
synthesis, and water absorption, **8:**212; 246
in excised organs, **8:**213–214; 222–223; 225–226; 227; 231; 236; 237–238; 243; 246
in grafts, **8:**214–217; 218–222; 224–227; 227–231; 231–233; 234; 237–238; 240–241
hereditary control, **8:**205
in transplantations, **8:**214–217; 218–222; 224–277
of tobacco, **8:**211; 217–224; 227–231; 234; 235–236; 240–241; 245–246
of uncertain origin, **8:**231–233

Alkaloids (*Continued*)
in asymmetric decomposition, **7:**97
in asymmetric synthesis, **7:**89–90
degradation in processing of tobacco, **10:** 429–433
fate in fermenting tobacco, **10:**386–394; 428
methods for determination, **10:**388–389
optical antipodes in, **20:**2
in plant oxidase studies, **16:**133
phenolic, metabolism of, **16:**141–143; 150
in plant respiration, **18:**283
Alkane dehydrogenase, **27:**531–534
Alkane dehydrogenase:olefin reductase, reversible, **27:**531; 534
Alkanediol hydrates, **65:**246
Alkane oxidase system, **27:**532; 537–538
Alkane-oxidizing enzymes, **27:**475–477; 485
Alkane oxygenase, **27:**533
n-Alkanes
microbial oxidation, **27:**480–493
Alkanes, fluorinated, **55:**257–264
Alkenes
fluorinated, **55:**265–268
microbial oxidation, **27:**481; 483–484
Alkoxyaryl hydroxylase, location of, **24:**308
Alkoxy radical, **58:**293
Alkylamine hydrates, **65:**243–245
Alkylamines, cellulose-bound, in synthesis of d-maldelic nitrile, **20:**7
Alkylamines in plant tissues, **9:**407
N-Alkyl-D-amino acids, **57:**91
α-Alkyl amino acids, **35:**86
Alkylarsonic acids, biological formation of, **12:**432
Alkylating agents
degradation of DNA by, **31:**14–15
for thiols, **11:**213–216; 226–227; 238–239
Alkylation, **39:**345, **42:**128
mutagenic effect, **27:**305; 320–326; 335–336; 338; 342
Alkylation of RNase, **24:**203; 214–220; 245
Alkylbenzenes, microbial oxidation even-numbered, **27:**516–519
odd-numbered, **27:**513–516
Alkyl benzenes, oxidation, **8:**348–349
n-Alkyl-*t*-butyldisulfide, exchanges, **27:**209
N-Alkylcarbamoyl pyridinium cations, **43:**385
Alkylfluorophosphate hydrolase, classification, **14:**248
7-Alkylguanines, formation of, **31:**14
s-Alkyl homocysteine sulfoximine inhibitors, glutathione synthesis, **73:**246–252
Alkyl hydroperoxide, **60:**73
Alkyl hydrophobic chains, **34:**489
ε-Alkyllysinase, **42:**245; 278
Alkyl pherases. *See also* Methylpherase and Transalkylation
and ethionine effects, **25:**387
Alkyl phosphates. *See also* specific compounds, eg, Diisopropyl fluorophosphate
in identification of active centers of esteraes, **20:**284; 287; 304–309; 311
in identification of imidazole, **20:**67; 71; 79–81; 89; 90
Alkyl sulfatase, in microorganisms, **22:**230
Alkyl sulfates, occurrence, **22:**206
Alkyl tosylates, structures, relative reactivities, **22:**88; 89
Allantoate, **56:**92
transport system for, **56:**126
Allantoicase, **12:**452–453, **56:**92
classification, **14:**246; 250
kinetic constant for, **56:**98
structural genes for, **56:**98; 99
Allantoicase in plants, **9:**442
Allantoin, **19:**204–206, **56:**118
control patterns for transport and catabolism of, **56:**132; 133
enantiomers of, **56:**92
excretion of, **17:**320–321; 323
fermentation of, **23:**327; 338
hydantoin effect on, **56:**125
production of, **56:**92
transport of, following repression, **56:** 121–123
transport system for, **56:**124–126
Allantoin and purine metabolism, **9:**463–465; 489
Allantoinase, **9:**442, **12:**452–453, **56:**92, **74:**72
classification, **14:**247; 250
kinetic constant for, **56:**98
structural genes for, **56:**98; 99
Allantoin degradation
biochemistry of, **56:**92–98
enzyme production control in, **56:**104–120
genetics of, **56:**109–117
by induction, **56:**104–109
by nitrogen catabolite repression, **56:** 117–120
enzymes of, kinetic constants for, **56:**98

as gene regulation and metabolic integration model, **56:**91–135
genetic organization of, **56:**98–101
homeostasis of, with adverse environment, **56:**129–132
nitrogen metabolism integration of, **56:** 132–134
physical organization of genes for, **56:** 101–104
reactions and genes associated with, **56:**93
transport systems for, **56:**121–128
Allantoin Degradation by Saccharomyces cerevisiae-A Model System for Gene Regulation and Metabolic Integration (Cooper), **56:**91
Allantoin permease, **56:**98; 99
Alleles, definition, **22:**4
Allicin, **15:**417–418, *See* Diallyldisulfoxide
antibacterial action of, **11:**380–381; 391; 398
constitution of, **11:**381
Alliin, **15:**417, **56:**34
chemical investigation of, **11:**377–400
therapeutic use of, **11:**397–399
Alliinase, **74:**134; 152
apoenzyme-holoenzyme equilibria of, **56:**42
classification, **14:**257
cysteine *S*-Conjugate β-lyase reactions, **72:**213–214
inactivation of, **56:**37
inhibition of, **56:**37
molecular parameters of, **56:**44
purification and properties of, **56:**34–35
as pyridoxal-P-dependent enzyme, **56:**35
quaternary structure of, **56:**42
replacing agents for, **56:**10
spectral absorption peak of, **56:**36
substrates for, primary, **56:**10
Alliinase in plants, **10:**258
Alliin lyase, classification, **14:**257
Allin, The Specific Principle of Garlic (Stoll and Seebeck), **11:**377
Allinase
action of alliin, **11:**390–393
preparation of, **11:**391
specificity of, **11:**393–397
and allicin, **15:**417–418
Allithiamine, and thiaminase, **15:**405; 417–418
Allium epidermis, cytological study, **18:**14–15
Allium sativum, **7:**467; 487; 489; 491; 493
Allocytochromes, **23:**267; 312
Alloheptulose, **15:**159
D-Alloisoleucine, conversion of β-methylglutamic acid to, **31:**200
L-Alloisoleucine, in transamination reactions, **16:**193; 197; 199
Allo isomer of L-leucine utilization of, **16:**250
Allophanate, **35:**401
in enzyme production stimulation, **56:**104
structure of, **56:**107
in urea catabolism, **56:**134
in urea transport, **56:**127
Allophanate hydrolase, **56:**94
functional half-life of, **56:**109
immunoprecipitation of activity of, **56:**97
induction of, **56:**104–106
sequence and timing of events for, **56:** 108–109
kinetic constant for, **56:**98
structural genes for, **56:**100
transport of, following repression, **56:** 121–123
urea concentration and activity of, **56:**104
Allophycocyanin, **46:**92; 93
Allopregnane-3β,21-diol-20-one, microbiological conversions, **20:**243
Allopregnane-3β,7β-diol-20-one, microbiological formation, **20:**243
Allopregnane-3,20-dione, microbiological conversions, **20:**273
Allopregnane-3α-ol-11,20-dione, microbiological formation, **20:**253
Allopregnane-11α-ol-3,20-dione, microbiological formation and conversions, **20:**257; 273
Allopregnane-3β-ol-20-one, microbiological conversions, **20:**243
Allopregnane-3,11,20-trione, microbiological conversions, **20:**253; 273
Allopurinol, **54:**68, **57:**150–151
D-Allose, in polysaccharide synthesis, **15:**306
D-Allose, presence of a single common carrier in translocation of, **29:**53
Allosteric activation, **28:**51–52; 58, **38:**309
Allosteric behavior, **39:**379
Allosteric-binding site, fibronectin, ligand-binding site model, **70:**14
Allosteric control, **37:**404; 405
mechanism of, **28:**56–60
Allosteric effector, **48:**288
Allosteric effectors, **50:**30
definition, **28:**49
desensitization to, **28:**52

Allosteric effects, adsorption at interfaces, **28:**30–33
Allosteric enzymes, aggregation state, **28:** 53–56; 68; 73; 83; 90; 105; 117; 129; 132; 138–142; 144
conformational changes, **28:**51–53; 57–59; 93; 105; 117; 120; 125; 136; 138; 142
kinetic properties, **28:**49–52; 72–73; 76; 80–86; 93; 103; 128–133; 136–137; 140–142
Allosteric inhibition, **28:**50–52; 58; 72–73
amidotransferases, **66:**222–223
Allosteric mechanism, **43:**348
Allosteric model, substrate binding and conformational changes, **68:**71–72
Allosteric protein, definition of, **37:**363; 371
Allosteric regulation, **43:**312, **57:**210–212. *See also* Control mechanisms; Feedback regulation; Inhibition; Regulation; Repression
of enzyme activity, **28:**41–154
glutamine PRPP amidotransferase
catalytic activities, **72:**102–104
feedback inhibition, **72:**100–102
nucleotide regulation, **72:**95
PRTase structures, **72:**100
Allosteric regulation of asparate β-decarboxylase, **35:**523
Allosteric site, of enzymes, **26:**105–107
Allosteric sites. *See also* Active sites; Binding sites; Regulatory sites
definition, **28:**49; 51
modification, **46:**371
of phosphofructokinase, **28:**80–83
in subunits of enzymes, **28:**53–58
Allosteric substances, effect on phosphofructokinase, **28:**78–86
Allostery, in proteins, **50:**30–33
D-Allothreonine, utilization of, **16:**250
Allotopy, **34:**457
Alloxan, **17:**311
effect of borate on action of, **12:**523
effect on glucose diphosphate concentration, **14:**212
effect on tryptohan peroxidase, **25:**377
reaction with cysteine, **16:**121
in thiol oxidation, **11:**207
Alloxan diabetes, prevention by thiols, **11:** 247–248; 251
Alloxazine, **15:**89
Allyglycine, **50:**165
Allyl epoxides, **46:**128
Allyl isopropylacetamide, **46:**106; 119; 120; 122; 124; 128; 129; 132; 136–137; 150; 151; 155
S-Allylmercaptocysteine, **15:**418
N-Allylnormorphine
morphine antagonism, **6:**41
Allylthiourea, **7:**207
Almond emulsin
enzyme constituents, **4:**46–48
β-glucosidase from (*See* β-Glucosidase)
hydrolyzing action on different glycosides, **4:**47; 48
purification of enzymes from, **4:**45–46
α-Structure of contractile proteins, **16:**318–319; 325–326
Alternaria species, enzymic transformations of steroids by, **20:**262
plant disease caused by, **20:**391
Alternaria tenuis, **23:**177
Alternative metabolic pathways for amino acids, **16:**239
Altitude, effect on blood glutathione, **11:** 252–253
D-Altroheptulose, **15:**161
Altropine, in schizophrenia, **29:**534
Aluminum
effect on pectic substances, **20:**351
effect on phosphoglucomutase activity, **14:**212
Aluminum and luminescence, **7:**253
Aluminum compounds, **17:**142; 147; 148
Aluminum-enzyme complex formation, **10:**201–202; 207–208
Aluminum hydride, in protein structure study, **14:**279; 286
Aluminum hydroxide, as adsorbent in isolation of prothrombin, **16:**28
Aluminum ion as decarboxylating agent, **12:**69–72; 80; 87–88
Aluminum ions, **15:**379
concentration in monolayer, **11:**68
model catalyses with, **21:**174–177
in nonenzymic transaminations, **16:**224; 225
Aluminum phthalocyanine, oxidation of, **15:**33; 39–40
Amadori rearrangement, **35:**50; 58
α-Amanitin, **47:**289
Amanitin in plants, **10:**285; 296

Amastatin, **68:**161
binding to leucine aminopeptidase, **68:** 184–185
complexes, **68:**177–178
interactions with leucine aminopeptidase, **68:**173
Amberlite, use in decalcification of plasma, **10:**15
Amberlite IRA-400, as adsorbent in isolation of clotting factors, **16:**27
Ameba, permeability studies, **13:**44–45
stratification by centrifuging, **13:**6; 12
Amethopterin, **38:**275, **43:**331, **67:**240
reduced analogs, **38:**274
synthesis and structure, **21:**375; 376
Amethopterin-resistant organisms, **27:**454; 455
Amicyanin, **59:**207
Amidase, interference in transamination study, **16:**200; 201
Amidase-esterase, cysteine-activated, **23:**188
Amidases, **13:**286; 295–296, **17:**176. *See also* Asparaginase and Glutaminase
classification, **14:**224; 242; 246; 250
avian liver, action on mixed disulfides, **14:**65
action on pantetheine, **14:**63
for muropeptides, **26:**211–215
penicillins as substrates, **28:**238; 247–249
soil, **28:**386
Amidases in plants, **9:**410; 417–420
Amidations, and ATP, **18:**229–231
Amide bonds
generation of, **17:**176–179
synthesis, in purine biosynthesis, **21:**233–240
Amide bond spectra, **42:**136
Amide-chain theory of energy transport in living cells, **14:**27–29; 41
Amide group, role in NADPH binding, **39:**381
Amide groups, **50:**2; 3; 13; 24; 27; 31; 32
iminol form, **50:**3
pK value of, **50:**2
polarity of, **50:**24
resonance in, **50:**2; 27; 31
tautomerization of, **50:**13
Amide groups of RNase, chemical modification of, **24:**209; 224–225
Amide linkage, planarity of, **50:**32
Amides. *See also* Asparagine, Glutamine, Glutathione
as agent for transfer across cell membrane, **13:**268–271
aliphatic, transglutaminase activity towards, **54:**12–17; 18
aminodicarboxylic acid, transglutaminase activity towards, **54:**17–19
ammonia of proteins, **13:**245–246
enzymic transformations in plants, **20:** 334–338
and esterase activity, **24:**462–464
cyclic, enzymes acting on, **24:**328
linear, enzymes acting on, **24:**327–328
kinetic studies, **17:**47; 59–60
linkage, synthesis by papain, **13:**307–308
use of term, **13:**241
metabolism in plants, **13:**263–266; 303
nitrogen of proteins, metabolic activity, **13:**266–267
in plant metabolism, **9:**392; 409–421; 432–433; 434
light-induced changes in, **9:**91–96
synthesis in plants, **9:**412–413
role in transamination reactions, **13:**265; 272; 284–285
as tranglutaminase substrate, **54:**51
Amides, Enzymic Transformations in Plants (Kretovich), **20:**319
Amides in preparation of dehydropeptides, **8:**124–128; 161–162
Amide synthetases, localization of, **24:**334
Amidinases. *See also* individual amidinases
classification, **14:**224; 247
Amidine group transfer, **15:**373–374
Amidines, **15:**379
cyclic, enzymes acting on, **24:**328
linear, enzymes acting on, **24:**328
inhibition of tumor growth by, **29:**336
N-Amidinoazetidine-2-carboxylic acid, **54:**387; 390
Amidinotransferase, liver, repression by creatine, **50:**202–204
ω-Amidodicarboxylate amid ohydrolase, **39:**186
Amidotransferase, allosteric regulation, **28:** 118–121
Amidotransferases, **66:**203–289
allosteric inhibition, **66:**222–223
amide transfer mechanism, **66:**208–210
arylamine synthetase, **66:**282–283
background research, **72:**88–89
catalysis, **66:**216–222

Amidotransferases (*Continued*)
chorismate conversion to anthranilate, **66:**208–210
covalent glutamyl thioester intermediate formation, **66:**217
cobyrinic acid *a,c*-diamide synthetase, **66:**284–287
cobyrinic acid synthetase, **66:**285–287
domain organization, **66:**219–221
evolution, **66:**287–289
examples, **66:**205
F-type, **66:**210; 212–214
asparagine synthetase, **66:**276–280
glucosamine 6-phosphate synthase, **66:**264–268
glutamate synthase, **66:**268–276
glutamine PRPP amidotransferase, **66:** 251–264
general information, **39:**91
reactivity with alkylating agents, albisiin, **39:**106; 111; 155
2-amino-4-oxo-chloropentanoic acid, **39:**106; 126; 135; 145; 167
D-azaserine, **39:**107
L-azaserine, **39:**106; 107; 120; 145; 155; 163; 166
6-diazo-5-oxo-L-norleucine (DON), **39:**97; 100; 106; 107; 120; 134; 135; 145; 146; 147; 155; 156; 161
5-diazo-4-oxo-L-norvaline, **39:**107; 135; 166
iodoacetamide, **39:**106; 111
iodoacetate, **39:**106; 111; 132; 165
sulfhydryl group at active site for glutamine reaction, **39:**108; 126; 132; 145
in synthesis of, amides, **39:**161
amines, **39:**152
aromatic and heterocyclic compounds, **39:**137
purine nucleotides, **39:**94
pyrimidine nucleotides, **39:**124
use of ^{14}C labeled antimetabolites in determination of glutamine active sites, **39:**93; 97; 106; 126
Gln-tRNA^{Gln}, **66:**281–282
glutamine affinity analogues, **66:**207–208
G-type, **66:**210–212
aminodeoxychorismate synthase, **66:** 223–225
anthranilate synthase, **66:**214–223
carbamyl phosphate synthetase, **66:** 228–238
CTP synthetase, **66:**243–247
formylglycinamidine synthetase, **66:** 238–243
GMP synthetase, **66:**247–249
imidazole glycerol phosphate synthase, **66:**249–251
missense mutants, **66:**219; 221
NAD synthetase, **66:**283–284
NH_3-dependent activity, **66:**208–210
NTN amidotransferases, **72:**93–119
asparagine synthetase, **72:**106–112
future research, **72:**117–119
glucosamine 6-phosphate synthase, **72:**104–106
glutamate synthase, **72:**112–117
glutamine PRPP amidotransferase, **72:** 93–104
structure-function relationship, **72:**90–93
structure-function relationship, **66:**206–208
structure-function relationships, **72:**89–93
subfamilies, **66:**210–214
triad amidotransferases, **72:**119–137
aminodeoxychorismate synthase, **72:** 136–138
ammonia transfer theories, **72:**138–139
anthranilate synthase, **72:**136–138
carbamyl phosphate synsthetase, **72:** 124–131
GMP synthetase, **72:**119–124
imidazole glycerol phosphate (IGP) sunthase, **72:**131–134
NAD synthetase, **72:**134–136
Amidotransferases, The (Buchanan), **39:**91
Aminases, classification, **14:**224; 246; 250
Amination, *See* Transamination, Transreamination
reductive, **15:**338; 363
in plants, **9:**402; 408; 436, **20:**323–328; 330
Amination and transamination, **1:**154; 155
Amination-deficient mutants of *Neurospora,* **16:**256
Amination of oxalacetate, **16:**226
of pyruvate, **16:**226
Aminations, and ATP, **18:**229–231
and citric acid cycle, **18:**251–256; 258
Aminco-Chance DW2 spectrophotometer, **51:**12; 25; 43

Amine dehydrogenases, **59:**191–192
Amine-glyoxylate adducts, **57:**110; 113; 122–123
Amine oxidase, **35:**115
action of tyrosinase on, **14:**147; 149
in beef serum, **36:**235
diamine oxidase, **36:**235
histochemical methods for, **9:**588
Amine oxidases, **7:**533–538; 542–543
and apoenzyme, **59:**194
and bovine plasma, **59:**193
and cofactor interaction, **59:**196
in electron transfer, **19:**201
and hydrazine formation, **59:**192; 201
and redox forms, **59:**199
and vitamins, **59:**209
Amines, **57:**102–105. *See also* and Spermine; Asparagine, Glutamine, Glutathione; 1,4-Diaminobutane; Spermidine
absorption spectra, **9:**77
acetylation of, **16:**1–2; 18
and pyruvic acid catabolism, **8:**388–389
reactions with quinones, **16:**108–121
action on thiaminase II, **15:**410
activators of thiaminase, **15:**399–400; 403–404
aliphatic, and thiaminase, **15:**402
aromatic, acetylation, **15:**186; 189; 196; 199; 211; 257
formation, **15:**376
heterocyclic, activators of thiaminase, **15:**400–404
reaction with peroxidase, **15:**17
aliphatic, transglutaminase intermediates with, **54:**33
antagonism by bezedrine, **6:**41–42
in exchange reactions, **13:**289–305; 305–310
fluorinated, **55:**227–240
inhibiting fibrin crosslinking, **54:**30
peroxidase action on, **14:**151
primary, catalysts for α-ketoacid decomposition, **14:**168–169; 170–171
pressor (*See* Pressor amines)
in transglutaminase interactions, **54:**30; 52
α-Aminoacetic acid, and glycine metabolism, **23:**456
Aminoacetone, **23:**456
α-Amino-β-acetylaminopropionic acid, **50:**121; 124
Amino acid activating enzymes, of nucleolus, **27:**268
of nucleus, **27:**275
Amino acid activation, **17:**184
by L-diaminobutyric acid, **33:**370
sequences, of gramicidin S, **33:**337–338
in gramicidin S biosynthesis, **33:**370–373
by L-leucine, **33:**370
by L-ornithine, **33:**370
by D-phenylalanine, **33:**370
by L-phenylalanine, **33:**370
by L-proline, **33:**370
by L-valine, **33:**370
Amino acid adenylate-forming enzymes, **25:**387–388
Amino acid adenylates, **23:**324
general structure, **22:**190
synthesis, **22:**190; 191
Amino acid amides
deamination, by metal ions, **21:**178; 179 194
in transpeptidation reactions, **19:**310
Amino acid analogs
effect on β-lactamase, **28:**274; 275
effect on nucleic acid synthesis, **29:**148
effect on sRNA, **29:**150
effects of, on formation of gramicidin S, **33:**340
isosteres, **29:**90
relation to protein synthesis, **25:**376
types of structural variations of, **29:**90
Amino acid analyses, **42:**88
of β-lactamase, **28:**270–272; 275–278; 280–284; 311
Amino acid anhydrides, and phrophosphorolysis, **18:**202; 203; 208–209
Amino acid biosynthesis
regulation by aminoacyl-tRNA synthetases, **40:**212; 213; 214; 215; 216; 217
Amino Acid Biosynthesis, Intermediates in (Davis), **16:**247
Amino acid chlorides, **19:**238; 239
Amino acid chromophores, and Cotton effect, **27:**46; 60; 95–96
Amino acid code, **47:**375–430
error-free use of, **47:**375
university of, **47:**379–380
Amino Acid Code (Jukes), **47:**375
Amino acid composition, **39:**347, **42:**86; 87; 88; 151; 294, **43:**144; 145
of contractile proteins, **16:**319; 330

Amino acid composition (*Continued*)
of histones, **16:**442; 444–445
of protamines, **16:**441–443
of prothrombin, **16:**29
of residual proteins, **16:**447
tyrosine hydroxylase, **70:**120–121
Amino acid content, of muscle phosphorylases, **24:**266–268
of proteins, effect of ionizing radiations, **24:**360; 381–383; 397
of RNase, **24:**175
Amino acid decarboxylase-pressor amine-amine oxidase system, **7:**533–538
Amino acid decarboxylases, **5:**67–5:83, **7:** 534–538, **16:**210; 219–220; 297
classification, **14:**256
coenzymes for, **14:**182
Amino Acid Decarboxylases, Bacterial (Gale), **6:**1
Amino Acid Decarboxylases of Mammalian Tissue (Blaschko), **5:**67
Amino acid dehydrogenases, **15:**118; 123; 125
PN-dependent, **26:**120
Amino acid derivatives, **19:**308; 309
Amino acid esters, **19:**237; 238
peptide synthesis, **19:**263
Amino acid fermentations, **58:**176
Amino acid hydroxylases, pterin-dependent, **67:**222
Amino acid interchanges in related proteins, **24:**417; 419; 420–425
permitted, **24:**422–436
Amino acid metabolism, genetic studies, **17:**395–422
D-Amino acid:O_2 oxidoreductase (deaminating), *See* D-Amino acid oxidase
Amino acid oxidase, **7:**374; 379; 393; 534
absorption spectrum, **4:**301; 305
in plants, **20:**322
D-Amino acid oxidase, **9:**563, **18:**77; 118; 306, **35:**94; 95, **37:**409; 414; 421; 424; 488; 508; 528; 557; 570
action on isomeric peptides, **8:**150–152
and dehydropeptidase, **8:**151
in nuclei, **8:**33
products from various substrates, **8:**149; 150–152
and amine-glyoxylate adducts, **57:**102–104
and amines to support O_2 uptake, **57:**102–104
catalytic efficiency of, **57:**108–109
catalytic reaction of, **57:**112–113
and cell growth control, **57:**94
characteristics of, **57:**91–92
classification, **14:**236; 253
in phenylalanine metabolism, **14:**99; 100
separation by chromatography, **14:**349
coenzyme binding, **22:**387; 389; 390
concentrations of, **57:**90
conversion oxidized enzyme-substrate complex to purple intermediate of, **34:**55–60
conversion of purple intermediate to fully reduced form of, **34:**60–63
electronic interaction involved in purple intermediate, **34:**63–67
interactions involved in oxidized enzyme-substrate complex of, **34:**52–55
mode of reaction of, with basic amino acids, **34:**71–75
occurrence, of oxidized enzyme-substrate complex of, **34:**48–51
of purple intermediate of, **34:**43–48
primary reaction product of, **34:**69–71
reaction of purple intermediate of, with oxygen, **34:**67–69
and cysteamine-glyoxylate adducts, **57:** 103–104; 106–107
and detoxification, **57:**92–93
discovery of, **57:**89
distributions of, **57:**90
effect of radiations, **24:**388
in elimination of optical antipodes, **20:**18; 19
functions of, physiological, **57:**92
inhibition by thiol reagents, **11:**237; 239; 241
inhibition of, **57:**92
inhibitors of, **57:**140–143
effects of, **57:**143–145
and plasma membrane enzymes, **57:**147
kinetic constants for, **57:**107–108; 110
kinetic mechanism of, **57:**89–91
limiting rate constant for, **57:**108
location in subcellular fraction, **24:**302; 305; 307; 340; 341; 343
Michaelis constant for, **57:**108; 111
and nucleophile-glyoxylate adducts, **57:** 99–101
and β-phenethylamine products, **57:**105
in plants, **10:**300
presence of, **57:**86
products of, **57:**135–137

and sodium-restricted diet, **57:**93; 138
specificity of, **57:**90–91
stabilization by coenzyme, **23:**47
substrate requirements for, **57:**99–101
in synthesis of *d*-alanine and *d*-serine, **20:**14
and thiazolidine-2-carboxylate products, **57:**104–108
and thiazoline-2-carboxylate products, **57:**105–106
in transamination reactions, **19:**352; 376; 377
turnover number of, **57:**108
as vestigial enzyme, **57:**92–93
debate of, **57:**93–94
L-Amino acid oxidase, **37:**428; 541; 570, **46:**451; 455
action on proline, **16:**259
distribution, **16:**230; 256
adsorption studies of, **14:**346
classification, **14:**253
imidazole in, **20:**100
isolation, **14:**397; 399; 400–401; 405
location in subcellular fraction, **24:**301; 340
in plants, **9:**402–403; 406
in snake venoms, **8:**476–484; 487–489
of kidney, **8:**355; 481–482
sources, **14:**377
in sperm, **9:**375
in transamination reactions, **19:**340
α-Amino acid oxidase
cacao, **25:**465; 466
D-Amino acid oxidase, EPR study, **27:** 132–133
in mechanism study, **27:**2–34
protein, physicochemical constants, **27:**25
L-amino acid oxidase, multiple forms, **25:**358
Amino acid oxidase, PN-dependent, **26:**121
D-*Amino Acid Oxidase, Reaction Mechanism of* (Yagi), **34:**41
Amino acid oxidases, **15:**123; 340
(D- and L-), EPR study, **27:**183
in electron transfer, **19:**201
properties, **19:**203
of *Proteus vulgaris,* **8:**481–482
Amino acid phosphate derivatives, as activated amino acids, **19:**249
Amino acid polymerase, **48:**7
Amino acid pools, in microorganisms, **22:**125
Amino acid racemases, **44:**71
inhibitors, **44:**106
pyridoxal phosphate enzymes, **44:**71
Amino acid racemization, **44:**99
nonenzymatic, **44:**99
Amino acid residues
chemical method for replacement of, **33:** 383–399
enzymic method for replacement of, **33:**399
fibronectin, **70:**2
Amino acid-RNA ligases, localization of, **24:**333–334
D-Amino acid r O_2 transhydrogenases, classification, **14:**236; 253
L-Amino acid r O_2 transhydrogenases, classification, **14:**253
Amino acids, **35:**480–483, **47:**46; 51; 53; 61; 62; 65; 67; 71; 79; 80; 81, **49:**42, **53:**163; 226. *See also* Diamino acids; Polypeptides; Proteins; individual amino acids
α, β-unsaturated, and imino acids, **8:**117–118
abiogenic syntheses, **27:**382–383
interactions with ploynucleotides, **27:**397
in meteorites, **27:**350
and origin of life, **27:**356–359; 363; 366–368; 374–375
sulfur-containing, determination of, **27:**231–232
thermal polymerization, **27:**392; 405
thermalsynthesis, **27:**358
absorption spectra, **4:**277; 279
absorption spectra of, **9:**77–81
acrylic acids from, **9:**408–409
deamination, **9:**84
dicarboxylic, in plant metabolism, **9:**394; 398–401; 402–409; 419; 421–430
effect on fertilizability, **9:**250–251
light-induced changes in, **9:**81–89
in plant metabolism, **9:**391–439
and pyrimidine synthesis, **9:**468
in semen, **9:**338–339; 343; 345
as sperm nutrients, **9:**375
of viruses, **9:**181–183; 231–234
and acetate metabolism, **23:**419–420; 426–429; 432–434
and glycolate metabolism, **23:**440–441
from labeled acetate, **23:**405–406
Stickland reaction, **23:**457–458
acetylation, **6:**262, **15:**196
composition of enzymic and non-enzymic proteins, **15:**36–37; 296–297

Amino acids (*Continued*)
dehydrogenation by model enzyme, **15:** 29–30
oxidation, and pyridine nucleotides, **15:**122–123
in rennin, **15:**429–430
sulfur-containing, **15:**124–126
in tobacco mosaic virus, **15:**461–464
in urea synthesis, **15:**330–331; 338 (N^{15}); 339–341; 343; 347; 374 (N^{15})
in virus (TMV) mutants, **15:**474; 479–480
action of carboxylase on, **14:**187
action of peroxidase on, **14:**152–153
activated, **18:**209
and anaerobiosis, **18:**271–274
from aspartic and glutamic, **18:**256–257
effect of fluoroacetate on levels, **18:**137
interconversions, **18:**226
labeled precursors of, **18:**267–269
in plant respiration, **18:**299; 300; 302–303
radioactive, in study of DNA function, **18:**183–184
sequence in protamines, **18:**173
source of nitrogen for, **18:**253–255
and vitamin B_{12}, **26:**254
metabolism, role of cobamide coenzymes, **26:**265
in mureins, **26:**196–204; 216–219
activation, in protein synthesis, **22:**129; 130
at active site, identification without covalent labeling, **22:**57–63
aliphatic, at active site, role and identification, **22:**56
auxiliary, **22:**68; 89
definition, **22:**48; 49
bond-changing functions in enzyme activity, **22:**69; 70
chromatography, **22:**419
in composition of active site, **22:**63–72
contact, **22:**56; 67; 89
definition, **22:**48
number, **22:**69; 89
contributing, **22:**89
definition, **22:**49
four-carbon, formation, **22:**259–261
free, role in enzyme formation, **22:**125
noncontributing, **22:**89; 90
definition, **22:**49
phosphoric anhydrides of, preparation by phosphorochloridate method, **22:**174
recurring sequences in enzymes, **22:**68; 69
specificity functions in enzymes, **22:**69; 70
transfer to soluble RNA, **22:**131; 132
activation *in vitro* and *in vivo,* **19:**235–266
activation of, **24:**2–4
labeled, in enucleation experiments, **24:**12
metabolism of, **24:**340
in sperm metabolism, **24:**62–66; 88–89
acylated, action of peptidase on, **12:**235–237
amides, action of leucine aminopeptidase on, **12:**200–204
biosynthesis of, **12:**43
effect on oxaloacetate carboxylase, **12:** 80–81
oxidase, **12:**451; 467
residues, effect on peptidase action, **12:**229–232; 247–248
role of biotin in metabolism of, **12:**81–85
unnatural derivatives, action of peptidases on, **12:**232–235
adsorption studies, **14:**326–327; 329–330; 343–344; 345
albumin-bound, **49:**92
and alkaloid synthesis, **8:**243–244
in allantoin transport, **56:**126
alliin, **56:**34
allosteric regulation by, **28:**140
effect on β-lactamase, **28:**263
amino group, **35:**481
analogs of sulfur-containing, **29:**134
aromatic
activation and incorporation into protein of, **29:**110
biosynthesis of, **29:**116
control of biosynthesis by analog, **29:**115
phenylalanine analogs, **29:**110
resistance to analogs of, **29:**122
structures, occurrence, and growth effects, **29:**108
synthesis of, **29:**116
benzyl esters of, **29:**400
tert-butyl, **29:**412–413; 442
esters of, **29:**400
tert-butyloxycarbonyl group of, **29:** 397–398; 437; 441
carbobenzoxy group of, **29:**394–395; 412–413
modified, **29:**395–397
concentration in neoplastic cells of, **29:**364
conservation of the twenty, **29:**155

containing selenium, **29:**141
control of biosynthetic pathways of, **29:**92
evaluation of synthetic methods of protecting groups, **29:**437–443
N-formyl, formation and cleavage of, **29:**399
formyl group of, **29:**399; 437; 441
growth inhibiting, analog of, **29:**91–92
growth-inhibiting properties of, **29:**90
incorporation into proteins of tumors, **29:**365
methylation of, **29:**399–401
nitro group, for masking the guanidino group of arginine, **29:**401–402
p-nitrophenyl esters of, **29:**406
nonprotein, of plants, **29:**91
phthalyl group of, **29:**399
derivatives of, **29:**399
structural analogs of, **29:**90
p-toluenesulfonyl group of (toysl), **29:**398; 438; 441
toxic, **29:**89–163
of *Lathyrus,* **29:**147
mechanism of action, **29:**148
protective mechanisms against, **29:**94
in *Vicia,* **29:**147
triphenylmethyl group of (trityl), **29:** 398–399
uptake of, inhibited by cyanide in tumors, **29:**365
inhibited by dinitrophenol in tumors, **29:**365
in antibiotic peptide biosynthesis, activation, **35:**17; 23
analogues, **35:**17
ATP-AMP exchange, **35:**6; 8
ATP-PP_i exchange, **35:**8
effect of analogues, **35:**18
polymerization, **35:**7; 11
racemization, **35:**5; 22
thioester-linked, **35:**10; 14
aromatic
biosynthesis, **16:**287–295
of arthropodin, **16:**157–158
basic, effect on blood clotting, **16:**88
4-carbon group, interrelations of, **16:**268–275
C^{14}-labeled, uptake of, **16:**491–492
deamination by *o*-quinones, **16:** 117–120
deuterium labeled, **16:**199; 218; 294
dicarboxylic, in transaminations, **16:**199; 203; 225–226
incorporation of labeled acetate, **16:**269; 276; 282; 283; 285; 286; 290; 294; 296–299; 301
intermediates in biosynthesis, **16:** 247–312
lebeled, in nucleic acid study, **16:** 450–462
uptake by cytoplasmic ribonucleoprotein, **16:**474–475
monocarboxylic, in transaminations, **16:**203–204
N^{15}-labeled, in transamination study, **16:**186; 193; 218; 223; 226–227
in nonenzymic transaminations, **16:** 223–226
poly-α, effect on blood clotting, **16:**88
preparation of pure isomers, **16:**238
reactions with quinones, **16:**112–117
role in transamination in metabolism of, **16:**226–237
in transamination systems, **16:**185–246
in energy transport studies, **14:**32–35; 36–37
arrangement in enzymes, **14:**190
asparagine, **56:**27; 33–34; 117
carboxyl group, **35:**481
catalytic oxidation of, **14:**166–168; 178; 179
cellular levels, **49:**43
fetal tissues in pregnancy, **49:**43
liver, elevated supply, **49:**44
liver during hyperplasia, **49:**43
neoplastic tissues, **49:**43
change or destruction during protein hydrolysis, **14:**265; 267–270
changes during fermentation of tobacco, **10:**340; 382–386; 396; 403; 429
charged, **47:**49; 53; 67; 79; 81; 103; 114–115; 131
chelates, in metal ion activation of enzymes, **21:**145
common, **14:**263–264; 290
complex formation with quinones, **10:**381; 382–386; 408; 429
as complexing agents, **17:**138; 149–151
labeled, in virus protein synthesis, **17:** 490–495
composition, **35:**448, **47:**47; 85; 139

Amino acids (*Continued*)
reactions, alkylation, **35:**93
C_β-C_γ isomerization, **35:**111
cleavage at C_α-C_β, **35:**93–97
decarboxylation, **35:**84–92
elimination-addition, **35:**107–115
racemization, **35:**106
transamination, **35:**97
transport, **35:**189–192
analogues, in study of, **35:**190
diversity of systems, **35:**192
energy requirement, **35:**190
gram-negative bacteria, **35:**189
gram-positive bacteria, **35:**189
role of peptides, **35:**194
composition of proteins, **14:**262–277
conformational potentials, **47:**51; 66; 68–71; 97; 135
for cyanoalanine synthase activity, **56:**32
cysteine (*See* Cysteine)
deamination, by metal ions, **21:**178; 179; 194
deamination of α-acids, **4:**75–76
decarboxylases, bacterial (*See* Bacterial amino acid decarboxylases)
dehydrogenase, diaminopimelate dehydrogenase, **72:**319–320
dehydrogenation, **8:**362–363; 384; 387–388
dipeptides, **47:**102–105; 130
distribution in proteins, **47:**406; 412
essential, effect of vitamin B_6 on synthesis, **19:**351
and ethionine effects, **25:**370–373; 376–380
photosynthesis of, **25:**47; 55; 62; 71
sequence in enzymes, **25:**332
fluorinated, **55:**227–240
helix formers, **47:**50–51; 67
homologous sequence, **47:**84; 125; 127; 130; 133; 136
hydrophobic residues, **47:**72; 81; 114–115; 131
and imino acids, **8:**117–118
incorporation into proteins, in mitochondria, **21:**44
inhibitory analogue of, **6:**133–134
isomeric peptides, **8:**147–152
and keto acids, **8:**117–118
interaction, **4:**77–79
in lipoxidase, **19:**99
and liver phosphatide, **7:**141
breakdown by kidney enzymes, **7:**533–536
oxidases, **7:**374; 379; 393; 534
sulfur-containing, and growth, **7:**376; 383
low lipid solubility, **49:**45–48
mercapto-, **56:**63–67
metal ion complexes of, model catalyses with, **21:**189
in mitochondrial lipoproteins, **21:**109
as model substrates, **49:**60
modification, **47:**76–77
movements among tissues, **49:**42
mutation, **47:**79–83; 85; 135
in myasthenia gravis, **6:**212
from nitrogen fixation, **13:**362–368
in nitrogen metabolism, **19:**335–386
^{15}N-labeled, **38:**434
optical activity of, **38:**229
noncompetitive inhibition, **49:**55
nonprotein, of plants, **50:**117–171
biological role, **50:**157
metabolic action of, **50:**157–171
nutritional significance of, **49:**44
optical activity, change during transamination, **4:**77–78
optical configuration, **35:**481
optical rotation and biological activity, **4:**9
during transamination, **4:**77–78
oxidation, **8:**412–414
in mitochondria, **21:**12
oxidation in peptides, **8:**150–152
in papain, **19:**282–285
phenylalanine hydroxylase activation, **67:**167–168
plasma levels, **49:**43
positional preferences, **47:**68–72; 97
in progressive muscular dystrophy, **6:**220
in proteins, **1:**1
pyridoxal-P-dependent enzyme transformation of, **56:**5
isotopic exchange of hydrogen atoms in, **56:**50
racemic, enzymatic resolution, **1:**94
racemization of, **14:**193; 195–196; 263; 265; 269–270
rare, **14:**264–266; 290
ratio
in insulin, **4:**3
in tobacco mosaic virus, **4:**7
reaction with sugars in aging of tobacco, **10:**408
regulated flows, **49:**42

regulation of aminoacyl-tRNA synthetase synthesis, **40:**219; 220; 221; 222
relation to cystein inhibition of *Aspergillus niger,* **10:**225
release by liver, **49:**42
residues, **1:**19–21; 64, **47:**65
zonal distribution, **47:**87
resolution of racemic mixtures by papain, **19:**308; 309
role in activation of enzymes, **10:**100; 103–104; 106–107
secondary, **67:**267–271
secondary structural predictions, **47:**130
sequence, **47:**46–67; 74; 77; 112; 127; 134
serine, **56:**18; 19; 22
β-sheet formers, **47:**49–51; 67; 79; 83
side chains, **35:**481–483
side chains, funtion of, **34:**1
regulation mechanisms, **34:**29–37
side chains of, as zinc ligands, **56:**315
of some crystalline enzymes, (table), **14:** 274–275
steady state in protein molecule, **4:**11
as substrates and activators, **67:**166–172
sulfur-containing, **11:**390–397
synthesis, **6:**271; 300, **8:**119
synthesis in animal tissues, **19:**354–386
terminal oxidation, **8:**412–414
tetrapeptides, **47:**52–53; 61–62; 105
transamination, **5:**154–155
translocation, transport systems mediating, **53:**208
as transport substrate, **49:**41–101
tripeptides, **47:**102–105; 130
t RNA, **35:**489
β-turn formers, **47:**71–72
tyrosine, **4:**100; 115; 144–145
unnatural or D-amino acids in Tyrocidine and Gramicidin, **4:**154; 163; 164; 165; 166; 168; 169; 183
D-Amino acids
in ACTH peptides, **29:**455
elimination in living organisms, **20:**18
as growth factors, **16:**249–250
transaminations of, **16:**206–207; 218–219
oxidative deamination of, **19:**349
D-α-Amino acids, **57:**90–91
L-Amino acids, **57:**98
for activation in liver mitochondria, **19:**261
oxidative deamination of, **19:**341; 348; 354; 378
cysteine *S*-conjugate β-lyase reactions, prodrug targeting, **72:**227–230
oxidative deamination, **16:**226
α-Amino acids
charge on side chain, **49:**48
relation to activity of *E. coli* β-glucuronidase, **16:**379
DL-Amino acids, resolution of, **53:**243
Amino Acids, Toxic, Action as Antimetabolites (Fowden, Lewis, and Tristram), **29:**89
Amino acids (C^{14}), use by sperm, **24:**88–89
Amino acids dicarboxylic
biosynthesis in plants, **20:**319–340
in chloroplasts, **20:**209
in chymotrypsinogen, **20:**285
in chytochrome *c,* **20:**193
effect on protein synthesis in plants, **20:**223
optical purity in living organisms, **20:**2
Amino acid sequence, **39:**364, **42:**86. *See also* Peptide sequences; Protein structure
analysis, **27:**229–231
genetic control, **27:**295
mutations, **27:**302; 304; 312
at receptor site, **27:**80–82; 432
aspartate transcarbamylase, **68:**81
of chymotrypsinogen, **24:**416; 418; 420–422; 432–435
collagen hydroxylases, **72:**330–334
cosubstrate binding, **72:**349–357
peptide substrates, **72:**342–345
in cytochrome *P*-450, **60:**49
of cytochromes *c,* **24:**416; 418; 419
and enzyme structure, **26:**89–114
of esterases and proteases, **24:**458–459; 465
evolutionary concept of, **24:**415–439
genetic control of, **24:**7–11
lysyl hydroxylase, **72:**380–382
and molecule conformation, **24:**23; 200–202
in papain and mercuripapain, **19:**294–301
of phosphoglucomutase, **24:**465
of pituitary hormones, **24:**418–419; 422–425; 430
prolyl 4-hydroxylase α-subunit structure, **72:**366–368
protein disulfide isomerase (PDI) prolyl 4-hydroxylase β subunit, **72:**370–375
and purines, **24:**9–11; 15

Amino acid sequence (*Continued*)
and pyrimidines, **24:**9–11; 15
of ribonuclease, **24:**173–176; 416; 418; 421–424; 426–427; 430; 432–435; 448
total, **39:**348
of trypsinogen, **24:**416; 421; 422; 432–435
Amino Acid Sequence, Prediction of the Secondary Structure of Proteins from (Chou and Fasman), **47:**45
Amino Acid Side Chains, Function of (Matheja and Degens), **34:**1
Amino Acids of Plants, Nonprotein (Fowden, Lea, and Bell), **50:**117
Amino acid structure, **49:**41
exploiting to learn about membrane transport, **49:**41–101
modifiability of, **49:**48
Amino Acid Structure, Exploiting to Learn About Membrane Transport (Christensen), **49:**187
Amino acid substitutions, **57:**400
Amino acid substitution studies, fibronectin, central cell-binding domain of, **70:**8; 10
Amino acid substrate, tyrosine hydroxylase, substrate specificity, **70:**116–119
Amino acid synthesis, allosteric regulation, **28:**44–45; 60–71
Amino acid transport, **49:**41–101
effect of sodium or potassium ions, on, **29:**365
in neoplastic cells, **29:**365
role of phospholipids in, **29:**376
neutral, **49:**54
of nucleus, **27:**260
overlapping systems for, **49:**54–60
tolerance for structural variation, **49:**55
Amino acid-tRNA synthetase, **43:**333
Aminoacrylate, cysteine *S*-conjugate β-lyase reactions, β-elimination reactions, **72:**218–219
Aminoacrylic acid, **49:**29
α-Aminoacrylic acid, in tryptophane synthesis, **16:**294
Aminoacyl adenylates, **19:**251–252, **35:**485; 487. *See also* Adenosine monophosphate and specific compounds, eg, Valyl adenylate
carbamoyl adenine, **35:**469
Aminoacyladenylates, intermediates of aminoacyl-tRNA synthetases, **40:**173; 174; 175; 177; 178; 179; 180
Aminoacylase, **33:**271
Aminoacyl mercaptans. *See also* Cystine, Cysteine, and specific compounds
as activated amino acids, **19:**248
effect on intramolecular aminolysis of, **19:**242–245
as models for biologically active amino acids, **19:**239
properties and transformations, **19:**241–249; 261
syntheses, **19:**239–241
Aminoacyl phosphates, **19:**250
Aminoacyl synthetases, binding to t-RNA, **40:**50
Aminoacyltransferases, localization of, **24:**311
Aminoacyl Transfer RNA Synthetases, Understanding the Recognition of Transfer RNAs by (Schimmel), **49:**187
Aminoacyl-tRNA, **38:**211, **53:**253
control of cognate synthetase levels, **40:**220
corepressor of amino acid biosynthesis, **40:**214; 215; 216; 217
deacylation, **40:**186; 187
donor substrate for aminoacyl-tRNA protein transferase, **40:**101; 102
for alanyl-tRNA phosphatidylglycerol transferase, **40:**115; 116
in the arginine-transfer reaction, **40:**102
in the biosynthesis of interpeptide bridges, **40:**121; 130; 131
in the leucine, phenylalanine-transfer reaction, **40:**102
for lysyl-tRNA phosphatidylglycerol transferase, **40:**115; 116
Aminoacyl-tRNA biosynthesis, regulatory role in amino acid biosynthesis, **40:**213; 214; 215; 216; 217
Aminoacyl-tRNA GlcNAc-MurNAc-(pentapeptide)-*P*-*P*-phospholipid, transferases, aminoacyl-tRNA specificity, **40:**130; 131
particulate, **40:**120; 121; 122; 123; 124; 125; 126; 127; 128; 129; 130
Aminoacyl-tRNA N-acetylmuramyl-pentapeptide transferases, **40:**118; 119; 120; 121; 122; 123; 124; 125; 126; 127; 128; 129; 130; 131; 132; 133; 134; 135; 136; 137
Aminoacyl-tRNA phosphatidylglycerol transferases, **40:**111; 112; 113; 114; 115; 116; 117; 118

acceptor specificity, **40:**116; 117
cellular funtion, **40:**117; 118
Aminoacyl-tRNA synthetase, **50:**135; 160; 161; 165
alterations in active site of, **50:**160–167
Amino Acyl-tRNA Synthetase, Structure and Activity of, Some Questions About (Mehler and Chakraburtty), **35:**443
Aminoacyl tRNA synthetases, **33:**371, **49:**187
enzyme-tRNA contact points, **49:**191–207
Aminoacyl-tRNA synthetases, **35:**444, **43:**5; 20; 43
Amino acyl-tRNA synthetases, **55:**340
Aminoacyl-tRNA synthetases
acylation by mutant synthetases, **40:**207; 208
amino acid composition, **40:**143
amino acid regulation of enzyme synthesis, **40:**219; 220; 221
association complexes, **40:**223; 224; 225
association constants, **40:**160; 161; 162; 163; 164; 165; 166; 167; 168
association with ribosomes, **40:**223; 224
ATP-PP_i, exchange, **40:**169; 170; 171; 172
binding affinity of mutants, **40:**201; 202; 203; 204
complexes with noncognate tRNA, **40:**184
control in bacterial cells, **40:**217; 218; 219; 220; 221; 222
correlation of number of binding sites to number of submits, **40:**169
crystalline, **40:**151; 152
deacylation of tRNA, **40:**186; 187
enzymological characteristics of mutants, **40:**201; 202; 203; 204
genetic mapping, **40:**208; 209; 210; 211
histidine, **40:**180; 181
interaction with tRNA, **40:**183; 184; 185; 186; 187; 188; 189; 190; 191; 192; 193; 194; 195; 196
internal duplication, **40:**150
isolation of mutants with defective synthetases, **40:**196; 197; 198; 199; 200; 201
isolation studies, **40:**153; 154; 155; 156
metabolic regulation of enzyme formation, **40:**221; 222
misacylation, **40:**184; 185; 186
molecular weights, **40:**145; 146; 147; 148; 149; 150
mutants, eukaryotic, **40:**200; 201
medium deficiency, **40:**198; 202
streptomycin-conditional, **40:**199
temperature sensitive, **40:**197; 198; 202; 203
wild-type, **40:**199
order of substrate binding, **40:**169; 170; 171; 172; 173
ping-pong mechanism, **40:**172; 173
primary structure near the active center, **40:**143; 144
products, **40:**142
prokarytoic, **40:**222; 223; 224; 225
quarternary structure, **40:**146; 147; 148; 149; 150; 151
rate-limiting stager, **40:**149; 150; 151; 181; 182; 183
reaction intermediates, **40:**173; 174; 175; 176; 177; 178; 179; 180; 181
regulatory role in protein synthesis, **40:**212; 213
repression of amino acid biosynthesis, **40:**213; 214
structural alterations of mutant synthetase, **40:**204; 205; 206; 207
structural genes, **40:**208; 209; 210; 211
tRNA recognition, **40:**194; 195; 196
tRNA specificity moiety, amino acid stem, **40:**191; 192; 193
anticodon loop, **40:**188; 189; 190; 191
stem of the dihydrouridine loop, **40:**193; 194
in vivo functioning, **40:**212; 213; 214; 215; 216; 217; 218; 219; 220; 221; 222
arginine
activating enzyme, **35:**449
enzyme, **35:**478; 479; 481; 482; 485; 489
molecular weight of, **35:**449
comparison with luciferase, **44:**64
cysteine enzyme, **35:**451–452
disulfide bond, **35:**452
glutamate enzyme, **35:**489
glutamic acid, **35:**451
glutamine enzyme, **35:**485
glycine enzymes, **35:**489
isoleucine
activating enzyme, **35:**477; 485
enzyme, **35:**450; 452; 482; 486; 488
isoleucyl-t RNA synthetase, **35:**455
molecular weight of, **35:**446

Aminoacyl-tRNA synthetases (*Continued*)
subunits, **35:**446
leucine, **35:**490
enzymes, **35:**487
leucyl-t RNA synthetase, **35:**449–450
molecular weight of, **35:**450
lysine enzyme, **35:**449–452; 482; 485
methionine
activating enzymes, **35:**484
enzyme, **35:**451; 483; 485
methionyl-t RNA synthetase, **35:**446
molecular weight of, **35:**446
subunits, **35:**446
phenylalanyl-t RNA synthetase, **35:**447
molecular weight of, **35:**447; 449
subunits, **35:**446
phenylaline
activating enzyme, **35:**491
enzymes, **35:**487; 492
proline enzyme, **35:**482; 488
prolyl-t RNA synthetase, **35:**449
molecular weight of, **35:**449
serine enzyme, **35:**449; 487
molecular weight of, **35:**449
sulfhydryl groups, **35:**451–453
thermal denaturation, **35:**482
threonine enzyme, **35:**449; 484; 487
molecular weight of, **35:**449
tryptophan
activating enzyme, **35:**450
enzyme, **35:**452; 488
subunits of, **35:**450
tyrosine enzyme, **35:**481; 484; 487
tyrosyl-t RNA synthetase, **35:**446–447
molecular weight of, **35:**446
valine, **35:**490
enzyme, **35:**478; 482; 487
system, **35:**477
valyl-t RNA synthetase, **35:**447
molecular weight of, **35:**447
yeast
tyrosine enzyme, **35:**477
valine enzyme, **35:**477
Amino-acyl-tRNA synthetases, **44:**38
Aminoacyl-tRNA Synthetases, Some Recent Results and Achievements (Kisselev and Favorova), **40:**141
Aminoacyl-tRNA synthetase-tRNA complex, sensitivity to nuclease digestion, **40:**155
Aminoacyl-tRNA Transferases (Soffer), **40:**91
L-α-Aminoadipate, metabolism in *Neurospora,* **16:**293; 295–296
α-Aminoadipate in lysine metabolism, **16:** 295–297; 299
in transamination reactions, **16:**192; 194; 232
α-Aminoadipic acid, **17:**403
α-Aminoadipic-δ-semialdehyde, in transaminations, **16:**197; 232
α-Amino alcohols, PN-dependent oxidations, **26:**118; 122
Aminoaldehyde, formation from spermine, **36:**236
Aminoalkylsilane glass, converted to aminoaryl derivative, **34:**450–453
converted to isothiocyanate derivative, **34:**450–451
N^{ω}-Amino-L-arginine, nitric oxide synthases inhibition, **65:**315
Aminobenzaldehyde, **49:**31; 33; 34
o-Aminobenzaldehyde, **17:**396, **40:**70
use in study of amino acid intermediates, **16:**258
p-Aminobenzoate, **50:**149
chorismate conversion to, **66:**223–225
microbial synthesis and breakdown, **7:**454
synthesis, **72:**136–138
p-Aminobenzoate synthetase, compound A, a possible intermediate of reaction, **39:**150
cross feeding, **39:**150
extracts, fractionation of enzymatic components, **39:**149
metabolic interlock, **39:**151
genetic mutants, studied with *B. subtilis,* **39:**150; 151
Neurospora crassa, **39:**149
Aminobenzoic acid, **7:**207; 418
p-Aminobenzoic acid, **21:**397, **53:**353
absorption spectrum, **4:**285; 304
in animal system, **6:**40
antagonism to sulfonamides, **6:**38–40
in dextran synthesis, **11:**312
in *Diplococcus pneumoniae,* **22:**17
in *E. coli,* **22:**17
as growth factor, **6:**38–39
inhibition of β-lactamase, **28:**295
as inhibitor, **6:**39; 51–57
inhibitory analogues, **6:**130; 132–133; 137; 139; 140
methyl synthesis, **25:**406

nonspecific antagonism, **6:**51–57
origin of, **18:**366
in protein, **6:**39–40
and purine synthesis, **9:**461
and resistance to sulfonamides, **6:**102–103
and thiaminase model, **15:**404
in yeast, **6:**40
p-Aminobenzoic acid (*p*-AB), action of sulfonamides, **27:**418; 426–431; 457
analogs, in folate metabolism, **27:**428–433
biosynthesis of, **27:**421–422
in folic acid biosynthesis, **27:**420–424; 426–428; 432; 459
p-Aminobenzoic acid (C^{14}), effect of sulfathiazole, **27:**427–428
p-Aminobenzoic acid in genetic studies, **13:**136–138
o-,m-, and *p*-Aminobenzoic acids, action on thiaminase, **15:**400; 410
action on thermostable factor, **15:**410
p-Aminobenzoic acid-utilizing enzyme, **27:**432
p-aminobenzoyl glutamate, eukaryotic dihydrofolate reductase, **70:**30
p-Aminobenzoyl glutamic acid
in folate biogenesis, **27:**422
folic acid from, **15:**404
Aminobenzoylpyruvic acid, **49:**28; 29
o-Aminobenzoylpyruvic acid, in transaminations, **16:**197; 235
p-aminobenzyl cellulose, **34:**453–456
b-Aminobenzylmethylthiazolium chloride, as inhibitory analogue, **6:**131
o-, m-, and *p*-Aminobenzyl-(3)-(4)-methylthiazolium salts, effect on thiaminase action, **15:**399
m-Aminobenzyloxymethyl ether of cellulose, **34:**453
α-Amino-benzyl-penicillin, action of β-lactamase, **28:**258
2-Amino-3-butenoic acid, **54:**224
N-(4-aminobutyl)-3-aminopropionaide-hyde, product of amine oxidase, **36:**235
N-(4-aminobutyl)-3-aminopropionic acid, in nervous tissues, **36:**236
γ-Aminobutyraldehyde, **36:**233
product of, diamine oxidase, **36:**224; 233
diamine transamination, **36:**223
putrescine oxidase, **36:**224
product of speridine dehydrogenase, **36:**225
α-Aminobutyrate, **35:**482
4-Aminobutyrate, **42:**259, **49:**48
α-Aminobutyrate, from threonine, **16:**270
origin of 6-carbon skeleton of isoleucine from, **16:**281–284
relation to valine metabolism, **16:**281
source, **16:**284–285
in transamination reactions, **16:**190; 192; 194; 201; 203–204; 208; 229; 234; 235
γ-Aminobutyrate, in transamination reactions, **16:**190; 194; 205; 237
4-Aminobutyrate aminotransferase, **74:**133
evolution of, rate of, **74:**158
species distribution of, **74:**158
α-Amino butyric acid, **35:**481
α-Aminobutyric acid
lack of effect on glutamate dehydrogenase-NADPH complexes, **39:**375
in pyrimidine biosynthesis, **21:**281
in synthetic ACTH, **29:**466
β-Aminobutyric acid, **7:**437
4-Aminobutyric acid, **49:**67
γ-Aminobutyric acid, in biosynthesis of glutamic acid in plants, **20:**327
α-Aminobutyric acid, thermodynamic data on ionizations, **20:**64
γ-Aminobutyric acid, transamination reactions of, **19:**343; 367
Aminobutyric acid production, **6:**4
α-Amino-γ-butyrothetin, *See* S-Methylmethionine
ε-Aminocaprioc acid methylester, **54:**40; 41
ε-Aminocaproic acid, **54:**14
ε-Aminocaproylamide, **54:**43
α-Aminocaprylate, in transaminations, **16:**201
3-Amino-3-carboxypropyldimethylsulfonium salt, *See* S-Methylmethionine
3-(3-Amino-3-carboxypropyl) uridine, **54:** 228–230
7-Aminocephalosporanic acid (ACA), action of β-lactamase, **28:**240; 249; 259; 262; 287
derivatives, structures, **28:**240
L-α-(α-S,5*s*)-α-Amino-3-chloro-4,5,dihydro-5-isoxazoleacetic acid, **59:**152
Aminocyclase, **56:**293
cis-3-Aminocyclohexane-1-carboxylic acid, **49:**48
1-Aminocyclopentanecarboxylic acid, growth-inhibitory activity of, **29:**143

1-Aminocyclopropane-1-carboxylate deaminase, **74:**134
phylogenetic distribution of, **74:**157
1-Aminocyclopropane-1-carboxylate synthase, **74:**134; 151
1-Aminocyclopropane-1-carboxylic acid, **54:**231–232; 234
Amino decarboxylases, **9:**407–408
3′-Amino-3′-deoxy ATP, **43:**25
4-Amino-4-deoxychorismate lyase, **74:**134; 141
Aminodeoxychorismate synthase, **72:**136–138
4-amino-4-deoxychorismate synthase relation to anthranilate synthase, **66:**225–227
amphibolic TrpG, **66:**227–228
chorismate conversion to *p*-aminobenzoate, **66:**223–225
4-Amino-4-deoxychorismate synthase, relation to anthranilate synthase, **66:**225–227
cis-1-Amino-1,3-dicarboxycyclohexane, as substrate for glutamine synthetase, **31:**204
1-Amino-1,3-dicarboxycyclohexane, as substrate for glutamine synthetase, **31:**203–205
α-Aminodicarboxylic acid, as substrate for glutamine synthetase, **31:**190–194
Aminodicarboxylic acid amide, **54:**17–19
3-Amino-4,5-dihydroxymethyl-2-methylpyridine, substrate for pyridoxal kinase, **16:**219
4-Amino-1,1-dimethylpiperidine-4-carboxylic acid, **49:**63
4-Amino-1-dimethylpiperidine-4-carboxylic acid, **49:**57
5-Amino-2,4-dioxy-6-ribitylamino-pyrimidine, **53:**349
α-Aminodipamic acid
carboxamide side chain of, **54:**25
peptide derivatives of, **54:**17
4-Amino-3-enolpyruvylcyclohexa-1,5-dienecarboxylic acid, **50:**149
Amino-enzyme, **53:**285
β-Aminoethanesulfinate, **39:**341
β-Amino-ethanethiosulfanate, **39:**341
Aminoethanol, in choline synthesis, **21:**427
Aminoethanol and choline, **9:**440
Aminoethanols, partial methylation of, **25:**417–420
S-Aminoethylcysteine, **50:**169; 170
3-β-Aminoethyl-4-methylthiazolium salt, **15:**399
2-Aminoethylphosphonic acid, **53:**169
Aminoethylphosphoric acid and phosphatides, **7:**151
Aminoexopeptidases, **68:**153–154. *See also* Leucine aminopeptidase
structural aspects, **68:**160; 162
Aminoglycoside acetyl transferase 3-1, antibiotic resistance, enzyme action, **73:**31–32
Amino group
binding energy, **43:**344
differentiation from imidazole, **20:**76
α-Amino group, metal binding by, **20:**56
use of blocking reagents with, **20:**68
ε-Amino group, metal binding by, **20:**56
a-Amino group, tendency to lose its hydrogen, **49:**65
Amino group and α-amylase activity, **12:**399
Amino group interchange, in plant metabolism, **9:**398–401
Amino group migration, **38:**434–437
mechanism of ^{15}N experiments, **38:** 436
1,2-Amino group migration, nonenzymatic model, **66:**14
Amino groups, **34:**453; 469
and enzyme activity, **14:**84; 181; 187; 189–190
effect of tyrosinase on, **14:**141
relation to activation of β-glucuronidase, **16:**376
role in sclerotization, **16:**160–161
source, **16:**255–256
relation to activity of alkaline phosphatase, **10:**97–98; 101
ω-Amino groups, in transaminations, **16:** 204–206
α-and ε-Amino groups of RNase, reactions, **24:**206–209; 220–224; 244; 245
α-Amino-γ-guanidinoxybutyric acid, *See* Canavanine
α-Amino-γ-guanidinoxylbutyric acid, **15:** 361–362
4-Amino-1-guanylpiperidine-1-carboxylic acid, **49:**62
4-Amino-1-guanylpiperidine-4-carboxylic acid, **49:**57; 59
α-Aminoheptylate, in transamination reactions, **16:**190; 194
2-Aminohex-4,5-dienoic acid, **50:**165

8-(6-aminohexyl)-amino-AMP, **46:**208
p-Aminohippurate
synthesis, in mitochondria, **21:**42
synthesis of, **24:**311; 334
p-Aminohippurate clearance, effect of fluoroacetate on, **18:**134
m- and *p*-Aminohippuric acid, action on thiaminase, **15:**400
p-Aminohippuric acid, enzyme system for, **13:**12
Amino hydrolase, allosteric regulation, **28:** 132–133
Aminohydrolase, AMP, **56:**304
Aminohydroquinone in sclerotization process, **16:**160
2-Amino-4-hydroxpteridine, inhibitory effect, **27:**424
2-Amino-4-hydroxy-6-hydroxymethyl-7,8-dihydropteridine, in folate biogenesis, **27:**423
2-Amino-4-hydroxy-6-hydroxymethylpteridine, in folic acid biogenesis, **27:**422
2-Amino-4-hydroxy-6-pteridine aldehyde, in folate biogenesis, **27:**422; 436
2-Amino-4-hydroxypteridine-6-carboxylate, folate metabolism, **27:**422; 436
5-Amino-4-hydroxyvalerate, **46:**84
5-Amino-4-hydroxyvaleric acid, **46:**91
α-Amino-δ-hydroxyvaleric acid, as growth factor for *Neurospora* mutant, **16:**264
α-Amino-γ-hydroxyvaleric acid, as possible leucine analog, **29:**144
α-Amino-β-hydroxyvaleric acid, effect on growth of *E. coli* by, **29:**144
inhibition of enzymes in *E. coli* by, **29:**145
mode of action of, **29:**145
4-Aminoimidazole, in purine degradation, **21:**401
5-Amino-4-imidazolecarboxamide, in purine biosynthesis, **21:**217–219; 242; 243; 265
Aminoimidazolecarboxamide (AICAR) ribotide transformylase, reaction, **27:**425
5-Amino-4-imidazolecarboxamide ribonucleotide, folate metabolism, **27:**459
4-Amino-5-imidazole carboxamide riboside, **18:**222
4-Amino-5-imidazolecarboxamide ribotide
folate metabolism, **27:**459
in histidine biosynthesis, **21:**403–405
in purine biosynthesis, **21:**399–401
5-Amino-4-imidazolecarboxamide ribotide, in purine biosynthesis, **21:**204; 206; 215–223; 225; 242; 252; 253; 265
4-Amino-5-imidazolecarboxylic acid, in purine degradation, **21:**401
5-Amino-4-imidazolecarboxylic acid ribotide, in purine biosynthesis, **21:**204; 213; 214; 216; 243; 252
5-Amino-4-imidazole-*N*-succinocarboxamide ribotide, in purine biosynthesis, **21:**204; 214–216; 252–254
4-Aminoimidazole ribonucleotide, **53:**364
5-Aminoimidazole ribotide
carboxylase, **21:**214; 216
in purine biosynthesis, **21:**204; 211–214; 216; 400
2-Aminoisobutyric acid, **49:**57; 60; 69
α-Amino-β-keto-adipate-pyridoxal enzyme, **46:**40
α-Amino-β-ketoadipic acid
and glycine metabolism, **23:**455
in synthesis of active formate, **21:**410; 411
S-Aminolaevulinate synthetase, localization of, **24:**335
Aminolevulinate dehydratase, **56:**293
inhibition of, by Pb(II), **56:**400–401
metal centers of, propertis of, **56:**399–400
metal content of, **56:**397–398
molecular properties of, **56:**398–399
δ-Aminolevulinate dehydratase, **37:**167–168
5-Aminolevulinate synthase, **74:**133
erythroid specific, mitochondrial, species distribution of, **74:**158
evolution of, rate of, **74:**158
nonspecific, mitochondrial, species distribution of, **74:**158
δ-Aminolevulinate synthetase, **37:**140
δ-Aminolevulinic acid, **35:**95; **46:**39; 40; 41; 42; 83; 95; 102; 111; 114–115; 144; 152; 157; 162
administration to *Cyanidium caldarium,* **46:**67
biosynthesis, in plants, **46:**81
conversion to porphobilinogen, **46:**42–44
in DR synthesis, **28:**423
formation, **46:**84; 94; 98
in fetal liver, **46:**106
induction, **46:**39; 126
in plants, **46:**41–42
synthesis, **46:**114; 128

δ-Aminolevulinic acid (*Continued*)
via δ-aminolevulinic acid synthetase, **46:**39–41
and glycine metabolism, **23:**455
metabolism, **16:**255
in synthesis of active formate, **21:**410; 411
δ-Aminolevulinic acid dehydrase, **46:**42; 43; 44; 45; 86; 88; 92; 103; 106; 111; 121; 176; 178
activity, **46:**87; 115
genetics of, **46:**44
inhibition, **46:**91
from mouse liver, **46:**44
δ-Aminolevulinic acid synthetase, **46:**39; 40; 87; 88; 91; 103; 106; 107; 112; 113; 115; 122; 124; 125; 126; 128; 129; 131; 132; 137; 139; 147; 157; 159; 162; 165
activity, **46:**104; 111–115; 134; 136; 138; 144; 164; 176–178
in cultured cells, **46:**121
of liver, half-life, **46:**124–125
in animals, **46:**83
in bacteria, **46:**83; 86
formation in plants, **46:**41–42
hemin inhibition site, **46:**127
induction of, **46:**39; 121–139; 151
by steroids, **46:**155–160
inhibition of activity by heme, **46:**87
reaction, **46:**41
repression, **46:**141–144
synthesis, **46:**122; 126–127; 141
transport of into mitochrondia, **46:**123
Aminolysis. *See also* Transacylation
alkaline, of activated amino acids, **19:**241
intramolecular, of activated amino acids, **19:**241–248; 250
in peptide biosynthesis, **19:**263
Amino malonate, **35:**95
Aminomalonate, **35:**505
Aminomalonic acid, in transaminations, **16:** 192; 194
a-Aminomalonic acid monoamide, **49:**52
2-Aminomalonic acid monoamide, **49:**77
Aminomethanesulfonate, **67:**42–43
α-Amino-δ-*N*-methyladipamic acid, in transaminations, **16:**194; 202
2-Amino-5-methyl-4-enoic acid, **50:**165
2-Amino-4-methylhex-4-enoic acid, **50:**126; 129; 130; 132; 164; 165
2-Amino-5-methyl-6-hydroxyhex-4-enoic acid, **50:**129
Aminomethylphosphonic acid, **53:**168
4-Amino-1-methylpiperidine-4-carboxylic acid, **49:**63; 65
2-Amino-2-methyl-1,3-propanol(DAMP),**38:**309
Aminomutase
adenosylcobalamin-dependent, **66:**26–32
pyridoxal phosphate role in reactions, **66:** 28–30
3-Amino-α-naphthoxindole, as model decarboxylase, **15:**29
ω-Amino nitrogen, **49:**65
2-Amino-3-nitropropionate, ionization of, **74:**303–304
2-Aminonorbornane-2-carboxylic acid, **49:**57; 58
5′-Amino nucleotides, **43:**7
α-Amino-β-oxalylaminopropionic acid, **50:**122; 123; 124
3-Aminooxindole as model decarboxylase, **15:**29
4-Amino-2-oxobutyrate, **50:**152
6-Amino-2-oxocaproate, **50:**125
L-2-Amino-4-oxo-5-chloropentanoic acid, **53:**220
8-Amino-7-oxononanoate synthase, **74:**133
phyogenetic distribution of, **74:**157
Amino-oxyacetate, **57:**265; 326–328
Aminooxyacetic acid, **56:**22
2-Amino-4-oxy-6-hydroxymethyl-7,8-dihydropteridine, **53:**353
6-Aminopenicillanic acid (6-APA), action of β-lactamase on, **28:**249; 258–262; 267; 271; 274; 287–293; 296; 306
derivatives, structures, **28:**239
Aminopeptidase, **56:**292; 348
action on collagen, **23:**205
bacterial, **23:**231
Aminopeptidase from yeast, **7:**529
Aminopeptidases, **2:**76–81, **45:**397
classification, **2:**51; 54
intestinal, **2:**54; 76–79
leucylpeptidase, **2:**79; 80
yeast polypeptidase, **2:**80
α-Aminopeptide amino acid hydrolases, localization of, **24:**324
p-Aminophenol, **15:**6
o-Aminophenol, action on thermostable factor, **15:**410

p-Aminophenylalanine, **50:**149
p-aminophenylalanine, **34:**453
4-Aminophenylalanine, inhibition of growth in *E. coli* by, **29:**114
o- and *m*-Aminophenylglucuronide, as substrate for β-glucuronidase, **16:**382
formation, **16:**403
γ-(*o*-Aminophenyl)-L-homoserine, **49:**31
Aminopherase, **4:**86, *See* Transaminase
specificity, **19:**343
in transamination reactions, **19:**341; 342; 346; 347; 349; 353–358; 365; 373
Aminopherases, coenzymes for, **14:**182
Aminophorases in plant metabolism, **9:** 398–401
Amino phosphonic acid, **53:**168
β-Aminophosphonic acids, **53:**169
2-Amino-3-phosphonopropionic acid, **53:**169
Aminophthalic acid, chemiluminescence, **25:**122–123; 128–129
4-Aminopipecolic acid, proline analog, **29:**95
Aminopolypeptidase, effect on glycyl-dehydrophenylalanine, **8:**134
Amino polypeptidase, separation by chromatography, **14:**350
Aminopolypeptidase, specificity of, **12:**193
Aminopolypeptidase in yeast, **9:**418
N-3-Aminopropyl-1,5-diaminopentane, in *E. coli* mutant, **36:**207
γ-aminopropyltriethoxysilane, **34:**453
Aminopterin, **35:**263, **67:**237
antifolic factor, **27:**424; 434–437; 441; 442; 456; 457
effect on pyrimidine biosynthesis, **21:**288; 289; 420
synthesis and structure, **21:**375
eukaryotic dihydrofolate reductase, ligand binding (side chain roles), **70:**62
reduced analogs, **38:**274
2-Aminopurine, mutagenic agent, **27:**302; 305–306; 318–319; 335–337; 339; 341
2-Aminopurine ribonucleoside triphosphate, **43:**35
2-Aminopyridine, action on thiaminase, **15:**401; 410
3-Aminopyridine, action on thiaminase, **15:**401; 410
Aminopyrine demethylase, **46:**134
Aminoquinone type pigments, formation, **16:**112
Aminoquione derivatives, E'_0 values for, **16:**113–114
3-Amino-ribose, in transamination, **16:**236
p-Aminosalicylic acid, and folate, **27:**428–429
Aminosäuren, Aktivierung von (Wieland and Pfleiderer), **19:**235
Amino sugars. *See also* Hexosamines and Sugar amines
in cell walls, **26:**198; 200–204; 207; 216–219; 223
linkages in glycoproteins, **25:**442–453
recent developments, **25:**433–456
in sugar nucleotides, **26:**55–56
Amino Sugars (Jeanloz), **25:**433
Amino-terminal, **42:**81
Amino terminal analysis, **46:**369
2-Aminothiazole, action on thiaminase, **15:**401
Aminothioacids, effect of carbon dioxide on, **19:**246
synthesis, **19:**238
Aminothiols, **56:**63–67
β-Aminothiols, **57:**141
o-Aminothiophenol, in disulfide exchanges, **27:**203
Aminotransferase, **49:**29
Aminotransferases, **50:**132; 156, **57:**88. *See also* Ntn amidotransferases; Triad Aminotransferases
localization of, **24:**312
β-3-Amino-1,2,4-triazol-1-ylalanine, as a histidine analog, **29:**124
Aminotripeptidase, **56:**292
classification, **14:**247
specificity of, **12:**206–210
Aminotripeptidase activity, **45:**393
δ-Amino-*n*-valeric acid, in biosynthesis of nicotinamide, **15:**103
δ-Aminovaleric acid, in transaminations, **16:**195; 206
Amitriptyline
anticholinergic activity of, **55:**117
structure-activity studies of, **55:**178
structure of, **55:**177
Amitrole, action on imidazole glycerol phosphate dehydrogenase, **29:**125
Amitryplyline, in depression, **29:**491
Ammonia, **38:**202, **50:**141; 142; 146; 147; 148, **53:**209; 211. *See also* Asparagine and Glutamine; Nitrogen compounds
allantoin degradation to, **56:**92

Ammonia (*Continued*)
conversion to amino groups, **16:**255–256; 268
in arginine synthesis, **16:**260; 264–265
from asparagine, **16:**202
in glutamine formation, **16:**257
incorporation into pyrimidine nucleus, **16:**450–451
(N^{15}), from glutamine, **16:**200
effect on cytochrome a_3, **20:**166
labeled, in plants, in studies of asparagine and glutamine biosynthesis, **20:**331–334
in plants
fixation, **20:**331–334
in biosynthesis of amino acids, **20:** 323–328; 330
in studies of glutamic acid metabolism, **20:**322
thermodynamic data on ionization, **20:**64
in exchange reactions, **13:**289–305; 312
labeled, in transpeptidation reactions, **19:**310; 311
in nitrogen metabolism, **19:**339
in ornithine cycle, **19:**373; 383
loss from tobacco during fermentation, **10:**340; 350; 377
as nitrogen source, **56:**118
in purine biosynthesis, **21:**203; 210; 211; 229; 250; 251
in pyrimidine biosynthesis, **21:**264; 266–268; 271; 272; 279
purine synthesis, **28:**121
role in nitrogen fixation, **13:**361–368; 369–373
sources in tobacco fermentation, **10:**395–398
in synthesis of urea, **15:**321; 323 (N^{15}); 332; 336; 338(N^{15}); 341 (N^{15}); 356; 362; 364; 364 (N^{15}); 373; 374 (N^{15}); 375–378; 380–382
urinary N^{15} in, **15:**341
and transferase activity, **13:**292; 294–295; 299
in tryptophan metabolism, **10:**61–62; 63–64; 68
urea degradation to, **56:**94
Ammonia (amide), **51:**217
Ammonia haemochrome *a*, absorption spectrum, **23:**289–297; 306–310
Ammonia-lyase family, **74:**301
Ammonia lyases, localization of, **24:**332–333
Ammonia metabolism
acinar glutamine metabolism, acid-base homeostasis, **72:**71–75
amidotransferases, **72:**89–93
asparagine synthetase, **72:**106–112
catalysis mechanisms, **72:**146–147
glutamine-dependent nitrogen transfer, **72:**169–175
nitrogen transfer, **72:**107–112
carbamyl phosphate synthetase, **72:**131
glucosamine 6-phosphate synthase, **72:** 105–106
glutaminase amplification system, **72:**69–70
glutamine synthase as perivenous scavenger, **72:**67–69
glutamine synthetase and hepatocyte heterogeneity, **72:**58–61
GMP synthetase, **72:**121–123
hepatic pathophysiology, **72:**75–77
long-term glutamine regulation, **72:**65; 67
short-term glutamine regulation, **72:**64–66
triad amidotransferase, **72:**138–139
tunnel model, **72:**119
Ammonia (N^{15}), in amination study, **18:**252–253
Ammonia secretion, **53:**227
Ammonium, **34:**509
ions, **34:**511
Ammonium assimilation, and citric acid cycle, **18:**251–254
Ammonium chloride, as inhibitor of *E. coli* β-glucuronidase, **16:**380
Ammonium group, binding of, **43:**333
Ammonium ion
competition with guanosine triphosphate in glutamate dehyreogenase reaction, **39:**389
and convulsions, **18:**149–150
for fructose-bisphosphatase activity, **54:**127; 129; 132
in fructose-bisphosphatase regulation, **54:**178
Ammonium ions, **15:**116; 246; 251; 257; 258, **38:**202
enzyme activation by, **21:**153; 158
properties, **21:**157
in purine biosynthesis, **21:**213; 229
Ammonium ions and ketone body formation, **8:**407
Ammonium maleate, conversion to Laspartic acid, **74:**299
Ammonium nitrogen, assimilation in plants, **9:**392–398; 413–414; 417

Ammonium salts
and plant respiration, **18:**284
radioactive, in studies of purine biosynthesis, **21:**202
Ammonium sulfate, **58:**105; 152
Amoeba, enucleation studies on, **16:**469–471; 473
Amoeba-enucleation experiments on, **24:**20
Amoebic dysentery, **11:**380
Amoeboid movements of cytoplasm, **8:**41–42; 56
AMP, **53:**49; 51; 72; 76, **58:**125, *See* Adenosine monophosphate; Adenosine monophosphate, Adenosine-5′-phosphate, and Adenylic Acid; Adenylic acid; Aminoacyl adenylates; Cyclic AMP; Fatty acid adenylates
activating effect on phosphorylase *b*, **28:** 104–110
in allosteric regulation, **28:**78–86; 111–112; 136–139
cyclic, in allosteric regulation, **28:**78–81; 99; 101; 102; 107–110
inhibition of fructose diphosphatase, **28:**93–94
reduction of, **28:**443; 445
acylation by aminoacyl adenylates, **19:**252
in amino acid activation, **19:**258
β-aspartyl-AMP intermediate, aspartic acid, **72:**158–159
and ATP, **18:**230; 231
in ATP system for energy transport, **14:**16–17
binding to LADH, **27:**66–70
effect on LADH complexes, **27:**61–70
effect on Zn exchange of LADH, **27:**64–65
biosynthesis of, **18:**217
cyclic, **55:**350
effect on PR-enzyme, **24:**280
in fatty acid activation, **19:**257; 258
in fructose-bisphosphatase regulation, **54:**178–179
effect of, on cations, **54:**128–132
in microbial metabolism, **54:**172
in fructose-bisphosphate regulation, **54:** 136–140
concentration necessary for, **54:**138
divalent cation effect on, **54:**137; 138
in microbial metabolism, **54:**169
glucagon effects and, **54:**150
glutamine synthesis expression, **72:**36–38
inhibition by, **53:**67–68
interaction with fructose 1,6-bisphosphatase, **53:**48; 67–68
labeled, in acetyl CoA biosynthesis, **19:**257
in pantothenic acid biosynthesis, **19:**259
and liver phosphorylase, **24:**282–283; 286
and luciferin action, **25:**129–131; 133–134; 135–137; 140; 143–144
in oxygen metabolism, **19:**127; 128; 189
and phosphorylases *a* and *b*, **24:**265; 266; 268–270; 274–275; 276; 286
phosphorylation of, **18:**196–197
in photosynthesis, **19:**7; 10; 37; 63
from pyrophosphorylase action, **18:**208–209
in urea synthesis, **15:**327–329; 334; 347–349; 380
[^{3}H] AMP, **49:**242
N^6-(6-aminohexyl)-AMP, **46:**208
3,5-AMP, cyclic, **46:**91
L-AMP, free energy of hydrolysis, **44:**40
AMP (adenosine monophosphate). *See also* Adenylic acid
and CO_2 fixation, **12:**51; 56–57; 62
in muscle relaxation, **16:**350
role in deamination, **12:**83–84; 90
AMP aminohydrolase, **43:**33
5′-AMP analogs, **46:**208–211
AMP-ATP exchange reaction, **33:**371
AMP (C^{14}), and light emission, **25:**135
AMP_cPP, **35:**483
AMP deaminase, **56:**293, **74:**72
localization of, **24:**328
relation to myosin, **16:**337
Ampelopsin, metabolism in plants, **16:**141
d-Amphetamine, action of, **29:**490–491
fall in brain norepinephrine by, **29:**487
inhibition of dopamine-β-hydroxylase by, **29:**491
and schizophrenia, **29:**534–535
Amphetamine, β-fluorinated, **55:**239
Amphibia
digestion, **1:**402–410
food, **1:**402
gastric digestion, **1:**403–407
intestinal canal, **1:**377
methylated lysine and arginine derivatives in, **42:**235
resorption of food, **1:**409; 410
secretion of gastric juice, **1:**402; 403
Amphipathic helices, **65:**207

Amphipathic molecules, **34:**483
substances, **34:**489–491
Amphipaths, **34:**491
Amphiphilic compounds, **38:**202, **51:**32; 43
Amphoterisin B, **34:**485
Ampicillin, action of β-lactamase, **28:**239; 258; 288–293
Ampicillinase, as β-lactamase, **28:**241
AMP nucleosidase, **37:**499
AMP-PCP, **35:**488, *See* Adenyl methylenediphosphonate
AMP phosphodiesterase, **54:**254
AMP-P(NH)P, inhibition of ATPase activity, **49:**246–249; 271
AMP-PNP, divalent metal complex, **43:**16
ionization constants, **43:**16
radioactive form, **43:**20
AMP-pyrophosphorylase, inhibition by nucleotides, **28:**124
AMPS, **55:**299–303
hydrolysis of, **55:**350
AMPSase, *See* Adenylosuccinate synthetase (AMPSase)
Amygdalin, **56:**199
formation in plants, **20:**20
hydrolysis of, **28:**345
Amylase, **5:**23, **33:**265, **36:**71, **43:**258, **56:**292
activation by amino acids, **10:**100
adsorption of on starch granules, **33:**259–260
in chromidia, **8:**31
in mitochondria, **8:**20; 33
endo- *vs.* exoenzyme form of, **56:**143–144
inhibition of, **13:**332
localization in cytoplasm, **13:**12
intracellular distribution, **16:**463; 466
in low vertebrate, **1:**381; 383; 389; 391; 392; 394; 407; 408; 411; 412
multiple forms, **25:**334–336
product inhibition of, **25:**224; 252
pancreatic, **23:**130–135
salivary, **23:**130
in plastids, **20:**211; 212
rate of labeling, **26:**108–110
structure, **26:**102–104
separation by chromatography, **14:**322; 329; 330; 334; 343; 349; 363–365
stabilization of, **14:**390
structure studies, **14:**308
use in isolation of enzymes, **14:**398
in sperm, **24:**88
α-Amylase, association of molecules, **24:**119
localization of, **24:**316; 322; 325
molecular weight from radiation studies, **24:**366
requirements for synthesis, **24:**24
role of metal, **24:**112–113
structural role of zinc in, **56:**316
taka, **56:**210
α-Amylase. *See also* Amylases
α-Amylase, **6:**396–407; 416–419, **18:**289, **23:**229, **34:**452, **37:**423; 429; 460; 541
of *B. subtilis,* **29:**135
ethionine incorporated into, **29:**135
of *Bacillus subtilis,* denaturation of, **23:** 32–35; 41; 50–58
bacterial
denaturation of, **23:**31–39; 41; 62
ethionine in, **22:**271
column chromatography, **22:**447
in *Pseudomonas saccarphila,* **22:**106
structure, noncovalent forces in, **22:**73
criteria of purity, **6:**405–406
dextrinization, **6:**396–399
effect on bacterial wall, **26:**225
effect on glycogen synthetase assay, **34:**319
pancreatic inhibitor from *Phaseolus vulgaris,* **34:**319–320
role in providing primer for glycogen synthesis, **34:**352
erosion of starch granules, **6:**407
limit of hydrolysis, **6:**403–405
measurement of activity, **6:**422
reaction stages, **6:**402–403
saccharification, **6:**399–401
stability of, **23:**44
β-Amylase. *See also* Amylases
β-Amylase
action of, **12:**383–384; 385–386; 391–395; 417–419
action on glycogen, **12:**371–372
occurrence of, **12:**395; 400–401
use in determining structure of amylopectin, **12:**356–357
use in purification of amylose, **12:**348–351; 358
pH dependence, **15:**283
β-Amylase, **6:**391–396; 416–419, **18:**289, **55:**247
active center of, **24:**115

addition to coacervate, **27:**368
location of active site in, **22:**63
mechanism of action of, **22:**81
measurement of activity, **6:**422–423
specific action, **6:**392–394
α-Amylase, action of, **12:**358; 383–384; 385–391
occurrence of, **12:**395; 396–400
α, β-Amylase, effects of polycations, **40:**41; 42
Amylase activity, assay for, **23:**35
α-Amylase (diastase) cacao, **25:**466; 467; 480
ethionine-containing, **25:**376
multiple forms, **25:**334
Amylase mold in industrial fermentations, **7:**591–595
Amylases, **6:**389–412; 415–463, **12:**382–409, **48:**49. *See also* α-Amylase; β-Amylase
adsorbed, **28:**3; 28
removal from laminarinase preparation, **28:**343
soil, **28:**375; 386; 387
soya bean, exohydrolysis, **28:**350
α - amylase, **48:**45
β - amylase, **48:**45
amylokinase activating, **6:**425
comparison of α and β enzymes, **12:**402–406
dextrinogenic, **11:**347–348
in pectic enzymes, **11:**292
in polysaccharide synthesis, **11:**298–299; 322–324
saccharogenic, **11:**344–347; 348
general features, **6:**391–407
heat stability, **6:**408–409
inhibitors of, **12:**408–409
ions affecting, **6:**407–408
kinetics of action, **6:**410–412
excess enzyme, **6:**410; 411
excess substrate, **6:**410; 411
liquefaction, **6:**402–403
measurement of activity, **6:**421–424
autolytic methods, **6:**423–424
in plant respiration, **18:**288; 289; 290
purity tests for, **12:**401–404
sisto-amylase inhibiting, **6:**425
specificity and affinity for substrate, **12:** 406–407
synthesis of starch by, **12:**382
of tobacco, **6:**367
of wheat, **6:**415–463 (*See* Wheat amylases)
α-Amylases. *See also* Amylases, dextrinogenic
α-Amylases
classification, **14:**242; 246
isolation, **14:**397–398; 399; 401; 405
peroxidase action on, **14:**155; 157–158
separation by chromatography, **14:**342; 343; 349; 365
as inhibitor of amylosucrase, **11:**318
β-Amylases. *See also* Amylases, saccharogenic
β-Amylases
classification, **14:**246
isolation, **14:**397; 405
peroxidase action on, **14:**155; 157–188
separation by chromatography, **14:**343; 348
degradation of amylose by, **11:**359
effect on primers, **11:**357–358
effect on synthetic starch, **11:**352–353
as inhibitor of phosphorylase, **11:**303
requiring —SH groups, **11:**236; 239; 241
α-Amylases, catalysis by, **50:**33
β-Amylases, multiple forms, **25:**335–336
product inhibition of, **25:**222
steps in reaction, **25:**14
Amylases (Hopkins), **6:**389
Amylases of Wheat and Their Significance in Milling and Baking Technology (Hopkins), **6:**415
Amylodextrins, as receptor molecules, **15:**308
Amylo-1,6-glucosidase, pK value, **22:**58; 59
Amyloglucosidases, **12:**409–410
(1r6)-Amyloglucosidases, **15:**309
Amylo-1,6-glucosidase transferase, **34:**352; 403
action of glycogen, **34:**403
assay of, **34:**404
deficiency, **34:**405
reversibility of, **34:**362
yeast, **34:**406
Amyloid protein, **36:**77
Amylokinase, **6:**425
Amylolysis, **11:**344–348
Amylomaltase
action of, **12:**411
in polysaccharide hydrolysis and synthesis, **12:**420–421
classification, **14:**251
enzymic adaptation, **13:**70; 82; 86; 106
in polysaccharide synthesis, **11:**298–299; 320–322; 330, **15:**306–307
properties of, **30:**118; 123

Amylomaltase (*Continued*)
reaction catalyzed by, **30:**117
in utilization of maltose by *E. coli,* **10:**124
reversibility of action, **10:**141
Amylopectin. *See also* Amylose, Glycogen, Polysaccharides, Starch; Starch
action of α-amylase on, **12:**388–391
action of β-amylase on, **12:**393–395
action of amylases on, **11:**345–346; 347–348
action of cyclodextrinase on, **12:**419–420
action of phosphorylase on, **12:**412–414
from amylose, **12:**422
branching of, Q-enzyme, **34:**356
synthesized in *vitro* from amylose, **34:**356
complex with iodine, **12:**363–365
constitution of, **12:**353–357
degradation with β-amylase, **12:**348–351
from glucose-1-phosphate, **12:**419; 422
molecular weight of, **12:**359–363
nature of, **12:**380–381
in polysaccharide synthesis, **11:**300; 328
reactivity of, **12:**357–358
separation from amylose, **12:**343–345
structures of, **11:**339–341
synthesis from amylose, **11:**362–365; 366–370
Amylopectinosis, **34:**362
Amylopectins, as receptor molecules, **15:**305; 308
enzymic synthesis, **15:**304; 307–309; 309–310
Schardinger dextrins from, **15:**307
Amylopectin-type polysaccharides. *See also* Amylopolysaccharides, branched, and Starch-glycogen class of polysaccharides
Amylopectin-type polysaccharides., synthesis of, **11:**298–299
Amylophosphatase, **6:**420
Amyloplasts, **18:**291
Amylopolysaccharides, branched. *See also* Starch-glycogen class of polysaccharides
formation from amylose-type chains, **11:** 324–329
Amylose, **56:**192. *See also* Amylopectin, Glycogen, Polysaccharides, Starch; Starch
action of α-amylase on, **12:**386–388
action of amylases on, **11:**346–347; 347–348; 359–361
action of branching enzyme, **15:**307–309
enzymic synthesis, **15:**304; 306–309
as receptor molecules, **15:**305
action of cyclodextrinase on, **12:**419–420
action of phosphorylase on, **12:**412–414
addition products of, **12:**353
aging, (*See* Aging)
behavior in water, **12:**351–353
complex with iodine, **12:**363–365
conversion to amylopectin, **11:**362–365; 366–370
molecular weight of, **12:**359–363
nature of, **12:**380–381
in plants, **19:**5
in polysaccharide synthesis, **11:**298–299; 300; 324–329
quantitative estimation of, **12:**345–348
rate of formation, **25:**246
separation from amylopectin, **12:**343–345
structure of, **11:**339–341
use of β-amylase in control of purity, **12:** 348–351
Amylose, A_2, preparation, **12:**345
Amylose A_1, preparation, **12:**344
Amylose isomerase, **12:**423, **15:**309
Amylose r amylopectin transglucosidase, classification, **14:**251
Amylose-type polysaccharides. *See also* Starch-glycogen class of polysaccharides
synthesis, **11:**298–299
Amylosucrase
action of, **12:**411; 421
classification, **14:**222; 251
in polysaccharide, synthesis, **11:**298–299; 316–320; 326; 327; 330; 372
in polysaccharide synthesis, **15:**306; 309–310
Amylosucrase in synthesis of sucrose to polysaccharide, **10:**141
Amylosynthease, **11:**326
β-Amyrin, in rubber biosynthesis, **29:**244
Amytal, **34:**87; 89; 96; 104; 123; 134; 146
effect on mitochondrial oxidative phosphorylation, **21:**31
effect on oxidative phosphorylation, **20:**172
. effect on reduced diphosphopyridine nucleotide dehydrogenases, **20:**171; 172
effect on succinic oxidase and reduced diphosphopyridine nucleotide oxidase, **20:**178
inhibition by, **34:**109

inhibition site of, **34:**101–111
inhibitor of oxidation, **23:**368–369; 383
Anabasine, and nicotine biosynthesis, **8:** 229–231
in *Nicotiana glauca,* **8:**219; 227
produced in roots, **8:**223–224
similarity to nornicotine, **8:**228–229
Anabasis aphylla in alkaloid studies, **8:**223
Anabena, **46:**67
Anabolism, **34:**478. *See also* Microbial assimilations
of acetyl phosphate, **6:**233; 242–250
microbial, **6:**269–305
Anacystis, photoinduced changes in, **25:**75
Anacystis, phycocyanin in, **30:**17
Anacystis nidalans, **54:**169
Anacystis nidulans
biopterin in, **21:**377
plastoquinone in, **30:**31
Anaerobic bacteria in cellulose decomposition, **2:**249–251; 255–256; 260–262
Anaerobic decarboxylation, **27:**535
Anaerobic flash, in luminous bacteria, **25:** 151–153
Anaerobic growth
on accetate, **23:**431–434
uncoupling agents inhibited by, **29:**353
Anaerobic heat production in spermatozoa, **24:**51–52
Anaerobic metabolism, **1:**132–139
Anaerobic metabolism of spermatozoa, **24:** 46–51
Anaerobic oxidation of aliphatic hydrocarbons, **27:**529–534
Anaerobic pathway, quinolinate synthesis, **73:**136–140
Anaerobic respiration, in plants, **18:**295–298; 311; 313
Anaerobic techniques, EPR, **27:**118
Anaerobic transhydrogenases, classification, **14:**227–228; 253–254; 255
Anaerobiosis, **58:**207
and citric acid cycle, **18:**270–274
8-Analinonaphthalene sulfonate, **46:**392; 393
Analogs, altered proteins from, **29:**153
amino acid, **29:**90 (*See also* Amino acid analogs)
affinity for acyl-sRNA synthetases by, **29:**93
arginine, **29:**131
characteristic of toxic amino acid, **29:**91
control of histidine synthesis by, **29:**126–127
histidine, phenotypic resistance to, **29:**130
resistance to, **29:**128
structures of, **29:**124
isoleucine, **29:**142
leucine, **29:**142; 144–145
methionine, **29:**134
miscellaneous, **29:**142
phenylalanine, **29:**110; 115; 122
proline, **29:**94–108
random incorporation of, **29:**153
resistance to arginine, **29:**133
selenium containing, **29:**140
of sulfur-containing amino acids, **29:**134
threonine, **29:**142
tryptophan, **29:**121
tyrosine, **29:**114–115; 122
valine, **29:**142
Analogues, inhibitory, **6:**129–144. *See also* Inhibitory action
Analogues of pyridoxal 5′-phosphate, **35:**512
Analogy
algebraic *versus* graphic, enzyme action, **73:**49–52
enzyme action, **73:**25–26
Analysis, graphical, of enzyme data, **29:**1–2
least squares method of statistical, **29:**2–4
statistical, of enzyme kinetic data, **29:**1–32
Anaphylactic reactions, from thrombin preparations, **16:**84
Anaphylactic shock, antithrombin in, **10:**36
Anaphylaxis-desensitization reaction, in schizophrenia, **29:**503–504
Anchimeric assistance, **43:**269
Anchor principle, **43:**296
Anderson test, for racemization, **29:**410–411
Androgamone I, effect on sperm, **24:**84
Androgens, **47:**298. *See also* Steroid metabolism
control of nerve growth factor content in mouse submaxillary gland by, **31:**54–56
effect on β-glucuronidase levels, **16:**388; 399–400
effect on mRNA coding for α 2u-globulin, **47:**298
induction of α 2u-globulin synthesis, **47:**298
as inhibitory analogues, **6:**135
and seminal citric acid, **9:**376–378
and seminal fructose, **9:**361–362; 364–367; 376–380

Androgens (*Continued*)
and seminal phosphatases, **9:**352
site of primary effect of, **29:**377
$\Delta^{1,4}$-Androstadine-3,17-dione, microbiological formation, **20:**261; 270
5α-Androstane, **46:**158
Androstane-3β,17β-diol, microbiological formation and conversions, **20:**257; 258
Androstane-3,17-dione, formula, **20:**240
microbiological formation and conversions, **20:**252; 258
Androstane-3α-ol-17-one (5-trans), microbiological conversions, **20:**272
Δ^4-Androstendione-3,17, tritium labeling, **26:**127–128
Δ^5-Androstene-3β-17deta-diol, microbiological conversions, **20:**258; 259
Δ^4-Androstene-3,17-dione, formula, **20:**240
microbiological formation and conversions, **20:**239; 242; 252; 255; 257; 258; 261; 273
Δ^5-Androstene-3β-ol-17-one, microbiological formation and conversions, **20:**239; 258; 273
Androsterone, absorption spectrum, **4:**288; 289
Anemia, **44:**221, **58:**815
iron deficiency, **46:**176
pernicious, metabolism of tyrosine in, **14:**99
sickle-cell, genetics, **13:**141–142
sideroblastic, **46:**177
and vitamin B_{12}, **26:**265; 267
Anesthesia, increase in brain serotonin by, **29:**487
lack of change in brain catecholamines levels during, **29:**487
Anesthesia of protoplasm, **7:**41; 52–54
Anesthetics, **55:**199
fluorinated, **55:**257–263
catalysis of, **55:**258
distribution properties of, **55:**258
fluoroxene, **55:**262–263
halothane, **55:**259–261
methoxyflurane, **55:**261
Aneurin, *See* Thiamine
Aneurinase, **15:**389; 391
Aneurin in molds, **9:**572
Angelicin, structure, **66:**87
Angiosperm, **46:**96
Angiosperms, β-1,3-glucanase of, **28:**342–343
β-1,3 glucans of, **28:**329; 334; 337; 340
Angiotensin-converting enzyme, **53:**283, **56:**292
action of, **56:**355
activity of, **57:**4–7
and carboxypeptidase-A, **57:**20–24
catalytic process of
generalizations of, **57:**16
kinetic work with, **57:**16–18
reactions of, **57:**14–15
zinc dissociation of, **57:**19–20
contributions of, **57:**73–74
discovery of, **57:**4–5
and enalaprilat
active-site conformation of, **57:**56–57
analogs of, **57:**53–55
apoenzyme, **57:**52–53
dissociation constant, **57:**52
inhibition, **57:**49–50
initial rates, **57:**48–49
lactam analogs of, **57:**58–66
other analogs of, **57:**55–56
rate constant, **57:**51–52
steady-state velocities, **57:**50–51
structure of, **57:**57–58
studies of, **57:**46–47
tight-binding, **57:**47–48
and enzymatic reaction mechanisms, **57:**20
inhibition of, **56:**355; 358
inhibitors
as antihypertensives, **57:**2
captopril, **57:**28–31
carboxylkyldipeptides, **57:**32–46
design of, **57:**24–25
naturally occurring peptide, **57:**25–26
peptide inhibitor analogs, **57:**26–28
localization of, **57:**5–6; 8
metal centers of, properties of, **56:**356–358
metal content of, **56:**356
molecular properties of, **56:**356
properties of
catalytic, **57:**14–20
characteristics of, **57:**10–14
important, **57:**2–3
isolation of, **57:**7–8
molecular weights of, **57:**10–14
purification of, **57:**8–10
in rabbit lung, **57:**12–13
role of, biological
in brain, **57:**6–7
as inhibitors, **57:**2
in RAA system, **57:**3–4

studies of, **57:**4–6
specificity of, **56:**355
and thermolysin-inhibitor complex
inhibitors of, **57:**67–70
peptide cleavage by, **57:**70–73
Angiotensin-I
in catalytic processes, **57:**3
conversion to angiotensin-II, **57:**5; 14
hydrolysis of, **57:**17
as substrate for brain converting enzyme, **57:**7
Angiotensin I and II, **56:**355
Angiotensin II, **53:**240
Angiotensin-II
conversion from angiotensin-I, **57:**5; 14
effects of, opposite insulin, **57:**164
properties of, biological, **57:**3–4
Angiotonin, *See* Hypertensin
Angle, bond, of proteins, **23:**5–11; 22–23
of internal rotation, of proteins, **23:**5–11; 22–23
Anhydride, **58:**5
2,2′-Anhydro-ara-fc, **54:**62
N^5,N^{10}-Anhydroformyltetrahydrofolic acid, in purine biosynthesis, **21:**204; 206; 207; 220; 222; 223
interconversion with N^5,N^{10}-methylenetetrahydrofolic acid, **21:**415
isomers, **21:**384; 385
relative free energy of hydrolysis, **21:**414
structure, **21:**219; 383
synthesis and metabolism, **21:**225–228; 383; 384; 412; 413
in urine, **21:**377
1,5-Anhydro-D-glucitol, presence of a single common carrier in translocation of, **29:**53
1,2-Anhydro-D-glucitol-6-P, **43:**512
1,5-Anhydroglucitol-6-P, **43:**502
1,2-Anhydro-L-iditol-6-P, **43:**512
Anhydroleucovorin, *See* N^5,N^{10}-Anhydroformyltetrahydrofolic acid
2,5-Anhydromannitol, **59:**367–368
1,2-Anhydro-D-mannitol-6-P, **43:**512
Δ^2-Anhydromevalonic lactone, **35:**153; 154
Anhydro-β-rotunol, **55:**53
Anil formation type reactions during sclerotization, **16:**162
Anilides, of amino acids, synthesis with papain, **19:**308; 309
Aniline
action with thermostable factor, **15:**410
activator of thiaminase, **15:**400; 403; 407; 409–413
inhibition of invertase, **15:**287
and thiaminase model, **15:**404
and indolylacetic acid oxidase, **19:**127
peroxidase action on, **19:**106
reaction with *o*-benzoquinone, **16:**110
reaction with protocatechuic acid, **16:**110
thermodynamic data on ionization, **20:**64
Aniline derivatives, peroxidase action on, **19:**106; 107
1,8-Anilinonaphtalenesulfonate (ANS), **51:**8; 9; 38
8-Anilinonaphthalene sulfonate, **46:**391
Anilinothiamine, **15:**403
Animal Collagenases, **34:**175; 187
action on collagen of, **34:**177; 178
from amphibian tissues, **34:**175
assay procedures for, **34:**166; 175; 176
cell origin of, **34:**176; 181
controversial results on, **34:**186
from invertebrata, **34:**187
from mammalian tissues, **34:**178
physiologic implications of, **34:**191; 195
Animal membrane, **34:**467
Animal metabolism, asparagine synthetase, **72:**149–151
Animals
carbohydrate metabolism, regulation, **28:**76–80; 83–86
pentose synthesis, tracer studies, **28:**413–419; 431–436
uronic acid pathway, **28:**398–401; 414; 418
laboratory, **53:**204; 205–208
Animal tissues
and metabolic regulation, **59:**369–372
and PFK 1, **59:**324–327
subcellular fractions in, **24:**291–358
Animal Tissues, Distribution of Enzymes Between Subcellular Fractions in (de Duve, Wattiaux, and Baudhuin), **24:**291
Animal Tissues, Molecular Properties and Transformations of Glycogen Phosphorylase in (Krebs and Fisher), **24:**263
Anion, **34:**489
content of normal mitochondria, **29:**261–263
uptake of, in mitochondria, **29:**277–278
effects, competitive, **39:**345

Anion exchangers, in protein chromatography, **22:**431; 433
Anionic buffers, **51:**10
Anionic polymers, effect on PR-enzyme, **24:**280
RNase, **24:**187–190; 192–193; 219; 238
Anions
binding by myosin and its subunits, **16:**332
depolymerizing effect on actin, **16:**335
in hydrocarbon oxidation, **27:**471; 485; 530
influence on properties of F_1, **49:**250–252
inorganic, **44:**213
in semen, **24:**80–82
as surrogates, **53:**332–333
Anion transport, in plants, **18:**50–52
Anisic acid, in lignin study, **18:**374
Anisole, chlorination
by cyclodextrin, **58:**14
by epichlorhydrin, **58:**15
Anisomycin, in treatment of plant diseases, **20:**390; 400; 402
Anisotropic inclusions, cholesterol ester hydrolases, **69:**59; 63
Ankistrodesmus, photoreduction by, **30:**24
Ankistrodesmus braunii, photosynthesis by, **19:**57; 58
Ankistrodesmus cells, photoinduced changes, **25:**78
Anlage theory of protein synthesis, **1:**17
Anlarge theory of protein synthesis, **9:**430–431
Annelids, luminescence of, **25:**160
Annexin-membrane complex, **71:**221–226
ion channel activity, **71:**224–226
lipids affinity, specificity, and binding mode, **71:**221–222
lipid state, **71:**224
structure of, **71:**222–224
Annexins. *See also* Ion channel model system annexin V
generally, **71:**214–216
V crystal structure, **71:**216–220
Anomerase activities, glucose-6-P-1-epimerase, **44:**146; 148
mutarotase, **44:**146; 148
phosphoglucose isomerase, **44:**146
yeast aldolase, **44:**146
Anomer distribution, **44:**141
Anomeric specificity, **53:**53–55; 75
activation
of phosphofructokinase, **44:**149
of yeast pyruvate kinase, **44:**152
determination of, **44:**142; 150
fructokinase, **44:**146
fructose biphosphatase, **44:**149
of fructose 1,6-bisphosphatase, **53:**53–55; 74–76
fructose bisphosphate, **44:**159
fructose bisphosphate aldolase, **44:**149
galactokinase, **44:**146
galactose dehydrogenase, **44:**147
glucokinase, **44:**146
glucose, oxidase, **44:**147
glucose dehydrogenase, **44:**147
glucose-6-P-dehydrogenase, **44:**147
D-glucose-6-P-phosphatase, **44:**147
glycogen phosphorylase, **44:**148
hexokinase, **44:**146
implications for mechanism, **44:**154
aldolase, **44:**156
fructose bisphosphatase, **44:**157
glucose-6-phosphatase, **44:**156
hexokinase, **44:**156
phosphofructokinase, **44:**157
phosphoglucose isomerase, **44:**154
implications for metabolic regulation, **44:**157
glycolytic flux, **44:**158; 159; 160
rate of isomerization, **44:**158
kinetic methods, **44:**143
phosphoglucomutase, **44:**147
phsophofructokinase, **44:**149; 150
phsophoglucose isomerase, **44:**148
in situ generation, **44:**142
substrate analogs, **44:**145
sucrose phosphorylase, **44:**148
Anomeric Specificity of Glycolytic Enzymes, The (Benkovic and Schray), **44:**139
Anomeric sulphur atom, effect on rate, **36:**102
Anomer specificity, **43:**501; 502; 503
Anoxia, **57:**145, **59:**369
Anserine
amino acids in, **14:**266
imidazole in, **20:**102
synthesis from carnosine, **22:**256
L-Anserine, effect on fructose 1,6-bisphosphatase, **42:**202
Antagonism, **7:**244; 462
among factors affecting alkaline phosphatase, **10:**102–104
of nicotinic acid to benzamide, **10:**69
of riboflavin to mepacrine, **10:**69

Antagonisms, Biological, Between Structurally Related Compounds (Woolley), **6:**129
Antagonisms between structurally related compounds, **6:**129–144. *See also* Inhibitory action, analogues
Antagonist-inhibitor actin in chemotherapy, **6:**40–51. *See also* Inhibitory action
Antagonists. *See also* Antimetabolites and Inhibitors
amino acid, activity in transamination, **16:**237
noncompetetive, of sulfonamide, **16:**300
of vitamin B_6, in transamination study, **16:**220–222
structural (*See also* Competitive inhibitors)
action of ethionine and methionine, **12:**478–479
effect on peptidase activity, **12:**242–248
Antagonists of glutamic acid, **13:**255–260; 263; 278–279
Antheraea mylitta, **45:**403; 407; 408; 409
Antheraea mylitta proenzymes, **45:**407
N-terminal sequence of, **45:**407
Antheraea pernyi, **45:**402; 403; 406; 407; 415
cocoonases of, **45:**410
Antheraea polyphemus, **45:**403; 405; 406; 407; 408; 409
active-site serine peptides from, **45:**412
cocoonases of, **45:**410; 412
molting fluid of, **45:**417
N-terminal sequence of, **45:**407
Anthocyanase, cacao, **25:**466
Anthocyanin. *See also* Phytochrome
Anthocyanin formation, **46:**100
role of phytochrome, **26:**24–26
Anthocyanins
from cacao, **25:**462–465; 478–479; 482–483; 484; 488
metabolism in plants, **16:**142; 144
Anthracene, microbial oxidation, **27:**525
Anthracyclines, collagen hydroxylase inhibitor mechanisms, **72:**363
Anthranilate, **49:**3, **50:**149
chorismate conversion to, **66:**208–210
fluorescence, **49:**3
Anthranilate synthase, **66:**214–223, **72:**136–138
catalysis, **66:**216–222
glutamine amide transfer, **66:**217–218
GAT and synthase domains, **66:**215–216
relation to 4-Amino-4-deoxy-chorismate synthase, **66:**225–227
thioacylenzyme intermediate, **72:**178–183
Anthranilate synthetase, **50:**169
A Aerogenes, **38:**11
B. subtilis, **38:**9
binding of glutamine, **38:**17
carrot root(Daucus Carota), **38:**35
catalytic facilitation, **38:**29
channeling, **38:**29
chorismate binding, **38:**26; 35
conformation probes, **38:**34
cooperative kinetics, **38:**27
enzyme aggregates, **38:**2; 4–5*t;* 10
evolutionary relationships, **38:**32
glutaminase activity, **38:**17
hydroxamate formation, **38:**17
inactivation by bromopyruvate, **38:**24
intermediate steps, **38:**24
mechanism, of glutamine utilization, **38:**15
of tryptophan inhibition, **38:**34
N. crassa, **38:**14
physiological function of multienzyme complexes, **38:**28
pseudomonas enzyme, **38:**7
reaction mechanism, **38:**17; 24; 33
relationship to other glutamine amido-transferases, **38:**29
S. *typhimurium,* **38:**11
Serratia marcescens, **38:**6
subunit composition, **38:**3; 4–5*t*
S. marcescens, **38:**7
S. typhimurium, **38:**11
subunit interactions, **38:**27
tobacco pith(Nicotiana tabacum), **38:**35
trypsin digestion, **38:**32
tryptophan binding, **38:**35
tryptophan inhibition, **38:**25; 35
type I enzyme, **38:**2; 3
type II enzyme, **38:**2; 10
type III enzyme, **38:**3; 13
aggregated forms of the enzyme, **39:**139
association with phosphoribosyltransferase, **39:**139
complementation of enzymes from extracts of, **39:**139
Escherichia coli, **39:**139
Salmonella typhimurium, **39:**139; 140
experiments with purified enzymes, **39:**140
separation from phosphoribosyltransferase, **39:**140
soluble form of enzyme, **39:**141

Anthranilate synthetase (*Continued*)
brom opyruvate, **39:**141
hydroxylamine, **39:**142; 143
inhibition of activity by tryptophan, **39:**141
methyl hydroxylamine, **39:**142; 143
molecular weights, **39:**141
properties, **39:**141
separation of soluble enzyme into two components, **39:**141
sulfhydryl reagents, **39:**141
incorporation of proton of water into products, **39:**141; 142
particulate form of the enzyme, **39:**143
glutaminase activity of, **39:**146
inhibition by tryptophan, **39:**143; 144
reaction with alkylating reagents, **39:**145
reactivity of subunits with amino donors, **39:**145
separation of subunits, **39:**144
solubilization of, with proteolytic enzymes, **39:**144
with thiol compounds, **39:**144
reaction catalyzed by, **39:**138
sources of the enzyme, **39:**139; 146
unaggregated form of enzyme, **39:**146
properties of the enzyme from, *Bacillus subtilis,* **39:**147
complementation of enzymes derived from mutant extracts, **39:**147; 148
mutants of strain NP100, **39:**147
subunits of enzyme, **39:**148
pseudomonas, **39:**148
gene fusion, possible role in origin or aggregate enzyme, **39:**149
reactivity of subunits with nitrogen donors, **39:**148
separation of subunits, **39:**148
tryptophan, effect on wild type and mutant strains, **39:**148
serratia marcescens, **39:**146
inhibition by tryptophan, **39:**147
molecular weight of, **39:**147
reaction with alkylating agents, **39:**147
subunit structure, **39:**147
inhibition by tryptophane, **28:**44
Anthranilate synthetase Component I, *E. coli,* **38:**10
molecular weight, *E. coli* enzyme, **38:**11
N. *crassa,* **38:**14
Pseudomonas, **38:**8
Serratia marcescens, **38:**6
Anthranilate synthetase Component II, B. *subtilis,* **38:**9
function in glutamine binding, **38:**13; 31
synthesis of 4-aminobenzoate, **38:**31
glutamine binding subunit from *S. marcescens,* **38:**35
N. crassa, **38:**15
proteolytic digestion, **38:**13
Pseudominas, **38:**8
S. typhimurium, **38:**11
Serratia marcescens, **38:**7
Anthranilate synthetase subunit X, **38:**9
Anthranilate Synthetase (Zalkin), **38:**1
Anthranilic acid, **49:**28; 29
metabolic pathway, **15:**184–185
precursor of nicotinic acid, **15:**101–102
in metabolism of aromatic amino acids, **16:**235; 288; 293–295
reaction with phosphoribosyl phrophosphate, **18:**218
in tryptophan synthesis, **10:**71
Anthrone reagent for glycogen in tissue, **12:**371
Anti-ADP-glucose pyrophosphorylase, **46:**376
effect on *E. coli* ADP-glucose pyrophosphorylase activity, **46:**376
effect on *S. typhimurium* ADP-glucose pyrophosphoryase activity, **46:**376
Antiazoproteins, **15:**70
Antibacterial action, of alkyl thiosulfinic esters, **11:**393
of allicin, **11:**380–381; 391; 398
of garlic, **11:**380
Antibacterial activity, methods for determining, **7:**494–495
strains of bacteria used, **7:**490–493; 494
Antibacterial agents, **6:**35–57, *See* Gramicidin; Tyrocidine; Tyrothricin
effect on sperm, **24:**85
requisites for action, **6:**36–37
sulfonamides, **6:**35–44; 58–67
Antibacterial index, **6:**138
Antibacterial substances, alphabetic list with chemical and biological properties. *See also under* individual substances
Antibacterial substances, αbetic list with chemical and biological properties, **7:**466–494
active groups in, **7:**503–504

biological properties, **7:**489
classified by active groups, **7:**503–504
deep culture or tank method, **7:**463
from fungi and green plants, **7:**461–511
general methods for preparing concentrates, **7:**465–466
growth-inhibiting concentrations, **7:**490–493
identification, **7:**495–501
measurement in mixtures, **7:**497–499
methods of isolation, **7:**464–466
methods of sterilization, **7:**489
mode of action, **7:**501–504
problems for the future, **7:**505
production, **7:**462–466
properties, **7:**486–487
Antibacterial Substances from Fungi and Green Plants (Kavanagh), **7:**461
Antibiosis
fungal suppression, **28:**355–356
in plants, **16:**153
Antibiotic activity of streptomycin, relation to structure, **10:**160–161
Antibiotic compounds, from *Streptomyces,* **10:**145–166
in plants, **10:**267–268
Antibiotic effects of aerobic, spore-forming bacteria, **4:**154
Antibiotic peptides, **35:**2
Antibiotic resistance, enzyme action, V/K *constant,* **73:**31–32
Antibiotics
adsorption on solids, **28:**29
and β-lactamase, **28:**249; 250; 308; 312–313
resistance to, **28:**269
amino sugars, **25:**433
effect on bacterial walls, **26:**216–219
effect on carotenogenesis, **21:**318; 319
effect on collagenase, **23:**194; 203
effect on *in vitro* biosynthesis of poly(glycerol phosphate) and poly (ribitol phosphate), **30:**244
effect on plants, **20:**385
in treatment of plant diseases, **20:**383–406
effect on RNase, **24:**238
inhibition of peptidoglycan biosynthesis, **40:**122
lipophilic, **46:**390
polyene, **34:**483
polypeptide, **34:**483
reactivity toward —SH groups, **11:**246–247
as unnatural translocators, **29:**48
Antibiotics, Streptomycin and Other Streptomyces (Brink and Folkers), **10:**145
Antibiotics and Plant Diseases (Tanner and Beesch), **20:**341
Antibiotic substances, activity against fungi, **7:**495
effects on luminous bacteria, **7:**252
Antibiotic treatment, *Pseudomonas aeruginosa,* **70:**223; 245
Antibiotin, in metabolite displacement, **6:**144
Antibodies, **36:**31; 33; 53; 75 81, **48:**57–60. *See also* Immunological studies
adenosine receptors, **69:**89; 97
antithrombins as, **16:**84
bivalency of precipitating, **30:**148
blood group, peroxidase action on, **14:**156–157
chemical modification and biological activity, **24:**105–160
partial degradation of, **24:**127–131
partial loss of active centers, **24:**120–121
chromatographic purification, **22:**447
labeling of active site in, **22:**54; 55
constraining, **37:**594; 595
stimulation of enzymic activity, **37:**595
Rh, inactivation by peroxidase, **14:**154–157
of tobacco mosaic virus, **15:**468; 470–472
thermochemical studies, **15:**49–51; 64–77
univalence of, **15:**68–70
to tyrosinase, **14:**93; 96
use in isolation of enzymes, **14:**398–399
Antibody, **34:**481, **43:**313
antienzyme, **38:**309
Antibody-antigen, adsorption studies, **14:**337–338; 343; 364
Antibody-antigen interaction, **41:**132–137
complementarity of the reactive sites, **41:**133–137
complementary, **41:**133
conformation, **41:**132
mechanism of, **41:**132
subcomplementary, **41:**133
theory, **41:**132
Antibody-enzyme complex, **37:**595
Antibody excess, effect on RNase, **24:**238
Antibody-hapten interactions, elementary steps in, **25:**13–15

Antibody(ies), and complement interdependence, **8:**92–93
complement and union with antigen, **8:** 97–101
data on molecule, **8:**81
estimation in weight units, **8:**72
isolation, **8:**72
production, **8:**54–55
reaction with antigens, **8:**55; 73
relation to complement and antigen, **8:**82–89
Antibody nitrogen and hemolysin titer, **8:**77–79
Antibody to enzymes, **46:**386
Antibody to transferases, **46:**389
Anticatalase, formation, **1:**167
Anticephalin activity, distribution in centrifugal field, **16:**88
of lipide inhibitor, **16:**85
Anticoagulant action, of β-1,3 glucans, **28:**356
Anticoagulant action of poly-α-amino acids, **16:**88
Anticoagulants, **10:**12; 15–17; 33–38. *See also* Inhibitors
effect on inhibitors of threone activity, **16:**56
effect on stability of plasma Acglobulin, **16:**62–63; 71
lipide, from brain tissue, **16:**85–86
of serum, **16:**86–87
synthetic, **10:**34–35
Anticocoonase, **45:**410
Anticodon, **34:**3, **35:**467
Anticodons, **47:**378; 388–391; 399–403
thiouridine in, **47:**384
Anticollagenases, **23:**194; 197; 198; 200
Antidepressants, tricyclic, **55:**178
chronic administration of, **55:**179
structure of, **55:**177
Antielastase, serum, **23:**236
Antifatty-liver activity, method of assay, **8:** 195–196
Antifatty-Liver Factor of Pancreas (Chaikoff and Entenman), **8:**171
Antifatty-liver factors of pancrease, **8:**171–202
of Entenman *et al.,* **8:**194–195
mechanism of action, **8:**196–200
preparation, **8:**193–196
relation of choline, **8:**191–193; 196; 197–199
Antifertilizin, chemical nature, **9:**264–266
effect on activation of egg, **9:**294
in egg, **9:**256
role in fertilization, **9:**266–268; 268–275; 314–315; 348
Antifibrinolysin, **16:**71; 76
Antifibrotic drugs, collagen hydroxylases and development of, **72:**357–364
Antifolic acids, methylation study, **25:**400
Antifolics, **27:**436–455
synergism with sulfonamides, **27:**455–456; 460
Antifreeze glycoprotein, **52:**37; 39; 40; 41; 43; 44; 47; 52; 53; 56; 59; 98; 112
Antigen, **41:**132; 134, *See* M-antigen
alcohol dehydrogenase, **41:**132
bovine-serum albumin, **41:**132
casein, **41:**134
derivatized, **41:**134
gelatin, **41:**134
horse globulin, **41:**134
insuline, **41:**134
Antigen-antibody complex, **15:**50
composition of, **30:**148
Antigen-antibody interaction, **12:**252; 513
Antigen-antibody reactions, **43:**313
chemical theories, **8:**84–89
role of –SH groups, in, **11:**245
Antigen-antibody specificity, relation to stereochemical configurations, **11:**66
Antigenic action of proteins, **8:**54–55
Antigenic activity, of tobacco mosaic virus, **15:**468; 470–472
Antigenic behavior of ribonucleases, **24:**184; 231–232
chemical modification and biological activity, **24:**105–160
partial degradation of, **24:**127–131
partial loss of active centers, **24:**120–121
valency of, **24:**118
Antigenic character, of bacterial walls, **26:** 223–226
of polysaccharides, **26:**67–78
Antigenicity, of collagenases, **23:**197–198
Antigenic specificity, and resistance, **6:**118–119
Antigens, **36:**31; 33
activity of films, **28:**24; 30
glucans as, **28:**336–337; 357
glutamate dehydrogenase as, **28:**142
β-lactamase as, **28:**253; 255–257; 298–299; 304–307
appearance in embryo, **9:**319

complement, and union with antibody, **8:** 97–101
estimation in weight units, **8:**72
reaction with antibodies, **8:**55; 73
relation to complement and antibody, **8:** 82–89
development, **16:**478
effect on antithrombin, **10:**36–37
thermochemical studies, **15:**49–51; 64–77
Antigens, A B H, **36:**132
Lewis, **36:**132
present on human erythrocytes, **36:**131
Antigens-antibody reaction, use in localizing proteins in cell nucleus, **16:**418–419
Antiglucuronidase, in tissues, **16:**375
Antihemophilic factor, **48:**278, *See* Platelet cofactor I
Antihemophilic globulin, properties of, **30:**262. *See also* Factor VIII
Antihemorrhagic vitamin, **2:**286; 287. *See also* Vitamin K
Antiheparin activity, *See* Platelet factor 4
Antihistamine compounds, in blood clotting, **16:**54; 71
Antihistamines, inhibition of hyaluronidase by, **13:**225
Antiinhibition of enzyme activity, **10:**99; 102–104
Antiketogenesis, **3:**246–247
Antiketogenesis and oxalacetic acid, **8:**408
Antilipemic action, of β-1,3 glucans, **28:**356
Antilymphoma activity, **39:**187–189
Antimalarial activity, of phyrimidine derivatives, **27:**442; 444; 446–447; 458
Antimetabolites. *See also* Antagonists and Inhibitors; Structural inhibitors
and chemotherapy, **27:**418; 420; 431; 436; 461
in study of metabolic relationships, **16:**249; 302
toxic amino acids as, **29:**89–163
Antimetabolites, Action of Toxic Amino Acids as (Fowden, Lewis, and Tristram), **29:**89
Antimony mercaptide, **11:**216
Antimuscarinic drugs, **55:**178
Antimycin, effect on oxidation of vitamin K_3, **20:**184
effect on oxidative phosphorylation, **20:**172
effect on respiratory chain enzymes, **20:**159; 169; 171; 177; 178; 186; 187; 193
in treatment of plant diseases, **20:**403
Antimycin A, **54:**340, **58:**239
cyclic phosphorylation mediated by ferredoxin inhibited by, **30:**41
noncyclic photophosphorylation inhibited by, **30:**34
effect on mitochondrial components and processes, **21:**18; 31 105–107; 112; 115; 116
effect on succinate oxidation, **18:**69; 104
inhibitor, **28:**160; 178; 190; 192; 194; 198; 199; 202; 210–212; 214; 217; 221–222
Antimycin A sensitivity, **24:**303–304
Antimycin-labile factor, in respiratory chain, **20:**171; 172; 175–178; 182; 185
Antineoplastic activity, **39:**186
Antinvasin, **13:**216
Antinvasin I, and snake venoms, **8:**474; 485–486; 487
inactivation, **8:**485
Antinvasin II and snake venoms, **8:**485–486
Antioxidant molecules, maintenance, **65:**119
Antioxidants, **58:**195
action of, **8:**439–441; 447–448; 453
and free-radical mechanism, **8:**439–440; 453
lipoxidase action on, **19:**99; 102; 103
Antiport, definition of, **29:**38–39
exchange-diffusion type of mechanism, **29:**49
glucose/galactose, effect of phlorzin on, **29:**52
glucose flux during, **29:**52
Na^+ and H^+ across the mitochondrial membrane, **29:**52–53
Antipressor factor of kidney, **7:**541
Antipsychotic drugs, **55:**179–182
antischizophrenic activity of, **55:**180
dopamine receptor block from, **55:**180
structure of, **55:**180
Anti-R-antibodies (rabbit γ-globulin), steps in reaction, **25:**14
Antirenin formation, **7:**540–541
Antiseptic action of hydrocarbons, **10:**451–452; 452–453
Antiseptics, effect on collagenase, **23:**194–195
Antisera
to glucosylated wall teichoic acid, **30:**226
precipitation of dextransucrase by dextran-reactive, **30:**190

Antisera (*Continued*)
of β-lactamase, effect on enzymic activity, **28:**306–307
Antiserum, **41:**134–136
to *O-B*-glucosidotyrosyl-protein, **41:**136
to gelatin-azobenzene arsonate, **41:**136
to *O*-β-glucosido-N-carbo-benzyloxy-tyrosyl-protein, **41:**136
to ovalbumin-azobenzene arsonate, **41:**136
to paba-albumin, **41:**135
to paba-casein, **41:**134
to paba-gelatin, **41:**134
to paba-globulin, **41:**135
to paba-insulin, **41:**134
Antiserum cross, **39:**225
Anti-silk fibroin antibody, combining site of, **24:**115
Antithiamine factor, in ferns, **15:**390; 394; 406–411
in plants, **15:**390; 391; 394; 406–411
Antithrombin, **16:**25; 76–85; 92
Antithrombin accelerator effect, *See* Antithrombin-IV
Antithrombin and blood clotting, **10:**12–13; 19; 33–38
Antithrombin-binding region, **60:**280
Antithrombin-I, **16:**77–79; 83; 92
Antithrombin-II, **16:**77; 80–82; 83
Antithrombin III, **46:**6, **48:**282
Antithrombin-III, **16:**76; 77; 79–81; 83; 92
Antithrombin-IV, **16:**77; 82–83; 92
Antithrombin reactions (other than by specific antithrombins), **16:**83–85
Antithromboplastic activity, of lipide inhibitor, **16:**85; 86
Antithromboplastic agent in hemophilic blood, **10:**12
Antithromboplastin, **16:**86–88; 92
effect on platelets activity, **16:**69
inhibitor effect, **16:**25; 54–56
Antitoxin inactivation, **7:**248
Antitumor activity of GPS, **39:**221
Anti-vitamin, **11:**408
Ants, attine, **45:**402
AOA, **57:**326–328
A5P, *See* Adenosine 5′-monophosphate
AP endonucleases, **60:**3–14
analysis method, **60:**8
classification, **60:**7
cleavage sites, **60:**9
cross reactivity, **60:**7
depurination, **60:**7
DNA specificity, **60:**12
β-elimination reaction, **60:**11
exonucleolytic activity, **60:**6; 13
glycosylase activities, **60:**6
heterogeneity, **60:**7
molecular weight, **60:**5
phosphodiester bond cleavage, **60:**10
phosphodiester cleavage, **60:**5
properties, **60:**4
reaction mechanism, **60:**9; 11
Aphanocapsa-6308, **46:**338
ADP-glucose pyrophosphorylase activity in, **46:**68
activation by 3-P-glycerate, **46:**338
Apis mellifera, **45:**395; 398, **71:**100
Aplysia limacina, **46:**68
Aplysioviolin, **46:**68
Apoalkaline phosphatase
cadmium binding to, **55:**398–400
magnesium binding to, **55:**405
zinc binding to, **55:**400–404
Apocarboxylase, activation of, **26:**331–334
Apocarboxylase-activating enzyme, **26:**332–334
Apocarboxypeptidase, activation of, **24:**127
Apocatalase, **46:**152; 155
Apocatalase in *E. coli* mutant, **24:**23–24
Apo-Cp, **44:**226
Apocytochromes, **46:**86; 88
*b*1, **46:**86
*b*5, **46:**151; 154
c, **46:**154
P450, **46:**132; 146; 153
Apodehydrogenase, binding to DPNH analogs, **27:**56–70
LADH, substrate binding, **27:**72
Apodehydrogenases, **5:**211–215
concentrations in source materials, **5:**214
Apodhydrogenase, **34:**474
Apoenzyme, **7:**309, **35:**337, **58:**77
activation by pyridoxal phosphate, **6:**27
of methylmalonyl-CoA mutase, **26:**345–346; 351
preparation of amino acid decarboxylases, **6:**19–29
of thiaminase, **15:**395–396; 403–404; 406
Apoenzyme, Cotton effect, **27:**52
in ternary complex, **27:**4–12
Apoenzyme-cofactor interaction, **24:**113–114

Apoenzyme conformers, with different affinity, eukaryotic dihydrofolate reductase, ligand binding, **70:**53–55
Apoenzymes, **34:**474. *See also* Zymogens and individuals
activation of, **24:**111–114
mode of action, **14:**181; 183–190
in soil microorganisms, **28:**367
Apo glutamic aspartic transaminase, **35:** 100–102
Apohemoglobin, **46:**152
Apolar mass, effect of on transport, **49:**47
Apoleghemaglobin, **46:**88
Apolipoprotein A_2, **58:**113
Apolysis, **45:**417
Apoperoxidase haemochrome *a,* absorption spectrum, **23:**289–297; 306–310
Apophosphorylases *a* and *b,* **24:**272
Apoprotein, **34:**470
Apoprotein-heme enzyme, **7:**333
Apoproteins, **7:**308–309; 312–313; 325–327, **46:**151–155
Apotransaminase, activation, **16:**213; 215–218
Apotranscarboxylase, attachment of biotin, **26:**333
Apotryptophan pyrrolase, **46:**153
Apotrytophanase, binding to PLP, **42:**301
dimeric form, **42:**301
monomeric form, **42:**304
sedimentation rate, **42:**302
monovalent cation, effect of, **42:**302
temperature, effect of, **42:**302
subunit interaction, **42:**305
hydrophobic, **42:**305
ionic forces, **42:**305
tetrameres form, **42:**301
Apo-vitamin B_{12} enzyme, **26:**267
Apozymase system, **7:**424–425; 427; 428–435
Apparatus, for column chromatography of enzymes, **22:**437–440
for particle size separation of ion exchangers, **22:**442
Apparent Michaelis constant, **34:**460–465; 502; 522–525
Apple, cytochromes and oxidases, in, **18:**19; 29; 45–46
Apple juice, oxidized, effect on polygalacturonases, **20:**365
Apples, antibiotic treatment of diseases, **20:**398; 399
Aproteon, **6:**162
AP site, **60:**3
APS kinase gene, in *Salmonella typhimurium,* **74:**189
[^{32}P] APT, **49:**238
Apyrase, **6:**149, **34:**449; 459
in adenylic acid formation, **6:**161
classification, **14:**248; 250
in myosin sediment, **6:**174
Aqueous media, cryoenzymology in, **51:**1–74
water-in-oil emulsions, **51:**5–30
applications, **51:**13–30
emulsion procedure, **51:**5–6
physical-chemical properties, **51:**6–12
solutions of proteins, **51:**12
water-in-oil microemulsions, **51:**30–70
micella formation (nonaqueous media), **51:**30–37
rapid mixing and related problems, **51:** 65–70
reversed micellas as catalysts, **51:**47–55
temperature and, **51:**55–65
water pools and, **51:**37–47
withstanding subzero temperatures (unfreezable water), **51:**2–5
Aqueous phase, **34:**468; 485
phase, homogeneous, **34:**456
solution, **34:**487
Aqueous solutions. *See also* Water
Aqueous solutions of enzymes, effect of ionizing radiations on, **24:**383–399
Aquo complexes, **51:**127
Aquo(hydroxo)-cobalamin. *See also* Corrinoids and Vitamin B_{12}
absorption spectrum, **26:**247–249
synthesis, **26:**242–243; 246; 251
Aquo(hydroxo)-corrinoids. *See also* Corrinoids and Vitamin B_{12}
properties, **26:**247–251; 255–256
Araban, linkage to pectin in plants, **20:**345
Arabidopsis, phosphoribulokinase from, **74:**100; 120; 121
^{14}C-D-Arabinitol-1,5-diphosphate, as competitive inhibitor of aldolase, **31:**146
Arabinofuranosidases, **56:**150; 151
9-β-D-Arabinofuranosyl adedine (ara-A), **51:**200
Arabinogalactan, **56:**185

Arabinokinase, enzymic adaptation, **13:**70; 99
L-Arabinopyranose, disaccharide formation, **18:**224
Arabinopyranosidases, **56:**150; 151
Arabinose
oxidation, **15:**149
in pectin, **20:**344
L-Arabinose
as degradation product of pectins, **20:**350; 351
origin, **14:**206–208
D-Arabinose, **43:**493
L-Arabinose, **43:**493
enzymes and their structural determinants, **30:**110–111
map of operon, **30:**110
system of utilization of, **30:**110–116
L-Arabinose, as inhibitors, **36:**118
D-Arabinose, in disaccharide formation, **15:**312
Arabinose, UDP ester of, **18:**201
Arabinose derivative of coenzyme I, **15:**31
L-Arabinose isomerase, reaction catalyzed by, **30:**110
L-Arabinose kinase, **46:**247
Arabinose oxidase, **19:**202
L-Arabinose permease, properties of, **30:**111
Arabinose-5-phosphate, pentose metabolism, **28:**394–395
Arabinosidases
assay of, **56:**152
in cancer chemotherapy, **56:**154
classes of, **56:**150–151
occurrence and localization of, **56:**151–152
properties of, **56:**153–154
purification of, **56:**152–153
role of, **56:**154
β-L-Arabinoside, **36:**102
Arabinosides, action of β-galactosidase on, **13:**101–104
Arabinoxylans, **56:**223
Arabonate-5-P, **43:**494
Arachidonate-5-lipoxygenase, **58:**258
purification, **58:**259
self-inactivation, **58:**259
stimulation by nucleotides, **58:**259
Arachidonate 12-lipoxygenase, **58:**256; 259
in epoxyleukotriene synthesis, **58:**260
in lipoxins synthesis, **58:**260
from lung cytosol, **58:**261
purification, **58:**260
from skin, **58:**262
Arachidonate 15-lipoxygenase, **58:**260
molecular weight, **58:**261
Arachidonic acid, **55:**99, **58:**194
metabolism inhibitors, **58:**215
metabolites, **58:**217
oxygenation, **58:**284
prochiral center, **58:**283
steric structure, **58:**274
Arachin
adsorption studies, **14:**365
sedimentation and electrophoretic experiments on, **30:**142–143
Arachis hypogaea, adsorption studies or proteins, **14:**365
Arachis oil injection, effect on live β-glucuronidase, **16:**400
Ara-FC, 1-1β-D-arabinofuranosyl-5-5fluorocytosine, **54:**58
Arbacia sperm, basic proteins of, **16:**446
Arbutin, **56:**199
role in browning of plant tissues, **16:**150
Archaebacterium, glutaredoxin, **66:**164–165
Arctium minus, **7:**467; 468; 487; 489; 491; 493
Arene oxide, **46:**394
Arene oxides, xenobiotic substrate specificity and mechanism, **69:**32
AR fraction of pancreas, antifatty-liver effect, **8:**192; 198
Arg-145, **57:**21; 24
Arg^{70}, eukaryotic dihydrofolate reductase, ligand binding (side chain roles), **70:** 63–64
Argenine kinase, mechanism of deduced from relaxivity data, **33:**43
Arg-Gly-Asp (RGD) cell-adhesion motif, **70:** 2–3. *See also* Fibronectin
Arginase, **10:**300, **15:**322; 325; 331–333; 347; 353; 356; 358–360; 372–376; 379, **16:**193; 266, **18:**254, **37:**140; 151–152, **40:**73, **50:**141, **56:**134
and alkaloid biogenesis, **8:**238–240
in belladonna, **8:**244
in chromidia, **8:**31
in nuclei, **8:**33
classification, **14:**235; 247
isolation, **14:**403
effect of borate on, **12:**518
effect of metal ions on, **12:**225
from erythrocytes, **39:**45

from human liver, **39:**45
from kidney, **39:**46
from rat liver, **39:**44
mechanism of activation by Mn^{2+}, **39:**44
properties of, **39:**44
intracellular distribution, **16:**463; 466–468
localization in cytoplasm, **13:**11
localization of, **24:**320; 328; 340
macrophage, **65:**311
multiple forms, **25:**344–345
product inhibition of, **25:**223; 242–243
in ornithine cycle, **19:**372
pK value, **22:**58
in plants, **9:**405; 444
titrimetric methods, **9:**601
reactivation, **10:**111
relation to structure of alkaline phosphatase, **10:**116–117
in tobacco, **6:**369
Arginases, among uricotelic species, **39:**46
from bovine liver, **39:**44
properties of, **39:**45
Arginine, **6:**4, **34:**33; 36, **35:**482, **47:**410–415, **48:**157–158, **49:**77; 80, **50:**143; 144; 145; 156; 157; 158; 193, **51:**217, **54:**384, **58:**81
activation of, **24:**2
activity in *Neurospora* mutant, **16:**256; 293
biosynthesis, **16:**259–268
metabolic relationship with other amino acids, **16:**255; 257; 259–268
in myosin, **16:**323
in nuclear proteins, **16:**441–442; 447
reaction with quinones, **16:**114
in salmine, **16:**126
in transamination reactions, **16:**190; 192; 193; 195; 201; 227; 228
and alkaloid biogenesis, **8:**242; 244
and biosynthesis of solanaceous alkaloids, **8:**236–240
in hyoscyamine synthesis, **8:**236–240
analog activation of, **29:**131
analog of, **29:**131
biological effect of, **29:**131
control of biosynthesis of, **29:**133
effect of canavanine on synthesis, **29:**151
masking of guanidino group of, **29:**401–403
tosyl group for, **29:**403
reactivity of guanidino group toward, **29:**402–403
resistance to analogs of, **29:**133
from argininosuccinate, **40:**73
bacterial decarboxylase, adoptive formation in *E. coli,* **6:**8; 14
age of culture and growth, **6:**12
apoenzyme, activation by pyridoxal phosphate, **6:**27
codecarboxylase, **6:**29
growth temperature effect, **6:**11
Michaelis constant, **6:**16
*p*H of growth medium, **6:**10; 16
purified, **6:**12–21
reversible resolution, **6:**21
as carboxyl-terminal in Tuftsin, **41:**162
in chloroplasts, **20:**209
structure, **20:**86
thermodynamic data on ionizations, **20:**64
use of blocking reagents with, **20:**68
and citric acid cycle, **18:**253–254; 256–257; 261; 271
compartmentalization, **36:**217
decarboxylase, **36:**211
deamination, by metal ions, **21:**178
synthesis, nitrogen transfer in, **21:**255
glutathione S-transferases, **69:**20
inhibitors affecting activity, **6:**18
as inhibitory analogue, **6:**133
integrin-binding sequences, **70:**6; 10; 11
in *Neurospora,* **22:**27; 28
in ornithine cycle, **19:**372
in papain and mercuripapain, **19:**298; 300; 301
oxidation, **65:**319
rate-limiting creatine precursor, **50:**194
residues, **39:**347
synthesis, in invertebrate species, **39:**77
metabolites of, **39:**46
role in enzyme activation, **34:**32
selection against, **47:**411
six codons, **47:**410
specificity of preparation, **6:**15
synthesis in liver, **13:**7
L-Arginine, **65:**298
methyl ester, **19:**307
synthesis from citrulline, **65:**310
D-Arginine, and arginine desimidase, **15:**379

Arginine, Enzymes of Biosynthesis and Their Repressive Control (Vogel and Vogel), **40:**65
Arginine, Metabolism (Ratner), **15:**319
L-Arginine, oxidation in plants, **20:**320
Arginine adenylate, synthesis by carbodiimide method, **22:**191
L-Arginine amidinases, classification, **14:**235; 247
Arginine analogs, nitric oxide synthases inhibition, **65:**314–317
Arginine biosynthesis
allosteric regulation, **28:**126
eight-step pathway, **40:**65
regulations by repression, **40:**73; 74; 75; 76; 77; 78; 79; 80; 81; 82; 83; 84; 85; 86; 87
Arginine biosynthesis enzymes, depression, **40:**75
differential repressor effectiveness, **40:**77
half-life of gene messages, **40:**81; 85; 86
regulatory gene, **40:**67
repression-depression ranges, **40:**83
repressive complex, **40:**78
structural genes, **40:**67
Arginine decarboxylase
electron microscopy of, **33:**106–107
from *Escherichia coli,* **74:**143
structural data on, **74:**134; 142
eukaryotic, **74:**143
structural data on, **74:**134; 142
inhibition, **16:**222
Arginine decarboxylases, biosynthetic, **36:**211
control mechanisms, **36:**216–217
feedback inhibition, **36:**216
repression, **36:**216
induced, **36:**212
sequence of active site, **36:**213
L-Arginine decarboxylases, classification, **14:**256
Arginine desimidase, **15:**376–379
conversion of arginine to citrulline, **16:**266
Arginine dihydrolase, **7:**416, **15:**376
classification, **14:**235; 247
Arginine-dihydrolase, soil, **28:**386
Arginine distribution in tissues, **15:**374 (N^{15})
metabolism, **15:**319–387
Arginine:glycine amidinotransferase, **50:**180; 182–190; 196; 199
assays for, **50:**185
biological distribution, **50:**180
creatine as corepressor, **50:**205
depression, **50:**209–210
effects of fasting, **50:**211–213
effects of hormones, **50:**213–214
indirect evidence for intermediate, **50:** 185–187
inhibition by ornithine, **50:**199
isolation of enzyme-amidine intermediate, **50:**190
repression by N-acetimidoylsarcosine, **50:** 204
repression in embryo, **50:**199
repression in growing chick, **50:**207–209
repression in rat kidney, pancreas, and decidual, **50:**210–211
subcellular localization, **50:**196
substrate specificity, **50:**183–185
sulfhydryl reagents, **50:**187–189
in yolk sac, **50:**207
Arginine kinase, **37:**447; 541; 571; 600, **50:**373, **54:**414
classification, **14:**251
isolation, **14:**405
Arginine methyl ester, hydrolysis by thrombin, **16:**34
Arginine monohydrolase, **15:**376
Arginine phosphate, **1:**109; 111; 112; 118; 119; 128, **9:**437, **23:**324
Arginine phosphate in semen, **24:**42
Arginine phosphokinase, activation by metal ions, kinetics, **21:**140
Arginine phosphoric acid, hydrolysis, by metal ions, **21:**176; 194
Arginine production, precursor for, **53:**228
Arginine racemase, **44:**100, **49:**24; 25; 26; 37
transamination by, **44:**111
Arginine repressor, **40:**75; 76
control sites, **40:**76; 77; 78; 79
differential repressor effectiveness, **40:**77
Arginine-rich histones, **18:**172; 175; 177
Arginine transport and catabolism, **56:**132–134
L-Argininic acid, **15:**379
Argininosuccinase, **21:**255
from bovine kidney, **39:**39
amino acid composition, **39:**33
primary structure of, **39:**33
from bovine liver
amino acid composition, **39:**33
primary structure of, **39:**33

in brain, **39:**41
from kidney, physical and catalytic properties of, **39:**40
from liver
allosteric regulation of, **39:**29
association of argininosuccinase, thermodynamic constants for, **39:**34
catalytic regulation of, **39:**29
cold lability, **39:**30
cooperative substrate effects on, **39:**29
β-elimination, **39:**27
β-elimination mechanism in, **39:**28
identity of subunits of, **39:**34
mechanism of C-N cleavage, **39:**27; 28
molecular weight of, **39:**34
molecular weight of subunits, **39:**35
negative cooperativity in, **39:**29
number of binding sites, **39:**29
number of catalytic sites, **39:**29; 30
number of substrate binding sites, **39:**30
paracrystal formation, **39:**36
physical and catalytic properties of, **39:**40
reaction equilibrium, **39:**26
reaction stoichiometry, **39:**26
SH groups
number of, **39:**35
relation to activity, **39:**35
relation to structure, **39:**35
stereospecificity of β-elimination, **39:**27
subunit association, **39:**31
subunit dissociation, reversible, **39:**30
subunit equilibria, **39:**31
subunits, geometric arrangement of, **39:**36
in *Neurospora,* **22:**29; 35
of *Neurospora*
mutant enzyme, **39:**42
physical and kinetic properties of, **39:**43
wild-type enzymes, **39:**42
Argininosuccinate, **40:**72; 73
and anhydrides I and II
estimation of, **39:**49
interconversions of, **39:**49–51
properties of, **39:**49; 52
related guanidino anhydrides, structures of, **39:**53
structures of, **39:**52
in arginine synthesis, **16:**260; 265–267
biological occurrence of, in lower species, **39:**54
in mammalian species, **39:**54
in rat serum, **39:**55
in rat tissue, **39:**55
mechanism of formation, **18:**211–212
phosphorolysis of, **23:**338
synthetase
aspartate analogs
affinities of, **39:**20
as substrates, **39:**19
aspartate analogues as inhibitors, **39:**19
from bovine liver, **39:**17
allosteric regulation by substrates, **39:**25
subunit composition, **39:**26
equilibrium, **39:**18
free-energy changes, **39:**18
from hog kidney, **39:**18
α-methylaspartate and ^{32}PPi-ATP exchange, **39:**24
^{32}PPi-ATP exchange, **39:**22; 24
PPi inhibition, **39:**21; 23
product inhibition, **39:**21
reaction mechanism, **39:**21
steric requirements at active site, **39:**19; 20
substrate-dependent cleavage of ATP, **39:**23
two partial reactions, evidence for, **39:**22
Argininosuccinate lyase, **40:**73, **74:**314; 315
homology with δ-crystallin, **74:**334
nonenzymatic function of, **74:**334
translational repression, **40:**82
Argininosuccinate synthetase, **40:**72; 73
Argininosuccinic acid, **15:**339; 346–363; 372–373; 379
Arginosuccinate, *See* Argininosuccinate
Arginyl adenylate, **35:**489
Arginyl-glycine bonds of fibrinogen, hydrolysis by thrombin, **16:**37; 42
Arginyl groups, modification, **53:**69
fructose 1,6-bisphosphatase, **53:**67
Arginyl-tRNA protein transferase, acceptor specificity, **40:**102; 103; 104
angiotensin II as acceptor, **40:**109
arginylation of albumin, **40:**96
arginylation of thyroglobulin, **40:**96
assay methods, **40:**95
cellular function, **40:**107; 108; 109
donor substrate specificity, **40:**101; 102
immunoglobulin acceptors, **40:**103
β-melanocyte stimulating hormone as acceptor, **40:**109
molecular weight, **40:**96
nonpeptide acceptors, **40:**104

Arginyl-tRNA protein transferase, acceptor specificity (*Continued*)
peptide acceptors, **40:**103
reaction requirements, **40:**96
sedimentation constant, **40:**96
Arginyl-tRNA synthetase, **39:**25, **50:**166; 167
^{32}PPi-ATP exchange, **39:**24
Arg-26 residue
asparagine synthetase, **72:**111–112
glucosamine 6-phosphate synthase, **72:** 105–106
Arg-73 residue
glucosamine 6-phosphate synthase, **72:** 105–106
glutamine PRPP amidotransferase, **72:** 103–104
Arg-30 residue, asparagine synthetase, **72:** 111–112
Arg-39 residue, dihydrodipicolinate reductase, **72:**295–297
Arg-138 residue, dihydrodipicolinate synthase (DHDPS), **72:**290
Arg-49 residue, glutamine-dependent nitrogen transfer, **72:**179–183
Arg-107 residue, γ-glutamyl transpeptidase substrate binding, **72:**252–254
Arg-110 residue, γ-glutamyl transpeptidase substrate binding, **72:**252–254
Arg-111 residue, γ-glutamyl transpeptidase substrate binding, **72:**252–254
Arg-112 residue, γ-glutamyl transpeptidase substrate binding, **72:**253–254
Arignyl-*t*RNA synthetase, **37:**571
βARK-related enzymes, adenosine receptors, **69:**112
Armillaria mellea, luminescence in, **25:**161
Armour fraction V, *See* Serum albumin fraction
Arochlors, *See* Polychlorinated biphenyls
Aromatase, **55:**263; 264
Aromatic acids, peroxidase action on, **14:**151
Aromatic acylamidases, classification, **14:**247
Aromatic 1-amino acid decarboxylase
evolution of, rate of, **74:**158
species distribution of, **74:**158
Aromatic amino acid hydroxylases. *See also* Phenylalanine hydroxylase; Tryptophan hydroxylase; Tyrosine hydroxylase
active site iron, **74:**239–241
amino acid hydroxylation, mechanism of, **74:**254–257; 255–257
bacterial, **74:**235–236
eukaryotic, **74:**235–236
structure of, **74:**241–252
ferrous forms of, **74:**239–240
hydroxylation of phenylalanine by, mechanism of, **74:**269
mechanism of action, **74:**251–270; 268–270; 269
NIH shift, **74:**267–268
physiological roles of, **74:**235–236
pterin substrates or products, structures of, **74:**237; 238
reactions catalyzed by, **74:**235–236; 236; 238
hydroxylating intermediate, identity of, **74:**251–254
regulation of, **74:**270–281
structures of, **74:**241–249
substrate specificity of, **74:**249–251
Aromatic amino acids, biosynthesis, allosteric regulation, **28:**62–63; 71
Aromatic carboxylic acids, **57:**142
Aromatic chromophores in glutamate dehydrogenase spectral perturbation, **39:**372; 374; 377; 385; 393
Aromatic compounds
anaerobic oxidation of, **27:**534–536
dihydroxylation, **27:**476; 506–508; 516; 523–525; 538
in meteorites, **27:**351
deoxygenases, **55:**251–254
fluorinated, **55:**248–256
liver nonspecific hydroxylation action on, **19:**174–179
nonspecific hydroxylation by model system, **19:**134–143
oxidation to acetyl CoA and succinate, **19:**82
o-aminobenzoate, **39:**138
p-aminobenzoate, **39:**138
anthranilate, **39:**138
chorismate, in synthesis of, **39:**137; 138
folate, **39:**138
tryptophan, **39:**138
ring hydroxylation of, **55:**198
Aromatic hydrocarbons, dissimilative metabolism, **27:**505–528
Aromatic iodonium compounds, nitric oxide synthases inhibition, **65:**317–318
Aromatic monomers, C^{14}-labeled, in lignin study, **18:**374

Aromatic nucleus, synthesis, **14:**120–122
Aromatic pathway, **49:**4; 36
Aromatic polyauxotrophs, in study of amino acid metabolism, **16:**287–290
Aromatic residues, at both sides of the splitting point, **41:**185
  at the carbonyl side of the splitting point, **41:**182
Aromatic ring, anaerobic cleavage, **27:**536
Aromatic ring formation, PN-dependent, **26:**127–128
Aromatic ring-splitting enzyme, **27:**509
Aromatization process, in lignin formation, **18:**361–370
Arrhenius activation energy, **33:**288
Arrhenius coefficient, *See* Activation energy
Arrhenius′ equation, **3:**36–37; 39–41; 48–49
Arrhenius plot
  breaks in, **37:**483
  for levansucrase, **30:**207
Arrhenius plots, **36:**112, **51:**26–27; 59
Arsanil-serum globulin as antigen, **15:**70
Arsenate, **15:**104; 118; 147; 324; 380–382
  effect on microbiological formation of cortisone, **20:**270
  effect on photosynthesis, **19:**22
  and phosphate in exchange reactions, **26:**314
  and phosphorylation, **23:**365–366; 386
  replacement of phosphate in presence of phosphorylase, **10:**127; 140
  in succinic dehydrogenase assay, **18:**88–90
Arsenic, inhibtor action, **15:**26
Arsenicals
  effect on sperm motility, **9:**345
  inhibition of enzyme action by, **11:**235; 238
    reaction with thiols, **11:**217–218; 220; 227; 230
  organic, in inhibitor studies, **15:**289
Arsenic and melanin pigmentation, **14:**94
Arsenic compounds, and biological methylation, **12:**430–432; 435; 437–438; 444; 447–449; 455–457
Arsenite, **15:**166–167; 219; 223; 324; 327; 379
  ameliorative effect on selenium poisoning, **48:**5
  effect on 16α-hydroxylase, **20:**270
    effect on microbiological formation of cortisone, **20:**270
  effect on mold metabolism, **9:**567–568
  effect on phenylalanine hydroxylase, **19:**182; 199
    effect on tryptophan oxidase, **19:**122
  and oxidative phosphorylation, **23:**334–335
Arsenolysis, **15:**114; 245–248; 381, **35:**330
Arsenoxide, **15:**244
Arsines, alkylated, biological formation of, **12:**432; 435; 437–438; 447–448; 455–457
Artemia salina, **59:**82–83
Arterial glutamine, utilization of, **53:**204–214
Arthrobacter, **23:**457
Arthrobacter crystallopoietes, amino-acyl-tRNA specificity, **40:**131
  biosynthesis of interpeptide bridges, **40:**127
  cell function, **40:**136
Arthrobacter JG-9, **35:**168–171; 177
Arthrobacter viscosus, **46:**328; 330; 331
Artificial donors, **34:**477
  enzyme membrane, **34:**447; 495
  membrane, **34:**447; 485
  model system, **34:**477
  polyelectrolyte carriers, **34:**456
Artificial Enzymes and Enzyme Models (Breslow), **58:**1
Artificial light-sensitive enzyme systems, **57:**232
Artificially immobilized enzyme, **34:**446
Artificial Michaelis complex, **27:**3; 4
Artificial models of enzymes, **27:**373
Artificial sound-sensitive enzyme systems, **57:**233
Arum maculatum
  anaerobic respiration, **18:**295
  coenzyme Q in, **25:**288
Arum spadix, cytochromes in, **18:**23; 25–28; 305–306; 308
  respiration in, **18:**299
  terminal oxidases of, **18:**46–47
Aryl, hydroxylation, **67:**208–209
Arylamine acetylase, **15:**199; 200; 211–213
Arylamine acetyltransferase, location in subcellular fraction, **24:**310
Arylamines, **6:**103–107, **44:**206
  ratio to growth, **6:**104
  resistance, **6:**105–107
  tryptophan in formation, **6:**103–104
Arylamine sulfokinase, **22:**213
  specificity, **22:**213
Arylamine synthetase, **66:**282–283

Arylazidocytochrome *c*, cytochrome *c* oxidase, **71:**100
Aryldialkylphosphatase, **74:**72
Aryldiamines, **44:**195
Aryl galactosaminide, **60:**247
Aryl-β-glucoside-permease, properties of, **30:**128
reaction catalyzed by, **30:**128
Aryl glycosaminide, **60:**247
Arylglycosides, **60:**238
Aryl hydrocarbon hydroxylase, **46:**139
activity, **46:**119
Aryl-4-hydroxylase, location of, **24:**308
Aryl hydroxylases, **67:**205
Aryl sulfamate, synthesis, **22:**215
Aryl sulfatase, **16:**381; 383
in mitochondria, **21:**40
Arylsulfatase A, **38:**306; 309; 310, **60:**167
affinity chromatography, **38:**305
amino acid composition, **60:**173
assay, **60:**169
carbohydrate content, **60:**173
cerebroside 3-sulfate substrate, **60:**257
genetics, **60:**176
inhibition, **60:**170
by anions, **60:**176
isoelectric point, **60:**172
kinetic data, **60:**173
molecular weight, **60:**172
monosaccharide sulfate hydrolysis, **60:**175
physiological role, **60:**176
properties, **60:**172
purification, **60:**168
reaction mechanism, **60:**173
self-sufation, **60:**174
substrate, **60:**169
subunits, **60:**172
Arylsulfatase B, **60:**171
Arylsulfatase C, **60:**169
Aryl sulfatases, **22:**215–226
activities, **22:**224
distribution, **22:**223; 224
and lysosomes, **22:**225; 226
occurrence, specificity, **22:**216
physiological role, **22:**224–226
properties, **22:**216; 218; 219
Type I, **22:**216; 217
of *Aerobacter,* **22:**216; 218
of *Alcaligenes,* **22:**216; 218
of *Aspergillus,* **22:**216; 218
function *in vivo,* **22:**225; 226
mammalian liver, C, **22:**216; 218
mode of action, **22:**217
properties, **22:**216–218
Type II, **22:**217–223
and chloride ions, **22:**222
function *in vivo,* **22:**225
of *Helix pomatia,* **22:**219
and Kinetics of NCS hydrolysis, **22:**221
mammalian liver A, **22:**219; 220
mammalian liver B, **22:**219; 220
mode of action, active center, **22:**222
occurrence, **22:**217; 220
of *Patella vulgata,* **22:**219
properties, **22:**219; 220
of *Proteus vulgaris,* **22:**219
separation, **22:**223
specificity, **22:**220; 222
Arylsulfatase substrate, **60:**257
Aryl sulfates, occurrence, **22:**205; 206
Arylsulfate sulfohydrolase, **60:**256
Aryl sulfur free radicals, **27:**206
Arylsulphatases, localization of, **24:**321–322; 323; 343
AS-15, composition, **20:**384
Asarum canadense, **7:**468; 486; 487; 489; 490; 491; 493
Ascaricruorin, catalytic activity, **14:**190
Ascaris megalocephala, fertilization study, **16:**427
Ascaris suum, **59:**87
Ascidians, DNA in cytoplasm of reproductive cells, **16:**423
Ascites cells, folate metabolism, **27:**435; 443
irradiation of nuclei, **27:**268
oxidation of nuclear DPNH, **27:**273
permeability, **27:**269
Ascites hepatoma cells, pyrimidine metabolism in, **28:**135
Ascites tumors, DR metabolism, **28:**436–444; 448
β-glucan effect, **28:**357
pentose synthesis in, **28:**417; 419; 433; 436
Asclepain, requiring –SH group, **11:**241
Asclepias, nicotine in, **8:**223
Ascochyta pisi, plant disease caused by, **20:**392
Ascorbalamic acid, **50:**147
Ascorbate, **34:**477; 484, **44:**196
activation of prolyl hydroxylase by, **41:**286
as cytochrome, c_1 electron donor, **54:**314

in cytochrome c_1 complex reduction, **54:**283; 284; 314; 325–326
in oxygen metabolism, **19:**89; 90; 126; 133; 135; 141; 142; 176; 184; 196–199
tyrosine hydroxylase, substrate specificity, **70:**110
Ascorbate enviroment, collagen hydroxylases, consubstrate binding, **72:**356–357
Ascorbate free radicals, in peroxidase action, **27:**122–125
Ascorbate oxidase, composition of copper in, **33:**196
EPR-nondetectable copper ions in, **33:**223
extinction coefficient of, **33:**198
forms of copper in, **33:**195
nonequivalent cupric ions in, **33:**203
oxygen, product of water reduction of, **33:**230–231
presence of broad component in CPR of, **33:**216
reaction with azide, **33:**221
Ascorbate oxidase activity of Cp, **44:**195
Ascorbates, **46:**65
Ascorbic acid, **46:**83, **59:**121, *See* Vitamin C
activation of acid phosphatase, **9:**591
in tests for peroxidases, **9:**649
activation of ATPase, **6:**167
in biosynthesis of folic acid coenzymes, **21:**394
in cholesterol biosynthesis, **22:**296; 297
in squalene biosynthesis, **22:**308; 319
effect on hyaluronic acid, **13:**202
effect on pectic substances, **20:**352
oxidation, **20:**166
in photosynthetic phosphorylation, **20:**228
in respiratory chain, **20:**172
enzymes in formation of, **24:**299
as inhibitor of β-glucuronidase, **16:**373
presence in nuclei, **16:**489
relation to oxidation of diphenols, **16:**134
inhibitory analogues of, **6:**132; 139
metal-catalyzed oxidation, **15:**40
oxidation of reduced pyridine nucleotides, **15:**128
reaction with peroxidase, **15:**17
peroxidase action on, **14:**151
in tyrosine metabolism, **14:**80; 93; 96; 97–99; 101; 102–104; 106; 107; 108–111; 123
in photophosphorylation, **18:**58–59
in photosynthesis, **19:**55; 64; 65
and PN-enzymes, **26:**123; 127
as a reducing agent in collagen synthesis, **41:**253
reduction of coenzyme Q, **25:**281
and transmethylation, **12:**469
effect on cathepsin III, **12:**205–206
reaction with borate, **12:**500; 518
reaction with peroxidase-peroxide complex, **12:**167; 169–171; 175
in treatment of tellurium poisoning, **12:**434
Ascorbic acid and coupled oxidations, **8:**452
Ascorbic acid in plants, relation to thiamin, **10:**247
Ascorbic acid oxidase, **19:**203
cacao, **25:**466
product inhibition of, **25:**224; 250
in plant respiration, **18:**37; 40–42
in plants, **18:**29; 37; 40–43; 45; 47; 309–310
N-terminal histidine in, **20:**71; 101
Ascorbic acid oxidase, EPR study, **27:**125
Ascorbic acid-2-sulfate, **60:**177
Ascorbic oxidase, as terminal oxidase, **16:**131
formation of complexes, **16:**130
Ascorbic oxidases, classification, **14:**254
isolation, **14:**400; 405
ASF, *See* Acetobacter stimulatory factor
Ashbya gossypii as source of LBF, **14:**50–52; 61
Asialo-α-acid glycoprotein, **52:**30; 34; 41; 107
Asialoceruloplasmin, **41:**[64Cu]-111; 100; 102–104; 108–111; 113; 122–124, **52:**30
in humans, **41:**100; 104
inhibitory power, **41:**122–126
in rat, **41:**104
in serium, **41:**102–103; 108–110
ziptake and Catobolism, **41:**104
Asialo-Cp, **44:**218; 226
Asialo-factor VIII/von Willebrand factor, **52:**114
Asialogangliosides, in schizophrenia, **29:**503
Asialoglycoprotein receptor, **66:**43
ligand binding, highly selective nature, **66:**49
ligand requirements, **66:**44–47
natural ligand, **66:**77–78
probing primary specificity with monosaccharide analogues, **66:**44–46
secondary specificity with neoglycopeptides, **66:**46–47
solubilized, binding characteristics, **66:**52

Asialoglycoprotein receptor (*Continued*)
subunit organization, **66:**52–61
binding site topography determination using tri-branched photoaffinity labelling reagents, **66:**54–57
photoaffinity labelling efficiency, **66:**57–58
stereospecific binding of tri-branched glycopeptides, **66:**60–61
stoichiometry of rat receptor, **66:**58–60
Asialoglycoproteins, binding of, **41:**119; 122
inhibition of, **41:**122–124
tritiation of, **41:**99–103
Asialo-G_{m2}, **38:**307; 308
Asialo-mucin, **52:**42
Asialoprothrombin, **46:**13
Asialotransferrin, **52:**37; 40; 41; 43; 53; 98; 102; 106
asn A and B genes, asparagine synthetase, **72:**106–107
Asn-74 residue
asparagine synthetase, **72:**111–112
glutamine-dependent nitrogen transfer, **72:**180–183
Asn-98 residue
glucosamine 6-phosphate synthase, **72:** 105–106
glutamine-dependent nitrogen transfer, **72:**179–183
Asn-248 residue, dihydrodipicolinate sunthase (DHDPS), **72:**290
Asn-80 residue, dihydrodipicolinate synthase (DHDPS), **72:**290
asn^- sublines, **39:**234
asn^+ Variants, **39:**234
Asochyta pisi, **55:**47
Asolectin, **34:**473
Asparaginase, **5:**23, **13:**282–289; 291; 294; 295–296; 303, **37:**460; 473; 482; 491; 508; 542; 543; 572
activity, in guinea pig serum, **39:**221
circular dichroism of, **39:**196
disappearance curves, **39:**228
α-keto acid-dependent, **39:**186
cacao, **25:**465; 466
in plant metabolism, **9:**417–420
in yeast, **9:**418
soil, **28:**386
L-Asparaginase, **36:**73, **39:**186
asparagine synthetase methabolism, **72:** 150–151
chemotherapeutic applications, **72:**151–153
glutamine-dependent asparagine synthetases, **72:**153–157
L-*Asparaginase: A Review* (Wriston and Yellin), **39:**185
Asparaginase activity, and *p*H, **8:**158
lability, **8:**160
Asparaginase in tobacco, **6:**369
Asparaginase-resistant variants, **39:**233
Asparaginases
amino acid composition of, **39:**216
Km values of, **39:**230
classification, **14:**230; 234; 247
Asparaginases I, **39:**191
Asparaginases II, **39:**191
Asparagine, **39:**221, **50:**143; 156; 157, **53:**216, **56:**117
activation of, **24:**2
and RNase activity, **24:**450
and adenylic acid metabolism, **9:**497
in plant metabolism, **9:**392; 409–421
anomalous stereoselectivity for, **49:**51
biosynthesis, **16:**268–270
relation to histidine synthesis, **16:**301
in transaminations, **16:**195; 199–204; 207; 225–227
biosynthesis of, **56:**27
and citric acid cycle, **18:**261–264
labeled, in plant respiration, **18:**300
configuration of, **23:**24–25
cyanoalanine synthase and, **56:**33–34
cyclic structure, **20:**337; 338
ionization constants, **20:**58
in phosphorylated peptides of esterases, **20:**288
in plants, biosynthesis and metabolism, **20:**331–336
deamidation, **8:**152; 158–164
deamination, by metal ions, **21:**179
growth inhibition, **7:**437
intermediary metabolism of, **13:**237–319
peptides, deamidation of, **13:**244–245
utilization by microorganisms, **13:** 244–245; 250–253
peroxidase action on, **14:**152
in protein synthesis, **39:**239
role in nitrogen transfer, **15:**340–341
synthetase, **39:**231; 232
transamination reactions of, **19:**365
D-Asparagine, **33:**367

L-Asparagine, **39:**222; 236, **49:**77
amidohydrolase, **39:**186
analogs, **39:**211
synthetase, **39:**231
L-[*amide*-^{13}N]Asparagine, **57:**323–324
Asparagine, Intermediatry Metabolism (Waelsch), **13:**237
^{14}-C and ^{15}N-L-Asparagine, utilization of, **39:**239
4-Asparagine decarboxylase, column chromatography, **22:**482; 484
Asparagine-α-keto acid reactions, **16:**199–203; 207–208
Asparagine-linked oligosaccharides, **52:**97
Asparagine peptides, mechanism of hydrolysis, **15:**293
Asparagine synthetase, **37:**572, **66:**276–280, **72:**106–112, *See* Aspartate amidotransferase
AS-B-form
binding site identification, **72:**162–165
bound conformation, **72:**159–162
pyrophosphate binding motif, **72:**165–168
vs, papain, **72:**175–183
catalysis mechanisms
acid activation, **72:**158–167
biosynthesis and metabolism, **72:**149–151
chemotheraphy through inhibition, **72:** 151–153
GAT-domain, **72:**190–191
genetic control, **72:**148–149
glutamine-dependent nitrogen transfer, **72:**167–190
glutamine-dependent structure, **72:**153–157
research background, **72:**146–148
E. coli asnA and asnB, **66:**278
GAT and synthetase domains, **66:**276–278
glutamine domain residues, **72:**111–112
nitrogen transfer, **72:**107–111
AS-B *vs.* papain, **72:**175–183
glutamine-dependent machanisms, **72:** 167–175
heavy-atom kinetic isotope effects, **72:** 183–186
imide intermediates, **72:**188–190
kinetic mechanisms, **72:**186–188
sequence alignment, **72:**162–163
vertebrate, **66:**278–280
Asparagine transaminase, **16:**226–227; 268–269
Asparaginylglycine, ionization constant, **20:**58
Asparaginyl-tRNA, **39:**238
ligases, **39:**232; 239
Asparagus, inhibition of, **29:**102
specificity of proline sRNA synthetase in, **29:**102
Asparate semialdehyde dehydrogenase, bacterial amino acid biosynthesis, **72:**285–286
Aspareaginyl-tRNA synthetase, **50:**165
Asparginase, **20:**331
Aspartaldehyde dehydrogenase, in ATP synthesis, **23:**335
Aspartase, **15:**337; 362, **20:**328, **46:**459
and citric acid cycle, **18:**251–253; 268
classification, **14:**229; 237; 257
formation of aspartic acid, **16:**268
in conversion of ammonia to α-amino groups, **16:**255–256
L-Aspartase
activation of
covalent, **74:**330
by divalent metal ions, **74:**302
noncovalent, **74:**327–330
proteolytic, **74:**330
activator site in, **74:**329–330
active site
functional groups, **74:**324–327; 326
residues, identification of, **74:**304–307
activity of, pH and, **74:**305–307; 306; 327–328; 328
amino acid sequence homology, **74:**314; 315; 324
in blood clotting, **74:**333–334
carboxyl terminus
amino acid sequence of, **74:**330
lysine, and plasminogen binding, **74:** 333–334
truncation mutants, **74:**330–334; 331; 332
truncation sites, **74:**330
characterization of, **74:**297–298
chemical modification studies of, **74:** 304–305
commercial applications of, **74:**299
cysteine residues, mutagenesis, **74:**315–316; 316
deamination reaction of, pH dependence of, **74:**305–306
from *Escherichia coli,* **74:**297–298; 302; 333–334
activity of, pH and, **74:**327–328; 328
aspA gene for, **74:**314

L-Aspartase (*Continued*)
chemical modification studies of, **74:** 304–305
cysteine residues, **74:**316–317
histidine residues, **74:**317
mechanism-based inactivator of, **74:** 307–308
modification of, mechanism of, **74:** 313–1314
pH study of, **74:**305–307
structure of, **74:**321; 323
gene for, **74:**314
from *Haemophilus influenzae,* **74:**333–334
from *Hafnia alvei,* **74:**297; 301–302; 303
chemical modification studies of, **74:** 304–305
pH study of, **74:**305–307
histidine residues, mutagenesis, **74:**316
inactivation of, **74:**307–314; 308
by fumaric acid semialdehyde, **74:** 307–314; 312; 313
mechanisms of, **74:**311–314; 312; 313
inactivation-resistant mutants, generation of, **74:**310; 311
inactivation sites, identification of, **74:** 307–311; 309; 310
inhibitors of, **74:**300–301
kinetics of, **74:**302–303
lysine residues
lysine to arginine mutant, recovery of catalytic activity, **74:**319–321; 320
mutagenesis, **74:**315; 316; 317–319
mechanism of, **74:**301; 302–304
metabolic significance of, **74:**298–299
metal ion requirements, **74:**301–302
mutagenesis, **74:**315–321
mutants
inactive, recovery of catalytic activity, **74:**319–321; 320; 321
kinetic parameters for, **74:**315; 316; 325
nonenzymatic function of, **74:**333–334
plasminogen binding activity of, **74:** 333–334
proteolytic digestion of, **74:**307; 309
from *Pseudomonas fluorescens,* **74:**308–309
characteristics of, **74:**297; 298
purification of, **74:**297–298
species distribution of, **74:**298
structural homology, **74:**324
structure of, **74:**321–324; 322; 323
studies of, historical perspective on, **74:** 296–297
substrate activation, **74:**327–330
substrate specificity of, **74:**299–301
truncation mutants
formation and characterization of, **74:** 331–333
kinetic parameters of, **74:**331–333; 332
Aspartase in plants, **9:**408–409
Aspartate, **40:**72, **49:**50, **50:**142; 143; 144; 149; 152; 154; 169, **56:**315
and acetate metabolism, **23:**405; 419–420; 426–427; 432–434
and glycolate metabolism, **23:**437
activation of, **18:**212
anomalous stereoselectivity for, **49:**51; 52
bacterial amino acid biosynthesis, **72:** 279–281
biosynthesis, **16:**268–272
in arginine synthesis, **16:**260; 265–268
in enzymic transaminations, **16:**186; 189; 190; 192; 195; 203; 204; 207–209; 211–213; 215–218; 221–223; 226–228
incorporation of labeled acetate, **16:**290
in nonenzymic transaminations, **16:**225
in purine synthesis, **16:**451
relation to tricarboxylic acid cycle, **16:**249; 255–256
in synthesis of isoleucine, **16:**282; 284
in synthesis of lysine, **16:**296; 298
and biotin, **12:**81–85
CO_2 fixation in, **12:**132
and oxalacetate synthesis, **12:**79; 81–85
in mitochondrial carbamylaspartate synthesis, **21:**52
reaction with carbamyl phosphate, **68:**68–69
L-Aspartate, L-alanine from, **16:**275
Aspartate-alanine transaminase, **16:**189; 207
Aspartate amidotransferase
characteristics of the reaction
β-asparty adenylate as an intermediate, **39:**164; 165
inhibition by asparagine, **39:**164; 167
kinetic analysis of the reaction, **39:**165
levels in tissues during embryogenesis, **39:**166
properties of the enzyme, **39:**165
molecular weight, **39:**165
reaction with alkylating agent, **39:**165; 166
stabilization by substrate, glycerol and mercaptoethanol, **39:**165

significance in suppression of lymphomas, **39:**165; 166
source of the enzyme, **39:**164
Aspartate aminotransferase, **37:**414; 429; 448; 461; 491; 508; 509; 566; 572, **50:**6; 10; 14–23, **67:**37–38; 53–54, **74:**142; 148; 149; 156. *See also* glutamic aspartic transaminase; Transaminases
active site of, **74:**167; 168
circular dichroism of, **50:**16; 23
conversion to aspartate β-decarboxylase, engineering of, **74:**167–170
cytosolic
evolution of, **74:**160–163; 162
species distribution of, **74:**158
dynamic model of transaminase action, **50:**19
enzyme-substrate complexes of, **50:**16
evolution of, rate of, **74:**158
groups *a-c,* **74:**152; 157–159
homologous heterotopic isoenzymes, evolution of, **74:**160–163
internal aldimine, **74:**145–146
localization of, **24:**312; 341; 343
mitochondrial
evolution of, **74:**160–163; 162
species distribution of, **74:**158
mutant and wild-type
active site, structure of, **74:**167; 168
aspartate β-decarboxylase activity of, **74:**167; 168
N-methylpyridoxal phosphate in, **50:**23
phosphonates and, **50:**22
properties of, **50:**14
proton exchange in, **50:**37
reaction catalyzed by, intermediates, **74:**191
structural data on, **74:**133; 135; 143
L-Aspartate ammonia-lyase, *See* L-Aspartase
L-Aspartate as inhibitor of pyruvate carboxylase, **42:**7
Aspartate carbamoyltransferase, **37:**448; 473; 474; 573
location in subcellular faction, **24:**309
Aspartate carbamylase
kinetic studies of, **56:**334
in pyrimidine biosynthesis, **56:**329
reaction catalyzed by, **56:**329
spectral features of, **56:**333
Aspartate-carbamyl transferase, **21:**270; 272; 274
Aspartate decarboxylase, **74:**176
L-Aspartate-β-decarboxylase, **35:**94; 503–541
activation of, by α-keto acid, **35:**504
Aspartate β-decarboxylase, **49:**13; 23; 26; 29; 30; 37
aspartate aminotransferase conversion to, engineering of, **74:**167–170
Aspartate-β-decarboxylase, electron microscopy of, **33:**110–112
Aspartate-α-decarboxylase, pyruvoyl group as prosthetic group, **53:**374
L-*Aspartate*-β-*D*ecarboxylase, Structure, Catalytic Activities, and Allosteric Regulation (Tate and Meister), **35:**503
Aspartate-α-ketoglutarate reaction, free energy change, **16:**204
Aspartate kinase, **37:**573, **50:**169
Aspartate metabolism, allosteric regulation, **28:**60–62; 70–71
in *E. coli,* control, **28:**60–62
D-Aspartate oxidase, **37:**491
activators for, **57:**118–120
and amine-glyoxylate adducts, **57:** 113–117
catalytic constants for, **57:**120
catalytic reaction of, **57:**95
concentrations of, **57:**94
and L-cysteinylglycine-glyoxylate adducts, **57:**115–116; 122
discovery of, **57:**94
distributions of, **57:**94
and drugs, **57:**96
FAD activity for, **57:**95
functions of, physiological, **57:**96
and glyoxyalate adduct products, **57:**116–117
and hormones, **57:**96
inhibition, **57:**96; 121–122
kinetic activity of, **57:**117–120
limiting rate constant for, **57:**120–121
in liver tissue, **57:**96
Michaelis constant for, **57:**119–120
molecular weight of, **57:**95
and nucleophile-glyoxylate adducts, **57:**99–101
as peroxisomal enzyme, **57:**94
presence of, **57:**86–87
purification of, **57:**94–95
specificity of, **57:**117
substrate requirements for, **57:**99–101
substrates for, **57:**114–116
turnover number of, **57:**120

Aspartate racemase, **44:**100, **74:**176
Aspartate residue, *Caulobacter* cell cycle, transcription regulation, **71:**18
Aspartate semialdehyde, **50:**143; 144
as mechanism-based inactivator of aspartase, **74:**307–311
sensitivity of aspartase mutants to, **74:** 309–310
Aspartate synthesis, and biotin enzymes, **26:**293
Aspartate transcarbamylase, **37:**526, **43:**29; 31; 39; 245; 352, **56:**292; 306, **68:**67–136
activity and regulation, **68:**73–77
allosteric effects, modified catalytic trimers, **68:**135
allosteric regulation, **28:**44; 51–53; 128–132; 135
asymmetry and negative cooperativity, **68:** 93–99
adenosine triphosphate binding, **68:**94; 96
cytidine triphosphate binding, **68:**94–95
r1-r6 pathway displacements, **68:**94; 98–99
T state asymmetry, **68:**94; 97
binding of N-phosphonacetyl-L-aspartate to active site, **68:**82–83
biochemical and mutant studies of
constraints, **68:**114–116
frozen R or T states, **68:**117–118
heterotropic effects, **68:**114–124
homotropic cooperativity disconnection from maximum rate, **68:**124
nucleotide perturbation mechanism, **68:**115–116
separation of adenosine triphosphate and cytidine triphosphate effects, **68:** 120–123
separation of cytidine triphosphate effects from cytidine triphosphate-uridine triphosphate synergism, **68:**123
separation of homotropic and heterotropic effects, **68:**116–120
uridine triphosphate activation, **68:**123–124
and biotin, **26:**295
domain closure and homotropic effects, **68:**87–93
intermediate quaternary structures, **68:**93
R form interactions, **68:**88; 90
salt links, **68:**90–91
site-specific mutants, **68:**91–92
T and R forms, **68:**87–88
T from interactions, **68:**88–89
hetertropic effects, **68:**130–135
low angle X-ray solution scattering, **68:** 118–119
metal centers of, properties of, **56:**332–334
metal content of, **56:**330
molecular properties of, **56:**330–332
mutant review, **68:**136
nucleotide effectors, **68:**68; 70
pathways, **68:**124–130
allosteric-zinc interface, **68:**128
residue movement in T state, **68:**124; 126
T-form changes with allosteric site binding, **68:**129; 131
T-state catalytic chain changes, **68:**129–130
T-state regulatory dimer changes, **68:**124; 127
pH effects, **68:**76–77
polypeptide structure, **68:**80
in pyrimidine pathway, **68:**68–69
regulatory-catalytic subunit, **68:**156–157
regulatory dimers and catalytic trimers, **68:**73
secondary structure, **68:**78–80
structural aspects of heterotropic
active site, **68:**100–101
c1-c4 interface, T form, **68:**105; 107
domain separation changes, **68:**105
effects, **68:**94; 98–114
global structural changes, **68:**105–107
heterotropic signal transduction interfaces, **68:**105–106; 108–109
interactions with adenosine triphosphate or cytidine triphosphate, **68:**101–104
interface aspects, **68:**107–114
polypeptide backbone of c1-r1 chain, **68:**106–107; 110
regulatory site, **68:**101–104
β-sheet of allosteric domains, **68:**111; 117
solvent accessibility, **68:**98–99
superposition of four T state structures, **68:**100; 102
superposition of T and R forms, **68:** 100–101
zinc domain movement, **68:**107; 115
structural results and catalysis, **68:**77–87
active site binding, **68:**82–83
amino acid sequence, **68:**81
carbamyl phosphate and succinate binding, **68:**82–83; 85–86
domain closure, **68:**82

malonate and phosphonoacetamide binding, **68:**82; 84
structural role of zinc in, **56:**316
substrate, relationship with *N*-phosphonacetyl-L-aspartate, **68:**74–75
subunit structure, **68:**71–73
T to R transition, **68:**75–76
X-ray diffraction, **68:**78
Aspartate transport, **52:**296–298
Aspartic acid, **34:**15; 16; 21, **51:**217, **53:**169. *See also* Asparagine
at active site of enzyme action, **22:**68; 71
in methionine, homoserine synthesis, **22:**264
asparagine synthetase and activation, **72:** 158–167
β-aspartyl-AMP intermediate, **72:**158–159
binding site identification, **72:**162–165
bound conformation in as-B, **72:**159–162
pyrophosphate binding motif in AS-B, **72:**165–167
bacterial decarboxylation of, **6:**5
blocking of β-carboxyl groups by, **29:**434
in chloroplasts, **20:**209
ionization constant, **20:**58
metal binding by, **20:**56
in phosphorylated peptides of esterases, **20:**288
in plants, biosynthesis and metabolism, **20:**324; 325; 327–330; 335; 336
occurrence, **20:**319
thermodynamic data on ionizations, **20:**64
and citric acid cycle, **18:**251–254; 256–257; 261–264; 266; 271–276
labeled, **18:**268–269; 272
decarboxylation, **7:**436–437
growth inhibition, **7:**437
in glycoprotein linkage, **25:**451–453
in histidine biosynthesis, **21:**404
in purine biosynthesis, **21:**202–204; 214–216; 228; 232
nitrogen transfer reactions of, **21:** 252–254
in pyrimidine biosynthesis, **21:**264; 267; 270–272
radioactive, in study of pyrimidine biosynthesis, **21:**266; 273
integrin-binding sequences, **70:**5–6
in mercuripapain, **19:**301
in ornithine cycle, **19:**372; 375; 376
in synthesis of homoserine, **19:**256
transamination reactions of, **19:**338–340; 342; 344
and RNase activity, **24:**451
role in nitrogen fixation, **13:**370
utilization by microorganisms, **13:**250–253
and virus mutation, **15:**477
in formation of urea, **15:**324–327; 330; 341 (N^{15}); 343–347; 350–352; 354–356; 362–363; 371–373; 379
liberation from proteins, **15:**293; 432
in oxidative processes, **15:**124; 261
L-Aspartic acid
commercial production of, **74:**299
reversible deamination of, catalyzed by aspartase, **74:**296–297
transamination reactions of, **19:**347–350; 367; 368
L-Aspartic acid, oxidation in plants, **20:**320; 322
Aspartic acid β-decarboxylase, **16:**220; 275
L-Aspartic acid decarboxylases, classification, **14:**256
Aspartic acid in plant metabolism, **9:**394; 398–401; 402; 407–409; 419; 421–430, **10:**258
Aspartic acid kinase, **19:**257
Aspartic acid residues, at the carbonyl side of the splitting point, **41:**183
Aspartic acid transport, **49:**52
Aspartic aminopherase, **5:**155
Aspartic amino transferase, complexes of, Cotton effect, **27:**59
Cotton effect, **27:**86–87
Aspartic deammoniacase, classification, **14:**229; 237; 257
Aspartic-β-decarboxylase, **35:**88; 92; 95
reaction with analogues, **35:**91–92
variety of reactions, **35:**88
Aspartic dehydrogenase, **15:**337
Aspartic diamide, ionization constant, **20:**58
Aspartic β-semialdehyde, in synthesis of 4-carbon amino acids, **16:**270; 271
Aspartic semialdehyde dehydrogenase, repression of, **28:**70
Aspartic transcarbamylase, feedback effect on, **29:**367
Aspartohydroxamic acid (AHA), from transferase action, **13:**291–292; 305

Aspartokinase, **50:**143; 144
bacterial lysine biosynthesis, **72:**283–285
multiple forms, **25:**361
repression by other amino acids, **29:**145
Aspartokinase 1, **37:**448
β-Aspartokinase, in ATP synthesis, **23:**335
Aspartokinases, concerted feedback inhibition of, **28:**65
of *E. coli,* **28:**60–62; 72
homoserine-sensitive, **28:**61–62; 72
lysine-sensitive, allosteric regulation, **28:**53; 61–62; 70–72
threonine-sensitive, allosteric regulation, **28:**53; 61–62; 73
Aspartotransferases (ATF), **13:**290–296; 301–305; 309
β-Aspartyl-AMP
aspartic acid activation, **72:**158–159
binding site identification, **72:**164–165
glutamine-dependent nitrogen transfer, **72:**173–175
Aspartyl-AMP, asparagine synthetase, **72:** 108–111
Aspartylglucosaminidase, **60:**161
assay, **60:**162
biological role, **60:**164
genetics, **60:**164
heat stability, **60:**163
isoelectric point, **60:**163
kinetic data, **60:**163
molecular weight, **60:**163
properties, **60:**163
purification, **60:**162
substrates, **60:**164
Aspartylglycosaminuria, **60:**161
β-Aspartylhydroxamate, **39:**208; 213
β-Aspartylhydroxamic acid, **39:**207
Aspartyl-P formation, regulation of, **28:**61–62
Aspartyl phosphate, **23:**324; 325, **50:**143
β-Aspartyl phosphate, **18:**226
as intermediate in homoserine biosynthesis, **19:**256
synthesis, **19:**250; 256
β-Aspartyl phosphate, formation, **16:**268; 270; 271
Aspartyl phosphate dehydrogenase, **50:**143
Aspartyl radical, transfer to amines and amino acids, **13:**289–305; 305–310
Aspartyl semialdehyde dehydrogenase, in ATP synthesis, **23:**335
Aspects of the Biosynthesis of Enzymes (Chantrenne), **24:**1
Aspergillic acid, **7:**469; 486; 489; 490; 492, **35:**155; 163–165
biosynthesis, **35:**163
Aspergillin, **7:**485. *See also* Gliotoxin
Aspergillus, **7:**485
as aerobic enzyme, **2:**185–187
alcoholic fermentation and Pasteur effect, **2:**212–216
anabolic respiration, **2:**206–212
carbon monoxide poisoning of growth, **2:**216–218; 220–222
C Q value and oxidation, reduction, **2:** 188–200
and dehydrases, **2:**223–235
economical coefficient, **2:**196–200; 208
elementary composition of *aspergillus oryzae,* **2:**190
heat balance of growth, **2:**201; 202; 205
and hydrocyanic acid, **2:**218; 220; 221
hyperquotient, **2:**190–199
hyphae, **2:**185–187; 217–219
hypoquotient, **2:**190–199
maintenance respiration, **2:**206–212
metabolism, **2:**188–200
methylene blue reaction, **2:**224–226; 229
respiration and metabolism, **2:**201
respiratory quotient, **2:**188–195
role of *aspergillus oryzae* in brewing, **2:**212
Rubner coefficient, **2:**198; 199
späneatmung, **2:**210–212
system of iron catalystor, **2:**216–223
α-aminoadipate in, **16:**297
mutants in study of amino acid metabolism, **16:**278
deaminase in, **15:**108
enzymic transformations of steroids by, **20:**245; 262
pectinesterase of, **20:**360
fat synthesis by, **9:**673–674; 691
glucose oxidase of, **19:**206
mutations, **27:**308
protein synthesis by, **9:**654
Aspergillus: Respiration and Fermentation (Tamiya), **2:**183
Aspergillus, sulphureus, proteinase from, **41:**190
Aspergillus bisporus, **59:**68–69
Aspergillus cloacae

β-lactamase of, **28:**268–269; 290; 295; 297
Aspergillus conidia, relation of DNA to ploidy, **16:**425
Aspergillus flavus
enzymic transformations of steroids by, **20:**262
proteinase from, **41:**186; 190; 191
Aspergillus flavus-oryzae, coenzyme Q derivative in, **25:**322
Aspergillus foetidus, polygalacturonase in, **20:**343; 365–367
Aspergillus fumigatus
transamination in, **16:**206
ubichromenol in, **25:**290
Aspergillus griseus, **23:**194; 203
Aspergillus melleus, proteinase from, **41:**221; 223; 227
Aspergillus nidulans, **46:**49, **51:**116, **59:**66
enzymic transformations of steroids by, **20:**242; 274
genetic studies of, **13:**125; 131; 136–138; 139–145; 147
Aspergillus Niger, **49:**10
acid formation in, **9:**537–577
enzymes in, **20:**265
enzymic transformations of steroids by, **20:**245; 249; 251
polygalacturonases of, **20:**343; 362–366; 367
genetic studies of, **13:**147
hydroxylation by, **19:**157, **35:**140
multimolecular forms, **36:**97
nitrite reductase from, **46:**63
oxidation of hydrocarbons, **27:**517
synthesis of carbohydrate polymers, **15:**312; 313; 314
use of acetate, **23:**406; 411–413; 415; 420
Aspergillus niger and Related Molds, Acid Formation in (Walker), **9:**537
Aspergillus ochraceus, enzymes in, **20:**265; 269; 270
enzymic transformations of steroids by, **20:**242; 245
Aspergillus oryzae, **54:**204
α-amylase composition, **25:**334–335
proteases, **25:**351
α-amylase of, **23:**41
and mevalonic acid, **23:**473
proteinase of, **23:**177
aryl sulfatase of, **22:**216–218
enzymes, adsorption studies, **14:**365
isolation, **14:**397; 405
in industrial fermentation, **7:**592; 595
β-lactamase of, **28:**243
polygalacturonases of, **20:**364; 366; 367
proteases from, **33:**385
proteinase from, **41:**180; 182; 184–187; 190–191; 193; 194; 227
Aspergillus oryzae action on carbohydrates, **15:**312; 314; 315
Aspergillus sojae, proteinase from, **41:**190; 221; 223; 227
Aspergillus suinus, enzymic transformations of steroids by, **20:**262
Aspergillus tamarii, separation of enzymes by chromatography, **14:**349; 363
Aspergillus terreus
coenzyme Q derivative in, **25:**322
effect of fluoroacetate on, **18:**136
Asphyxia and snake venoms, **8:**490–491
Aspirin, **54:**139
Asp-123 residue, glucosamine 6-phosphate synthase, **72:**105–106
Asp-422 residue, γ-glutamyl transpeptidase substrate binding, **72:**253–254
Asp-238 residue, pyrophosphate binding motif, **72:**167–168
Asp-14 residue, triad amidotransferase, **72:**139
Assay, **39:**187; 332
growth hormone, **42:**74
methods, **39:**332
Assay substrates, **46:**394
Assembly/disassembly turnover, microtubule dynamics, **69:**132–136
Assimilating system in plants, reduced state, **1:**202; 242–255
Assimilation, Heterotrophic, of Carbon Dioxide (Werkman and Wood), **2:**135
Assimilation et dissimilation de l'azote chez les animaux (Braunstein), **19:**335
Assimilations, microbial, *See* Microbial assimilations
Assimilations, Microbial (Clifton), **6:**269
Association, rates of, **30:**144–145
Association constants, *See* Affinity constants
Association-dissociation changes, of allosteric enzymes, **28:**53–56; 68; 73; 83; 90; 105; 117; 129; 132; 138–142; 144
Association-dissociation of proteins, **1:**17; 18
Astasia, RuDP carboxylase in, **19:**32

Astaxanthin, biosynthesis, **21:**349
phototactic function of protein complex of, **21:**355
Aster yellows virus, transmission of, **9:**174
Astragalus, **48:**5
accumulation of selenium in, **29:**140
Astringenin, **55:**94
Asymmetrically selective nature of some catalysts, **15:**27
Asymmetric catalysis, by enzymes, **7:**83; 89; 91
interrupted, **7:**80; 83
kinetics, **7:**101
Asymmetric catalytic decomposition, **7:**80; 97–99
Asymmetric catalytic racemization, **7:**74–75
Asymmetric catalytic synthesis, **7:**80; 89–94
Asymmetric decomposition, absolute, **7:**80; 94
alkaloids, **7:**97
catalytic, **7:**80; 97–99
effect of polarized light on, **7:**94
eliminative, **7:**80; 95–97
by enzymes, **7:**98–99
noneliminative, **7:**80; 97
Asymmetric enzyme membrane, **34:**455
Asymmetric induction, **7:**65; 66–69; 70; 77; 82, **35:**85; 101; 119
and mutarotation, **7:**68–69
Asymmetric phytochemical reduction, **7:**91–92
Asymmetric synthesis, **7:**65–110
absolute, **7:**84–85; 100
alkaloids, **7:**89–90
and asymmetric transformation, **7:**76–79
catalytic, **7:**80; 89–94
effect of polarized light on, **7:**84–85
eliminative, **7:**80; 85–86
by enzymes, **7:**91–94, **27:**38
and origin of life, **27:**358
internal, noneliminative, **7:**80; 86–89
mechanism, **7:**67–69
Asymmetric Synthesis and Related Processes (Ritchie), **7:**65
Asymmetric transformations, and asymmetric synthesis, **7:**76–79
borderline cases, **7:**74–75
first-order, **7:**69–71; 74–76
second-order, **7:**70; 71–73; 74; 75; 80; 82; 84; 100
Asymmetry
induced, lability, **7:**76
of olefinic system, **7:**77–78
optical evidence for, **7:**68
in proteins, **27:**38–41; 46–49; 52–53; 55–56; 61
Asymmetry of protein molecules, **1:**23–28
Asymmetry regulation
polarity and, Caulobacter cell cycle (*See also* Caulobacter cell cycle)
polarity and, *Caulobacter* cell cycle, **71:**1–39
Atabrine, *See* Mepacrine
ATF, *See* Aspartotransferases
Athiorhodaceae, **1:**236; 280; 281; 289; 290; 302, **39:**338
carotenoids in, **21:**303; 304
metabolism in, **74:**7; 11
use of acetate, **23:**420–422; 430
Athropod cuticle, sclerotization, **16:**153–163
Athropodin, changes during sclerotization, **16:**160–163
properties, **16:**156–158
Atmungsferment, **18:**11, **23:**266–269; 284; 298; 304; 312
Atmungsferment, CO reaction, **28:**228–229
Atomic absorption spectroscopy, tyrosine hydroxylase, **70:**202
Atomic spectroscopy, **56:**286
ATP, **34:**520; 526, **35:**483; 487; 488, **38:**197, **43:**252; 344; 361, **49:**103; 106; 243
α, γ anhydride, **50:**376
βs, **50:**374; 378
in acetylation of sulfanilamide, **16:**2
in acetyl CoA formation, **19:**257; 258
achiral phosphate group of, **54:**417
in allosteric regulation of enzymes, **28:**76–89; 92–94; 104–110; 138–141; 143; 144
in DR metabolism, **28:**435–436; 438–440; 442; 444; 446; 448; 453
in amino acid activation, **19:**251–256; 258; 259
asparagine synthetase
catalysis mechanisms, **72:**146
glutamine-dependent nitrogen transfer, **72:**169–175
kinetic isotope effects, **72:**187–188
β-Aspartyl-AMP intermediate, aspartic acid, **72:**158–159
ATPase thermodynamics, **72:**13–15
and bioluminescence, **25:**129; 131–137; 142
borohydride trapping of acyl-P intermediate of glutamine synthetase, **72:**27–28
cleavage of, by glutamine synthetase, **31:**188

as competitive inhibitor of rabbit muscle aldolase, **31:**166
role of, in bonding of glutamate, **31:**188
and CO_2 fixation, **12:**51–52; 54–60; 62–63; 63–67; 68–69; 74; 76–77; 83; 90; 101–102; 104
in acetylation reactions, **12:**96; 98
energy source for acetylation, **12:**312–315; 317–318
role in choline acetylase action, **12:** 327–331
role in transmethylation, **12:**465
in conversion of pantetheine to CoA, **16:**6; 11; 16
in creative kinase reaction
activation, **54:**404
inversion at phosphorus, **54:**417–418; 420
cyclic and noncyclic phosphorylation site for formation of, **30:**68
cytochrome *c* oxidase, **71:**102; 116; 165; 166; 186
-deaminase, **34:**459
effect of cardiac glucosides on, **13:**182
deprotonation rate, **25:**22; 26
in diphosphoglyceric acid synthesis, **14:** 214–215
in galactokinase reaction, **14:**197
in glucose diphosphate synthesis, **14:**211
in synthesis of UDPG, **14:**203
-driven DPN reduction, **34:**142
effect of, on rubber biosynthesis, **29:**251–252
phosphofructokinase inhibited by, **29:**355
in schizophrenia, **29:**512
energy donor in glutamine formation, **16:**257
and enzyme inhibition, **25:**171; 223; 242–249
enzymic breakdown in muscle, **13:**163–169
and ethionine effects, **25:**375; 379–388
exchange reactions, tracer experiments, **26:**311–316
in carboxylations, **26:**307; 311–316; 320–327; 331–334; 345
in propionic acid metabolism, **26:** 284–286
in racemization of methylmalonyl-CoA(a), **26:**334–337
in fatty acid activation, **19:**257; 258
formation, **34:**476–477; 481
in formation of adenosyl-methionine, **16:**275
in formation of β-aspartyl phosphate, **16:**268
in fructose-2, 6–6bisphosphate synthesis, **54:**154–155
in fructose-bisphosphatase inhibition, **54:**140–141
in fructose-bisphosphatase regulation, **54:**178; 181
glutamine-dependent nitrogen transfer, kinetic isotope effects, **72:**186
glutamine synthetase activity, **72:**12–13
hydrolysis, **65:**33–35
effect of metal, **25:**18–19
hydrolysis of, **43:**355
interaction with actomyosin gels, **13:**173–178
interaction with actomyosin sols, **13:**171–173
interaction with myosin and actin, **13:**169–180
K_m values for, **35:**40
level in enucleated amoebae, **16:**473
and muscle contraction, **13:**151–197
in myosin
breakdown to adenylic acid, **6:**161–162
effect of ADP, **6:**160
flow birefringence, **6:**160
hydrolysis mechanism, **6:**158
myosin action on, **6:**158
myosin stability, **6:**162–163
rate of splitting, **6:**187
spectral changes, **6:**163
thread extensibility, **6:**175
triphosphate effect, **6:**160
NMP kinase catalytic mechanism, **73:** 114–117
NMP kinase primary structure, **73:**107
NMP kinase tertiary structure, **73:**108–110
N-1 oxide of, **43:**31
nuclear, **27:**270; 277–282
nuclear synthesis, **27:**272–274
oxidation of hydrocarbons, **27:**485; 486
polyphosphate substitute, **27:**361
in oxygen metabolism, **19:**175–177; 189; 192; 199
in pantothenic acid activation, **19:**257; 258
periplasmic permeases, **65:**25–28
from phosphocreatinine supply, **54:**368
and photosynthesis, **25:**42–44; 53–62; 65–66; 68–71; 85–95; 100; 102; 103; 105; 106
in photosynthesis, **19:**2; 3; 17; 18; 21; 24; 29; 30; 38–45; 52; 53; 64; 65
positional isotope exchange, **72:**28–29
in protein biosynthesis, **19:**260

ATP (*Continued*)
and proteins of special types of muscle, **13:**180–184
reactions, **49:**106; 120
application of mechanistic criteria to, **49:**110
mechanisms of, **49:**106–110
role of metals in, **49:**120
relation to myosin and actomyosin, **16:**314; 333–354
relation to polymerization of actin, **16:** 333–335
in *rigor mortis,* **13:**179–180
RNA polymerase II transcription, reversible phosphorylation of CTD during transcription cycle, **71:**57
role
N_2 fixation, **67:**307
nitrogenase, **67:**301–307
role in methyl biosynthesis, **25:**402–408
role in muscle contraction, **13:**187–192; 192–193
role in transmethylation, **25:**412–415; 418; 426
in spermatozoa, **6:**184
substitutions in, **49:**103–126
in synthesis, of amino acids, **16:**260; 264–267; 270; 271; 273
of pyridoxal phosphate, **16:**211; 219
synthesis of, **49:**243
and thioanalog, steady-state kinetic constants for, **54:**407
in transamination reactions, **19:**354; 372; 374
turnover studies, **7:**113–126
tyrosine hydroxylase regulation, **70:**130–131; 134; 141
[^{3}H] ATP, **49:**242
β-γ-^{32}P-ATP, as substrate for glutamine synthetase, **31:**210
γ-^{32}P-ATP, as substrate for succinyl thiokinase, **31:**189
ATP: *N*-Acetylgultamate-5-phosphotransferase, **37:**496; 510
ATP, Role of Metals in Enzyme-Catalyzed Substitutions at Each of the Phosphorus Atoms of (Mildvan), **49:**103
ATP (ADP) analogs, **46:**212
ATP-ADP exchanging system, of mitochondrion, **49:**75
ATP/ADP ratio, in anaerobic cells, **29:**357
in controlling glycolytic intermediates in a cell-free system, **29:**358
ATP-AMP exchange, **35:**6; 8
ATP analogs
6-6chloropurine, **43:**41
fluorophosphate, **43:**22
fused ring compounds, **43:**33
hemoglobin, **43:**35
metal binding, **43:**5
α-β methylene, **43:**7
polyphosphate, **43:**23
spin-labeled derivatives, **43:**35
sulfate, **43:**22
6-thiocyanato, **43:**39
ATP Analogs (Yount), **43:**1
ATPase, **34:**457; 474–475; 485, **37:**415; 429; 439; 448; 461; 474; 496; 509; 543; 565, **40:**39; 40, **49:**73; 224; 261; 265, **55:**367–368. *See also* Adenosine triphosphatase
activity
and fibril contraction, **13:**174
of muscle during embryonic development, **13:**182–183
of the sarcoplasm, **13:**167–168
actomyosin as, **16:**336; 338–339; 344; 345
intracellular distribution, **16:**465; 472
myosin as, **16:**314; 320–322; 326; 335–349; 351–352
alkaline phosphatase, **55:**386
borohydride trapping of acyl-P intermediate of glutamine synthetases, **72:**27–28
CaATP as a competitive inhibitor of, **29:**366
and carboxylations, **26:**325
classification, **14:**248; 250
in oxidative phosphorylation, **14:**13
and CO_2 fixation, **12:**51
conformational changes in energy transducing, **49:**261–267
membrane-bound ATPases, **49:**263–265
soluble ATPase, **49:**265–267
diagram of $2H^+$ translocator, **29:**75
electrical potential across mitochondrial membrane from, **29:**47
energization by, **49:**69
in energy control, **29:**346
in heart muscle preparations, **28:**164; 165
H^+ translocator, **29:**73–78
localization in cytoplasm, **13:**11
mitochondrial, **29:**73–74
multiple forms, **25:**363

product inhibition of, **25:**224; 229–233; 242
in myosin, **6:**147–189
activators, **6:**159; 164–168
activity values, **6:**168
and ADPase specificity, **6:**161
characteristics, **6:**155; 158–170
crystallization, **6:**151–155; 170–171
cyanide effect, **6:**165–166; 167
electrophoretic measurements, **6:**172–173
fractionation, **6:**170
identify with myosin, **6:**170–174
inhibitors, **6:**160; 164–168; 176; 185; 187
poisons, **6:**165–166
iodoacetamide effect, **6:**167
phase rule test, **6:**171–172
*p*H dependence, **6:**163–164
in progressive muscular dystrophy, **6:**223; 224
purification, **6:**150–158
sedimentation in ultracentrifuge, **6:**172
solubility test, **6:**171–172
soluble, **6:**155–156; 171–172
specificity, **6:**158–162
stabilization, **6:**162–163
sulfhydryl groups, effect, **6:**166–168
thermolability, **6:**162–163; 173
myosin activity as, **13:**157; 163–167; 183
Na^+/K^+ antiporter, properties of, **29:**69–73
Na^+/K^+ translocator, diagrams of, **29:**73
oligomycin-sensitive, composition of, **49:**226
of oxidative phosphorylation, **49:**224–225
rate of glycolysis dependent on balance of, **29:**335–336
in retina, **6:**185
and rhodopsin solution, **6:**185
reversibility of, **29:**74–75
sodium-potassium activated and membrane function of, **29:**505–506
in spermatozoa, **6:**184–185
thermodynamics, **72:**13–15
in yeast, **6:**186
ATPase activity
and muscle contraction, **16:**351–352
of myosin, SH-group, **27:**214–215
of relaxed muscle, **16:**347–349
ATP-binding pocket, biochemical and genetic analysis, **65:**28–29
ATP-binding site, formation requirements, **65:**29–30
ATP (C^{14}), in cobamide coenzyme biosynthesis, **26:**366–367
ATP-^{1} ^{4}C-ADP exchange, **35:**329
ATP citrate lyase, **43:**90. *See also* citrate-cleavage enzyme
equilibrium constant, **43:**92
mechanism, **43:**86; 91
metabolic rate, **43:**86
in rat brain, **43:**87
from rat liver, **43:**85
stereochemistry, **43:**86
subunits, **43:**85
ATP Co(III) phenanthroline, **43:**45
ATP derivative, 6–6thioether, **43:**40
ATP diphosphatases, classification, **14:**248; 250
ATP γ F, **43:**16
fumarase, **43:**22
inhibition, **43:**22
metal-binding properties, **43:**22
ATP hydrolysis, **60:**24
ATP inhibition, **48:**217–218
ATP-manganese complex, EPR, **27:**140
ATP-Mg-creatine, enzyme complex, Cotton effect, **27:**59
ATP/Mg^{2+} system, relaxation spectra, **25:**9; 14
ATP monophosphatases, classification, **14:**248; 250
ATP-^{1} ^{8} O, **50:**374; 376; 378; 379
ATP (P^{32}), in carboxylation experiments, **26:**322–326; 333
ATP-^{3} $^{2}P_1$ exchange, **35:**329
ATP-PFK, **59:**332
ATP-^{3} $^{2}P_i$-ADP exchange, **35:**329
ATP-PPase, GMP synthetase, **72:**122–123
ATP-PP_i exchange, **35:**4; 8
aminoacyl-tRNA synthetase, **40:**169; 170; 171; 172
ATP r arginine transphosphatases, classification, **14:**251
ATP r creatine transphosphatases, classification, **14:**251
ATP r DPN transphosphatases, classification, **14:**251
ATP r fructose-6-phosphate transphosphatases, classification, **14:**251
ATP r fructose transphosphatases, classification, **14:**251
ATP r galactose-(1) transphosphatases, classification, **14:**251

ATP r glucose-1-phosphate transphosphatases, classification, classification, **14:**251
ATP r glucose transphosphatases, classification, classification, **14:**251
ATP r nicotinamide mononucleotide transadenylase, classification, **14:**252
ATP r riboflavin phosphate transadenylase, classification, **14:**252
ATPαS, **43:**9; 10, **55:**303–306
  diastereoismers, **43:**9; 10
  inorganic thiophosphate synthesis with, **55:**319
  polynucleotides, thio-substituted, **43:**10
ATPβS, **43:**9; 10, **55:**303–306; 307
  NMR spectroscopy of, **55:**330; 331
ATPγS, **43:**9, **55:**299–303
  disulfide analog, **43:**10
ATP-sulfurylase, **48:**7, **55:**347–348, **74:**186–187
  depression of, **48:**8
  gene, in *Salmonella typhimurium,* **74:**189
ATP synthesis, **55:**313
  by respiratory chain, **28:**156; 159–161
ATP synthetases, by respiration, **28:**161
ATP system in energy transport, **14:**7–19
ATP-trapping action of ethionine, **25:**381–382; 386–388
ATP urea amidolyase, **35:**328; 400–401
Atractyloside, ATP hydrolysis and calcium ion uptake inhibited by, **29:**271
  calcium ions return in respiration inhibited by, **29:**270
  external ATP entry to mitochondrial internal sites inhibited by, **29:**282
  inhibition of ADP and ATP, **29:**66
  inhibition of oxidative phosphorylation by, **29:**66
Atropa belladonna in alkaloid studies, **8:** 224–226; 236–240; 241; 242; 243
Atropine, **55:**139. *See also* Hyoscyamine
  in phosphatidylinositol breakdown, **55:**146
  produced in roots, **8:**224–225
  structure of, **55:**141
  from tropine, **8:**240–241
Atropine esterase, classification, **14:**244
Atropine triphosphate, heterotropic effects, aspartate transcarbamylase, **68:**120–123
Auberginone, **55:**73
  biosynthesis of, **55:**75
  structure of, **55:**53
Aurantiogliocladin, relation to coenzyme Q, **25:**278–281; 296–298
Aureomycin, activity, **10:**164
  structure, **10:**164
Aurofusarin, **11:**403
Aurovertin, **49:**249; 258
  inhibition of ATPase activity of soluble and membrane-bound F_1, **49:**249–250
  mitochondrial ATPase activity inhibited by, **29:**271
Autoacceleration, phenomenon of, in exonuclease, **29:**170
Autoactivation, **45:**405
Autocatalysis
  of blood clotting, **10:**7–13; 25–28
  mechanism, **14:**172
  of protoplasm, **7:**58–59
Autocatalytic activation of prothrombin, study by electrophoresis, **16:**45
Autocatalytic enzyme reactions, **7:**4
Autocatalytic nature of enzyme synthesis, **13:**92–93
Autodigestion, **34:**455
Autoimmunity, **24:**114
Autokinase activity, α-crystallin, **69:**165
Autolysis
  of bacteria, agents for, **26:**213–214; 216
  changes in cell structure during, **8:**35–36
  effect on, pepsin, **24:**125–126
    trypsin, **24:**125
  role in soil microorganisms, **28:**367
Autolytic enzymes, phage-induced, **26:**213
Autolytic systems, localization of, **24:**326–327
Autones, **13:**189
  units of myosin, **16:**351–352
Autoperfused jejunal segment preparation *in vivo,* **53:**208
Autoperfused rat jejunal segments *in situ,* **53:**206
Autoperfused rat jejunal segments *in vivo,* **53:**225
Autoprothrombin I, *See* Proconvertin
Autoprothrombin II, *See* Factor IX
Autoprothrombin III, *See* Stuart power factor
Autoradiographic method, methaneoxidizing bacteria, **27:**478–479
Autoretardation, of DNase, **29:**172–174; 186
  phenomenon of, **29:**170
Autosynthetic molecules, **4:**4–8
Autosynthetic process, **4:**8–11; 19
Autotrophism, **2:**138–144, **74:**6–12
  definition of, **74:**6
  facultative, **2:**143

and heterotrophism, differentiation of, **74:** 8–9; 10–11
intermediate forms of, **74:**11
Autotrophs, carbon dioxide fixation by, **12:** 41–151
Autotrophs, Fixation of Carbon Dioxide by (Utter and Wood), **12:**41
Autoxidation
destruction of vitamin A, **8:**440; 441; 443
and double-bond conjugation, **8:**434–439
of fats, and lipoxidase, **8:**425–457
of fatty acids with conjugated double bonds, **8:**428; 434–439; 449–450
in fermenting tobacco, **10:**345–349; 373–375; 382–386; 417
of glutathione, **10:**294–295
and free-radical mechanism, **8:**431–439; 450–451; 453
and hemin, **8:**427; 440; 443; 452
hydroquinones, **65:**102–103; 121–123
of linoleate, and spectral changes, **8:**435–437
thermochemical studies, **8:**437–438
mechanism, **15:**2; 10; 11; 19–21; 23; 38; 41–42
of leucoriboflavin, **15:**89
of monoethenoid fatty acids, **8:**428–434; 437–438
primary steps, **8:**434–439
secondary rearrangements in products of, **8:**429–434; 437
semiquinones, **65:**121
of thio ethers, **8:**440
of unsaturated compounds, **8:**425–428; 428–439
and epoxides, **8:**428; 432; 433–434
Autoxidizability, of cytochrome c, **23:**63; 64
Autoxidizable cytochromes, **18:**36–37; 48
Auxin, **9:**438
aerobic requirements, **18:**39
indolepyruvate as precursor, **16:**235
Avena, **20:**371
Aviden, **43:**331; 332
Avidin, **35:**411–415, **36:**67–69; 75, **37:**461, **56:**94; 95
assay, **35:**415
biotin binding sites, **35:**412
cause of egg white injury, **3:**305–307
column chromatography, **22:**477
effect of, on acetate utilization in latex, **29:**232
interaction with enzyme-bound biotin, **35:**414
in metabolite displacement, **6:**144
molecular properties, **35:**412
purification, **35:**415
source, **35:**412
structure studies, **14:**283
Avidin, DNA in, **16:**423
Avin fibronectin receptor, **59:**40–46
Axes, of polypeptide chain, **23:**13–18
Axonal membrane, **34:**493
Axonemes
chlamydomonas reinhardtii, **53:**136
KCL-Extracted, **53:**147
8-Azaadenosine, **54:**210
Aza analog, **57:**42–45
8-Aza-9-deaza-adenosine, **54:**210
8-Azaguanine
effect on enzyme synthesis, **24:**16–18
folate metabolism, **27:**460
as a substrate-for hypoxanthine-quanine phosphoribosyl-transferase, **41:**14
Azaguanine, and resistance to aza-1-mutant of *S. pombe,* **41:**12
S-Azaguanine, effect on β-lactamase, **28:** 275–278
Azaguanine deoxyriboside, **18:**222
Azaguanine riboside, **18:**222
Azapeptide inhibitors, **57:**42; 46
Azaserine, **54:**68, **57:**265
mutagen, **27:**331
Azeleucine, **49:**63; 65
Azetidine carboxylic acid, **35:**482
Azetidine-2-carboxylic acid, **50:**119; 149; 152; 154; 157; 161–164; 181, **54:**225–227
E. coli and mung bean growth inhibited by, **29:**95–96
effect on growth of *E. coli* by, **29:**97
enzymatic degradation by *Agrobacterium* of, **29:**107
found in Liliacease, **29:**95
in plants, **29:**94
proline analog, **29:**94
in synthesis of actinomycin, **29:**104
L-Azetidine-2-carboxylic acid, **50:**163
Δ′-Azetine-2-carboxylic acid, **50:**152
Azide
complex with carbonic anhydrase, **47:**186; 225–226; 228; 231; 236–239
effect on carotenogenesis, **21:**348
effect on enzymes of *Cunning-hamella blakesleeana,* **20:**270

Azide (*Continued*)
effect on 16α-hydroxylase, **20:**270
effect on transformations of antibiotics in plants, **20:**387
effect on β-lactamases, **28:**297
effect on oxidases, **19:**118; 122; 124–128; 182; 199
effect on plant respiration, **18:**33–34; 290
inhibition of *o*-aminophenyl-glucuronide formation, **16:**403
Azide ions, **15:**17
Azide-sensitive iron-oxidizing activity, **44:**203
8-Azido-ATP, **43:**42
5-[(3-Azido-6-nitrobenzylidene)amino]-2,6-diamino-4 pyrimidone, **67:**223–225
Azlactone formation, **8:**119–124
p-Azobenzene arsonate, as hapten for antigens, **41:**134
Azobenzene reductase, location in subcellular fractions, **24:**304
Azo bonds, **34:**453
Azo-dye carcinogen, **46:**389; 409
Azo dye induced hepatoma, β-glucuronidase activity, **16:**396
Azo group, reductive cleavage, PN-dependent, **26:**129
δ-Azo-guanine of tobacco mosaic virus, **15:**474
Azol esterases, classification, **14:**244
Azospirillum brasilense, glutamate synthase, **66:**275–276
Azote, assimilation et dissimilation chex les animaux (Braunstein), **19:**335
Azotobacter
coenzyme Q in, **21:**111
nitrogen fixation by, **13:**345; 346–347; 351; 358; 360; 361–368; 371–372
oxidation of hydrocarbons, **27:**515
propionic acid metabolism, **26:**295
Azotobacter agile, **15:**193
citric acid cycle, **18:**247–248
Azotobacter agilis, **34:**466
Azotobacter chroococcum, oxidase of, **23:**267
Azotobacter ferredoxin, in nitrogen fixation, **33:**146–147
Azotobacter vinelandii, **51:**112, **70:**222; 235; 236
and citric acid cycle, **18:**250; 270
polynucleotide phosphorylase in, **18:**228
coenzyme Q in, **25:**288
cytochrome c_4 of, **20:**163
electron transport particle from, **21:**83
EPR spectra, **27:**156–158; 183
oxidation of hydrocarbons, **27:**515
use of acetate, **23:**411
Azotoflavin, in nitrogen fixation, **33:**146
Aztobacter vinelandii oxidase, double-headed, **23:**314–315
Azure A, use in staining nuclei, **16:**419
Azurin
composition of copper in, **33:**196
extinction coefficient of, **33:**195; 198
forms of copper in, **33:**195
ORD studies of, **33:**201
oxidation-reduction potential of, **33:**207
pH dependent reduction of, **33:**208
cytochrome *c* oxidase, **71:**129
Azurin EPR study, **27:**138

B

Babcock model, cytochrome *c* oxidase, proton pumping, **71:**143–146
Bacillarophyceae, carotenoids, in, **21:**298; 301–303
Bacillus amyloliquefaciens, **23:**179
Bacillus aneurinolyticus, 390, **15:**394; 397; 398; 410–415
Bacillus anthracis, **23:**179
amino acids in, **14:**266
polyglutamate of, **14:**278
β-lactamase of, **28:**242; 264
Bacillus anthracoides, **23:**184
Bacillus brevis, **23:**179
bactericidal substances from, biological properties, **4:**169–182
in biosynthesis of gramicidin S, **33:**339–343
by cell-free system from, **33:**343–350
mode of action, **182–194**
as producer of gramicidin S, **33:**337
purification and chemistry, **4:**156; 157–169
strains of
ATCC 8185, **33:**338; 350; 367
ATCC 9999, **33:**338; 339; 343; 353
ATCC 10068, **33:**338
Nagano, **33:**338; 340; 343; 349; 351; 353; 357–358; 369; 370; 374; 375
Bacillus butacone in industrial fermentations, **7:**563
Bacillus cadaveris enzymic adaptation in, **13:**71

Bacillus casei, reduction of hexose monophosphates, **15:**167–168
Bacillus cellulosae dissolvens action on cellulose, **8:**267
Bacillus cereus, **23:**179; 183, **44:**175, **71:**187
enzyme synthesis in, **24:**17–18
enzymic adaptation in, **13:**70; 95–98
enzymic transformations of steroids by, **20:**247; 248
β-lactamase from, **28:**241; 245; 247; 251; 253–257; 260; 262; 264; 266–267; 269; 271–281; 283–288; 291; 294–305; 312
penicillinase in, **22:**106; 109; 137
sensitivity to RNA antagonists, **22:**138
penicillinase induction in, **22:**119
kinetics, **22:**116; 117
and nature of precursor, **22:**124
ultraviolet sensitivity of, **22:**138
penicillinase mutants, **22:**21
constitutive, **22:**103
phospholipase C from, **74:**71
proteinase, **25:**338, **41:**183
protein turnover in nongrowing cells, **22:**126
Bacillus circulans, **54:**2
laminarinase from, **28:**344
yeast-lysing isolate of, **28:**348
Bacillus coagulans, **39:**338
Bacillus emersonii, **59:**71
Bacillus fossicularum action on cellulose, **8:**267
Bacillus licheniformis, **33:**370, **34:**481, **54:**169; 172
alkali hydrolysis of cell wall teichoic acid of, **30:**228
binding of antibody in the walls of, **30:**247
biosynthesis of poly(glycerol phosphate) in, **30:**235–236
CDP-glycerol pyrophosphorylase in the cells of, **30:**236
poly(glycerol phosphate) and poly(galactosylglycerol phosphate) synthesis in, **30:**237
allosteric enzymes, **28:**69
β-lactamase of, **28:**252; 264–265; 267; 270–274; 276; 279–288; 291; 294; 296; 303; 305; 306; 310–312
liberation, **28:**265–266
Bacillus licheniformis proteinase, **23:**179
Bacillus macerans
amylase in, **11:**298–299; 322–324, **14:**246
enzyme for polysaccharide synthesis, **15:**307
enzymes from, **14:**405
Bacillus megaterium, **23:**180, **34:**466, **49:**5
enzyme synthesis in, **24:**12
β-galactosidase induction in, **22:**119
location of teichoic acids in, **30:**245
nucleic acids from, **18:**163
proteinase from, **41:**183
quinones in, **25:**289
source of LBF, **14:**61
Bacillus mesentericus, **23:**179
polygalacturonases of, **20:**366
Bacillus methanicus, oxidation of hydrocarbons, **27:**477
Bacillus methanigenes action on cellulose, **8:**266–267
Bacillus mycoides, collagenase in, **23:**179
metabolism of glycine, **23:**453–454
Bacillus naphthalinicum nonliquefaciens, oxidation of hydrocarbons, **27:**519; 520
Bacillus paratypi, allosteric enzymes, **28:**111
Bacillus polymyxa, allosteric enzyme of, **28:**65; 72
Bacillus polymyxa in industrial fermentations, **7:**603; 605–607
Bacillus popilliae, superoxide dismutase from, **41:**81
Bacillus putrificus, enzymic transformations of steroids by, **20:**257
Bacillus pyocyaneus, **15:**378
Bacillus R-4, **59:**90
Bacillus sphaericus, enzymic transformations of steroids by, **20:**262
Bacillus Stearothermophilus, **39:**338
Bacillus stearothermophilus, **59:**328, **71:**187
wall teichoic acid, of, **30:**227
Bacillus sublilis, amylase of, **26:**103–104
cell walls, **26:**198; 223
PN-dependent enzymes, **26:**131
rate of protein synthesis in, **26:**108
Bacillus subtilis, **15:**416, **39:**338, **44:**165; 168; 169, **54:**172, **71:**3; 17; 187; 189
allosteric enzymes, **28:**71–73; 123–124
β-lactamase of, **28:**244–245; 254; 264; 288; 296
amino acids in, **14:**266
enzymes from, **14:**405
proteolytic enzyme of, **14:**300–301
as source of LBF, **14:**50; 52–53
α-amylase composition, **25:**335
ethionine effects in, **25:**376

Bacillus subtilis (*Continued*)
quinones in, **25:**289
α-amylase of, **23:**32–35; 41; 50–58, **24:**112; 119
collagenase of, **23:**179–181; 194; 203
cytochromes of, **23:**266; 269; 270; 287; 308; 309
elastase from, **23:**227; 229; 230; 232
enzyme formation in, **24:**24
proteinase of, **23:**31; 72–74
structure of subtilisin, **24:**449
and biosynthesis of ornithine, **33:**374
biosynthesis of poly(glycerol phosphate) in, **30:**235–236
CDP-glycerol pyrophosphorylase in the cells of, **30:**236
compounds found in the wall of, **30:**242
effect of phosphate deficiency on teichoic acid synthesis on, **30:**250
enzyme effect on *in vitro* synthesis of poly(glycerol phosphate) in, **30:**236–237
enzymic transfer of glucose from UDP-glucose to poly(ribotol phosphate) in, **30:**242
membrane teichoic acid of, **30:**225
Michaelis constant for glucosylation in, **30:**242
Michaelis constant for poly(ribitol phosphate) in, **30:**242
ribitol teichoic acid of, **30:**230
uptake of antigen by walls of, **30:**247
and citric acid cycle, **18:**250; 264
CO effect on, **18:**31
effect of acidity on growth, **18:**264–267
nitrogen metabolism of, **18:**251–255
syntheses of, **18:**255–256
folate metabolism, **27:**457
mutation, **27:**311
glutamine PRPP amidotransferase, **66:** 261–263, **72:**93–95; 97–99
mutant, in study of amino acid metabolism, **16:**269
transamination reactions in, **16:**192; 206-207; 237
proteases from, **33:**385
Bacillus subtilis, L-alanine dehydrogenase in, **19:**340
Bacillus subtilis, glutamyl peptides in, **13:**301
Bacillus subtilis in industrial fermentations, **7:**593–594
Bacillus subtilis-mesentericus group, β-lactamase of, **28:**242
Bacillus subtilis neutral proteinase, **53:**283
Bacillus Subtilis RNA Polymerase and Its Modification in Sporulating and Phage-Infected Bacteria (Losick and Pero), **44:**165
Bacillus subtilis, superoxide dismutase from, **41:**85; 86
proteinase from, **41:**182; 183; 187
Bacillus thermoproteolyticus, proteinase from, **41:**183
Bacillus thiaminolyticus, 390, **15:**394; 397; 398; 400–405; 407; 411–415
Bacillus Thuringiensis, **44:**175
Bacitracin, **33:**367
effect on synthesis of poly(glycerol phosphate) in *Bacillus,* **30:**244
in treatment of plant diseases, **20:**392
Bacitracin A, effect on murein synthesis, **26:**218
Backbone folding, NMP kinase tertiary structure, **73:**108
Bacteria, **34:**456; 479; 485, **46:**335, **59:**328. *See also under* specific organism
acid-fast, and branched-chain fatty acids, **8:**321
synthesis of waxes in, **8:**333–336
adenylosuccinate synthetase (AMPSase) gene structure, **73:**86–90
aerobic, spore-forming (*See also* bacillus brevis)
bacteriolytic effects of filtrates, **4:**153–157
autotrophic (*See* Autotrophism)
blue-green, **46:**335
ADP-glucose pyrophosphorylase of, **46:**335
chemo-autotrophic, **74:**6–7; 10
coliform, fixation of carbon dioxide by, **74:**16
cysteine *S*-conjugate β-lyases in, **72:**214–215
denitryfying, **46:**58
effect of fluoroacetate on, **18:**135–136; 142
photosynthetic, **18:**56
enteric
O-acetylserine sulfhydrylase expression in, **74:**189–190
O-acetylserine sulfhydrylase isozyme A in, **74:**189–190; 197
O-acetylserine sulfhydrylase isozyme B in, **74:**189–190; 197
L-cysteine biosynthesis in, **74:**186–191; 189
genetic mapping in, **22:**2

glycogen accumulation, **46:**319
Gram-negative, **4:**154; 156; 170; 171; 172; 173; 174; 175
Gram-positive, **4:**154; 156; 158; 170; 171; 172; 173; 174; 185
enzymes of teichoic acid synthesis in, **30:**234–235
muramic acid phosphate in, **30:**248
teichoic acids in, **30:**243
growth, effect of rare earth metal ions on, **21:**192
nonphotosynthetic, carotenogenesis in, **21:**311; 318–322; 329; 330
carotenoids in, **21:**306
photosynthetic, carotenogenesis in, **21:**308; 310; 311; 313–316; 319–321; 324; 339 344–349
carotenoid functions in, **21:**352–354; 357; 358
carotenoids in, **21:**298; 303–305
heterotrophic (*See* Heterotrophism)
nitrifying, **74:**6
photosynthetic, **74:**7
sulfur
non-photosynthetic, **74:**6
photosynthetic, **74:**7
and H^+-translocator oxidoreductases, **29:**76
location of markers of sugar metabolism in, **30:**86
nitrogen metabolism in, **9:**407–409
non-photosynthetic, **46:**39; 42
pathogenic, **58:**100
photosynthetic, **25:**41–43; 45–46; 55–58; 65; 76–77; 80–81; 85; 93–94, **46:**39; 42; 54; 58; 69; 89–91
photosynthesis in, **30:**2
synthesis of vitamin B_{12}, **46:**54
Bacteria, Gram-negative enteric, biosynthesis of cell wall lipo-polysaccharide in, **31:** 77–124
Gram-positive, cell wall and protein synthesis in, **31:**116
Bacteria, P450 expression, **68:**52–60
Bacteria in CO_2 fixation, **2:**144–169
and vitamin K, **2:**297; 317; 318
Bacterial amino acid decarboxylases, **6:**1–31
activity of codecarboxylases in growth medium, **6:**8–9
adaptation to substrate, **6:**7–8
in bacilli, **6:**7
biological function, **6:**29–31
in clostridia, **6:**7
codecarboxylase (*See* Codecarboxylase)
in *E. coli,* **6:**6
effect of age of culture, **6:**12
of growth temperature, **6:**2; 11
of substrate concentration, **6:**16
inhibitors, **6:**17–21
*p*H of growth medium, **6:**10–11; 15–16
in *Porteus,* **6:**7
properties, **6:**14–19
protein apoenzyme moieties, **6:**19–21
protein coenzyme moieties, **6:**19–21
purified, **6:**12–14
requisites for formation, **6:**4; 5
in streptococci, **6:**6
Bacterial cell disruption, with deoxyribonuclease, **33:**350
with French pressure cell, **33:**348
with glass beads, **33:**346; 347
by grinding with alumina, **33:**348
with lysozyme, **33:**350
by press grinding with alumina at freezing point, **33:**347
by sonic disintegration, **33:**343–344; 349–350
by ultrasonic disintegration, **33:**347
Bacterial cell-envelope, **34:**475–478
membrane, **34:**467
Bacterial cell protein, inhibitors on the synthesis of, **33:**340–343
Bacterial cell walls, and macromolecules, **26:**193–232
synthesis of, **26:**36; 66; 68–72
Bacterial Cell Walls-Bagshaped Macromolecules (Wiedel and Pelzer), **26:**193
Bacterial citrate synthetase, **43:**83; 84
Bacterial collagenases
from Aeromonas (*See also* Clostridiopeptidase A; Collagenase 2)
from *Aeromonas,* **34:**173
from *Bacteroides melaninogenicus,* **34:**173
from Clostridium chauvoei, **34:**172
from Clostridium histolyticum, **34:**156
application, **34:**156; 157; 169
from Clostridium integumentum, **34:**172
from Clostridium perfringens, **34:**171
from Clostridium septicum, **34:**172

Bacterial collagenases (*Continued*)
from Clostridium tetani, **34:**172
from *Mycobacterium tuberculosis,* **34:**173
from pigmented pseudomonas, **34:**173
from *Staphylococcus aureus,* **34:**173
from *Streptomyces* (Nocardia) *madurae,* **34:**172
Bacterial cytochrome P_{450}, **51:**42; 43; 64
oxyferrous compound of, **51:**14–16
Bacterial dihydrofolate reductase, vertebrate DHFR structures compared, **70:**42–44
Bacterial DNA, study of, **16:**431–433; 459–460; 481
Bacterial endotoxins, **58:**100
Bacterial enzymes, action on collagen, **23:** 177–183
Bacterial exotoxins, **58:**100
pertussis toxin, **58:**131
of *Pseudomonas,* **58:**115
Bacterial extracts, light emission in, **25:**144–151
Bacterial folate transport, **47:**315–327
energy requirement, **47:**320–322
general characteristics, **47:**316
inhibitors, **47:**322
regulation, **47:**326–327
substrate specificity, **47:**318
Bacterial formate dehydrogenase, **48:**16–20
Bacterial β-glucuronidases, **16:**377–381; 384
Bacterial glutaminases, **53:**227
Bacterial glutamine synthetase, regulation mechanisms, **72:**30–38
casecade control, **72:**31–32
cumulative end-product inhibition, **72:** 32–35
glutamine synthesis expression, **72:**35–36
Bacterial infection, trimethoprim, eukaryotic dihydrofolate reductase, ligand binding (of inhibitors), **70:**55–56
Bacterial luciferase, **51:**63–64
long-lived intermediate of, **51:**16–19
Bacterial luminescence, **25:**124; 125; 128; 144–156, *See* Luminescence, bacterial
effect of pressure on, **9:**68–69
Bacterial lysine biosynthesis
aspartokinase and aspartate semialdehyde dehydrogenase, **72:**285–286
diaminopimelate epimerase, **72:**315–320
dihydrodipicolinate reductase, **72:**291–298
dihydrodipicolinate synthase (DHDPS), **72:**285–291
L,L-diaminopimelate decarboxylase, **72:** 310–314
L,L-diaminopimelate epimerase, **72:**307–310
research backgound, **72:**279–283
N-succinyl-L,L-DAP desuccinylase, **72:** 303–307
N-succinyl-L,L-diaminopimelate aminotransferase, **72:**302–303
tetrahydropdipicolinate *N*-succinyltransferase, **72:**298–302
yeast-L-lysine pathway, **72:**283–284
Bacterial membrane fraction, as site of protein synthesis, **22:**132–134
Bacterial metabolism and acetyl phosphate, **6:**232–250
anabolism, **6:**233; 242–250
catabolism, **6:**233–241
Bacterial phosphoryl transfer protein, phosphohistidine in, **36:**191
Bacterial photosyntheses, **1:**236; 263–328
energetics, **1:**314–319
kinetics, **1:**289–301
Bacterial pigments, **1:**235; 292; 293; 307–314
Bacterial proteins, **16:**460; 491
Bacterial Protein Toxins with Latent ADP-Ribosyl Transferases Activities (Lai), **58:**61
Bacterial resistance, to penicillin, **28:**274–275
Bacterial superoxide dismutases, **41:**35; 70–73
from *E. coli,* **41:**70
the ferrientyme, **41:**71
intracellular localization, **41:**73
the manganese enzyme, **41:**70
from *streptococcus mutans,* **41:**71
Bacterial synthesis of methionine, **25:**404–408
Bacterial terminal oxidases, cytochrome *c* oxidase, **71:**186–191
Bacterial toxins, **58:**99; 102
functions domains, **58:**103
mode of action, **58:**101
protein synthesis inhibition, **58:**101
Bacterial transport systems, classified, **44:**241
Bacterial virus, **2:**6–17; 17–21. *See also* Host-virus relation
distribution in nature, **2:**4; 5
general properties, **2:**2–4
life-cycle in sensitive host, **2:**6–17
lysines, **2:**22–24
methods of assay, **2:**6; 15–17
mutations, **2:**21–22

size, homogeneity, **2:**24–27
size of particle, **2:**27
and x-ray, **2:**24–27
Bacterial Viruses (Bacteriophages) (Delbrück), **2:**1
Bactericidal action, **4:**154; 155; 169; 185.
Bactericidal action of soaps and fatty acid chain length, **8:**385–386
Bactericidal effects of fresh sera, relation to complement, **8:**111
Bacteriochlorin, **1:**309
Bacteriochlorophyll, **1:**235; 292; 309; 313
fluorescence of, **11:**131–139
Bacteriochlorophyllide, **46:**76
Bacteriochlorophylls, **46:**38; 72; 75
d, **46:**80
a, **46:**77; 80
b, **46:**76–77; 80
c, **46:**80
e, **46:**80
synthesis, **46:**89
Bacterioides ruminicola, **46:**102
Bacteriolysis, **4:**169; 192. *See also* Tyrocidine
Bacteriophage, **2:**2–4, **15:**482, **36:**237–238, **58:**104. *See also* Escherichia coli; λ phage
action on cell walls, **26:**196; 200; 213–214; 225
activity and particle identity, **9:**191–195
chemical composition, **9:**229
cofactors, **9:**178–179; 183
effect of light on, **9:**142; 155
growth curves, **9:**180
histochemical methods, **9:**598
infection by, **9:**177–179
mutations, **9:**183–184
nucleic acid of, **9:**479
physical constants, **9:**187; 188
purification, **9:**187
alterations of structure of O side chains by conversion of, **31:**113
chemical change and self-reproduction, **13:**124; 145–146
enzymic adaptation, **13:**79–80
effect of amines on T4rII infection, **36:**237
effect on DR metabolism, **28:**421; 423–428; 435–436
effect on β-lactamase, **28:**269–271; 280
genetic material of, **24:**9–11
inactivation, **7:**248
in industrial fermentations, **7:**563–564
inhibiting effect of amines, **36:**237
methylation, **25:**398
polyamines in, **36:**208; 242
protective effect of amines, **36:**237
T-even, **36:**218
polyamines in, **36:**218
Bacteriophage, DNA in, **16:**412–413; 432; 453; 480
5-hydroxymethylcytosine in DNA of, **16:**432
Bacteriophage coat protein, **47:**85; 137
Bacteriophage DNA, **18:**164; 168; 169
radioactive, DNA duplication study, **18:**183
and sugar nucleotides, **26:**57; 60; 61; 64; 75
Bacteriophage nucleic acids, **24:**17
Bacteriophage epsilon^{34} properties of, **31:**100
Bacteriophage protein, role of messenger RNA, **24:**19
Bacteriophage T4, mutation study, **27:**297–301; 306; 310; 313–318; 320; 323–326; 330; 331; 336–337; 339–340; 342
Bacteriophage Tz, effect on bacterial RNA, **24:**17
Bacteriopheophytin, **46:**81
Bacteriostatic action of gramicidin, **4:**169; 185; 188; 189
Bacterioviridin, **1:**313
Bacterium aliphaticum liquefaciens, oxidation of hydrocarbons, **27:**534
Bacterium aroideae, polygalaturonases of, **20:**365; 366; 368; 370
Bacterium cadaveris, *See* Hafnia alvei
Bacterium cadaveris, lysine decarboxylation in, **22:**109; 110
Bacterium linens, α-ketopantoic acid synthesis in, **21:**419
Bacterium paraffinicum, oxidation of hydrocarbons, **27:**472
Bacterium stewartti, plant disease caused by, **20:**393; 394
Bacterium subtilis, enzymic transformations of steroids by, **20:**262
Bacteroides melaninogenicus, **23:**182
Bacteroides symbiosus, **45:**108
pyruvate, phosphate dikinase in, **45:**108
Baculovirus, P450 expression, **68:**51
BAEE, **34:**508–511
Bagasse, lignin from, **18:**354–357
Baicalinase
classification, **14:**246; 249
as a β-glucuronidase, **16:**381

Bakers yeast, PP-nibose-P anidotrans ferase from, **41:**10
BAL. *See also* 2,3-Dimercaptopropanol
inhibition of melanin formation by, **14:**93; 94
inhibitor, **28:**159; 160
Balanoglassus biminiensis, luminescence in, **25:**160
A-Band, of muscle, **16:**352–353
H-Band, of muscle, **16:**352
I-Band, of muscle, **16:**352–353
Banding. *See also* Periodicity
of chromosomes, **16:**460–461
of light-meromyosin, **16:**325
of structural proteins, **16:**320
Bandorwski's base, **44:**212
Bangosomes, **29:**291
Barabourin, fibrinogen binding inhibition, **70:**7
Barium
and fat metabolism, **8:**398
Barium carbonate, as absorbent for clotting factors, **16:**64; 65; 83; 87
Barium ions, **15:**258
chelate formation with amino acids, **21:**145
model catalyses with, **21:**177
properties, **21:**157
in decarboxylation, **12:**80
in β-glucan synthesis, **28:**340
and isolation of enzymes, **14:**396
precipitation of argininosuccinic acid, **15:**353–354; 362
precipitation of compound X, **15:**369
promotion of citric acid formation, **10:**215
Barium salts, use in obtaining CoA from liver, **16:**2
Barium stearate films, **28:**15; 19
Barium sulfate
as adsorbing agent in isolation of clotting factors, **16:**26–28; 65; 73; 74; 86; 87
for prothrombin adsorption, **22:**456
Barley, **46:**75; 100; 336
Barley etioplasts, **46:**71; 74
Barry degradation procedure, applied to laminarin, **28:**332
Base analogs, effect on β-lactamase, **28:** 275–278
Base catalysis, of esterases, **24:**461–463; 464
model for RNase, **24:**446–448; 455
Base displacements, in disulfied exchanges, **27:**202–204
Base exchange reaction, by action of thiaminase, **15:**403–404; 411–414; 416
Base groups, **34:**492
Base pairing, **47:**378, **50:**36
Bases, **34:**446
Base Y, **35:**463; 468
Basic amino acids, reaction of D-amino acid oxidase with, **34:**71
Basic excitation unit, **43:**202
Basic FGF, cholesterol ester hydrolases regulation, **69:**73
Basidiomycete, hydrolysis of laminarin, **28:**350
Basophils, **58:**258
Bats, fruit, **46:**54
induction of vitamin B_{12} deficiency in, **46:**54
B blood group, **52:**61
galactosyltransferase, **52:**62; 67; 74
B-Cell dysfunction, decreased Ecto-5′-Nucleotidase, **51:**201–202
BCNC, **57:**336–340
BCNU, **57:**336–340
B_{12} Coenzyme, **38:**416
synthetase, **38:**428
B_{12} coenzymes, EPR study, **27:**142–143
B_{12} cofactor, **58:**179
β-B-Crystallin, *See* entries under Crystallin
Bean leaves, **46:**75; 76
Bean mosaic virus, **20:**402
Beans. *See also* Lima beans
antibiotic treatment of diseases, **20:**388–390; 401; 402
cytochromes in, **18:**17; 19; 29
respiration in, **18:**286
Bed, **34:**523–524
Beef heart F_1, **49:**230; 232; 258; 266
Beef heart mitochondria, **34:**474
Beef kidney enzyme, **1:**82; 83
Beef spleen cathepsin, **1:**81–84
Beef spleen enzyme preparation, **1:**82; 84
Beets, antibiotic treatment of diseases, **20:**396
Beggiatoa, metabolism in, **74:**6
Behavior, correlation of brain amines and changes of, **29:**485–489
norepinephrine and changes of, **29:**486
relation of neurotransmitters to, **29:**482
serotonin and changes of, **29:**486
Behrens procedure for isolation of cell nuclei, **16:**413; 415–417; 423
Benadryl, substitute for platelet cofactor I, **16:**54

Benoxaprofen, **58:**220
Bentonite clay, **33:**266
Benzacridine, mutation study, **27:**309
Benzaldehyde, **15:**257; 258
effect on contraction, **16:**339
Benzamidine, **48:**290
Benzanthracene, mutation study, **27:**309
Benzedrine, antagonism to amines, **6:**41–42
Benzene, **51:**33
biological oxidation to polyphenols, **10:**469
microbial oxidation, **27:**506–513
quinone formation, **65:**108–109
Benzenehexacarboxylate, **51:**236
Benzenepentacarboxylate, **51:**236
Benzene ring, biosynthesis, **16:**287–291
Benzenesulfonate, **39:**345
1,2,4,5-Benzenetetracarboxylate, **51:**236
Benzene 1,2,3-tricarboxylate, **43:**93
1,2,3-Benzenetricarboxylate, **51:**236
1,2,4-Benzenetricarboxylate, **51:**236
1,3,5-Benzenetricarboxylate, **51:**236
Benzenoid compounds, synthesis by plants, **18:**361–370
Benzetimide, **55:**148
muscarinic site binding of, **55:**150
Benzfuran systems in native lignins, **16:**147
Benzidine sulfate, **60:**171
ω-*N*-Benzilidene derivatives, of amino acids, **16:**232
Benzimidazole
as inhibitory analogue, **6:**134; 137; 138; 144
reaction with benzoyl chloride, **20:**83
Benzimidazolyl cobamide (BC) coenzyme, affinities of, **26:**346–347
characteristics, **26:**348; 360–362
effect on diol dehydrase, **26:**366
Benzo[*a*]pyrene, quinone formation, **65:** 109–110
Benzo(a)pyrenes, **60:**15
diol-epoxide derivative, **60:**47
Benzoate
anaerobic oxidation, **27:**534–536
as substrate substitute, in ternary complex, **27:**4–12
labeled, in plant respiration, **18:**303
Benzoate (^{14}C), degradation, **27:**514; 535
Benzoate oxidase, **27:**515
Benzohydroquinone, laccase action on, **27:** 125–126
Benzoic acid, **15:**184–185; 249
in biosynthesis of hippuric-acid, **19:**252
effect on ornithine cycle, **19:**376; 377
inhibition of β-lactamase, **28:**295
in lignin study, **18:**374
thermodynamic data on ionization, **20:**64
Benzoic acid as inhibitor of β-glucuronidase, **16:**374
thio ester with CoA, **16:**18
Benzoin condensation, **58:**2
Benzonoid class of phenolic compounds, metabolism in plants, **16:**140–141
Benzophenone photochemistry, **58:**48
Benzo [α] pyrene 4, 5–5oxide, **46:**394; 396
Benzo [α] pyrene-3-monoxycenase, **46:**171
Benzoquinone
as artificial electron acceptors, **30:**23
as inhibitory analogue, **6:**142
p-Benzoquinone, **65:**108–109. *See also* Quinones
o-Benzoquinone. *See also* Quinones
p-Benzoquinone, as tanning substrate, **16:**156
formation from hydroquinone, **16:**140
nuclear substitutions in, **16:**108
reaction, with amino acids, **16:**112–114; 121–122
with cysteine, **16:**121–122
with heterocyclic amines, **16:**110–112
with proteins, **16:**123–126
reduction by DPNH and TPNH, **16:**134
o-Benzoquinone, from catechol, **27:**128
o-Benzoquinone, inaction with collagen, **16:**125
metabolism in plants, **16:**140
reactions, with amino acids, **16:**112–115; 117–120
with aniline, **16:**110
with proteins, **16:**126; 161
reactive centers of, **16:**122
p-Benzoquinoneacetic acid, effect on homogentisate oxidase, **19:**92; 126
Benzoquinone free radicals, in peroxidase action, **27:**122
Benzoquinone ring, origin of, **25:**298–299
Benzoquinones, and coenzyme Q, **25:**305; 313
S-1,4-benzoquinone-2-ylcysteine, formation, **16:**121–122
S-Benzoyl-*N*-acetylcystine, intramolecular aminolysis of, **19:**243

8-Benzoylamino-2-naphthylglucuronide, as substrate for β-glucuronidase, **16:**366; 368
Benzoyl-L-arginine
amide, **34:**460; 501; 505–511
ethyl ester, **34:**463–465; 502
methyl ester, **19:**306
Benzoylarginine amine, as substrate for thrombin action, **16:**34
Benzoylarginine esters
hydrolysis by thrombin, **16:**34
hydrolysis by trypsin, **15:**292
Benzoyl arginine ethyl ester, **48:**288
Benzoylation, biological, role of coenzyme A, **13:**280
Benzoyl chloride, reaction with imidazole, **20:**73; 83
Benzoylcholinesterase, localization of, **24:**317
S-Benzoyl CoA, in formation of hippuric acid, **19:**249
Benzoyl CoA r glycine transbenzoylase, classification, **14:**251
Benzoyl-CoA synthetase, localization of, **24:**334
Benzoyl coenzyme A, **18:**208
N-Benzoylcystine, synthesis, **19:**243
Benzoylglycinamide, papain action on, **19:**307; 310
Benzoylglycine, esters, papain action on, **19:**307
Benzoylglycine ethyl ester, **34:**460; 509–511; 522
N-Benzoyl-L-phenylalanyl-L-valyl-L-arginine-paranitroanilide, **48:**288
Benzoyl-prolyl-phenyl-arginine-paranitro-anilide, **48:**291
Benzoyl-L-tyrosylglycine amide, hydrolysis of, **43:**314
Benzphetamine, **60:**72
Benzpyran systems, in native lignins, **16:**147
Benzpyrene hydroxylase, ethionine effect on, **25:**380
Benzpyrene-sarcoma, treatment of, **11:**398
Benztropolones, formation in browning of plant tissues, **16:**152
in plant metabolism, **16:**136–138; 152
5-Benzylamino-1,6-diamino-4-oxo-pyrimidine, **67:**196–198
Benzyl carbamates, rate constants and half-lives for fission of, **29:**396
Benzylcarbonyl glycine, rates of cleavage of, **29:**396–397
Benzylcephalosporin, action of β-lactamase, **28:**240; 245; 249; 252; 259; 271; 288; 291; 294
S-Benzylcysteine, in transaminations, **16:**195
5-Benzyl-2,4-diaminopyrimidines, folate antagonists, **27:**445–447
N′-Benzyl dihydronicotinamide, **15:**93
Benzyldisulfide, in disulfide exchanges, **27:**230
Benzylglutaric acid, **57:**67–68
Benzyl group, binding energy, **43:**334
Benzyl groups, as nucleoside protectants, **22:**162
removal from phosphate, in nucleotide coenzyme synthesis, **22:**160; 161
Benzylhomocysteine sulfoxide (BzSO), antagonist of glutamic acid, **13:**256–257
S-Benzyl-β-mercaptopyruvic acid, in transaminations, **16:**195
Benzyl mustard oil, occurrence and properties, **10:**265
Benzyloxy carbonyl-L-alanyl-glycyl-L-phenylalanine Chloromethyl Ketone (ZAGPCK), **41:**182; 225
Benzyloxy carbonyl-L-alanyl-L-phenylalanine chloromethyl Ketone (ZAPCK), **41:**182; 225
Benzyloxycarbonyl-α-aminoadipamylglycine, **54:**25
Benzyloxycarbonyl-D-glutaminylglycine, **54:**21; 23
Benzyloxycarbonyl-L-glutaminylglycine, **54:**17
as transglutaminase substrate, **54:**23; 25–26
Benzyloxycarbonyl-α-methyl-D-glutaminylglycine, **54:**21; 22; 25
Benzyloxy carbonyl phenylalanine chloromethyl Ketone (ZPCK), **41:**225
N-Benzyloxycarbonyl phenylalanyl chloromethylketone, **43:**341
Benzyloylcholine, **55:**143
Benzylpenicillin, as substrate for β-lactamase, **28:**239; 241; 244; 245; 249–253; 257–260; 262; 267–269; 271–272; 285; 287–294; 300–302; 306; 310–311
5-Benzylpyrimidines, inhibitor, structure-activity relation, **27:**446–447; 449–451
Benzylsuccinic acid, **57:**24; 40
L-Benzylsuccinic acid, **43:**354
Benzylsulfoxide, **43:**287
Benzyl thiocyanate, **46:**399
2-Benzylthioethylamine, in synthesis of pantetheine, **16:**6

Benzyl viologen, as artificial electron acceptors, **30:**23
B^6 enzymes, *See* Pyridoxal-5′-phosphate-dependent enzymes
Berberine
biosynthesis, **8:**242
distribution in *Berberis darwinit,* **8:**233
metabolism in plants, **16:**142
Berezin, **51:**48
Bergmann ratios, **4:**2; 3; 10
Bestatin, **68:**161
binding to leucine aminopeptidase, **68:** 184–185
complexes, **68:**177–178
interactions with leucine aminopeptidase, **68:**172
P_1 hydroxyl oxygen, **68:**180; 182
β/α barrel, **67:**17–18; 48–49
Betabacterium, in cacao fermentation, **25:**472
β-Bulge, **50:**35
Betagarin, **55:**91
Betain, **15:**127
Betainaldehyde, **15:**118; 127
Betaine
in air-cured tobacco, **6:**348
aldehyde, oxidation, in mitochondria, **21:**22; 24
and ethionine effects, **25:**372; 374
as methyl donor, **25:**423–426
in plants, **8:**234
lipotropic effect, **8:**192
Betaine aldehyde dehydrogenase
classification, **14:**254
and ethionine metabloism, **25:**372
localization of, **24:**298
in mitochondria, **21:**22
Betaine aldehyde r DPN transhydrogenase, classification, **14:**254
Betaine-homocysteine methyltransferase, **25:**424–426
Betaine r homocysteine transmethylase, **14:**250
Betaines. *See also* Methylation, biological
and biological methylation, **12:**444; 446; 449–450; 453–455; 458–469
biological oxidation of, **12:**481
in plants, **9:**439–442; 444
N-substituted, formation, **16:**112
Betanechol, **55:**176
β-Replacement reaction, **42:**288
β-*Replacement-Specific Pyridoxal-P-Dependent Lyases, The* (Braunstein and Goryachenkova), **56:**1
β-sheet, **47:**49; 50; 53; 59–60; 64; 65; 66; 81; 117
antiparallel, **47:**55; 59–60; 117; 133
assignments, **47:**51; 55; 66
boundary, **47:**50; 53; 67; 69–70; 99; 101; 113; 116
breakers, **47:**50–51; 53–54; 67; 83; 101; 111; 113
carbonic anhydrase, **47:**153; 158; 161
content, **47:**64; 117
crossover, **47:**134
formers, **47:**49–51; 54; 66–67; 79; 83
length, **47:**53–55; 59–60; 100
nucleation, **47:**53; 54; 101; 105; 121
parallel, **47:**59; 87
potential, **47:**50–51; 110; 115
prediction, **47:**49
propagation, **47:**53
in proteins, **47:**75–79; 81–85; 110; 118–119
termination, **47:**53
α(β transition, **47:**76; 79; 82–83; 135–136
β(α transition, **47:**74; 136
β-Sheets, **50:**23
β-Sheet structure, hydrogen bonds in, **50:**25
β-Structure, **42:**103
dipoles in, **50:**25
β-turn, **47:**55; 60–64; 65; 70; 71; 88; 120; 127; 130; 131
bend profile, **47:**126–127
bend types, **47:**61–64; 134
classification, **47:**134
computer predictions, **47:**124
content, **47:**64–65
cutoff value, **47:**106–107; 109
environmental analysis, **47:**72–74
frequencies, **47:**70–71; 125
probability, **47:**108–109; 121
in proteins, **47:**61–63; 85; 120; 122–127; 130
Betavulgarin, **55:**91
Beta vulgaris, **55:**91
BGH, **42:**74
BH_4
dependent activity, lysolecithin-mediated, **67:**102–104
6-diastereoisomers, **67:**174

BH_4 (*Continued*)
and phenylalanine hydroxylase inhibition, **67:**218–219
phenylalanine hydroxylase interconversion effects, **67:**141–145
tyrosine hydroxylase, **70:**163–165; 181; 198; 200
BH_{4-}, **39:**342; 362
Biacetyl, *See* Diacetyl
Bi Bi mechanism
ordered, **46:**401
for PP-ribose-P synthetase, **41:**4
Bicarbonate, **35:**343, **51:**243. *See also* Carbon dioxide
carbonic anhydrase, binding to, **47:**186–187; 236–237
dehydration catalyzed by, **47:**178–186; 224
exchange in gastric mucosa, **13:**57–58
labeled, and acetate metabolism, **23:**426–428
in propionic acid metabolism, **26:**284; 285; 312
in purine biosynthesis, **21:**203; 204; 213; 215; 216; 232
radioactive, in study of purine biosynthesis, **21:**214
Bicarbonate (C^{14}), in aspartate synthesis, **26:**293
in propionic acid metabolism, **26:**286; 301–302; 314; 320–331; 334–337; 342–343; 345
Bicarbonate (^{14}C), labeling of pentoses by, **28:**414–417; 431–433
Bicarbonate diffusion, **34:**512
ions, **34:**512
Bicarbonate ion
and amination of pyruvate, **19:**345
effect on aminoacyl mercaptans, **19:**241; 244; 245
influence on citric acid synthesis, **10:**215
Bicarbonate (O^{18}), ATP in carboxylations, **26:**316
Bicarbonate (O^{16})(O^{18}), photosynthesis, **25:**42
Biformin, **7:**469; 487; 489; 491; 493
Bifunctional enzyme, **59:**341–344
Bifunctional reagent, **34:**453
Bifurcated hydrogen bonds, **65:**224–226
Bilayer leaflet, **34:**470
membrane, **34:**467–468; 483–485
Bile, **46:**107; 142
Bile acids, **60:**48
Bile acids and absorption of vitamin K, **2:**309
Bile-free diet and K-avitaminoses, **2:**290
Bile pigments, **7:**321; 338–339, **19:**195–198, **46:**38; 39; 105; 114
destruction, **7:**337–338
glycine as precursor, **18:**258
from iron protoporphin, **7:**333–337
Bile salts. *See also* Cholate; Deoxycholate
binding of colipase to, **48:**349; 354; 356; 358
binding of lipase to, **48:**356; 358; 359
effect on interfacial denaturation of lipase, **48:**350
effect on lipase adsorption, **48:**347–348; 349; 359; 364
inhibition of lipase by, **48:**341; 343–347
interaction between lipase and, **48:**356
role of, in lipase inactivation by diethyl p-nitrophenylphosphate, **48:**360
cholesterol ester hydrolases, **69:**55
in cytochrome isolation, **18:**10
in preparation of succinic dehydrogenase, **18:**70–71; 108
effect on mitochondrial oxidative phosphorylation, **21:**32
effect on succinic oxidase system, **20:**175
and enzyme isolation, **14:**380
role in lipolysis, **23:**150–152
solubilizing effect on fatty acids, **2:**270; 274
Bilins, **46:**64–68
Bilirubin, **46:**38; 64; 65; 102; 107; 119; 139; 140; 142; 146; 148; 162; 171; 384; 392; 393; 407
conjugation, **46:**407
as ligand, **46:**391
peroxidase action on, **14:**151
serum levels in cyanate treated sickle-cell patients, **40:**24; 25
storage proteins, **46:**407
structure, **46:**65
Bilirubin diglucuronide, **46:**392; 407
Bilirubin glucuronide transesterification, **46:**407
Bilirubin monoglucuronide, **46:**392; 407
transesterification, **46:**407
Bilirubinoids, **7:**362–363
Bilirubin-transferase B complex, **46:**391
Biliverdin, **19:**195, **46:**38; 64; 88; 107; 146; 174; 184
formation, **46:**65
precursor from catalase, **1:**167; 171

Bimolecular reaction, diffusion-controlled, **39:**354
Binding
of chitin, **59:**61
of fibroblastic cells
cells in suspension, **59:**21–23; 25
hepatocytes, **59:**25
monolayers, **59:**16–21
platelets, **59:**25–26
of quinoprotein oxidoreductases, **59:**194–195
wrong-way, **39:**354
Binding affinity, oligosaccharides, and solution conformation, **66:**76–77
Binding capacity, **34:**521
Binding constants, **58:**41
effect of glutathione, **46:**393
Binding energy, **43:**222; 329
estimation, **43:**330
of glucose, **43:**233
of glucose-1-phosphate, **43:**233
relation to catalytic efficiency, **33:**436–437
small groups, **43:**334
Binding enthalpies, **43:**320
Binding proteins, **49:**59
essentiality in transport, **65:**31–33
interaction of conserved component, **65:** 30–31
periplasmic, **65:**3
Binding sites. *See also* Active centers and Active sites; Active sites; Allosteric sites; Regulatory sites
of allosteric enzymes, **28:**50–51; 53–56; 58–60; 68; 73; 75; 80–83; 120–121; 123; 129; 130; 133; 136–137; 140–141
in antigen-antibody action, **24:**115
of enzyme-substrate complex, **24:**116–117
pyridine coenzymes, **22:**369
TPD, **22:**340; 341; 345
yeast, ADH, **22:**354
Binding site topography, **46:**256
Binding specificity, **33:**403–404; 430, **53:**263
of pepsin, **33:**413
Binding Specificity, of Hypoxanthine-guanine phospho-ribosyltransferase, **41:**23
Binuclear cu_A models, mononuclear *versus*, cytochrome *c* oxidase, structural biochemistry, **71:**93–99
Bio A, **35:**336
Bioadsorption, **46:**233–236
Bio B, **35:**336
Biochanin A, **55:**5
aryl migration in formation of, **55:**31
Biochemical accompaniments, ADA deficiency, **51:**170–174
Biochemical changes, effect of ionizing radiations, **24:**381–383
Biochemical organization, **24:**291–358
Biochemical similarity principle, **27:**390; 394
Biocytin, **35:**324; 338; 349
in carboxylations, **26:**332
and CO_2 fixation, **12:**55
molecular measurements, **35:**334
Biocytin (C^{14} and C^{12}), in carboxylations, **26:**331
Biodegradability of alkylbenzene sulfonates, **27:**498
of hydrocarbons, relation to structure, **27:**473–474
Bioelution, **46:**236–239
abortive complex, **46:**239
adduct, **46:**239
binary complex, **46:**237
coenzyme inhibitor, **46:**237
ternary complex, **46:**239
Biogenesis, alkaloid, **8:**203–251
and diamine oxidase, **8:**238; 239
Biogenesis, Alkaloid (Dawson), **8:**203
Biogenetic isoprene rule, **22:**321
Biogenic amines, **44:**228; 231
Biological activity, **42:**83
Biological activity of soil, enzyme methods, **28:**365–390
Biological activity of tobacco mosaic virus, **15:**468–472
Biological Antagonisms Between Structurally Related Compounds (Woolley), **6:**129
Biological control. *See also* Product inhibition of enzymes
mechanism of, **25:**170; 173
Biological distribution, **39:**335–339
Biological function, **44:**213
Biological membrane, **34:**456–457; 467; 483; 492–494
Biological Methylation (Greenberg), **25:**395
Biological oxidation of carbohydrates, *See* Carbohydrate metabolism
Biological regulation of lactate dehydrogenases, **37:**115–124
Biological specificity, **37:**607

Biological systems
complex, **57:**193–195
state and function of copper in, **33:**177–244
Biological Systems, State and Function of Copper in (Malkin and Malmström), **33:**177
Bioluminescence, **7:**215; 216; 225; 229, **25:**131–161. *See also* Chemiluminescence, Light emission, and Luminescence
bacterial, **25:**144–156
source of oxygen in CO_2 release, **44:**42
use of ATP system, **14:**14–17
Bioluminescent spectra, effect of zinc on, **44:**45
Bioluminscent process, **51:**18–19
Biomass, bacteria action on hydrocarbons, **27:**472–473
Bioperin binding, eukaryotic dihydrofolate reductase, **70:**32–33
Biopolymers
mechanism of growth of, **30:**179–221
propagation reactions of growth of, **30:**181
structure of, **30:**181
origin of, **38:**223; 226
Biopterin, **35:**36; 53
biosynthesis of, **35:**69
dihydrofolate reductase, **70:**28
eukaryotic dihydrofolate reductase ligand binding, **70:**50–52; 60
structural changes due to derivatization with, **70:**46–47
isolation, structure, synthesis and properties, **21:**377
H_4-Biopterin
mechanistic role, **65:**326
NO synthesis enhancement, **65:**302–304; 319
redox role, **65:**328
synthesis, **65:**312–313
Bioreductive alkylating agents, **65:**118–119
Biospecific sorption, **57:**209
Biosyn Process of protein synthesis, **9:**654
Biosynthese induite des enzymes (Monid and Cohn), **13:**67
Biosyntheses, Industrial, Fats (Hesse), **9:**653
Biosyntheses, using ATP system, **14:**14–17
Biosynthesis
calabolic role, **42:**291
induction by tryptophan, **42:**289
repression by glucose, **42:**289
synthetic action, **42:**291
by cell-free system from *B. brevis,* **33:** 343–350
directed by protein molecules, **7:**103
optically selective, **7:**84
and genic control, **8:**205
of alkaloids, **8:**204–206
of berberine, **8:**242
of glucose-1-phosphate, **34:**294
of glycogen
from glucose, **34:**294
by glycogen synthetase, **34:**316
of gramicidin S, **33:**337–378
by *B. brevis,* **33:**339–343
of β-lactamase, control of, **28:**272–274
mechanism of, **33:**367–375
oligosaccharides, **52:**108–113
of ornithine, **33:**373–375
arginase in, **33:**374; 375
arginine desiminase in, **33:**374
ornithine transcarbamylase in, **33:**374
via glutamic-γ-semialdehyde, **33:**374
via *N*-acetylated derivatives, **33:**374
by plants, role of phenolase complex in, **16:**136–153
of amino acids, **16:**247–312
procollagen, **42:**179
of starch, **34:**356
of teichoic acids, **30:**234–245
Biosynthesis
of Enzymes, Aspects of the (Chantrenne), **24:**1
of Gramicidin S *(Saito, Otani and Otani),* **33:**337
Biosynthesis by acyl phosphates, **1:**152–154
Biosynthesis of Arginine (Vogel and Vogel), **40:**65
Biosynthesis of ATP, **23:**327–387
Biosynthesis of coenzyme Q, **25:**295–304
Biosynthesis of enzymes, **24:**1–33
induced, **13:**67–119
Biosynthesis of fats, **9:**653–704
Biosynthesis of Riboflavin, Folic Acid, Thiamine, and Pantothenic Acid (Brown and Williamson), **53:**345
Biosynthesis of Rubber (Archer and Audley), **29:**221
Biosynthetic activities
carbamyl phosphate synthetase, **72:**131
regulation of bacterial form glutamine synthetase, **72:**33–35

Biosynthetic capacity of nucleus, **27:**272–275
Biosynthetic pathway for biotin, **38:**321
based on isotopic studies on glutamic fermentation, **38:**321
based on isotopic studies on mutant technique, **38:**321
Biosynthetic pathways, regulation of, **28:**44–46; 143
regulation of branching, **28:**45–46
Biosynthetic reactions, pyrophosphorylases and phosphorylases in, **18:**191–240
Biosynthetic role of the tricarboxylic acid cycle, **18:**241–280
Bio-thrombin, from prothrombin, **16:**51–52; 62
Biotin, **35:**332–342; 347–357, **43:**315; 331; 332, **46:**114, **48:**10, **56:**94
absorption spectrum, **3:**299
active carbon, **26:**326–330
carboxylation of, reaction mechanism, **26:**316–320
formation of holocarboxylase, **26:**331–334
mode of action, **26:**305–307
in propionic acid metabolism, **26:**284; 291; 292
in propionyl-CoA carboxylase, **26:**287
antibiotin or avidin, **3:**305–307
assay procedures, **3:**291–292
attachment to apoenzyme, **35:**334
avidin and, **56:**95
bibliography, **3:**311–313
biochemistry, **3:**289–292; 305–311
biosynthesis, **35:**334–337
and cancer, **3:**309–311
in cancer and normal tissues, **4:**240; 248
N-carboxybiotin, **35:**347
carboxylase (BC), **35:**363; 369
carboxylation, of free biotin
avian liver acetyl CoA carboxylase, **35:**348
biotin carboxylase, **35:**348
carboxyl transferase, **35:**348
β-methylcrotonyl CoA carboxylase, **35:**347
specificity patterns, **35:**349
ureido nitrogen or oxygen as carboxylation site, **35:**351
carboxylation of, **23:**338
chemistry, **3:**294–304
and CO_2 fixation, **12:**55; 81–85
crystallization, **3:**294
deficiency, **3:**307–308, **35:**337
dependent enzymes, **35:**325–332
classification, **35:**325–326
reactions, **35:**325–326
carboxylases, first partial, **35:**328–329
second partial, **35:**330
decarboxylases, **35:**331–332
enzyme-CO_2, formation and reactions, **35:**327
transcarboxylases, **35:**331
derivatives, absorption spectra, **4:**278–279
effect on purine biosynthesis, **21:**216
elementary composition, **3:**294
enzymatic attachment to the apoenzyme, **35:**337–338
and fat formation, **8:**310
destruction during autoxidation, **8:**440
formation of, **54:**227–228
free and bound biotin, **3:**304
functional groups, **3:**295–296
in genetic studies, **13:**136–138
as growth factor, **6:**38
inhibitory analogues of, **6:**132; 144
identification of carboxylation site for free biotin, **35:**347
identity with vitamin H, **3:**289–291
in induction of allantoin enzyme production, **56:**109
isolation, **3:**292–294
loci on *E. coli* DNA for biosynthetic enzymes, **35:**336
occurrence in nature, **3:**304–305
optical activity, **3:**294
by ordered mechanism, **35:**363
pathway from 2-C units, **35:**335
side chains, **3:**296–298
site carboxylation enzyme-bound biotin, **35:**348
solubility, **3:**294
stereochemistry, **35:**332
structure, **3:**301–304, **35:**332
sulfur ring, **3:**300
translocation between catalytic sites, **35:**328; 364
transport system, **35:**337
urea ring, **3:**298–299
and urea synthesis, **15:**342
x-ray crystallographic analysis of, **35:**332
Biotin: Biogenesis, Transport, and Their Regulation (Eisenburg), **38:**317
Biotin, Chemistry and Biochemistry (Hofmann), **3:**289

Biotin action in yeast, **7:**454
Biotin-activating enzyme, in carboxylations, **26:**333–334
Biotin (C^{14}), in carboxylation experiments, **26:**326–334
Biotin-containing carboxylases, allosteric enzymes, **28:**55–56; 89–92
Biotin-containing enzymes, **45:**379
  kinetic mechanism of, **45:**291
Biotin enzymes
  and acetoacetate, **26:**296–297; 303–304; 333
    and carboxylation, **26:**289; 292–311
    carboxylation of, **26:**320–330
    reaction mechanism, **26:**311–330
    tracer mechanism, **26:**311–316
  and acid catalyzed proton exchange, **59:**299
  and base catalysis of proton exchange, **59:**299
  and enolization, **59:**298
  and enzymatic mechanism of carboxylation, **59:**301
  and exchange of NH protons, **59:**298–299
  model reactions, **59:**297–298
  and non-concerted carboxylation and proton transfer, **59:**301–302
  polypeptide composition of, **42:**21
  and sulfur, **59:**301
  and transannular interaction of sulfur, **59:**298–299
Biotinidase, **35:**415–416
  release of biotin, **26:**325–326; 328; 332
  source, **35:**415
  substrate specificity, **35:**415
Biotin mutants, classification of, **38:**324
  based on crossfeeding experiments, **38:**326
  based on excretion pattern of the biotin vitamers, **38:**326
Biotin oxidase, **19:**203
Biotins, in plant metabolism, **10:**227; 247–250
  relation to cysteine in plants, **10:**269
Biotin vitamers, and related compounds, **38:**319
Biotinyl adenylate, in holocarboxylase formation, **26:**332–334
Biotinyl-5′-AMP, **35:**340
Biotinyl hydroxamate, and carboxylations, **26:**333
Biotransducers, of membrane-mediated information, **50:**303–320
Biotransducers of Membrane-Mediated Information (Brady and Fishman), **50:**303
BiP chaperone, protein disulfide isomerase (PDI) prolyl 4-hydroxylase β subunit function, **72:**378–380
Biphasic increase, in absorbance at 550 m μ, **34:**46
Biphenyl, effect on carotenogenesis, **21:**320
Biphenyl ether structure, **34:**213
Biphenylglucuronide, as substrate for β-glucuronidase, **16:**370
Biphenyl-type linkages, **34:**211
2,2′-Bipyridyl, **54:**133
Birds, methylated lysine and arginine derivatives in, **42:**235
Birefringence, streaming, of tobacco mosaic virus, **15:**455
Birefringence of protoplasm, **7:**38
Birefringence of various cell structures, **8:**8–14; 18–19; 21–23; 35–36
Birefringence of virus solutions, **9:**196–197
Bis-β-carboxyethyl disulfide, exchanges, **27:**214; 215
Bis-(3-carboxy-4-nitrophenyl)-disulfide, measurement of thiols, **27:**231–232
Bischloroethylnitrosourea, **55:**273–274
1,3 Bis(2-chloroethyl)-1-nitrosourea, *See* BCNU
Bismuth and melanin pigmentation, **14:**94
Bismuth mercaptide, **11:**216
Bis(*p*-nitrophenyl) phosphate, velocity of hydrolysis at 37°C of, **31:**43
Bis-Noryangonin, **55:**22
2,3-Bisphosphoglycerate (2,3DPG), **51:**211–253
  activators of the hydrolysis of, **51:**214
  enzymatic reactions, **51:**212–215
    mutase, **51:**212–213
    synthase, **51:**213–215
  introduction to, **51:**211–212
  kinetic properties, **51:**230–244
  overview of reaction mechanism, **51:**244–248
  phosphorylated intermediate, **51:**219–223
  purification and properties of, **51:**215–219
  rate-limiting steps, **51:**223–226
  red blood cell and, **51:**248–249
  structural studies, **51:**226–230
    active-site phosphopeptides, **51:**228–229
    chemical modification, **51:**229–230
    molecular structure, **51:**226–228
Bisphosphoglycerate synthase (DPGM), **51:**213–215; 249
  active-site peptides, **51:**228
  amino acid composition, **51:**217

purification from red cells, **51:**217
Bis-Tris, **51:**10
Bistrispropane, **51:**10
Bisulfite, effect on yeast metabolism, **9:**693
Bisulfite reaction of pyridine nucleotides, **15:**85; 93; 94–95
Biuret reaction, as denaturation test, **23:**37
Bixin destruction by lipoxidase, **8:**440; 452–453
BKA, *See* Bacillus aneurinolyticus
Blackman period, **1:**206–209; 213; 215
Blackman reaction, **1:**205; 320
Blackman's master reaction, **7:**6
Blanching conditions, effect on flavors of frozen vegetable tissue, **9:**614–624; 631–637; 637–649
Blasticidin S, **55:**44
Blastocladiella, **59:**76
Blastocrithidia culicis, **46:**102
Bleaching effect, in algae, **46:**99
Bleomycin, **58:**231
Blind staggers, **48:**4
S-blocked glutathiones, **59:**110; 130
Blocking reagents, in identification of imidazole, **20:**68; 71; 72–82; 90; 101
Blood
anticoagulants of, **5:**47–55
coagulation, **16:**23–103
effect of rare earth metal ions on, **21:**192
coagulation of, **5:**31–5:65
activator system in, **5:**34–41
and calcium, **5:**33–34
classical theory, **5:**33
controlling factors in, **5:**32
and fibrin, **5:**33; 43–45
and fibrinogen, **5:**32; 33; 43–45; 49
and heparin, **5:**32; 49–54
inhibition of, **5:**47–55
and lipide inhibitors, **5:**54
mechanism of, **5:**55
and prothrombin, **5:**33; 34; 41–42
role of —SH groups, **11:**245
substances with coagulating properties, **5:**45–47
and thrombin, **5:**32; 33; 34; 42–43
codehydrogenases in, **5:**230
hemophilic, **5:**35; 36
oxidation-reduction in, **5:**140
plasma, **5:**33; 35
platelets, **5:**34
pyruvate in, **5:**139–143
water exchange with intestinal fluid, **13:**47–48
whole, clotting mechanism in, **30:**303
Blood, Coagulation (Chargaff), **5:**31
Blood, Coagulation (Seers), **16:**23
Blood and phosphatide turnover, **7:**146–147
Blood-brain barrier, **29:**483, **38:**310, **49:**91; 92
penetration by 3,4-dihydroxyphenylalanine and 5-hydroxytryptophan of, **29:**483
prevention of uptake of amines by, **29:**483
in schizophrenia, **29:**508; 519
Blood cells, enzyme synthesis in, induction of, **27:**455
Blood changes, in familial periodic paralysis, **6:**215–218
in myasthenia gravis, **6:**212
in myotonia, **6:**213
in nutritional muscular dystrophy, **6:**206–207
in progressive muscular dystrophy, **6:**222–223
Blood clotting, **10:**1–49
effect of egg jelly coat on, **9:**263; 313
effect of sulfhydryl groups, **9:**278–279
effect of peroxidase on, **14:**158
effect of tyrosinase on, **14:**142–144; 147–150
factors *(See also Fibrin, Fibrinogen, Heparin, Prothrombin),* adsorption studies, **14:**346
mechanism of, and enzymology, **30:**255–315 (*See also* Fibrinogen, Factors, and Cascade)
and snake venoms, **8:**489–490
effect of concentrations of components, **8:**108
role of enzymes, **8:**461; 489–490
Blood Clotting and Related Processes (Astrup), **10:**1
Blood Clotting Mechanisms, Enzymology and the (Esnouf and Macfarlane), **30:**255
Blood coagulation
contact system of, **48:**281
initiation of, **48:**281–283
intrinsic pathway of, **48:**279–280
vitamin *K*-dependent proteins in, **46:**3–7
Blood coagulation factors, nomenclature, **48:**278
Blood flow, **57:**292–294
Blood group, A_1, 144, A_2, **36:**144
Blood group H antigen, **60:**160
Blood Groups, **41:**136
type A determinants, **41:**136
type B determinants, **41:**136

Blood groups, **52:**46
agglutination studies, **15:**55–56; 58–64
galactose in, **20:**119
Blood groups in man, **36:**131–147
Blood Groups in Man, Enzymatic Basis for (Ginsburg), **36:**131
Blood group specificity, glycoprotein linkages, **25:**446–450
Blood group splitter, **23:**186
Blood group substances
reaction with borate, **12:**509–513
synthesis of, **26:**76–78
Blood lipids, *See* Lipides of blood
Blood plasma, and vitamin E synergies, **6:** 498–503
Blood platelets, *See* Platelets
Blood pressure
homeostasis, **65:**310; 318
NO role, **65:**310
Blood proteins, in schizophrenia, **29:**501–515
Blood sugar control, **25:**173
Blood sugar level, regulation, **28:**95–98
Blood thromboplastin, use of term, **16:**29
Blood-type specificity, **48:**56
Blowfly, *See* Calliphora erythrocephala
Blue chromophore, **44:**202
Blue-copper proteins, cytochrome *c* oxidase, **71:**129
Blue cupric ion, *See* Cupric ion, blue
Blue proteins, EPR study, **27:**138; 144; 147
Blue radical, **34:**64
Blue-shift, **50:**308
Blue substance from catalase, **1:**170–172; 174
B lymphocytes, **70:**5
B-Lytic protease, from *Sorangium,* **41:**207; 237
BMM, *See* Bacillus thiaminolyticus
Body temperature, effect of fluoroacetate on, **18:**139
Body water, and fatty acid synthesis, **8:**324–330
and sterol synthesis, **8:**331
Bohr effect, **7:**271; 278
hemoglobin-catalyzed translocation of oxygen, **29:**44
translocation of protons due to, **29:**44
Bohr magneton numbers, of cytochrome *c* peroxidase, **33:**320
Boltzmann constant, **34:**461
Bombardia, genetic studies of, **13:**126
Bombyx mori, **45:**402; 403; 405; 407
galeae of, **45:**406
phenoloxidase in, **45:**418
prephenoloxidase-activating enzyme from, **45:**419
Bond angles, and protein structure, **23:**5–11; 22–23
Bond-changing function, of amino acids in enzymes, **22:**69; 70; 90; 91
Bond energies, **11:**13
Bond energies in proteins, **9:**81
Bond energy, and protein structure, **23:**5–11
Bonding, ionic type, **34:**452
Bond length, and protein structure, **23:**5–11; 22–23
Bond sensitivity and peptidase action, **12:** 228–229
Bond strength, effect on peptidase action, **12:**235–238
of carbon isotopes, **12:**70–71
Bond type, in enzyme-substrate compounds, **12:**157–159; 178–181
in metal-protein complex, **12:**198–200; 205; 212–216.
Bonellia viridis, **46:**68
Bonellin, **46:**68–69
Bone marrow, **46:**39; 142; 158
human, **47:**357
terminal deoxynucleotidyl transferase, **47:**357
Bone marrow cells, lipoxygenase mRNA in, **58:**250
Bone marrow lymphocytes, **47:**361
Bone marrow transplantation, ADA deficiency and, **51:**196–197
Bone tissue, radiophosphorus turnover, **7:**113
Borate
γ-glutamyl transpeptidase catalytic sites, **72:**255–262
reaction with substances of biological interest, **12:**493–527
Borate, Reaction with Substances of Biological Interest (Zittle), **12:**493
Bordetella pertussis, **58:**131
Born salvation energies, **34:**493
Bornyl glucuronide, as substrate for β-glucuronidase, **16:**368–369; 379; 380
Borohydride, **35:**462; 473; 474, **39:**328; 343
hydrogen transfer model, **26:**137–139; 141
Borohydride trappong, glutamine synthetase acyl-P intermediate, **72:**26–28

Boron, **51:**147
  and nitrogen metabolism in plants, **9:**437–438
  in sperm, **24:**80
Botrytis cinerea, polygalacturonases of, **20:**366; 368
Botrytris cinera, **55:**45
  capsidiol metabolism of, **55:**83
Bottom layer, **5:**10
Boundaries, protein, **34:**454
Boundary conditions, **34:**496–500
Boundary stabilization model, microtubule dynamics, **69:**124
Bound ligand motion, eukaryotic dihydrofolate reductase, **70:**64–65
Bound water, eukaryotic dihydrofolate reductase, chemical transformation step (proton donation), **70:**89–90
Bovine factor XII_a (activated Hageman factor), **48:**284
Bovine growth hormone, **35:**267
Bovine kidney enzyme, crystalline, **39:**339; 350
Bovine lactalbumin, **57:**184–185
Bovine liver enzyme, **39:**334; 364
  crystalline, **39:**339; 350
Bovine liver rhodanese, crystalline, **39:**336; 348
  molecular weight of, **39:**350
Bovine plasma amine oxidase, **59:**193
Bovine prolactin mRNA, methylation, **65:** 262–264
Bovine RNase, physicochemical parameters of, **24:**177–181
Bovine serum albumin, **34:**475. *See also* Serum albumin
  adsorption at interfaces, **28:**13; 15; 20
  effects of chemical modification, **24:**141–152
    ionizing radiations, **24:**381–383
  rate constant for enzyme-substrate reaction, **25:**14
    rate of change in, **25:**20–21
  reaction with mercaptans, **27:**210; 230–231
  SH-groups, **27:**212; 221–226
    and viscosity, **27:**222
Bovine Serum albumin, stimulation of prolyl hydroxylase by, **41:**255
Bovine submaxillary gland mucoprotein (BSM), composition and structure of prosthetic group, **20:**138–144
BPF, **57:**25
BPP_{5a}, **57:**25–27; 35
BPP_{9a}, **57:**25
B.R. factor, **18:**320
Brachiose from amylopectin, **12:**389
Bradykinin, **23:**199, **56:**355, **57:**5; 17; 25
  analogs, hydroxylation of by prolyl hydroxylase, **41:**261; 262
    inhibition of prolyl hydroxylase, **41:**268
  peptide sequences in, **24:**425–430
Bradykinin in blood, **10:**39
Bradykinin potentiating factor, **57:**25
Brain
  cholinesterase in, **8:**463–464
  phosphorus metabolism, **7:**126–127; 143–145
  tyrosine hydroxylase, **70:**106–108
Brain ammonia extraction
  and blood flow, **57:**292–294
  entry mechanism of, **57:**295–296
  results of, general, **57:**290–292
Brain ammonia metabolism
  compartmentation of, **57:**301
  and hypercapnia, **57:**301–302
  inference of, **57:**296–297
  investigation of, **57:**297–299
  routes of, **57:**299–301
Brain hexokinase, and glycolysis, **39:**307
  kinetic mechanism, **39:**294
Brain phosphatides, **7:**165; 170
Brain tissue, glutamine and glutamic acid in, **13:**246–248; 278
  glutohydroxamic acid formation in, **13:** 298–299
  permeability, **13:**62–63; 271
Brain waves, **7:**47
Branched-chain amino acid aminotransferase, **74:**134; 141
  phylogenetic distribution of, **74:**157
Branched reaction scheme, **57:**379
Branch formation, in polysaccharide synthesis, **11:**298–299; 301; 304–306; 310; 314; 318; 321; 324–329; 340–341; 361–370
Branching
  in β-1,3 glucans, **28:**327; 328; 331–332; 334–337; 340
  of polyribonucleotide, **16:**437
  of polysaccharide chains, **3:**120; 126
  in polysaccharides, effect on enzymic action, **11:**345–346; 347–348; 359–360
    factor, in polysaccharide synthesis, **11:** 298–299; 304–306; 326; 350; 361–370
Branching enzyme, in polysaccharide synthesis, **15:**304; 307–309; 310; 311

Branching enzymes, **18:**287, **34:**351
action, on amylopectin, **34:**357
on amylose, **34:**357
on branched molecules, **34:**361
algae, **34:**361
chain segment transferred by, **34:**356–357
properties, **34:**354
redistribution of 1 → 6 linkages by, **34:**361
Branching factor, action of, **12:**423
Brassica oleracea, **55:**19
Bray, Robert C., **51:**107–165
Breadmaking, and wheat amylases, **6:**443–463
Brevibacterium flavum, use of acetate, **23:**411; 412
Bridase, elastase from, **23:**173
Brilliant cresyl blue
effect on mitochondrial oxidative phosphorylation, **21:**31
and reduced pyridine nucleotides, **15:**89; 98
Broccoli, antibiotic treatment of diseases, **20:**391
Bromelain, **34:**449, **53:**279, **56:**210
action on collagen, **23:**176
action on, collagen, **23:**176
elastin, **23:**226; 228; 229
specificity of, **33:**433
Bromelain inhibitor, **47:**127–130
Bromelin, **10:**300
classification, **14:**247
requiring –SH group, **11:**241
Bromide, **51:**243
complex with carbonic anhydrase, **47:** 225–226; 233; 236; 238
effect on fluorescence of riboflavin, **20:**32
permeability of gastic mucosa, **13:**56
Bromide content in blood, in myotonia, **6:** 213–214
Bromine, effect on pectin substances, **20:**352
Bromoacetate
effect on muscle, **13:**162
reaction with thiols, **11:**213–214
Bromoacetic acid, in labeling active site of ribonuclease, **22:**53
N-Bromoacetylethanolamine, **51:**229
m-Bromobenzaldoxime, **43:**380
Bromobenzene, **7:**194
5-Bromodeoxyuridine
dehalogenation of, **54:**74
mutagenic agent, **27:**315–318
2-Bromoethylamine, **67:**44–45
2-Bromo-*d*-lysergic acid diethylamide (BOL148) in schizophrenia, **29:**519
6-Bromo-2-naphthyl-β-D-glucopyruronoside, for tissue localization of β-glucuronidase, **16:**385–386
Bromoproteins, **7:**196–197
Bromopyruvate, **55:**214
Bromosuccinamide, **56:**209
N-Bromosuccinimide, **39:**346, **58:**151
Bromosulfophthalein, **46:**385; 394; 396; 397
binding, **46:**405
dissociation constant, **46:**391
2-Bromotetradecanoic acid, CoA thioesters, **67:**401
5-Bromouracil
mutagenic agent, **27:**296; 302; 305–306; 315–318; 331; 335–342
replacement of thymine, **18:**164
5-bromouridine, synthesis of gramicidin S, inhibitor of, **33:**340
Bronsted-Bjerrum equation, **43:**388
Brönsted equation, disulfide exchanges, **27:** 202–203
Bronsted value, **59:**134
Browning, in plants, **16:**150–153
Browning reaction, imidazole in, **20:**93–95
Brown rot, in lignin study, **18:**351–354; 362–366
Brucella, in adsorption studies, **14:**364
Brucella abortus
and citric acid cycle, **18:**252; 253
transamination in, **16:**204
Brucellus abortus, metabolism in, **74:**10
Bryophyllum calcynum, heptulose formation in, **15:**162
Bryophyta, carotenoids in, **21:**300
BSM, *See* Bovine submaxillary gland mucoprotein
BSO, cancer chemotherapy, glutathione synthesis, **73:**252–255
Buffer effect, on luminescence, **7:**226; 258
on respiration, **7:**258
Buffering system of tobacco, **10:**352–354
Buffer ions, effect on LADH complexes, **27:** 73–75
Buffers, **35:**492
effect on stability of pyridine nucleotides, **15:**90
effect on succinic dehydrogenase, **18:**87–90
pH values, **51:**10

Bufo marinus oocytes, **46:**149
Bufotenin, in schizophrenia, **29:**527
in urine of schizophrenics, **29:**529
Bumblebee, **42:**222
Burning properties of tobacco, **10:**356–357
Burn Treatment, by collagenase, **23:**219–221
Burst, in transient-state kinetics of glutamate dehydrogenase, **39:**397; 409
Burton's equation, **7:**6; 20
Busycon, polygalactan production by, **20:**119
Butan-2,3-diol, cleavage, with metal ions, **21:**177
2,3-Butanedione, **54:**132; 133; 139
t-Butanol, **43:**292
Butanol-acetone fermentation, **7:**560–573; 578
maintenance of sterility, **7:**566–568
of materials other than molasses and grains, **7:**564–566
protection against phage, **7:**563–564
and vitamin-rich by-products, **7:**569–573
n-Butanthiol-S^{35}, in disulfide exchanges, **27:** 201
Butter yellow, **46:**389
Butylamine, reaction with quinones, **16:**114
Butylbenzene, tertiary, on hapten, **15:**67
4-*t*-Butylcatechol, **58:**31
2,3-Butylene glycol by fermentation, **7:**578; 601–611
tert-Butyl group, for protecting carboxyl groups of amino acids, **29:**442
t-Butylmercaptan, in disulfide exchanges, **27:**204
sec-Butyl mustard oil, optically active, **10:**264
tert-Butyloxycarbonyl group, method of introducing into amino acids of, **29:**397
resistance to hydrogenolysis of, **29:**398
susceptibility to acid catalysis of, **29:**398
tert-Butyloxycarbonyl hydrazine, preparation of, **29:**397
sodium salts reacted with, **29:**397
Butylphenacyl alcohol, **58:**28
Butylphenyl acetate, **58:**18
Butylphenyl groups, **58:**19
Butyl thiophosphate, catalyzed hydrolysis, **22:**83
Butyraldehyde, oxidation, **15:**256; 257; 258
Butyrate, **15:**238; 249
and acetate metabolism, **23:**431–432
Butyrate activation, **18:**208
Butyribacterium rettgeri, **15:**214
and lipoic acid, **18:**320; 342
Butyribacterium rettgeri, superoxide dismutase from, **41:**79
Butyric acid, **34:**489
biosynthesis of, **12:**109
fermentation, hydrogen inhibition of, **6:**247
synthesis, **6:**271; 300
metal ion complexes of, model catalyses with, **21:**189
Butyrivibrio fibrisolvens, superoxide dismutase from, **41:**80
λ-Butyroβine hydroxylase, **41:**270
Butyrophenones, **55:**179
extrapyramidal side effects of, **55:**181
Butyryl adenylate, **19:**258
Butyryl CoA dehydrogenase, in electron transfer, **19:**201
Butyryl-CoA dehydrogenase, in soluble fraction of liver, **21:**49
Butyryl coenzyme A, **15:**233; 236; 237
Butyryl dehydrogenase, spectra, **22:**390; 391
and substrate addition, **22:**391; 392
Butyryl phosphate, and acetyl phosphate, **6:**238
Butyryl-S-CoA dehydrogenase, **23:**423
Byssochlamys fulva, polygalacturonases of, **20:**366

C

C. perfringens, allosteric enzymes, **28:**79
Ca^{2+}, effect on rate of poly(glycerol phosphate) synthesis, **30:**236
Cabbage butterfly, *See* Pieris brassica
Ca^{2+}/calmodulin, **65:**307–308
Cacao, browning during drying, **25:**482–483
curing, mechanism of, **25:**457–492
enzymes in, **25:**465–468
factors affecting quality, **25:**484–488
fat content, **25:**461; 469
flavors, **25:**458–460; 479; 480; 484–488
fractionation of components, **25:**461–462
microbiology of pulp, **25:**470–475
seed, fermentation death, **25:**476
unfermented, biochemistry of, **25:**460–469
Cacao Curing (Forsyth and Quesnel), **25:**457
Cachexia, cancer, **54:**167
C_4 acids, and CO_2 fixation, **12:**46; 47; 49–50
from acetate, **12:**117
Cacodylate, **51:**10
Cadaverin, **2:**93; 95; 96; 98; 102–108

Cadaverine, **54:**42–43, *See* 1,5-Diaminopentane
oxidation, in mitochondria, **21:**22
Cadaverine production, **6:**4
^{14}C-Adenosine, use by tumor cells, **28:**433
^{14}C-Adenylic acid, DR synthesis, **28:**420
Cadmium ($^{115}Cd^{2+}$), binding to LADH, **27:** 79–80
Cadmium ion, effect on peptidases, **12:**222
in decarboxylation, **12:**80
Cadmium ion in creatine kinase activation, **54:**404; 405; 406
Cadmium ions
chelate formation with amino acids, **21:**145
effect on mitochondrial oxidative phosphorylation, **21:**31
enolase activation by, **21:**152
model catalyses with, **21:**174
properties, **21:**157
effect on elastases, **23:**234
Cadmium-LADH, absorption spectrum, **27:** 79–80
Cadmium mercaptide, **11:**216
Cadmium-mercaptide chromophore, Cotton effect, **27:**80
CAD protein
carbamyl phosphate synthetase, **72:**125–131
thioacylenzyme intermediate, **72:**178–179
Caenorhabditis elegans, **70:**3
Caeruloplasmine, multiple forms, **25:**359–360
Caffeic acid
esters of, **55:**9
formation of, **55:**39
in plant oxidase studies, **16:**132; 133
metabolism in plants, **16:**141; 144; 145; 150
Caffeic acid *O*-methyl transferase, **55:**39
Caffeine
cacao content, **25:**464; 469
mutagen, **27:**331
Ca-fibrin, *See* Calcium fibrin
Calciferol. *See also* Vitamins D
Calcification
effect of elastase, **23:**247
during hardening of cuticle, **16:**153
mitochondrial participation in, **29:**313
renal, administration of parathyroid extracts in, **29:**313
Calcineurin (Klee, Draetta and Hubbard), **61:**149
Calcium, **34:**490–492
activation of adenosinetri-phosphatase, **8:**45
and bactericidal activity of fresh serum, **8:**111
and fat formation, **8:**308–309; 315
and functioning of cytoplasm, **8:**7–8; 17; 19–20; 24; 45; 47; 50
and lysis, **8:**94–95
and opsonic activity, **8:**111
and activation of egg, **9:**292–297; 308–310; 313; 317–318
and cytoplasmic structure of egg, **9:** 292–294
effect on oxygen consumption of egg, **9:**308–310
and fertilizability, **9:**248; 249; 263–264 279
in plant metabolism, **9:**397
and plasmolysis of fertilized egg, **9:** 282–283
as activator of enzymes, **15:**39–40; 258
as inhibitor of enzymes, **15:**347
requirement for clotting of milk, **15:**430; 432; 434; 436; 438–445
adenosine receptors, **69:**99
annexins, **71:**214–215
antagonism with Mg, **25:**18–20
effect on luminescence, **25:**148–149
and product inhibition, **25:**231–232
in blood coagulation, **5:**33–34; 42; 45
inhibition of action in, **5:**48
Caulobacter cell cycle, flagellar structure and assembly, **71:**11
complex with actin, **11:**231
ions, effect on membrane permeability, **11:**71
formation of complexes, **11:**71
exopolysaccharide alginate synthesis, **70:**235
ion channel model system annexin V, **71:**221; 222 (*See also* Ion channel model system annexin V)
in nucleus, **27:**256
and phosphoribulokinase activity, **19:**21
in transamination reactions, **19:**354; 370
and stability of trypsin, **20:**285
binding, by pectin substances, **20:**355; 356
by protein groups, **20:**56
stimulating agent for ATP breakdown, **4:**217; 220; 221; 223
storage, ATP driven, **34:**475
transport, **34:**474

tyrosine hydroxylase regulation, **70:**140–142; 143; 144; 145; 150; 162; 178
Calcium and luminescence, **7:**253
Calcium ATPase, **65:**185–210
amino acid sequence, **65:**194
ATP utilization mechanism, **65:**192
calcium binding localization, **65:**202–208
calcium binding, **65:**202–206
vectorial translocation, **65:**206–208
catalytic domain localization, **65:**196–202
nucleotide binding, **65:**196–198
phosphorylated enzyme intermediate, **65:**198–200
spatial relationships, **65:**200–202
conservative mutations, **65:**199
enzyme activation, **65:**192
hydropathy plots, **65:**194–195
linked functions, **65:**208–210
sarcoplasmic reticulum, **65:**186–196
catalytic and transport cycle partial reactions, **65:**192–194
experimental systems, **65:**186–192
topology, **65:**194–196
stable expression, **65:**189; 191
transient expression, **65:**189–190
Calcium binding-protein inhibitors, **65:**308
Calcium (bis-*p*-nitrophenyl) phosphate, as substrate for acid DNase, **31:**33
Calcium caseinate, **15:**435–436
Calcium channel, **65:**207–209
Calcium chloride, effect on actomyosin dissociation, **16:**345
effect on relaxation of muscle, **16:**348–351
Calcium dichloride, phosphomannomutase/phosphoglucomutase (PMM/PGM), **70:**230
Calcium fibrin, relation to thrombin fibrin, **16:**43–44; 92
Calcium hydroxyapatite, in liver and heart mitochondria, **29:**275
Calcium in semen, **24:**79–80
Calcium ion
complexed with ADP, **54:**373
in creatine kinase activation, **54:**403
in fructose-bisphosphatase regulation, **54:**163
in transglutaminase reactions, **54:**5; 6
with *p*-nitrophenyl acetate, **54:**8
Calcium ions, **48:**278
activation of alkaline phosphatase by, **10:**99; 102–104; 110
binding to proteins, **10:**14
and blood clotting, **10:**2; 10; 11; 12; 13–15; 20; 24; 25; 26
in citric acid cycle, **10:**201; 215
effect on activation of trypsinogen, **10:**12
effect on tryptophanase action, **10:**71
in aldehyde condensation, **14:**172
protection of α-amylase by, **14:**399
alkalinization of mitochondria during uptake of, **29:**280–281
efflux from mitochondria, **29:**273
mitochondrial uptake of, **29:**264–267
and amylase activity, **12:**400
and choline acetylase activity, **12:**318–322
in decarboxylation, **12:**80
effect on peptidases, **12:**217; 220; 222
chelate formation with amino acids and organic phosphates, **21:**145–148
effect on magnesium ion-activated enzymes, **21:**153; 155; 156
enzyme activation by, **21:**153; 159
in mitochondria, **21:**10
model catalyses with, **21:**175
properties, **21:**157
effect on, collagenase, **23:**195–196; 202
denaturation, **23:**39–41; 51–58
elastases, **23:**234
elastomucase, **23:**233
lipase action, **23:**152–153
mitochondria, **23:**352; 360; 366; 368
effect on action of myosin as an enzyme, **16:**336–337; 338; 339
effect on muscle proteins, **16:**332; 335; 352
effect on permeability of nuclear membrane, **16:**487
in isolation of cell nuclei, **16:**415
role, in blood clotting, **16:**33; 43–44; 87
in prothrombin activation, **16:**25; 49–55; 61–62; 66; 69; 91–92
and fluorocitrate, **18:**146–148
with glycogen synthetase, **28:**99; 101; 102
in phosphorylase conversions, **28:**107–109
muscarinic receptors and, **55:**145; 147
and muscle contraction, **13:**193
effect on cell permeability, **13:**60
on action, **13:**152–153; 155
on actomyosin and ATP interaction in gels, **13:**175–176
on actomyosin and ATP interaction in sols, **13:**171–173

Calcium ions (*Continued*)
on actomyosin solutions, **13:**182
on ITPase, **13:**168
on myosin activity, **13:**164–166
on sarcoplasm ATPase, **13:**167–168
and myofibrillar structure, **13:**186–187
myosin binding of, **13:**157
role of, in transglutaminases, **38:**133
transport of, **55:**387
Calcium metabolism, effect on ATPase, **6:**159; 164–165
in familial periodic paralysis, **6:**217; 218
in muscular dystrophy, **6:**205; 222
in myasthenia gravis, **6:**212
in myotonia, **6:**213
Calcium oxalate, **57:**131
precipitation, **57:**129–130
Calcium pantothenate, and oxidation of hydrocarbons, **27:**477
Calcium phosphate
as adsorbent in isolation of clotting factors, **16:**27
for column chromatography of enzymes, **22:**445; 446; 457
effects on bacterial luminescence, **7:**258
in mitochondria, **29:**275
Calcium phosphocaseinate micelles in milk, **15:**437–440; 443
Calcium salts, effect on gelation of pectin substances, **20:**357
Calcium uptake, accumulation of ADP (ATP) during, in presence of phosphate, **29:**282
effect of phosphate on, **29:**269–270
ejection of K^+ during, **29:**281
H^+ ejection during, **29:**278–280
movements and effects of ions accompanying respiration-linked, **29:**277–285
oxidative phosphorylation and, as alternate processes, **29:**270–271
during resting respiration, **29:**272–273
stoichiometry between ATP hydrolysis and, **29:**271
stoichiometry in the absence of phosphate, ATP, and magnesium ions, **29:**268–269
superstoichiometry, **29:**282–285
Calcofluor White M2R New, **59:**73–74
Calf thymus polyuncleotide kinase, **48:**271
Calliphora erythrocephala, **45:**400
acid protease in, **45:**401
Callose, as β-1,3 glucan, **28:**329; 337; 338; 340–343
Calmodulin, **55:**147, **58:**258
annexin crystal structure, **71:**217
Calmodulin antagonists, **65:**308
Calonectria decora, enzymic transformations of steroids by, **20:**246–261; 262
Calories, relation of production to life span, **20:**22–24
Calorimetric determinations, **51:**6–8
Calorimetric techniques, **34:**468
Calorimetry, and thermodynamic functions, **11:**12–13
Calvin-Bassham cycle, **54:**169
Calvin cycle, **48:**163; 172, **54:**175; 176
in photosynthesis, **25:**95, **30:**3
cAMP, **46:**124; 125; 131; 159, *See* Adenosine 3,5′-cyclic monophosphate
cAMP-CAP complex, *Caulobacter* cell cycle, flagellar structure and assembly, **71:**9
cAMP (cylic adenosine monophosphate) system, **59:**213–214. *See also* Cyclic adenosine monophosphate-dependent regulation of 2′,5′-oligoadenylate system
cAMP dependent protein kinase, *See* Cyclic adenosine monophosphate-dependent protein kinase
cAMP-dependent protein kinase, *See* Protein kinase, cAMP-dependent
cAMP-dependent protein kinases, **36:**69
Camphor, **43:**331
Camphor binding, **47:**31; 32
D-and L-Camphocarbonic acid, decarboxylation by model enzyme system, **15:**27
CaM-PKII, tyrosine hydroxylase regulation, **70:**142–143; 144; 145; 147–148; 152; 153; 154; 155; 159–160; 163; 177; 182
CAMP receptor, **43:**42
Canaline, **50:**141; 142; 168
Canavalia ensiformis, **55:**37
Canavanine, **15:**361–362; 375; 379, **35:**482, **50:**121; 136; 137; 139; 141; 142; 157–160; 166; 168; 183
as amidine donor, *in vivo,* **50:**201
as amidine donor *in vitro,* **50:**183
effect on arginine synthesis, **29:**151
inhibition of bacterial growth, **29:**131
naturally occurring arginine analog, **29:**131
as substrate for arginine sRNA synthetase, **29:**132

effect on β-lactamase, **28:**275
Canavaninosuccinate, **50:**141
Canavaninosuccinic acid, **15:**361–362
Cancer. *See also* Cancer virus; Carcinogenic agents; Carcinoma, Hepatoma, Malignant cells, and Tumor; Carcinomas and Tumors; Enzymes; Enzymes and enzyme components; Growth control; Tumor enzymology; Tumors
abnormal proteins as cause, of, **4:**223; 228
alkylation of nucleic acids, **27:**320
chemotherapy, **27:**418
and biological energy transformation, **4:** 201–250
and biotin, possible relationship to, **3:** 309–311
cachexia with, **54:**167
control mechanisms, **4:**202; 203; 209; 218–227
Cori cycle in, **54:**167
correlation with optical activity, **20:**24
fluorinated pyrimidines in treatment of, **54:**58
fluorouracil treatment, **25:**400
fructose-bisphosphatase in, **54:**166–168; 181
glucose metabolism in patients with, **54:**168
and growth factors, **57:**163–164
and GSH transferase, **57:**399
nature of, five postulates on, **4:**202
and nitrogen-13 ammonia, **57:**313–314; 329–332
polyamine levels in blood and urine, **36:**228
prostatic, and acid phosphatase, **9:**352
relation to β-glucuronidase, **16:**394; 397; 399–400
a theory of, **4:**228–232
and vitamin E, **6:**507
Cancer chemotherapy, **47:**340. *See also* Glutathione synthesis
BSO use in, glutathione synthesis, **73:** 252–255
cystamine, inactivation by, glutathione synthesis, **73:**255–257
relation to folate transport, **47:**340
resistance, glutathione synthesis, **73:**233–235
Cancerous growth
and changes in blood and serum, **3:**341–344
diagnosis of, **3:**340–345
Cancerous tissue. *See also* Melanomas
amino acids in, **14:**265
Cancerous tissue, protein synthesis in, **8:**61
Cancer Problem, Biological Energy Transformations and (Potter), **4:**201
Cancer theory, free radicals, **15:**6
Cancer therapy, glutathione synthesis and, **72:**241–242
Cancer virus, **4:**228; 229; 230; 231
Candicidin, in treatment of plant diseases, **20:**399; 403
Candida, oxidation of hydrocarbons, **27:**472; 474
Candida albicans, **54:**63
methylthioadenosine isolation in, **54:**199
Candida albicans:
and chitinase activity, **59:**91
and chitin synthetase, **59:**66; 68; 72; 75; 79
Candida guillermondia in riboflavin production, **7:**572
Candida guilliermondii, oxidation of hydrocarbons, **27:**472
Candida intermedia, oxidation of hydrocarbons, **27:**472
Candida lipolytica, oxidation of hydrocarbons, **27:**472; 499; 501; 502
Candida parapsilosis, oxidation of hydrocarbons, **27:**472
Candida tropicalis, oxidation of hydrocarbons, **27:**472
Candida utilis, **37:**465
Candida utilis, **46:**60, **54:**169
pentose metabolism, **28:**405
tracer studies, **28:**408–409; 428–429
Candida utilis, PN-dependent enzymes, **26:**119
Candida utilis, transaldolase from, **31:**138
Capacitance, **34:**467
Capillary endothelium, **49:**91
Cap methylation, **65:**257–259
Caproate, and acetate metabolism, **23:**431–432
Caproic acid synthesis, **6:**300
ε-Caprolactam, configuration of, **23:**14–15
Capsanthin, structure, **21:**305
Capsicum annum, **55:**69
Capsidiol
biosynthesis of, **55:**74
formation of, **55:**69; 73
metabolism of, **55:**83
structure of, **55:**54
Capsinone, **55:**54; 83
Capsular polyglutamate synthesis, **44:**110
L-glutamate racemization, **44:**111

Capsular polysaccharides, adaptation of enzymes for, **13:**70
Captan, compatibility with antibiotics, **20:**386; 396; 400
Captopril
design of, **57:**28–31
and hypertension, **57:**7
as inhibitor of angiotensin-converting enzyme, **57:**2; 9; 46
proline residue in, **57:**31
and purification procedures, **57:**9
and SHR, **57:**7
structure of, **57:**2
Captoten(r), *See* Captopril
Carbamate kinase, **37:**365, **39:**7–8
location in subcellular fraction, **24:**309; 313
metabolic function of, **39:**7
properties of, **39:**7
reaction equilibrium of, **39:**7
Carbamatse formation, role in intramolecular aminolysis, **19:**245–247
Carbaminohemoglobin, formation of, **74:**37–38
Carbamoyl-β alanine, **50:**155
Carbamoylation, **58:**80
Carbamoyl phosphate synthetase, **53:**221; 222; 223
Carbamoylphosphate synthetase
glutaminase ammonia amplification system, **72:**69–70
glutamine synthetase and hepatocyte heterogeneity, **72:**60–61
Carbamoyltransferases, location in subcellular fraction, **24:**309
Carbamyl-β-alanine, in pyrimidine biosynthesis, **21:**283; 284
Carbamyl-β-alanine ribotide, in pyrimidine biosynthesis, **21:**284
Carbamyl arsentate, **15:**382
Carbamylaspartate, *See* Ureidosuccinate
in pyrimidine biosynthesis, **21:**264; 266–274; 277; 283
radioactive, in study of pyrimidine biosynthesis, **21:**266; 267
synthesis and metabolism, in mitochondria, **21:**52; 266; 272
Carbamylaspartate decarboxylase, **21:**284
Carbamylaspartic acid, cyclization of, **28:**135
Carbamylated red cells, biological function of, **40:**15
Carbamylation, as phosphorylase reactions, **18:**227–228
Carbamylcholine, **55:**139
muscarinc binding of, **55:**162
structure of, **55:**140
tyrosine hydroxylase induction, **70:**168
Carbamylglutamate, **15:**N^{15}); 339; 342; 364; 364 (C^{14}; 370–372; 382
in arginine synthesis, **16:**265–267
Carbamyl oxamate, phosphorolysis of, **23:**338
Carbamyl phosphate, **15:**382, **18:**231, **42:**29, **50:**141
in ATP synthesis, **23:**324; 325; 338; 373
binding to aspartate
binding sites, **68:**93–94
transcarbamylase, **68:**82–83; 85–86
biosynthesis, allosteric regulation, **28:**66–69
decomposition of, **39:**47; 48
in arginine synthesis, **39:**5
equilibrium of, **39:**47
formation of, **39:**47
mechanism of reaction, **39:**129
occurrence of, **39:**48
partial reactions catalyzed by, **39:**126
properties of, **39:**47
in pyrimidine synthesis, **39:**5
sources, **39:**124; 125
subunit structure of enzyme, **39:**126
synthetase I, **39:**125
synthetase II, **39:**125
alkylation of, by glutamine analogs, **39:**126
conformational states of enzyme, **39:**127
ligands, effect on structure, **39:**128; 129
monomeric forms, **39:**128
sulfhydryl groups, titration of, **39:**128
control of enzymatic activity, by nucleotides, **39:**125
by ornithine, **39:**125
from glutamine, **40:**66
in arginine biosynthesis, **40:**66
substrate for ornithine carbamoyltransferase, **40:**72
in mitochondrial carbamylaspartate synthesis, **21:**52
in pyrimidine biosynthesis, **21:**264; 267–273; 284
as a phosphate donor for phosphotransferase activity of glucose 6-phosphatase, **34:**308
reaction with aspartate, **68:**68–69

Carbamyl phosphate synthetase, **50:**384–389, **66:**228–238, **72:**124–131
amide transfer, **66:**230–231
catalytic mechanism, **66:**228–233
column chromatography, **22:**484
domain map, **66:**232
E. coli, relation to
carbamyl phosphate synthetase I of urea cycle, **66:**236–238
other monofunction enzymes and CAD, **66:**233–236
relationship to anthranilate synthetase, **38:**30
repressibility, **40:**75; 76
structural relationships with different forms, **72:**133
structure-function, **66:**228–233
type II enzymes, **72:**125–131
type III enzymes, **72:**125–131
Carbamylphosphate synthetase, inactivation of, **27:**216
Carbamyl phosphate synthetases, pathway specificity of, **39:**6
Carbamyl phosphokinase, **39:**124
N-Carbamylputrescine, **36:**228. *See also* 1,4-Diaminobutane
Carbamylputrescine, in biosynthesis of putrescine in plants, **36:**230. *See also* 1, 4-4Diaminobutane, carbamylation of
Carbamyl transfer, **18:**192
in pyrimidine biosynthesis, **21:**273; 274
Carbanion, in pyruvate decarboxylation, **18:**335–336
Carbanion-DPT coordination complex, in α-keto acid decarboxylation, **15:**218–219
Carbanionic intermediates, oxygenation reactions, **67:**10–11
Carbanyl phosphate synthetase
from *E. coli,* **39:**10
allosteric regulation, **39:**13
molecular weight, **39:**11
partial reactions, evidence for, **39:**11
of subunits, **39:**12
rule of glutamine, **39:**12
SH groups, relation to activity, **39:**13
subunit composition, **39:**11
from liver
conformational requirements, **39:**8
from frog liver, **39:**8
mechanism of action of, **39:**9–10
partial reactions, evidence for, **39:**9
from rat liver, **39:**8
role of *N*-acetylglutamate in, **39:**8–9
Carbazole, lack of reactivity with *p*-benzoquinone, **16:**112
Carbenolamine, **54:**75–76
Carbethoxylation, **60:**260
Carbinolamine adducts, **57:**100–101; 105
Carboanhydratase, multiple forms, **25:**362
Carbobenzoxyamino acid derivatives, in synthesis of activated amino acids, **19:**249; 251
Carbobenzoxyamino acids. *See also* specific compounds, eg, Carbobenzoxyglycine
in anilide and phenylhydrazide syntheses, **19:**308
in synthesis of activated amino acids, **19:** 240
Carbobenzoxy-L-aspartic acid, effect on papain, **19:**319; 320
Carbobenzoxy chloride, carbonic anhydrase catalyzed hydrolysis, **47:**189
Carbobenzoxy-L-glutamic acid, effect on papain, **19:**306; 318–320
Carbobenzoxy-L-glutamic acid diamide, papain action on, **19:**304–306
Carbobenzoxy-α-L-glutamyl-L-tyrosine, as substrate for pepsin, **30:**147
Carbobenzoxyglycine, in anilide synthesis, **19:**308
Carbobenzoxyglycylglycine, papain action on, **19:**325
Carbobenzoxyglycyl-L-phenylalanine, in Anderson test for racemization, **29:** 410–411
Carbobenzoxy group, in peptide synthesis, **29:**395–397
Carbobenzoxy-L-histidinamide, papain action on, **19:**304; 325
Carbobenzoxy-L-isoglutamine, papain action on, **19:**304–306
Carbobenzoxylation, without racemization, **29:**394
Carbobenzoxy-L-leucinamide, papain action on, **19:**304–306
δ-*N*-Carbobenzoxyornithine, in transaminations, **16:**195
Carbobenzyloxy-β-alanyl azide, in synthesis of pantetheine, **16:**6
N-Carbobenzyloxy-*O*-tosyl-L-serine methyl esther, **33:**384

N-Carbobenzyloxy-*S*-trityl-DL-cysteine methyl ester, **33:**384
Carbodiimide method, for adenosine-5′-sulfatophosphates synthesis, **22:**192
for amino acid adenylates synthesis, **22:** 190–191
for cytidine diphosphate choline synthesis, **22:**177
for cytidine diphosphate glycerol synthesis, **22:**189
for cytidine diphosphate ribitol synthesis, **22:**189–190
for DPN and TPN synthesis, **22:**188
for FAD synthesis, **22:**188
for nucleoside-5′-polyphosphates, **22:**186; 187
for uridine diphosphate glucose synthesis, **22:**188
Carbodiimides, **34:**450–453
in acid anhydride synthesis, **18:**212
general structure, **22:**175
reactions with phosphates, **22:**177–184
ε-*N*-Carboenzoxylysine, in transaminations, **16:**195
Carbohydrase, **18:**288; 289
mechanism of action, **15:**3
multiple forms, **25:**350
Carbohydrase activities of algae, **28:**341
Carbohydrase isozymes, **28:**352
Carbohydrases. *See also* Glycosidases
acting on barley β-glucan, **28:**341–343
classification, **4:**69–72, **14:**220; 224
column chromatography, **22:**472–477; 486
nomenclature, **4:**72
titrimetric methods, **9:**601
Carbohydrases in tobacco, **6:**366–368
Carbohydrate
breakdown, mechanism in *Fusaria,* **5:** 174–176
phase sequence in, **5:**179–181
degradation, **5:**180; 181
metabolism, oxidation, **5:**129; 130; 136; 137
phosphorylating and nonphosphorylating mechanisms, **5:**136; 137
relationship to protein and fat metabolism, **5:**154–159
Carbohydrate, Biological Oxidation (Kerbs), **3:**191
Carbohydrate anabolism, *See* Carbohydrate metabolism
Carbohydrate and Fat Catabolism (Witzemann), **2:**265
Carbohydrate catabolism, *See* Carbohydrate metabolism; Fat and carbohydrate catabolism
Carbohydrate chains of glycoproteins, heterogeniety among, **36:**134; 147
Carbohydrate degradation, distribution of tracer, **28:**402–404
Carbohydrate fermentation, **3:**151–163. *See also* Carbohydrate metabolism
adenosine and guanidine phosphates, role in, **3:**154–156
equilibrium of enzymatic glycogen hydrolysis, **3:**153–154
glycogen, breakdown to pyruvate, **3:**151–160
phosphorylative and non-phosphorylative fermentation, **3:**156–160
pyruvate decomposition, anaerobic, **3:** 161–163
pyruvate formation as first phase of, **3:** 151–160
Carbohydrate ligands, **66:**42
Carbohydrate mesogens, **65:**246
Carbohydrate metabloism, in luminous bacteria, **25:**151
Carbohydrate metabolism. *See also* Carbohydrate fermentation; Carbohydrate metabolism, theory; Carbohydrate oxidation
allosteric regulation, **28:**71–118
mammalian cells, **28:**76–80; 83–86
yeast cells, **28:**77–83
alternative pathways, **28:**116–118
anabolism, **3:**174–176
antiketogenesis, **3:**246–247
basic principle of experimentation, **3:**192
bibliography, **3:**184–189; 248–252
catabolism, **3:**151–163
enzymatic reactions *in vitro* and in living cells, **3:**172–174
and fat metabolism, **3:**246–247
glucogenesis from carbohydrates, **3:**245–246
glutathione, role in, **3:**176–178
inhibitors of, **3:**194–196
intermediary stages, **3:**191–252
interrelationship of pyridine nucleotides and, **29:**341
mechanism, **3:**149–189
metabolic activities in animal tissues, **3:**218

metabolic quotient, definition, **3:**196
methods of experimentation, **3:**192–217
muscle as an experimental material, **3:**193
in muscle tissue, **3:**199–217
orientation of reactions, **3:**180–183
oxidative breakdown, **3:**163–174
Pasteur effect, **3:**180–183
in plants, **19:**1–79
and protein metabolism, **3:**245–246
reaction schemes, **3:**217–244
regulatory mechanisms, **3:**178–180
related problems, **3:**244–248
role of pantothenic acid in, **3:**266
role of vitamin B_{12}, **26:**270
summary, **26:**37–39
synthesis of carbohydrates, **3:**174–176
theory of (*See also* Tricarboxylic acid cycle)
citric acid cycle, **3:**223–235
earlier theories, **3:**221–223
general principles, **3:**217–221
oxidation *via* acetate, **3:**242–244
Szent-Cyörgyi cycle and tricarboxylic acid cycle, **3:**236–238
Carbohydrate metabolism and adrenal cortical hormones, **2:**344
Carbohydrate metabolism and cysteine, **7:**392
Carbohydrate Metabolism (Barron), **3:**149
Carbohydrate metabolism in muscle tissue
biochemical reactions of metabolites, **3:** 199–217
of acetate, **3:**214–215
of acetoacetate, **3:**215
of citrate, **3:**200–202
of α-keto glutarate, **3:**202–205
of oxaloacetate, **3:**205–208
catalytic effects, **3:**216
metabolites, survey, **3:**196–199
transaminations, **3:**215–216
Carbohydrate oxidase, **19:**202
Carbohydrate oxidation, **3:**163–174. *See also* Carbohydrate metabolism
enzymatic reactions *in vitro* and in the living cell, **3:**172–174
phosphorylation, aerobic, **3:**170–172
pyruvate oxidation and reversible red. ox. systems, **3:**167–169
via pyruvate, **3:**164–169
without previous fermentation, **3:**163–164
Carbohydrate pool, of plants, **18:**288–291
Carbohydrate-protein interactions, in cells, **20:**93; 94
Carbohydrate-protein linkage in glycoproteins, **25:**450–453
Carbohydrate racemases, **44:**71
Carbohydrate resynthesis, **1:**143–148
Carbohydrates, **57:**12. *See also* Amino sugars and Photosynthesis; individual carbohydrates
abiogenic syntheses, **27:**385–386
and origin of life, **27:**356–357; 359–361; 363; 368
absorption in depancreatized dogs, **8:**180–181
and fat formation, **8:**300–301; 302–305; 306–314; 322; 325–330
synthesis, and cell structure, **8:**56
and chloroplasts, **8:**56
terminal oxidation, **8:**412–414
in air-cured tobacco, **6:**327–330; 349–353
concentration of esters in myasthenia gravis, **6:**212
in flue-cured tobacco, **6:**364
metabolism in progressive muscular dystrophy, **6:**222
oxidation in tobacco leaves, **6:**358–362
synthesis, **6:**271
polysaccharides, **6:**299–300
as antigens, **15:**67
citric acid cycle, **15:**183–270
metabolism, alternate pathways, **15:** 141–182
role of pyridine nucleotides, **15:**113–122
synthesis of polysaccharides, **15:**301–318
biological oxidation of, **3:**191–252 (*See also* Carbohydrate metabolism)
cacao, **25:**468; 477; 479; 481–482
in carbon reduction pathway, **25:**94–103
changes during fermentation of tobacco, **10:**340; 363–364; 396; 397; 401–406
combination with lignins, **16:**147
in blood-clotting factors, **16:**29; 30; 46–48; 51
link to amino acid metabolism, **16:** 226–227
metabolism, in cell nuclei, **16:**488–490
complex, biosynthesis, **28:**110; 113–114
effect on amino acids during hydrolysis, **14:**269
effect on protein synthesis, **9:**421–423; 425–426; 431–433

Carbohydrates (*Continued*)
fermentable, inhibition of indole production by bacteria, **10:**74–81
fluorinated, **55:**240–248
formation of aromatic compounds from, **18:**362–370
in plant respiration, **18:**282–291; 302–303
relation to lignification, **18:**351; 355; 359–370; 375
free energy of formation, **25:**46–48; 53–54; 102–106
metabolism, and fat synthesis, **9:**692–700
in nucleic acid, **9:**470; 484–485
organic acids from, **9:**540–580
phosphorus-containing, turnover studies, **7:**113–127
in plant metabolism, **10:**224–225
respiration in plants, **9:**618–621
synthesis
genetic control, **26:**67–78
role of sugar nucleotides, **26:**35–88
of viruses, **9:**230–231
Carbohydrates, Surface, **41:**99–127
in hepatic recognition, **41:**99–127
in transport of circulating glycoproteins, **41:**99–127
Carbohydrate sulfates, occurrence, **22:**206
Carbohydrate Synthesis, and Sugar Nucleotides (Ginsberg), **26:**35
Carboligase, **7:**93
Carboligase reaction, mechanism of catalysis, **14:**170
Carboligases, classification, **14:**256
Carbolyases, classification, **14:**228; 256
1′-*N*-Carbomethoxybiocytin (C^{14}), hydrolysis by biotinidase, **26:**328
N-Carbomethoxybiotin methyl ester, synthesis of, **26:**318
Carbomoylphosphate synthase, **37:**424; 449; 491; 510; 543; 574
Carbon
isotopic, retention by DNA, **16:**481
radioactive, in study of methionine synthesis, **16:**272
labeled (*See also* compounds)
in determination of galatose-l-phosphate uridyl transferase activity, **20:**114
in identification of active centers of esterases, **20:**288
in studies of amino acid metabolism in plants, **20:**321
in study of uridine diphosphoglucose pyrophosphorylase, **20:**122
in photosynthesis, **25:**42
Carbon-14
in citric acid cycle, **18:**267–269
in lignification studies, **18:**360–361; 364; 366–370; 372–374
labeling of acetate, in cholesterol biosynthesis, **22:**282–285
Carbon 13. Entries will be found under specific compounds.
Carbon 14. Entries will be found under specific compounds.
Carbon (14), in *E. coli* metabolism, **16:**460
in study of amino acid metabolism, **16:** 227–228; 269; 290; 301
in transamination study, **16:**191; 223
Carbon (13), in transamination study, **16:**223
Carbon-13, labeling of acetate, in cholesterol biosynthesis, **22:**282; 283
Carbonaceous chondrites, composition of, **27:**350–352
Carbonaceous material, organized, in meteorites, **27:**393
Carbon acids (C_3 and C_4), resynthesis in plants, **18:**302–303
Carbon Anhydrase: Structure, Catalytic Versatility, and Inhibition (Pocker and Sarkanen), **47:**149
Carbon assimilation, **6:**272–287
Carbonate activation, **18:**210
Carbonate esters, carbonic anhydrase catalyzed hydrolysis, **47:**188; 206–207; 209
Carbonate hydrolyase, *See* Carbonic anhydrase
Carbon (C^{14}), in glyoxylate cycle, **23:**404–409
Carbon-carbon condensation, in pentose synthesis, **28:**395–397; 412; 428
Carbon-carbon hydrolases, classification, **14:**225; 249
Carbon (^{14}C) compounds, folate metabolism, **27:**457
Carbon chain, cleavage, with metal ions, **21:**177; 182
Carbon (C^{14})-labeled, acetate, use by sperm, **24:**89
amino acids, use by sperm, **24:**88–89
carboxymethylribonuclease, **24:**214
fatty acids, in sperm, **24:**65

galactose, in sperm metabolism, **24:**62
glucose, in sperm study, **24:**48; 74
glycerol, utilization by sperm, **24:**54; 74; 91
glycine, use by sperm, **24:**91
lecithin, sperm metabolism, **24:**58
octanoate, sperm metabolism of, **24:**65
Carbon compounds, evolution of, **27:** 349–354
4-Carbon compounds, from acetate, **23:** 413–415; 425–426; 429–430
2-Carbon compounds, metabolism by microorganisms, **23:**401–470
Carbon (C_1) pool, in coenzyme Q synthesis, **25:**296–297
4-Carbon dicarboxylic acids, inhibition of isocitratase, **23:**415–420; 422
in synthesis, **23:**403–404; 413–415
Carbon-dihalogenases, classification, **14:**248
Carbon dioxide, **43:**315. *See also* Active carbon dioxide; Bicarbonate
and acetate metabolism, **23:**421; 423–429; 432–434
labeled, and acetate metabolism, **23:** 426–429; 432–434
and glycine, **23:**458–460
and glycolate, **23:**441
and oxalate, **23:**449–450
activation by ATP, **19:**42
and amination of pyruvate, **19:**345
in arginine synthesis, **16:**260; 264–265
incorporation into pyrimidine nucleus, **16:**450–451
assimilation, **6:**298–299
fixation in lactic acid, **6:**248–250
in pyruvate carboxyl, **6:**246
synthesis, **6:**300
assimilation of by ferredoxins, **33:**138–143
bacterial assimilation (fixation) of
in acetate, **74:**34–36
C_3 and C_1 addition in, **74:**14–34
by carbon to carbon linkage, with both components from carbon dioxide, **74:** 36–37
and formation of formic acid, **74:**12–14
and formation of methane, **74:**14
involving carbon to carbon linkage, **74:** 14–37
in lactate, **74:**34–36
miscellaneous reactions in, **74:**34–37
not involving carbon to carbon linkage, **74:**12–14
phosphoribulokinase in, **74:**96
and biological methylation, **12:**469–474; 480–481
exchange in acetic acid, **12:**114–115
fixation, **12:**41–151
by chloroplasts, **12:**32–37
carbonic anhydrase, binding to, **47:**186–187; 236–237
hydration catalyzed by, **47:**176–177; 181–186; 194–195; 206–207
in chemosynthesis, **74:**6
dark fixation, **26:**27
as degradation product of pectins, **20:**349
in biosynthesis of glutamic acid in plants, **20:**327
labeled, in studies of biosynthesis of pectic substances, **20:**369
effect on activated amino acids, **19:** 245–247
effect on glycyl phosphate, **19:**250
effect on sperm metabolism, **24:**49–51; 83
eukaryotic assimilation of, phosphoribulokinase in, **74:**96
evolution by *Chlorella,* **19:**59; 60
exchange, **15:**156 (C^{14}); 208 (C^{14}); 213 (C^{14}); 218–219; 230
from C^{14} of nicotinamide and nicotinic acid, **15:**110
fixation, **15:**144; 172
from glucose labeled with C^{14}, **15:**176
in isocitrate, **15:**126
photochemical fixation, in heptulose, **15:**160–161 (C^{14})
in urea synthesis, **15:;** (C^{14}); 321; 323 (C^{13}); 328; 339; 342; 356; 357; 364; 364 (C^{14}); 372; 375–377; 380; 382 (C^{14}); 383 (C^{14})
fixation by animal tissues
by C_3 and C_1 addition, **74:**38–44
in dissimilation of pyruvate by liver tissue, **74:**38–44; 41
in glycogen, by liver tissue, **74:**44–46
involving carbon to carbon linkage, **74:** 38–47
mechanism of, **74:**37–47
not involving carbon to carbon linkage, **74:**37–38
fixation by *Trypanosoma lewisi,* **74:**46–47

Carbon dioxide (*Continued*)
fixation in photosynthesis, **19:**2; 3; 18–45; 51–54; 56; 58–61; 63–66
fixation on α-keto-glutaric acid, **10:**208–213
production in microbial attack on hydrocarbons, **10:**447–448; 465–466
functions of, in cellular physiology, **74:**11–12
heterotrophic assimilation, **2:**135–179
photosynthetic utilization, **2:**138–139
heterotrophic assimilation of
definition of, **74:**6
studies of, historical perspective on, **74:**3–6; 5
heterotrophic fixation of, mechanism of, **74:**12–37
labeled
in Hill reaction, **19:**49
in photosynthesis, **19:**19–21; 28–33; 35; 37–40; 42–45; 61
in sucrose synthesis in plants, **19:**5
in hydrocarbon oxidation study, **27:**478; 484
in photosynthesis, **74:**6
in purine biosynthesis, **21:**201; 202; 204; 214
in pyrimidine biosynthesis, **21:**267; 268; 271; 272
radioactive, in study of carotenogenesis, **21:**337–340
in study of pyrimidine biosynthesis, **21:**266
radioactive, lignin formation from, **18:** 360–361
role in syntheses by citric acid cycle, **18:**259–264; 265–267
reduction, **1:**242–259; 279
in dark reaction, **74:**8
and RuDP carboxylase, **19:**33
unidentified fixation of, **74:**47
Carbon Dioxide, Fixation by Heterotrophs and Autotrophs (Utter and Wood), **12:**41
Carbon Dioxide, Heterophic Assimilation (Werkman and Wood), **2:**135
Carbon dioxide-biotin enzyme formation, tracer experiments, **26:**311–316
Carbon dioxide (C^{13})
in carboxylation, **26:**293
effect on Hill reaction, **25:**88
Carbon dioxide (C^{14})
and biotin enzymes, **26:**295; 296; 299; 301
in carboxylation, of biotin, **26:**316–320
of enzyme, **26:**320–330
in propionic acid metabolism, **26:**287; 291; 343; 368
in citric acid cycle, **18:**264; 268–269; 274
exchange with pyruvate, **18:**334; 336
in photosynthesis study, **18:**57; 340–342
in plant respiration, **18:**290; 299; 300; 302–303
in photosynthesis, **25:**61–62; 96; 101–103
Carbon dioxide (^{14}C), methane oxidation study, **27:**478
Carbon dioxide fixation, **18:**290–291; 300–303, **23:**450, **42:**5, *See* Photosynthesis
and ATP, **18:**229–230
by hydrocarbon-oxidizing microbes, **27:**478; 484
Carbon dioxide fixation by animal tissue, C_3 and C_1 addition, **2:**170–176
in glycogen, **2:**176–179
involving C-C linkage, **2:**170–179
in lactates, **2:**175–178
mechanism, **2:**169–179
not involving C-C linkage, **2:**169–170
in pyruvates, **2:**171–174; 178
in succinates, **2:**174–175; 178
Carbon dioxide fixation by bacteria, in acetates, **2:**166–168
by C_3 and C_1 addition, **2:**146–169
and cocarboxylase, **2:**158–160
involving C-C linkage, **2:**146–169
in lactates, **2:**166–168
mechanism, **2:**144–169
not involving C-C linkage, **2:**144–146
and propionic acid, **2:**160–166
and pyruvic acid, **2:**154–158
and succinic acid, **2:**146–153
Carbon dioxide fixation in molds, **9:**564–577
Carbon dioxide fixing enzymes, location, **25:** 67–69
Carbon dioxide (O^{16})(O^{18}), in photosynthesis, **25:**42
Carbon dioxide tension, and sperm motility, **24:**39–40; 44; 50
Carbon dioxide uptake, in photosynthesis, **11:**153; 155–157; 189–195
Carbon electrophilic, **46:**398
Carbon-flourine bond, stability of, **18:**138–139
Carbon-flourine compounds, toxicity of, **18:**117
Carbonic acid anhydrides, and pyrophosphorolysis, **18:**202; 203; 209–212

Carbonic anhydrase, **34:**512–513, **36:**74; 75, **41:**53, **46:**486–487, **47:**149–265, **50:**367, **57:**138, **58:**9
acetate complex, **47:**239–240
acetazolamide complex, **47:**216–218
activation, **21:**134
activation parameters, **47:**196–198
active-site-directed covalent modification, **47:**244–252
active site of, **56:**390
active site structure
bovine isozyme B, **47:**161
human isozyme B, **47:**159–160
human isozyme C, **47:**154–157
alkylated with
bromoacetate, **47:**219; 244–246; 251; 260
bromoacetazolamide, **47:**211; 250–251
N-bromoacetylacetazolamide, **47:**250–251
bromopyruvate, **47:**194–195; 248–249; 260
N-chloroacetyl chlorothiazide, **47:**247; 251
N-chloroacetyl cyclothiazide, **47:**247
^{13}C NMR, **47:**245–246
fluoro-2,4-dinitrobenzene, **47:**247
iodoacetamide, **47:**246
iodoacetate, **47:**194–195; 244; 260
amino acids, active site, **47:**155–156; 159–160
aromatic clusters, **47:**155; 159; 161
anion complex, NMR of, **47:**227; 232–236; 238–239
structure of, **47:**186–187; 226; 236–240
apoenzyme, binding with carboxylate esters, **47:**199
binding with sulfonamides, **47:**203; 215–216
binding with zinc ions, **47:**162; 174
catalytic activity, **47:**203; 209–210
arginyl residues, **47:**155
aromatic amino acid clusters, **47:**155; 159; 161
azide complex, **47:**186; 222; 226; 228; 231; 236–239
bicarbonate complex, **47:**186–187; 236–237
binding with, anions, **47:**226–227; 232–236
substrates, **47:**173–174; 186–187; 199; 201–203
sulfonamides, **47:**217; 219–224
bovine erythrocyte isozymes, **47:**152–153; 161; 175
bromide complex, **47:**225–226; 233; 236; 238
cadmium enzyme, **47:**163; 214–215
NMR of, **47:**214
catalytic activity, **47:**175–212
high pH, **47:**208–212
ionizations governing, **47:**210–212
linear free energy correlations, **47:**199; 201
temperature effect, **47:**196–198; 209–210
catalytic efficacy, **47:**205–208
catalytic role of zinc in, **56:**316
chloride complex, **47:**225–226; 232–236
circular dichroism of human isozymes, **47:**161–162
cobalt enzyme, **47:**163–165; 168; 171–172; 213; 223; 231
bicarbonate complex, NMR, **47:**186–187
cyanide complex, EPR, **47:**238–240
fluoroacetate complexes, NMR, **47:**227
formate complex, NMR, **47:**227; 236
magnetic circular dichroism of complexes, **47:**262–263
proton exchange, **47:**171–172
complex with anions, NMR, **47:**227; 232–236; 238–239
structure, **47:**186–187; 226; 236–240
complex with sulfonamides, EPR, **47:**222
NMR, **47:**219
structure, **47:**216–218; 251
conformational changes
pH induced, **47:**162
substrate induced, **47:**253
sulfonamide induced, **47:**219
copper enzyme, **47:**163; 165
cyanide complex, EPR, **47:**238–240
crystal structure of human isozyme B, **47:**158–160
crystal structure of human isozyme C, **47:**153–157
cyanide complex, **47:**222; 226; 229–230; 232; 238–240
cysteinyl residue, **47:**155; 159
deuterium isotope effects, **47:**176–177; 194–196; 257–259
distribution, in nature, **47:**150; 175
in tissues, **47:**175–176
entatic state of, **56:**314
evolution of isozymes, **47:**160–161
in eye, **16:**172
fluoroacetate complex, **47:**227
formate complex, **47:**227; 236
function of, **56:**388

Carbonic anhydrase (*Continued*)
helical structure in human isozymes, **47:**154; 159; 161
histidyl residues, **47:**155; 159; 167; 170
titration, **47:**167–171; 254
human erythrocyte isozymes, **47:**152–161
hydrophobic locus in active site, **47:**155–156; 173–175; 216; 220; 222
inhibition, **13:**332, **47:**212–243
anions, **47:**225–240
insecticides, **47:**241
localization in cytoplasm, **13:**7
substrate, **47:**229; 240–243
substrate analogs, **47:**240–243
sulfonamides, **47:**150; 202–204; 208; 212–215; 223–224; 227–229
iodide complex, **47:**225–226; 228; 233
ionic strength effect on catalysis, **47:**201–202
kinetic schemes, **47:**257–259
kinetics of metal reconstitution of, **56:** 304–306
ligand geometry at metal ion, **47:**164–165; 216–218; 222; 234–235; 262–263
linear free energy relationships, **47:**199; 201
lysyl residues, **47:**155; 159; 167
macrocyclic metal complexes as models for, **47:**215; 262; 265
manganese enzyme, **47:**163; 173
acetate complex, NMR, **47:**240
proton exchange, **47:**173
mechanisms proposed, **47:**252–265
mercury enzyme, **47:**163
metal centers of, properties of, **56:**389–397
metal complex formation, **25:**19
relaxation methods for, **25:**6; 7; 35
turnover number, **25:**32; 33
metal content of, **56:**389
metal ion
ligand geometry, **47:**164–165; 216–218; 222; 234–235; 262–263
specificity, **47:**163
metal removal from, **56:**303
Michaelis-Menten parameters, **47:**176–179; 191–195; 206–207; 257–259
effect of pH on, **47:**176–179; 191–195; 257–259
models, macrocyclic metal complexes, **47:**215; 262; 265
molecular properties, **47:**152–175
pH effect, **47:**162; 165–171; 254
molecular properties of, **56:**389
natural substrates, **47:**175–187; 194–195; 236–237
nickel enzyme, **47:**163
non-natural
aldehydes, **47:**174; 187; 190–191; 194; 206–207
carbobenzoxy chloride, **47:**189
carbonate esters, **47:**188; 206–207; 209
carboxylate esters, **47:**188; 191–194; 196–200; 206–215; 223; 227–229
fluoro-2,4-dinitrobenzene, **47:**189–190
phenyl *N*-methylacetimidate, **47:**189–190
phosphate triesters, **47:**190–191; 193; 206–207
pyruvate esters, **47:**188
pyruvic acid, **47:**187
sulfonate esters, **47:**189; 205–207
non-natural substrates, **47:**174; 187–194; 196–200; 205–215; 224; 227–229
nuclear magnetic resonance of, **56:**311
nucleophilicity, **47:**264–265
oxygen exchange catalysis, bicarbonate-water, **47:**180–184
inter-carbon dioxide, **47:**181–184
pH, effect on, amino acid residues, **47:** 165–171; 254
conformation, **47:**162
ligand geometry of metal ion, **47:** 165–167
Michaelis-Menten parameters, **47:** 176–179; 191–195; 257–259
pH effect on, **56:**394; 395
phenylalanyl residues, **47:**155; 159
pre-steady state transient, search for, **47:** 198–200
proton exchange at active site, **47:**171–173
mechanism, **47:**263
proton transfer, buffer participation, **47:** 179–182
recent information, **47:**265–266
reverse burst, **47:**212–213
role and source of, **56:**293
role of zinc in, **56:**409
spectral features of, **56:**309; 391–392; 393
in sperm, **24:**49; 80
localization of, **24:**331
role of metal, **24:**112–113
structure
active site, **47:**154; 155–157; 159–161

anion complexes, **47:**186–187; 226; 236–240
human isozyme B, **47:**158–160
human isozyme C, **47:**153–157
sulfonamide complex, **47:**216–218
β-structure in human isozymes, **47:**153; 158; 161
substrate
bicarbonate, **47:**178–187; 224; 236–237
carbon dioxide, **47:**176–177; 181–187; 194–195; 236–237
natural, **47:**175–187; 194–195; 236–237
substrate bindings locus, **47:**173–174; 186–187; 199; 201–203
sulfide complex, **47:**225–226; 229–230; 232
sulfonamide complex, **47:**216–219; 222–223; 251
EPR of, **47:**222–223
PMR of, **47:**219
tryptophanyl residues, **47:**155; 159; 167
tyrosyl residues, **47:**155; 159; 165–167
vanadyl enzyme, **47:**165
water in active site, **47:**156
as zinc enzyme, **56:**290
zinc ion, active site, **47:**150; 154–155; 159; 162
Carbonic anhydrase B, **47:**152–153; 158–161
Carbonic anhydrase C, **47:**58; 124; 153–157
Carbonic anhydrase in blood, **9:**374
Carbonic anhydratases, classification, **14:**228; 237; 256
^{14}C-Carbon dioxide, from labeled glucose, **28:**408
incorporation into chrysolaminarin, **28:**338
Carbonic esterases, classification, **14:**224–225; 244; 250
Carbon ions, inactivation of enzymes, **24:**365; 372; 375; 376
Carbon isotopes, effect on bond strength, **12:** 70–71
effect on reaction rates, **12:**70–71
in photosynthesis study, **12:**135–138; 138–143
in transmethylation studies, **12:**460–462; 469–474; 479; 480–482
Carbon isotopes as tracers for fixation of CO_2, **2:**137; 146–148; 152–154; 160–164; 166; 168; 170; 172; 176
Carbonium ion
in biological oxidations, **15:**12
as product of biological oxidations, **19:**205; 206
Carbon-11 labeled compounds, **57:**253
Carbon monoxide, **46:**65
in detection of denatured-protein hemochromogen, **20:**173; 174
reaction with cytochromes, **20:**150; 151; 164–166
effect on cytochrome *c,* **28:**221
effect on cytochrome *c*-cytochrome oxidase complex, **28:**204–206
effect on cytochromes, **18:**6; 7–11; 12; 16; 30–33
reaction with hematin derivative, **18:**5
effect on dihydroxyfumaric acid oxidase, **19:**118; 120; 127; 199
effect on tryptophan oxidase, **19:**122
effect on plant respiration, **16:**135–136
enzyme inhibitor, **15:**3; 26
influence on growth and respiration of *aspergillus,* **2:**216–218; 220–222
and Pasteur effect, **2:**216
and phosphatide formation, **7:**149; 151
effects on thyroxine and diiodotyrosine formation, **7:**206
release, **46:**67
Carbon monoxide-cytochrome a_3, reversible photodissociation, **28:**228–229
Carbon monoxide-cytochrome oxidase, photochemical change, **26:**13
Carbon monoxide haemochromes a, **23:** 289–297; 304; 306–312
Carbon monoxide heme linkage, **15:**20; 33
Carbon monoxide spectra, of cytochromes, **23:**306–312
of oxidase, **23:**298–301; 303–304
Carbon monoxyhemoglobin, rates of dissociation, **29:**36
Carbon-nitrogen bonds, formation, in purine biosynthesis, **21:**229–255
Carbon-nitrogen lyases and syntheases, classification, **14:**229; 257
1-Carbon particle. *See also* Formate and Methyl group
relation to glycine and serine metabolism, **16:**275–277
relation to histidine metabolism, **16:** 300–301
relation to methionine metabolism, **16:**270; 274

Carbon path, in plant respiration, **18:**282–303
1-Carbon pool, role of folate, **27:**426
Carbon reduction cycle, in photosynthesis, **25:**42–44; 47; 61–62; 94–103
energetics of, **25:**48–51; 102–106
products of, **25:**53–56
Carbon reduction pathways, **25:**94–103
Carbon reduction reactions in photosynthesis, **25:**94–106
Carbon sources, effect on carotenogenesis, **21:**310; 311
Carbon suboxide, reaction with thiols, **11:**227
Carbon-sulfur hydrolases, classification, **14:**225; 249
Carbon-sulfur lyases, classification, **14:** 228–229; 257
Carbon tetrachloride, **51:**33
Carbon tetrachloride-induced liver cirrhosis, **42:**274
Carbon-to-carbon condensations, in squalene biosynthesis, **22:**309; 310
1-Carbon transferases, in folic acid system, **27:**419
Carbonyl compounds, as inhibitors for diamine-oxidase, **2:**100
Carbonyl compounds reduction of, mechanisms, **24:**474–475
Carbonyl cyanide *p*-trifluoromethoxy-phenylhydrazone (*p*-CF_3O-CCP), as uncoupler of oxidative and photophosphorylation, **30:**69
Carbonyl cyanide-*p*-trifluromethoxy-phenylhydrazone, catalysis of equilibration of the electrochemical activity of protons across natural membranes, **29:**65
Carbonyl group, **38:**215; 216, **60:**3
entropy for dehydration, **43:**301
polarization, **43:**509
Carbonyl oxide, **67:**196
Carbonyl phosphate, **42:**29
Carbonyl-polarizing enzyme reactions
and glyoxalase I, **59:**288–290; 292–293
mechanisms of, **59:**287
and model reactions of biotin enzymes, **59:**297–299; 301–302
and polarization process, **59:**287–288
and yeast aldolase, **59:**293–297
Carbonyl sulfide, **67:**11–12
Carbosyntheases, classification, **14:**228; 256
Carboxamide group, in transport, **49:**76
Carboxamide ribotide transformylase, **21:** 220–222; 400
Carboxyalkanoylproline derivatives, **57:**29–30
N-Carboxyalkyldipeptide design, **57:**39
Carboxyamidopeptidase, **53:**273
2-Carboxyarabinitol bisphosphate, **67:**7
active site
interactions, **67:**21–22
orientation, **67:**23
site-directed mutagenesis, **67:**34–35
Carboxybiotin, **43:**315
mechanism of formation, **26:**316–320
N-Carboxybiotin, **35:**352
decarboxylation of, analogues, **35:**354
covalent-linked compound, **35:**354
Carboxybiotin-enzyme complex
methylation, **26:**327–328
reaction mechanism, **26:**320–330
cis-α-(Carboxycyclopropyl) glycine, **50:**128; 134; 135
trans-α-(Carboxycyclopropyl) glycine, **50:**128; 129; 134
2-Carboxy-5,6-dihydroxyindole, from dopachrome, **16:**115
Carboxydismutase, **23:**430; 441; 450
in plastids, **20:**228
Carboxydismutase system, in photosynthesis, **25:**95; 96; 98; 99
Carboxydomonas oligocarbophilia, metabolism in, **74:**6
Carboxyendopeptidases, **41:**236
Co-β-carboxyethyl-cobalamin, **26:**245
N^5-(1-*Carboxyethyl) ornithine and Related [N-Carboxyalkyl]-Amino Acids: Structure, Biosynthesis, and Function* (Thompson and Miller), **64:**317
S-Carboxyethylcysteine, **50:**121; 123; 124
S-Carboxyethylcysteine sulfoxide, **50:**121; 124
5-Carboxyfurfural, as degradation product of pectins, **20:**350
γ-Carboxyglutamic acid, **46:**2; 17; 20, **48:**277
acid hydrolysis, **46:**21
alkaline hydrolysis, **46:**18; 21
amino acid analysis, **46:**17
in bone protein, **46:**25
chemical synthesis, **46:**22
decarboxylation, **46:**18; 21
electrophoretic mobility, **46:**17; 18; 22
^{3}H-diborane reduction, **46:**18; 22
mass spectrometry, **46:**17; 18; 22

proton nuclear magnetic resonance spectroscopy, **46:**17
sequenator degradation, **46:**18
thiohydantoin derivatives, **46:**18
urinary excretion, **46:**23
Carboxyhemoglobin, structure studies, **14:**305
Carboxyhemoglobin structure, **7:**276
3-(3-Carboxy-4-hydroxyphenyl)alanine, **50:**148
(3-Carboxy-4-hydroxyphenyl)glycine, **50:**149
3′-Carboxy-4′-hydroxyphenylpyruvic acid, **50:**148
2-Carboxyindole-5,6-quinone, relation to *Sepia* melanin, **16:**172
S-Carboxyisopropylcysteine, **50:**121; 124
Carboxyl, **36:**114
carrier protein, **35:**334; 363; 364; 369
transferase (CT), **35:**369
Carboxylalkyldipeptides
and azopeptide inhibitors, **57:**42–44
and enalapril, **57:**39–40
and enalaprilat, **57:**32; 39–40
I_{50} values of, **57:**33–37
and lisinopril, **57:**35
and 2-methylglutarylproline, **57:**32–33
and NH, **57:**40–43
and proline, **57:**35–38
Carboxylase, **9:**571; 624; 678; 693, **15:**158; 177; 208. *See also* specific carbodylases
complex formation, **10:**204–208
oxalosuccinic acid in citric acid cycle, **10:**202–208; 209–213
in plants, **10:**300
reactivation, **10:**111
relation to structure of alkaline phosphatase, **10:**116–117
formation of holocarboxylase, **26:**331–334
inhibition, **13:**332
mechanism of blockade in *Fusaria,* **5:** 171–174
and mechanisms of catalysis, **24:**467–469
role of metal, **24:**112–113
models, **15:**27; 29
in *Neurospora,* **22:**25
requiring –SH group, **11:**241
specificity, rubisco, **67:**56–60
stabilization by coenzyme, **23:**47
vitamin K-dependent, **46:**2
β-Carboxylase, and CO_2 fixation, **12:**54–60
α-Carboxylase, isolation, **14:**405
α-Carboxylase, TPP in, **22:**401
Carboxylases, **35:**328–331
allosteric regulation, **28:**55–56; 88–92
classification, **14:**228; 230; 256
model for study of catalysis by, **14:** 177–178; 187
Carboxylated acceptor-^{14}C-labeled acceptor exchange, **35:**330
Carboxylate esters, carbonic anhydrase binding to, **47:**174; 199–203
hydrolysis catalyzed by, **47:**188; 191–194; 196–200; 206–215; 223; 227–229
3-acetoxy-2,6-dinitropyridine, **47:**191; 198–200; 206–207; 211–212
3-acetoxy-2-nitropyridine, **47:**198–200; 206–209; 211–212; 223
2,4-dinitrophenyl acetate, **47:**206–207; 212–213
p-nitrophenyl acetate, **47:**164; 191–194; 196–198; 206–207; 209–210; 214–215; 227–229
p-nitrophenyl propionate, **47:**206–207; 211; 223
Carboxylate group, binding of, **43:**333
Carboxylate groups, **34:**465
Carboxylating enzymes
homologous mechanisms for, **33:**49
mechanisms of action, **24:**443
Carboxylation
and biotin enzymes, **26:**292–311
exchange reactions, tracer experiments, **26:**311–316
of propionyl-CoA, **26:**284; 286–334
and citric acid cycle, **18:**261; 274; 276
reductive, of acetyl phosphate to pyruvate, **6:**248
rubisco, **67:**5–10
α-Carboxylation, and CO_2 fixation, **12:**92–106
in photosynthesis, **12:**140
β-Carboxylation, and CO_2 fixation, **12:**50–92
γ-Carboxylation, geranyl CoA carboxylase, **35:**389
β-methylcrotonyl CoA carboxylase, **35:**389
β-Carboxylation, in plants, **18:**301
Carboxylation-decarboxylation reactions, **50:**272; 281–283
Carboxylation enzyme, *See* Ribulose 1,5,diphosphate carboxylase
Carboxylation enzymes, biotin-linked, **23:**358
Carboxylation of α-ketoglutaric acid to oxalosuccinic acid, **10:**208–213
Carboxylation reactions, **12:**46–106
in mitochondria, **21:**46; 47

Carboxylations, and ATP synthesis, **23:**338
Carboxyl carriers, **34:**453
group, **34:**453
methyl bromelain, **34:**464
methyl cellulose bromelain, **34:**464–465
Carboxylesterase, **37:**462; 491
localization of, **24:**316
Carboxyl gorups of RNase, chemical modification of, **24:**209; 224–225
Carboxyl group
binding energy, **43:**344; 347
differentiation from imidazole, **20:**76
in sialic acid, **60:**141
as site of amino acid activation, **19:**236
α-Carboxyl group, metal binding by, **20:**56
γ-Carboxyl group, metal binding by, **20:**56
Carboxyl group activating enzymes, protein ligase problem, **72:**38–40
Carboxyl groups
modification, **53:**69–70
on fructose 1,6-bisphosphatase, **53:**69
relation to activation of β-glucuronidase, **16:**376
Carboxylic acid esters, hydrolysis, by metal ions, **21:**179; 180; 191; 192
Carboxylic acid residue, cytochrome *c* oxidase, polypeptide composition, **71:**85
Carboxylic acids, formation of thio esters with CoA, **16:**18
Carboxylic ester hydrolases, localization of, **24:**316–318
Carboxylic esters, **34:**526
Carboxylic group, **58:**278
α-Carboxyl ketene, **59:**125
S-(Carboxylmethyl)glutathione, **59:**116
Carboxyl terminal analysis, **46:**368
Carboxyl-terminus, **42:**81
Carboxyltransferases, location in subcellular fraction, **24:**309
Carboxylyases, localization of, **24:**330
Carboxymethyl alcohol dehydrogenase, Cotton effect, **27:**58
Carboxymethylation, **58:**123
of RNase, **24:**203–206; 214–220; 245
Carboxymethylation (^{14}C) of LADH, **27:** 80–82
Carboxy-β-methylcrotonyl-CoA carboxylase, stability, **26:**329–330
Carboxymethyl cysteinyl peptide (^{14}C), from LADH, **27:**81–82
Carboxymethyl pachyman, as substrate for laminarinase, **28:**344; 352
Carboxymethylribonclease (C^{14}), **24:**214
3-Carboxymuconate lactonizing enzyme, **74:**314; 315
β-Carboxymuconic acid, **19:**98; 99
2-Carboxypentitol bisphosphates, diastereoisomers, **67:**7; 8
Carboxypeptidase, **18:**173, **34:**449–451, **41:**197, **43:**238; 312; 336, **50:**35, **53:**249, **55:**356, **56:**293
action on
muscle phosphorylases, **24:**267
RNase, **24:**133; 229–230; 244; 246; 250
action on *a*-keto acid peptides, **8:**129
effect on glycyldehydrophenyl-alanine, **8:**134
susceptible configuration, **8:**166
action on collagen, **23:**205
action on hypertensin, **7:**256
and proteinases from kidney, **7:**529
in aldolase study, **25:**344
in glycoprotein linkage study, **25:**451
turnover number, **25:**32
amino acid content, **15:**296
kinetics, **15:**273; 292; 293
in virus protein study, **15:**463; 474
anion effect on, **56:**365–366
catalytic role of zinc in, **56:**316
decomposition of glutathione by, **10:**295
effect of, on aldolases, **31:**130
effect of DFP on, **13:**332
effect of radiation on, **24:**388; 394–395
entatic state of, **56:**314
esterase activity of, **19:**306
kinetic constants, **19:**313
specificity, **19:**329
hydrogen bonds in, **50:**29
isolation of, **56:**298
kinetics of metal reconstitution of, **56:** 304–306
mechanism of action of, **56:**369–370; 373–376
mechanism of deduced from relaxivity data, **33:**56–59
x-ray crystallographic studies of, **33:**56–57
metal center of, properties of, **56:**360–376
metal content of, **56:**298; 299; 359
metalloenzyme substitutions of, **56:**304
metal removal from, **56:**303

molecular properties of, **56:**359–360
oxidation of, **56:**368
partial degradation of protein by, **24:**125; 128
pH effects on, **56:**364
role of zinc in, **56:**409
spectral features of, **56:**309; 361; 362
thermolysin and, **56:**381
types of, **56:**358–359
valence of, **24:**118
Carboxypeptidase A, **36:**62, **37:**410; 415; 439; 449; 462; 510; 526, **43:**249; 354, **47:**54; 60; 62; 124, **53:**268, **58:**5, **59:**111
anhydride mechanism, **58:**7
bovine, substrate binding, **27:**72
carboxylate group, **58:**8
column chromatography, **22:**451
imidazole in, **58:**9
metal binding group, **58:**8
peptide hydrolysis by, **58:**6
sequential response, **37:**588
substrate model, **58:**8
zinc role in, **58:**8
Carboxypeptidase-A, **45:**359
activity, **45:**396
and benzylsuccinic acid, **57:**40
and carboxypeptidase-B, **57:**14–15
and Gly-L-Tyr, **57:**21
inhibition of, **57:**23–24
inhibitors for, **57:**20–21
kinetic mechanism of, **45:**359
mechanism of, **57:**22–24
product inhibition of, **25:**181–182; 222 (data)
structure of, **57:**21
X-ray studies of, **57:**67
Carboxypeptidase A., **52:**185; 193; 198; 211; 222
Carboxypeptidase A and B, localization of, **24:**324; 325
Carboxypeptidase activity, **45:**393
Carboxypeptidase A inhibitor, **68:**198
Carboxypeptidase-B, **45:**396, **57:**14–15
kinetic mechanism of, **45:**359
Carboxypeptidase B, treatment of γ-globulin, **41:**162
Carboxypeptidase (pancreatic), requirement for —SH group, **11:**241–242
Carboxypeptidases, **45:**397, **57:**15; 184–185
chemical mechanism of, **45:**367
classification, **14:**238; 247
in protein structure studies, **14:**280; 286–288
from *I. bisselliella,* **45:**399
Carboxypeptidases A and B
LADH study, **27:**81–82
of pancreas, **23:**130; 132–133
Carboxypeptidase Y, **53:**260; 272–273; 290, **59:**153–154
α-Carboxypeptide amino acidhydrolases, localization of, **24:**324
3-(3-Carboxyphenyl)alanine, **50:**148
(3-Carboxyphenyl)glycine, **50:**149
N-[1(S)-Carboxy-3-phenylpropyl]-L-leucyl-L-tryptophan, **57:**67–70
3′-Carboxyphenylpyruvic acid, **50:**148
Carboxypolypeptidase, specificity of, **12:**193; 206; 224; 225–251
Carboxytransphosphorylase, *See* P-Enolpyruvate carboxytransphosphorylase
mechanism of deduced from relaxivity data, **33:**49
Carboxytransphosphorylase, Properties of, and Role in the Metabolism of Inorganic Pyrophosphate (Wood, O'Brien, and Michaels), **45:**85
4-Carboxyuracil. *See also* Orotic acid
in pyrimidine synthesis, **15:**356–357
Carboxyurea, **56:**94
Carcinogenesis
and mutations, **27:**309; 314
Warburg theory of, **29:**324
Carcinogenic agents, **4:**223; 228; 248–250
Carcinogens
as alkylating agents, **46:**406
azo dyes, **46:**409; 489
as inducers, **46:**406
as substrates, **46:**406
cleavage, **15:**127
deacylating enzyme for, **24:**327
detoxification, **69:**6
effect on mitochondria, **25:**377
effect on mitochondrial oxidative phosphorylation, **21:**32
inactivation of, **55:**254–256
4-4nitroquinoline *N*-oxide, **27:**331
transformation in body, **24:**308
Carcinoma, **54:**102. *See also* Cancer; Cancer Hepatoma, Malignant cells, and Tumor; Tumors
DNA synthesis in, **16:**453
Flexner-Jobling, glycolysis in, **24:**335
β-glucuronidase activity, **16:**396; 397
human, RNA ribose synthesis, **28:**417

Carcinosarcoma, Walker, **24:**256; 325
  enzymes in, **24:**313; 329
Carcinosarcoma and phosphatide turnover, **7:**138; 146; 171
*Cardiac Cytachrome C*1 (King), **54:**267
Cardiolipin, **34:**466
  cytochrome *c* oxidase, electron transfer, **71:**118
  in lipopolysaccharide synthesis, **31:**90
Cardiovascular effects, of adenosine, **69:**84
Carnations, antibiotic treatment of diseases, **20:**401
Carnitine, **42:**259
  hydroxylation of γ-buty-roβine to, **41:**270; 271
Carnitine acetyltransferase, **37:**415; 544, **43:**351, **58:**160
L-Carnosinase, classification, **14:**247
Carnosinase, specificity of, **12:**221–225
Carnosine, **20:**91; 102; 103
  amino acids in, **14:**266
  in decarboxylation of oxalacetate, **12:**81
  ethyl derivative of, **25:**383
L-Carnosine, effect on fructose 1,6-bisphosphatase, **42:**202
l-Carnosine and β-alanine, **7:**437
Carotene, **58:**210
β-Carotene
  biosynthesis, **21:**337–340; 343–346
    degradation products, **21:**331–333
    effect o phototropism, **21:**358–359
    structure, **21:**297
  in chloroplast lamellae, **30:**9; 50
γ-Carotene, phototactic function of, **21:**356
  structure, **21:**297
α-Carotene, structure, **21:**297
ζ-Carotene, structure, **21:**342
Carotene oxidase in plants, **8:**442
Carotenes
  in air-cured tobacco, **6:**333–334
  definition, **21:**297
    in algae, **21:**301
  destruction during autoxidation, **8:**440; 443; 452–453
  in photosynthesis, **19:**62; 63
  in tobacco, **10:**372
Carotenoid, **58:**66
Carotenoid oxidase, and lipoxidase, **8:**441
Carotenoid Pigments, Biosynthesis and Function (Goodwin), **21:**295
Carotenoid pigments from microbial action on hydrocarbons, **10:**470; 479
Carotenoids
  absorption spectra, **4:**281–283
  biosynthesis, **21:**307–349
    cis isomers, synthesis, **21:**297
    distribution, **21:**298–306
    functions, **21:**349–361
    in mitochondrial fragments, **21:**113
  and coupled oxidation, **8:**452–453
  degradation, **46:**100
  effect on gametes, **9:**257–258
  in photochemical systems, **30:**50
  and photosynthesis, **1:**293; 308; 311; 313
  in plastids, **20:**204
  role in photosynthesis, **25:**71–75
  spectral changes, **6:**163
Carotenoproteins, **21:**299; 300; 306; 350
Carp muscle, ATPase from, **13:**168
Carrier hypothesis of gene mutation, **13:**133
Carrier mobility, **49:**75
Carriers. *See also* Oxygen transferring agents
  artificial polyelectrolyte, **34:**456
  carboxyl, **34:**453
  catalytic, **29:**33–39
    center, **29:**43–44
    mobile *vs.* fixed, **29:**43
  for column chromatography of enzymes, **22:**447
  electrically neutral, **34:**453
  high molecular weight, **34:**447
  inorganic, **34:**453
  in plant tissues, **10:**374–375
  polyelectrolyte, **34:**453
  water soluble, **34:**452
Carrier systems for active transfer in cells, **8:**51
Carrots, antibiotic treatment of diseases, **20:**392
Carr-Purcell measurement of relaxation rates, *See* Relaxation rates, nuclear, Carr-Purcell pulsed method of measurement
Cartesian diver manometry, in histochemistry, **9:**603
Cartilage, hyaluronidase action on, **13:**229
Cartilage matrix, proteolytic resorption, **60:**227
Casbene, **55:**78; 79
  structure of, **55:**55
  synthesis of, **55:**78–79
Casbene synthetase, **55:**79
Cascade, evidence for an enzyme, in blood clotting, **30:**268

implications of hypothesis of, **30:**297–305
kinetics of the coagulation, **30:**299–305
Cascade reactions, **48:**279
Cascase control mechanisms, regulation of bacterial form glutamine synthetase, **72:**31–32
Casein. *See also* Milk
δ, **15:**432
action of peroxidase on, **14:**158
action of tyrosinase on, **14:**138
adsorption studies of, **14:**336; 360; 361
amino acids in, **14:**263
oxidation of tyrosine in, **14:**86
action of rennin on, **15:**431–434
condition in milk, **15:**437–439
isotope studies of biosynthesis, **19:**260
optimal pH for papain action on, **19:**303
mechanism of clotting, **15:**440–442
milk, D-leucine in, **20:**25
phosphorylation, in mitochondria, **21:**45
properties, **15:**435–437
β-Casein, **54:**27; 29–30
Casein kinase, RNA polymerase II transcription, C-terminal domain, characteristics of, **71:**44
Caseinogen, amino acid content, **3:**70
Casein-trypsin units, **23:**33
C^{14} as label in photosynthesis studies, **12:** 135–138; 138–143
use in transmethylation studies, **12:**460–462; 469–474; 479; 480–482
Cassava, **39:**359
Cassic acid, **7:**469; 486; 489; 490; 492
Castorbean endosperm, **54:**175
Castor bean phytoalexins, **55:**78–79
Catabolic activity of the cell, **8:**30–37
Catabolic enzymes, **60:**90
Catabolin, **60:**226
Catabolism, **34:**478, *See* Depot fats; Fat and carbohydrate catabolism
of serum proteins, **41:**104; 121; 126
Catabolite inactivation, **59:**373–374
Catabolite repression, **44:**253, **46:**86
Catalase, **5:**23, **15:**274; 276; 287, **18:**90; 118, **33:**265, **34:**449–451, **35:**171; 172; 265, **41:**79, **45:**435–437, **46:**59; 105; 142; 170, **52:**179, **57:**87; 116, **67:**94–95. *See also* Hemoproteins, Iron prophins, Porphins, and Tetrapyrroles
absorption bonds, **1:**168
action, mechanism, **1:**168–170
action of perturbators on, **23:**45–46
action on hydrogen peroxide, **19:**109; 113–116; 119
complexes with hydrogen peroxide, **19:**111; 114–116
effect on oxidases, **19:**118; 120–122; 124; 125; 127; 199
kinetic constants, **19:**313
peroxidase activity of, **19:**104
substrates for, **19:**104; 105; 109
activity, and flavor changes in vegetable tissue, **9:**615; 618–624; 628–631; 631–637; 646–647
chemistry of, **9:**632
effect of freezing on, **9:**626–627
and semen metabolism, **9:**374–375
titrimetric methods, **9:**601
activity of, **24:**123
attachment of porphyrin, **24:**23–24
binding sites, **24:**116
inactivation by radiations, **24:**370; 371; 374; 376; 394–395; 401; 403
location in subcellular fraction, **24:**302; 305; 307; 340; 341; 343
molecular weight from radiation studies, **24:**366
in sperm, **24:**66
adsorbed
activity of, **28:**25; 26
radiosensitivity, **28:**34–35
adsorption, **1:**165
adsorption at interfaces, **28:**3
amino acid content, **14:**277, **15:**296
anticatalase, **1:**167
cacao, **25:**466; 480
and light emission, **25:**143
multiple forms, **25:**334
turnover number, **25:**32; 33
on cellulose anion exchanger, **33:**271
chemical nature, **1:**163–176
in chromidia, **8:**31
classification, **14:**227–228; 236; 254
and copper, **23:**304
Cotton effect, **27:**47; 59; 90–92
crystalline, **1:**172–174
denaturation of, **23:**31–33; 36; 44–46; 62
effect of disulfide, **27:**216–217
effect of tyrosinase on, **14:**147

Catalase (*Continued*)
effect on D-amino acid oxidase complex, **27:**22
effect on melanin formation, **16:**167
intracellular distribution, **16:**463; 466; 467; 492
oxidation of phenols to pigments, **16:**152
effect on myosin monolayer, **6:**184
in tobacco, **6:**375–376
EPR study, **27:**118; 138
and glycolate metabolism, **23:**436
inactivation, **1:**165
inactivation of, **23:**73
inhibition, **13:**332
localization in cytoplasm, **13:**7; 8
inhibition by thiol reagent, **11:**237
isolation, **14:**379; 396; 397; 398; 405
mechanism of action, **15:**3; 16–18; 23; 27–28; 32–33
in mitochondria, **21:**5; 40
mode of action, **14:**182
and nuclear purity, **27:**249; 252–253
occurrence, **1:**164
and L-ophioamino acid oxidase, **8:**476–477
and oxidation of hydrocarbons, **27:**491
as peroxidase, **14:**151; 152; 158
in peroxide measurement, **14:**153
in plastids, **20:**214
preparations, **1:**172–173
prosthetic groups, **1:**170
rate of synthesis, **26:**108
of *S. aureus,* sensitivity to RNA antagonists, **22:**138
in schizophrenics, **29:**507
separation by chromatography, **14:**325; 342; 346; 349; 352; 362
SH-groups, **27:**228
soil, **28:**387
stability, **1:**165
stimulation of prolyl hydroxylase by, **41:**255
in thiaminase-producing bacteria, **15:**413
in tobacco, changes during processing, **10:**414; 415; 421; 423
turnover number, **12:**171–174
urea treatment, **23:**44–45
Catalase, Chemical Nature (Summer), **1:**161
Catalase-H_2O_2, rate constant for, **25:**14
Catalase-H_2O_2 complex, **15:**3; 17
Catalase-H_2O_2 complexes (I and II), SH-group, **27:**217
Catalase-positive coccus, oxidation of hydrocarbons, **27:**490
Catalase s, **7:**227; 232; 266–269; 295–301; 308–309; 310; 312; 317
active groups in, **7:**295
amino acid analyses, **7:**296
and coupled oxidations, **7:**300–301
derivatives and inhibitory action, **7:**298–299
in industrial fermentation, **7:**571
and iron protoporphin IX, **7:**316
magnetometric determinations, **7:**297–298
mechanism of action, **7:**299–301
Catalase-substrate compounds, **12:**153–190; 195
Catalatic oxidations, mechanisms, **19:**104–117
Catalyses, enzymic theory of, **21:**230–254
model, with metal ions, **21:**160; 161; 169–198
Catalysis, **34:**522, **46:**394–404, **58:**11, *See* Asymmetric catalysis
acid-base, general, **43:**154; 165; 167; 169; 170; 172; 196; 221
covalent, **43:**326; 364
acid-base cooperative, **58:**34
anthranilate synthase, **66:**216–222
assay substrates, **46:**394
base-catalyzed reversal, psoralens, **66:**89–90
binding and, enzyme action, algebraic *versus* graphic analogies, **73:**51–52
chemical mechanism, **46:**402
electrophilicity, **46:**398
enzyme, mechanisms related to, **24:**441–482
general mechanism of, **29:**41–44
through lipoprotein membranes, **29:**39–41
translocation, **29:**37
glutathione S-transferases, **69:**5–7; 18–20
heterogeneous, **15:**2; 22; 24–26
kinetic mechanisms, **46:**401
mechanism
carbamyl phosphate synthetase, **66:**228–233
thioltransferases, **66:**172–174
organic, chemistry of, **14:**163–192
polyfunctional, **50:**11
tautomeric, **50:**11–14
proton and electron transfer, **25:**12; 21–32
Catalysis, Mechanisms Related to Enzyme (Westheimer), **24:**443
Catalysis in reverse micellas, **51:**47–55
enzyme, **51:**48–55
ionic surfactants, **51:**51–54
limitations of data, **51:**54–55
partitioning of substrates, **51:**49–51

at subzero temperatures, **51:**58–65
water-to-surfactant ratio, **51:**48–49
miscellar enzymatic activity, **51:**47–48
Catalyst concentration, effect on model catalyses by metal ions, **21:**172; 173
Catalysts
specific active grouping, **7:**91
synthetic, **58:**11
Catalytic activity, **34:**472–474; 480–482; 495; 507–509; 521
conformational control, **37:**601
conformational transitions, **37:**599
protein, **34:**472
and protein structure, **24:**105–160
rate, **34:**487
reaction, **34:**452; 499
unit, **34:**471
Catalytic Aspects of Enzymatic Racemization (Adams), **44:**69
Catalytic asymmetric decomposition, **7:**80; 97–99
Catalytic asymmetric synthesis, **7:**80; 89–94
Catalytic cycle, **39:**349; 352
Catalytic cycle (cytochrome *c* oxidase), **71:** 110–116
dioxygen chemistry, **71:**110–114
thermodynamics, **71:**114–116
Catalytic diversity, evolution of, glutathione S-transferases, **69:**36–40
Catalytic efficiency, relation to binding energy, **33:**347–436
Catalytic groups, multiple, **43:**309
Catalytic mechanisms, **39:**339; 364
α-complex, **69:**28; 30; 35
electrophilic catalysis, **69:**34
glutathione S-transferases, **69:**2; 20; 21–22
glutathione synthesis, **73:**236–245 (*See also* Glutathione synthesis)
nucleophilic aromatic substitution, **69:**21; 27; 31–32; 35
nucleoside monophosphate (NMP), kinases, **73:**110–119 (*See also* Nucleoside monophosphate (NMP) kinases)
rate-limiting step, **69:**34–35
stereospecificity, **69:**26
thiolate anion, **69:**21
tyrosine, **69:**21
xenobiotic substrate, **69:**23–36
Catalytic oxidation
N-acetyl-D-glactosamine, **36:**162
N-acetyl-D-glucosamine, **36:**162
L-arabinose, **36:**162
D-fucose, **36:**162
D-galactose, **36:**162
D-galacturonic acid, **36:**162
D-glucuronic acid, **36:**162
D-mannose, **36:**162
Catalytic power
definition, enzyme action, **73:**34–39
origins, enzyme action, **73:**43–49 (*See also* Enzyme action)
pyrophosphoryl transferring enzymes, MutT pyrophosphohydrolase, **73:**187
Catalytic proficiency, enzyme action, V/K constant, **73:**33
Catalytic properties and the adsorbed state of proteins, **14:**332–333
Catalytic reactions, **50:**66–81
effect on diffusional constraints, **52:**190
Catalytic residues, **33:**403
Catalytic sites
glutamine PRPP amidotransferase, **72:** 102–103; 118–119
γ-glutamyl transpeptidase catalytic sites, **72:**254–262
modification, **46:**371
Catalytic specificity of pepsin, **33:**413
Catalytic systems and luminescence, **7:**232
Cataracts
glutathione in, **11:**248
and pancreatectomy, **8:**182
Catatonia, urinary pigments in, **29:**529–530
epi-Catechin
of cacao, **25:**461–462; 467; 469; 479; 481–484
and flavandiol dimer, **25:**463–464
from tea leaves, **16:**151; 152
Catechin gallate
from tea leaves, **16:**151
epi-Catechin gallate, from tea leaves, **16:**151; 152
Catechins
from cacao, **25:**461–465; 478; 481–483
metabolism in plants, **16:**142; 151–152
Catechol
absorption spectrum, **4:**285
effect on indolylacetic acid oxidase, **19:**124
effect on phenylpyruvic acid oxidase, **19:**125
oxidation, **19:**81–83; 85–87

Catechol (*Continued*)
effect on tyrosinase, **14:**81
oxidation by serum, **14:**87
in oxidation of phenol of Vibrio, **14:**112
in peroxidase action, **14:**154
in tyrosinase action, **14:**140; 141; 144–150
metabolic pathway, **15:**184–185
in oxidation of hydrocarbons, **27:**506–511; 513–515; 520; 522; 525; 526; 535
oxidation mechanisms, **27:**128
Catecholamines
from adenrochromes, **29:**516
in the brain, **29:**482–483; 489
depletion by reserpine, **29:**485–486
depletion of brain stores by α-methyl-*m*-tyrosine, **29:**485
effect of increases in brain of, **29:**487
effect on newly hatched chicks of, **29:**488
function of, **29:**482
hypothesis of, in affective disorders, **29:**489–494
in manic-depressive insanity, **29:**492
relation to sedation, **29:**488
biosynthesis of, **74:**236
tyrosine hydroxylase regulation, **70:**127; 136; 156
Catecholase, **18:**372, **19:**134; 203
activity in phenolase complex, **19:**132–134
in human serum, **14:**87
Catecholase activity
and oxidation of coniferyl alcohol, **16:**149
of tyrosinase, rate measurements
chronometric, **4:**122–125
colorimetric, **4:**115; 116; 118–119
manometric, **4:**116; 117–119; 122; 125
in presence of reductant, **4:**119–120; 120–125; 128
Catechol derivatives, *N*- and *O*-methylation of, **25:**415–416
Catechol melanin, formation, **16:**140
Catechol-*O*-methyltransferase (COMT), **25:**416, **46:**484, **50:**289
action on extraneuronal space, **29:**490
in rat liver, **29:**524
in schizophrenia, **29:**482; 517
steric requirement for, **25:**385
Catechol oxidase
in oxidation of coniferyl alcohol, **16:**148
of belladonna, **16:**117
in plant respiration, **18:**18–20; 37
in tobacco fermentation, **10:**383
Catechol oxidation by tyrosinase
to *o*-benzoquinone, **4:**110–112; 122; 124
to hydroquinone, **4:**113; 114
inactivation of tyrosinase during, **4:**126–132
mechanism, **4:**113–114
peroxide formation, **4:**102; 110–112
protective action of gelatin, **4:**116; 118; 119; 120; 121; 146
rate measurements, **4:**113; 114; 117–125
Catechol-2,3-oxygenase (metapyrocatechase), **27:**510
Catechol-1,2-oxygenase (pyrocatechase), **27:** 509
Catechols, **67:**216–217; 237–238. *See also* Diphenols, Phenols, and Polyphenols
in curing of cacao, **25:**483
as tanning substrate, **16:**156
deamination of glycine in presence of phenolase complex, **16:**117; 118
enzymic oxidation of, **16:**110; 112; 127; 140
formation of *o*-quinones from, **16:**128–131; 138–140; 161
in phenolase system with thiols, **16:**122
in plant metabolism, **16:**132; 141; 150; 151
reaction with oxygen and amino acids, **16:**114
substituted, *o*-quinone formation in plants, **16:**138–140
tyrosine hydroxylase, **70:**205
Cathecoyl dipeptide, leucine aminopeptidase inhibition, **68:**178–179
Cathepsin, **19:**286, **53:**255; 261, **54:**146
adsorption studies, **14:**327
classification, **14:**232; 238; 247
isolation, **14:**388
amide formation by, **13:**309
localization of, **24:**326
in mitochondria and lysosmes, **21:**39–41
in particles of liver cells, **16:**367
Cathepsin, D, **44:**1; 6; 8
Cathepsin A, localization of, **24:**324; 325
partial degradation by, **24:**128
Cathepsin B, **53:**261; 280
localization of, **24:**326
Cathepsin B1, **53:**280
Cathepsin B2, **53:**280
Cathepsin C, **53:**261; 280

isolation, **14:**396
localization of, **24:**325
product inhibition of, data, **25:**222; 225
in transpeptidation reactions, **19:**310
Cathepsin D, **33:**405
hydrolytic action on insulin of, **33:**423
localization of, **24:**326
precursor, **60:**90
Cathepsin in chromidia, **8:**31
Cathepsins
action on, collagen, **23:**171; 172; 176; 219
elastin, **23:**225
and disulfide exchanges, **27:**211
kidney, **7:**522; 529–530
requirement for —SH group, **11:**239; 241
specificity of, **12:**194; 205–206
Cathespin B_1, effect on fructose 1,6 bisphosphatase, **42:**217
Cation exchangers, in protein chromatography, **22:**428–430; 442; 443
Cationic, center, **39:**345
group, **39:**341; 346
site, **39:**355; 357
Cationic acid, **34:**492
Cationic buffers, **51:**10
Cation radicals, **66:**31
Cation requirement, for PP-ribose-P synthetase, **41:**9
Cations, **34:**490, *See* Metal ions
bivalent, effect on DNase, **29:**180
content of normal mitochondria, **29:** 261–263
divalent, energy-dependent accumulation of, **29:**264–277
monovalent, induction of, uptake by valinomycin and gramicidin, **29:** 287–292
divalent
effect on phosphorylase *b,* **24:**264–265; 269; 274; 276; 277–279; 286
effect on sperm, **24:**78–80
interchangeability for alkaline phosphatase action, **10:**113–114
and lipoic acid, **18:**340
monovalent, activation of pyruvate carboxylase by, **42:**32
Caulobacter cell cycle, **71:**1–39
background, **71:**1–2
cell-cycle, **71:**3–5
flagellar structure and assembly, **71:**7–14
polar differentiation and asymmetric cell division, **71:**30–33
protein localization, **71:**25–30
chemotaxis proteins, targeting of, to swarmer pole, **71:**26–28
flagellar assembly and, **71:**28–29
generally, **71:**25–26
non-flagellar proteins, **71:**29–30
stalk formation, **71:**5–7
transcription regulation, **71:**15–25
asymmetric expression of non-flagellar genes, **71:**24
cell-cycle regulation of flagellar gene expression, **71:**15–21
polarized expression of flagellar genes, **71:**21–23
polarized transcription basis, **71:**24–25
Caulobacterium crescentis, **58:**73
Cbz-Phe-His-Leu, **57:**26
C=C binding, reduction of, PN-dependent, **26:**127–128
C-Cephalosporinase, as β-lactamase, **28:**241
^{14}C-CMP, reduction of, **28:**444–445
CD curve, of azurin, **33:**201
of laccase, **33:**201–202
C_4-dicarboxylic acids. *See also under* individual acids
and fat metabolism, **8:**392–401; 401–409; 412–414
cDNA
cloning of, **57:**400; 403; 405
enzyme and transcripts, **72:**263–264
γ-glutamyl transpeptidase, **72:**243–245
in mouse enzymes, **57:**396–397
pig liver thioltransferase gene, **66:**169
in rat enzymes, **57:**393–396
sequences of, **57:**400–404
cDNA clones
cholesterol ester hydrolases regulation, **69:**66; 105–107
isoenzymes, **69:**2
cDNA cloning, tyrosine hydroxylase transcriptional regulation, **70:**172
CDP, *See* Cytidine diphosphate; Cytidine diphosphate choline
agarose, **52:**34; 39; 42
ethanolamine-agarose, **52:**149
hexanolamine-agarose, **52:**149
CDP, DR metabolism, **28:**438–449; 453
radioactive, reduction of, **28:**444

CDPabequose, role of, in LPS synthesis, **31:**92
CDPabequose synthetase, reaction catalyzed by, **31:**107
CDP-choline, in synthesis of lipids, **26:**63
CDP-3,6-dideoxyhexoses, accumulation of, in mutants, **31:**116
CDP-ethanolamine, in synthesis of lipids, **26:**63
CDPglucose oxidoreductase, reaction catalyzed by, **31:**107
CDP-D-glucose pyrophosphorylase
  action of, **26:**52
  allosteric regulation, **28:**111; 113
CDPglucose pyrophosphorylase, reaction catalyzed by, **31:**107
CDP-glycerol, as precursor of teichoic acid, **30:**234–235; 237
CDP-L-glycerol pyrophosphorylase, action of, **26:**54
CDP-glycerol pyrophosphorylase, in the cells of *Bacillus licheniformis* or *Bacillus subtilis,* **30:**236
  in the membrane fraction of *Staphylococcus lactis* or *Lactobacillus buchneri,* **30:**236
^{3}H-CDP, in DR metabolism, **28:**446
CDP-paratose-2′-epimerase, **44:**81
CDP reductase, in DR metabolism, **28:**436–449; 453
CDP-ribitol, as precursor of teichoic acid, **30:**234–235
CDP-L-ribitol purophosphorylase, action of, **26:**55
CDP sugars, **26:**47–48; 51; 61; 63; 66
CD spectrum, of purple intermediate, **34:**63
Ce^{++++}, in decarboxylation, **12:**102; 104
Celerio euphorbiae, **45:**419
Celery, antibiotic treatment of diseases, **20:**392; 393; 398
Celite, adsorbent in separation of clotting factors, **16:**87
Cell adhesion, fibronectin and, **70:**1–21. *See also* Fibronectin
Cell-bound forms, **34:**482
  -envelope, bacterial, **34:**475–478
  -envelope membrane, **34:**456; 467–470
  epithelial, **34:**479
  junction membrane, **34:**470
  structure, **34:**494
Cell culture models, ADA defiency, **51:**182–184
Cell cycle, **42:**267
Cell differentiation, **58:**246
Cell division
  polyamines in, **36:**219
  role of thiols in, **11:**253–255; 257
Cell functions, **58:**103
Cell growth factors, **57:**87; 162–164
Cell lysates, RNA polymerase II transcription, reversible phosphorylation of CTD during transcription cycle, **71:**51
Cell membrane, **44:**228
Cell membranes, adsorption of enzymes, **28:**31
Cell metabolism and structural arrangements of constituents, **13:**1–20
Cell nuclei, phosphatides, **7:**168–169
  phosphorus turnover in, **7:**182–186
Cell Nucleus, Chemistry (Allfrey, Mirsky, and Stern), **16:**419
Cellobiase, **5:**120, **8:**268–269; 285
  separation by chromatography, **14:**349
Cellobiose
  from cellulose, **8:**256–258
  in β-glucan synthesizing system, **28:**339
    hydrolysis of, **28:**353
    in laminarinase digests, **28:**351
  in polysaccharide synthesis, **15:**306; 315
Cellobiose-2′ epimerase, **44:**91
3-*O*-β-Cellobiosyl D-glucose, from barley β-glucan, **28:**345
  in β-glucan synthesis, **28:**340
Cellophane, **34:**456
Cell organization, **8:**24–26
Cellotriose, in β-glucan synthesizing system, **28:**339
Cellotriose from cellulose, **8:**258
3-*O*-β-Cellotriosyl D-glucose, from barley β-glucan, **28:**345
Cell population, **36:**33
Cell proliferation, **42:**255
Cell proteins, dynamic state, **37:**137–138
Cell receptors, and drug resistance, **27:**432; 445–450
  sulfur compounds, **27:**226–228
Cell respiration, **58:**246
Cells
  division, relation to DNA metabolism, **16:**429; 452–454; 458
    germ (*See* Egg cell and Sperm cell)
    nondividing, DNA in, **16:**454–457
    nucleus, chemistry of, **16:**411–500
  epidermal, **45:**404
    duct, **45:**404
    zymogen, **45:**404

irradiation effects, **24:**401–403
transformed, **48:**82–83
treatment with sialidases, **48:**66–68
Cell separation, bone marrow lymphocytes, **47:**361
thymocytes, **47:**358–359
Cell spreading assay, fibronectin, central cell-binding domain of, **70:**11
Cell surface, localization of enzymes in, **13:**16
Cell surface enzymes, **33:**287–288
Cell-surface interactions, **40:**51; 52; 53; 54; 55
Cell-surface receptors, integrin, fibronectin and, **70:**1
Cellular concentration of enzymes, **7:**456–457
Cellular levels, and growth, **49:**43
Cellular particles, enzymes of, **27:**251; 252
Cellular surfaces, as catalysts, **28:**32
Cellular topology of enzymes, **28:**263–269
Cellular water, need to refer uptake to, **49:**69
Cellulase, **8:**268–269; 271; 285
action of, **26:**58
cacao, **25:**466
multiple forms, **25:**350
column chromatography, **22:**447; 473–475
kinetics, **28:**345–347
soil, **28:**386
Cellulases, action on dialyzing membranes, **14:**399
classification, **14:**246
isolation, **14:**401
separation by chromatography, **14:**348
Cellulodextrins, endwise degradation, **28:**353
Cellulose, **5:**198, **34:**453; 521, **38:**200; 210
action of acid on, **8:**259–262
action of wood-destroying molds on, **8:**269–279
crystalline structure, **8:**263–264; 283–284
degradation, and lignin, **8:**279–295
dissimilation by wood-destroying molds, **8:**278–279
enzymic degradation, **8:**265–279
linkage between glucose units, **8:**255–259
linkage to lignin, **8:**283–288
microbiological degradation, **8:**253–298
oxidative degradation, **8:**265–269
preliminary hydrolysis, **8:**265; 266–269; 272–274; 279
products, **8:**266–268; 278–279
relation to glucose, **8:**255–264; 267–269; 271–273
structure, **8:**254–264; 283–284
Tollens-Hiller-Pacsu concept of structure, **8:**259–262
x-ray studies, **8:**262–264
as cell-wall constituent, **2:**242; 243
chain, diagram, **2:**241
changes during fermentation of tobacco, **10:**369–370; 396; 402
oxidized, action on thrombin, **10:**23
chemistry of, **2:**240–243
decomposing organisms, **2:**245; 251–262
digestion in plants, **18:**288
formation from $C^{14}O_2$, **18:**361
removal from lignin, **18:**351–352; 355
effect on properdin system, **28:**357
β-1,3 glucan, **28:**327; 337; 338
enzyme action on, **33:**260–261
linkage to pectin in plants, **20:**345
micellae, **2:**242; 243
primers for, **26:**58
synthesis of, sugar nucleotide donor, **26:**66
properties of, **30:**180
structure of, **30:**180
synthesis, **15:**315
weak bonds in, **15:**304
Cellulose, Decomposition by Microorganisms (Norman and Fuller), **2:**23
Cellulose, Microbiological Degradation (Nord and Vitucci), **8:**25
Cellulose bacteria, classification, **2:**256–259
Cellulose decomposition, **2:**244–260
biochemistry of, **2:**244–251
cultural methods, **2:**252; 253
optimal temperature and *p*H, **2:**246; 250
and oxycelluloses, **2:**247; 248
through actinomycetes, **2:**260
through aerobic bacteria, **2:**245; 249–255; 260–262
through anaerobic bacteria, **2:**249–251; 255–256; 260–262
through fungi, **2:**251; 259–262
Cellulose depolymerization, inhibition, **28:**348
Cellulose exchangers, in protein chromatography, **22:**431–436; 443–445; 447
Cellulose-hydrolyzing enzyme system, **18:**351
Cellulose in flue-cured tobacco, **6:**364
Cellulosetrisulfuric acid, effect on blood clotting, **10:**35

Cell wall
bacterial
interpeptide bridges, **40:**118
structure and biosynthesis, **40:**118
yeast (*See* yeast cell wall)
Cell wall biosynthesis, *Caulobacter* cell cycle, **71:**6–7
Cell-wall lytic enzymes, in β-lactamase study, **28:**253
Cell walls, **59:**60
bacterial, and bagshaped macromolecules, **26:**193–232
synthesis of, **26:**36; 66; 68–72
biosynthesis, **28:**110
β-1,3 glucans of, **28:**327; 330; 335; 336; 341–343
lysis of, **28:**253; 348–349; 355–356
of mating yeast cells, enzyme for softening, **28:**348–349
molecular sieving, **35:**235
Centers of enzyme formation, **24:**4–7
Central pore mutations, ion channel model system annexin V, mutants analysis, **71:**226–229
Centrifugal fractionation, enzyme distribution, **24:**291–358
Centrifugation, density gradient, in isolation of mitochondria, **21:**4; 75
differential, in isolation of mitochondrial fragments, **21:**4; 6; 75–77
in isolation of mitochondrial fragments, **21:**102
Cephalin, **5:**32; 39; 40; 45; 46. *See also* Phosphatides
from brain thromboplastin, **16:**50–51
speed of formation, **7:**137–138
Cephalin and blood clotting, **10:**4
Cephalocethium roseum, steroid hydroxylation by, **19:**191
Cephaloridine, action of β-lactamase, **28:**240; 247
Cephalosporin amidase, and β-lactamase, **28:**248
Cephalosporinase. *See also* β-Lactamase; Penicillinase
biochemistry of, **28:**238; 252–254
occurrence of, **28:**241–244
Cephalosporin C
action of β-lactamase, **28:**240; 245–247; 250; 252; 288–292; 294; 297
in penicillinase induction, **22:**119
in penicillinase synthesis, **22:**113
Cephalosporin C_a (pyridine), action of β-lactamase, **28:**240; 259
Cephalosporins
definition, **28:**241
hydrolysis by microorganisms, **28:**242–244
β-lactamase action, effect of, nucleus, **28:**260–261; 304
side chain, **28:**259; 260; 304
structures, **28:**240
as substrate of β-lactamase, **28:**244; 245; 247–254; 262; 268; 272; 273; 287–294; 306–309; 312–313
fluorinated, **55:**274
Cephalosporum subverticillatum, enzymic transformations of steroids by, **20:**262
Cephalothecium roseum, enzymin transformations of steroids by, **20:**242; 245; 249; 274
Cephalothin, **46:**392; 393
action of β-lactamase, **28:**240; 243; 245–247; 252; 259; 290
Cepholin, methylation of, **25:**417; 419
Ceramide, **34:**466
Ceramidelactoside, **38:**295; 308, **48:**73
Ceramide lactoside-β-galactosidase, **38:**302
Ceramidelactoside lipidosis, **38:**294; 295; 298
Ceramidetrihexosidase, **38:**306; 307
human-placental tissue, **38:**306; 307
rat-intestinal tissue, **38:**307
Ceramide trihexoside, **38:**295; 298; 299
Ceramide trihexoside-α-galactosidase, **38:**299
Ceratostomella ulmi, polygalacturonases of, **20:**366
Cercospora beticola, **55:**91
Cereal embryos and seedlings, terminal oxidases in, **18:**39–43
Cereal β-glucan, **28:**327; 345; 353
Cereal laminarinases, patterns of hydrolysis by, **28:**350
Cereals
β-glucanases of, **28:**341–342; 344–346; 353
respiration in, **18:**286; 294
roots, cytochromes in, **18:**20–22; 33; 42; 49–52
seed, cytochromes and oxidases in, **18:**9; 19
Cerebrocuprein, **41:**52
extinction coefficient of, **33:**212
Cerebroside, **34:**466

requirement for —SH groups, **11:**241
sulfate, **34:**466
Cerebroside sulfatase activator, **60:**177
Cerium
effect on phosphoglucomutase activity, **14:**212
effect on tryptophan metabolism, **10:**68
Cerium ion, model catalyses with, **21:**171; 172; 175–180; 182–188; 190
physiological effect, **21:**193
Ceruloplasmin, **41:**100; 102–104; 106–114; 117; 118; 122; 123, **44:**187, **48:**67
asialo, **41:**100; 102; 103; 104; 109; 110; 111; 112; 113; 122; 123
broad component in EPR of, **33:**214; 216
composition of copper in, **33:**196
EPR-nondetectable copper ions in, **33:**223
extinction coefficient of, **33:**198; 200
forms of copper in, **33:**195
mechanism of deduced from relaxivity data, **33:**63–64
nonequivalent cupric ions in, **33:**203
oxidation-reduction potential of, **33:**206
oxygen, product of water reduction of, **33:**230–231
reaction with azide, **33:**221
[64Cu]-, **41:**106
modified rabbit, **41:**102
radioactive preparation of, **41:**100
in schizophrenia, **29:**506–507
in serum, **41:**102–103; 108–110
tritiatron of, **41:**117–118
Ceruloplasmin, Cotton effect in, **27:**60; 92–94
EPR study, **27:**110; 126–128; 138; 144–148; 150
laccase reaction, **27:**126–127
Ceruloplasmin: The Copper Transport Protein with Essential Oxidase Activity (Frieden and Hsieh), **44:**187
Cesium ions, **15:**257
properties, **21:**157
C factor, and contraction, **23:**385
C-Factor, SH-group, **27:**228
CFTR, function, **65:**38; 40
CGD, classification, **65:**66–67
Chaetomin, **7:**469; 470; 487; 489; 491; 493
Chain, paraffin, **34:**483
reactions, **34:**479
Chain branching, in β-1,3 glucans, **28:**327–328; 331–332; 334–337; 340
Chain growth, in polysaccharide synthesis, **11:**301–302; 310–312; 314–315; 319; 321–322; 323; 354–358
Chain length. *See also* Intermolecular forces, Particle size, and Van der Waals′ dispersion forces
and enzyme activity, **12:**174; 201–202; 230; 247–248; 265–268; 325
Chain lengthening, in β-glucan synthesis, **28:**338–340
Chain length in polysaccharides, effect on enzyme action, **11:**346–347
Chain reaction
in clotting of milk, **15:**445
in enzymic oxidoreductions, **15:**2–11; 13; 16–18; 32
and ionizing radiations, **24:**385
α-chains, collagen hydroxylase occurrence, **72:**328; 330
Chains, subunit structure, **44:**191
Chain termination, **47:**407–409
Chalcone
enzymic formation of, **55:**19–29
flavonone interconversions, **55:**20
enzymes in, **55:**27
initial reactions catalyzed by, **55:**25
isoflavones from, **55:**29
naringenin, **55:**19; 20; 22
as precursor of quercetin in plants, **16:**145
Chalcone cyclase, **55:**29
Chalcone isomerase, **55:**15; 20
mechanism of action, **55:**27–29
in phaseollin accumulation, **55:**108
properties of, **55:**29; 30
substrate specificity of, **55:**29
Chalcone synthase
abiotic and biotic elicitors for, **55:**26–27
discovery of, **55:**19
fatty acid synthase and, **55:**22
formation of release products by, **55:**22; 23
induction of, **55:**106; 107
mechanism of action of, **55:**22; 25
properties of, **55:**20; 21
regulation of, **55:**111; 112; 113
in response to wounding, **55:**27
substrate specificity of, **55:**20
Chan CU_A model, cytochrome *c* oxidase, proton pumping, **71:**146–149; 170–176
Chan CU_A model (alternate), cytochrome *c* oxidase, proton pumping, **71:**150–152

Chan CU_B model, cytochrome *c* oxidase, proton pumping, **71:**152–154
Channeling, **34:**521
Chara ceratophylla, **7:**18
Characterization, procollagen, **42:**171
Charcoal, in model enzyme system, **15:**3; 21–24; 39
Charcoal particles, **34:**452
Charge, role in transport, **49:**53
Charged residue mutation, ion channel model system annexin V, mutants analysis, **71:**230
Charge groups, **34:**486
  induced associations, **34:**485
Charge isomers, **46:**388
Charge rearrangements, in isoalloxazine, on light absorption, **27:**15
Charge relay system, **43:**164; 196
Charge-relay system, **41:**239, **50:**21; 22
  of serine proteases, **50:**2; 12
Charges, pairing of, **50:**20
  in hemoglobin, **50:**32
Charge-transfer complex, **43:**177; 178; 181; 196
Charge transfer complex, of cytochrome *f* and P-700, **30:**37–38
Charge transfer complexes, **22:**348, **39:**346
Charge transfer transition, copper, **27:**147
Charonia lampas, as source of enzymes involved in sulfate ester formation, **22:**214; 215
  as source of glucosulfatase, **22:**228
Chastek paralysis, **15:**390–391
α-C-H bond, of substrate, cleavage of, **34:**55; 58; 66
ChE, *See* Cholinesterase
Cheirolin (γ-thiocarbimidopropylmethylsulfone), occurrence and properties, **10:**264–265
Chelate configuration, of metal ions, effect on enzyme activation, **21:**157; 158; 162
Chelate formation, by metal ions, in enzyme activation, **21:**143–148
Chelate ring, in vitamin B_6-catalyzed reactions, **16:**224
Chelate ring in enzyme-substrate complex, **12:**205; 214–215; 218; 240; 253
Chelate rings of metals with proteins, **15:**39–40
Chelates, *See* Metal complexes
  of copper, EPR study, **27:**147
    metal, as electron carrier, **27:**179
    or iron, EPR study, **27:**140; 159
Chelating agent
  effect on activity of β-glucuronidase, **16:**379
    effect on myosin ATPase, **16:**335
  ethylenediamine tetraacetic acid, **15:**184; 223; 250
Chelating agents, **58:**85
  alkaline phosphatase and, **56:**341
  in allosteric regulation, **28:**140
    effect on laminarinase, **28:**347
  and ATP synthesis, **23:**325–326
    effect on, collagenase, **23:**195–196
      elastases, **23:**234; 242
      oxidase activity, **23:**305
  carbonic anhydrase and, **56:**389
  carboxypeptidase and, **56:**359
  effect on, dehydrogenase action, **24:**470
    sperm, **24:**71–72
  effect on hyaluronidase activity, **13:**214
  metalloenzymes and, **56:**300–303
  polymerases and, **56:**338–341
  and succinic dehydrogenase, **18:**75; 88; 94–99
  thermolysin and, **56:**379
Chelation, **43:**281
  in alcohol dehydrogenase study, **27:**49–50; 53–56; 58; 66
    and Cotton effect, **27:**88
    cuprous, cytochrome oxidase, **27:**148
    and metal-flavin contact, **27:**184–186
    of YADH, and Cotton effect, **27:**83–85
Chelators, iron, effect on succinate dehydrogenase, **28:**225–228
Chemical evolution, **27:**388
Chemical genetics, **4:**1; 2
Chemical kinetics, of photosynthesis, **1:**209–216
Chemical mechanism of lactate dehydrogenase, **37:**102–114
Chemical mechanisms, **39:**352; 355
  of acetyl CoA carboxylase, **45:**381
    of aconitase, **45:**354
    of alcohol dehydrogenase, **45:**372
    of alkaline phosphatase, **45:**358
    of biotin carboxylation from MgATP and bicarbonate, **45:**381
    of carboxypeptidase, **45:**367
    of dehydrogenases, **45:**371; 373
    of enolase, **45:**353
    of fructokinase, **45:**370
    of fumarase, **45:**352
    of glutamate dehydrogenase, **45:**377

of glyceraldehyde-P dehydrogenases, **45:**372; 374
of β-hydroxy-β-methylglutaryl CoA reductase, **45:**373
of isocitrate dehydrogenase, **45:**374
of lactate dehydrogenase, **45:**377
of malic enzyme, **45:**374
of ribonuclease, **45:**359
of thymidylate synthetase, **45:**370
of transcarboxylase, **45:**379–380
of triose phosphate isomerase, **45:**356
Chemical Mechanisms of Enzyme-Catalyzed Reactions, Determining by Kinetic Studies (Cleland), **45:**273
Chemical modification
affinity labeling, **67:**27–30
cholesterol ester hydrolases regulation, **69:**68
general reagents, **67:**25–27
Chemical modifications of proteins, **24:**105–160; 202–227; 244; 248
Chemical Modifications of Proteins and Their Significance in Enzymology, Immunochemistry, and Related Subjects (Sri Ram, Bier, and Maurer), **24:**101
Chemical mutation, *See* Mutation, chemical
Chemical phenotypes, **4:**21–26
Chemical rescue, rubisco, **67:**37–40; 59–60
Chemical transformation, maximum speed of, **25:**9–12
Chemiluminescence, **7:**216; 227, **41:**44; 53. *See also* Bioluminescence, Light emission, and Luminescence
and oxidation, **25:**120–131
Chemiluminescence of LH_2 AMP, **44:**43
Chemiluminescent reactions, mechanisms of, **25:**120–131
Chemiluminescent spectra of lumines cent substances, **7:**256
Chemiosmotic hypotheses, **34:**481
Chemiosmotic system, **51:**99; 104–105
Chemiosmotic theory, **44:**256
Chemisorption, of enzymes, **28:**20; 28
Chemistry of ferredoxins, **33:**149–165
Chemo-autotrophic bacteria, **1:**265; 266
Chemo-autotrophs, **74:**6–7; 10
Chemosensory apparatus, *Caulobacter* cell cycle, **71:**3; 4; 26–27
Chemosynthesis, carbon dioxide in, **74:**6
Chemosynthesis and acetyl phosphate anabolism, **6:**248–250
Chemosynthesis in assimilation of CO_2, **2:**138; 139; 143
Chemotaxis, **44:**241
Chemotaxis proteins, targeting of, to swarmer pole, protein localization, *Caulobacter* cell cycle, **71:**26–28
Chemotherapeutic compounds, effect on luminescence, **7:**252
Chemotherapy
antifolic agents, **27:**436–450
antimetabolites, **27:**461
and folate transport, **27:**458
and inhibition of folate biosynthesis, **27:**417–468
synergistic effects, **27:**456
and enzyme problems, **6:**34–122
Chemotherapy, and Enzyme Problems (Sevag), **6:**33
Chemotherapy, Inhibition of Folate Biosynthesis and Function as a Basis for (Hitchings and Burchall), **27:**417
Chemotherapy (methotrexate)
dihydrofolate reductase studies, **70:**25
eukaryotic dihydrofolate reductase, ligand binding (of inhibitors), **70:**55–56
Cherries, antibiotic treatment of diseases, **20:**400
N^5,N^{10}-CH_2-FAH_4 reductase, reaction, **27:**425
5,10-CH_2 H_4 PteGlu, **54:**58; 83
in assessing antitumor activity of FU, **54:**105
in binary complex with FdUMP, **54:**82
methylene bridge joining, to dUMP, **54:**76
in potentiation of fluoropyrimidine cytotoxicity, **54:**97
in ternary complex with FdUMP and TS, **54:**81–86
thymidylate synthetase binding to, **54:**78
5,10-CH_2H_4 PteGlu binding to, **54:**78
in ternary complex with FdUMP and 5 10–10CH_2H_4PteGlu, **54:**80; 81–86
dissociation of, **54:**83
Chick embryo, cholesterol in, **7:**168
fatty acids in, **7:**168
phosphatides in, **7:**165–166
Chick-embryo enzyme, amino acid composition of prolyl hydroxylase from, **41:**248
Chick embryo extracts, DR metabolism, **28:**436; 448–449
Chick embryonic tissue, synthesis of pentoses, **28:**434–436

Chick embryos, rich source of prolyl hydroxylase, **41:**248
used to determine requirement for molecular oxygen in hydroxylation of proline, **41:**247
Chicken liver phosphofructokinase, **48:**208
Chicks, ribose synthesis, **28:**414; 418–419
Chimeric enzymes, glutathione S-transferases, **69:**31–34
Chimeric G-protein-coupled receptors (GPCRs), adenosine receptors, **69:**109
Chimeric isoenzymes, glutathione S-transferases, **69:**33
Chinchona alkaloids, and quinuclidine nucleus, **8:**235
biosynthesis, **8:**242
origin, **8:**231–233
Chinese hamster ovary cell culture, **42:**244; 272
Chinese restaurant syndrome, **53:**216
Chiral acetate, configurational assay of, **50:**251
synthesis of, **50:**257; 258; 259; 260
Chiral methyl groups, **50:**243–298
analysis of, **50:**247–256
biochemical applications of, **50:**263; 298
synthesis of, **50:**256–263
Chiral Methyl Groups (Floss and Tsai), **50:**243
Chiral Phosphorothioates: Enzymatic Substitution at Phosphorus, Stereochemical Analysis of (Frey), **62:**119
Chironomus tentans, **45:**414
Chitin. *See also* Chitin synthetase
chain length of, **59:**60–61
colloidal, **59:**84–85
degradation
action, **59:**90–93
assay of, **59:**84–86
distribution, intracellular, **59:**93–94
properties of, **59:**86–87; 90
purification of, **59:**86–87; 90
regulation, **59:**93
description of, **59:**59–60
distribution of, **59:**61–62
enzyme action on, **33:**261
and hydrogen bonding, **59:**61
nascent, **59:**90–93
orientation of crystallites of, **16:**160
polysaccharide of arthropod cuticle, **16:**153
overview of, **59:**94–95
preformed, **59:**90–93
regenerated, **59:**84
structure of, **59:**60–61
synthesis of, **26:**59; 66
in yeast cell wall, **14:**204
Chitin, The Synthesis and Degradation of (Cabib), **59:**59
Chitinase, lytic action, **28:**355–356
Chitinase activity, **45:**417
assay of, **59:**84–86
distribution of, intracellular, **59:**93–94
on nascent chitin, **59:**90–93
on preformed chitin, **59:**90–93
properties of, **59:**86–87; 90
purification of, **59:**86
regulation of, **59:**93
Chitinases, *See* Chitinase activity
classification, **14:**246
isolation, **14:**401
separation by chromatography, **14:**349
structure studies, **14:**303
Chitin deacetylase, **59:**83
Chitindisulfuric acid, effect on blood clotting, **10:**35
Chitin-protein complex, tanning of, **16:**154
Chitin synthetase. *See also* Chitin
and *N*-acetylglucosamine, **59:**62
in animals, **59:**81–83
assay of, **59:**62–63
and cloning of gene, **59:**80
cooperativity of, **59:**65
in crustaceans, **59:**81–82
digitonin effect of, **59:**70
divalent cation requirement of, **59:**63
in fungi
conclusions of, **59:**79–81
distribution of, intracellular, **59:**74–76
inhibitors of, **59:**71–74
particulate preparations, **59:**63; 65
properties of particulate preparations, **59:**64
purified preparations, **59:**68–71
vectorial, by membranes, **59:**76; 78–79
zymogen form of, **59:**65–68
inhibitors of
in animals, **59:**82–83
in fungi, **59:**71–74
in insects, **59:**81
molecular weight, **59:**69
and *Neurospora crassia,* **59:**63; 68; 76
phospholipid requirement, **59:**69
proteolytic activation, **59:**65

solubilization, **59:**68
stoichiometry of, **59:**63
Chitobiase, classification, **14:**246
separation by chromatography, **14:**349
Chitodextrinase, classification, **14:**246
Chitodextrins, **59:**81–82
Chitosan, **55:**101, **59:**83–84
activation of β-glucuronidase, **40:**41
as activator of β-glucuronidase, **16:**376
Chitosomes, **59:**75
Chlamydomonas
carotenoid-deficient mutant of, **19:**62; 63
effect of light on, **25:**74
phosphoribulokinase from, **74:**100; 114; 117; 120; 121
in multienzyme complex, **74:**104–105
Chlamydomonas eugametos, biogenesis of flavonoids in, **16:**143–146
Chlamydomonas reinhardi, light-induced absorbancy changes in chloroplast fragments of, **30:**35; 39
Chlamydomonas reinhardii, **46:**100; 101; 337
ADP-glucose pyrophosphorylases of, **46:**337
Chlamydomonas reinhardtii axonemes, **53:**136
Chlamyudomonas reinhardi, photoelectron transport in, **25:**66
Chloracetate, **51:**10
Chloracetic acid, thermodynamic data on ionization, **20:**64
Chloracetyldehydrophenylalanine, absorption spectrum, **8:**132
and dehydropeptidases, **8:**134; 135; 139; 140; 144; 151
preparation, **8:**123
Chloral hydrate, **17:**138; 151
Chloramine-T, action on fibrinogen, **10:**33
Chloramphenical, effect on murein synthesis, **26:**218
Chloramphenicol, **17:**313, *See* Chloromycetin
absorption by *Nitella clavata,* **20:**387
in treatment of plant diseases, **20:**394; 402
accumulation of intermediates, **27:**476; 487; 492; 494–495; 499–500
effect of, on synthesis of O antigen, **31:**116
effect on carotenogenesis, **21:**319
effect on β-lactamase, **28:**266
synthesis of gramicidin S, inhibited by, **33:**340–342; 345; 347; 348
tyrocidine synthesis, effect on, **33:**350
synthesis of 70S ribosomes inhibited by, **30:**26
Chloramphenicol acetyltransferase (CAT), tyrosine hydroxylase transcriptional regulation, **70:**173; 175–176
Chloramphenicol-sensitivity, **42:**279
Chloranil (tetrachlorobenzoquinone), P/O ratio in pseudocyclic phosphorylation with, **30:**65
Chlorella, **46:**37; 42; 46; 70; 76; 77; 98; 100
absorbancy changes of chlorophyll *b* in, **30:**31–32
absorption bands of, **30:**11
action spectrum for the Emerson enhancement effect in, **30:**61
chlorophyll *b* in, **30:**17
difference spectrum for, **30:**32
fast light-induced reactions of cytochrome *f* in, **30:**37
manganese in, **30:**25
photosynthetic unit in, **30:**18
plastocyanin from, **30:**14
quantum yield of photosynthesis in, **30:**57
amino acids in, **14:**266
biosynthesis of glutamic acid in, **20:**327
carbon dioxide fixation by, **19:**21; 33
cytochrome *f* in, **19:**50
glycolytic enzymes in, **19:**7–10
Hill reaction in, **19:**50
methemoglobin reducing factor in, **19:**50
oxidative phosphorylation by, **19:**17–18
photosynthetic studies with, **19:**54–56; 59; 60; 62
cytochromes in, **18:**25
lipoic acid study, **18:**341
photoxidizable pigment of, **18:**54
mutants, **46:**69
splitting enzyme in, **15:**353; 359; 361; 362; 375
Chlorella, ESR measurements in, **25:**83
photosyntheis, **25:**52; 53; 56; 60; 61; 63–64; 97
Chlorella fusca, **46:**49
Chlorella pyrenoidosa, **46:**340
ADP-glucose pyrophosphorylase, **46:**337
ADP-glucose synthesis, **46:**340
allosteric enzymes, **28:**69
starch synthesis, **46:**340
Chlorella regularis, **46:**74
Chlorella vulgaris, **46:**98; 337
regulation of chlorophyll synthesis in, **46:**99

Chlorenchyma, dehydrogenases in, **20:**219
Chlorethanol, **47:**74
Chlorguanide (proguanil), folate antagonist, **27:**438; 444
Chloride, **51:**243
complex with carbonic anhydrase, **47:**222; 226; 232–236
effect on dissociation constant of liver alcohol dehydrogenase system, **20:**41; 42
effect on fluorescence of riboflavin, **20:**32
muscle permeability to, **13:**58–59
permeability of gastric mucosa to, **13:** 56–58
uptake by frog skin, **13:**48–50
Chloride concentration, in familial periodic paralysis, **6:**215
in nutritional muscular dystrophy, **6:**204; 205; 206; 207
Chloride ions, and α-amylase, **12:**397; 398; 400
and β-amylase activity, **12:**400
role in chloroplast activity, **12:**13
Chlorin a_2, **23:**287–289
Chlorination, **58:**14
free radical, **58:**52
template directed, **58:**52
Chlorin *e,* carbon content, **19:**60
Chlorine, effect on pectic substances, **20:**352
Chlorins, **46:**80
Chloroacetate, reaction with thiols, **11:**213–214
Chloroacetylalanine, absorption spectrum, **8:**131
Chloroacetyldehydroalanine, absorption spectrum, **8:**131
and dehydropeptidases, **8:**139–142; 144; 145; 147; 148; 151
preparation, **8:**125; 128
Chloroacetylglutamate, **15:**371
in arginine synthesis, **16:**265
Chloroacetylglycyldehydroalanine and dehydropeptidases, **8:**140–142; 144; 145
Chloroacetylglycyldehydrophenylalanine and dehydropeptidases, **8:**140–141
ε-*N*-Chloroacetyllysine, in transaminations, **16:**195
δ*N*-Chloroacetylornithine, **49:**90
δ-*N*-Chloroacetylornithine, in transaminations, **16:**195
Chloroacetylphenylalanine, absorption spectrum, **8:**132
Chloroacetylsarcosyldehydroalanine and dehydropeptidases, **8:**140; 141; 144
Chloroacteyl-DL-alanyldehydroalanine and dehydropeptidases, **8:**140–142; 145
Chloroalanine, **53:**169, **56:**27; 28
β-Chloro-L-alanine, **35:**505
β-Chloroalanine, cysteine *S*-conjugate β-lyase reactions, β-elimination reactions, **72:**218–219
Chloroamphenicol, glucuronide as substrate of β-glucuronidase, **16:**368; 383
Chlorobacteriaceae, **39:**338, **46:**79
p-Chlorobenzaldehyde, **43:**294
addition of methoxyamine, **43:**294
p-Chlorobenzoyl, virus, **17:**478
Chlorobium chlorophylls, **46:**79; 80; 81
Chlorobium limicola, cytochromes in, **19:**57
Chlorobium limicola f. thiosulfatophilum, **46:**332
Chlorobium thiosulfatophilum, **58:**73
Chlorobium thiosulfatophilum, propionic acid metabolism, **26:**291
Chlorocruorin, **46:**63
Chlorocruorohaem, **23:**269; 312–313
Chlorocruorohaemochrome, **23:**281; 295; 313
Chlorocruoroheme, *See* Heme s
Chlorocruoroporphyrin, **23:**280–281; 284; 285
1-Chloro-2,4-dinitrobenzene, **46:**394; 395; 396; 397; 409
as alkylating agent, **46:**409
Chloroform, action in blood clotting, **10:**12–13
Chlorogenic acid, **55:**9
in oxidation of amino acids in plants, **20:**321; 322
in plant oxidase studies, **16:**132; 133; 150; 151
Chlorogenic acid from tyrosine, **14:**87
Chlorogloeopsis fritschii, phosphoribulokinase from, **74:**98
β-Chloroglutamate, **35:**118
Chlorohydrolase, **74:**72
β-Chloro-β′-hydroxydiethyl sulfide, mutagenic action, **27:**323
Chloroketone, inactivation by, glutathione synthesis, **73:**257–258
α-Chloroketones, **68:**189–190
p-Chloromercuribenzenesulfonate, in determination of sulfhydryl groups in papain, **19:**291–294
p-Chloromercuribenzoate
action on sulfhydryl enzymes, **15:**286; 289

reaction with thiols, **11**:217; 222; 225; 227; 230
p-Chloromercuribenzoate, **39**:349, **54**:400, **60**:243
action on dehydrogenases, **23**:46–48
in determination of sulfhydryl groups in papain, **19**:291–294
effect on oxidases, **19**:90; 97; 126; 176; 184; 199
effect on photosynthesis, **19**:64
inhibition of isopentenyl pyrophosphate into rubber in latex by, **29**:243
inhibition of reactions involving flavine coenzymes, **22**:381
as inhibitor of metalloflavoproteins, **17**:240; 254; 258; 259; 265; 311; 323
in cytochrome study, **17**:378
p-Chloromercuribenzoate, effect on mitochondrial oxidative phosphorylation, **21**:31
Chloromercuribenzoate, inhibition of ATPase activity, **6**:167
p-chloromercuribenzoate (-*p*-CMB) inactivation of prolyl hydroxylase by, **41**:256
inhibitor of proteinases, **41**:183
p-Chloromercuribenzoate (PCMB), effect on hydroxysteroid dehydrogenases, **20**:269
effect on muscle phosphorlase, **20**:70
in identification of imidazole, **20**:75; 95
p-Chloromercuribenzoic acid, effect on β-glucuronidase, **16**:379
p-Chloromercuriphenyl sulfonate, **41**:120
Chloromethylketone, **48**:291
Chloromethyl ketone derivatives, of peptide substrates, **41**:182; 225–227
Chloromethylurea, in photosynthesis, **25**:58; 86
Chloromycetin, **18**:211
activity, **10**:162–163
structure, **10**:163
in treatment of plant diseases, **20**:392
4-Chloro-7-nitrobenzofurazan, **49**:252; 253; 259
Chloroperbenzoic acid, **56**:368
5-*p*-Chlorophenoxypyrimidines, antifolate properties, **27**:442–443
4-Chlorophenylalanine, **35**:286
5-*p*-Chlorophenyl-6-ethylpyrimidines, structure-activity relationships, **27**: 440–443; 447–448
p-Chlorophenylglucuronide, as substrate for β-glucuronidase, **16**:365; 368–369; 374; 382
Chlorophyceae, carotenoids in, **21**:208; 300–302
Chlorophyll, **17**:436. *See also* Chloroplasts; Fluorescence, Photosynthesis
absorption bands, of, **30**:10
in chloroplast lamellae, **30**:9
fluorescence microscopy of, **30**:5
fluorescence of, **30**:10–11
forms of, **30**:10–11
P-700 in, **30**:11–12; 22
bacterial, **27**:536
in coacervate, **27**:370–371
fluorescence, **25**:45–46
molecules per quantasome, **25**:69
orientation in chloroplast, **25**:71–73
semiconduction in crystal, **25**:81–82
changes during fermentation of tobacco, **10**:340
sulfur in formation of, **10**:224
in energy transport, **14**:31; 35
excitation energy of, **11**:95; 97
fluorescence of, **11**:91–99
in photosynthesis, **19**:58–62
in plant respiration, **18**:305
in plastids, **20**:204–207
-protein compound, **12**:11–12
role of magnesium in, **12**:18–20
Chlorophyll a_2, fluorescence yield from, **30**: 27–28
Chlorophyll *a*, in photosynthesis, **19**:58; 61
Chlorophyll *a* (chl *a*), absorption band of, **30**:3
content of, in chloroplast fractions, **30**:50
Chlorophyll and coupled oxidations, **8**:452
Chlorophyllase, **5**:23, **46**:74; 75; 100
action of, **14**:223; 244
in plastids, **20**:213
Chlorophyll *b*, absorbancy bands of, **30**:31–32
absorbancy changes in, **30**:31–33
content of, in chloroplast fractions, **30**:50
as indicator of the high energy state of the chloroplast, **30**:33
Chlorophyll fluorescence, **1**:216
sensitizer, **1**:232
Chlorophyll Fluorescence and Photosynthesis (Wassink), **11**:91
Chlorophyllide, **46**:74; 75; 76
a, **46**:80; 94
Chlorophyllide-holochrome complex, **46**:75

Chlorophyllin, **58:**211
Chlorophyll in air-cured tobacco, **6:**333; 334
Chlorophyll-lipoprotein, isolation, **20:**206
Chlorophyll-nitrogen ratio, in chloroplasts, **20:**205
Chlorophyll-protein complex, in chloroplasts, **20:**206
Chlorophyll-protein complexes, separation on polyacrylamide gel electrophoresis of, **30:**54–55
Chlorophylls, **7:**307; 320–321; 330, **46:**37; 69–81; 93; 94; 95; 98; 101; 102. *See also* Magnesium tetrapyrroles and Tetrapyrroles
  a, **46:**72; 76; 77; 79
  a', **46:**76
  b, **46:**76; 77
  c, **46:**78; 79
  *c*1, **46:**79
  *c*2, **46:**79
  d, **46:**79
  degradation, **46:**100
  formation, **46:**76
  light energy transfer to, **21:**350–354
  synthesis, **46:**89; 99
    in barley, **46:**96
    in *Chlorella vulgaris*, **46:**99
    inhibition by 8-hydroxyquinoline, **46:**70
Chlorophylls, Hemes, and Related Compounds: Biosynthesis and Metabolic Regulation (Granick and Beale), **46:**33
2-Chloro-3-phytyl-1,4-naphthoquinone, **46:**25
Chloroplast, **34:**479
  membrane, **34:**470; 492
Chloroplast ATPase, **43:**36, **49:**229; 231; 233; 241; 258; 262; 263; 264; 265; 269
Chloroplast F_1, **49:**234; 238
Chloroplast fragments, CO_2 fixation by, **25:** 60–62; 66
Chloroplastin, **1:**311
Chloroplast preparations, in photosynthesis, **19:**45–65
Chloroplasts, **7:**39; 326–327; 329, **33:**294, **41:**41; 89, **46:**84; 94; 98; 99, **49:**272, **58:**244. *See also* Photosynthesis
  absorbancy changes of, in ultraviolet region, **30:**30 (*See also* Lamellae)
    absorption spectra of fractions of, **30:**45
    carotenoid content of, **30:**50
    chlorophyll content of centrifugal fractions of, **30:**44
  classes of, **30:**6
  CO_2 fixation by, **30:**6
  copper composition in fractions of, **30:**47
  differential centrifugation of digitonin-treated, **30:**43
  electron microscopy of, **30:**5
  fluorescence emission spectra of fractions of, **30:**52
  iron composition in fractions of, **30:**47
  light-induced electron flow in, **30:**23
  magnesium in fractions of, **30:**47
  manganese in fractions of, **30:**47
  mesophyll, characteristics of, **30:**5
  molar ratios of chlorophyll/cytochrome and chlorophyll/P-700 for fractions of, **30:**48
  $NADP^+$ reduction by, **30:**6
  phase-contrast microscopy of, **30:**6
  photochemical activities of fractions of, **30:**46
  photoinduced electron flow in, **30:**21
  photooxidation of Mn^{2+} to Mn^{3+} by, **30:**25
  quantum requirements of ferrocytochrome *c* photooxidation and reduction by, **30:**60
  70S ribosomes, in, **30:**26
  structural proteins of lamellae of, **30:**12
  structure of, **30:**4–7
  thylakoids of, **30:**5
  time course of fluorescence emission from, **30:**28
  trace metal and P-700 contents of centrifugal fractions of, **30:**47–48
  trace metal composition of fractions of, **30:**47
  treatment with sodium dodecyl sulfate (SDS), **30:**54–55
  types of, **30:**4–5
adsorption of phenolase on, **16:**152
  effect on enucleated *Acetabularia*, **16:**470
and carbohydrate synthesis, **8:**56
  and fat synthesis, **8:**56
  and functioning of cytoplasm, **8:**20–23
  structure, **8:**23
chemiosmotic coupling hypothesis of, **29:**309
  and H^+-translocator oxidoreductases, **29:**76
  and Na^+/K^+ antiporter ATPase, **29:**73
cytochrome content, **18:**24; 52–57; 305
  and lipoic acid, **18:**341

photosynthesis by, **18:**58–59
separation from leaves, **18:**43–44
starch formation in, **18:**291
formation and structure, **20:**203; 204
nucleic acids in, **20:**210; 211
proteins in, **20:**206–209
fractionation of, **25:**44; 67–69
illuminated, **1:**219–220
isolated, photoreduction in, **25:**42–44; 53–55; 59; 60–62; 67–69
lamellar system, **25:**70–71
oxidoreduction in, **12:**1–39
rhodanese activity of, **39:**339
Chloroplasts, ATP synthesis in, **23:**385–387
Chloroplasts, Cotton effect, **27:**59
Chloroplasts, Oxidoreduction in (Hill), **12:**1
Chloroplasts, TPN specificity of, **19:**32; 38; 54
Chloroplast structure, **25:**67–73
Chloropromazine
effect on mitochondrial oxidative phosphorylation, **21:**32
inhibitor of oxidation, **23:**368–369
α-Chloropropionyl chloride, rotational isomers of, **23:**18
a-Chloropropionyldehydroalanine, absorption spectrum, **8:**131
and dihydropeptidases, **8:**139; 140; 144
preparation, **8:**128
6-Chloropyridoxine, **50:**6
dissociation constant, **50:**5
Chloropyruvate, **55:**214
Chloroquine, inhibition of β-lactamase, **28:**295
Chlorozotocin, **55:**272
Chlor-phenylmagnesium bromide, chemiluminescence, **25:**121
Chlorpormazine, tyrosine hydroxylase regulation, **70:**151
Chlorpromazine, **55:**180; 182
in depression, **29:**492
in schizophrenia, **29:**532; 535
sedation due to, **29:**493
oxidation by peroxidase-H_2O_2, **27:**129–130
Chlortetracycline, in treatment of plant diseases, **20:**392; 395; 396; 398; 402
CHO cells, **52:**79
N^{10}-CHO-FAH_4 deacylase, reaction, **27:**425
N^{10}-CHO-FAH_4 isomerase, reaction, **27:**425
Cholangiosarcomas, loss of glucokinase in, **29:**334
Cholase, **23:**310
Cholate, **34:**474
in fragmentation of respiratory chain, **28:**190; 192; 196; 202; 204–210
and haem *a* spectra, **23:**290; 292; 293; 306–310
and haematin *a* spectra, **23:**291; 293; 296
in preparation of succinic dehydrogenase, **18:**70–71
sodium, as solubilizing agent, **17:**346; 348; 354–355; 358; 360
Cholate-haem-CO model compound, **23:**292; 293; 304; 306
Cholate inactivation, of cytochrome oxidase, **23:**310–311
Choleglobin, formation, **19:**196; 197
Choleinic acids, **17:**153
Cholera, **58:**118
Choleragen, **58:**119
subunit reduction of, **58:**123
Choleragenoid, **50:**305
Cholera toxin, **50:**305–313; 314–318, **58:**119
ADP-ribosyltransferase, **50:**313
binding to cell surface, **50:**306–307
conformational change, **50:**309
interaction with ganglioside, **58:**125
interchain disulfide bond, **50:**310
molecular weight, **58:**122
multivalent binding to the cell surface, **50:**307–308
purification, **58:**121
receptors, **58:**125
subunit A
active sites, **58:**126
ADP-ribosyltransferase activity, **58:**126
amino acid sequence, **58:**126
arginine, **58:**128
cysteine, **58:**128
subunit B, **58:**129
subunits, **58:**122
structure, **58:**124
subunit structure, **50:**305
B component (choleragenoid), **50:**305
A component, **50:**305
Cholestanol, **58:**53
Cholesterol, **17:**362, **34:**466–471, **38:**374, **50:**275
and adsorption of enzymes, **28:**26; 31
association with heparin, **16:**81
of cell nuclei, **16:**449
ovarian, **16:**394

Cholesterol (*Continued*)
biosynthesis, **22:**281–335, **38:**374–375
in developing brain, **38:**405–406
scheme of intermediates in, **22:**284
site of, **38:**376
site of regulation, **38:**375; 391–392
for column chromatography of lipase, **22:**447
distribution of acetate carbons in, **22:**282–285
effect on liver phosphatide, **7:**139
in chick embryo, **7:**168
hydroxylation, in mitochondria, **21:**26
in mitochondrial lipoproteins, **21:**109
and irritability of protoplasm, **8:**3
in Golgi bodies, **8:**21–23
in livers of depancreatized dogs, **8:**177
in mitochondria, **8:**20–23
labeled, synthesis from labeled mevalonate, **22:**290
metabolism of, **69:**45
microbiological conversions, **20:**242; 243; 267
serum, and elastase activity, **23:**246
steroidogenic pathway, **68:**44–46
synthesis, from lanosterol, **22:**324–327
quantitative, from mevalonic acid, **22:**289
and vitamin B_{12}, **26:**270
Cholesterol biosynthesis, and mevalonic acid, **23:**482–483
Cholesterol changes in blood, in nutritional muscular dystrophy, **6:**205; 207
Cholesterol dehydrogenase, **19:**203, **20:**267; 270
Cholesterol α-epoxide, **57:**376
Cholesterol esterase
localization of, **24:**317
of pancreas, **23:**130
Cholesterol esterases, classification, **14:**244
Cholesterol ester hydrolases regulation, **69:** 45–82
biochemical characterization, **69:**47–65
generally, **69:**47–50
pH optima, **69:**50–55
substrate-cofactor interactions, **69:**55–57
substrate molecular motions and thermophase behavior, cell activation effects, **69:**60–65
substrate physical state, **69:**57–60
generally, **69:**45–47
molecular characterization, **69:**65–69
Cholesterol from acetate, **12:**43; 312
Cholesterol inhibitors, in phytochrome study, **26:**28
Cholesterol levels, and β-1,3 glucans, **28:**356
Cholesterol metabolism of, **24:**296; 310; 331; 334; 338–339
Cholesterol α-oxide, **57:**377
Δ^7-Cholesterol reductase, localization of, **24:**299
Cholesterol synthesis, **15:**127
and coenzyme Q, **25:**298–299; 303
Cholesteryl arachidonate, cholesterol ester hydrolases regulation, **69:**48
Cholesteryl linoleate, cholesterol ester hydrolases regulation, **69:**48
Cholesteryl oleate, cholesterol ester hydrolases regulation, **69:**48
Cholesteryl palmitate, cholesterol ester hydrolases regulation, **69:**48
Cholesteryl stearate, cholesterol ester hydrolases regulation, **69:**48
Cholestrol, control of biosynthesis of, **29:**250
negative feedback control of synthesis of, in neoplastic cells, **29:**368
Cholestyramine, **60:**39
Choline, **4:**278–279, **17:**408; 411; 435–436, **34:**490. *See also* Methylation, biological
acetylation, **6:**260–262, **10:**221, **15:**186; 196; 199–201; 205; 206; 210
in air-cured tobacco, **6:**348
dehydrogenation, **15:**118; 127
acetylation of, **16:**1–2
and biological methylation, **12:**444; 449–450; 453; 458–469; 469–474; 478–479
biological oxidation of, **12:**481
biosynthesis, **21:**416; 427
in mitochondrial lipoproteins, **21:**100
oxidation, in mitochondria, **21:**22; 24
cleavage by microorganisms, **23:**327
oxidation of, **23:**350
and ethionine metabloism, **25:**370–372; 374
tracer studies, **25:**375; 379–380; 384
and formate, **12:**48
formation of, **24:**338
and methyl transfer, **7:**376–377
effect on liver phosphatide, **7:**138–141
and plasma phosphatides, **7:**139–140
phosphatidyl, biosynthesis, **25:**417–420
and phospholipide content of blood, **8:**180; 197–198

of blood plasma, **8:**197–198
effect on fatty livers, **8:**191–193
lipotropic action, **8:**172
in plants, **8:**234
relation to antifatty-liver factors of pancreas, **8:**191–193; 196; 197–199
synthesis, and methionine, **8:**198
and methyl donors, **8:**198
in reactivation of phosphorylated esterases, **20:**305; 306
and semen metabolism, **9:**353
in plants, **9:**439–440
source of methyl groups, **25:**383
synthesis, from dimethylaminoethanol, **22:**256
Choline acetylase, **12:**312–331, **15:**199; 200; 205; 206; 210
activity of –SH group in, **11:**243
in mitochondria, **21:**40
Choline acetylase system, location in subcellular fraction, **24:**310; 334
Choline-acetyl transferase, **55:**168; 169
Choline acetyltransferase, location in subcellular fraction, **24:**310
Choline analogs in phosphatide turnover, **7:** 172–173
Choline-containing ^{15}N, **7:**172
Choline-containing phosphatides, separation, **7:**134
Choline-containing phospholipides, in livers of depancreatized dogs, **8:**177
of plasma, **8:**180
Choline deficiency, effects, **7:**140–141
Choline dehydrogenase, **15:**118, **17:**383–384, **18:**100
inhibition by fluorophosphates, **13:**332
location in subcellular fraction, **24:**298
Choline dehydrogenase, PN-dependent, **26:**121
Choline esterase
hydrolysis by, activation energy, **23:**150
requirements for –SH group, **11:**241
Choline esterases, **9:**20
gasometric measurement, **9:**603
histochemical methods for, **9:**593
titrimetric methods, **9:**601
Cholineglycerophosphatide, **58:**187
Choline oxidase, **12:**466, **17:**383–384, **19:**203
activity of –SH group, **11:**241
and ethionine metabolism, **25:**372; 378; 380; 381
in mitochondria, **21:**22
Choline oxidase activity, location of, **24:**298
Choline oxidase in nuclei, **8:**33
Choline oxidase in plants, **9:**440
Choline phosphate, and inorganic phosphate, **7:**188–189
and liver phosphatides, **7:**141
Cholinergic agonists, **55:**138
Cholinergic antagonists, **55:**138
Cholinergic receptors, **55:**138
Cholinergic synapse, **55:**139
Choline r guanidoacetic transmethylase, classification, **14:**250
Choline r homocysteine transmethylase, classification, **14:**250
Choline r noradrenaline transmethylase, classification, **14:**250
Cholinesterase, **18:**118; 150
active groups in, **15:**288–289
kinetic studies, **15:**273; 284; 292
active sites, **24:**459
inactivation of, **24:**117
inhibitors, from insecticides, **24:**308
localization of, **24:**317
adsorption studies, **14:**337
classification, **14:**244
isolation, **14:**400
column chromatography, **22:**467
serum, abnormalities in man, **22:**15–17
and alteration in substrate affinity, **22:**19
human, use of inhibitors for differentiation of enzyme types, **22:**15; 16
pK value, **22:**58
effect of borate on, **12:**518
effect of van der Waals' forces on action of, **12:**252
specificity of, **12:**262; 264–270
immobilization of on urethan foam pad, **33:**279
kinetic studies, **17:**45–46
in muscle, effect of prostigmine, **6:**209; 210; 211
in myasthenia gravis, **6:**209; 210; 211
and nerve permeability, **13:**61
inhibition, **13:**321; 332; 333–336
Cholinesterase, Hydrolysis by (Davies and Green), **20:**283

Cholinesterases
active surface of, **20:**69
imidazole in, **20:**76; 77; 289; 290; 293; 300; 302; 305; 309–312; 314; 315
mechanism of hydrolysis by, **20:**283–318
phosphorylated, amino acids in, **20:**288
and nerve impulse transmission, **8:**468
from brains, **8:**463–464
distribution, **8:**467–469
from erythrocytes, **8:**462–467
in erythrocytes, **8:**462–467
inhibition mechanism, **8:**464
inhibitors, **8:**470–471
measurement, **8:**467
methods of differentiation, **8:**462; 464–467
multiple nature, **8:**462–467
properties, **8:**462–464; 466–467; 470–471
pseudo, **8:**463–464; 467
from serum, **8:**462–467
and sex hormones, **8:**468
of snake venoms, **8:**466–467; 469–471; 490–491
true, **8:**463–464; 465–467
types, **8:**461–471
Cholinestrase, **34:**452
Cholinestrases, plasma, in schizophrenia, **29:**505
Choline sulfuric acid ester in algae, **10:**234
Cholorophyllase in tobacco, **6:**369
Choloyl-CoA synthetase, localization of, **24:**310; 334
Cholyl coenzyme A, **18:**208
Chondoritin sulfate, ammonium salt, **17:**149
Chondrococcus columnaris, **23:**180
Chondrocytes, **60:**227
Chondroitin ABC lyase, **60:**285
Chondroitin AC lyase, **60:**285
Chondroitinase, activity of, **25:**438–440
Chondroitin B lyase, **60:**288
Chondroitin C lyase, **60:**288
Chondroitin sulfatases, **23:**228
classification, **14:**245
Chondroitin sulfate, **52:**27; 81; 83, **60:**110; 273
activation of lipoprotein lipase, **40:**33
activation of pepsinogen, **40:**37; 38
complexed with lipoproteins, **40:**56
effect on crude lipoprotein lipase, **40:**34
effect on oxidative phosphorylation system, **40:**39
effect on pepsin, **40:**56
effect on trehalose phosphate synthetase, **40:**36
interaction with polylysine, **40:**57
core protein B-xylosyltransferase, **52:**27
degradation by β-galactosidase, **60:**252
endocytosis, **60:**222
glycosyltransferases, **52:**81–87
sequential degradation, **60:**274
structure, **60:**274
synthesis, **22:**214
Chondroitin 4-sulfate, chemical structure, **25:**437–438
Chondroitin 6-sulfate, chemical structure, **25:**438
Chondroitin sulfate A, *See* Chondroitin 4-sulfate
Chondroitin sulfate B, *See* Dermatan sulfate
Chondroitin sulfate C, *See* Chondroitin 6-sulfate
Chondroitin sulfates, amino sugars in, **25:** 437–440
biological degradation and synthesis, **25:** 438–440
Chondrosine, from chondroitin 4-sulfate, **25:**437
Chondrosulfatase, **22:**229; 230
classification, **14:**245
occurrence, **22:**229; 230
properties of, **22:**229
in *Proteus vulgaris,* **22:**229
true substrate for, **22:**229
Chondrosulfatase in plants, **10:**310–311
Chondrosulfatases, **25:**439
Chorionic gonadotropin, **52:**32
human, **50:**314
Chorismate, **50:**148; 149; 169
conversion to
p-aminobenzoate, **66:**223–225
anthranilate, **66:**208–210
Chorismate mutase, **36:**69–71, **50:**170
reaction catalyzed, **29:**119
Chorismic acid
in *p*-AB biosynthesis, **27:**422
branch point in aromatic amino acid biosynthesis, **29:**119
Chorismic acid mutase, allosteric regulation, **28:**71
Christmas disease, hemostatic efficiency in, **30:**265
Christmas factor, **48:**278, *See* Factor IX
as component of thromboplastin, **16:**30; 73–74
Chromanols of coenzymes Q, **25:**284–285

Chromatic adaptation, **46:**92
Chromatic transients, in photosynthesis, **25:** 63–64
Chromatin, **47:**278
binding of enzymes, **27:**278; 282–283
and sodium concentration, **27:**261
Caulobacter cell cycle, flagellar structure and assembly, **71:**9
cellular differentiation and, **38:**82–85
control of DNA transcription, **38:**82–88
enzymology of, **38:**70–81
fractionation, **38:**47–49
occurrence, **38:**42
preparation, **38:**45–47
structure of, **38:**54–70
chemistry, **4:**11–17
experimental procedures in chromatin chemistry, **4:**12–14
metal content, **4:**14; 15
molecular weight, **4:**13
structure, **4:**13–14
synthesis during cell division, **16:**429
Chromatin, Structure and Function of (Simpson), **38:**41
Chromatin and hepatic tissue, **7:**187–189
Chromatin-bound enzymes, **27:**266–268
Chromatin space, of nucleus, **27:**265–268
Chromatin threads, **16:**417
isolation, **9:**597
Chromatium, **39:**361
phosphoribulokinase from, **74:**97
use of acetate, **23:**411; 412; 420; 430
Chromatium cells, photosynthesis in, **25:**-77; 55; 57; 76; 80–81
Chromatium high-potential iron protein, **47:**58; 59; 124
Chromatium vinosum, **46:**332
Chromatofocusing, **57:**375
Chromatographic adsorption of borate-carbohydrate complex, **12:**501
Chromatographic analysis of oxalacetate, **12:**76–77
Chromatographic analysis of RNase, **24:** 167–168
Chromatography, **46:**179; 208; 259; 262. *See also* Adsorption
affinity, **48:**36; 37; 51; 61, **57:**8–10; 373
column, **48:**36
gel emulsion, **48:**38
ion exchange, **48:**38
of amino acids and peptides in papain and mercuri-papain, **19:**294–300
of effect of vitamin B_6 on transamination, **19:**350
in evaluation of protein purity, **19:**280
of intermediates in photosynthesis, **19:**20; 21
of Kerbs cycle compounds in plants, **19:**13
of mercuripapain, **19:**281; 282
of papain, **19:**281; 282; 284
column
of enzymes, **22:**417–488
in isolation of folic acid compounds, **21:**386; 390; 413
in isolation of mitochondria, **21:**4
in isolation of purine precursors, **21:**205; 209–211; 214; 217; 218; 221; 228; 229
in isolation of THFA derivatives, **21:**225
composition of proteins, **14:**263; 271–272; 281; 286; 295–302; 303–309
effect on enzyme activity, **28:**28
in enzyme isolation, **14:**396
in enzyme purity test, **14:**389
gas, **57:**259
hydrophobic interaction, **57:**373–374
in identification of imidazole, **20:**74
in identification of imidazole phosphate, **20:**85
in immunization studies with uridine disphosphogalactose-4-epimerase,**20:**120
in preparation of cytochrome *c,* **20:**193
in separation of polygalacturonases, **20:**367
in studies of *N*-acetylneuraminic acid, **20:**140
in studies of amino acid biosyntheses and metabolism in plants, **20:**324; 327–329
in studies of enzymic transformations of steroids, **20:**238
in studies of uridine diphosphogalactose-4-epimerase action, **20:**115
in study of browning reaction, **20:**93
in study of pectic substances, **20:**342; 343; 345; 370
immunoaffinity, **57:**9
of LBF, **14:**50–53; 54–55; 61–62
paper
for coenzymes Q, **25:**293–294
in isolation and study of folic acid and folic acid compounds, **21:**225; 372; 381; 393; 409

Chromatography (*Continued*)
in isolation of purine precursors, **21:**205; 218; 221
and purification procedures, **57:**8–10
in studies of enzymes and other proteins, **14:**334–338; 359–365
in study of hydroxymethyl transfer, **21:**417
in study of pterins, **21:**377
in study of UDPG, **14:**199; 201
of thiol-alcalase, **33:**391
of thiol enzymes, **33:**390–391
of thiol-subtilisin, **33:**390–391
Chromatography applied to porphins, **7:**355
Chromatography in methylthioadenosine analysis, **54:**206–208
HPLC, **54:**207–208
ion exchange, **54:**206
paper, **54:**206–207
Chromatography on carboxymethyl cellulose, **42:**81
Chromatophores, and H^+-translocator oxidoreductases, **29:**76
Na^+/K^+-antiporter ATPase, **29:**73
Chromenols of coenzymes Q, **25:**284–285; 290
Chromidia, *See* Microsomes
and respiratory enzymes, **8:**7; 31–37
chemical compounds in, **8:**7–8
mutation, **8:**60–62
similarity to chromomeres, **8:**16–17
and structure of cytoplasm, **8:**4–14
Chromidial mutations and differentiation of cells, **8:**60–62
Chromium, effect on tryptophan metabolism, **10:**68
Chromium adenosine triphosphate, **46:**370
Chromium and phosphoglucomutase activity, **14:**212
Chromium (III) nucleotides, **45:**285; 289; 293; 294
Chromium salts, **17:**137; 138; 142
Chromobacterium violaceum
phenylalanine hydroxylase from, **74:**235; 240–241; 253
kinetic mechanism of, **74:**259
tryptophan hydroxylase from, **74:**236
α-Chromoglobulin, isolation from plastids, **20:**207; 208
β-Chromoglobulin, isolation from plastids, **20:**207; 208
Chromomeres, **3:**2
and nucleic acid, **8:**3
and structure of chromosomes, **8:**16–17
Chromones, **55:**84
Chromopeptides from cytochrome *c,* **24:**418; 419
Chromophore, ionized-SeH group of selenocysteine residue, **48:**23
oxygen-sensitive, **48:**23
Chromophores, amino acid, Cotton effect, **27:**46; 60; 95–96
of enzyme complexes, Cotton effects, **27:** 53–56; 58–60; 66
LADH-DPNH, **27:**50–54; 56; 61–70
metal-mercaptide, **27:**78–80; 82
and protein asymmetry, **27:**40–41; 46–49; 52–53; 61
Chromophoric changes, **4:**272–276
bathochromic effect, **4:**273–275; 280; 290; 296
hyperchromic effect, **4:**272; 275; 276; 280; 290; 297
hypochromic effect, **4:**273
hypsochromic effect, **4:**274; 276
Chromoplasm, in chloroplasts, **30:**4–5
Chromoplasts, formation, **20:**203
Chromosomal nonhistone protein, **42:**254
Chromosomal structure and DNA molecules, **17:**464–465
Chromosome breakage, **17:**453–455
Chromosome replication, *Caulobacter* cell cycle, **71:**3–4
Chromosomes. *See also* Cell nucleus and DNA; Chromatin; Nucleic acids; Nucleoproteins; Nucleus
bibliography, **3:**32–34
and cell differentiation, **16:**476; 478–479
DNA in, **16:**412; 423; 425–430
duplication in dividing cells, **16:**429; 453
isolated, composition, **16:**412; 423; 429–430; 437; 444–450
isolation, **16:**414; 417; 423
mechanism of action, **16:**486–491
morphological integrity, **16:**447–448
plasticity in metabolism, **16:**457–461
of *Rhynchosciara angelae,* **16:**460–461; 485
role in genetic transmission, **16:**475–486
of salivary gland of *Drosophila,* **16:**422; 476
composition, **18:**178–181

duplication of, **18:**181–183
constituents, **3:**13–19
cytochemical studies on, **9:**593
and genetic mapping, **22:**3
historical, **3:**1–4
mutation through ultraviolet irradiation, **3:** 22–23
nucleic acid content, **3:**13–15
nucleoprotein content, **3:**15–23
protein content, **3:**12–23
as self-perpetuating fibrils, **8:**4
comparison with cytoplasmic fibrils, **8:**11; 14; 17
motility, **8:**38–39
structural periodicity, **8:**14–17
—SH groups in, **11:**257
splitting of, **3:**18
structure of, **3:**18
and viruses, **4:**4–8; 14
Chromosomes (Mirsky), **3:**1
Chromosomin, **16:**449–450
Chronic granulomatous disease, **41:**39, **65:**51
Chrysanthemums, antibiotic treatment of diseases, **20:**400
Chrysolaminarin, as β-1,3, glucan, **28:**329; 334; 338
Chrysophyceae, carotenoids in, **21:**301
ChTr, *See* Chymotrypsin
Chymopapain, **53:**278
effect of pH on, **19:**315
isolation, **19:**302
preparation, **19:**280
separation from papain, **14:**383
Chymosin, **44:**6, **53:**261; 286
Chymotrypsin, **23:**130; 136, **33:**263; 266, **34:**10; 13; 21; 31; 448–449; 457–462; 523, **37:**416; 429; 430; 439; 463; 511; 557, **41:**179; 180; 186; 237, **43:**162; 164; 165; 166; 173; 196; 223; 253; 298; 341; 369; 371; 374, **46:**481–483, **50:**24, **52:**211, **53:**243; 246; 251; 252; 255; 261; 263; 265; 269–272, **60:**7
χ-Chymotrypsin, **36:**62
acetylation, **58:**17
action on
collagen, **23:**173; 205
degraded glyceraldehyde-3-P, dehydrogenase, **23:**331–332
denaturated protins, **23:**32–36
native proteins, **23:**32–36
phosphorylase, **24:**273
P^{32}-labeled phosphorylase *a,* **24:**271
RNase, **24:**168; 174–176; 220; 232–235; 244; 247; 250
action on amino acid esters, **19:**237; 238.
action on insulin, **19:**302
esterase activity of, **19:**238; 306
mechanism of action, **19:**313
in transpeptidation reactions, **19:**310
action on insulin, **14:**149
classification, **14:**232; 247
inactivation by oxidation, **14:**131
peroxidase action on, **14:**155; 157–158
separation from trypsin, **14:**383
structure studies, **14:**301
tyrosinase action on, **14:**138; 141; 144; 146–147
action on myosin, **16:**323
as activator of β-glucuronidase, **16:** 376–377
action with synthetic peptides, **13:**305; 309–310
reactive group, **13:**338–342
stoichiometric inhibition, **13:**321–343
activation by indole, **43:**266
active center, **23:**233
active sites, **24:**459; 465
acylation, **43:**308
acyl enzyme, **43:**304
intermediate, **43:**223
adsorbed, activity of, **28:**3; 19; 22; 27; 30
aggregation of, **27:**224
effect on disulfide exchange, **27:**310
sequence analysis, **27:**229–230
amino acid composition, **34:**13
amino acid composition near active site, **22:**65
column chromatography, **22:**424; 425; 443; 448–450
labeling of active site, **22:**52; 66; 67
and model reactions, **22:**81
pK value, **22:**58
reversible denaturation, **22:**76
structure, **22:**73
anionic hole, **43:**252
change upon complex formation, **25:**20
multiple forms, **25:**350–351; 363
product inhibition of, **25:**221–225; 226
relaxation spectrometry, **25:**35
in R Nase study, **25:**337
turnover number, **25:**32

χ-chymotrypsin (*Continued*)
chemical modification of, **24:**134; 137; 214
conformation change, **43:**235
crystallographic studies, **34:**21–29
deacylation, **43:**306; 308
denaturation of, **23:**34; 37–38
dilatometric methods, **9:**605
photochemistry of, **9:**139; 150–151
effect of irradiation of, **24:**369–370; 376; 379–380; 382; 385–388; 392–394; 395
effect on copper-protein, **23:**305
effect on oxygen exchange between acetate and water, **12:**304
specificity of, **12:**194; 206; 228–229
and elastase, **23:**235; 238
ester substrates of, **33:**431
homogeneity of, **24:**418
hydrogen bonding to amide hydrogen atom, **43:**310
hydrolysis of amides, **43:**370
hydrophobic interaction of, **33:**431
imidazole in, **20:**62; 74; 76–82; 86; 87; 89
imidazole in polymerization of, **20:**101
mechanism of hydrolysis by, **20:**283–318
phosphorylated, amino acids in, **20:**288; 290
serine in, **20:**88
immobilization of in acrylamide gels, **33:**278
inactivation of, **24:**117
effect on absorption spectrum, **24:**369–370
information change in, **45:**321
kinetic mechanism of, **45:**337
of *vespa orientalis,* **45:**396
instability and ionization, **43:**371
kinetic studies, **17:**41; 46–49, **24:**234–235; 459–460
localization of, **24:**325
mechanism for action of, **24:**463–464
methyl acetate, **43:**306
methyl hippurate, **43:**305; 306
molecular weight from radiation studies, **24:**366
N-acetyl-L-tryptophan, **43:**305
N-acetyl-L-tyrosine, **43:**304
ethyl ester, **43:**306
negative specificity, **43:**266
partial degradation by, **24:**128–131
peptide bond cleavage, **67:**221
phenylalanine hydroxylase conversion, **67:**219–220
phosphomannose isomerase-guanosine diphospho-D-mannose pyrophosphorylase (PMI-GMP), **70:**227; 229
in prepn. of DNA, **18:**163; 168
and rennin, **15:**434
inhibitor studies, **15:**288–289
mechanism of action, **15:**292
radioactive tracer studies on active centers, **15:**276
in schizophrenia, **29:**507
Sepharose conjugates, **34:**522
side-chain specificity of, **33:**410
specificity, **1:**73–75
specificity of, **33:**431
spectra, **34:**471
steric structure, **26:**93
subsites, **43:**298
substrate binding, **43:**335
tyrosine hydroxylase, **70:**182
valence of, **24:**118–119
d-Chymotrypsin, **41:**193; 221; 227
α-Chymotrypsin, **37:**411; 478; 524; 544, **47:**49; 124; 134, **52:**192, **53:**264, **57:**184–185
active center of, **24:**434
effect of radiation on, **24:**388
kinetics of action, **25:**178; 225
δ-Chymotrypsin, **53:**254–255
α-Chymotrypsin, inactivation by oxidation, **14:**132; 146
structure studies, **14:**285; 310
union with DFP, **14:**283; 284; 287; 310
Chymotrypsin, Stoichiometric Inhibition (Balls and Jansen), **13:**321
Chymotrypsin-A, amino acid sequence in, **34:**32
Chymotrypsin action on hypertensin, **7:**526
Chymotrypsinagen, **43:**254
Chymotrypsin-B, amino acid sequence in, **34:**32
Chymotrypsin-like digestive proteases, **45:**398
Chymotrypsinogen, **23:**131–133; 148; 238, **34:**32
action of tyrosinase on, **14:**146
structure studies, **14:**274; 279; 283; 285; 287
activation of, **24:**24; 126–127, **50:**30
amino acid sequence, **24:**416; 418; 420–422; 432–435
effect of irradiation, **24:**379–380; 388; 394
amino acids in, **20:**285
behavior with DFP, **13:**327

column chromatography, **22:**424; 433; 448–450
reversible denaturation, **22:**77
enzymic activation, **34:**35
sequence analysis, **27:**230
Chymotrypsinogen A
column chromatography, **22:**451
rate of biosynthesis, **23:**134
Chymotrypsinogen A and B, localization of, **24:**325
Chymotrypsinogen activation, **45:**409
Chymotrypsinogen B
column chromatography, **22:**451; 452
in DNase, **29:**170
Chymotrypsin substrate, **23:**240
Chymotryptic digestion, LADH study, **27:**81–82
Chymotzypsin C, **41:**191
Cibacron blue, **46:**233; 239, **65:**112
Cibacron Blue F3GA, **59:**131
Cicer arietinum, **55:**5
chalcone isomerase of, **55:**27
isoflavone 4′-*O*-methyl transferase of, **55:**36; 39
CI Esterase inhibitor, **48:**288; 291
Cigarette tobacco, aging, **10:**399–409
Cigar leaf tobacco, sweat processes for, **10:** 358–399
Ciliapterin, **35:**72
Ciliary movement, effect of inhibitors on, **12:**281
Cinchonidine, origin, **8:**232
Cinnamic acid, **58:**21
coverted to 4-coumaric acid, **55:**16
mechanism of action of, **55:**16
modulation of, **55:**17
phenylalanine ammonia lyase and, **55:**12
relation to lignin, **18:**373; 374
Cinnamic acid 4-hydroxylase, **55:**16
abiotic elicitors of, **55:**17
Cinnamic acids, hydrotropic effects, **17:**153; 154
Cinnamoyl chymotrypsin, hydrolysis of, **24:**460
Cinnamyl alcohols, hydroxy-, in lignin formation, **18:**363; 371–372; 375
Circe effect, **43:**223
Circe Effect, Binding Energy, Specificity, and Enzyme Catalysts, The (Jencks), **43:**219
Circular dichroic measurements, **50:**333
Circular dichroic spectra, **50:**309
Circular dichroic spectrum, **39:**197
Circular dichroism, **27:**42; 46–47, **35:**477, **42:**102; 103; 105; 106; 108; 141; 143, **47:**50; 74; 76–77; 84–85; 136, *See* CD curve
of aspartate amino-transferase, **50:**16; 23
spectra from superoxide dismutase, **41:**60
Circulin, **33:**367
Cirrhosis, hepatic glutamine regulation, **72:** 75–77
Cirrhosis of liver and pancreatectomy, **8:**182–183
cis-trans Isomerase
in oxidation of alkylbenzene, **27:**518
in oxidation of naphthalene, **27:**522
cis-trans Isomerization, **46:**403
Cistron, defined, **22:**4
and lactose utilization in *E. coli,* **22:**102; 103
Citramalate lyase, **43:**64
Citrase, product inhibition, of, **25:**224; 243–244
Citrate, **35:**365, **51:**236, **53:**208; 210; 211
acetate in formation of, **12:**98
and CO_2 fixation, **12:**47; 53–54; 91; 104–105
and acetate metabolism, **23:**405–406; 408–410; 414; 419–420; 425; 427–428; 430; 463
and glycolate metabolism, **23:**435; 439; 440–441; 463
from labeled acetate, **23:**405–406; 408–409
allosteric control by, **28:**59–60; 78; 79; 84; 86; 90–92; 136–139
effect on fatty acid synthesis, **26:**300–303
effect on 3-hydroxyanthranilate oxidase, **19:**97; 126
and kynurenine 3-hydroxylase, activity, **19:**194
and papain activity, **19:**286
in photosynthesis, **19:**13–17; 52; 53
inactivity in *E. coli* mutant, **16:**255
inhibition of aspartate β-decarboxylase activity by, **35:**539
as inhibitor of phosphofructo-kinase activity, **48:**218
polyfluorinated, **55:**224
and pyridine nucleotides, **15:**121–122
and reactivation of choline, **6:**261
stereochemistry, **43:**58; 59
citrate synthases, **43:**73
parent numbering system, **43:**60
re-citrate synthases, **43:**58
si-citrate synthases, **43:**58; 66
synthesis and oxidation, in mitochondria, **21:**20; 34

Citrate, The Enzymology of the Formation and Breakdown of (Srere), **43:**57
Citrate buffer, inhibition of β-glucuronidase, **16:**370; 382
Citrate-1,5-C^{14}, effect on acetyl-CoA carboxylase, **26:**302
Citrate-cleavage enzyme, in fatty acid synthesis, **43:**88. *See also* ATP citrate lyase
in human adipose tissue, **43:**89
in mammary tissue, **43:**88
in Penicillium spiculosporum, **43:**87
in Rhodotorula gracilis, **43:**87
Citrate cleavage enzyme, phosphohistidine in, **36:**191
Citrate condensing enzyme
effect of deficiency in *E. coli* mutant, **16:**268
relation to requirement for glutamate, **16:**255
genetic studies, **17:**426
Citrate in fructose-bisphosphatase regulation, **54:**142
Citrate-isocitrate transformation, enzymic, **14:**194; 197
Citrate lyase, **43:**60, **55:**219; 220
enolase, **43:**61
equilibrium constant, **43:**71
mechanism of deduced from relaxivity data, **33:**51
from Phodopseudomonas gelatinosa, **43:**67
reaction enactivation, **43:**61
from S. diacetilactis, **43:**64
suicidal substrates of, **55:**226
Citrate lyase ACP, **43:**63
Citrate lyase enzymes, evolutionary relations, **43:**95
Citrate metabolism, **43:**92
Citrate permease, **43:**60
Citrates, solubilization study, **17:**143; 144; 148; 149; 152
Citrate synthase, **37:**449; 511; 544; 545, **43:**72, **46:**247; 260; 464, **55:**206; 207; 219, **58:**144
aldosterone, **43:**82
from azobacter vinelander, **43:**73
from beef heart, **43:**72
from *E. coli,* **43:**73; 83
effectors, **43:**79
equilibrium constant, **43:**74
in fluorocitrate generation, **55:**220
from Halobacterium cutirubrum, **43:**84
in human fibroblasts, **43:**81
in human liver, **43:**81
in human muscle, **43:**81
kinetics, steady state, **43:**75
localization of, **24:**331
in *P.* aeruginosa, **43:**85
from pigeon-breast muscle, **43:**72
from pig heart, **43:**72
from rat brain, **43:**72
from rat kidney, **43:**72
from rat liver, **43:**72
regulation, **43:**80
subunits, **43:**83
from Thiobacillus denitrificans, **43:**84
from Thiobacillus neopolitanus, **43:**84
vs. malate synthase, **55:**208; 209
from yeast, **43:**72
Citrate-thrombin, from prothrombin, **16:**51
Citrate transport, **43:**94
Citric acid
asymmetric degradation, **10:**196–199
changes during fermentation of tobacco, **10:**340; 364–366; 396; 402
degradation in yeast, **10:**173–176
equilibrium with isocitric acid and *cis*-aconitic acid, **10:**194–196
influence of ions on formation of, **10:** 214–216
into α-ketoglutaric acid, **10:**167–168
transformation into isocitric acid, **10:** 193–194
in bones, **18:**128
and the convulsive state, **18:**146–150
determination of, **18:**117
effect of fluoroacetate, **18:**118–146
effect of fluoro fatty acids, **18:**140–141
in carbohydrate oxidation, **6:**359; 361
effect on carotenogenesis, **21:**318
in isolation of mitochondria, **21:**5; 16
metal ion complexes of, model catalyses with, **21:**188
in mitochondria, **21:**8; 10
effect on pectic substances, **20:**368
and fatty acid catabolism, **8:**392–401
cycle of Krebs and Johnson, **8:**395–396
isolation, **8:**394
as inhibitor of β-glucuronidase, **16:**373
synthesis, **16:**1–2
use in nuclear isolation, **16:**414–415; 422

and semen metabolism, **9:**342; 348; 367; 376–378
degradation by molds, **9:**550–554
formation in molds, **9:**537–545; 547–559; 564–576; 673
use in rennin extraction, **15:**426
Citric acid and choline acetylation, **12:**319; 322; 327–329
Citric acid cycle, **3:**223–235, **4:**209; 210, **5:**153–154, **10:**167–222. *See also* compounds in; Respiration, Tricarboxylic acid cycle, and compounds of the cycle; Tricarboxylic acid cycle
in aromatization process, **18:**364–365
ATP role in, **12:**315
and coenzyme Q, **25:**305–306
multiple forms of enzymes, **25:**333; 346–347
enzymes, in nuclei, **27:**263
enzymes effective in, **10:**213–217
enzymes for, in sperm, **24:**86
enzymic mechanisms in, **15:**183–270
mechanisms of condensation reactions, **10:**177–190
phosphoenolpyruvate carboxykinase (GTP), **69:**205
in plants, **18:**298–303; 308
and propionic acid metabolism, **26:**368–369
role of glutamine in, **13:**268; 271
scheme, **10:**198
and urea synthesis, **15:**320; 330; 333–342
Citric acid cycle enzyme, **58:**146
Citric acid cycle in plants, **9:**417; 441–412
Citric Acid Cycle (Martius and Lynen), **10:**167
Citric acid cycle metabolites, ethionine effects, **25:**380–381
Citric Acid Cycle (Ochoa), **15:**183
Citric acid desmolase, **15:**201
Citric dehydrogenase, in *Chlorenchyma,* **20:**219
Citrinin, **7:**470; 486; 489; 490; 492
Citrobacter freundii, ADP-glucose pyrophosphorylase of, **46:**341
Citrogenase, **5:**156; 157
activity of –SH group, **11:**243–244
Citrogenase and choline acetylation, **12:**328
Citrogenase and fat metabolism, **8:**398–400; 401; 410
Citrovorum Factor, *See* N^4-Formyltetrahydrofolic acid
Citrovorum factor, from prefolic A, **25:**408
Citrovorum Factor cyclodehydrase, **21:**413
Citrovorum factor (folinic acid, leucovorin), **27:**426; 436–437; 439; 441–442; 454; 455; 457; 458
Citrullinase, **15:**376
Citrulline, **40:**72, **50:**141, **53:**208; 209; 210; 211; 222
activation of, **18:**211–212
in arginine-requiring *Neurospora* mutants, **22:**27
effect on acetylarginine uptake, **40:**77
genetic studies, **17:**398–399
in ornithine cycle, **19:**372–375; 377; 379; 380
phosphorolysis of, **23:**327; 338
in pyrimidine biosynthesis, **21:**267; 270–272
synthesis, carbamyl phosphate in, **21:**267; 270; 272
in mitochondria, **21:**43; 44
released by isolated rat livers, **53:**229
released by small intestine, **53:**228
in transaminations, **16:**196; 227; 231
metabolic relationships with other amino acids, **16:**259–261; 264–268
uptake rate, kidneys of rats, **53:**229
Citrulline, Metabolism (Ratner), **15:**319
Citrulline decomposing system, site of cleavage by, **17:**163; 178
Citrulline formation, and ATP, **18:**227–228; 231
Citrulline metabolism, **15:**319–387
Citrulline phosphorylase, **15:**370; 376–377; 379–383, **17:**162; 175
conversion to ornithine, **16:**266; 267
Citrulline synthesis, **17:**180–181
Citrulline ureidase, **15:**376
Citrus limonum, effect of fluoroacetate on, **18:**138
Citrus paradisi, **55:**29
Citrus phosphatase, **17:**165; 167; 170
Citryl CoA, **43:**64; 66
Cltryl coenzyme A, **15:**196; 201
C-1 Kinases, in monosaccharide synthesis, **26:**41
CKL, *See* Clostridium thiaminolyticum
Cladosporium herbarum, **55:**48
Claisen condensations, **50:**271–280
Clarase
action on penicillin, **28:**243
inhibition of penicillin, **10:**253
Clarase-papain action on CoA, **16:**3
Classification of enzymes, **14:**219–260

Classification of Enzymes (Hoffmann-Ostenhof), **14:**219
Classification of organisms on basis of nutritional requirements, **2:**143
Clathrate hydrates
hydrogen bonding, **65:**240–245
bond lengths and angels, **65:**242–243
gas hydrates, **65:**242
pentagonal dodecahedron, **65:**242–243
polyhedra, **65:**246–247
proton-deficient and proton-excess, **65:**244; 246
Clathrins, adenosine receptors, **69:**113
Clavacin, *See* Patulin
as wilting toxin, **11:**403; 407; 412; 422; 423
Clavatin, *See* Patulin
Claviceps purpurea, succinate dehydrogenase from, **28:**179–180
Clavicin, in treatment of plant diseases, **20:**402
Claviformin, *See* Patulin
Clay
soil enzyme study, **28:**367; 371
use in fermentations, **7:**563
Clays, digestion of proteins adsorbed on, **33:**266
properties of enzymes adsorbed on, **33:** 270–272
Clearance, rate, **39:**226
times, **39:**227
Clearing factor, lipolytic action, **23:**247
Cleavage, **34:**524
Cleaving enzyme, for cystathionine, **16:**272
Clinical manifestations, ADA deficiency, **51:**169–170
Clinical significance of β-lactamase, **28:** 307–308
Clitocybine, **7:**485
Clofibrate, **57:**93, **60:**39
Clomiphene, **55:**162
Cloning, **59:**80
Clostridia
and acetyl phosphate formation, **6:**237
reversibility, **6:**246–248
and amino acid decarboxylase formation, **6:**7
effect of age of culture, **6:**12
purification of histidine in, **6:**13
purification of ornithine in, **6:**14
Clostridia, **15:**118
Clostridial collagenase, **41:**194
Clostridial glycine reductase, **48:**20–24
Clostridial glycine reductase selenoprotein, **48:**12
essential selenocysteine residue, **48:**12
Clostridin, **23:**199
Clostridiopeptidase A (from Cl. histolyticum), **34:**161
action on collagen of, **34:**162
application, **34:**169
assay procedures for, **34:**166
peptide substrates, **34:**163
properties, **34:**167–169
purification, **34:**157
specificity requirements, **34:**164
substrates, **34:**159; 161; 163
subunits, **34:**161; 168
Clostridium, **46:**103
amination in, **18:**253
and lipoic acid, **18:**342
Clostridium aceticum
glycine metabolism in, **23:**460
metabolism in, **74:**9–10
Clostridium acetobutylicum
amylase of, **14:**343
racemase in, **14:**197
folate metabolism, **27:**457
metabolism in, **74:**34; 36
Clostridium acetobutylicum in industrial fermentation, **7:**571
Clostridium acetobutylicum, superoxide dismutase from, **41:**80
Clostridium acidiurici, **48:**19
Clostridium acidi-urici
glycine metabolism in, **23:**457–458
metabolism in, **74:**9–10; 36
Clostridium acidi-urici, purine degradation in, **21:**227; 401; 402; 408
synthesis of active formaldehyde in, **21:**412
Clostridium barkeri, superoxide dismutase from, **41:**80
Clostridium bifermentans, proteolytic activity, **23:**178; 181; 184
Clostridium butylicum, **23:**337
amylosucrase in, **15:**310
citric acid cycle, **15:**194; 245; 246
Clostridium butylicum, PN-dependent enzymes, **26:**132
Clostridium butylicum in industrial fermentation, **7:**565
Clostridium butyricum, polygalacturonases of, **20:**366

Clostridium capitovale, action on collagen, **23:**178; 183; 186
Clostridium cellobio parum, superoxide dismutase from, **41:**80
Clostridium chauvoci, **23:**184; 185
Clostridium cylindrosporum, **23:**458, **48:**19
folinic acid conjugates in, **21:**393
purine degradation in, **21:**227; 401; 402
tetrahydrofolic formylase in, **21:**407; 408
formimino transferase of, **27:**425
Clostridium felsineum, polygalacturonases of, **20:**366; 367
Clostridium formicoaceticum, formate dehydrogenase activity in, **48:**18
Clostridium HF, **23:**461
Clostridium histolyticum, bacterial collagenase from, **41:**260
clostripain from, **41:**181; 183; 197
Clostridium hoffmani, β-glucuronidase in, **16:**377
Clostridium, hyaluronidases in, **13:**200; 205; 207; 221–222
nitrogen fixation by, **13:**347; 360; 362; 365; 367–368; 372
Clostridium, inactivation of aldolase from, with borohydride, **31:**140
Clostridium kluyveri
and citric acid cycle, **18:**273
fermentation in, **23:**431–434
oxidative phosphorylation in, **23:**335
in preparation of *N*-acetylimidazole, **20:**84
Clostridium kluyveri, PN-dependent enzymes, **26:**132
propionic acid metabolism, **26:**367–370
Clostridium kluyveri, citric acid cycle, **15:**188; 207; 234; 237; 245–246; 249; 254–256; 258
enzymes in, **15:**116; 118
Clostridium lentoputrescens, **23:**178
Clostridium lentoputrescens, superoxide dismutase from, **41:**80
Clostridium oedematiens, **23:**178
Clostridium pasteurianum, **52:**2
extract, EPR spectrum, **27:**112
isotope, studies of ammonia fixation by, **20:**334
metalloflavoprotein of, **17:**248; 269; 281–282
Clostridium pasteurianum phosphofructokinase, **48:**214–215
Clostridium pasteurianum, superoxide dismutase from, **41:**80
Clostridium perfringens, **60:**284
Clostridium perfringens, **15:**377–379, **34:**490
Clostridium perfringens, AC coenzyme from, **26:**366
vitamin B_{12}, **26:**265–266
Clostridium perfringens, hyaluronate from, **13:**208
Clostridium perfringens neuraminidase, treatment, **46:**388
Clostridium perfringens, neuraminidase activity from, **41:**108; 109; 118; 120
Clostridium propionicum, propionic acid metabolism, **26:**367–370
Clostridium putrificum, **23:**184
Clostridium septicum, proteolytic activity, **23:**178; 184; 185
Clostridium septicum, enzymic adaptation in, **13:**71
Clostridium sp., superoxide dismutase from, **41:**80
Clostridium sporogenes
glycine metabolism in, **23:**457; 461
proteolytic activity, **23:**178; 184; 195
quinones in, **25:**289
Clostridium sticklandii, **38:**414, **48:**4
action on glycine, **23:**461
oxidative phosphorylation in, **23:**336
dihydrofolate reductase, **27:**434
Clostridium sticklandii, folic reductase in, **21:**395
reduction of folic acid compounds in, **21:**378
serine hydroxymethylase in, **21:**393
Clostridium tetani, glutamine and glycolysis in, **13:**261
Clostridium tetanomorphum, **46:**51; 53; 55; 103
cobamide coenzyme synthesis, **26:**366
methylaspartate mutase of, **26:**346; 359–365
vitamin B_{12}, **26:**263; 264
glutamate degradation in, **23:**434
Clostridium tetramorphum, folate use, **27:**427
Clostridium thermoaceticum, **23:**460
formate dehydrogenase activity, **48:**16; 18
Clostridium thiaminolyticum, **15:**390; 394; 397; 411; 412; 413
Clostridium welchii, **23:**382
amino acid metabolism in, **16:**220; 275
metabolism in, **74:**34–36
neuraminidase of, **20:**144

Clostridium welchii, 376, **15:**378
Clostridium welchii, enzymic adaptation in, **13:**71
mucinase in, **13:**200
Clostridium welchii toxin, in methylation study, **25:**419
Clostridum histolyticum, collagenase of, **23:**167; 176; 178–179; 182–186; 188–217; 219–221
Clostridum perfringens, collagenase of, **23:**167; 176; 178–179; 181; 183–188; 190–202; 205; 208; 210; 219
lecithinase from, **23:**367
Clostridum sticklandii, superoxide dismutase from, **41:**80
Clostripain, **23:**188, **41:**179; 180; 181; 197–199, **53:**260; 261
Clothes moth, *See* Tineola bisselliella
Clot retraction, **16:**44; 68; 72
Clottable material, of platelets, **16:**71–72
Clotting, blood, **10:**1–49
Clotting factors, purification, **16:**25–32
Clotting factors, Roman numeral designation of, **48:**278
Clotting of blood, *See* Blood clotting; Coagulation of the blood
Clotting of milk, **15:**423–448
Clotting of milk by chymotrypsin, **24:**376
Clotting system within protoplasm, **8:**28
Clotting time
effect of heparin on, **16:**38–39
in disease, **16:**72–75; 84
relation to fibrinogen concentration, **16:**38
relation to thrombin concentration, **16:**33; 38
and thromboplastin, **10:**4–5; 9–10
effect of ion concentration, **10:**30–32
use of silicone technique, **10:**15–17
Cloxacillin, action of β-lactamase, **28:**239; 289
Clozapine, **55:**181
potency of, **55:**182
Clupein, amino acid composition, **16:**442–444
Clupeine, structure studies, **14:**284
Clyceraldehyde-phosphate dehydrogenase, **37:**411
CMC-azide, **34:**449–451
CM-cellulose, **34:**452–453; 522
-cellulose-bromelain, **34:**460
-cellulose-ficin, **34:**456; 460
CMP, *See* Cytidylic acid
CMP, DR metabolism, **28:**438–440; 442; 445–446; 448
^{3}H-CMP, in DR metabolism, **28:**438
^{32}P-CMP, in Dr metabolism, **28:**438
CMP-*N*-acetylneuraminic acid
in carbohydrate synthesis, **26:**51
synthesis of, **28:**114–115
CMP-*N*-acetylneuraminic acid pyrophosphorylase, **26:**41; 56
CMP-*N*-glycolylneuraminic acid pyrophosphorylase, **26:**56
CMP-2-keto-3-deoxyoctonic acid pyrophosphorylase, **26:**51; 56
CMP kinase, in DR metabolism, **28:**438; 442
CMP-sialic acid, **52:**28
CMP-sugars, **26:**65; 66
CMU, *See* Chloromethylurea
CN, **39:**339; 341; 343; 344; 362
CO_2, **34:**513, **35:**343–347
diffusion, **34:**512
ions, **34:**512
$^{14}CO_2$, incorporation of, **34:**216
CoA, **57:**143, *See* Coenzyme A
Coacervate droplets, and origin of life, **27:** 366–371; 374; 392
Coacervates, **34:**447
enzyme action on, **33:**265
enzyme digestion, **33:**265
Co-acetyl-cobalamin, **26:**245; 253–254
Coactivation with respect to alkaline phosphatase, **10:**102–104
CoADP in creatine kinase activation, **54:**412
Coagulase-negative staphylococci, glucuronidase in, **16:**378
Coagulation
accelerators inhibitors of, **30:**303–305
activating surfaces for, **30:**287–288
basis for current theories of, **30:**263–270
of blood, **30:**303–305
evidence for an enzyme cascade in, **30:**268–270
Factor XIII in, **30:**276
fibrinogen in, **30:**271
inhibitors of, **30:**306
kinetics of, **30:**302
kinetics of the cascade of, **30:**299–305
properties of factors of, **30:**262
sequential theory of, **30:**266–268

thrombin-fibrinogen interaction in, **30:** 274–276
thrombin in, **30:**272–274
heat, **17:**154
of plant plasma, **11:**405–406; 408
role of —SH groups in, **11:**245–246
of proteins, **9:**131–132; 135–136
of semen, **9:**346–348; 378
Coagulation anomaly and vitamin K, **2:**237
Coagulation factor, transglutaminase as, **54:**5
Coagulation of blood, **10:**1–49
activation, **10:**3–28
inhibition, **10:**33–38
Coagulation of proteins, SH-groups, **27:**223–225
Coagulation of the blood, **16:**23–103
Coagulation properties, of pectic substances, **20:**356
Co-alkyl-cobalamines, synthesis, **26:**245
CoA—SH, *See* Coenzyme A
CoA transferase, and propionic acid metabolism, **26:**308–309; 356
Cobalamin, *See* Vitamin B_{12}
as cofactor, **58:**180
deficiency, **58:**185
Cobalamin analogs, alkanolamine, **26:**240–241; 271–273
competitive action, **26:**270–274
synthesis of, **26:**239–241
Cobalamin-Co-carbonic acid ethyl ester, **26:**245
Cobalamin coenzyme, absorption spectrum, **26:**247–249
cathode potential, **26:**255
cyanide cleavage, **26:**252–253
electrophoresis, **26:**250
in isomerization, **26:**264–265
light sensitivity, **26:**242; 251–252; 256
partial synthesis, **26:**244
structure, **26:**237–238
transfer reactions, **26:**253–255
Cobalamin-Co-enzyme, and corrin ring, **26:**257
Cobalamin-Co-sulfonate, properties, **26:**252; 254; 255
Cobalamin hydride, absorption spectrum, **26:**247–249
reactions, **26:**242–243
synthesis, **26:**242–245; 257
Cobalamins. *See also* Cobalamin coenzymes, Corrinoids, Corrinoid coenzymes, and Vitamin B_{12}
absorption spectra, **26:**247–249
biosynthesis, **26:**239
molecular biology, **26:**268–276
partial synthesis, **26:**240–241
phosphorylation, **26:**238–241
structural specificity in man, **26:**275–276
structure, **26:**236–237
transfers
acyl group, **26:**253–254
alkyl group, **26:**253
sulfonate group, **26:**254–255
Cobalamin sulfonate, cathode potential, **26:**255
Cobalamin-sulfur compounds properties, **26:**252; 254; 255
Cobalt, **46:**53
binding, by histidine, **20:**98
by protein groups, **20:**56; 97
catalysis by, **15:**22; 25; 32–33; 379; 380
coordination linkages, **26:**236–238
effect of corrin ring on, **26:**256–258
in corrins, **46:**38
effect on, collagenase, **23:**195–196
elastases, **23:**234
lysine 2,3-aminomutase, **66:**17–26
in melanin, **16:**172
in phosphorylation of pyridoxal, **16:**219
and model hydroxylating system, **19:**135
and RuDP carboxylase, **19:**22
in ornithine metabolism, **17:**397
and peptidase action, **12:**195; 210; 212–216; 220; 222
in decarboxylation, **12:**80; 90
effect on transmethylation, **12:**468–469; 474
reduced, in corrinoids, **26:**242–243; 247–249
in tetrapyrroles, **46:**81; 82
in vitamin B_{12} structure, **26:**234–238
Cobalt acetate, **46:**51
Cobalt-carbon binding, in corrinoids, **26:**243; 245–247; 250–257
Cobalt (Co^{2+}), in corrinoids, spectrum, **26:** 247–249
Cobalt (Co^{60}) γ-rays. *See also* γ-Rays
inactivation of enzymes by, **24:**365; 374–376; 379-380; 399
Cobalt complexes, in thiol oxidation, **11:** 204–205
Cobalt compounds, EPR study, **27:**142–143
Cobalt compounds, solubilization of, **17:**138; 142; 147; 150

Cobalt ethylenediamine complex, as model enzyme, **15:**27
Cobaltic ions, chelate formation with amino acids and organic phosphates, **21:**146; 148
properties, **21:**157
Cobalt(III) nucleotides, **45:**285
Cobalt ion, as probe of carbonic anhydrase, **47:**163–166; 186–187; 223; 231; 233; 236; 238–240; 244–245; 246–247; 251; 260; 262
Cobalt ion in creatine kinase activation, **54:**404; 412
Cobalt ions. *See also* Cobaltic ions; Cobaltous ions
activation of alkaline phophatase by, **10:**99; 110
effect on penicillin action, **10:**254
on tryptophan metabolism, **10:**68–69
model catalyses with, **21:**177
Cobalt mercaptide, **11:**216; 217
Cobalt-methyl derivatives, of tertiary alcohols, **26:**245
Cobaltocyanide, reducing properties, **15:**32–33
Cobaltous chloride, effect on enzymes of *Cunninghamella blakesleeana,* **20:**270
Cobaltous ions
chelate formation with amino acids and peptides, **21:**145–147
enzyme activation by, **21:**148; 152; 153; 155
in hydroxymethyl transfer, **21:**419
properties, **21:**157
in DR metabolism, **28:**440
effect on laminarinases, **28:**347
Cobalt-sulfur binding in corrinoids, **26:**246–250; 252–255; 258
Cobamide coenzyme, in DR metabolism, **28:**444–447
Cobamide coenzymes. *See also* Corrinoid coenzymes and Vitamin B_{12}
biosynthesis of, **26:**366–367
characteristics, **26:**346–348; 351; 360–363; 366
in diol dehydrase reaction, **26:**265–266; 359; 365–366
enzyme reactions using, **26:**264–266; 359–365
in-β-methylaspartate mutase reaction, **26:**359–365
in isomerization of methylmalonyl-CoA(b), **26:**343
mode of action, **26:**358–359
in propionic acid metabolism, **26:**265; 285; 335; 343–367
Cobamide conjugate, and methylmalonyl-CoA mutase, **26:**347
Cobamide enzymes, coenzyme affinities of, **26:**346–347
effect of AC coenzyme, **26:**346–347; 359–360; 366
Cobamide-requiring diol dehydrases, **28:**447
Cobamides. *See also* Cobamide coenzymes, Corrinoids, Corrinoid coeznymes, and Vitamin B_{12}
activity in DR metabolism, **28:**444–447
alkanolamine analogs, synthesis, **26:**240–241
biogenesis, **26:**260
and methyl synthesis, **25:**405; 410
nucleotide residue, **26:**240; 260–262; 275
partial synthesis, **26:**240–241
structures, **26:**235–236
Cobamide specificity, for ribonucleotide reductase, **28:**446–448
Cobinamide. *See also* Cobinamide coenzymes, Corrinoids, Corrinoid coenzymes, and Vitamin B_{12}
absorption spectra, **26:**247–249
alkanolamine analogs, molecular biology, **26:**268–271
biosynthesis, **26:**239
of *E. coli,* **26:**239
structure, **26:**235–236
in synthesis of cyano-corrinoids, **26:**238–241
P_1-DL-Cobinamide-P_2-adenosine-5′-pyrophosphate, synthesis of, **26:**238–240
Cobinamide analogs, **26:**238–241; 260–262; 268–271; 274
Cobinamide coenzymes, absorption spectrum, **26:**247–249
cathode potential, **26:**255
cyanide cleavage, **26:**252–253
electrophoresis, **26:**250
synthesis, **26:**244
Cobinamide conjugate, and methylmalonyl-CoA mutase, **26:**352
P_1-DL-Cobinamide-P_2-guanosine-5′-pyrophosphate, **26:**238–240
Cobinamide hydride, absorption spectrum, **26:**247–249
Cobinamide phosphate, and cobinamide analogs, **26:**239–241; 260–262

Cobinamide-r, absorption spectrum, **26:**247–249
Cobinic acid, biogenesis, **26:**260
structure, **26:**235
Cobra venom, *See* Snake venom
Co-butyl-cobalamin, **26:**257
Co-butyl derivatives of tertiary alcohols, **26:**245
Cobyric acid (Cobyrinic acid-a,b,c,d,e,g-hexamide, Factor V_{1a}), in vitamin B_{12} synthesis, **26:**238–240
structure, **26:**235
Cobyrinic acid, **46:**51; 52; 53
biogenesis, **26:**259
structure, **26:**235
Cobyrinic acid *a,c*-diamide synthetase, **66:**284–287
Cobyrinic acid synthetase, **66:**285–287
Cocaine
effect on Krebs cycle, **19:**360
effect on transamination, **19:**363; 364
Cocaine esterase, classification, **14:**244
Co-β-carboxyethyl-cobalamin, **26:**245
Cocarboxylase, **15:**157; 158. *See also* Diphosphothiamin; Diphosphothiamine and Thiamine pyrophosphate; Thiamine pyrophosphate
and bacterial fixation of carbon dioxide, **74:**23–24; 26–28
effect of ionizing radiations, **24:**388; 395
effect on pyruvate cleavage, **6:**239
fluorometric assay, **20:**33
in *Fusaria,* **5:**168
gasometric measurement, **9:**603
mode of action, **14:**174; 180–183; 187
in plant respiration, **18:**293; 296
in plants, **10:**227
and urea synthesis, **15:**342
Cocarcinogen, **4:**228; 229
Coccomyces hiemalis, plant disease caused by, **20:**400
CO-chlorocruorin, **23:**269
Cocoa butter, fresh cotyledon content, **25:**461
Cocoonases, **45:**402–411; 412
active-site peptide of, **45:**411
from *B. mori* galeae, **45:**406
chemistry of, **45:**409
inhibitors, **45:**409
map positions, **45:**410
occurrence, **45:**405
purification, **45:**406
substrate activation, **45:**409
substrate specificity, **45:**409
Cocoon-softening fluid, **45:**406
Codecarboxylase, **5:**77
preparation from yeast, **16:**211
Codecarboxylases, **6:**21–29
activity in growth mediums, **6:**8–9
chemical nature, **6:**25–29
distribution, **6:**21–23
function of pyridoxal phosphate, **6:**27
identity, **6:**28–29
preparation of concentrate, **6:**24–25
properties, **6:**23
protein reactivation, **6:**19
salt formation with metals, **6:**23
solubility in organic solvents, **6:**23
stabilities, **6:**23; 28
Codehydrogenase, in the carboxylation of α-ketoglutaric acid, **10:**208–213
Codehydrogenase I, *See* Coenzyme I; Diphosphopyridine nucleotide
Codehydrogenase II, *See* Triphosphopyridine nucleotide
Codehydrogenases, **5:**208–211
action of phosphatase on, **5:**212
destruction by tissues, **5:**230
enzymatic decomposition of, **5:**227; 232
methods for quantitative determination, **5:**216; 217
mode of action, **5:**215–216, **14:**168; 180–183
stability in blood, **5:**230
Co-5-deoxyadenosyl-cobinamide, **26:**244
Co-5′-deoxyguanosyl-cobalamin, **26:**244; 252
Co-5′-deoxyinosyl-cobalamin, **26:**244; 252
Co-5-deoxyinosyl-cobinamide, **26:**244; 252
Codes
amino acid, **47:**375–430
frozen accident, **47:**386
two-letter, **47:**382
Code without commas, **24:**10–11; 21
Coding, in mechanism of protein synthesis, **22:**135; 136
Coding properties, of RNA, effect of ultraviolet, **27:**327
Coding signals. *Information carrier,* and *Information transmission,* in protein synthesis, **24:**10–11; 21; 427; 437
Codons, **34:**3, **47:**378
anticodon, **47:**378; 388–391; 399–403
anticodon pairing, **47:**381–382

Codons (*Continued*)
assignments, **34:**2–17, **47:**392–399
chain-terminator, **47:**407
in DNA bacterial virus phiX174, **47:**399
initiator, and the ribosome binding site, **47:**403–407
internal AUG, **47:**405
in MS2 viral proteins, **47:**399
pyrimidine-restricted, **47:**383
chain-terminating, **47:**384
single-base changes in, **47:**385
terminator, **47:**380
Codon table, **38:**214
Codosomes, **40:**225
Coefficient of friction, and shape of molecules, **16:**324
Coefficients
lipid/water partition, **29:**39–40
permeability, **29:**39–40
Coenzyme, A Structure (Baddiley), **16:**1
Coenzyme, R, **35:**323
repression, **35:**342
succinyl-A, **35:**95
Coenzyme A, **15:**85; 115; 116; 121–123; 129; 183–213; 217; 244; 262, **17:**181–182; 410, **48:**10, **57:**128; 154–156; 159. *See also* Acetylation; Acetyl-CoA
and acetate, **23:**407–409; 411–413; 423–424; 432
acetylation, **18:**203–205; 208; 230
in acetylation reactions, **12:**98–99
in acetylations, **13:**280; 311–312
activity for various organisms, **14:**65–69
and ATP synthesis, **23:**325; 334; 335; 337; 338; 341; 348; 387; 388
in benzoylations, **13:**280
and bioluminescence, **25:**134–135
in carotenogenesis, **21:**346
in cholesterol biosynthesis, **22:**283; 284; 295
and choline acetylase, **12:**326; 329
in citric acid cycle, **10:**190; 221–222
classification in enzyme systems, **14:**241; 252
effect of radiations on, **24:**401
and energy transport, **14:**13–14; 17
in enzyme synthesis, **24:**23
folate metabolism, **27:**434
in folic acid biosynthesis, **21:**394
formation of β-aspartyl phosphate, **16:**271
and glycolate metabolism, **23:**437–440; 445–446
HMG ester, in cholesterol biosynthesis, **22:**284; 289
and lipoic acid, **18:**333–338; 342; 344
localization, **27:**276
in mitochondria, **21:**8; 23
in mitochondrial reactions, **21:**23; 25; 48; 50; 51
mode of action, **14:**180–183
and oxalate metabolism, **23:**447–452
and oxidation of hydrocarbons, **27:**485; 486; 490–491; 495; 502; 512–513
to pantethine, **14:**62–65
to pantothenic acid, **14:**63–65; 70
in peptide biosynthesis, **19:**249
and phosphotransacetylase, **18:**226
and photosynthesis, **25:**99
in photosynthesis, **19:**18; 41–45
in plants, **18:**299
in plastids, **20:**217
product inhibition, **25:**223; 242; 244–246
in reduction of diphosphopridine nucleotide, **20:**151
relationship to LBF, **14:**53–55; 61–62; 62–65
role in phytochrome action, **26:**24–29
role in propionic acid metabolism, **26:** 284–334
structure, **16:**1–21
synthesis, by mixed anhydride method, **22:**174
by phosphoramidate method, **22:**200; 201
and transferases, **13:**299
Coenzyme activation of apotransaminase, **16:**215–218
Coenzyme activity, mechanism, **7:**417–418
Coenzyme A deacylase, **15:**210; 224–229
Coenzyme A derivatives. *See also* specific compounds
acylation of glycine by, **19:**249
synthesis, **19:**240; 257; 258
Coenzyme affinities of cobamide coenzymes, **26:**346–347
Coenzyme analogs, and Cotton effect, **27:**56–70
Coenzyme A phosphate, **15:**232
Coenzyme A-pyrophosphate, production of, **18:**199
Coenzyme A transferase, **15:**234; 237–242; 252
Coenzyme A transphorase, **15:**234; 236; 252
Coenzyme B_{12}, **58:**1
EPR study, **27:**142–143
model study, **58:**3

Coenzyme binding, **22:**338–415
and changes in conformation, **22:**362; 363
of enzyme, **27:**52; 61–64
and enzyme activity, **22:**364–366
by LADH, native configuration, **27:**77; 81–82
by metal, **27:**68–71; 72
sites of, **22:**364; 365
Coenzyme Binding (Shifrin and Kaplan), **22:**337
Coenzyme-enzyme complex, sterochemistry of, **26:**156–165
Coenzyme factor, *See* Diaphorase
Coenzyme factors for tryptophanase, **10:**60–62
Coenzyme forms of corrinoids
biogenesis, **26:**258–263
partial synthesis, **26:**242–246
properties of, **26:**247–258
structures, **26:**234–238
Coenzyme forms of lipoic acid, **15:**222–223, **18:**338–340
Coenzyme I. *See also* Coenzymes; Diphosphopyridine nucleotide
conversion to coenzyme II, **7:**414; 425–427; 433–435
in deamination, **9:**402
effect on chloroplast system, **12:**5; 39
inactivation, **7:**410–414
pyocyanin system, **7:**427–435
in mechanism of enzymic oxidoreductions, **15:**5; 9; 14–15; 30–32
and nicotinamide, synergism, **7:**421
in normal and cancer tissues, **4:**208; 212; 226; 234; 245; 248
reactivation, **7:**430–432; 434
role in metabolism, **8:**414
separation by adsorption, **14:**332–333
and stimulation of metabolism, **7:**411–412
and V factor, **7:**414–415
Coenzyme I-cytochrome c reductase in normal and cancer tissues, **4:**230; 235; 237; 238; 239–240; 241; 245; 247; 249
Coenzyme II, **4:**208; 212; 232; 239; 245, **15:**4. *See also* Coenzymes; TPN; Triphosphopyridine nucleotide
conversion to coenzyme I, **7:**414; 419–420; 425; 433–435
effect on chloroplast system, **12:**5; 39
nature of, **7:**411
role in metabolism, **8:**413; 414
and stimulation of metabolism, **7:**411–412
and V factor, **7:**414–415
Coenzyme III
enzymes acting with, **15:**98; 114–115; 117; 118; 131
structure, **15:**83–87
in sulfur metabolism, **15:**124–125
Coenzyme inactivation, **7:**412–413; 414; 415; 427–435
Coenzyme models, **15:**30–32
Coenzyme-needing reductases, disulfides, **27:**218
Coenzyme nucleotides, **9:**487–490; 503–507
Coenzyme-oxidizing systems, **24:**337
Coenzyme Q, **25:**275–328, **34:**85; 122–129; 476–477; 480. *See also* Ubiquinone
anti-oxidant role, **65:**124–128
biosynthesis of, **25:**295–304
compounds related to, **25:**286–288; 322
detection, by chromatography, **25:**293–394
by color reactions, **25:**292–293
by enzymic reactions, **25:**294
by spectroscopic methods, **25:**291–292
homologs, **21:**111
isolation and purification of, **25:**294–295
labeled, in biosynthesis, **25:**300–302
in mitochondria and mitochondrial fragments, **21:**108–112; 115; 116; 118
in mitochondrial systems, **25:**305–319
natural occurrence of, **25:**288–291
nutritional aspects, **25:**301–304
and photosynthesis, **25:**75
position in mitochondrial electron transfer chain, **21:**119; 120; 122
in respiratory chain, **28:**157; 159; 160; 162; 190; 192; 194; 202; 209–214; 217; 219; 220
in sequence of carriers, **29:**77
structural studies, **25:**277–281
Coenzyme Q_1, in respiratory chain, **28:**212
Coenzyme Q_2, in respiratory chain, **28:**194; 212; 213
Coenzyme Q_6, in respiratory chain, **28:**194; 201
Coenzyme Q_{10}, in respiratory chain, **28:**211; 212
Coenzyme Q compounds
naturally occurring, **25:**277
properties of, **25:**281–284
Coenzyme Q group, physical constants, **25:**282
Coenzyme Q (Hatefi), **25:**275
Coenzyme Q hydroquinones, occurrence, **25:**291
separation, **25:**293–294

Coenzyme Q isoprenologs, DPNH-Q reductase action on, **25:**313
Coenzyme Q oxidase, in mitochondria, **21:**74
in mitochondrial fragments, **21:**115; 116
Coenzyme Q oxidase system, **25:**318
Coenzyme Q (QH_2)-cytochrome c reductase, **25:**312; 315–319
Coenzyme Q reductase
in electron transport system, **25:**310–312; 318
EPR study, **27:**180
in mitochondria, **21:**74
Coenzymes, **46:**262–270; 268, **58:**35. *See also* Enzyme binding; Enzyme-coenzyme complex; Enzyme complex; Ternary complex
A, **46:**75
absorption spectra, **4:**301; 302; 304 (*See also under* specific coenzyme)
assimilation of, **7:**413–414
asymmetric binding to alcohol dehydrogenase, **26:**164–165
B_{12}, **46:**53
classification, **14:**240–243
and coliform organisms, **7:**416–423
and Cotton effect, **27:**48; 86–87
dehydrogenase specificity, **7:**413–414
effect on enzyme denaturation, **23:**46–50
and enzyme catalysis mechanisms, **24:** 442–443; 467–475
in enzyme electrode, **46:**265–268
in enzyme reactor, **46:**265–268
and enzymic adaptation, **13:**75
FAH_4-containing, specific, **27:**460
folic acid containing, **27:**425; 426; 458–460
hydrogen bonds to, **50:**28
I and II reaction with borate, **12:**514
immobilized active preparations, **46:**264–272
inactivation of, **7:**410; 412–415; 427–435
influence on growth of influenza bacilli, **7:**415–416
interaction with enzyme, **15:**37–42
of thiaminase, **15:**395–396; 403–404; 406
kinetic studies, **17:**14–19; 42–45
mechanisms of interactions, **17:**272–281
mode of action, **14:**174; 180–183; 183–190
of NAD, NADP, and ATP analogs, **46:** 270–271
nicotinamide, abbreviations for, **24:**294
nuclear, **27:**276–278
and origin of life, **27:**373–374
phosphoric acid anhydride, formation of, **18:**206; 208
in plants, **18:**303–304; 306–310; 312
polyhydroxy, reaction with borate, **12:** 516–518
products of metabolism, **7:**414
and radiophosphate, **7:**190
regeneration methods, **46:**264–265
of respiratory chain, **23:**341
retention methods, **46:**264–265
and soil microorganisms, **28:**367
stereospecificity, for hydrogen transfer, **26:**169–171
synthesis, **18:**196; 206–208
in ternary complex, **27:**4–12
theory, **1:**85; 88; 95
in yeast, synthesis and degradation, **7:** 423–432
Coenzyme transhydrogenase reaction, **26:**167
Cofactor-apoenzyme interaction, **24:**113–114
Cofactor E/E^+, in photosynthesis, **25:**79–80
Cofactors, **36:**95
of bacteriophages, **9:**178–179; 183
classification, **14:**240–243
coenzyme I, **4:**245 (*See also* Coenzyme I)
coenzyme II, **4:**245 (*See also* Coenzyme II)
of dihydrofolate reductases, **27:**434–436
flavin, **4:**247
in normal and cancer tissues, biotin, **4:**248
for PN-dependent enzymes, **26:**120–132
riboflavin, **4:**246; 247
thiamine, phosphorylated, **4:**248
Cofactor X, in photosynthesis, **25:**79–80
Cofactor Z, in photosynthesis, **25:**79
Coffea sp., multimolecular forms, **36:**97
Cofibrin, **16:**25; 36–37; 41–42; 44; 92. *See also* Fibrinopeptide
Coformycin, **51:**182
Co-glucosidase, **60:**122
binding studies, **60:**123
molecular weight, **60:**123
physiological role, **60:**124
purification, **60:**123
Co-*S*-glutathionyl-cobalamin, **26:**246
Co-glycyl-cobalamin, **26:**245; 253–254
CO-haem-cholate model compound, **23:**292; 293; 304; 306
CO-haem-Emasol model compound, **23:**292; 293; 304; 306–312
Co-hydroxybutyl-cobalamin, **26:**245

Co-β-hydroxyethyl-cobalamin, **26:**245
Co(III) AMP, bound at AMP allosteric site, **43:**45
  with glycogen phosphorylase, **43:**45
Coil conformation, **47:**60; 64; 65
Colchicine, **11:**416
  effect on β-glucuronidase, **16:**400
  GTPase reaction microtubule dynamics, **69:**146
  metabolism in plants, **16:**137; 138; 142
Cold
  fall in brain norepinephrine by, **29:**487
  and thyroid stimulation, **7:**199
Cold inactivation, **35:**449
Cold-insoluble globulin, **52:**33
Cold-labile enzymes, **23:**381–382
Cold lability, pyruvate carboxylase, **35:**396
Cold sensitivity, of allosteric enzymes, **28:**53; 56; 72; 90
Cold shock, **24:**74
Cold stress, tyrosine hydroxylase induction, **70:**166
Colfibrate, **57:**88
Coliform organisms, coenzymes, and nicotinic acid derivatives, **7:**416–423
Colinearity hypothesis of genetic control, **24:** 9–10
Colipase
  activity, measurement of, **48:**346
  interaction between lipase and, **48:**357
  molecular properties of, **48:**341
  purification of, **48:**341
Colipase-micelle complex, **48:**359
Coliphages S13 and thetaX174, mutations, **27:**340
Collagen, **1:**11; 12, **3:**72; 73; 91–96, **33:**276, **34:**156, **41:**245; 264; 265; 277; 279; 281, **52:**27; 87; 88; 92, **56:**382
  agarose, **52:**92
  alteration in tissues, **23:**171–173; 180–183; 185; 202–203; 218–221
  amino acids in, **14:**265; 266
  ascorbic acid collagen formation, **37:**282–283
  banding of, **16:**320
  binding of *p*-benzoquinone by, **16:**123–125
  breakdown *in vivo,* **34:**191
  breakdown mechanism, **23:**213–217
  chemistry, **42:**182
  concentration in progressive muscular dystrophy, **6:**223
  cross striations of, **23:**213–217
  degradation of animals, **37:**294
  denaturation temperature, **41:**264
  depolymerization of, **23:**169; 173
  different types, **42:**182
  effect of γ-glutamyl-β-amino-propionitrile on, **29:**147
  enzymic digestion, **15:**292
  glycosylation of, **37:**292
  glycosyltransferases, **52:**87–93; 146
  helical structure, **23:**169; 213; 217–218
  under hydroxylated, **41:**265; 276; 281
  metabolism of, **37:**282–312
  microstructure, **23:**217
  molecular aging, **37:**275
  native, as collagenase substrate, **23:**166–170; 209–217
  phenylalanine and tyrosine content, **14:**119
  in plasma, **37:**311
  precursor forms, **42:**183
  pro-d-chains, **41:**279
  rate of solubilization, **23:**170–171
  relation to elastin, **23:**238
  scheme of biosynthesis, **37:**287; 289
  secretion, **41:**277
  shrinkage temperature, **23:**168–169; 172; 203; 204
  trypsin action on, **23:**164; 167–170; 172–174; 180; 182; 202; 204; 205; 209; 212
Collagen A, **23:**167–168; 181
Collagenase, **33:**262, **56:**293
  characteristics of, **56:**382
  column chromatography, **22:**453; 454
  distribution in animals, **37:**295
    mechanism of action, **37:**295
    in microganisms, **37:**296
    precursor, **37:**332
    properties of, **37:**296
  metal centers of, properties of, **56:**384
  metal content of, **56:**382–383
  molecular properties of, **56:**382–384
β-Collagenase, **23:**202
Collagenase 2, action on collagen of, **34:**162
  properties, **34:**169
  purification, **34:**163
  substrates, **34:**159; 161
Collagenase-like enzyme, from *aspergillus oryzae,* **41:**194; 238

Collagenase-like proteinase
from *Pseudomonas aeruginosa,* **41:**205
from *Serratia marcescens,* **41:**205
Collagenases, **23:**163–221. *See also* Animal Collagenases; Bacterial Collagenases; Clostridiopeptidase A; Collagenase 2; Mold Collagenases
action on elastin, **23:**222; 228; 237; 238
applications of, **23:**217–221
assay procedures for, **34:**166; 196
from bacteria, **23:**177–186
from *Cl. histolyticum,* **23:**240
definition, by substrate, **23:**166
definition for, **34:**156; 195
kinetics of action, **23:**209–210
mode of action, **23:**209–217
occurrence, **23:**171–183
pancreatic, **23:**172–173
production of, **23:**187–190
properties of, **23:**190–209
reaction products, **23:**210–213
specificity of, **23:**201–209
synthetic substrates for, **23:**171; 205–207
toxicity of, **23:**199–201
type A, **23:**192; 200
Collagenases and Elastases (Mandl), **23:**163
Collagen BrO-hydroxylysyl galactosyltransferase, **52:**90–91
Collagen hydroxylases
cosubstrates and reaction mechanisms, **72:**345–357
hydroxyprolines and hydroxylysine occurrence, **72:**328–334
inhibitors and inactivators, antifibrotic drug potential, **72:**357–364
intracellular hydroxylation sites, **72:**338–340
lysyl hydroxylase
gene mutation in Ehlers-Danlos syndrome, **72:**383–384
subunit structure, **72:**380–382
molecular properties, **72:**334–338
lysyl hydroxylase, **72:**336–338
prolyl 3-hydroxylase, **72:**338
prolyl 4-hydroxylases, **72:**334–336
peptide substrates, **72:**339; 341–345
prolyl 4-hydroxylases, a subunits, **72:**364–368
protein disulfide isomerase/β subunit polypeptide
functions, **72:**375–380
structure, **72:**368–375
regulation mechanisms, **72:**384–386
reseach backgound, **72:**326–327
Collagenmucoproteinase, **23:**239
Collagenolytic enzymes
action of on collagen, **34:**189; 190
from *Aspergillus oryzae,* **34:**190
nonspecific, **34:**188
Collagenolytic Enzymes (Nordwig), **34:**155
Collagenous tissues, hydroxyproline in, **16:**259
Collateral specificity patterns of β-lactamase, **28:**255–257
Colletotrichum antirrhini, enzymic transformations of steroids by, **20:**248
Colletotrichum coccodes, **55:**45; 47
Colletotrichum coffeanum, **55:**45
Colletotrichum lindemuthianum, **55:**45; 46; 47
plant disease caused by, **20:**390
Colletotrichum species, plant disease caused by, **20:**398
Colletotrichum truncatum, plant disease caused by, **20:**390
Colloidal elemental sulfur, **39:**362; 364
Colloidal structure, and succinic oxidase, **18:**104–108
Colloidal surfaces, adsorption of enzymes, **28:**367
Colloids of egg, changes following fertilization, **9:**281–286; 318
Collybia velutipes, luminescence in, **25:**161
Colominic acid, **52:**33
Color changes in bacteria, **1:**308
Color reactions. *See also* Melanin; Pigments
of proteins with quinones, **16:**122–123
Colostrum, **52:**30
Combination chemotherapy, use of adenosine deaminase inhibitors in, **51:**200
Combined immuno deficiency disease, **41:**1; 26–27
Combing ratio, **52:**10
Combustion, heat of, changes during tobacco fermentation, **10:**337–342
Co-methyl-cobalamin, cathode potential, **26:**255
electrophoresis, **26:**250
enzyme reactions using, **26:**266–267
in methane synthesis, **26:**267
photolysis, **26:**251; 256
synthesis, **26:**243; 245; 246
in TOH, **26:**257
transfer reactions, **26:**253–255

Comets, organic matter from, **27:**350; 353
Commission on Enzymes of the International Union of Biochemistry, **24:**294
Common precursors, in protein synthesis, **22:**137
Comparative biochemistry, of phenolase complex, **16:**105–184
Comparative specificity, of microbial proteinases, **41:**179–239
Competition. *See also* Inhibitors
  and enzyme adaptation, **13:**81–83; 100–105; 109–112
  in enzyme systems, **17:**12–13
Competitive action. *See also* Inhibition and Inhibitors
  of cobalamin analogs, **26:**270–274
  of cobinamide analogs, **26:**268–269; 274
Competitive inhibition, **35:**479, **46:**391
  components, **43:**160
  studies, RGD sequence, **70:**3
Competitive inhibitors, **36:**31. *See also* Antagonists, structural
  of acetylcholinesterase, **12:**286–290; 292–293; 305–311
  of carboxypeptidase, **12:**242–248
Competitive reactions, effect of nucleate complexes, **17:**141
Complement, **15:**58
  activity and calcium ions, **8:**95
    and magnesium ions, **8:**95
  ammonia-treated, **8:**104–111
  and antibody interdependence, **8:**92–93
  effect of heat on component, **8:**4; 107
  enzymic, **14:**241–242
  estimation of components, **8:**75
  fixation data, **8:**83
  fixation test, **8:**97–101
  heat inactivation, **8:**76
  hemolytic activity, **8:**76; 89–97
  inorganic, in enzyme systems, **14:**240–243
  lytic activity and electrolytes, **8:**93–96
  and mass law, **8:**84–89
  method of analysis, **8:**72–76
  microestimation in weight units, **8:**72–89
  molecular weight, **8:**80
  nitrogen in guinea pig serum, **8:**73
  nitrogen values, **8:**76
  participation in union of antigen and antibody, **8:**97–101
  properties of components, **8:**102
  purification and isolation of components, **8:**101–103
  quantitative estimations of fixation, **8:**97–101
  quantitative studies, **8:**71–112
  relation to antigen and antibody, **8:**82–89
  relation to bactericidal and opsonic effects of fresh sera, **8:**111
  relation to hemolysin and red cells, **8:**79–82
  specifically inactivated, **8:**103; 109
  stability, **8:**96
  stabilization, **8:**74
  titration, **8:**98
    of components, **8:**103–111
    reagents, **8:**108–109
  uptake in antigen-antibody combination, **8:**73
  zymosan-treated, **8:**104–111
Complement, Quantitative Studies (Heidelberger and Mayer), **8:**71
Complementation
  interallele, in enzyme formation, **22:**34–37
  intralocus *vs.* interlocus, **22:**37
  map, **22:**34; 35
Complex 1, **34:**82; 87; 99; 104; 106
  nonheme iron in, **34:**87
Complex catalysis, **14:**173–176; 179–190
Complexes
  and coenzyme action, **17:**14–19; 42–45
  enzyme-PNH, fluorescence, **26:**157–163; 165
  enzyme-substrate, **17:**9–10; 42–45; 55–58; 221–224; 308–309
  loose, **43:**286
    charge transfer, **43:**286
    hydrogen bonded, **43:**286
    racemization of, **43:**287
  of metal ions, **17:**25–26; 59; 60
Complex formation
  in enzyme action, **10:**73–74; 96–97; 102–104; 111; 113–117; 201–202; 204–208
  of enzymes, **26:**156–165
  and inhibition of alkaline phosphatase, **10:** 94–97
  in monolayers, **11:**82–85
  with polydentate ligands, **53:**309
    chemical equation, **53:**310–311
    order of bond formation, **53:**314
    role of solvated species, **53:**314–315
  stopped-flow trace of, between oxidized enzyme and *o*-aminobenzoate, **34:**46

Complex II (succinate-coenzyme Q reductase), **37:**191; 192; 193; 194
Complexing agents, **39:**348
anionic, **17:**275–276
Complexing groups of nucleotides, **17:**139–141
Complex systems, distribution of, **24:**335–342
Composition, **34:**446
Compound 64, **57:**64–66
Compound 66, **57:**64–66
Compound V, in metabolism of aromatic amino acids, **16:**291
Compound X, in arginine synthesis, **16:**260; 264–265; 267
Compound X in urea synthesis, **15:**339–342; 366–372; 382
Computer analysis, **47:**50; 97; 104–109; 124–125; 132
Computerized predictive methods, protein structure, **47:**97–130
Computer systems, in enzyme kinetic studies, **25:**173; 174; 181–182
Conalbumin, **17:**151; 273
chelating properties, **21:**143
column chromatography, **22:**477
Cotton effect, **27:**60; 93
electrophoresis of, **30:**168
Conalbumin-copper, EPR, **27:**111; 137
Conarachin, adsorption studies, **14:**365
Concanavalin A, **34:**341, **47:**49–50; 57; 74; 79; 124, **48:**69; 76; 84, **60:**90
use in purification of glycogen synthetase, **34:**318
Concanavalin A-agarose, **52:**92
Concentration, **34:**492–494; 499; 504; 522–525
gradient, **34:**502–503; 507
of outer substrate, **34:**463
of product, **34:**502
profiles, **34:**495
of reactant at an interface, **11:**37–69
Concerted feedback inhibition, of enzymes, **28:**46; 65
Conchiolin, mineralization processes, **34:**3
Conconavalin A, interaction with heparin, **40:**56; 57
Condensation enzyme, citrate splitting by, **12:**47
Condensation equilibrium model, microtubule dynamics, **69:**132–133
Condensation reaction to form argininosuccinic acid, **15:**344–353; 356; 362–363; 373–374
Condensing acetyl CoA r acetyl CoA transacetylase, **14:**251
Condensing acetyl CoA r oxalacetate transacetylase, **14:**251
Condensing enzyme, **15:**121; 183; 188–189; 191–201; 202–207; 210; 213; 235; 237; 241; 247–249; 259; 262; 270, **23:**408; 409; 413; 419–420; 435; 437; 438; 440; 445. *See also* Citrate condensing enzyme
adsorption studies, **14:**346
in arginine synthesis, **17:**399
citrate, **17:**426
classification, **14:**222; 251
cystathionine formation by, **16:**272
deficiency in mutants, **18:**269–270
and fluorocitric acid, **18:**128; 129–131
in folate system, **27:**429; 430
localization of, **24:**339
for oxalacetate and acetate, **12:**327–328
in pantothenic acid synthesis, **17:**442–445
and phytochrome action, **26:**25
in plants, **19:**14
sources, **14:**377; 405
Condensing reaction, between two-carbon compound and oxalacetic acid, **15:** 186–201
Conditional gene replacement studies, *Caulobacter* cell cycle, transcription regulation, **71:**16
Conductance, membrane, **34:**485
Conducting tissue, acetylcholinesterase in, **12:**270–271; 271–274; 312–315
and enzyme specificity, **12:**265–270
Conduction in nerve tissue, metabolic processes in, **12:**312–317; 331–334
relation to inhibition of acetylcholinesterase, **12:**279–290
Conduction of energy in living cells, **14:**3; 19–42
Conductivity. *See also* Semiconduction
thermal, in photosynthetic tissues, **25:**82
Conductivity of cytoplasm, **8:**42–47
Conduritol B, **48:**46
Conduritol B epoxide, **48:**47
Conferin, **56:**201
Configurations, polypeptide, **23:**1–27
Conformation, **42:**98; 122; 123; 137; 157; 158; 160, **59:**245
changes, **42:**110
entropy, **42:**134

of enzyme proteins
and oxidation state, **23:**58–70
polypeptides, **23:**1–27
and solubility, **23:**70–72
of β-glucans, **28:**327
Conformational boundaries, **47:**67; 99; 100; 106–108; 109; 113; 130
Conformational change
effect of coenzyme binding, **27:**52
in enzyme in slow change, **27:**23–29; 33–34
in toxin structure, **50:**308
Conformational change and enzyme processes, **25:**20–21
Conformational change during catalysis, **59:**139
Conformational change in glutamate dehydrogenase, **39:**387
Conformational changes, **36:**112, **39:**356; 357; 364, **47:**74; 76; 79; 81–83; 135; 136
in adsorption, **28:**4–5; 7; 8; 11–12; 17–21; 23–25; 31–32
of allosteric enzymes, **28:**51–53; 57–59; 93; 105; 117; 120; 125; 136; 138; 142
of β-lactamase, **28:**251; 254–257; 262; 267; 272; 297–304; 306–307
trypsin test for, **28:**99; 107; 108; 125
Conformational flexibility, **59:**130
Conformational flexibility of enzymes, *See* Enzymes, conformational flexibility of
Conformational homology, **47:**125–129; 136
Conformational isomerism, **50:**30
Conformational isomers, anti and syn, **43:**28; 32
Conformational parameters, **47:**50; 64; 103–104; 135
helix, **47:**50–51; 64; 66
helix boundary, **47:**67–68; 106–108
β-sheet, **47:**50–51; 64; 66
β-sheet boundary, **47:**69–70; 106–108
β-turn, **47:**64; 70–72
Conformational profiles, **47:**88; 97; 110; 114–115; 131
Conformational regions, **47:**58–63
Conformational states, of proteins, **50:**30
Conformational transition, **39:**349
Conformational transitions, rates of, **37:**557
Conformation change, **43:**247; 312; 345; 348; 354; 356; 370
Conformation of enzymes, **26:**89–114
Conformation of proteins, Cotton effect, **27:** 46–48
and enzyme activity, **27:**403
SH-groups, **27:**222–223; 226–227
Conformation of RNase molecule, **24:**200–202; 246–248
Conformations of enzyme-bound glutathione, **59:**293
Conformative response, **37:**404; 405
activation, **37:**479; 481
amplification, **37:**591–592
antagonistic effects, **37:**500; 566–586; 594
in antibodies, **37:**598
assumptions, **37:**406–407
biological role of, **37:**597–606
in catalysis, **37:**600
catalytic function, **37:**599; 602
changes in hydrodynamic properties, **37:**422
chemical inactivation studies, **37:**502–503
chemical modification, **37:**444–459
competitive inhibition, **37:**480
concept of, **28:**299–302
of β-lactamase, **28:**255–257; 301
constraint, **37:**593–597
correlation among criteria, **37:**504–527
crystallographic evidence, **37:**408–422; 526
definition of parameter, F_{CR}, **37:**533
K_{CR}, **37:**533
R, **37:**533
definitions of, **37:**404
K, **37:**531
K_D, **37:**531
K_{DM}, **37:**531
R, **37:**531
deformers, **37:**486
derivation of parameters, **37:**534–537
determinants, **37:**586–588; 594
diagnostic tests, **37:**560–562
difference criteria, **37:**505–523
difference sedimentation, **37:**422
difference spectroscopy, **37:**422–423
and fluorescence, **37:**428–437
differential inactivation, **37:**444; 479; 500; 502
by conformation disrupting agents, **37:**503
as a tentative criterion, **37:**500
differential modification, **37:**444; 459; 472
drug receptor, **37:**597
elimination, **37:**593–597
evidence based on, differential inactivation, **37:**486
enzyme function, **37:**478–485

Conformative response (*Continued*)
fluorescence techniques, **37:**423
fluorescent probes, **37:**438; 444; 598
function of muscle, **37:**598
to the hapten, **37:**599
hormone action, **37:**597
hydrodynamic properties, **37:**424–427
hydrogen exchange, **37:**459–469
hydrogen exchange studies, **37:**470–471
inhibition, **37:**479–481
isomerization, **37:**485
kinetic analysis, **37:**560
kinetic consequences, **37:**556–562
kinetic diagnosis, **37:**479–481
kinetic evidence, **37:**483; 484
labilization by ligands, **37:**486
ligand-induced stabilization, **37:**488; 489
meaning of, **37:**606
membrane systems, **37:**598
methodology, **37:**407–408
of differential inactivation, **37:**486
moderation, **37:**591–592
modifications, **37:**591–597
mutational constraint, **37:**596
nervous system, **37:**598
NMR spectroscopy, **37:**421
noncompetitive inhibition, **37:**479; 480
ordered binding, **37:**564
parameter persistence, **37:**562
parameter R, **37:**600
parameters, **37:**531
persistence, **37:**562
polarimetric evidence, **37:**414–420
polarimetric methods, **37:**421
proteolytic inactivation, **37:**533
studies of, **37:**500–502
proteolytic probe, **37:**469–478
quantitative parameters, **37:**527–556
rapid reaction techniques, **37:**484
reactivation, **37:**481–483
reporter groups, **37:**423; 438; 598
and fluorescent probes, **37:**439–443
repressor, **37:**597
resilience, **37:**591; 592
role of, in binding, **37:**605
role of ligand, **37:**608
sequential response, **37:**563–590
in ordered binding, **37:**605
to a single ligand, **37:**589–591
and specificity, **37:**602–606
structural details, **37:**590
synergistic and antagonistic effects, **37:** 568–585
synergistic effects, **37:**500; 566–586
thermal inactivation, **37:**533
studies of, **37:**487
time scale, **37:**492; 556–562
transport system, **37:**598
Conformative response patterns, β-lactamase, **28:**302–304
Congenital X-linked agammaglobulinemia, **51:**201–202
Congential nonsperocytic hemolytic a anemia, **41:**1; 26; 28
AMP kinase deficiency, **41:**28
PP-ribose-P deficiency, **41:**28
Conglutin, **56:**210
Congo red, **59:**73–74
effect on myosin, **16:**345
Congo rubin, effect on myosin, **16:**345
Congruency kinetics, enzyme action, **73:**41–43
Congruent thermodynamic cycle, enzyme action, **73:**39–41
Coniferase, **18:**371
in lignin synthesis, **16:**147
Coniferin, labeled, **18:**372
relation to lignin, **18:**371–373
D-Coniferin, radioactive, incorporation into lignin, **16:**149
Coniferin in plant tissues, **16:**148
Coniferyl alcohol
metabolism in plants, **16:**141; 148–149
relation to lignin, **18:**371–373; 375
Coniferyl aldehyde, **18:**371
Coniferyl aldehyde and alcohol, **55:**97
Coniferyl carbinol C-14, incorporation into lignin, **16:**149
Coniferyl type lignin, **16:**147
Coniophora cerebella, action on cellulose, **8:**271; 272; 273; 278–279
degradation of glucose, **8:**272–274
Coniothyrium species, enzymic transformations of steroids by, **20:**247
Conjugase, **15:**222–223
activity of –SH group, **11:**241
Conjugases, **21:**391
classification, **14:**247
and use of folic acid, **27:**428
Conjugated double bonds and autoxidation, **8:**428; 434–439; 449–450

Conjugated pterins, **35:**35
Conjugates, **34:**453
Conjugation
quinones, **65:**119–120
yeast, cell wall-softening enzyme, **28:** 348–349
Connective tissue, amino sugars from, **25:** 434–441
Consecutive membrane enzyme reaction, **34:**471
reaction, **34:**513
Constant
dissociation
influence of $[Na^+]$ on K_A of carrier for sugar, **29:**59–60
of translocation of D-glucose, D-mannose, and D-arabinose, **29:**45
of translocation of L-leucine, L-phenylalanine, L-methionine, and L-valine, **29:**45
kinetic, standard errors of, **29:**6–7
nonlinear, plot of a function against the value of, **29:**5
variances of fitted, **29:**6
Constitutive mutants, **44:**247
Contact system of blood coagulation, **48:**281
Continuous linkage, **34:**452
Continuous spectrophotometric assay, protein tyrosine phosphatases, **68:**21–23
Continuum theory of energy transport in living cells, **14:**25–29
Contractile processes, imidazole in, **20:**91
Contractile protein, in respiratory chain, **28:**161
Contractile structure, of mitochondria, **23:** 383–385
Contractile substance in spermatozoa, **24:**42
Contractile system, precipitation in, **16:**331–333
Contractility, relation to ATPase activity, **16:**339–343
Contractility of protoplasm, **7:**42–43, **8:**37–42; 50; 56
Contractin, in muscle contraction, **13:**162
Contraction
of actomyosin, —SH group, **27:**215
muscle, **24:**285–286
muscular, **11:**23–25; 304
of myosin threads, **11:**230–231
non-specific role of actin, **16:**345
of actomyosin, relation to ATP, **16:**337–339
of muscle, relation to actomyosin, **16:** 346–347
relation to myosin, **16:**351–354
of myosin, relation to meromyosin components, **16:**326
relation to precipitation of proteins, **16:**331
Contraction cycle, efficiency, **16:**342
Contraction of muscle, mechanism of, **13:**174; 187–192; 192–193
Control, **38:**354–365
coordinate repression, **38:**356
enzyme repression, **38:**355
specificity of biotin, **38:**355
Control functions, of nuclear enzymes, **27:** 275–283
Control mechanisms, **25:**170; 173. *See also* Allosteric regulation; Feedback regulation; Product inhibition of enzymes; Regulation; Regulatory mechanisms; Repression
of cancer cells, **29:**322
in cell wall growth, **26:**215–222
coupled reactions, **28:**47; 87; 89; 93; 122
DNA synthesis, **28:**436; 443
in DR metabolism, **28:**447–449
enzyme, and protein structure, **26:**90; 105–107; 110–111
of enzyme activity, **28:**41–154
enzyme multiplicity, **28:**45; 60–64
of β-lactamase synthesis, **28:**272–274
metabolic patterns, **28:**60–71
Convallaria majalis, imino acid in, **29:**95
Convergence, hypothesis of, **29:**371
Convergent evolution, **43:**298
Conversion, **34:**482
Conversion factor, adenosine deaminase deficiency, **51:**176–177
Conversion procollagen to collagen, **42:**180
procollagen peptidase, **42:**180
Conversion rate, apoenzyme, conformers with different affinity, eukaryotic dihydrofolate reductase, **70:**54
Convertibility factor, relation to prothrombin activation, **16:**59; 60
Convertin. *See also* Factor VII
term for a blood clotting factor, **16:**60
Converzyme, **7:**593
Convulsive state and citric acid, **18:**146–150
Cooperative feedback inhibition, of enzymes, **28:**46; 62; 65–66
Cooperative interactions, of allosteric enzymes, **28:**49–50

Cooperativity, **39:**379; 380; 387; 393; 410
eukaryotic dihydrofolate reductase, ligand binding, **70:**52–53
negative, **46:**402
Coordination, complexes of enzyme, metal, and substrate, **49:**105–106
Coordination chemistry of copper, *See* Copper, coordination chemistry of
Coordination complex, of non-heme iron, in respiratory chain, **28:**226–228
Coordination complexes, and catalytic activity, **15:**26; 32–33
metalloporphyrin, **15:**16–21; 32–33; 38–39
Coordination complexes, Co(III), **43:**43
Cr(III), **43:**43
Cr(III) ATP, **43:**43
Cr(III) nucleotides, **43:**44
Coordination compounds. *See also* Corrinoids and Vitamin B_{12}
of cobalt, **26:**236–238; 256–257
Coordination compounds as catalysts, **7:**98. *See also* Iron porphins
Coordination number, of metal ions, effect on enzyme activation, **21:**157; 158; 162
Coordinative complexes, **9:**1; 2; 9–15; 17–20
Co-oxidation, of alkanes, **27:**181; 480
CO-photolysis studies (cytochrome *c* oxidase)
electron transfer, **71:**119–121
proton pumping, **71:**137–138
Copper, **44:**188
action with cytochromes, **23:**297
in cytochrome oxidase, **23:**304–306
effect on, collagenase, **23:**196
elastases, **23:**234
in oxidase, **23:**314–315
role in haemin *a* biosynthesis, **23:**315
in beef serum amine oxidase, **36:**235
in diamine oxidase, **36:**233
as component of respiratory chain, **20:**166; 167; 170; 185; 188
binding, by pectic substances, **20:**356
by protein groups, **20:**56; 71; 97
effect on facilitated diffusion of glycerol, **20:**95
effect on intestinal maltase and oligo-1,6-glucosidase, **20:**94
effect on pectic substances, **20:**351
porphyrin compounds, **20:**167
reaction with diisopropylfluoro-phosphate, **20:**97
in respiratory chain preparations, **20:**153; 188
synergism with Agrimycin, **20:**390; 500
coordination chemistry of, **33:**181–183
electron configuration of complexes of, **33:**184–186
EPR properties of, **33:**188; 191–195
extinctions of complexes, **33:**182–183
intensities of absorption bands, **33:**187
ions of, **33:**182–183
properties of Cu^+ complexes, **33:**183
properties of Cu^{2+} complexes, **33:** 182–183
visible spectra of, **33:**186–188
EPR-nondetectable, definition of, **33:**222
relation to diamagnetic copper, **33:**222
state and function of in biological systems, **33:**177–244
effect on amino acid decarboxylases, **6:**17; 18
effect on ATPase, **6:**165
effect on cacao polyphenol oxidase, **25:** 467–468; 482
in photophosphorylation, **25:**69–70
effects on enzyme systems, **15:**2; 42; 379; 398; 410
in β-keto acid oxidation, **15:**234
and electron transfer oxidases, **19:**202; 203; 205
and mixed function oxidases, **19:**132–134; 185; 199
in model hydroxylating system, **19:**135
and oxygen transferases, **19:**81; 118; 122; 126; 127
fourth type, **44:**231
inhibition of xanthine oxidase, **17:**311
in metalloflavoproteins, **17:**261; 263; 265; 266; 268; 273; 274; 276; 283; 285; 294
redox function, **17:**77; 82
in mitochondria, EPR study, **27:**148–155; 187–190
in submitochondrial particles, EPR study, **27:**153–154
in respiration, **28:**157; 160; 162; 174; 192
role of in iron mobilization, **44:**214
state of, **44:**193
type 1 Cu(II), **44:**193
type 2 Cu(II), **44:**193
type 3 Cu(II), **44:**193
in tyrosinase, **4:**105–110; 148
utiliztion of, **44:**221

Copper, State and Function of in Biological Systems (Malkin and Malmstrm), **33:**177
Copper and autoxidation of fats, **8:**441
Copper atoms, numbers of, **44:**190; 191
Copper-binding substances, and oxidase activity, **23:**305
Copper carboxypeptidase, lack of Cotton effect, **27:**93
Copper complexes, inorganic, relation to oxidation-reduction, **17:**280
Copper-conalbumin, EPR spectrum, **27:**111; 137
Copper-containing oxidase, EPR study, **27:** 125–128
Copper-containing proteins, classes of (table), **33:**179
  composition of copper in (table), **33:**196
  functions of, **33:**178–180
Copper enzymes, in seedlings, **18:**41–42
Copper-histidine complex, EPR spectrum, **27:**110–111
Copper(II)-albumin, **44:**224
Copper (II) complexes, EPR spectrum, **27:**109; 111; 116; 117; 119; 137–138
Copper(II)-histidine, **44:**224
Copper in semen, **9:**346
Copper ions. *See also* Cupric ions
  and amylase activity, **12:**400; 401
    and choline acetylase, **12:**319
    combination with porphyrin, **12:**18
    complex with oxalacetic, **12:**72
    as decarboxylating agent, **12:**80; 87
  concentration in monolayer, **11:**68
    effect on membrane of permeability, **11:**71; 247
    effect on monolayer photochemical reactions, **11:**67
    in thiol oxidation, **11:**203–204; 206–210; 211; 216; 235–236; 249
  effect on β-glucuronidase, **16:**379
    in melanoprotein, **16:**171–172
    in nonenzymic transaminations, **16:**224; 225
  effect on tryptophan metabolism, **10:**68–69; 71
    in citric acid cycle, **10:**201
  in mitochondria, **21:**10; 100
  in mitochondrial fragments, **21:**100 101; 106; 113; 114
  in mitochondrial reactions, **21:**27
  model catalyses with, **21:**177
  and tyrosinase activity, **14:**78–84; 85–86; 87; 88; 91; 92; 93–94; 133; 140; 144; 146
    in isolation of enzymes, **14:**397; 398
    in oxidation by quinones, **14:**132
Copper-ligand covalency, **27:**149
Copper oxidase, multiple forms, **25:**359–360
  product inhibition of, **25:**183; 250
Copper oxidases, EPR study, **27:**145
Copper phthalocyanine, EPR spectrum, **27:**146
Copper porphins and yeast, **7:**320
Copper protein, **39:**364
Copper-protein, relation to phenolase complex, **16:**106; 128–131
Copper-protein enzyme, oxidation state of copper, **16:**145–146
Copper protein enzymes, **1:**183; 185
Copper-protein enzymes, inhibitors of, **18:**36–37
Copper-proteins, **44:**226
Copper proteins, and Cotton effects, **27:**92–94
  EPR study, **27:**144–155
  oxidase reactions of, **27:**128
Copper salts, solubilization of, **17:**137; 138; 142; 147; 150
Copper sulfate, effect on enzymes of *Cunninghamella blakesleeana,* **20:**270
Copper system, in plant respiration, **18:**309–310
Copper transport, **44:**188; 224
Copper transport protein oxidase, **44:**187
Copper-zinc superoxide dismutases, **41:**35; 51–70
  activity and mechanism, **41:**58–63
  apoenzyme, **41:**60
  circular dichroism spectra, **41:**60
  effects of metals, **41:**61
  historical background, **41:**51–56
  Inhibition by cyanide, **41:**67
  Isoenzymes, **41:**67
  Isolation procedures, **41:**56–58
  roles of Cu^{++} and Zn^{++}, **41:**60
  stability, **41:**63
  structure, **41:**64
  ultraviolent absorbance, **41:**60
  X-ray-crystallography, **41:**67
Coprinus atramentarius, quinones in, **25:**288
Coprinus cinereus, **59:**66; 68
Coprinus enzyme, **59:**68–69
*Coprinus radia*tus, **49:**12
Coprogen, **35:**167; 168; 177
Coproporphin I and yeast, **7:**337
Coproporphins II and III, **7:**319

Coproporphyria, **46:**161
hereditary, **46:**163
Coproporphyrin, **46:**86; 102
Coproporphyrin, III, **46:**92
Coproporphyrinogen, **46:**86; 102; 138
decarboxylation, **46:**57
Coproporphyrinogen decarboxylase, **46:**54; 56; 57; 86
Coproporphyrinogen I, **46:**165
Coproporphyrinogen III, **46:**54–58; 81; 87; 91–92
metabolism of, **24:**330; 342
Coproporphyrinogen IV, **46:**57
Coproporphyrinogen oxidase, **46:**164
Coproporphyrinogen oxidase (decarboxylating), location of, **24:**300
Co Q_1, **34:**87; 91; 111; 141; 148
depleted particles, **34:**128
depletion of, **34:**123
heterogeneity of the pool, **34:**129
homologs, **34:**129
low molecular weight form of, **34:**124
reductase activity of, **34:**124
resolution of membrane preparation with respect to, **34:**123
reversible removal of, **34:**126
unphysiological reduction of external, **34:**123
Co Q_6, **34:**11; 87; 96
Co Q_{10}, **34:**87; 91; 96; 104; 111; 126; 139
content, **34:**82
interaction of, with dehydrogenases and the cytochrome system, **34:**128
physiological reduction of, **34:**82
reductase activity, **34:**93
reincorporation, **34:**126
$CoQH_2$, **34:**477
Cordycepin, **43:**26
Coreuloplasmin oxidase, isolation of, **14:**405
Cori cycle, **54:**167
Cori fraction, **15:**168
Corn, antibiotic treatment of diseases, **20:**393; 394; 401
Corneal opacity, **60:**261
Corn oil, **51:**5
Corn proteins, amino acids in, **14:**266
Cornybacterium xerose, β-glucuronidase in, **16:**377
Corpuscular proteins
chemical and x-ray evidence on structure, **3:**96–104
egg albumin, **3:**67; 68; 101–102
hemoglobin, **3:**68–70; 102–103
insulin, **3:**96–99
lactoglobulin, **3:**99–101
tobacco mosaic virus, **3:**103–104
Correlation time, dipolar interaction for, **33:**6
scalar interaction for, **33:**7
Corrinoid-cobalt-sulfonic acids, partial synthesis, **26:**246
properties, **26:**247–250; 252–256; 258
Corrinoid coenzyme analogs, partial synthesis, **26:**242–246
Corrinoid coenzymes. *See also* Cobalamin coenzymes, Cobamide coenzymes, Cobinamide coenzymes, and Vitamin B_{12}
biogenesis, **26:**258–263
nucleoside ligands, **26:**252–253; 275
partial synthesis, **26:**242–246
properties of, **26:**247–258
structures, **26:**234–238
Corrinoid derivatives, spectra of, **26:**247–249
Corrinoid hydrides, reactions, **26:**242–243
properties, **26:**247–249; 255; 257
Corrinoids, **46:**87; 103. *See also* Cobalamins, Cobamide, Cobinamides, Coenzymes, Corrinoid coenzymes, and Vitamin B_{12}
acid-amide groups, **26:**273–275
biochemical functions, **26:**264–270
biogenesis of, **26:**258–263
with cobalt-carbon binding, **26:**243; 245–247; 250–257
complete, synthesis, **26:**240–241
coordination compounds, **26:**236–238; 256–257
with Co-S binding, **26:**246–250; 252–255; 258
cyanide cleavage, **26:**252–253
effect of reducing agents, **26:**255–256
electrophoresis, **26:**249–250
incomplete, structure, **26:**236–237
synthesis, **26:**238–240
iodine cleavage, **26:**253
nucleotides, **26:**236; 239–241; 243; 244; 248; 250; 256; 260–262
paper chromatography, **26:**249–250
photolysis, **26:**251–252
polarography, **26:**242; 255–256
reduced, reactions of, **26:**242–243
structure, **26:**234–238

trans effect, **26:**256
transfer reactions, **26:**253–255
ultraviolet absorption spectra, **26:**247–249
Corrinoids (C^{14}), in coenzyme formation, **26:**263
Corrin ring. *See also* Corrinoids
biogenesis, **26:**258–260
effect on coordination linkages of cobalt, **26:**256–258
in vitamin B_{12}, **26:**234–236
Corrins, **46:**38
Corriphyrin, **46:**51
Cortex, brain, Na^+-amino acid symporter in, **29:**61
Cortex and plasma membrane of cells, **8:**17–20
Cortex hormones, *See* Adrenal cortical hormones
Cortex of egg, **9:**245–246
changes following fertilization, **9:**275–286
Cortexolone, tyrosine hydroxylase induction, **70:**172
Cortexone, effect on *Neurospora crassa,* **20:**264
formula, **20:**240
microbiological conversions, **20:**242–244; 246–249; 254; 263; 264; 270; 273
Cortical cells, **47:**368
Cortical granules of egg, **9:**245–246
and fertilization, **9:**275–286
Corticium sasakii, enzymic transformations of steroids by, **20:**249
Corticoid compounds E and F, effect on β-glucuronidase, **16:**388
Corticoids, glucuronides as substrates for β-glucuronidase, **16:**368; 380; 384
Corticosteroid hormone, **49:**28
Corticosteroids, **46:**124
in affective disorders, **29:**498
Corticosterone, **19:**189; 191
formation, in mitochondria, **21:**27
formula, **20:**241
microbiological formation, **20:**246; 270
Corticotropin
structure studies, **14:**294; 299; 305–306
variation of amino acid composition, **22:**65
Corticotropin-releasing factor, disulfides, **27:**220
Corticotropins, amino acid sequence, **24:**417; 418; 422–425; 430; 435
lower peptide sequences in, **24:**420–425
Cortin
activity (*See* Adrenal cortical hormones)
nomenclature, **2:**326
Cortin active adrenal hormones, amorphous fraction, **2:**327–345
corticosterone, **2:**327–347
dehydrocorticosterone, **2:**329–344
desoxycorticosterone, **2:**330–349
17-hydroxycorticosterone, **2:**329–347
17-hydroxydehydrocorticosterone, **2:**330–344
17-hydroxydesoxycorticosterone, **2:**330–340
Cortisol, **67:**227, *See* Hydrocortisone
dissociation constant, **46:**391
steroidogenic pathway, **68:**44–46
Cortisone
effect on β-glucuronidase, **16:**388
relation to transamination, **16:**228
effect on hyaluronidase inhibition, **13:**218
effect on rate of phenolsulfophthalein excretion, **13:**229
and fluoroacetate, **18:**145
formula, **20:**241
microbiological formation and conversions, **20:**251; 252; 255; 261; 263–265; 270; 273
tryptophane pyrrollase induction by, **29:** 375–376
4,5-Cortisone reductase, location in subcellular fraction, **24:**299
20-Cortisone reductase, location in subcellular fractions, **24:**298
Corynebacteria, enzymes and reactions of pyrimidine biosynthesis in, **21:**274; 275
Corynebacteriophage, **58:**107
Corynebacterium, diphtheriae, endonuclease of, **29:**200
Corynebacterium, oxidation of hydrocarbons, **27:**471; 475; 493
source of pteridines, **27:**423
Corynebacterium bovis, in lipoic acid assay, **18:**332; 342
Corynebacterium creatinovorans, use of acetate, **23:**412
Corynebacterium diphteriae, quinones in, **25:**289
Corynebacterium diphtheriae
amino acids in, **14:**266
cytochromes of, **23:**270
succinic dehydrogenase of, **18:**71

Corynebacterium, folic acid conjugate in, **21:**371
Corynebacterium michiganese, **20:**396
Corynebacterium simplex, enzymic transformations of steroids by, **20:**254; 255; 261; 262
Corynebacterium species, plant disease caused by, **20:**396
Coryneform bacteria, glycine metabolism in, **23:**457
Corynephage, **58:**104
COS cells, P450 expression, **68:**42–49
COSMO algorithm, glutamine-dependent nitrogen transfer, AS-B *vs.* papain, **72:**176–183
Cosubstates, collagen hydroxylases, **72:**345–357
Cosubstrates, **1:**95; 96
  for papain, **19:**311; 312
Cothromboplastin, *See* Factor VII
Co-translational reactions, **67:**274–276
Cotransmethylase, **9:**492
Co-transmethylase, **12:**463
Cotransport, with Na^+, **49:**67
Cotton effect, **35:**509
  extrinsic
    designation for proteins, **27:**41; 47–49
    and mechanism of enzyme action, **27:** 37–104
  intrinsic, designation for proteins, **27:**41; 46–47; 51–52
  in proteins, terminology, **27:**41
Cotton Effects, Extrinsic, and the Mechanism of Enzyme Action (Ulmer and Vallee), **27:**37
Cottonseed allergen picrate, separation by chromatography, **14:**350
Coulomb forces, and hapten-antibody combination, **15:**67
Coumalic acid, collagen hydroxylase inhibitor mechanisms, **72:**363
4-Coumaric acid
  from cinnamic acid, **55:**16
  phenylalanine ammonia lyase and, **55:**14
4-Coumaric esters, **55:**9
Coumarin
  7–7hydroxylase activity, **68:**48
  as inhibitory analogue, **6:**134; 138; 139; 142
Coumarin anticoagulants, **46:**7–14
Coumarin drugs, **48:**293
Coumestrol, **55:**5
  chalcone isomerase and, **55:**29
  daidzein and, **55:**29
  structure of, **55:**8
Countercurrent distribution, **42:**78
  coefficients of peptide intermediates in toluene system, **29:**423
    of peptide intermediates, **29:**423–426
  of RNase, **24:**168
Countercurrent technique in amino acid determination, **14:**272–273; 299; 300; 303; 304
Coupled oxidations, in plants, **16:**133–135
Coupled oxidations and catalase, **7:**300–301
Coupled reactions, and enzymic adaptation, **13:**77–80
Coupled spectrophorometric assay, tyrosine hydroxylase, **70:**110
Coupled systems, effect of radiations on, **24:**399
Coupled vectorial processes, utilization of binding energy and, **51:**75–106
  introduction to, **51:**75–76
  ion transport (calcium-transporting ATPase), **51:**89–98
    equilibria and utlization of binding energies, **51:**92–98
    rules, **51:**90–92
  muscle contraction, **51:**79–89
    rules, **51:**83–84
    utilization of binding energies (steps A, B, and C), **51:**84–89
  phosphorylation coupled to proton transport, **51:**99–105
    binding and equilibria, **51:**101–105
    rules, **51:**100–101
  properties of, **51:**77–78
  sodium-potassium ATPase, **51:**98–99
Coupling, of oxidation and phosphorylation, **23:**369–372; 375; 378–385
Coupling factors
  in ATP synthesis, **28:**161
  role in oxidative phosphorylation, **23:**378–382
Coupling of molybdenum to ^{33}S, **51:**146
Coupling of reactions, control, **28:**47; 87; 89; 93; 122
Coupling procedures, in affinity chromatography, **46:**228–229
Coupling site 1, **34:**146
  and DPNH dehydrogenase, **34:**145
  localization of, **34:**143
  role of non-heme iron, **34:**144

Co-5-uridyl-cobalamin, synthesis, **26**:244
Covalent affinity labeling, **43**:157
Covalent bond cleavage, ligand reactions associated with, **53**:336–338
Covalent bond formation, **34**:452
Covalent enzyme-substrate intermediates, **43**:308
Covalent Inhibitors of the Gelation of Sickle Cell Hemoglobin and Their Effects on Function (Manning), **64**:55
Covalently attached cofactors, **67**:280–281
Covalent structure of ribonuclease, **24**:173–177; 248–251
Covitamin E, **6**:480
Cow colostrum trisachharide, **20**:143
Cozymase, **4**:304–305, **5**:208–211; 221; 227; 232. *See also* Coenzyme I; Diphosphopyridine nucleotide
  in biosynthesis of glutamic acid in plants, **20**:323; 325
  methods for quantitative determination, **5**:216
Cozymase in sperm, **9**:363
Cozymase-splitting enzyme, classification, **14**:248
CP, *See* Phosphocreatine
C5P, *See* Cytidine 5′-monophosphate
Cp, catalyzed oxidation of catecholamines, **44**:210
  and iron mobilization, **44**:206
  physiological role, **44**:218
  substrates, **44**:289
2C-particles. *See also* Acetaldehyde and Acetic acid
  activation, **10**:180–182
  in citric acid cycle, **10**:188–190
Crabtree effect, in minimal deviation tumors, **29**:330–332
  requirement of endogenous substrate for, **29**:334
Crassulacean acid metabolism, role of phytochrome, **26**:27
Creatinase, enzymic adaptation, **13**:70
Creatine, **15**:354; 356; 361; 373–374; 379, **50**:177–237. *See also* Methylation, biological
  affinity label structurally related to, **54**: 391–395
  analogs, as probes, **50**:224
    structure, **50**:205
  analogs of
    acyclic, **54**:384–386
    conformationally restricted, **54**:388–389
    monocyclic, **54**:386–391
  balance studies, **50**:193
  binding of
    to creatine kinase, **54**:391
    pH in, **54**:378
  and biological methylation, **12**:453; 458–469; 469–474; 478–479
  biosynthesis, **50**:180
    regulation of, **50**:192–215
    subcellular sites of, **50**:196
  chemical structure of, **54**:386
  cleavage by microorganisms, **23**:327
  concentration, in cardiac muscle hypertrophy, **6**:202; 203
  conformational preferences of, **54**:378–379
  constant excretion of, **50**:192–193
  corepressor of arginine: glycine amidinotransferase, **50**:205
  deprotonation of, via histidine, **54**:395
  effect of dietary precursors, **50**:194
  effects on biosynthesis of myosin and actin, **50**:223–234
  and ethionine metabolism, tracer studies, **25**:375; 379–380
    ethyl derivative of, **25**:383
    source of methyl group, **25**:383; 384
  in familial periodic paralysis, **6**:216
  free and phosphorylated, **1**:118
  functions, **50**:215
  guanidinium group in, **54**:379
    deprotonation of, **54**:396
  metabolic costs, **50**:198
  in MgADP-enzyme complex, **54**:377; 378
  in muscular atrophy, **6**:196
  in muscular dystrophy, **6**:204; 206; 207; 220; 221; 222; 223; 505
  in myasthenia gravis, **6**:212
  in myotonia, **6**:214
  phosphokinase, activity of –SH group in, **11**:243
  phosphylation of, conformation for, **54**:420
  repression of liver amidinotransferase by, **50**:202–204
  sparing effect on arginine and glycine, **50**:193
  stability of methyl group, **12**:481–482
  synthesis from quanidinoacetic acid, **22**:255
  vs. glycocyamine, **54**:386

Creatine: Biosynthesis, Regulation and Function (Walker), **50:**177
Creatine amidinase, classification, **14:**247
Creatine and methyl transfer, **7:**376–377
Creatine-ATP-transphosphorylase (creatine kinase), Cotton effects, **27:**59; 94
Creatine in testes, **9:**342
Creatine kinae, kinetic mechanism of, **45:**285; 289; 308
Creatine kinase, **34:**449; 523, **37:**409; 416; 421; 450; 474; 492; 501; 511; 512; 524; 545; 546; 566; 574; 575; 587; 600, **43:**36; 44, **46:**247, **55:**359
activation of, by metal ions, **54:**403–413
active site of, **55:**372; 373
affinity labeling of, **54:**378; 391–395
alkylation of, via iodoalkanes, **54:**402
arginyl side chain of, **54:**419
assay and isolation of, **54:**369–370
and ATP relationship, **54:**368
bound to creatine, representation of, **54:**391
circular electron flow in, **50:**13
cytosolic, **50:**220
mitochrondrial M-line, **50:**220–223
classification, **14:**251
competitive inhibition of, **54:**398–399
composition of, subunit, **54:**370–371
and conformational preference of creatine, **54:**378–379
and conformational preference of phosphocreatinine, **54:**378; 379–384
Cr (III) nucleotides with, **54:**404–405; 406
cysteine residues of, chemical modification studies of, **54:**400–403
differential inactivation, **37:**565
enzyme-bound substrate studies of, **54:** 413–414
epoxycreatine inactivation of, **54:**392; 393
epuilibrium constant for reaction of, **54:** 371–372
forms of, **54:**401–403
histidine residues of
at active site, **54:**395–400
pH effect on, **54:**395–396; 397
inactivation of, by sulfhydryl blocking reagents, **54:**400
intermediates in reaction of, **54:**417–418
isoenzymes of, **54:**368–369
kinestics of, steady-state, **54:**372–378
lysyl side chain of, **54:**419
mechanism of deduced from relaxivity data, **33:**39–43
with methoxycarbonylmethyl disulfide, **54:**402
with *N*-methylcreatine, **54:**386
with methylmercury chloride, **54:**403
with methyl methanethiosulfonate labeling, **54:**400–402
in MgADP complex, **54:**378
molecular weight of, **54:**370–371
in myrocardial infarction diagnosis, **54:**369
occurrence of, **54:**368
photoaffinity labeling of, **54:**395
product-inhibition patterns in reactions of, **54:**373; 375; 376; 403
protein-substract interaction studies of, **54:**414–417
rate constants in reaction of, **54:**419
rate-limiting step in reaction of, **54:**372
specificity of, for nucleotide analogs, **54:**384
stereochemistry of reaction of, **54:**417–418
structure-activity relationships of, **54:**367; 426
structure of, primary, **54:**391–392
in substrate specificity for creatine analogs, **54:**384–391
sulfhydryl groups of, blocking of, **54:**402
transition state-analog complexes of, **54:**403
activation metal ion in, **54:**407
ADP in, **54:**414; 415–416
anion coordination with metal ions in, **54:**409
EPR spectra for, **54:**408
formate in, **54:**414–415
free energies in, **54:**420
infrared spectroscopy of, **54:**407–408
metal ion coordination scheme of, **54:**409
nitrate effects on, **54:**413; 414
Creatine kinase, Cotton effects, **27:**59; 94
EPR study, **27:**140
Creatine Kinase: Structure-Activity Relationships (Kenyon and Reed), **54:**367
Creatine kinase in myosin preparations, **16:**315; 318
relation to muscle relaxation, **16:**350–351
removal from actin, **16:**334
Creatine kinase-MnADP-fromate-creatine complex, **54:**410
Creatine phosphate, **1:**100; 109; 111; 113; 117; 118; 128, **23:**324, **50:**215, *See* Phosphocreatine

analogs as probes, **50:**224
effects of starvation, **50:**197
as energy messenger, **50:**222
functions, **50:**215; 221
modulation of glycolysis, **50:**219–220
in muscular dystrophies, **50:**197
pH-buffering action, **50:**218
reservoir of high-energy phosphate, **50:** 216–218
source of inorganic phosphate, **50:**218–219
in tumor cells, **50:**232
Creatine phosphate in energy transport, **14:** 14–17
Creatine phosphate in semen, **24:**42
Creatine phosphokinase
effect of radiations, **24:**388; 395
effect on Q_{o2}, **17:**122
Creatine phosphoric acid, hydrolysis, by metal ions, **21:**176; 194
Creatinephosphoric acid, rate of renewal, **7:** 113–120
Creatininase, **56:**293
apoenzyme-holoenzyme equilibria of, **56:**43
cation effect on, **56:**43; 46
cosubstrate specificities of, **56:**47
in elimination of amino acid substituents, **56:**77
isotopic exchange of hydrogen atoms with substituents of, **56:**53
Michaelis and affinity constants for, **56:**48
molecular parameters of, **56:**44
steady state kinetics of, **56:**74–75
sulfhydryl reagent reactions of, **56:**60
Creatinine, **15:**354; 361
decomposition, as adaptive process, **6:**80–81
in muscular dystrophy, **6:**207; 220; 221; 222
in myasthenia gravis, **6:**212
Creatinine amidinase, classification, **14:**247
Creatinine desimidase, column chromatography, **22:**461; 462
Creatinine formation, **1:**118
Creatinuria, **50:**197–198
and diabetes, **8:**402
and ketonuria, **8:**402; 407
CRE-binding protein, tyrosine hydroxylase transcriptional regulation, **70:**178–179
Crenothrix, metabolism in, **74:**6
Crepis, **7:**470; 471
chromosome duplication, **18:**182–183
m- and *p*-Cresol, enzymic oxidation, **16:**110
p-Cresol, peroxidase action on, **19:**106; 143
p-Cresol, use in plant oxidase studies, **16:**132
Cresolase, **18:**37, *See* Monohydric Phenols; Tyrosinase
isolation, **14:**397
Cresolase classification, **19:**134
mechanism of action in phenolase complex, **19:**131–134
o-Cresolindophenol as hydrogen acceptor, **7:**543
p-Cresol in tyrosinase action, **14:**145; 146; 149
Crick's code without commas, **24:**10–11; 21
Cr (III) nucleotides, **54:**404–405; 406
in nmr studies of creatine kinase, **54:**413
Crithidia, **46:**62; 102
Crithidia fasciculata, growth factor for, **21:**377
Crithidia oncopelti, **46:**102
Critical size, **34:**502
CRM 197, **58:**113
CRM-producing mutants (muteins), of β-lactamase, **28:**272; 305
Crocin, **58:**66; 210
Crop, **45:**391; 400; 401; 407
Cross and Beavan reaction, **34:**208
Crossed immunoelectrophoresis, **46:**9
Crossing-over of genes, **13:**127–129; 129–134; 134–138; 145–147
Cross-linked polymeric, matrices, **34:**452
networks, **34:**447
Crosslinked polymerization in sclerotization process, **16:**161–162
Crosslinking
between DNA fibers, **18:**168
in melanin formation, **16:**166
of fibrin strands in gel formation, **16:**38
Cross-linking, **34:**451–454
and actin-troponin-tropomyosin complex, **57:**207–208
bifunctional, **57:**233–234
and bifunctional agents, **57:**187–188; 193–195
and cleaving agents, **57:**189
and conformation of enzymes, **57:**206–207
drawbacks of, **57:**194–195
glutaraldehyde, **57:**207
and heterobifunctional agents, **57:**187; 194
intermolecular, **34:**453
method of, **57:**187–189
and penicillinase, **57:**208
proteins, **57:**211–212
and quaternary structure of proteins, **57:** 189–193

Cross-linking (*Continued*)
and tropomyosin, **57:**207
and troponin, **57:**207
Cross-linking agent, bisdiazobenzidine 2,2′-disulfonic acid, **34:**451
dimethyl adipimidate, **34:**451
diphenyl-4,4′-diisothiocyanate 2,2′-disulfonic acid, **34:**451
glutaraldehyde, **34:**451–454
Cross-links, **67:**278–280
rubisco, **67:**32
Crosslinks, epsilon(γ-glutamyl) lysine, formation by transglutaminases, **38:**187
Crossover point analysis in ordered mechanisms, **45:**283
Crossover points, in respiratory chain, **17:**93–97; 126; 128–129, **23:**350–352
Crossover technique, **34:**102
Crossover theorem
allosteric control, **28:**84–85
CTP, allosteric regulator, **28:**54; 67–69; 128–130
reduction of, **28:**438–440; 445–447
for electron transport system, **17:**101–106
Cross-reacting material, **22:**11
in tryptophan synthetase mutants, **22:**31
Crossreacting protein, separated from prolyl hydroxylase, **41:**285
Cross-reaction protein, **38:**306
Cross sections of enzymes, by ionizing radiations, **24:**365–367
Cross striation, of action fibers, **13:**156
of myofibrils, **13:**184–187
Cross striations in fibrin clot, **16:**35
Crotalidae, nerve growth factor in venom of, **31:**57
Crotonaldehyde, **15:**258
Crotonase, **15:**234
in *R. rubrum,* **23:**423
in soluble fraction of liver, **21:**49
Crotonase (crystalline), in photosynthesis, **19:**44; 45
Crotonate, **15:**238; 261
Crotonyl coenzyme A, **15:**233
Crotonyl mustard oil, occurrence and properties, **10:**264
Crotonyl-s. CoA, and phosphorylation, **23:**432
Crotylglycine, **50:**165
Crustacea, thiaminase in, **15:**389; 391; 393*t;* 394
Cryoenzymology, **52:**178–187
Cryoenzymology in Aqueous Media (Douzou), **51:**21
Cryosolvents, **52:**194–202
Cryptates, **43:**281
Crypticity, definition, **22:**109
Cryptococcus neoformans, **70:**25
Cryptohaem, **23:**284
Cryptohaemin *a,* **23:**271
Cryptohaemins, **23:**284–287
Cryptohaemins P, **23:**285; 286
Cryptohaemochrome *a,* **23:**287
Cryptomonads, **46:**67
Cryptoporphyrin *a,* **23:**271; 272; 280; 282; 284–287
Cryptostegia grandiflora, rubber-producing vine, **29:**222
Crystal habit, of polymers, **27:**402–403
Crystal lattice, **34:**501
α-Crystallin
autokinase activity, **69:**165
gene expression, **69:**165–188
gene structure and evolution, **69:**158
molecular chaperone activity, **69:**163–165
non-lens expression, **69:**160–162
small heat shock proteins, relationship to, **69:**157–160
δ-Crystallin, **74:**314; 315
homology with argininosuccinate lyase, **74:**334
α-A-Crystallin
gene expression, **69:**165–182
chicken gene, **69:**175–182
5′ flanking sequences, **69:**169; 175–177
functional redundancy, **69:**173–174
mouse gene, **69:**169–175
transgenic mice, **69:**165; 173–174
non-lens expression, **69:**161–162
transcriptional regulatory elements
αA-CRYBP1, **69:**169–172; 173–174; 177–179
αCE1, **69:**181–182
αCE2, **69:**181–182
αCE3, **69:**181–182
DE1, **69:**172–174
DE1A/DE1B, **69:**179–180
DE2A/DE2B, **69:**180–181
PE1, **69:**175
PE2, **69:**175
α-B-Crystallin
disease expression, **69:**163

experimentally induced expression, **69:** 162–163; 188
gene expression, **69:**182
alternative transcription initiation, **69:** 187–188
enhancer region, **69:**185–186
5′ flanking sequence, **69:**182–183
stress induction, **69:**188
transgenic mice, **69:**183
non-lens expression, **69:**160–161; 183; 185–188
transcriptional regulatory elements
α-BE1, **69:**185–186
αBE2, **69:**185–186
αBE3, **69:**185–186
LSR, lens-specificity, **69:**186–187
MRF, **69:**185–186
Crystalline enzymes, reactivity, **52:**185–186
Crystalline form, **36:**97
Crystalline nature of tobacco mosaic virus, **15:**456
Crystallins
diversity of types, **69:**156–157
lens S-crystallins, glutathione S-transferases, **69:**3
Crystallization, of enzymes, **15:**114–115
of proteins, **15:**427–428
Crystallization and purity of proteins, **14:**302; 401–404
Crystallographic analysis of RNase, **24:**181–182
Crystallography, **34:**501, **43:**514, **65:**218
eukaryotic dihydrofolate reductase, **70:**42; 44; 53; 54; 61; 64
neutron diffraction analysis, **65:**221–222; 225; 237; 239
Crystals, glutathione S-transferases, **69:**7
Crystal structure
annexin V, **71:**216–220
of cytochrome *P*-450, **60:**65–66
glutamine PRPP amidotransferase, **72:**96
Crystal transfer, **52:**197–200
Crystal violet, inhibition of poly-(glycerol phosphate) synthetase activity by, **30:**244
Crythrocytes, **34:**457
Csaky test, **35:**170
CSAT antigen, **59:**40–41; 43–46
C splittings, xanthine oxidase, **51:**143–145
Ctalase, **37:**140; 168–172
C-Terminal amino acids, of arthropodin, **16:**157
of myosin, **16:**319
C-Terminal compounds, *See* Terminal
C-terminal domain (RNA polymerase II transcription), **71:**44–77. *See also* RNA polymerase II transcription
DNA binding, **71:**50–51
extended structure, **71:**50
generally, **71:**44–46
posttranslational modifications, **71:**49–50
promotor specificity, **71:**46–49
unique and essential qualities, **71:**46
C-terminal domains, adenosine receptors, **69:**104
C-terminal Phe-Lys-Phe, **57:**395
C-terminal residues, *See* Terminal
^{3}H-CTP, in DR metabolism, **28:**446
CTP reductase, in DR metabolism, **28:**446–447
CTP, *See* Cytidine triphosphate
CTP synthetase, **66:**243–247
E. coli, properties, **66:**243–244
gene cloning, **66:**245–246
mammalian, **66:**245
synthetase mechanism, **66:**246–247
C_6-tricarboxylic acid cycle, **8:**392–401
and ketone body oxidation, **8:**401–409; 409–412
C_6-tricarboxylic acids and fatty acid catabolism, **8:**392–401
Cu_A ligands
cytochrome c oxidase, structural biochemistry, subunit II (*See also* cytochrome c oxidase)
cytochrome *c* oxidase, structural biochemistry, subunit II, **71:**91–93
Cu_A models, mononuclear *versus* binuclear
cytochrome c oxidase, structural biochemistry (*See also* Cytochrome c oxidase)
cytochrome *c* oxidase, structural biochemistry, **71:**93–99
Cucumber mosaic virus, mutations, **9:**180–181
Cucumbers, antibiotic treatment of diseases, **20:**392; 393; 398
Cucurbitacins, inhibition of tumor growth by, **29:**336
Cucurbita pepo, **46:**49; 70
Cumarines, **17:**154
Cumeme hydroperoxide, **60:**74
Cumulative feedback inhibition, regulation of bacterial form glutamine synthetase, **72:**32–35

Cumulative feedback inhibition of enzymes, **28:**46; 62; 66–69
Cumulus dispersing factor, **9:**349–352
Cumyl alcohol, **60:**76
C_2 units, in biosynthesis, **12:**49–50; 93–101; 140–143
Cunninghamella blakesleana, steroid hydroxylation by, **19:**191
Cunninghamella blakesleeana, enzymes in, **20:**270
enzymic transformations of steroids by, **20:**246; 247; 251; 252; 270; 274
Cupric enzyme, conversion to cuprous, **16:** 145–146
Cupric ion
blue, **33:**195; 200–211
effect of pH on, **33:**200
EPR properties of (table), **33:**204
extinction coefficient of, **33:**200
optical properties of, **33:**195; 198*t;* 200
oxidation-reduction potentials of (table), **33:**207
oxidation-reduction properties of, **33:** 206–209
structural models for, **33:**209–211
effect on RNase, **24:**451
nonblue, in blue copper containing oxidases, **33:**214–222
component of blue oxidases, **33:**218
EPR properties of (table), **33:**213
occurrence of, **33:**211–212
optical properties of, **33:**212
reactions with anions, **33:**218
Cupric ions
chelate configuration, **21:**158
chelate formation with amino acids and peptides, **21:**145–147
effect on mitochondrial oxidative phosphorylation, **21:**31
model catalyses with, **21:**160
properties, **21:**157
effect on hyaluronic acid, **13:**202
effect on hyaluronidase, **13:**214
effect on myosin ATPase, **13:**167
effect on transferases, **13:**291; 297; 302–303
Cuproflavoproteins, **17:**274
Cuprous ion, **44:**194
Curare compounds, effect on choline acetylase, **12:**331
effect on esterases, **12:**308
Curing of cacao, mechanism of, **25:**457–492
Curing of tobacco, flue, **10:**327; 353; 399–409; 427
processes used, **10:**326–327
shed, **10:**326–327; 353; 372; 426–427
Curve-fitting, in differentiation, **49:**55
Curvularia, falcata, PN-dependent enzymes, **26:**168
Curvularia falcata, enzymic transformations of steroids by, **20:**271
Curvularia lunata, enzymes in, **20:**265
enzymic transformations of steroids by, **20:**243; 246; 247; 250; 251; 254; 255; 264
plant disease caused by, **20:**401
Curvularia pallescens, enzymic transformations of steroids by, **20:**243; 244
Curvularia species, enzymes in, **20:**265
enzymic transformations of steroids by, **20:**243; 247
Cuticle
arthropod, sclerotization, **16:**153–163
insect, **45:**416
detachment from epidermal cells (apolysis), **45:**417
formation of, **45:**416
shedding of, **45:**411
Cuticular enzyme, oxidation of tanning reaction substrate, **16:**160
Cuticulin, **16:**153; 156
Cyanamide, in nucleotide coenzymes synthesis, **22:**182–183
Cyanase, soil, **28:**387
Cyanate, and deoxyhemoglobin S solubility, **40:**2
antisickling property, **40:**8
assay in whole blood, **40:**22
carbamylation, inhibition by carbon dioxide, **40:**14
in proteins, **40:**5
of serum proteins, *in vivo,* **40:**12
of whole blood, *in vivo,* **40:**12; 13
carbamylation of amino groups, ε-amino of lysine, **40:**10; 11; 12
irreversible, **40:**8
mechanism of, **40:**5
carbamylation of hemoglobin, oxy- and deoxy-, **40:**14
effect on Bohr protons, **40:**17

effect on carbon dioxide capacity of blood, **40:**16
effect on 2,3-diphosphoglycerate binding, **40:**16
site, **40:**9
in vivo, **40:**12; 13
as compared with urea, effects on deoxyhemoglobin S gelling, **40:**3; 4
in prevention of sickling, **40:**3
effect on metabolites and enzymes within the red cell, **40:**15
effect on red cell metabolism and function, **40:**17
effect on the sickling of erythrocytes, **40:**2
inhibition of deoxyhemoglobin S gel formation, **40:**3
LD_{50} in mice and rats, **40:**22
mechanism of antisickling effect, **40:**18; 19; 20; 21
oral therapy, **40:**12; 13
oxygen affinity in red cells, **40:**17
physiological studies with, **40:**21; 22; 23; 24; 25
reaction with hemoglobin, specificity, **40:**2; 4
reactive tautomer, **40:**4
stability of the carbamyl bond, **40:**13
in vivo studies, **40:**22; 23; 24
Cyandie, effect on oxidation of vitamin K_3, **20:**184
effect on respiratory chain enzymes, **20:**155; 157–159
reaction with cytochromes, **20:**150; 151; 164; 165
Cyanidase, enzymic adaptation, **13:**74
Cyanide, **7:**206; 207, **39:**328; 343; 354, **58:**2; 67. *See also* Hydrogen cyanide, Potassium cyanide
citrate accumulation in mitochondria inhibited by, **29:**312
uptake of amino acids in tumors inhibited by, **29:**365
clearance, **39:**359
complex with carbonic anhydrase, **47:** 225–226; 229–230; 232; 238–240
cytochrome *c* oxidase, polypeptide composition, **71:**85
detoxication, **39:**335; 359
effect of pH on inhibition by, **18:**34–35
effect on amino acid decarboxylases, **6:**17; 18
effect on ATPase, **6:**165–166; 167
effect on enzyme systems, **15:**3; 17; 20; 23; 26; 144; 287; 324; 327
effect on mitochondrial coenzyme Q, **21:**112
effect on mixed function oxidases, **19:**182; 198; 199
effect on mold metabolism, **9:**568–570
effect on oxygen transferases, **19:**97; 118; 122; 124; 125–128
effect on respiratory chain, **28:**174; 205–206; 208; 221; 225–228
effect on uricase, **19:**205
in hydrogen transfer model, **26:**138–140
inhibition of superoxide dismutase, **41:**45; 67
as inhibitor of metalloflavoproteins, **17:**240; 242; 246; 247; 250; 252; 253; 265; 323
effect of incubation, with aldehyde oxidase, **17:**247
with xanthine oxidase, **17:**240; 309–310; 313; 317–318; 323
in mutation studies, **17:**453; 455
as inhibitory analogue, **6:**142
and papain activity, **19:**269
and phosphatide formation, **7:**149–150
photooxidation of Mn^{2+} inhibited by, **30:**25
poisoning, **39:**359
sensitive electrode, **39:**334
and yeast, **7:**189
Cyanide adduct of pyridine salts, association constants for, **26:**155
Cyanide cleavage, of cobalamin coenzymes, **26:**252–253
Cyanide formation, **46:**398
Cyanide haem *a* compound, **23:**289–297; 306–310
^{14}C-Cyanide, incorporation of ^{14}C into succinate dehydrogenase, **28:**225; 227
Cyanide in cytochrome c_1 reactions, **54:**293; 311–312; 323–324; 325
Cyanide-inhibited mitochondria, formation of ATP^{32}, **17:**107
Cyanide inhibition of *o*-aminophenylglucuronide formation, **16:**403
Cyanide-insensitive respiration, **17:**439–440
Cyanide ions, effect on disulfides, **27:** 211–212
Cyanide-labile sulfur atom, **51:**154
Cyanide particle, preparation of, **28:**173–174
in reconstitution of succinate oxidase, **28:**180–181
with succinate dehydrogenase, **28:**180–181

Cyanide reaction of pyridine nucleotides, **15:**84; 85; 90; 94–95; 97
Cyanide sensitivity of cytochromes, **17:** 334–335; 339–341; 345; 361–363; 371
Cyanide spectrum, of cytochrome oxidase, **23:**301–302
Cyanidin, **17:**433–434, **55:**34
metabolism in plants, **16:**142
Cyanidium caldarium, **46:**42; 67; 92; 93
Cyano-adenine-cobamide, cathode potential, **26:**255
β-Cyanoalanine, **50:**146; 165
β-Cyano-L-alanine, **35:**521, **39:**203; 204
β-Cyanoalanine synthase
amino acid substrate specificity of, **56:**32
characteristics of, **56:**28
cystine and, **56:**30; 31
function of, **56:**28–30
isolation of, **56:**27–28
purification of, from lupine seedlings, **56:**29
β-replacement reactions, **74:**197
replacing agents for, **56:**10
spectral absorption peak of, **56:**31–32
substrates for, **56:**10; 30
from white *vs.* blue lupine seedlings, **56:**33
6′-Cyano-ATP, epimer of, **43:**25
Cyanocobalamin, **46:**53; 54
cathode potential, **26:**255
from cyanide cleavage, **26:**252
structure, **26:**234–237
synthesis, **26:**242
in DR metabolism, **28:**447; 448
Cyano-cobinamide, absorption spectrum, **26:**247–249
cathode potential, **26:**255
cyanide cleavage, **26:**252–253
Cyano-corrinoids. *See also* Corrinoids
partial synthesis, **26:**238–241
properties, **26:**247–249; 252–253; 255–256
trans effect, **26:**256
Cyanoethylated, **35:**464
Cyanogen, in comets, **27:**350
Cyanogen bromide, **35:**476
Cyanogen bromide activation, **34:**449–450
Cyanogen Bromide cleavge, of orosomucoid, **41:**124
of rat-skin collagen, **41:**265
Cyanogen bromide method for thiothiamine, **15:**405
Cyanogenic glycosides, **39:**339; 359
Cyanolysis, **39:**340; 351
2-Cyanophenanthroline, **58:**7
Cyanophoric plant, **39:**359
Cyanophyceae, carotenoids in, **21:**298; 301; 303
2-Cyanopyridine, **58:**7
Cyanuric chloride activation, **34:**449–450
Cycasin, as β-1,3 glucan, **28:**329
Cyclic adenosine monophosphate, **59:**213–214; 373–374. *See also* Cyclic adenosine monophosphate-dependent protein kinase, Cyclic adenosine monophosphate-dependent regulation of 2′, 5′-oligoadenylase system
Cyclic adenosine monophosphate-dependent protein kinase
conformation of bound
$Co(NH_3)_4ATP$, **59:**268–269
heptapeptide substrate, **59:**269; 272
coordination scheme of, **59:**267–268
mechanism of phosphoryl transfer, **59:**272
and nucleophilic substitutions on phosphorus, **59:**267–269; 272–274
phosphorylation by, **59:**337–338
reaction coordinate distance, **59:**272–273
regulatory dimer, **59:**274
relation to *src* tyrosine kinase, **59:**274
sequence homology with adenylate kinase, **59:**273
Cyclic adenosine monophosphate-dependent regulation of 2′,5′-oligoadenylate system
behavior of, coordinated, **59:**230–235
generalization about, **59:**235–238
induction of, **59:**221–222; 224
inhibition of 2′-phosphodiesterase
experiments with cell cultures, **59:** 224–225
phosphorylation involvement, **59:**225–230
Cyclic adenosine 3′,5′-phosphate (cAMP), tyrosine hydroxylase regulation, **70:**129; 138–139; 145–146; 150; 156; 157; 159; 164; 168; 170–171
Cyclic AMP, **34:**328; 384; 398
activation of kinase by, **34:**329
analogs, effect on glycogen metabolism, in rat liver, **34:**344
a to *b* interconversion of mammalian liver glycogen synthesis by, **34:**342
cholesterol ester hydrolases, **69:**46–47
dependence of phosphorylase *b* kinase-kinase on, **34:**381

effect, on I → D interconversion of glycogen synthetase, **34:**331
on synthetase *b* kinase, **34:**336
increase by glucagon, **34:**342
phosphoenolpyruvate carboxykinase (GTP), **69:**219; 235–240
regulation of folate transport, **47:**337
Cyclic amp, **36:**69
Cyclic AMP-dependent protein kinase, cholesterol ester hydrolases, **69:**53
Cyclic AMP-independent effects, adenosine receptors, **69:**104–105
Cyclic dextrins, coupling with linear, **17:** 210–211
Cyclic disulfides, **39:**342
and lipoic acid, **18:**340–341
Cyclic metaphosphate, structure, **22:**181
Cyclic nucleotides, **17:**171–173
Cyclic peptides
effect on β-lactamase, **28:**263
fibrinogen binding inhibition, **70:**6–8; 14–15
Cyclic phosphatase activity of RNase, **24:**173; 214
Cyclic phosphate, formation in carbodiimide method of nucleotide synthesis, **22:**165; 176; 177; 179; 180; 188–190
Cyclic phosphate derivatives of RNA, **24:**164; 165; 171; 172; 173
Cyclic phosphate formation, **18:**198; 200
Cyclic phosphates, and RNase activity, **24:** 187–188; 235–243; 443–448; 452–455
Cyclic units, of myosin, **16:**327
3′,5′-Cyclid AMP-stimulated protein kinase, **53:**150–153
Cyclindrocarpon radicicola, enzymic transformations of steroids by, **20:**262; 263
Cyclization
of γ- and δ-aminoaldehydes, **16:**258–259
prevention of, in lysine precursors, **16:** 298–299
of squalene, **22:**320–324
hypothetical scheme, **22:**286–288
Cycloalkylalkanes, microbial oxidation, **27:**495–498
Cycloamidases, classification, **14:**224; 247; 250
8,5′-Cyclo-AMP, epimer of, **43:**33
Cycloamylose. *See also* Schardinger dextrins
in polysaccharide synthesis, **11:**298–299; 322–324
Cycloamylose-forming amylase, classification, **14:**246
8,5-Cyclo ATP, **43:**34
Cyclocreatine phosphate, **50:**226–227
delay of rigor, **50:**230
kinetics *in vivo,* **50:**234
in muscle, heart, and brain, **50:**229
thermodynamics *in vivo,* **50:**234
in tumor cells, **50:**231
Cyclodeaminase, reaction, **27:**425
Cyclodextrin, **58:**12
acylation substrates for, **58:**22
amisole chlorination by, **58:**14
derivatives, **58:**13
Diels-Alder reaction catalyzing, **58:**16
as enzyme mimics, **58:**13
ethyl capped, **58:**20
ferrocene substrate, **58:**24
glucoses in, **58:**30
hydroxyl group
acylation, **58:**17
geometry, **58:**23
imidazole attachment to, **58:**30
metal binding group, **58:**13
methylation, **58:**19
methyl capped, **58:**20
modification of, **58:**19
nitrophenyl acetate substrate, **58:**18
pyridoxamine attachment, **58:**35
thiamine attachment, **58:**46
tosylate derivative, **58:**37
Cyclodextrinase, action of, **12:**411; 419–420
Cyclodextrin heptatosylate, **58:**19
Cyclodextrin hydrates, hydrogen bonding, **65:**237; 239–240
Cyclodextrin phosphates, **58:**28
Cycloheptaamylose, **58:**28
2,3,5,6-Cyclohexadiene, in aromatic ring synthesis, **17:**414
3,5-Cyclohexadiene-1,2-diol dehydrogenase, **27:**520
Cyclohexane, **58:**25
1,4-Cyclohexanedione, use in hypertension, **7:**544
Cyclohexanol, **60:**72
Cyclohexanone, **60:**72
Cycloheximide, **46:**15, **47:**289, **55:**172, **56:**150, *See* Actidione
in allantoin transport, **56:**126
effect on carotenogenesis, **21:**319

Cycloheximide (*Continued*)
effect on Hill reaction of, **30:**25–26
in galactosidase inhibition, **56:**179
inhibition of mannan biosynthesis, **40:**264
mannosidases and, **56:**211
Cyclohexylalanine, in transaminations, **16:**196; 201
Cyclohexylglycine, in transaminations, **16:**196
Cyclohydrolase, **21:**222; 223; 225; 228; 402; 413
reaction, **27:**425
Cyclol-cage structure, **1:**29–33
Cycloleucine, **49:**60, **56:**264
Cyclooxygenase, **58:**215
antarafacial character, **58:**285
cholesterol ester hydrolases regulation, **69:**73
as lipoxygenase, **58:**289
prostaglandin, **55:**199
Cyclooxygenase metabolites, cholesterol ester hydrolases regulation, **69:**71
Cycloparaffins, microbial oxidation of, **27:** 503–504
Cyclopentadiene, **58:**16
Cyclopentane, **58:**25
1,2,3,4-*cis, cis, cis, cis*-Cyclopentane-tetracarboxylate, **51:**236
Cyclopentenylmethylpenicillin, action of β-lactamase, **28:**258
Cyclopeptides, in myosin, **13:**158
Cyclophorase, *See* Krebs cycle enzymes
Cyclophorase complex, **21:**12; 124; 125
Cyclophorase preparation, proline oxidase in, **16:**259
Cyclophorases, **15:**36
Cyclophorase system, **13:**12; 14
classification, **14:**233
Cyclophosphamide, fluorinated, **55:**278–279
Cyclopropyl carbinyl radical, rearrangement, **66:**14
Cycloserine, **35:**106; 117, **56:**61–63
D-Cyclo-serine (Oxamycin), effect on murein synthesis, **26:**218
Cyclotron, **57:**252; 254–255
Cylic AMP, **43:**43; 331
Cypridina, **7:**216; 229; 232; 234; 255; 256
luciferase in, **19:**198
Cypridina hilgendorfii, light emission by, **25:**156–158
Cypridina luciferin, **25:**121
Cys-1 residue
asparagine synthetase, **72:**107–111
glutamine-dependent nitrogen transfer, **72:**169–175
glutamine-dependent structure, **72:**153–157
vs. papain, **72:**176–183
glucosamine 6-phosphate synthase, **72:**104
glutamine PRPP amidotransferase active sites, **72:**97–99
Ntn amidotransferases, **72:**117–119
Cys-135 residue, aspartate semialdehyde dehydrogenase synthesis, **72:**285–286
Cys-248 residue, carbamyl phosphate synthetase, **72:**129–131
Cys-86 residue, GMP synthetase, **72:**120–123
Cystamine
in disulfide exchanges, **27:**201; 203; 204; 207; 209; 212–214; 216; 218–219
inactivation by, glutathione synthesis, **73:**255–257
Cystaminemonosulfoxide, exchanges, **27:**203; 205; 213
Cystamine (^{35}S), reaction with proteins, **27:**212
Cystathionase, **22:**264, **35:**111, **39:**360
estimation of sulfur amino acids, **27:**232
isolation, **14:**405
linkage with pyridoxal phosphate, **22:**395
and vitamin B_6, **24:**273
β-Cystathionase
bacteria and fungi, **72:**215
cysteine *S*-conjugate β-lyase reactions, **72:**212–214
γ-Cystathionase
in cysteine biosynthesis, **56:**24
isotopic exchange of hydrogen atoms with substituents of, **56:**53
molecular parameters of, **56:**45
purification of, **56:**38
pyridoxal-P-binding sites of, **56:**40
quaternary structure of, **56:**42
replacing agents for, **56:**11
spectral absorption peak of, **56:**39
substrates for, primary, **56:**11
sulfhydryl reagent reactions of, **56:**60
tetrameric structure of, **56:**40
γ-Cystathionase, cysteine *S*-conjugate
bacteria and fungi, **72:**215
β-lyase reactions, **72:**212–214
Cystathionine, **7:**381–382, **22:**243, **56:**22; 24; 39
in biosynthesis of methionine, **16:**270; 272–274

labeled, conversion to methionine, **16:**272
and ethionine effects, **25:**371
and methyl synthesis, **25:**402
genetic studies, **17:**408–410
in methionine synthesis, **21:**424
pyridoxal phosphate-dependent conversion to methionine, **29:**147
Cystathionine hydrolase, classification, **14:**249
Cystathionine in plant metabolism, **10:**268–284
L-Cystathionine in the synthesis of methionine, **10:**274–275
Cystathionine β-lyase, **74:**152
Cystathionine (S^{35}), and ethionine metabolism, **25:**375
Cystathionine β-synthase, **74:**134; 141
coenzyme analogs and, **56:**55
cosubstrate specificities of, **56:**47
in elimination of amino acid substituents, **56:**77
Michaelis and affinity constants for, **56:**48
molecular parameters of, **56:**44
purification of, **56:**24
β-replacement reactions, **74:**197
replacing agents for, **56:**10
serine sulfhydrase homology with, **56:**25–26
substrates for, primary, **56:**10
Cystathionine γ-synthase, **74:**134; 149
Cystathionine-γ-synthetase, **35:**110; 111
Cysteamine, **39:**342
and D-amino acid oxidase, **57:**103
in animals, **57:**154
anti-insulin effects of, **57:**156
biosynthesis of, **57:**154
in disulfide exchanges, **27:**202; 212
formation of, **57:**154–155
and glycolysis, **57:**156
and glyoxylate, **57:**112
and insulin, **57:**158
metabolism of, **57:**153–156
effects of, **57:**156–158
in tissues, **57:**112
Cysteamine-glyoxylate adducts, **57:**103–104
Cysteate, as inhibitor of succinic dehydrogenase, **18:**94
Cysteate-inhibitable glutamate uptake, **49:**82
Cysteic acid, **7:**385–390; 401–402, **15:**84; 118; 125–126, **49:**50; 57; 59; 62, **56:**18
from cysteine, **7:**385; 388–390
in transamination reactions, **16:**189; 190; 192; 196; 201; 233
transamination reactions of, **19:**367
Cysteic acid decarboxylase, **5:**75–76; 239, **7:**389; 401–402
activity of various liver extracts, **5:**75
inhibition of action by cyanide, **5:**76
of action by *dl*-homocysteic acid, **5:**76
and vitamin B_1 deficiency, **5:**76
L-Cysteic decarboxylases, classification, **14:**256
Cysteinase, **11:**390
L-Cysteinase, occurrence and action, **10:**307
Cysteine, **15:**93; 219; 241; 257, **34:**453, **35:**447, **39:**342; 343; 361; 362, **48:**159, **49:**55, **50:**125; 146; 147; 164. *See also* Cystine; Sulfhydryl groups; Sulfhydryl groups and specific aminoacyl derivatives; Sulfhydryl groups and Thiol groups
acetylation, **19:**242
action on allithiamine, **15:**417–418
action on thermolabile factor, **15:**410
action on thiaminase II, **15:**410
action on thiamine disulfide, **15:**416–417
activation of ATPase, **6:**167
activation of peptidase, **12:**198; 206
activator of plant enzymes, **10:**278
and blood clotting, **10:**23
complexes with heavy metals, **10:**276
inhibition of bacteriostatic action of penicillin by, **10:**254
in plant metabolism, **10:**225; 226; 268–284
reaction with glucose at room temperature, **10:**276–277
at active site, of cytochrome *c*, **22:**64; 65
of succinic dehydrogenase, **22:**63
S-aminoacyl derivatives, intramolecular aminolysis of, **19:**241; 242
synthesis, **19:**240; 256
and biological methylation, **12:**467–469
biosynthesis of, by transsulfuration, **56:**24
and carbohydrate metabolism, **7:**392
chelates, in metal ion activation of enzymes, **21:**145; 146
β-cyanolalanine synthase and, **56:**28
cystathionase and, **56:**39
cysteine lyase and, **56:**16
from cystine, **7:**395
degradation by desulfurase, **8:**146–147
desulfhydrase, **39:**360
desulfuration, **7:**390–392
in disulfide exchanges, **27:**202; 204; 209; 211; 216–218; 222; 229–232

Cysteine (*Continued*)
and oxidation of hydrocarbons, **27:**485
effect of irradiation, **24:**377–379
effect on elastases, **23:**234–235
effect on β-glucuronidase, **16:**379
in CoA formation, **16:**17
in cystine biosynthesis, **16:**278–279
in methionine synthesis, **16:**270; 272–274
reaction with alloxan, **16:**121
reaction with quinones, **16:**121–122; 134
in transamination reactions, **16:**190; 192; 196; 201; 232–234
effect on homogentisate oxidase, **19:**126
effect on mercuripapain, **19:**278; 279
effect on phosphoglucomutase activity, **14:**212
in enzyme isolation, **14:**380; 381
formation of, **39:**361
genetic studies, **17:**407–412; 445; 458
iron-catalyzed oxidation, **17:**280
oxidation by cytochrome *c*, **17:**350–351
as hydrogen donor, **7:**386
inhibition of melanin formation by, **14:**93; 94
and kidney cathepsin, **19:**286
mercaptide formation, **11:**216
metabolism, **7:**383–393
metal-catalyzed oxidation, **15:**40
of metallothionein, **56:**402
oxidation of, by oxygen, **11:**203–206
by various agents, **11:**206–207; 210
oxidative deamination, **7:**393
oxidative metabolism, **15:**124–126
and papain, **19:**270; 272–274; 276–279; 286–291; 293; 304; 305; 310
in photosynthesis, **19:**8; 22; 45
photosynthesis of, **25:**47
reaction with alkylating agents, **11:**213–216
reaction with metal electrodes, **11:**211–212
reactivation of choline, **6:**261
reactivation of phosphoglyceraldehyde dehydrogenase by, **12:**206
in reactivation of thiols, **20:**66
ionization constants, **20:**58
metal binding by, **20:**56
use of blocking reagents with, **20:**68
relation to coenzyme A, **14:**65
relation to LBF, **14:**55
reversal of enzyme inhibition by, **15:**286
serine sulfhydrase and, **56:**19
sulfhydrase, **35:**107
sulfur, **39:**360
synthesis of, **74:**186–191
in enteric bacteria
pathways of, **74:**186; 188–189
regulation of, **74:**189–191
pathways of, **74:**186–189; 187; 188
of tobacco mosaic virus, **15:**462–463
of tobacco mosaic virus mutants, **15:**479
tyrosine hydroxylase, substrate specificity, **70:**110
and yeast proteinase, **19:**286
zinc ligands of, **56:**315
L-Cysteine, **57:**113–116; 122
metabolism, in mitochondria, **21:**26
transamination reactions of, **19:**347; 367
Cysteine-22, apparent pK_a, **66:**178–179
sulfhydryl group, **66:**160
Cysteine-activated amidase-esterase, **23:**188
Cysteine-activated (γ) enzyme, **23:**188; 195; 198; 199
Cysteine *S*-Conjugate β-lyase reactions
bacteria and fungi, **72:**214–215
β-elimination catalysis, **72:**212–214
in human liver, **72:**215–216
prodrug targets in kidney, **72:**226–230
rat kidney
cysteine *S*-conjugate β-lyase/cytosolic glutamine transaminase K, **72:**219–222
cysteine *S*-conjugate β-lyase/kynurenine pyruvate aminotransferase, **72:**222–223
high M_r cysteine *S*-conjugate β-lyase, **72:**223–224
rat liver, cysteine *S*-conjugate β-lyase/kynureninase, **72:**216–219
thiols and thiomethyl products, **72:**201–203
toxicity, **72:**207–212
unstable thiols, **72:**203–206
in vivo control, **72:**224–226
Cysteine *S*-conjugates, early research, **72:** 200–201
Cysteine derivatives, cleavage with alliinase, **11:**393–395
Cysteine desulfhydrase, **7:**379–380; 381; 382; 385; 391–392; 396; 398
classification, **14:**237; 257
effect of dialysis on, **16:**211; 213
in transamination study, **16:**233
and sperm maturation, **24:**57; 59
localization of, **24:**333; 341
in sperm, **24:**82

Cysteine desulfurase, **10:**307, **16:**278, **74:**134; 152
Cysteine groups, modification, on fructose 1,6-bisphosphatase, **53:**71–72
L-Cysteine-HCl, action on thiaminase, **15:**401
Cysteine-inhibited (δ) proteinase, **23:**188; 195; 198
Cysteine lyase
  apoenzyme-holoenzyme equilibria of, **56:**42
  borohydride reduction of, **56:**18
  catalytic mechanism of, **56:**14–15
  cosubstrate specificities of, **56:**47
  in elimination of amino acid substituents, **56:**77
  historical perspectives on, **56:**14
  inactivation of, **56:**18
  inhibitors of, competitive, **56:**18
  isotopic exchange of hydrogen atoms with substituents of, **56:**51
  Michaelis and affinity constants for, **56:**48
  molecular parameters of, **56:**44
  physical characteristics of, **56:**15
  reaction products of, **56:**19
  β-replacement reactions, **74:**197
  replacing agents for, **56:**10; 16; 18
  spectral absorption peak of, **56:**18
  substrates for
    primary, **56:**10; 16
    relative activities based on, **56:**17
L-Cysteine monohydrochloride in plant metabolism, **10:**281
Cysteine of creative kinase, **54:**400–403
Cysteine oxidase, **7:**379; 386; 388; 395
Cysteine oxidation, **44:**231
Cysteine oxidation system, reconstitution of, **28:**167–168
Cysteine proteins. *See also* Sulfhydryl enzymes
  oxidation by model system, **15:**23
Cysteines, cytochrome *c* oxidase, **71:**96
Cysteine (S^{35}), from ethionine, **25:**375
L-Cysteinesuccinic acid, formation, **10:**277
Cysteinesulfenic acid, **7:**385–388; 395. *See also* Sulfinic acids
Cysteine sulfinate
  and ATP, **23:**382
  in cystine metabolism, **16:**278
    in transamination reactions, **16:**189; 196; 233–234
L-Cysteinesulfinate, **35:**505
Cysteinesulfinate desulfinase, **49:**24
Cysteinesulfinic acid, **7:**385–389, **10:**282, **15:**83–85; 118; 123; 124–125. *See also* Sulfinic acids
Cysteinesulfinic dehydrogenase, **15:**83–85
Cysteinesulfinic dehydrogenase I, **15:**118
Cysteinesulfinic dehydrogenase II, **15:**118
Cysteinesulfinic desulfinase, classification, **14:**257
Cysteine sulfur, oxidation, **7:**383–385
  to cysteic acid, **7:**385; 388–390
  to cystine, **7:**383–384; 385–386
  mechanisms, **7:**385–390
  to sulfate, **7:**384; 386
  to sulfenic and sulfinic acids, **7:**386–388
  to taurine, **7:**385
  to thiosulfate, **7:**384
Cysteine synthetase, **56:**25
Cysteine thiol studies, cytochrome *c* oxidase, **71:**92
Cysteine trisulfide, **46:**89
Cysteinic acid, **10:**282
Cysteinylasparagine, ionization constants, **20:**58
Cysteinylcysteine, ionization constants, **20:**58
L-Cysteinylglycinase, **43:**523
Cysteinylglycinase, column chromatography, **22:**457; 458
Cysteinyl-glycine dipeptidase, lacalization of, **24:**324
Cysteinyl Proteinases and Their Selective Inactivation (Shaw), **63:**271
Cysteinyl-tRNA synthetase, **48:**7
L-Cysteinyl-tRNA synthetase, from *Phaseolus aureus,* **48:**9
5-Cysteinyluracil, **54:**74
Cysterine, **41:**42
Cysthathionin, **33:**517
Cystic fibrosis
  mutations, **65:**39–40
  Pseudomonas aeruginosa, **70:**222; 223; 235; 236
Cystic fibrosis, Pseudomonas aeruginosa. *See also* Exopolysaccharide alginate synthesis
Cystic fibrosis transmembrane conductance regulator, **65:**7–9
Cystinase, occurrence and action, **10:**307
Cystine, **5:**238–246, **15:**124–126, **17:**349, **34:**21, **42:**128; 137, **49:**63, **50:**125, **56:**30; 31. *See also* Cysteine; Disulfides; specific aminoacyl derivatives

Cystine (*Continued*)
aminoacyl derivatives, intramolecular aminolysis of, **19:**241–245; 250; 260
synthesis, **19:**240; 250; 251
biosynthesis, **16:**277–279
in CoA formation, **16:**17
relation to other sulfur-containing amino acids, **16:**270–273
serine as precursor of, **16:**277–279
in transamination reactions, **16:**190
in chloroplasts, **20:**209
in phosphorylated peptides of esterases, **20:**288
conversion to hydrogen sulfide, pyruvic acid, and ammonia, **5:**240; 246
conversion to keto acid, **7:**397
degradation by desulfurase, **8:**146
synthesis, **8:**134
desulfuration, **7:**396–397
in disulfide exchanges, **27:**202–204; 209; 210; 212; 216–217; 223; 227–232
effect of irradiation, **24:**377–378; 393; 400
and ethionine effects, **25:**371; 372
formation from methionine, **5:**238
γ-glutamyl transpeptidase and, **72:**246
metabolism, **7:**393–397
from methionine, **7:**380–383
oxidation of, **14:**132; 133
peroxidase action on, **14:**151–154
relation to coenzyme A, **14:**65
relation to LBF, **14:**55
in papain, **19:**284
in plant metabolism, **10:**225; 268–284
in plants (tables), **10:**282–284
reaction with thiols, **11:**222; 223; 225–226
synthetis, **50:**146
D-Cystine, **33368**
L-Cystine, oxidation in plants, **20:**320
Cystine auxotrophs, of *E. coli*, **16:**273
Cystine desulfhydrase activity, in tumors, **25:**377–378
Cystine dimethyl ester, in disulfide exchanges, **27:**219
Cystine disulfoxide, **7:**395–396
Cystine free radicals, **27:**206
Cystine groups, in enzyme inactivation, **24:**367; 370; 382–383
Cystine oxidase, **7:**395
Cystine peptide dimer, of vasopressin, **27:**220
Cystine peptides, **33:**494; 498; 504; 509
insulin, **27:**204; 229
rate of desulfuration, **8:**135
enzymic degradation, **8:**145–147; 148
Cystine reductase, location in subcellular fraction, **24:**304; 341
Cystine reductase, $NADH_2$-dependent, **27:** 217–218
Cystine (^{35}S), reaction with proteins, **27:**212
Cystine sulfur, oxidation, direct, **7:**395–396
mechanisms, **7:**394–396
to sulfate, **7:**393–394
to taurine, **7:**394
Cystine-thioglycolic acid reaction, pH dependence, **27:**201
Cystinuria, **7:**379; 380; 383; 393, **17:**407
study of, **16:**271–272; 278
L-Cystinyl-bis-valine, disulfied exchanges, **27:**204
Cystinyldidiglycine, ionization constants, **20:**58
Cystinyldiglycine, ionization constants, **20:**58
Cytases, classification, **14:**246
cytAspAT, cysteine *S*-conjugate β-lyase reactions, β-elimination reactions, **72:**218–219
Cytichrome, **7:**267
Cytidine, **35:**465, **38:**215
in DR metabolism, **28:**435
as phosphate acceptor, **17:**172
in genetic studies, **17:**429–430
in pyrimidine biosynthesis, **21:**285–287
^{15}N-Cytidine, incorporation into nucleic acids, **28:**420–421
Cytidine aminase, classification, **14:**246
Cytidine cyclic phosphate. *See also* Cyclic phosphates and specific cytidine cyclic phosphates
enzymic opening, rate, **24:**445–447; 452–455
Cytidine-2′,3′-cyclic phosphate
in assay of RNase, **24:**172–173; 236–239; 241; 242
and product inhibition, **25:**227–229
Cytidine deaminase, **9:**498
^{14}C-Cytidine deoxyriboside, formation of, **28:**434; 444–445
Cytidine derivatives, in allosteric regulation, **28:**135
Cytidine diphosphate, in mitochondrial lipid synthesis, **21:**52
phosphorylation, in mitochondria, **21:**53
Cytidine diphosphate choline

in phospholipid synthesis, **18:**207–208
structure, **22:**177
synthesis by carbodiimide method, **22:**177; 181; 188; 189
Cytidine-diphosphate choline synthesizing enzyme, **17:**185
Cytidine diphosphate ethanolamine, synthesis by carbodiimide method, **22:**188
Cytidine diphosphate glycerol, synthesis by carbodiimide method, **22:**189
Cytidine diphosphate glycerol structure of, **30:**224; 234
Cytidine diphosphate ribitol
structure of, **30:**224
synthesis by carbodiimide method, **22:**189; 190
Cytidine-labeled RNA, in enzyme synthesis, **24:**16
^{14}C-Cytidine, labeling of nucleic acids, **28:**420–421; 423
use by tumor cells, **28:**433
Cytidine 5′-monophosphate, phosphorylation of, **18:**197
Cytidine monophosphosialic acid, **52:**25
Cytidine nucleotides, reduction of, **28:**436–444
Cytidine-2′-phosphate
effect on RNase, **24:**237; 241
Cytidine-3′-phosphate
effect on RNase, **24:**237; 241
Cytidine-5′-phosphate
and nucleic acid synthesis, **25:**170–171
Cytidine phosphates, in DR metabolism, **28:**436–440; 442–449; 453
Cytidine-5′-pyrophosphate, synthesis by dimethylcyanamide method, **22:**187
Cytidine triphosphate, **68:**70
aspartate transcarbamylase
asymmetry of binding, **68:**94–95
inhibition, **68:**73; 75
and ATP, **18:**230; 231
in pyrophosphorylase reactions, **18:** 207–208
binding
to aspartate transcarbamylase, domain separation changes, **68:**105
sites, **68:**73–74
in T form, **68:**124–125
heterotropic effects
aspartate transcarbamylase, **68:**120–123; 130–135
separation from cytidine triphosphate-uridine triphosphate synergism, **68:**123
interaction with aspartate transcarbamylase, **68:**101–103
in mitochondria lipid synthesis, **21:**50–52
in purine biosynthesis, **21:**216
in pyrimidine biosynthesis, **21:**264; 265; 280
Cytidine triphosphate synthetase, **53:**223, *See* Uridine-5′-triphosphate amidotransferase
relationship to anthranilate synthetase, **38:**30
Cytidine-U-^{14}C, in DR synthesis, **28:**434–435
Cytidylate kinase, NMP kinase primary structure, **73:**107
Cytidylic acid, biosynthesis, **21:**264; 280; 281
allosteric regulation, **28:**66–69
from glucose-2-^{14}C, **28:**409
hydrolysis, by metal ions, **21:**184
in mitochondrial lipid synthesis, **21:**52
in pyrimidine biosynthesis, **21:**287
^{14}C-Cytidylic acid, DNA synthesis, **28:** 420–421
Cytidylic acid, RNA, P^{32} uptake, **16:**462
Cytidylic nucleotides, reduction to deoxy-, **28:**135
Cytochemical methods
for enzymes, **8:**30
in virus infection study, **17:**494–495
Cytochemistry
effect of nucleates on dyes, **17:**137
enzymic, **9:**585–611
use of absorption spectra, **9:**485–487
Cytochromase, **17:**438
Cytochrome *f*
absorbancy changes due to, **30:**35–36
absorbancy ratio of, **30:**13
chromatography of, on Sephadex, **30:**13
difference spectrum for oxidized minus reduced, **30:**37
as electron carrier between pigment systems, **30:**35–38
in *Euglena,* reduction of, **30:**33–34
kinetics of oxidation of, **30:**35–36
molar extinction coefficients of, **30:**13
oxidation at - 150C, **30:**37–38
P-700 accepting electron from, **30:**22
properties of, **30:**13
sedimentation coefficient of, **30:**13
in spinach chloroplasts, **30:**33
in chloroplast, **18:**52–53; 55–57

Cytochrome *f* (*Continued*)
in plants, **18:**9; 16; 22–24; 28
and hematin compounds of leaves, **12:**21–25; 33
in photosynthesis, **19:**48–50; 57
Cytochrome 559, in chloroplasts, **30:**13
reduction of, **30:**34–35
in spinach chloroplasts, **30:**33
Cytochrome *556,* in mitochondria, **21:**19
Cytochrome a_3. *See also* Cytochrome oxidase
Cytochrome *a,* **23:**297–306
absorption spectra, **20:**164; 165; 192, **21:**93; 95; 96
as component of respiratory chain, **20:** 154–156; 164–167; 175; 176; 185–187
identification, **20:**150
in mitochondria, **21:**74; 97
in mitochondrial fragments, **21:**94; 97; 106; 107
in mitochondrial oxidations, **21:**115
position in mitochondrial electron transfer chain, **21:**119; 120; 122
in respiratory chain preparations, **20:**153; 189; 191; 192
rates of oxidation and reduction, **20:** 157–159; 165
absorption spectrum, **23:**306–311
addition of glucose causing oxidation of, **29:**352
Ehrlich ascites tumor cells oxidized by, **29:**352
characteristics of, **23:**266–270
and coenzyme Q, **25:**317–319
haemin of, **23:**271; 283–284; 286
models for, **23:**289–312
and oxidative phosphorylation, **23:**350; 352
in plants, **18:**3; 6; 10–12; 16; 20–21; 24; 26–28; 42; 49; 52
relation to cytochrome oxidases, **18:** 10–11
in respiratory chain, **18:**11–13
in respiratory chain, **17:**67–69; 72–85; 89–97; 100–108; 120; 121; 124–129; 261; 334–337; 341–344; 375; 439–442, **28:**160; 162–163
structure and function, **23:**312–314
Cytochrome a_1, **23:**297–306
absorption spectrum, **23:**306–310
characteristics of, **23:**266–270; 306–311
models for, **23:**289–312
Cytochrome a_2, **17:**68, **18:**11; 15
Cytochrome a_3, **17:**67; 68; 71–82; 84; 85; 89; 92; 93; 100–105; 111; 112; 121; 124; 128; 261; 335–337; 342; 375; 440, **23:**297–306
absorption spectra, **20:**164; 165; 192
as component of respiratory chain, **20:** 154–156; 164–167; 181; 185–187
identification, **20:**150; 164
in respiratory chain preparations, **20:**153; 189; 191; 192
rate of oxidation, **20:**157; 165
absorption spectrum, **23:**306–311
characteristics of, **23:**266; 268–270
and cytochrome *a,* **18:**10–11
as cytochrome oxidases, **18:**38–46
in plants, **18:**17; 20; 36
in respiratory chain, **18:**11–13
effect of Emasol on, **23:**298
haemin of, **23:**271; 283–284; 286
models for, **23:**289–312
and pyruvic oxidase activity, **19:**202
in respiratory chain, **28:**160; 162–163; 174; 206; 228; 229
Cytochrome a_7, **23:**297
Cytochrome $(a + a_3)$
absorption spectra, **23:**298; 307–310
preparation of, **18:**10
redox potentials, **23:**312–314
in respiratory chain, **23:**342; 347; 350
Cytochrome a_2 (d), **23:**266–270; 287; 311–312
Cytochrome $a_4(d_1)$, **23:**267–270; 311–312
Cytochrome *b,* **17:**67; 69–84; 88–110; 115; 116; 120–128; 334–337; 352–357; 364–382, **23:**311; 313, **54:**269
absorption spectra, **20:**162; 191, **21:**93; 95; 96
as component of respiratory chain, **20:** 154–160; 171; 173–178
identification, **20:**150
isolation and composition, **20:**188; 190; 191
in mitochondria, **21:**15; 74
in mitochondrial fragment, **21:**94; 97; 98; 104; 105
in mitochondrial oxidations, **21:**33
position in mitochondrial electron transfer chain, **21:**119; 120; 122
and coenzyme Q, **25:**311–312; 315–317; 319
in *Euglena,* reduction of, **30:**33–34

redox potential of, **30:**13
role of, **30:**33–35
slow reduction of, **30:**34
genetic studies, **17:**438–442
in metalloflavoproteins, **17:**246; 254–256; 260; 261; 275; 281–282; 284–285; 323
and oxidative phosphorylation, **23:**349–352; 378; 383
in plants, **18:**3; 6; 10; 12; 15; 16; 20–22; 24–28; 42; 47; 49; 51; 52
in respiratory chain, **18:**11–13
in reconstitution of respiratory chain, **28:**158; 160; 162; 177; 181; 190; 192; 194–199; 201–204; 208–210; 213; 214; 217; 220; 222–224; 229
in respiratory chain, **23:**343; 347; 354
Cytochrome b_2, **17:**69; 70; 71; 77; 244, **23:** 314–315
Cytochrome b_3
in leaves, **12:**21–23
in plants, **18:**9; 15; 16; 23–25; 27–28; 48; 53
as terminal oxidase, **18:**37
Cytochrome b_5, **17:**69; 71; 76; 77; 79; 262; 336; 352; 354; 358–359; 361–362; 378, **18:**48, **23:**311; 343, **46:**117; 119; 133; 146; 154, **47:**62; 72; 124; 135, **60:**70
in microsomes, **21:**18; 19
in mitochondria, **21:**19
in nuclear preparations, **27:**254
reductase, in microsomes, **21:**19
synthesis, **46:**91
Cytochrome b_6
and photosynthesis, **18:**55–56
in plants, **18:**16; 24–26; 55–56
in photosynthesis, **19:**57
Cytochrome $^{b}558$
gp91^{phox}, **65:**75–78
patients with CGD, **65:**66
Cytochrome b', in plants, **18:**4
Cytochrome b_7, in plants, **18:**15; 16; 25–28; 39; 47
as terminal oxidase, **18:**37
Cytochrome *b-c*, complex, **54:**273; 274
Cytochrome *b*-c_1 particle, properties of, **28:** 194–195
in reconstitution of respiratory chain, **28:**190; 196–199; 204; 208–209; 223; 229
variable adsorption of cytochrome *c*, **28:**204
Cytochrome b_5 reductase, **46:**474–476
Cytochrome *c*, **17:**67–110; 115; 116; 120–128; 331–333; 348–353; 363–367; 369–384; 437–442, **33:**209; 312, **34:**469; 477–484, **35:**167, **42:**258, **46:**88; 102; 133; 135; 154, **47:**48; 59; 124; 135, **60:**7. *See also* Cytochromes; Cytochromes, Electron transfer and Respiratory chain; Hemoproteins; Iron porphins; Porphins, and Tetrapyrroles; Succinoxidase system
absorption spectra, **20:**162, **21:**93; 95; 99, **23:**64–66; 69–70; 308
action of proteinase on, **23:**62–64
autoxidizability of, **23:**63; 64
changes during oxidation and reduction, **23:**58–70
as component of respiratory chain, **20:** 154–156; 159–166; 173–178; 181–183; 185–187
conformation charges, **23:**58–70
and cytochrome *o*, **23:**267
and cytochrome oxidase, **23:**304
effect of, heating, **23:**66
purification procedures, **23:**58–62
identification, **20:**150
imidazole in partial hydrolyzates of, **20:**69; 98; 99
and lipids, **23:**310–311
lipoprotein complex, **21:**108
in mitochondria, **21:**15; 58
in mitochondrial fragments, **21:**94; 96; 97
in mitochondrial oxidations, **21:**19; 20; 30; 82; 112; 115
modified, **23:**66–69
molecular weight, **21:**95
native, **23:**66–69
oxidation, in mitochondria, **21:**12; 15; 17; 33–35
in mitochondrial fragments, **21:**38; 39; 116; 117; 118
in oxidation of reduced triphosphopyridine nucleotide, **20:**172
and oxidative phosphorylation, **23:** 348–354; 373
physicochemical characteristics of, **23:** 68–69
position in mitochondrial electron transfer chain, **21:**119; 120; 122
redox potentials, **23:**312; 313
reduction, in microsomes, **21:**19
in respiratory chain, **23:**341–343; 347; 368

Cytochrome *c* (*Continued*)
in respiratory chain preparations, **20:**153; 192
isolation and composition, **20:**188; 193
rates of oxidation and reduction, **20:**157; 158; 160
valine incorporation into, in mitochondria, **21:**44; 45
yeast, absorption spectra, **23:**64–66
chromatographic species of, **23:**58–66
effect of digestion on, **23:**64–66
absorption spectra of, **54:**277
absorption spectra of pigments responsible for reduction of, **30:**62
properties of, **30:**13
active site, amino acid sequence, **22:**64; 65; 68; 69
labeling of, **22:**55
addition of glucose causing oxidation of, **29:**352
Ehrlich ascites tumor cells oxidized by, **29:**352
amino acid sequences, **24:**416; 418; 419
effect of radiation on, **24:**395; 397
chromatographic evaluation of purity, **19:**280
in photosynthesis, **19:**48; 49; 51; 57
and coenzyme Q, **25:**305–319
interaction with peroxidase, rate constant, **25:**14
and multiple enzyme forms, **25:**362; 364
column chromatography, **22:**424; 429; 430; 478
contamination, **58:**68
Cotton effect, **27:**60; 92; 95
criteria of purity, **7:**278–279
in cyanide reaction, **54:**323–324; 325
in cytochrome *b*-c_1 particle, **28:**194–199; 204; 208–209; 223; 229
as electron acceptor, **28:**172; 177; 211–215
interaction with cytochrome oxidase, **28:**203–208
range in metazoa, **28:**311
reconstitution of, cysteine oxidation system, **28:**167–168
respiratory chain, **28:**157; 160–163; 174; 176; 190; 192–194; 196–197; 216–220; 222; 229
succinate-cytochrome *c* reductase, **28:**197–203; 220–222
with cytochrome c_1, **54:**317–318
in cytochrome oxidase test, **18:**28–30
in leaves, **18:**53–54
in plants, **18:**3; 6; 7–13; 15; 16; 18; 20–30; 36; 38–40; 42; 44; 49; 52–54; 57
porphyrin peptides from, **18:**79
in respiratory chain, **18:**11–13
and succinate oxidation, **18:**66–71; 84–87; 100; 104; 106–108
deficiency in nucleus, **16:**489–490
effect in hypertension, **7:**544
electron transfer between, and cytochrome c_1, **54:**296; 315
in energy transport studies, **14:**38–42
separation by chromatography, **14:**326; 331; 339; 351; 354–355; 361; 362; 367
structure studies, **14:**305
study of catalytic action, **14:**188
enzyme adaptation, **13:**71; 75; 89–90
localization in cytoplasm, **13:**3
in ferricyanide reaction, **54:**312
and heme-linked groups, **7:**278–286
and hypertensin destruction, **7:**533
mechanism of deduced from relaxivity data, **33:**63
methionine ligand of, **54:**310
oxidase, **34:**472; 477–480
oxidation-reduction potential, **7:**283–284
photoreduction of, **54:**298
rate constant for the reduction by O_2-, **41:**45
reactions between, and cytochrome c_1, **54:**315–317
reaction with redogenic substrate, **27:**121–125
in respiratory chain, position of, **54:**270–271
and spectrophotometric investigations, **7:** 280–282
Cytochrome c_1, **17:**71–73; 76; 79–82; 84; 92; 94; 97; 336; 355; 358–361; 364–367; 373–374, **18:**10; 12; 16
absorption spectra, **21:**96
lipoprotein complex, **21:**108; 110
in mitochondria, **21:**15; 19; 74; 94
in mitochondrial fragments, **21:**94; 96
molecular weight, **21:**95
position in mitochondrial electron transfer chain, **21:**119–122
and coenzyme Q, **25:**310; 315–317; 319
in respiratory chain, **23:**343; 347, **28:**160–161; 190–192; 201–203; 213; 217; 229
in respiratory chain preparations, **28:** 162–163

Cytochrome c_2, **47:**58; 62; 124
  in photosynthesis, **19:**49; 57
Cytochrome c_5, **17:**71
Cytochrome c_1, cardiac, **54:**268–366
  absorption spectra of, **54:**277–279
  amino acid composition of, **54:**300–301; 305
  amino acid sequence of, **54:**301–307
  as an apoprotein, **54:**303
  artificial donors and acceptors in, **54:**312–315
  in ascorbate reaction, **54:**283; 284; 311; 314
  autooxidation of, **54:**284; 285; 296
    from guanidine-HCI, **54:**289
    from SDS, **54:**291
    from urea, **54:**288
  axial ligand of, **54:**308–310
  chromatogram of, gel permeation, **54:**347
  circular dichroic spectra of, **54:**279–283
  complexes of, **54:**317–344
  in cyanide reaction, **54:**283–291; 311–312; 323
  with cyrochrome c (*See* Cytochrome c_1-cytochrome c complex)
  with cytochrome b, **54:**273
  with cytochrome c and cytochrome oxidase, **54:**336–342
  development of, **54:**268–271
  electron transfer of, **54:**312–314
  isolation of, **54:**271–275
  methionine and lysine ligands of, **54:**310
  molecular configuration of, **54:**310–312
  molecular weight of, **54:**276
  one-band, **54:**344–356
    amino acid composition of, **54:**301; 307
    preparation of, **54:**344–349
    properties of, **54:**349–350
    role of, **54:**350–356
  pH changes affecting, **54:**283–286
    conformational changes with, **54:**310
  in phospholipid vesicles, **54:**342–344
    oxidation-reduction potential of, **54:** 343–344
  photoreduction of, **54:**297–300
  properties of, **54:**275–300
  in quenching of *ans* fluorescence, **54:**294–297
  reactions between, and cytochrome c, **54:**315–317
  reduction of, by ferrocyanide, **54:**312
  in solution *vs.* liposome form, **54:**345
    oxidation-reduction of, **54:**343
  structure of
    primary, **54:**300–307
    secondary, **54:**300; 307–308
    stability of, **54:**310–312
    tertiary, **54:**300; 308–312
  two-band
    amino acid composition of, **54:**307
    amino acid sequence of, **54:**356–357
    circular dichroic spectra of, **54:**308
    role of colorless band in, **54:**350–356
    secondary structure of, **54:**308
  urea, guanidine, and SDS effect on, **54:** 286–293
  yeast, **54:**274–275; 276
Cytochrome c-ascorbate reaction, kinetics of, **27:**125
Cytochrome c_1-cytochrome c complex, **54:** 318–329
  absorption spectra for, **54:**319
  autooxidation of, **54:**322
  cyanide reaction to, **54:**324–325
  dissociation of, **54:**327
  induced cotton effect increase in, **54:**350–353
  molar ellipticity increases of, **54:**355
  photoreduction of, **54:**326
  in quenching of ANS, fluorescence, **54:** 319–320
  rate of formation of, **54:**326
  reducibility of, **54:**325–326
  structure of, **54:**328–329
Cytochrome *c*-cytochrome oxidase, dissociation rate, **27:**149
Cytochrome *c*-cytochrome oxidase complex, **54:**317–318
Cytochrome c_1-cytochrome oxidase complex, **54:**329–335
  cyanide reaction to, **54:**328
  with cytochrome c, **54:**336–342
  electron transfer in, **54:**334–335
  pH change effect on, **54:**340
Cytochrome *c*-cytochrome oxidase complex, absorption spectra, **28:**204–207
  properties, **28:**206–208
  in reconstitution of respiratory chain, **28:**197; 203–208; 220–222
Cytochrome components, distribution in plants, **18:**15–28
  isolation of, **18:**7–10
  in plants, **18:**3–15
Cytochrome components and semen metabolism, **9:**337; 346; 370–372

Cytochrome *c* oxidase, **23:**266; 297; 304, **44:**227, **58:**242, **71:**79–208. *See also* Cytochrome *a;* Cytochrome a_3; Cytochrome *c*
background, **71:**80–83
bacterial terminal oxidases, **71:**186–191
catalytic cycle, **71:**110–116
dioxygen chemistry, **71:**110–114
thermodynamics, **71:**114–116
complete turnover cycle model, **71:**180–186
definition, **20:**149
electron transfer kinetics, **71:**116–125
fully reduced enzyme, **71:**124–125
generally, **71:**116
steady-state kinetics, **71:**116–118
transient kinetics, **71:**118–124
enzyme turnover control and H + /e - ratio, **71:**162–167
future research, **71:**191–192
inhibition, **20:**175; 176; 181; 183
non-redox-active metal centers, **71:**101–103
preparation and composition, **20:**188; 191; 192
product inhibition of, **25:**224; 264–265
proton pumping, **71:**125–162
chemical models, **71:**135–138
Babcock model, **71:**143–146
Chan CU_A model, **71:**146–149
Chan CU_A model (alternate), **71:** 150–152
Chan CU_B model, **71:**152–154
generally, **71:**135
Mitchell copper loop model, **71:**158–160
Rousseau models, **71:**138–143
Wilkström model, **71:**154–158
Woodruff model, **71:**135–138
problem definition, **71:**125–127
restrictions on potential models, **71:** 127–134
reaction, **20:**165–167
redox linkage, extended model of, **71:** 167–180
structural biochemistry, **71:**83–110
polypeptide composition, **71:**83–85
subunit I, **71:**85–94
subunit II, **71:**91–101
subunit II (Cu_A ligands), **71:**91–93
subunit II (cu_A models, mononuclear *versus* binuclear), **71:**93–99
subunit II (cytochrome *c* binding site), **71:**99–101
subunit III, **71:**103–108
three-dimensional structure, **71:**108–110
Cytochrome *c* oxidation, model for, **15:**23
Cytochrome *c* peroxidase, **33:**309–335, **65:** 179–180
activity, **14:**151
apoenzyme moiety of, **33:**311
complex ES of, **33:**329; 331–332
compulsory order mechanism of, **33:**312
crystalline, heavy-atom derivatives of, **33:**311
space group of, **33:**311
unit-cell dimensions of, **33:**311
x-ray diffraction analysis of, **33:**311
diffusion content of, **33:**311
magnetic susceptibility of, **33:**319
molecular activity of, **33:**312
molecular characteristics of, **33:**311–312
molecular weight of, **33:**311
optical absorption spectra of, effect of pH on, **33:**317
low temperature spectra of, **33:**317
porphyrin side chains of, effect on enzymatic activity, **33:**315
preparation of, **33:**310–311
protocheme moiety of, **33:**311
sedimentation content of, **33:**311
x-ray diffraction analysis of, **33:**312
Cytochrome c Peroxidase (Yonetani), **33:**309
Cytochrome *c* reductase, **14:**227; 254; 398; 405, **17:**69; 260, **18:**118; 308, **34:**97, **58:**65
antimycin-insensitive, **34:**84; 92
composition of, **34:**96
deficiency in nucleus, **16:**489
PCMB inhibition of, **22:**381
turnover number of, **34:**96
Cytochrome *c* (*Rh-*) and photosynthesis, **18:** 56–57
Cytochrome-cytochrome oxidase system, in sperm, **24:**87
Cytochrome *d,* **46:**63
Cytochrome d_1 *c,* **46:**63
Cytochrome *dh,* **18:**16; 22; 51
Cytochrome *e,* **17:**71; 225; 257; 336; 355; 358–361; 364–367; 373–374; 440, **18:**12; 16. *See also* Cytochrome c_1
Cytochrome *h,* in snails, **46:**63
Cytochrome in Higher Plants (Hartree), **18:**1
Cytochrome-linking (SC) factor, **17:**255–257
Cytochrome *m, See* Cytochrome b_5
Cytochrome *o,* as terminal oxidase, **23:**267; 312

Cytochrome *m* or b_5, location of, **24:**303; 304; 337
Cytochrome oxidase, **2:**116–119, **7:**24; 206; 227; 232; 316; 324; 385, **17:**68; 78; 339–349; 364; 365; 368–370; 373; 375–382, **44:**226. *See also* Cytochrome a_3; Succinic; Succinoxidase system
absorption spectra, **28:**193
composition and activity, **28:**192
interaction with cytochrome *c,* **28:**203–208; 229
isolation, **28:**203
in reconstitution of succinate oxidase, **28:**197
in respiratory chain, **28:**160; 163; 168; 173–174; 181; 190; 196; 214–223
activation of, **54:**334
adsorption studies, **14:**346
classification, **14:**254; 255
in energy transport studies, **14:**32–33; 37–42
and aldehyde oxidase, **17:**246; 285; 323
and *aspergillus,* **2:**219
biosynthesis of, **44:**230
cholate inactivation of, **23:**310–311
copper in, **23:**304–306
effect of, CO on, **23:**298–301; 303–304
cyanide on, **23:**301–302
haemin *a* from, **23:**270–271
inhibition of, **23:**369
as multiheaded enzyme, **23:**314–315
NO spectrum, **23:**302
reduction of, **23:**286
relation to cytochromes *a,* **23:**297
in respiratory chain, **23:**342
and S-S bridges, **23:**312
in chromidia, **8:**31
in nuclei, **8:**33
in complex with cytochrome c_1, **54:**317–318
and cytochrome *a,* **18:**10–11
and cytochrome *f,* **18:**23–24; 57
detection and estimation of, **18:**28–35
and fluoroacetate, **18:**118
in plant respiration, **18:**13–15; 35–52
in plants, **18:**6; 7–10; 12; 16; 305; 307–309
preparation of, **18:**244
and succinic dehydrogenase, **18:**66; 104; 106
and cytoplasmic transmission of heredity, **22:**28
disulfide bridges, **27:**211
EPR study, **27:**117; 126; 128; 138; 139; 148–155
kinetics of oxidation-reduction, **27:**153
in mitochondria, EPR study, **27:**187–190
and nuclear purity, **27:**254
in DPNH oxidase complex, **17:**260–261
effect on cysteine, **10:**278
enzymic adaptation, **13:**71; 85
inhibition, **13:**321; 332
localization in cytoplasm, **13:**9; 11
genetic studies, **17:**437–442; 453–455
histochemical methods for, **9:**588
inhibition by thiol reagent, **11:**237
isolation and properties, **21:**106; 107
isolation of, **54:**268
location in cell, **24:**300–302; 306; 307
mechanism of action, **19:**210; 211
in photosynthesis, **19:**49
properties, **19:**203
in mitochondria, **21:**12; 15; 21; 58; 59; 74
effect of sonic vibrations on, **21:**57
in mitochondrial oxidations, **21:**92; 112; 114; 115
oxidation of coenzyme Q, **25:**305; 315–319
in plastids, **20:**215–217; 230; 231
of respiratory chain, **54:**268
secondary structure of, **54:**307
and tea oxidase, **2:**119
as terminal oxidase, **16:**131
inhibition, **16:**129; 135
intracellular distribution, **16:**415; 468; 489
in melanin granules, **16:**169; 172
Cytochrome oxidase, EPR-nondetectable copper ions in, **33:**223
EPR properties of, **33:**203
oxygen, product of water reduction of, **33:**230–231
Cytochrome oxidase-carbon monoxide, photochemical change, **26:**13
Cytochrome oxidase complex, effect of CO, **28:**204–206
Cytochrome oxidase (complex IV), in reconstitution of respiratory chain, **28:**214–220
Cytochrome oxidase-cytochrome system, in sperm, **24:**87
Cytochrome oxidase preparations, absorption spectra, **23:**290–292; 298–302; 307–308
comparison of, **23:**298–302

Cytochrome oxidase (reduced)-ferricytochrome c complex, dissociation rate, **27:**153
Cytochrome *P*-448, **46:**117; 139
Cytochrome *P*-450, **43:**331, **46:**36; 105; 106; 116; 117; 118; 119; 122; 128; 134; 135; 138; 142; 153; 156; 164; 166; 171; 174; 244; 409, **47:**15–31, **51:**50; 51, **58:**231, **60:**35–87
active site, **60:**59
distal heme ligand, **60:**38
membrane topology, **60:**61
thiolate ligand, **60:**37
aggregation number, **60:**44
amino acid composition, **60:**48
amino acid sequence, **60:**49
alignments, **60:**52
microheterogeneity, **60:**52
amino acyl residue, **60:**63
bacterial, **60:**39
substrate hydroxylation, **60:**46
carbon-carbon bond cleavage, **60:**46
chiral orientation of heme, **60:**63
with cinnamic acid 4-hydroxylase, **55:**16; 17
CO_2H terminal, **60:**48
complete primary structure, **60:**49
corresponding forms, **60:**59
crystal structure, **60:**65
hydrogen bonding, **60:**66
cumyl alcohol formation, **60:**76
cystein residue, **60:**61
deamination, **60:**46
dehalogenation, **60:**46
demethylation reaction, **60:**72
denatured form, P-420, **60:**40
desulfurization, **60:**46
distribution, **60:**38
drug metabolizing form, **60:**47
electrophoretic properties, **60:**44
energy transfer study, **60:**63
epoxidation, **60:**46
ferrous dioxygen complex, **60:**70
gene subfamilies, **60:**61
genetic regulation, **60:**39
halt transfer signal, **60:**48
hexacoordinate, **60:**38
homolytic mechanism, **60:**74
hydrogen peroxide formation, **60:**71
hydroxylation reactions, **60:**66
inducers, **60:**39
in infested plant cells, **55:**17
isozymes structure, **60:**48–66
membrane topology, **60:**63
mercaptide ligand, **60:**61
metabolic significance, **60:**46
metal cofactor, **65:**175–176
molecular weight, **60:**41
monooxygenase system, **60:**40
bacterial, **60:**45
mitochondrial, **60:**45; 48
mocrosomal, **60:**44
reconstitution, **60:**72
multiple forms, **60:**40; 52
NH_2-terminal, **60:**48
nomenclature, **60:**36
oligosaccharide acceptor sites, **60:**59
oxidation-reduction potential, **47:**25
oxygen reduction and substrate hydroxylation, **60:**75
oxygen utilizing reactions, **60:**66
NADPH oxidase reaction, **60:**67
pentacoordinate, **60:**38
pentacoordinated heme position, **60:**65
peroxide utilizing reactions, **60:**73
as peroxygenase, **60:**73
peroxygenation, **60:**46
phenylalanine, nonenzymatic hydroxylation, **67:**210
properties, **60:**36–48
purification, **60:**39
detergents, **60:**40
quasi lipoxygenase activity, **58:**234
rapid proteolysis site, **60:**59
reaction cycle, **60:**68
carbon hydroxylation, **60:**69
dioxygen binding, **60:**69
electron transfer, **60:**69
hydrogen abstraction, **60:**69
oxygen bond cleavage, **60:**69
oxygen insertion, **60:**69
rate determining step, **60:**70
stoichiometry, **60:**72
substrate binding, **60:**69
reaction mechanism, **60:**66–76
ethanol oxidation, **60:**67
steps in reaction cycle, **60:**68
reactions catalyzed by, **60:**46
reaction scheme, **60:**74
role of cytochrome b_5, **60:**70
signal peptide, **60:**48
spectral properties, **60:**36

high-spin, **60:**38
low-spin, **60:**38
spectral properties of, **47:**16
spin equilibrium, **47:**24; 26
steroid hydroxylation, **60:**48
stoichiometry, **60:**67
substrate hydroxylation, **60:**71
substrate specificity, **60:**46
superoxide release, **60:**75
synthesis, **46:**91
three dimensional structure, **60:**66
Cytochrome peroxidase, **17:**284
in plastids, **20:**215
reaction with hydrogen peroxide, **12:**159; 178; 181; 187
Cytochrome P_{450} heme-monooxygenase, **55:**249
Cytochrome P_{450} monooxygenase, **55:**278
fluorinated steroids and, **55:**263
in vitamin D_3 conversion of, **55:**276
Cytochrome P_{450} monooxygenases, **46:**118
Cytochrome poisons, mutagenic effects, **17:**453
Cytochrome P-450 oxidoreductase, **65:**328
Cytochrome reactions, **19:**57; 58
Cytochrome reductase, **7:**251, **15:**89; 97–98
effect of borate on, **12:**517
EPR study, **27:**131; 178–179; 183
localization in cytoplasm, **13:**10; 11
microsomal, **22:**372; 373
PN-dependent, **26:**125; 130–131
in tobacco, **6:**372
Cytochrome reductases I and II, in plants, **18:**307
Cytochromes, **7:**24; 266–269; 278–286; 308; 312; 316–317; 324; 385, **34:**36, **41:**201, **46:**39; 58; 86; 91, **65:**54–57; 61. *See also* Coenzyme Q and Electron transfer; Electron transfer and Respiratory chain; specific compounds
algal chloroplast, **30:**13
in the higher plant chloroplast, **30:**12
molecular weights of algal chloroplast, **30:**13
in photochemical systems, **30:**48–49
types of, **30:**12–13
autoxidizable, **18:**36–37; 48
b group, and cyanide-resistant respiration, **18:**46–48
cyanide-resident, **18:**48
in higher plants, **18:**1–64
and photosynthesis, **18:**52–59
properties of, **18:**16–17
spectroscopic differentiation, **18:**5–7
as terminal oxidases in plants, **18:**35–48
b_5, **37:**140
b_1, **17:**357–359
b_5, **20:**171; 174
biosynthesis of, **23:**315
differentiation of, **23:**266–270
of group A, **23:**265–321
of group D, **23:**267; 268; 311–313; 315
models for, **23:**289–312
and photophosphorylation, **23:**386
prosthetic groups, **23:**265–321
unitarian concept, **23:**297; 303–304
b_5 reductase, in direct hydrogen transfer, **26:**166
c, **15:**327; 334
c_4, **20:**163
c_1, absorption spectra, **20:**162; 173–175
as component of respiratory chain, **20:**171; 173–175; 177; 185–187
isolation, **20:**191
in respiratory chain preparations, **20:**153; 189; 193
in chromidia, **8:**31
and coenzyme Q, **25:**276; 277; 305–319
and copper in cytochrome oxidase, **27:** 148–155
and nuclear purity, **27:**253–254
separation of a and a_3, **27:**154
c reductase, **37:**140
definition, **20:**151
differentiation from denatured-protein hemochromogens, **20:**173; 174
in plastids, **20:**227
electron transfers between, **15:**14; 15; 16; 33; 38
enzymes using, **24:**298; 299; 303–304
mechanisms of action, **24:**443
and enzymic adaptation, **13:**75; 83; 84–85; 93; 106
estimation, **21:**94; 95
evolutionary changes, **34:**29
in green leaves, **12:**14; 16; 21–25; 33
in heart muscle preparations, **28:**159–165; 224
in metabolic scheme, **8:**414
methionine content, **10:**270
effect on tryptophanase activity, **10:**61

Cytochromes (*Continued*)
oxidation, in mitochondria, **21:**92; 94
and phosphate turnover, **17:**175
in DPNH oxidase complex, **17:**260–261
mammalian, some controversial aspects, **17:**329–392
spectra, **17:**71–77
and phospholipide synthesis, **8:**331
in photosynthesis, **25:**58–60; 66; 75–81; 87; 89–93
and PN-dependent enzymes, **26:**121; 125; 128; 130–131
reduction, in mitochondria, **21:**92; 94
study of catalytic action, **14:**173–176; 182
system, production of melanin by, **14:**137
Cytochromes, Mammalian, Controversial Aspects (Wainio and Cooperstein), **17:**329
Cytochromes D, **23:**267; 268; 311–313; 315
Cytochromes of Group A and Their Prosthetic Groups (Lemberg), **23:**265
c-Cytochromes, redox potential, **67:**311–312
Cytochrome-system preparation, species specificity, **28:**179–180
Cytochrome systems, **15:**120; 144
and ATP formation, **15:**130
and bioluminescence, **25:**150–151; 153–154
components of, **17:**335–362
genetic studies, **17:**437–442
cyanide particle as, **28:**180–181
reaction with, NADH dehydrogenase, **28:**208–209
in reconstitution of succinate oxidase, **28:**174–179; 186–190; 223; 225
in reversible dissociation of succinate oxidase, **28:**186–190; 225
succinic dehydrogenase, **28:**174–179; 223
genetic studies of, **13:**143
oxidation of reduced pyridine nucleotides by, **15:**128
in plants, **18:**304–308
and succinate oxidation, **18:**67; 69–70; 71; 100; 104–108
role of glutathione in, **11:**249
Cytodeuteroporphyrin, **23:**272; 276–280
Cytohaemin, **23:**271; 284
Cytokines, cholesterol ester hydrolases regulation, **69:**69
Cytokinin, **35:**469
Cytolysis
and birefringence of cytoplasm, **8:**35–36
osmotic, of luminous bacteria, **7:**253
and rate of respiration, **8:**44
role of lecithinases in, **8:**54
separation of lipides from proteins in, **8:**44
Cytolysis quotient, **17:**332
Cytophilic γ-Globulin, Binding to arterial endothelrim, **41:**172
Cytoplasm. *See also* Protoplasm
amino acids of, **16:**447
DNA in, **16:**412; 423; 425; 427
RNA in, **16:**423; 438; 439; 461–462; 473–475; 491–492
enzyme formation in, **24:**4–5; 11–14
source of RNA in, **24:**14–20
functioning, **8:**1–69
and chloroplasts, **8:**20–23
and defense mechanism, **8:**51–55
interaction with cell nucleus, **8:**56–65
irritability, **8:**3; 42–47; 56
isolation from nucleus, **3:**13–17
localization of enzymes in, **13:**1–20
nucleoprotein content, **3:**6; 23–29
permeability, **8:**20; 48–51; 53; 56
phosphatides in, **7:**168
specific functions, **8:**55–65
structure, **8:**4–26
and chromidia, **8:**4–14
ultraviolet absorption, **3:**25; 26
Cytoplasm, Functioning (Monn,), **8:**1
Cytoplasm, Localization of Enzymes in (Holter), **13:**1
Cytoplasmic fibrils. *See also* Fibrils
and birefringence phenomena, **8:**8–14
characteristics, **8:**11–14
and contraction of cell, **8:**37–42
effect of hydration on, **8:**37–42
orientation, **8:**10; 25
stretching, **8:**10
and structure of protoplasm, **8:**4–14
Cytoplasmic fractions, reference in isolation of nuclei, **27:**242–244
Cytoplasmic fragments, functioning, **8:**55–56
Cytoplasmic malate dehydrogenase, **25:**346–347
Cytoplasmic maturation, and activation of egg, **9:**290–292; 296–297
causes, **9:**244; 249–252
changes occurring, **9:**246–249; 319
Cytoplasmic membrane, relation to cell wall, **26:**227

Cytoplasmic particles. *See also* Granules, subcellular; Mitochondria
oxidation of D-lactate, **14:**197
transaminase in, **14:**105
tyrosinase attachment to, **14:**85; 133
Cytoplasmic particulates, separation, **9:**597–598
Cytoplasmic region, adenosine receptors, **69:**109
Cytoplasmic structure of egg and fertilization, **9:**250
Cytoplasmic transmission of heredity, **22:**3
Cytoplasm of egg, changes following fertilization, **9:**280–286
Cytosine, **17:**429, **35:**464; 476
and DNA structure, **18:**164
in DR synthesis, **28:**423
formula, **21:**263
in pyrimidine biosynthesis, **21:**285
of tobacco mosaic virus, **15:**460
in transamination, **16:**237
of DNA, **16:**431–433
of nuclear RNA, **16:**437; 439
Cytosine aminase, classification, **14:**246
Cytosine deaminase, **74:**72
Cytosine deoxyriboside, synthesis of, **28:**423
Cytosine-5-methyltransferase, tRNA, **54:**89; 90
Cytosine nucleotides, in mitochondria, **21:**10
Cytosine ribonucleotide reductase, substrates for, **28:**446–447
Cytosol, **39:**360, **58:**180
Cytosolic enzymes
evolution of, **69:**36–37
glutathione S-transferases, **69:**3; 4
Cytosolic glutamine transaminase K (cytGTK)
cysteine *S*-conjugate β-lyase reactions, **72:**219–222
M_r enzyme, **72:**223–224
Cytosolic glutathione S-transferases, **69:**3
Cytosolic pathways, asparagine synthetase, **72:**149–151

D

Dactylium dendroides, enzymic transformations of steroids by, **20:**245; 249
plant disease caused by, **20:**402
dADP
in allosteric regulation, **28:**137
NMR spectroscopy of, **55:**323; 324; 325; 326
synthesis of, **55:**346
Dahlias, antibiotic treatment of diseases, **20:**401
DAHP, **34:**521
DAHP synthetase
allosteric enzyme, **28:**71
inhibition by analogs of phenylalanine and tyrosine, **29:**117–118
Daidzein, **55:**29; 39
degradation of, **55:**42
Dalbergioidin, **55:**36
45,000-Dalton fibronectin receptor, **59:**46–47
Dalziel relationships, in mechanisms of product inhibition, **25:**211
Dam-Karrer test, for coenzyme Q, **25:**292; 293
dAMP
configuration analysis of, **55:**334; 335
in DR metabolism, **28:**423; 448
effect on aspartate transcarbamylase, **28:**132
^{17}O NMR spectroscopy of, **55:**337; 338; 339
synthesis of, **55:**313; 314
Danielli-Davson model, **34:**470–471; 483
DAP auxotrophs, of *E. coli,* **16:**297; 298
dapB gene sequence, dihydrodicolinate reductase, **72:**290–297
dapC gene, *N*-succinyl-L,L-diaminopimelate aminotransferase, **72:**302–303
DAP decarboxylase, **16:**297
dapD gene
N-succinyl-L, L-diaminopimelate aminotransferase, **72:**302–303
tetrahydrodipicolinate *N*-succinyltransferase, **72:**298–302
DAP (α,α′-diaminopimelic acid). *See also* α,α′-Diaminopimelic acid
dapE gene, *N*-succinyl-L,L-diaminopimelate descuccinylase, **72:**303–307
dapF gene
diaminopimelate epimerase, stereochemistry, **72:**281–283
L,L-diaminopimelate epimerase, **72:**307–310
Daphnetin, role in browning of plant tissues, **16:**150
Dark fixation of CO_2, **26:**27
Dark inhibition, of cytochrome oxidase, by CO, **18:**30–31
Dark metabolism of purple bacteria, **1:**301–307
Dark oxidase, in photosynthetic bacteria, **18:** 56–57

Dark reaction, and chemiluminescence, **25:**123; 141; 148
in photosynthesis, **25:**42–44; 48; 50–51; 56; 59; 64–66; 79; 83–89
Dark reversion of phytochrome isomers, **26:** 21–23
Darwin, **38:**202; 207
Data, kinetic, FORTRAN program for fitting, **29:**23–31
preliminary examination and choice of equation for fitting, **29:**19–22
statistical analysis of enzyme kinetic, **29:**1–32
dATP, **35:**483
in allosteric regulation, **28:**134; 135
Datura stramonium, **55:**73
Datura stramonium in alkaloid studies, **8:**219; 224–226; 238–240; 241; 243
Daucus carota, polygalacturonases of, **20:**306
dCDP, in allosteric regulation, **28:**134
DR metabolism, **28:**435–440; 442; 446; 448; 453
DCIP, as electron acceptor, **28:**159; 177; 181; 183; 190; 192; 194; 197–203; 210; 220; 222; 223
dCMP, in allosteric regulation, **28:**132–133
DR metabolism, **28:**423; 438–439; 446; 448
dCMP transhydroxy methylase, reaction, **27:**425
DCMU, *See* Dichloromethylurea
dCTP, DR metabolism, **28:**438; 446
DDE, effect on carbonic anhydrase, **47:**242
ddh gene, diaminopimelate dehydrogenase, **72:**318–319
DDT, effect on carbonic anhydrase, **47:**242
Deacetylation, **58:**13, **60:**252
by diphtheria toxin, **58:**104
Deacetylcephalosporin C, action of β-lactamase, **28:**240; 245; 250; 259
N-Deacylase, localization of, **24:**327
Deacylases, classification, **14:**249
Deacylating principle, in oxidation of hydrocarbons, **27:**485
Dead-end complex, **39:**354
DEAE-cellulose, **34:**452; 459; 523
lactic dehydrogenase, **34:**456
Sephadex, **34:**452
DEAE-Sephadex chromatography, for separation of lysyl and prolyl hydroxylases, **41:**275
for separation of prolyl hydroxylase from crossreacting protein, **41:**287
Dealdolation, **53:**164
pyridoxal-catalyzed, **53:**185–189
Dealkylation, **60:**46
Deamidated products, **42:**83; 103
Deamidation, **42:**80; 81
of asparagine, **16:**199–203; 226–227
of glutamine, **16:**199–203; 226–227
of α-ketoglutaramate, **16:**257
of α-ketosuccinamate, **16:**269
of asparagine and glutamine, **13:**241–243; 282–284
of asparagine and its peptides, **13:**244–245; 251–252
of glutamine and its peptides, **13:**244–245; 261–262
of asparagines, **8:**152; 158–164
effect of *a*-keto acids, **8:**158–164
of glutamine, **8:**152; 158–164
of various amides, **8:**158–159
Deaminase, AMP, **56:**293
Deaminase, ATP, **34:**459
free, **34:**459
Deaminases, **53:**349. *See also* specific deaminases
action on DPN, **15:**106–108
from takadiastase, **15:**80
allosteric regulation, **28:**132–133
associated with myosin, **16:**337
determination of adenosine by, **16:**4
and cardiac glucosides, **13:**182
column chromatography, **22:**461; 462
and nucleic acid metabolism, **9:**495–499
for threonine and serine, adsorption studies, **14:**346
Deamination, **60:**46. *See also* Transamination, Transdeamination
of adenosine, **9:**472; 487–490
of adenosine triphosphate, **9:**495–499
of amino acids, **9:**84
of nucleic acid derivatives, **9:**495–499
oxidative, in plants, **9:**402–403; 406–408; 412; 422; 434–439
of amino acids, oxidative, **15:**338; 363
of pyridine nucleotides, **15:**106–108
of amino acids by quinone-amino acid complex, **10:**382–386; 429
of amino acids during hydrolysis, **14:**268
of A5P, **18:**231
oxidative, **18:**251–253
of glycine, **12:**449–450

role of AMP in, **12:**83–84; 90
role of biotin in, **12:**82–85
by metal ions, **21:**178; 179; 194
nonenzymic, **16:**117–120
of 3,4-dihydroxyphenylalanine, relation to protein tanning, **16:**155–156
of ornithine, **16:**262
oxidative, of L-amino acids, **16:**226
of nucleic acids, **27:**309–311
oxidative, **7:**374–375; 389; 393; 397
of glycine, **23:**453–454
in plants, with glutamine, **20:**334
in sperm, **24:**66
Deamino-3-acetylpyridine-DPNH, enzyme-bound, Cotton effect, **27:**57; 58
Deamino-DPNH, enzyme-bound, Cotton effect, **27:**57; 58
Deamino pyridine-3-aldehyde-DPNH, enzyme-bound, Cotton effect, **27:**57; 58
Deammoniacases. *See also* individual enzymes
classification, **14:**257
7-Deasa ATP-HgOAc, **43:**46
7-Deazaadenosine, **54:**210
7-Deazadenosine (tubercidin), **43:**46
5-Deazafolate, eukaryotic dihydrofolate reductase, ligand binding (of inhibitors), **70:**60
5-Deaza-6-methyltetrahydropterin, **67:**218
Debaryomyces guilliermondii, β-lactamase of, **28:**243
Debranching enzyme, **11:**348; 368–369, **34:**401
specificities of, **34:**409
Debye-Hückel theory, **33:**251
Decalcification of plasma, by ion exchanging resins, **10:**15
by metaphosphates, **10:**15
by pyrophosphates, **10:**15
n-Decane (^{14}C), oxidation of, **27:**488
Decapryn, substitute for platelet cofactor I, **16:**54
Decapsulation of microorganisms by hyaluronidase, **13:**226; 230
Decarboxylases, **35:**331–332, **46:**476–478, **55:**227; 228, *See* Adnosylmethionine decarboxylase; Arginine decarboxylases; Ornithine decarboxylase
activation by metal ions, **21:**160
amino acid, **7:**534–538
cysteic acid, **7:**389; 401–402
pyridoxal phosphate as coenzyme for, **10:**62; 70
reactivation, **10:**111
tyrosine, **10:**62–63; 65
amino acid, activation by pyridoxal phosphate, **16:**219–220
in pantetheine formation, **16:**17
bacterial (*See* Bacterial amino acid decarboxylases)
bacterial, product inhibition of, **25:**224
classification, **14:**221; 228; 256
column chromatography, **22:**483–484; 486–487
enzymic adaptation, **13:**71; 79
in formation of amines, **14:**118
in microbiological assays, **14:**273
in oxidation of naphthalene, **27:**519; 522
Decarboxylases of mammalian tissue, **5:**67–5:83
cysteic acid, **5:**75–76
dopa, **5:**72–75
histidine, **5:**69–71
tryptophan, **5:**71
tyrosine, **5:**71–72
Decarboxylating malic r DPN transhydrogenase, **14:**254
Decarboxylating malic r TPN transhydrogenase, **14:**238; 243; 254
Decarboxylation, **43:**250; 251; 384, **53:**164, **58:**176. *See also* Carbon dioxide fixation, α-Carboxylation, and β-Carboxylation; Carboxylation
of aminomalonate to glycine, **35:**518
anaerobic, **27:**535
D-aspartate, **35:**516
in carbohydrate synthesis, **26:**43
condensative, **23:**442; 445
of glyoxylate, **23:**454
of oxalate, **23:**447–449
in fermenting tobacco, **10:**349; 365–366
of α-keto acids, **18:**333–338
oxidative, **18:**273
in plants, **18:**295–298
of α-ketoglutarate, **12:**92; 103–106
by light, **9:**86; 88
in melanin formation, **16:**166–167
of amino acids, **16:**119
of oxalacetate, **16:**223
relation to decolorization of pigments, **16:**114–117
relation to phenolase activity, **16:**110

Decarboxylation (*Continued*)
relation to transamination, **16:**219–220
in tricarboxylic acid cycle, **16:**254–255
by model enzyme system, **15:**27; 29
of α-keto acids, **15:**218–219; 222; 224–230; 230–232
of α-ketoglutarate, **15:**115
of oxalosuccinic acid, **15:**117
oxidative, of glucose-6-phosphate, **15:** 164–166
of phosphogluconate, **15:**153; 156
in plants, **15:**172
and pyridineoproteins, **15:**122
of pyruvate, **15:**207–208; 218–219; 223
by transketolase, **15:**158
of oxalosuccinic acid, **12:**71–72; 86–88
and oxidative deamination, **4:**76–82
and pyridoxine, **7:**454
reversibility of, **12:**46; 49–50; 51
role of amino groups, **12:**81
role of metal ions in, **12:**69–72; 80; 87–88; 90; 102; 104
of succinyl-CoA, **26:**292
α-Decarboxylation, **35:**507
β-Decarboxylation, **35:**506
γ-Decarboxylation, **53:**163; 164; 191–192
δ-Decarboxylation, **53:**196
Decarboxylative fatty acid degradation systems, **27:**486; 487
Decarboxylative rearrangement of dopachrome, **16:**115–117
Decorsin, fibrinogen binding inhibition, **70:**7; 8
Decreased Ecto-5′-nucleotidase, B-Cell dysfunction associated with, **51:** 201–202
n-Decyl citrate synthase, **43:**82
Defective anthranilate synthetase Component II, **38:**9; 31
Defense
molecules of, **41:**130
organs of, **41:**130
Defense mechanisms and functioning of cytoplasm, **8:**51–55
Deficiency syndromes, **48:**2
exudative diathesis, **48:**2
necrotic liver degeneration, **48:**2
white muscle disease, **48:**2
Degeneracy, **47:**382
Deglycosylation, **60:**91
Degradation
chitin
action of chitinases, **59:**90–93
assay of, **59:**84–85
distribution of chitinases, intracelluar, **59:**93–94
purification of chitinases, **59:**86–87; 90
regulation of chitanases, **59:**93
of heparan sulfate, **60:**277
of heparin, **60:**277
of keratan sulfate, **60:**281
of polysaccharides, in soils, **28:**354
of proteins by light, **9:**132–133; 141
RNA, **59:**215
Degradation methods, for laminarin, **28:** 332–333; 335
for tracer distribution, **28:**396; 402–404
Degradation patterns, of polysaccharides, **28:**348–351; 353
Degradations, microbiological, for tracer distribution, **28:**403–404
Dehalogenation, **60:**46
light-induced, **9:**134
Dehydrases
and *aspergillus oryzae,* **2:**223–235
diamine-oxidase, **2:**94
and nitrate reductase, **2:**229
requiring co-ferments, **2:**225
genetic studies, **17:**401
threonine, in tumors, **29:**364
Dehydratases
aminolevulinate, **56:**203; 397–401
classification, **14:**221; 228; 243; 256
homoserine, **56:**38
serine (*See* Serine dehydratase)
threonine, **56:**43
Dehydroalanine, **33:**384, **50:**146
Dehydroascorbate, protein disulfide formation, **66:**181–182
Dehydroascorbate reductase, **66:**180–181
activity
intrinsic, **66:**179–181
protein disulfide isomerase, **66:**181–182
thiol-disulfide exchange, **66:**187
Dehydroascorbic acid
and active ribonuclease, **26:**103
complex formation with glutathione, **10:**295
in model hydroxylating system, **19:**135
reaction with borate, **12:**516
reduction, electron transport pathway, **66:**180

Dehydroascorbic reductase, in plants, **18:**310
24-Dehydrocholesterol, reduction, PN-dependent, **26:**127; 128
Dehydroconiferyl alcohol, formation from coniferyl alcohol, **16:**148
11-Dehydrocorticosterone, microbiological conversions, **20:**249
1-Dehydrocortisone, *See* Prednisone
Dehydrodiveratric acids, **34:**215
Dehydroepiandrosterone, **46:**163
Dehydroepiandrosterone sulfate, inaction of β-glucuronidase on, **16:**380
Dehydrogenase, **34:**474
  alcohol, **34:**448
    in plants, **10:**300
      in carboxylation of α-ketoglutaric acid, **10:**208; 212–213
      choline, in plants, **10:**300
      isocitric, **10:**215; 216–217
      in microbial action of hydrocarbons, **10:**448
      pyroracemic acid, **10:**180–181
      pyruvic, action of arsenic oxide on, **10:**216
      succinic, in plants, **10:**300; 309
      selective inhibition, **10:**174–175
  and cell adaptation, **17:**456
  DEAE, **34:**456
  in *Fusaria,* **5:**181
    of phosphoglyceraldehyde, **5:**133
  glucose-6-phosphate, **34:**456; 520
  glutamic acid, auxotropic mutants of, of *Neurospora crassa,* **30:**149
  hydroxybutyric, **34:**474
  lactic, **34:**448–452; 523
  for phenolic hydroxyl groups, **16:**148
    in plant oxidations, **16:**133
  soil, **28:**387
Δ^1-Dehydrogenase, **20:**268
Δ^4-5α-Dehydrogenase, **20:**268
Δ^4-5β-Dehydrogenase, **20:**268
Dehydrogenase activity, of xanthine oxidase, **17:**312–313
Dehydrogenase flavoprotein, and coenzyme Q, **25:**277; 309; 311–312; 314; 315; 317–319
Dehydrogenase-Ligand Complexes, A Unifying Model of the Thermodynamics of Formation of (Fisher), **61:**1
Dehydrogenase model, **15:**29–30
Dehydrogenase protein, in respiratory chain, **28:**209
Dehydrogenases, **15:**80; 84, **46:**468–474. *See also* Hydrogen transfer and specific dehydrogenases; individuals; specific compounds; specific dehydrogenases; specific enzyme
  carbon-14, **26:**173
  chemical mechanism of, **45:**371; 373
  classification, **14:**228
  coenzyme stereospecific, **26:**169–171
  column chromatography, **22:**479–481
  in control mechanisms, **29:**360 (*See also* Neoplastic tissues)
    hexose monophosphate shunt, induction of, **29:**374
    isocitric, in neoplasms and normal tissues, **29:**362–363
    lactic, in normal and tumor cells, **29:** 361–362
    malic, in neoplasms, **29:**361
  definition, **20:**149
    in plastids, **20:**219; 227
  denaturation of, **23:**36; 46–50; 62; 64; 71; 72
    double-headed, **23:**330; 331; 335
    DPN-linked, **23:**46–50
    in α-glycerophosphate oxidation cycle, **23:**345–346
    inhibitors for, **23:**351
    and swelling agents, **23:**346
    TPN-linked, **23:**346
  enzyme-coenzyme complex, **26:**156–165
  enzymic adaptation, **13:**71; 76; 106
  FAD-and NAD-linked, **24:**338
  fluorscence spectra of complexes, **26:**158–163
  hydrogen acceptors for, **18:**13
    in lignin formation, **18:**371
    plant, **18:**296–301; 303–304; 306–308
    relation to particle respiratory activity, **18:**13–15
  inhibitor I activity toward, **17:**107
  and mechanisms for catalysis, **24:**442–443; 469–471
  NAD-dependent, forms of, **25:**331
  nucleotide binding, **50:**29; 33
  or pentose-phosphate oxidative cycle, **24:**337
  oxidation of hydrocarbons, **27:**476; 529–534
  PN-depedent, **26:**117–121
    research models, **26:**132–156

Dehydrogenases (*Continued*)
and stereochemistry of H-transfers, **26:**156–171
principle of catalytic action, **14:**168; 181
product inhibition of, **25:**223; 224; 250–266
purification of, **46:**240–245
and pyridine alkaloids, **8:**246–247
and pyridine nucleotides, **15:**112–131
quantitative estimation of, **25:**332
relaxation study, **25:**35
specificity and coenzymes, **7:**413–414
in sperm, **24:**87
substrates, **26:**62
ternary complexes, **26:**164; 165
in tobacco, **6:**371–372
turnover number, **25:**32
Dehydrogenase substrates, photochemical reduction of, **18:**58
Dehydrogenase systems
and cytochrome b_3, **18:**23
effect of bile salts on, **18:**10
effect of urethan, **18:**6
electron transfer from, **18:**11–13
in leaves, **18:**43–44
effect on tyrosinase activity, **17:**448
kinetics of enzyme-substrate compounds in, **12:**184–185
Dehydrogenase theory, of respiration, **28:**156
Dehydrogenation
of *a*-amino acids, **8:**362–363; 384; 387–388
of fatty acids, **8:**360–373; 375–376; 380–387
of glucose 6-phosphate in nuclei, **16:**488
microbial, of hydrocarbons, **10:**469
and nicotinic acid derivatives, **7:**416–417
of steroids by microorganisms, **20:**259–263; 273; 275
studies with wood molds, **8:**273–278
Dehydrogenation in *Fusaria,* **5:**181–189
course of dehydrogenation of alcohols, **5:**182–187
Dehydroluciferin
emission spectra of mixture with luciferase, **44:**59
fluorescent emission spectra, **44:**57
fluorescent ionization states, **44:**58
structure of, **44:**39
and light emission, **25:**134–142
structure, **25:**131–132
Dehydroluciferyl-adenylate, **44:**40
Dehydroluciferyl-AMP-enzyme complex, **25:**135–137
Dehydronicotinamide, binding effect, **27:**70
Dehydropeptidase activity, **8:**134–139; 145; 163
Dehydropeptidase I
sources, **14:**376
in tumors and pathological sera, **8:**164
purification, **8:**143–145
specific action, **8:**152
substrates, **8:**141–143
susceptible configuration, **8:**166
Dehydropeptidase II
and exocystine desulfurase, **8:**147; 148
purification, **8:**143–145
substrates, **8:**141–143
susceptible configuration, **8:**166
Dehydropeptidases, **8:**117–167, **13:**284
and D-amino acid oxidase, **8:**151
characteristics of substrates, **8:**166–167
and chloroacetyldehydroalanine, **8:**139–142; 144; 145; 147; 148; 151
and chloroacetyldehydrophenylalanine, **8:**134; 135; 139; 140; 144; 151
and chloroacetylglycyldehydroalanine, **8:**140–142; 144; 145
and chloroacetylglycyldehydrophenylalanine, **8:**140–141
and chloroacetylsarcosyldehydroalanine, **8:**140; 141; 144
classification, **14:**247
isolation, **14:**398; 405
and dialanylcystine, **8:**146
and dibromopropionylcystine, **8:**146
discovery, **8:**134–135
distribution, **8:**134–143
and glycyldehydroalanine, **8:**139–145; 148; 151; 152; 157; 159; 163
and glycyldehydrophenylalanine, **8:**134; 135; 137–140; 143–145; 151; 163; 166
inhibition, **8:**145
linkage attacked by, **8:**165
in normal and pathological human sera, **8:**165
peptide metabolism, **8:**151–152
purified preparations, **8:**143–145
reversibility of action, **8:**167
specific action, **8:**118–119; 134–143
structural requirements of substrates, **8:** 141–142
substrates (table), **8:**140
Dehydropeptidases (Greenstein), **8:**117

Dehydropeptides, absorption spectra, **8:** 130–133; 136; 137; 138; 139; 152
chemical reactivity, **8:**133–134
from cystine peptides, **8:**146–147
and diacylaminopropionic acids, **8:**119; 124–128; 152
enzymic degradation of precursors, **8:** 145–164
enzymic hydrolysis, **8:**134–145
formation in tissues, **8:**152
preparation, and properties, **8:**119–134; 161–162
from amides and *a*-keto acids, **8:**124–128; 161–162
from nitriles and *a*-keto acids, **8:**128–129
produced by action of D-amino acid oxidase, **8:**150–152
products of enzymic hydrolysis, **8:** 142–143
4,5-Dehydro-L-pipecolic acid, proline analog, **29:**95
16-Dehydroprogesterone, microbiological conversions, **20:**259
1-Dehydroprogesterone, microbiological formation, **20:**261; 263
3,4-Dehydroproline, **50:**162
effect on growth of *E. coli,* **29:**96
proline analog, **29:**95
Dehydroquinase, **16:**289
Dehydroquinate dehydratase, **34:**220
Dehydroquinic acid, **17:**413; 420
5-Dehydroquinic acid
and aromatic amino acids, **18:**362
5-Dehydroshikimic acid, **17:**412–414; 421
from phosphorylated sugars, **18:**362
Dehydroshikimic acid, metabolism in *Neurospora* mutants, **22:**21
Dehydroshikimic acid dehydrase, formation in *Neurospora* mutants, **22:**21
Dehydrosqualene, formation, in squalene biosynthesis, **22:**317
1-Dehydro-substance S, microbiological conversions, **20:**245
$\Delta^{9,11}$-Dehydro-substance S, microbiological conversions, **20:**250; 251; 254; 274
Δ^{14}-Dehydro-substance S, microbiological conversions, **20:**250; 251; 274
1-Dehydrotestololactone, microbiological formation, **20:**263
1-Dehydrotestosterone, microbiological formation, **20:**263
Dehydroxypeptidase, **7:**397
Deiodinating activity, location of, **24:**308
Delacroixia coronata, enzymic transformations of steroids by, **20:**245
Delactonizing enzyme, **27:**511–512; 535
Deletion mutagenesis studies, tyrosine hydroxylase, **70:**183; 184; 186; 189–193
Deletion mutation, *See* Mutation, deletion
Demethylation
of methionine, **7:**375–378
of nicotine, **8:**234
oxidative, of N-methyl compounds, **12:** 450–451; 461–462
Demolybdo xanthine oxidase, **51:**112
Denaturants, **42:**96; 127; 128; 144; 145
Denaturases, **5:**57, **23:**171; 209
Denaturation, **1:**35–40, **34:**452–455, **42:**83; 104; 110; 112, **52:**201; 202; 203; 208, **57:**216–219
by acid, **23:**51–58
of α-amylase, **23:**50–58
of collagen, **23:**164–171
degree of definition, **23:**31
detection of, **23:**30
effect of, calcium ions, **23:**39–41; 51–58
coenzymes on, **23:**46–50
substrates on, **23:**48–50
of elastin, **23:**165
of enzyme proteins, **23:**29–82
by heat, **23:**39; 48–58
kinetics of, **23:**51
and loss of activity, **23:**31; 39–44; 46–48; 50–58
and oxidative phosphorylation, **23:**382
protection against, **23:**46–50
ratio of, **23:**31–44
and solubility of proteins, **23:**70–72
thermodynamic aspects, **23:**50–58
by urea, **23:**35; 37–41; 43–58
from adsorption, **28:**28
of albumin, **7:**248
of egg albumin, **7:**248
of enzyme, irreversible, **7:**258
reversible, **7:**236; 237; 242; 249; 258
of luminescence, effect of *p*H, **7:**245–248
irreversible, **7:**251
reversible, **7:**235; 251

Denaturation (*Continued*)
of copper proteins, effect on EPR signal, **27:**149–150
and EPR, **27:**159
of proteins, by sulfhydryl reagents, **27:** 77–78; 82
Cotton effect, **27:**88
SH-groups, **27:**222–226; 230
YADH, and Cotton effect, **27:**84–85
of cytoplasmic proteins during mitosis, **8:**38
effect of pressure and temperature, **7:**248–251
irreversible, **7:**251
reversible, **7:**235; 251
and milk clotting, **15:**440–442; 445
of proteins, **24:**137–140
by heat, **9:**31–39; 66–69; 103; 127; 146
and chemical agents, **9:**136
by light, **9:**101–108; 112; 124–126; 127; 129–131; 133–136; 141; 143; 148
mechanism, **9:**35–39; 66; 98–99; 102–103; 113; 128–131; 136–137; 146–148; 159–162
of viruses, **9:**220–228
by ionizing radiations, **24:**360
of ribonuclease, **24:**184–197
effect of solvent, **24:**194–197
and enzymic activity, **24:**187–194
by heat, and low pH, **24:**188–193; 196
transition temperature, **24:**188–193; 249–251
and helical content, **24:**191–192; 194
and hydrogen bonding, **24:**186; 192–193
reversibility of, **24:**185–193
and protein structure, **26:**91–98; 103–105
interpretations of, **26:**94–98
reversible, of non-disulfide enzymes, **26:**103–105
of ribonuclease, **26:**91; 93; 100–103
by radiations, **14:**36–37
effect adsorption on, **14:**329; 332–333; 365–366
of serum globulin, **7:**248
and —SH groups, **11:**220–221; 221–223; 224; 226; 255
of tobacco mosaic virus, **7:**248, **15:**459–460; 480–482
Denaturation and Inactivation of Enzyme Proteins (Okunuki), **23:**29
Denaturation equilibrium, **7:**240; 242
of invertase, **7:**259
Denaturation temperature, for hydroxylated collagen, **41:**264
Denatured proteins, **42:**146
with haem *a,* **23:**294–295
Denervation, muscle, *See* Muscle diseases and altered metabolism, atrophy
Denitrification
exchange reaction in, **57:**278–279
and nitrogen-13, **57:**274–276
and nitrogen-15, **57:**276–278
and nitrogen pathway, **57:**276–278
process of, **57:**273–274
rates of, **57:**274–275
and rates of conversion, **57:**274
and trapping of ^{15}NO, **57:**275–276
Density centrifugation, *Caulobacter* cell cycle, **71:**3; 4
Dental disease, and β-glucuronidase, **16:**397
Deoxyadenosine, **51:**170
3′-Deoxyadenosine, **54:**210
5′-Deoxyadenosine, **46:**53
from *S*-adenosylmethionine, **66:**19; 22
3′-Deoxyadenosine (cordecepin), **51:**200
2′-Deoxyadenosine 3,′5′-cyclic phosphate, **55:**312; 313
Deoxyadenosine kinase (dAdoK), **51:**181; 190–194
S-(5′-Deoxyadenosine-5′)-L-methionine, *See* S-Adenosylmethionine
Deoxyadenosine triphosphate (APPP), incorporation into DNA, **18:**213–215
5′-Deoxyadenosyl part of corrinoid coenzyme
biogenesis of, **26:**262–263
photolysis, **26:**251
trans effect, **26:**256
transfer reactions, **26:**253
Deoxyadenosyl radical, **66:**29
S-adenosylmethionine cleavage to, **66:**35
2-Deoxyaldoses, from DR-aldolase action, **28:**451
Deoxy-AMP, and LADH complex formation, **27:**63–64
3-Deoxy-D-arabino-heptulosonic acid 7-phosphate, formation of, **29:**115
3-Deoxy-D-*arabino*-heptulosonic acid-7-P synthetase, (DAHP synthetase), allosteric control, **28:**71
3-Deoxy-D-*arabino*-heptulosonic acid synthetases, allosteric enzymes, **28:**63

Deoxyaspergillic acid, **35:**165
3′-Deoxy-ATP, **43:**25
2′-Deoxy-ATP, substrate for pyruvate carboxylase, **42:**30
6-Deoxychalcone synthase, **55:**24; 26
Deoxycholate, **34:**487, **57:**8
in fragmentation of respiratory chain, **28:**209; 214–215
in preparation of succinic dehydrogenase, **18:**68; 70–71
Deoxycoformycin (covidarabine), **51:**182
11-Deoxycorticosterol, isotope studies of hydroxylation, **19:**190
11-Deoxycorticosterone, **19:**189; 191; 192
Deoxycorticosterone, hydroxylation, in mitochondria, **21:**27
Deoxycytidine, in thymine synthesis, **21:**423
Deoxycytidine deaminase, **21:**423
Deoxycytidine-3′-phosphates, as substrates for acid DNase, **31:**33
Deoxycytidine triphosphatase, regulating effect, **27:**283
Deoxycytidine triphosphate (CPPP), incorporation into DNA, **18:**213–215
Deoxycytidine triphosphate deaminase, **38:**263
Deoxycytidine-U-^{14}C, in chick embryo, **28:**434
Deoxycytidylate, deaminase, **38:**264
hydroxymethyl derivative, enzymatic synthesis of, **38:**244
5-methyl derivative, **38:**265
Deoxycytidylate deaminase, allosteric regulation, **28:**54; 55; 132–133
Deoxycytidylate hydroxymethylase, **54:**73
Deoxycytidylate kinase, allosteric regulation, **28:**133–134
Deoxycytidylic acid, biosynthesis, **21:**264; 286; 287
in pyrimidine biosynthesis, **21:**290; 421
Deoxycytidylic nucleotides, allosteric regulation, **28:**135
5-Deoxy derivatives of vitamin B_6, action as antagonists, **16:**221
5′-Deoxy-5′-(dimethylsulfonium) adenosine, **54:**210–211
6-Deoxy-6-fluoro-D-glucose, **18:**152
presence of a single common carrier in translocation, **29:**53
2-Deoxy-2-fluoro-(±)-glyceraldehyde, **18:**116
2-Deoxy-2-fluorotetritols, **18:**116
6-Deoxy-D-galactose, presence of a single common carrier in translocation, **29:**53
7-Deoxy-D-glucoheptose, presence of a single common carrier in translocation, **29:**53
3-Deoxy-D-glucose, presence of a single common carrier in translocation, **29:**53
6-Deoxy-D-glucose, presence of a single common carrier in translocation, **29:**53
2-Deoxyglucose-6-phosphate, **44:**243
Deoxyguanosine, **18:**222, **51:**198; 201
labeling, **28:**423
Deoxyguanosine-*p*-nitrophenyl ester, as substrate for acid DNase, **31:**33
Deoxyguanosine triphophate (GPPP), incorporation into DNA, **18:**213–215
Deoxyhemoglobin, viscosity, **40:**2
Deoxyhemoglobin A, salt bridges, **40:**19; 20
Deoxyhemoglobin S, aggregation of tetramers and oxygen affinity, **40:**19
carbamylated and solubility, **40:**21
complementary stacking, **40:**2
gel formation, **40:**2
three-dimensional structure, **40:**2
Deoxyhexoses, formation of, **26:**43–44
Deoxyinosine, **18:**222, **54:**100
5-Deoxyisoflavonoid phytoalexins, **55:**38
3-Deoxy-D-*manno*-octulosonic acid (KDO), in lipopolysaccharide, **31:**78; 79
5′-Deoxy-5′-methylthioadenosine, **54:**197
5′-Deoxy-5′-methylthio-7-deazaadenosine, **54:**212; 213
5′-Deoxy-5′-methylthioinosine, **54:**210–211; 248
5-Deoxy-5-methylthioribose, **54:**245
Deoxynucleoside diphosphates, in allosteric regulation, **28:**134
Deoxynucleoside phosphorylase, **18:**216; 224–225
Deoxynucleosides, and transfer reactions, **18:**201
Deoxynucleoside therapy, **51:**198
Deoxynucleoside triphosphatases, regulating effect, **27:**283
Deoxynucleoside triphosphates, **60:**27
effect on DNA synthesis, **27:**283
labeled, incorporation into DNA, **18:**213–215
from monophosphates, and ATP, **18:**197
Deoxynucleotide monophosphates, reaction with ATP, **18:**197
Deoxynucleotide phosphodiester bond synthesis, **18:**193

Deoxynucleotides
and LADH Complex formation, **27:**68
phosphorylation, in mitochondria, **21:**53
Deoxynucleotidyl transferase
replicative, **29:**178
terminal, **29:**178
terminal, **47:**347–372
biology of, **47:**356–369
immune fluorescence, **47:**366–369
terminal transferase function, **47:** 369–371
tissue distribution of enzyme activity, **47:**356–358
bone marrow, **47:**357; 361
cytoplasmic, **47:**368
enzyme assays, **47:**354–356
initiators, **47:**349–352
mechanism of polymerization, **47:**352–353
molecular weight, **47:**354
murine lymphomas, **47:**364
murine plasmacytomas, **47:**365
nuclear, **47:**368
protein structure, **47:**353–354
spleen, **47:**368
substrates, **47:**349–352
thymus, **47:**357
thymus cortex, **47:**368
viral leukemias, **47:**366
Deoxynucleotidyl Transferase, Terminal: Biological Studies (Bollum), **47:**347
Deoxyorotidine-5′-phosphate, in pyrimidine biosynthesis, **21:**285
5′-Deoxy-5′-*S*-*n*-propylthioadenosine, **54:**209
5-Deoxypyridoxal, **50:**7–9
dissociation constant, **50:**5
and phosphorylase activity, **24:**272
Deoxypyridoxine, 5–5phosphate, preparation, **16:**214–215
4-Deoxypyridoxine, substrate for pyridoxol kinase, **16:**219
vitamin B_6 antagonist, **16:**221
4-Deoxypyridoxine 5-phosphate, enzyme studies, **16:**221–222
Deoxypyridoxine phosphate-enzyme complex, **16:**217
Deoxyribonuclease
adsorption at interfaces, **28:**3
in bacterial cell disruption, **33:**350
effect on peptide synthesis, **33:**345–346
column chromatography, **22:**451
effect of ionizing radiations, **24:**370; 373–374; 381; 388; 391; 392; 395–396; 399–400
effect on enzyme synthesis, **24:**12
localization of, **24:**321; 343
molecular weight from radiation studies, **24:**366
neutral, location of, **24:**321
effect on β-glucuronidase, **40:**40
in mitochondria and lysosomes, **21:**39–41; 58
of pancreas, **23:**130; 132–134
uncoupling action, **23:**367
Deoxyribonuclease I
from different tissues, **27:**246
effect of radiations on, **24:**395–396; 397; 404
Deoxyribonuclease II
column chromatography, **22:**467
Deoxyribonucleic acid, *See* DNA
Deoxyribonucleic acid pyrimidines, biosynthesis, **21:**285–290
Deoxyribonucleohistone, *See* Chromatin
Deoxyribonucleoprotein. *See also* Chromatin
a genetic material, **18:**161–190
Deoxyribonucleoprotein, a Genetic Material (Butler and Davison), **18:**161
Deoxyribonucleosides
from DR-5-P, **28:**421–422
formation in cell-free systems, **28:**435
in growth of *L. leichmannii,* **28:**429–431
excess, effect on mutations, **27:**341–342
synthesis, vitamin B_{12} role, **26:**267
Deoxyribonucleotides
bases of, **27:**290–295
biosynthesis, **28:**447
allosteric regulation, **28:**132–133
formation in cell-free systems, **28:**435–436; 438
synthesis of, allosteric control in, **29:**368
physiological regulation of, **29:**369
Deoxyribose, **60:**3
biosynthesis of, **28:**419–435
in animals, tracer studies, **28:**431–435
from bound ribose, **28:**420–421
from ^{14}C-glucose, **28:**419; 426–433
de novo, **28:**421–423; 435
from lactate-1-^{14}C, **28:**426
pathways, **28:**420–423
degradation of, **28:**396; 403–404
distribution of ^{14}C in, **28:**407

metabolism, enzymes of, **28:**435–451
oxidation of, **28:**399
purine-bound
from glucose-1-^{14}C, **28:**425
liability, **28:**433
pyrimidine-bound, **28:**433
from glucose-1-^{14}C, **28:**425
Deoxyribose, Biosynthesis of Ribose and (Sable), **28:**391
Deoxyribose aldolase. *See also* Aldolase; Deoxyribose-5-P aldolase
function, **28:**403; 422; 424–425; 435
substrate specificity, **28:**450–451
Deoxyribose derivatives, origin, **28:**420–421
Deoxyribose nucleoside diphosphate, **23:**324
Deoxyribose nucleoside triphosphate, **23:**324
Deoxyribose-5-phosphate, **60:**11
in pyrimidine biosynthesis, **21:**285
Deoxyribose-5-phosphate aldolase, in DR metabolism, **28:**449–452
effect on DR-5-P, **28:**393
2-Deoxy-D-ribose-5-phosphate aldolase (DRP aldolase), from *Lactobacillus plantarum,* **31:**139
from *Salmonella typhimurium,* **31:**139
Deoxyribose-5-phosphate aldolase pathway, deoxyribose synthesis, **28:**421–422
Deoxyriboses, synthesis, vitamin B_{12} in, **21:**431
Deoxyribosyl-transferring enzymes, **28:**435
Deoxysugars, synthesis, **26:**40; 43–44; 47–48; 51; 54
2-Deoxy-2-sulfamate-glucose-3-sulfate 3-sulfase, **60:**267
2-Deoxy-2-sulfamate-D-glucose-6-sulfate-6-sulfatase, **60:**265
Deoxythymidine, in hexosamine metabolism, **26:**50
deoxyThymidine diphosphate-D-glucose oxidoreductase, **37:**436; 485
Deoxythymidine kinase, allosteric regulation, **28:**133–134
Deoxythymidylate, biosynthesis of, **28:**133–134
Deoxyuridine, **18:**222
in DNA formation, **25:**397
in pyrimidine biosynthesis, **21:**289; 420–423
Deoxyuridine kinases, methylation study, **25:**397
2′-Deoxyuridine-5′-monophosphate (dUMP), **54:**58
Deoxyuridine triphosphatase, regulating effect, **27:**283
Deoxyuridylate, **54:**70
5–5formyl analog, **38:**274
hydroxy methyl analog, **38:**274
5-mercapto analog, **38:**274
nucleophilic attack at carbon-6, **38:**247
synthesis, **38:**264
tritium labeled, **38:**243
Deoxyuridylic acid, biosynthesis, **21:**264; 286
in pyrimidine biosynthesis, **21:**264; 265; 288–290; 421; 422
Depancreatized dogs, absorption of carbohydrates in, **8:**180–181
absorption of fat in, **8:**180–181
cholesterol in livers, **8:**177
choline-containing phospholipides in livers, **8:**177
effect of raw pancreas on, **8:**183–186
fat in liver, **8:**172; 178; 179
fatty liver in, **8:**175–179
insulin-treated, diet for maintenance, **8:** 173–175
pathological changes on special diet, **8:** 175–183
survival, **8:**172–175
Dephospho CoA, **44:**41
Dephospho-CoA, **57:**143
3′-Dephospho-CoA, in enzymic synthesis of CoA, **16:**16; 17
Dephospho coenzyme A, **18:**202
3′-Dephospho-coenzyme A, synthesis by phosphoramidate method, **22:**200
Dephosphonylation, **53:**163
pyridoxal-catalyzed, **53:**192–195
γ-Dephosphonylation, **53:**164
Dephosphopepsin A, *See* Pepsin D
Dephosphophosphorylase, **24:**282–283
Dephosphorylation, **58:**150, *See* Phosphorylation-dephosphorylation
enzymic, in ribose synthesis, **28:**395
Dephosphorylation of phosphopyruvate, **5:**133
Dephosphorylations, of hexose monophosphates, **15:**151–153
Depigmentation, **14:**94–96
Deploymerase action of RNase, **24:**173; 214
Depolarization of adenylic system, **6:**186
Depolymerases, **23:**208
product inhibition of, **25:**184

Depolymerization. *See also* Hyaluronidases and Polymerization
of collagen, **23:**169; 173
of myosin, **23:**382
by hyaluronidases, **13:**203; 226
of light-meromyosin, **16:**332
of myosin by urea, **16:**326–330
Depot fats
catabolism, **2:**271–283
transport as phospholipids and colic acid-cholesterol complexes, **2:**271–275
Depression, **74:**236
catecholamine hypothesis of, **29:**489; 491
drugs and their mechanism of action on, **29:**483
endogenous, **29:**490
Depressor substance in oils, **7:**542
Derepression, **40:**75; 77; 79; 80; 82
Dermatan sulfate, **52:**73; 82, **60:**273
of connective tissue, **25:**439–442
degradation by β-galactosidase, **60:**252
faulty degradation, **60:**241
sequential degradation, **60:**275; 276
structure, **60:**275; 276
Dermatitis from products of microbial action on hydrocarbons, **10:**474
Dermatophytes, β-1,3 glucan of, **28:**336–337
Dermatosparaxis, **42:**174
deficiency of procollagen peptidase in, **37:**332
Dermoitin sulfate, *See* Dermatan sulfate
Dermonecrosis, penicillin treatment, **28:**274
Dermonecrotic toxin, **58:**131
Desamidase, classification, **14:**234
Desamidonicotinamide adenine dinucleotide amidotransferase
adenyl desamido NAD, a possible intermediate of reaction, **39:**163
amino donors, **39:**163
cation requirements, **39:**163
characteristics of reaction, **39:**162
inhibition of azaserine, **39:**163
pathways for synthesis of NAD, **39:**161; 162
enzymes involves in reaction starting with nicotinamide, **39:**161
enzymes involves in reaction starting with nicotinic acid, **39:**162
Desamine CoA, **44:**41
Desaminocozymase, **7:**415
Desamino DPN, **15:**80–81; 98; 106–108; 117
Desaspidin, breakdown products of, effects on inhibition of electron flow and phosphorylation, **30:**69
effects on cyclic and noncyclic phosphorylation, **30:**68–69
Desaturation of fatty acids, enzymes for, **8:** 316–317; 329
Descarboxyenalaprilat, **57:**53–55
Desensitizaion of enzymes, **26:**106–107
Desensitization, in allosteric enzymes, **37:**595
Desensitization (adenosine receptors), **69:** 99–105; 111–113
A_1AR, **69:**99–101
A_2AR, **69:**102–105
Desferriferrichrome, **35:**145; 175; 176
aluminum chelate, **35:**143; 177
hydrogen exchange, **35:**143
Desferrioxamine, **46:**115
Desialylation, of ceruloplasmin, **41:**106–109; 114
of glycoproteins, **41:**106–115; 119
with neuraminidase, **41:**118
Desiccation, and protein synthesis in plants, **9:**424–426
Design and Properties of N-Carboxyalkylpeptide Inhibitors of Angiotensin Converting Enzymes (Patchett and Cordes), **57:**1
Desipiramine
anticholinergic activities of, **55:**177
structure-activity studies of, **55:**178
Deslysine ribonuclease, **24:**209; 221
Deslysyl carboxyl-ribonuclease, **24:**171
Desmoenzyme, in conversion of pyroracemic acid to citric acid, **10:**213
Desmoenzymes, isolation, **14:**378
Desmoglycogen, **3:**133
Desmolases
classification, **14:**221
mechanism of catalysis by, **14:**172–173; 183
in tobacco, **6:**369
Desmolysis, auxiliary enzymes of, **14:**221
Desmolysis of hexoses, **7:**16; 17
Desmosterol, in cholesterol biosynthesis, **22:**326; 327
structure, **22:**325
Desorption, **34:**452
Desoxyadenylic acid, in DNA, **16:**431
Desoxycholate
effect on DPNH oxidase I, **20:**163; 164

effect on DPNH oxidation, **17:**379–380
solubilization of cytochromes, **17:**343; 354–358
Desoxycorticosterone, **7:**524, **15:**127
11-Desoxycorticosterone, *See* Cortexone
Desoxycorticosterone acetate, effect on rate of excretion of phenolsulfophthalein, **13:**229
Desoxycytidylic acid, in DNA, **16:**431
Desoxyglucose, **15:**145; 146
Desoxyguanylic acid, in DNA, **16:**431
Desoxyhemigossypol, **55:**80
structure of, **55:**56
4-Desoxyheptulose phosphate, synthesis of, **15:**160
2-Desoxyketohexoses, **15:**154
Desoxy-6-methoxyhemigossypol, **55:**80
structure of, **55:**56
Desoxypentosenucleic acid, *See* DNA
Desoxyribonuclease, **9:**517; 519
effect on phage DNA, **17:**494; 496
genetic effects, **17:**456; 458; 464
Desoxyribonucleases
classification, **14:**245
isolation, **14:**405
hydrolysis of DNA, **16:**434–435; 447
intracellular distribution, **16:**463; 466; 467
Desoxyribonucleic acid. *See also* DNA; Nucleic acids
and phage, **17:**477–478; 483–484; 493–496
genetic effects, **17:**455–457; 459–465
solubilizing power, **17:**137
and sarcoma, **7:**175–179
turnover, **7:**173–179; 188–189; 192
in sperm, **9:**338–340
Desoxyribonucleic acid (DNA), in plastids, **20:**210; 211
Desoxyribose. *See also* Carbohydrates and Pentose; Nucleic acids
of DNA, **16:**431–432
metabolism, **9:**474
Desoxyribose nucleotides, **9:**507–508. *See also* Nucleotides
elementary composition, **9:**521
Desoxyribose phosphate, **15:**172
Desoxyribose-5-phosphate, **15:**154; 158; 160; 167; 171
Desoxyribose-5-phosphate aldolase, **15:**171
Desoxyribosides, as B_{12} replacement in lactic acid bacteria, **16:**274
Desoxyribosyl transfer, **17:**220–221
Destabilization, **43:**239; 327, **51:**94
desolvation, **43:**249
electrostatic, **43:**249; 264
geometric, **43:**240
induced, **43:**254
mechanism, **43:**360; 376
tight binding with, **43:**351
by micelles, **43:**384
nonlinear, **43:**259
unfavorable interaction between two substrate molecules, **43:**245
Destabilization energy, muscle contraction and, **51:**84–85
Desthiobenzylpenicillin, effect on β-lactamase, **28:**262
Desthiobiotin in plant metabolism, **10:**249
Destructors, definition, **14:**242
Desulfation, **60:**256
Desulfatomaculum nigrificans, **39:**338
Desulfhydrases. *See also* specific compounds
classification, **14:**228–229; 257
Desulfinases, classification, **14:**229; 257
Desulfination, **35:**506–507
Desulfination reaction, **16:**234
Desulfinicase, **7:**388; 401
Desulfinicase in the liver of higher animals, **10:**308
Desulfo enzyme, EPR signal, **51:**138
Desulfo Inhibited spectrum, **51:**138
Desulfotomaculum nigrificans, **46:**49
Desulfovibrio, oxidation of hydrocarbons, **27:**530
Desulfovibrio desulfuricans, **46:**49
Desulfovibrio gigus, **46:**49
Desulfovibrio vulgaris, **46:**49
Desulfo xanthine oxidase, **51:**135–138; 150
reactivity toward alcohol substrates, **51:**138
redox potential of molybdenum in, **51:** 137–138
slow signal, **51:**136–137
Desulfurase, **11:**390
action on cysteine, **8:**146–147
exocystine, **8:**147; 148
Desulfuration, **7:**379–383; 390–392; 394; 396–397
Desulfuration of cystine peptides, **8:**135
Desulfurization, **60:**246
Detegents, effect on mitochrondrial oxidative phosphorylation, **21:**32
Detergent, effect on lipase activity, **48:**352

Detergent-covered interface, **48:**359
Detergents, **34:**469; 482–485, **60:**40; 96
action on toxicity of hexyl-resorcinol, **11:**85
effect on —SH groups of proteins, **11:** 221–223; 224
inactivation of pectinesterase by, **11:**284
anionic
effect on β-lactamases, **28:**295
as lytic agents, **26:**214
cytochrome *c* oxidase, polypeptide composition, **71:**85
effect on sperm, **24:**70
effects on cell structures, **8:**64
Detergents as inhibitors of hyaluronidase activity, **13:**214
Detergent solutions, of haem *a,* **23:**289–297; 306–310
of haematin *a,* **23:**289–297; 306–310
Dethiobiotin, **35:**336, **38:**340–343
conversion of, to biotin, **38:**343
effective sulfur donors, **38:**347
effect of the antibiotic actithiazic acid, **38:**348
in growing cultures of A. *niger,* **38:**344
DAPA as a precursor, **38:**340
kinetic studies, **38:**342
mechanism for dethiobiotin synthesis, **38:**343
purification and properties of the enzyme, **38:**341
requirements for HCO_{3^-}, ATP and Mg^{2+}, **38:**340
in resting cells, of *E. coli,* **38:**347
of *S. cereviseae,* **38:**345
synthetase, **35:**336
Detoxication, **39:**361
of aromatic hydrocarbons, **10:**477–478
of nicotine, **10:**429
role of glutathione in, **13:**273–274
Detoxication and biological methylation, **12:**451
Detoxification, **46:**408, **57:**92–93, **59:**131–132
carcinogens, **69:**6
glutathione S-transferases, **69:**5–7
xenobiotics, **69:**6
xenobiotic substrate specificity and mechanism, **69:**23–36
Detoxifying function, **39:**359
Detrite rocks, age, **27:**354–355
Deuterated ethane, **27:**481; 533
Deuterium
and citric acid cycle, **18:**243
in cobamide coenzyme reaction, **26:**358; 364–366
in diols to aldehyde conversion, **26:** 265–266
in exchange reaction, **26:**340
in hydrogen transfer study, **26:**137–139; 166–167; 170
in glycerides, **7:**156
in phosphatides, **7:**170
in studies of biosynthesis of pectic substances, **20:**369
in studies of steroid enzymes, **20:**273; 274
in study of pyrimidine biosynthesis, **21:**275
as tracer in enzyme kinetics, **17:**55
in transamination study, **16:**199; 218
labeling of serine, **16:**294
use as tracer in transamination mechanism, **4:**80–83
α-Deuterium, of amino acid, labilization, **16:**218
Deuterium-containing fatty acids, **7:**169–170
Deuterium exchange, **42:**104, **50:**345–346
Deuterium-hydrogen exchange, **24:**449; 468–469
and hydrogen bonding in RNase, **24:**192–193
in oxidized RNase, **24:**200
of PIR, **24:**228
of ribonucleases, **24:**170; 181; 232; 247; 248
Deuterium in transmethylation studies, **12:** 458–464; 469–474; 479; 480–481
Deuterium isotope effect, in reduction of carbonyl compounds, **24:**474
on RNase activity, **24:**193
Deuterium-labeled succinate, oxidation of, **18:**100–101
Deuterium oxide
effect on ESR of *Chlorella,* **25:**83
in EPR study, **27:**133–136
in hydrocarbon oxidation study, **27:**483; 533
mutagenic effect, **27:**330
in mechanism studies, **24:**448; 461; 468–469
Deuterium oxide isotope effects, **43:**173
Deuterium transfer from reduced DPN, **15:**88
Deuterium transfer to olefin, **24:**473
α-Deuteroalanine, in transaminations, **16:**218
Deuteroheme, **46:**102
Deuterons, and enzyme in activation, **24:**373; 376
LET from, **24:**365; 371
Deuteroporphyrin, **46:**102
iron chelates of, **46:**101

x-Deuteroglutamate, in transaminations, **16:**218
Deuteroporphyrins, **23:**278; 285
Dexamethasone, **67:**228
Dexamethasone treatment, adenosine receptors, **69:**110–111
Dexanethasome, tyrosine hydroxylase induction, **70:**169; 170
Dextran
acceptor constant K_a of (*See also* Mechanisms)
acceptor constant K_a of, **30:**194
action of the acceptors in, **30:**198–200
activation energy of, **30:**194
from bacteria, **30:**190
from dextransucrase of *Leuconostoc mesenteroides* containing branches, **30:**197
formation of, **30:**190–202
glycosidic bonds in, **30:**182
as a helix, **30:**196
hypothesis of the formation of, **30:**218
initial rates of formation of, at various substrate concentrations and temperatures, **30:**193
kinetics of formation of, **30:**192–195
maximum rate V_m of, **30:**194
mechanism of formation of, **30:**179–221
Michaelis constant K_m of, **30:**194
molecular weights of, **30:**190; 195–197
number average molecular weights of, with additions of acceptors, **30:**199
origin of branches in, **30:**197–198
schematic picture of the formation of, **30:**213
structure of, **30:**195–197
summary of formation of, **30:**201–202
verification of formation of, **30:**195
antibodies against, **24:**115
enzymic synthesis from sucrose, **10:**141
inhibition of threone activity, **16:**55
in isolation of mitochondria, **21:**5
synthesis, **6:**300
Dextran-antidextran system, **24:**130
Dextranase, **34:**522
Dextran-dextrinase
classification, **14:**251
in polysaccharide synthesis, **11:**298–299; 327–329; 330
Dextrans, **17:**208; 213; 222; 224
adsorption studies, **14:**331
formation of, **11:**298–299; 330
from amylose-type chains, **11:**327–329
from sucrose, **11:**308–312
hydrolysis of, **11:**340
synthesis, **15:**310–312
Dextransucrace, classification, **14:**222; 251
Dextransucrase, **5:**119, **11:**371, **17:**191; 195; 207–208; 222; 225–228
action of, **30:**190
molecular weight of, **30:**191
optimum pH of, **30:**191
properties of, **30:**191–192
purification of, **30:**191
specificity of, **30:**192
stability of, **30:**191
action on maltose after heat denaturation, **22:**70
in polysaccharide synthesis, **11:**298–299; 308–312; 327; 330
synthetic action of, **12:**421
Dextran sucrases, **15:**310–311
Dextran sulfate, activation of lipoprotein lipase, **40:**33
Dextran synthesis factor, **11:**311
Dextrantriaose, **17:**211–213; 227
O-α-Dextrantriaosyl-(1→4)-D-glucose, **17:** 212–213
Dextrinase, limit, isolation, **14:**397; 405
Dextrin-dextranase, **17:**213
Dextrinizing amylase, *See* α-Amylase
Dextrin r dextran transglucosidase, classification, **14:**251
Dextrins. *See also* Oligosaccharides; Polysaccharides
coupling of linear with cyclic, **17:**210–211
cyclic, **11:**300; 322–324; 328; 354, **12:** 419–420, **15:**305; 307
fermentation by thiaminase-producing bacteria, **15:**413
formation in plants, **18:**289; 290
limit action of isophosphorylase on, **12:** 417–419
from amylopectin, **12:**389; 393–394
from phosphorylase action, **12:**413
structure of, **12:**356
linear, **11:**298–299
in polysaccharide synthesis, **11:**298–299; 300; 327–329
properties and structures, **12:**372–374
Schardinger, reaction with iodine, **12:**365

α-Dextrins, resulting from α-amylolysis, **11:**347–348
Dextrose, utilization in progressive muscular dystrophy, **6:**222
DFP, *See* Diisopropyl fluorophosphate; Diisopropylfluorophosphate
effect on tyrosinase inactivation of chymotrypsin, **14:**146
in enzyme crystallization, **14:**310
DFP-α-chymotrypsin, structure studies, **14:**283; 284; 287; 310
DFP (diisopropyl fluorophosphate) effect on choline acetylase, **12:**323
inactivation of esterases, **12:**281–290; 308–311
DFP-sensitive proteinase, from *staphylococcus aureus,* **41:**195
DFP-trypsin, structure studies, **14:**283
dGMP, in allosteric regulation, **28:**132–133
in DR metabolism, **28:**423; 448
effect on aspartate transcarbamylase, **28:**132
dGTP, in allosteric regulation, **28:**133; 135
DHA, *See* Dihydroxyacetone
DHAP, *See* Dihydroxyacetone phosphate
DHAS, *See* Potassium dehydroepiandrosterone sulfate
DHI, *See* α, p-Dihydroxy-isovaleric acid
DHQ (5-dehydroquinic acid), in metabolism of aromatic amino acids, **16:**287–291
DHS (5-dehydroshikimic acid), in metabolism of aromatic amino acids, **16:**287–291
DHS reductase, **16:**289
DHV, *See* α, β-Dihydroxy-β-methylvaleric acid
Diabetes, **54:**164–165, **59:**365–366
accumulation of fatty acyl CoA, **28:**117
alloxan, effect on ribose synthesis, **28:**417
and creatinuria, **8:**402
and oxidation of ketone bodies, **8:**408; 412
and phosphorylation of glucose, **8:**403
and familial periodic paralysis, **6:**216
glucose diphosphate in, **14:**212
phenylalanine hydroxylase activation and, **67:**137–138
phosphoenolpyruvate carboxykinase (GTP), **69:**252–255
and –SH groups, **11:**247–248; 251
sodium fluoroacetate, **18:**144–146
Diabetes mellitus, **33:**446
Diabetic liver
glucose-6-phosphatase from, **25:**235
glucose-6-phosphatase of, **28:**98
Diabetogenic hormone of hypophysis, **8:**414
a,a-Diacetaminopropionic acids, preparation, **8:**124–128
Diacetyl
dismutation of, **15:**115
effect on sulfanilamide acetylation, **6:**258
formation in plant tissues, **9:**623; 624
Diacetyl and fat metabolism, **8:**389–391; 400
Diacetylchitobiose, **60:**161
Diacetylcystamine, S-exchanges, **27:**209
Diacetyl mutase and fat metabolism, **8:**391
O,N-Diacetylneuraminic acid, **20:**135; 138; 140; 141; 143
S, N-Diacetylthioethanolamine, **15:**190
Diacylaminopropionic acids, absorption spectra, **8:**153
and dehydropeptidases, **8:**152–158
and dehydropeptides, **8:**119; 124–128; 152
and pyruvic acid, **8:**153–158
stability, **8:**125–126
Diacylimide bond, role in peptide biosynthesis, **19:**263
Diacylimide compounds, formation, **19:**243–245
Diacylimino compound, *See* Pepsin, diacylimino compound intermediate of
P′,P^5-Di(adenosine-5′) pentaphosphate, **43:**353
Di-adenosine polyphosphate, **1:**114
Di-adenosine tetraphosphate, **1:**114
Diaffinity theory of enzyme-substrate complex, **12:**193
Diagnostic laboratories, use of β-lactamase, **28:**307
Diagonal diagrams, **47:**133–134
Dialanylcystine and dehydropeptidases, **8:**146
Dialdehyde starch, **34:**453
Dialkyl amino acid
decarboxylase
mechanism of action, **35:**87
stereochemistry of transamination, **35:**102
transaminase, **35:**104
α-Diaklyl amino acid transaminase, **35:**87
Dialkylbenzanthracenes, **60:**15
Dialkyl disulfides, and biological methylation, **12:**436–437; 441–442; 443–445; 457
2,2-Dialkylglycine decarboxylase, **74:**133; 148; 156
Dialkyl phosphates, alkali stability, **14:**214
Diallylbarbituric acid, **46:**130
Diallyl disulfoxide, synthesis in plants, **10:**258

Diallyl sulfide in plants, **10:**242
Diallyl sulfides, **11:**379; 380; 381
Dialuric acid, **41:**43
formation, **16:**121
Dialysis, effect on pyrocatechase, **19:**126
Dialysis tubing, **34:**452
Diamagnetic species, **34:**44; 63
Diamine, chi ketoglutarate aminotrans-ferase, **36:**223
Diamine oxidase, **36:**233–235, **46:**455
and alkaloid biogenesis, **8:**238; 239
and inhibition, **8:**480
classification, **14:**253
inhibition by taraxein of, **29:**509
product inhibition of, **25:**224
properties, **19:**203
in spermatozoa, **24:**55
location in subcellular fraction, **24:**302
Diamine-oxidase, **2:**93–112
and activators, **2:**102; 103
affinity to substrates, **2:**98; 99
definition, **2:**93; 94
determination, **2:**94; 95
enzymatic disintegration, **2:**103–105
extraction, **2:**95; 96
and hormones, **2:**107
and inhibitors, **2:**99–102
occurrence in nature, **2:**105–107
physiological value, **2:**107–109
specificity, **2:**96–98
therapeutical use, **2:**109–110
Diamine oxidase (hog kidney), **59:**193
Diamine oxidase in semen, **9:**367
Diamine r O_2 transhydrogenases, classification, **14:**253
Diamines, **5:**82, *See* 1,4-Diaminobutane
Diamino acid dehydrogenases, **38:**440–445
2,5-diaminohexanoate dehydrogenase-(2,4-diaminopentanoate dehydrogenase), **38:**441
DPN, activity with, **38:**441
equilibrium constant, **38:**441
L-erythro-3,5-diaminohexanoate dehydrogenase, **38:**440
3-keto-5-aminohexanote, **38:**440
Δ'-pyrroline-2-methyl-5-carboxylic acid, **38:**441
TPN binding sites, **38:**441
Diamino acids
paradoxical transport by neutral systems, **49:**63–75
relation to protein absorption maximum, **16:**421
1,9-Diamino-5-azanonane, **54:**221
2,6-Diamino-8-azopurine, FAH_2-reductase inhibitor, **27:**444–445
2,5-Diamino-1,4-benzoquinone, formation of derivatives, **16:**112
1,4-Diaminobutane, **36:**203–251, **54:**214
acetylation, **36:**222–223; 225
agmatine ureohydrolase, **36:**213
amine oxidases, **36:**232–236
analytical procedures, **36:**206
automated ion-exchange methods, **36:**206
chromatography, **36:**206
dansyl chloride, **36:**206
fluorescent, **36:**206
gas chromatography, **36:**206
ion-exchange chromatography, **36:**206
paper, **36:**206
serum amine oxidase, **36:**206
spermidine dehydrogenase, **36:**206
thin-layer electrophoresis, **36:**206
arginine decarboxylase, **36:**211
binding to nucleic acids, **36:**242–243
biosynthesis, **36:**220–221
biosynthesis of, **36:**215–218; 228
in animals, **36:**229
ornithine decarboxylase, **36:**229
in *E. coli,* **36:**208
in *Neurospora Crassa,* **36:**220
in plants, **36:**230–231
agmatine iminohydrolase, **36:**230
arginine decarboxylase, **36:**230
carbamylputrescine pathway, **36:**230
carbamylation of, **36:**224–225
cell division and inhibition of microcyst formation, **36:**240–241
compartmentalization of arginine, **36:**217
concentration, **36:**206–208; 227–228
animals, **36:**227
bacteria, **36:**206
bacteriophages, **36:**208
fungi, **36:**206
in gram-negative bacteria, **36:**206
in gram-positive bacteria, **36:**207
plants, **36:**228
viruses, **36:**208
control mechanisms, **36:**215

1,4-Diaminobutane (*Continued*)
diamine oxidase, animal, **36:**233
plant, **36:**235
diamine oxidation, **36:**224
Aspergillus niger, **36:**224
distribution, **36:**206–208
effect on bacteriphage T4rII, **36:**237
growth effects, **36:**238
growth requirements for, **36:**218
growth requirements of mutant strains of *E. coli,* **36:**218
hydroxylation of, **36:**225
inhibitory effect on bacteriophages, **36:**237
localization, **36:**207
metabolism of, **36:**224–226
in animals and plants, **36:**232–237
in *E. coli,* **36:**221–223
mutants lacking, **36:**218–219
ornithine decarboxylase, **36:**211
protection of bacteriophage T5, **36:**237
redistribution, **36:**207
stabilization of nucleic acids, **36:**242–243
transamination, **36:**223
uptake by *E. coli,* **36:**226
1,4-Diaminobutane, Spermidine, and Related Amines, Biosynthesis and Metabolism (Tabor and Tabor), **36:**203
2,4-Diaminobutyrate, **50:**152
2,4-Diaminobutyric acid, **49:**64; 67, **54:**26
1,4-Diamino-[^{14} C] butane, metabolism of, *in vivo* on animals, **36:**237
cis-1,4-Diaminocyclohexanecarboxylic acid, **49:**63
1,10-Diamino-*n*-decane, as activator of β-glucuronidase, **16:**376
Diaminodioxypyrimidine (III), **67:**201
2,4-Diamino-6,7-disubstituted pteridines, folic antagonists, **27:**438
Diaminodurol, as artificial electron donor, **30:**23
α,γ-Diaminoglutaric acid, in transaminations, **16:**196; 206
3,5-Diaminohexanoate, **38:**426
acid ninhydrin color, **38:**427
formation of, from sorbic acid and ammonia, **38:**426
2,5-Diaminohexanoate dehydrogenase, **38:**444
inactivation of, **38:**444
kinetic constants of, **38:**445
properties of, **38:**444
Diaminohexanoates, chromatographic mobilities, **38:**421
electrophoretic mobilities, **38:**421
3,6-Diaminohexanoic acid, **38:**417
Diaminohexanoic acid isomers, **38:**420
α,ω-Diaminomonocarboxylic acid derivatives, **54:**33–36
Diaminopamelic decarboxylase, **17:**404
7,8-Diaminopelargonate, **35:**336
7,8 Diaminopelargonic acid, S-adenosyl-L-methionine as amino donor, **38:**338
enzyme purification, **38:**339
7-KAP substrate inhibition, **38:**337
specificity of aminotransferase, **38:**338
synthesis in cell-free systems, **38:**336
1,5-Diaminopentane, **36:**205. *See also* 1,4-Diaminobutane
biosynthesis in *E. coli,* **36:**208
in *E. coli mutant,* **36:**207
L-lysine dearboxylase, **36:**220
putrescine oxidase, **36:**224
substrate of propylamine transferase, **36:**215
L,L-Diaminopimelate decarboxylase
bacterial lysine biosynthesis, **72:**310–314
chemical mechanism, **72:**312; 314
Diaminopimelate decarboxylase, structural data on, **74:**134; 142
Diaminopimelate dehydrogenase
bacterial lysine biosynthesis, **72:**315–320
chemical mechanism, **72:**316–317
chemical structure, **72:**281
three-dimensional structure, **72:**316; 318–319
L′,L-Diaminopimelate epimerase
bacterial lysine biosynthesis, **72:**307–310
chemical mechanism, **72:**308–310
fluorinated diaminopimelate reaction, **72:**309; 311
Diaminopimelic acid, **50:**120
α, α′-Diaminopimelic acid, *See* DAP
meso-Diaminopimelic acid, incorporation in nucleotide, **26:**50
2,6-Diaminopimelic acid (C^{14}), in murein structure, **26:**218–219
α, α′-Diaminopimelic acid (DAP), **17:**403–405
2,6-Diaminopimelic acid (DAP), in mureins, **26:**198–203; 205–207; 212; 217–219
2,6-Diaminopimelic acid (H^3), in *E. coli* growth, **26:**220–221
Diaminopimelique, and citric acid cycle, **18:**257

α,ω-Diaminopolymethylenes, as activators of β-glucuronidase, **16:**376
1,3-Diaminopropane, **36:**205, **54:**222. *See also* 1, 4–4Diaminobutane
analytical procedures, **36:**206
automated ion-exchange methods, **36:**206
chromatography, **36:**206
dansyl chloride, **36:**206
fluorescent, **36:**206
gas chromatography, **36:**206
ion-exchange chromatography, **36:**206
paper, **36:**206
serum amine oxidase, **36:**206
spermidine dehydrogenase, **36:**206
thin-layer electrophoresis, **36:**206
concentration, **36:**206–208
bacteria, **36:**206
bacteriophages, **36:**208
fungi, **36:**206
in gram-negative bacteria, **36:**206
in gram-positive bacteria, **36:**207
virus, **36:**208
distribution, **36:**206–208
product, of spermidine dehydrogenase, **36:**225
of spermidine oxidation, **36:**226
2,3-Diaminopropionic acid, **49:**64
α,β-Diaminopropionic acid, **50:**122; 123; 124; 125
β-Diaminopropionic acid, **50:**146
2,4-Diaminopteridine ring, eukaryotic dihydrofolate reductase, **70:**34–35; 37
Diaminopteridines, potentiators of sulfonamides, **27:**455
2,6-Diaminopurine
effect on virus multiplication, **17:**480
genetic studies, **17:**427–428
folate inhibition, **27:**444–445
mutagenic action, **27:**318–319
4,5-Diaminopyrimidine, from purine, **27:**423
2,4-Diaminopyrimidines, as folate antagonists, **27:**438–440
models, as dihydrofolate reductase inhibitors, **27:**448–449
Diaminopyrimidines, synergism with sulfonamides, **27:**455; 460
Diamin-Oxydase (Zeller), **2:**93
4,5-Dianilino-1,2-benzoquinone, formation, **16:**110
Dianisidine, **58:**64
Diaphorase
in electron transfer, **19:**201
inhibition of, **13:**332, **22:**380; 381
liver, and coenzyme Q, **25:**315
in mitochondrial fragments, **21:**100
in plastids, **20:**219
isolation and composition, **20:**168; 169; 188; 192
in respiratory chain, **20:**154; 159; 160; 168–172; 185–187
in respiratory chain, **17:**67; 69; 81; 82; 376–377; 380
relation to DPNH-cytochrome reductase, **17:**258–263; 284
xanthine oxidase as, **17:**296; 299
in tobacco, **6:**372
Diaphorase activity, PN-dependent, **26:**132
Diaphorase-Fe-riboflavin complexes with anions and chealators, **17:**276
Diaphorases, **15:**89; 97–98
classification, **14:**255
separation by chromatography, **14:**335; 350
location in subcellular fraction, **24:**304; 305; 337; 343
Diaphorse, and ATP synthesis, **23:**333; 352
Diaphragm, Na^+-glucose symporter in, **29:**61
Diaphragm muscle, effect of myosin ATPase, **13:**167
Diarydisulfide, photolysis of, **27:**206
Diarylsulfide, **58:**54
Diastase, **7:**590, *See* α-Amylase
secretion in progressive muscular dystrophy, **6:**222
Diastase (amylase), **34:**448
Diastase in tobacco, **10:**423
Diastereoisomerides, **7:**67–110
interrupted formation, **7:**80; 83
in resolution of racemates, **7:**82–83
Diastereomer, **59:**138
Diathesis in chick, and vitamin E, **6:**485–486
Diatomic molecules, data for, **23:**5
Diatoms
photosynthesis in, **11:**142–147
polysaccharides of, **28:**334
Diaveridine, dihydrofolic reductase inhibitor, **27:**440
synergism with sulfonamides, **27:**455
Diazoacetyl-D, L-norleucine methyl ester, **41:**185
O-Diazoacetyl-L-serine, **59:**152

Diazo arsanilic acid, coupled with serum proteins as antigen, **15:**64–65
Diazobenzene sulfonic acid, **54:**139
Diazomethane, alkylation of nucleic acids, **27:**320–322; 330
Diazonium cation, **58:**15
Diazonium coupling, in identification of imidazole, **20:**68; 72; 75; 77; 90; 99
  with histidine and tyrosine, **20:**59; 68
6-Diazo-5-oxonocleucine (DON), inactivation of anthranilate synthetase, **38:**16
  inhibition of anthranilate synthetase, **38:**16
6-Diazo-5-oxo-L-norleucine, **54:**101, **59:**152
6-Diazo-5-oxo-L-norleucine (DON), **53:**220
5-Diazo-oxo-norleucine (DON), asparagine synthetase catalysis, **72:**153–157
Diazosulfanilic acid, **41:**66
Diazotization, **60:**260
Dibasic acids, kinetics of dissociation, **9:**1; 4–6
  titration curves, **9:**8–9
1,2,5,6-Dibenzanthracene, binding to protein, **27:**527
Dibenzylphosphoric acid and derivatives, **19:**249; 250
Dibromopropionylcystine and dehydropeptidases, **8:**146
3,6-Dibromosulfophthalein, **46:**392; 393
Dibucaine number, use in cholinesterase assay, **22:**15
Di-*n*-butanedisulfide, in disulfide exchanges, **27:**201
Di-*t*-butyldisulfide, exchanges, **27:**204; 209
Dibutyryl-cyclic adenosine 3′,5′-phosphate (cAMP), tyrosine hydroxylase regulation, **70:**129
3,5-Dicarbethoxy-1,4-dihydrocollidine, **46:**106; 122; 124; 127; 129; 130; 137; 147; 154
Dicarboxyethyldisulfide, exchanges, **27:**204
Dicarboxylic acid cycle
  in bacteria, **18:**246–247; 249
  and glycolate metabolism, **23:**437–440; 445–446; 462
Dicarboxylic acid metabolism, **17:**183; 423–426
Dicarboxylic acid oxidation, **1:**142
Dicarboxylic acids. *See also* individual acids; individual acids and Dibasic acids
  from acetate, **12:**117
  CO_2 fixation in, **12:**46; 47; 49–50; 50–86
  and fat metabolism, **8:**392–409; 412–414
    from fat catabolism, **8:**374–380
    and β oxidation, **8:**349
  hydrolysis of thiol esters, **26:**350
  in mold metabolism, **9:**553; 564; 576–577
  in plant metabolism, **9:**394–395; 398–401; 402–409; 410–411; 419; 421–430
  and propionic acid metabolism, **26:**284–286
  synthesis, **15:**144
  synthesis by CO_2 fixation, **26:**293–294; 296
  in transaminations, **16:**199; 203; 225–226
Dicarboxylic acid shuttle, pyruvate carboxylase, **35:**398
Dicarboxylic amino acid β-lyase
  active site, structure of, **74:**164; 165
  tyrosine phenol-lyase conversion to, **74:** 163–167
Dicarboxylic amino acids, **49:**49; 50; 82–83. *See also* Aspartic acid and Glutamic acid
  in actin, **13:**152
  transport by neutral systems, **49:**50
    distinction from anionic systems, **49:**50
Dicarboxylic Amino Acids, Biosynthesis in Plants (Kretovich), **20:**319
Dichapetalum cymosum, Toxicity (Peters), **18:**113
Dichapetalum cymosum, toxicity of active constituent of, **18:**113–159
Dichapetalum toxicarium, toxic component of, **18:**143–144
Dichlone, compatibility with antibiotics, **20:**386; 396
Dichloroacetyl chloride, ratational isomers of, **23:**17
Dichloroacetylene, cysteine *S*-conjugate β-lyase reactions, **72:**212
2,6-Dichlorobenzenoneindo-3′-chlorophenol, **17:**366; 373
2,6-Dichloro-1,4-benzoquinone, as electron acceptor, **16:**134
Di-(chloroethyl) ethylamine, alkylation of nucleic acids, **27:**321
Di-(2-chloroethyl) methylamine, alkylation of nucleic acids, **27:**321
β, β′-Dichloroethyl sulfide, effect on pectic substances, **20:**348
Di-(chloroethyl) sulfide, mutagenic agent, **27:**324
2,6-Dichloroindophenol, **57:**140
Dichloromethyl urea, photosynthesis study, **25:**65; 66; 75; 86

2,3-Dichloronaphthoquinone, inhibitory action on yeast, **6:**135
1,2-Dichloro-4-nitrobenzene, **46:**395; 396; 397; 406, **69:**2
2,4-Dichlorophenol
effect on liver nonspecific hydroxylase, **19:**176; 199
effect on microbiological formation of hydrocortisone, **20:**270
3-(3,4-Dichlorophenol)-1,1-dimethylurea (DCMU), photooxidation of Mn^{2+} inhibited by, **30:**25
2,6-Dichlorophenolindophenol, **50:**333, **58:**210
as artificial electron donor, **30:**23
as electron acceptor, **17:**242; 246; 251; 253; 254; 257; 258; 282; 299–301
and *p*-hydroxyphenylpyruvate oxidase, **19:**184–186
as indicator in chloroplast system, **12:**10
reaction with thiols, **11:**218; 222; 226; 240
2,6-Dichlorophenol-indophenol, **39:**333
Dichlorophenolindophenol, and succinic dehydrogenase, **18:**68; 70; 71; 84–87; 108
2,6-6 Dichlorophenol-indophenol, assay for pyridinoprotein dehydrogenases, **15:**98
assay for pyruvic oxidase, **15:**212
2,6-Dichlorophenol indophenol, velocity constant for PN reaction, **26:**155
Dichlorophenolindophenol reductase activity, location of, **24:**304–305
2,4-Dichlorophenoxyacetic acid, effect on pectic substances, **20:**370
5-(3,4-Dichlorophenyl)-pyrimidines, activity against plasmodia, **27:**444
Dichlorophosphoric acid and derivatives, **19:**249; 250
Dichloro-*s*-triazinyl dye, **34:**449
S-(1,2-Dichlorovinyl)-L-cysteine (DCVC), cysteine *S*-conjugate β-lyase reactions, **72:**204–207
glutamine transaminase K reactions, **72:** 219–222
high M_r enzyme, **72:**223–224
kynureninase β-elimination, **72:**216–219
prodrug targeting, **72:**228–230
toxic effects, **72:**208–212
S-(1,2-Dichlorovinyl)-L-homocysteine (DCVHC), cysteine *S*-conjugate β-lyase reactions, prodrug targeting, **72:**228–230
Dichroism, **42:**107; 110
circular, **39:**349
Dicoumarol, **7:**471; 486; 489; 490; 492, **46:**2; 7; 9
and blood clotting, **10:**24; 27
and prothrombin concentration, **10:**19–20
effect on oxidative phosphorylation, **20:**182; 183
effect on prothrombin activation, **16:**58; 60–61; 63; 64; 66; 83
quinone toxicity enhancement, **65:**117
in respiratory chain studies, **17:**99; 110; 117–119
Dictyostelium, adenylosuccinate synthetase (AMPSase) gene structure, **73:**91
Dictyostelium discoideum, **59:**34–35, **71:**83; 84; 215
Dicumarol
addition to cyanide-treated mitochondria, **29:**353
effect on mitochondrial oxidative phosphorylation, **21:**32
Dicyanide, mechanism of catalytic conversion to oxamide, **14:**169–170; 172–173
Dicyclohexylcarbodiimide, **49:**226; 227; 268
in peptide bond formation of ACTH, **29:**427
in peptide synthesis, **29:**405; 412
Dicyclohexylcarbodiimide (DCCD), cytochrome *c* oxidase, **71:**103; 105
Dicyclohexylcarbodiimide-inhibited particles, **49:**254
a,a-Dicyclopropylglycine, **49:**62; 86
3,6-Dideoxy-D-galactose (abequose), in lipopolysaccharide from *Salmonella,* 79, **31:**92
3,6-Dideoxy-D-glucose (paratose), in lipopolysaccharide from *Salmonella,* **31:**79
Dideoxyhexoses, in *Salmonellae,* **26:**47–48
3,6-Dideoxy-D-mannose (tyvelose), in lipopolysaccharide from *Salmonella,* **31:**79
Dideoxynucleotide chain termination method, **67:**446–448
Di-DNP-cystine, exchanges, **27:**204; 206; 231
Didymella lycopersici, enzymic transformations of steroids by, **20:**262; 265
Didymella vodakii, enzymic transformations of steroids by, **20:**248; 249
Didymohelix, metabolism in, **74:**6

Dieldrin, effect on carbonic anhydrase, **47:**242
Dielectric breakdown, **34:**467
constant, **34:**446; 456; 486; 492–494; 527
Dielectric constant, **39:**345
effect on papain, **19:**317; 320–322; 326
Dielectric constants, **51:**37–38
Diels-Alder reaction, **43:**278, **58:**16
in water, **58:**17
Diepoxybutane, effect on pectic substances, **20:**348
Diesterase, and phosphorylase action, **24:**286
Diet
effect on thiosulfate metabolism, **7:**399
in hypertension, **7:**544–545
relation to xanthine oxidase levels, **17:** 304–306; 313; 318–322
Dietary regulation, phosphoenolpyruvate carboxykinase (GTP), **69:**249–252
Dietary state of animal and liver phosphorus turnover, **7:**120–125
Diethylacetothetin, **22:**272
Diethyl-*N*-acetyl-β-glutamate, cleavage of, by chymotrypsin, **31:**196
Diethylamine, reaction with quinones, **16:**114
Diethyldisulfide, exchanges, **27:**209
Diethyl dithiocarbamate, **58:**77
Diethyldithiocarbamate, effect on oxidases, **19:**91; 124; 126; 127; 185; 199
Diethyldithiocarbamate (DDC), effect on plant oxidases, **18:**41–42
Diethyl ester, **34:**491
Diethyl fluorooxalacetate, **18:**116
a,a-Diethylglycine, **49:**62
Diethyl-*p*-nitrophenylphosphate, **48:**360. *See also* E600
Diethyl-*p*-nitrophenyl phosphate, inhibition of lipase by, **23:**148–149
Diethylpyrocarbonate, **59:**126
Diethyl pyrocarbonate, **65:**202
Diethylstilbesterol, in allosteric regulation, **28:**140; 141
Diethylstilbestrol, **44:**272
effect on melanin formation, **14:**90
effect on phosphatide turnover, **7:**160–161
Diethyltryptamine, increase in brain serotonin level produced by, **29:**518–519
in schizophrenia, **29:**534
Difference absorption spectra, **42:**93; 111; 112
Difference spectra, *See* Spectra, difference
Differential inactivation, by group specific reagents, **37:**460–468
by proteolytic enzymes, **37:**473–477
quantitative studies, **37:**538–555
by thiol reagents, **37:**446–458
urea, **37:**496–498
Differential microbial assay, true and total biotin, **38:**321
Differential modification, by group specific reagents, **37:**460–468
by proteolytic enzymes, **37:**473–477
quantitative studies, **37:**538–555
by thiol reagent, **37:**446–458
Differential scanning calorimetry, **34:**469
cholesterol ester hydrolases, **69:**58
Differential scanning calorimetry (DSC), **51:**7
Differentiation, from phagocytosis, **41:**112–113
Differentiation in developing egg and enzyme localization, **13:**15
Differentiation of cells and chromidial mutations, **8:**60–62
Differentiation processes, at intracellular, level, **16:**169–170
Diffusion, **34:**496; 503; 509–512; 519–523, **42:**148, **52:**189; 218; 219; 220; 225; 226; 240. *See also* Permeability; Permeation
bicarbonate, **34:**512
carbon monoxide, facilitated by hemoglobin or myoglobin, **29:**36
facilitated, **29:**34
of amino acid analogs affected by [Na^+], **29:**62
definition of, **29:**36
facilitation by catalytic carriers, **29:**33–39
of oxygen, **29:**34
theory of translational facilitated, **29:**35–36
into cells, **7:**9–13
CO_2, **34:**512
coefficients, **34:**496–504
entropy changes in, **11:**19–23
rates, at interfaces, **11:**67–69
exchange, **13:**29–33
free, **13:**29–33
Fick equation for, **7:**9–10
layer, **34:**504
on oxygen, influence on respiration, **7:**8
and reaction rate, **25:**9–13; 15; 18; 23; 25–26; 28; 35
relation to location of enzyme in nucleus, **16:**418

Diffusional constraints, **52:**216; 217–222
Diffusion and energy transport in living cells, **14:**7; 18–19; 40
Diffusion coefficient, of cytochrome *c*, **17:**353
Diffusion-controlled bimolecular reaction rate, **43:**162; 173
Diffusion distances of radiation products in biological systems, **24:**384; 387; 400–402
Diffusion of RNase, **24:**169
Diffusion rates, effect of nucleate complexes, **17:**140
Diffusion studies, of papain, **19:**227
Diflubenzuron, **59:**82
Difluoroalanine, **55:**231
Difluoroalcohol, **55:**267; 268
Difluoroandrost-4-ene-3, 19–19dione, **55:**199
Difluoroaspartate, **55:**234
1,5-Difluoro-2,4-dinitrobenzene, **34:**451
Difluoromethyl ornithine, **55:**199
Difluoromethylornithine, **56:**274
Difluorooxalacetate, **55:**234
2,2′-Difluorosuccinate, **58:**156
Difluoro vitamin D_3, **55:**276–277
N^5,N^{10}-Diformyltetrahydrofolic acid, synthesis and properties, **21:**385
Digalactosylceramide, **52:**62, **60:**97
Digalactosyldiglyceride, **52:**64
  in chloroplast lamellae, **30:**9
m-Digallic acid, from tea leaves, **16:**151
Digestibility, of proteins, and denaturation, **23:**38–39; 70–71
Digestion, **42:**132; 146
  amphibia, **1:**402–410
  effect of pectin on, **20:**357
  fish, **1:**385–402
  low vertebrata, **1:**371–417
  rate of, **33:**262
  reptiles, **1:**410–412
  vertebrata and invertebrata, **1:**371–374
Digestion of food, nonintestinal, **45:**401
Digestive enzymes, insect, **45:**394
Digestive proteases, *See* Proteases, digestive
Digitonin, **55:**164; 165
  action of, in rat liver mitochondria, **29:** 294–295
  Hill reaction inhibited by, **30:**23
  quantum requirements of ferrocytochrome *c* photooxidation and reduction by particles treated with, **30:**60
  treatment of chloroplasts with, **30:**55
Digitonin preparations, of respiratory chain, **28:**164
Digitoxigenin, microbiological conversion, **20:**247
Digitoxin, effect on heart muscle protein, **13:**182
Diglucose disaccharide, synthesis, **15:**314–315
3,6-Di-*O*-β-glucosyl-D-glucose, from laminarin, **28:**332; 350
Diglyceridase, **23:**153
Diglycinamide, synthesis, **19:**263
Diglycylaminopropionic acid, and pyruvic acid, **8:**153–158
  hydrolysis, **8:**153–158
  substrate for enzymes, **8:**127
Diglycylcystine, ionization constants, **20:**58
Diglycylimide, intramolecular aminolysis of, **19:**263
1,2-Diguaiacylpropane-1,3-diol structure, **34:**211
Diguanidines, **15:**379
Diguanylspermidine, **36:**227
Dihedral angle, **59:**148
Dihedral angles (phi, psi), **47:**48; 58–63; 64; 71; 83; 87; 121; 133
12.20-diHETE, cholesterol ester hydrolases regulation, **69:**71–72
Dihydroalliin, **11:**385; 389
5,6-Dihydro-β-(2-aminopyrimid-4-yl) alanine, **50:**154
7,8-Dihydrobiopterin, complex with active site of tyrosine hydroxylase, **74:**244–245; 245
Dihydrobioterin, **35:**261
Dihydrocarvone epimerization, **44:**96
Dihydrocozymase, **7:**281
Dihydrodiols, formation by liver nonspecific hydroxylase, **19:**177–179
Dihydrodipicolinate reductase
  chemical mechanism, **72:**293–294
  lysine biosynthesis, **72:**290–297
  2,6-pyridine dicarboxylate interaction, **72:**297
  three-dimensional structure, **72:**293; 295
Dihydrodipicolinate synthase (DHDPS), **50:**169
  bacterial lysine biosynthesis, **72:**285; 287–291
  chemical mechanism, **72:**287–288
  three-dimensional structure, **72:**289–290
1,5-Dihydroepilubimin, **55:**52

Dihydrofolate, **54:**58; 70
biogenesis, **27:**424
enzymes reducing, **27:**434–436
isomeric, **27:**433–434
Dihydrofolate dehydrogenase, **37:**463
localization of, **24:**302
Dihydrofolate reductase, **27:**433–455, **35:**262; 264, **36:**71, **43:**331, **54:**82, **67:**83–84; 92–93
cofactors, **27:**434–436
generally, **70:**24–25 (*See also* Eukaryotic dihydrofolate reductase)
inhibition of, **27:**458–460
inhibitors, **27:**436–453
and chemotherapy, **27:**419–420
small molecules, **27:**437–450
structural analogs, **27:**436–437
synergism with sulfonamides, **27:**455–456
isolated, inhibition, of, **27:**450–455
in methylation, **25:**397; 400; 401
reaction, **27:**425; 426
resistance to inhibitors of, **27:**454–455
species differences, **27:**450–453
substrate specificity, **27:**434–436
and transport mechanisms, **27:**457–458
Dihydrofolic acid, **53:**353–355
enzymic synthesis of, **35:**37–48
stereospecific reduction, **38:**240
synthesis and structure, **21:**378; 379; 395
Dihydrofolic reductase, **21:**395; 423, **37:**503
α, β-Dihydrogenation, in metabolism of leucine, **16:**199
Dihydrogeranylgeraniol, **46:**75
Dihydrogossypetin, **55:**48
1,2-Dihydro-2-hydroxy-1-naphthol, identification of its glucuronide, **16:**383
Dihydrohydroxywyerol, **55:**90
Dihydrokaempferol, **55:**34
Dihydro-α-lipic acid, as electron donor, **18:** 342–343
and pyruvate oxidation, **18:**335–338
Dihydrolipoamide dehydrogenase, and DR metabolism, **28:**443
Dihydrolipoate, **39:**352–354; 362; 363
Dihydrolipoate-glyoxylate adduct, **57:** 127–128
Dihydrolipoic acid dehydrogenase, PN-dependent, **26:**119; 126
Dihydrolipoic dehydrogenase, **18:**337–338, **23:**334
Dihydrolipoic transacetylase, **18:**336–337
Dihydrolipolyl transacetylase, electron microscopy of, **33:**105
Dihydrolubimin, **55:**52; 69
Dihydromethylnicotinamide, autoxidation, **15:**41
Dihydroneoptein aldolase, **53:**359
Dihydroneopterin aldolase, K_m for, **35:**63
molecular weight of, **35:**62
properties of, **35:**61–66
purification of, **35:**61–66
Dihydroneopterin triphosphate, **53:**357
Dihydroneopterin triphosphate pyrophosphohydrolase, **53:**359
Dihydrooratate, **55:**271
Dihydroorotase, **21:**274, **74:**72
Dihydroorotic acid, **15:**357, **43:**536; 541
from orotic acid, **26:**127
in pyrimidine biosynthesis, **21:**264; 274–277
in pyrimidine metabolism, **28:**135
Dihydroorotic acid dehydrogenase, PN-dependent, **26:**119
Dihydroorotic dehydrogenase, **21:**274–277
Dihydro orotic dehydrogenase, **41:**40
Dihydroorotic dehydrogenase, EPR spectrum, **27:**117; 136–137; 172–175; 180; 183
nonheme iron, **27:**89
Dihydroortase, **56:**293
Dihydrophenanthrene phytoalexins, **55:**9
biosynthesis of, **55:**95; 96
Dihydroporophyrin, **46:**63
Dihydroporphins, absorption spectra, **7:**345–348
Dihydroporphyrin, **23:**269; 288
Dihydropteridine reductase, **35:**260, **65:**302; 305, **67:**87–88; 231–243
amino acid composition, **67:**242–243
assay, **67:**232–233
inhibitors, **67:**237–238
mechanism, **67:**238–239
molecular biology, **67:**242–243
molecular properties, **67:**231–232
regulation *in vivo,* **67:**238–242
aminopterin, **67:**240
hormones, **67:**241–242
substrate specificity, **67:**233–237
kinetic data, **67:**234
NADH *versus* NADPH, **67:**233–235
pyrimidines as substrates, **67:**235–236
Dihydropteridine reductase, NADP-dependent, **26:**125

Dihydropteridine reductase (DHPR), tyrosinase/ tyrosine hydroxylation contrasted, **70:**106
7,8-Dihydropterin, **74:**238
structure of, **74:**237
Dihydropterin, quinonoid tautomers of, **74:**238
structure of, **74:**237
Dihydropteroate, antagonists to, **27:**460
in *E. coli,* **27:**422–424; 426
model, **27:**447; 448
Dihydropteroate-glutamate coupling enzyme, **27:**430–431
Dihydropteroate synthetase, **35:**40–44
characteristics of (table), **35:**42
K_m for *p*-AB, **35:**41
molecular weight of, **35:**41
pH optimum, **35:**41
purification of (table), **35:**41
Dihydropteroic acid, **53:**353–355
Dihydropteroyl glutamic acid, *See* Folate-H_2
Dihydropyridines, oxidation of, one-electron, **26:**143–144
two-electron, **26:**144–156
Dihydropyrimidinase, localization of, **24:**328
Dihydropyrimidine aminohydrolase, **56:**293
Dihydroquercetin, **55:**34
Dihydroquinic acid, in *p*-AB biosynthesis, **27:**421
Dihydroquinoline, thyminyl derivative, **38:**255
Dihydrosphingosine, biosynthesis of, **35:**95; 96
Dihydrostreptomycin
activity against *Mycobacterium tuberculosis,* **10:**160
synthesis of gramicidin S, inhibited by, **33:**340–342
Dihydrothiazine, cyclic, formation, **16:**122
1,2-Dihydro-*N*-thyminylquinolone, **54:**77
Dihydrotriazines, inhibitor, species differences, **27:**450–451
synergism with sulfonamides, **27:**455
Dihydrouracil, **35:**473–474
loop, **35:**466; 468
in pyrimidine biosynthesis, **21:**282–284
5,6-Dihydrouracil, **50:**155
Dihydro-uracil dehydrogenase, localiazation of, **24:**299
Dihydrouridine, **54:**89
Dihydrouridylic acid, in pyrimidine biosynthesis, **21:**284
Dihydrowyerone, **55:**85
Dihydrowyerone acid, **55:**85
N-(2,3-Dihydroxpropyl) glycocyamidine, **54:**394
Dihydroxyacetone, **15:**116; 120
in photosynthesis, **25:**95; 101; 102; 104
Dihydroxyacetone-P, **43:**493
Dihydroxyacetone-P fluorinated analogs, **55:**242–244
Dihydroxyacetone phosphate, **15:**114; 163; 164; 169–170; 171
oxidation, **19:**118–120
in photosynthesis, **19:**24–26
in pentose synthesis, **28:**396
in photosynthesis, **25:**101; 102; 104
Dihydroxyacetone phosphate-3-glyceraldehyde phosphate, **14:**195
Dihydroxyacetone zymase, as adaptive enzyme, **6:**78–79
2,3-Dihydroxyanisole, oxidation, **16:**137
3,4-Dihydroxyanthranilic acid, **15:**100
3,4-Dihydroxybenzoate, collagen hydroxylase inhibitor mechanisms, **72:**361–364
3,4-Dihydroxybenzoic acid, **17:**414
role in cuticle hardening, **16:**154–155
o-Dihydroxybenzonid substances, relation to phenolase complex, **16:**140–146
Dihydroxycinnamyl alcohol
in lignin formation, **18:**372
transformation to lignin-like material, **16:**148
5β-Dihydroxy-coprostane, **46:**156
6β,17α-Dihydroxycortexone, proof of structure, **20:**276
17α, 19Dihydroxycortexone, proof of structure, **20:**276
Dihydroxyeicosatetraenoic acid, stereochemistry, **58:**295
Dihydroxyeicosatrienoic acid, **58:**297
Dihydroxyfumarate, EPR study, **27:**124
Dihydroxyfumaric acid, and peroxidase, hydroxylating system, **19:**143; 174; 199
Dihydroxyfumaric acid oxidase, **19:**117–121; 127
2,3-Dihydroxygermacene, **55:**54; 73
Dihydroxyiminoacetone, in reactivation of phosphorylated esterases, **20:**307; 308
5,6-Dihydroxyindole, from dopachrome, **16:**115–116
in melanin formation, **16:**164–168
5,6-Dihydroxyindole and tyrosine metaboilsm, **14:**76–77; 89

5,6-Dihydroxyindole-2-carboxylic acid, in melanin study, **16:**165
Dihydroxyisoleucine, *See* α, β′-Dihydroxy-β-methylvaleric acid
α, β-Dihydroxyisovaleric acid (DHI), **17:**400–402
α,β-Dihydroxyisovaleric acid (dihydroxyvaline), in biosynthesis of valine, **16:**279–281
Dihydroxylation, microbial, of aromatic ring, **27:**476; 506–508; 516; 523–525; 538
  of olefins, **27:**501
Dihydroxymaleic acid
  action of peroxidase on, **12:**175; 181–182
  peroxidase action on, **14:**151
  in ribose synthesis, **28:**397
4,5-Dihydroxymethyl-2-methylpyridine, substrate for pyridoxal kinase, **16:**219
α, β-Dihydroxy-β-methylvaleric acid (DHV), **17:**400–402
α,β-Dihydroxy-β-methylvaleric acid (dihydroxyisoleucine), in biosynthesis of isoleucine, **16:**279–281; 283
2,5-Dihydroxyphenylacetic acid, *See* Homogentisic acid
3,4-Dihydroxyphenylacetic acid, action of phenolase on, **16:**110
  role in cuticle hardening, **16:**154–155; 161
Dihydroxyphenylacetic acid, metabolism in plants, **16:**144
L-3,4-Dihydroxyphenylalanine, **50:**157
3,4-Dihydroxyphenylalanine, *See* Dopa
  in per-oxidase-catalyzed reaction, **19:**143
  as precursor of melanin, **16:**163–169; 172
    metabolism in plants, **16:**132; 133; 141; 144; 145; 150
    role in hardening of cuticle, **16:**154–156; 158; 160
    in transaminations, **16:**196
Dihydroxyphenyl-L-alanine, phenolase complex specificity for, **19:**131
D- and L-3,4-4 Dihydroxyphenylalanine, oxidation by model enzyme, **15:**27
Dihydroxyphenylalanine, pressor effect, **7:**536
L-Dihydroxyphenylalanine decarboxyase, classification, **14:**256
3,4-Dihydroxyphenylalanine (DOPA), in the depression hypothesis, **29:**492–493
  effect on reserpinized mice, **29:**487; 488
  function of neurotransmitter on, **29:**483
  sedation with, **29:**489
3,4-Dihydroxyphenyl-DL-(carboxy-C-14)-alanine, in melanin studies, **16:**166–167
o-Dihydroxyphenyl compound, role in sclerotization of cuticle, **16:**154–156
Dihydroxyphenyl compounds. *See also* specific compounds
  and homogentisic acid formation, **14:**102–104
  in hydroxyphenyluria, **14:**111
2-(3,4-Dihydroxyphenyl)-ethylamine, formation of melanin, **16:**167
3,4-Dihydroxyphenyllactic acid, role in cuticle hardening, **16:**154–155; 161
Dihydroxyphenylpyruvase in tyrosine metabolism, **14:**106
2,5-Dihydroxyphenylpyruvate, metabolic path, **15:**184
3,4-Dihydroxyphenylpyruvic acid, in transaminations, **16:**196
2,5-Dihydroxyphenylpyruvic acid, in tyrosine metabolism, **14:**98; 106; 110; 115
Dihydroxyphenylpyruvic decarboxylase, in tyrosine metabolism, **14:**98; 106
3,4-Dihydroxyphenylserine, effect of, on reserpinized mice, **29:**488
3,4-Dihydroxyphenylserine in noradrenaline formation, **14:**117–118
6β-11α-Dihydroxyprogesterone, microbiological formation, **20:**245
11α-17α-Dihydroxyprogesterone, microbiological formation, **20:**245
Dihydroxyproline, in diatoms, **37:**278
ε-*N*-β-(1,3-Dihydroxypropyl) lysine, structure of, **31:**135
3,4-Dihydroxypyridine, from mimosine, **29:**109
  toxicity to mung bean seedlings, **29:**109
4,4′-Dihydroxystilbene, effect on succinic oxidase, **20:**176; 177
Dihydroxytartrate, peroxidase action on, **19:**125
Dihydroxytartrate oxidase, **19:**123; 125
2,5-Dihydroxyterephthalic acid, as inhibitor of β-glucuronidase, **16:**374
Dihydroxyvaline, *See* α, β-Dihydroxy-isovaleric acid
3,5-Diimino-1,2,4-dithiazoline, in disulfide exchanges, **27:**202
3,5-Diiodo-4-hydroxyphenylpyruvic acid, in transaminations, **16:**196
Diiodotyrosine
  thermodynamic data on ionizations, **20:**64
  in thyroxine formation, **14:**112–115

peroxidase action on, **14:**151; 152
3,5-Diiodotyrosine, in transamination reactions, **16:**192; 196; 237
Diiodotyrosine formation, effect of carbon monoxide, **7:**206
in tissue slices, **7:**205–208
Diiodotyrosine-pepsin, activity, **14:**190
Diiodotyrosine turnover, **7:**199–209
Diisopropoxy [P^{32}] phosphinyl elastase, **23:**231
Diisopropyl fluorophosphate, **17:**48
Diisopropylfluorophosphate, **55:**175, *See* DFP
action on lipase, **23:**148
complex with proteinase, **23:**72–74
(DFP), radioactive, in inhibitor studies, **15:**288
use as quasi-substrate, **22:**52
Diisopropyl fluorophosphate (DFP), and nerve permeability, **13:**61
inhibition of chymotrypsin by, **13:**321–343
Diisopropylfluorophosphate (DFP), effect on esterases, **20:**287; 288; 305; 306
effect on serum complement activity, **20:**89
in identification of imidazole, **20:**78–81
labeled, in identification of active centers of esterases, **20:**284; 288; 290
in study of imidazole in enzymes, **20:**81
reactions with imidazole, in enzymes, **20:**89; 90; 97
Diisopropyl fluorophosphate (P^{32}), active center label, **24:**131
Diisopropyl phosphofluoridate, **41:**181
Diisopropylphosphofluoridate (DFP), effect on elastases, **23:**233; 235
Diisopropylphosphorofluoridate, **33:**386
Diis opropyl phosphorofluoridate, **59:**123
Diisopropyl phosphorofluoridate (DEP), **48:**290
2,5-Diketogluconic acid, **15:**157; 177
α γ-Diketoglutaric acid, in transaminations, **16:**196
3,5-Diketohexanoic acid, lactone of, **15:**155
Diketopiperazine, configuration of, **23:**14–15
Diketopiperazine hypothesis of protein synthesis, **9:**430–431
Diketopiperazines
from amino acid esters, **19:**237
in protein structure studies, **14:**279–280; 292–293
Diketosuccinate, in peroxidase-catalyzed reaction, **19:**118–120
Dilatometric methods in histochemistry, **9:** 605–606
Dimedon, **5:**182; 183; 187; 188
Dimefox, change in body, **24:**308
Dimer, **39:**347; 350; 364
stable, **39:**350
6,8-Dimercaptooctanoic acid, in citric acid cycle, **15:**184; 215–216; 220–221; 230
2,3-Dimercaptopropanol, **15:**219. *See also* Dithiols
basis for antilewisite action of, **11:**218
in electron transport studies, **17:**69; 79; 253; 323; 368–370; 374; 376–377; 384
estimation of, **11:**217
inactivation of phosphorylase by, **11:**304
oxidation of, **11:**207–210
2,3-Dimercaptopropanol(BAL), **39:**342
2,3-Dimercaptopropanol (BAL)
effect on blood clotting, **10:**23
effect on phenolase complex, **19:**199
in studies of papain, **19:**279; 290; 314; 317–319
effect on respiratory chain enzymes, **20:** 171–176–180
2,3-Dimercaptopropanol-labile factor, in respiratory chain, **20:**175–180; 185
Dimeric molecule, **39:**349
Dimerization
mediated
calculations on, include conditions corresponding to sedimentation of proteins, **30:**174
conservation equations for sedimentation of, **30:**172–173
definition of, **30:**171
theoretical sedimentation patterns computed for, **30:**173
theory for kinetically controlled, **30:** 144–145
Dimerization of polypeptide chains, **24:**23
7-(2,6-Dimethoxybenzamido)-cephalosporanic acid, action of β-lactamase, **28:**240; 259
6,6′-Dimethoxygossypol, **55:**80
structure of, **55:**56
3,4-Dimethoxyphenylethylamine (DMPE), dietary origin of, **29:**524
no hallucinogenic potency in, **29:**522–523
in schizophrenia, **29:**500; 524
4,4-Dimethylacetoacetic acid, **58:**182

Dimethylacetothetin, as methyl donor, **22:**240
Dimethylallyl pyrophosphate, **22:**297; 304; 305; 309–311
elimination from, **22:**310
enzymic condensation of, **55:**66
in formation of rubber, **29:**246
from mevalonic acid, **55:**64; 65
structure, **22:**305
Dimethylallyl transferase, in latex, **29:**245
Dimethylaminoazobenzene, **5:**219, **15:**6; 127
metabolism of, **5:**219
4-Dimethylaminoazo benzene, *See* Butter yellow
p-Dimethylaminoazobenzene, as carcinogenic agent, **4:**243; 248; 249
p-Dimethylaminoazobenzene, demethylation of, **12:**461–462
4-Dimethylaminoazobenzene demethylase, ethionine effect on, **25:**380
4-Dimethylaminoazobenzene reductase, ethionine effect on, **25:**380
Dimethylaminoazobenzene reductase, repression of, **29:**380
N,N-Dimethylaminoethylamine, **34:**453
cis-(β-Dimethylamino-L-alanine), **49:**63
1-Dimethylaminonaphthalene 5-sulfonyl chloride, in molecular weight determination of actin, **16:**334; 345
1-Dimethylaminonaphthalene-5-sulfonyl chloride, reaction with chymotrypsin, **20:**286
1-Dimethylaminonapthalene-5-sulfonyl chloride, **54:**400
Dimethylaniline, **60:**72
N^G,N'^G-Dimethyl-L-arginine, **42:**263
N^G,N^G-Dimethyl-L-arginine, **42:**263
5,6-Dimethylbenzimidazole, **46:**53
Dimethylbenzimidazolylcobamide, methylation, **25:**407–408; 409
Dimethylbenzimidazolyl cobamide (DBC) coenzyme, characteristics, **26:**348–349; 360–362
effect on, diol dehydrase, **26:**366
methylmalonyl-CoA mutase, **26:**343–344; 346–352
4,5 Dimethyl-*o*-benzoquinone, lack of reactivity with indoles, **16:**111
1,3-Dimethyl-5-benzylleukolumiflavin, semiquinone cation of, **27:**185–186
Dimethyl biacridinium nitrate (DBA) chemiluminescence, **25:**121; 125; 126; 148
4,5-Dimethylcatechol, quinones from, **16:**122
3,4-Dimethylcatechol, use in plant oxidase studies, **16:**132
Dimethylcyanamide, in cytidine-5′-pyrophosphate synthesis, **22:**187
in nucleotide coenzymes synthesis, **22:**182; 183
5′-*S*-Dimethyl-5′ deoxyadenosine, **54:**223
Dimethyldichlorosilane, effect on pectic substances, **20:**346
Dimethyldiselenide, accumulation of, by *Astragalus* and *Stanleya,* **48:**5
Dimethyldisulfide, exchange, **27:**209
Dimethyl disulfide in mushrooms, **10:**240; 257
4,6-Dimethylglucose, oxidation of, **15:**148
Dimethylglycine as methylating agent, **12:**464; 466
6,7-Dimethylletrahydropterin, **35:**249
Dimethylluciferin, **44:**43
5–5′-Dimethylluciferin, **44:**44
ε-*N*-Dimethyllysine, ratio to ε-N-monomethyllysine, **42:**256
N-Dimethyl-β-mercaptoethylamine, in synthesis of activated amino acids, **19:**239
Dimethylnapthoquinone, **17:**118–119
Dimethylnitrosoamine, **42:**236; 256
Dimethyl-*p*-phenylenediamine, oxidation of, in patients with mental diseases, **29:**506
Dimethyl-β-propiothetin, biosynthesis, **22:**245; 246
discovery, **22:**242
as methyldonor, **22:**240
stability to alkali, acid, enzyme action, **22:**248
6,7-Dimethyl-8-ribityllumazine, **53:**351–352
Dimethylselenide, accumulation of, by *Astragalus* and *Stanleya,* **48:**5
Dimethyl suberimidate, **39:**202, **50:**335
Dimethylsuberimidate, **58:**147
Dimethylsulfate, alkylation of nucleic acids, **27:**320–322
Dimethyl sulfide, biological formation of 441–446, **12:**454–455
Dimethylsulfone, in splitting of sulfonium compounds, **22:**261
Dimethyl sulfoxide, **35:**492
Dimethylthetin-homocysteine methyltransferase, **25:**424–426
Dimethylthetin methyltransferase, **25:**424–426
Dimethylthetin r homocysteine transmethylase, classification, **14:**250

Dimethylthetins as methyl donors, **25:**423–426
2,4-Dimethylthiazolium salt, **15:**399
N,N'-Dimethyltryptamine, hallucinogenic effects prevented by, **29:**518
increases in brain serotonin level produced by, **29:**518–519
in rabbit lung, **29:**527
γ,γ-Dimethylvaleramide, **54:**13
Dinactin, properties of, **29:**289
2,5-Dinicotinylornithine, relation to nicotinamide, **15:**109–111
2,4-Dinitrochlorobenzene, **43:**380
2,4-Dinitrofluorobenzene, **43:**380, **54:**400
Dinitrofluorobenzene (DNFB), in identification of imidazole, **20:**68; 72; 75; 77; 78; 81; 90
reaction with imidazole, **20:**61; 98; 286
Dinitrofluorobenzene method, with cytochrome *c,* **17:**349
Dinitrogenase, **52:**5–13
reductase, **52:**5–13
saturation curve for, **52:**16
Dinitrogen fixation
in blue-green algae, **57:**264–267
in soybean nodules, **57:**267–268
Dinitrophenol, **15:**141; 166–167; 225, **46:**138
effect on chromosome breaks, **17:**455
effect on mitochondrial enzymic functions, **21:**31; 32; 37; 46
effect on valine incorporation into cytochrome *c,* **21:**45
effect on translocation of antibiotics in plants, **20:**387
as inhibitor of *o*-amino-phenylglucuronide formation, **16:**403
mode of action, **23:**364–365
in study of virus protein, **15:**463–464
2,4-Dinitrophenol, **17:**241, **49:**251
action of, in rat liver mitochondria, **29:** 294–295
addition to mitochondria, **29:**280
amino acid uptake in tumors inhibited by, **29:**365
catalyzed equilibration of the electrochemical activity of protons across natural membranes, **29:**65
citrate accumulation in mitochondria inhibited by, **29:**312
mechanism for uncoupling oxidative phosphorylation by, **29:**299
net uptake in K^+ in mitochondria inhibited by, **29:**285
respiration-dependent Ca^{++}-H^+ exchange reversed by, **29:**279
stimulation of glucose oxidation by, **29:**345
effect on photosynthesis, **19:**55; 64
effect on steroid 11β-hydroxylase, **19:**189
effect on synthesis of coenzyme Q, **25:**304; 322
as uncoupling agent in respiratory chain, **17:**99; 108–109; 112–117; 129; 174–176
2-4-Dinitrophenol, in enzymic adaptation studies, **13:**77–79; 82
Dinitrophenol (DNP), **18:**290
2,4-Dinitrophenyl acetate, carbonic anhydrase catalyzed hydrolysis, **47:**206–207; 212–213
Dinitrophenylation of RNase, **24:**207; 223–224; 245
N-Dinitrophenyl-[1–^{14}C] carboxymethyl-Se-cysteine, **48:**23
2,4-Dinitrophenylhydrazine, **58:**176
2,4-Dinitrophenylsulfenyl chloride, disulfide exchanges, **27:**205
Dinophyceae, carotenoids in, **21:**391
Dinucleoside monophosphate, **55:**374
Dinucleoside pyrophosphates, synthesis by phosphoramidate method, **22:**195–197
Dinucleotide, flavin adenine and phosphorylase, **11:**304
Dinucleotide fold, **59:**131
Dinucleotides, in protein synthesis, **17:**464
Dioldehydrase, cobamide coenzyme dependence, **26:**359; 365–366
Diol dehydrases, cobamide-requiring, **28:**447
Diol dehydrogenase, column chromatography, **22:**479; 481
Dioleyl lecithin, **58:**245
Diols, complexes with borate, **12:**495; 496–499; 501–502
1,2-Diols, conversion to aldehyde, role of cobamide coenzyme, **26:**265–266
γ,δ-Dioxovalerate, **46:**42; 84
Dioxygen, cytochrome *c* oxidase, **71:**80–81; 84; 110–114; 122–123; 126; 128; 137; 152; 154; 163; 183; 185
Dioxygenase, **46:**166, **47:**2. *See also* Tryptophan pyrrolase
Dioxygenases, **55:**251–254

Dioxygenation, **58:**193
positional specificity, **58:**202
stereospecificity, **58:**202
Dioxypropyl-fructoside, formation of, **30:**205
1,2-Dipalmitin, **55:**316
Dipalmitoyl-L-α-glycerophosphoryl-DL-serine-1-C^{14}, decarboxylation of, **25:**418
Dipalmityl phosphatidylethanolamine, **55:** 316–317
phospholipase D and, **55:**348
Dipeptidase, **45:**420, **56:**292
separation by chromatography, **14:**350
Dipeptidase in yeast, **9:**418
Dipeptidases, **2:**81–83
and DL-*a*-chloropropionyldehydroalanine, **8:**139; 140; 144
in chromidia, **8:**31
effect on glycyldehydrophenylalanine, **8:**134
in nuclei, **8:**33
susceptible configuration, **8:**166
and synthetic processes, **8:**56
classification, **2:**51
intestinal, **2:**82; 83
localization of, **24:**324
in yeast, **2:**81; 82
Dipeptide hydrolases, localization of, **24:**324
Dipeptides, **38:**204, **47:**102–105; 130
action of leucine aminopeptidase on, **12:**200–204
acylated, action of carboxypeptidase on, **12:**233–235
amides, action of leucine aminopeptidase on, **12:**204–205
inhibition of β-lactamase by, **28:**262
and tripeptide interactions, **47:**130
Z-Dipeptides, as synthetic substrates for proteinases, **41:**205
Dipeptide sequences. *See also* Peptide sequences
in protein chains, **24:**420–425
Dipeptide transport, **35:**198–200
competition phenomena, **35:**221–222
energy requirement, **35:**225–227
growth dependence on, **35:**198
oligopeptide inhibition, **35:**224–225
stereospecificity, **35:**222–224
Dipeptidyl aminopeptidase, **53:**255
Dipeptidyl carboxypeptidase, **57:**12
Dipeptidyl transferase, **53:**255
Diphdropteroic acid, enzymic synthesis of, **35:**37–48
o-Diphenolase activity, estimation, **16:**134
o-Diphenolases, in plants, **16:**132
o-Diphenol dehydrogenase, *See* Catecholase
Diphenols, **44:**195. *See also* Phenolase complex, Phenols, Polyphenols, and Quinones
enzymes acting on, **24:**307
quinones from, **16:**135; 136
relation to oxidation of ascorbic acid, **16:**134
o-Diphenols, and phenolase complex activity, **19:**199
o-Diphenols, dehydrogenation, **16:**106; 130
effect on phenolase, **16:**129; 130
formation in plants, **16:**131; 133; 136
role in sclerotization, **16:**154–156
terminal oxidation during synthesis, **16:**146
Diphenylamine, effect on carotenogenesis, **21:**319–321; 343–346
Diphenyleneiodonium, **65:**80
Diphenylmethane-4,4′-disulfonyl chloride, **58:**30
Diphenylphosphoric acid, in nucleotide synthesis, **22:**165
Diphosphoadenosine, from TPN, **15:**82–83; 106
3′,5′-Diphosphoadenosine (PAP), and light emission, **25:**158–159
1,3-Diphosphoglycerate, **15:**129; 171; 172; 258
and ATP, **23:**324; 328–333; 338
2,3-Diphosphoglycerate
allosteric regulation, **28:**111
binding to hemoglobin after carbamylation, **40:**16
in salt bridges of hemoglobin, **40:**20
Diphosphoglyceric acid, **1:**121
2,3-Diphosphoglyceric acid, **15:**172
classification as coenzyme, **14:**241
role in phosphoglyceric acid mutase action, **14:**213–215
1,3-Diphosphoglyceric acid, in photosynthesis, **19:**10; 24
Diphosphoglyceric acid mutase, **14:**214–215
1,3-Diphosphoglyceric acid r ADP transphosphatases, **14:**251
1,3-Diphosphoglyceric mutase, **15:**172
Diphosphoinostide, **54:**144
5-Diphosphomevalonate, **22:**301; 302
formation, **22:**301
in squalene biosynthesis, **22:**297
structure, **22:**303

5-Diphosphomevalonic anhydrodecarboxylase, in squalene biosynthesis, **22:**297; 304
Diphosphonate kinase, **43:**11
5-Diphospho-3-phosphomevalonate, in squalene biosynthesis, **22:**304
Diphosphopyridine dinucleotide, synthesis by carbodiimide method, **22:**177; 181
Diphosphopyridine nucleotidase, **15:**80; 84–85; 88 See also Coenzyme I
in *Neurospora,* **20:**116
in sperm, **24:**88
Diphosphopyridine nucleotidase transglycosylation reaction, **18:**206–207
Diphosphopyridine nucleotide, **5:**220; 221, **15:**80–98; 103–108; 112–131; 296. *See also* Coenzyme I; DPN; DPN-cytochrome reductase; Pyridine nucleotides
acetylpyridine analogue, binding to TPD, **22:**341; 342
complex with liver LDH, **22:**367
complex with muscle LDH, **22:**366
complex with TPD, **22:**346; 347
complex with yeast ADH, **22:**356; 357
acetylpyridine deamino analogue, complex with TPD, **22:**346; 347
in alternate pathways of glucose and fructose metabolism, **15:**141; 145; 148–150; 154–157; 167; 171; 177–178
analogues, in complexes with TPD, **22:**346
as coenzymes of TPD, **22:**347
and ATP synthesis, **23:**328–335; 338; 344–347; 388
effect on denaturation of dehydrogenases, **23:**46–48; 64; 71–72
extra, **23:**376–377
and glycine, **23:**457; 460
and glycolate, **23:**439; 442–444
and oxalate metabolism, **23:**447; 449; 451–452
reduced, effect on denaturation, **23:**49–50; 64; 71–72
cytochrome *c* reductase, **23:**343–367
in respiratory chain, **23:**344–357; 363; 368–370; 372–373; 375–377; 379–380; 382–383
binding, with GAD, sites, **22:**369
and spectral changes, **22:**373
with TPD *in vivo,* **22:**375
biosynthesis, **15:**104–105
in citric acid cycle, **15:**184; 185; 194–195; 197–212; 216–230; 233–235; 243; 251; 255–259
complex with liver ADH, dissociation constant, **22:**350; 351
complex with TDP, **22:**340
effect of pH on, **22:**344; 345
tryptic digestion, **22:**345; 346
deamino analogue, complex with TPD, **22:**346; 347
effect of various compounds on fermentation of, **5:**220
effect on mitochondrial coenzyme Q, **21:**112
in mitochondria and mitochondrial fragments, **21:**15; 100
in mitochondrial reactions, **21:**24; 25; 33; 48
in purine biosynthesis, **21:**204; 229; 250
in pyrimidine biosynthesis, **21:**264
reduced, in carotenogenesis, **21:**346
in mevalonic acid biosynthesis, **21:**336; 337
in mitochondria, **21:**15; 80; 81; 83
in mitochondrial oxidative phosphorylation, **21:**123; 124
oxidation, in mitochondria and mitochondrial fragments, **21:**19; 20; 29–31; 34; 35; 39; 81–83; 101; 103–106; 111; 115; 117; 119–122
in pyrimidine biosynthesis, **21:**274; 275; 289; 422; 423
in THFA biosynthesis, **21:**395
synthesis, **21:**208
in synthesis of active formate, **21:**411
enzymes using, **15:**114–119*t*
and epimerase activity, **22:**371
evidence for structure, **15:**80–82
α form, **18:**225
formation of, **18:**193
functional groups, in TPD binding, **22:**346; 347
gasometric measurement, **9:**603
inactive form, **18:**206
inhibition of fermentation, **5:**221
inhibition of *p*-phenylenediamine, **6:**42
and lipoic acid, **18:**333–335; 337; 342
in mitochondria, **22:**375
in oxidation-reduction systems, **11:**241–244; 249
protection of TPD, **22:**402

Diphosphopyridine nucleotide (*Continued*)
pyridine-3-aldehyde analogue, complex with liver LDH, **22:**367
complex with muscle LDH, **22:**366
complex with TPD, **22:**346; 347
complex with yeast ADH, **22:**355
and pyrophosphorolysis, **18:**202–203; 206–207
reduced, acetylpyridine analogue, **22:**350; 351; 358; 359; 361; 372; 402
binding in photosynthetic organisms, **22:**373; 374
binding to GAD, **22:**367
binding to α-glycerophosphate dehydrogenase, **22:**370; 371
in cholesterol biosynthesis, **22:**296; 297
complex with heart LDH, **22:**358; 359; 366
complex with liver ADH, **22:**348–354
complex with MDH, **22:**370
complex with muscle LDH, **22:**367
complex with TPD, **22:**341; 346; 347
complex with yeast ADH, **22:**354; 358
and cytochrome reductase, **22:**372; 373
deamino analogue, **22:**372
effect on dissociation of GAD, **22:**368; 369
fluorescence changes in presence of LDH, **22:**363
fluorescence shift with isocitric dehydrogenase, **22:**371
fluorescence spectrum of acetylpyridine analogue, **22:**350; 361
nature of binding to ADH, **22:**353; 354
pyridine-3-aldehyde analogue, absorption in presence of liver ADH, **22:**351
structure, **22:**177
and succinic dehydrogenase, **18:**67–69; 84
sulfide complex with liver LDH, **22:**367
synthesis, by carbodiimide method, **22:**188
by mixed anhydride method, **22:**172; 173
and TPD esterolytic activity, **22:**347; 348
and urea synthesis, **15:**327; 331; 333–334; 338.
Diphosphopyridine nucleotide cytochrome *c* reductase, and fluoropyruvate, **18:**152
in plants, **18:**12; 44; 57
Diphosphopyridine nucleotide cytochrome reductase, in oxidation of succinate, **18:**69; 106
Diphosphopyridine nucleotide dehydrogenase, and succinic dehydrogenase, **18:**69; 105
Diphosphopyridine nucleotide (DPN)
and activity of hydroxysteroid dehydrogenases, **20:**266; 268; 269; 273
in biosynthesis of glutamic acid in plants, **20:**324
cleavege, **20:**102
effect of liver alcohol dehydrogenase on, **20:**43
effect on triose phosphate dehydrogenase on, **20:**43
fluorometric assay, **20:**33
labeled, in studies of uridine diphosphogalactose-4-epimerase action, **20:**115
in liver alcohol dehydrogenase system, **20:**31–48
in oxidation of reduced triphosphopyridine nucleotide, **20:**172
in phosphorylating respiratory chain, **20:**183
reduced (DPNH), in liver alcohol dehydrogenase system, **20:**31–48
labeled, in studies of uridine diphosphogalactose-4-epimerase action, **20:**115
oxidation in respiratory chain, **20:** 154–156; 159–161; 163; 166; 168–172; 184–187
in uridine diphosphogalactose metabolism, **20:**115–117
reduction, by α-ketoglutarate or pyruvate, **20:**151
in plastide, **20:**219; 227
rate in liver mitochondria, **20:**157; 158
in respiratory chain preparations, **20:**153
system, oxidation reduction potential, **20:**38; 44
in uridine diphosphogalactose metabolism, **20:**113; 115–117; 122
in photosynthesis, **19:**10; 11; 14; 16; 18; 23; 32; 34; 37–39; 48; 51–55; 65
reduced (DPNH), in oxygen metabolism, **19:**179–183; 194; 199
in photosynthesis, **19:**2; 9; 16; 23; 24; 29; 30; 34; 39; 56; 65
in transamination reactions, **19:**346–348
Diphosphopyridine nucleotide in tryptophanase action, **10:**61–62
Diphosphopyridine nucleotide oxidase, effect of fluoropyruvate, **18:**152

Diphosphopyridine nucleotides (DPN), and mechanisms of catalysis, **24:**442; 469–475
in sperm, **24:**87–88
Diphosphopyridine nucleotide specific reductase, in *Arum,* **18:**306
Diphosphopyridinoprotein enzymes, **1:**180; 181
Diphosphothiamin, **5:**143; 147; 154. *See also* Cocarboxylase
Diphosphothiamine. *See also* Cocarboxylase, and Thiamine pyrophosphate; Thiamine
in citric acid cycle, **15:**122; 184; 202–204; 207–210; 213; 215; 217–219; 222–223; 229–230
and CO_2 fixation, **12:**55
in formate fixation, **12:**99
and *p*-hydroxyphenylpyruvate oxidase, **19:**184
in plants, **18:**296
and propionyl CoA carboxylase, **19:**41
in synthesis of active formate, **21:**411
Diphosphothiaminoprotein enzymes, **1:**184; 185; 186
Diphthamide, **58:**112
Diphtheria antitoxin, **15:**74–75, **58:**112
Diphtheria toxin, **50:**313
ADP-ribosylation, **58:**105
ADP-ribosyltransferase activity, **58:**110
amino acide sequence, **58:**106
arginine residue, **58:**106
binding sites, **58:**113
binding to cell surface, **58:**113
chemical structure, **58:**106
commercial production, **58:**104
CRM 197, **58:**113
cysteine residue, **58:**106
cytoplasm entrance, **58:**114
disease caused by, **58:**103
disulfide reduction of, **58:**110
DNA nucleotide sequence, **58:**107
fragment A, **58:**101
fragment B, **58:**112
functional regions, **58:**113
hydrophobicity, **58:**113
glycine residue, **58:**110
historical overview, **58:**103
inactivation, **58:**105
interaction with cell membrane, **58:**112
lethal dose, **58:**105
mechanism of entry, **58:**114
mode of action, **58:**101; 105
protein synthesis inhibition, **58:**105
purification, **58:**105
resistant cells, **58:**112
serotonin residue, **58:**107
sulfhydryl group, **58:**111
trypsinolysis, **58:**106
tyrosine nitration, **58:**111
uptake by endocytosis, **58:**114
Diphtheria vaccine, **58:**104
Diphtheroid bacilli, neuramidase of, **20:**144
Diplococcus, folate metabolism, **27:**457
Diplococcus glycinophilus, glycine metabolism in, **23:**457–462
Diplococcus glycinophilus, synthesis of active formaldehyde in, **21:**412
Diplococcus pneumoniae
folate metabolism, **27:**435; 453
genetic studies, **17:**455–457
sulfonamide drug-resistant mutants, **22:**17
Diploid cells, *See* Polyploid cells
Diplotomic mechanism, definition of, **31:**8
Dipotassium phosphate, effect on foliar absorption of streptomycin, **20:**390; 399
Diptera, genetic studies of, **13:**127–128
Diptheria toxin, **46:**249
2,2′-Dipyridine, **35:**282
α′α′-Dipyridyl, **17:**240
Dipyridyl derivatives, as artificial electron acceptors, **30:**23
α, α-Dipyridyl, effect on oxidases, **19:**91; 97; 126–128; 176; 182; 185; 199
Dipyrrole compounds, **7:**338–339
absorption spectra, **7:**358–362
Dipyrroles, tripyrroles, and tetrapyrroles, interrelationships, **7:**330–332
Dipyrroles and acute muscular dystrophy, **7:**338
Direct hydrogen transfer phenomenon (molybdenum centers in enzymes), **51:**143
Directional flow of metabolites, **28:**46–47
Disaccharide, **5:**88; 90
Disaccharide epimerases, **44:**91
Disaccharides, **4:**57; 58. *See also* Carbohydrates; Oligosaccharides
biosynthesis, **28:**110
enzyme synthesis, **10:**123–143
from fructose transfers, **17:**195; 196; 198–205; 221; 224

Disaccharides (*Continued*)
from glucose transfers, **17:**207–209; 212–214; 227
β-linked, **17:**217–218
hydrogen bonding, **65:**234–236
β-Disaccharides, order of hydrolysis by β-glucosidase, **28:**345
Disaccharides, Synthesis with Bacterial Enzymes (Hassid and Duodoroff), **10:**123
Disaccharide units, in polysaccharides, **26:**63
Disaccharidozymase, **5:**91
Disagregin
fibrinogen binding inhibition, **70:**7
fibronectin and, **70:**4
Discadenine, **54:**230–231
Discovery and Chemistry of Mevalonic Acid (Wagner and Folkers), **23:**471
Diseases, hereditary fructose intolerance, **31:**167–168
Disintegration of viruses, **9:**220–228
Disintegrins, fibrinogen binding inhibition, **70:**6–8; 14
Dismutation, **60:**75
Dismutation between α-ketoglutaric and oxalic acids, **10:**190
Dismutation reactions in CO_2 fixation, **12:**92; 101–103
Dismutations
ketoaldehyde mutase, **7:**92
in urea synthesis, **15:**331–332
of acetaldehyde, **15:**254–257
of aldehyde, **15:**14
of diacetyl, **15:**115
of α-ketoglutarate, **15:**224–225; 229; 235
of pyruvate, **15:**202–207; 210–211; 214; 216–218; 221
Dismutation systems, **18:**333; 339–340
Disodium tetrathionate, **54:**400
Disopropylfluorophosphate (DFP), **34:**9; 10
Dispersion constants, of bacterial proteinase, **23:**73
Dispersion energy, **43:**247
Disproportionating enzyme (D-enzyme), **17:**214–217; 225; 226; 229
Dissimilation et assimilation de l'azote chex les animaux (Braunstein), **19:**335
Dissimilative metabolism, hydrocarbons, aliphatic, **27:**477–505
aromatic, **27:**505–528
Dissociation, **34:**494, **39:**193; 194; 197
constant, **34:**461; 492–493
rates of, **30:**144–145
reversible
of alkaline phosphatase, **10:**108–110
of enzyme system, **10:**117
of β-glucuronidase, **16:**370–371; 375–377
Dissociation constant
of DPN-ADH complex, **22:**350; 351; 358
of DPNH-ADH complex, **22:**350; 351; 358
of TPD complexes with DPN analogues, **22:**347
of TPD-DPN complex, **22:**340; 341
of TPD-DPNH complex, **22:**341
of liver alcohol dehydrogenase system, **20:**36–46
Dissociation constants, **46:**391, **50:**3, **58:**148. *See also* pK value
determination, **46:**391
by fluorescence, **46:**391; 392
by inhibition, **46:**391
3-hydroxypyridines, **50:**3
microscopic dissociation constants, **50:**3
in redox systems, **9:**15–16
of enzyme-substrate complexes, **9:**16–22
of micellar complexes, **9:**9–15
of polyvalent acids, **9:**4–6
of proteins, **9:**14
Dissociation of proteins
light-induced, **9:**123–124
symmetry relation to, **33:**95
Dissociation of t RNA, **35:**488
Dissymmetric configuration, **7:**67–70; 72; 75; 76
Dissymmetric surface of enzyme, **27:**38–39
Dissymmetry, and α-carbonyl group, **7:**76
centers of inactivation of, **7:**80; 94–99
centers of synthesis of, **7:**80; 84–94
induced, **7:**67; 76
and olefinic group, **7:**76
Distance measurements by NMR-EPR, **53:** 72–74
Distortion, **43:**239
Distribution, in protein chromatography, **22:**436; 437
Distribution energy model, **43:**247
Distribution of Enzymes Between Subcellular Fractions in Animal Tissues (Duve, Wattiaux, and Baudhuin), **24:**291
Distribution of mammalian hexokinase and, latency, **39:**299
synaptosomal mitochondria, **39:**299

Disulfidaustausches, Chemie und Biochemie des (Lumper and Zahn), **28:**199
Disulfide, **39:**341; 362, **59:**147–148
bond, **39:**351; 352
bridge, **39:**343; 344; 347
formation, **39:**345
intramolecular, **66:**156
mixed, **39:**345
Disulfide bond, **58:**106
in succinyl-CoA synthetase, **58:**151
Disulfide bonds, **42:**105; 107; 126; 127; 128; 130; 134; 137; 143, **43:**146; 147; 153; 154; 155
and denaturation, **23:**43–44
of Taka-α-amylase, **23:**74–76
effect of light on, **9:**78; 91; 147; 157; 161; 162
in enzyme structure, **22:**73; 74
formation in reduced RNase, **24:**200–202
reactions of, **24:**197–202; 206; 212
of RNase, accessibility of, **24:**229
and activity, **24:**198–199
positions, **24:**174–175
intersubunit bonding, **43:**154
Disulfide bonds, Cotton effect, **27:**96
splitting by nucleophilic reagent, **27:**202
Disulfide bridges, **45:**410
and active site, **24:**448–449; 451–452
and denaturation, **24:**185; 193; 195–197, **26:**95
from polypeptide chains, **26:**100–103
of globular proteins, **27:**45–46
in β-lactamase, **28:**295; 299
of RNase, **24:**174–176; 244–246; 250; 251
Disulfide bridges in proteins, **10:**278; 286–288
Disulfide compound as acceptor for oxidoreductases, **24:**304
Disulfide crosslinks, **42:**177
Disulfide dihedral angle, **59:**148
Disulfide exchange, myosin inactivation, **43:**38
Disulfide exchanges, acid-catalyzed, **27:** 204–206
base-catalyzes, **27:**200–204
biochemical significance, **27:**210–229
chain reactions, **27:**205
chemistry of, **27:**199–237
equilibrium, **27:**207–209
free radical mechanism, **27:**206–207
induction period, **27:**205
mechanism, **27:**200–207
of proteins, **27:**220–229
Disulfide forms of CoA, **16:**2–3
Disulfide interchange reaction, catalysis of, **46:**403
Disulfide link, fission of, **12:**436–437
Disulfide linkages, **47:**117; 129
relation to thrombin production, **16:**29
Disulfides, **57:**155. *See also* Cystine; Mercaplans, Sulfhydryl groups, and Thiols; Sulfhydryl; Sulfide
aromatic, as phosphorylase inhibitors, **24:**269
from coenzyme A, **15:**187
effect of ionizing radiations, **24:**362–364; 367; 375; 382–383; 391; 393; 397–398; 400; 404
and enzyme activity, **27:**212–217
enzymic reduction of, **27:**217–219
of glutathione, **15:**221
as growth factors for *Lactobacillus bulgaricus*, **14:**49–71
mixed, from pantetheine, **14:**61–62; 65–69; 70
in coenzyme A, **14:**62–65
and linkages among respiratory components, **28:**227–228
reduction by thioredoxin reductase, **28:**443
lipoic acid, **15:**215–216
mixed, aryl-alkyl type, **27:**206
exchanges, **27:**206; 209–211
Prot-S-S-R type, **27:**212
RSA-S-S^{35}-G type, **27:**210–212
nonnitrogenous alicyclic, in plant metabolism, **10:**257–259
reactions with proteins, **27:**212–217
reduction of, **11:**228–229
reduction of, PN-dependent, **26:**129
symmetrical, exchange, **27:**201
Disulfides (S^{35}), reactions with proteins, **27:**212
3,5-Disulfonic acid, as inhibitor of β-glucuronidase, **16:**374
Disulfotan, **55:**175
Disulphide bonds, adenosine receptors, **69:**89
Diterpenoid phytoalexins, **55:**49; 78–80
Dithane, compatibility with antibiotics, **20:**396
Dithane D-14, in treatment of plant diseases, **20:**401
1,2-Dithiane-3-butyric acid, **18:**328; 332
1,2-Dithiepane-3-propionic acid, **18:**328; 332
5,5′-Dithiobis (2-nitrobenzoic acid), **54:**400, **59:**123
Dithioerythrytol, **39:**342

Dithioglycolic acid, disulfide exchanges, **27:**204; 214–215; 231
DL-1,2-Dithiolane-3-butyric acid, **18:**332
DL-1,2-Dithiolane-3-caproic acid, **18:**332
1,2-Dithiolane properties, **18:**341–342
Dithiolane ring, **39:**342; 358
1,2-Dithiolane ring, strain energy, **18:**329–330
1,2-Dithiolane-3-valeric acid, **18:**320; 328
Dithiol dehydrogenase, **18:**342
Dithiolipothiamine pyrophosphate, **15:**119
DL-7,8-Dithioloctanoic acid, **18:**332
6,8-Dithioloctanoic acid, **18:**332
Dithiols, **39:**328; 333; 342; 362
  in DR metabolism, **28:**446
  oxidation by irradiation, **11:**256–257
    reaction with arsenic, **11:**218
    relation of structure to ease of oxidation, **11:**207–210; 220–221
    stability of mercaptide of, **11:**240
  in pyrimidine biosynthesis, **21:**286
  synthetic, **10:**300–301
Dithionite, **51:**134
  in two-electron reduction models, **26:**135–137
Dithionite($S_2\ O_4^{2-}$), **39:**342
6,8-Dithiooctanoic acid cyclic disulfide, *See* α-Lipoic acid
2,3-Dithiopropanol, *See* BAL; 2,3-Dimercaptopropanol
Dithiothreitol, **39:**342, **42:**128, **58:**150, **67:** 94–95
  and binding by plasma membranes, **41:**120
    and prolyl hydroxylase, **41:**251; 256; 286; 287; 289
  phosphoenolpyruvate carboxykinase (GTP), **69:**208
Dithiothreitol (DDT), phosphomannose isomerase-guanosine diphospho-D-mannose pyrophosphorylase (PMI-GMP), **70:**226
Dithymidylate dinucleotides (C^{14}), **18:**214
Ditran, in schizophrenia, **29:**534
Diuretics, **57:**137–138
Divalent cation, **35:**483
Divalent cations, **39:**348
  interchangeability for alkaline phosphatase action, **10:**113–114
  pyrophosphoryl transferring enzymes, MutT pyrophosphohydrolase, **73:**187–191
Divalent metal ions, **39:**349
Divalent metals, **68:**163. *See also* Leucine aminopeptidase
  phosphomannomutase/phosphoglucomutase (PMM/PGM), **70:**230
  phosphomannose isomerase-guanosine diphospho-D-mannose pyrophosphorylase (PMI-GMP), **70:**226–227
Divergent transcription as a model for control based on, biotin escape synthesis in λ lysogens, **38:**360
  biotin the corepressor, **38:**362
  complementation analysis, **38:**360
  the fusion of the two operons, **38:**362
  hybridization technique, **38:**358
Divicine, **67:**201
Djenkolic acid, **7:**370, **50:**121; 124; 125, **56:**39
Djenkolic acid in plants, **10:**243
Djenkolic acid sulfoxide, **50:**122; 124
$DMPH_4$, tyrosine hydroxylase, **70:**122; 126; 127; 138; 200
DNA, **18:**162–169, **20:**100, **38:**225; 226; 228; 229, **44:**168, **59:**10–11. *See also* Mutation; Nucleic acids; RNA
  by acid DNase digestion
    oligonucleotides obtained from, **31:**18–23
    relationship between M_w, R_z, and [eta] of fragments of, **31:**17–18
  and amino acid sequence in proteins, **22:**38; 39
  association with proteins, **16:**441; 444; 446–448; 450; 485
  and ATPase, **40:**39; 40
    complexes with polypeptides, **40:**45
    effect on hyaluronidase, **40:**41
  in bacteriophage, **16:**412–413
  bacteriophage, and DR metabolism, **28:** 424–428
  base pairing, **50:**36
  base sequence, **27:**290–295; 341
  biosynthesis of, **18:**212–215
  biosynthesis of, allosteric regulation, **28:** 132–133
    control mechanisms, **28:**436; 443
    from ^{14}C-thymidine, **28:**429
    from ^{14}C-uridine, **28:**420–421; 423
    from ^{14}C-uridylic acid, **28:**420–421
    from labeled cytidine, **28:**420–421; 423
    vitamin B_{12} in, **28:**423; 429–431
  biosynthesis of deoxyribose, **28:**392–394; 419–454

and carbohydrate synthesis, **26:**67; 73; 75
in cell nuclei, **16:**412; 415–417; 422–440
chelates, in metal ion activation of enzymes, **21:**148
hydrolysis, by metal ions, **21:**173; 174; 186
composition, **16:**430–436
conformation in chromatin, **38:**55–59
containing incorporated Fu, **54:**58; 61
cross-linkage, **27:**309–311
double helix, **27:**291–295
double-stranded, photoreactivity with spin-labeled psoralens, **66:**134–135
dynamics study, spin-labeled psoralen application, **66:**128–130; 139–141
effect of amino acid analogs on synthesis of, **29:**148
effect of ethionine on synthesis of, **29:**151
effect of histidine analogs on synthesis of, **29:**151
effects of nitrous acid, **27:**309–311
and enzyme synthesis, **24:**11–14
ethyl derivative of, **25:**383; 389
methylation, **25:**397–398; 414–415
product inhibition of, **25:**222
and the extragenic information carrier, **24:** 14–20
in extranuclear material, **16:**412; 423; 425; 427
FdUMP in inhibiting synthesis of, **54:**105
fluoropyrimidine incorporation into, **54:**68–69
5-fluorouracil effect on, **54:**98–99; 106
furan-side monoadduct, protein-DNA cross-linking, **66:**120–122
as genetic material, **24:**9–11
in β-glucuronidase assay, **16:**365; 370–371; 374–376; 400; 402
heat denaturation of intact and partially digested, **31:**14
hydrogen-bonding in, **27:**291; 311; 318–319; 324–325; 335
labeling, **28:**419
and β-lactamases, **28:**311
metabolism, **16:**454–461
native, degradation of
by acid DNase, **31:**3–4
statistics of, **31:**15–16
native calf thymus
digestion of, by hog spleen acid DNase in presence of different actinomycin concentrations, **31:**32
$S_{20,w}$ of, **31:**4
native chicken erythrocyte
degradation of, **31:**12
preparation of, **31:**4
$S_{20,w}$ of, **31:**4
in non-dividing cells, **16:**454–457
and nuclear purity, **27:**242–244; 248–250
origin of, **38:**225
and origin of enzymes, **27:**374
λ phage, degradation of, **31:**12
of pneumococcus, and sulfonamide resistance, **27:**432
polymerized, **16:**423; 432; 447–448
predecessor of, **38:**226
protein cross-linking, via psoralens, **66:** 119–126
quantity in erythrocytes, **16:**482
quantity per chromosome, **16:**412; 413; 427–428; 429–430; 435; 481
quantity per nucleus, **16:**415; 419; 422; 424–429
relation to physiology of cell nucleus, **16:**480–483; 492
repair synthesis, **67:**436–437
dideoxynucleotide chain termination method, **67:**446–448
plus-and-minus method coupled to improved gel system, **67:**443–446
polymerase chain reaction, **67:**449
recombinant DNA procedures and M13 vectors, **67:**448
scheme of degradation of, by difunctional alkylating agents, **31:**15
single-stranded, sensitivity, **27:**340; 460
single-stranded as template, **67:**443
sonication of, **31:**17
in sperm, **24:**90
and sperm fertility, **24:**37
and studies with superoxide dismutases, **41:**42
synthesis in cell division, **16:**428–429; 452–454
thymine glycol containing, **60:**14
transfer from *E. coli* to bacteriophage, **28:**424
variation of intrinsic viscosity and sedimentation constants with weight-average molecular weight for samples of, **31:**18
variation of radius of gyration with weight-average molecular weight for samples of, **31:**17
velocity of hydrolysis at 37°C of, **31:**43

186p DNA, cohesive end sequence analysis, **67:**440
lambda DNA, cohesive end sequence analysis, **67:**438–439
DNAase, **18:**163; 169; 214
decomposition of DNA in cell nuclei, **16:**492
solubilization of enzymes, **27:**266
DNAase II, intracellular distribution, **16:**465
DNA bases, enolization of, **27:**315–318
DNA binding, RNA polymerase II transcription, C-terminal domain characteristics, **71:**50–51
DNA cloning, simplification
with DNA ligase and synthetic oligodeoxynucleotide, **67:**452–453
with synthetic oligodeoxynucleotide adaptors, **67:**453–455
DNA damaging agents, **60:**15
DNA degradation
biological methods of detection, **31:**4
chemical methods of detection of, **31:**4
kinetics of initial, **31:**5–15
physical methods of detection of, **31:**3
DNA-degrading enzyme, chromatin space, **27:**268
DNA (deoxyribonucleic acid), **59:**10–11
DNA deoxyribosyl, from guanosine, **28:**431
DNA-dependent RNA polymerase, **38:**70–73., **43:**5; 10; 35
electron microscopy of, **33:**105–106
fluorescent properties, **43:**32
DNA glycosylases, **60:**2; 6
in DNA damage recognition, **60:**14
DNA helix, and deletion mutation, **27:**302
DNA ligase, **67:**451–452
DNA cloning simplification, **67:**452–453
DNA methylation, γ-glutamyl transpeptidase gene expression, **72:**269–270
DNA-nucleotidyl transferase, **38:**75
DNA of bacteriophage, D-glycosyl residues, **26:**60; 61
DNA (^{32}P), folate inhibitor study, **27:**456
from TTP, **27:**241
DNA phosphorus, comparison with RNA in cell nuclei, **16:**436
DNA polymerase, **38:**73–75, **43:**26; 35, **56:**292, **60:**26
inhibition of, **56:**339
localization of, **27:**241; 250; 274
and mutations, **27:**342
terminal, chromatin space, **27:**268
metal content of, **56:**335–336
molecular properties of, **56:**336–337
phosphatase-exonuclease accompanied by, **29:**197
reaction, **49:**113
reaction mechanism in excision-repair process, **60:**27
regulation, **28:**447; 448
DNA Polymerase-Catalyzed Reactions, The Dynamics of (Mizrahi and Benkovic), **61:**437
DNA polymerase I
and abortive pol I-Mn^{2+}-dTTP complex, **59:**276
and amino acids
at substrate site, **59:**282–283
at template site, **59:**283
conformations of
bound substrates, **59:**277–278; 280
bound template, **59:**280
and discrimination against incorrect nucleotides, **59:**276
divalent cation in, **59:**275
and enzyme-bound divalent cation, **59:**276
from *Escherichia coli,* **74:**71
and mechanism of Mn^{2+} mutagenesis, **59:**277
reaction, **59:**274–275
DNA polymerases, **33:**269, **36:**82; 83
secondary interactions of, **33:**436
DNA primer, **27:**295
DNA-protein interactions, **47:**79; 81
DNA repair, **60:**1
of alkylation damage, **60:**14
base excision repair, **60:**2
of bulky adducts, **60:**14–29
demethylation, **60:**3
and DNA polymerases, **60:**26
nucleotide excision repair, **60:**2
oligonucleotide excision mechanism, **60:**25
of purine modification, **60:**14
recombination repair, **60:**3
RNA polymerase II transcription, potential roles, **71:**68–69
DNA replication
Caulobacter cell cycle, **71:**4–5
transcription regulation, **71:**16–17
in vitro, **27:**290; 293–295; 316
and nuclear enzymes, **27:**275–283

and origin of life, **27:**396
DNA-RNA hybrid duplex, **60:**12
DNA-RNA hybridization, **38:**83
DNA:RNA ratio, in nuclear preparation, **27:**245–247
DNA-RNA relations in chromosomes of *Rhynchosciara angelae,* **16:**460–461
DNase, **23:**186, **46:**249, **55:**375. *See also* Acid DNase
adsorbed, radiosensitivity, **28:**33–35
in HeLa cells infected with cowpox virus, **29:**201
hog spleen, amino acid analysis of, **29:**185
hydrolysis of dinucleotides, **29:**169
from λ infections, **29:**201
in invertebrates, **29:**182–183
from phage SP3 infection of *Bacillus subtilis* SB19, **29:**201
in pleuropneumonia-like organisms, **29:**199
rat liver, localization of, **29:**181
properties of, **29:**181
studies of structure of nucleic acids and the uses of, **29:**165–220
terminology and classification of, **29:** 167–170
from T_2 infection of *E. coli,* **29:**202
pancreatic, digestion of DNA sample with, **31:**6; 7
DNase I
attack on native DNA, **29:**169
mode of action of, **29:**210
pancreatic, autoretardation of, **29:**172–174
homogeneity of, **29:**170–171
ionic requirements of, **29:**177–180
location of inhibitor for, **29:**171–172
naturally occurring protein inhibitor for, **29:**171–172
physical and chemical characteristics of, **29:**171
polyelectrolytes as inhibitors of, **29:**172
preparation and inhibition of, **29:**170–172
ribonuclease contamination in, **29:**170
specificity of, **29:**175–177
time-activity curve of crystalline, **29:**173
physical and chemical characteristics of, **29:**209
specificity for strandedness, **29:**175
DNase II
cleavages by, **29:**175
difference between DNase I and, **29:**187
and DNase II-like enzymes, **29:**180–187
effects of radiation on, **29:**181
homogeneity of, **29:**184
hydrolysis of dinucleotides by, **29:** 169–170
location of enzymes like, **29:**180–181
mechanism of action of endonuclease I resembles, **29:**198
mode of action, **29:**184–187; 210
physical and chemical characteristics of, **29:**209
preparation of, **29:**183–184
properties of, **29:**180
in regenerating rat liver, **29:**181
specificity for strandedness, **29:**177
splenic, **29:**170
temperature sensitivity of, **29:**183
in vertebrate organs, **29:**180–181
DNaseI footprint, T7 RNA polymerase, **66:** 105–106
dynamics, **66:**110–112
DNA sequence analysis, **67:**431–464. *See also* DNA polymerase I; Genome project
cytochrome *c* oxidase polypeptide composition, **71:**83–84
E. coli, **67:**435
exonuclease III1, **67:**449–450
general plan for method development, **67:**433–434
importance of developing new methods, **67:**432–433
progress made between 1967 and 1990, **67:**455–459
as tool for labeled nucleotide incorporation into specific DNA regions, **67:**435–449
basic idea, **67:**435–436
chemically synthesized primers, **67:** 441–442
lambda DNA, cohesive end sequence analysis, **67:**438–439
mobility-shift method, **67:**442–443
186p DNA, cohesive end sequence analysis, **67:**440
primers for repair synthesis, **67:**440–441
the problem, **67:**435–436
single-stranded DNA as template, **67:**443
test system, **67:**436–437
DNases and Their Use in the Studies of Primary Structure of Nucleic Acids (Laskowski), **29:**165

DNA synthesis
enzyme system for, **27:**293–295
temporal relationship between and histone methylation, **42:**253
DNFB, *See* Dinitrofluorobenzene
DNP, *See* Dinitrophenol; 2,4-Dinitrophenol
DNP-cysteinic acid, disulfide exchanges, **27:**212
Docking, *See* Molecular docking
Dodecadepsipeptide, **34:**483
Dodecane, **51:**33
Dodocanoic acid esters, **34:**488
D_2O exchange, near 4Fe-4S cluster, **67:**315
Dogfish lactate dehydrogenase, **46:**255
Dolichol, **52:**28
phosphate, **52:**94
relation to coenzyme Q, **25:**287–288
Domains, of proteins, **50:**24
Domain structure, glutathione S-transferases, **69:**8–11
DON, **39:**207
Donnan equilibrium, **11:**37–47
activity of adsorbed enzymes, **28:**29–30
Donnan relationship, **34:**463
Donnan's theory, **33:**251; 290
Donor substrate, **39:**341
DONV, **39:**194; 198; 203; 205; 206
DOPA, *See* 3,4-Dihydroxyphenylalanine
in transmination reactions, **16:**192
Dopa, **5:**72–75; 77; 79–81, **9:**86; 87
action of peroxidase on, **14:**152
role in tyrosine metabolism, **14:**75–96; 103; 109–111; 117–119; 123
in tyrosinase action, **14:**133; 137; 139–140; 142; 145; 146; 148–149
formation in animal body, **5:**80
inhibition by, **67:**217
relation to melanin, **5:**80
tyrosine hydroxylase, substrate specificity, **70:**110
D- or L-Dopa, **44:**196
L-DOPA, **49:**90
interaction between histidine and, for brain, **49:**90
L-dopa
L-tyrosine conversion, **70:**105
tyrosine hydroxylase, **70:**109
Dopachrome, role in tyrosine metabolism, **14:**76–78
Dopachrome (2-carboxy-2,3-dihydroindole-5,6-quinone), **16:**113
decarboxylation, **16:**115–117
in melanin formation, **16:**164; 168
Dopa decarboxylase, **5:**72–75; 77; 79–81, **55:**235
distribution in different tissues, **5:**73
inhibition of action by cyanide, **5:**75
by hydroxytyramine, **5:**75
by sodium sulfide, **5:**75
by sucramine, **5:**75
by Trypan Blue, **5:**75
DOPA decarboxylase, localization of, **24:**330
Dopa melanin, from dihydroxyphenylalanine, **16:**163–169
Dopamin-(3,4-dihydroxyphenylethylamine) hydroxylase, mechanism of deduced from relaxivity data, **33:**64
Dopamine, **44:**211
biosynthesis of, **74:**236
location of, in brain, **29:**482
uptake of, by hypothalamus, **29:**488
tyrosine hydroxylase regulation, **70:**158; 163
Dopamine β-hydroxylase, **57:**135–136, **67:**85, **70:**105
amphetamine inhibited by, **29:**491
Dopa oxidase, **9:**87
adsorption studies of, **14:**356
in tyrosine metabolism, **14:**80; 81
cytochemical method for, **8:**30
histochemical methods for, **9:**588
Dopaquinone, **57:**376
Dopa quinone
in melanin formation, **16:**164
in tyrosinase action, **14:**140
role in tyrosine metabolism, **14:**75–96
Dothichiza species, enzymic transformations of steroids by, **20:**147
Double bond reduction, mechanisms, **24:** 472–474
Double bonds, PN-reduction, **26:**127–128
Double-displacement, form, **39:**364
formal mechanism, **39:**353
formulation, **39:**354
mechanisms, **39:**328; 352
Double-headed enzyme, **23:**314–315; 330; 331; 335
Double helix, of DNA, **27:**291–295
Double refraction of protoplasm, **7:**60
Double-strand scission, in λ phage DNA, **29:**170

Doublets, in mammalian RNA's, **47:**422
in MS2 RNA, **47:**420
in nucleic acid strands, **47:**419
in phi X174 DNA, **47:**422
selection against, **47:**422
in vertebrate DNA, **47:**419
Douzou, Pierre, **51:**1–74
Dow M-137, compatibility with antibiotics, **20:**396
Down regulation, **50:**315
adenosine receptors, **69:**103
DPA, *See* Diphenylamine
DPHN, *See* Diphosphopyridine nucleotide, reduced
DPN. *See also* Diphosphopyridine nucleotide; NAD; Pyridine nucleotides and Respiratory chain
enzymic synthesis of, **14:**203
in energy transport, **14:**10–12; 13; 17
in enzyme assays, **14:**382
in oxidation of phenylalanine, **14:**97–98; 101; 106
in racemization of lactic acid, **14:**193; 197
interaction with dehydrogenase, rate constant, **25:**14
in neoplastic tissue, **29:**363
of nucleus, **27:**272–274
and phosphate turnover, **17:**175; 184; 185
effect on choline oxidation, **17:**383
kinetics of formation from nicotinamide, **17:**36
in kinetic studies of alcohol dehydrogenase, **17:**42–45
maintenance of oxidized state, **17:**66
reduction by metalloflavoprotein, **17:**249
synthesis of, effect of ethionine on, **25:**382
DPN-adducts, **37:**90
DPN-analogs, **37:**83
DPNase, product inhibition of, **25:**197; 222; 229; 237–240
DPN (C^{14}), enzyme inhibitor study, **25:**237
DPN cytochrome c reductase classification, **14:**254
DPN-cytochrome reductase, **15:**86
localization in cytoplasm, **13:**10
DPN-dehydrogenase, in pre-cytochrome system, **17:**363–365
DPN-dependent enzyme, in histidine metabolism, **17:**406
DPN (diphosphopyridine nucleotide). *See also* Coenzyme I
and CO_2 fixation, **12:**51; 55; 61–62; 65; 67; 90–91
in enzyme-substrate complex, **12:**184–185
as hydrogen carrier for glutamic dehydrogenase, **16:**255
in cell nuclei, **16:**468; 488
in plant oxidations, **16:**133–135
synthesis, **16:**474
in synthesis of amino acids, **16:**270; 271; 289; 299
reduced, as hydrogen donor for peroxidase-peroxide complex, **12:**175; 187
DPN formation, in nucleus, **27:**269–271
DPNH, **34:**476–477, **35:**246. *See also* Diphosphopyridine nucleotide, reduced; NADH; NADH ($NADH_2$); Pyridine nucleolides
and bacterial luminescence, **25:**144–145; 149–151; 154; 155; 160–161
fluorescence, **25:**124
reduction of coenzyme Q by, **25:**305; 312–315
tumor content, **25:**378
in DR metabolism, **28:**436–438; 443; 448
effect of radiations, **24:**394
enzyme binding moiety of, **27:**55–70
and EPR spectra, **27:**124; 136; 173–175
and liver alcohol dehydrogenase, **27:**49–83
oxidation, EPR data, **27:**119
oxidation of, and aerobic glycolysis, **29:** 339–346
rotatory dispersion of, **27:**55
and xanthine oxidase, **17:**296–301; 309–313; 317–318; 322; 323
function in respiratory chain, **17:**67; 105
oxidation of, **17:**238–241; 334
reduction of cytochrome b_5, **17:**361–362
DPNH analogs, complexes with LADH, **27:**68
enzyme binding of, **27:**56–70
DPNH-coenzyme Q reductase, **25:**294; 312–319
DPNH-coenzyme Q reductase, EPR study, **27:**179
DPNH-CoQ, reductase, **34:**106
DPNH-CoQ_{10}, reductase, inhibition of, by Amytal and rotenone, **34:**109
DPNH-cytochrome b_5 reductase, **20:**169; 171

DPNH-cytochrome c reductase, **25:**312–315; 318–319
inactiviation, **20:**179; 181
isolation, preparation and composition, **20:**168; 169; 188; 192; 193
in respiratory chain, **20:**168–172
DPNH-cytochrome *c* reductase, **17:**69–71; 77; 79; 349–352; 362–365; 376–380, **34:**84; 109; 133; 480
different from DPNH-ubiquinone reductase, **34:**84
extraction of, **34:**92
in mitochondria, **21:**15–19; 21; 35; 58; 59
effect of sonic vibrations on, **21:**57
in mitochondrial fragments, **21:**100; 113
molecular weight, **21:**100
DPNH-cytochrome c reductase from, **20:**192; 193
in study of respiratory chain, **20:**152; 153; 161–164
DPNH-cytochrome reductase, **19:**201
as metalloflavoprotein, **17:**257–264; 267–270; 274–277; 283; 284; 316
DPNH dehydrogenase. *See also* Diaphorase and DPNH-cytochrome c reductase; DPHN-cytochrome c reductase and Diaphorase
action of inhibitors on, **34:**101–122
activity determination of, **34:**88
aerobic baker's yeast, **34:**145
aggregates, **34:**83
in anaerobically grown *S.* cerevisiae cells, **34:**146
anaerobic fragmentation of, **34:**94
assay of, **34:**87–88
as a binding site of rotenone, **34:**108
C. utilis, **34:**145–146
centrifugation in sucrose gradients, **34:**83
composition of, **34:**86
concentration in mitochondria, **34:**147
conditioning effect on, **34:**113
conformation changes in, **34:**99
conformation of, **34:**98
conversion to cytochrome reductase, **34:**91
damage to, by lyophilization, **34:**125
difference spectrum of, **34:**90
different forms in, **34:**82–86
DPNH binding of, **34:**98
effect of proteolytic enzymes to urea or thiourea, **34:**85
effect on, by mercurial, **34:**117
by prolonged contact with substrates, **34:**94
by subtilisin, **34:**93
by thiourea, **34:**93
by urea, **34:**93
electron acceptors of, **34:**87–89
electron transport from flavin to nonheme iron in, **34:**113
endogenous CoQ step, **34:**142
EPR and optical spectra of, **34:**89–90
EPR signal of, **34:**89; 90; 92
EPR studies of, **34:**89; 90; 92
extraction by digestion with phospholipase A at 37, **34:**85
FAD in, **34:**86
flavin content in, **34:**86
FMN content of, **34:**86
FMN in, **34:**97
fragmentation of, **34:**92; 100
freeze-thawing of, **34:**91
heat-acid-ethanol effect on, **34:**84
high molecular weight, form of, **34:**86–95
preparation, isolation of, **34:**81
soluble form of, **34:**82–84
high molecular weight form, conversion to low molecular weight form, **34:**84
inactivating agents of, **34:**90–95
inactivation by detergents, **34:**94
intramolecular electron transport, sequence in, **34:**132–134
irreversibly bleached chromophore of, **34:**140; 145
K_m, value of, **34:**87
linkage to the respiratory chain, **34:** 129–132
low molecular weight, form of, **34:**84; 86; 95; 96
isolation of, **34:**80
origin of, **34:**81
preparation, differences vary, **34:**81
lyophilization of, **34:**91
modification of, by Triton, **34:**84
molecular transformation of, **34:**90–95
molecular weight of, **34:**86
as a multienzyme complex, **34:**99
multiple forms of, **34:**83
nonheme iron, components of, **34:**103
nonheme iron in, **34:**87
nonheme iron-liable sulfide components of, **34:**89

nonheme irons of, **34:**108; 145
optical absorbance changes, **34:**135–141
origin of different forms of, **34:**98–101
particulate preparations, **34:**82
phospholipase extraction of, **34:**130
preconditioned state of, **34:**115
preparation of, **34:**83
low molecular weight form, **34:**84
properties of, from baker's yeast, **34:**146
from *C. utilis*, **34:**146
prosthetic group of, **34:**86
protection of, **34:**92
rate limiting step in the catalytic cycle of, **34:**90
reactivity with ferricyanide, **34:**96
redox cycle of, **34:**117
redox equilibrium of, **34:**139
redox state of, **34:**135
redox state of the flavin, **34:**132–134
relation to energy conservation site 1, **34:**142–145
reversible effect of mercurials on, **34:**119
role of SH groups in intramolecular electron transport, **34:**119
SH group of, **34:**97
SH inhibitors and, **34:**111–120
specificity, for electron carriers of high molecular weight form of, **34:**87
of substrates, **34:**87–89
stability of, **34:**90–95
thermal modification of, **34:**91
transhydrogenase activity of, **34:**112
tryptic action on, **34:**94
turnover number, **34:**87–89
type I, SH groups of, **34:**111; 116
type II, SH groups of, **34:**111; 115
type III, SH groups in, **34:**113; 115
type IV, DTT on, **34:**119
type IV, SH groups on, **34:**116
type V, SH groups on, **34:**116; 117
column chromatography, **22:**478
lipoprotein complex, **21:**108
in mitochondria, **21:**74
in mitochondrial fragments, nonheme iron in, **21:**114
position in mitochondrial electron transfer chain, **21:**120–122
DPNH -dehydrogenase, **17:**240
DPNH dehydrogenase, EPR study, **27:**126; 177–183
DPNH-dehydrogenase flavoprotein, and coenzyme Q, **25:**309; 315; 319
DPNH-dehydrogenase interaction, rate constant, **25:**14
DPNH-flavin oxidase, and luminescense, **25:**150
DPN(H)-LADH, effect of inhibitors, mechanism, **27:**73–75
DPNH oxidase, **19:**182; 183; 201; 203, **34:**480. *See also* Diaphorase and DPNH-cytochrome c reductase
activity, inhibition, **20:**170; 176; 178; 179
in respiratory chain preparations, **20:**153; 184
and coenzyme Q, **25:**307–308; 310; 318
interference from, **26:**350–351
isolation and properties, **21:**102; 103
restoration activity of, **34:**124
DPNH oxidase complex, as metalloflavoprotein, **17:**258; 260–263; 270; 275; 281; 284–285
DPNH oxidase I, in study of respiratory chain, **20:**152; 153; 161–164
preparation, **20:**161
DPNH oxidase II, composition, **20:**188
DPNH oxidase system, **17:**81; 82; 84; 128; 375–380
$DPNH_2$ r acetaldehyde transhydrogenases, **14:**253
$DPNH_2$ r cytochrome c transelectronases, **14:**254
DPNH-reductases, in inorganic nitrogen metabolism, **17:**250–252
$DPNH_2$ r O_2 transhydrogenase, classification, **14:**253; 255
$DPNH_2$ r pyruvic transhydrogenases, **14:**253
DPNH-ubiquinone reductase, different from DPNH-cytochrome *c* reductase, **34:**84
DPNH-X, **15:**93–94
DPN-isocitric dehydrogenase, in mitochondria, **21:**16; 17; 21; 58; 74
DPN kinase, TPN forming, **27:**270
DPN-linked dehydrogenases, and phosphorylations, **17:**78
DPN metabolism, of nucleus, **27:**276–278
DPN-nucleosidase, reaction pattern of, **15:** 107–108
DPN pyrophosphorylase
chromatinbound, **27:**267–268
nuclear, **27:**276–278

DPN pyrophosphorylase (*Continued*)
and nuclear purity, **27:**248–250; 252–253
solubilization of, **27:**265–266
in sucrose nuclei, **16:**474
DPN-requiring dehydrogenases, **15:**84–85; 97–98; 123
DPN ribose, labeling, **28:**416
DPN-specific isocitrate dehydrogenase, **28:**135–139
DPN-synthesizing enzyme, **15:**107
and phosphate transfer, **17:**184–186
DPN-TPN transhydrogenase, in mitochondria, **21:**15; 17; 35
DPT, *See* Diphosphothiamine
Dried enzyme preparations, effects of ionizing radiations on, **24:**360; 364–383
Drohoche's integrator, analysis of EEG records, **29:**533
Drosophila, mutation rate, **24:**90
Drosophila, genetic studies of, **13:**130–131; 135–136; 138
Drosophila larvae, origin of cytoplasmic RNA, **24:**16
Drosophila melanogaster, **45:**419, **70:**3; 11
genetic studies, **17:**420–422; 447–448; 453
xanthine dehydrogenase-deficient mutants, **22:**29
Drosophila melanogaster, biopterin in, **21:**377
Drosophila, mutations, **27:**331
nuclear membrane, **27:**269
origin of pteridines in, **27:**422
Drosophila virilis, **17:**447
Drosophilia, **71:**46; 48; 63; 65
Drosophilla melanogaster, **51:**111
Drude equation, and Cotton effect, **27:**44–45; 52
Drug action, and electrical changes, **7:**47
mechanism, **7:**251
Drug detoxification, **47:**3
Drug-protein-coenzyme complex of respiratory enzymes, **6:**37
Drug resistance
and cell receptors, **27:**432; 445–450
genetic studies, **17:**455–457
Drugs
antagonisms between, **6:**140
antagonists, **6:**40–51; 140
enzyme role, **6:**40
antidepressant, **29:**484
hallucinogenic, changes produced in EEG by, **29:**534 (*See also* LSD 25, Mescaline, Bufotenin, and N,N′-dimethyltryptamine)
methylation and, **29:**527
in psychiatry, **29:**481
in schizophrenia, **29:**515
effect on enzymes, **1:**194; 196
glucuronic acid conjugates of, **16:**364
hydrotropic solubility of, **17:**154
solubilization and transport, **17:**137; 138; 143
oxidation of, **24:**308
pharmacological activity, **1:**193
photobiological sensitization to, **11:**67
resistance to, **6:**68
Drug tolerance, **60:**47
Drying oils and double-bond conjugation, **8:**439
DRY motif, G-protein-coupled receptros (GPCRs), adenosine receptors, **69:**87
dsDNA, photoreactivity with spin-labeled psoralens, **66:**134–135
DT-Diaphorase, **46:**26
DT-diaphorase, **65:**110–119; 128–129
antioxidant molecule maintenance, **65:**119
concerted antioxidant system, **65:**120–124
induction, **65:**113–115
occurrence, **65:**112
primary structure, **65:**112–113
quinone
detoxification roles, **65:**117
metabolism, **65:**116–119
structure and kinetics, **65:**111–113
substrate specificity, **65:**113
vitamin K-dependent protein carboxylation, **65:**115–116
dTDP-*N*-acetyl-D-glucosamine-4-epimerase, **26:**55
dTDP-D-galactose pyrophosphorylase, **26:**53
dTDP-D-glucosamine phrophosphorylase, **26:**55
dTDP-D-glucose, enzymes for reduction, **26:**54
dTDP-D-glucose-4-epimerase, **26:**53
dTDP-glucose oxidoreductase reaction, **50:**297
dTDP-D-glucose pyrophosphorylase, **26:**52; 62
dTDP-L-rhamnose synthetase, **44:**85
sequential enediol steps, **44:**85
dTDP-sugars, **26:**43–50; 61–62; 69–70
dTMP, **55:**268
in allosteric regulation, **28:**132–133
labeling of DR, **28:**423
dTMP (Thyminedeoxyribo-5,-phosphate), in methyl synthesis, **25:**395; 397; 399–401

DTNB, **35:**452
dTPD-D-glucose pyrophosphorylases, allosteric regulation, **28:**111; 113
DTT, **42:**135; 136
dTTP, in allosteric regulation, **28:**132–135
DR metabolism, **28:**448
Duboisia hopwoodii, in alkaloid studies, **8:**224
Duboisia myoporoides in alkaloid studies, **8:**223; 224; 226; 240–241
dUDP, in DR metabolism, **28:**443
dUDP-D-galactose pyrophosphorylase, **26:**53
dUDP-D-glucose, enzymes for reduction, **26:**54
dUDP-D-glucose pyrophosphorylase, **26:**52
dUDP sugars, **26:**61
dUMP, **55:**268; 270; 271
5, 10-10CH_2H_4PteGlu binding with, **54:**84
accumulation of, in tumor responsiveness to fluoropyrimidines, **54:**105
activation of C-5 position of, **54:**71–75
biosynthesis of, **28:**132–133
in DR metabolism, **28:**443; 447
conversion of, to dTMP, **54:**70
from dUTP, **54:**69
in FdUMP inhibition of thymidylate synthetase, **54:**93
levels of, following FU treatment, **54:**96
methylene bridge joining, to 5, 10–10CH_2H_4PteGlu, **54:**76
in TS protection from inactivation, **54:**84
dUMP (P^{32}), methylation study, **25:**399
dUMP (tritium-labeled), methylation study, **25:**399
dUMP (Uracildeoxyribo-5,-phosphate), in methyl synthesis, **25:**395; 397–400
5-F-dUMP, **55:**199
Duplex films, **28:**9; 11; 23
Duponol, effect on –SH groups of proteins, **11:**221–222; 223–224
Dupuytren's contracture, and vitamin E, **6:**484; 485
Durenol, peroxidase action on, **19:**108
Duroquinone, oxidation of thiols by, **16:**121
Dutch elm disease, **11:**416; 417
dUTP
cleavage of, to dUMP, **54:**69
in DR metabolism, **28:**443; 447
dUTP pyrophosphatase, and DNA, **28:**443; 447
Dye-binding study on nucleoproteins, **18:** 176–177
Dye-protein interactions, steps in, **25:**13–15
Dye reduction system, **39:**333
Dyes
in assay for succinic dehydrogenase, **18:**72; 75; 84–87
basic, binding to histones, **16:**446
binding to protein-bound cations, **17:**274
effect on xanthine oxidase, **17:**309
nucleates, **17:**137
oxidation-reduction, as electron acceptors, **17:**239–242; 246; 251; 253; 256–259; 263; 282; 285; 298; 380
chemiluminescence, **25:**123; 127; 149; 150
effect on coenzyme Q, **25:**306; 312–313
in photosynthesis study, **25:**65–66
indophenol, as artificial electron acceptors, **30:**22
polyglutamic acid-bound, Cotton effect, **27:**47–48
in study of reactions at surfaces, **11:**42–43; 45–47
Dynamic affinity chromatography, **46:**259
Dynamic instability, microtubule dynamic, **69:**134–136
Dynamic state, **7:**191; 192
Dynein adenosine 5′-triphosphatase, **53:** 144–150
brain ATPase affinity, **53:**150
interaction with microtubules, **53:**146–150
isolation and molecular properties, **53:** 145–146
Dysentery, amoebic, **11:**380
Dystonia, DOPA-responsive, **74:**236
Dystrophy, **7:**338
muscular, **6:**195–201
progressive, **6:**218–225
vitamin E in, **6:**483–485

E

E. coli, See Escherichia coli
E. coli alkaline phosphatase, **43:**17
E. coli B, mutation study, **27:**297–301
E. coli code, **47:**379
E. coli K12, mutation study, **27:**300–301
E. coli phosphatase gene, **26:**102
E. coli protein R2, **65:**155–158
dimer structure, **65:**156–158
iron center, **65:**158–177
binuclear, **65:**162–165

E. coli protein R2 (*Continued*)
conserved amino-acid residues, **65:** 159–160
diferric-peroxide intermediate, **65:**172–173
function, **65:**176–177
hydroxo-bridged diferrous model complex, **65:**170–171
iron coordination, **65:**158–160
metal cofactors, **65:**175–176
model complexes, **65:**162–165
oxo-bridge exchange reactions, **65:**167
reaction with low molecular weight exogenous ligands, **65:**167
reduction, **65:**167–170
spectroscopic properties, **65:**160–162
stability, **65:**165–166
structure and stability of reduced R2, **65:**170–176
structure, **65:**155–156
tyrosyl radical
reactivity, **65:**178–180
structure, **65:**177–178
E. coli-T4 system, mutation study, **27:**297–301
E, E-Farnesol, **46:**79
E600, effect on esterases, **20:**289; 305; 309
E'_0, of water-oxygen couple, **30:**22
Eadie-Hofstee plot, cytochrome *c* oxidase, electron transfer, **71:**117–118
Eberthella typhi, metabolism in, **74:**13–14
Eburicoic acid, biosynthesis from acetate, **22:**329
structure, **22:**328
EC-1, **39:**191
EC-2, **39:**191
Ecdysone, **45:**420
Ecdysone pulse, **45:**415; 420
Echinochrome in egg, **9:**257–258
Echinoderms, **46:**63
Echis carinatus venum, **46:**11; 15; 16
Ectoenzymes, **57:**159–162
Edatrexate, eukaryotic dihydrofolate reductase, ligand binding (side chain roles), **70:**62
Edeine, **36:**226
Edestein, structure studies, **14:**284
Edestin
amino acid content, **3:**66; 67
dicarboxylic amino acids in, **20:**319
EDTA, **53:**51, **57:**4; 19–20; 24, *See* Ethylenediaminetetraacetic acid; Versene; Versene-etheylene-diaminetetraacetic acid
effect of, on cations activating fructose-bisphosphatase, **54:**129
in fructose-bisphosphatase stimulation, **54:**133
E,E,E-Geranylgeraniol, **46:**76
Effective concentration, **43:**272
Effective molarity (EM), **59:**135
Effectiveness factor, **34:**501–502
Effect of local field, on electron flow, in D-amino acid oxidase reaction, **34:**74
Effect of pH, on rate of conversion of purple intermediate to fully reduced enzyme, **34:**61
on rate of formation of purple intermediate, **34:**58
Effector, in enzyme systems, **14:**240–243
Effectors, **34:**524
Efficiency of plants for the reduction of CO_2, **1:**203
E-F hand, **65:**204
EF-hand proteins, annexins and, **71:**215
Egasyn, **60:**237
deficiency, **60:**238
Egg albumin, *See* Albumin; Albumin, egg, and Egg white proteins; Albumins
adsorption at interfaces, **28:**13–15; 19–20; 22
amino acid content, **3:**67; 68
and labeled phosphorus, **7:**159–160
denaturation, **7:**248
mol. weight, **1:**14; 15; 22; 23
shape of molecule, **3:**101–102
Egg albumin, SH-groups, **27:**221–226
Egg cells, DNA in, **16:**412; 423; 427
fertilization, **16:**427; 479–480
Egg formation, **7:**157–160
Eggplant, phytoalexins of, **55:**73
Eggs
fertilization and species specificity, **9:** 268–275
mammalian, artificial activation, **9:**296
metabolic changes following fertilization, **9:**297–312
parthenogenesis, **9:**286–297
physiological state and fertilizability, **9:**244–255
structural changes following fertilization, **9:**275–286

structural changes in hypertonic solution, **9:**292–294
structure, **9:**245–246
substances produced by, **9:**255–268
theory for activation, **9:**312–320
marine, localization of enzymes in, **13:**6; 7; 8; 9–10; 12; 16
trout, permeability to water, **13:**43
Egg shell and labeled phosphorus, **7:**159–160
Egg water, **9:**246
effect on cross fertilizations, **9:**269–270; 272–275
on sperm, **9:**252; 255–268; 271; 314; 369
Egg white
hen, ovomucin of, **20:**137
proteins, **7:**196
Egg white injury, **3:**289–291; 305–307
Egg white protein. *See also* Albumins, Microsomes
nonenzymic reactions, **19:**261
Egg white proteins. *See also* specific proteins
separation by chromatography, **14:**339; 345; 350; 354; 358–359
Egg yolk, phosphorus compounds in, **7:** 157–160; 165–166
vitellin, **7:**163–165
Ehlers-Danlos syndrome, lysyl hydroxylase gene mutations, **72:**383–384
Ehrlich ascites, thymidylate synthetase from, **25:**398
liver, ethionine effect, **25:**373–375
Ehrlich ascites cells
enzymes of, **24:**301; 306; 323; 328
glycine translocation in, **29:**61
Na^+-amino acid symporter in, **29:**61
Ehrlich ascites tumor
aerobic glycolysis in, **29:**344
pentose synthesis in, **28:**417; 419; 433; 436
Ehrlich ascites tumor cells
difference spectra of the 6C3HED lymphoma cells and the cytochromes of, **29:**326; 328; 329
electron transfer in, **29:**352
Na^+-glucose symporter in, **29:**61
uptake of glycine in, **29:**365
proteins of, **18:**175
Ehrlich ascites tumor cells, PP-ribose-P synthetase from, **41:**4
Ehrlich cell, H^+ cotransport in, **49:**69
Ehrlich-Lettré carcinoma cells, **59:**372
E-I complex, **57:**48–49; 51–52
Eicosanoids, cholesterol ester hydrolases regulation, **69:**69
Eicosapentaenoic acid, **55:**99
Eigen-Tamm mechanism, formation of E-Mn-S complexes for, **33:**54
EILDV-conjugated substrates, **70:**5
Elaidic acid as tracer, **7:**171
Elapsidae, nerve growth factor in venom of, **31:**57
Elastase, **34:**10; 13; 21, **43:**244; 368, **46:**482, **47:**49; 74; 124, **52:**181; 186; 198; 203; 206; 208; 209; 210; 211; 212; 213; 214; 215; 216; 221; 223; 224; 230–234; 235; 237–238; 239, **53:**261; 264; 265, **56:**293
active center of, **24:**434
amino acid composition near active site, **22:**65
column chromatography, **22:**454
amino acid sequence in, **34:**22
oligopeptide inhibitors, **43:**255
peptide aldehydes, **43:**299
from *Pseudomonas aeruginosa,* **41:**201; 238
Thiol proteinase from, **41:**186
specificity of, **33:**432–433
Elastases, **23:**163–165; 221–248
applications, **23:**243–248
assays, **23:**222–223
effect of metal ions on, **23:**234
flavobacterium, mode of action, **23:**243
preparation of, **23:**230–231
properties of, **23:**232–238; 240
microbial, **23:**165; 201; 225; 227–231
occurrence, **23:**223–227
pancreatic, **23:**165; 172–173; 222–226; 228–232
mode of action, **23:**240–243
preparation of, **23:**229–230
properties of, **23:**231–239
specificity of, **23:**237–239
peptidase-free, **23:**240
properties of, **23:**231–240
synthetic substrates for, **23:**238
Elastases, Collagenases and (Mandl), **23:**163
Elasticity of protoplasm, **7:**40–43; 60
Elastin, **1:**12; 13
collagen hydroxylases, **72:**331–334
enzyme action on, **33:**257–258
fine structure of, **23:**244
nature of, **23:**221–222

Elastolytic activity, microbial, **23:**245
Elastolytic preparation, **23:**179
Elastomucase, **23:**232; 241; 244
Elastomucoproteinase, **23:**239
Elastoproteinase, **23:**230; 237
ElcB pathway, **59:**125
Electrical changes and drug action, **7:**47
Electrical processes in nerve fibers, **12:** 312–315
Electrical stimulation, tyrosine hydroxylase regulation, **70:**128–129; 151; 156
Electrical stimuli, effect on mitochondrial oxidative phosphorylation, **21:**31
Electric charge, **34:**446; 456
  conductance, **34:**485
  field, **34:**446
  potential gradient, **34:**492
  tissue, **34:**525
Electric convulsion theory, effect on catecholamines, **29:**490
Electric current, **57:**232–233
Electric dichroism in chlorophyll, **25:**72–73
Electric discharges, and origin of life, **27:**356; 359; 360; 382–386
Electric field pulse, relaxation methods, **25:**2–9
Electric organ, use of ATP system, **14:**14–17
Electric polarizability, photoinduced, **25:**82
Electric potentials, in organisms, **7:**45–48
  in traumatized tissue, **7:**48
Electric shock, fall in brain epinephrine by, **29:**487
Electrochemical activities of a substance, relation to flux, **13:**33–42
Electrochemical activities of ions, relation to flux, **13:**48–63
Electrochemical fields, **34:**492
Electrochemical gradient, Na^+, **49:**66; 67
Electrode potential. *See also* Redox potential
  and toxicity of metal ions, **11:**69–74
Electrodes
  ion-specific, **39:**334
  noble metal, effect on thiols, **11:**211–212
Electrodialysis, **42:**78
  of glycogen, **3:**113
Electroencephalographic patterns, and citrate levels, **18:**148
Electrofocusing, **42:**271
Electrolyte, distribution in affective disorders, **29:**494
Electrolyte behavior, in nucleus, **27:**256–261
Electrolytes
  concentration in nucleus, **27:**256
  effect on agglutination, **15:**51
    effect on immunologic precipitation, **15:**68
  effect on mitochondria, **21:**5
  effect on pectinesterases, **20:**360
    effect on polygalacturonases, **20:**365
    effect on properties of pectic substances, **20:**354; 356
Electromagnetic energy, conversion to chemical, **25:**48–51
Electrometric titration, **42:**90; 91
Electron, microscope image trilaminar, **34:**467
  microscopy, **34:**473
  shuttle compounds, **34:**477; 484
  spin resonance spectroscopy, **34:**468
  transfer activity, **34:**477
  transfer chain, **34:**470; 477
  transport, **34:**480
  transport particles, **34:**477
Electron acceptor pigment P_{450}, of microsomes, **27:**190–191
Electron acceptors
  artificial, reactions of succinate dehydrogenase, **28:**176–178; 181–184
    cytochrome *c* as, **28:**172; 177; 211–215
    DCIP, **28:**159; 177; 181; 183; 190; 192; 194; 197–203; 210; 220; 222; 223
    ferricyanide as, **28:**169–172; 181; 183; 194; 197; 210–214; 225–227
    menadione as, **28:**211
    methylene blue, **28:**172; 173; 177; 210; 222; 225
    phenazine methosulfate, **28:**168–173; 176; 181–184; 194; 197; 210; 225; 227
    Wurster's blue, **28:**159; 172; 177; 178; 194
  in holoenzyme, **17:**273–281
    with metalloflavoproteins, **17:**239–242; 246–248; 251–254; 259; 264; 277–288
  for SDC, **18:**68–70; 86
Electron acceptor specifcity, for xanthine oxidose substrates, **17:**297–301
Electron and proton transport of energy in living cells, **14:**19–42
Electron carriers
  of aldehyde oxidase, EPR study, **27:**171–172
    iron as, **27:**89; 181
    metal chelates, **27:**179
  for succinic dehydrogenase, **18:**72; 84–87
Electron crystallography, ion channels and membrane proteins, **71:**212
Electron donors

dithiols as, **18:**342
in kinetic studies of peroxidase, **17:**55–58
β-Electronegative groups, elimination, **53:**163; 178–184
π-electrons, in imidazole, pyridine and pyrrole, **20:**53; 54; 57
Electronegativity
of metal ions, effect on enzyme activation, **21:**157–159; 162; 163
relation to toxicity of metal ions, **11:**71–74
Electron flow in tissues, **7:**46
Electronic displacement, during substrate binding, **50:**34
Electronic interaction, involved in purple intermediate, **34:**63–67
Electronic interpretations, of acetylcholinesterase action, **12:**290–311
of methyl group transfer, **12:**455–457
Electronic rearrangements, and reaction rate, **25:**12
Electronic rearrangements in irradiated enzymes, **24:**377–381
Electronic structure, and catalytic activity, **15:**19–20; 25–26; 29–30; 38–39
in activating group of enzyme, **15:**29–30; 40; 296–297
Electron ladder, and anion transport in plants, **18:**50–52
Electron microcopy, **48:**205
Electron microscope analysis
annexin-membrane complex, ion channel model system annexin V, **71:**222
Caulobacter cell cycle, flagellar structure and assembly, **71:**10
immunogod electronmicroscopy, *Caulobacter* cell cycle, **71:**26–27
Electron microscopy, **57:**234
applications of, **33:**98–113
astigmatic image, **33:**83
astigmatism, correction of, **33:**84
calibration methods for, **33:**88
contour phenomena, **33:**82–83
contrast considerations, **33:**86
with darkfield microscopy, **33:**114
defocus, relation of to resolution, **33:**85
determination of molecular weight, **33:** 71–118 (*See also* Subunit stoichiometry of enzymes)
history of, **33:**73–74
to examine structure of prolyl hydroxylase, **41:**251
experimental procedure for, **33:**78–82
future role of, **33:**113–115
instrumental and operating requirements, **33:**82–88
of lysosomes, **21:**41
of mitochondria, **21:**3–6; 83–86
of mitochondrial fragments, **21:**86–93
molecular dimensions, determination of, **33:**87–88
with phase-contrast microscopy, **33:**114
phase grain (*See* Electron microscopy, phase image)
phase image, **33:**83
resolution requirements for, **33:**84–86
results with small proteins, **33:**112–113
signal-to-noise considerations, **33:**86–87
specimen preparation, **33:**75–82 (*See also* Negative staining; shadow casting)
subunit numerology, information from, **33:**91
support films, preparation of, **33:**78–79 (*See also* Negative staining)
titling stage, **33:**97
Electron Microscopy of Enzymes (Haschemeyer), **33:**71
Electron microscopy of pyruvate carboxylase, **42:**19
Electron nuclear double resonance (ENDOR) spectroscopy, cytochrome *c* oxidase, **71:**84; 87; 92
Electron nuclear double resonance spectra, **59:**179; 195
Electron paramagnetic resonance, *See* EPR
cytochrome *c* oxidase, **71:**87; 93; 99; 108–109
Electron-paramagnetic-resonance (EPR) spectroscopy, **41:**38; 63
rapid-freeze, **41:**45
Electron paramagnetic resonance spectroscopy, **50:**339, *See* EPR spectroscopy
Electron paramagnetic resonance studies, of coenzyme Q, **25:**311; 314–316
Electron pooling of proteins, **15:**34–36; 37; 38
Electrons
fast, effect on enzymes, **24:**365; 388–390
low voltage, effect on enzymes, **24:** 402–403
high energy, and origin of life, **27:**352; 359; 384
unpaired, number from EPR spectra, **27:**114–116
unpairing for transfer, **17:**279–281
Electron sink, **49:**33; 35

Electron spin resonance, *See* ESR spectroscopy
bound ligand motion, eukaryotic dihydrofolate reductase, **70:**65
of enzymes, **22:**378
of flavines, **22:**377; 378
of Michaelis complex, **27:**20–21
in photosynthetic tissues, **25:**82–84
in study of corrinoids, **26:**242; 248; 254
Electron spin resonance and mechanisms of catalysis, **24:**471
Electron spin resonance signals from irradiated enzymes, **24:**377–379
Electron-spin-resonance spectroscopy, to detect O_2 radical, **41:**37
Electron transfer, **59:**149. *See also* Coenzyme Q, Cytochromes, and Photosynthesis; Cytochromes, Metalloflavoproteins and Respiratory chain
by cytochrome *c,* **17:**349–351
free energy changes, **17:**333–334; 347
mechanism of, **17:**109–112; 337–339; 382
effect of ionizing radiations, **24:**362; 373; 376; 378; 382
flavoproteins in, **19:**201
in photosynthesis, **19:**48; 49; 56
in reduction of oxygen to water, **19:**200; 201
in homogeneous catalysis, **25:**12; 21–32
introgenase, order of events associated with, **67:**355–356
and lipoic acid, **18:**341
in plants, **18:**11–13; 22; 303–311
by pyridine nucleotides, **15:**82; 84; 87; 113; 130
in oxidoreductions, **15:**9–16; 32–39; 41
rates, **25:**31
during respiration in animal tissues, **20:** 147–199
in plastids, **20:**227
Electron transfer, EPR spectroscopy, **27:**107; 166; 172; 183–188; 190–191
in mercaptan-disulfide exchanges, **27:**200
and origin of life, **27:**369–371
single, **27:**120; 125; 127
in slow reaction, **27:**3; 14; 33–34
Electron transfer chain, lipoproteins in, **21:**118; 119
order of components and structural pattern, **21:**119–123
Electron transfer (ET) (cytochrome *c* oxidase), **71:**82–83; 84; 114
complete turnover cycle model, **71:**180–186
enzyme turnover control and H + /e - ratio, **71:**164–165
fully reduced enzyme, **71:**124–125
generally, **71:**116
proton pumping, **71:**129; 131; 134; 141; 144–146; 154; 156; 158; 160
redox linkage, **71:**168; 174; 175
steady-state kinetics, **71:**116–118
transient kinetics, **71:**118–124
Electron-transfering flavoprotein, **17:**264–266; 285; 383
Electron-transfer models, absorption spectra, **26:**134–140; 143; 156–157
PN-dependent, **26:**132–157
Electron transfer oxidases, classification, **19:**80
in oxygen metabolism, **19:**201–211
properties, **19:**202; 203
Electron transfer particle (ETP), **18:**67–68; 106–107
Electron transfer path, in xanthine oxidase, **27:**166–168
Electron transfer reactions, **60:**66
Electron-transferring enzymes, product inhibition of, **25:**249–266
Electron transfers, in oxidation-reduction reactions involving quinones, **16:**108
in plants, relation to phenolase activity, **16:**131–136
Electron transfer sequence, **18:**99
Electron transfer systems, xanthine oxidase, EPR study, **27:**160–168
Electron transport
and ATP synthesis, **23:**327; 340–385
and copper, **23:**305–306
and phosphorylation, sequence, **23:** 373–374
and photophosphorylation, **23:**386
ferredoxins in, **33:**133–136
inhibition of, **17:**93–112
metalloflavoproteins in, **17:**284–285
monphosphorylating systems, **17:**81–82
possible mechanisms, **17:**84–87
and interenzymic organization, **18:**104–108
in mitochondria and mitochondria and mitochondrial fragments, **21:**73–129
sequence, **28:**160–161
Electron transport and oxidative phosphorylation, **24:**337

Electron Transport and Oxidative Phosphorylation (Green), **21:**73
Electron transport apparatus in photosynthesis, **25:**67–73
Electron transportases, classification, **14:**227
Electron transport chain
crossover theorem, **28:**84–85
and the elementary particle, **28:**159; 216; 219; 230
lipids of, **28:**160; 162; 170; 192–194; 202; 207; 220–223
reconstitution of, **28:**155–236
distribution in cell, **24:**300
Electron transport components, in mitochondria, **21:**12–21; 81; 94; 97–100
in mitochondrial fragments, **21:**81–83; 91–116
notation system for, **21:**101; 102
Electron transport particle, **25:**43; 69–70
Electron transport particles, components, **21:** 91–112; 114–116
isolation, **21:**102
properties, **21:**59; 60; 80; 83; 102; 103; 114
Electron transport particles (ETP), analysis and enzymic activities, **20:**152–153
composition, **20:**188
cytochromes in, **20:**162; 163; 168; 192
Electron transport pathway, dehydroascorbic acid reduction, **66:**180
Electron-transport processes, and light emission, **25:**145; 153–154
Electron transport system, **23:**341–354
role of coenzyme Q, **25:**283; 289; 294; 304–319
Electrophile, **39:**355
Electrophiles, **59:**132
Electrophilic, catalysis, **39:**356
catalysts, **39:**331
group, **39:**357
Electrophilic and nucleophilic, attack, concomitant, **39:**355
catalysis, concomitant, **39:**364
groups, **39:**357
Electrophilic atom, **46:**398
Electrophilic carbon, **46:**398
Electrophilic cataysis, catalytic mechanisms, **69:**34
Electrophilicity, as criterion for substrates, **46:**398; 400
Electrophilic replacement reactions, at methyl groups, **46:**464–465
Electrophilic substrate, xenobiotic substrate, **69:**24
Electrophoresis, **33:**251, **42:**79; 80, **44:**191, **48:**33; 36; 65
capabilities of, for detection and characterization of macromolecular interactions, **30:**140
convective disturbances, of, **30:**160
mobility, of enzyme-substrate complex, **30:**146
moving-boundary, for complex formation between pepsin and bovine serum albumin, **30:**146
computation of, **30:**145
non-ideal effects complication to, **30:**140
results of computations for, **30:** 154–163
time of development of patterns of theoretical, **30:**155; 157–161; 163–165; 170–171
transport equations for, **30:**148
multiple zones due to reversible protein-buffer interaction, **30:**168
protein-buffer interaction, **30:**162
single macromolecule showing two peaks in, **30:**163
zone, immune to non-ideal effects, **30:**140
results of computations for, **30:** 163–171
free boundary, **42:**79
paper, in identification of imidazole, **20:**74
in separation of polygalacturonases, **20:**367
on polyacrylamide gels, **42:**79; 81
of RNase, **24:**168–169
zone, **42:**79
Electrophoresis, SDS gel, **46:**387
Electrophoresis/chromatography, tyrosine hydroxylase regulation, **70:**145
Electrophoresis of tyrosinase preparations, **4:**146–148
Electrophoretic heterogeneity, of, hypoxanthine-guanine phosphoribosyl transferase, **41:**17
Electrophoretic mobility, **34:**489–490, **42:**87, **48:**59; 64; 67–69; 76; 78

Electrophoretic mobility, RNA polymerase II transcription
C-terminal domain, characteristics of, **71:**44; 50
reversible phosphorylation of CTD during transcription cycle, **71:**51
Electrophoretic patterns, light-induced changes in, **9:**126–127
Electrophoretic studies, of mercuripapain, **19:**278; 279
of papain, **19:**271; 274; 275
Electrophorus electricus, **43:**106; 133
Electrophorus electricus, **34:**525
Electrostatic effects, **47:**131. *See also* Buffer effects; Ionic strength; pH
glutathione S-transferases, **69:**21–22
Electrostatic effects on ezymatic site, **17:**37–40; 54
Electrostatic forces, in proteins, **50:**24
Electrostatic interaction, **34:**455
glutathione S-transferases, **69:**16; 17; 20
parameter, **34:**464
potential, **34:**461–462
Electrostatic interaction in 3-hydroxypyridines, **50:**4
Electrostatic interactions, rubisco, **67:**55–56
Electrostatic stabilization, **43:**378
Electrostriction effects, **39:**356
Elementary particle, in reconstitution of respiratory chain, **28:**159; 216; 219; 230
Elementary steps in enzyme reactions, **25:**1–38
Elements, abundance of, **27:**349
Eleocharis tuberosa, **7:**485
α, β-Elimination, **53:**164; 196
α-β-Elimination, **42:**288; 308
β, γ-Elimination, **53:**164
β-Elimination, **35:**507
β-elimination, **67:**15; 46
δ-Elimination, **53:**196
γ-Elimination, **53:**163; 189–191
Elimination of β-electronegative groups, **53:**163; 178–184
β-elimination reactions
cysteine *S*-conjugate β-lyase reactions, **72:**212–214
rat liver with kynureninase, **72:**216–219
Ellipticity, **42:**108
Ellman procedure, **43:**117
Elmulsin β-glucosidase, **17:**217
Elodea, hexokinase in, **19:**8
Elongation process, RNA polymerase II transcription, **71:**44
potential roles, **71:**64–69
El Tor Vibrio, **23:**183
Eluate, analysis in protein chromatography, **22:**447; 448
Eluate factor, **15:**124–125
Elution, negative, **46:**254
EMA, **34:**453; 458
chymotrypsin, **34:**453; 461
derivatives, **34:**455
EMA-papain, **34:**453; 460
Emasol, effect on, cytochrome a_3, **23:**298
haem *a,* **23:**290; 292; 293; 306–310
haematin *a,* **23:**291; 293; 296; 306–310
Emasol-haem-CO model compound, **23:**292; 293; 304; 306–312
Emasol reactivation, of cytochrome oxidase, **23:**310–311
EMA-trypsin, **34:**460–463
Embden-Meyerhof enzymes, and thiol group, **27:**213–214
Embden-Meyerhof pathway, **34:**219
non-use by *P. saccharophila,* **29:**409
and plant metabolism, **18:**294
Embryological studies and enzyme localization, **13:**15
Embryologic differentiation, causes, **8:**61–65
Embryonic development, muscle proteins and ATPase activity during, **13:**182–183
Embryonic tissue, effect of hyaluronidase on, **13:**229
Embryonic tissues
lactate dehydrogenase, **25:**353
pentose synthesis, **28:**418–419; 434–436; 448–449
Embryos, plant, respiration in, **18:**39–43; 286; 294
EM (effective molarity), **59:**135
Emerson enhancement effect, in photosynthesis, **30:**3; 17
Emission spectra. *See also* Light emission
luminescensce, **25:**124; 129; 130; 133; 136–142
Emission spectrum, of luminous bacteria, **7:**227
of *Photobacterium phosphoreum,* **7:**257
of *Proteus splendidum,* **7:**257
Emulsin, **48:**31; 51
almond, **4:**45–61 (*See also* Almond emulsin)
action on β-1,6 linkage, **28:**345

definition, **4:**61; 72
from sources other than almond, **4:**61–68
cacao, **25:**466
multiple forms of enzymes, **25:**350
fractionation of, **14:**401
separation by chromatography, **14:**349
as a β-glucuronidase, **16:**381
in synthesis of *d*-mandelic nitrile, **20:**7; 14
in tobacco, **6:**368
Emulsion stabilizer, effect on enzyme activity, **28:**26–27
Enalapril, **57:**37; 39–40
Enalaprilat
active-site conformation of, **57:**56–57
analogs of
aza, **57:**42–45
descarboxyenalaprilat, **57:**53–55
lactam, **57:**58–66
other, **57:**55–56
carbon of, **57:**40
conformation of, **57:**74
definition of, **57:**2
design of, **57:**32
as inhibitor of angiotensin-converting enzyme, **57:**9–10
interaction with angiotensin-converting enzyme
and apoenzyme, **57:**52–53
dissociation constant, **57:**52
inhibition development, **57:**49–50
initial rates, **57:**48–49
objectives of studies of, **57:**46–47
rate constant, **57:**51–52
slow-binding inhibitors, **57:**47–48
steady-state velocities, **57:**50–51
tight-binding inhibitors, **57:**47–48
lactam analogs of
bicyclic, **57:**62–66
computations for, **57:**59–60
seven-membered, **57:**61–62
simple, **57:**59–60
studies of, **57:**58–59
structure of, **57:**2; 44–45; 57–58
Enantimorphs, in competitive inhibition, **17:**49
Enantio-ferrichrome, **35:**145
Encephalitogenic basic protein, **42:**242; 264
Encephalomalacia, in chick, and vitamin E, **6:**485–486
Encephalopathy, hepatic, **53:**228
Endamoeba histolytica, ethionine studies, **25:**386
Endergonic reactions using ATP system, **14:** 14–17
End-group determination of virus (TMV) protein, **15:**463–464; 474
Endocytosis, **58:**103
of chondroitin sulfate, **60:**222
of glycosaminoglycans, **60:**222
of heparin, **60:**222
of hyaluronate, **60:**223
of proteodermatan sulfate, **60:**224
of proteoglycans, **60:**222
of proteoheparan sulfate, **60:**225
receptor mediated, **60:**222; 224
Endoenzymes, **56:**143
Endo-β-galactosidase, **60:**291
Endogenous proteases, adenosine receptors, **69:**96
Endogenous substrate protein, **42:**264
Endo-β-glucanase, **28:**341–342; 345–347; 349–353
Endoglucosaminidases, **60:**165
Endoglucuronidase, **60:**232
Endo-β-glucuronidase, **60:**284
Endoglycosidases, **60:**227–235
Endohydrolases, action on cell walls, **28:**342
Endohydrolysis, of polysaccharides, **28:** 349–351
Endo-laminarinase, **28:**342
Endomyces (Endomycopsis) vernalis, fat synthesis by, **9:**656–658; 658–661; 662–663; 663–664; 666; 667–670; 686; 689–692; 693–696; 699
protein synthesis by, **9:**659–661; 669
Endomycin, in treatment of plant diseases, **20:**392
Endonexin fold, annexin crystal structure, **71:**217
Endonuclease, of *Corynebacterium diphtheriae,* **29:**200
micrococcal nuclease as, **29:**191
from *Mustelus canis* liver, **29:**182
from *Neurospora arassa,* **29:**197
properties of, **29:**182
Streptodornase as, **29:**199
venom, **29:**175; 209; 211
Endonuclease I, nuclease of *E. coli,* **29:**197–198
Endonuclease mapping, tyrosine hydroxylase, **70:**179

Endonucleases
of *E. coli,* **60:**18
effect on single-stranded DNA, **27:**460
Endopeptidase, classification, **2:**55; 56
Endopeptidases, **23:**77; 240, **41:**237
classification, **14:**224–225; 232; 247
column chromatography, **22:**448–459; 484–485
disulfide exchanges, **27:**219
Endoplasmic reticulum, **58:**200
smooth, **46:**116; 117; 118; 119; 132; 135; 144; 149
Endoplasm of egg, **9:**245–246
changes following fertilization, **9:**280–286
Endorcrine control, *See* Hormones
Endo-α1r6-mannanase, **40:**264
ENDOR spectroscopy, **59:**179; 195
Endotoxin, **65:**309
extraction of, **31:**78
Endoxylanases, **56:**222
Endpiece of complement, **8:**101–111
Endpoint, **34:**500
Endproduct inhibition, in threonine deaminases biosynthesis, **37:**370
End product regulation, of enzyme activity, **28:**43; 70–71; 100; 128; 134; 135
Enduracididine, **50:**139
Endwise hydrolysis, of β-glucans, **28:**348–351; 353
Enediol, **59:**114
cis-Enediol, **43:**500
Enediol formation, in photosynthesis, **19:**23; 24
Enediol intermediate, **43:**492
Enediol mechanism, **43:**494
Enediols, and homogentisate oxidase, **19:**126
and model hydroxylating system, **19:**135; 199
Energetics and Related Topis (Bassham), **25:**39
Energetics of action, of Michaelis complex, **27:**29–32
Energetics of bacterial photosyntheses, **1:** 314–319
Energetics of carbon reduction cycle, free-energy changes, **25:**102–106
Energetics of photosynthesis, **25:**39–117
Energietransports innerhalb labender Zellen, Probleme des (Bücher), **14:**1
Energization, by cotransport with Na^+ flows, adequacy, **49:**69
Energized carriers, in oxidative phosphorylation, **23:**375–377
Energized membrane state, **44:**256
Energy barrier, and internal rotation, **23:**10
Energy barriers, to reversal of reactions, **28:** 46–47; 86–88
Energy changes, in interconversions of carbohydrates, **11:**329–330; 369–370
Energy conduction by proteins, **15:**33–37
Energy conservation. *See also* Oxidative phosphorylation
sites, in respiratory chain, **17:**105–106
localization of, **17:**88–112
Energy conservation site 1 and DPNH dehydrogenase, **34:**142
Energy coupling to transport, membranes, **44:**254
whole cells, **44:**255
Energy efficiency, photosynthetic, of carbon reduction cycle, **25:**48–51
enhancement of, **25:**62–64
of phosphorylation, **25:**57–58
Energy functions, of citric acid cycle, **18:**256; 259
Energy-linked DPN reduction, **34:**107
Energy-linked Ion Movements in Mitochondrial Systems (Lehninger, Carafoli and Rossi), **29:**259
Energy-linked reactions, **28:**157; 164
Energy metabolism, allosteric regulation, **28:**48; 64; 65
Energy migration, in irradiated enzymes, **24:**400
Energy minimization, **47:**47; 87; 132; 135
Energy of activation, *See* Activation of enzymes
Energy requirements, of photosynthesis, **25:** 46–48
Energy-rich bonds. *See also* Phosphate bonds, energy rich
of *S*-acyl derivatives of coenzyme A, **15:** 198–199; 201; 232
and protein synthesis, **9:**436–437
Energy rich phosphates, **43:**355
Energy sinks, effect of ionizing radiations, **24:**381–382
Energy source, glutamine as, **53:**224–226
Energy source for synthesis of di- and polysaccharides, **10:**128–129; 141
Energy sources, and origin of life, **27:**352; 355; 356; 382–386
Energy sources of protoplasm, **7:**54
Energy transfer
in muscle, relation to ATPase, **16:**339–343

in photosynthesis, **11:**117–119; 127–130; 133–134; 137; 145; 148–149; 150; 154–156; 166–167; 182–189; 191–195
Energy transfer in radiobiological processes, **24:**360–363; 365; 368; 370–371; 380–381; 387–390
Energy transfer mechanisms, of hydrolytic enzymes, **15:**295–297
Energy transfer mechanisms in normal and cancer tissues, **4:**206–232
control mechanisms, **4:**218–232
glycolytic mechanisms, metabolic pathways, **4:**206–209
oxidative hydrogen transport, **4:**212–213
oxidative mechanisms, metabolic pathways, **4:**209–212
phosphorylative mechanisms, **4:**213–218
Energy Transformations, Biological, and Cancer Problem (Potter), **4:**201
Energy transport, in living cells, **14:**1–48
in myoglobin system, **14:**185
Energy-yielding reactions, in nuclear metabolism, **16:**468; 471–473; 488–490
Enhancement of relation rates in paramagnetic macromolecular systems, *See* Relaxation rates, nuclear
Enkephalins, **53:**240; 271; 276
Enniatin, **11:**402–403
Enolase, **5:**133, **15:**172; 348, **18:**118; 293, **23:**339–340, **34:**448, **37:**482, **45:**352; 353, **46:**458
action of fluoride on, **5:**135
activation by metal ions, **21:**142–144; 150; 152; 153; 155; 156; 158–160; 162
kinetics, **21:**136; 139–141
active site, **22:**63; 78
column chromatography, **22:**483
reversible denaturation, **22:**76
structure, noncovalent forces in, **22:**73
chemical mechanism of, **45:**353
classification, **14:**220; 237; 256
isolation, **14:**397
effect of magnesium on, **12:**240
intracellular distribution, **16:**468; 488
kinetic mechanism of, **45:**352
mechanism of deduced from relaxivity data, **33:**49–50
in photosynthesis, **19:**10
residual activity, **21:**135
reversible denaturation, **26:**105
in sperm, **24:**63
Enolase, EPR study, **27:**140
Enolase of citrate enzymes, **43:**61
Enolases. *See also* specific enolases
multiple forms, **25:**343–344; 352
turnover number, **25:**32
Enolase synthesis, effect of nucleus, **24:**12; 20
Enolization, **43:**494; 510, **58:**28
of oxalacetic acid, **12:**68–73
of oxalosuccinic acid, **12:**88–89
Enol of pyruvate, **57:**147
P-Enolpyruvate carboxytransphosphorylase, **45:**88–108
electron micrographs of, **45:**92
formation of, **45:**88
ligands and subunit interactions, effect of, **45:**94–100
mechanism of reaction, **45:**102–108
CO_2 as active species, **45:**107–108
oxygen studies, **45:**103–105
pyruvate, irreversible formation of, **45:**105–106
reactions, partial, **45:**105
stereospecificity of reactions, **45:**106
metals and thiols, role of in reaction, **45:** 100–102
subunit structure of, **45:**90–94
5-Enolpyruvylshikimate 3-phosphate, **50:**148
Enoyl ACP reductase, **36:**301
Enoyl-CoA hydratase
localization of, **24:**332; 338
and propionic acid metabolism, **26:**304; 367; 369
Enoyl-CoA redyctase, localization of, **24:**299
Enoyl hydrase, in mitochondria, **21:**74
Entamoeba histolytica, **45:**108; 140, **59:**329–330
fermentation scheme, **45:**143
metabolism of, **45:**140–146
pyruvate, phosphate dikinase in, **45:**108
Entasis, **56:**313; 317
Entatic state, **43:**248
Enteric Bacteria, Gram-Negative, Biosynthesis of Cell Wall Polysaccharide in (Nikaido), **31:**77
Enterobacter cloacae, ADP-glucose pyrophosphorylases of, **46:**341
effect of energy charge on, **46:**350
Enterobacter cloacae, β-lactamase from, **28:**245; 290

Enterobacteriaceae, ADP-glucose pyrophosphorylases of, **46:**341–350
Enterobacteriaceae, L-cysteine biosynthesis in, **74:**186–189
pathways of, **74:**186–189; 187; 188
Enterobacteriaceae, immunologic polysaccharides, toxicity of, **26:**225
Enterochrome-556, *See* Cytochrome h
Enterococci, enzymes from, **23:**183
Enterocytes, **53:**217, **59:**372
glutamine utilization rate, **53:**218
Enterokinase, **53:**265
and elastase, **23:**225
Enteromorpha, photosynthesis in, **25:**59
Enterotoxin
ADP-ribosyl transferase assay, **58:**120
animal model for, **58:**118
assays for, **58:**119
cell-biological assay, **58:**120
functional domains, **58:**122
immunology, **58:**129
purification, **58:**119; 121
subunit structures, **58:**122
vaccines, **58:**130
vascular permeability assay, **58:**120
Enthalpy, **53:**252
cholesterol ester hydrolases regulation, **69:**61
Enthalpy change, of immunological precipitation reaction, **15:**70
of isohemagglutination, **15:**57; 58–60; 61–63
Enthalpy of activation, **43:**323
Entner-Doudoroff pathway, **48:**100; 123; 171
use by *P. saccharophila,* **29:**409
Entrapping method, **34:**452
Entropy, **38:**196, **39:**356, **43:**268; 283, **51:**85; 94. *See also* Thermodynamic constants
activation, **43:**283; 288; 289; 293
in biochemical processes, **11:**1–33
compensating changes, **43:**291
of enzyme activation, **9:**43–48; 58; 60–67
of enzyme inactivation, **9:**32–38
fusion, **43:**286
hydrogen bonds, **43:**317
internal, **43:**277; 286; 290; 321; 346; 359; 363
intrinsic, **43:**288; 290
ionization, **43:**290
loss with destabilization, **43:**351
moments of inertia, **43:**322
observed, **43:**287
peptide chain, **43:**317
problems, **43:**283
of protein denaturation, **9:**32
radical reactions, **43:**289
requirements, bimolecular reactions, **43:**275
rotational, **43:**275; 284; 322; 359
solids, **43:**286
solution, **43:**277; 283
solvent effects, **43:**288
substrates, two or more, **43:**363
translational, **43:**276; 322; 359
utilization, **43:**323
vaporization, **43:**284
vibrations, **43:**286
water, **43:**307
Entropy, Nature and Role in Biochemical Processes (Gutfreund), **11:**1
Entropy change
of immunological precipitation reaction, **15:**70; 72; 75
in isohemagglutination, **15:**61–63
for ionization of carboxyl ion, phenol ion, imidazole and amino groups, **20:**64–66
Entropy-enthalpy compensation, **39:**383
Entropy of activation of RNase, **24:**455–458
Entropy strain, **59:**148
Enucleated cells, metabolism, **16:**413; 469–474; 477; 485; 491–492
Enucleation experiments on enzyme synthesis, **24:**11–14; 19–20
Environment, **34:**527
lipid, intermediate, **34:**477
Environmental temperature, **34:**469
Enzymatic activity
and external fields, **57:**229
fish bile, **1:**393; 395
and mechanochemical phenomena, **57:** 229–232
of normal and cancerous animals, **3:**335–340
regulation of, **57:**232–233
of tumor-bearing and normal tissues, **3:** 315–334
Enzymatic assay, **42:**247
Enzymatic breakdown of glycogen, **3:**121; 124–126
Enzymatic Catalysis, Tautomerism in Pyridoxal Phosphate and (Metzler), **50:**1
Enzymatic catalysis, tautomerism of, **50:**1–37
Enzymatic digestion, **42:**131; 133
Enzymatic isomerization, **59:**138
Enzymatic processes, multistep, **50:**34–37

Enzymatic reactions
influence of hormones, **4:**257–264
mechanisms of hydrolysis, **4:**43
study of
fixation of substrate-induced conformations, **57:**204–209
preparation of enzyme intermediates, **57:**209
prevention of protein-protein interactions, **57:**209–210
in vitro and in the living cell, **3:**172–174
Enzymatic synthesis, **1:**332–334; 347
of glycogen, **3:**129
D-Enzyme, *See* Disproportionating enzyme
κ-enzyme, **23:**186; 188; 191; 198
θ-enzyme, **23:**186
Enzyme 205–2, methyl synthesis, **25:**405–408
Enzyme A, in *B-coli,* **25:**405–406
Enzyme action, **73:**25–55
on adsorbed substrates, **33:**266–269
on chitin, **33:**261
on coacervates, **33:**265
on elastin, **33:**257–258
on fats, **33:**262
freezing effects on, **33:**286
in frozen foods, **33:**285–286
on gelatin, **33:**257
on gels, **33:**255–259
in ice, **33:**285–286
of insoluble enzymes of soluble and insoluble substrates, **33:**269–283
on insoluble substrates, **33:**255–269
on lipids, **33:**263–264
at low humidity, **33:**283–284
at low temperatures, **33:**285–286
pH optima of (table), **33:**267
on polymeric esters, **33:**264–265
on proteins, **33:**255–259
in solid enzymes, **33:**284–285
on starch, **33:**260
algebraic *versus* graphic analogies, **73:**49–52
catalysis and binding, **73:**51–52
ground state binding, **73:**51
mutant enzymes, **73:**49–50
transition state binding, **73:**50–51
analogy, **73:**25–26
catalytic power definition, **73:**34–39
catalytic power origins, **73:**43–49
ground and transition states reciprocal stabilization, **73:**47–49
reactant ground state stabilization, **73:** 43–46
transition state stabilization, **73:**46–47
conflict resolution, **73:**52–53
congruency kinetics, **73:**41–43
steady-state hypothesis, **73:**26–29
thermodynamic cycle, **73:**39–41
V/K *constant* definition, **73:**29–31
V/K *constant* significance, **73:**31–33
antibiotic resistance, **73:**31–32
catalytic proficiency, **73:**33
perfect enzymes, **73:**32–33
Enzyme Action, Active Site and (Kosland), **22:**45
Enzyme activation, **53:**322
activators, **53:**377
by Cat ligands, **53:**337
by displacing inhibitory ligands, **53:**330–331
Enzyme activators, **7:**39
Enzyme activity
adsorption, effect of, **28:**3; 28–30
soil, qualitative, **28:**371–372
quantitative determination, **28:**372–385
and amino acid modification, kinetic analysis, **22:**60–63
and catalysis, **22:**81–83
effect of shape alterations on, **22:**75; 76
factors influencing velocity of, **22:**84; 90
increase, as a function of total protein synthesis, **22:**114; 115
induced fit model of, **22:**90
pH dependence, **22:**91
proximity and orientation effects in, **22:** 83–89
and side chain modification, **22:**60–63
and structure modification, **22:**90; 91
and conformation, **27:**403
effect of disulfides, **27:**212–217
in evaluation of mitochondrial purity, **21:**76; 77
and heme-linked groups, **7:**290
and temperature, **7:**259
imidazole in, **20:**61–82; 100
Enzyme Activity, Allosteric Regulation of (Stadtman), **28:**41
Enzyme Activity, Genetically Controlled Differences in (Fincham), **22:**1
Enzyme Activity, Irreversible Modification of Kinetics of Substrate Reaction During (Tsou), **61:**381

Enzyme Activity in Frozen Vegetable Tissue (Joslyn), **9:**613
Enzyme adaptation, definition, **22:**101
Enzyme amino acyl adenylate complexes, **35:**484
Enzyme-antibody complex, **37:**594
Enzyme-antibody interactions, β-lactamase, **28:**304–307
Enzyme B_1, in DR metabolism, **28:**442
Enzyme B_2, in DR metabolism, **28:**442
Enzyme binding. *See also* Enzyme complex, Enzyme-coenzyme complex, and Ternary complex
by chromatin, **27:**266–268; 278; 282–283
of inhibitors, **27:**450–453
Enzyme-binding moiety of DPNH, **27:**56–70
Enzyme biosynthesis, **24:**1–33
Enzyme-biotin-CO_2, **35:**325; 375
Enzyme-biotin-CO_2 complex, **42:**5
Enzyme-bridge complexes, **33:**30
Enzyme catalysis, **51:**48–55, *See* NMR studies of mechanisms of enzyme action
mechanisms related to, **24:**441–482
subzero temperatures, **52:**179
Enzyme Catalysis, Mechanisms Related to (Westheimer), **24:**441
Enzyme-Catalyzed Chemical Modification, Regulation of Enzymes by (Holzor), **32:**297
Enzyme-catalyzed O-exchange, **50:**344
Enzyme-catalyzed reactions, **3:**45–62
mechanisms of, **45:**273–387
selectivity, **58:**10
velocity of, **58:**10
Enzyme-Catalyzed Reactions, Determining the Chemical Mechanisms of, by Kinetic Studies (Cleland), **45:**273
Enzyme-Catalyzed Substitutions at Each of the Phosphorus Atoms of ATP, Role of Metals in (Mildvan), **49:**103
Enzyme-CO_2, **35:**329
Enzyme-coenzyme complex. *See also* Coenzyme, Enzyme complex, and Ternary complex
chromophoric, **27:**58–60; 98
Cotton effect, **27:**50–54; 56–70
effect of substrate homologs, **27:**71–72
stereochemistry of, **26:**156–165
Enzyme-coenzyme substrate, role of metal ion, **27:**72
Enzyme complex. *See also* Coenzymes, Enzyme binding, Enzyme-coenzyme complex, Michaelis complex, and Ternary complex
binding site on DPNH, **27:**56–70
crystallography, **27:**72
extrinsic Cotton effects, **27:**56–70
formation, **10:**73–74; 95–97; 102–104; 111; 113–117; 201–202; 204–208
LADH, dissociation constants, **27:**63–64
sulfate reducing, **27:**215–216
YADH, **27:**83–85
Enzyme concentration
effect of substrate, **8:**108
and rate of reaction, **7:**4
Enzyme content of soils, effect of added substances, **28:**368
Enzyme coupling, definition, **28:**167
Enzyme determination, in soils, principles, **28:**369–371
Enzyme distribution
in cell, from radiation study, **24:**402–403
in subcellular fractions of animal tissues, **24:**291–358
methods for study, **8:**30
Enzyme efficiency
calculation, **4:**42
of glycosidases from almond emulsin, **4:**47; 48; 50; 52; 54; 57; 58; 59; 61; 63
Enzyme extracts, oxidation of catechol, **27:**510
Enzyme formation, **37:**607
in nongrowing cells, **22:**124–127
Enzyme function, **37:**607
Enzyme induction
effects of mutation in, **22:**19–24
nucleic acids in, **22:**137–140
precursor in, nature, **22:**123; 124
and repression, theory of mechanism, **22:**24
specificity, **22:**117–121
in mammalian system, **29:**375–379
membranes and, **29:**376–379
of TNP-linked enzymes and TPNH utilization, **29:**374–375
mechanisms for, **44:**252
Enzyme inhibition and lipides, **8:**44
Enzyme-inhibitor complexes, and Cotton effect, **27:**87
Enzyme inhibitors, *See* Inhibitors
Enzyme Inhibitors, Slow-Binding, The Behavior and Significance of (Morrison and Walsh), **61:**201

Enzyme inhibitors and immune hemolysis, **8:**97
Enzyme kinetics. *See also* Kinetics of enzyme reactions
activation energy
of enzyme systems, **3:**46; 47; 51–53; 59; 60
of extracted and intracellular enzymes, **3:**55–57
and nature of substrate, **3:**49; 53–55
activation entropy, **3:**58
bibliography, **3:**60–62
covalent bond cleave study, **53:**341
dissociation of enzyme-substrate complex, **3:**57
inactivations of enzymes by heat, **3:**41–45
with ligand substitution reactions, **53:** 316–336
ligand substitution reactions study, **53:**341
mixtures of surrogates, **53:**327–330
modified Briggs-Haldane formulation, **53:**316–317
modified Michaelis-Menton rate equation, **53:**318
product dissociation as possible rate-determining step, **53:**311–312
temperature dependence of enzyme-catalyzed reactions, **3:**45–62
temperature dependence of inactivation of enzymes, **3:**41–45
Enzyme Kinetics, Effects of Temperature on (Sizer), **3:**35
Enzyme Kinetics (Alberty), **17:**1
Enzyme levels
drug-induced changes, **37:**163–167
genetic regulation of, **37:**167–172
Enzyme Levels, Mammalian Tissues (Schimke), **37:**135
Enzyme-linked immunosorbent assays, RNA polymerase II transcription, reversible phosphorylation of CTD during transcription cycle, **71:**53
Enzyme mechanisms
nuclear spin relaxation induced by paramagnetic probes and, **33:**1–70
positional isotope exchange studies, **50:**361–393
Enzyme Mechanisms, Positional Isotope Exchange Studies of (Rose), **50:**361
Enzyme Mechanisms, Study of, by Nuclear Spin Relaxation Induced by Paramagnetic Probes (Mildvan and Cohn), **33:**1
Enzyme memory mechanism, **57:**383
Enzyme-metal complexes, in metal ion activation of enzymes, **21:**138–144; 155; 156
Enzyme-metal substrate interactions, EPR study, **27:**140
Enzyme models, **58:**1
and oxidation, **58:**47
for zinc binding sites, **58:**9
Enzyme multiplicity
basis of, **28:**352
as control mechanism, **28:**45; 60–64
Enzymen, Multiple Formen (Wieland und Pfleiderer), **25:**329
Enzymen, Saulenchromatographie von (Turba), **22:**417
Enzyme-oxygen-substrate complex
of epoxide, **27:**519–520
oxidation of hydrocarbons, **27:**501; 538
Enzyme P, tumor, in DR metabolism, **28:**440; 441
Enzyme Problems in Relation to Chemotherapy, Adaptation, Mutations, Resistance, and Immunity (Sevag), **6:**33
Enzyme-product complex, stability, **27:**128
Enzyme Proteins, Denaturation and Inactivation of (Okunuki), **23:**29
Enzyme purity, **25:**330–333
Enzyme reactions, electric charges in, **33:**263
in heterogeneous systems, **33:**245–308
locus effects in, **33:**250–254
and permeability, **7:**1–34
pH dependence of, **33:**288
pH effect on, **33:**249–250
rate equations for, **33:**247–250
surface tension in, **33:**263
water effect on, **33:**249–250
Enzyme Reactions, Elementary Steps as Studied by Relaxation Spectrometry (Eigen and Hammes), **25:**1
Enzyme Reactions in Heterogeneous Systems (McLaren and Packer), **33:**245
Enzyme regulation, molecular mechanisms of, **37:**172–179
Enzyme repression, **22:**121; 122
definition, **22:**122, **40:**74
in arginine biosynthesis, **40:**73

Enzymes, **34:**445; 452–457; 467; 472; 478–479; 485–487; 491–495; 499; 513–524, **44:**189, **48:**59–60. *See also* Adaptive enzymes; Enzymatic activity: Enzyme reactions; Enzymes of 2′,5′-oligoadenylate metabolism; Enzyme synthesis; Genetic factors in enzyme synthesis; Glutathione-dependent enzymes; Hydrolytic enzymes; Immobilized enzymes; NMR studies of mechanisms of enzymes; Proteins; Proteolytic enzymes
absorption spectra, **4:**277; 278; 304; 306; 307
acid invertase, **56:**155–163
activated by polyelectrolytes, **40:**32
activated by polynucleotides, **40:**47
of arginine biosynthesis, **40:**65
containing bound polynucleotides, **40:**47
inhibition by polyelectrolytes, **40:**50
specific interaction with polynucleotides, **40:**44
yeast mannan degradating, **40:**262
activation, **46:**40
affinity, bound to effector site, **46:**236
concentration in tissues, **46:**405
freezing, **46:**263–264
isoelectric point, **46:**386
mechanism, **46:**260
purification of, **46:**247–249; 254–256
technology, **46:**265
active site of, serine residue as part of, **33:**385
bound, in columns, **33:**278–279
comparison of with solubilized enzymes, **33:**286–294
in films and membranes, **33:**279–283
in suspension, **33:**272–278
cell surface (*See* Cell surface enzymes)
conformational flexibility of, **33:**436
covalently bound, **33:**272–279
electron microscopy of (*See* Electron microscopy, of enzymes)
flexibility of, **33:**433
insoluble (*See* Insoluble enzymes)
kinetic constants for soluble and insoluble, comparison of (table), **33:**270
particulate (*See* Particulate enzymes)
solid (*See* Solid enzymes)
subcellular (*See* Subcellular enzymes)
activity, **34:**446; 475; 480; 491–493; 500; 509–512
adenosylmethionine decarboxylase, **56:** 265–275
adsorption, **48:**335
alcohol dehydrogenase, **56:**292; 318–326
aldolase, **56:**385–388
alkaline invertase, **56:**155; 156; 158; 162
alkaline phosphatase, **56:**341–350
allantoicase, **56:**92; 95; 99
allantoinase, **56:**92; 98; 99
allantoin permease, **56:**98; 99
alliinase, **56:**10; 34–35; 37; 42
allophanate hydrolase (*See* Allophanate hydrolase)
allosteric, **57:**204; 210
aminolevulinic acid dehydrogenase, **56:** 397–401
aminopeptidase, **56:**292; 348
amylases, **56:**143–144; 292; 316
angiotensin-converting enzyme, **56:**356–358
arabinofuranosidase, **56:**150; 151
arabinopyranosidase, **56:**150; 151
arabinosidases, **56:**150–154
of arginine synthesis
activity in extrahepatic tissue, **39:**60
activity in kidney, **39:**59
intracellular concentrations of, **39:**56
artificial, **58:**1
artificially immobilized, **34:**448
artificial membrane, **34:**447; 495
aspartate carbamylase, **56:**329; 333; 334
aspartate transcarbamylase, **56:**330–334
aspergilli as aerobic enzymes, **2:**185–187
assay, **4:**232–240
detection of absolute amounts, **4:**239–240
detection of catalytic effect, **4:**235–238
homogenates as test materials, **4:**232–235
slice technique, **4:**204; 205; 232
assays, **47:**354–356
asymmetric membrane, **34:**455
autosynthesis, **4:**4; 8
B gene, **36:**145–147
carbonic anhydrase (*See* Carbonic anhydrase)
carboxypeptidase (*See* Carboxypeptidase)
catalysis, **34:**461
catalytic efficiency of, **57:**108
characterization, **22:**105–107
column chromatography, **22:**417–488
chemistry of, **34:**2
classification of, principles of, **56:**2–3
collagenase, **56:**282–284

columns, **34:**447
complex formation, **58:**11
concentrations, **34:**494; 512
copper-containing, **57:**136
creatininase (*See* Creatininase)
crystalline, **39:**334
labeled, **39:**352
crystals, **34:**501–502
cyanoalanine synthase (*See* Cyanoalanine synthase)
cystathionase (*See* Cystathionase)
cystathionine β-synthase (*See* Cystathionine β-synthase)
cysteine lyase (*See* Cysteine lyase)
cysteine synthetase, **56:**25
deaminase, AMP, **56:**293
definition, **1:**190
definition of, **29:**78
feedback effects and repression of, **29:** 367–371
FORTRAN program for fitting kinetic data, **29:**23–31
statistical analysis of kinetic data of, **29:** 1–32
degradation of, **37:**176–179
half-lives, **37:**141; 149
levels in different tissues, **37:**156
synthesis, **37:**172–176
denaturation, **48:**328
denaturation of, **57:**216–219
detoxification, **57:**92–93
digestive, **1:**371–374; 378–383; 390–397; 407; 408; 411; 412
dihydroortase, **56:**293
dihydropyrimidine aminohydrolase, **56:**293
dipeptidase, **56:**292
effect of drugs, **1:**194–196
elastase, **56:**293
esterase, **34:**4; 9
evolution of, **38:**205–207
exogeneous administration, **38:**310
mutant, **38:**309
origin of, **38:**203–210; 227
therapy, **38:**308–310
FMN requiring, **57:**97
fructofuranosidases, **56:**155–165
fucosidase, **56:**220
galactosidases (*See* Galactosidases)
gel polymerize, **34:**452
A gene, **36:**142–144
and genes, **4:**14; 16; 19; 21; 24–26; 28
glucosidase (*See* Glucosidase)
glucuronidase, **56:**220–221
glycosidases (*See* Glycosidases)
H gene, **36:**141
and hormones, **1:**190–192
hydrolases (*See* Hydrolases)
hydrolysis, **34:**488–489
identification, **1:**186; 187; 190; 192
immobilized, **34:**455
inhibitors, **1:**194*t;* 195; 196
and insulin, **57:**157–158
intermediates, **57:**209
invertase (*See* Invertase)
iron-containing, **57:**136
isomerases, **56:**293; 401
lactate dehydrogenase, **56:**292
Le gene, **36:**141–142
leucine aminopeptidase, **56:**350–355
lignases, **56:**293
in the living cell, **1:**330; 331; 334–342; 346; 347; 366; 367
lyases (*See* Lyases)
mammalian, **57:**86; 130–133
mannosidase (*See* Mannosidases)
membrane, **34:**447; 495; 500–505; 509
artificial enzyme, **34:**447
-bound, **34:**446–447; 472; 478–481; 494; 501
preparation and structure, **34:**454
in membrane system, **57:**219
metal coordination, **58:**11
metalloenzymes (*See* Metalloenzymes)
metallothionein, **56:**401–407
methionine adenosyltransferase, **56:**251–265
mouse, **57:**396–397
N-acetylglucosaminidases, **56:**145–149
native, **34:**462; 504–508
native particulate, **34:**447
nuclease and nucleotidase, **56:**292
one-sided and reversible actions, **1:**330
ornithine decarboxylase, **56:**273
ornithine transaminase, **56:**134
oxidoreductases, **56:**318–329
particle, **34:**462
peroxide-producing, **57:**87
peroxisomal, **57:**94
phosphodiesterase, **56:**292
phosphorylating, **34:**4
phosphoserine phospholase, **56:**19–20

Enzymes (*Continued*)
physiological role, **1:**347–365
polyelectrolyte conjugate, **34:**462
polymerase, **56:**335–341
procarboxypeptidases, **56:**292; 293
and production of oxalate, **57:**150
proteases, **34:**12–14, **56:**293; 304
proteinases, **56:**2
proteolytic, **4:**2, **34:**4; 9; 455
purification, **48:**38
pyridoxal-P-dependent (*See* Pyridoxal-P-dependent enzymes)
pyrophosphatase, **56:**293
rat, **57:**393–396
reaction, **34:**447; 476; 494; 502; 509; 513; 526–527
reactivation of, **57:**217–219
regioselectivity, **58:**10
resolution, **1:**94
respiratory (*See* Respiratory enzymes)
rhamnosidase, **56:**221
role in wine making, **3:**381–384
other enzymes, **3:**384
oxidases of the grape, **3:**381–383
pectic enzymes, **3:**383–384
and sampling of soil, **28:**368–369
secretion, **34:**482
selectivity, **58:**10
soluble, **34:**447; 481
specificity, **48:**31
specificity of prosthetic group and protein, **1:**177; 178
stabilization of, **57:**219
stereoselectivity, **58:**11
structure, **1:**193
substrate activity, **34:**489
substrate binding, **58:**11
substrate geometry, **58:**27
substrate-induced conformation of, **57:** 204–209
substrate interface, **34:**491
substrate specificities of, **57:**361
sulfhydryl (*See* Sulfhydryl enzymes)
synthesis, repression-induction in, **50:**327
synthesis of peptide bonds, **1:**90–93; 94; 96
synthetic, **34:**494
in tea-fermentation, **2:**115–120
thermolysin, **56:**376–382
thio-disulfide regulation, **66:**184–187
three-layer membrane, **34:**454
thrombin and blood coagulation, **2:**287
and trace substances, **1:**177–198
transferases, **56:**292; 329–341
transfer of phosphate, **34:**9
trehalases, **56:**215–219
tryptophanase, **56:**11; 41; 42; 43; 45
turnover number, **1:**178
urea amidolyase, **56:**94; 95; 96
urea carboxylase, **56:**94; 97; 98
of urea synthesis
activity in liver, **39:**58
activity measurements of, **39:**57
intracellular concentrations of, **39:**56
ureidoglycollate hydrolase, **56:**92; 98; 99
utilization of S and Se compounds as substrates, **48:**7
vestigial, **57:**92–93
water-insoluble derivatives, **34:**447–448
xylosidases, **56:**221–223
yellow, amino acid content, **3:**70
zinc-containing, **57:**136
zinc's role in, **56:**285–286; 290
Enzymes, Absorption Spectra (Brode), **4:**269
Enzymes, Adsorption Studies (Zittle), **14:**319
Enzymes, Aspects of the Biosynthesis of (Chantrenne), **24:**1
Enzymes, Bacterial, Synthesis of Disaccharides with (Hassid and Duodorff), **10:**123
Enzymes, Biosynthese induite des (Monod and Cohn), **13:**67
Enzymes, Biotin-Dependent (Moss and Lane), **35:**321
Enzymes, Classification and Nomenclature (Hoffman-Ostenhof), **14:**219
Enzymes, Conformational Adaptability (Citri), **37:**397
Enzymes, Distribution of, Between Subcellular Fractions in Animal Tissues (Duve, Wattinaux, and Baudhuin), **24:**291
Enzymes, Effects of Ionizing Radiation on (Augenstine), **24:**359
Enzymes, Electron Microscopy of (Haschemeyer), **33:**71
Enzymes, Formation of Secondary and Tertiary Structure of (Straub), **26:**89
Enzymes, Hydrolytic, Action (Lindley), **15:**271
Enzymes, Hydrolytic, Kinetics (Van Slyke), **2:**33
Enzymes, Interactions with Polyelectrolytes and Polynucleotides (Elbein), **40:**29

Enzymes, Isolation (Schwimmer and Pardee), **14**:375
Enzymes, Kidney, and Essential Hypertension (Schales), **1**:513
Enzymes, Metal Ion Activation of (Malmström and Rosenberg), **21**:131
Enzymes, Molecular Size Determination of, by Radiation Inactivation (Kempner), **61**:107
Enzymes, Origin of, and Origin of Life (Oparin), **27**:347
Enzymes, Oxidative, Contributions of EPR Spectroscopy to Our Knowledge of (Beinert and Palmer), **27**:105
Enzymes, Photochemistry (McLaren), **9**:75
Enzymes, Product Inhibition (Walter and Frieden), **25**:167
Enzymes, Proteolytic, Classification (Bergmann), **2**:49
Enzyme S, tumor, in DR metabolism, **28**:440; 441
Enzymes and enzyme components, relative activity of
  in *normal tissues* of normal and of tumor-bearing animals, **3**:335–340
    in blood and serum, **3**:340–344
    in kidney, spleen and adrenals, **3**:337–338
    in liver, **3**:335–337
    in muscle, **3**:338
  in *tumors* and normal tissues, **3**:315–334
    adenocarcinomas mucosa, **3**:329–330
    carcinoma of prostate, **3**:330–332
    hepatomas and liver, **3**:318–327
    lymphomas and lymph nodes, **3**:327–328
    mammary tumors and breasts, **3**:328–329
    osteogenic sarcoma and bone, **3**:332–333
    rhabdomyosarcoma and muscle, **3**:329–330
Enzymes and Trace Substances (Green), **1**:177
Enzymes of Arginine and Urea Synthesis (Ratner), **39**:1
Enzymes of Arginine Biosynthesis and Their Repressive Control (Vogel and Vogel), **40**:65
Enzymes of Cytoplasm, Localization (Holter), **13**:1
Enzymes of 2′,5′-oligoadenylate metabolism
  mechanism of action, **59**:218–221
  2′,5′-oligoadenylate synthetase, **59**:216–217
    induction of, by cAMP, **59**:221–222; 224
    inhibition of, **59**:224–230
  2′-phosphodiesterase regulation by cAMP involvement, **59**:225–230
Enzymes of Snake Venoms (Zeller), **8**:459
Enzymes of Starch Degradation and Synthesis (Bernfeld), **12**:379
Enzyme specificity, **33**:429–438, **37**:605. *See also* Specificity
  history of, **33**:401–403
  induced fit model of, **22**:77; 78
    template model of, **22**:77
  operational definitions of, **33**:403–404
  in regulation of fatty acid chain length, **36**:296–298
Enzyme structures, secondary and tertiary, **26**:89–114
Enzyme-substrate complexes, **35**:486. *See also* Kinetics, enzyme; Michaelis constant
  affinity constants, **9**:2; 4; 9–15; 16–22; 56–60
    effect on activation energy, **9**:46
    mechanism of formation, **9**:60
    spectroscopic evidence for, **9**:58
  of aspartate aminotransferase, **50**:16; 17
    crystalline, **50**:16; 17
  crystallization of, **27**:2–4; 9–10; 16–23
    mechanism study, **27**:1–36
    slow reaction, **27**:12–16
  in deaminations, **16**:223
  effect of film formation, **28**:25
  effect on mechanism of action, **19**:245
    nature of combination and cleavage, **19**:313
  and entropy of activation, **24**:455–458
    binding sites for, **24**:116–117
  of gelatin, **33**:256; 258–259
    in rate equations, **33**:247–249
  glycosidases, mechanism of hydrolysis, **4**:42–44
    specificity of aglycon group, **4**:53–55
  influence of thiol groups, **11**:232–233
  mechanisms of product inhibition, **25**:183–220
  rate constants, **25**:14
  relaxation spectrometry studies, **25**:8; 12–15
  stability, **26**:93
  in transglycosylation, **17**:221–224
    Michaelis-Menten mechanism with, **17**: 9–10
    spectrophotometric measurements of, **17**:42–45; 55–58
    xanthine oxidase, **17**:308–309
  tyrosinase, *o*-dihydric phenols, **4**:131; 132
    monohydric phenols, **4**:136–139

Enzyme-substrate complex model, **27:**3–12; 23–28
Enzyme-substrate compounds, **12:**153–190. *See also* Peroxidase kinetics
effect of copper on formation, **14:**140; 144
peroxidase kinetics, **73:**3–23
Enzyme-substrate concentration and reaction rate, **7:**3–4
Enzyme-substrate interaction, electron unpairing, **27:**176
Enzyme-substrate interactions, β-lactamase, **28:**297–304
Enzyme-substrate intermediates, **51:**61–65
Enzyme-substrate-metal complexes, in metal ion activation of enzymes, **21:**139; 140; 149–151; 162
Enzyme surface, dissymmetry of, **27:**38–39
and inhibitors, **27:**442–453
Enzyme synthesis
of chitin
in animals, **59:**81–83
assay of, **59:**62–63
in fungi, **59:**63; 65–76; 78–81
of chitosan, **59:**83–84
control of, in neoplastic cells, **29:**371–379
convergence hypothesis of, **29:**371
dietary enzyme induction of, **29:**372–373
hormonal induction and possible mechanisms for enzyme alterations in, **29:**374–379
lack of specificity of inhibitors of, **29:**376
relationship of transport phenomena to the control of, in neoplastic cells, **29:** 379–381
repression by amino acid analogs, **29:**93
induced, kinetics of, **22:**107–117
induction of, **27:**455
repression of, **28:**61; 70; 72–73
Enzyme systems, complex, distribution of, **24:**335–342
effect of radiations, **24:**399
simple, distribution of, **24:**295–335
Enzyme-thiol ester intermediate, **23:**329
Enzyme-thiosulfate, complex, **39:**345
Enzyme-tRNA complex, **35:**477; 482
Enzyme tryptophan, fluorescence, **43:**175
Enzyme turnover control, H + /e - ratio and, cytochrome *c* oxidase, **71:**162–167
Enzyme X, **4:**249; 250
and cancer virus, **4:**228; 229; 230; 231
Enzymic cleavage, of alliin, **11:**381; 390–393; 398
of sulfoxides, **11:**393–395
Enzymic metal ion, **39:**358
Enzymic nucleophile, **39:**356; 357
Enzymic Reactions, Periodic, and Their Possible Applications (Christiansen), **23:**83
Enzymic reactions as chain reactions, **8:** 450–451; 453
Enzymic sulfhydryl group, essential, **39:**358
Enzymoids, **24:**121–123
Enzymology. *See also* Mixed solvents; Subzero temperatures
and the blood clotting mechanism, **30:** 255–315
kinetic isotope effects in, **46:**415–494
problems of, **57:**196; 204
at subzero temperatures, **45:**157–271
Enzymology, 2,3-Bisphosphoglycerate (Rose), **51:**211
Enzymology, Immunochemistry, and Related Subjects, Chemical Modifications of Proteins and Their Significance in (Sri Ram, Bier, and Maurer), **24:**105
Enzymology at Subzero Temperatures (Douzou), **45:**157
Enzymology of Human Alcohol Metabolism (Li, Ting-Kai), **45:**427
Enzymology of Quinoproteins (Duine, Jzn and Jongejan), **59:**169
E_{ox} S, *See* Oxidized enzyme-substrate complex
EP, *See* Elementary particle
Ephedrine, **55:**308–310
antagonism to adrenaline, **6:**40–41
Epichlorhydrin, **58:**15
effect on pectic substances, **20:**346
Epichlorohydrin, **55:**243
Epicoccum oryzae, enzymic transformations of steroids by, **20:**254
Epicorticosterone, **19:**191
11-Epicorticosterone, formula, **20:**241
microbiological formation, **20:**246
Epicuticle, definition, **16:**153
Epidermal cells, *See* Cells, epidermal
Epidermal growth factor (EGF), **47:**84; 137
effect on ornithine decarboxylase levels, **36:**246
tyrosine hydroxylase regulation, **70:**146; 150
Epidermal proteins, binding of pigments, **16:**172

Epidermin, spatial arrangements of polypeptide chain, **13:**159
Epidermis
of arthropod integument, **16:**153
elimination of optical antipodes in, **20:**18
Epiglycanin, **52:**32
11-Epihydrocortisone, microbiological formation, **20:**246; 249; 271
Epilepsy, glutamic acid treatment, **13:**258
Epilubimin, **55:**69
biosynthesis of, **55:**71
Epimerase, **22:**371; 372
mammalian, DPN requirement of, **22:**371
yeast, and DPN, **22:**371
4-Epimerase, complex with diphosphopyridine nucleotide, **20:**40
Epimerase action, in ribose synthesis, **28:** 396–397
Epimerases, **35:**66–69
action of, **26:**41–42; 49–50; 52–57
classification, **44:**70
definition, **44:**70
localization of, **24:**333
Epimerization, **59:**123
exopolysaccharide alginate synthesis, **70:**235
and hydrogen transfer, **26:**173
of nucleotide-bound sugars, **26:**41–42; 46; 49–50; 52–57
in pentose synthesis, **28:**398
Epinephrine, **41:**42; 49, **46:**165, **54:**154, **57:**164, **58:**62, **67:**146–147
adnamine formed by, **29:**491
changes in the hypothalamus by, **29:**514
excretion patterns, **29:**517
Michaelis-Menten, **29:**40–41
neurotransmitter function of, **29:**482
oxidized to adrenochrome, **29:**516
sedative effects of, **29:**488
for translocator-catalyzed reactions, **29:**41
biosynthesis of, **74:**236
effect on sperm, **24:**84
formation of cyclic AMP, **28:**107; 109
in glycogen synthesis, **28:**102
methylation of, **22:**257
and phosphorylase, **24:**279; 283; 286
source of methyl group, **25:**415; 416
tyrosine hydroxylase, **70:**104; 145; 169
Epinine, in tyrosinase action, **14:**145; 146
Epistasis studies, *Caulobacter* cell cycle, flagellar structure and assembly, **71:**8
Epithelial cell, **34:**479
Epithelioma, uterine, enzymes in, **24:**296; 297; 301; 312
Epitope tagging, adenosine receptors, **69:** 107–108; 109
Epoxidation, **58:**56, **60:**46
of aromatic double bond, **27:**519
of double bonds, **58:**57
by liver nonspecific hydroxylase, **19:**177–179
of steroids by microorganisms, **20:**250; 251; 274
Epoxide, **58:**37
ring opening, **58:**38
xenobiotic substrate, **69:**25; 35
Epoxide hydrolase, **60:**47
1,2-Epoxide reductase, in oxidation of hydrocarbons, **27:**533
Epoxides, **43:**511–513, **46:**128; 132; 173; 174; 392, **57:**376–377
from microbial oxidation of olefins, **27:** 499–503
Epoxides and autoxidation of unsaturated fatty acids, **8:**428; 432; 433–434
Epoxidetransferase, **57:**361
Epoxidizing enzyme, **27:**499
Epoxycreatine
in creatine kinase inactivation, **54:**392; 393
in creatine kinase structure analysis, **54:**392
inactivation of, **54:**392; 393
phosphorylated, **54:**392–393
Epoxyeicosatrienoic acids, cholesterol ester hydrolases regulation, **69:**71
Epoxyhydroxy acids, **58:**196
1,2-Epoxy-3-(*p*-nitrophenoxy) propane, **46:**394; 396; 406
N-(2,3-Epoxypropyl)-*N*-amidinoglycine, **54:**392
EPR-detectable iron, **34:**103
EPR-nondetectable copper, *See* Copper, EPR-nondetectable
EPR signal, **34:**96; 132–134
EPR spectra, **52:**3; 4
information contained in, **27:**109–117
of transition metal ions, **27:**107; 118–120; 137–144
lysine 2,3-aminomutase, **66:**9–11
high-spin cobalt(II), **66:**17–18
for pyruvate, phosphate dikinase, **45:**127; 128
spin-labeled psoralens, **66:**137–139

EPR spectroscopy, **33:**322–325, **47:**22; 23; 26, **51:**108; 110; 155
anaerobic techniques, **27:**118
applied to oxidative enzymes, **27:**105–198
aqueous standards, **27:**114–116
complex systems, **27:**187–197
flavin free radicals, **27:**133–137
liquid flow techniques, **27:**120–130
optical measurements, **27:**117
quantitative information, **27:**114–116
rapid freezing, **27:**117
symbols, **27:**108; 109
techniques, **27:**109–118
temperature effects, **27:**115–117
of azurin, **33:**203
basic resonance condition of, **33:**189
of binuclear complexes, **33:**194
of Cu^{2+} complexes, **33:**191–194
effect of anisotropy on, **33:**191
fundamental principles of, **33:**188–191
g1, **33:**191; 193; 194
g11, **33:**191; 193
g value of, **33:**190
hyperfine structure in, **33:**190–191
quantitation of signals of, **33:**194
super-hyperfine structure in, **33:**191
of blue Cu^{2+}, **33:**203–206
of cytochrome oxidase, **33:**203
deuterium substitution exchange and transfer, **51:**138–143
general features, **51:**117–121
generation of the spectra, **51:**121–122
high-pH forms, **51:**128–130
inhibitory side products, **51:**131–135
intermediates, **51:**130–131
of nonblue cupric ion (table), **33:**213
nonfunctional degradation products, **51:** 135–138
of plastocyanin, **33:**203
prospects of, **51:**159
of stellacyanin, **33:**203
for studying molybdenum in enzymes, **51:**113–117
substitution with other isotopes, **51:** 143–147
coupling of molybdenum to ^{33}S, **51:**146
C splittings, **51:**143–145
N splitting, **51:**146
O splittings, **51:**145–146
substrate product, and anion complexes, **51:**122–128
Equations, nonlinear, by iterative methods, **29:**5–6
rate, directly fitting nonlinear, **29:**11–14
in linear form, **29:**7–11
Equilenin, absorption spectrum, **4:**288; 289
Equilibrium, **34:**494
adenylosuccinate synthetase (AMPSase), **73:**72–77
enzyme action, steady-state hypothesis, **73:**26–29
peroxidase kinetics, **73:**9–11
and product inhibition of enzyme, **25:** 179–182; 188–189; 195–196; 202–220
Equilibrium constant, **58:**145
for argininosuccinic acid splitting, **15:**360; 363
in citric acid, cycle, **15:**198; 232; 239; 242–243; 248; 250; 255–256
for DPN synthesis, **15:**104
for immunological reactions, **15:**51–54; 60–61; 64; 70; 72; 74; 76
for synthesis of nicotinamide riboside, **15:**104
ATPase thermodynamics, **72:**15
of disulfide exchanges, **27:**207–209
of α-ketoglutarate-*d*-isocitrate conversion, **19:**36
of pyruvate-L-malate conversion, **19:**35
of Ru-5-P-6-phosphogluconate conversion, **19:**38
for liver alcohol dehydrogenase system, **20:**35–38; 43; 45–47
in simultaneous reactions, **9:**54–56
of enzyme activation, **9:**28; 29
of enzyme inactivation, **9:**33
of enzyme reactions, **9:**17–23
of enzyme-substrate association, **9:**57–60
of step-wise reactions, **9:**3–16; 49–53
for sucrose synthesis, **10:**128
and thermodynamic functions, **11:**11–12
Equilibrium dialysis, **46:**370; 393
Equilibrium isotope effects, **46:**427
Equilibrium processes, entropy of, **11:**4–7
Equilibrium shifts, enzyme reaction steps, **25:** 1–38
Equilin, absorption spectrum, **4:**288; 289
Equilin, PN-oxidation, **26:**127
Equine encephalomyelitis virus

activity and particle identity, **9:**194
photochemistry of, **9:**105; 115; 129; 142; 154–155
purification, **9:**187
Equine influenza virus, **52:**137
Equisetum, nicotine in, **8:**223
Erabutoxin b, **47:**85; 137
Eremothecium ashbyii
pterin synthesis in, **21:**393
in riboflavin production, **7:**572
Erepsin in low vertebrata, **1:**391; 392; 407; 412
Ergastoplasmic material, in nuclear preparations, **27:**244; 252
Ergosterol, **46:**103
absorption spectrum, **4:**290; 291
biosynthesis, **50:**291
methylation of, **25:**423
synthesis of, **25:**383
in microorganisms, **20:**264
structure, **22:**328
synthesis, from acetate, formate, methionine, **22:**327; 328
S-adenosylmethionine in, **22:**257
Ergothioneine
imidazole in, **20:**102
occurrence in plants, **10:**261–262
Ergothionine, **7:**370
in disulfide exchanges, **27:**203
in semen, **24:**64; 70
and transglycosylation, **18:**206
Ergotropic system, increased activity of, **29:**487
norephinephrin effect on, **29:**486
sedation due to, **29:**489
Eriodictyol, **55:**22
from naringenin, **55:**34
ERKs, tyrosine hydroxylase regulation, **70:** 151–152; 160
ERp72 polypeptide, protein disulfide isomerase (PDI) prolyl 4-hydroxylase β subunit structure, **72:**374–375
ERp60 polypeptides, protein disulfide isomerase (PDI) prolyl 4-hydroxylase β subunit structure, **72:**373–375
Error, envelopes of expected, **29:**8
standard, **29:**6–7
of fitted constants, **29:**7
Errors, **48:**207
Erthrityl tetranitrate, **46:**399
Erwinia amylovora, plant disease caused by, **20:**398
Erwinia aroideae, polygalacturonases of, **20:**365; 367
Erwinia asparaginase, **39:**213; 219
Erwinia atroseptica, plant disease caused by, **20:**395
Erwinia carotovora, **71:**11
plant disease caused by, **20:**394
polygalacturonases of, **20:**365–367
Erwinia carotovora asparaginase, substrate specificity of, **39:**208–209
Erwinia chrysanthemi, plant disease caused by, **20:**388; 400
Erwinia species, plant disease caused by, **20:**401
streptomycin-resistant strains, **20:**387
Erwinia tracheiphila, plant disease caused by, **20:**393
Erysiphe polygoni, plant disease caused by, **20:**389
Erysoline (δ-thiocarbimidobutylmetyl-sulfone), occurrence, **10:**265
Erythritol, **56:**194
Erythritol permeability, **34:**467
Erythroblasts, **46:**176, **58:**249
δ-aminolevulinic acid synthesis in, **46:** 103–104
marrow, **46:**103
Erythrocuprein, **41:**52
extinction coefficient of, **33:**212
Erythrocuprein, EPR study, **27:**137; 150
Erythrocuprein transport, **44:**224
Erythrocyte enzymes, fermentation of ribose, **28:**395
Erythrocyte isozymes, bovine, **47:**152–153; 161
human, **47:**152–153; 161
Erythrocyte lysate, **58:**72
Erythrocyte membranes, **58:**239
Erythrocyte mucopolysaccharide, as substrate for neuraminidase and receptor destorying enzyme, **20:**136; 137; 144
Erythrocyte phosphofructokinase, **48:**209–210
Erythrocytes, **34:**466, **41:**136–138; 145; 150; 178, **42:**270, **46:**42; 55; 57; 389, **48:**56; 76; 81; 82. *See also* Red cells
absorption of hemolysin by, **15:**54
agglutination studies on, **15:**54–64
in hapten-antibody studies, **15:**64–67
heptulose phosphates in, **15:**163
hexokinases in, **15:**146
hexose phosphate formation in, **15:** 163–164

Erythrocytes (*Continued*)
synthesis of nicotinamide mononucleotide, **15:**103
A-cells, **41:**136
aged, **48:**77
avian, **46:**42; 55; 57
nuclei of, **27:**245; 255
B-cells, **41:**136
chicken, **46:**57
choline acetylase in, **12:**324–325
esterase, effect on choline esters, **12:** 264–270
laking by borax, **12:**511
cholinesterase in, **8:**462–467
duck, **46:**59
effect of pectin on, **20:**357
γ-globulin binding *in vitro,* **41:**138
γ-globulin binding *in vivo,* **41:**145
γ-globulin binding to, **41:**178
glucose oxidation in, **28:**117
regulation of hexokinase, **28:**76–77; 85
goat, **48:**75
half-life, **41:**150
horse, **48:**68; 75
human, **46:**63; 385, **48:**56; 61; 65; 66; 69; 77; 80, **59:**371
enzyme from, **46:**389
mammalian, **46:**42; 43
nucleated, DNA content, **16:**424; 430
O-cells, **41:**137
or reproducible lytic behavior, **8:**92
rabbit, **48:**61
rhodanese activity of, **39:**336
sheep, **48:**75; 76
sialidase-treated, **48:**77
treatment with β-galactosidase, **48:**81
treatment with β-N-acetylhexosaminidase, **48:**81
Erythrocytes, L-leucine uniporter of, **29:**52
sugar efflux from human, **29:**52
Erythro-β-hydroxyasparatate, **35:**505
Erythro-9-(2-hydroxy-3-nonyl) adenine (EHNA), **51:**182–183
Erythroid cells, **46:**147; 158
Erythro-L_8-isocitrate, allosteric effect, **28:** 59–60; 137
Erythromycin, in treatment of plant diseases, **20:**398; 401
Erythrophilic γ-Globulin, **41:**140–144; 147; 149; 152
amino acid composition, **41:**143
chemical characteristics, **41:**143
chromatographic characteristics, **41:**141
effect of lysis of red cells, **41:**144
effect on the shape of red cells, **41:**144
effect on viability of red cells, **41:**147
fraction II, **41:**152
fraction III, **41:**149; 152
fraction IV, **41:**149
immuno chemical characteristics, **41:**140
physiological function, **41:**147
reduction after splenectomy, **41:**149
Erythropoietic porphyria, **46:**165
Erythropoietic protoporphyria, **46:**165
Erythropoietin, **41:**115, **46:**158; 159
Erythrose, **15:**154; 163
Erythrose-4-P, **43:**494; 502
and ATP synthesis, **23:**337
Erythrose phosphate, **15:**166
Erythrose-4-phosphate, **48:**171
in *p*-AB biosynthesis, **27:**421
allosteric regulation, **28:**118
D-Erythrotrihydroxypropyl-H_2-pterin, **35:**53
Erythrulose, **15:**157; 158
D-Erythrulose, transketolase action on, **19:**26
Erythrulose phosphate, **15:**171
Erythrulose-4-phosphate, **19:**26; 27
synthesis, **21:**420
Erythrulose-phosphate aldolase, **15:**171
Erythrytol-P, **43:**494
Escherichia coli, **17:**249; 264, **33:**101–102, **34:**424; 466, **35:**38, **41:**5; 15; 42; 70; 71; 79–82; 84–87, **42:**230, **44:**171, **46:**49; 58; 169, **48:**4, **51:**112, **54:**169, **55:**383; 384–386, **71:**7; 8; 9; 27; 87; 95; 96; 97; 108; 143; 189; 191; 212–213. *See also* Aspartate transcarbamylase; Glutamine synthetase
and acetyl phosphate formation, **6:**238
in adenylosuccinase, **22:**33
amino acid activation in, **22:**130
chromosome structure, **22:**3
constitutive galactose permease in, **22:**110
free amino acid pools in, **22:**125
galactokinase-deficient mutants, **22:**29
β-galactosidase in, **22:**106; 144
dominance relationships, **22:**143; 144
genetic loci, **22:**102–104
inhibition in constitutive mutants, **22:**121

sensitivity to RNA antagonists, thiouracil, **22:**138; 139
β-galactosidase induction in, **22:**19–21; 107; 111; 112; 115
inducers, substrates of, **22:**119; 120
kinetics, **22:**111
linkage relationships in, **22:**20; 21
and melibiose, **22:**120
in starved cells, **22:**125
β-galactoside concentration in regulation of, **22:**109
β-galactoside permease in, **22:**103
inducible, **22:**110
and genetic mapping, **22:**2–3
genetic resistance to *p*-aminobenzoic acid analogues in, **22:**17
histidine synthesis in, regulation of, **22:**17
isoleucine + valine-requiring mutants, **22:**33
methionine mutants, **22:**269
methionine synthetase in, **22:**122
mutants, raffinose utilization in, **22:**22–24
organic phosphatase-deficient mutants, **22:**17; 18; 39
ornithine transcarbamylase synthesis in, **22:**122
pantothenic acid synthetase mutants, **22:**4; 5
protein turnover in, and β-galactosidase synthesis, **22:**127
in growing cells, **22:**126; 127
in nongrowing cells, **22:**126
L-serine deaminase induction in, **22:**119
2-thiazolealanine-resistant mutant, and histidine synthesis, **22:**17
tryptophan synthesis in, control of, **22:**122
tryptophan synthetase mutants, **22:**12–15
genetic mapping of, **22:**38
and interlocus complementation, **22:**37
and suppressor mutations, **22:**30; 32
ADP-glucose pyrophosphorylase of, **46:**341; 376
covalent binding of Pyridoxal-P to, **46:**373
alanine racemization by, **14:**195
amino acids in, **14:**266
glucose diphosphate synthesis by, **14:**212
alkaline phosphatase, **46:**388
alkaline phosphatase from, **74:**71
and amino acid decarboxylase formation, **6:**6; 8; 11; 13
arginine-biosynthetic system, **40:**65
arginine decarboxylase from, **74:**143
structural data on, **74:**134; 142
aromatic amino acid biosynthesis in, **29:**115
control of aromatic amino acid biosynthesis in, **29:**115
DAHP synthetase, **29:**115
DNP catalysis of galactosides through membrane of, **29:**65
effect of addition of L-azetidine-2-carboxylic acid and 3,4-dehydro-DL-proline to, **29:**97
effect of amino acid analogs on nucleic acid synthesis, **29:**148
effect of *O*-methylserine on, **29:**145
4-fluoroprolines as substrates for protein synthesis, **29:**103
growth inhibition by α-methylhistidine, **29:**130
homoarginine inhibition of growth, **29:**91
inhibition of, **29:**91; 95; 96
inhibition of DAHP synthetase by analogs of phenylalanine and tyrosine, **29:**118
inhibition of growth, by mimosine, **29:**108
by norleucine, **29:**136
by tyrosine analogs, **29:**114
methionine sRNA in, **29:**154
mutant strains resistant to proline analogs, **29:**106
nucleases of, **29:**197–199
proline permease in, **29:**105
repression of tryptophan biosynthetic enzymes, **29:**121
resistance to analogs of aromatic amino acids, **29:**122
strains resistant to azetidine-2-carboxylic acid, **29:**107
strains which overproduce proline, **29:**107
tryptophan synthesis, **29:**121
L-aspartase from, **74:**297–298; 302; 333–334
activity of, pH and, **74:**327–328; 328
aspA gene for, **74:**314
chemical modification studies of, **74:** 304–305
cysteine residues, **74:**316–317
histidine residues, **74:**317
mechanism-based inactivator of, **74:** 307–308
modification of, mechanism of, **74:**313–314
pH study of, **74:**305–307

Escherichia coli (*Continued*)
structure of, **74:**321; 323; 324
and biosynthesis of ornithine, **33:**374
and carbon dioxide fixation in pyruvate carboxyl, **6:**246
and citric acid cycle, **18:**246–251; 257; 258; 260–261; 264; 267; 269–270
DNA polymerization by, **18:**213
effect of fluoro compounds on, **18:**135; 151
formation of aromatic compounds by, **18:**362; 366
and lipoic acid, **18:**332–339
mutants, **18:**269–270
nucleic acid metabolism of, **18:**217–218; 228–229
phosphorylation in, **18:**197; 227
reversible oxidation of succinate, **18:**81
condensation of acetate and CO_2 by, **16:**268
in amino acid studies, **16:**255; 276; 284–286; 297–299
DNA of, **16:**432; 433
β-galactosidase of, **16:**460
β-glucuronidase of, **16:**377–381; 384
mutants, amino acid metabolism of, **16:**258–261; 263–274; 278–282; 284; 287–292; 295–300
arginine biosynthesis by, **16:**259; 261; 263–266
proline biosynthesis by, **16:**258–259
reductases in, **16:**134–135
transminations in, **16:**190–193; 203–209; 219; 222; 227; 229–231; 234; 237; 239
conversion of D-glucuronic acid by, **20:**371
galactose-negative mutants, **20:**128–131
galactose-1-phosphate uridyl transferase of, **20:**114
uridine diphosphogalactose-4-epimerase of, **20:**119
L-cysteine biosynthesis in, **74:**188–191
cysteine synthesis, pathways of, **74:**186; 188–189
cytochrome a_2 of, **23:**266; 267
oxidative phosphorylation in, **23:**335
phosphorylating system from, **23:**353
use of, acetate, **23:**402; 406; 411–413; 415; 420
glycine, **23:**453–454
glycolate, **23:**435; 437; 440
pyruvate, **23:**337
DNA polymerase I from, **74:**71
endonucleases, **60:**18
enterotoxins, **58:**118
enzyme system for metabolizing sugars of, **30:**81–137
induction of β-galactosidase by isopropyl thiogalacataside in, **30:**149
enzymic adaptation in, **13:**70; 71; 72; 81–82; 85–88; 89–90; 94–95; 97–98; 99–100; 101–105; 105–108; 109–113
genetic studies of, **13:**138; 139–145; 146
in nitrogen fixation studies, **13:**351; 358; 359–360
folate metabolism, **27:**422–424; 426–432; 435; 441; 447; 451–452; 456–457; 459
and formate condensation, **6:**244–246
fumarase from, structure of, **74:**324
genetic studies, **17:**396–410; 412–414; 417–419; 422; 426–431; 444–445
glycogen reserve, **34:**291
glycogen synthetase mutants, **34:**350
hypoxanthine-quanine phosphoribosyl-transferase from, **41:**5; 15
metabolism in, **74:**21
methylthioadenosine isolation from, **54:** 199; 206
mutagens for, **27:**330; 331
nuclear RNA, **27:**247
nucleoside triphosphatases, **27:**283
phosphorylase in, **34:**388
phosphotriesterase homology protein of, **74:**71–72
purification of lysine in, **6:**13
pyrophosphorylase, **34:**315
succinic acid fermentation by, **74:**15; 17
succinyl-CoA synthetase, **58:**147
sulfate permease from, **74:**186
superoxide dismutase from, **41:**42; 70; 71; 79–82; 84–87
transferring enzymes in, **17:**208–209; 218–219; 224
and vitamin B_{12}, **26:**239; 260; 268–269; 272–275
carbohydrate synthesis, **26:**48; 62; 68–69; 75
mureinases, **26:**211–215
murein sacculus, **26:**196–197; 200–211; 224–228
growth, **26:**216–222
murein structure, **26:**207–211
muropeptide hydrolases, **26:**211–215

PN-dependent enzymes, **26:**119; 129–131
roughness, carbohydrate synthesis, **26:**69
Escherichia coli:
adenosylmethionine decarboxylase from, **56:**266–268
alginate biosynthesis, **70:**224; 231; 235; 246; 248
asnA and asnB, asparagine synthetase, **66:**278
carbamyl phosphate sunthetase, **72:**125–131
carbamyl phosphate synthetase, **66:**228–233
relation to carbamyl phosphate synthetase I of urea cycle, **66:**236–238
relation to other monofunction enzymes and CAD, **66:**233–236
CTP synthetase properties, **66:**243–244
DNA sequence analysis, mode of action, **67:**435
eukaryotic dihydrofolate reductase, **70:**37; 57; 79; 83–84; 91
exonuclease III, **67:**449–450
glucosamine 6-phosphate synthase, **66:** 265–267
glutamate synthase, **66:**271–275
glutamine PRPP amidotransferase, **66:** 257–261, **72:**93–95; 97–99
glutaredoxin, **66:**157–158; 160
high-level expression in, pig liver thioltransferases, **66:**169–171
17-α hydroxylase and 17, 20–20lyase activities in, **68:**57–58
imidazole glycerol phosphate synthase, **72:**131–134
methionine adenosyltransferase from, **56:**255–257
mutants, thioredoxin, **66:**171
mutants deficient in glutaredoxin, **66:**168–169
P45017α expression, **68:**52–59
protein N-myristoylation, **67:**382–383
ribonucleotide reductase, inhibition by cisplatin, **66:**183
thioredoxin, **66:**163–164
tyrosine hydroxylase, **70:**181; 187
Escherichia coli, L-alanine dehydrogenase in, **19:**340
leucine activating enzyme in, **19:**258
malic dehydrogenase in, **19:**40
malic enzyme in, **19:**35; 40
pantoic acid activating enzyme in, **19:**259
RuDP carboxylase in, **19:**33
Escherichia coli, The Hexose Phosphates Transport System of (Dietz), **44:**237
Escherichia coli, allosteric enzymes of, **28:**43; 54; 60–64; 66–72; 78–79; 111–112; 122; 125; 128; 131; 133
cytidine-diphosphate reductase of, **28:** 436–440; 448–449; 453
β-lactamase of, **28:**242; 244; 260; 265; 268; 277; 285; 286; 290; 293; 294; 297; 305; 310
pentose metabolism, **28:**393; 395–396; 413; 436–446
tracer studies, **28:**405–408; 420–421; 423–428; 435–436; 453
thioredoxin reductase, **28:**441–443; 446
Escherichia coli, arginine-requiring mutant, **15:**356
citric acid cycle, **15:**188; 191; 194; 201–213; 216–223; 229; 230; 245; 246; 249; 253; 256
enzymes, **15:**116; 119; 122; 306; 375; 416
galactan synthesis, **15:**314
glucose-6-phosphate oxidation shunt, **15:** 153–157; 175
glucose-1-phosphate transphosphorylase, **15:**152
reduction of hexose monophosphates, **15:**167
Escherichia coli asparaginases, **39:**192
Escherichia coli B AC70RI ADP-glucose pyrophosphorylase, **46:**366
amino acid composition, **46:**367
molecular weight, **46:**366–368
subunit studies, **46:**366–368
Escherichia coli B ADP-glucose pyrophosphorylase, **46:**350; 370; 373
activity, **46:**350
amino and carboxyl terminal sequence, **46:**368
binding, **46:**370
equilibrium dialysis, **46:**370
Escherichia coli B enzyme, pyridoxal-P linkage to, **46:**344
Escherichia coli B, glycogen accumulation in, **46:**356
Escherichia coli, effect of pressure on metabolism, **9:**67
Escherichia coli, effect of radiations on, **24:**403
protein synthesis in, **24:**6; 9; 14; 17; 18; 19; 23–24
thiamine compound in, **24:**468

Escherichia coli enterotoxin, **50:**313
Escherichia coli enzyme, **46:**344
purification by affinity chromatography, **46:**365
structural requirements for an activator of, **46:**344
Escherichia coli, enzymes and reactions of purine biosynthesis in, **21:**217; 253; 254; 399
enzymes and reactions of pyrimidine biosynthesis in, **21:**266; 272; 277; 280; 285; 289; 290; 421
histidine biosynthesis in, **21:**403
α-ketopantoic acid synthesis in, **21:**419
methionine synthesis in, **21:**430
nitrogen transfer reactions in, **21:**252
Escherichia coli-galactosidase, **48:**45; 46
inhibitor of, **48:**49
Escherichia Coli Glutamine Synthetase, Regulation of (Rhee, Chock and Stadtman), **62:**37
Escherichia coli mutant SG14 ADP-Glucose pyrophosphorylase, **46:**363–365
Escherichia coli phage, adsorption studies of, **14:**356
Escherichia coli-phosphofructokinase, **48:**215
Escherichia coli protease I, **41:**197
Escherichia coli SG5 and CL1136 ADP-Glucose pyrophosphorylases, **46:**355
Escherichia freundii, 3α-hydroxysteroid dehydrogenase in, **20:**268
Eschericia coli, mechanism of utilization of maltose, **10:**124; 141
tryptophan metabolism, **10:**51–82
Escherischia coli, acid phosphatases, **25:**348
amino acid synthesis, **25:**361
β-galactosidases, **25:**336
and luminescence, **25:**150
methionine synthesis, **25:**404–409
methylation of RNA, **25:**414–415
product inhibition in, **25:**173
quinones in, **25:**289
thymidilate synthetase, **25:**398
Esculetin, role in browning of plant tissues, **16:**150
Esculin, metabolism in plants, **16:**141
Eserine
effect on cholinesterase, **20:**300; 301
inhibitor of acetylcholinesterase, **12:**280–281; 284–290; 292–293; 294; 299; 318
and nerve permeability, **13:**61
effect on esterases, **13:**334–335
ES-5-P, **34:**521
ESR measurements, in photosynthetic tissues, **25:**83–84
ESR spectroscopy
and pyrrolo-quinoline quinone, **59:**171–172; 178; 183
and quinoprotein oxidoreductases, **59:** 195–196; 198; 202
ESR spectroscopy of alkaline phosphatase, **55:**404; 407–409
Essential hypertension and fever, **7:**540
Essential oils, changes during fermentation of tobacco, **10:**370–372
Ester, hydrophobic, **48:**327
Esterase, inhibition, **13:**321–343
localization in cytoplasm, **13:**7; 10
Esterase activity
of carboxypeptidase, **12:**228–229, **19:**306
of chymotrypsin, **19:**238; 306
of papain, **19:**306–308
of proteinases, **12:**206
of trypsin, **19:**306
of imidazole, **20:**62; 70; 74; 76–82; 87–90; 92; 289; 293; 300; 302; 305; 309–312; 314; 315
of proteolytic enzymes, **15:**432–434
Esterase activity of carboxypeptidase, classification, **14:**238
Esterase (pancreatic), requirement for –SH group, **11:**241
Esterases, **34:**487. *See also* Acetylcholinesterase and under other esterases; individual esterases; specific esterases
action of fluoroacetylcholine, **18:**139
action on acetylcholine, **12:**264–270
and activation energy, **23:**150
of *Cl. perfringens,* **23:**208
horse liver, **23:**146
pancreatic, **23:**145–148
use of term, **23:**135
active centers, **24:**434; 459
effect of radiations, **24:**388; 392–394
mechanism of action, **24:**243; 458–464
amino acid composition near active site, **22:**65–70
labeling of active site, **22:**52
classification, **14:**224–225; 230; 244–245

for decomposition of aromatic hydrocarbons, **10:**463
vegetable, relation of *p*H to activity, **10:**84
intracellular distribution, **16:**464; 466–468
relation to estrogen-stimulated growth, **16:**387; 399
kinetic, **14:**173
multiple forms of, **25:**331; 348–349
in muscle, effect of prostigmine, **6:**210
in plastids, **20:**213; 214; 229
mechanism of action, **20:**311–315
from sera, **8:**461–463; 465
in nuclei, **8:**33
in snake venoms, **8:**486–487
synthesizing and hydrolizing activity in animal tissues, **1:**345
titrimetric methods, **9:**601
in tobacco, **6:**368–369
Esterase synthesis, effect of nucleus, **24:**20
Esteratic site, and esterase activity, **12:**295–302; 302–311
Ester bond, **34:**486
groups, **34:**489
orientation of, **34:**489
β-Ester bond, **34:**475
Ester bonds, enzymes acting on, **24:**316–322
Ester hydrolysis, **50:**364
Esterification, **58:**182
of amino acids, **29:**399–401
Esterolytic activity, of pancreas, **23:**145–148
Ester phosphate, group potential, **1:**103–105
Ester phosphorus, **7:**124; 167
Esters
digestion by pancreatin, **11:**76–77
emulsified, action of lipase on, **23:**129; 142–149
hydrolysis of, influence of alcohol, **23:**155
insoluble, and lipase, Michaelis curves for, **23:**143–145
soluble, and lipase, Michaelis curves for, **23:**145–148
hydrolysis, by metal complexes, **21:**161
hydrolysis by α-chymotrypsin, kinetic studies, **17:**46–49
from oxidation, of alkanes, **27:**489–491
of olefins, **27:**502
Estradiol, **18:**134, **36:**41
effect on β-glucuronidase, **16:**388; 391
tyrosine hydroxylase induction, **70:**172
α-Estradiol, **15:**127
17 β-Estradiol, effect on β-glucuronidase, **16:**387; 388
Estradiol 17 β, effect on protein methylase I, **42:**267
17β-Estradiol, microbiological formation and conversion, **20:**253; 259
Estradiol benzoate, effect on β-glucuronidase, **16:**388
Estradiol-binding proteins, **36:**41; 56; 76
Estradiol dipropionate, effect on β-glucuronidase, **16:**391
Estradiol receptors, **55:**162
Estriol, effect on β-glucuronidase, **16:**391
Estriol glucuronide, as substrate for β-glucuronidase, **16:**368–369; 380; 383; 385
Estrogen, **44:**221, **56:**275
effect on acid phosphatase, **9:**352
effect on uterus muscle proteins, **13:**181
Estrogenic compounds as inhibitors of thrombin-fibrinogen interaction, **16:**85
Estrogenized roosters, **44:**221
Estrogens. *See also* Steroid metabolism
depression of α 2u-globulin mRNA synthesis, **47:**298; 300
effect on β-glucuronidase levels, **16:**386–388; 390–391; 394; 398–399
glucuronides as substrate for β-glucuronidase, **16:**380; 384–385
as inhibitory analogues, **6:**135
in sperm cells, **24:**84
Estrone
absorption spectrum, **4:**288; 289
effect on tissue β-glucuronidase, **16:**387; 391; 399
identification of its glucuronide, **16:**383
microbiological formation and conversions, **20:**253; 255; 259; 263
Estrone sulfate, inaction of β-glucuronidase on, **16:**380
Estrone sulfokinase, **22:**212
Ethacrynic acid, **46:**396; 406, **57:**137–138
Ethallyglycine, **50:**165
Ethane
(D_6), in cooxidation experiment, **27:**481
deuterated, oxidation of, **27:**533
Ethane dehydrogenase, **27:**481
1,2-Ethanedisulfonate, **18:**94
Ethane-oxidizing system, **27:**475; 476

Ethanethiosulfanate, **39:**341
1,N[6]-Ethano-ATP (epsilon ATP), **43:**29
allosteric effects, **43:**30
kinetic properties, **43:**30
Ethanol, **15:**255; 257, **35:**492, **59:**367, *See* Alcohol; Alcohols
and acetate metabolism, **23:**431–432
in ADH complex, **22:**357
catalase action on, **19:**109; 115; 116
effect on papain, **19:**292; 295
in cellulose synthesis, **15:**315 (C^{14})
effect on papain action, **15:**292
formation in plant tissues, **9:**617; 621; 623–624
mold formation of acids from, **9:**545–546; 547–550; 553–554
from glucose, **15:**176 (C^{14})
mechanism of enzymic oxidation, **15:**7; 9–10; 11; 13; 15 (deuterium labeled); 23
oxidation, coupled to reduction of *o*-quinones, **16:**134
Ethanolamine, **5:**83, **17:**436, **57:**103; 112
biological origin of, **12:**472
formation from serine, **5:**83
metabolism of, **24:**338
Ethanolamine ammonia lyase, **46:**451, **50:**364
Ethanolamine ammonia-lyase, **74:**301
Ethanolamine as transglutaminase substrate, **54:**30
Ethanolamine (C^{14}), methylation of, **25:**417; 418
Ethanolamine deaminase, **66:**29–30
oxycation radicals, **66:**31
Ethanolamine (N^{15}), methylation of, **25:** 417–420
Ethanol (C^{14}), and citric acid cycle, **18:**272
Ethanol dehydrogenase, **59:**188–189; 196
Ethanol in gluconeogenic enzyme regulation, **54:**160; 161
Ethanol oxidation, **60:**67
Ethanolysis, of dextrans, **30:**195
Ether, effect on mitochondrial oxidative phosphorylation, **21:**31
Ethionine, **22:**244, **50:**169, **56:**253. *See also* Methionine and Methylation
activation of, **29:**136
competition for SAM-synthetase, **29:**137
effect on DNA synthesis, **29:**151
growth inhibition by, **29:**135
incorporation into protein, **29:**135
methylation reactions blocked by, **29:** 93–94
as a natural product in bacteria, **29:**135
antagonist to methionine, **12:**478–479
ATP-trapping action, **25:**381–382; 386–388
bacterial synthesis of, **25:**370
effect in biological systems, **22:**271–273
effects on tissues, **25:**369–393
and inhibition of protein synthesis, **22:**271
in transamination reactions, **16:**190; 196; 201; 237
and transethylation, **22:**272
DL-Ethionine, effects, **25:**371
Ethionine acyl adenylate, enzymic synthesis, **25:**382–388
Ethionine (C^{14}), metabolism of, **25:**375–380
Ethionine Effects on Tissues, Biochemical Basis (Stekol), **25:**369
Ethionine metabolism, **25:**369–380
tracer studies, **25:**375–380
Ethionine (S^{35}), metabolism of, **25:**375
Ethionine studies *in vitro,* **25:**380–388
Ethionine sulfonium compounds, effect in biological systems, **22:**271–273
Ethopromazine, **55:**181
Ethoxyacetylene, in conversion of acids to anhydrides, **22:**182
7-Ethoxycoumarin, deethylation, **60:**73
Ethoxyformic anhydride, **58:**159
Ethoxyformyl group, **58:**160
β-Ethoxy-α-ketobutyraldehyde, **59:**130
α-Ethoxypropioguaiacone, **34:**213
Ethoxzolamide, inhibition of carbonic anhydrase by, **47:**212–213
Ethyl acetate, hydrolysis, by metal ions, **21:**180
Ethyl alcohol
dehydrogenation by yeast, **10:**190
and fat formation, **8:**315; 328–330
and wood molds, **8:**273–279
Ethyl alcohol fermentation, **7:**573–601
of molasses, **7:**585–587
of wheat, **7:**587–591
of wood, **7:**574–585
Ethylamine, **58:**19
Ethyl aminoacetate, hydrolysis, by metal ions, **21:**180
Ethylation, **42:**254
Ethylbenzene, microbial oxidation, **27:**516–519
Ethyl-*p*-benzoquinone, relation to cuticle hardening, **16:**155

2-Ethyl citrate tricarballylate, **43:**93
Ethylcobalamin, methyl transfer, **25:**407
Ethyl dimethylaminoacetate, hydrolysis, by metal ions, **21:**180
Ethylene, **54:**231
formation of, **54:**231–232; 233
methionine as precursor of, **54:**231
Ethylenediamenetetracetic acid, phosphomannomutase/ phosphoglucomutase (PMM/PGM), **70:**229
Ethylene diamide, solubility of pectic substances in, **20:**353
Ethylenediamine, **57:**103–104
Ethylenediamine-glyoxylate adducts, **57:** 111–112
Ethylenediamine tetraacetate (EDTA), **41:**183; 184; 200; 204; 208; 252
inhibitor of proline hydroxylation, **41:**252
inhibitor of Proteinases, **41:**183; 184; 200; 204; 208
Ethylenediaminetetraacetic acid, **17:**240; 242
and collagenase action, **23:**182; 195–196; 202
in denaturation study, **23:**40–41
effect on elastases, **23:**234; 242
in isolation of mitochondria, **21:**5
in stabilization of mitochondrial oxidative phosphorylation, **21:**124
Ethylenediamine tetraacetic acid, chelating agent, **15:**184; 223; 250
Ethylenediaminetetraacetic acid (EDTA), effect on pectic substances, **20:**355; 368
in analysis of pectin, **20:**345
Ethylenediaminetetracetic acid, *See* Versene
Ethylene glycol, **51:**23–25; 134; 138
2-P-Ethyleneglycol, **43:**494
Ethylene glycol-bis N,N,N′-tetraacetic acid (EGTA), tyrosine hydroxylase regulation, **70:**142; 144
Ethylene glycol dinitrate, **46:**399
Ethylene imine, **67:**41
Ethyleneimine in synthesis of pantetheine, **16:**6
Ethylene-malcic acid (EMA), **34:**461
Ethylene reductase, **15:**234
in soluble fraction of liver, **21:**49
Ethyl ethanesulfonate, mutagenic action, **27:**320; 324–326; 338–340
Ethyl 7-fluoroheptanoate, **18:**115
Ethyl 16-fluorohexadecanoate, **18:**115
Ethyl fluoromalate, **18:**116
Ethyl 9-fluorononanoate, **18:**115
N-Ethyl-DL-glutamate, **15:**371
N-Ethylglycocyamine, **54:**384
Ethyl group, effect on oxidation of cinnamic alcohol, **8:**366
effect on β oxidation of fatty acids, **8:**352
Ethyl group transfer, **25:**382–388
γ-Ethylidene glutamic acid, **50:**118
Ethyl indole-3-acetate, in treatment of plant diseases, **20:**399
5-Ethyl-5-isoamylbarbiturate, *See* Amytal
β-Ethyl-α-ketobutyrate, oxidation of, **18:**338
N-Ethylmaleimide, **35:**110; 452, **39:**344, **54:**400, **59:**123; 134; 138, **60:**14
effect on *p*-hydroxyphenylpyruvate oxidase, **19:**184–199
N-ethylmaleimide, inactivation of prolyl hydroxylase by, **41:**256
N-Ethylmaleimide, phenylalanine hydroxylase activation, **67:**108–110
N-Ethylmalemide, in identification of imidazole, **20:**75
Ethyl malonate, hydrolysis, by metal ions, **21:**179
Ethylmethanesulfonate
effect on β-lactamase, **28:**280
mutagenesis, **67:**406–407
Ethyl methanesulfonate, mutagenic action, **27:**320; 323–326; 335–336; 338–340
Ethyl morphine, **46:**119
Ethyl morphine demethylase, **46:**134
Ethyl nicotinate as inhibitor of acetylcholinesterase, **12:**306
Ethyl oxalate, hydrolysis, by metal ions, **21:**179
Ethyl pantothenyl carbonate, in synthesis of pantetheine, **16:**6
Ethyl phenylacetate, hydrolysis, by metal ions, **21:**180
N-ethyl-5-phenylisoxazolium-3’sulfonate, **34:**453
N-Ethyl-5-phenyl-isoxazolium-3′ sulfonate, in peptide synthesis, **29:**406
Ethyl succinate, hydrolysis, by metal ions, **21:**180
Ethylthioadenosine, **54:**209; 223
sulfoxides of, **54:**211
5′-Ethylthioadenosine, in ethionine study, **25:**385–386
Ethyl thiocyanate, **46:**397; 399

Ethylthiocyanate, **46:**397
Ethyl xanthate, **17:**252. *See also* Potassium ethyl xanthate
effect on phenolase complex, **19:**199
Etiocholanolone, **46:**157; 158
5β-Etiocholanolone, **46:**163
Etiocholanolone-17β-ol-3-one, **46:**158
Etiolation. *See also* Phytochrome
prevention of, **26:**7–9
Etioplasts, **46:**94; 95
barley, **46:**71
ETP, *See* Electron transport particle; Electron transport particles
in respiratory chain, **28:**159; 162; 164; 218
ETPH, mitochondrial particles, EPR study, **27:**187–188
Eubacterium, methylated lysine and arginine derivatives in, **42:**234
Eucaryote phosphofructokinase, **48:**201–214
Euchromatin, **38:**47
Eugenin, **55:**84–85
Eugenol, in lignin formation, **18:**373–374
Euglena, **46:**62; 69; 74; 98
reduction of cytochrome b_6 in fragments of chloroplasts of, **30:**33–34
Euglena, ATP in, **19:**18
Krebs cycle in, **19:**13
RuDP carboxylase in, **19:**32
triosephosphate dehydrogenase in, **19:**32
Euglena, photosynthesis in, **25:**72
Euglena garcilis, **46:**57
Euglena gracilis, vitamin B_{12}, **26:**273
Euglena gracilus, **59:**329
Euglena laminarinase, kinetics, **28:**346–347
Euglena transferase, activation of, **28:**339
Euglenids, β-1,3 glucan of, **28:**327; 334; 354
laminarinases of, **28:**341
Euglenoids, **46:**54
Euglinenae, carotenoids in, **21:**298; 301; 302
Euglobin and complement, **8:**101–102
Euglobulin fraction from hemophilic plasma, therapeutic effect for hemophilia, **16:**56
Eukaryotes, **46:**54
Eukaryotic carriers, traffic ATPases relevance, **65:**34; 36–40
Eukaryotic cells, RNA polymerases, **71:**42
Eukaryotic dihydrofolate reductase, **70:**23–102
alternate substrate, **70:**92–97
folate as, **70:**93–96
generally, **70:**92–93
short side chains, **70:**96–97
bound ligand motion, **70:**64–65
chemical transformation step, **70:**73–92
chemical transformation step (generally), **70:**73–74
chemical transformation step (isomerization component), **70:**90–92
chemical transformation step (mutation effects), **70:**92
chemical transformation step (proton donation), **70:**81–90
bound water, **70:**89–90
preprotonation, **70:**81–83
proton donation, **70:**83–89
chemical transformation step (reactant activation), **70:**74–81
facilitation of entry, **70:**76–78
NADPH activation by hydride donor, **70:**75
pteridine activation as hydride acceptor, **70:**78–81
generally, **70:**24–25
kinetic scheme, **70:**65–73
generally, **70:**65
kinetic scheme (steady-state results), **70:**65–66
kinetic scheme (transient state results), **70:**66–73
human DHFR, **70:**67–69
murine DHFR, **70:**69–70
mutation effects, **70:**71–73
P. carinii DHFR, **70:**70–71
ligand binding, **70:**47–65
ligand binding (generally), **70:**47–48
ligand binding (of inhibitors), **70:**55–60
7-hydroxymethotrexate, methotrexate polyglutamates and other inhibitors, **70:**58–60
ternary inhibitor complexes, isomerization of, **70:**56–58
trimethoprim and methotrexate, **70:**55–56
ligand binding (physiological ligands)
apoenzyme conformers with different affinity, **70:**53–55
cooperativity of, **70:**52–53
NADPH and NADP binding, **70:**48–50
substrate binding and tetrahydrofolate, **70:**50–52
ligand binding (side chain roles), **70:**60–64
Arg^{70} role in, **70:**63–64
Glu^{30} role in, **70:**60–61

hydrophobic residues lining active site, **70:**61–63
Lys^{54} role in, **70:**64
primary structure, **70:**25–27
secondary and tertiary structure, **70:**28–47
bioperin binding in substrate site, **70:**32–33
generally, **70:**28
methotrexate binding, **70:**34–35
NADPH binding, structural changes induced by, **70:**44–45
NADPH-binding site and NAD-binding site compared, **70:**41–42
nucleotide-binding site, **70:**40–41
organomercurials, structural changes due to derivatization with, **70:**46–47
phenyltriazine inhibitor binding, **70:**35–37
substrate binding, structural changes induced by, **70:**45–46
substrate-binding site, **70:**28–31
trimethoprim binding, **70:**37–40
vertebrate and bacterial structures compared, **70:**42–44
structural evidenced from nuclear magnetic resonance spectroscopy, **70:**47
structure, **70:**25–47
Eukaryotic expression systems, p450 enzymes, **68:**39–40
Eukaryotic folate transport, **47:**327–342
exploitation for cancer chemotherapy, **47:**340–342
general characteristics, **47:**327
inhibitors, **47:**333
regulation by cyclic nucleotides, **47:**337
substrate specificity, **47:**331
two separate systems for, **47:**328–331
Eukaryotic mRNA, methylation, **65:**256–260
cap methylation, **65:**257–259
capping enzymes, **65:**258
DHFR mRNA, **65:**271–273
function, **65:**265–270
inhibitors, **65:**266–267
internal methylation, **65:**259–260
intron D, **65:**271
in vitro, **65:**270–273
Eukaryotic transporters, **65:**9–10
Eurotium chevalieri, enzymic transformations of steroids by, **20:**245
Evolution, **44:**188; 189
of catalytic diversity, glutathione S-transferases, **69:**36–40
driving force of, **38:**195–198; 199; 204; 232
of proline sRNA synthetase, **29:**102
of protein molecules, **47:**417
of carbonic anhydrase isozymes, **47:** 160–161
Evolutionary concept or protein structure, **24:**415–439
Evolution of Glutathione Metabolism (Fahey and Sundquist), **64:**1
EXAFS (X-ray absorption extended fine structure), **51:**108; 112; 148; 150; 151; 155
information from, **51:**147–149
prospects of, **51:**159
for studying molybdenum in enzymes, **51:**113–117
Exchange, K^+/H^+, across human red cell membranes, **29:**47
Exchangeable guanine nucleotide site, microtubule dynamics, **69:**123
Exchange constant, definition, **13:**25
Exchange enzyme, of Chiga and Plaut, **26:**312
Exchange-incorporation reactions, **22:**129
in *S. aureus,* **22:**128
Exchange of CO_2 in acetic acid molecules, **12:**114–115
Exchange of glycoside linkages in synthesis of di- and polysaccharides, **10:**124; 138–141
Exchange reactions
biotin enzymes, mechanism, **26:**311–316
methylmalonyl-CoA racemization, **26:**340
in pentose biosynthesis, **28:**400–402; 407
in peptide bond synthesis, **13:**289–305; 305–310; 311–315
Exchange reactions, ADP^{32}-ATP, **23:**333; 355; 356; 358–359; 381; 385; 386
ADP-C^{14}-ATP, **23:**356; 358–359
H_2O^{18}-ATP, **23:**327; 356; 359–360; 368; 386
H_2O^{18}-Pi, **23:**356; 359–360; 364; 373
O^{18}, **23:**327; 355; 356; 359–360; 364; 368; 373; 378; 386
and oxidative phosphorylation, **23:**355–360
Pi^{32} and 1,3-di-P-glycerate, **23:**332–333
Pi^{32}-ATP, **23:**333; 348; 356–358; 365; 366; 368; 370; 371; 374; 375; 382; 386
Excitation energies and enzyme in-activation, **24:**367–371
Excitation energy in energy transport in living cells, **14:**29–31

Excitation process, in light emission, **25:** 143–144
Excitations, sub-ionizing, and enzyme inactivation, **24:**367–368
Exclusion chromatography, **42:**76; 77; 81; 128; 148
Excreting kidney cells and permeability, **8:** 50–51
Excretion, kidney, effect of fluoroacetate, **18:**134
Exercise, fall in brain norepinephrine from, **29:**487
Exocystine desulfhydrase, **7:**396–397
 classification, **14:**257
Exocystine desulfurase, and dehydropeptidase II, **8:**147; 148
 distribution, **8:**147; 148
Exocytosis, **40:**281; 283
Exoenzymes, **56:**143
Exo-enzymes, cellulose decomposing, **2:** 244–249
Exo-form, **34:**482
Exoglucanases, **28:**341–342; 345–353
Exoglycosidases, **60:**235–253
Exohydrolysis, of polysaccharides, **28:**348–351; 353
Exo-β-lactamase, isotope study, **28:**254
Exolaminarinase, hydrolysis by, **28:**349–351
Exo-α-mannanase, **40:**243; 245; 247; 251; 263
Exon-intron boundaries, catalytic diversity evolution, **69:**37; 39–40
Exonuclease, **56:**292
 bovine spleen, **55:**352–353
 commercial, and RNA, **29:**203
  in elucidating sequence of terminus in tobacco mosaic virus, **29:**204
  preference for denatured DNA, **29:**195
  spleen, **29:**207
  venom, **29:**203; 207; 209; 211
 spleen
  bis(*p*-nitro-phenyl) phosphate as substrate for, **31:**35
  chromatographic properties of, **31:**35
  inhibitors of, **31:**35
  p-nitrophenyl esters as substrate for, **31:**35
  pH optimum of, **31:**35
  sedimentation coefficient of, **31:**35
  substrates for, **31:**35
  thermal inactivation of, **31:**35
Exonuclease I, of *E. coli,* **29:**170; 209; 211
 hydrolysis of native DNA by, **29:**197
Exonuclease II, of *E. coli,* **29:**197
Exonuclease III, *See* Phosphatase exonuclease
 E. coli, **67:**449–450
 sequence analysis of duplex DNA, **67:** 450–451
 S1 nuclease and, progressive deletion to facilitate DNA sequence analysis, **67:**451
Exonuclease IV A, mode of action of, **29:**211
 physical and chemical characteristics of, **29:**209
Exonucleases, **60:**12
Exonucleotidases, oxidation of oligonucleotides by, **29:**202
Exopeptidase, classification, **2:**54; 56
Exopeptidases, **41:**237
 classification, **14:**224–225; 247
 column chromatography, **22:**448–459; 484–485
 specificity of, **12:**191–257
Exopolysaccharide alginate synthesis, **70:** 221–254
 alginate gene expression regulation, **70:** 236–238
 algR2 gene role in TCA cycle operation, energy metabolism, and alginate synthesis, **70:**238–245
 background, **70:**222–224
 biosynthetic enzymes, **70:**224–234
  guanosine diphospho-D-mannose dehydrogenase, **70:**232–234
  phosphomannomutase/ phosphoglucomutase (PMM/PGM), **70:**229–232
  phosphomannose isomerse-guanosine diphospho-D-mannose pyrophosphorylase (PMI-GMP), **70:**224–229
 modifying enzymes, **70:**234–236
 spontaneous interconversion between mucoidy and nonmucoidy, **70:**245–251
Exoxylanases, **56:**222
 effect of heat, **28:**346
Expansine, *See* Patulin
Experimental observations, **34:**474
Experiments, kinetic, design and evaluation of, **29:**14–23
Extensibility of protoplasm, **7:**41
Extensin, (hydroxy proline containing protein in cell wall), **41:**291–292

Extension, in cell elongation, **37:**324; 325
structure of, **37:**322
Extinction coefficients, **43:**142
Extrinsic system, *See* System, extrinsic
Exudative diathesis, **48:**2
Eye, aqueous humor, permeability studies, **13:**37
Eye pigments, genetic studies, **17:**420; 421

F

F-, **35:**286
F-17, in treatment of plant diseases, **20:**394
Fabrin, covalent crosslinking of, by transglutaminases, **38:**164
Fabry's disease, **38:**294; 295; 298; 299
therapy, **38:**310
Fabry's diseases, **60:**92
Facilitated diffusion, imidazole in, **20:**95; 96
F-Actin, **16:**345
F-actin, **6:**157
F-Actin, SH-groups, **27:**215
Factor
blood coagulation, properties of, **30:**262
bovine clotting, antisera to, **30:**266
clotting, interaction of, **30:**270–276
catalysis, **29:**36–37
Factor 5 and blood clotting, **10:**3; 9; 10; 11; 13; 19; 20; 23–28
Factor 6 and blood clotting, **10:**24–28
Factor I, fibrinogen, **48:**278
Factor II
factors concerned with the activation of, **30:**269 (*See also* Prothrombin)
mechanism of conversion of, **30:**281–283
properties of, **30:**277–278
prothrombin, **48:**278
Factor III, tissue factor, **48:**278
Factor IV, calcium ions, **48:**278
Factor IX, activation of, **30:**292. *See also* Christmas factor
interactions of, with Factor VIII, **30:**293–295
with Factors VIII and X, **30:**295–296
properties of, **30:**262; 291
Factor IX, Christmas factor, **48:**278
Factors, in respiratory chain, defined, **28:**160
Factor $(SH)_2$ dehydrogenase, in DR metabolism, **28:**440
Factor U, **21:**372
Factor V
discovery of, **30:**259 (*See also* Ac-globulin)
properties and interactions of, **30:**278–279
proaccelerin, **48:**278
term for blood clotting factor, **16:**59; 60; 61
Factor V_{1a}, *See* Cobyric acid
Factor VI, term for blood clotting factor, **16:**60; 61
Factor VII
discovery of, **30:**259 (*See also* Proconvertin)
interaction of tissue factor and, **30:**286–287
properties of, **30:**285–286
proconvertin, **48:**278
Factor VIII
antihemophilic factor, **48:**278
interaction of, with Factor IX, **30:**293–295 (*See also* Antihemophilic globulin)
with Factors IX and X, **30:**295–296
properties of, **30:**292
Factor VII (SPCA, cothromboplastin, stable conversion factor, convertin), blood clotting factor, **16:**29–31; 61; 65–67; 76; 91
Factor X, **46:**1; 3; 5. *See also* Stuart power factor
activated (X_a)
esterolytic and coagulant activities of, **30:**280
formation of, **30:**280
activation of, **30:**283–296
by the extrinsic system, **30:**284–287
calcium ion activation of, **30:**280
esterase from, **30:**261
interactions with Factors IX and VIII, **30:**295–296
interactions with tissue factor, **30:**286–287
intrinsic system activation of, **30:**287–296
IX, **46:**3; 5
properties, of, **30:**279–280
role in clotting, **30:**267
Russell's viper venom activation of, **30:**280
Factor X, Stuart factor, **48:**278
Factor XI
interaction of Factor XII and, **30:**291 (*See also* PTA)
properties of, **30:**290
plasma thromboplastin antecedent (PTA), **48:**278
Factor XII, activation of, **30:**289–290. *See also* Hageman factor
analysis of, **30:**289
interaction with Factor XI, **30:**291

Factor XII, activation of (*Continued*)
properties of, **30:**288–289
requirements for activation of, **30:**288
Factor XII, Hageman factor, **48:**278
Factor XIII
fibrin-stabilizing factor, **48:**278
properties of, **30:**262; 276
FAD, **57:**92; 95, **65:**61. *See also* Flavin adenine dinucleotide; Flavin Adeninenucleotide; Flavin adeninenucleotide; Flavin adenine nucleotide; Flavine adenine dinucleotide; Flavin nucleotides
absorption spectrum, effect of intermolecular bindings, **27:**3–16; 30
fluorescence changes, **27:**3; 9; 22; 29–31
localization, **27:**276
in mechanism study, **27:**3–16; 20–23; 28
in oxidation of hydrocarbons, **27:**485; 515; 537
reduced, EPR signal, **27:**162
brain, nitric oxide synthases, **65:**297
in methyl synthesis, **25:**405–406; 408; 410
nitric oxide synthases, isoforms, **65:**300
in putrescine oxidase, **36:**224
in spermidine dehydrogenase, **36:**225
respiratory burst oxidase cofactor, **65:**54
in respiratory chain, **28:**159; 162; 194
in thioredoxin reductase, **28:**443
tyrosine hydroxylase, substrate specificity, **70:**110
F_1 adenosine triphosphate, **59:**262–263; 265–267
FAD-enzyme, redox chain, SH-group, **27:**216
$FADH_2$, *See* Flavin adenine nucleotide, reduced
FAD-linked dehydrogenases, **24:**338
FAD pyrophosphorylase, localization of, **24:**314
FAICAR, *See* 4-Formamido-5-imidazolecarboxamide ribotide
Falcarindiol, **55:**90
Falcarinol, **55:**90
Familial periodic paralysis, **6:**214–218. *See also* Muscle diseases and altered metabolism
Faradic stimulation, effect on respiratory chain, **17:**113
FARCE, **43:**269
FARCE (freezing at reaction center on enzymes), **59:**286
Farnesoic acid, inhibitory action on sterol synthesis, **22:**320
role in squalene synthesis, **22:**308
structure, **22:**319
Farnesol, **46:**60
in biosynthesis of rubber, **29:**244
structure, **22:**306
Farnesyl-condensing enzyme, in squalene biosynthesis, **22:**297
Farnesyl isomerase, in squalene biosynthesis, **22:**297
Farnesyl pyrophosphate, **22:**317; 318, **55:**66
biosynthesis, **22:**319
condensation with nerolidyl pyrophosphate, **22:**316; 317
inhibition from, **55:**67
from IPP, **55:**78
isomerization to nerolidyl pyrophosphate, **22:**315; 316
in squalene biosynthesis, **22:**297; 305–309
condensation, **22:**313
synthesis, **22:**309–313
stereochemistry of formation of, **29:**247–248
structure, **22:**311
Fast green, use in staining nuclei, **16:**419
Fasting, **42:**277
and blood and urine changes in muscular dystrophy, **6:**207
and blood changes in progressive muscular dystrophy, **6:**222
effect on DNA metabolism, **16:**458–459
effect on enzyme distribution, **16:**467–468
and muscle weight loss, **6:**200
and vitamin E, **6:**507
Fasting and muscle phosphatides, **7:**142
Fast reactions, methods of studying, **25:**2–9
Fat, *See* Lipide
anabolism of, **5:**158–159
catabolism of, **5:**155–158
formation in *Fusaria,* **5:**196–198
Fat absorption and phosphatide turnover, **7:**137
Fat and carbohydrate catabolism, fundamental facts, **2:**266–271
general discussion, **2:**281–283
hypothesis of their reciprocal integration, **2:**266–283
mobilization of depot fats, **2:**271–283
Fat and Carbohydrate Catabolism (Witzemann), **2:**265
Fat coefficient, **9:**661; 668; 673; 674; 681; 685–686; 694
Fat-deficient diet and muscle phosphatides, **7:**142

Fat ingestion and phosphatide, metabolism, **7:**135–137
Fat metabolism, **17:**422–426
and carbohydrate metabolism, **3:**246–247
relation to carbohydrate metabolism, **9:** 692–700
Fats, **34:**486. *See also* Fatty acids, Lipides, Phosphatides, and Phospholipides
absorption in depancreatized dogs, **8:**180–181
autoxidation, **8:**425–457
chemistry of biosynthesis, **9:**692–700
industrial biosynthesis, **9:**653–704
from microorganisms, conditions governing formation, **9:**658–661
industrial methods, **9:**661–666
properties, **9:**686–692
recovery, **9:**689–692
composition and factors involved, **8:**318–324
enzyme action on, **33:**262
formation, and body water, **8:**324–330
and calcium, **8:**308–309; 315
from carbohydrate, conditions of formation, **8:**306–312; 322; 329–330
efficiency of conversion, **8:**302–305
and ethyl alcohol, **8:**315; 328–330
from noncarbohydrate sources, **8:**314–316; 325–326
and phosphate, **8:**307–308; 328; 330
and protein synthesis, **8:**310–312
and pyridoxine, **8:**310; 316; 330
and pyruvic acid, **8:**315; 327–330
and salt concentrations, **8:**308–309
site of, **8:**318; 319
and thiamine, **8:**310; 322; 330
and variety of carbohydrates, **8:**312–314; 328
and vitamins, **8:**310; 316; 322; 330
in liver of depancreatized dog, **8:**172; 178; 179
metabolism
and acetoin, **8:**389–391; 400
and acetyl phosphate, **8:**400–401
and aconitase, **8:**393
and barium, **8:**398
and diacetyl, **8:**389–391; 400
and diacetyl mutase, **8:**391
and dicarboxylic acids, **8:**392–409; 412–414
and external secretion of pancreas, **8:**186–191
role of vitamin B_{12}, **26:**265; 266; 270
in plant respiration, **18:**283; 288
synthesis, and chloroplasts, **8:**56
from fatty acids, **8:**317–318
and respiratory quotient, **8:**300; 302–305; 307; 310
and tricarboxylic acid cycle, **8:**328
synthesis from acetate, **12:**43
synthetic, biochemistry of, **8:**405–406
composition, **8:**405–406
from yeasts, **8:**320–322; 325; 328
Fats, Industrial Biosyntheses (Hesse), **9:**653
Fats in milk, precursors of, **7:**167–168
Fat synthesis, and photosynthesis, **25:**47–48; 62
Fat synthesis by acyl phosphates, **1:**153; 154
Fatty acid, **54:**181
in fructose-bisphosphatase regulation, **54:**143
synthesis in latex, **29:**232
Fatty acid adenylates, **19:**257; 258
Fatty acid anhydrides, and pyrophosphorolysis, **18:**202; 203; 208–209
Fatty Acid Catabolism (Breusch), **8:**343
Fatty acid cycle, pantetheine activity, **18:**207
Fatty acid esters
for column chromatography of lipase, **22:**447
effect on microbiological formation of hydrocortisone, **20:**246
in mitochondrial lipoproteins, **21:**109
Fatty acid oxidase, **19:**123; 125
Fatty acid oxidases, in electron transport, **14:**40
Fatty acid oxidation, enzyme system for, **13:**12
ω-Fatty acid oxygenase, **27:**493
Fatty acid radical, rearrangement, **58:**279
Fatty acids, **17:**182; 362, **34:**475–478; 486–488, **57:**88, **58:**185. *See also* Fats and Lipids; β-Keto acids; Lipids
absorption spectra, **7:**172, **9:**77
change in unsaturation, **7:**170–171
conjugated, **7:**172
deuterium-containing, **7:**169–170
iodinated, **7:**171
in plasma phosphatide, **7:**171
and acetate metabolism, **23:**431–432
long chain, as uncouplers, **23:**366–367; 387
activation by AMP, **19:**258
peroxidase action on, **19:**125
in photosynthesis, **19:**42; 43
acyl CoA derivatives, allosteric regulation, **28:**117

Fatty acids (*Continued*)
allosteric regulation, **28:**92
biosynthesis, **28:**138–139
effect on glycolysis, **28:**84
alternating β oxidation, **8:**358–360; 404; 409–412
aromatic, hydrotropic effects, **17:**153
branched chain, **8:**321; 369; 377–379; 406
(C^{14}), in sperm metabolism, **24:**65
from carbohydrate, **8:**300–301; 302–305 306–8:312; 312–314; 322; 325–330
catabolism, **8:**343–423
and citric acid, **8:**392–401
and C_6-tricarboxylic acids, **8:**392–401
and oxalacetic acid, **8:**392–401
and pyruvic acid, **8:**392–401
in various organs, **8:**409–412
from cellulose, **8:**266–268
chains, **34:**469; 471
in cholesterol biosynthesis, **22:**283–284
cyclohexyl, metabolism, **8:**356–357
degradation, **10:**189–190
oxidation, **10:**186–188
relation to citric acid cycle, **10:**213–214
α, β dehydrogenation, **8:**360–369; 380–387
relation to β oxidation, **8:**380–387
9,10 dehydrogenation, **8:**369–373; 375–376; 380–387
relation to omega oxidation, **8:**380–387
double bonds, **58:**275
enzymes for desaturation, **8:**316–317; 329
enzymes for oxidation, **17:**264–266; 382–383
epoxyhydroxy derivatives, **58:**303
essential and double-bond position, **8:** 370–371; 385
even-numbered, metabolism of, **8:**349; 401–409; 410
α γ oxidation, **8:**354
α,β hydrogenation, **8:**352–353; 365–369
9,10 hydrogenation, **8:**369–373
hydroxy, from alkanes, **27:**493
from isoparaffins, **27:**494
from microbial oxidation, of *n*-alkanes, **27:**484–493
of ring-substituted paraffins, **27:** 496–498
and mitochondrial swelling, **27:**229
interconvertibility, **8:**316–317
in livers of depancreatized dogs, **8:**177–178
long-chain fluoro, **18:**115; 130; 140–142
low molecular weight, **8:**387–391
mechanism of synthesis, **8:**324–330
mechanisms of enzyme action, **8:**380–387
metabolism, **15:**185–186; 201; 232–244; 251–253
monoethenoid, autoxidation of, **8:**428–434; 437–438
naturally occurring, **8:**343–345
odd-numbered, **8:**349; 355; 356–357; 370; 401–409; 410 (*See also* Propionic acid)
incorporation of propionic acid, **26:**370
metabolism of, **26:**369–370
oxidation, **15:**122–123
in mitochondria, **21:**12; 21–23; 57
synthesis, in mitochondria, **21:**47–49
α oxidation, **8:**354–355; 404
α,β oxidation, **8:**362
β oxidation, **8:**317; 336; 346; 347–357; 360; 364; 365; 369; 375; 376; 377; 380–387; 404; 411; 412
relation to α, β dehydrogenation, **8:** 380–387
δ oxidation, **8:**358–360
omega oxidation, **8:**336; 346; 365; 372; 374–380; 380–387; 406
relation to 9,10 dehydrogenation, **8:** 380–387
oxidation by thiols, **11:**249
oxygenation, **58:**192
in plastids oxidation, **20:**217; 218
synthesis, **20:**223; 224
polyethenoid, methylene-interrupted, autoxidation of, **8:**434–439
polyunsaturated dioxygenation, **58:**276
from protein, **8:**316
role of CoA in metabolism, **16:**18
salts of, solubilizing effects, **17:**138; 142; 153
saturated, rate of synthesis, **8:**324–330
spatial arrangement and enzyme action, **8:**380–387
in sperm metabolism, **24:**58; 65
spin label, **34:**468
synthesis, **8:**56; 300–318; 324–330
synthesis by acetyl coenzyme A, **15:**122
terminal oxidation, **8:**412–414
unsaturated
absorption spectra, **4:**280–281
and lipoxidase, **8:**425–457
autoxidation, **8:**425–457

metabolism, **8:**360–369
rate of synthesis, **8:**324–330
in respiratory chain, **20:**179; 180
α,β unsaturated, characterization, **8:**361
Fatty Acids, Unsaturated, Lipoxidase and Autoxidation (Bergstrom and Holman), **8:**425
Fatty acids (unsaturated) r O_2 transhydrogenases, **14:**253
Fatty acid synthase, **55:**22
Fatty acid synthesis, **43:**89, **45:**2–8
and acetyl-CoA carboxylase, **26:**297–303
effect of biotin, **26:**296
enzyme source, identity, **45:**8–9
Fatty Acid Synthesis in Mycobacterium smegmatis, Control Mechanisms for (Bloch), **45:**1
Fatty acid synthesizing system, location of, **24:**296; 299; 310; 311; 332; 334; 338
Fatty acid synthetase, **35:**365, **45:**292
allosteric regulation, **28:**139
kinetic mechanism of, **45:**292
and propionic acid metabolism, **26:**298–300; 302; 370
subunit structure of, **45:**58–62
Fatty acid synthetase complex, **67:**409–410
Fatty acid synthetases, **46:**248; 259
Fatty acid thiokinase, in mitochondria, **21:**74
Fatty acyl CoA dehydrogenase. *See also* Yellow enzymes
mechanism of action, **19:**209; 210
properties, **19:**202
Fatty acyl CoA dehydrogenases, **17:**364; 382–383, **18:**106
Fatty acyl CoA derivatives, and glucose oxidation, **28:**117
Fatty-acyl CoA synthetase, **44:**40
Fatty acyl coenzyme A dehydrogenase, **15:**85
Fatty acyl dehydrogenase, in photosynthesis, **19:**44; 45
Fatty livers
and ethionine metabolism, **25:**369–380
and high protein diet, **8:**173–183; 199–200
development in rats, **8:**196
and hydrolyzed protein, **8:**199–200
in insulin-treated depancreatized dog, **8:**175–179
and ligation of pancreatic ducts, **8:**187–189
prevention by pancreatic juice, **8:**190–191
time interval for development, **8:**195–196
FBPase 2
and animal tissues, **59:**353–354
and fungi, **59:**354–356
and plants, **59:**356
and protists, **59:**357
and rat liver bisfunctional enzyme
discovery of, **59:**341
evidence for, **59:**341–344
and histidyl-P enzyme, **59:**352–353
kinetic properties, **59:**349–350
other reactions catalyzed by, **59:**350; 352
and phosphorylation by cAMP-dependent protein kinase, **59:**344–346
FC, **54:**58
FCdR, **54:**58
FCR, **54:**58
FDP, *See* Fructose diphosphate
FdTTP, **54:**58
FdUDP, **54:**58
FdUMP, **54:**58
5, 6-6dihydro, **54:**79; 83
in assessing tumor responsiveness to Fu, **54:**103–105
in binary complex with 5, 10–10CH_2H_4PteGlu, **54:**82
in binary complex with TS, **54:**81; 83
synthesis of, **54:**64
in ternary complex with TS and 5 10–10CH_2H_4PteGlu, **54:**81–86
dissociation from, **54:**85
in thymidylate synthetase inhibition, **54:**70; 92; 93–94
in thymidylate synthetase interactions, **54:** 77–78; 79–80
Fe^{2+} binding, collagen hydroxylases
consubstrate binding, **72:**349–557
inhibitor mechanisms, **72:**360–364
Feces, human, thiaminase production in, **15:**390; 394; 414–415
Feedback, in cholesterol synthesis, **29:**368
in deoxyribonucleotide synthesis, **29:**368–370
effects of TTP on Tdr kinase, **29:**367
effects on aspartic transcarbamylase, **29:**367
effects on proline inhibition of pyrolline carboxylate reductase, **29:**367
enzyme repression and effects of, **29:** 367–371

Feedback inhibition. *See also* Product inhibition of enzymes
of enzymes, **25:**169–172
glutamine PRPP amidotransferase
binding sites, **72:**100–102
crystal structure, **72:**96
nucleotide regulation, **72:**95
glutathione synthesis, **73:**245–246
in synthesis of DR-5-P, **28:**393
Feedback inhibition of enzymes, explanation, **26:**105–107
Feed back mechanism, of aspartokinases, **25:**361
Feedback regulation. *See also* Allosteric regulation; Control mechanisms; Inhibition; Regulation; Repression
of enzymes, **28:**43–46; 61–63; 65–69; 72–73; 100; 103; 113–116; 124; 128
Fe-enzyme, **35:**282
Fe(II)-complexing agents, **44:**205
FeMoco, reactivation, **67:**330
Fenton's reaction, **15:**6–8; 10; 11; 16–17; 23
Fenton's reagent, **41:**43
mechanism of action, **19:**137–140
FeP-MoFeP interaction, **67:**343–344
during catalysis, **67:**353–356
models, **67:**354–355
FeP protein, **67:**340–349
amino acid sequence, **67:**340–342
conserved hydroxyl-containing group, **67:**347–348
Fe-S center, **67:**340–343
making competent, **67:**349–352
MgATP-binding, **67:**344–347
nucleotide interaction, **67:**344–349
$FeP_R \cdot 2MgATP$, regeneration, **67:**356–358
Ferment, *See* Verdoperoxidase
Fermentation, **7:**57, **34:**525. *See also* Fermentation and Yeast
activating agents in, **7:**563
alcoholic, **5:**213
direct, Willstätter theory, **5:**91–94; 107; 108; 113
in *Fusaria,* **5:**167–179
indirect, Fischer theory, **5:**90–91; 107; 108
of oligosaccharides, **5:**87–5:127
rôle of cell organization, **5:**119–124
specific catalysts and inhibitors, **5:**114–116
and alkali metal ions, **7:**27–30
of cacao, **25:**458; 466–467; 469–484; 486 487
changes in cotyledon, **25:**476–482
death of seed, **25:**476
chemical mechanism, **9:**682
of Douglas-fir hydrolyzates, **5:**201
effects of ethylene, **7:**21
effects of pressure, **7:**249
ethyl alcohol (*See* Ethyl alcohol fermentation)
gluten recovery, **7:**590
and glycogen synthesis, **7:**27–28
industrial, **7:**557–615
acetone, **7:**560–573
of agricultural wastes, **7:**578–579
biological stability of systems, **7:**560–562
butanol-acetone, **7:**560–573
2,3-butylene glycol, **7:**578; 601–611
ethyl alcohol (*See* Ethyl alcohol fermentation)
glycerol, **7:**587; 596–600
microorganisms used, **7:**563; 565; 570–572; 578; 579; 581; 585; 592; 593; 602–607
molasses, **7:**585–587
sluggishness, **7:**263–264
value of residues, **7:**569–573; 589–590
and ion exchange, **7:**28–29
by luminous bacteria and salts, **7:**253–255
by luminous bacteria and substrates, **7:**253–255
microbiology, **25:**470–475
in microorganisms, effects of pantothenate deficiency, **7:**442–444
and phosphorylation, **7:**27–30
potassium ions, **7:**27–30
products of luminous bacteria, **7:**255
and pyridine nucleotides, **15:**120
of sugars by thiaminase-producing bacteria, **15:**413
reaction rate, **7:**18–20
relation to tryptophanase activity, **10:**78–81
of sulfite liquor, **7:**565; 575–576; 582–585
of sulfite waste liquors, **5:**198
temperature effects, **25:**474–476; 479
of tobacco
aging of cigarette tobacco, **10:**399–409
causative agents, **10:**409–433
changes during sweating, **10:**361–399
changes in water retention, **10:**354–355
changes of *p*H, **10:**352–354
enzymic theory, **10:**412–415; 415–425
gas exchange during, **10:**342–352

improvement of burn, **10:**356–357
loss of weight, **10:**335–336; 361–363
qualitative changes, **10:**333–335
sweat techniques, **10:**358–361
temperature increase, **10:**336–342
of tobacco leaf, **10:**325–441
types of processing, **10:**327–333
use of acetate, **23:**431–434
of wheat hydrolyzates, **5:**198
wheat in, **7:**587–591
of wood hydrolyzates, **5:**198
wood in, **7:**574–585
by yeasts, **7:**13–20
Fermentation, Industrial, Recent Progress (Hildebrandt), **7:**557
Fermentation, Tea (Roberts), **2:**113
Fermentation, Tobacco (Frankenburg), **10:**325
Fermentation and phosphate bond energy, **1:**132
and phosphorylation, **1:**120
Fermentation pathways, in analysis of tracer distribution, **28:**402–404
Fermentation reaction, of glycine, **23:**457–462
Fermentative metabolism, ferredoxins in, *See* Ferredoxins, in fermentative metabolism
Ferment fibrinolytique, **23:**184
Ferns, thiaminase in, **15:**390–391; 394; 397; 398; 403
Ferredoxin, **39:**361, **47:**58; 60; 124, **50:**142; 157, **68:**40. *See also* PPNR
in chloroplasts, **25:**55–56; 85
in Hill reaction of spinach chloroplasts, **30:**46–47
properties of, **30:**15–16
purification of, of chloroplasts, **30:**15–16
purification, **67:**306
role in nitrogenase, **67:**301–307
Ferredoxin, Cotton effects, **27:**59; 87–89
EPR study, **27:**139–140
labile sulfide in, **27:**159
Ferredoxin-NADP reductase, **33:**124–125
transhydrogenase activity, **33:**124–125
Ferredoxin-NADP reductase, NADPH-cytochrome *f* reductase activity exhibited by, **30:**41
properties of, **30:**16
reduction of $NADP^+$ mediated by, **30:**22
Ferredoxin-reducing factor (FRS), **33:**127
Ferredoxin reductase, **60:**46
Ferredoxins, **41:**38; 201, **46:**82; 166; 167, **48:**10
absorption spectra of, **33:**149
activation of fructose diphosphatase by reduced, **33:**136–138
active sites of, **33:**165
amino acid composition and sequence of, **33:**157; 160
apoprotein recombination with iron and sulfide in, **33:**163
in assimilation of carbon dioxide, **33:** 138–143
bacterial, amino acid composition of (table), **33:**158
amino acid sequence of (table), **33:**159
homology of, **33:**160
biological synthesis of, **33:**160
chemical evidence for iron-sulfur bonding in, **33:**163
chemical properties of (table), **33:**156
chemical synthesis of, **33:**160
chemistry of, **33:**149–165
circular-dichroism (CD) spectra of, **33:**155; 157
classification of, **33:**166
electron-paramagnetic resonance (EPR) properties of, **33:**163–164
electrons transferred by, **33:**154
in electron transport, **33:**133–136
evolutionary aspects of, **33:**168–169
in fermentative metabolism, **33:**119–176
and breakdown of pyruvate, **33:**147
and carbon dioxide assimilation, **33:**149
as electron donor for reduction, of bisulfite to thiosulfate, **33:**149
of hydroxylamine, **33:**149
of nicotinamide adenine dinucleotides, **33:**149
of sulfite to sulfide, **33:**149
as electron donors, **33:**149
and evolution of hydrogen gas, from acetaldehyde, **33:**147
from formate, **33:**147
from hypoxanthine, **33:**147
from α-ketoglutarate, **33:**147
history of, **33:**121–123
iron in, **33:**153–154
nature of, **33:**160–164
oxidation state of, **33:**160; 162
iron-sulfur bonding in, chemical evidence for, **33:**163
physical evidence for, **33:**163–164
labile sulfide in, **33:**162–163

Ferredoxins (*Continued*)
labile sulfur in, **33:**154
magnetic-susceptibility measurements of, **33:**154
models of, **33:**165
molecular weight of, **33:**151–152
Mössbauer effect in, **33:**165
in NADP reduction, by chloroplasts, **33:**121–125
transhydrogenase activity, **33:**123
in nitrogen fixation, **33:**119–176
illuminated chloroplasts as reductant in, **33:**145
nitrogenase and its components in, **33:**145–146
pyruvate as reductant in, **33:**145
nomenclature of, **33:**166; 167*t*
optical-rotatory-dispersion (ORD) spectra of, **33:**155; 157
other names for, heme-reducing factor, **33:**121
methemoglobin-reducing factor, **33:**121
NADP-reducing factor, **33:**121
photosynthetic pyridine nucleotide reductase, **33:**122
red enzyme, **33:**122–123
and other reactions in photosynthetic cells, **33:**144
NAD and NADP reduction by photosynthetic bacteria, **33:**144
nitrite reduction, **33:**144
sulfite reduction, **33:**144
oxidation-reduction potential of, **33:** 152–153
in oxygen evolution, **33:**125–127
stoichiometry between ferredoxin and oxygen, **33:**126
in photophosphorylation, **33:**127–133
antimycin A as inhibitor, **33:**131; 132
cyclic, **33:**128–129
cyclic electron transport as system I reaction, **33:**135
cyclic process and the red rise, **33:**130
3-(3,4-dichlorophenyl)-1,1-dimethyllurea (DCMU) as inhibitor, **33:**130
dinitrophenol as inhibitor, **33:**132
effect of inhibitors on, **33:**130–133
noncyclic, **33:**127–128
noncyclic electron transport as system II reaction, **33:**135
oligomycin as inhibitor, **33:**131; 132
oxygen-linked noncyclic, **33:**128
o-phenanthroline as inhibitor, **33:**130
pseudocyclic (*See* Ferredoxins, in photophosphorylation, oxygen-linked noncyclic)
separation of cyclic and noncyclic, **33:**130–133
sites of photophosphorylation, **33:**132; 133–135
system I, **33:**135
system II, **33:**135
in photosynthesis, **33:**119–176
physical evidence for iron-sulfur bonding in, **33:**163–164
plant, amino acid sequences of (table), **33:**161
homology of, **33:**160
reaction with PCMB, **33:**162
o-phenanthroline, **33:**160
sodium mersalyl, **33:**162
Tiron, **33:**162
reconstitution of, **33:**163
in regulation of photosynthetic carbohydrate synthesis, **33:**137–138
spectral characteristics of, **33:**149–151
x-ray crystallographic analysis of, **33:**165
Ferredoxins, Chemistry and Function of in Photosynthesis, Nitrogen Fixation, and Fermentative Metabolism (Buchanan and Arnon), **33:**119
Ferredoxin type iron, in respiratory chain, **28:**209
Ferric. *See also* Iron and Metal ions
Ferric chloride, in determination of activated amino acids, **19:**236
Ferric complex, **39:**332
Ferric compounds, EPR spectrum, standards, **27:**116
Ferric cytochromes *a,* absorption spectra, **23:**307–309
Ferric haem *a* compounds, **23:**289–297; 306–310
Ferric horse radish peroxydase, absorption spectra of, **51:**45
Ferrichrome, **35:**136; 140–145; 158–163; 172; 174
absorption, **35:**142
spectrum, **35:**141
acetate, **35:**159
active transport, **35:**174; 177
acyl groups, **35:**139

amino acid, incorporation, **35:**159
sequence, **35:**141
biological activity, **35:**144; 148; 150; 151–152
biosynthesis, **35:**158
chemical synthesis, **35:**143–145
as coenzyme, **35:**171
conformation and active transport, **35:**176
conformation of, **35:**141–143
as growth factor, **35:**168
in heme synthesis, **35:**172
hydrogen, bonds, **35:**143
exchange, **35:**142–143
hydrolysis, **35:**141
periodic, acid, **35:**141
reaction, **35:**141
reduction of, **35:**141
stereochemistry, **35:**142
Ferrichrome A, **35:**145–146; 150; 175; 176
absorption spectrum, **35:**145
acyl groups, **35:**139
amino acid sequence, **35:**146
biological activity, **35:**146
biosynthesis, **35:**158; 160
conformation, **35:**146
hydrogen, bonds, **35:**146
exchange, **35:**146
hydrogenation, **35:**145
x-ray analysis, **35:**146
Ferrichrysin, **35:**148
Ferric ion. *See also* Iron
in model hydroxylating system, **19:**135
Ferric ions. *See also* Ferrous ions and Iron; Iron
chelate formation with organic phosphates, **21:**147
model catalyses with, **21:**176
properties, **21:**157
in citric acid cycle, **10:**201
as decarboxylating agent, **12:**80
effect on hyaluronidase, **13:**214
effect on laminarinases, **28:**347
effect on phosphoglucomutase activity, **14:**212
effect on transmethylation, **12:**468
formation of complex with oxalacetic acid, **12:**72
in tyrosine metabolism, **14:**107
Ferric oxidase, effect of cyanide on, **23:**301–302
Ferric porphin, **7:**306; 310
Ferric protoporphyrin IX, **33:**311
Ferric reagent, **39:**332
Ferricrocin, **35:**148
Ferric superoxide complex, **60:**75
Ferric thiocyanates, **39:**332
Ferric triethylenetetramine, as model for catalase-peroxide complex, **19:**115; 116
Ferricyanide, **15:**38; 88; 89, **51:**134
as artificial electron acceptors, **30:**22; 23
assay for pyruvic oxidase, **15:**212; 216–218; 223
as electron acceptor, **17:**239–242; 254; 256; 266; 275; 298; 316; 323; 366–369; 372, **23:**386, **28:**169–172; 181; 183; 194; 197; 210–214; 225–227
as hydrogen acceptor in photosynthesis, **15:**126
in inhibitor studies, **15:**289
luminescence, **25:**149; 150
reaction with thiols, **11:**207; 219–220; 222; 223–224
and succinic dehydrogenase, **18:**67; 68; 71; 72; 75; 76; 84–87; 99–100
Ferricyanide ions, **34:**490
Ferricyanide reductase activity, EPR study, **27:**180
Ferricytochrome
c_1, **54:**277
autooxidation of, **54:**286
CD spectra of, **54:**285; 286
with cyanide, **54:**323
pH for reduction of, **54:**297
in phospholipid vesicles, **54:**342
electron transfer with, **54:**342–343
quenching ANS fluorescence, **54:**294
reduction of, **54:**325
structural stability of, **54:**310; 311
urea effect on, **54:**286–287; 288
effect of radiation on, **24:**395
Ferricytochrome *a,* and phosphorylation, **23:**352
Ferricytochrome a_3, effect of cyanide on, **23:**301–302
Ferricytochrome c, **7:**267; 280–285
Ferricytochrome *C,* **41:**38; 90
Ferricytochrome *c,* **34:**477, **58:**65
in photosynthesis, **19:**51
Ferricytochrome c, Cotton effect in, **27:**47; 92; 95
Ferricytochrome c monomer, EPR study, **27:**138
Ferridoxin, **39:**339
Ferrienzyme, superoxide dismutase, **41:**86
Ferrihaemochromes, absorption characteristics, **23:**291; 294
Ferrihemoglobin, Cotton effect, **27:**60; 90–92

Ferrihemoglobin, enzymic inactivity of, **12:** 178–181
stability of the peroxide compound, **12:**160; 161–163; 165–166
Ferrihemoproteins, EPR study, **27:**138–140
Ferrihodin, **35:**148
Ferrimycin, **35:**144; 172; 177
Ferrimycins, **35:**151–152
biological activity, **35:**151–152
Ferrimyoglobin
complexes with hydrogen peroxide, **19:**111; 114; 116; 117; 138
enzymic inactivity of, **12:**178–181
stability of the peroxide compound, **12:**160; 161–163; 165–166
Ferrimyoglobin, Cotton effects, **27:**60; 90–92
Ferrimyoglobin azide, EPR study, **27:**138
Ferrioxamine, **35:**168; 170; 175
Ferrioxamines, **35:**148–151
biological activity, **35:**149
hydrogen exchange, **35:**150
stereochemistry, **35:**149
synthesis, **35:**150
Ferriperoxidase, **7:**290; 292–294
Ferriprotoporphyrin, **7:**267
Ferrireductase, **44:**222
Ferrirhodin, **35:**150; 153; 176
Ferrirubin, **35:**148; 150; 176
Ferritin, **17:**306
effect of chemical modification, **24:**145
electron microscopy of, **33:**103–104
histochemical methods for, **9:**598
Ferritin Family of Iron Storage Proteins, The (Theil), **63:**421
Ferri-transferrin formation, **44:**205; 215
Ferrocene, **58:**24
Ferrochelatase, **35:**171; 172, **46:**58; 59; 60; 86; 90; 107; 129; 137; 163; 165; 175; 176; 179
Ferrochlorin *d,* **23:**312
Ferrocyanide, electronic structure, **15:**20
Ferrocytochrome c, **7:**267; 281–285
autooxidation of, **54:**321; 322
from cytochrome c or c_1 reduction, **54:**313
effect on cytochrome oxidase, **27:**151–152
as hydrogen donor, **12:**170–171; 175; 187
oxidation of, **54:**334; 335
in phospholipid vesicles, **54:**342
electron transfer with oxidation of, **54:** 342–343
structural stability of, **54:**310; 311
Ferrocytochrome *c*
cytochrome *c* oxidase, **71:**80–81; 82; 86; 111; 117; 125; 181
oxidation of, **23:**349; 352; 354; 367; 382
Ferrocytochrome c in energy transport, **14:** 39–41
Ferrocytochrome c_1-oxidase, **54:**331
Ferrodoxin, **54:**175; 177
Ferrodoxin-thioredoxin reductase, **54:**176
Ferroflavoprotein, succinic dehydrogenase as, **18:**80
Ferroflavoprotein DPNH-cytochrome reductase, **17:**256–257
Ferroflavoproteins, **17:**256; 270; 274. *See also* Metalloflavoproteins
Ferrohaemochrome spectra, **23:**294
Ferromolydboflavoproteins, **17:**270. *See also* Metalloflavoproteins
Ferro-O_2 oxidoreductase, **44:**197
Ferroperoxidase, **7:**289; 292; 293, **23:**290; 296
Ferroprotoporphyrin, **7:**267
Ferrous compounds, EPR study, **27:**140
Ferrous cytochrome a_2, **23:**312
Ferrous cytochrome a_4, **23:**312
Ferrous cytochrome *a,* absorption spectrum, **23:**300–302; 306–309
Ferrous cytochrome a_1, absorption spectrum, **23:**306–307
Ferrous cytochrome a_3, absorption spectrum, **23:**301; 302; 306–309
Ferrous (Ferric) ion, requirement for prolyl hydroxylase, **41:**252
Ferrous haem *a* compounds, **23:**289–297; 306–310
Ferrous haemoproteins, **23:**290; 295–296
Ferrous ion. *See also* Heme, Iron, Iron porphyrin; Iron
action on hydrogen peroxide, **19:**137–139
effect on degradation of pectic substances, **20:**352
in model hydroxylating system, **19:**135; 140–143; 199
in oxygen metabolism, **19:**81; 85; 89–91; 97; 98; 176; 177; 179; 182; 199
in phosphorylation of pyridoxal, **16:**219
Ferrous ions. *See also* Iron
activation of alkaline phosphatase by, **10:**99; 102–104

chelate formation with organic phosphates, **21:**147
enolase activation by, **21:**152
properties, **21:**157
as decarboxylating agent, **12:**72; 80; 87
in DR metabolism, **28:**440
effect on collagenase, **23:**196
effect on hyaluronidase, **13:**214
in glutathione splitting, **13:**287
in oxidation of tyrosine, **14:**80; 97–98; 104–105; 107; 109–111
and peptidase action, **12:**195; 206; 222
Ferrous oxidase, effect of cyanide on, **23:** 301–302
Ferrous P450, oxygenated, formation and breakdown, **47:**33–34
Ferrous porphin, **7:**306; 310
Ferrous protoporphin IX, **7:**306–307
Ferrous protoporphyrin, *See* Protoheme
Ferroxidae activity, alternative, **44:**203
of human Cp, **44:**197
true, **44:**203
Ferroxidase, **44:**197
Ferroxidase and iron mobilization, **44:**215
Ferroxidase II, **44:**204
Ferryl cytochrome a_3, **23:**302
Ferryl ion
in heme enzyme action, **15:**7; 18
as hydroxylating intermediate in oxygen metabolism, **19:**113; 116; 117; 142; 143; 174; 177; 179; 197; 211
Fertilizability, dependence on physiological state of sex cells, **9:**244–255
effect of albumin on, **9:**250–251; 272–274; 278; 281
of amino acids on, **9:**250–251
of calcium on, **9:**248; 249; 262–264; 279
of magnesium on, **9:**248
Fertilization
comparison of DNA in two germ cells, **16:**427
effect on sperm nucleus, **16:**479–480
and glutamine metabolism, **13:**273
role of hyaluronidase in, **13:**226–227
and semen metabolism, **9:**348–352
mechanism in metazoa, **9:**241–327
Fertilization in Metazoa (Runnstrm), **9:**241
Fertilization membrane of egg, **9:**245; 247; 251–252; 254–255
effect of jelly coat on, **9:**313–314
formation, **9:**275–286; 311
variation with ripeness of egg, **9:**247–248
Fertilization of egg, and glutathione content, **11:**253–255
Fertilization reactions, **9:**255–268
Fertilizin, **9:**246
and antifertilizin, **9:**264; 267–268
chemical nature, **9:**262–263
in egg jelly coat, **9:**258–259; 313–315
in egg water, **9:**255–259
role in fertilization, **9:**266–268; 268–275; 313–315; 348
Ferulic acid
esters of, **55:**9
formation of, **55:**39
as lignin precursor, **18:**373–374
2Fe-2S, from parsley, **48:**10
Fe-S center, **67:**340–343
[Fe-S] cluster
glutamate synthase, **72:**114–116
glutamine PRPP amidotransferase, **72:**93–95
Fe_2Se_2 centers, **48:**10
Fe_2SSe centers, **48:**10
Fetuin, **41:**110–111; 116; 122; 123, **52:**32; 59; 99; 105
separation by chromatography, **14:**351; 358
Feulgen reaction, for DNA, **16:**412; 417; 419; 422; 423; 426; 427; 429; 460
Feulgen's nucleal reaction, **3:**4–6; 10; 23 of viruses; 30
Fever and essential hypertension, **7:**540
FGAM, *See* N-Formylglycinamidine ribotide
FGAR, *See* N-Formylglycinamide ribotide
Fiber proteins, *See* Proteins fibrous
Fibers of polymerized actin, **16:**334
Fibrillar aggregates, from β-1,3 glucans, **28:**327; 335
Fibrillar structure of ground cytoplasm, **8:**8–14
Fibrillation, and muscular atrophy, **6:**197; 200
Fibrils. *See also* Cytoplasmic fibrils and cell organization, 8:24–8:26,8:33
comparison, **8:**14–17
in cortex of cell, **8:**19–20
motility and contractility, **8:**29
and thixotropy of protoplasm, **8:**27–28
Fibrin, **53:**258. *See also* Fibrinogen
as antithrombin, **16:**78
banding, **16:**320
formation, **16:**25; 32–44; 68; 91–92
autocatalytic formation, **10:**8
crosslinking of, **54:**30

Fibrin (*Continued*)
optimal pH for papain action on, **19:**303
oxidation of tyrosine in, **14:**86
structure studies, **14:**284; 285
transflutaminase reactivity to, **54:**26; 29
Fibrinase, *See* Factor XIII
Fibrin clot
dissolving by various agents, **16:**35–37; 39; 42–44
imidazole in, **20:**101
α- and β-Fibrin derivatives, from action of fibrinolysin, **16:**42
Fibrinogen, **5:**32; 33; 43–45; 46; 57–59, **43:**315, **48:**278. *See also* Clotting, blood; Fibrin
action of various enzymes on, **10:**12–13; 33
clotting, **10:**30–33
distribution in centrifugal field, **16:**89
interaction with thrombin, **16:**25; 35–42; 92
platelet materials behaving as, **16:**71–72
purification, **16:**26
inhibition of action in blood coagulation, **5:**49
oxidation of tyrosine in, **14:**86
structure studies, **14:**284; 285
properties, **10:**29–30
properties of, **30:**262
interactions of, **30:**271
Fibrinogenase, in semen, **9:**348
α- and β-Fibrinogen derivatives, from action of fibrinolysin, **16:**42
Fibrinogen-fibrin, effect of tyrosinase on, **14:**142–144; 147–150
peroxidase action on, **14:**158
Fibrinogen-fibrin transformation, **16:**32–44; 91–92
Fibrinolysin, **10:**12; 22; 32
comparison of action with thrombin action, **16:**42–43
effect on thrombin, **16:**78
equilibrium conditions among trace proteins, **16:**76
mechanisms, **16:**24
in separation of fibrinogen from blood, **16:**26
in semen, **9:**348
Fibrinolytic enzyme, neutralization by platelet factor, **16:**71
Fibrinolytic factor, **23:**200; 201
Fibrinopeptide. *See also* Cofibrin
removal from fibrinogen by thrombin, **16:** 36–37; 41–42; 44; 92
Fibrinoplastic action, of acacia, **16:**33; 39
Fibrinoplastic activity, of platelet factor, **16:**2; 70
Fibrins, disulfide bonds, **27:**220
Fibrin stabilizing factor, *See* Factor XIII
Fibrin-stabilizing factor, **48:**278
Fibrin-stabilizing factor (Serum factor), in formation of Ca-fibrin from T-fibrin, **16:**43–44
Fibroblastic cells
binding of
cells in suspension, **59:**21–23; 25
hepatocytes, **59:**25
monolayers of, **59:**19–21
platelets, **59:**25–26
and fibronectin synthetic peptides, **59:**26–31
Fibroblasts, **46:**121
from patients with gout, **41:**13
Fibroin
properties of, **30:**180
structure studies, **14:**302
Fibronectin, **52:**158, **70:**1–21
and adhesion, **59:**2
binding of
fibroblastic cells in suspension, **59:** 21–23; 25
fibronectin-coated beads, **59:**16; 18–19
hepatocytes, **59:**25
monolayers of fibroblastic cells, **59:**19–21
platelets, **59:**25–26
bovine, **59:**7; 10; 13
cell recognition pentapeptide, **59:**15
cellular, **59:**3–6
central cell-binding domain of, **70:**8–12
chicken cellular, **59:**12–13
development of, **59:**35–36
disintegrins and cyclic peptides, **70:**6–8
extra domain, **59:**14
features of, **59:**3–6
generally, **70:**1–3
human, **59:**7; 13
integrin cell-surface receptors, **70:**3
ligand-binding site model, **70:**13–15
metastasis of, **59:**36–38
overview of, **59:**49–50
physical description of
molecular biology, **59:**7; 9–12
primary structure, **59:**6–7
plasma, **59:**3–6; 8
rat, **59:**7; 13
receptors

analyzation, **59:**23
avian, **59:**40–46
classes of, **59:**38–39
and CSAT antigen, **59:**40–41; 43–46
45,000-dalton, **59:**46–47
gangliosides, **59:**46–48
glycoprotein IIb/IIIa, **59:**48–49
hamster, **59:**46
human, **59:**39–40
140 k complex, **59:**40–44
model, **59:**22–24
platelet, **59:**48–49
progress in, **59:**38–39
reviews on, **59:**3
schematic structure of, **70:**9
short integrin-recognition sequences, **70:**3–6
synthetic peptides as probes of
development of, **59:**35–36
fibroblastic cells, **59:**26–27; 30–31
macrophages, **59:**34
metastasis of, **59:**36–38
platelets, **59:**31–32; 34
slime mold cells, **59:**34–35
type III connecting sequence (IIICS) region, **70:**12–13
Fibronectin (Akiyama and Yamada), **59:**1
Fibronectin-coated beads, **59:**16; 18–19
Fibronectin receptors
analyzation, **59:**23
avian, **59:**40–46
classes of, **59:**38–39
and CSAT antigen, **59:**40–41; 43–46
45,000-dalton, **59:**46–47
gangliosides, **59:**47–48
glycoprotein IIb/IIIa complex, **59:**48–49
hamster, **59:**46
human, **59:**39–40
140 k complex, **59:**40–44
model, **59:**22–24
platelet, **59:**48–49
progress in, **59:**38–39
Fibrous forms, from nuclei, **18:**162; 165; 180
Fibrous proteins
chemical and x-ray evidence on structure, **3:**71–96
collagen group, **3:**72; 73; 91–96
distribution of side chains, **3:**81–88
gelatin, **3:**91–94
keratin-myosin group, **3:**72; 73; 91–96
rabbit myosin, **3:**78–81
silk fibroin, **3:**73–76
wool keratin, **3:**77–81
x-ray classification, **3:**71–73
α-keratin in porcupine quill, **3:**88–91
structure of, **23:**2–4; 22; 25
Fibrous proteins and enzyme activity, **14:** 184–185
Ficin, **10:**300, **19:**287; 289; 327, **34:**449; 459, **37:**450, **53:**261; 278
action on, collagen, **23:**176; 219
elastin, **23:**226; 229
chemical attachment to carboxymethylcellulose, **33:**277
specificity of, **33:**433
classification, **14:**247
effect of acylation on, **24:**138
pK value, **22:**58
specificity of, **12:**194
synthesis of amide linkage by, **13:**309
Fick's equations, **7:**9–10
Fick's law, **29:**41, **34:**496
FIG, *See* Formiminoglycine
FIGLU, *See* Formimino-L-glutamic acid
Filamentous fungi. *See also under* individual genera, eg, Aspergillus
genetic studies of, **13:**143; 144
relation of nitrogen to growth of, **13:**347–348
Filipin, in treatment of plant diseases, **20:**394; 395; 403
Film compression, hysteresis, **28:**8; 13
Film formation, effect on biological activity, **28:**3; 21–33
effect on protein structure, **28:**4–5; 7; 8; 11–12; 24; 31–32
by enzymes, **28:**1–40
Film formation of amylose esters, **12:**353
Film pressure, and concentration, **28:**7; 10
Films, *See* Surfaces
compressibility of, **28:**9–12
concentrated, **28:**8–12
definition of terms, **28:**6–8
dilute, **28:**7–8
gaseous type, protein structures, **28:**4; 7; 8
nature of, **28:**6–21
spatial arrangements, **28:**4; 9–12; 16–17
Filters, **34:**452
Firefly bioluminescence, **25:**123; 124; 131–144. *See also* Luciferin
Firefly luciferase, **44:**37
activation and oxidation reactions, **44:**39

Firefly luciferase (*Continued*)
mechanism of, light-emitting steps, **44:**43
oxidative reaction, **44:**41; 43
reactions catalyzed by, **44:**39
Firefly Luciferase (DeLuca), **44:**37
Firefly luciferin, **25:**121; 124–126; 129–144
Fire worms, **25:**160
First-order kinetics, **34:**497; 520
Fischer lock and key, **65:**249–250
Fish
digestion, **1:**385–402
intestinal canal, **1:**375–377
secretion of gastric and pancreatic juice, **1:**385–387
thiaminase in, **15:**390; 391–392*t;* 395–404
Fishman unit, for activity of β-glucuronidase, **16:**385
Fission, enzymic, mechanism, **10:**62–64
Fission fragments, LET from, **24:**365
Fixed charge groups, **34:**463
Fixin, requirement for —SH group, **11:**241
Flagella protein, **42:**230
Salmonella typhimurium, **42:**229
Flagellar structure and assembly, *Caulobacter* cell cycler, **71:**7–14
Flash-flow studies, cytochrome *c* oxidase, **71:**113; 122
Flash photolysis, relaxation methods, **25:**2; 3; 31
Flash photolysis techniques, **51:**20–28
Flash techniques for enzyme reactions, **25:**2–9
Flavandiol, and epicatechin dimer, **25:**463–464
Flavicidin, *See* Penicillin
Flavicin, *See* Penicillin
Flavin, **7:**223; 375. *See also* individual flavins
acid extractable, in respiratory chain, **28:**162; 190; 193–194; 212; 214; 217
acid nonextractable, in respiratory chain, **28:**162; 170–174; 176; 178; 189; 192; 201; 202; 210; 217
in respiratory chain preparations, **28:**162; 169–172; 214; 223; 226; 227
in double-headed enzyme, **23:**314–315
of fumaric reductase, **18:**102
and succinic dehydrogenase, **18:**69; 70; 71; 73; 75–79; 86; 94
in mitochondrial fragments, **21:**77; 101; 104
spectral changes, **6:**163
Flavin adenine dinucleotide, **15:**82; 89, **18:**202, **60:**44
in ATP synthesis, **23:**334; 337
in citric acid cycle, **15:**184; 234; 237–238; 251
as component of metalloflavoproteins, **17:**235–243; 246; 248–260; 263–267; 273–274; 277; 282; 285; 301–308; 314–318
effect on choline oxidation, **17:**384
terminal oxidase, **17:**454
in glucose oxidation, **15:**145; 148–150
in plant respiration, **18:**306
Flavin-adenine dinucleotide, action of alkali on, **14:**200–201
in ATP system, **14:**17
Flavin-adenine-dinucleotide, comparison with CoA, **16:**15
Flavin adenine dinucleotide and phosphorylase, **11:**304
Flavin adenine mononucleotide, in phytoflavin, **30:**16
Flavin adenine nucleotide, in mitochondria, **21:**8
in mitochondrial oxidations, **21:**33
reduced, in pyrimidine biosynthesis, **21:** 276–277
in succinic and DPHN dehydrogenases, **21:**98–100
Flavin adeninenucleotide (FAD), as prosthetic group of diaphorase, **20:**168
effect on fluorescence of flavin mononucleotide, **20:**32
fluorometric assay, **20:**32
Flavin adeninenucleotide (FAD) mechanism of action, **19:**209; 210
in oxygen metabolism, **19:**202; 203; 206–210
Flavin adrenine dinucleotide, **58:**178
Flavin coenzyme. *See also* Flavins, Flavin enzymes, Flavin nucleotides, Flavoproteins, and Metalloflavoproteins
in methylation, **25:**400; 404–407; 410
and PN-dependent reactions, **26:**121; 127–129; 144
Flavin coenzymes, mode of action, **14:**180–183; 188
Flavin component, glutamate synthase, **72:** 114–116
Flavin-containing enzyme, in DR metabolism, **28:**442
Flavine, in respiratory chain preparations, **20:**153; 188
Flavine adenine dinucleotide, absorption spectrum of, **22:**376

in active site of xanthine oxidase, **22**:382; 383
adenine amino group and binding of, **22**:389; 390
analogues, synthesis by mixed anhydride method, **22**:172
changes in fluorescence with binding, **22**:376
and cytochrome reductase, **22**:372
isoalloxazine ring in, **22**:383
in mitochondrial lipoflavoprotein, **22**:386
reduction of iron by, **22**:383
ribityl, ribose moieties, and binding, **22**:387
structure, **22**:172
synthesis, by carbodiimide method, **22**:172; 188
by mixed anhydride method, **22**:171–173
by phosphoramidate method, **22**:199; 200
Flavine coenzymes, binding, **22**:376–393
metals in, **22**:379; 380
Flavine inosine dinucleotide, synthesis by carbodiimide method, **22**:188
Flavine mononucleotide, association with apoenzyme of old yellow enzyme, **22**:383; 384; 387–390
changes in fluorescence with binding, **22**:376
isoalloxazine ring in, **22**:383
Flavine nucleotides, molecular structure, **22**:339
Flavin enzyme, methyl synthesis, **25**:407
multiple forms, **25**:358–359
Flavin enzymes. *See also* Flavins, Flavin coenzyme, Flavin nucleotides, Flavoprotein, and Metalloflavoproteins
free-radical mechanism, **27**:20–21; 30–31; 33
and PN-dependent reactions, **26**:121–127; 131; 132; 166
Flavin free radicals, distribution of unpaired electron in, **27**:133–137
in succinic dehydrogenase, **27**:176
in xanthine oxidase action, **27**:162
Flavin intermediates, EPR study, **27**:130–137
Flavin leucoflavin, **26**:121
Flavin mononucleotide, **15**:89, **60**:44
adenyl transfer from ATP, **17**:184
and aldehyde oxidase, **17**:243
and inorganic nitrogen metabolism, **17**:250–252
in lactic oxidas, **17**:244
in citric acid cycle, **15**:184; 251
in photophosphorylation, **18**:58–59
Flavin mononucleotide (FMN), **51**:17
dihydrodipicolinate reductase, **72**:290–291
fluorescence, effect of flavin adeninenucleotide on, **20**:32
in old yellow enzyme, **20**:33
fluorometric assay, **20**:32
localization, **27**:276
in oxygen metabolism, **19**:202; 203
in photosynthesis, **19**:52; 53; 55; 64; 65
in photosynthetic phosphorylation, **20**:228
Flavin nucleotide, **15**:131
and glycolate metabolism, **23**:436
and lipoic dehydrogenase, **23**:333–334
Flavin nucleotides. *See also* Flavins, Flavin coenzyme, Flavin enzymes, Flavoproteins, and Metalloflavoproteins
in cobamide coenzyme synthesis, **26**:366–367
and PN-dependent reactions, **26**:121; 126–129; 131
in respiratory chain, **28**:156; 159; 162; 194; 214; 216
and succinic dehydrogenase, **18**:77–79; 82–84; 84–87; 94; 95; 98; 99
Flavin nucleotides, EPR study, **27**:183
Flavin peptides, from succinic dehydrogenase, **18**:79
Flavin phenol hydroxylase, **74**:251
Flavin reductase, **65**:170
Flavin ring, hydrogen bonds to, **50**:29
Flavins, **41**:38, **59**:113; 120; 147; 149. *See also* FAD, Flavoproteins, and FMN; FAD, Flavoproteins, and Riboflavin; Flavoproteins and specific flavins
of dihydroorotic dehydrogenase, EPR study, **27**:172–175
EPR study, **27**:117; 130–137; 183
and luminescence, **25**:123–125; 128; 145–151; 154; 160–161
and nuclear purity, **27**:254
in oxidation of hydrocarbons, **27**:485
and PN-dependent enzymes, **26**:120–121; 125–132
reaction of oxygen with, **74**:251
reduction of, **15**:4; 11; 15–16
triplet states, EPR study, **27**:191
in xanthine oxidase, EPR study, **27**:160–168
Flavin semiquinone, **58**:65
Flavin semiquinone-metal interaction, EPR study, **27**:183–187
Flavin-type coenzymes, **15**:11
Flavin types, **28**:172

Flavobacteria, degradation of dermatan sulfate, **25:**440–441
Flavobacterium
enzymic transformations of steroids by, **20:**258
gene for phosphotriesterase, **74:**53
hydrolysis of organophosphates, **74:**52
Flavobacterium enzymes, and elastin, **23:**183; 227; 229–238; 240
Flavobacterium heparinum, **60:**285–290
chondrosulfatase activity from, **25:**439
Flavobacterium keratolyticus, **60:**291
Flavobacterium, oxidation of hydrocarbons, **27:**503; 525
Flavobacterium thermophilum, **54:**169; 172
Flavocol, **35:**164
Flavodoxin, **47:**58; 124; 133
as substitute for ferredoxin in nitrogen fixation, **33:**145
Flavodoxins, hydrogen bonds in, **50:**29
Flavoenzymes, flavin component of, **28:**172
Flavokinase, **36:**72
β-D-fucosidase, inhibition by D-fucono, **36:**172
inhibition, by galactonolactone, **36:**172
by glucono-, **36:**172
Flavone glycosides, **55:**15
Flavones, **17:**434
glucuronides of, **16:**381
metabolism in plants, **16:**141; 144
peroxidase action on, **14:**151
and photophosphorylation, **25:**53; 58; 61; 69
reaction with borate, **12:**500; 521
Flavonoid, **55:**9
biosynthesis of
hydroxycinnamic acid CoA thiol esters in, **55:**17
phenylalanine ammonia lyase in, **55:**15
Flavonoids, as inhibitors of β-glucuronidase, **16:**381
in lignin synthesis, **16:**147
metabolism in plants, **16:**133; 136; 140–147; 150–153
Flavonols, **17:**434
substrates of tea leaf oxidase, **16:**151
Flavonols in tobacco, **10:**347–349; 373–375
Flavonone, chalcone interconversions with, **55:**20
enzymes in, **55:**27
Flavonone synthase, **55:**19
Flavooxidase, **57:**88; 118. *See also* D-Amino acid oxidase; D-Aspartate oxidase; L-α-Hydroxy acid oxidase
Flavoprotein, **47:**12, **55:**202
and Luminal, **7:**24
and luminescence, **7:**227; 232
and urethan, **7:**22
oxidases, **55:**230
rates of oxidation and reduction in liver mitochondria, **20:**157; 158
separation by chromatography, **14:**350
Flavoprotein aldehyde oxidase, in nicotinamide metabolism, **15:**112
Flavoprotein catalysis, **17:**271–273
Flavoprotein dehydrogenases, application of EPR spectroscopy, **27:**107; 117; 131
nonheme iron, **27:**89
Flavoprotein electron transfer, EPR study, **27:**183
Flavoprotein enzymes, **1:**181; 182; 184
in nitrogen transfer, **15:**340
mechanism of action, **15:**4; 8–16
Flavoprotein-Fe-*o*-phenanthroline, **18:**98
Flavoprotein intermediates, EPR study, **27:** 130–133
Flavoprotein monooxygenases, **67:**210–211
Flavoprotein NADPH-cytochrome *P*-450 reductase, **60:**40
Flavoprotein oxidase, in photosynthesis, **19:**16
Flavoprotein oxidase system, of Arum spadix, **18:**47
Flavoprotein-oxidoreductases, multiple forms, **25:**358–359
Flavoprotein reductase, **47:**10; 12–13, **60:**44
Flavoprotein reductases, in plants, **18:**306; 310
Flavoproteins, **59:**146. *See also* Flavins, Flavin coenzyme, Flavin enzymes, Flavin nucleotides, and Metalloflavoproteins; Metalloflavoproteins and Respiratory chain
autoxidizable, in plant respiration, **18:**36–37
and succinic dehydrogenase, **18:**69; 71; 73; 80–81; 86; 98; 100
as terminal enzymes in seedlings, **18:**41–42
and choline oxidation, **17:**384
DPNH-cytochrome *c* reductase as, **17:**79; 376–377
and fatty acyl CoA dehydrogenases, **17:**382–383

metallo-, nature and function of, **17:**233–291
oxidation by cytochrome *c*, **17:**349–350; 363–365; 376–378
and oxidative phosphorylation, **17:**334
pure, redox studies, **17:**277
in respiratory chain, **17:**69–70
spectra, **17:**72–77
and coenzyme Q, **25:**276; 277; 309; 311–312; 314; 315; 317–319
column chromatography, **22:**477
metals in, **22:**392
deficiency in nucleus, **16:**489–490
in plants, **16:**131
in electron transfer, **19:**201
mechanism of action, **19:**209; 210
in oxidation-reduction systems, **11:**241–244
in photophosphorylation, **23:**386
in respiratory chain, **23:**341; 343–344; 347; 350–352; 357; 363; 368–369; 374
and PN-dependent reactions, **26:**125; 126; 128–130
and resistance, **6:**113–118
in respiratory chain, **28:**156; 159–160
in tobacco, **6:**372
Flavoproteins, EPR spectra, **27:**117; 119–120; 159
in mechanism study, **27:**3–34
metal, EPR study, **27:**160–187
semiquinone, **27:**183–187
metal-free, EPR study, **27:**130–137; 183–184
Flavoprotein spectrum, of dihydro-orotic dehydrogenase, **27:**173
Flavoprotein terminal oxidase, genetic studies, **17:**453
Flavor
chocolate
components of, **25:**458–460; 479; 480
factors affecting, **25:**484–488
compounds in cacao, **25:**458–460; 479; 480; 484
Flavor changes in frozen vegetable tissue, **9:**613; 614–624; 627–631
Flavorportein, **15:**112–113
and amino acid oxidation, **15:**123
in glucose oxidation, **15:**148
and pyridine nucleotides, **15:**89; 98
Flavosemiquinone, **34:**64
and PN-dependent reactions, **26:**121
Flexibility of enzyme structure, **22:**76–78
under denaturation stress, **22:**76; 77
Flexibility of hydrocarbon chains, effect on rate of interfacial reaction, **11:**58–64; 64–66
Flexible linkages, identification in complex oligosaccharides, **66:**61–62
fliLM gene, *Caulobacter* cell cycle, **71:**10; 15
Floor process for growth of microorganisms, **9:**657; 662; 663–664; 668–670
Florigen, and phytochrome action, **26:**28
Flourescence method, in measurement of DPNH, **17:**44
Flow, **34:**499; 504
rate, **34:**523
Flowering. *See also* Phytochrome
biochemical control, **26:**28–29
Flow methods, for transient state kinetic, **17:**4; 44
Flow techniques, in enzyme reactions, **25:**2; 3; 9; 33
Fludrocortisone, **55:**198; 199
Fluidity, **34:**468–469
Fluidity of protoplasm, **7:**60
Fluorescein
chemiluminescence, **25:**123; 127
in localizing proteins in nuclei, **16:**418
Fluorescein isothiocyanate, **65:**197
Fluorescein mercuric acetate, **39:**346
Fluorescence, **7:**227, **39:**197; 346, **47:**74. *See also* Absorption spectra and Luminescence; Light emission
bacteriochlorophyll, **1:**313
of bilirubinoids, **7:**363
of *Cypridina hilgendorfii*, **25:**157
depolarization, **43:**30
of DPNH-GAD complex, **22:**367
of DPNH, TPNH complexes with GAD, **22:**369
of DPNH-liver LDH complex, **22:**367
of DPNH-MDH complex, **22:**370
of DPNH-muscle LDH complex, **22:**367
of epimerase preparations, **22:**371
of flavine coenzymes, **22:**376; 377
of free, bound DPNH, **22:**373
of LDH-DPNH complex, effect of carboxylic acids on, **22:**362
of LDH solutions, **22:**360; 361
measurements of changes in coenzyme conformation on binding, **22:**362; 363
of mitochondrial pyridine nucleotides, **22:**374; 375
of riboflavin, effect of various compounds on, **22:**384

Fluorescence (*Continued*)
quenching of, **22:**377
of TPNH with isocitric dehydrogenase, **22:**371
effect on thiols, **11:**210
of bacteriochlorophyll, **11:**131–139
of chlorophyll, **11:**91–199
changes with age of organism, **11:**130; 164; 174
with CO_2 concentration, **11:**97; 105; 115; 134–136; 144–145; 150–155; 162–163; 171; 180
with light intensity, **11:**109–112; 126; 142–143; 146; 147; 153; 159–160; 161; 165; 174; 181
with narcotics, **11:**96–97; 98; 99; 102; 112–113; 120–121; 125–126; 136; 166; 174–175
with oxygen tension, **11:**97–99; 101–104; 106–109; 113–115; 121; 124–127; 153–154; 160
with temperature, **11:**96; 101; 105; 109–112; 113–114; 126; 138; 142; 160; 163; 166
with time, **11:**95; 103; 105; 109–111; 113; 124; 142–144; 152; 159–160; 163; 165; 174
green plants, **1:**216–219
of open-chain tetrapyrroles, **7:**362
phenylalanine hydroxylase, **67:**117–119
alkaline pH effect, **67:**127
and photosynthesis, **1:**216; 223–231
polarization, **39:**350
of porphins, **7:**318–319
quenching, **39:**346
reduced nucleotide in presence of glutamate dehydrogenase, **39:**371
Fluorescence changes, during enzyme action, **27:**3; 9; 22; 29–31
Fluorescence energy transfer
N-linked oligosaccharide solution conformation, **66:**62–65
time-resolved, oligosaccharides, **66:**71–73
Fluorescence maxima, of heart LDH complexes, **22:**358; 359
Fluorescence methods, **46:**178; 391; 392
Fluorescence microscopy, microtubule dynamics, **69:**140
Fluorescence of porphyrins, **12:**18
Fluorescence polarization
of DPNH-heart LDH complex, **22:**366
and LADH complex formation, **27:**66
Fluorescence probing, **51:**8–9
Fluorescence spectra, **42:**112; 113
of acetylpyridine analogue of DPNH, **22:**350
of ADH-DPNH complex, **22:**348–351
of DPNH, changes on binding to TPD, **22:**343
of DPNH-yeast ADH complex, **22:**354
of photosynthetic organisms, **22:**373; 374
in enzyme complex study, **26:**157–163*t;* 165
in photosynthesis, **25:**45–46; 71–73
tryptophanyl emission, **42:**113
tyrosine emission, **42:**113
Fluorescence spectrum, of lactoflavin, **7:**227
of luminescent substances, **7:**256
of porphins, **7:**348–350; 351; 357
Fluorescent analogs, **43:**29
Fluorescent labels, calcium ATPase, **65:**200–201
Fluorescent ligands, **43:**139; 177; 183; 191
Fluorescent organic compounds, conjugated, and chemiluminescence, **25:**120–131
Fluorescent probes, **49:**263–264
Fluoride, **15:**144; 250; 251; 279; 347; 379; 380; 383, **17:**242
effect of pH on inhibition by, **18:**34–35
effect on succinic dehydrogenase, **18:**92; 94
of teeth, effect of fluorovalerate, **18:** 138–139
effect on ATPase, **6:**166
effect on dihydroxyfumaric acid oxidase, **19:**118; 127
effect on photosynthesis, **19:**7; 59; 60
effect on tryptophan oxidase, **19:**122; 127
effect on lytic activity, **8:**95
inhibition of *o*-aminophenyl-glucuronide formation, **16:**403
inhibitory action, **6:**99–101
as inhibitory analogue, **6:**142
inhibitory rôle in glycolysis, **5:**135; 137; 138
Fluoride and phosphorus turnover, **7:**123; 124; 151
Fluoride inhibition of fermentation, **1:**120
Fluorides, solubilization of, **17:**138; 142; 144; 145; 149
Fluorimetry
eukaryotic dihydrofolate reductase, ligand binding, **70:**47–48
in study of kinetics and equilibria of enzymic reactions, **20:**32–34

in study of liver alcohol dehydrogenase system, **20:**35; 36; 38–43
Fluorinated amines, **55:**238–240
Fluorinated amino acids, **55:**227–240
Fluorinated amphetamines, **55:**239
Fluorinated aromatic compounds, **55:**248–256
dioxygenases, **55:**251–254
monooxygenases, **55:**248–251
Fluorinated carbohydrates, **55:**240–248
Fluorinated cephalosporins, **55:**274
Fluorinated cyclophosphamides, **55:**278–279
Fluorinated dicarboxylates, **55:**216–219
Fluorinated hexoses, **55:**245–248
Fluorinated norepinephrine, **55:**239; 240
Fluorinated organic acids, **55:**203–226
dicarboxylates, **55:**216–219
fumarates, **55:**217–219
succinates, **55:**216–217
monocarboxylates, **55:**203–216
α-fluorocarboxylates, **55:**203–210
β-fluoromonocarboxylic acids, **55:**210–215
ω-fluoromonocarboxylic acids, **55:**215–216
tricarboxylates, **55:**219–226
citrates, **55:**219–226
Fluorinated pentoses, **55:**244
Fluorinated purines, **55:**271–272
Fluorinated pyrimidines, **55:**268–271
Fluorinated Pyrimidines and Their Nucleosides (Heidelberger, Danenberg, and Moran), **54:**57
Fluorinated steroids, **55:**263–264
Fluorinated substrate analogs, **55:**197–284
amino acids and amines, **55:**227–240
carbohydrates, **55:**240–256
cephalosporins, **55:**274
cyclophosphamides, **55:**278–279
difluoro vitamin D_3, **55:**276–278
in enzymic reactions, **55:**203–279
fluoroalkanes and fluoroalkenes, **55:**257–268
fluoroketones, **55:**275–276
halogenated nitrosoureas, **55:**272–274
organic acids, **55:**203–226
purines and pyrimidines, **55:**268–272
Fluorinated Substrate Analogs: Routes of Metabolism and Selective Toxicity (Walsh), **55:**197
Fluorinated succinates, **55:**216–217
Fluorinated trioses, **55:**240–244
Fluorine, **55:**200
size and electronegativity of, **55:**201
Fluoroacetaldehyde, **55:**214
Fluoroacetamide, **18:**142
Fluoroacetate, **15:**198, **55:**198
bacterial and yeast resistance to, **55:**209
biochemistry and mode of action, **18:** 118–132
estimation of, **18:**117
protection against, **18:**141–142; 148
and glycolate metabolism, **23:**435; 439–440
processing of, **55:**281; 282
reaction with thiols, **11:**213–214
2-Fluoroacetate
fluoroacetyl CoA, **55:**204; 208
2-fluorocitrate from, **55:**223
to glycolate, **55:**209
occurrence of, **55:**203–204
toxicity of, **55:**204
Fluoroacetic acid
effect on Krebs cycle, **19:**359
effect on transamination, **19:**359–363; 367; 368
toxicity of, **18:**113–159
Fluoroacetoacetic acid, **18:**151
Fluoroacetylcholine, **18:**139; 150
Fluoroacetyl CoA, **55:**204; 208
Fluoroacetyl-CoA, **18:**128–132; 140
and glycolate metabolism, **23:**439
Fluoroacetyl coenzyme A, **15:**198
2-Fluoroadenosine, **54:**210
Fluoroalanine, **55:**199; 213; 227–234
antibacterial activity of, **55:**229
metabolism of, **55:**281; 283
in racemase inactivation, **55:**227
in transaminase inactivation, **55:**227
turnover/inactivation with, **55:**228; 229
ω-Fluoroalcohols, **18:**115
ω-Fluoroaldehydes, **18:**115
Fluoroalkanes, **55:**257–264
Fluoroalkenes, **55:**265–268
ω-Fluoroalkylamines, **18:**115
α-Fluoro amino acids, **55:**227
β-Fluoro amino acids, **55:**234–235
β-Fluoroaminobutyrate diastereomers, **55:**234
threo-β-Fluoroaspartate, **55:**234–235
β-Fluoroaspartate diastereomers, **55:**234
Fluorobenzoate, **55:**252
p-Fluorobenzylpenicillin, **10:**255
Fluorobutyrate, **18:**140–141
ω-Fluorobutyrate, effect on bacteria, **18:**135
α-Fluorocarboxylate, **55:**203–210

ω-Fluorocarboxylate series, toxicity of, **18:**115
Fluorocatechol, **55:**252
Fluorocitrate, **43:**60; 93; 95
  and the convulsive state, **18:**147
    effect of, **18:**153–154
    synthesis of, **18:**128–132; 151–152
  effect on acetyl-CoA carboxylase, **26:**302
2-Fluorocitrate, **55:**204–205; 219–224
  aconitase action with, **55:**222; 223
  from fluoroacetate, **55:**223
  isomers of, **55:**224
3-Fluorocitrate, **55:**224–226
  enzyme inactivation by, **55:**225
  isomers of, **55:**207
Fluorocitrate synthesis, **15:**197–198
Fluorocitric acid, effect on aconitase, **18:** 123–128, **19:**359
  effects on enzymes, **18:**128
  isomers of, **18:**116
  protection against, **18:**141–142; 148
  synthesis of, **18:**121–123; 128–132
Fluorocitric acids, **43:**78
3-FluorocitrylCoA, **55:**225
ω-Fluoro compounds, toxicity of, **18:**115–116
Fluorocortisone, **18:**152
2-Fluorocytidine (FCR), **54:**58; 89; 90
5-Fluorocytosine (FC), **54:**58
  conversion of, to nucleotides, **54:**62
  FU conversion to, **54:**91
  for yeast and fungal infections, **54:**59; 63
Fluorodecanoate, **18:**140–141
5-Fluoro-2′-deoxycytidine (FCdR), **54:**58; 61
5-Fluoro-2′-deoxyUMP, **55:**268
5-Fluoro-2′-deoxyuridine-5-diphosphate (FdUDP), **54:**58
5-Fluoro-2′-deoxyuridine (FUdR), **54:**58; 61
  in cancer treatment, **54:**59
5-Fluoro-2′-deoxyuridine-5-monophosphate (FdUMP), **54:**58; 61
  chemotherapeutic activity of, **54:**62
  in thymidylate synthetase inhibition, **54:**59
5-Fluoro-2′-deoxyuridine-5′-triphosphate, **54:**58
Fluorodeoxyuridine triphosphate (FdUTP), **54:**68–69
Fluorodeoxyuridylate, **38:**253; 271
11-Fluoro-1,2-dihydroxyundecanol, **18:**115
Fluorodinitrobenzene
  estimation of end groups in myosin by, **13:**157–158
  in protein structure study, **14:**278–281; 283–285; 299–302; 309–310
1-Fluoro-2,4-dinitrobenzene
  carbonic anhydrase catalyzed hydrolysis, **47:**189; 190
  in papain and mercuripapain structure, **19:**294; 296; 301
    in peptide synthesis, **19:**244
Fluorodinitrobenzene method, in histone study, **18:**173
α-Fluoroethanethiols, cysteine *S*-conjugate β-lyase reactions, **72:**206–207
2-Fluoroethanol, **18:**117
Fluoro fatty acids, long chain, **18:**115; 130; 140–142
  protection against, **18:**140–142; 148
Fluorofumarate, **55:**202; 216; 217–219
3-Fluoro-D-glucitol, **55:**245
2-Fluoroglucose, **55:**245
3-Fluoroglucose, **55:**245
3-Fluoroglucose-6-phosphate, **55:**245
α-Fluoroglycine, **55:**227
3-Fluoroglyoxal, **55:**214–215
Fluorohexanoate, **18:**140
Fluorohydrins, **55:**218
  enzyme-catalyzed routes to, **55:**202
9α-Fluorohydrocortisone, microbiological conversion, **20:**239
1-Fluoro-3-hydroxyacetone-P_1, **55:**243
Fluoroisocitrate, **55:**223
Fluoroketones, **55:**275–276
Fluorokinase, action on propionyl CoA carboxylase system, **19:**59
3-Fluoro-L-lactate, **55:**212–213
Fluorolactone, **55:**252
ω-Fluoroleate, **55:**215
Fluoromalamide, **18:**116; 154
3-Fluoromalate, **55:**212; 217
Fluoromalonate, **55:**216
2-Fluoromannose, **55:**245
Fluoromethyl acetophenone, **55:**275
α-Fluoromethyl-α-amino acids, **55:**235–236
Fluoromethylglyoxal, **59:**113
β-FluoromethylmalonylCoA, **55:**210; 211
β-Fluoromonocarboxylic acids, **55:**210; 215
ω-Fluoromonocarboxylic acids, **55:**215–216
2-Fluoromuconate, **55:**253
Fluoronicotinic acid, **18:**152
ω-Fluoronitriles, **18:**115
ω-Fluoro-omega′-nitroalkananes, **18:**115

Fluoronitrobenzene, **54:**139
5-Fluoroorotic acid (FO), **54:**58
nuclear RNA incorporation of, **54:**88
in rRNA inhibition of maturation, **54:**86
3-Fluorooxalacetate, **55:**205; 211; 212; 218–219
Fluorooxaloacetate, **18:**152–154
Fluorophenylalanine, **55:**395
p-Fluorophenylalanine, **50:**169
2-Fluorophenylalanine, **35:**286
3-Fluorophenylalanine, **35:**286
4-Fluorophenylalanine, **35:**286
effect on DNA synthesis, **29:**151
incorporation into protein of *E. coli,* **29:**110
kinetics of incorporation into protein, **29:**112
resistance to, **29:**122
p-Fluorophenylalanine, effect on β-lactamase, **28:**275
p-Fluorophenylalanine, effects of on gramicidin S formation, **33:**340
effects of on synthesis of gramicidin S, **33:**347
Fluorophosphatases, **20:**77; 90; 92
Fluorophosphates, action as inhibitors, **13:** 321–343
4-Fluoroprolines, *cis*- and *trans*-, substrates for protein synthesis, **29:**103
3-Fluoropropionate, **55:**210
β-FluoropropionylCoA, **55:**210–211; 216
Fluoropyrimidine, *See* Pyrimidine, fluorinated
Fluoropyruvate, **55:**211–214
from fluoroalanine, **55:**230
processing and carboxylation of, **55:**281
reduction of, **55:**212–213
Fluoropyruvic acid, **18:**115; 116; 150–154
Fluorosuccinate, **55:**252
5′-Fluorosulfonylbenzoyladenosine, **58:**158
α-Fluorothiolethers, **55:**236
Fluorotricarboxylates, **55:**219–226
Fluorotryptophan, **55:**395
5-Fluorotryptophan, synthesis of gramicidin S, effect on, **33:**340
Fluorotyrosine, **55:**395
3-Fluorotyrosine
competitive inhibitor of melanin formation, **14:**93–95
as a substrate for tyrosine sRNA synthetase, **29:**114
as a tyrosine analog, **29:**114
2′-Fluoro-UDP, **55:**199
5-Fluorouracil
effect on enzyme synthesis, **24:**16–17
effect on murein synthesis, **26:**218
effect on mutants, **27:**337
synthesis of gramicidin S, inhibitor of, **33:**340
Fluorouracil, cancer treatment, **25:**400
5-Fluorouracil (FU), **54:**58
activation of, **54:**65–68
parameters assessing, **54:**104–105
anabolic metabolism of, **54:**61
in assessing tumor responsiveness, **54:**103
analogs of, **54:**109
as antimetabolite in RNA, **54:**59
antitumor activity of, **54:**99–100
in cancer treatment, **54:**59
catabolism of, **54:**61
cellular uptake of, rate of, **54:**63
chemotherapeutic activity of, **54:**59; 61
in DNA, **54:**58; 61
cytotoxicity of, **54:**98–99
effect of, on transcription, **54:**88
with effectors of pyrimidine metabolism, **54:**100–102
fluorocytosine from, **54:**91
5-formyltetrahydrofolate effect on, **54:**97
properties of, **54:**59
in *rna,* **54:**58; 61; 86; 92; 93
cytotoxicity of, **54:**99
nuclear, **54:**88
with TdR, **54:**100
5-Fluorouridine, **55:**268
5-Fluorouridine-5′-diphosphate (FUDP), **54:**58
identification of, **54:**86
5-Fluorouridine (FUR), **54:**58; 61
5-Fluorouridine-5′-monophosphate (FUMP), **54:**58
hexadecyl ester of, **54:**62
5-Fluorouridine-5′-triphosphate, **54:**58; 61
identification of, **54:**86
Fluorovalerate, **18:**115; 138–139
Fluoroxene, **55:**262–263
Fluphenazine, **55:**180; 181
DL-3-Flurotyrosine, resolution by papain, **19:**309
Flux, components, **13:**29–33
determination of values from tracer experiments, **13:**25–29
ratio of a substance and its electrochemical activities in inside and outside solutions, **13:**33–42
use of term, **13:**23–25

α-F-Methyl amino acids, **55:**227–228
FMN. *See also* Flavin adenine mononucleotide; Flavin mononucleotide; Flavin nucleotides
as artificial electron acceptors, **30:**23
photooxidation of Mn^{2+} stimulated by, **30:**25
and bioluminescence, **25:**124; 145–151; 153–155; 160–161
and photophosphorylation, **25:**53; 58; 61
brain, nitric oxide synthases, **65:**297
nitric oxide synthases, isoforms, **65:**300
in respiratory chain, **28:**159; 162; 194; 214; 216
FMN interaction with yellow enzyme, rate constant, **25:**14
FO, **54:**58
Folate
as alternate substrate, eukaryotic dihydrofolate reductase, **70:**93–96
dihydrofolate reductase, **70:**28
eukaryotic dihydrofolate reductase
ligand binding, **70:**50–52
ligand binding (side chain roles), **70:**62
glutamate moiety of, eukaryotic dihydrofolate reductase, **70:**30; 37
Folate analogs, biosynthesis, of, **27:**428–431; 457
as inhibitors of dihydrofolate reductase, **27:**436–437
Folate antagonists, small molecule, **27:**437–451; 455; 458
Folate-binding proteins, **47:**323; 335–336
Folate biosynthesis, inhibition, and chemotherapy, **27:**417–468
Folate Biosynthesis and Function, Inhibition of as a Basis for Chemotherapy (Hitchings and Burchall), **27:**417
Folate cofactor synthesis, pathways, **27:**426
Folate Compounds, Transport of, in Bacterial and Mammalian Cells (Huennekens, Vitols, and Henderson), **47:**313
Folate (dihydrofolate) reductase, **27:**426
Folate-H_4, in *de novo* methyl synthesis, **25:** 396–409; 420
tritium-labeled, in methyl synthesis, **25:**401
Folate-H_2 (dihydropteroyl glutamic acid), in methylation, **25:**395; 397; 399–401
Folate-reducing enzymes, **27:**434–436
Folate-reducing system, of *Cl. sticklandii*, **27:**434
Folate reductase, **37:**166–167. *See also* Dihydrofolate reductase
inhibitors, **27:**437
binding sites, **27:**450
in vivo behavior, **27:**456–458
Folates, from natural sources, **27:**420–421
Folate synthesis, inhibition of, **27:**427–428
Folate-synthesizing enzymes, and sulfonamides, **27:**432
Folate-synthesizing enzymic system, folate analogs, **27:**428–431
Folate transport, **47:**313–343
in bacterial cells, **47:**313
in *Diplococcus pneumoniae,* **47:**318
in *Lactobacillus casei,* **47:**317; 318
in mammalian cells, **47:**313; 327
Ehrlich cells, **47:**328
intestine, **47:**328
L1210 cells, **47:**328
L5178Y cells, **47:**328
rabbit choroid plexus, **47:**328
S180 cells, **47:**328
WI-L2 cells, **47:**328
mechanism of transport, **47:**342–343
in *Pediococcus cerevisiae,* **47:**316; 318
in *Streptococcus faecalis,* **47:**315; 318
Folding of the peptide chain, **1:**9; 29
Folic acid, **9:**495
biogenesis of, **27:**420–426
natural sources, **27:**420–421
biosynthesis, **21:**393; 394
estimation and occurrence, **21:**390–392
isolation, structure, synthesis and properties, **21:**371–374
biosynthesis of, **53:**353–363
p-aminobenzoic acid, as precursor, **53:** 353
2-amino-4-oxy-6-hydroxymethyl-7,8-dihydropteridine, **53:**353
deficiency, **38:**237
halogenated derivatives, **38:**281
effect of RNase, **24:**238
and indole oxidase activity, **19:**127; 128
pteridine portion of, **53:**355–359
and purine synthesis, **9:**461
and pyrimidine synthesis, **9:**468
in synthesis of histidine, **16:**300
relation to glycine and serine metabolism, **16:**277
in synthesis of methionine, **16:**270; 274

and thiaminase exchange reaction, **15:**404
in tyrosine metabolism, **14:**98; 104; 108–110
linkages in, **14:**278
Folic acid analogs, **21:**375; 376
Folic acid and formate utilization, **12:**143
Folic acid coenzymes
biogenesis of, **27:**421
in methyl synthesis, **25:**396–409
Folic Acid Coenzymes and One-Carbon Metabolism (Huennekens and Osborn), **21:**369
Folic acid compounds. *See also* specific substances
absorption spectra, **21:**388–390
isolation and identification, **21:**390
notation system for, **21:**380
in serine biosynthesis, **21:**224
Folic acid conjugates, degradation, **21:**391
occurrence and physiological action, **21:**371; 372
Folic acid cycle, and vitamin B_{12}, **26:**266–267
Folic acid deficiency, effect on xanthine oxidase level, **17:**310; 332
Folic acid derivatives, as formylating enzymes, **18:**211
Folic acid production, **7:**454
Folic acid reductase, and folate analogs, **27:**430
Folic reductase, **21:**395
Folin-Ciocalteu reagent for phenols, **16:**365
Folinic acid, *See* N^5-Formyltetrahydrofolic acid
Folinic acid (citrovorum factor, leucovorin), **27:**426; 436–437; 439; 441–442; 454; 455; 457; 458
Follicle-stimulating hormone, **50:**314
Follicle-Stimulating hormone (FSH), **41:**113; 118
Follicular fiber, effect on myosin, **16:**345
Fomes annosus, degradation of cellulose, **8:** 272–273; 278–279
degradation of glucose, **8:**272–274
Food
fish, **1:**385
utilization and resorption in low vertebrata, **1:**397–401; 409–411
Food constituents, scheme for metabolism, **8:**412–414
Food utilization, and lipoic acid, **18:**343
Football protein, *See* Leucine aminopeptidase
Force constants, of molecules, bending, **23:**6
repulsive, **23:**7
stretching, **23:**5
Foregut (Stomodeum), **45:**391; 401
Formaldehyde, **38:**202, **51:**133, **58:**35, **60:**72
activation, **21:**418; 419
in choline biosynthesis, **21:**427
metabolism, formyl derivatives of THFA in, **21:**223; 225
in pyrimidine biosynthesis, **21:**289; 290; 420–422
synthesis, in mitochondria, **21:**22; 25; 26; 428; 429
active, in methyl group synthesis, **25:**396–410
and transmethylation, **25:**418
and biological methylation, **12:**448–449; 449–453; 471–474; 481
fixation in propionic acid fermentation, **12:**127–128
role in biosynthesis, **12:**43; 48–49; 95; 142–143
catalase action on, **19:**109
in photosynthesis, **19:**19; 20
effect on milk clotting, **15:**442
in formation of erythrulose, **15:**171–172
oxidation, **15:**257; 258
oxidation by chain mechanism, **15:**7
effect on pectic substances, **20:**345
effect on pectinesterases, **20:**360
effect on polygalacturonases, **20:**365
and glycine metabolism, **23:**456–457
mutagen, **27:**331
reaction with virus, **17:**479
in ribose synthesis, **28:**397
Formaldehyde activating enzyme, **21:**417–419
Formaldehyde and nitrogen metabolism in plants, **9:**395–397
Formaldehyde (C^{14}), in methyl synthesis, **25:**402–404
Formaldehyde dehydrogenase, **59:**127–131
Formaldehyde dehydrogenase, NAD-linked, **27:**479
Formaldehyde effect on color formation from protein-quinone reaction, **16:**123
Formaldehyde incorporation in methionine, **22:**266
N′-Formalkynurenin, metabolic pathway, **15:**184–185
Formal mechanism, **39:**351; 354
Formamidase, localization of, **24:**327
Formamide, **54:**12
solubility of pectic substances in, **20:**353

Formamidinesulfinate, **39:**362
Formamidinethiosulfonate, **39:**362
4-Formamidoimidazole-5-carboxamide, **17:**427
4-Formamido-5-imidazolecarboxamide ribotide, in purine biosynthesis, **21:**399; 401
5-Formamido-4-imidazolecarboxamide ribotide, in purine biosynthesis, **21:**204; 220–223
Formanilide, configuration of, **23:**15
Formate, **41:**44, **54:**414. *See also* 1-Carbon particle and Methyl group
  activation and deacylation, **21:**225; 226; 407–410
    in biosynthesis of folic acid coenzymes, **21:**394
    in histidine biosynthesis, **21:**403
    metabolism, formyl derivatives of THFA in, **21:**223; 225
    in purine biosynthesis, **21:**201–203; 232; 399; 400
    in pyrimidine biosynthesis, **21:**288; 420
    in pyruvic acid biosynthesis, **21:**431
    radioactive, in study of carotenogenesis, **21:**339
      in studies of purine biosynthesis, **21:**202; 205; 206; 220
    in serine biosynthesis, **21:**417
  and biological methylation, **12:**450; 469–474
    reaction with catalase-peroxide complex, **12:**167; 169; 175–176; 186
    role in biosynthesis, **12:**43; 48–49; 95–101; 142–143
    in uric acid metabolism, **12:**121–127
  in biosynthesis of pectic substances, **20:**369
  condensation in *E. coli,* **6:**244–246
  effect of homocysteine on incorporation into compounds, **16:**274
    labeled, incorporation into imidazole ring, **16:**300
      uptake by chromosomes, **16:**477
  and glycine, **23:**453–455
    and glycolate, **23:**435–437
    labeled, **23:**436; 449–450
    and oxalate, **23:**447–452
  heavy formate, **6:**245
  labeled, in plant respiration, **18:**303
  in methylatiom, **25:**397; 401–402; 418
  and pyruvate equilibrium, **6:**245
  from pyruvate fermentation, **6:**238
  role of folic acid, **27:**426
Formate activating enzyme, *See* Telrahydrofolic formylase
Formate-activating enzyme, FAH_4-coenzyme, **27:**425
Formate activation, **17:**184
Formate (C^{14})
  in coenzyme Q synthesis, **25:**296–299
  in folate biogenesis, **27:**422
Formate dehydrogenase, **48:**3; 4, **51:**109
  activity in *Clostridium formicoaceticum,* **48:**18
  activity in *Clostridium Thermoaceticum,* **48:**18
  bacterial, **48:**16–20
  of *M. Vannielli,* **48:**18
  oxygen sensitivity, **48:**19
  tungsten incorporation in, **48:**19
Formate dehydrogenase-$NADP^+$-oxidoreductase, **48:**19
Formate dehydrogenase-nitrate reductase enzyme complex, **48:**15
C^{14}-Formate, incorporation into DNA, **16:** 434–435; 455–459; 482–483
  incorporation into RNA, **16:**455–456; 461
  in nucleic acid study, **16:**434–435; 450–451; 455–459; 461; 482–483
$^{1\ 4}C$-Formate, incorporation into pteridines, **35:**48
^{14}C-Formate, ribose synthesis, **28:**418–419
Formic acid, **15:**26; 101–102; 184–186; 204
  formation of, bacterial fixation of carbon dioxide and, **74:**12–14
  hydrogen bond data, **23:**10–11
  product of action of GTP cyclohydrolase, **35:**54
Formic acid, PN-dependent oxidation, **26:**119; 123
Formic acid in mold metabolism, **9:**559–560
Formic acid through reduction of CO_2 by bacteria, **2:**144
Formic dehydrase in plants, **10:**309
Formic dehydrogenase, **7:**317–318, **15:**118, **23:**449; 454
  classification, **14:**254; 255
Formic hydrogenlyase
  as adaptive enzyme, **6:**79–80
  classification, **14:**257
Formicodehydrase, **9:**560
Formic r DPN transhydrogenases, **14:**254
Formiminoaspartic acid, formation in histidine metabolism, **19:**179; 180

Formimino-L-glutamic acid transferase, **21:**406; 407
Formiminoglycine, in purine degradation, **21:**401; 402
Formiminoglycine formiminotransferase, **21:**227; 402; 406
Formiminotetrahydrofolic acid, in histidine degradation, **21:**406; 414
in purine degradation, **21:**402
N^5-Formiminotetrahydrofolic acid, isolation, structure and synthesis, **21:**385; 386
Formiminotetrahydrofolic cyclodeaminase, **21:**227; 402
Formononetin, **55:**5, **56:**204
degradation of, **55:**42
formation of, **55:**31
isoflavone, **55:**5
in medicarpin, **55:**29
Formotransacetylase, classification, **14:**251
Formycin A, **54:**210
Formycin nucleotides, **43:**32
Formycin triphosphate, **43:**16
N-Formyl amino acids, formation and cleavage of, **29:**399
Formylanthranilic acid, **19:**127; 128
Formylase, **15:**102
enzymic adaptation, **13:**73
Formylated RNase, **24:**211; 226; 244; 248
Formylating conenzymes, **18:**211
Formyl-coenzyme A, **23:**448–449; 451–452
N-Formylcysteine, synthesis in mitochondria, **21:**26
N^{10}-formylfolate, as substrate of dihydrofolate reductase, **27:**436
N^{10}-Formylfolic acid, isolation and synthesis, **21:**373; 375
Formyl-L-glutamate, in urea synthesis, **15:**371
Formylglutamic acid, **15:**186
N-Formylglutamic acid
in histidine degradation, **21:**407
synthesis, **21:**412; 414
in synthesis of active formate, **21:**411; 412
Formyl-L-glutamine, **15:**371
S-(Formyl)glutathione hydrolase, **59:**127
Formylgly cinamide ribonucleotide amido transferase, **39:**102–119
active site for glutamine, **39:**106
active site for nucleotide substrates, binding of FGAR and MgATP, **39:**113
albizziin, **39:**110
azaserine, **39:**107
binding of glutamine to, **39:**111
factors influencing reactivity of, **39:**111
hydrolytic reactions at with other substrates, **39:**111
glutaminase activity, **39:**111
γ glutamyl derivatives, **39:**111
kinetic analysis of addition of substrates, **39:**114
proposed mechanism of reaction, **39:**116
structural determination, by use of, **39:**107
assay of enzyme, **39:**104; 118
properties of purified enzyme, from chicken liver, **39:**102
from Ehrlich ascites tumor cells, **39:**102
Formylgly cinamide ribonucleotide amidotransferase, relationship to anthranilate synthetase, **38:**30
N-Formylglycinamide ribotide
in purine biosynthesis, **21:**204–207; 211–213; 222; 223; 240; 247–249; 400
Formylglycinamidine ribonucleotide synthetase, *See* Formylglycinamidine ribonucleotide amidotransferase
Formylglycinamidine synthetase, **66:**238–243
GAT and synthetase domain organization, **66:**238–239
GAT function, **66:**239–240
synthetase function, **66:**240–243
Formylglycine, **35:**157
Formyl group, of pyridoxal phosphate, **22:**394; 395; 398
4-Formyl group, requirement in transaminations, **16:**220–222
Formylkynurenine, **15:**102, **17:**418; 420–422
formation in tryptophan metabolism, **19:** 121–123
from peroxidase action, **14:**152
Formylmethanofuran dehydrogenase, **74:**72
N-Formylmethionine, and chain initiation, **29:**154
N^{10}-Formylpteroic acid, *See* Rhizopterin
N^5-Formyltetrahydrahydrofolic acid conjugates, **21:**381; 392; 393
5-Formyltetrahydrofolate, **54:**58
effect of, on FU cytotoxicity, **54:**97
Formyltetrahydrofolate synthetase, **43:**7
N^{10}-Formyltetrahydrofolic acid
and ATP, **18:**230

N^{10}-Formyltetrahydrofolic acid (*Continued*)
biosynthesis and metabolism, **21:**225–228; 383; 394; 407–413
in histidine degradation, **21:**406
in purine biosynthesis and degradation, **21:**204; 220; 222; 223; 400; 402
relative free energy of hydrolysis, **21:**414
structure, **21:**383
and purine metabolism, **27:**459
5-Formyltetrahydrofolic acid, **27:**428; 435
N^5-Formyltetrahydrofolic acid, biosynthesis, **21:**394; 397
estimation, **21:**392
in formyl transfer reactions, **21:**222; 412–414
isolation, synthesis and degradation, **21:** 380–383
in mitochondria, **21:**8
in purine biosynthesis and degradation, **21:**399; 400
relative free energy of hydrolysis, **21:**414
structure, **21:**219; 381
5-Formyl-5,6,7,8-tetrahydrofolic acid, Citrovorum factor, **27:**436
N^{10}-Formyltetrahydrofolic deacylase, **21:**409; 410
N^5-Formyltetrahydrofolic isomerase, **21:**413
Formyl transfer, folic acid coenzymes in, **21:**396–415
Formyltransferase
location in subcellular fraction, **24:**309
reaction, **27:**425
N-Formyltyrosine, action of tyrosinase on, **14:**134–135; 138
competitive inhibitor in tyrosine oxidation, **14:**81; 84; 93–95
N-Formyltyrosine ethyl ester, action of tyrosinase on, **14:**134–135; 138–139
Forssman glycolipid, **52:**71, **60:**101
FORTRAN, application of program for enzyme kinetic data, **29:**31–32
programs for enzyme kinetic data, **29:**23–31
Four-center exchange reaction, **33:**429
Fourneau 883, **7:**548
Fourneau 933, **7:**548
Fowl plague virus, **52:**138
Fox's microspheres, and origin of life, **27:**358; 366; 392
F6P, *See* Fructose-6-phosphate
Fractional membrane, **34:**446
Fractionation
of enzyme system for gramicidin S formation, **33:**356–367
by DEAE-cellulose column chromatography, **33:**365; 366
by DEAE-Sephadex, **33:**365
by DEAE-Sephadex, A-25 column, **33:**359
by DEAE-Sephadex, A-50 column, **33:**366
by a hydroxylapatite column, **33:**358
by Sephadex, G-200, **33:**360–361; 366
by Sephadex, G-200 gel filtration, **33:** 363–364
subcellular, enzyme distribution, **24:**291–358
Fractionation factors, **46:**429; 438
Fractionation of respiratory chain, **28:**157–158; 196
Fragmentation, of nuclei, **16:**417
of unicellular organisms, **16:**469
Fragmentation of respiratory chain, by bile salts, **28:**190–192; 196; 202; 204–210; 214–215
sequential, **28:**190–197
Fragments, subunit structure, **44:**191
Free-boundary method, **42:**87
Free deaminase, **34:**459
radicals, **34:**452
Freeedoxin, glutamate synthase, **72:**112–117
Free energy
of activation, **43:**323
of binding, **43:**319
specificity, **43:**264
determination of changes in, **11:**11
of metabolic processes, **11:**8–9
of photosynthetic processes, **25:**46–48; 53–54
Free energy change
for ionization of carboxyl ion, phenol ion, imidazole and amino groups, **20:**64; 65
and occurrence of reaction, **15:**41–42
in citric acid cycle, **15:**198; 232; 248; 250; 255–256
of hapten-antibody combination, **15:**65–67; 76
of immunological precipitation reaction, **15:**70; 74; 75
of isohemagglutination, **15:**55; 61–63
in model enzyme systems, **15:**23
or argininosuccinic acid splitting, **15:**363
of urea formation, **15:**372

Free energy changes, in carbon reduction cycle, **25:**102–106
in photophosphorylation, **25:**53; 54
Free-energy changes, microtubule dynamics, **69:**129
Free energy increase, for Hill reaction, **19:**47
for photochemical reaction, **19:**48
Free energy levels, **38:**200; 204; 207
Free energy of enzyme reactions, *See* Thermodynamic constants
Free energy of hydrolysis, of acetyl imidazole, **20:**85
Free energy of the ATP system, **14:**7–9; 11; 15–17; 18
Free energy of transfer, **43:**249; 390; 394
methylene group, **43:**392
Free-radical EPR signals, **27:**156–157; 162; 168; 174–177
Free-radical formation, in mitochondria, **27:**187–188
Free-radical intermediates, and flavoprotein function, **27:**130–137
Free-radical mechanism, of disulfide exchanges, **27:**206–207
of flavin enzymes, **27:**20–21; 30–31; 33; 133–137; 162; 176
Free-radical reactions and enzyme catalysis mechanisms, **24:**470–472
Free radicals, **59:**104–105. *See also* Metastable species and Radical species
aryl sulfur, **27:**206
EPR spectra, **27:**109–112; 119–130
and origin of life, **27:**349
in succinic dehydrogenase system, **27:**176
and enzyme action, **12:**168–169; 176
in enzymic oxidoreductions, **15:**2–18; 21; 23–26; 32; 38; 42
formation, effect of ionizing radiation, **24:**362; 370; 377–379; 383–384; 395; 401–403
formation in oxidation-reductions, **11:** 242–243
in oxidation of thiols by iodine, **11:**225
and luminescence, **25:**125–126; 147
and mechanism of trypto-phanase action, **10:**62–64; 69
Free-radical signals, from DPNH dehydrogenase, **27:**177
Freeze-stop technique, in isolation of nuclei, **27:**242; 244
Freezing
effect on enzyme activity, **9:**625–627
effect on radiosensitivity of enzymes, **24:**392
Freezing and respiration, **7:**57
Freezing at reaction center on enzymes (FARCE), **59:**286
Freezing technique, rapid, for EPR study, **27:**161; 168; 173; 177; 179–180; 188
Fremyella diplosiphon, **46:**92
FR enzyme, cytochrome *c* oxidase, **71:**122–124; 125; 185
Freundlich adsorption isotherm, **33:**259
Freundlich isotherm, relation to chromatography, **14:**320–325
Frey-Wyssling complexes, in rubber latex, **29:**223–225; 234
Friedkin intermediate, **54:**71
modification of, **54:**75–76
Friend leukemia cells, **46:**115; 158
Frisch isotherm, **33:**258
Frog muscle, relation of ATP to contraction, **13:**178
x-ray studies of, **13:**186; 188
Frogs, effect of rare earth metal ions on, **21:**192; 193
Frog skin, permeability studies, **13:**45–46; 48–56
transfer of sodium across, **13:**28
water permeability, **13:**36–37
Frohman factor, in schizophrenia, **29:**512–514
Frohman test, racial differences in, **29:**513
in schizophrenia, **29:**512–513
values in monkeys reared in isolation, **29:**501
values in relatives of schizophrenics, **29:** 500–501
Frozen foods, enzyme action in, **33:**285–286
Frozen vegetable tissue, enzyme activity, **9:** 613–652
FRS, *See* Ferredoxin-reducing factor
β-Fructofuranosidase, **15:**313
β-Fructofuranosidase, **17:**198
β-Fructofuranosidase (invertase), cacao, **25:**466
β-Fructofuranosidases
assay of, **56:**156–157
occurrence and localization of, **56:**155–156
properties of, **56:**159–162
purification of, **56:**157–159
regulation of, **56:**163–165
role of, **56:**162–163

O-β-D-Fructofuranosyl-(2→6)-D-fructofuranose, **17:**204
O-β-D-Fructofuranosyl-(2→1)-D-fructofuranose, **17:**204
O-β-D-Fructofuranosyl-(2→6)-D-glucopyranose, **17:**204
Fructokinase, **14:**209, **45:**324
chemical mechanism of, **45:**370
classification, **14:**251
kinetic mechanism of, **45:**284; 308; 311; 324; 370
product inhibition of, **25:**223
C-1-Fructokinase, phosphorylation by, **15:** 145–148
Fructolysis, and semen metabolism, **9:**344; 355–367; 372; 373; 378–379; 381
Fructosans, **18:**288
synthesis, **15:**313–314
Fructose
fermentation by thiaminase-producing bacteria, **15:**413
metabolism by alternate pathways, **15:** 141–182
in β-1,3 glucans, **28:**336
inhibition of phosphoglucose isomerase, **52:**253
labeled, in plant respiration, **18:**290
monophosphate, and saccharose synthesis in plants, **1:**358–360
in plants, **19:**5
and semen metabolism, **9:**342; 344; 355–367; 372–374; 376–382
effect on fertilizing power of sperm, **9:**253; 255
as sperm nutrient, **9:**332–333
as solubilizing agent, **17:**148
in transglycosylations, **17:**192–205; 208; 219; 222; 224; 226; 229
in sperm metabolism, **24:**52; 60–62; 64; 66–68
transformation into glucose, **1:**152
use in synthesis of sucrose, **10:**123–132; 138–141
Fructose-1, **15:**6–6diphosphate; 141; 147; 151–153; 163; 167; 170; 171–713.. *See also* Hexose diphosphate
Fructose, 1,6-Bisphosphatase: Properties of the Neutral Enzyme and its Modification by Proteolytic Enzymes (Horecker, Melloni, and Pontremoli), **42:**193
Fructose-2, 6-bisphosphate
biosynthesis of, **54:**154–155
enzymatic regulation of, **54:**156
epinephrine effect on, **54:**154
fructose-bisphosphatase regulation by, **54:**153–157
glucagon effect on, **54:**154; 155; 180
glucose effect on, **54:**154
identification of, **54:**153
insulin effect on, **54:**154
ionophore A 23187 effect on, **54:**154
occurrence of, **54:**154; 155
as PFK activator, **54:**154; 180
phenylephrine effect on, **54:**154
recycling of, **54:**161; 163; 164
vasopressin effect on, **54:**154
in yeast, **54:**174
D-Fructose, from D-sorbitol, **15:**118
in polysaccharide synthesis, **15:**306; 310–314
D-Fructose, pathway of metabolism of, in enteric bacteria, **31:**81
Fructose-1,6-biphosphatase, **56:**292
Fructose-bisphosphatase, **54:**155–157
activation of, **54:**123; 126
adenosine nucelotide regulation of, **54:** 136–142
AMP inhibition, **54:**136–140
ATP and ADP inhibition, **54:**140–141
by energy charge, **54:**141–142
alkaline, **54:**146; 148
AMP binding sites on, **54:**137; 139–140
AMP inhibition of, **54:**123
in ATP hydrolysis, **54:**122
in bumblebee flight muscle, **54:**162–163
in Calvin cycle, **54:**176
in cancer, **54:**166–168; 181
chelating agents in stimulation of, **54:**133
chloroplast, **54:**175–177
as constitutive protein, **54:**159–160; 179
deficiency of, **54:**165–166
in diabetes, **54:**164; 165
divalent cation requirement for activity of, **54:**123; 127–129
AMP effect on, **54:**128; 137
EDTA effect on, **54:**127; 129
kinetics of, **54:**128–129
magnesium ion, **54:**127
manganese ion, **54:**127
effectors of, **54:**126–145

kinetic and binding constants for, **54:** 130–131
endosperm, **54:**175
fatty acids and phospholipids in regulation of, **54:**143–145; 181
ferrodoxin affinity of, **54:**176
fructose-2, 6–6bisphosphate regulation by, **54:**154; 155
fructose-6-phosphate regulation by, **54:** 142–143
glucagon effect on, **54:**150; 152; 180
in heat production, **54:**164; 178
hormonal regulation of, **54:**150–160
via fructose-2, 6-6bisphosphate, **54:** 153–157
via phosphorylation, **54:**150–153
inhibitors of, **54:**122
magnesium ion binding sites on, **54:**127
in malignant tissues, **54:**180
in microbial metabolism, **54:**168–174; 180; 181
AMP regulation of, **54:**169
of autotrophs, **54:**169; 181
catabolite inactivation of, **54:**173
divalent ion activation of, **54:**168
of heterotrophs, **54:**169; 181
monovalent ion regulation of, **54:**169
phosphoenolpyruvate activation of, **54:**169–172
properties of, **54:**170–171
protease and, **54:**172–174
of yeast, **54:**173–174
molecular weight of, **54:**123
monovalent cation effect on, **54:**129–133
ammonium ion, **54:**129; 132
potassium ion, **54:**129; 132
sodium ion, **54:**129; 132
temperature effect on, **54:**132
native *vs.* proteolyzed, **54:**146
occurrence of
and general properties of, **54:**123–125
mammalian, **54:**124; 125; 177
or corn, **54:**177
in PFK regulation, **54:**160; 178
pH for activity of, **54:**123
phosphorylation of, **54:**151–153
in photosynthesis, **54:**175; 176; 178
plant, **54:**174–177
castorbean endosperm, **54:**175
spinach chloroplast, **54:**175–177; 181
protease effect on, **54:**180
proteolytic modification of, **54:**145–148
by cathepsins, **54:**146
regulation of, **54:**122; 178–179
spinach cytosolic, **54:**177
starvation effect on, **54:**158–159; 179
thioredoxin affinity of, **54:**176
tryptophan in, **54:**148–150
of yeast, **54:**180
zinc ion as activator and inhibitor of, **54:** 133–135
physiologic significance of, **54:**135–136
zinc ion binding sites on, **54:**134
Fructose 1,6-bisphosphatase, **53:**45–76, *See* Fru-1,6-P_2
activation of, **53:**49
adenosine monophosphate
distance to, **53:**72–73
inhibition by, **53:**48
aldolase, interaction with, **53:**49; 50
alkaline type, **53:**46–47
amino acid composition of, **42:**197
Ca^{2+}, inhibition by, **42:**200
cold, effect of, **42:**208
conversion of neutral to alkaline enzyme, **42:**204
digestion with proteolytic enzyme, **42:**204
divalent cation, effect of, **42:**199
EDTA, activation by, **42:**200
fasting, effect of, **42:**208
histidine, effect of, **42:**201
isolation of, **42:**195
kidney, **42:**221
liver, **42:**221
lysosomes, effect of, **42:**208
muscle, **42:**222
natural chelators, effect of, **42:**201
pH, effect of, **42:**199
proteolytic modifications, **42:**215
regulatory properties, **42:**214
seasonal variation, **42:**208
subunit structure, **42:**196
Zn^{2+}, inhibition by, **42:**201
amino acid compositions, **53:**50
aminotransferase, **53:**223
AMP, **53:**49; 51; 72; 76
analogs, substrate, **53:**51–57; 61
anomeric specificity, **53:**53–55; 75
apoenzyme, **53:**58
arginyl groups modification, **53:**69

Fructose 1,6-bisphosphatase (*Continued*)
binding properties, **53:**51; 56
carboxyl groups modification, **53:**69–70
catalytically active unit, **53:**47
chemical modification, **53:**69–72
comparison from different sources, **53:**47; 48
complex with ligands, **53:**72
discovery, **53:**46
distance measurements, **53:**72–73
EDTA, **53:**51
equilibrium, **53:**66–67
fructose 6 phosphate, binding, **53:**51–52
fru-6-P analogs, **53:**63
histidyl groups modification, **53:**70
inhibition by AMP, **53:**67–68
inhibition of, **53:**53–55; 67–68
kinetic model, **53:**60; 61; 62; 63; 64; 65; 66; 68; 74
of reaction, **53:**59–63; 65; 74–76
kinetic studies, **53:**59–63
lysyl groups modification, **53:**70
manganese cofactor, **53:**51–52; 55–59; 72–74
metal ion, **53:**55–59
methyl-D-fructo-furanoside 1,6 P_2, **53:**51–57; 61
Mg^{2+}, **53:**52; 54; 55; 56; 57; 58; 59; 61; 63; 67; 75
Mn^{2+}, **53:**51; 52; 54; 55; 56; 57; 58; 59; 61; 63; 67; 72
model for reaction, **53:**75
modification, chemical, **53:**69–72
molecular weights from different sources, **53:**47–48
monovalent cation activation, **53:**58–59
neutral form, **53:**46–47
^{18}O exchange from phosphate, catalyzed, **53:**63–64
phosphofructokinase, interaction with, **53:**49
protein chemical modification, **53:**69–72
rabbit liver enzyme, **53:**46
radioimmunologic assays, **53:**47
reviews on, **53:**46–47
structural properties, **53:**47–49
substrate analogs, **53:**51–57; 61; 62
subtilisin cleavage, **53:**50–51
subunits of, **53:**46–49
sulfhydryl (cysteine) groups modification, **53:**71–72
tyrosyl residues modification, **53:**70–71
zinc cofactor, **53:**51; 55–59
Zn^{2+}, **53:**51; 52; 55; 56; 57; 58; 61; 75
Fructose 2,6-bisphosphatase, *See* FBPase 2
Fructose Bisphosphatase Activity, Regulation of (Tejwani), **54:**121
Fructose 2,6-bisphosphate, *See* Fru-2,6-P_2
Fructose-1,6-bisphosphate aldolase, **56:**293; 385, **74:**1765
Fructose 2,6-Bisphosphate (Van Schaftingen), **59:**315
Fructose (C^{14}), in fructose transfer study, **17:**196; 198–199
Fructose dephosphatase, correlation of growth rate of hepatomas with, **29:**331
Fructose diphosphatase, **24:**125, **37:**459; 526, **52:**272–277
activation of by reduced ferredoxin, **33:** 136–138
allosteric regulation, **28:**92–95; 122
in ribose synthesis, **28:**411
change in activity after glucagon, **52:**288–289
congenital deficiency, **52:**291
effect of AMP, **52:**289
effect of Ca^{2+}, **52:**274
effect of fatty acid, **52:**275
effect of Zn^{2+}, Ca^{2+}, **52:**273
general properties, **52:**272–275
inhibition of AMP, **52:**273; 277
inhibition of fructose diphosphatase, **52:**272
interconvertible enzyme, **52:**275–276
change in activity, **52:**275
phosphorylation, **52:**275
kinetic mechanism of, **45:**285
in photosynthesis, **19:**25
proteolysis of, **52:**276
starvation, diabetes, corticoid, **52:**276
stereospecificity, **52:**273
stimulation of phosphofructokinase, **52:**278
Fructose-1,6-diphosphatase, **43:**7
ethionine effects, **25:**381
Fructose diphosphate
allosteric control, **28:**78–85; 87; 92–95; 111–112; 118
cleavage of, **28:**396; 397
effect on pyruvate kinase, **52:**311–312
in photosynthesis, **25:**95–96; 100; 102; 105
in schizophrenia, **29:**512
separation from glucose diphosphate, **14:**211
Fructose 1,6-diphosphate, **19:**7; 9; 15; 25; 26; 29; 30; 52; 53

Fructose-1,6-diphosphate, aldolase action, **25:**344
Fructose-1,6-diphosphate adolase, **37:**483
Fructose diphosphate aldolase
electron microscopy of, **33:**100–101
in photosynthesis, **25:**96
protozoa and green algae in, **31:**132
Fructose diphosphate (FDP), effect of, on conformation of aldolase, **31:**151
Fructose diphosphate-1-phosphatase, in photosynthesis, **25:**96
Fructose isomerase, in β-glucan synthesis, **28:**340
Fructose Metabolism (Racker), **15:**141
Fructose monophosphatase, **24:**125
Fructose-P_2, **46:**366; 373
as activator, **46:**332; 333
concentration and energy charge, relation to glycogen accumulation, **46:**359–362
Fructose-6-P, **43:**493
phosphorolysis of, **23:**324; 327; 337
Fructose 6-P, as activator, **46:**328–333
Fructose-1-phosphatase, **15:**153
Fructose-1-phosphate, **15:**141; 145; 147; 151–153; 154; 168–170
aldolase action, **25:**344
Fructose-6-phosphate, **15:**151–152; 154; 155; 162; 163–164; 165; 167; 168–170, *See* Hexose phosphates
allosteric regulation, **28:**74–75; 78–87; 92; 94; 111–112; 114–118
in pentose metabolism, **28:**395–397; 402; 411
in fructose-bisphosphatase regulation, **54:**142–143; 156
fructose-bisphosphate from, **54:**154–155
hydrolysis, by metal ions, **21:**184
2-2kinase, **54:**154–155; 156
in yeast, **54:**174
in photosynthesis, **25:**95; 98–102; 105
recycling of, **54:**161
Fructose 1-phosphate, **60:**255
aldolase action on, **19:**9
Fructose 6-phosphate, **19:**5; 6; 24–27; 29–31
alginate biosynthesis, **70:**224
Fructose phosphate, in plants, **19:**5
D-Fructose-6-phosphate amidotransferase
characteristics of reaction, **39:**153
properties of enzyme, **39:**153
fructose-6-phosphate as stabilizing agent, **39:**153
molecular weight, **39:**153
regulation of activity *in vitro*
alkylating agents, reaction with, **39:**155
AMP and glucose 6-phosphate, relationship to, **39:**154
feedback inhibition by UDP-*N*-acetyl-D-glucosamine, **39:**153; 155
glutamine, relationship to, **39:**154
UTP relationship to, **39:**154
regulation of activity *in vivo,* **39:**155
sources of enzyme, **39:**152; 153
Fructose 6-phosphate cycle
cori cycle, **52:**284
effect of hormones, glucagon, phenylephrine, **52:**286–288
in intact animals, **52:**284–286
fed state, **52:**285
starved state, **52:**285
in isolated liver preparation, **52:**281–284
measurement by [3-^{3}H] glucose, **52:**284
measurement by randomization of ^{14}C, **52:**285
usefulness, **52:**290–291
in fed conditions, **52:**290
in starvation, **52:**290
Fructose 6 phosphate/fructose diphosphatase recycle
extent of recycling, **52:**283
measurement, **52:**283
occurrence, **52:**283
Fructose-6-phosphate-glucose-6-phosphate transformation, **14:**195; 205–206
Fructose-1-phosphate kinase, **15:**151–152
Fructose-6-phosphate kinase, **15:**151–152
Fructose 6-phosphate kinase, in plants, **18:**286
Fructose-6-phosphate reductase, **15:**167
Fructose phosphates
isomerization of, **14:**195; 205–206; 209
in plant respiration, **18:**285–287; 289; 292–294; 312
in solubility studies, **17:**145–147; 149; 151
Fructose-T, as acceptor in dextran formation, **30:**199
Fructose transfer, oligosaccharides from, **17:**192–205; 222; 224; 226; 229
β-Fructosidase, *See* Invertase
β-Fructosidase, **48:**47
β-*h*-Fructosidase, in tobacco, **6:**367–368
β-Fructosidase (inulase), cacao, **25:**466

β-h-Fructosidases, classification, **14:**246
Fructosyl-fructoside, **17:**199; 202; 203
1-Fructosyl-1-kestose, **17:**210
Fructosyl-methylfructoside, **17:**205
1-Fructosyl-sucrose-F, **17:**203
6-Fructosyl-sucrose-G, **17:**204
Fruit, carotenogenesis in, **21:**308; 317; 325–328; 333
  carotenoids in, **21:**305
Fru-1,6-P_2
  inhibition of, by Fru-2,6-P_2, **59:**334–335
    mechanism of, **59:**335–337
  occurrence of, **59:**334
  and phosphorylation by cAMP-dependent protein kinase, **59:**337–338
  properties of, **59:**334
Fru-2,6-P_2. *See also* Metabolic regulation, and Fru-2,6-P_2
  anomeric configuration of, **59:**323–324
  assay procedure of, **59:**331
  concentration, **59:**374–375
  conclusions of, **59:**380–381
  description of, **59:**317–318
  discovery of
    and Fru-6-P/Fru-1,6-P_2 cycle, **59:**319–320
    and gluconeogenesis in liver, **59:**318–319
    and glycolysis in liver, **59:**318–319
  effect of, on PP_i-PFK, **59:**330–331
  and enzyme activities, regulation by
    Fru-1,6-P_2, **59:**334–338
    other enzymes, **59:**339–340
    PFK 1, **59:**324–329
    PP_i-PFK, **59:**329–334
  and FBPase 2
    and liver bifunctional enzyme, **59:**341–350; 352–353
    other, **59:**353–357
  and glycolytic signal, **59:**380–381
  identification of, **59:**321
  inhibition by, **59:**334–335
    mechanisms of, **59:**335–337
  and low molecular weight stimulator of liver PFK 1, **59:**320–323
    anomeric configuration of, **59:**323–324
    integration of, in metabolic reaction, **59:**357
  and PFK 2
    other, **59:**353–357
    rat liver bifunctional enzyme, **59:**341–350; 352–353
    properties, **59:**321
    role of, **59:**375–377
    stabilizing effects of, on liver, **59:**328–329
    synthesis, **59:**322; 323
Ft, **54:**58
FTdR, **54:**59
  in DNA incorporation, **54:**61–62
  synthesis of, **54:**61
Ftorafur, **54:**58
  synthesis of, **54:**62
FtsZ protein, *Caulobacter* cell cycle, **71:**7
FU, *See* 5-Fluorouracil
Fucogly colipids, **38:**303
N-Fucopentaose, **52:**50
Fucose, **14:**207
  in β-1,3 glucans, **28:**336
D-Fucose
  induction of galactose system by, **30:**104
    inhibition of arabinose induction by, **30:**112
  as inhibitor, **36:**118
Fucose polysaccharides, synthesis, **28:**113–114
Fucosidase, **38:**308
α-L-Fucosidase, **38:**298; 303, **48:**39; 55–56; 83; 84
  amino acid composition, **60:**159
  assay, **60:**158
  biological role, **60:**160
  catalytic mechanism, **60:**159
  genetics, **60:**160
  inhibitions, **60:**160
  kinetic data, **60:**159
  Michaelis constant, **60:**159
  molecular weight, **60:**158
  occurrence, **60:**157
  properties, **60:**158
  purification, **60:**157
  substrates, **60:**160
  subunits, **60:**158
Fucosidases, **56:**220
χ-D-Fucoside, **36:**102
Fucosidosis, **38:**294; 295; 298; 303
2′-Fucosyllactose, **52:**46; 55; 62; 70
Fucosyl α1r2 galactoside, **52:**46
(Fucosyl α1r2) galactoside α1r3 *N*-acetylgalactosaminyltransferase, **52:**27; 60; 68; 70; 73–75; 98; 105; 110; 112; 155
(Fucosyl α1r2) galactoside α1r3 galactosyltransferase, **52:**27; 67–68

(α1r2 Fucosyl) galactoside α1r3 *N*-acetylgalactosaminyltransferase, **52:**155
α 1r2 Fucosyltransferase, **52:**39
Fucosyltransferases, **52:**25; 44–56
Fucoxanthin, **11:**145–146
light absorption by protein complex of, **21:**350
Fuculose 1-phosphate aldolase, catalysis of carbon-carbon cleavage/condensation reactions by, **31:**131
FU-DNA, **54:**58
FUDP, **54:**58; 86
FuDR, **54:**58
activation of, **54:**64
anabolism of, **54:**103
cytotoxity of, **54:**97
with 5-fluorouracil, **54:**101–102
growth-inhibitory potency of, **54:**97–98
transport system of, **54:**64
Fuel cell, microbial, **27:**531
Fully reduced enzyme, activation energy for conversion of purple intermediate to, **34:**61
conversion of purple intermediate to, **34:**60
effect of pH, on rate of conversion of purple intermediate to, **34:**61
reversible reaction between purple intermediate and, **34:**63
Fumagillin, in treatment of plant diseases, **20:**401
Fumarase, **15:**186; 191; 259–262; 287; 332; 344–345; 353; 358–360, **37:**482, **46:**456–458, **50:**367, **74:**303
and acetate metabolism, **23:**428
digestion of, **23:**70–71; 78
action on *trans* acid, **8:**353
and citric acid cycle, **18:**244; 272–273
and fluoroacetate, **18:**118
in plants, **18:**298–299
classification, **14:**237; 256
isolation, **14:**404; 405
class II, **74:**314
elementary steps in reaction, **25:**14
kinetics, and product inhibition, **25:** 180–181
multiple forms, **25:**347
product inhibition of, **25:**224
turnover number, **25:**30; 32
from *Escherichia coli,* structure of, **74:**324
imidazole reactivity in, **20:**70; 100
kinetic mechanism of, **45:**318; 349–352
kinetic studies, **17:**11; 24; 26; 31; 35; 49–55
localization of, **24:**331; 332
in mitochondria, **21:**13; 20; 21; 58; 74
in photosynthesis, **19:**14
pK value, **22:**58
Fumarase-aspartase family
amino acid sequence homology within, **74:**324
conserved regions among, **74:**314; 315; 316–319
histidine residues, **74:**317
lysine residues, **74:**318–319
structure of, **74:**321–324
Fumarase in plants, **9:**409
Fumarate, **34:**476, **40:**73, **50:**142; 149, **51:**236
from acetate, **23:**405; 406; 414; 425; 428
effect of streptomycin on oxidation, **10:**161
as electron acceptor, **46:**58
formation in phenylalanine metabolism, **19:**87; 88; 183
and kynurenine 3-hydroxylase activity, **19:**194
in ornithine cycle, **19:**372; 380
and steroid 11β-hydroxylase activity, **19:**189
inhibitor, **28:**186; 190; 199–200; 210; 225
kinetics of succinic dehydrogenase action, **18:**91–92
reduction by leucosuccinic dehydrogenase, **18:**81–84
as succinic dehydrogenase inhibitor, **18:**94
in mitochondrial hydroxylations, **21:**27
in oxidation of reduced diphosphopyridine nucleotide and succinate, **20:**157–160
relation to amino acid synthesis, **16:**255–256; 260; 265; 268
and urea synthesis, **15:**327; 332; 335–337; 339–340; 343–345; 358–362; 372–373; 379
in citric acid cycle, **15:**125; 184–185; 259–262
Fumarate and CO_2 fixation, **12:**51; 132
Fumarate and pantothenate, **7:**443
Fumarate from phenylalanine and tyrosine, **14:**74; 96–111; 112
Fumarate hydratase, **37:**424
localization of, **24:**331; 337; 341
Fumarate-inhibitor complexes, **15:**261
Fumarate reductase, **37:**210; 212; 215; 217; 218; 239; 246; 262
K_m for $FMNH_2$, **37:**213

Fumarates, as solubilizing agents, **17:**149
Fumarhydrase, classification, **14:**255
Fumaric acid. *See also* Citric acid cycle
in biosynthesis of aspartic acid in plants, **20:**328
from ethylbenzene, **27:**517; 518
labeled, in plant respiration, **18:**299; 300; 303
in plant metabolism, **10:**258
in plants, **19:**13; 52; 53
in yeast metabolism, **10:**174–176
Fumaric acid formation in molds, **9:**538; 540; 545–546; 547–550; 558–560; 566; 576
Fumaric acid semialdehyde, inactivation of aspartase, **74:**307–311
mechanism of, **74:**311–314; 312; 313
Fumaric hydratases, classification, **14:**237; 256
Fumaric hydrogenase, classification, **14:**255
Fumaric reductase, classification, **14:**255
Fumaric reductase assay, of succinic dehydrogenase, **18:**95; 98
Fumaric reductases, **18:**83
unidirectional, **18:**101–103
Fumarylacetoacetase in tyrosine metabolism, **14:**97–98; 107
Fumaryl acetoacetate, metabolic path, **15:**184; 241
4-Fumarylacetoacetic acid, formation in phenylalanine metabolism, **19:**89
Fumarylacetoacetic acid in tyrosine metabolism, **14:**97–99; 104–105; 107; 110
Fumigacin, *See* Helvolic acid
Fumigatin, **7:**471; 472; 486; 489; 490; 492
FUMP, **54:**58
identification of, **54:**86
synthesis of, from FU, **54:**64
in uracil replacement, **54:**86
Functional domains, of fibronectin
amino-terminal, **59:**9
cell-binding domain, **59:**17; 22
cell-binding region, **59:**10–11
collagen-binding fragment, **59:**9–10
DNA-binding fragment, **59:**10–11
fibrin-binding domain, **59:**12
globular, **59:**69
heparin-binding domain, **59:**11–12
Functional group, **34:**452
lipoprotein mosaics, **34:**479; 481
membrane, **34:**481
subunits, **34:**471
Functional groups, of DPNH involved in coenzyme binding, **22:**363–366
of yeast ADH, **22:**354–358
Functional groups of proteins, **24:**105–160
Functional redundancy, α-crystallin/small heat-stock protein, **69:**173–174
Fungal laminarinases, components, **28:**351–353
types of, **28:**350
Fungal proteinase, action on RNase, **24:**230
Fungi, **46:**42, **59:**327–328; 354–355. *See also* Chitin synthetase; specific fungi
action on wood, **18:**351–354
antibacterial substances from, **7:**461–511
carotenogenesis in, **21:**307–311; 317; 318; 321; 328–330; 333; 335–339; 343–345
inhibition, **21:**319–321
carotenoid function in, **21:**358–360
carotenoids in, **21:**306
cysteine *S*-conjugate β-lyases in, **72:**214–215
ethionine-resistant mutants of, **29:**138
filamentous, β-1,3 glucans of, **28:**328; 336–337
β-1,3 glucans of, **28:**328; 334; 338; 340
β-lactamase of, **28:**243; 312–313
laminarinases, **28:**338; 341–343; 346–352
microbial lysis, **28:**355–356
types, action on laminarin, **28:**354
in isoflavonoid phytoalexin metabolism, **55:**44–49
luminous, **25:**161
oxidation of hydrocarbons, **27:**476; 499
in pathological wilting, **11:**403; 426–432
in terpenoid phytoalexin metabolism, **55:**83
Fungi, Antibacterial Substances from (Kavanagh), **7:**461
Fungichromin, in treatment of plant diseases, **20:**403
Fungicidal action of sulfur, **10:**238
Fungicidin, *See* Mycostatin
Fungus, methylated lysine and arginine derivatives in, **42:**234
FUR, **54:**58
activation of, **54:**64
nuclear RNA incorporation of, **54:**88
Furacin, **17:**298
oxidation of reduced pyridine nucleotides by, **15:**128
Furanacrylolyl-Phe-Gly-Gly
and enalaprilat concentrations, **57:**46–47
hydrolysis of, **57:**16–17; 19

Furanocoumarins, **55:**41
Furanose, **60:**3
Furanose-pyranose, equilibrium mixture of D-fructose, **10:**132
Furanoside acceptors, **18:**224–225
Furanoterpenoids, **55:**76
Furfural, as degradation product of pectins, **20:**349; 350
FU-RNA, **54:**58
Furosemide, **57:**137–138
Fusaria, **5:**165–5:205
  alcoholic fermentation in, **5:**167–179
  dehydrogenations in, **5:**181–189
    course of dehydrogenations of alcohols, **5:**182–187
    role of sulfur in, **5:**189–192
    role of xanthones in, **5:**189–192
  fat formation in, **5:**196–198
  influence
    of nicotinic acid, **5:**169; 170
    of nitrate concentration, **5:**170
    of thiamin, **5:**168; 169
  mechanism
    of carbohydrate breakdown, **5:**174–176
    of carboxylase blockade, **5:**171–174
    of pentose breakdown, **5:**170–171
  nutritive value of, **5:**202–203
  phase sequence of carbohydrate breakdown in, **5:**179–181
  phosphorylation, **5:**179–181
  possible mode of action of xanthones, **5:** 195–196
  rôle of nitrogen source, **5:**168
  sulfur as hydrogen acceptor, **5:**176–178
  technical applications, **5:**198–203
  training of, **5:**187–188
Fusaria Biochemistry (Nord and Mull), **5:**165
Fusaria, glycolaldehyde in, **15:**177
Fusarinine, **35:**152–154; 160; 168
  instability, **35:**153
  structure of, **35:**152
Fusarium, fat synthesis by, **9:**666; 681–689; 699–700
Fusarium caucasicum, enzymic transformations of steroids by, **20:**262
Fusarium culmorum, enzymic transformations of steroids by, **20:**248
  plant disease caused by, **20:**394
Fusarium lini, enzymic transformations of steroids by, **20:**247
Fusarium lini in industrial fermentation, **7:**585
Fusarium moniliforme, polygalacturonases of, **20:**366
Fusarium oxysporum, **55:**47
  capsidiol metabolism, **55:**83
Fusarium oxysporum f. lycopersicum, pectinesterase of, **20:**360; 368
  polygalacturonases of, **20:**365; 368
Fusarium proliferatum, **55:**47
Fusarium roseum, **35:**152; 163
Fusarium sambucinumo, plant disease caused by, **20:**396
Fusarium solani, **55:**46
  enzymic transformations of steroids by, **20:**254; 255; 261–263; 273
Fusarium species, enzymic transformations of steroids by, **20:**248
Futile cycle, **48:**230
Futile cycles, **59:**318–319
  amplification mechanism, **52:**250
  in glycolysis and gluconeogesis, **52:**249
  measurement of, general considerations, **52:**250–251
  measurement of interpretation, **52:**258
  metabolic crossroads, **52:**250; 259; 272
  themogensis, **52:**249
FUTP, **54:**58
  identification of, **54:**86
Fycosidosis, **52:**50; 51

G

GABA transaminase, **55:**234
G-actin, **6:**157
GAD, *See* Glutamic dehydrogenase
GADH, *See* Glutamic acid dehydrogenase
Galactal, **48:**50
D-Galactal, as inhibitor, **36:**119
Galactans
  linkage to pectin plants, **20:**345
    lung, galactose in, **20:**119
  synthesis, **15:**314–315
Galactocerebrosidase, **38:**298; 301; 305; 307; 309, **60:**111
  rat-intestinal tissue, **38:**307
Galactocerebroside, **52:**63, **60:**111
D-Galactoflavin, **17:**220

Galactokinase
classification, **14:**251
deficiency in *E. coli,* **22:**29
enzymic adaptation, **13:**70
galactose metabolism in absence of, **20:** 127–129
levels in normal and galactosemic subjects, **20:**125; 126
in phosphorylation of galactose, **20:**112
kinetic mechanism of, **45:**308
properties of, **30:**101
reaction catalyzed by, **30:**101
C-1-Galactokinase, phosphorylation by, **15:**145; 147–148
Galactokinase reaction, **14:**197–198; 201–203
Galactolipid, in lamellae, **30:**9
Galactolipids, **36:**122, **60:**167
brain, deposition in mammals, **20:**119; 124; 125
in photosynthetic structure, **25:**70
Galactomannan, **56:**214
Galactomannans, **36:**103
Galactonic acid, formation and conversions in galactose metabolism, **20:**119
D-Galactono-γ-lactone, **60:**104
Galacton permease, and galactose utilization in mutant yeast, **22:**21; 22
Galactosamine, **15:**147. *See also* Amino sugars
in detection of galactokinase, **14:**198
in glycoproteins, **25:**442–443; 446–453
linkage in glycoproteins, **25:**446–453
in transamination, **16:**236
D-Galactosamine, from connective tissue, **25:**434; 437–439; 441
α-Galactosaminidase, **48:**57
endo-α (1→4), **48:**55
Galactose, **41:**99–101; 103; 104; 106–109, **52:**25
in biosynthesis of pectic substances, **20:**369
labeled, in studies of biosynthesis of pectic substances, **20:**369
in study of galactose-negative *Escheridia coli,* **20:**129–130
metabolism, **20:**111–134
in pectin, **20:**344
enzymatic hydrolysis of, **41:**104
fermentation by thiaminase-producing bacteria, **15:**413
phosphorylation of, **15:**145; 147–149
genetic studies, **17:**457
in transglycosylations, **17:**196; 208; 218–219; 227
inhibitor of constitutive β-galactosidase synthesis, **22:**121
quantitation of the requirement for, **41:**99; 106–109
specificity of, **41:**99; 103–104
in sperm metabolism, **24:**61
as terminal group by action of neuraminidase, **41:**100; 101
[1-^{14}C] Galactose, **52:**255
D-Galactose
conversions in nature, **14:**206–208
in polysaccharide synthesis, **15:**306
D-Galactose, **60:**96
enzymes and their structural determinants, **30:**101–103
location of markers on the bacterial chromosome, **30:**86
system of utilization of, **30:**101–110
as inhibitor, **36:**109
in lipopolysaccharide from *Salmonella,* **31:**79; 92
pathway of metabolism of, in enteric bacteria, **31:**81
D-Galactose, presence of a single common carrier in translocation, **29:**53
L-Galactose, origin, **14:**207
Galactose, UDP ester of, **18:**201
Galactose (C^{14}), **17:**219
in sperm metabolism, **24:**62
D-Galactose configuration, **36:**118
Galactose fermentation, **6:**69
hexose diphosphate formation by oxygen treatment, **6:**75–77
induction periods, **6:**71–75
abolition by hexose diphosphate, **6:**72–74
abolition by oxygen, **6:**74–77
products, **6:**69–71
Galactose-glucose disaccharide, synthesis, **15:**314–315
Galactose-glucose interconversion, **18:**208
D-Galactose kinase, **46:**247
Galactose linkage in sialic acid, **25:**444–446
Galactosemia
allosteric enzymes, **28:**112–113
cellular damage in, **20:**127
enzymic deficiency in, **20:**124–127
test for, **20:**114

D-Galactose nucleotides of, *See* Sugar nucleotides
Galactose oxidase, **38:**301, **41:**103, **50:**306
Galactose pathway, UDP galactose-4-epimerase of tumor cells represent rate-limiting step in, **29:**367
Galactose permease, constitutive, in *E. coli,* **22:**110
Galactose-permease, induction of, **30:**107–110
properties of, **30:**103
Galactose-1-phosphate, **15:**141; 145; 147–148; 168; 169 (P^{32})
allosteric regulation, **28:**110–113
effect on growth of galactose-negative *Escherichia coli,* **20:**131
effect on phosphoglucomutase, **20:**127
formation and conversions in living organisms, **20:**112; 113; 122; 125; 126
labeled, in studies of galactosemia, **20:**125
metabolism in galactosemic subjects, **20:**124; 125
isomerases, classification, **14:**237; 241; 258
Galactose-6-phosphate, hydrolysis, by metal ions, **21:**184
Galactose-1-phosphate-glucose-1-phosphate transformation, **14:**195; 197–204
Galactose-1-phosphate uridyltransferase, **38:**311
localization of, **24:**314
properties of, **30:**101–102
reaction catalyzed by, **30:**101–102
Galactose-1-*P*-uridyl transferase, **26:**53
Galactose-6-sulfate sulfatase, **60:**261
Galactose transfer enzyme, **34:**476
Galactose uridyl transferase, *See* Galactose-1-phosphate uridyl transferase
α1r4 galactoside β-*N*-acetylgalactosaminyltransferase, **52:**72
α1r4 galactoside β1r3 *N*-acetylgalactosaminyltransferase, **52:**72
β1r4 galactoside β1r4 *N*-acetylgalactosaminyltransferase, **52:**71
Galactosidase, **34:**517; 523
α-Galactosidase, **60:**92
properties of, **30:**125
purification, **60:**93
in utilization of raffinose by *E. coli,* **22:**22–24
β-Galactosidase
of *E. coli,* **16:**460
enzymic adaptation of, **13:**70; 72; 78–80; 82–83; 85–88; 89–90; 94–95; 97–98; 99–100; 101–105; 105–108; 109–113
β-Galactosidase, **34:**449, **36:**40; 58–62, **38:**302; 303; 308, **41:**103; 105; 116, **46:**479–480, **60:**21; 102; 106; 250
acid, **60:**103
activation by metal ions, **21:**153; 154
activity, **60:**106
assay, **60:**105
binding to chondroitin sulfate, **60:**110
biological role, **60:**110; 252
biosynthesis, **60:**107
catalytic mechanism, **60:**107
chonroitin sulfate degradation, **60:**252
column chromatography, **22:**474; 475
in *E. coli,* **22:**106; 110; 144
dominance relationships in synthesis, **22:**143; 144
effect of thiouracil on synthesis, **22:**139
genetic loci, **22:**102–104
inducers and substrates, **22:**119; 120
inhibition in constitutive mutants, **22:**121
sensitivity to RNA antagonists, **22:**138
synthesis and protein turnover, **22:**127
synthesis inhibited by glucose, **22:**112
induction, in *E. coli,* **22:**19–21; 107; 111; 112; 115; 120; 125
nature of precursor in, **22:**123; 124
linkage relationships in mutants of *E. coli,* **22:**19–21
production in raffinose, utilizing mutants of *E. coli,* **22:**22–24
in *S. aureus,* sensitivity to RNA antagonists, **22:**138
thiogalactosides in, **22:**113
complex form, **60:**104
deficiency, **60:**103
dermatan sulfate degradation, **60:**252
effect of analogs on structure, **29:**110
synthesis in presence of phenylalanine analogs, **29:**110
effect of radiations on, **24:**403
localization of, **24:**322
molecular weight from radiation studies, **24:**366
structural gene for, **24:**13
electron microscopy of, **33:**107–108

β-Galactosidase (*Continued*)
genetic inactivity, **60:**251
genetics, **60:**110
glycosaminoglycan degrading activity, **60:**251
heparan sulfate degradation, **60:**252
inhibition of
by D-fucono-, **36:**172
by galactonolactone, **36:**172
by glucono-, **36:**172
inhibitor, **60:**104; 109
inhibitory power of galactone-1:4-lactone, **36:**157
interaction with glycolipid substrate, **60:**109
isoelectric point, **60:**107
keratan sulfate degradation, **60:**251
kinetic data, **60:**108
Michaelis constant, **60:**251
molecular weight, **60:**103
natural substrate, **60:**251
oligosaccharide substrate, **60:**252
oliomeric form, **60:**107
optimum pH, **60:**108; 251
properties, **60:**107
protective protein, **60:**103; 137
purification, **60:**104; 251
specificity of induction of, **30:**87–88
substrates, **26:**62, **60:**103; 108
subunits, **60:**107
β-D-Galactosidase
inhibited by galactal, **36:**177
regulation of activity, **25:**173
α-Galactosidase A, **60:**93–97
activator protein, **60:**96
amino acid composition, **60:**94
assay of activity, **60:**94
binding sites, **60:**97
biological significance, **60:**97
biosynthesis, **60:**95
DNA nucleotide sequence, **60:**95
genetics, **60:**97
heterogeneity, **60:**95
homodimeric structure, **60:**94
hydroxyl group, **60:**97
inhibitors, **60:**96
isoelectric focusing, **60:**95
kinetics, **60:**95
molecular weight, **60:**94
pH and stability, **60:**95
properties, **60:**94
purification, **60:**93
substrate interaction, **60:**96
α-Galactosidase B, **60:**98
assay, **60:**99
binding to concanavalin A, **60:**100
biological function, **60:**102
biosynthesis, **60:**99
genetics, **60:**102
half life, **60:**100
heat stability, **60:**100
inhibitors, **60:**102
kinetics, **60:**100
Michaelis constant, **60:**100
molecular weight, **60:**99
properties, **60:**99
purification, **60:**99
substrate hydrolysis, **60:**101
α-Galactosidases, **48:**39; 40; 56; 59
in agglutination of RBC, **56:**175; 177
amino acid analysis of, **56:**174
assay of, **56:**171
inhibition of, **56:**175
mechanism of action of, **56:**175–178
molecular forms of, **56:**167–170
occurrence and localization of, **56:**166–167
origin of multiple forms of, **56:**224–225
pH effect on, **56:**174; 178
properties of, **56:**173–178
purification of, **56:**171–173
regulation of, **56:**179
role of, **56:**178–179
β-Galactosidases, **14:**246
β-Galactosidases, **48:**38; 39; 47; 51; 55; 77; 79; 81; 82
active sites of, **56:**184
assay of, **56:**182
of *Escherichia coli,* **48:**8; 30; 56
inhibition by galactal, **48:**50; 62
inhibition of, **56:**184
molecular forms of, **56:**181
occurrence and localization of, **56:**180
pH effect on, **56:**183–184
properties of, **56:**183–185
purification of, **56:**182–183
reaction with conduritol B epoxide, **48:**47
regulation of, **56:**186
role of, **56:**185–186
x-Galactosidases, Biochemistry of (Dey and Pridham), **36:**91
α-Galactosidases, classification, **14:**246
separation by chromatography, **14:**349

Galactosidases (α and β), in soil, **28:**375
β-Galactosidases (lactase, anthocyanase), cacao, **25:**466; 467; 480; 485–487
sources, **25:**336–337
α-Galactosidases (melibiase, raffinase), cacao, **25:**466
multiple forms, **25:**350
χ-Galactosidase, stability of, **36:**111
χ-Galactosidase II (1), **36:**97
χ-Galactosidase II (2), **36:**97
χ-Galactosidases, **36:**91–123
activity, measurement of, **36:**95
assay, methods of, **36:**95
histochemical detection, **36:**95
chemical analysis, **36:**101
induction, **36:**94
isolation, **36:**96–97
kinetic properties, **36:**109–119
effect of inhibitors, **36:**114–119
group specific reagents, **36:**116
metal ions, **36:**116–118
sugars, **36:**118–119
effect of pH, **36:**112–114
effect of substrate concentration, **36:** 109–111
effect of temperature, **36:**111–112
mechanism of action, **36:**119–121
occurrence, **36:**92–94
physical properties, **36:**97–99
molecular weights, **36:**99–100
multimolecular forms, **36:**97–99
physiological significance, **36:**121–123
specificity, **36:**101–109
De Novo synthesis, **36:**108–109
hydrolase activity, **36:**101–105
transgalactosylation, **36:**105–108
χ-Galactosidases I, **36:**97
multimolecular forms, **36:**97
χ-Galactosidases II, **36:**97
multimolecular forms, **36:**97
χ-D-Galactoside, **36:**102
β-Galactoside, concentration in *E. coli*, regulation of, **22:**109
β-Galactoside binding protein, **52:**116
Galactoside diglycerides, **52:**64
χ-D-Galactoside galactohydrolases, **36:**92
β-Galactoside permease, function of, **30:**84
α-Galactoside-permease, induction of, **30:**126
properties of, **30:**125
β-Galactoside permease in *E. coli*, **22:**103; 109–111
in *E. coli* mutants, **22:**19–21
Galactoside α1r2 fucosyltransferase, **52:**98
β-Galactoside α1r2 fucosyltransferase, **52:**27; 47; 48; 52–54; 105; 110; 118; 155; 156
β-Galactoside β1r3 glucuronyltransferase, **52:**83
Galactoside α2r6 sialytransferase, **52:**98
β-Galactoside α2r3 sialytransferase, **52:**27; 31; 32; 38; 42; 44; 98; 105; 112; 127; 130; 142; 149; 156
β-Galactoside α2r6 sialytransferase, **52:**27; 30; 34–38; 42; 98; 105; 114; 137; 149; 151; 155; 156
Galactosides
2, 4-4dinitrophenol inhibition of, **29:**66
action of β-galactosidase on, **13:**101–105; 107; 110–113
biosynthesis, **20:**118; 119
Galactosides, **36:**102
α-Galactosides, characteristics of induction of, **30:**126
system of utilization of, **30:**125–127
β-Galactosides, function of, **30:**84
system of utilization of, **30:**84–101
Galactosylceramidase, **60:**111
activators, **60:**114
assay, **60:**112
biological significance, **60:**114
genetics, **60:**114
inhibitors, **60:**114
isoelectric focusing, **60:**113
kinetic data, **60:**113
molecular weight, **60:**113
optimum pH, **60:**113
properties, **60:**113
purification, **60:**112
substrate, **60:**113
Galactosylceramide, **52:**63, **60:**102; 111
Galactosyldiglyceride, **60:**102
o-Galactosylglycerol phosphate, in teichoic acid, **30:**227
Galactosylhydroxylsine, **60:**108
Galactosyl hydroxylsine α1r2 glucosyltransferase, **52:**27
Galactosylhydroxylysyl α1r2 glucosyltransferase, **52:**91–93
Galactosylmethyl carbonium ion, **60:**108
Galactosylmethyl-*p*-nitrophenyltriazene, **60:**108

Galactosyl residues, exposed by sialicase treatment, **48:**61
Galactosylsphingosine, **52:**93, **60:**113
Galactosyl transferase, **43:**412
 chemical properties, **43:**448
 enzymatic assay of, **43:**415
 formation of complexes, α-lactal-albumin, **43:**468
 physical and chemical properties, **43:**448
 purification of, **43:**446
Galactosyl transferase reaction, **34:**476
Galactosyltransferases, **52:**56–68
Galactosyl transfer reactions, mechanism of, **43:**455
Galactotransferase, **15:**314–315
Galactowaidenase, enzymic adaptation, **13:**70
Galactowaldenase, **15:**164; 168–169
 reaction catalyzed by, **14:**201
Galacto-waldenase, *See* Uridine diphosphogalactose-4-epimerase
Galactozymase
 as adaptive enzyme, **6:**69–78
 effect of oxygen, **6:**74–77
 enzymic adaptation, **13:**81–83; 91–93
 and hexokinase, **6:**77
Galacturonate, **17:**147
D-Galacturonic acid
 conversions in nature, **14:**206–208
 in pectin, **20:**342–344; 349–351
  optical rotation, **20:**353
Galacturonic acid, determination, **20:**370
Galatose oxidase, **48:**64; 84
Galatose-1-phosphate uridyl transferase, absence in *Escherichia coli* mutant, **20:**129
 in formation of uridine diphosphogalactose, **20:**113; 114
 induction in galactokinase-deficient *Saccharomyces cerevisiae,* **20:**128
 levels in normal and galactosemic subjects, **20:**124–127
 occurrence and properties, **20:**114; 119; 120
Galeae, **45:**107; 403; 406
Galleria larvae, **59:**81
Gallic acid, in plant metabolism, **16:**132; 137–138; 150; 152
Gallin, amino acid composition, **16:**442; 444
epi-Gallocatechin, from tea leaves, **16:**151; 152
epi-Gallocatechin gallate, from tea leaves, **16:**151; 152
Gallocyanine, **59:**117
Gallotannin, metabolism in plants, **16:**142
Galloyl esters of flavonols, substrates of tea leaf oxidase, **16:**151
Galls, pigments in, **16:**137
Gal-1-P, *See* Galactose-1-phosphate
Gamones, effect on sperm, **24:**84
Ganglioside, **58:**119
Ganglioside *N*-acetylgalactosaminyltransferase, **52:**73
Ganglioside G_{M1}, **60:**105
Ganglioside G_{M2}, **60:**126; 133
Ganglioside G_{M3}, **60:**148
Ganglioside G_{m1}, **38:**295
Ganglioside G_{m2}, **38:**300
Gangliosides, **38:**297, **48:**60; 74, **50:**306; 318, **52:**31, **59:**47–48
 GD_2, **52:**61
 G_{D3}, **52:**33
 G_{D1a}, **52:**31; 41
 G_{M1}, **52:**31; 40
 G_{M2}, **52:**59; 61; 71
 G_{M3}, **52:**31; 32; 33
 G_{T1a}, **52:**33
 interferon interaction with, **50:**318–320
 in schizophrenia, **29:**503
 tetanus toxin bound to, **50:**318
GAR, *See* Glycinamide ribotide
Garlic, **56:**34
 antibacterial action of, **11:**380
  specific principle of, **11:**377–400
  therapeutic use of, **11:**397–399
Garlic, Specific Principle, Allin (Stoll and Seebeck), **11:**377
Garlic oil, composition of, **11:**379
Gas chromatography, **57:**259
 disulfide exchanges, **27:**208
 of yeast polysaccharides, **40:**247
Gaseous environment, effect on inactivation of enzymes by radiation, **24:**375–376
Gases, effect on sperm, **24:**82–83
Gas exchange in microbial attack on hydrocarbons, **10:**446–448; 465–466
Gastric digestion, in low vertebrate, **1:**387–390; 403–407; 410; 411
Gastricin, column chromatography, **22:**456
Gastric juice
 and milk clotting, **15:**424; 425
 secretion in low vertebrata, **1:**384–386; 402; 403

Gastric mucosa, permeability studies, **13:**56–58
Gastricsin, **33:**405
hydrolytic action on insulin of, **33:**422
Gastro-intestinal proteinases, specificity, **1:**69–78
Gastrointestinal tract, and plasma phosphatides, **7:**154
labeled phosphatides of, **7:**129
Gaucher's disease, **38:**294; 295; 296; 298; 303, **60:**115
appearance of, **29:**480
enzymatic abnormality, **38:**296
therapy, **38:**310
GDH, *See* Glycerinaldehyde-3-phosphate dehydrogenase
GDP, *See* Guanosine diphosphate; Guanosine diphosphate mannose
fucose, **52:**46; 47
hexanolamine-agarose, **52:**52; 54; 150
reduction of, **28:**449
6-6thiol analog, **43:**39
GDP-Colitose, formation of, **26:**44; 45
GDPfucose, synthesis of, **31:**115
GDP-L-fucose synthetase, **44:**85
GDP-D-glucose pyrophosphorylase, **26:**52
GDP-D-*glycero*-D-*manno*-heptose pyrophosphorylase, **26:**53
GDP-D-mannose, enzymes for reduction, **26:**54
metabolism, **26:**45; 46
GDPmannose, role of, in lipopolysaccharide synthesis, **31:**92
GDP-D-mannose dehydrogenase, **26:**54
GDP-D-mannose pyrophosphorylase, allosteric regulation, **28:**111; 113
GDPmannose pyrophosphorylase, control of, **31:**116
reaction catalyzed by, **31:**107
GDP-D-Mannose pyrophosphorylases, occurrence of, **26:**45; 53
GDP-mannuronate, alginate biosynthesis, **70:**224
GDP-sugars, **26:**43–49; 61; 65
Gel, **34:**520–521
acrylamide, **34:**452
muscle, studies of, **13:**173–178
polyacrylamide, **34:**448
silica, **34:**452; 521
starch, **34:**448
Gelatin
action of tyrosinase on, **14:**150
adsorption studies, **14:**322–323; 325–328; 338
structure studies, **14:**265; 308
as activator of β-glucuronidase, **16:**376
antigenicity of, **23:**219
effect of chemical modification, **24:**145
enzyme action on, **33:**257
optimal pH for papain action on, **19:**303
Gelatinase, **33:**405
Gelatinases, of bacteria, **23:**177–179; 184; 185; 188; 190; 191; 193–195; 198
of protozoa, **23:**176
Gelatin liquefaction by thiaminase-producing bacteria, **15:**413
Gelation of protoplasm and coagulation, **7:**40
Gelation properties, of pectic substances, **20:**356; 357
Gel chromatography, **58:**146
Gel electrophoresis, **57:**11; 14; 187
Gel filtration chromatography, alginate biosynthesis, **70:**226
Gel filtration chromatography, T7 RNA polymerase, **66:**114–117
Gel formation
of amylose, **12:**352–353
of borate-polyhydroxy complex, **12:**502–509; 510–511; 514
of starch, **12:**367–371
protein, and SH-groups, **27:**221–222
Gels, enzyme action on, **33:**255–259
Gel-sol transformation, in myosin study, **16:**331; 338
Gene
definition, **13:**123–129, **22:**4
positional aspects of action, **13:**134–138
relationship to enzymes, **13:**138–145
size of, **13:**124; 129–134
structure and gene action, **13:**121–149
regulatory, for lipopolysaccharide synthesis, **31:**111
Gene cloning
CTP synthetase, **66:**245–246
glucosamine 6-phosphate synthase, **66:**264–265
glutamate synthase, **66:**268–271
Gene conversion, catalytic diversity evolution, **69:**37
Gene deficiencies, effect on enzyme activity, **17:**394

Gene-enzyme relationships, **17:**437–452
Gene expression, **44:**166; 170
asparagine synthetase, **72:**148–149
glutamine synthetase and hepatocyte heterogeneity, **72:**60–61
γ-glutamyl transpeptidase, **72:**264–270
DNA methylation, **72:**269–270
gene structure and mechanism, **72:**264–265
human gene expression, **72:**267–269
multiple promoters and tissue-specific expression, **72:**265–267
hepatic glutamine transport, hepatocyte hydration, **72:**52–53
regulation of, adenosine receptors, **69:** 110–111
Gene families catalytic diversity evolution, **69:**37
Gene interaction, **36:**145
Gene-protein relationship, in folate synthesis, **27:**432
General base catalysis, **67:**51; 53
Generalized gangliosidosis, **38:**294; 295; 298; 302
General transcription factors (GTF), RNA polymerase II transcription, **71:**43–44
Genes. *See also* Chromomeres; Mutants and Cell nucleus
action and structure, **4:**1–40
and biological evolution, **4:**27–32
as chemical mechanism, **4:**17–21; 30
control of cystine metabolism, **16:**278
control of β-glucuronidase, **16:**389–396; 401
importance of DNA in, **16:**412–413; 429–430; 434; 453–455
influence on phenolase activity, **16:**172b
relation to number to amount of DNA, **16:**430
and cytoplasm, **4:**19–21; 26–27
and enzymes, **4:**4; 8; 14–16; 19; 21; 24–26; 28
enzymes of nucleotide-sugar synthesis determined by, **31:**106–107
and hormones, **4:**20
as megamolecules, **4:**18; 19
mutation by ultraviolet irradiation, **3:**21; 22
mutation of, **11:**31–32
mutations, **8:**62
and nucleoproteins, **3:**21
origin of, **38:**227
and paleontological concepts, **4:**32; 33
and pigmentation, **4:**21–26
polyurgic, **4:**29; 30
regulator
C
in arabinose system, **30:**113–116
function of, **30:**82
i, definition and properties of, **30:**92–93
for lactose operon, **30:**85
in maltose system, **30:**122
Rgal, properties of, **30:**105
relation to viruses, **9:**182–184
and specific reactions in synthesis of vitamins and amino acids, **8:**205; 235
transmission of structural information from, **24:**11–20
and viruses, **3:**3; 4
Genes for tyrosine metabolism, **14:**92
Gene sharing, α-crystallin/small heat-stock protein, **69:**157
Genes (phi X174), translation of, **47:** 415–419
reinitiation during, **47:**415
Gene structure, adenylosuccinate synthetase (AMPSase), **73:**86–94. *See also* Adenylosuccinate synthetase (AMPSase)
Gene Structure and Action, Chemical Formulation (Gulick), **4:**1
Gene Structure and Action, Genetic Formulation (Pontecorvo), **13:**121
Genetically Controlled Differences in Enzyme Activity (Fincham), **22:**1
Genetic alternations, of PP-ribose-P-synthetase, **41:**7
Genetic basis for structurally related proteins, **24:**419; 423
Genetic blocks, **17:**394–395
Genetic code, **27:**295; 304, **38:**213; 221; 224, **47:**379; 419
during evolution, **47:**427
interpretation of, **38:**214
origin of, **38:**210
primitive mechanism, **38:**231
(table), **38:**214
universality of, **38:**230
Genetic coding, and enzyme structure, **26:** 89–114
Genetic control
of carbohydrate synthesis, **26:**67–78
of enzyme activity, **28:**43

Genetic control of enzyme structure, **24:**7–11; 13
Genetic Control of Yeast Mannans (Ballou), **40:**239
Genetic determination of induced biosynthesis of enzymes, **13:**105–108
Genetic engineering, **38:**310–311, **58:**131
Genetic factors in enzyme synthesis, **6:**86–91; 118
  antigenic specificity, **6:**118–119
  artificially colored light, **6:**95
  irradiation, **6:**82–84; 92–93
  orthoquinoid dyes, **6:**95; 114
  in paramecia, **6:**89–90
  in pneumococcal types, **6:**86–89
  resistance to inhibitors, **6:**91–118
  sulfonamides, **6:**94
  toxic chemicals, **6:**93–94
  in trypanosomes, **6:**90–91; 95; 113; 118
  in yeast, **6:**90
Genetic function of nuclear enzymes, **13:**15
Genetic locus, and complementation maps, **22:**34; 35
  concept of, **22:**4
  single *vs.* multilocus control of enzyme activity, **22:**29; 30
Genetic map, of arabinose operon, **30:**110
  of bacterial chromosomes, **30:**86
  of galactose operon, **30:**105
  of maltose system, **30:**119
  of rhamnose region, **30:**116
Genetic mapping, and relation to function, **22:**37; 38
Genetic material, deoxyribonucleoprotein as, **18:**161–190
Genetic material in sperm, **24:**90
Genetic mechanisms and nuclear enzymes, **27:**240; 275–283
Genetic polymorphism, **60:**246
  adenosine deaminase, **51:**174–176
Genetics
  chemical, metabolic aspects, **17:**393–476
  chemical modification of virus, **24:**109
  exopolysaccharide alginate synthesis, **70:**221–254 (*See also* Exopolysaccharide alginate synthesis)
  of β-lactamase, **28:**269–272; 280; 311–313
  molecular, survey, **27:**291–295
  tyrosine hydroxylase induction, **70:**172–182
Genetics, Chemical Metabolic Aspects (DeBusk), **17:**393
Genetic studies, of carotenogenesis, **21:**322–330
Genetic transmission, relation to nucleus, **16:**475–486
Gene transfer, **58:**74
Genic significance of deoxyribonucleoprotein, **18:**181–186
Genistein, **55:**5
  prenyl group of, **55:**41
  structure of, **55:**6
Genome, origin of, **38:**227
Genome project
  approaches, **67:**460–462
  basic requirements for success, **67:**459–460
  need for new methods, **67:**462–463
  progress up to 1990, **67:**459
Genomic clones, adenosine receptors, **69:**111
Gentiobiose
  from β-1,3 glucans, **28:**331; 335; 350–353
    hydrolysis of, **28:**353
  isomaltose from, **11:**341
  in stabilization of *d*-mandelic nitrile, **20:**20
  in transglycosylations, **17:**214; 217–218; 220
Gentiobiosyl, in teichoic acids, **30:**225
3-*O*-β-Gentiobiosyl-D-glucose, and β-1,3 glucans, **28:**331; 340; 350
Gentiotriose, from β-1,3 glucans, **28:**335
Gentisic acid
  formation in peroxidase-catalyzed reaction, **19:**143
  from toluene, **27:**515
Gentistic acid, **59:**137
Geotrichum lactis, **59:**73–74
Geotrichum species, plant disease caused by, **20:**402
Geraniol, **22:**306; 307
  structure, **22:**306
Geranium, antibiotic treatment of disease, **20:**401
Geranyl CoA carboxylase, **35:**388–390
Geranyl-condensing enzyme, in squalene biosynthesis, **22:**297
Geranylgeraniol, **46:**77
Geranylgeraniol pyrophosphate, **46:**75
Geranylgeranyl pyrophosphate, **55:**78
Geranyl pyrophosphate, **55:**66
  biosynthesis, **22:**319
    chemical synthesis, **22:**317
    in squalene biosynthesis, **22:**297; 305–309
      synthesis, **22:**311; 312

Geranyl pyrophosphate (*Continued*)
structure, **22:**311
in biosynthesis of nonrubber terpenoids in latex, **29:**243
Germicidal action of hydrocarbons, **10:**451–454
Germination, **36:**122, **44:**175
Germination-promoting properties of thiosinamine, **10:**260–261
GHA, *See* Glutamohydroxamic acid
Gibberella baccata, enzymic transformations of steroids by, **20:**248; 273
Gibberella species, enzymic transformations of steroids by, **20:**248
Gibberellic acid, **56:**179
Gibberellin, effect on β-glucanases, **28:**342
Gibberellins, **55:**79
Gibbs-Donnan equilibrium, **7:**46
Gibbs energy, **53:**244
Gibbs energy, ATPase thermodynamics, **72:** 14–15
Gibbs free energies, **43:**222
Gilbert theory, with sedimentation, **30:**144
Gladiolic acid, **7:**472; 486; 489
Glandular activity, relation to cytochromes, **17:**335
Glass, **34:**521
adsorption of enzymes on, **28:**298; 299
beads, **34:**452
coagulant effect of, **30:**287–288
ground, effect on platelet-poor plasma, **16:**68
properties of enzymes adsorbed on, **33:**270
Glass surfaces, effect on clotting, **16:**69
thrombin adsorption on, **16:**78–79
GlcNAc-MurNAc(-pentapeptide)-*P*-*P*-phospholipid, acceptor substrate in biosynthesis of interpeptide bridges, **40:**133
Gliadin
adsorption at interfaces, **28:**19
adsorption studies, **14:**322; 330
structure studies, **14:**263; 284
dicarboxylic amino acids in, **20:**319
N-terminal histidine in, **20:**71
Gliocladium catenulatum, enzymic transformations of steroids by, **20:**262; 263
Gliocladium roseum, **23:**177
methoxyl carbon labeling, **25:**296
Gliotoxin, **7:**472; 473; 487; 489; 490; 493
Gliotoxin in plant metabolism, **10:**257
glmS gene, glucosamine 6-phosphate synthase, **72:**104
glnA gene, glutamine synthesis expression, **72:**35–36
Gln-59 residue, triad amidotransferase, **72:** 138–139
Gln-90 residue, triad amidotransferase, **72:** 138–139
Gln-179 residue, triad amidotransferase, **72:**139
Gln-tRNA^{Gln} amidotransferase, **66:**281–282
Globin, **46:**141; 152; 159
synthesis, **46:**106; 149
synthesis by reticulocytes, **16:**491–492
Globin-carboxyheme interaction, elementary steps in, **25:**14
Globin haemochrome *a,* absorption spectrum, **23:**289–297; 306–310
Globin-heme, conversion to biliverdin, **19:**195
Globins, structure studies, **14:**277; 281; 284; 285
Globin stability, **6:**162
Globoid cell leukodystrophy, **60:**111
Globoside, **38:**295; 297; 307; 310, **52:**71
Globotriaosylceramide, **60:**92
Globotriglycosylceramide, **52:**62; 72
Globular molecules, in air-water interface, **28:**22–24
in oil-water interface, **28:**16
Globular proteins, *See* Proteins globular
active centers of, **24:**128–131
Globular proteins, Cotton effect, **27:**45–46
Globular proteins and enzyme activity, **14:** 184–185
Globulin
activation of prothrombin, **16:**70
in cell nucleus, **16:**449
cold-insoluble, **52:**66
column chromatography, **22:**423; 430
nature of thromboplastin, **10:**9–10
α Globulin, **46:**150
α_2-Globulin, **46:**160
α 2u-Globulin, **47:**291; 294; 303
biosynthesis, **47:**291
effect of androgens, **47:**291; 298
effect of estrogens, **47:**291; 298; 300
effect of glucocorticoids, **47:**291
effect of growth hormones, **47:**291
effect of thyroid hormones, **47:**291; 294; 295
multihormonal control, **47:**291
functional α 2u-globulin mRNA, **47:**291

host livers and hepatomas, **47:**304
sex horm one control of, **47:**297
α 2u-globulin cDNA, **47:**303
wheat-germ translation system for measurement of functional level of, **47:**294
γ-Globulin, **41:**129–174; 178
bovine, **41:**131; 178
effect of chemical modification, **24:**145
canine, **41:**131
chromatographic evaluation of purity, **19:**281
disulfide exchanges, **27:**221; 231
effect of pH on viscosity and optical rotation, **22:**426; 427
γG-, **41:**142
human, **41:**131
γM-, **41:**142
Physiological role of, **41:**129–174
β-Globulin, action of peroxidase on, **14:**158
Ac-globulin, action of tyrosinase on, **14:**147; 150
β_2-Globulin, association of platelet cofactors with, **16:**32; 73
α_2-Globulin, comparison with prothrombin, **16:**28
β_1-Globulin, effect on succinic dehydrogenase, **18:**95–97
γ-*Globulin, Physiological Role of* (Najjar), **41:**129
β-Globulin haemochrome *a,* absorption spectrum, **23:**289–297; 306–310
α_2 Globulin human serum oxidase, **44:**187
γ-Globulin inhibition of complement activity, **8:**95
Globulin nature of complement, **8:**88
Globulins. *See also* Egg white proteins; Immune globulin; Lactoglobulin; Serum proteins; Thyroglobulin
action of tyrosinase on, **14:**150
pea seed, separation by chromatography, **14:**351; 358
serum, separation by chromatography, **14:**332; 335; 345; 348; 350; 361; 362; 363
γ-Globulins, adsorpion studies on, **14:**360
structure studies, **14:**284
Gloloid leukodystrophy, **38:**294
Glomerella, genetic studies, **17:**447
Glomerella cingulata, **55:**83
mutants, reduction of melanin production in, **22:**24
Glomerella, enzymic adaptation in, **13:**106
Glossina morsitans, **45:**393
engorgement of crop in, **45:**400
Glu-143, **57:**23
Glu-270, **57:**21–24
Glu^{30}, eukaryotic dihydrofolate reductase, ligand binding (side chain roles), **70:**60–61
Glu-183, GMP synthetase, **72:**120–123
Glucagon, **36:**84, **46:**159, **47:**75–77; 136
activation of liver phosphorylase by, **34:**399
effect of mammalian liver glycogen synthetase, **34:**341
increase of cyclic AMP by, in liver, **34:**342
amino acid sequence, **24:**422; 430–431
and phosphorylase, **24:**283; 286
in diabetes, **54:**164
effect of, **54:**180
effect on cyclic AMP, **28:**107; 110
effect on ribose synthesis, **28:**417
effect on enzymes, **37:**158
effect on fructose diphosphatase, **52:**275–276
effect on fructose 6P cycle, **52:**286; 312
effect on gluconeogenesis, **52:**307–314
effect on glutamate, **52:**298
effect on glycolytic flux, **52:**306
effect on lipolysis, **52:**308
effect on mitochondria respiration, **52:**293; 312
effect on phosphoenolypyruvate carboxykinase, **52:**298; 299
effect on phosphofructokinase, **52:**280; 289
effect on pyruvate carboxylase, **52:**296
effect on pyruvate dehydrogenase activity, **52:**294
effect on pyruvate kinase, **52:**303
flux, **52:**305; 306; 308
effect on pyruvate transport, **52:**312
effect on pyruvate transport and metabolism, **52:**292
in fructose-2, 6–6bisphosphate regulation, **54:**154; 155
in fructose-bisphosphatase phosphorylation, **54:**152
in fructose-6-phosphate regulation, **54:**153
in gluconeogenesis, **54:**150; 154
mechanism of action, **54:**150–151
phenylalanine hydroxylase activation, **67:**135–137
in vivo, **67:**138–139
release from pancreatic islets, **49:**59

Glucagon (*Continued*)
secondary structure in, **47:**75
starvation effect on, **54:**157
N-terminal histidine in, **20:**71; 94; 95; 97
zinc binding in, **20:**97
tertiary structure, **47:**76
x-ray resolution of, **47:**77
Glucagon-agarose, **36:**40
β-Glucanase, effect on barley glucan, **28:**345; 355
effect on mating yeast cells, **28:**348–349
of higher plants, **28:**341; 342
β-1,6-Glucanase, effect on β-1,3 glucans, **28:**335
induction of, **28:**348–349
kinetics, **28:**346
preparation of, **28:**343
β-Glucanase multiplicity, **28:**351–352
β-1,4 Glucanases, action on cellulodextrins, **28:**353
in *Aspergillus niger,* **29:**351
inhibition of, **28:**347
isozyme concept, **28:**352
kinetics, **28:**346
β-1,2 Glucanases, kinetics, **28:**346
β-1,3 Glucanases (β-1,3 glucan 3-glucanohydrolase), **28:**341–353
effect on β-1,3 glucans, **28:**335
endo-, from laminarinase, **28:**341–342; 351–353
kinetics, **28:**345–347
patterns of hydrolysis, **28:**349–350
exo-, from laminarinase, **28:**341–342; 351–353
kinetics, **28:**345–348
patterns of hydrolysis, **28:**349–351; 353
inhibitors, **28:**347–348
isozyme concept, **28:**352
preparation of, **28:**343
β-1,3 Glucanase specificity, **28:**345
α-Glucan phosphorylase, **37:**431; 450; 463; 474; 575
Glucan phosphorylases, **74:**174
β-1,6 Glucans, action of β-glucanase, **28:**349
β-1,3 Glucans, antigenic properties, **28:** 336–337
biosynthesis, **28:**337–340
degree of polymerization, **28:**332–337
distribution of, **28:**328–329
hydrolytic enzymes for, **28:**341–353
1,4 linkages, **28:**336
occurrence and properties, **28:**326–330
other than laminarin, **28:**334–337
β-Glucans, branching, **28:**327–328; 331–332; 334–337; 340
cereal, degradation sequence, **28:**353
conformation of, **28:**327
extracellular, **28:**335–336
insoluble, **28:**327; 328; 331–333; 339
soluble, **28:**328; 331–333; 336; 339
as substrates for laminarinases, **28:**342
x-ray patterns, **28:**327
Glucans, mixed, **28:**341
physiological activity, **28:**356–358
β-1,3 Glucan synthesis, kinetics of, **28:**339
β-D-1,3 Glucan synthesizing systems, **28:** 337–340
Gluceraldehyde phosphate dehydrogenase, in plastids, **20:**228
Glucidases, enzymic adaptation, **13:**105; 107
Glucoascorbic acid, as inhibitory analogue, **6:**132; 138; 139
Glucocerebrosidase, **38:**297; 298; 304; 307, **60:**115
beef spleen, **38:**304
human-placental tissue, **38:**304; 307
Glucocerebroside, **38:**295; 297
Glucocheiroline, occurrence, **10:**266
Glucocorticoid
effect on glucose/glucose 6P cycle, **52:**268
effect on glycogen metabolism, **52:**268
effect on phosphoenolpyruvate carboxykinase, **52:**298
Glucocorticoids, **47:**277; 281, **54:**157, **60:**39, **67:**227
acceleration of synthesis of tryptophan oxygenase, **47:**277
adenoisne receptors, **69:**110
control of α 2u-globulin biosynthesis, **47:**300; 302
increase in nucleolar RNA polymerase 1 activity, **47:**289
modulation of gene transcription, **47:**277
phosphoenolpyruvate carboxykinase (GTP), **69:**243–244
stimulation in RNA metabolism, **47:**285
Glucocorticosteroids, tyrosine hydroxylase induction, **70:**169; 172
Glucodecafructose, **18:**288
Glucodifructose, **18:**288
Glucoerysoline, occurrence, **10:**266–267

Glucofructosane B, **17:**203
Glucogenesis, **57:**88
allosteric regulation, **28:**86–98
from carbohydrates, **3:**245–246
Glucoheptulose, **17:**210
D-Glucoheptulose, presence of a single common carrier in translocation, **29:**53
Glucoinvertase in honey, **15:**310
Glucokinase, **34:**291; 305; 307, **37:**140, **46:**246, **54:**158, **55:**360
activities, effects of fasting and refeeding with hexoses on, **39:**313
diabetes and insuline effects on, **39:**312
allosteric regulation, **28:**95; 98
correlation of hepatoma growth rate with, **29:**331
in malignant cells, **29:**334
in diabetes, **54:**164
distribution in species, **52:**259
effect of *N*-acetylglucosamine, **52:**263
effect of K^+ and Na^+ ions, **52:**261
exclusive role in glucose phosphorylation, **52:**263–264
in liver extracts, **52:**261–262
measurement of activity, **52:**260
in newborns, **52:**260
properties, **52:**259
in starvation and diabetes, **52:**260; 267
6-Glucokinase, **18:**286
C-6-Glucokinase, phosphorylation by, **15:** 145–147; 152
α-Glucomelizitase, in yeast, **22:**22
Gluconasturtin, occurrence, **10:**266
Gluconate, in carbohydrate metabolism, **15:**148; 149–150 (C^{14})
1-^{14}C-Gluconate, labeling of pentoses, **28:** 430–435
6-^{14}C-Gluconate, labeling of pentoses, **28:**410; 430
Gluconate-6-phosphate, in oxidation shunt, **15:**154–157
Gluconate-6-phosphate dehydrogenase, **15:**154; 156–157
D-Gluconate-6-phosphate dehydrogenase, PN-dependent, **26:**119
Gluconegenesis, dependence on pyridine nucleotides by, **29:**360
Gluconeogenesis, **59:**318–319
with diabetes, **54:**164
effect of glucagon
in diabetes and starvation, **52:**311
on FGP cycle, **52:**312
on mitochondria, **52:**312
hormonal control of
effect of adrenaline, **52:**314
molecular mechanism, **52:**315
effect of angiotensin, **52:**314
effect of glucagon, **52:**307–314
cross-overs, **52:**307
role of phosphoenolpyruvate, **52:**310
role of pyruvate kinase, **52:**308; 313
via cyclic AMP, **52:**307
effect of oxytocin, **52:**314
effect of phenylephrine, **52:**314
molecular mechanism, **52:**315
role of Ca^{2+}, **52:**316
effect of vasopressin, **52:**314
molecular mechanism, **52:**315
role of Ca^{2+}, **52:**316
hormonal regulation of, **54:**150; 154
inhibition by hypoglycin, **52:**271
phosphoenolpyruvate carboxykinase (GTP), **69:**205; 207; 214–218
in plants, **54:**175
regulation of glucose/glucose 6P cycle, **52:**269
role of fructose 1,6-bisphosphatase, **42:**221
starvation effect on, **54:**157
Gluconeogenic substrates, **59:**367–368
Gluconic acid
formation in molds, **9:**538–541; 544; 547–550; 558–559; 560–563; 571–572; 574–576
salts, **17:**148–149
Gluconokinase, enzymic adaptation, **13:**70
Gluconolactonase
localization of, **24:**317
in photosynthesis, **19:**15
D-δ-Gluconolactone, **15:**116
δ-Gluconolactone, **15:**145; 148–149
1,5-Gluconolactone, **48:**48; 50
δ-Gluconolactone, formation, **19:**206–209
D-δ-Gluconolactone-6-phosphate, **15:**116; 121
Gluconolactones, as β-glucanase inhibitors, **28:**347–349
Gluconyl phosphate, **15:**155
Glucopeptonic acid, slats of, **17:**149
Glucoprotein, isolation from plastids, **20:**207; 208
β-Glucopyranose, oxidation, **19:**206–209

Glucopyranose units, β-1,3 linked, cholesterol control, **28:**356
α-D-Glucopyranosidases, multiple forms, **25:**349
α-Glucopyranosyl, in teichoic acid, **30:**225
α - D - Glucopyranosyl - β - D - xyloketo-furanoside, synthesis by sucrose phosphorylase, **10:**129–131; 134–135; 138–141
β-Glucopyranosylepoxyalkanes, is inhibitors, **48:**48
O-D-Glucopyranosyl-(1→5)-D-fructofuranose, **17:**207–208
O-α-D-Glucopyranosyl-(1→4)-D-fructose, **17:**220
O-α-D-Glucopyranosyl-(1→3)-D-glucopyranose, **17:**214
O-α-D-Glucopyranosyl-(1→3)-D-glucose, **17:**220
3-α-D-Glucopyranosyl-D-glucose, **15:**313
1-*O*-α-D-Glucopyranosylglycerol, in teichoic acid hydrolysates, **30:**227
β-Glucopyranosyl isothiocyanate, as inhibitor, **48:**48
β-(2-β-D-Glucopyranosyl-3-isoxazolin-5-on-4-yl-1-alanine, **50:**147
3 [α - D - Glucopyranosyl] - L - arabino-pyranose, synthesis by sucrose phosphorylase, **10:**129–131; 136–138; 138–141
α - D - Glucopyranosyl - α - L - arabo-ketofuranoside, *See* D-Glucosyl-L-araboketoside
α - D - Glucopyranosyl - α - L - sorbo-furanoside, synthesis by sucrose phosphorylase, **10:**129–131; 132–134; 138–141
Glucopyruronic acid 1-β-phosphate, hydrolysis, **16:**372–373
β-D-Glucopyruronides, specificity of β-glucuronidase for, **16:**372–373
β-glucopyruronosidase, in rumen microorganisms, **16:**377
Glucosaccharase, **17:**198
  classification, **14:**245
Glucosaccharic acid, **17:**149
D-Glucosaccharic acid, as inhibitor of β-glucuronidase, **16:**373
β-D-Glucosaminase in snake venoms, **8:**473; 475; 476
Glucosamine, **15:**145; 146, **45:**407; 410. *See also* Amino sugars
  acetylation, **60:**271
  in glucolipids, **25:**442–443; 445–450
  linkages in glycoproteins, **25:**446–453
  linkage with galactosamine, **25:**446–450
  in prothrombin molecule, **16:**29
    in transmination, **16:**236
D-Glucosamine
  from connective tissue, **25:**434–437
  sugar nucleotides, from, **26:**49–50
L-Glucosamine, from streptomycin, **25:**433
D-Glucosamine, in formation of pyrrole-2-carboxylic acid, **20:**139; 140
Glucosamine acetyltransferase, **60:**270
  active site, **60:**271
  deficiency, **60:**271
Glucosamine deaminase, column chromatography, **22:**460; 462
Glucosamine (GN), in mureins, **26:**198; 200–204
Glucosamine in viruses, **9:**230–231
D-Glucosamine-6-P, **43:**493
Glucosamine-6-P, biosynthesis, allosteric regulation, **28:**66–69
  in carbohydrate metabolism, **28:**114–115
Glucosamine-6-phosphate acetyltransferase, **60:**270
Glucosamine 6-phosphate synthase
  acceptor domains and glutamine, **72:**104–105
  domain structure of glutamine, **72:**105–106
  structure and function, **72:**103
Glucosamine-6-phosphate synthetase, *See* D-Fructose-6-phosphate amidotransferase
Glucosamine-6-P-isomerase, **43:**494; 508–509
Glucosaminidase, action on murein, **26:**205; 210; 212; 219
Glucosaminidases, effect on hyaluronic acid, **13:**203; 210
Glucosamino 6-phosphate synthase, **66:**264–268
  E. coli, **66:**265–267
  gene cloning, **66:**264–265
  Rhizobium nodM, **66:**267–268
β-Glucosanase, **18:**288
Glucosans, in plant wilting, **11:**416–422
Glucose, **34:**525, **38:**200, **43:**223, **52:**25, **53:**210; 225. *See also* Lactate; Pyruvate
  absorption and phosphorylation, **7:**119–120
    effect on phosphorus turnover, **7:**118–125
    intermediary metabolism, **7:**127
    metabolism by yeast, **7:**15–20

permeation into yeasts, **7:**13–20
and phosphorus in yeast, **7:**189
allosteric control, **28:**86–98; 106
carbons occurring in pentose, **28:**400–401
metabolism, effect of adenine on, **28:** 408–409
biosynthesis of, **12:**43
conversion to acetate, **12:**114–116
blood, control of, **25:**173
carrier system, **34:**480
conversions in nature, **14:**206–208
metabolism and energy transport, **14:**13–17
in desoxyribonucleic acid, **17:**463
in transglycosylations, **17:**196–200; 202; 204–220; 222–229
dissimilation by molds, **8:**272–274; 294–295
linkage in cellulose, **8:**255–259
phosphorylation, **8:**51; 403
relation to cellulose, **8:**255–264; 267–269; 271–273
effect of superoxide dismutase production, **58:**83
effect on glucose/glucose 6P cycle, **52:**268
effect on glycogen metabolism, **52:**268
and erythrocyte synthesis of nicotinamide mononucleotide, **15:**103
in ATP trapping agent, **15:**381–382
fermentation by thiaminase-producing bacteria, **15:**413
in α-ketoglutarate oxidation system, **15:**225; 229; 230–231
metabolism by alternate pathways, **15:**144; 175–177 (C^{14})
in phosphogluconic fermentation shunt, **15:**166–167 (C^{14})
in polysaccharide synthesis, **15:**305; 306; 307; 310–315
facilitated diffusion of, **20:**96
in biosynthesis of glutamic acid in plants, **20:**325
in biosynthesis of pectic substances, **20:**369
formation and conversions in living organisms, **20:**118; 120; 121
optical antipodes in living organisms, **20:**2
in studies of biosynthesis of pectic substances, **20:**369
in familial periodic paralysis, **6:**217
formation from fructose, **1:**152
in fructose-2, 6-6bisphosphate regulation, **54:**154
in gluconeogenesis and glycolysis regulation, **54:**154
as inducer and complexant of β-glucosidase, **22:**118
inhibitor of β-galactosidase synthesis in *E. coli,* **22:**112
mutarotation by *O*-hydroxyproline, **22:**82; 83
inhibition of tryptophanase, **10:**69; 74–81
reaction with cysteine, **10:**276–277
and K^+ permeability, **13:**62–63
labeled, in plant respiration, **18:**290
UDP ester of, **18:**201
metabolism, formyl derivatives of THFA in, **21:**223; 224
in muscular dystrophy, **6:**207; 222
oxidase, **34:**448–449; 452; 456; 480
phosphorylatin
glucose 6-phosphate, necessary step, **52:**260
role of glucose 6-phosphatase, **52:**260
presence of a single common carrier in translocation, **29:**53
tolerance in depression, **29:**497–498
utilization in manic and depressed patients, **29:**494
rate of production of, **54:**160
in sprem metabolism, **24:**61–63; 66–68
and steroid 11β-hydroxylase activity, **19:**189
in synthesis of proline by *E. coli,* **16:**259
labeled, in synthesis of aromatic amino acids, **16:**290–291; 294
in synthesis of leucine, **16:**286
oxidation in cell nuclei, **16:**474
Glucose-1, **15:**6–6diphosphate; 141; 151–152; 169. *See also* Hexose diphosphate
6-6diphosphate, and phosphoglucomutase activity, **20:**127
D-Glucose
in lipopolysaccharide from *Salmonella,* **31:**79
pathway of metabolism of, in enteric bacteria, **31:**81
nucleotides of, **26:**35–88
sugar nucleotides from, **26:**44–49
α-Glucose, action of mutarotase on, **15:**145
oxidation, **15:**148–149; 150–151
β-Glucose, action of mutarotase on, **15:**145
oxidation, **15:**148–149; 150–151
Glucose acceptors, in di- and polysaccharide synthesis, **10:**138–141

Glucose-ATP phosphotransferase, in neoplasia, **29:**334
Glucose balance, **59:**359–360
D-Glucose-1-^{14}C, **34:**217
D-Glucose-6-^{14}C, **34:**217
Glucose (C^{14})
and citric acid cycle, **18:**272–274
in lignin study, **18:**364–366
in coenzyme Q biosyntheis, **25:**298
in product inhibition study, **25:**234
in genetic studies, **17:**425
in transglycosylations, **17:**196; 198–200; 204; 209; 210; 212–214; 219; 224
in sperm metabolism, **24:**48; 74
in synthesis of starch, **26:**62
Glucose (C-14), as source of carbon for amino acids, **16:**227–228
[2-^{3}H] Glucose, **52:**251–253
detritiation, **52:**251; 252
in hypoglycin poisoning, **52:**271
measurement of glucose cycle, **52:**251
measurement of glucose phosphate, **52:**251
measurement of glucose phosphorylation, **52:**262; 282
phosphoglucose isomerase, **52:**251
in type I glycogenosis, **52:**269–272
[3-^{3}H] Glucose, **52:**253–255
detritiation and enzymes, **52:**253
measurement of F6P cycle, **52:**254; 283
measurement of glycolysis, **52:**282
measurement of phosphorylation of Fructose 6P, **52:**282
[4-^{3}H] Glucose, **52:**253–255
detritiation and enzymes, **52:**253
[5-^{3}H] Glucose, **52:**253–255
detritiation and enzymes, **52:**253
measurement of fructose 6P cycle, **52:**253
measurement of 6P phosphonyl, **52:**253
transaldolase role in detritiation, **52:**253
[6-^{3}H] Glucose
complete recycling of glucose, **52:**255
Cori cycle, **52:**255
detritiation, **52:**251; 257
detritiation and enzymes involved, **52:**255
measurement of glycolysis, **52:**282
measurement of glycolytic flux, **52:**255; 258
measurement of pyruvate kinase flux, **52:**258
[1-^{3}H] Glucose, detritiation, **52:**251
Glucose-1-C^{14}, distribution of isotope in amino acids, **16:**269; 290; 301
Glucose (^{14}C-1 and ^{14}C-6), in plants, **18:**295
Glucose catabolism, calculation of pathways, **28:**424
Glucose dehydrogenase, **7:**418, **9:**561, **15:**116; 145; 148–150; 200, **35:**264, **37:**496
classification, **14:**253; 254
and pyrrolo-quinoline quinone, **59:**173; 185; 194
and quinoprotein oxidoreductases, **59:** 189–190; 197; 200; 207–208
Glucose derivative of coenzyme I, **15:**31
Glucose 1,6-diphosphate, and phosphoglucomutase, **19:**8
Glucose-1,6-diphosphate, classification as coenzyme, **14:**241
role in phosphate migrations, **14:**208–213
Glucose donors, for sucrose phosphorylase, **10:**138–140
in di- and polysaccharide synthesis, **10:** 138–141
Glucose effects, **22:**121
Glucose fermentation, and lipoic acid, **18:**342
Glucose-galactose interconversion, **18:**208
D-Glucose-G-^{14}C, **34:**216
Glucose-glucose oxidase system, H_2O_2-generating, **27:**479
Glucose/glucose 6-phosphate cycle
extent of recycling, **52:**266
in hapatocytes, **52:**266
in perfused liver, **52:**266
in sheep, **52:**267
starvation, **52:**266
in type I glycogenosis, **52:**267; 269–272
in vivo, **52:**266
measurement by
[2-^{3}H] glucose, **52:**265
[5-^{3}H] glucose, **52:**265
occurrence in different species, **52:**265–267
physiological significance, **52:**267–269
effect of glucocorticoid, **52:**268
effect of glucose, **52:**268
effect of K^{2}-ions in hepatocytes, **52:**268
in gluconeogenesis, **52:**269
in glycogenolysis, **52:**269
in glycogen synthesis, **52:**268
in pentose phosphate pathway, **52:**269
role in glucose 6 phosphate, **52:**268
in synthesis of glycogen, **52:**267

1-^{14}C-Glucose, in DR synthesis, **28:** 423–426
labeling of pentoses, **28:**426–431; 433–435; 454
in ribose synthesis, **28:**405–406; 408–411; 414
Glucose inhibition, in acetate-dependent *Neurospora,* **17:**423–425
2-^{14}C-Glucose, labeling of pentoses, **28:** 426–431; 434–435
ribose synthesis, **28:**407–411; 414–417
3,4-^{14}C-Glucose, labeling of pentoses, **28:**405; 414–416; 426
6-^{14}C-Glucose, labeling of pentoses, **28:**409; 411; 414; 426–431; 433–435
^{14}C-Glucose, labeling of pentoses, **28:**419; 426–431; 434–435
oxidation, **28:**117
pentose synthesis, in tumor-bearing rats, **28:**431–433
phosphorylation of, **28:**95–96
Glucose metabolism, and fluoroacetate, **18:** 144–146
Glucose Metabolism (Racker), **15:**141
Glucosemonophosphatase, **5:**23
Glucose output, **52:**266
Glucose oxidase, **7:**300, **9:**356; 560–563; 572; 574–576, **15:**145; 148; 150–151; 306, **17:**77, **37:**424; 474; 512, **46:**475
in electron transfer, **19:**201
mechanism of action, **19:**207–210
properties, **19:**202; 206
in laminarinase assay, **28:**344
peroxidase action on, **14:**155; 157–158
Glucoseoxidase, chemical modification of, **24:**140
Glucose oxidase, EPR study, **27:**118; 183
Glucose oxidase, PN-dependent, **26:**121
Glucose oxidation
and lipoic acid, **18:**342
stimulation by dinitrophenol, **29:**345
^{14}C-Glucose precursors, in ribose synthesis, **28:**416
Glucose-1,6P_2, **51:**219
Glucose-1-P, **51:**219
Glucose-6-P, **51:**219
D-Glucose-6-P, **43:**493
Glucose-1-P concentration, **46:**360
Glucose-6-phosphatase, **15:**153, **53:**221, **54:**10
allosteric regulation, **28:**77; 95–98; 122
and blood sugar control, **25:**173
effect of ethionine on, **25:**380
product inhibition of, **25:**197; 224; 232–235; 237; 239
classification, **14:**245
correlation of hepatoma growth rate with, **29:**331
deficiency of, **54:**166
in diabetes, **54:**164
kinetic mechanism of, **45:**304
localization of, **24:**316; 320; 322; 338
nuclear, **27:**251–253
starvation effect on, **54:**158
Glucose 6-phosphatase, **34:**298–311; 474–475
activities, of glycogen synthetase by, **34:** 324–325
of mammalian liver glycogen synthetase by, **34:**337–338
after glucocorticoids, **52:**265
deficiency of, **34:**309
discussion of multiple activities associated with, **34:**301–304
distribution of, **34:**301–304
effect, of detergents on, **34:**300
of Mg^{2+} on phosphotransferase activities of, **34:**308
in fatal liver, **52:**265
fate in liver-slices, **34:**301
in glycogen biosynthesis, **34:**294
in G6P synthesis, **34:**304–309
identity with pyrophosphate phosphotransferase, **34:**302
inhibition by ATP, **34:**304
inhibition in hypoglycin poisoning, **52:**271
inhibitors, **52:**265
inhibitors of, **34:**309–310
intestinal, **34:**311
kidney, **34:**310–311
membrane-bound nature of, **34:**299
phospholipid requirement, **34:**299
pools, **34:**295; 296
properties, **52:**264
in pseudo type I glycogenosis, **52:**271
purification and assay, **34:**299
reactions catalyzed by, **34:**299
requirements of glycogen synthetase, **34:**317; 322
role of glucose 6 phosphatase in synthesis of, **34:**304

Glucose 6-phosphatase (*Continued*)
in starvation and diabetes, **52:**265
thyroxin, **52:**265
as a transferase, **34:**298; 302
two components, **52:**264
in type I bis glycogenosis, **52:**271
in type I glycogenosis, **52:**269–272
Glucose-6-Phosphatase: Two Concepts of Membrane-Function Relationship (Sukalski and Nordlie), **62:**93
Glucose-6-Phosphatase Dehydrogenases (Levy), **48:**97
Glucose-L-phosphate, **34:**480
Glucose-1-phosphate, **5:**131, **15:**141; 147–148; 151–152; 167; 168–170; 245, **43:**226. *See also* Phosphate, Phosphorylase, and Phosphorylated sugars
allosteric regulation, **28:**104–106; 110–112
chelates, in metal ion activation of enzymes, **21:**148
conversions in living organisms, **20:**112; 122; 123
conversion in plastids, **20:**213
dismutase, **34:**294
kinase, **34:**294
synthesis in yeast, **34:**298
enzymic hydrolysis, **10:**89
use in synthesis of sucrose, **10:**123–143
as phosphoryl donor, **17:**167; 169
in polysaccharide synthesis, **11:**298–299; 299–306; 309; 316; 317–318; 321; 324–326; 329; 330; 348–358; 366–367; 370–373, **15:**309 (C^{14})
Glucose-6-phosphate, **5:**131, **15:**145–147; 163; 167; 174, **55:**366–367, *See* Hexose phosphates
allosteric regulation, **28:**48; 74–78; 87; 95–103; 106; 108–110; 116–118
in pentose metabolism, **28:**394–395
and control of hexokinase action, **25:**170
formation of, **25:**246–249
in photosynthesis, **25:**102
enzyme for formation, **27:**361
formation, in living organisms, **20:**113
in plastids, **20:**213
hydrolysis, by metal ions, **21:**184
oxidation cycle, **15:**149–150; 164–166
oxidation of, **14:**207–208
oxidation shunt, **15:**153–166; 175–178
from phosphorylating enzyme, **15:**230–231
and pyridine nucleotides, **15:**113; 121
relations to other hexose monophosphates, **15:**168–170
β-Glucose-1-phosphate, **17:**223
Glucose 1-phosphate
phosphomannomutase/phosphoglucomutase (PMM/PGM), **70:**230
in plants, **19:**7
Glucose 6-phosphate
and liver nonspecific hydroxylase, **19:**175
in photosynthesis, **19:**5; 7; 14; 15; 63
metabolism, **16:**488
phosphomannomutase/phosphoglucomutase (PMM/PGM), **70:**230
Glucose 1-phosphate (C^{14}), **18:**287
Glucose-6-phosphate concentrations, **44:**249; 250
Glucose-6-phosphate dehydrogenase, **15:**82; 97; 116; 154–156; 165–167, **17:**449, **37:**424, **46:**233; 240; 271
allosteric regulation, **28:**116–117
in DR metabolism, **28:**443; 444
enzyme coupling, **28:**167
ribose synthesis, **28:**411
column chromatography, **22:**474; 475
ethionine effects, **25:**380–381
in formation of $NADPH^3$, **25:**401
in luminous bacteria, **25:**151
multiple forms, **25:**347
levels in cyanate treated red cells, **40:**17
localization of, **24:**297; 304; 308
nuclear-cytoplasmic ratios, **27:**262
and pyruvate carboxylase, **26:**306
PN-dependent, **26:**119; 160
Glucose-6-phosphate-dehydrogenase, **34:**456; 520
correlation of hepatoma growth rate with, **29:**331
induction of, **29:**372
in latex, **29:**225–226
Glucose 6-phosphate dehydrogenase
in cell nuclei, **16:**468
in photosynthesis, **19:**15
in plants, **18:**293–295
Glucose-phosphate dehydrogenase, and CO_2 fixation, **12:**55
Glucose-6-phosphate dehydrogenases, **48:** 97–191
acetyl-CoA, effect of, **48:**170

acetyl-CoA thioesters, long-chain, effect of, **48:**172–174
ACTH, effect of, **48:**177
adaptive hyperlipogenesis and, **48:**99
adenocarcinoma and, **48:**127; 128
ADP, effect of, **48:**166; 167; 169
aggregation to tetramers and larger oligomers, **48:**117–118
aging and, **48:**130; 177
aging and erythroid cell differentiation and, **48:**129
amino acid composition, **48:**108; 157–159
  C-terminal amino acids, **48:**110
  N-terminal amino acids, **48:**110
AMP, effect of, **48:**166; 167; 168; 169
ascorbate, effect of, **48:**177
ATP, effect of, **48:**160; 165–170
binding constants, **48:**145–146
borate, effect of, **48:**152; 154
catalytic features, **48:**131–159
cellular location, **48:**104
classification, **14:**235; 254
clinical diagnostic procedures, **48:**101
coenzyme A, effect of, **48:**170
coenzyme concentrations and ratios, effects of, **48:**161–165
conformational isomers, **48:**127
deficiency of, **48:**100
deinhibition mechanisms, **48:**160; 161; 162; 166
diabetes and, **48:**99
dissociation to monomers, **48:**118–119
divalent cation requirement, **48:**138–140
electron microscopy, **48:**122
enzyme immunoassays, **48:**101
equilibrium constant, **48:**131
erythrose-4-phosphate, effect of, **48:**171
forms of, **48:**123
  genetic variants, **48:**123; 126
glutathione, effect of, **48:**161; 162; 166
glyceraldehyde-3-phosphate, effect of, **48:**171
GTP, effect of, **48:**167
hemoglobin, effect of, **48:**177
instability of, **48:**104
interconversion of dimers and tetramers, **48:**117
ionic strength, effects of, **48:**152–155
isolation, **48:**102–108
kinetic constants, **48:**140–145
kinetic mechanisms, **48:**147–152
lactation and, **48:**99; 127; 160; 177
metabolic intermediates, **48:**171–172
Mg^{2+}, effect of, **48:**138; 165; 166; 167; 168; 169; 170
modification of
  chemical, **48:**157
  post-translational, **48:**129
modifying factor, **48:**104; 130
molecular weights of dimers and tetramers, **48:**121–122
NADH, effect of, **48:**160; 164; 165
$NADP^+$, structural in, **48:**120
NADPH, effect of, **48:**160; 161; 162; 163; 164; 166
NADPH/$NADP^+$ ratio, effect of, **48:**160; 162; 163; 164
non-amino acid components, **48:**108–110
nucleotides, effect of, **48:**165–170
palmitoyl-CoA, effect of, **48:**173; 174
pH, effects of, **48:**152–155
phosphates, effects of, **48:**177
phosphoenolpyruvate, effect of, **48:**160; 171
physical properities, **48:**124–125
polymers of single subunit, **48:**127
pregnancy and, **48:**99; 127; 160; 177
preneoplastic hyperplastic alveolar nodules and, **48:**127
regulatory features, **48:**159–177
reversible inactivation, **48:**119
ribulose-1,5-diphosphatem effect of, **48:**171; 172
sigmoid kinetics, **48:**151; 152; 161
source, **48:**102–104
specificity, **48:**132–138
  anomeric, **48:**132
  coenzyme, **48:**137
  substrate, **48:**133–137
stereospecificity of hydrogen transfer, **48:**137
steroids, effect of, **48:**174–177
stoichiometry, **48:**145–146
structure
  forces involved in maintaining, **48:** 119–120
  minimum, for catalysis, **48:**116–117
  models for, **48:**122–123
  primary, **48:**108
  quaternary, **48:**116–123
  secondary, **48:**115–116
  tertiary, **48:**116–123
subunit, identity, **48:**111–112

Glucose-6-phosphate
dehydrogenases (*Continued*)
subunit molecular weight, **48:**112–115
temperature, effects of, **48:**152–155
thermodynamics, **48:**131–132
thyroid hormones, effects of, **48:**177
X-chromosome and, **48:**100; 126
β-D-Glucose-1-phosphate from phosphorylation of maltose, **15:**315
Glucose-6-phosphate-fructose-6-phosphate, transformation, **14:**205–206
Glucose-1-phosphate-galactose-1-phosphate transformation, **14:**195; 197–204
Glucose-1-phosphate-glucose-6-phosphate transformation, **14:**195; 209–213
α-D-Glucose-1-phosphate in polysaccharide synthesis, **15:**304–306; 307; 309; 311
D-Glucose-6-phosphate isomease, carbohydrate synthesis, **26:**68
Glucosephosphate isomerase, **37:**546
localization of, **24:**320; 333; 335
Glucose-6-phosphate isomerases, classification, **14:**237; 258
Glucose-1-phosphate kinase, **15:**151–152
Glucose-6-phosphate mutarotose, **44:**90
Glucose-1-phosphate r amylose transglucosidase, **14:**251; 252
Glucose-1-phosphate r glucose-1-phosphate-(6) transphosphatase, **14:**251
Glucose-6-phosphate r TPN transhydrogenases, **14:**235; 254
Glucose phosphates
isomerization of, **14:**195; 197–213
in plant respiration, **18:**285–287; 289; 292–295; 312
Glucose phosphates and insulin, **8:**413–414
Glucose phosphates in citric acid cycle, **10:** 211–212
Glucose-1-phosphate transphosphorylase, **15:**151–153
Glucose-1-phosphate-transphosphorylase, classification, **14:**251
Glucose-1-phosphate uptake, **44:**242
Glucose phosphorylation
correlation of growth rate of hepatomas with, **29:**331
exclusive role of glucokinase, **52:**263–264
in extracts of hepatocytes, **52:**261
glucokinase, **52:**259
hexokinase, **52:**259
in intact hepatocytes, **52:**262–263
effects of ions, **52:**262
kinetic characteristics, **52:**262
in liver extracts, **52:**261–262
role of glucose 6-phosphatase, **52:**263[
role of hexakinases, **52:**261
Glucose phosphotransferase system, **55:**362–364
Glucose P isomerase, **43:**491; 497; 502
mutarotation by, **43:**504–508
Glucose-pyruvate-lactate relations, **5:**139
Glucose r DPN transhydrogenases, **14:**254
Glucose r O_2 transhydrogenases, **14:**253
Glucose sulfatases, classification, **14:**245
Glucose-T, as acceptor in dextran formation, **30:**199
Glucose transfer, oligosaccharide formation from, **17:**205–220; 224–229
Glucose transferring agent in di- and polysaccharide synthesis, **10:**138–141
Glucose transporters, adenosine receptors, **69:**114
Glucose uptake, **52:**266
Glucose uptake system, *Caulobacter* cell cycle, **71:**5–6
α-Glucosidase, **17:**198, **37:**496, **40:**242
assay of, **56:**188
glucoamylase and, **56:**192
induction in yeast, **22:**113; 117
and protein turnover rate, **22:**127
ultraviolet sensitivity, **22:**138
inhibition of, **56:**194
mechanism of action of, **56:**194
occurrence and localization of, **56:**187–188
pH effect on, **56:**191
properties of, **56:**189–194
purification of, **56:**188–189
regulation of, **56:**195
role of, **56:**194–195
in soil, **28:**375
in test for cell damage, **28:**265–266
substrate specificity of, **56:**193
in yeast, **22:**106; 107
β-Glucosidase, **8:**271
β-Glucosidase, **17:**219–220, **60:**115
amino sugar study, **25:**437
multiple forms, **25:**350
assay of, **56:**197
column chromatography, **22:**472–474
induction in yeast, kinetics, **22:**114–116
stereospecificity of inducer, **22:**119

in yeast, **22:**106
constitutive mutants, **22:**103
constitutive synthesis, **22:**103
inducers, complexants of, **22:**118; 119
inhibition of synthesis, **22:**115; 121
properties, **22:**104
ribosomal-bound material, **22:**146
induction of, **30:**128
properties of, **30:**128
reaction catalyzed by, **30:**128
inhibition of
by D-fucono-, **36:**172
by galactonolactone, **36:**172
by glucal, **36:**178
by glucono-, **36:**172
by glucono-1:5-lactam, **36:**179
by glucono-1:4-lactone, **36:**161
by glucono-1:5-lactone, **36:**161
by L-idono-1:5-lactam, **36:**179
by nojirimycin, **36:**179
mechanism of action of, **56:**201–203
occurrence and localization of, **56:**196–197
pH effect on, **56:**199–201
properties of, **56:**198–203
protein cofactors, **60:**185
purification of, **56:**197–198
regulation of, **56:**205
relative activity of, based on substrates, **56:**200
role of, **56:**203–205
specificity of, **56:**199
β-xylosidase activity, **60:**276
χ-Glucosidase, inhibtion of, by glucal, **36:**178
inhibitory power, of glucono-1:4-lactone, **36:**156
of glucono-1:5-lactone, **36:**156
Glucosidase, in tobacco, **6:**368
α-D-Glucosidase, isolation, **14:**401
separation by chromatography, **14:**349
β-D-Glucosidase, isolation, **14:**401
separation by chromatography, **14:**349
β-1,6 Glucosidase, of laminarinase complex, **28:**345
β-Glucosidase (emulsin), cacao, **25:**466
Glucosidase fractionation experiments, **28:**352
Glucosidases, **4:**47–64; 70–71
α-1,4-Glucosidases, **12:**409
α-1,6-Glucosidases, **12:**409–410
α-Glucosidases, **48:**42
β-Glucosidases, **48:**43; 44; 47; 50; 51; 58
from almond emulsin, **4:**48–61
from other sources, **4:**62–64
α-Glucosidases, classification, **14:**245
β-Glucosidases, classification, **14:**232; 246
structure studies, **14:**303
α- and β-Glucosidases, in sperm, **24:**53
β-Glucosidases (β-D-glucoside glucohydrolases), **28:**341
acting on barley β-glucan, **28:**341–342
adsorption of, **28:**31
inhibitors, **28:**348
from laminarinase system, **28:**353
role in laminarinase action, **28:**345
in soil, methods, **28:**374–375; 379–380
results, **28:**387
α-Glucosidases (maltase, trehalase), cacao, **25:**466
β-1,2 Glucoside, interchain, **28:**335
β-1,6 Glucoside linkages, in β-glucans, **28:** 331–338
β-glucosidase for, **28:**345
α-Glucoside permease, in yeast, **22:**106; 109
Glucosides, **56:**202; 204
hydrolysis of, **12:**381–382
β-Glucosides
classification, **4:**70–71
distinction from α-glucosides, **4:**47
hydrolysis
and chain length and structure of aglycon group, **4:**54–61
and substitution of aromatic aglycon group, **4:**59–61
and substitution of glycosyl radical, **4:** 48–53
as inducers and complexants of β-glucosidase, **22:**118; 119
β-D-Glucosides
characteristics of induction of, **30:**128–129
enzymes and their structural determinants, **30:**128
system of utilization of, **30:**127–129
β-1,6 Glucosides, action of β-glucanase, **28:**349
of laminarin, **28:**345
β-D-Glucosides, in polysaccharide synthesis, **15:**306
α-D-Glucosides in polysaccharide synthesis, **15:**306
α-Glucosides in yeast, **22:**106
Glucosidic bonds, priming agents for synthesis of, **12:**413–414; 416; 417
splitting and synthesis by phosphorylase, **12:**413–414

α-1,4-Glucosidic linkage, action of α-amylase on, **12:**386; 388–389
action of β-amylase on, **12:**391; 394
α-1,6-Glucosidic linkage, action of α-amylase on, **12:**391
action of isophosphorylase on, **12:**417–419
(1r3)-α-Glucosidic linkages, **15:**310; 312–313
Glucosidic linkages, mixed, formation of, **28:**340
α-D-Glucosidoarabinose, **18:**222
α-D-Glucosidoketoarabinoside, **18:**222
α-D-Glucosidoketoxylose, **18:**222
Glucosidoronic acid, *See* Glucuronide
α-D-Glucosidosorbofuranoside, **18:**222
α-D-Glucosidoxylose, **18:**222
Glucosucrase, enzymic adaptation, **13:**108
Glucosulfatase
classification, **14:**245
effect of borate on, **12:**518
function, **22:**229
occurrence, **22:**228
specificity, **22:**228
in plants, **10:**311
relation of *p*H to activity, **10:**84
Glucosyl acceptors, stereochemical relationships, **18:**223–224
D-Glucosyl-L-araboketoside, synthesis by sucrose phosphorylase, **10:**129–131; 135; 138–141
Glucosylating enzymes, of phage-infected *E. coli,* **26:**75
Glucosyl disaccharides, in plants, **18:**287
Glucosyl donors, allosteric regulation, **28:** 110–114
Glucosyl-fructose (1→3), **17:**227
1-*O*-β-Glucosylmannitol, and laminarinase action, **28:**345; 352
β-Glucosyl mannitol, hydrolysis by β-glucosidase, **28:**345
from laminarin, **28:**350
Glucosylsphingosine, **60:**125
Glucosyltransferase, **52:**146
Glucosyltransferases, in polysaccharide synthesis, **28:**340
α-Glucosyl transfers, **17:**219–220; 226–227
β-Glucosyl transfers, **17:**217–218; 227
Glucotropaolin, occurrence, **10:**266
D-Glucovanillin, **18:**372
incorporation into lignin, **16:**149
Glucozymase, enzymic adaptation, **13:**81
β-Glucurinidase, inhibtory action of galactaric acid, **36:**155
Glucuronate, **17:**147–148
Glucuronate pathway to ribose, origin and numbering of carbon atoms, **28:**399
Glucuronate-2-sulfate sulfatase, **60:**256
Glucuronate-xylulose pathway, in ribose synthesis, **28:**394
D-Glucurone, as inhibitor of β-glucuronidase, **16:**373; 374
Glucuronic acid, **46:**65, **52:**25, **60:**47; 226
formation of pentose, **28:**398–399
as inhibitor of β-glucuronidase, **16:**373; 374; 403
UDP ester of, **18:**201
D-Glucuronic acid
conversion by *Escherichia coli,* **20:**371
conversions in nature, **14:**206–208
D-Glucuronic acid, of connective tissue, **25:** 434–438; 440; 441
Glucuronic acid conjugates. *See also* β-Glucuronidase and β-Glucuronides
of drugs, hormones, etc., **16:**364
Glucuronic acid-1-phosphate
in formation of glucuronides, **16:**403
preparation of, **14:**207
α- and β-Glucuronic acid 1-phosphates, action of β-glucuronidase on, **16:**372
Glucuronic acid transferase activity of β-glucuronidase, **16:**402; 403
β-Glucuronidase, **16:**361–409
intracellular distribution, **16:**464; 466–468
β-Glucuronidase, **17:**148; 149, **40:**40, **48:**82, **60:**219; 235
activity at interfaces, **28:**3; 27
amino acid composition, **60:**236
assay, **60:**238
biological role, **60:**240
biosynthesis, **60:**237
carbohydrate composition, **60:**236
catalyzed reactions, **60:**238
fixation by protein, **60:**237
genetics, **60:**240
from human placenta, **60:**236
hyaluronate substrate, **60:**239
in hydrolysis, **60:**240
inactivity, **60:**240
inhibition of
by glucaric acid, **36:**152

by glucaro-1:4-lactone, **36:**155
isoelectric point, **60:**236
kinetic forms, **60:**238
localization of, **24:**323; 344
lysosomal, **60:**236
maturation, **60:**237
microsomal, **60:**237
in mitochondria, **21:**39
molecular size heterogeneity, **60:**236
molecular weight, **60:**236
purification, **60:**236
substrates, **60:**239
subunits, **60:**236
in transglycosylation, **60:**239
β-D-Glucuronidase, **48:**62
β-*Glucuronidase* (Fishman), **16:**361
β-Glucuronidase induction, in *E. coli,* **22:**119
β-Glucuronidases, **56:**220–221
β-Glucuronidases, classification, **14:**246; 249
sources, **14:**376; 405
Glucuronidases, relation to hyaluronidase action, **13:**203; 208; 210; 216
Glucuronide, biosynthesis, **20:**122
labeled, in studies of galactosemia, **20:**125
β-Glucuronide-*N*-acetylglucosamine, in action of β-glucuronidase on, **16:**371
β-Glucuronide-6-galactose, in action of β-glucuronidase, **16:**371
β-Glucuronides, characterization by β-glucuronidase, **16:**383
Glucuronolactonase activity, localization of, **24:**318
Glucuronolactone, as inhibitor of glucuronide formation, **16:**403
Glucuronolactone reductase, localization of, **24:**295; 299
Glucuronosylation, **57:**258
Glucuronte-xylulose cycle, **26:**39
Glucuronyl transferase, **26:**59–60
Glucuronyltransferase, **52:**83
Glu-Ile-Leu-Asp-Val-Pro-Ser (EILDVPS) sequence, fibronectin, **70:**12–13
Glu-841 residue, carbamyl phosphate synthetase, **72:**129–131
Glu-38 residue, dihydrodipicolinate reductase, **72:**295–297
Glu-100 residue, glutamine-dependent nitrogen transfer, **72:**179–183
Glutamate, **15:**124–125, **35:**489, **46:**42, **49:**50, **50:**145; 149; 152; 156; 157, **56:**315, **58:**81; 178
and acetate, **23:**405; 419–420; 426–427; 429; 432–434
and glycine, **23:**457
and glycolate, **23:**435; 437; 440
structure studies, **23:**24
active site, **43:**512
conjugation with hydropteroic acids, **27:**424; 426
conversion to β-methylaspartate, mechanism of reaction, **26:**353–354; 363–365
decarboxylation of, **35:**84
enzymic addition to dihydropteroate, **35:** 47–48
and methionine derivatives, relationships between, **31:**209
oxidation of, **24:**340
in photosynthesis, **19:**13; 52–53
in racemization of alanine, **14:**195
racemization of, **14:**195; 196
reactivation of choline, **6:**261
substrate for glutamate acetyltransferase, **40:**69
as substrate for glutaniate dehydrogenase, **41:**24
in synthesis of active formate, **21:**411
radioactive, in study of carotenogenesis, **21:**340
D-Glutamate, **43:**549; 552
in synthesis of glutamine, **19:**253
in transaminations, **16:**206
synthesis of D-glutamine from, **16:**257
L-Glutamate
attachment to active site of glutamine synthetase, **31:**212
stereochemical structure of, **31:**215
conformation of, **39:**389
effect on glutamate dehydrogenase-NADPH complex, **39:**374; 389
L-[U-^{14}C] Glutamate (6mM), **53:**215
Glutamate, CO_2 fixation in, **12:**132
Glutamate-alanine transaminase preparations, **16:**189; 207–209; 211–213; 222–223; 228
Glutamate-amino acid reactions, **16:**203–204; 207–209; 222–223; 228; 229; 231
L-Glutamate and brain tissue permeability, **13:**62–63

Glutamate antagonists, glutathione synthesis, **73:**258–259
Glutamate as inhibitor of threonine formation, **16:**271
biosynthesis of related amino acids, **16:** 257–268
effect, on blood coagulation, **16:**88
on incorporation of acetate into amino acids, **16:**286
in enzymic transaminations, **16:**186; 189; 191–193; 199–201; 203–209; 211–213; 215–218; 221–223; 226–229; 231; 233–236
formation, **16:**254–257
in histidine metabolism, **16:**299; 301
in nonenzymic transaminations, **16:**225
N-terminal residues of fibrinogen, **16:**37; 41–42
in purine synthesis, **16:**451
racemase for, **16:**250
reaction with quinones, **16:**114
from tricarboxylic acid cycle, **16:**249; 252; 254–255
Glutamate-aspartate-alanine transamination, **16:**207
Glutamate-aspartate transaminase, **16:**189; 207–209; 211–213; 215–218; 221–223; 225; 228
Glutamate (1-C^{13}), in transamination, **16:**223
Glutamate (4-C^{14}), methylaspartate mutase reaction, **26:**363–364
Glutamate carboxyl, **58:**5
D-Glutamate cyclotransferase, **43:**523; 549; 550
physiological function of, **43:**551
L-Glutamate cyclotransferase, **43:**523
Glutamate (γ-D), methylaspartate mutase reaction, **26:**364–365
Glutamate decarboxylase, **46:**477, **49:**37, **50:**6, **55:**235, **74:**133; 148; 298
evolution of, rate of, **74:**158; 159
groups *a* and *b*, **74:**152; 159
species distribution of, **74:**158–159
Glutamate dehydrogenase, **37:**409; 459; 557, **41:**1; 2; 24, **46:**241; 472–474, **47:**84–85; 136; 138, **50:**157, **53:**224, **74:**298–299
allosteric regulation, **28:**54; 55; 65; 139–144
chemical mechanism of, **45:**377
kinetic mechanism of, **45:**316; 318; 320
connectivity scheme for ligand interactions, **39:**390
-NADPH binding site, **39:**372
steady-state kinetic features of the reaction, **39:**395
transient-state kinetic features of the reaction, **39:**396; 400
deficiency, **41:**2
diaminopimelate dehydrogenase, **72:**319–320
electron microscopy of, **33:**102–103
location in subcellular fraction, **24:**301; 306
and nuclear purity, **27:**249; 252–253
L-Glutamate dehydrogenase, **43:**523, **46:**271
Glutamate dehydrogenase complexes, **39:**371
-3-acetylpyridine-ADH, **39:**373; 381; 404
-adenosine diphosphate, **39:**382; 383; 384; 387; 405; 411
-adenosine diphosphate-L-leucine ternary, **39:**387
effects of ligands on, **39:**391
enzyme reduced coenzyme, **39:**371; 382
-L-glutamate, **39:**374
-glutarate, NADP, **39:**393
involving oxidized coenzyme, **39:**392
-α-ketoglutarate binary, **39:**376
-L-leucine binary, **39:**385
-NADH binary (regulatory site), **39:**384; 409
-NADP binary, **39:**393
NADPH, **39:**371; 372; 374; 375; 381; 385; 387; 388; 405; 414
-NADPH-L-glutamate ternary, **39:**388; 401; 403; 408
-NADPH-guanosine triphosphate, **39:**387; 413
-NADP-α-ketoglutarate, **39:**393
Glutamate Dehydrogenase-Ligand Complexes and Their Relationship to the Mechanism of the Reaction (Fisher), **39:**369
Glutamate dehydrogenase NADP, **37:**451; 492
Glutamate dehydrogenase NAD(P), **37:**417; 425; 431; 440; 464; 470; 475; 492; 513; 547; 575; 576
Glutamate-dehydrogenases, multiple forms, **25:**360–361; 363–364
Glutamate-dihydropteroate condensing enzyme, **27:**424; 426
Glutamate-isoleucine-leucine-valine, system of transamination, **16:**229
Glutamate moiety of folate, eukaryotic dihydrofolate reductase, **70:**30; 37

Glutamate (N^{15}), in transamination, **16:**193; 223; 226
Glutamate-oxalate-transaminase, multiple forms, **25:**360
Glutamate/proline auxotrophs of *E. coli,* **16:**259; 264
Glutamate-pyruvate transaminase (alanine aminotransferase), cobamide enzyme study, **26:**364
Glutamate racemase, **44:**100, **74:**176
  and cobamide coenzyme, **26:**364
  hydrogen exchange, **44:**111
Glutamate semialdehyde, **56:**134
Glutamate synthase, **50:**156; 157, **66:**268–276, **72:**112–117, **74:**299
  Azospirillum brasilense, **66:**275–276
  E. coli, **66:**271–275
  [Fe–S] and flavin components, **72:**114–116
  gene cloning, **66:**268–271
  glutamine domain, **72:**116–117
  Klebsiella aerogenes, **66:**271–275
Glutamate synthetase, characteristics of reaction, **39:**156
  control of synthesis of, **39:**156; 159
    relationship to levels, of glutamate dehydrogenase, **39:**157
      of glutamine synthetase, **39:**157
  nitrogen fixation, role in, **39:**159
  properties of enzyme, flavin, content, **39:**160
    inhibitors of enzyme, **39:**160; 161
    iron, content of non-heme, **39:**160
    reaction with alkylating agents, **39:**161
    specificity of substrates, **39:**160
    subunit structure, **39:**160
  sources of enzyme, **39:**156; 157; 159; 161
L-Glutamate ternary complex with glutamate dehydrogenase and NADPH, **39:**388
Glutamate transaminases, **16:**203–204
Glutamate-valine system, for transamination, **16:**204
Glutamic acid, **51:**217. *See also* Glutamine
  activation of, **24:**2
  adaptive formation in *E. coli,* **6:**8
    effect of age of culture on growth, **6:**12
    *p*H of growth medium, **6:**10; 16
    inhibitors, **6:**18
    Michaelis constant, **6:**16
    purified, **6:**12–21
    specificity, **6:**15
  bacterial decarboxylase, **6:**4
  in biosynthesis of glutamine, **19:**252–254
    role in nitrogen metabolism, **19:**337–354
  in chloroplasts, **20:**209
    labeled, in studies of amino acid metabolism in plants, **20:**321
      in studies of protein synthesis in plants, **20:**220; 221
    optical antipodes in living organisms, **20:**2
    in phosphoylated peptides of esterases, **20:**288
    in plants, biosynthesis and metabolism, **20:**322–330; 335; 336
      occurrence, **20:**319
    thermodynamic data on ionizations, **20:**64
  and citric acid cycle, **18:**251–258; 261–264; 268–269; 271–276
    labeled, **18:**268–269; 272
  dehydrogenation of, **26:**117
  as equivocal model, **49:**59
  and ethionine effects, **25:**372
    in glycoprotein linkage, **25:**451; 452
  genetic studies, **17:**400; 426
    isomers, in synthetase reaction, **17:**178
  as inhibitory analogue, **6:**139
  metabolism of, **13:**254–260; 263; 278–279
  and nucleic acid metabolism, **9:**497
    in plant metabolism, **9:**394; 398–401; 402–409; 419; 420; 421–430
  in purine biosynthesis, **21:**203
    oxidation, in mitochondria and mitochondrial fragments, **21:**21; 34; 38
  and pyridine ring, **8:**236
    and water absorption, **8:**246
  role in nitrogen fixation, **13:**369–373
  as substrate for glutamine synthetase, **31:** 189–198
  in transamination, **14:**100; 102
  and virus mutation, **15:**477
    to citric acid cycle, **15:**185–186; 214–224
    in formation of urea, **15:**324–342; 364; 370–371; 373
    in glutamine synthesis, **15:**254
    from rennin action, **15:**432
D-Glutamic acid, **33:**367
  in mureins, **26:**199–201; 205; 207; 210–211; 216–219
L-Glutamic acid
  formylation, **21:**222; 412; 414
  transamination reactions of, **19:**367; 368

D-Glutamic acid, occurrence, in malignant tumors, **20:**24; 26
in native proteins, **20:**25
L-Glutamic acid, oxidation in plants, **20:** 320–322
DL-Glutamic acid, resolution by papain, **19:**309
L-Glutamic acid and choline acetylation, **12:**319; 322; 325–326
L-Glutamic acid decarboxylase, determination of glutamic acid, **16:**191
Glutamic acid dehydrogenase
auxotropic mutants of, of *Neurospora crassa,* **30:**149
effect of radiations, **24:**388; 393
Glutamic acid derivatives, genetic studies, **17:**395–399
L-Glutamic acid-γ-ethyl ester, inactivity of, **16:**201
Glutamic acid-glutamine system, and translocation of materials in mammalian tissues, **13:**268–271
Glutamic acid in relation to penicillin, **10:** 253–254
Glutamic acid isomerase, role of cobamide coenzyme, **26:**264
Glutamic acid residues, at the carbonyl side of the splitting point, **41:**183
Glutamic acid γ-semialdehyde, **53:**222
Glutamic acid transport, **49:**52
Glutamic-alanine transaminase, **37:**140
Glutamic aminopherase, **5:**155
Glutamic aspartic transaminase, **35:**104
mechanism of action, **35:**99–105
stereochemistry of transmenation, **35:**99–105
Glutamic-aspartic transaminase, **15:**330–331; 334; 337; 339
aldehydic, steps in reaction, **25:**14
aminic, steps in reaction, **25:**14
conformational change upon complex formation, **25:**20
mechanism for action, **25:**33–35; 243
effect of pH on, **22:**395
pyridoxal phosphate in, **22:**395
and vitamin B_6, **24:**273
L-Glutamic-L-aspartic transaminase, **35:**99
γ-Glutamic carboxypeptidase, classification, **14:**247
L-Glutamic decaboxylases, classification, **14:**256
Glutamic decarboxylase, **35:**86
Glutamic dehydrogenase, **15:**90; 114; 123; 224; 229; 331; 334; 335; 337–338; 340, **22:**367–370
activation, **26:**302
PN-dependent, **26:**119; 120; 160
active, activating sites of, **22:**369
activity of –SH group, **11:**241
in *Chlorenchyma,* **20:**219
and citric acid cycle, **18:**251–254
denaturation of, **23:**48–50; 64; 71
solubility of, **23:**72
dissociation, **22:**368; 369
and DPNH binding, **22:**369; 370
importance in metabolism, **16:**226
in formation of glutamate, **16:**255–256
inhibition by 1,10-phenanthroline, **22:**368
in mitochondria, **21:**21; 22; 58; 74
effect of sonic vibrations on, **21:**57
molecular association of, **24:**119
role of metal, **24:**112–113
Neurospora mutants, **22:**8–11; 29
and heat activation, **22:**19
and interallele complementation, **22:**34; 36
properties of, **22:**9–11
and phosphorylations, **17:**78; 80
genetic studies, **17:**448–450
in photosynthesis, **19:**52; 53
in transamination reactions, **19:**340–342; 346; 347; 354; 375
product inhibition of, **25:**223; 265–266
role of sulfhydryl groups in, **22:**367; 368
zinc content and structural integrity of, **22:**368
L-Glutamic dehydrogenase
inhibition by thiol reagents, **11:**239
isolation, **14:**405
L-Glutamic dehydrogenase, in plants, **9:**402–403
Glutamic dehydrogenases, classification, **14:**254
Glutamic dehydrogenase system, redox potentials in, **17:**120
Glutamic γ-semialdehyde, **40:**70
in amino acid synthesis, **16:**258–259; 262–264
in transaminations, **16:**197; 205; 231
Glutamic-oxalacetic transaminase, product inhibition of, **25:**222–223; 243
Glutamic r oxalacetic transaminases, **14:**252
Glutamic r pyridine nucleotide transhydrogenases, **14:**254
Glutamic r pyruvic transaminases, **14:**252

Glutaminase, **13:**282–289; 291; 293; 294; 295–296; 297; 303–304, **15:**342, **20:**331, **39:**214; 226, **41:**1; 2; 24
activation, **8:**164
activity, **8:**158; 160
amidotransferases, **72:**89–93
deficiency, **41:**2
hepatic glutamine transport, **72:**54
breakdown regulation, **72:**63–65
hepatocyte heterogeneity and enzyme distribution, **72:**56–58
Ntn amidotransferases, glutamine PRPP amidotransferase active sites, **72:**96–99
and plant metabolism, **9:**403; 417–420
product inhibition of, **25:**222; 241
L-Glutaminase, **39:**223, **43:**523
Glutaminase A and B, *E. coli,* **38:**31
Glutaminase-asparaginase, **39:**215; 226
enzymes, **39:**213; 214
Glutaminase I, phosphate-catalyzed, **16:**199
Glutaminase II, deamidation by, **16:**199
Glutaminases, **53:**221; 258
bacterial, **53:**227
classification, **14:**234; 247
inactivated, **53:**220
localization in rat gastrointestinal tract, **53:**219
phosphate-dependent, **53:**224
in small intestine, **53:**218–221
phosphate-independent, maleate-stimulated, **53:**223
properties of, **53:**219
Glutaminases I & II, localization of, **24:**327; 343
Glutamine, **35:**489, **46:**42, **50:**141; 152; 156; 157, **53:**211; 250; 258, **56:**117
and adenylic acid metabolism, **9:**497
and fertilization, **9:**312
and glutathione, **9:**420
in plant metabolism, **9:**392; 403; 409–421; 433
as amino group donor, **16:**255
assay, **16:**257–258
in purine synthesis, **16:**451
relation to amino aid biosynthesis, **16:** 257–268
in transaminations, **16:**196; 199–204; 207; 211; 225–227; 229
arsenolysis of, **31:**186
mechanism of enzymic synthesis of, **31:**183–184
arterial, utilization of, **53:**204–214
and citric acid cycle, **18:**261–264
labeled, in plant respiration, **18:**300
reaction with phosphoribosyl pyrophosphate, **18:**217–218
configuration of, **23:**24–25
phosphorolysis of, **23:**338
deamidation, **8:**152; 158–164
by phenylpyruvic acid, **8:**160–164
as energy source, **53:**224–226
in histidine biosynthesis, **21:**403–405
in purine biosynthesis, **21:**201–205; 208; 210; 211; 229; 281; 401
nitrogen transfer reactions of, **21:** 240–252
in pyrimidine biosynthesis, **21:**281
synthesis, **21:**234–240
intermediary metabolism of, **13:**237–319
synthesis, **13:**275–280
synthesis in liver, **13:**7
utilization of peptides by microorganisms, **13:**244–245; 261–262
luminal, utilization of, **53:**214–218
metabolism of, **24:**340
peptide derivatives of *vs.* lysine peptide derivatives, **54:**37
transglutaminase activity towards, **54:** 17–19; 37
in phosphorylated peptides of esterases, **20:**288
in plants, biosynthesis and metabolism, **20:**331–336
role in nitrogen transfer, **15:**339; 340–342; 371
synthesis, **15:**254; 349; 351; 363
stereospecificity of, for transglutaminase activity, **54:**21–25
as substrate for glutaminase, **41:**24
tissue levels, in intestinal mucosa, **53:**220
transamination reactions of, **19:**365
transglutaminase effect on, **54:**2
transglutaminase specificity for, **54:**45; 52
as liver enzyme *vs.* blood coagulation factor, **54:**51
as transglutaminase substrate, **54:**51
uptake, **53:**204; 205
uptake rate, concentration dependency, **53:**206
as urea precursor in liver, **53:**203
utilization, hepatic, **53:**212
utilization rate, by enterocytes, **53:**218
utilization sources, **53:**230

L-[*amide*-^{13}N]Glutamine, **57:**331–332
L-Glutamine, **49:**77
biosynthesis, **19:**252–254
^{15}*N*-L-Glutamine, and $^{15}NH_4Cl$, protein and nucleic acid synthesis, **39:**239
D-Glutamine, inactivity, **16:**201
synthesis, **16:**257
Glutamine, Intermediary Metabolism (Waelsch), **13:**237
β-Glutamine, synthesis of, **31:**190
Glutamine amide transfer
anthranilate synthase catalysis, **66:**217–218
glutamine PRPP amidotransferase, **66:** 258–260
Glutamine amide transferase (GAT)
amidotransferases, **72:**89–93
asparagine synthetase catalysis, **72:**147–148
glutamine-dependent structure, **72:**154–157
Glutamine amido hydrolase, classification, **14:**234
Glutamine amidotransferases, **53:**223, *See* Amidotransferases
effect of glutamine analogs, **38:**15
evolutionary relationships, **38:**32
Glutamine analogs, **53:**220
in study of nitrogen transfer reactions, **21:**249; 250
Glutamine analogues in transamination study, **16:**201–202
Glutamine-asparagine transaminase, column chromatography, **22:**458–460
Glutamine cyclotransferase, **54:**2
L-Glutamine cyclotransferase, **43:**523
Glutamine decarboxylase, inactivation of, **23:**382
Glutamine-Dependent Carbomyl Phosphate Synthetase of Escherichia Coli, Mechanism and Regulation of (Meister), **62:**315
Glutamine domain
asparagine synthetase, **72:**111–112
structural studies, **72:**153–157
glutamate synthase, **72:**116–117
GMP synthetase, **72:**120–123
Ntn amidotransferases, **72:**96–99
Glutamine formation, and ATP, **18:**230–231
O^{18} studies, **18:**231
L-Glutamine-D-fructose-6-phosphate transaminase, allosteric regulation, **28:**114–115
Glutamine in transamination, **14:**100
Glutamine-keto acid activity, in vitamin B_6 deficient rats, **16:**211
Glutamine-α-keto acid reactions, **16:**199–203; 207
Glutamine metabolism
divergent pathways, **28:**48; 66–69
in intestinal transport, **53:**227
products of intestinal, as urea precursors, **53:**228
Glutamine (N^{15}), amide-labeled, in transamination study, **16:**200
Glutamine phosphoribosyl-pyrophophate amidotransferase, allosteric regulation, **28:**65–66; 118–121
Glutamine phosphoribosylpyro phosphate aminotransferase, **37:**440
Glutamine PRPP amidotransferase, **66:**251–264
acceptor domain, **72:**99–100
allosteric transition and coordination, **72:**102–104
feedback inhibitor binding sites, **72:** 100–103
avian and human, **66:**263–264
B. subtilis, **66:**261–263
cloning and sequence information, **66:** 255–257
crystal structures, **72:**96
E. coli, **66:**257–261
enzyme classification, **72:**93–95
glutamine amide transfer, **66:**258–260
glutamine-dependent nitrogen transfer, **72:**183–186
glutamine domain, **72:**96–99
nucleotide regulation and allosteric control, **72:**95
properties, **66:**251–255
size and molecular weight, **66:**256–257
structure-function relationship, **72:**89–93
structure-function relationships, **72:**90–93
Glutamine residue, and RNase activity, **24:**450
Glutamine synthesis
biochemistry developments, **72:**12–38
ATPase thermodynamics, **72:**13–15
bacterial glutamine synthetase atomic structure, **72:**36–38
bacterial glutamine synthetase regulation, **72:**30–36
borohydride trapping of acyl-P intermediates, **72:**26–28

glutamine synthetase activity, **72:**12–13
γ-glutamyl-phosphate intermediacy, **72:** 16–18
isotope exchange positional experiments, **72:**28–29
isotope trapping for cycle intermediate detection, **72:**25–26
isotopic tracer studies, glutamine synthetase reaction, **72:**15–16
kinetic reaction mechanism, glutamine synthetase, **72:**22–24
ligand interactions, sterochemical mapping, **72:**20–22
methionine sulfoximine inhibition, **72:**17; 19–20
enzymic catalysis, background, **72:**10
hepatic transport and metabolism, **72:**44–78
acid-base homeostasis, **72:**71–75
acinar-level integration, **72:**67–75
enzyme distribution, **72:**56–58
enzymes, **72:**53–56
functional consequences of swelling, **72:**49–53
glutaminase, **72:**54
glutaminase ammonia amplification, **72:**69–70
glutamine breadown regulation, **72:**63–65
glutamine synthase as ammonia scavenger, **72:**67–69
glutamine synthetase, **72:**55–56
gutamine transaminases, **72:**55
heterogeneity, **72:**56–63
heterogeneous glutamine synthetase distribution, **72:**58–61
hetrogeneous transport, **72:**62–63
hydration mechanism, **72:**48–49
intercellular glutamine cycling, **72:**70–71
long-term glutamine regulation, **72:**65
mitochondrial transport, **72:**47–48
pathophysiology, **72:**75–78
plasma membrane transport, **72:**45–47
regulation of, **72:**66–67
research history, **72:**44–45
short-term glutamine regulation, **72:**64–65
Meister's theory, **72:**10–12
protein ligase problem of carboxy group versatility, **72:**38–40
Glutamine synthetase, **21:**234–237, **36:**80, **39:**171–177, **43:**31; 526; 552, **50:**156; 365; 380; 381; 382; 383, **53:**224
activation of glutamate by, **31:**187
activity of, toward 1-amino-1,3-dicarboxycyclohexane, **31:**203–205
toward glutamic acid and other α-aminodicarboxylic acids, **31:**190–198
toward β-hydroxyglutamic acid and γ-hydroxyglutamic acids, **31:**201–203
toward methylglutamic acids, **31:** 194–195; 198–201
toward monocarboxylic amino acids, **31:**192
cis-1-amino-1,3-dicarboxycyclohexane as substrate for, **31:**204
D-and L-glutamate as substrates of, **31:**189
binding site for amino group of, **31:** 193–194
and binding sites for amino acid, ammonia and nucleotide, relationship between, **31:**205–208
^{14}C-methyl-L-methionine sulfoximine as substrate for, **31:**210
conformation of the amino acid substrate at active site of, **31:**190–205
dissociation of, **31:**216
D-glutamine synthesized with, **31:**186
γ-glutamyl transfer action of, **31:**186
mechanism of action of, **31:**185–190
general scheme for, **31:**190; 191
mechanism of inhibition by methionine sulfoximine, **31:**208–212
molecular weight for, **31:**216
optical specificity of, **31:**215
β-γ-^{32}P-ATP as substrate for, **31:**210
reaction catalyzed by, **31:**183
relationship to substrate conformation at the active site, **31:**183–218
relative activities of substrates of, with ammonia and hydroxylamine, **31:**206
schematic representation of orientations of glutamate, ammonia, and methionine sulfoximine on, **31:**211
specificity of, **31:**183–218
for optical isomers, **31:**184
studies with ^{18}O-labeled glutamate in, **31:**188–190
substrates for, **31:**189–206; 210
substrate specificity for, **31:**213
subunits of, **31:**216
summary of effects of methyl and hydroxyl substitutions on susceptibility of, **31:**214

Glutamine synthetase (*Continued*)
synthesis of hydroxymates from α-methylglutamic acid and glutamic acid by, **31:**195
allosteric regulation, **28:**54; 66–69
in biosynthesis of glutamine, **19:**252–254
in biosynthesis of glutathione, **19:**255; 256
characteristics of reaction, **39:**130; 156; 170
γ-glutamyl phosphate as an intermediate, **39:**172
evidence for, **39:**172
reaction of enzyme with pseudosubstrates
3–3aminoglutaric acid, **39:**173
cycloglutamate(cis-L-amino-1,3,-dicarboxy-cyclohexane), **39:**173
methionine sulfoxamine, **39:**173
reactions with hydroxylamine, **39:**172; 173
from *E. coli,* electron microscopy of, **33:** 101–102
ovine, electron microscopy of, **33:**108
kinetic analysis by isotopic exchange method
ordered for ovine brain enzyme, **39:**173
order of binding of substrates, **39:**173
random for adenylated *E. coli* enzyme, **39:**173
localization of, **24:**334; 340
partial reactions, **39:**173
product inhibition of, **25:**224; 241
regulation of, in liver, **39:**82
role of, in NH_3 utilization, **39:**82
site of cleavage by, **17:**163; 176–179
source of enzyme, **39:**172
L-Glutamine synthetase, **43:**523
Glutamine Synthetase, Specificity of, and its Relationship to Substrate Conformation at the Active Site (Meister), **31:**183
Glutamine-synthetase-deadenylating enzyme, polynucleotide stimulation, **40:**48
Glutamine synthetase-glutamate synthase reaction
in blue-green algae, **57:**264–267
in soybean nodules, **57:**267–268
Glutamine transaminase, **16:**268
broad specificity, **16:**201–202; 203; 226–227
Glutamine transaminases, cysteine *S*-conjugate β-lyase reactions
glutamine transaminase K, **72:**219–222
hepatic glutamine transport, **72:**55
Glutamine transamination, irreversibility, **16:**257
Glutamine Utilization by the Small Intestine (Windmueller), **53:**201
Glutaminylglycine, ionization constant, **20:**58
Glutamohydroxamic acid (GHA), from transferase action, **13:**290–299; 305
Glutamo-transferase reaction, **17:**176–179
Glutamotransferases (GTF), **13:**290–296; 296–299; 300–301; 301–305; 309
Glutamoyl bonds, **41:**196
γ-Glutamyl amino acids, **43:**524
γ-Glutamyl-α-aminobutyrylglycine, as substrate for tripeptide synthetase, **31:**188
γ-Glutamyl-α-aminobutyrylphosphate, as substrate for tripeptide synthetase, **31:**188
Synthesis of, by yeast tripeptide synthetase, **31:**188
L-γ-Glutamyl compounds, γ-glutamyl transpeptidase, **72:**249–250
γ-Glutamyl cycle, **43:**521; 536; 543
modified in 5-oxoprolinuria, **43:**546
γ-Glutamyl cyclotransferase, **43:**521; 522; 523; 532; 536; 545
distribution of, **43:**522
specificity of, **43:**524
Glutamyl cysteine, **17:**179
γ-Glutamyl cysteine, **43:**546
γ-Glutamylcysteine, biosynthesis, **19:** 254–256
γ-Glutamylcysteine formation, and ATP, **18:**230
γ-Glutamylcysteine-glycine-ATP enzyme, **15:**254
L-γ-Glutamyl-L-cysteine synthetase, **43:**523
γ-Glutamyl cysteine synthetase, **43:**526; 546; 552
γ-L-Glutamyl-L-cysteine synthetase, reaction catalyzed by, **31:**215
γ-Glutamyldimethylamide, in transaminations, **16:**196; 202
γ-Glutamyldjenkolic acid, **50:**122; 124
γ-Glutamyl-γ-glutamyl-P nitroanilide, specificity of, **43:**523
S-Glutamylglutathione, synthesis, **19:**247
γ-Glutamylhydroxamate, synthetase, of, by glutamine synthetase, **31:**185–186
γ-Glutamyl lactamase, **43:**521
localization of, **24:**311
γ-Glutamylmethylamide, in transaminations, **16:**196; 202

γ-Glutamyl-α-(methylenecyclopropyl) glycine, **50:**126
γ-Glutamyl-*p*-nitroanilide, **43:**524
γ-Glutamyl peptides. *See also* Glutamine and Glutathione
in the transferase system (GTF)(GSH), **13:**300–301; 312–315
γ-Glutamyl-L-phenylalanine, **45:**419
Glutamyl phosphate, **17:**178, **50:**145; 381
γ-Glutamyl phosphate, **43:**526
as intermediate in glutamine biosynthesis, **19:**253; 254
as intermediate in glutathione biosynthesis, **19:**256
synthesis, **19:**250
γ-Glutamyl-phosphate, glutamine synthetase reaction, **72:**16–17
Glutamyl radical, transfer to amines and amino acids, **13:**289–305; 305–310
Glutamyl-γ-semialdehyde, **50:**144; 145
D-Glutamyltransferase, **37:**576
γ-Glutamyl transferase, glutamine synthesis, **72:**13
γ-Glutamyl transpeptidase, **43:**521; 522; 523; 532; 537; 546, **53:**223; 258, **54:**10, **57:**159, **59:**151–154
active sites
catalytic and active site chemistry, **72:** 254–262
light subunit, **72:**250–251; 254–262
substrate binding sites, **72:**251–254
biosynthesis and processing, **72:**244–245
cancer-assisted changes, **72:**270–271
catalytic reactions and substrate specificity, **72:**247–250
gene structure and expression, **72:**264–270
DNA methylation status, **72:**269–270
human gene expression, **72:**267–269
multiple promoters and tissue specificity, **72:**265–267
glutathione metabolism, **72:**245–247
human lung enzymes and transcripts, **72:** 262–264
localization of, **24:**311
research background, **72:**240–242
structure and topology, **72:**242–244
γ-Glutamyl transpeptidase-related enzyme, structure and transcripts, **72:**262–264
Glutamyl-tRNA^{gln} amidotransferase
characteristics of reaction, **39:**167; 168
amino donor, **39:**167
asparagine, **39:**167
conditions for binding of substrates, **39:**169; 170
exchange reactions, **39:**167
glutamine, **39:**167
glutaminyl-tRNA, metabolic pathways for synthesis of, **39:**167
nucleoside triphosphate, **39:**167
phosphorylation of enzyme-substrate complex, **39:**168; 169
reaction with methylene analog of ATP, **39:**169
sources of enzyme, **39:**167
specificity of tRNA acceptor, **39:**167; 168
stability of phosphorylated complex, **39:**169; 170
Glutaraldehyde, **33:**298, **57:**207
in preparation of insoluble trypsin, **33:**276
in preparation of water-insoluble enzymes, **33:**278
Glutarate, **51:**236
ternary complex with glutamate dehydrogenase and NADP, **39:**393
Glutaredoxin, **59:**146, **65:**148–150, **66:**150–152
Archaebacterium, **66:**164–165
calf thymus, **66:**156–158; 165–167
E. coli, **66:**157–158; 160; 163–164
mutants deficient in, **66:**168–169
glutathione S-transferases, **69:**15
in mammalian ribonucleotide reductase, **66:**167–168
rabbit bone marrow, **66:**167
vaccinia virus, **66:**165
Glutaric acid
effect on glutamate dehydrogenase-NADP complex, **39:**385; 393
effect on glutamate dehydrogenase-NADPH complex, **39:**375
from lysine metabolism, **16:**206
Glutaric semialdehyde, in transaminations, **16:**195; 206
(Glutathiomethyl)glyoxal, **59:**110
Glutathione, **7:**393; 396, **10:**229–231, **11:** 248–257, **15:**93; 111; 126, **39:**342; 343; 362; 363, **43:**521, **46:**383; 384, **48:**161; 162; 166, **53:**250; 258, **58:**63, **60:**47. *See also* Glutathione-dependent enzymes; Glutathionylspermidine; Sulfhydryl groups; Thiol groups

Glutathione (*Continued*)
and acetylornithine deacetylase, **40:**71; 72
action on thiaminase, **15:**401; 410
activation of ATPase, **6:**167
S-acyl and *S*-aminoacyl derivatives, glyoxalase II action on, **19:**248
synthesis, **19:**240; 241; 247
and ATP, **18:**230
glycine as precursor, **18:**258
in triose phosphate oxidation, **18:**226
autoxidation, **10:**294–295, **15:**42
biosynthesis, **19:**254–256
and cathepsin activity, **19:**286
in cell nuclei, **16:**489
reaction with quinones, **16:**121–122; 134
reduction of CoA by, **16:**3
in citric acid cycle, **15:**184; 219; 221–222; 241; 248; 257–259
complex formation with dehydro-ascorbic acid, **10:**295
in conversion of disulfides to mercaptans, **10:**240
in corrinoid, **26:**246
crystalline form, **59:**116
cysteine *S*-conjugate β-lyase reactions, **72:**201–203
cysteinyl derivative, **72:**200–201
description of, **59:**103–106
and detoxification of biomembranes, **59:**143
in disulfide exchanges, **27:**199; 201–204; 210–213; 215; 217–219; 228–229
formaldehyde dehydrogenase action on, **27:**479
and oxidation of hydrocarbons, **27:**485; 509; 515; 518
disulfide formation, **46:**398; 399
domains, **59:**154
effect on blood clotting, **10:**23
effect on oxalacetate cocarboxylase, **12:**56
effect on tyrosinase, **14:**90; 93; 94
and glutamine, **9:**420
γ-glutamyl transpeptidase
metabolic role, of, **72:**245–247
synthesis mechanisms, **72:**240–242
and γ-glutomyl transpeptides, **59:**153
and GSH, **59:**155
hydrolysis of, **5:**243; 244; 245
at interfaces, **11:**49–54; 56
intermediary metabolism of, **13:**237–319
synthesis, **13:**280–282
linkages in, **14:**278
mercaptide formation, **11:**216
in muscular atrophy, **6:**198
N,S-modified, **59:**111
as nucleophile, **46:**384; 403
and nucleophilic attack of GSH, **59:**139
^{18}O experiments with tripeptide synthetase on, **31:**188
oxidation by cytochrome *c*, **17:**350–351
reversal of inhibition, **17:**254; 265; 323
oxidation by oxygen, **11:**203–206
by various agents, **11:**206–207
in oxidation of reduced triphosphopyridine nucleotide, **20:**172
in reactivation of thiols, **20:**66
in oxygen metabolism, **19:**81; 89; 90; 126–128; 185; 198; 199
and papain activity, **19:**286; 312
peptide formation by, **19:**238
in photosynthesis, **19:**22; 45
in plant metabolism, **10:**225; 226; 284; 288–295
in production of coenzyme A, **14:**65
in purine biosynthesis, **21:**229
synthesis, **21:**234–236; 239
reaction with alkylating agents, **11:**214–216
with metal electrodes, **11:**211–212
in redox systems in plants, **10:**289–295
reversal of enzyme inhibition, **15:**286
role in carbohydrate metabolism, **3:**176–178
specificity for, **46:**398
in squalene biosynthesis, **22:**319
and sulfenyl nitrite, **59:**132
synthesis, **15:**254; 349; 363
and thioperoxide, **59:**133
in TPD, **22:**343; 344
tyrosine hydroxylase, substrate specificity, **70:**110
Y conformation, **59:**116
and yeast proteinase activity, **19:**286
L-Glutathione, inactivity, **16:**201–202
Glutathione and Glutathione S-Transferases in Mercapturic Acid Biosynthesis, Role of (Boyland and Chasseaud), **32:**173
Glutathione-*S*-conjugates, γ-glutamyl transpeptidase, **72:**241–242
Glutathione-cystine, mixtures, iodine titration of, **5:**243–244
Glutathione dehydrogenase, **11:**249, **17:**351

Glutathione-dependent enzymes. *See also* Glutathione
description of, **59:**103–106
formaldehyde dehydrogenase, **59:**127–131
γ-glutamyl transpeptidase, **59:**151–154
glutathione peroxidase, **59:**143–146
glutathione reductase, **59:**146–151
glutathione *S*-transferases, **59:**131–136
glyoxalase I
description of, **59:**106–108
as hydride, **59:**111–114
inhibitors of, **59:**114–118
mechanisms of, advanced, **59:**118–122
nature of, **59:**108–111
as proton transfer, **59:**111–114
glyoxalase II, **59:**122–127
maleyl isomerases
description of, **59:**136–139
GS group rotation, **59:**139; 141–143
maleylacetone moiety rotation, **59:**139
overview of, **59:**154–156; 158
Glutathione-Dependent Enzymes, Mechanism of Action of (Douglas), **59:**103
Glutathione epoxidase, **46:**173
Glutathione (GS^{35}H), reaction with serum albumin, **27:**210
Glutathione-homocysteine, transhydrogenase, **27:**218
Glutathionc in mitochondria, **8:**34
Glutathione peroxidase, **27:**228, **46:**170, **48:**3; 12–15, **58:**202; 257, **59:**143–146, **67:**95
glutathione S-transferases, **69:**15
non-selenium dependent, **48:**13
Glutathione reductase, **15:**155; 222, **37:**432, **41:**1; 2; 23, **46:**241
action on disulfides, **27:**217–219
catalytic mechanisms for, **59:**150–151
charge-transfer complex, **59:**147
chemical modification studies, **59:**149–150
and DR metabolism, **28:**443
and flavin, **59:**147; 149
group of enzymes in, **59:**146–147
increased, **41:**2
localization of, **24:**304
and photophosphorylation, **23:**387
in plants, **18:**310
PN-dependent, **26:**119; 126
and protein disulfides, **59:**147–148
Glutathione S-alkenetransferase, **46:**384
Glutathione S-alkyltransferase, **46:**384
Glutathione S-aralkyltransferase, **46:**384
Glutathione S-aryltransferase, **46:**146; 384
Glutathione S-epoxide transferase, **46:**384
Glutathione S-transferase B, **48:**14
Glutathione S-Transferases: A Group of Multifunctional Detoxification Proteins, The (Jakoby), **46:**381
Glutathione sulfenyl nitrate, **46:**399
Glutathione synthesis, **17:**179–180, **73:**209–267
background, **73:**209–214
biosynthesis overview, **73:**214–216
cancer chemotherapy resistance, **73:**233–235
catalytic mechanism, **73:**236–245
chemical mechanisms, **73:**243–245
reactions catalyzed, **73:**240–241
steady-state kinetics, **73:**241–243
substrate specificity, **73:**236–240
cloning of γ-GCS genes and regulation of expression, **73:**229–233
human catalytic subunit (γ-GCS$_h$) gene, **73:**230–231
human regulatory subunit (γ-GCS$_h$) gene, **73:**231–233
yeast γ-GCS genes, **73:**229–230
γ-GCS subunit cloning, sequencing and expression, **73:**220–229
eukaryotic γ-GCS catalytic subunit, **73:**221–224
γ-GCS$_h$ and γ-GCS$_l$ mRNA expression, **73:**225–226
γ-GCS regulatory subunit, **73:**224
recombinant expression γ-GCS$_h$ and γ-GCS$_l$ proteins in *E. coli,* **73:**228–229
recombinant expression γ-GCS$_h$ and γ-GCS$_l$ proteins in mammalian cells, **73:**226–228
inhibition, **73:**245–260
BSO use in cancer patients, **73:**252–255
chloroketone and related compounds inactivation, **73:**257–258
cystamine inactivation, **73:**255–257
feedback inhibition, **73:**245–246
glutamate antagonists, **73:**258–259
nitric oxide donors, **73:**259–260
s-alkyl homocysteine sulfoximine inhibitors, **73:**246–252
perspectives on, **73:**260–261
posttranslateral regulation, **73:**236
purification and protein chemistry, **73:** 216–220

Glutathione synthetase, **21:**235; 237, **23:**358, **43:**523
  localization of, **24:**335
  product inhibition of, **25:**223
Glutathione transferases, **57:**357–358. *See also* GSH transferase
Glutathione *S*-transferases, **46:**383–414, **69:**1–44. *See also* Transferases
  ab initio calculations, **69:**19–20; 22
  acquired drug resistance, **69:**6–7
  active site, **69:**10; 23; 30
  active site tyrosine, **69:**16; 17–18
  affinity chromatography, **46:**385
  amino acid composition, **46:**387–389
  binding domain, **69:**10; 15
  binding site, **69:**15–18
  Bronsted value for, **59:**134
  catalysis
    detoxification and, **69:**5–7
    role of conserved tyrosine in, **69:**18–20
  catalytic efficiency, **46:**404
  catalytic efficiency of, **59:**135
  catalytic mechanisms, **69:**20; 21–22
  catalytic role, **46:**407
  catalyzation of, **59:**132–133
  chemical mechanism, **46:**402–404
  chimeric enzymes, **69:**31–34
  chimeric isoenzymes, **69:**33
  crystals, **69:**7; 28
  cytosolic, **69:**3
  cytosolic enzymes, **69:**3; 4
  deamidation, **46:**388
  and detoxification, **59:**131–132
  in detoxification, **46:**409
  dissociation constants, **46:**392
  effective molarity (EM) of, **59:**135
  electrostatic interaction, **69:**16; 17; 20
  evolution, **69:**7; 36–40
  fluorescence, **46:**391
  general base catalyst, **59:**134
  and GSH, **59:**134
  histidine, **69:**19
  historical perspective on, **69:**1–2
  human, **46:**385; 392; 399
    dissociation constants, **46:**392
      bilirubin, **46:**392
      diglucuronide, **46:**392
      bilirubin monoglucuronide, **46:**392
      hematin, **46:**392
      indocyanine green, **46:**392
  inhibitors, **69:**6
  isoelectric point, **46:**386
  isoenzymes, **69:**12; 13
  kinetic mechanism, **46:**401
  kinetic parameters, **46:**397
  lens S-crystallins, **69:**3
  leucine, **69:**17
  leukotriene C_4 synthase, **69:**4; 5; 6
  liver concentration, **46:**405
  macrophage migration inhibitory factors (MIF), **69:**4
  microsomal forms, **69:**4; 6
  molecular weight, **46:**386
    subunit, **46:**386
  multigene families of, **69:**3–7
  multiple bonding, **46:**387
  mutagenesis, **69:**3
  nomenclature, **69:**5
  nomenclature, AA, A, B, C, D, E, M, α, β, γ, δ, **46:**385
  pH-rate profiles, **69:**20
  physical properties, **46:**386
  physiological roles, **46:**404–409
    catalytic role, **46:**407
    covalent binding, **46:**408
    role as ligandins, **46:**407
  pK_a of GSH, **59:**134
  plants, activity in, **46:**405
  primary structures, **69:**27
  product complexes, **69:**23
  protonate leaving group, **59:**134–135
  rat, **46:**385; 392; 398
    dissociation constants, **46:**392
      8-analinonaphthalene sulfonate, **46:**392
      bilirubin, **46:**392
      bilirubin diglucuronide, **46:**392
      bilirubin monoglucuronide, **46:**392
      cephalothin, **46:**392
      3,6-dibromosulfophthalein, **46:**392
      hematin, **46:**362
      indocyanine green, **46:**392
      rose bengal, **46:**392
  roles of, **69:**2
  sequence alignments, **69:**3–4; 14; 15; 26
  specific activities, **46:**396
    with organic thiocyanates and nitrate esters, **46:**399
    with selected substrates, **46:**396
  as storage proteins, **46:**408
  structure solutions, **69:**7–8

subunit structure, **46:**387
thiolate anion, **69:**18; 19
three-dimensional structures, **69:**7–15; 19
comparisons of α, *mu,* and *pi* enzymes, **69:**12–15
quaternary, tertiary, and domain structure, **69:**8–11
tryptophan, **69:**17
from wax moth, **46:**396
xenobiotic substrate binding site, **69:**12; 23
Glutathionylspermidine, **36:**221–222
Gluten proteins, **10:**299–300
Gluten recovery in fermentation, **7:**590
Glutens, disulfide bonding, **27:**221
Glutinosin, **7:**473; 487; 489
Glutinosone, **55:**73
Glycagonase, localization of, **24:**326
α-Glycan phosphorylase, localization of, **24:**311
Glycemic curve, in muscular dystrophy, **6:**207
Glyceollin, **55:**15
inhibition of synthesis of, **55:**44
metabolism of, **55:**43
prenyl groups of, **55:**41
structure of, **55:**8
P-Glycerald-3-P, **43:**493
Glyceraldehyde, **15:**154; 158–160; 169–170; 258
formation and conversions in galactose metabolism, **20:**119; 120
reserving inhibition of hexokinase, **6:**77
Glyceraldehyde kinase, correlation of hepatoma growth rate with, **29:**331
Glyceraldehyde P dehydrogenase, **43:**501
Glyceraldehyde-P-dehydrogenase, **37:**482
Glyceraldehyde 3-P-dehydrogenase, **46:**242
Glyceraldehyde-P dehydrogenases, chemical mechanism of, **45:**372; 374
Glyceraldehyde-3-phosphate, **15:**129; 143; 158–160; 167; 170; 171; 174, **48:**171
in allosteric regulation, **28:**111–112; 116
in deoxyribose synthesis, **28:**421–422; 424; 450
in ribose synthesis, **28:**396–397; 402; 410; 411
in ATP synthesis, **23:**328–333; 337; 338; 372–376
Glyceraldehyde 3-phosphate, **19:**7; 23–27; 29–31
Glyceraldehyde phosphate, dehydrogenase, **17:**100; 122; 163
Glyceraldehyde phosphate dehydrogenase, **18:**273
Glyceraldehydephosphate dehy drogenase, **37:**417; 418; 425; 426; 432; 441; 452; 464; 470; 475; 492; 514; 515; 548; 577
Glyceraldehyde phosphate dehydrogenase, **58:**211
Glyceraldehydephosphate dehydrogenase
adsorption studies, **14:**333
isolation, **14:**380; 393
location of, **24:**298; 335
Glyceraldehyde-3-phosphate dehydrogenase, **15:**93–94; 114; 120; 129; 143; 171; 289–290, **37:**409; 557
in ATP synthesis, **23:**326; 328–333; 338; 355; 374; 379–381
kinetic properties of, **23:**332
in the extramitochondrial glycolysis system, **29:**351
in sperm, **24:**87
Glyceraldehyde 3-phosphate dehydrogenase, **59:**107; 128
Glyceraldehyde phosphate dehydrogenase, in neoplastic cells, **29:**360
Glyceraldehydephosphate dehydrogenase denaturation process, **26:**105
Glyceraldehyde phosphate dehydrogenases, multiple forms, **25:**343
3-P-Glycerate, **46:**335; 339–340
as activator, **46:**335; 336; 337; 338; 339
effect of concentration on phosphate inhibition, **46:**338
physiological significance of activation and phosphate inhibition, **46:**339
Glycerate, and glycolate metabolism, **23:**436; 441; 443–446
and oxalate metabolism, **23:**452
Glycerate kinase, **23:**444
Glycerate-1,3-P_2, **51:**219; 222
Glycerate-3-P, **51:**222; 229; 231; 232; 233; 242; 246
Glycerate-1,3P_2 (1,3_DPG), **51:**249
Glycerate-2-P (2-PGA), **51:**249
Glycerate-3-P (3-PGA), **51:**249
Glyceric acid, **15:**167
conversion to serine, **16:**234
free energy of formation, **25:**47
in plants, **19:**16

Glyceric acid dehydrogenase (Tartronsemialdehydreductase), PN-dependent, **26:**118
D-Glyceric dehydrogenase, **23:**443
in photosynthesis, **19:**23
Glycerides, **34:**488
action of lipase on, **23:**152–158
synthetic, **23:**155
deuterium in, **7:**156
synthesis, **8:**317–318
trisaturated, **34:**488
water interface, **34:**487
Glyceride synthesis, role of CoA, **16:**18
Glycerinaldehyde-3-phosphate dehydrogenase, denaturation of, **23:**33; 46–49; 62
Glycerinaldehyde phosphate dehydrogenases, multiple forms, **25:**333; 345
α-Glycerin-1-phosphate dehydrogenase, PN-dependent, **26:**118; 161
Glycerin-T, as acceptor in dextran formation, **30:**199
Glycerokinase, **43:**44, **46:**271
kinetic mechamism of, **45:**284; 293; 294
Glycerol, **15:**120, **34:**486, **56:**194
and citric acid cycle, **18:**253; 255–256; 264–267
cytidine diphosphate, structure of, **30:**224; 234
effect on reaction rate of levan formation, **30:**218
in teichoic acid, **30:**226
effect on foliar absorption of streptomycin, **20:**386; 390
effect on gelation of pectic substances, **20:**357
imidazole in facilitated diffusion of, **20:**95; 96
solubility of pectic substances in, **20:**353
as fructose receptor, **17:**196
in kinetic studies, **17:**423
as solubilizing agent, **17:**148
permeability, **34:**467
phosphate, **34:**524
phosphorylation by *Mycobacterium,* **19:**17
Glycerol (C^{14}), utilization by sperm, **24:**54; 74; 91
Glycerol dehydrase, **55:**242
Glycerol dehydrogenase, **15:**116; 120, **18:**255, **37:**433; 516, **59:**194
α-Glycerol dehydrogenase, in photosynthesis, **19:**9
Glycerol esters, lipase action on, **23:**152–158
Glycerol fermentation, **7:**587; 596–600
Glycerol fluorinated analogs, **55:**240–242
Glycerol kinase, **55:**240–242
in phosphoryl transfer reactions, **55:**357–358
Glycerol monoacetate, as protection against fluoroacetate, **18:**140; 141; 144–145; 148
Glycerol oxidase, activity of —SH group, **11:**241
inhibition by oxidation, **11:**236
Glycerol-3-P dehydrogenase, **36:**72
Glycerol-3-P-dehydrogenase, **55:**242
β-Glycerol phosphate, as phosphoryl donor, **17:**167
Glycerol phosphate, in teichoic acid, **30:**232
Glycerolphosphate acyltransferase, localization of, **24:**311
Glycerolphosphate dehydrogenase, **37:**493
localization of, **24:**295; 298; 335; 337
Glycerol-1-phosphate dehydrogenase, in normal and neoplastic tissues, **29:**343
Glycerol phosphate dehydrogenase, in photosynthesis, **19:**9
Glycerolphosphorylcholine, in semen, **24:**64–65
L-Glycero-D-*manno*-heptose, in lipopolysaccharide, **31:**78; 79
Glyceroneogenesis, phosphoenolpyruvate carboxykinase (GTP), **69:**218; 223
Glycerophosphatase, effect of adsorption, **28:**29
α-Glycerophosphatase from yeast, **10:**87
Glycerophosphatase of cacao, **25:**465–466
α-Glycerophosphate, **15:**171–172
in lipid synthesis, **15:**120
Glycerophosphate, action of rennin on, **15:**433
α-Glycerophosphate, in mitochondrial lipid synthesis, **21:**50
oxidation, in mitochondria, **21:**22; 24
sn-Glycero-3-phosphate acyltransferase, **36:**303–304
α-Glycerophosphate cycle, advantage of, **29:**343
to oxidize DPNH by mitochondria, **29:**342
α-Glycerophosphate dehydrogenase, **15:**114; 120; 171
classification, **14:**254
α-Glycerophosphate dehydrogenase, **22:**370
in tumor cells, **29:**343
Glycerophosphate dehydrogenase, in isolation of enzymes, **14:**382
isolation, **14:**402; 405

L-α-Glycerophosphate dehydrogenase, PN-dependent, **26:**121
Glycerophosphate dehydrogenases, multiple forms, **25:**347
α-Glycerophosphate oxidase, in mitochondria, **21:**22
α-Glycerophosphate oxidation cycle, **23:** 345–346
α-Glycerophosphate r DPN transhydrogenases, **14:**254
Glycerophosphates
and phosphatides, **7:**152
in liver, **7:**152
solubilization by, **17:**143; 146; 149
Glycerophosphokinase, in mitochondrial lipid synthesis, **21:**50; 51
α-Glycerophosphoric acid, hydrolysis, by metal ions, **21:**180–184; 186
β-Glycerophosphoric acid, hydrolysis, by metal ions, **21:**182–184; 186
D-Glycerotetrulose-1-phosphate, in ribose synthesis, **28:**397
Glychocholate, formation of, **24:**310
D-Glycidol-P, **43:**511
Glycinamide, **54:**30, **59:**156
Glycinamide ribotide
phosphorolysis of, **23:**338
in purine biosynthesis, **21:**204–211; 222; 223; 234–240; 400; 401
Glycinamide-ribotide (GAR) transformylase, reaction, **27:**425
Glycinamide ribotide kinosynthase, **21:**208; 234–237
Glycinamide ribotide transformylase, **21:**206; 207; 222; 400
Glycine, **46:**39; 40; 91; 103; 111–113, **50:**193, **51:**217, **57:**91; 148–150, **58:**35
and acetate, **23:**432–433
and acyl coenzyme A, **23:**455–456
aerobic metabolism, **23:**453–457
anaerobic metabolism, **23:**457–462
cleavage by microorganisms, **23:**327
C-transfer reactions, **23:**456–457; 463
and glycolate, **23:**435–436; 441
labeled, **23:**457; 458–462
metabolism by microorganisms, **23:**401; 404; 453–462
and oxalate, **23:**450
action of tyrosinase on, **12:**449
and biological methylation, **12:**451; 469–474; 478–479
CO_2 fixation in, **12:**142
deamination of, **12:**449–450
in decarboxylation of oxalacetate, **12:**81
fermentation, and CO_2 fixation, **12:** 119–121
role in biosynthesis, **12:**43; 48–49; 95; 116
in uric acid metabolism, **12:**121–127
biosynthesis, **16:**275
in glutamate formation, **16:**255
inactivity in *Neurospora* mutant, **16:**256
labeled, uptake by chromosomes, **16:**477
N-terminal residues of fibrin, **16:**37; 40; 41
in nucleic acid synthesis, **16:**450–451; 455–458; 461
reaction with quinones, **16:**112; 114; 117–120
in residual protein, **16:**447
in transamination reactions, **16:**190; 192; 196; 201; 204; 225; 228; 234–235
two-carbon precursor of, **16:**276–277
biosynthesis and metabolism, **21:**223–225; 227; 416; 418
chelates, in metal ion activation of enzymes, **21:**146
metabolism, in mitochondria, **21:**26
methyl derivatives, metabolism, in mitochondria, **21:**26; 429
in purine biosynthesis degradation, **21:** 201–205; 208; 211; 223; 232; 234; 235; 238; 239; 401; 402
in pyrimidine biosynthesis, **21:**288
radioactive, in study of carotenogenesis, **21:**339
in studies of purine biosynthesis, **21:**205; 206
in synthesis of active formate and active formaldehyde, **21:**410–412
in biosynthesis of glutathione, **19:**254; 255
in biosynthesis of hippuric acid, **19:**249; 252
deamination of, **19:**379; 380
labeled, in biosyntheses, **19:**260
methyl ester, condensation, **19:**237
in papain, **19:**284
in transacylation reactions of CoA derivatives, **19:**249
in biosynthesis of pectic substances, **20:**369

Glycine (*Continued*)
in chloroplasts, **20:**209
ionization constant, **20:**58
labeled, in studies of protein synthesis in plants, **20:**221–223
metal binding by, **20:**56
in phosphorylated peptides of esterases, **20:**288
thermodynamic data on ionizations, **20:**64
effect on muscle atrophy, **6:**200; 221
effect on papain action, **15:**292
inhibitor of fumarase, **15:**261
in synthesis, of glutathione, **15:**254
guanidino-acetic acid, **15:**374
nicotinamide, **15:**103
ribose, **15:**159; 175–176
urea, **15:**338 (N^{15}); 341 (N^{15})
in tobacco mosaic virus, **15:**462–464
and ethionine effects, **25:**372; 376
genetic studies, **17:**422; 442
as inhibitory analogue, **6:**133; 139
metabolism, **46:**113
in mureins, **26:**199; 202–204; 209–212; 219; 222
reductive deamination of, **48:**20
as synthetic unit, **18:**257–258; 271
Glycine-2-t_1, **35:**94; 95
(R)-Glycine-2-t_1, **35:**92; 95
(S)-Glycine-2-t_1, **35:**92; 93; 95
Glycine-*N*-acylase, product inhibition of, **25:**223; 245–246
Glycine acyltransferace, localization of, **24:**310; 311
Glycine adenylate, synthesis by carbodiimide method, **22:**191
Glycine and pyrrole formation, **7:**329
Glycine-2-C^{14}
incorporation into DNA and RNA, **16:** 455–456
synthetic use by sperm, **24:**91
Glycine(C^{14}), in folate biogenesis, **27:**422
Glycine-C^{14}, in study of nuclear RNA, **16:**440
Glycine (C^{14}), rate of labeling amylase, **26:** 108–110
Glycine carbamate, **57:**115
Glycine dehydrogenase, decarboxylating
evolution of, rate of, **74:**158
species distribution of, **74:**158–159
structural data on, **74:**133
Glycine depletion, asparagines synthetase metabolism, **72:**150–151
Glycinedesaminase, soil, method for, **28:**386
Glycine esters, effect of thrombin on, **16:**34
Glycine formation, and ATP, **18:**230
Glycine-glycine, **41:**194
$^{1\ 4}$C-Glycine, incorporation into pteridines, **35:**48
^{14}C-Glycine, in DR synthesis, **28:**423
ribose synthesis, **28:**418–419
Glycine-α-ketoglutarate reaction, free energy change, **16:**204
1-^{14}C-Glycine, labeling of pentoses, **28:**433
Glycine max, **55:**4, **71:**83
chalcone isomerase of, **55:**27
chalcone synthase of, **55:**19
hydroxycinnamoyl-CoA ligase of, **55:**17; 18
phenylalanine ammonia lyase and, **55:**15
Glycine metabolism, role of glutamine, **28:** 66–69
Glycine-N^{15}, incorporation into DNA, **16:**455; 458
incorporation into histone and protamine, **16:**480
incorporation into histones of somatic cells, **16:**446
incorporation into nuclear proteins, **16:**449; 458; 484
Glycine oxidase, **15:**340
in transamination reactions, **19:**340; 376
Glycine reductase, **48:**3
Glycine/serine auxotrophs, **16:**275–277
Glycine (serine) hydroxymethyltransferase, **74:**144–145; 152
phylogenetic distribution of, **74:**157
structural data on, **74:**133
Glycocoll, in plants, biosynthesis, **20:**329
oxidation, **20:**320; 322
Glycocyamine, **15:**361–362, **54:**384; 386
Glycogen, **3:**109–135, **46:**159; 318; 319; 359. *See also* Polysaccharides
accumulation, **46:**359
in bacteria, **46:**319
in *E. Coli* B, **46:**356
action of β-amylase on, **12:**394
action of phosphorylase on, **12:**412–414
from glucose-1-phosphate, **12:**419
similarity to amylopectin, **12:**371–372
activation of phosphorytion of phosphorylase *b* kinase by, **34:**382

activities of enzymes, involving metabolism of, in rat liver, **34:**313
and albumin, **3:**133
anabolism, **5:**131
α and β particles, **34:**290–292
association with protein, **34:**290
bibliography, **3:**133–135
binding of phosphorylase to, **34:**378
biosynthesis, in bacteria, **46:**318
in biosynthesis of uridine diphosphoglucose, **20:**122
biosynthetic pathways from glucose, **34:**294
branching, **3:**120; 126
breakdown, **52:**268
in digestive tract, **3:**126–127
enzymatic, **3:**121; 124–126
in liver, **3:**130
in muscle and yeast, **3:**130–131
phosphorolysis, **3:**127–130
carbon dioxide fixation in, by liver tissue, **74:**44–46
characterization of, **34:**289
chemistry of, **3:**109–133
constitution, **3:**119–122
control of deposition in tumor cells by, **29:**359
in tumor cells, **29:**359
control of glycogen synthetase activity by, **34:**334
debranching of, **34:**401
definition, **3:**109
de nova synthesis of, **34:**223
distribution, **34:**289
in dog, **34:**293
effect on muscle atrophy, **6:**195; 196; 197; 198
effect on phosphorylase, **28:**107
labeling, by 6-^{14}C-glucose, **28:**427
in ribose study, **28:**407; 414–415; 418
role of UDP-glucose, **28:**110
synthesis, **28:**338
in embryo and fetus, **34:**291–292
fermentation by thiaminase-producing bacteria, **15:**413
synthesis, **15:**305; 307–309; 310
fractionation, **3:**113
in guinea pig, **34:**293
incorporation, of 2-deoxyglucose into, **34:**317
of glucosamine into, **34:**318
incorporation of D-galactose, **26:**61
primers for, **26:**58
synthesis of, activators for, **26:**59
sugar nucleotide donor, **26:**66
iodine reaction, **3:**110; 120
isolation, **3:**111–113
loss of, during starvation, **34:**323
mammalian, **34:**292–294
in man, **34:**293
metabolism, by sperm, **24:**53; 62
enzyme localization, **24:**311; 338
metabolism of, **34:**285–417
methylation, **3:**120–124
molecular shape, **3:**119
molecular size, **3:**116–118
molecular weight
osmotic, **3:**114–116
viscosimetric, **3:**118–119
in mouse, **34:**293
in muscular dyastrophy, **6:**205; 223
nonmammalian, **34:**291
in organs, **3:**110–111; 132–133
osmotic pressure of solutions, **3:**116–118
α-particle, **34:**358
β-particle, **34:**358
phosphorolysis, **3:**127–130
phosphorus and nitrogen content, **3:**111; 113
polymerization degree, **3:**120; 133
as primer for glycogen synthetase, **34:**322
qualitative detection, **3:**110; 122
quantitative determination, **3:**110; 111
in rat, **34:**292–293
resynthesis from lactic acid, **6:**270; 271
role, **34:**289
role in fermentation, **5:**108–110; 113
and starch, **3:**119–120
storage, **5:**131
subcellular distribution and association with membranes, **34:**289
in sweet corn, **34:**356
synthase, **52:**24; 28
synthesis, **6:**299–300, **46:**159; 355, **52:**268
biological, **3:**131
enzymatic, **3:**129–130
in Type IV glycogen storage disease, **34:**362
type of linkage in, **11:**340
viscosity of solutions, **3:**118–119
in vitro synthesis of, **34:**358
and water storage, **34:**289
Glycogen *(α)*, synthesis of, **18:**222
Glycogen and cell structure, **8:**34

Glycogenase, **5:**120
Glycogen biosynthesis, in *Aerobacter aerogenes,* **34:**323
in bacteria, regulation of, **34:**314
branching enzymes in, **34:**351
by concerted action of glycogen synthetase and branching enzyme, **34:**358
from glucose, **34:**294
in liver homogenates, **34:**296
Novikoff ascites-hepatoma cells, **34:**296
primer for, **34:**351
role of glucose 6-phosphatase as a pyrophosphate phosphotransferase in, **34:**305
from UDPG, **34:**316
Glycogen breakdown and fertilization, **9:**307
Glycogen Chemistry (Meyer), **3:**109
Glycogen derivatives, molecular weight, **3:** 115–116
Glycogenesis, allosteric regulation, **28:**98–103
Glycogen-excess mutants, **46:**356
Glycogen-glucose shunt in liver, **15:**153
Glycogen in nucleus, **4:**14
Glycogen metabolism, **34:**285–417
hormonal regulation, **34:**412–413
importance of adrenaline and cyclic AMP, **34:**398
in perfused heart, **34:**394
rate-limiting factors in, **29:**358–360
of tumors, **29:**358–359
role of phosphotransferase activities of glucose 6-phosphatase in control of, **34:**305
Glycogen Metabolism, New Aspects of (Ryman and Whelan), **34:**285
Glycogen of liver and ketone bodies, **8:**408
Glycogenolysis, **3:**131, **57:**156
allosteric regulation, **28:**104–110
role of phosphorylase in regulation of, **24:**283–286
Glycogenosis Type I
analogy with hypoglycin poisoning, **52:**271
clinical features, **52:**271
different forms, **52:**271
glucose 6-phosphatase, **52:**271
glucose recycling, **52:**271
Glycogen phosphorylase, **46:**249, **50:**10; 370, **74:**174, *See* Phosphorylase
activation of, **26:**302
allosteric regulation, **28:**47; 54; 55; 104–106
tautomerism of, **50:**5
Glycogen Phosphorylase, Molecular Properties and Transformations of, in Animal Tissues (Krebs and Fischer), **24:**263
Glycogen phosphorylase *b,* **46:**236, **50:**58
allosteric properties, **28:**104–106
allosteric regulation, **28:**54; 55; 104–106
Glycogen phosphorylase in animal tissues, **24:**263–290
Glycogen-starch class of polysaccharides. *See also* Amylopectin-type polysaccharides and Amylopolysaccharides, branched
formation of, **11:**299–306; 315–324; 330
Glycogen storage disease, **34:**346
type I, **34:**302; 309; 311
type II, **34:**416
type III, **34:**405
type IV, **34:**362; 416
diagnosis leucocyte phosphorylase in, **34:**397
type V, **34:**290; 398; 403–404; 416
type VI, **34:**397
type VII, **34:**335
Glycogen storage disease, Type I, **41:**1; 26–27
Glycogen synthesis
hepatic glutamine transport, hepatocyte hydration, **72:**51–53
and potassium ions, **7:**27–28
in fermentation, **7:**27–28
Glycogen synthesizing enzyme, localization of, **24:**311
Glycogen synthetase, **36:**71; 73
acceptor requirements, **34:**321
activation of, **26:**106
assay of, **34:**319
D and I forms
control of interconversion
by hormones, **34:**331
by melabolites, **34:**334
in leucoytes, **34:**346
properties and interconversion of, **34:**317; 328
D phosphatase, **34:**317
effect
of hormones on, **34:**320
of insulin on, **34:**331; 332
of sex hormones on, **34:**344
glucosyl, donors for, **34:**316
I kinase, **34:**317
interference of α-amylase in assay of, **34:**319

liver, *a* and *b* forms, control of interconversion by glucose, **34:**342
liver location, **34:**338
in mammalian blood cells, **34:**346
mammalian liver, **34:**337
a and *b* forms, **34:**339
control by metabolites, **34:**339
effect of hormones on, **34:**341
mammalian muscle, **34:**326
control by metabolites, **34:**326
D and I forms, properties and interconversion, **34:**326
effect of ions on, **34:**327
metabolite control of, *in vivo,* **34:**337
muscle, dephosphorytion of, **34:**382
in *Neurospora crassa,* **34:**349
preparation, **34:**320
properties of, **34:**316
purification of, use of concanavalin A in, **34:**318; 320
similarity to phosphorylase *b* kinase, **34:**411
stimulation by glucose, **34:**385
yeast, **34:**324
allosteric regulation of, **34:**325
effect of glucose 6-phosphate on, **34:**324
Glycogen synthetase, ADP-glucose dependent regulation, **28:**75
allosteric regulation, **28:**47; 97–104
interconvertible forms, **28:**99–102; 107–109
Glycogen synthetase I, phosphorylation by phosphorylase *b* kinase-kinase, **34:**381
Glycogen synthetase kinase, **34:**341
Glycogen synthetase phosphorylase phosphatase, activation of by hormones, **34:**415
Glycogen-type polysaccharides, synthesis of, **11:**298–299
2-P-Glycohydroxamate, **43:**494
Glycolaldehyde
oxidation, **15:**256; 257; 258
role in carbohydrate metabolism, **15:** 154–155; 157–158; 163; 165; 177
in pentose synthesis, **28:**396; 397
as product of action of dihydroneopterin aldolase, **35:**63
relation to amino acid metabolism, **16:** 275–276
significance of identification in *Fusaria* fermentations, **5:**188
Glycolase, in tobacco, **6:**370
Glycolate, **55:**209
labeled, **23:**435; 441–445
metabolism by microorganisms, **23:**401; 404; 434–446; 464
oxidation of, **23:**434–440
syntheses from, **23:**440–446
(−)-2-d_1-Glycolate, **35:**95
2-t_1-Glycolate, **35:**94
Glycolate-glyoxylate oxidation cycle, **23:**345
Glycolate oxidase, **35:**94, **57:**88; 95; 97–98. *See also* L-α-Hydroxy acid oxidase
Glycolate-2-P, **51:**213; 214; 215; 222; 231; 232; 233; 242; 243; 246
Glycolates, **17:**149
Glycolic acid, **17:**422
conversion and function in photosynthesis, **19:**16
conversion to glycine, **16:**234
metabolic transformations in plants, **20:**218; 219
Glycolic acid dehydrogenase, PN-dependent, **26:**118
Glycolic acid from acetic acid, **9:**546; 547–550; 551–554; 558
Glycolic acid oxidase, in plastids, **20:**218
Glycolic acid oxidase, PN-dependent, **26:**121
Glycolic oxidase, **23:**435–436; 441; 445
in electron transfer, **19:**201
in photosynthesis, **19:**12
properties, **19:**202
Glycolipid, degradation enzymes, **60:**92
Glycolipid and Glycoprotein Degradation (Conzelmann and Sandhoff), **60:**89
Glycolipids, **48:**56; 60; 63; 64; 65; 66; 73; 74; 81; 83, **52:**25
amino sugars in, **25:**442
blood-group active, **48:**55
endogenous, **48:**82
Forssman, **52:**71
human intestinal, **52:**27
hydrolysis, **48:**40; 52
membrane, **48:**73
metabolism of, **24:**311–312; 338
sialoglycolipids, **48:**73
Glycol-lactone accumulation of, in mutants, **16:**283
Glycollate, relation to amino acid metabolism, **16:**275–276
Glycols, **17:**138
aliphatic, reaction with borate, **12:**494–495

N-Glycolylneuraminic acid, as product of neuraminidase action, **20:**135; 144
Glycolysis, **5:**93; 130–138; 213, **15:**151; 171–173, **48:**227, **57:**144–145; 156, **59:**318–319. *See also* Glycosidases
adenine nucleotide regulation of, **29:**347–348
aerobic, DPNH oxidation and, **29:** 339–346
anaerobic, rates of, **29:**326–327
effect of, **29:**352
of malignant cells, **29:**337–339
multisite control of tumors by, **29:**357–358
in oncogenesis, **29:**334
sites of regulation in, **29:**355
stimulation of, by external potassium, **29:**346–347
suppression by oxygen of, **29:**325
allosteric regulation, **28:**74–86
and ATP synthesis, **23:**324–325; 327–334; 339–340
and glycolate metabolism, **23:**444; 463
and citrulline phosphorylase action, **15:**382–383
creatine kinase in, **54:**368
distribution of, **24:**330; 332; 333; 335–336
in sperm, aerobic, **24:**52–55
effect of, pH, **24:**76
of, temperature, **24:**74
and motility, **24:**46–51
energy supply from, **15:**142–144; 152–153; 330
enzymes from different sources, **25:**333
in luminous bacteria, **25:**151
enzyme system of, **4:**41; 72; 202; 206; 207; 208; 209
glucagon effects on, **54:**150; 154
and inactivation of pantothenate, **7:**445–448; 450–453
streptococcal, **7:**410; 445–448; 450–453
inhibition, by mitochondria, **21:**56
in mitochondria, **21:**54–56
inhibitors, **5:**134–135; 138
localization of enzymes, **13:**7
muscle, effect of fluorophosphates on, **13:**332
effect of rate of ATP turnover, **13:**179
relation to deamidation of asparagine and its peptides, **13:**251–252
relation to deamidation of glutamine and its peptides, **13:**261–262
and nitrogen metabolism in plants, **9:** 411–412; 431–433
of semen, **9:**336–337; 354; 355–367; 369; 372–374; 378–379; 381
nonphosphorylating, **5:**136
in normal and cancer tissues, **4:**203; 211; 213; 218–232; 239; 240; 243; 244; 245; 249
and phosphate bond energy, **1:**132
phosphorylating, **5:**136; 137
phosphorylative, **4:**206; 207; 208; 209
in plants, **18:**291–293, **19:**5–12
and pyridine nucleotides, **15:**120
relation to tryptophanase action, **10:**78–81
and sorbitol, **15:**150
Glycolytic activity, and DPN level, **17:**66
Glycolytic activity of cell nuclei, **16:**468
Glycolytic enzymes
in mitochondria, **21:**54–56
multiple forms, **25:**333; 351–358
of nucleus, **27:**247; 250–253; 261–263; 272–274
thiol groups, **27:**213–214
Glycolytic enzymes in sperm, **24:**86
Glycolytic flux, **52:**251
effect of glucagon, phenylephrine and vasopressin, **52:**306
Glycolytic pathway, barriers to reversal, **28:**86–87
Glycolytic signal, **59:**380–381
Glycolytic system, classification, **14:**233
Glycon, configuration of, **36:**101
Glycone, **56:**143
Glycopeptide, **34:**457; 478
complex, **34:**478
Glycopeptides, **48:**42; 55; 57; 65
blood group N, **48:**56
chemical coupling to proteins of, **41:**99; 115–117
hydrolysis, by *endo*-N-acetyl-β-glucosaminidases, **48:**40
rates of, **48:**41
isomeric, defining terminal linkage specificity, **66:**50–51
sensitivity to *endo*-β-N-acetylglucosaminidase, **48:**52
from sialoglycopeptides, **48:**68
structure of, **48:**39
triantennary, structure, **66:**64–65
tri-branched, stereospecific binding, **66:** 60–61

Glycopetide, linkages of, **25:**451–452
Glycophorin, **52:**27; 32; 99; 105; 122; 124; 126; 135
Glycoprotein, **43:**144, **45:**407
biosynthesis, **28:**114–115
catabolism, **60:**91
degradation
by β-*N*-acetylhexosaminidase, **60:**247
by β-galactosidase, **60:**250
degradation enzymes, **60:**250
endocytosis, **60:**222
transport to degradation site, **60:**222
Glycoprotein hormones, **41:**113–115; 118
Glycoprotein IIb/IIIa complex, **59:**48–49
Glycoproteins, **36:**132, **41:**99–127, **47:**84; 88; 137, **48:**33; 38; 39; 42; 57; 60–67; 76; 77; 83; 84, **52:**25
α_1-acid, **48:**61
alteration by sialidase, **48:**59
antifreeze, **52:**37; 39; 40; 41; 43; 44; 47; 52; 53; 56; 59; 98; 112
asiolo-α1-acid glycoprotein, **52:**30; 34; 41; 47
biologically active, **48:**57
blood-group active, **48:**54; 55
M and N, **48:**56
with blood group specificity, **25:**447–450
carbohydrate-protein linkage, **25:**450–453
linkage of amino sugars in, **25:**442–453
bovine submandibular, **48:**53
endogenous, **48:**83
enzymes, **48:**59
hydrolysis, **48:**40; 42; 52
new method of labeling, **41:**99; 105; 117–118
oligosaccharide units in, **26:**58–59
saccharides of, **26:**49
rabbit Tamm-Horsfall, **48:**52
in the regulation of copper metabolism, **41:**100
secreted, **48:**56
sensitivity to glycosidases, **48:**52
sialidase activity on, **48:**74
sialoglycoprotein, **48:**56
sublingual, **52:**34
surface, **48:**69
transport of circulating, **41:**99–127
tritiation of terminal galactose, **41:**100; 101
Glycoproteins, Role of Surface Carbohydrates in the Hepatic Recognition and Transport of Circulating (Ashwell and Morell), **41:**99
Glycoprotein synthesis, **39:**236
Glycosaminoglycan Degradation (Kresse and Glossl), **60:**217
Glycosaminoglycans, **52:**81
role in cell aggregation, **40:**51
sequential degradation, **60:**271–283
Glycosaminoglycuronoglycans, of connective tissue, **25:**434–441
Glycoseen, in DR metabolism, **28:**438
Glycosidase, **46:**480
β-Glycosidase, column chromatography, **22:**437
Glycosidases, **4:**41–72, **23:**130, **48:**29–95
N-acetylglucosaminidases, **56:**145–150
assay of, **56:**145–146
occurrence and localization of, **56:**145
properties of, **56:**147–149
purification of, **56:**146–147
regulation of, **56:**150
role of, **56:**149
action on coil surfaces, **48:**67
action on complex carbohydrates, **48:**51–63
action on human erythrocytes, **48:**81
action on membranes and cells, **48:**63–85
activation of substrate, **4:**43
activity of, **56:**143
from almond emulsin (*See* Almond emulsin)
from almond emulsion, **48:**39
application to analysis of complex carbohdrates, **48:**54
arabinosidases, **56:**150–154
assay of, **56:**152
occurrence and localization of, **56:**151–152
properties of, **56:**153–154
purification of, **56:**152–153
role of, **56:**154
assay of, **48:**51
chemistry of, **56:**143–145
classification, **4:**69–72, **14:**220; 223–224; 245–246
complex formation with substrate, **4:**43
endogeneous, **48:**66
endoglycosidases, **48:**30; 32; 40–42; 45; 46; 54; 55
exoglycosidases, **48:**30; 31; 38; 39; 45; 46; 53; 54; 57; 62; 77; 80
purification, **48:**34
extracellular, **48:**30
extrinsic, **48:**66
β-fructofuranosidases, **56:**155–165
assay of, **56:**156–157
occurrence and localization of, **56:**155–156

Glycosidases (*Continued*)
properties of, **56:**159–162
purification of, **56:**157–159
regulation of, **56:**163–165
role of, **56:**162–163
fucosidases, **56:**220
α-galactosidases, **4:**46; 47; 51; 52; 65–66, **56:**166–179
assay of, **56:**171
occurrence and localization of, **56:**166–171
properties of, **56:**173–178
purification of, **56:**171–173
regulation of, **56:**179
role of, **56:**178–179
β-galactosidases, **4:**46; 47; 51; 52; 58; 64–65, **56:**180–186
assay of, **56:**182
occurrence and localization of, **56:**180
properties of, **56:**183–185
purification of, **56:**182–183
regulation of, **56:**186
role of, **56:**185–186
β-glucosidase (*See* β-Glucosidase)
α-glucosidases, **56:**186–195
assay of, **56:**188
occurrence and localization of, **56:**187–188
properties of, **56:**189–194
purification of, **56:**188–189
regulation of, **56:**195
role of, **56:**194–195
β-glucosidases, **56:**195–205
assay of, **56:**197
occurrence and localization of, **56:**196–197
properties of, **56:**198–203
purification of, **56:**197–198
regulation of, **56:**205
role of, **56:**203–205
β-glucuronidases, **56:**220–221
hydrolysis of pentosides, hexosides heptosides, **4:**50; 51; 52
inactivation with conduritol B epoxide, **48:**46
inhibition by aldonolactones, **36:**152, **48:**49
intracellular, **48:**30
invertases and inulases, **4:**66–68
isolation and purification, **48:**33–38
lyosomal, **48:**62
α-mannosidases, **56:**205–211
assay of, **56:**206
occurrence and localization of, **56:**205–206
properties of, **56:**208–210
purification of, **56:**206–208
regulation of, **56:**210–211
role of, **56:**210
β-mannosidases, **56:**211–214
assay of, **56:**211–212
occurrence and localization of, **56:**211
properties of, **56:**212–213
purification of, **56:**212
role of, **56:**213–214
mechanism of action, **4:**42–45, **48:**43
multiple forms, **25:**336–337
product inhibition, **25:**190
multiple forms of, **56:**145
origin of, **56:**223–228
occurrence as isozymes, **48:**33
occurrence of, **56:**143
purification of, **48:**36; 37; 38; 53; 56; 157
affinity chromatography, **48:**36; 37; 51; 65
column chromatography, **48:**36
electrophoresis, **48:**33; 36; 65
gel emulsion chromatography, **48:**38
ion exchange chromatography, **48:**38
isoelectric focusing, **48:**33; 36; 38; 65
purified, table of, **48:**34
rhamnosidases, **56:**221
sources other than almond, **4:**61–68
specificity of, **48:**31; 32; 38–43; 46; 65, **56:**144
substrates for, **56:**143
in transglycosylation reaction, **56:**144
trehalases, **56:**214–219
assay of, **56:**215–219
occurrence and localization of, **56:**214–215
properties of, **56:**217–218
purification of, **56:**216
role of, **56:**218–219
Weidenhagen theory, **4:**68–69
β-xylosidases, **56:**221–223
N-Glycosidases, classification, **14:**224; 225; 247
Glycosidases, Inhibition by, by Aldonolactones (Levvy and Snaith), **36:**151
β-Glycosidases, multiple forms, **25:**350
Glycosidases, Specificity, Classification, and Action (Pigman), **4:**41
Glycosidases (α- and β-), soil, **28:**386
Glycosidases-Properties and Application to the Study of Complex Carbohydrates and Cell Surfaces (Flowers and Sharon), **48:**29
Glycoside, **37:**14

Glycoside cluster effect, **66:**46–47
Glycoside formation, microbial, from alkanes, **27:**493
Glycoside hydrolases, localization of, **24:** 322–323
Glycoside linkage exchange in synthesis of di- and polysaccharides, **10:**124; 138–141
Glycoside phosphorylases, **18:**226
C-Glycoside phosphorylases, **18:**224
Glycosides, **17:**432, **48:**43; 45, **56:**143. *See also* β-Glycosides; Sugar nucleotides
  hydrolyzability of pentoses, hexoses, and heptoses, **4:**50–53
  in plant respiration, **18:**283
    synthesis of, **18:**221–225
  substrates for hydrolytic enzymes, **4:**41
  synthesis of, **26:**63; 64
N-Glycosides, pyrophosphorolysis of, **18:**194
  synthesis of, **18:**215–219
Glycosidic bond, inversion by phosphorylases, **18:**224–225
Glycosidic bonds, **56:**143
1,2-Glycosidic linkage in polysaccharides, **11:**340
1,3-Glycosidic linkage in polysaccharides, **11:**340
1,4-Glycosidic linkage in polysaccharides, **11:**301; 304; 306; 310; 317; 319; 324–326; 327–329; 340; 345–346; 348; 365; 366; 368–369; 369–370
1,6-Glycosidic linkage in polysaccharides, **11:**301; 304; 307; 310; 324–326; 327–329; 340–341; 348; 363–365; 366; 368–369; 369–370
2,6-Glycosidic linkage in polysaccharides, **11:**307; 314
Glycosidic linkage with galactose in glycoprotein, anomery of, **25:**445–446
Glycosine hydrolysis, mechanism of, by enzymes, **36:**174–175
Glycoslylation, RNA polymerase II transcription, C-terminal domain characteristics, posttranslational modifications, **71:**49–50
Glycosphingolipids, **52:**117
Glycosylase-endonucleases, **60:**13
Glycosylases, specificity, **60:**15
Glycosylasparagines, **60:**161
Glycosylation, **58:**200
N-Glycosyl compounds
  enzymes hydrolyzing, **24:**323
Glycosyl compounds, enzymes acting on, **24:**322–323
Glycosyl donors. *See also* Sugar nucleotides
  in synthesis of polysaccharides, **26:**36; 63–78
Glycosyl donors (C^{14}), in synthesis of starch, **26:**62
Glycosyl-enzyme complex, **17:**221–224
Glycosyl enzymes, product inhibition, **25:**190
Glycosyl fluorides, **55:**247
Glycosylglycerol, in teichoic acid, **30:**226
Glycosylic bond, hydrolysis, **60:**2
Glycosyl phosphorylases, **18:**223–225
C-Glycosyl transfer, **18:**200–201
N-Glycosyl transfer, **18:**200–201
Glycosyl transfer, from sugar nucleotides, **26:**57–78
Glycosyltransferase, **52:**24; 25; 27; 28; 88
  purification, **52:**144–158
β-glycosyl transferase enzyme, exopolysaccharide alginate synthesis, **70:**234
Glycosyltransferases, **36:**74
Glycosyl transferases, **46:**407
Glycosyltransferases
  γ-glutamyl transpeptidase, cancer-associated changes, **72:**270–271
  localization of, **24:**311–312
  specified by genes, that determine blood groups, **36:**141
Glycosyltransferases and Their Use in Assessing Oligosaccharide Structure and Structure-Function Relationships (Beyer, Sadler, Rearick, Paulson, and Hill), **52:**23
Glycosyl transfer reaction, phospholipid-LPS complex in, **31:**90–91
D-Glycuronic acid in mold metabolism, **9:**544; 571
Glycylalanine
  absorption spectrum, **8:**131
  hydrolysis, by metal ions, **21:**178
Glycylaminobenzoic acid, transamination in, **16:**239
2-*N*-Glycyl-*p*-benzoquinone, $E_{o'}$ value for, **16:**114
4-*N*-Glycyl-*o*-benzoquinone, $E_{o'}$ value for, **16:**114
S-Glycylcystine, intramolecular aminolysis of, **19:**241

Glycyldehydroalanine, absorption spectrum, **8:**131; 152
  and dehydropeptidases, **8:**139–145; 148; 151; 152; 157; 159; 163
  from diglycylaminopropionic acid, **8:**127
  preparation, **8:**128
Glycyldehydrophenylalanine, absorption spectrum, **8:**132; 137; 138
  and dehydropeptidases, **8:**134; 135; 137–140; 143–145; 151; 163; 166
  effect of various enzymes, **8:**134
  preparation, **8:**123
Glycylglycinase, column chromatography, **22:**458; 459
Glycylglycine
  chelates, in metal ion activation of enzymes, **21:**146; 147
  configuration of, **23:**14–16
  effect on succinic dehydrogenase, **18:**88–90
  inactivity in transamination, **16:**239
    reaction with quinones, **16:**114
Glycylglycine dipeptidase
  activation by metal ions, kinetics, **21:**140; 148
  specificity of, **12:**211–216
Glycyl-glycine hydrolase, localization of, **24:**324
Glycylglycine peptidase, classification, **14:**247
Glycylglycyldehydroalanine, and dehydropeptidases, **8:**140; 144
Glycylglycylglycine, reaction with quinones, **16:**114
Glycyl-L-leucine, peptidase, classification, **14:**247
Glycyl-L-leucine dipeptidase, specificity of, **12:**216–218; 224
Glycyl peptides
  rate of deamination, **16:**117
  steric factor in enzymic hydrolysis, **15:**292
O-Glycylphenol, effect of carbon dioxide on, **19:**247
Glycylphenylalanine, absorption spectrum, **8:**132
Glycyl phosphate, effect of carbon dioxide on, **19:**247; 250
  synthesis, **19:**250
Glycylthiophenol, alkaline aminolysis of, **19:**241
Glycyl-tRNA GlcNAc-MurNAc(-pentape-ptide)-*P-P*-lipid transferase, **40:**124
N-Glycyl-L-tyrosine, action of tyrosinase on, **14:**139
Glyogenosis, *See* Glycogen storage disease
Glyoxal, **59:**130
  effect on pectic substances, **20:**346
Glyoxalase, **15:**143, **23:**339; 456
  activation by glutathione, **11:**251
Glyoxalase, I, **19:**241; 249
Glyoxalase I, **55:**214, **56:**293
  action of, **59:**118
  active site of, **59:**292
  bound substrate conformation, **59:**292
  CD studies, **59:**115
  cis-enediol, **59:**109
  conformation of inhibitor, **59:**117
  and coordination geometry, **59:**289
  Co^{2+}-product distances, **59:**290
  Co^{2+}-substituted, **59:**114
  description of, **59:**106–108
  detoxification by, **59:**288–289
  and diastereomers, **59:**108
  effective molarity of, **59:**136
  EPR analyses, **59:**114
  and EXAFS, **59:**114; 289–290
  flexibility in GSH backbone, **59:**117
  fluorescence, **59:**117
  as hydride, **59:**111–114
  inhibitors of, **59:**114–118
  mechanisms of, advanced, **59:**118–122
  Mn(II)-enzyme, **59:**116
  Mn^{2+}-product distances, **59:**290
  nature of, **59:**108–111
  and NMR spectroscopy, **59:**108; 110; 116
  product conformation on enzyme, **59:**290
  as proton transfer, **59:**111–114
  proton transfers, **59:**118–119
  rate
    of advantage of, **59:**136
    of product dissociation, **59:**292
  reversibility, **59:**289
  second sphere product complex, **59:**290
  trans-enediol, **59:**109
  tyrosine groups, **59:**115
  water immobilized, **59:**118
  and Zn-ligands, **59:**290
Glyoxalase II, **59:**122–127
  action on glutathione derivatives, **19:**248
    biological role and specificity, **19:**249
Glyoxalase reaction, **23:**328
Glyoxalase use in estimation of glutathione, **13:**248–249; 280–281
Glyoxaline, *See* Imidazole

Glyoxylase, **46:**447
Glyoxylate
and cysteamine, **57:**112
and L-cysteine, **57:**113–114
from deamination of glycine, **16:**117
relation to amino acid metabolism, **16:** 275–276
in transaminations, **16:**196; 203–205; 225; 234
formation of, **57:**148–152
and glycine metabolism, **23:**453–454; 457
and glycolate metabolism, **23:**435–446
labeled, **23:**442; 443
and oxalate metabolism, **23:**450–452
as inhibitor of D-aspartate oxidase, **57:**121
as inhibitor of oxidative metabolism, **57:**151
and α-ketoglutarate dehydrogenases, **57:**151
metabolism of, **57:**87; 148–150
effects of, **57:**151–152
and pyruvate, **57:**151
in tissues, **57:**112
Glyoxylate adducts
of L-cysteine, **57:**115–116; 122
of L-cysteinylglycine, **57:**115–116; 122
as substrates for mammalian enzymes, **57:**86
Glyoxylate carboligase, and glycine, **23:**454
and glycolate, **23:**442; 443; 445
and oxalate, **23:**452
Glyoxylate cycle, **57:**88; 148
control of, **23:**415–420
of microorganisms, **23:**404–420; 425; 428; 430–431; 434; 463
and ribose synthesis, **28:**412; 413
Glyoxylate thiohemiacetals, **57:**128
Glyoxylate transaminase, **16:**225
Glyoxylic acid, **15:**128, **17:**422
amination of, **18:**258
and biological methylation, **12:**471–474
from purines, **12:**451–453
formation and conversion in photosynthesis, **19:**16
in synthesis of active formate, **21:**411; 412
transamination with, in plants, **20:**329
Glyoxylic acid in mold metabolism, **9:**550–554; 558–560; 570–571
Glyoxylic acid oxime, **35:**157
Glyoxylic acid reductase, **15:**116
in photosynthesis, **19:**16
Glyoxylic dehydrogenase, **23:**451
Gly-Pro, **57:**64
Gly-186 residue, dihydrodipicolinate synthase (DHDPS), **72:**290
Gly-L-Tyr, **57:**21
G_{m2}, **38:**300
labeled, **38:**300
GM_2 activator protein
assay, **60:**179
genetics, **60:**181
glycolipid specificity, **60:**180
kinetics, **60:**181
molecular weight, **60:**180
physiological role, **60:**182
properties, **60:**180
purification, **60:**179
stability, **60:**181
G_{m1}-β-Galactosidase, **38:**298; 302
GM_1 gangliosidosis, **60:**103; 165
GM_2 gangliosidosis, **60:**165
variant AB, **60:**126
variant B, **60:**126
G_{m1} Gangliosidosis, **38:**303
G_{m2}-Hexosaminidase, **38:**298; 300
GMP, **55:**146
in muscarinic receptor binding, **55:**173; 182
reduction of, **28:**443; 445; 448
GMP-dehydrogenase, allosteric regulation, **28:**122
GMP-PNP, **43:**15
GMP-pyrophosphorylase, inhibition by nucleotides, **28:**124
GMP-reductase, allosteric regulation, **28:** 121–123
GMP synthetase, **37:**476; 493; 497; 516; 548; 563; 577; 596, **66:**247–249, **72:**119–124
GAT and synthetase domain organization, **66:**247–248
GAT function, **66:**248–249
human enzymes, **72:**123–124
pyrophosphate binding motif, **72:**166–167
sequential response, **37:**587; 597
structure and metabolism, **72:**120–123
synthetase function, **66:**249
triad amidotransferase and, **72:**138–139
G_{m2}-sialidase, **38:**300
Goitrogenic compounds, **7:**207
Gold, and tyrosinase activity, **14:**86; 93; 94
Goldacre bubbles, and origin of life, **27:**365
Goldblatt method, hypertension, **7:**515
Gold mercaptide, **11:**216
Gold sol reaction with irradiated proteins, **9:**132

Golenkinea, **46:**98
Golgi apparatus, **58:**200, **60:**90
change in mass upon injury, **8:**47
Golgi bodies, and defense mechanisms, **8:**52–55
and functioning of cytoplasm, **8:**20–23
Gomori-Takatmatsu technique for alkaline phosphatase, **16:**418
Gonyaulax polyedra, luminescence in, **25:**161
Gorgonines, amino acids in, **14:**266
Gossypium, **55:**80
Gossypol, **55:**80
biosynthesis of, **55:**82
structure of, **55:**55
Gout, **41:**1–28
enzyme abnormalities in, **41:**2–25
enzymology of, **41:**1–28
Gout, Enzymology of (Kelley and Wyngaarden), **41:**1
G-1-P, *See* Glucose-1-phosphate
G5P, *See* Guanosine 5′-monophosphate
G6P, *See* Glucose-6-phosphate
gp91^{phox}, **65:**75–78
hydropathy and glycosylation sites, **65:**77
G protein, **55:**161, **65:**58
G-protein-coupled receptors (GPCRs), adenosine receptors, **69:**84; 85–89; 94–99; 100; 108–110; 114
G-proteins, states of, **69:**122
Gradient affinity chromatography technique, **46:**262
Gradients, **34:**524
Grafting, influence on disease, resistance, **11:**430–432
Gramacidin S, amino acid sequence of, **33:** 337–338
biosynthesis of, **33:**337–378
activation of amino acids in, **33:**370–373
by *B. brevis,* **33:**339–343
by cell-free system from *B. brevis,* **33:** 343–350
mechanism of, **33:**367–375
effect of streptomycin on synthesis of, **33:**345
enzyme system for formation of, fractionation of, **33:**356–367
fractionation of by, DEAE-cellulose column chromatography, **33:**365; 366
DEAE-Sephadex, **33:**365
DEAE-Sephadex A-25 column, **33:**359
DEAE-Sephadex A-50 column, **33:**366
hydroxylapatite column, **33:**358
Sephadex G-200, **33:**366
Sephadex G-200 column, **33:**360–361
Sephadex G-200 gel filtration, **33:** 363–364
formation of, by different enzyme fractions (table), **33:**343
effects of, amino acid analogs on, **33:**340
p-fluorophenylalanine on, **33:**340
norleucine on, **33:**340
norvaline on, **33:**340
fractionation of by differential centrifugation, **33:**344
inhibited by 5-bromouridine, **33:**340
inhibitors of (table), **33:**344
fractionation of enzymes of by differential centrifugation, **33:**347
intracellular, **33:**339
small peptides related to, **33:**351–356
HCO-D-Phe-L-Pro-L-Val-L-Orn-L-Leu-D-Phe-L-Pro-L-Val-L-Orn-L-Leu-$NHCH_2CH_2OH$, **33:**356
D-phenylalanyl-L-proline diketopiperazine, **33:**351
D-phenylalanyl-L-prolyl-L-valine, **33:**351
D-phe-L-Pro-DKP, **33:**359
tetrapeptides, **33:**353
synthesis of, effect of *p*-fluorophenylalanine on, **33:**347
effect of
5–5fluorotryptophan on, **33:**340
norleucine on, **33:**347
DL-β-2-thienylserine on, **33:**347
inhibited by, actinomycin D, **33:**342–343
chloramphenicol, **33:**340–342; 345; 347; 348
dihydrostreptomycin, **33:**340–342
5-fluorouracil, **33:**340
mitomycin C, **33:**342–343
puromycin, **33:**340–342; 345; 347; 348
ribonuclease, **33:**348
inhibitors on, **33:**340–343
pH optima of, **33:**348
Gramacidins, straight chain, with formyl group on N-terminal amino acid, **33:**354–355
Gramicidin. *See also* Tyrothricin
action of tyrosinase on, **14:**150
adsorption studies, **14:**347
structure study, **14:**266; 278; 279; 302–303
alkali metal cation induction by, **29:**288

induction of monovalent cation uptake by, in mitochondria, **29:**287–292
potassium ion efflux from phospholipid micellar systems stimulated by, **29:**291
carbohydrate and phosphate storage, interference with, **4:**185–194
chemical nature, **4:**161–166
color reactions, **4:**161
concentration and biological activity, **4:**170; 171; 172
constituents, **4:**163; 164; 165
crystalline, **4:**161
effect on bacteria, **4:**154; 169–173; 185
on bacterial infections, **4:**175; 176
on cellular metabolism, **4:**184–185
on metabolism of staphylococci, **4:**186; 187
effect on mitochondrial oxidative phosphorylation, **21:**31
inhibition of, **4:**174; 175
ion channels, **71:**214
isolation, **4:**154
mode of action, **4:**185–192
molecular weight, **4:**161; 162; 163
physical properties, **4:**159
resistance to enzymatic hydrolysis, **4:**169
respiration of bacteria and cells in presence of, **4:**184–185
solubility, **4:**158; 159; 160; 161
straight chain structure, **35:**25
structure and biological activity, **4:**165; 182–184
toxicity for animal cells and tissues, **4:** 178–182
in treatment of plant diseases, **20:**392
Gramicidin (Hotchkiss), **4:**153
Gramicidin S, **35:**162, **36:**304
biosynthesis, **35:**3–20
activation of amino acids, **35:**11
ATP-AMP exchange, **35:**6; 8
ATP-PP_i exchange, **35:**4; 8
complementary enzyme fractions, **35:**4
elongation, **35:**15
initiation, **35:**8; 14; 15; 29
effect on K^+, **29:**46
homolog of, effect on β-lactamase, **28:**263
structure, **35:**15
Fig. 1, **35:**2
Fig. 2, **35:**2
in treatment of plant diseases, **20:**392
Gramicidins, Biosynthesis of (Saito, Otani, and Otani), **33:**337
Gramicidins (A, B, and C), as cation conductor, **29:**46
effect on K^+, **29:**46
effect on permeability of mitochondria to choline, **29:**46–47
structure of, **29:**48
Gramidicin D, **55:**173
Gramine, synthesis of, **25:**383
Gram-negative bacteria, **41:**197
β-lactamase of, **28:**242–243; 248; 249; 265–270; 274; 287; 290; 293; 295
Gram-negative β-lactamase, properties of, **28:**309–311
Gram-negative organisms, cell walls, **26:**196; 201–203; 209; 213; 216; 218; 221–222; 224–228
Gram-positive bacteria, β-lactamase of, **28:**242; 249; 263–266; 272–273; 286; 287; 303
Gram-positive β-lactamases, properties of, **28:**309–311
Gram-positive organisms, cell walls, **26:**196; 201–203; 209; 211; 218; 221–228
Granular flour, use in fermentation, **7:**590
Granules
cytoplasmic (*See* Cytoplasm, Microsomes, Mitochondria, Particulate material and Particles)
and oxidation, **18:**244–246
pigment, and tyrosinase action, **14:**133; 137; 145–146
subcellular *(See also Cytoplasmic particles, Mitochondria),* adsorption studies, **14:**356
Granulomatosa, chronic, **46:**170
Graphic analogies, algebraic analogies *versus,* enzyme action, **73:**49–52
Grasses, antibiotic treatment of diseases, **20:**401; 402
Grass lignin, **34:**213
Gratuity principle, **22:**112–114
Green electron-transport particle, **23:**305
Green particle fraction, of mitochondria, evaluation of purity, **21:**76
properties, **21:**103
Green pigment
green sulfur bacteria, **1:**312; 313
purple bacteria, **1:**292; 293; 308; 309; 310; 312

Green plant photosynthesis, a photoelectron transport scheme, **25:**89–93
Green plants, antibacterial substances from, **7:**461–511
Green sulfur bacteria
metabolism, **1:**236; 269; 281*t*
pigments, **1:**312; 313
Grinding, effect on pectic substances, **20:**353
Grisein, **35:**147
Griseofulvin, **46:**137
in treatment of plant diseases, **20:**384; 390–392; 394; 402; 403
translocation in plants, **20:**387
Ground plasm, localization of enzymes in, **13:**5–8
Ground state
binding, enzyme action, algebraic *versus* graphic analogies, **73:**51
reciprocal stabilization, enzyme action, **73:**47–49
stabilization, enzyme action, **73:**43–46
Group potential
definition, **1:**102; 103
phospho-organic compounds, **1:**103–110
Group-specific reagents, differential modification and inactivation, **37:** 460–468
Group transfer, as metabolic reaction, **1:** 154–158
Group-transferring enzymes and kinases, **18:**195–201
Group-translocation mold, and amino acid transport, **49:**56
Growling point, **34:**482
Growth
effects on carotenogenesis, **21:**307; 308
and folate-containing enzymes, **27:**458–460
plant, and phytochrome, **26:**1–33
of murein sacculi, **26:**215–222
promotion by α-keto analogues of amino acids, **16:**187; 211
relation to β-glucuronidase, **16:**386–394; 397–401
steroid hormone induced, **16:**386–394; 397–401
propagation reactions of biopolymer, **30:**181
role of thiols in, **11:**253–255; 257
and sulfur-containing amino acids, **7:**376; 383
of coliform organisms, and nicotinic acid and derivatives, **7:**421–423
of influenza bacilli, effect of coenzymes, **7:**415–416
of luminous bacteria, and luminescence, **7:**251–253
and salts, **7:**253–255
and substrates, **7:**253–255
of microorganisms, action of pantothenate, **7:**448–453
and metabolism of vitamin-like substances, **7:**409–460
and pantothenic acid metabolism, **7:** 435–454
of organisms, and side chains of heme, **7:**324–325
and synthetic function of citric acid cycle, **18:**259–264
Growth and Metabolism of Vitamin-like Substances, Interrelations in Microorganisms (Mellwain), **7:**409
Growth control mechanism, **4:**202; 203; 209; 218–232
and the ATP system, direct control, **4:** 221–223
growth control, **4:**223; 225–227
pharmacological control, **4:**223–225
Growth factor activity, as criterion for normal intermediate, **16:**254
Growth factor for *Lactobacillus,* **9:**468; 494–495
Growth factors, **57:**87; 162–164. *See also p-*Aminobenzoic acid; Biotin; Thiamin
bacterial separation by chromatography, **14:**350
for *Lactobacillus bulgaricus,* **14:**49–71
for various *Lactobacilli,* **14:**49–50; 65–69
for various organisms, **14:**65–69
cholesterol ester hydrolases regulation, **69:**69
effect of pH on activity, **18:**34–35
inhibition of (*See* Inhibitory action, analogues)
inhibitor-antagonist ratio, **6:**44–51
Growth hormone
relation to transamination, **16:**228
tyrosine hydroxylase regulation, **70:**146
Growth hormone receptor, fibronectin, central cell-binding domain of, **70:**11
Growth hormones, **47:**136
Growth in animals, and glutamine utilization, **13:**272
Growth-inhibiting concentrations of antibacterial substances, **7:**490–493

Growth-promoting activity, **42:**75; 120; 127; 128; 133; 138; 139; 140; 144; 145; 148; 149; 152; 156
Growth-promoting factors, **23:**471–475
Growth rate, and lipoic acid, **18:**343
Growth rates of bacteria, **7:**250
Growth zone, in *Streptococcus pyogenenes,* **26:**221–222
Gruppenstoffe, use of term, **18:**358
GSH, **59:**132–134; 143; 155, *See* Glutathione
  oxidation, **65:**105
GSH-binding cavity, **57:**384
GSH-independent formaldehyde dehydrogenase, **59:**128
GSH peroxidase activity
  for basic human GSH transferase, **57:**372
  for cytosolic protective protein, **57:**377–378
  for microsomal rat GSH transferase, **57:**372
  for rat GSH transferase 2–2, **57:**369
GSH S-aryltransferases, **57:**370
GSH transferase
  acidic, **57:**378
  active site of, **57:**383–387
  binding of, **57:**382–386
  and carcinogenesis in rodents, **57:**399
  control of, biological, **57:**404–405
  evolution of, biological, **57:**375–378
  gene coding for, **57:**399–404
  human
    acidic, **57:**373
    amino acid composition of, **57:**367
    basic, **57:**372
    characteristics of, **57:**367
    inhibition characteristics of, **57:**368
    near-neutral, **57:**372–373
    nomenclature of, **57:**361; 363
    substrate specificities for, **57:**368
  identification of, **57:**357–360
  importance of, **57:**358
  induction of, **57:**404–405
  isoenzymes of
    characterization of, **57:**363–373
    nomenclature of, **57:**360–363
    occurrence of, **57:**360
    purification of, **57:**373–375
  and kinetic reaction scheme
    branched reaction scheme, **57:**379
    discrimination between models, **57:**380
    enzyme memory mechanism, **57:**383
    hybrid mechanism, **57:**379
    ping-pong mechanism, **57:**379–382
    sequential mechanism, **57:**379; 381–383
    single displacement mechanism, **57:**382
    substrate-rate saturation curve, **57:**378–379
  placental, **57:**391
  protein structure of
    comparison of rat and mouse, **57:**398
    mouse, **57:**396–397
    rat, **57:**393–396
  rat
    1-1, **57:**369
    1-2, **57:**370
    2-2, **57:**369–370
    3-3, **57:**370
    3-4, **57:**371
    4-4, **57:**370–371
    5-5, **57:**371
    6-6, **57:**371
    amino acid compositions of, **57:**364
    characteristics of, **57:**362
    inhibition characteristics of, **57:**366
    liver, **57:**361
    microsomal, **57:**372
    nomenclature for, **57:**362
    substrate specificities of, **57:**365
  rhesus monkey, **57:**367
  specificity for, **57:**384–386
  stereospecificity in catalyzed reaction, **57:**386–387
  substrates of, **57:**375–378
  tissue distribution of, **57:**387–388
    human, **57:**391–393
    rat, **57:**388–390
  tripeptide structures for, **57:**384
  in vitro translation of, **57:**399–404
g = 1.94 signal, **34:**103; 119; 146; 147
G-site, **57:**384–385
GS-SG, *See* Glutathione
GST_1, **57:**391–393
GST_2, **57:**391–392
GST_3, **57:**391; 393
GTF, *See* Glutamotransferases
GTF(G), *See* Glutamotransferases
GTF(GSH), *See* Glutamotransferases
GTP, **50:**311, **55:**159, *See* Guanosine triphosphate
  in allosteric regulation, **28:**65; 74; 81; 88; 89; 92; 140–141
  in muscarinic site binding, **55:**155; 161
  as precursor of folic acid, **53:**355

GTP (*Continued*)
as precursor of riboflavin, **53:**346–348
tubulin binding sites for, properties of, **53:**119–124
tyrosine hydroxylase regulation, **70:**163
GTPase, **55:**367–368
GTPase activity, **50:**311
GTPase activity of tubulin, intrinsic, **53:** 119–139
GTPase reaction, pharmacologic implications of, microtubule dynamics, **69:**146–147
GTP cyclohydrolase, **35:**55, **53:**347; 357
K_m for, **35:**56
molecular weight of, **35:**56
properties of, **35:**54–61
purification and properties, **53:**358
purification of, **35:**54–61
GTP cyclohydrolase II, **53:**348
GTP/(GDP) cycle, *Caulobacter* cell cycle, **71:**7
GTP hydrolysis, analog interactions during assembly, **53:**124–130
Guaiacol
peroxidase action on, **19:**107
in plant metabolism, **16:**141
Guaiacylacetone, **34:**213
Guaiacylglycerol-α-aryl ether, **34:**211
Guaiacylglycerol-β-aryl ether, **34:**211
Guaiacyl residues, in lignin formation, **18:** 360–361; 366; 373; 374
Guamine, **35:**469; 470
Guanase, **9:**442; 463; 489–490; 498
classification, **14:**246; 250
intracellular distribution, **16:**466
Guanidination, of RNase, **24:**207; 209; 220–224; 245; 247; 248
effect on immunological properties, **24:**184; 221
Guanidine, **15:**346; 355–356; 361; 363; 374; 379, **58:**106
actomyosin dissociation by, **13:**161
as inhibitor of clot formation, **16:**40
as inhibitor of oxidations, **23:**367–369
Guanidine-HC1, **54:**286; 289–290
Guanidine hydrochloride
dissolving of fibrin, **16:**36
effect on carotenogenesis, **21:**319
effect on RNase, **24:**185; 186–187
effect on –SH groups of proteins, **11:**222; 223–224; 229
Guanidines
inhibition of tumor growth by, **29:**336
substituted, **15:**379
Guanidinium chloride, **46:**387
Guanidinium ion, **49:**78
Guanidinium salts, **17:**151
Guanidinoacetate, **50:**194; 196
effect on liver fat, **50:**194
Guanidinoacetate methylpherase, steric requirements for, **25:**385
Guanidinoacetate methyl-transferase, **25:**144
Guanidinoacetic acid, **15:**373–374; 379
Guanidino carbon of arginine, origin, **16:**265
Guanidino group, **15:**329; 355–357; 362–364
arginine residues, **42:**230; 262
3-Guanidinopropionate phosphate, **50:**225
as synthetic phosphagen, **50:**225–226
Guanidinosuccinic acid, **15:**361–362
Guanidoacetic acid and methyl transfer, **7:**377
Guanine, **35:**461, **38:**215
alkylation of, **31:**14
and DNA structure, **18:**164
as folate precursor, **27:**424
genetic studies, **17:**426–428
and glycine metabolism, **23:**455
in histidine biosynthesis, **21:**403
inhibitory analogue of, **6:**134
nucleoside, precursor of pteridines, **35:**52
nucleotide, precursor of pteridines, **35:**52
-specific ribonuclease, **35:**463
of tobacco mosaic virus, **15:**460
Guanine aminase, classification, **14:**246; 250
Guanine catabolism, **56:**92
Guanine deaminase, localization of, **24:**328; 341
Guanine incorporation of C^{14} formate, **16:**457; 458
of DNA, **16:**431–433
of nuclear RNA, **16:**437; 439
in transamination, **16:**237
Guanine nucleotide, **58:**128
tubulin interactions with microtubule dynamics, **69:**123–124
Guanine nucleotides
adenosine receptors, **69:**96–98
GTPase activity, **53:**12; 120; 121
in mitochondria, **21:**10
reduction of, **28:**444
regulatory effects, **53:**10; 11
other systems, **53:**36–37
receptor, **53:**27

regulatory protein, **53:**11–15
GTPase activity, **53:**12
nucleotide exchange, **53:**14
substrate for cholera toxin, **53:**12
Guanosine, **18:**222, **35:**466, **51:**210
in folate synthesis, **27:**424
phosphorolysis of, **15:**104
ribosyl to deoxyribosyl of DNA, **28:**431
Guanosine aminase, classification, **14:**246
^{14}C-Guanosine, and DR synthesis, **28:**423
Guanosine diphosphate
in mitochondrial reactions, **21:**33; 35; 47
phosphorylation, in mitochondria, **21:**53
regulation of bacterial form glutamine synthetase, **72:**33–35
role in protein biosynthesis, **19:**264
Guanosine diphosphate choline, synthesis by carbodiimide method, **22:**189
Guanosine diphosphate mannose, **18:**201; 202; 207
Guanosine diphospho-D-mannose dehydrogenase, exopolysaccharide alginate synthesis, biosynthetic enzymes, **70:**232–234
Guanosine diphosphosugars, **52:**25
Guanosine 5′-monophosphate, and ATP, **18:**230
phosphorylation of, **18:**197
Guanosine-5′-phosphate synthetase, *See* Xanthylate amidotransferase
Guanosine-5-phosphate transaminase, column chromatography, **22:**459–461
Guanosine triphosphate
competition with NH_4^+ in glutamate dehydrogenase reaction, **39:**389
effect on glutamate dehydrogenase-NADPH complex, **39:**377; 405
exclusion of adenosine diphosphate from glutamate dehydrogenase surface by, **39:**387; 411
formation of ternary complex with glutamate dehydrogenase-NADPH, **39:**387; 413
as phosphate donor, **18:**195
in pyrophosphorylase reactions, **18:**207
in polypeptide synthesis, **24:**3
in purine biosynthesis, **21:**204; 253; 254
role in protein biosynthesis, **19:**264
Guanosine 5′-triphosphate (GTP), phosphomannose isomerase-guanosine diphospho-D-mannose pyrophosphorylase (PMI-GMP), **70:**227
Guanosine triphosphate (GTP), regulation of bacterial form glutamine synthetase, **72:**33–35
Guanylate cyclase, **55:**146
in muscarinic receptor binding, **55:**173
Guanylate kinase, NMP kinase primary structure, **73:**104–107
Guanylic acid
biosynthesis, **21:**204; 229; 232; 240; 241; 250–252; 281
hydrolysis, by metal ions, **21:**176
solubilizing power, **17:**137
Guanylic acid deaminase, **18:**118
^{14}C-Guanylic acid, deoxyribose synthesis, **28:**420
Guanylic aminase, classification, **14:**246
Guanyl Nucleotide-Binding Regulatory Proteins, ADP-Ribosylation of, by Bacterial Toxins (Moss and Vaughn), **61:**303
Guanyl spermidine, in edeine, **36:**226
Guanyly cyclase, **65:**290
Guinea-pig granuloma slices, for studying the hydroxylation of proline, **41:**256; 257
Guinea pig serum, **39:**221
asparaginase activity in, **39:**221
Gulonic acid, **60:**233
Gulonolactonase, localization of, **24:**317–318
Gulonolactone dehydrogenase, localization of, **24:**295; 299
Gultamine synthetase
borohydride trapping of acyl-P intermediate, **72:**26–28
γ-glutamyl-phosphate intermediacy, **72:**16–17
hepatic glutamine transport, **72:**55–56
hepatic regulation
long-term regulation, **72:**67
short-term regulation, **72:**66–67
hepatocyte heterogeneity and enzyme distribution, **72:**56–58
dynamics and regulation, **72:**58–61
isotopic tracer studies, **72:**15–16
kinetic reaction mechanism, **72:**22–24
ligand interactions, stereochemical mapping, **72:**20–22
regulation of bacterial form, **72:**30–38
cascade control, **72:**31–32
cumulative end-product inhibition, **72:**32–35

Gultamine synthetase (*Continued*)
glutamine synthesis expression, **72:**35–36
research on enzyme activity, **72:**12–13
Gultimate synthetase, **37:**451; 452; 492; 500; 547; 576
Gums, **17:**147
vegetable, reaction with borax, **12:**502–509
Gutta, isomer of rubber, **29:**222
Guvacine, as precursor of nicotinamide, **15:**103
Gymnosperms, **46:**76
β-1,3 glucans of, **28:**329

H

Habu, **17:**173–174
Hadacidin, **35:**156–158; 168
biosynthesis, **35:**156
Haem. *See also* Heme proteins
in double-headed enzyme, **23:**314–315
Haem a_2
biosynthesis of, **23:**315
of cytochromes, **23:**269; 287–289
and redox potential of D group cytochromes, **23:**313
compounds, **23:**311–312
Haem *a,* absorption spectrum, **23:**289–297
autoxidation of, **23:**302
biosynthesis of, **23:**315
concentration in tissues, **23:**282–284
of cytochromes, **23:**269; 270; 281; 293
Haem *a*-apomyoglobin, autoxidation of, **23:**302
Haem *a*-CO compounds, absorption spectra, **23:**292
Haem *a* compounds, as models for cytochromes *a*, a_1, a_3, **23:**289–312
Haem *a* (ferrous) compounds, absorption spectra, **23:**290
Haem *a* model compounds, absorption spectra, **23:**289–297; 306–310
comparison with cytochromes, **23:**306–312
Haem *a* proteins, absorption spectra, **23:** 306–310
Haem *a* synthesis, and copper, **23:**304–306
Haematin *a,* **23:**271; 293; 294
Haematin *a*-apomyoglobin, **23:**291; 295
Haematin *a*-apoperoxidase, **23:**291; 295
Haematin *a* compounds, ionic character of, **23:**296
Haematin *a* (ferric) compounds, absorption spectra, **23:**291
Haematin *a*-globin, **23:**291; 295
Haematin compounds, **23:**265–321
Haematohaemochrome, **23:**313
Haem-CO-cholate model compound, **23:**292; 293; 304; 306
Haem-CO-Emasol model compound, **23:**292; 293; 304; 306–312
Haemin *a,* **23:**270–284; 286; 288; 289
and lipids, **23:**310–311
Haemin a_2, **23:**288; 289
Haemochromes, **23:**265–321
absorption spectra, **23:**306–310
Haemochromes *a,* absorption spectra, **23:** 289–297; 306–310
Haemoglobin, and copper, **23:**304
Haemoglobin *a,* absorption spectrum, **23:**289; 293; 295
Haemophilus influenzae, **46:**59
Haemophilus influenzae, L-aspartase from, **74:**333–334
Haemoproteins, separation from cytochromes *a,* **23:**303
Haemoproteins *a,* absorption spectra, **23:** 295–296; 306–310
CO reaction, **23:**306
Hafnia alvei, aspartase from, **74:**297; 301–303
chemical modification studies of, **74:**304–305
pH study of, **74:**305–307
Hageman factor, **48:**278; 284
properties of, **30:**262 (*See also* Factor XII)
sensitivity to glass, **30:**261
Hair
elasticity, stretching, **1:**8
elimination of D-leucine in, **20:**25
elimination of optical antipodes in, **20:**18
Haldane equation, **17:**35–50
Haldane relationship, **43:**225
Half-cystine, **42:**118; 120; 188, **51:**217
Halide ions, effect on polypeptidase, **12:**210
Hallachrome, and action of peroxidase on amino acids, **14:**152
and action of tyrosinase on proteins, **14:** 139–140
use of term, **14:**78
Hallachrome analog from tyrosine derivative, **14:**137–138; 140
Halla parthenopala, red pigment in, **14:**78
Hallucinations, in schizophrenia, **29:**534

Halobacterium halobium, **34:**469–470
Halobacterium salinarium, superoxide dismutase from, **41:**79
Haloenzyme, **44:**228
Halogenases, classification, **14:**225; 248
Halogen compounds, organic, poisonous effect, **10:**277–278
Halogenocyclohexanes, configurations of, **23:**23
Haloperidol, tyrosine hydroxylase regulation, **70:**128
Haloperiodol
potency of, **55:**182
side effects of, **55:**181
structure of, **55:**180
Halophenols, effect on sperm respiration, **9:**369
Halophilic organism, **36:**240
Halothane, **42:**223, **55:**199
metabolism of, **55:**258
aerobic, **55:**259–260
anaerobic, **55:**260–261
Hammett constants, **46:**400; 403
Hammett equation, **50:**6; 9
Hammett values, **59:**121
Hamster fibronectin receptor, **59:**46
Haplopappus gracilis, **55:**19; 20
Haplotomic mechanism, definition of, **31:**8
L_1-Hapten, definition of, **31:**106
Haptens, **41:**134
combination with antibodies, **15:**50; 64–67; 76
Haptoglobin, **41:**110–111, **46:**139
Harden-Young fermentation, **6:**186
Harderian glands, rat, **46:**56
rodents, **46:**103
Harderoporphyrinogen, *See* 2-Vinylporphyrinogen III
Harmaline, in schizophrenia, **29:**520; 522
structure of, **29:**521
Harmine, in schizophrenia, **29:**520; 522
structure of, **29:**521
Hartley-Roe equation, **33:**251; 252
Hartley's rule, **43:**388
Hartridge-Roughton flow method, peroxidase kinetics, **73:**5
H blood group, **52:**46; 52
HCFU, **54:**58
HCO_3, **34:**513
H^+ contransport, in Ehrlich cell, **49:**69
HCS (human chorionic somatommotropin), **42:**113; 117; 118; 119; 120; 125
HDP, *See* Hexose diphosphate
HDP triosephosphate lyases, classification, **14:**256
Head-to-tail arrangement, microtubule dynamics, **69:**123
Heart block, by fluoroacetate, **18:**138
Heart muscle phosphofructokinase, **48:**206–207
Heart muscle preparation. *See also* Keilin-Hartree heart muscle preparation
alkali-treated, in reconstituted respiratory chain, **28:**173–180; 186–190; 208–209; 223; 224
for respiration study, **28:**159–165; 168–170; 192; 196; 200–203; 207
sequential fragmentation, scheme, **28:** 190–197
Heart muscle proteins, **13:**182
Heart tissue, effect of fluoroacetate on, **18:**137
Heat capacity of adenosine diphosphate binding to glutamate dehydrogenase, **39:**384
Heat conduction in energy transport in living cells, **14:**3–6
Heat content, determination of changes in, **11:**11–13
Heat denaturation, **39:**352. *See also* Temperature
of RNase, **24:**188–193; 249–251
Heat denaturation of proteins, **27:**226
Heat inactivation
of β-glucuronidase, **16:**377
of β-lactamase, **28:**302; 303
Heat of neutralization, of carboxyl ion, phenol ion, imidazole and amino groups, **20:**64; 65
Heat of reaction, *See* Thermodynamic constants
of carbon dioxide fixation in 3-phosphoglyceric acid, **19:**22
Heat production, annaerobic, in sperm, **24:** 51–52
Heat sensitivity, of adsorbed enzymes, **28:**5–6
Heat shock proteins
Caulobacter cell cycle, flagellar structure and assembly, **71:**9
RNA polymerase II transcription, potential roles, **71:**63
Heat stability, of allosteric enzymes, **28:**53; 72
Heavy-atom derivative, **39:**349
Heavy atom effects, **46:**443–446
Heavy atom kinetic isotope effects, glutamine-dependent nitrogen transfer, **72:** 183–186

Heavy metal ions, effect on laminarinases, **28:**347
Heavy metal tolerance, adenylsuccinate synthetase (AMPSase), **73:**94
Heavy water. *See also* Water, labeled
  effect on papain, **19:**318
  in Hill reaction, **19:**53
  in 11β-hydroxylation of 11-deoxycortisol, **19:**190
  in squalene cyclization, **19:**192; 193
HeLa cell, **58:**112
  N^6-methyladenosine, **65:**270–271
HeLa cells, lack of Pasteur effect in, **29:**327
HeLa S-3, **42:**253
HeLa tumor cells, pentose labeling in, **28:**417; 433
Helical content of RNase, **24:**180–182; 191–192; 194; 248
Helical structure, of collagen, **23:**169; 213; 217–218
  of proteins, and specific rotation, **23:**73–74; 76
Helical structure of ribonuclease, **24:**448–449; 457
Helical wheel representation, adenosine receptors, **69:**93
Helicostylum piriforme, enzymic transformations of steroids by, **20:**243; 244; 247; 251
Helix
  carbonic anhydrase, **47:**154; 159; 161
    content, **47:**47; 64; 74; 85; 117
    formers, **47:**47–48; 51–52; 54; 66
    growth parameter, **47:**85
    helical wheel, **47:**48; 68; 89
    helix-coil, **47:**48–49
    indifferent, **47:**48; 51
    initiation parameter, sigma, **47:**48–49; 85
    inner, **47:**50
    length, **47:**53–54; 59–60; 100
    non-helix, **47:**49
    nucleation, **47:**52; 86; 101; 105; 111; 131
    permissive, **47:**48; 89; 96
    potential, **47:**50–51; 110; 115
    propagation, **47:**52; 55; 59; 77
    termination, **47:**52
  properties of, **30:**181
α-Helix, **38:**224, **50:**23; 25
  chains of hydrogen bonds in, **50:**26
  and denaturation, **27:**225
    of proteins, and the Cotton effect, **27:**45–48
  as dipole, **50:**25
  and enzyme structure, **26:**91–92; 99
  interaction of other groups with, **50:**25
  of proteins, **23:**2–4; 18–23; 25
Helix breakers, **50:**27
α-Helix content, **42:**102; 103; 110; 140; 148
Helix (α,αII, ^{3}10, epsilon), **47:**58–59; 61; 65; 134
  assignments, **47:**51; 55; 66; 100
  boundary, **47:**50–54; 67–68; 86; 98–101; 105; 114; 116; 119; 130–131
  breakers, **47:**47–48; 51; 52; 54; 99; 102; 113
  α → β transition, **47:**76; 79; 82–83; 135–136
  β → α transition, **47:**74; 136
Helixin, in treatment of plant diseases, **20:**403
Helix pomatia, **17:**217, **48:**71
  aryl sulfatase of, **22:**219; 220
    steroid sulfatase of, **22:**227; 228
  enzymes in, **8:**285
  polygalactoside production in, **20:**119
  separation of enzymes by chromatography, **14:**349
Helix-random coil transition, rate of, **25:**20
Helminthosporium dictyoides, plant disease caused by, **20:**402
Helminthosporium sativum, plant disease caused by, **20:**394
Helminthosporium species, plant disease caused by, **20:**393; 401
Helminthosporium vagaus, plant disease caused by, **20:**402
Helvolic acid, **7:**473; 474; 486; 489; 490; 492
Hemagglutinin, soy bean, chemical modification of, **24:**144
Hematin, **15:**18, **46:**392
  as inhibitor of succinic dehydrogenase, **18:**94
Hematin compound, in respiratory chain, **20:**179
Hematin compounds. *See also* Cytochrome
  nomenclature, **7:**267
  in plants, **18:**3; 4; 22
Hematin compounds of leaves, **12:**21–25
Hematocrit, **44:**221
Hematoheme, **46:**102
Hematological diseases, and autoimmunity, **24:**114
Hematopoiesis, effect of glutathione on, **11:**253
Hematoporphyrin, **46:**63; 102
  iron chelates of, **46:**101
Hematoporphyrin IX, **17:**436
Hematoside, **38:**297, **52:**33

Hematoxylin, **58:**63
Heme, **15:**18; 37–38, **46:**37; 40; 51; 59; 83; 86; 88; 101; 103; 105; 107; 118; 126; 128; 132; 135; 143; 144; 152; 157; 171. *See also* Hemoproteins, Iron Porphyrin; Iron; Tetrapyrrolic compounds
  biosynthesis, **46:**86
  b-type, **48:**17
  definition, **20:**151
    in cytochrome *c,* **20:**188; 193
  degradation, **46:**38; 64
  free, **46:**142; 151
  inhibition of activity of ALA synthetase, **46:**87
  linkages and enzymic activity of hemoprotein, **12:**178–181
  in mitochondrial fragments, **21:**101; 104; 105
  oxidation by, **46:**103
  oxygenase, **46:**67; 107; 128; 132; 150; 162; 174
  in oxygen metabolism, **19:**120; 122; 127; 143; 174; 186; 199; 202; 203; 211
  respiratory burst oxidase cofactor, **65:**54–55
  in respiratory chain, **28:**162; 193; 194
  synthesis, **46:**106; 141
    by bacteria, **46:**89
    in barley, **46:**96
  turnover, **46:**147
  varieties of, **46:**59–63
Heme *d,* **46:**63
Heme *s,* **46:**63
Heme, EPR spectrum, **27:**117
Heme *a,* **46:**60; 120
  in respiratory chain, **28:**192; 204; 206; 214–217; 219
Heme alkylation, **60:**70
Heme b, *See* Photoheme
Heme *c,* **46:**60
Heme chromophores, cytochrome *c* oxidase, subunit I, **71:**86
Heme compounds
  enzymes for metabolism, **24:**341–342
Heme-copper ratio, in cytochrome oxidase, **27:**148–155
Heme enzymes, **44:**189
  active centers of, **15:**275
    mechanism of action, **15:**16–21
Heme enzymes, Cotton effect, **27:**89–92
Heme enzyme-substrate complex, kinetics of, **15:**278; 280
Heme *h,* **46:**63
Heme-linked groups, action of hemoproteins, **7:**265–303
  and catalases, **7:**295–301
  and cytochrome c, **7:**278–286
  and enzyme activity, **7:**290
  and hemoglobin, **7:**265–278
  and peroxidases, **7:**266–269; 274; 286–295
Heme-Linked Groups and Mode of Action of Some Hemoproteins (Theorell), **7:**265
Heme pigments, and nuclear purity, **27:**253–254
Heme protein derivatives, EPR study, **27:** 138–140
Hemeprotein hydroxylases, **55:**202
Heme proteins, **60:**37
  relationship among, **15:**18–21; 37–38; 42
  side chains and activity, **7:**322–325
  susceptibility to pancreatin, **23:**64
Heme proteins, Cotton effect in, **27:**47–48; 59; 60
Hemerythin, diffusion of oxygen facilitated by, **29:**34
Hemerythrin, **47:**84; 118, **65:**162–164
Hemerythrin, Cotton effect, **27:**60; 88
Hemerythryn, **46:**82
Hemes, Chlorophylls, and Related Compounds: Biosynthesis and Metabolic Regulation (Granick and Beale), **46:**33
Hemes (or iron porphins), **7:**267; 306; 307; 308–318
  catalytic oxidation of, **7:**334
  methene link in, **7:**334–337
  occurrence, **7:**308–309
Heme spin equilibria, **47:**30
Heme synthesis, **35:**171–173
Hemicellulases, **56:**222
Hemicellulose, **56:**222
Hemicelluloses, changes during fermentation, **10:**366–368; 402
Hemiglobin, **7:**267
Hemigossypol, **55:**80
Hemimetabola, **45:**411
Hemin, **7:**267, **15:**22; 41; 42, **46:**86; 92; 114; 120; 124; 129; 131; 135; 139; 140; 142; 146; 149; 163; 174
  biosynthesis of, **12:**43
  from catalase, **1:**170; 171; 174
  clearance from bloodstream, **46:**140
  derivatives, **15:**22
  dissociation constant, **46:**391

Hemin (*Continued*)
effect on protein synthesis, **46:**149–151
in enzyme system, **15:**22; 27–28; 32
inhibition of synthesis of ALA synthetase, **46:**126–127
movement of in cell, **46:**145–146
and succinic dehydrogenase, **18:**69; 70; 71; 74
in thiol oxidation, **11:**206
transport, **46:**139–141
Hemin and autoxidation, **8:**427; 440; 443; 452
Hemin compounds, **51:**40
Hemin-hemopexin, **46:**140
Hemin-proteins, in energy transport, **14:**31–35; 37–42
structure studies, **14:**307
Hemins
catalytic action of, **14:**173–176; 182–183; 185; 190
mode of action as coenzymes, **14:**180–183; 188
of cytochromes, **17:**343–345; 357; 382
Hemithiolacetal, **59:**107; 119; 128
Hemoagglutinin, **58:**132
Hemochromes, constitution, **7:**272–273
Hemochromogen, denatured-protein, detection, **20:**173; 174
Hemochromogens, **7:**267; 306; 310; 311; 312; 417
and cytochrome, **18:**3–5; 7
mechanism of oxido-reductions, **15:**16; 18–19; 42
properties of, **12:**15–20
reversible oxidation, **9:**2
in thiol oxidation, **11:**206
Hemocuprein, **41:**51
Hemocyanin, **7:**308, **46:**82
Hemocyanin, Cotton effect, **27:**93–94
Hemocyanin, EPR-nondetectable copper ions in, **33:**223
Hemocyanine as antigen, **15:**75
Hemo-ferroflavoproteins, **17:**270
Hemoflavoproteins, **17:**244–245; 254–256; 261–262
Hemoglobin, **7:**265–278; 313; 322, **17:**30, **34:**36, **36:**72, **43:**247; 312, **44:**218, **46:**59; 65; 139; 146; 157; 159, **47:**47; 49; 83; 124, **50:**36, **54:**307. *See also* Hemoproteins, Iron porphins, Porphins, and Tetrapyrroles; Tetrapyrrolic compounds
action of perturbators on, **23:**45
action of rennin on, **15:**431
mechanism of action, **15:**18–21; 37–38
molecule shape, **15:**458
variation with age, **15:**274
action of tyrosinase on, **14:**150
peroxidase activity of, **14:**190
separation by chromatography, **14:**322; 325; 328; 332; 350; 358
acyl-trypsin activity towards, **20:**75
imidazole reactivity in, **20:**70; 98; 99
amino acid content, **3:**68–70
amino terminal groups, effects of partial carbamylation, **40:**15
levels in cyanate treated patients, **40:**24; 25
salt bridges, **40:**19; 20
X-ray diffraction studies, **40:**19; 20
autoxidation of, **41:**89–90
biosynthesis, **44:**214
breakdown to bile pigments, **7:**333–337
change in acidity on oxygenation, **7:**271–277
charge pairing in, **50:**32
column chromatography, **22:**443
effect of pH on structure, **22:**137; 426; 428
synthesis, precursor material of, **22:**146
conduction of oxygen through solutions of, **29:**33–34
dried, effect of irradiation, **24:**376–378; 381
formation, from polypeptide, **24:**22–25
in reticulocytes, **24:**12; 14; 19–20; 21
human, structure, **24:**416; 419
lower peptide sequences in, **24:**421–422
x-ray studies, **24:**448
ferrohemoglobin, **58:**234
frog, **44:**222
genetic studies of, **13:**141–142
in nitrogen fixation, **13:**370–371
in root nodules, **13:**346
glycera, **47:**58–59; 62; 124
and heme-linked groups, **7:**265–278
horse, **47:**47; 49; 124
human, **47:**81–83
hydrogen bonds in, **50:**27; 28
inertness of, **16:**454; 460
intracellular distribution, **16:**463; 466
synthesis in sickle-cell anemia, **16:**475
lamprey, **47:**52; 57; 62; 124
midge larva, **47:**58; 124
mol. weight, **1:**19
molecular weight, **7:**270

and nuclear purity, **27:**254–255
reactions with disulfides, **27:**212; 217
sickle cell, **27:**47; 60
oxygen equilibrium curve, **7:**270–271
in plants, **7:**313
quasi-lipoxygenase activity, **58:**233
sickle cell, **47:**81–83; 135–136
structure, **3:**102–103, **7:**327–328
structure of, **26:**92; 98–100; 107
subunits, **50:**24
tadpole, **44:**222
use in determining oxygen from photosynthetic process, **12:**3–8; 30–31
Hemoglobin, Cotton effects, **27:**47; 90–92
mutations, **27:**302
Hemoglobin A, affinity for carbon dioxide, **40:**8
carbamino form, **40:**13
pK_a of the amino terminal residues of the α-chain, **40:**8
Hemoglobin films, enzymic activity on, **28:**29
Hemoglobin mutants, cytochrome *c* oxidase, **71:**138–139
Hemoglobin-oxygen combination, rate constant for, **25:**14
Hemoglobin-oxygen complex, affinity constant, **9:**13
Hemoglobin preparations, semiconduction current of, **24:**381
Hemoglobin S
carbamylated, **40:**2
oxygen affinity, **40:**18
salt linkages, **40:**18
carbamylation, kinetic studies of, **40:**10; 11; 12; 13
β-chain substitution, **40:**1
solubility, **40:**1; 2
Hemoglobins, **37:**525; 594; 596
in assay of papain activity, **19:**270
chromatographic evaluation of purity, **19:**281
optimal pH for papain action on, **19:**303
peroxidase action of, **19:**196; 197
Hemoglobin S. Carbamylated, Biochemical and Physiological Properties of (Manning, Cerami, Gillette, De Furia, and Miller), **40:**1
Hemolysin
absorption by red cells, **15:**54
from *Cl. histolyticum,* **23:**188–189; 199
requirements for —SH group, **11:**241
Hemolysins, data on antisheep cell, **8:**77
estimation by titer, **8:**76
microestimation in weight units, **8:**76–79
molecular weight, **8:**80
relation to complement and red cells, **8:**79–82
Hemolysis, **41:**43
and lysolecithin, **8:**460; 487; 490
effect of temperature, **8:**95
immune, mechanism of, **8:**96–97
role of enzymes, **8:**461; 490
by snake venom, **11:**78–79
by thiaminase-producing bacteria, **15:**413
Hemolytic activity, of complement, **8:**76; 89–97
β-Hemolytic streptococci, inactivation of pantothenate by, **7:**444–448
Hemopexin, **46:**141
Hemophilia
blood factors in, **30:**261
factor deficiencies in, **30:**265
hemostatic efficiency in, **30:**265
clotting time of mixed blood, **16:**72–75
possible use of substitutes for platelet cofactor I, **16:**54
relation to platelet cofactor II, **16:**72–74
relation to threone activity, **16:**56–57
and factor 5, **10:**25; 28
in experimental animals, **10:**17
prothrombin conversion rates, **10:**16–17
thromboplastic activity in, **10:**6; 7; 9; 10–13
Hemophilic plasma, **16:**31
and serum, inhibitor in, **16:**86
Hemophilus influenzae, **7:**317; 324–326; 333, **17:**172
Hemophilus influenzae, in estimation of DPN and TPN, **15:**97
Hemophilus parainfluenzae, **15:**97
Hemophilus parainfluenzae, and nicotinic acid derivatives, **7:**411–416
Hemoprotein, **34:**482
and succinic dehydrogenase, **18:**69; 71
Hemoproteins, **46:**92; 93; 102; 105; 151, **58:**233
absorption spectra of, **12:**154–157
activity as enzymes, **12:**171–174; 178–181
equations for reactions of, **12:**161–163
kinetic studies of compounds of, **12:** 163–171
substrate and donor specificity of, **12:** 174–175

Hemoproteins (*Continued*)
titration with peroxide, **12:**159–160
catalytic properties, **58:**237
enzyme-substrate combinations in, **19:**313
in protocatechuic acid oxidase, **19:**98
reactions with oxygen, hydrogen peroxide, and electron donors, **19:**117
heme iron, 3d electrons of, **33:**333:320
heme-linked groups and mode of action, **7:**265–303
magnetic moment, **7:**269; 275–278
magnetometric measurements, **7:**269; 275–278; 283; 291–293; 297–298
titrimetric determinations, **7:**269; 273–274
imidazole reactivity in, **20:**61; 98; 99
optical absorption spectra of, **33:**315
Hemoprotein-substrate complexes, magnetic properties of, **12:**157–159
Hemoserine, **49:**80
Hendersonia acicola, enzymic transformations of steroids by, **20:**251
Heparanase, **60:**233
Heparanoglucuronidase, **60:**232
biological role, **60:**234
in growth regulation, **60:**235
isoelectric point, **60:**233
optimum pH, **60:**234
properties, **60:**233
purification, **60:**233
reaction
with heparan sulfate, **60:**233
with heparin, **60:**233
reaction mechanism, **60:**233
substrate specificity, **60:**234
Heparan sulfate, **52:**81; 82
in cell membranes, **40:**51
degradation, **60:**277
faulty, **60:**241
by β-galactosidase, **60:**252
structure, **60:**278
Heparan sulfate lyase, **60:**290
Heparin, **5:**32–54, **48:**288, **52:**81; 82, **59:**11–12, **60:**223; 233
activation of pepsinogen, **40:**38
activation of trehalose phosphate synthetase, **40:**35; 36
effect on hyaluronidase, **40:**41
effect on lipoprotein lipase, **40:**35
effect on oxidative phosphorylation system, **40:**39
effect on pepsin, **40:**56
effect on trehalose phosphate synthetase, **40:**58
interaction with Concanavalin A, **40:**56; 57
anticoagulant action on blood, **5:**51–54
binding to antithrombin, **60:**280
and blood coagulation, **5:**32; 49–54
cofactor activity, **16:**80–82
effect on clotting time, **16:**38–39; 66
effect on platelet-low plasma, **16:**68
inhibitor, of β-glucuronidase, **16:**373–375
of prothrombin activation, **16:**25; 92
neutralization of activity, **16:**70–71; 88
production in response to labeled inorganic sulfate, **16:**471
in prothrombin preparation, **16:**29
relation to antithrombin, **16:**77; 80–84
degradation, **60:**277
effect on blood plasma, **14:**361
effect on phosphorylase, **28:**107
effect on RNase activity, **24:**238
in isolation of mitochondria, **21:**5
reviews on, **25:**434
tyrosine hydroxylase regulation, **70:**122–127
virus crystallizing agent, **9:**196; 198; 199
β-Heparin, *See* Dermatan sulfate
Heparin activation, of collagenase, **23:**195
Heparin activity, laminarin sulfate, **28:**356
Heparin and blood clotting, **10:**5; 33–38
Heparin and snake venom, **8:**476
Heparinase, **10:**34, **60:**289
classification, **14:**246
Heparin lyase, **60:**289
Heparitan (heparitin) sulfate, reviews on, **25:**434
Heparitinase, **60:**233; 290
Hepatectomy, effect on ribose synthesis, **28:**417
Hepatic encephalopathy, **53:**228
Hepatic glucocorticoid receptor, **47:**278–284
activation, **47:**279
by calcium, **47:**281
binding to anionic substances, **47:**282
binding to phosphocellulose, **47:**282
purification of, **47:**282
sensitivity to protease, **47:**281
Hepatic gluconeogenesis, phosphoenolpyruvate carboxykinase (GTP), **69:**214–216; 223–226; 245–249; 256; 264–265
Hepatic 3-Hydroxy-3-Methylglutaryl-Coenzyme A Reductase, Regulation of (Rodwell, McNamara, and Shapiro), **38:**373

Hepatic Recognition, and the role of surface carbohydrates, **41:**99–127
Hepatic regeneration of adult rat, **42:**256
Hepatic tissue and chromatin, **7:**187–189
Hepatic transport
  cysteine *S*-conjugate β-lyase reactions
    human liver, **72:**215–216
    rat liver, **72:**216–219
  glutamine synthesis, **72:**44–78
    acid-base homeostasis, **72:**71–75
    acinar-level integration, **72:**67–75
    enzyme distribution, **72:**56–58
    enzymes, **72:**53–56
    functional consequences of swelling, **72:**46–53
    glutaminase, **72:**54
    glutaminase ammonia amplification, **72:**69–70
    glutamine breakdown regulation, **72:**63–65
    glutamine synthase as ammonia scavenger, **72:**67–69
    glutamine synthetase, **72:**55–56
    gutamine transaminases, **72:**55
    heterogeneity, **72:**56–63
    heterogeneous glutamine synthetase distribution, **72:**58–61
    htetrogeneous transport, **72:**62–63
    hydration mechanism, **72:**48–49
    intercellular glutamine cycling, **72:**70–71
    long-term glutamine regulation, **72:**65
    mitochondrial transport, **72:**47–48
    pathophysiology, **72:**75–78
    plasma membrane transport, **72:**45–47
    regulation of, **72:**66–67
    research history, **72:**44–45
    short-term glutamine regulation, **72:**64–65
Hepatic uptake of protein, **41:**104
Hepatitis B virus, annexins, **71:**215
Hepatocarcinogen, ethionine as, **25:**373–375; 377–378
Hepatocellular carcinoma
  γ-glutamyl transpeptidase, **72:**244–245
    cancer-associated changes, **72:**270–271
  glutathione synthesis and, **72:**242
Hepatocuprein, **41:**51; 56
Hepatocyte, plasma membrane of, **49:**69
Hepatocytes, **46:**140, **59:**25
  glutamine transport and hydration, **72:**48–53
    hepatocyte swelling, **72:**49–53
    mechanisms, **72:**48–49
  heterogeneity in glutamine metabolism, **72:**56–63
    enzyme distribution, **72:**56–58
    glutamine synthetase dynamics and regulation, **72:**58–61
    transport systems, **72:**61–63
  normal liver ALA synthesis in, **46:**105
Hepatoma, **46:**119. *See also* Cancer, Carcinoma, Ethionine, Malignant cells, Tumor
  azodye, enzymes of, **24:**300; 306; 313; 314; 315; 318; 319; 321; 324; 326; 329
    mouse, enzymes of, **24:**307
    Novikoff, enzymes of, **24:**299–301; 303; 305; 312–316; 318–320; 323; 326; 328; 329; 335
  azo dye induced, β-glucuronidase activity, **16:**396
  phosphatide content, **7:**168
  production by ethionine, **25:**370; 373–375
Hepatoma BW 7756, EPR study, **27:**190–191
Hepatoma cells, pentose metabolism, **28:**135; 436
Hepatomas, **39:**336
  anaerobic glucolysis *vs.* growth rate in, **29:**333
    correlation of intermediary metabolic factors and enzymes with growth rates of, **29:**331
    incorporation of amino acids into protein in, **29:**364–365
    mitochondrial content of, **29:**336
    Morris, **29:**330
    rates of glycolysis of minimal deviation, **29:**330–331
    undifferentiated, **29:**332
Hepatosplenomegaly, **60:**240; 261
HepG2 cells
  γ-glutamyl transpeptidase, **72:**245
  human gene expression, **72:**268–269
Heptane, **34:**477
n-Heptane, **51:**33
n-Heptanol, **34:**489
Heptapeptide repeats, RNA polymease II transcription, C-terminal domain, characteristics of, **71:**46
6-Heptenoic acid, **18:**331
Heptolenticular disease, **44:**230
Heptoses, from glucose, **16:**249
Heptulose phosphates, **15:**154; 157–160; 162–167
Heptuloses, metabolism, **15:**160–163; 165

Heptulosonic acid, derivative of, **41:**118
Heptyl alcohol, **34:**489
n-Heptylpenicillin
action of β-lactamase on, **28:**257
H + /e - ratio, enzyme turnover control and, cytochrome *c* oxidase, **71:**162–167
Hereditary mechanisms, and nuclear enzymes, **27:**240; 275–283
Herellea sp., β-lactamase of, **28:**243; 252
Herpes keratitis, **54:**59; 61
Herpes simplex virus (HSV), cholesterol ester hydrolases regulation, **69:**60–65; 66; 67
Herpes virus infection, **54:**63
Heterochromatin, **4:**20, **38:**47
Heterocycles, effect on RNase activity, **24:**446–448
Heterodimer arrangements, microtubule dynamics, **69:**123
Heteroenzymes, definition, **25:**330
multiple forms of, **25:**333–345; 363; 364
Heterogeneity
of bovine serum albumin, **30:**166; 167
inherent, definition of, **30:**141
multiple electrophoretic zones and, **30:**141
index of, antibodies, **15:**64–67
Heterogeneous system, **34:**519
Heterogeneous systems, enzyme reactions in, **33:**245–308
Heterogeneous Systems, Enzyme Reactions in (McLaren and Packer), **33:**245
Heterologous, reactions, **35:**468
reaction systems, **35:**472
Heterophosphatase, and hexokinase, **6:**70
Heteropolyamino acids, volcanic regions, **27:**389
Heteropyrithiamine, **15:**399; 402–403; 407
Heterosaccharide transformations, **17:**219–221; 229
Heterosigma carterae, phosphoribulokinase, **74:**99–100
Heterotrophism, **2:**138–144, **74:**6–12. *See also* Carbon dioxide, heterotrophic assimilation of
and autotrophism, differentiation of, **74:**8–9; 10–11
facultative, **2:**142; 143, **74:**10–11
intermediate forms of, **74:**11
Heterotrophs, carbon dioxide fixation by, **12:**41–151
Heterotrophs, Fixation of Carbon Dioxide by (Utter and Wood), **12:**41
Heterozygotes for ADA deficiency, **51:**170
Heterozygous carriers, **38:**308
Hevea brasiliansis, **55:**59
Hevea brasiliensis, cytology of latex vessels of, **29:**222–225
Hexachlorobenzene, **46:**55; 138
α-Hexachlorocyclohexane, **46:**138
Hexacoordinate zinc ion, **59:**122
Hexadecane ($^{14}C_1$-labeled), oxidation of, **27:**490
Hexametaphosphate, effect on lytic activity, **8:**95
Hexamethylene glycol, activation of fibrinogen in, **16:**39; 44
as inhibitor of clot formation, **16:**40
n-Hexane, **51:**33
Hexanediol-P_2, **46:**366; 373
Hexapeptide repeat, tetrahydrodipicolinate *N*-succinyltransferase, **72:**301–302
N-Hexarbamoyl-FU, **54:**58; 62
H-5 exchange of pyrimidines, **49:**208; 212; 213
H-5 exchange of uridines in *E.* coli $tRNA^{Ile}$ affected by Ile-tRNA synthetase, **49:**212
rate constants for H-5 exchange of uridines in *E.* coli $tRNA^{Tyr}2$ affected by Ile- and Tyr-tRNA synthetase, **49:**213
H-8 exchange reaction of purines, **49:**202
Hexitol phosphate, **15:**167–168
Hexobarbital, **46:**119
Hexobarbital narcosis, and LSD 25, **29:**518
Hexodiphosphatase from the kidney and liver of mammals, **10:**87
Hexokinase, **5:**112; 131, **15:**121; 155; 167; 174; 381–382, **34:**291; 296; 305; 325; 520, **37:**433; 501; 516; 549; 565; 587, **43:**22; 28; 32; 35; 44; 223; 231; 330; 331; 413, **46:**271, **50:**391, **52:**192; 193, **55:**358–359
activity in mitochondria, **33:**290–291
allosteric regulation, **28:**75–76; 85; 86; 95; 122
biological significance, **28:**76–77; 84
in ascites tumor cells, **29:**347
cytoplasmic ATP utilization upon glucose addition by, **29:**348
glucose-6-phosphate inhibited by, **29:**355
influence of the phosphate potential on, **29:**354–357
K_i values of, **29:**356
in malignant cell, **29:**334
product inhibition of, **29:**356

release of glucose-6-phosphate inhibition by P_i of, **29:**355
and ATP synthesis, **23:**326; 338; 355; 379
denaturation of, **23:**48
in avian liver, **52:**263
in carboxylation study, **26:**306; 307
classification, **14:**235–236; 251
isolation, **14:**376; 401; 402
yeast, **14:**211
competition with ATP-ase, **17:**114
effect on Q_{o2}, **17:**122
increase of ADP by, **17:**125–127
phosphoryl transfer by, **17:**163; 164; 168
control of action, **25:**170
effect of metal ion, **25:**19
multiple forms, **25:**351–352
product inhibition of, **25:**224; 246–249
steps in reaction, **25:**14
destabilization mechanism, **43:**265
determination of activity, **52:**260
distribution, **52:**263
enzymic adaptation, **13:**70; 106
localization in cytoplasm, **13:**7
yeast, inhibition of, **13:**332
and galactozymase, **6:**77
and heterophosphatase, **6:**70
in α-ketoglutarate oxidation system, **15:**225; 229; 230–231
kinetic mechanism of, **45:**285; 293; 299; 305; 308; 310
in liver extracts, **52:**261
localization, **52:**261; 263
localization of, **24:**312; 335
sperm, **24:**40; 54
mitochondrial binding of, **39:**298–301
and phosphorylation, **18:**195
in plants, **18:**286; 311
in photosynthesis, **19:**8; 63
properties, **52:**260
reversible inhibition by glyceraldehyde, **6:**77
sperm, **9:**362
Hexokinase, SH-group, **27:**214
Hexokinase activity, and fluoroacetate, **18:**145
Hexokinase reaction, enzyme coupling, **28:**167
C-1-Hexokinases, phosphorylation by, **15:** 147–148
C-6-Hexokinases, phosphorylation by, **15:** 145–147; 148; 152
Hexokinases, reversibility studies of, **39:**294; 295
Hexokinases, The: Kinetic, Physical, and Regulatory Properties (Purich, Fromm, and Rudolph), **39:**249
Hexokinase (yeast), activity of –SH group in, **11:**243
resistance to action of ionizing radiations, **11:**256
Hexosamine metabolism, **14:**204
Hexosamines. *See also* Amino sugars and Sugar amines
in bacterial cell walls, **26:**198; 200–207; 210; 212; 223
in carbohydrate metabolism, **26:**38–40; 46; 49–50; 55–56; 58–59; 66; 70; 73–78
Hexosaminidase, **38:**298; 300
Hexosaminidase A, **38:**301; 307; 310, **48:**60
human, **38:**310
liver, **38:**306; 307
placenta tissue, **38:**306; 307
urine, **38:**305; 306; 307
Hexosaminidase B, **38:**301, **48:**60
Hexosaminidase C, **60:**127
Hexosaminidase isozymes, **38:**301
Hexosaminidases, **60:**125
assay, **60:**128
genetics, **60:**133
glycolipid hydrolysis, **60:**129
heat stability, **60:**130
inactivity, **60:**129
inhibition, **60:**133
interaction with substrate, **60:**132
isoelectric point, **60:**130
isoenzymes, **60:**125
kinetic data, **60:**131
molecular weight, **60:**129
physiological role, **60:**133
properties, **60:**129
purification, **60:**127
subunits, **60:**126
Hexose, **34:**524
alcoholic fermentation of, **5:**167–179; 199
from labeled acetate, **14:**122
and phosphatide formation, **7:**148–149
desmolysis, **7:**16; 17
Hexosediphospatase, localization of, **24:**320; 335
Hexosediphosphatase, **37:**442; 453; 464; 516; 549; 578
Hexose diphosphatase, in photosynthesis, **19:**15
Hexose diphosphatases, classification, **14:**245

Hexose diphosphate. *See also* Fructose-1,6-phosphate, and Glucose-1,6-phosphate
and arginine synthesis, **15:**330–331; 333
fermentation and ATPase activity, **6:**186
formation by oxygen treatment, **6:**75–77
metabolism to pyruvate in cell nuclei, **16:**488
from pentose phosphates, **15:**154
Hexosediphosphate, synthesis in withered leaves, **1:**365
Hexose diphosphate aldolase, **15:**171
Hexose diphosphate phosphatase, **15:**170
Hexose diphosphates, *See* Hexose phosphates
3,4-^{14}C-Hexose, from labeled acetate, **28:**407
Hexose isomerase, allosteric regulation, **28:** 117–118
Hexosemonophosphatase, **1:**360
Hexose monophosphate dehydrogenase, **5:**138
Hexose monophosphates. *See also* Hexose phosphates; individual phosphates
and dehydrogenase, **7:**413
metabolism, **15:**151–170; 171; 176–177
Hexose monophosphate shunt, **4:**211–212, **5:**139, **48:**99; 100; 123; 159; 160; 163; 172, *See* Pentose phosphate cycle
control of, **48:**99
for deoxyribonucleotide formation from TPNH, **29:**354
in latex, **29:**225–226
and pyridine nucleotides, **15:**120–121
Hexose-phosphatase activity, multiple forms, **25:**348
Hexose phosphate
direct oxidation of, **5:**137
metabolism in familial periodic paralysis, **6:**217
Hexose-6-phosphate, production, **15:**145–147
Hexose phosphate dehydrogenase, **48:**126
Hexose phosphate isomerase, **14:**205, **15:**153; 168, **19:**15
allosteric regulation, **28:**75
Hexose phosphate isomerization, enzymic, **14:**193–195; 197–213
Hexose phosphate mutases, **18:**198
Hexose phosphates, **7:**15–16; 413–414; 424–432
in Pasteur effect, **15:**173–174
in plant oxidations, **16:**133; 134
in plant respiration, **18:**285–287; 289; 292–295; 312
turnover studies, **7:**117–120; 124
Hexose phosphate transport system distribution, **44:**239
Hexose phosphate uptake, K_m for, **44:**250
Hexose-1-phosphate uridyltransferase, **67:**38; 40
Hexoses. *See also* individual hexoses
fluorinated, **55:**245–248
metabolism, **15:**145–151
in sperm metabolism, **24:**39–41; 46–51; 52–68
in transglycosylations, **17:**196–200; 202; 219
uronic acids of, in ribose synthesis, **28:** 398–399
Hexoses (C^{14}), in transglycosylations, **17:** 198–200
Hexosyltransferases, localization of, **24:** 311–312
H_2-Folate, enzymic synthesis from glutamate and H_2-pteroate, **35:**47
5,6,7,8-H^3-Folate-H_4, in methyl synthesis, **25:**401
H_2 folate synthetase, **53:**355
HGH dimer, **42:**148; 149
HGH (human pituitary growth hormone), **42:**113; 117; 118; 119; 120
human plasma treated, **42:**151
isolation of, **42:**76
purity criteria, **42:**77
reoxidized, **42:**135
stability of, **42:**83
synthesis of, **42:**155
High-density lipoprotein, interaction with lipoprotein lipase, **40:**34
High energy bonds, **23:**323–399
High-energy bonds, nucleotidyl cleavage, **18:**220
High-energy compounds. *See also* ATP, Bonds, high energy, Energy conservation and Oxidative phosphorylation
from respiratory chain, **17:**114–116
High energy compounds, in absence of ADP, **23:**372–373
table, **23:**324
High-energy state, coenzyme Q, **25:**320
High-energy state cytochromes, **28:**161
High-energy (X) intermediates, in ATP synthesis, **28:**159
Highly conserved/strictly conserved residues, cytochrome *c* oxidase, **71:**91
High molecular weight carrier, **34:**447

High-performance liquid chromatography, tyrosine hydroxylase, **70:**110
High performance liquid chromatography methods, **59:**185
High-pH forms, EPR and, **51:**128–130
High spin-low spin conversion, **51:**13–14
High temperature, mutagenic effect, **27:**331
Hill coefficient
glutamate dehydrogenase-NADPH, **39:**372
simple relationship to modifier concentration, **53:**319–320
Hill coefficients, of PP-ribose-P anudo transferase, **41:**10
Hill factor, *See* Methemoglobin reducing factor
Hillman Nutrition Clinic experiments, **6:** 501–503
Hill reaction, **19:**45–54; 57; 62; 63
and cytochromes, **18:**23; 54–57
and lipoic acid, **18:**341
inhibition of, **30:**23
phenylureas (CMU, DCMU) inhibitor in, **30:**40
in photosynthesis, **30:**3
simazin inhibitor in, **30:**40
in plastids, **20:**227; 228
Hill reaction, CO_2 effect on, **25:**88
Hill reagents, **19:**46–54
Hindgut (Protodeum), **45:**391
Hiochic acid, and mevalonic acid, **23:**473
Hip-His-Leu, **57:**16
Hippurate, synthesis, in mitochondria, **21:**42
Hippurate formation, **24:**311; 334
Hippurate synthesis, and benzoyl CoA, **18:**208
Hippuric acid, biosynthesis, **19:**249
role of ATP in, **19:**252
Hippuric acid synthesis, enzymic system for, **13:**12
L-Hippuricases, classification, **14:**247
Hippurylamide, papain action on, **19:**304–306; 325
α-Hippuryl amino acid amides, **54:**33; 36
Hirudin, **46:**15, **48:**297
Hirudinone, **36:**227
HisJ, **65:**18–19
His-Leu, **57:**27
His-Leu dipeptide, **57:**3
HisM, **65:**11–14
HisP, **65:**14–16
analysis, **65:**28–29
ATP-binding activity, **65:**29–30
structural model, **65:**21–22; 24
HisQ, **65:**11–14
ATP-binding activity, **65:**29–30
His-195 residue, dihydrodipicolinate reductase, **72:**297
His-181 residue, GMP synthetase, **72:**120–123
His-143 residue, triad amidotransferase, **72:** 138–139
Histamin, **2:**93–99; 103; 104; 107–109
Histaminase, **2:**93; 94; 96; 97; 101, **5:**68; 69; 70; 71, *See* Diamine oxidase
location in subcellular fraction, **24:**302
Histamine, **5:**67; 68; 69; 70; 71; 78, **20:**101; 102, **58:**291
effect on capillary permeability, **13:**227–228
effect on tryptophan peroxidase, **25:**377
formation from histidine, **5:**78–79
methylation of, **22:**255; 256
in mitochondria, **21:**8; 10
oxidation, **21:**22
relation to blood clotting, **16:**54; 71
Histamine-hemin, as catalase model, **15:**28
Histamine phosphate, in transaminations, **16:**196
Histamine production, **6:**2; 4
effect of temperature, **6:**2
Histidase, **9:**404
imidazole in, **20:**100
kinetic studies, **17:**40
pH dependence studies, **15:**282
in tobacco, **6:**369
L-Histidase, classification, **14:**247
Histidinal, relation to histidine metabolism, **16:**300
Histidine, **34:**16; 21; 453; 462, **48:**158, **50:**156; 157; 164, **51:**10; 217, **56:**315. *See also* Imidazole
active residue of aminoacyl-tRNA synthetases, **40:**180; 181
active site, of β-lactamase, **28:**260
biosynthesis, allosteric regulation, **28:**44; 66–69; 118–119; 124–126
at active site of enzyme action, **22:**53; 56; 57; 64–69; 71; 80–82; 88
role in enzyme action, **22:**66; 67; 69
synthesis in *E. coli,* regulation of, **22:**17
5-amino-4-imidazole ribonucleotide, **39:**138
analogs of, **29:**124
biosynthetic pathway, **29:**127

Histidine (*Continued*)
control of synthesis by analogs, **29:** 126–127
phenotypic resistance to analogs of, **29:**130
resistance to, **29:**128
binding to hemin in enzymes, **14:**188
in electron transfer system, **14:**40–41
oxidation of, **14:**132
biosynthesis, **16:**299–301, **21:**208; 396; 398; 403–405
chelates, in metal ion activation of enzymes, **21:**145; 146
degradation, **21:**223; 228; 405–407
inactivity in *Neurospora* mutant, **16:**256
possible function in activated fibrinogen molecule, **16:**42
in protamines, **16:**442
relation to base binding capacity of myosin, **16:**320
relation to glutamate metabolism, **16:**252; 255; 299; 301
in transamination reactions, **16:**190; 192; 196; 235–237; 239
biosynthesis of, **39:**138; 151
complex with iron and porphyrin, **15:**38
metabolic pathway, **15:**185–186
of tobacco mosaic virus mutants, **15:**479
in creatine kinase active site, **54:**395–400
in creatine kinase reaction, **54:**376; 384
cytochrome *c* oxidase, **71:**87; 92; 96; 105; 138–139; 141
in decarboxylation of oxalacetate, **12:**81
effect of ionizing radiations, **24:**382; 397
effect on succinic dehydrogenase, **18:**88–90
and enzymic catalysis, **24:**203–218; 241–243; 248; 250; 450–454; 457; 460; 467
esterase activity of, **20:**284
ethyl derivative of, **25:**383
in fructose-bisphosphatase regulation, **54:**133; 178
genetic studies, **17:**405–407; 418; 452
glutamine or ammonia, **39:**151
glutathione S-transferases, **69:**19
in identification of imidazole, **20:**71; 75
imidazole glycerol phosphate, **39:**151; 152
from imidazolyllactate, **19:**350
metabolism, **19:**179–181
in papain, **19:**284
ionization constant, **20:**58
metal binding by, **20:**56
in peptides and proteins, **20:**58–61
phosphoribulosyl forminino-aminoimidazole carboxamate ribonucleotide, **39:**151; 152
in phosphorylated peptides of esterases, **20:**288
role of ATP, **39:**138
structure, **20:**53; 86
thermodynamic data on ionizations, **20:**64
in zinc ion effect on fructose-bisphosphatase, **54:**135; 136
L-Histidine, deamination, **19:**344; 378; 379
L-Histidine, oxidation in plants, **20:**320; 322
Histidine ammonia lyase, **43:**244, **46:**459–460, **55:**238
Histidine auxotrophs, of *E. coli,* **16:**299
of *Neurospora,* **16:**299
Histidine carboxylase and inhibition, **8:**480
Histidine complex with metals, **27:**96; 110–111
Histidine deaminase, mechanism of deduced from relaxivity data, **33:**50–51
Histidine decarboxylase, **5:**69–71; 76; 77, **55:**235
bacterial, **6:**4
adaptive formation in *E. coli,* **6:**8
effect of age of culture on growth, **6:**12
of growth temperature, **6:**2; 11
of inhibitors on activity, **6:**18
*p*H of growth medium, **6:**10; 16
Michaelis constant, **6:**16
purified, **6:**12–21
specificity, **6:**15
distribution in tissues, **5:**69
purification by absorbants, **5:**70
pyruvoyldependent, **74:**175–176; 176
specific inhibitors of, **5:**69–70
L-Histidine decarboxylases, classification, **14:**256
L-Histidine-HC1, action on thiaminase, **15:**401
1-Histidine-hemin as catalase model, **15:**28
Histidine imidazoiium group, **36:**114
Histidine kinase, *Caulobacter* cell cycle, **71:**9; 18
Histidine link in hemoglobin, **7:**273–274
Histidine methyl ester, reaction with benzoyl chloride, **20:**83
Histidine permease, **65:**3–4; 32
membrane-bound proteins, **65:**16–17
Histidine residue
adenosine receptors, **69:**92; 93; 107–108
in luciferase binding, **25:**142
Histidine residues, collagen hydroxylases, cosubstrate binding, **72:**349–357

L-Histidinol, **17:**405–407
relation to histidine metabolism, **16:**299–300
Histidinol phosphate, **16:**236
relation to histidine metabolism, **16:**299–300
Histidinol-phosphate aminotransferase, **74:**133
phylogenetic distribution of, **74:**157
Histidyl group modification, **49:**172
by diethylpyrocarbonate, **49:**172
by photooxidation sensitized by pyridoxal phosphate, **49:**172
of the β_2 subunit of tryptophan synthase, **49:**172
Histidyl groups, modification, **53:**70
on fructose 1,6-bisphosphatase, **53:**70
Histidylhistidine, ionization constants, **20:**58
Histidyl-P enzyme, **59:**352–353
Histidyl riboflavin, **37:**205
8α-(*N*-Histidyl)-riboflavin, **37:**204
Histo- and Cytochemistry, Enzymic (Glick), **9:**585
Histochemical detection of –SH groups, **11:**217
Histochemical localization of RNase, **24:** 165–166
Histochemical methods, use in sclerotization process, **16:**156
Histochemistry, **17:**324
enzymic, **9:**585–611
of β-glucuronidase, **16:**385–386
Histochemistry in the detection of alkaline phosphatase in animal tissues, **10:**89
Histograms, **47:**94
Histological studies of RNA synthesis, **24:**15
Histological study, of ethionine effects, **25:** 372–375
Histone, **38:**49; 84, **42:**230
acetyl transferase, **38:**79
deacetylase, **38:**79
effects on phosphorylase b, **40:**44
effects on sialyl transferase c, **40:**44
interaction with DNA, **40:**45
Histones, **47:**46; 131; 138, **60:**7
amino acid composition, **16:**442; 444–445
association with DNA, **16:**441
in cell nucleus, **16:**415; 420; 444–446; 479–480
change to protamines in maturation of sperm, **16:**443
in chromosomes, **16:**413; 447–448
incorporation of glycine-N^{15}, **16:**458
use in chromatography of nucleic acids, **16:**434
arginine-rich, **18:**172; 175; 177
associated with DNA, **18:**162; 169; 170–177; 185–186
definition, **18:**175
lysine-rich, **18:**169; 170; 172; 175; 177
nucleo-, structure, **18:**176–177
test for, **18:**176
enzymic activities, **27:**266
with RNA, coacervate, **27:**367; 369
HIV-1, elongation factors, **71:**44
HMG, *See* 3-Hydroxy-3-methylglutaric acid
HMG-CoA, *See* 3-Hydroxy-3-methylglutaric acid, coenzyme A ester
HMG-CoA reductase, **55:**108
and andrenal corticosteroids, **38:**401–402
assay, **38:**377
and bile acids, **38:**398
and bile inhibitory proteins, **38:**399–400
and cancer, **38:**402–404
clofibrate, **38:**402
cyclic rhythm, **38:**378–384
effect, of fasting, **38:**378; 381
of light, **38:**379
and enzyme synthesis, **38:**381–383
fine structure, **38:**379–380
in developing brain, **38:**405–406
developmental pattern, **38:**384–391
and cholesterol levels, **38:**387
and enzyme synthesis, **38:**389–390
and hormones, **38:**388
and weaning, **38:**388–389
and fasting, **38:**401
intestinal, **38:**404
mechanism of cholesterol action, **38:** 394–398
preparation, **38:**376
properties, **38:**377–378
reaction, **38:**375
regulation by cholesterol, **38:**393–398
structure of regulatory steroids, **38:** 393–394
and thyroxine, **38:**401
and triton, **38:**401
x-ray, **38:**402
cholesterol ester hydrolases, **69:**45
HNB-Br, **42:**138
HNB-HGH, **42:**139

H_2-Neopterin, monophosphate, **35:**72
triphosphate, **35:**57; 72
HNI, alkylation of nucleic acids, **27:**321
H-NS protein, *Caulobacter* cell cycle, flagellar structure and assembly, **71:**9
H_2 O_2, **35:**265
H_2O_2, **39:**340, **57:**116; 148
Hofmeister series, **39:**345
Hogeboom's gradient method, **18:**105
Holo-ACP synthetase, **36:**284
Holocarboxylase, formation of, **26:**331–334
release of biotin from, **26:**325–326
Holochrome, **46:**72; 75; 95
Holoenzyme
eukaryotic dihydrofolate reductase, NADPH binding, structural changes induced by, **70:**44–45
of methylmalonyl-CoA mutase, **26:**345–349
role of metal, **17:**273–281
Holoenzyme synthetase, **35:**337
biotinyl-5′-AMP, intermediate, **35:**340
partial reactions, **35:**338–340
ATP-^{3} ^{2}P^{3} ^{2}P$_i$ exchange, **35:**338
regulation of, **35:**342
substrate specificity, **35:**339; 340
Holometabola, **45:**411; 413
Holophosphorylase, **11:**303–304
Holotoxin, **58:**113
enterotoxicity, **58:**130
Holotryptophanase, monovalent cations, effect of, **42:**302
sedimentation rate, **42:**302
temperature, effect of, **42:**302
Homaline, **36:**228
Homoarginine, **29:**131, **49:**59; 62, **50:**152
inhibition of bacterial growth, **29:**131
as substrate for aginine sRNA synthetase, **29:**131
L-Homoarginine, **65:**298; 307
Homoarginine exodus, **49:**66
Homocitrate, **67:**317
Homocitrate synthase, **67:**332–333
Homocitric acid, **15:**197
in lysine synthesis, **16:**297
Homocysteate, **49:**62
Homocysteic acid, **49:**59
in transamination reactions, **16:**192; 196
Homocysteine, **17:**408–411, **51:**181, **66:**152. *See also* Homocystine
S-alkyl derivatives, **25:**384
de novo methylation of, **25:**396; 401–410
and ethionine metabolism, **25:**370; 372
and transmethylation, **25:**418; 423–426
biological synthesis in microbial cells, reaction scheme, **22:**265
in biosynthesis of *S*-adenosylmethionine, **22:**244; 245
conversion to methionine, **22:**265–268
in mammalian cells, **22:**264
in microbial cells, **22:**264; 265
origin, in mammalian cells, **22:**264
in microbial cells, **22:**264; 265
in transmethylations, **22:**252; 253
with *S*-adenosylmethionine, **22:**257
biosynthesis of, **56:**24
in biosynthesis of folic acid coenzymes, **21:**394
in methionine biosynthesis, **21:**423–426; 430
in purine biosynthesis, **21:**223
in biosynthesis of methionine, **16:**270–274
reaction with quinones, **16:**121
relation to cystine metabolism, **16:**278–279
cystathionase and, **56:**39
desulfuration, **7:**379–383
in disulfide exchanges, **27:**202; 218; 232
fate of sulfur in, **7:**378–383
oxidation, to homocysteic acid, **7:**379
to homocystine, **7:**378–379
Homocysteine desulfhydrase, **7:**379–380; 382
classification, **14:**257
L-Homocysteine (HC), **51:**194
Homocysteine methyltransferases, **25:**424–427
Homocysteine (S^{35}), from ethionine, **25:**375
Homocysteine sulfur oxidation, **7:**378–379
Homocystine. *See also* Homocysteine
in cystine metabolism, **16:**278
use by *E. coli* mutants, **16:**272
disulfide exchanges, **27:**212
and ethionine effects, **25:**371
metabolism, **7:**378–379
reduction, **21:**430
Homocystine reductase, **21:**430
Homocystinuria, **56:**26
Homoeriodictyol, **55:**22
Homofolic acid, **38:**276
Homogenate, whole, **42:**247
Homogeneity, of enzymes at the cellular level, **22:**107
Homogeneity of viruses, **9:**215–220

Homogeneous solutions, **34:**520
systems, **34:**446; 516–519
Homogentisase in tyrosine metabolism, **14:** 97–99; 107; 109–111
Homogentisate, formation, **16:**230
Homogentisate 1,2-dioxygenase activity, **59:**137
Homogentisate oxidase, **19:**87–92; 126
Homogentisate oxygenase, **27:**518
Homogentisic acid, **17:**415; 416; 421, **59:**137. *See also* Alcaptonuria
from enthyl-benzene, **27:**517–518
metabolic path, **15:**184
in phenylalanine metabolism, **19:**87–92; 183
in tyrosine metabolism, **14:**89; 90; 97–100; 102–104; 104–105; 106; 107; 108–111
Homoglutamine, in transaminations, **16:**196; 202
Homoglutathione, **46:**398, **59:**129
Homoisoleucine, **50:**126; 130; 132
Homologies sequence
of adenylate kinase, **59:**262–263; 265–267
and fibronectin, **59:**9; 14
Homologous mechanisms for carboxylating enzymes, **33:**49
Homologous sequences, **47:**130
definition of, **74:**132
Homology, **42:**118; 119; 123
internal, **42:**126
Homology scaning, fibronectin, central cell-binding domain of, **70:**10
Homopepsins, classification, **14:**247
Homoprotocatechuate oxygenase, **27:**518
Homoserine, **17:**402; 409–411, **22:**241; 260; 264, **49:**63, **50:**120; 142; 143; 144; 152; 168, **56:**39
from adenosylmethionine enzymatic decomposition, **54:**222–227
azetidine-2-carboxylic acid from, **54:**227
biosynthesis, **19:**256
and citric acid cycle, **18:**256–257
formation in splitting of *S*-adenosylmethionine, **22:**259
in isoleucine synthesis, **16:**282; 284; 285
labeled, conversion to methionine, **16:**272
relation to methionine, **16:**272–274
relation to threonine, **16:**269–271
in study of lysine metabolism, **16:**298
in methylthiadenosine formation, **54:**213
Homoserine deaminase-cystathionase, absorption spectrum, **22:**396
pyridoxal phosphate in, **22:**396
Homoserine dehydrase, **35:**109; 111
Homoserine dehydratase, **56:**38
Homoserine dehy drogenase, **37:**418; 453; 476; 493; 517; 527; 549; 578
Homoserine dehydrogenase, **50:**143; 144
allosteric regulation, **28:**53; 54; 73
repression of synthesis, **28:**70–71
Homoserine kinase, product inhibition of, **25:**172
Homoserine-*O*-phosphate, **54:**226–227
Homoserine synthesis, allosteric regulation, **28:**60–62; 70–72
Homoserline, **50:**141
Homospermidine, **36:**228
Homosulfanilamide, action on thiaminase, **15:**400
for thiaminase disease, **15:**415
Homotrypsins, classification, **14:**247
Honeybee, *See* Apis mellifera
Honey-invertase, transglucosylase activity of, **17:**206–207
Hopkins phenomenon, **18:**93
H_2O_2 production, respiratory burst oxidase, **65:**53–54
Hordeine, synthesis of, **25:**383
Hordenine distribution in germinating barley seed, **8:**233
Hormodendrum viride, enzymic transformations of steroids by, **20:**248
Hormonal control, of phosphorylase interconversion, **28:**107; 110
Hormonal factors, in conversion of glycogen synthetase I to D, **28:**102
Hormonal Modulation of Specific Messenger RNA Species in Normal and Neoplastic Rat Liver (Feigelson and Kurtz), **47:**275
Hormonal regulation of phosphorylase activity, **24:**286
Hormone, **36:**83
parathyroid, attributes of valinomycin shown by, **29:**292
effect on cations and mitochondria by, **29:**46
effect on permeability of mitochondrial membrane by, **29:**378
K^+ and Mg^{++} accumulation induced by, **29:**301
required for Mg^{++} uptake, **29:**275
thyrotropic (*See* Thyrotropic hormone)

Hormone action, and nucleoside triphosphatases, **27:**283
Hormone Inactivation, mechanism of, **41:**99; 113–115
Hormone-inhibitor complex, effect of adsorption, **28:**28
Hormone receptors, **36:**40
glucagon, **36:**51
insulin, **36:**51
Hormones, **53:**212. *See also* individual hormones
absorption spectra, **4:**277; 278; 285–289
action on phosphorylase, **34:**398
in control, on interconversion of D and I forms of glycogen synthetase, **34:**328; 331
of mammalian liver glycogen synthetase activity, **34:**341
effect, on glucose 6-phosphatase, **34:** 302–303
in phosphorylase in heart, **34:**395
and glycogen level in fetal liver, **34:**292
in the regulation of glycogen metabolism, **34:**413
sex, effect on glycogen synthesis and glycogen synthetase, **34:**344
activity concentration, **1:**187–190
adenohypophyseal, synthesis of peptides related to, **29:**392
adrenal cortical, **262–265**
adrenocortical, effect on β-glucuronidase, **16:**389
anterior pituitary growth, relation to transamination, **16:**228
pituitary, **16:**48
pupation, relation to sclerotization, **16:** 159–160
sex, of *Chlamydomonas eugametos,* **16:**144
steroid, determination in blood and urine, **16:**383–385
effect on β-glucuronidase, **16:**363; 386–394; 401
adsorption studies, **14:**346
and D-amino acid oxidase, **57:**93
and D-aspartate oxidase, **57:**96
as cofactors, **24:**113
chemical modification of, **24:**107
effect on sperm, **24:**40; 83–85
peptide, dispensable part, **24:**117–118
structures, **24:**416–419
pituitary, amino acid sequences, **24:** 418–419; 422–425; 430
definition, **1:**190
effect of tyrosinase on, **14:**142; 147; 149
effect on enzymes, **37:**157–163
and enzymatic reactions, **4:**257; 264
enzyme relation, **1:**190–192; 196
epinephrine and glycogen metabolism, **4:**262 (*See also* Adrenaline)
gonadotropic, separation by chromatography, **14:**340; 342; 350
and hypothesis of insulin action
cell growth, **57:**162–164
effects of, opposite insulin, **57:**164
growth factors, **57:**162–164
insulin (*See* Insulin)
mechanisms of, **57:**87
in oxidative phosphorylation, **14:**13
parathyroid hormone, **4:**262
peptide, **57:**150
peptide, SH-groups, **27:**226–228
peroxidase action on, **14:**155
phenylalanine hydroxylase activation effect, **67:**146–150
pituitary, and melanin formation, **14:**90–91; 96
pituitary hormones, **4:**258
plant, absorption spectra, **4:**287
and semen metabolism, **9:**343–344; 379–380
light-induced changes in, **9:**138; 142; 156
sex hormones, **4:**265
steroid, and melanin formation, **14:**90–91
steroid hormones, absorption spectra, **4:** 288–289
and thiols, **11:**247–248
thyroid hormone, **4:**261–262
phosphoenolpyruvate carboxykinase (GTP), **69:**240
thyroid-stimulating, **14:**113–116
tyrosinase action on, **14:**144–146
and vitamin E, **6:**503–504
Hormones, Absorption Spectra (Brode), **4:**269
Hormones, Influence on Enzymic Reactions (Jensen and Tenebaum), **4:**257
Hormones and diamine-oxidase, **2:**107
Hormones-sensitive lipase, mRNA for, cholesterol ester hydrolases regulation, **69:**69
Hormone stimulation, for discrimination of system, **49:**85

Hornet, *See* Vespa orientalis
Horse chestnut leaves, **46:**75
Horsehair, spatial arrangements of polypeptide chain, **13:**159
Horse liver alcohol dehydrogenase, **46:** 471–472
Horse liver alcohol dehydrogenase, Cotton effects, **27:**49–82
Horseradish peroxidase, **33:**284; 309–310; 312, **58:**67, **60:**37
Compound II of, **33:**329
Compound I of, **33:**329
peroxidase kinetics, **73:**4 (*See also* Peroxidase kinetics)
Horse-radish peroxidase, **52:**179; 210, *See* Peroxidase
Horse radish peroxidase (HRP), **51:**25; 46; 61–63
rate constant values for formation of, **51:**69
Host-virus relation, **2:**6–21. *See also* Bacterial viruses
adsorption
capacity, **2:**8
irreversibility, **2:**8
rates, **2:**6–7
residual fraction, **2:**7
growth of virus alone, **2:**27
lysis of cell, **2:**8–9
methods of assay, **2:**6; 15–17
and serological specificity, **2:**17–21
virus liberation, **2:**11–15
Housefly, *See* Musca domestica
HPLC, **59:**185
h-Protein, mouse, **46:**409
HPT, *See* Heteropyrithiamine
H_4PteGlu, **54:**59
condensation of, **54:**75
H_2-Pterin-CH_2 OH, as product of action of dihydroneopterin aldolase, **35:**63
H_2 pteroate synthetase, **53:**354–355
HS^-, **39:**363
HS-CoA, *See* Coenzyme A
Hückel, *See* Debye-Hückel theory
Human Alcohol Metabolism, Enzymology of (Li, Ting-Kai), **45:**427
Human brain ganglioside, resistance to neuraminidase, **20:**144
Human cancers, **39:**222
Human cervical mucus, as substrate for neuraminidase and receptor destroying enzyme, **20:**144
Human Chlorionic gonadotropin (HCG), **41:**113–115; 118
Human chorinoic gonadotropin, **50:**314
Human erythrocyte enzyme, **47:**152
Human erythrocytes, **59:**371
Human erythroeytes, Hypoxanthine-guanine phospho-ribosyl transferase from, **41:**14
Human fibronectin receptor, **59:**39–40
Human GMP synthetase, **72:**123–124
Human intestinal glycolipid, **52:**27
Human lactoferrin, **52:**27
Human lung, γ-glutamyl transpeptidase enzyme and transcripts, **72:**262–264
Human lymphoblasts, PP-ribose-P amidotransferase from, **41:**10
Human neutrophil cathepsin G, **57:**4
Human Pituitary Growth Hormone, The Chemistry of (Bewley and Li), **42:**73
Human placenta, PP-ribose-P amido transferase from, **41:**10
Human poisoning, by fluoroacetate, **18:**138
Humans, adenylosuccinate synthetase (AMPSase) gene structure, **73:**93–94
Humates, **17:**147; 152
Humidity, low, effect of on enzyme action, **33:**283–284
Humus
formation in browning of plant tissues, **16:**153
importance in soil, **28:**365; 371
Hunter disease, **60:**253
Huntington's chorea, **42:**277
appearance of, **29:**480
Huntington's disease, **44:**230
Hurler disease, **60:**241
Hyaline, degeneration in muscular dystrophy, **6:**219
Hyalobiuronic acid, synthesis of, **25:**435–436
Hyaloplasm, localization of enzymes in, **13:**5–8
Hyaluronate, **41:**42, **60:**223; 271
catabolism, **60:**240
composition, **60:**272
β-glucuronidase action, **60:**273
mechanism of degradation by hyaluronidase, **13:**208–212
Hyaluronate lyase, **60:**228; 283
activity, **60:**284
bacterial, **60:**284
Hyaluronate mucodextrinase, classification, **14:**246

Hyaluronate synthase, **60:**272
Hyaluronectin, **60:**271
Hyaluronglucuronidase, **60:**228; 284
Hyaluronic acid, **18:**201, **25:**434–439, **52:**81; 82. *See also* Hyaluronidases
chemistry, **13:**201–202; 203
monosulfuric acid ester, **10:**236
oligosaccharide from, **16:**371
preparation, **13:**200–201
synthesis, **26:**59
virus crystallizing agent, **9:**196
Hyaluronic acid and snake venoms, **8:**472; 487
Hyaluronic acid oligomucase, classification, **14:**246
Hyaluronidase, **23:**186; 188; 200; 201; 208; 218, **48:**84, **60:**227
column chromatography, **22:**475; 476
effect on egg jelly coat, **9:**263
in sperm, **9:**268; 333; 348–352; 378
use in histochemistry, **9:**593
effect on semen, **24:**86
inactivation by radiations, **24:**373
molecular weight from radiation studies, **24:**366
effects of polycations, **40:**40; 41
inhibition, **8:**473–475
measurement, **8:**473–475
multiple nature, **8:**472–473
properties, **8:**473–475
of snake venoms, **8:**471–476; 484–486; 487
Hyaluronidase hydrolysate, action of β-glucuronidase on, **16:**371
Hyaluronidases, **13:**199–236, **25:**435; 436
classification, **14:**246
Hyaluronidases (Meyer and Rapport), **13:**199
Hyaluronoglucosaminidase, **60:**227
activity
with chondroitin-4-sulfate, **60:**229
with glycosaminoglycan, **60:**229
characterization, **60:**228
inhibition, **60:**229
lysosomal, **60:**230
biosynthesis, **60:**232
characterization, **60:**231
medical applications, **60:**232
molecular weight, **60:**231
optimum pH, **60:**228
purification, **60:**228
substrate specificity, **60:**231
subunits, **60:**228
testicular, **60:**228
transglycosylation, **60:**229
Hybrid anthranilate synthetase, **38:**36
Hybrid enzymes, **24:**23
Hybrid insulin, **33:**515
Hybridizable messages, **40:**85
Hybridization, of hypoxanthine-quanine phospho ribosyl transferase, **41:**20
Hybridization experiments, **39:**194
Hybridization of the dimeric apoenzymes from *Alcaligene faecalis* and *Pseudomonas dacunhae,* **35:**536
Hybridization techniques, and transcriptional repression, **40:**79
Hybrid mechanism, **57:**379
Hybrids, **57:**202–204
Hybrid transfer, eukaryotic dihydrofolate reductase, **70:**73. *See also* Eukaryotic dihydrofolate reductase
Hydantoin, **56:**125
Hydantoinase, **74:**72
3-Hydorxyproline, bacterial degradation, **37:**316
distribution in nature, **37:**278
formation of, **37:**294
Hydratases, classification, **14:**228; 256
Hydrated electron, EPR spectrum, **27:**119
Hydrate inclusion compounds, **65:**241
Hydrates
channel and layer structures, **65:**246–248
hydrogen bonding, **65:**234–236
Hydration, **34:**486
Hydration and contractility of protoplasm, **8:**37–42
Hydration energies, **43:**249
Hydration of tobacco mosaic virus, **15:**456
Hydration processes, catalyzed by metal ions, **21:**177
Hydra vulgaris, **71:**215
Hydrazides, hydrolysis by *a*-chymotrypsin, kinetic studies, **17:**46–49
Hydrazine, **60:**252
effect on degradation of pectic substances, **20:**352
in exchange reactions, **13:**289–295; 312
inhibition of amino acid decarboxylases, **6:**18
mutagenic agent, **27:**314–315; 335–336; 340
Hydrazine derivatives, labeled, in determination of activated amino acids, **19:**237
Hydrazone, **58:**176

Hydride acceptor, pteridine activation, eukaryotic dihydrofolate reductase, chemical transformation step (reactant activation), **70:**78–81
Hydride mechanism, hydrogen transfer, **26:** 137–139; 141; 144; 145; 155
Hydride transfer, **59:**146
  in mechanisms of enzyme catalysis, **24:** 472–475
  in oxidation of NADH, **26:**143–156
Hydride transfer mechanism, **43:**494
Hydride transfer step in the glutamate dehydrogenase reaction, **39:**401
Hydrobenzoin trimesters, **55:**310–312
Hydrocarbon chains, **34:**468
Hydrocarbon chains at interfaces, **11:**58–64; 64–69; 76–82
Hydrocarbon-oxidizing enzymes, adaptive nature, **27:**475–477; 494–495; 499; 519
Hydrocarbon-oxidizing extracts, microbial, **27:**356–537
Hydrocarbon-oxidizing organisms, action, **10:**454–456; 456–462
  distribution, **10:**449–451
Hydrocarbon polymer, abiogenic synthesis, **27:**386
Hydrocarbon radicals, in stars, **27:**349
Hydrocarbons, **38:**203
  aliphatic
    anaerobic oxidation, **27:**529–534
    dissimilative metabolism, **27:**477–505
  aromatic, dissimilative metabolism, **27:** 505–528
  assimilation by animals, **10:**476–478
    assimilation by microorganisms, **10:** 443–486
    bacterial dehydrogenation, **10:**469
    criteria used for microbial attack on, **10:**445–448; 465
    derivatives, action of microorganisms on, **10:**462–464
    germicidal action, **10:**451–454
    oxidizability by microorganisms, **10:** 454–462; 467–469
    polycyclic, microbial oxidizability, **10:**461–462
    production by microorganisms, **10:** 478–479
      by plants, **10:**478
    products of microbial oxidation, **10:** 444–445; 464–470
    susceptibility to attack by micro-organisms, **10:**456–462
      difference between paraffinic and cyclic compounds, **10:**456–457
      variation with branching, **10:**460–461
      within chain length, **10:**449; 454–456; 457–459
      with degree of unsaturation, **10:**459–460
      variety attacked, **10:**444–445
  binding to protein in oxidation, **27:**527–528
  carcinogenic, **15:**6; 34
    mechanism of autoxidation, **15:**10
  evolution of, **27:**349–354
  oxidation
    adaptive *vs.* constitutive, **27:**474–477
    initial reactions, **27:**528–538
  oxidizability, and structure, **27:**473–474
  petroleum, microbial oxidations, **27:**469–546
  synthesis by electron bombardment, **27:**352
Hydrocarbons, Assimilation by Microorganisms (ZoBell), **10:**443
Hydrocellulose from cellulose, **8:**260–262
Hydrochloric acid from gastric mucosa, **13:** 57–58
Hydroclastic splitting of pyruvate, **6:**236; 244
Hydrocortisone
  effect on enzymes, **37:**158
  formula, **20:**241
    microbiological formation and conversions, **20:**246; 249; 251–255; 270; 271; 274; 275
  induction of tryptophan peroxidase by, **25:**377
  level of glycogen enhanced by, **29:**378
Hydrodisulfide, **39:**329
Hydrodisulfides(persulfides), **39:**330
Hydrodynamic studies, **34:**503
Hydrogen
  inhibition of butyric acid fermentation, **6:**247
  ion (*See* pH)
  ions, properties, **21:**157
  molecular, and pyridine nucleotides, **15:**113
  phosphate effect on rate of evolution, **6:**237
  radioactive (*See* Deulerium)
Hydrogen acceptors. *See also* Electron acceptors and Electron transfer
  with xanthine oxidase, **17:**295
Hydrogen acceptors and chloroplast activity, **12:**8–9; 26–28; 28–31

Hydrogenase, **15:**118, **46:**166
activation of, **11:**149
from *M. lactilyticus,* **17:**242; 248–249
and nitrogen fixation, **13:**358–360
oxidation of hydrocarbons, **27:**534
relation to sulfur metabolism, **10:**308–310
Hydrogenase for fatty acids, **8:**317
Hydrogenases
classification, **14:**229; 257
in electron transfer, **19:**201
as metalloflavoproteins, **17:**247–250; 267–270; 275; 277; 281–282
Hydrogenation, of steroids, by microorganisms, **20:**253–259
Hydrogenation of fatty acids, **8:**352–353; 365–373
α-Hydrogen atoms, effect of thiol ester formation, **16:**19
Hydrogen bond, in energy transport, **14:**19–25; 36
in organic catalysis, **14:**185–188; 189–190
Hydrogen-bonded network, **50:**24
of proteins, **50:**24
Hydrogen bond formation, in mixed monolayers, **11:**84
Hydrogen-bond formation, rate of, **25:**20
in protolytic reactions, **25:**21–23
Hydrogen bond in enzyme-substrate complex, **10:**73–74
Hydrogen bonding, **34:**469, **38:**213; 216, **43:**254, **47:**49; 51; 59; 60; 62; 64; 79; 81; 83, **58:**52, **65:**217–250
biological polymers, **43:**316
bond lengths and angles, **65:**221–223; 228; 242–243
chain stoppers, **65:**231
clathrate hydrates, **65:**240–245
cooperativity, **65:**226–241
cyclic systems, **65:**237
cyclodextrin hydrates, **65:**29–240; 237
in cytochrome P-450, **60:**66
and denaturation, **23:**47; 56
and protein structure, **23:**1–27
and denaturation of RNase, **24:**186; 192–193
and enzyme activity, **24:**460
disaccharides, **65:**234–236
disorder, **65:**230–231
in DNA, **27:**291; 311; 318–319; 324–325; 335
and protein aggregation, **27:**224; 225
electronic structure, **65:**226–227
electrostatic, **65:**220–221
and enzyme action, **12:**236–237; 299–302; 305
Fischer lock and key, **65:**249–250
flip-flop disorder, **65:**231
force constants, **65:**221
α-helix, **47:**49; 58–59, **65:**229
hydrates, **65:**234–236
infinite and finite chains, **65:**231–233
intramolecular interresidue, **65:**234
ions, **34:**460; 508–510
locating hydrogen atoms, **65:**223–224
molecular recognition, **65:**247–250
monosaccharides, **65:**231; 233; 235–236
N-methylacetamide, **43:**316
nonadditivity, **65:**228–229
nucleic acids, **65:**218; 224
of nucleotides, **17:**459–464
solubilization effects, **17:**140
patterns, **65:**231–232
pleated sheets, **65:**229–230
in polysaccharides, **28:**327
properties, **65:**220
proteins, **65:**218; 224
between purine and pyrimidine, **16:**435
in sclerotization, **16:**162
resonance-assisted, **65:**229
semiclathrates, **65:**240–245
β-sheet, **47:**49; 59–60; 81
strong *versus* weak, **65:**219–220
sugar alcohols, **65:**231; 233
tetrasaccharides, **65:**234; 238
three-center, **65:**224–226
trisaccharides, **65:**234; 238
tunneling, **65:**230
β-turn, **47:**61–64
Hydrogen-bonding, of DNA, **18:**166–167
Hydrogen bonds, **38:**216; 219, **43:**317
in activity of esterases, **20:**299; 313
in gelation of pectic substances, **20:**357
in imidazole, **20:**60; 92–96; 99; 101
among myosin subunits, **16:**330
in gel formation of fibrin, **16:**40; 42
in catalysis, **50:**12
to coenzymes, **50:**28
competition among, **50:**34
in hemoglobin, **50:**27; 28
network in, **50:**21
in proteins, **50:**3; 19; 23; 24

cooperativity in hydrogen bond formation, **50:**24
dependence of energy on interatomic distance, **43:**247
effect of radiations on, **24:**367; 369; 393; 404
entropy of formation, **43:**317
and enzyme activation, **9:**35–39
in enzyme-catalyzed reactions, **9:**69
and fertilization membrane, **9:**279
in enzyme structure, **22:**73; 74
formation, in coenzyme binding, **22:**384
glutathione S-transferases, **69:**9; 15–16; 19; 20; 22; 29–30; 36
in hydrolytic enzyme action, **15:**293; 296–297
in activation transmission, **15:**30; 36–37
in nucleoproteins, **15:**461
of proteins, and semiconductive properties, **15:**34–37
in sulfhydryl groups of papain, **19:**294
Hydrogen bonds in proteins, **13:**158–159; 161
Hydrogen cyanide. *See also* Cyanide, Potassium cyanide
effect on cytochromes, **18:**6; 7; 10; 12
and papain, **19:**285–287; 290; 302
Hydrogen dehydrogenase, PN-dependent, **26:**132
Hydrogen-deuterium exchange, *See* Deuterium-hydrogen exchange
Hydrogen donors
ferrocytochrome c, **12:**170–171; 175; 187
from the light reaction of photosynthesis, **12:**132–135
reaction with enzyme-substrate complex, **12:**166–168; 168–171; 181–182; 182–184
specificity for catalases and peroxidases, **12:**174–175
in photosynthesis, **11:**131–133; 139; 150; 183; 184; 185; 194
Hydrogen donors and fatty acid metabolism, **8:**391
Hydrogen gas, and acetate metabolism, **23:**422; 423
Hydrogen-ion concentration, and denaturation, **24:**188–193; 196
and effect of radiations, **24:**391–392
effect on sperm, **24:**75–76
Hydrogen ions, **38:**202
role as inhibitor of hydrolytic enzymes, **15:**281–282; 284; 286 (*See also* pH)
Hydrogen isotope effects, primary, *See* Isotope effects
Hydrogenlyase, **7:**317
Hydrogenlyases, classification, **14:**229; 257
Hydrogenolysis, liability of carbobenzoxy group to, **29:**394–395
Hydrogenomonas, RuDP carboxylase of, **19:**32; 33
Hydrogenomonas entropha, **54:**169
Hydrogenomonas eutropha, phosphoribulokinase from, regulation of, **74:**110
Hydrogenomonas eutropha, PN-dependent enzymes, **26:**132
Hydrogenomonas facilis, phosphoribulokinase from, regulation of, **74:**109
Hydrogenomonas, oxidation of hydrocarbons, **27:**534
Hydrogenomonas ruhlandii, PN-dependent enzymes, **26:**132
Hydrogenomonas ruklandii, **17:**249; 268
Hydrogen peroxide, **41:**36–37, **57:**141–142, **58:**61. *See also* Peroxides
in acetyl phosphate formation, **6:**233
in biosynthesis of nicotinic acid, **15:**102
in enzymic oxidoreductions, **15:**3; 6–7; 12; 16–18; 23; 25–28; 33; 38
catalase action on, **19:**109; 113–116; 119
decomposition, **19:**137–140
effect on natural melanins, **16:**169
formation in catechol autoxidation, **16:**140
in melanin formation, **16:**167
reaction of quinones with, **16:**127
effect on papain, **19:**287
effect on pectic substances, **20:**352
as hydrogen donor for catalase action, **12:**182–184
physiological sources of, **12:**185–188
inhibition of ATPase, **6:**167
as inhibitor of xanthine oxidase, **17:**309–310
complex with peroxidase, **17:**55–56
in oxidation of thiols, **11:**206–207
oxidation-reduction potentials for, **33:** 231–232
product of oxygen reduction, **33:**231
in oxygen metabolism, **19:**97; 126; 127; 135; 136; 141–143; 191; 196–198
peroxidase kinetics, **73:**4 (*See also* Peroxidase kinetics)
and semen metabolism, **9:**374–375

Hydrogen peroxide (*Continued*)
formation in sperm, **9:**302
tyrosine hydroxylase, **70:**200–201
Hydrogen peroxide as acceptor, oxidoreductases, **24:**307
Hydrogen peroxide complexes, with catalase, **19:**111; 114–116
with ferrimyoglobin, **19:**111; 114; 116; 117; 138
with peroxidase, **19:**105; 109–114; 118–121; 123–125
Hydrogen peroxide formation, and irradiation of enzymes, **24:**394–397
in sperm, **24:**66
Hydrogen sulfide, **5:**242. *See also* Sulfide
effect on cytochrome a_3, **20:**166
effect on phenolase complex, **19:**199
enzymatically formed in animal tissues, **5:**242
formation
from cysteine, **5:**242; 246
from homocysteine, **5:**242
formation in nature, **10:**231–232
from microbial action on hydrocarbons, **10:**448; 473
in plant metabolism, **10:**232–234
hydrogen donor, **7:**399
from colloidal sulfur, **7:**385–386
form cysteine, **7:**390–392; 397
form cystine, **7:**396–397
form digestive tract, **7:**397
form homocysteine, **7:**379–380; 397
metabolism, **7:**397–399
and papain, **19:**285–287; 290
Hydrogen transfer. *See also* Pyridinc nucleolide coenzymes
adenosylcobalamin, **66:**27–28
S-adenosylmethionine, **66:**6–9
by glucose oxidase, **15:**148
mechanism, **15:**13–16; 32; 40; 41
and origin of life, **27:**369–371
stereospecificity of, **27:**56
and oxidative phosphorylation, **15:**177
in photosynthesis, **19:**49
by pyridine nucleotides, **15:**87–90, **26:** 115–191
direct, **26:**165–166
PN binding to dehydrogenases, **26:**156–171
stereochemistry of, **26:**165–171
stereospecific with respect to, coenzyme, **26:**169–171
substrate, **26:**166–169
reconstitution of chain, **28:**155–236
during respiration in animal tissues, **20:** 147–199
in microbiological conversions of steroids, **20:**273
in plastids, **20:**219
reversal of, **23:**382–383
Hydrogen transfer, B_{12} coenzyme-mediated, **38:**436
Hydrogen transfer and luminescent system, **7:**231–233
Hydrogen transport, by C_4-dicarboxylic acids, **8:**412–414
Hydroglucan, **28:**335
Hydrolase, **39:**362, *See* Aspergillus oryzae
Hydrolase activity of soil, **28:**367; 368
Hydrolases, **5:**92–94, **56:**292
acid, localization of, **24:**315; 322
amino acid sequences in, **24:**422
chemical modification of, **24:**214
localization of, **24:**299; 316–330; 340
alkaline phosphatase, **56:**341–350
angiotensin converting enzyme, **56:**355–358
carboxypeptidase A, **56:**358–376
classification, **14:**221–225; 234–235; 244–250
mode of action, **14:**172–173; 183
collagenases, **56:**382–384
as distinguished from transferases, **33:**430
enzymic adaptation, **13:**70
imidazole identification in, **20:**72
leucine aminopeptidase, **56:**350–355
nonfunctional, **18:**250
specificity of, **17:**169; 170
thermolysin, **56:**376–382
in tobacco, **6:**366–369
Hydrolase systems, in relation to redox potential, **10:**289
Hydro-lyases, localization of, **24:**331–332
Hydrolysis, **34:**475–480; 488; 490–491; 501; 508–511; 524–526, **48:**42; 44; 48; 49; 50
acid, **48:**53
mild, **48:**66; 68
by amylases, **48:**45
of ATP, labeled oxygen study, **16:**337
by α-chymotrypsin, kinetic studies, **17:**46–49
O^{18} in water, **17:**162
of cinnamoyl-thiol-subtilsin, kinetics of, **33:**394–395

of NPA, kinetic constants of (table), **33:**393
of complex carbohydrates, **48:**41; 54
controlled, of complex glycopeptides, **48:**57
$EGFR_{988-988}$ Ala-scan peptides, **68:**26–28
enzymatic, **48:**44; 45; 46; 53
enzyme, **34:**488–489
of glycosidase, **48:**47
of glycosids, **48:**43; 45
at interfaces, **11:**43–45; 58; 59–66
of pectic substances, enzymic, **11:**270–276
of membrane teichoic acid of Lactobacillus arabinosus, **30:**226
metal catalyzed, **58:**6
patterns of, **28:**349–351
phosphate monoesters, **68:**14
phosphorylated peptides, **68:**25
protein tyrosine phosphatase catalyzed, **68:**12–13; 23
reaction, **34:**487
Hydrolysis by Cholinesterase and Related Enzymes (Davies and Green), **20:**283
Hydrolysis products, in detection of soil enzymes, **28:**371–372
Hydrolytic activity, **34:**485, **49:**239–241
Hydrolytic enzymes, **2:**33–48, **51:**58–61
activity measurement at varying substrate concentration, **2:**36; 43–47
binding site, **23:**76
and cell organization, **17:**191
carbohydrases, group-transfers by, **17:** 190–232
control of action, **2:**41–43
decomposition and monomolecular curve, **2:**33–36; 45
kinetics, **2:**33–48
location in cell, **16:**474
mechanism of action, **15:**271–299
in mitochondria, **21:**39–41
multiple forms, **25:**348–351
product inhibition of, **25:**221–241
for penicillins, **28:**238; 242; 243; 247–253
of soil, **28:**366–367; 370; 388
reaction velocity, **2:**36; 37
two-phase reaction, mechanism and formulation, **2:**37–40
Hydrolytic enzymes requiring –SH, **11:** 240–242; 244
Hydrolytic site, in mitochondria, **23:**379–380
Hydrolyzates, protein, analysis of, **6:**15
Hydroperoxides
and antioxidants, **8:**439–440
from autoxidation of unsaturated compounds, **8:**427–439; 449–451
in oxidation of hydrocarbons, **27:**491–492; 537
Hydroperoxyallene, **58:**215
Hydroperoxy fatty acid radical, **58:**278
Hydroperoxy group, **58:**290
Hydroperoxylinoleic acid, **58:**196; 206
Hydroperoxyl radical, **41:**37
Hydrophilic amino acids, RNA polymerase II transcription, C-terminal domain characteristics, extended structure, **71:**50
Hydrophilic carrier, **34:**455
Hydrophilic-lipophilic balance (H-L-B), **51:**32
Hydrophobic amino acids, **38:**200; 202; 213; 215; 219
Hydrophobic association, **39:**358
Hydrophobic binding, **34:**486, **38:**217, **43:**335; 339, **46:**383
bonding, **34:**469
film, **34:**513
group, **34:**455
interaction, **34:**469–470
regions, **34:**468
solvent, **34:**455
structures, **34:**471
substances, **34:**478
Hydrophobic components, periplasmic permeases, **65:**11–14
Hydrophobic effect, **58:**16
Hydrophobic groups, **42:**118; 120
Hydrophobic interaction, **34:**52; 55, **43:**327
Hydrophobic interaction chromatography, **57:**373–374
alginate biosynthesis, **70:**226
Hydrophobic interactions, of E-S complex, **33:**418
Hydrophobicity, **38:**213; 230
Hydrophobic membrane proteins, *Caulobacter* cell cycle, flagellar structure and assembly, **71:**11
Hydrophobic region, **39:**346; 350
Hydrophobic residues, **41:**182; 183; 185, **47:**49; 72; 81; 104; 114–115; 131
at the amino side of the splitting point, **41:**183
arcs, **47:**48
at both sides of the splitting point, **41:**185
at the carbonyl side of the splitting point, **41:**182

Hydrophobic residues (*Continued*)
lining active site, eukaryotic dihydrofolate reductase, ligand binding (side chain roles), **70:**61–63
nucleus, **47:**79
triplets, **47:**88; 132
Hydroquinone
in cytochrome study, **17:**356; 360; 367; 369
mechanism of oxidation, **15:**14
in oxygen metabolism, **19:**124; 125; 127; 184
peroxidase action on, **19:**106
in peroxidase action, **27:**123
Hydroquinones. *See also* Quinones
autoxidation, **65:**102–103; 121–123
chromatographic separation, **25:**293–294
redox transitions, **65:**101–104
as tanning substrate, **16:**156
autoxidation, **16:**128
binding to quinones, **16:**125
equilibrium with quinones in plants, **16:**131
oxidation by phenolase, **16:**140
role in browning of plant tissues, **16:**150
stabilization, **16:**107
Hydroquinone SO_2 complex, **65:**241
Hydroquinone thioether derivative, **65:**106
Hydrosulfide, **39:**329
Hydrosulfite (dithionite), **39:**328
Hydrotropic effects, solubilization by, **17:**136; 137; 146; 147; 152–156
Hydroxamate, in identification of pectin, **20:**370
Hydroxamates, amino acid, **29:**100
Hydroxamic acids, **35:**137–140, **58:**218
absorption, **35:**138
as antibiotics, **35:**170–171
association constant, **35:**139
biological activity, **35:**167–178
biosynthesis, **35:**156–167; 158
control of, **35:**165
dissociation, **35:**137
geometry, **35:**138
in heme synthesis, **35:**167
in iron transport, **35:**173–178
metal exchange, **35:**139
periodate oxidation, **35:**139–140
periodic reaction, **35:**153
in reactivation of phosphorylated esterases, **20:**305–308
solubility, **35:**139
stability, **35:**138
tautomerism, **35:**138
Hydroxamic Acids of Natural Origin (Emergy), **35:**135
Hydroxamides, hydrolysis by α-chymotrypsin, kinetic studies, **17:**46–49
Hydroxo (aquo)-cobalamin. *See also* Corrinoids and Vitamin B_{12}
absorption spectrum, **26:**247–249
synthesis, **26:**242–243; 246; 251
Hydroxo (aquo)-corrinoids. *See also* Corrinoids and Vitamin B_{12}
properties, **26:**247–251; 255–256
Hydroxo-cobinamide, absorption spectrum, **26:**247–249
7-Hydroxy, 2′, 4′-dimethoxy isoflavone, **55:**35
α-Hydroxy-*N*-acetyl-ε-amino-caproic acid, genetic studies, **17:**403
D-α-Hydroxy acid dihydrogenase, product inhibition of, **25:**223; 260
L-α-Hydroxy acid oxidase. *See also* Glycolate oxidase
L-α-Hydroxyacid oxidase, **46:**475
L-α-Hydroxy acid oxidase
and alcohol-glyoxylate adducts, **57:**122
and amine-glyoxylate adducts, **57:**122–123
catalytic reaction of, **57:**96–97
concentrations of, **57:**97
functions of, physiological, **57:**98
inhibition of, **57:**98; 139–141
kinetic constants for, **57:**126–127
limiting rate constant for, **57:**127
long-chain, **57:**97–98
Michaelis constants for, **57:**127
and nucleophile-glyoxylate adducts, **57:**99–100
and oxalyl thioesters, **57:**124–125; 128
presence of, **57:**86–87
short-chain, **57:**97
specificity for thiol products, **57:**124
substrate requirements for, **57:**99–101
and thiol-carbonyl adducts, **57:**123
and thiol-glyoxylate adducts, **57:**122; 124; 128
turnover number of, **57:**127
α-Hydroxyacid oxidase, **19:**202
L-*a*-Hydroxy acid oxidase in kidney, **8:**355
and synthesis of fats, **8:**336
Hydroxy acids
reaction with borate, **12:**495
salts of, as solubilizing agents, **17:**144; 146–149

β-Hydroxyacids, activation, **15:**249; 251–252
α-Hydroxy acids, PN-dependent oxidations, **26:**118; 122
α-Hydroxy acids, racemization of, **14:**193; 196–197
β-Hydroxyacyl ACP dehydrases, **36:**298–301
β-Hydroxyacyl-CoA dehydrogenase, in soluble fraction of liver, **21:**49
β-Hydroxyacyl-CoA dehydrogenase, PN-dependent, **26:**118
β-Hydroxyacyl coenzyme A, and pyridine nucleotides, **15:**123
β-Hydroxyacyl coenzyme A dehydrogenase, **15:**115
β-Hydroxyacyl dehydrogenase, in mitochondria, **21:**74
Hydroxyacylglutathione, **59:**105
α-*N*-Hydroxyalanine, **35:**155
Hydroxyaldehyde, in glucose-6-phosphate oxidation shunt, **15:**154
4-Hydroxyalkenals, **57:**377
2-Hydroxy-3-alkyl-1, 4–4 napththoquinone, effect on succinic oxidase, **20:**176; 177
N-Hydroxyamino acids, **35:**165
α-Hydroxy-ε-aminocaproic acid, **54:**40; 41–42
ε-Hydroxy-α-aminocaproic acid, in transaminations, **16:**196
11β-Hydroxy-Δ^4-androstene-3,17-dione, identification of its glucuronide, **16:**383
Hydroxyanthranilate, **49:**36
3-Hydroxyanthranilate, **49:**28
3-Hydroxyanthranilate oxidase, **19:**92–97; 126
Hydroxyanthranilic acid, **17:**421
3-Hydroxyanthranilic acid
  in biosyntheis of nicotinic acid, **15:**99; 100 (N^{15})
  in tryptophan metabolism, **19:**92–96
N^{ω}-Hydroxy-L-arginine, **65:**321; 323; 328–329
  NADPH intermediate required, **65:**323–324
threo-β-Hydroxyaspartate, **35:**505
β-Hydroxy aspartate, **35:**89
β-Hydroxyaspartate, **50:**17
p-Hydroxybenzaldehyde, **17:**312, **34:**213
  relation to lignin, **18:**355–358; 361; 373; 374
p-Hydroxybenzaldehyde, in oxidation products of lignins, **16:**149
p-Hydroxybenzoate, in synthesis of lysine, **16:**298
  use by aromatic polyauxotrophs, **16:**287
ρ-Hydroxybenzoate hydrolase, **37:**418; 426; 433; 517; 578
p-Hydroxybenzoate hydroxylase, **74:**251
p-Hydroxybenzoate hydroxylases, **55:**249
p-Hydroxybenzoic acid, **15:**184–185
p-Hydroxybenzoic acid, genetic studies, **17:**405; 412; 416; 421
p-Hydroxybenzoic acid (C^{14}), as lignin precursor, **18:**374
p-Hydroxybenzoyl methyl ketone, **34:**213
p-Hydroxy-benzyl mustard oil, occurrence and properties, **10:**265
p-Hydroxybenzylpenicillin, activity of β-lactamase on, **28:**257
p-Hydroxybiphenylglucuronide, as substrate for β-glucuronidase, **16:**365
Hydroxybutyrate, **57:**126–127
β-Hydroxybutyrate, **15:**238
β-Hydroxybutyrate
  and acetate metabolism, **23:**423–424; 429
    oxidation of, and phosphorylation, **23:** 349–350; 354; 357; 361; 362; 368; 370; 371; 383; 385
  glycolysis in 6C3HED lymphoma inhibited by, **29:**360
    oxidation of, **29:**77–78
3-Hydroxybutyrate, **53:**225
β-Hydroxybutyrate-acetoacetate-system, in respiratory chain studies, **17:**73; 78; 88; 100; 105; 107–108; 114
β-Hydroxybutyrate cycle, to oxidize DPNH by mitochondria, **29:**342
D-3-Hydroxybutyrate dehydrogenase, **46:**242
β-Hydroxybutyrate dehydrogenase, **23:**345–346
  adsorption at interface, **28:**3
    lipid requirement, **28:**31
  increase of concentration of DPNH produced by, **29:**360
  localization of, **24:**296
β-Hydroxybutyrate oxidation, and phosphate turnover, **17:**175
γ-hydroxybutyrate, tyrosine hydroxylase regulation, **70:**161–162
β-Hydroxybutyric acid. *See also* Ketone bodies
β-Hydroxybutyric acid, acetoacetate from, **15:**122–123
β-Hydroxybutyric acid, oxidation, in mitochondria and mitochondrial fragments, **21:**34; 38; 39

β-Hydroxybutyric acid from methionine, **12:**478
β-Hydroxybutyric acid oxidase system, **17:**83
β-Hydroxybutyric dehydrogenase, **15:**118
activity of —SH group, **11:**241
classification, **14:**253
β-Hydroxybutyric dehydrogenase, **17:**78; 80
in mitochondria, **21:**26; 27
β-Hydroxybutyric r DPN transhydrogenases, **14:**253
β-Hydroxybutyrl CoA epimerization, **44:**98
S-β-Hydroxybutyryglutathione, synthesis, **19:**240
S-β-Hydroxybutyryl CoA, synthesis, **19:**240
β-Hydroxybutyryl coenzyme A, **15:**233; 243
β-Hydroxybutyryl-S-CoA dehydrogenase, **23:**423
4-(4′-Hydroxy-2′-carbethoxy-1′-pyrrolidyl)-5-methyl-1,2-benzoquinone, formation, **16:**112
Hydroxycarboxylic acids, as inhibitors of β-glucuronidase, **16:**373
p-Hydroxycinnamic acid, **18:**366
Hydroxycinnamic acid coenzyme A thiol estres, **55:**9; 17
Hydroxycinnamic acids, **55:**9
Hydroxycinnamoyl-CoA ligases, **55:**17
biotic elicitors of, **55:**18
isomers of, **55:**17; 18
substrate specificity of, **55:**18
p-Hydroxycinnamyl alcohol, in lignin formation, **18:**363; 371–372; 375
p-Hydroxycinnamyl alcohol-glucoside, labeled, prepn. of, **18:**372–373
Hydroxycitrate, **43:**60; 79; 90; 91
inhibition of lipagensis, **43:**91
Hydroxycitric acid, **43:**78
14α-Hydroxycortexone, microbiological formation, **20:**270
15α-Hydroxycortexone, microbiological formation, **20:**248; 271
15β-Hydroxycortexone, microbiological formation, **20:**248; 271
17-Hydroxycorticosteroids, glucuronide of, **16:**384
6β-Hydroxycorticosterone, proof of structure, **20:**276
19-Hydroxycorticosterone, proof of structure, **20:**276
6β-Hydroxy-11-dehydrocorticosterone, proof of structure, **20:**276
6β-Hydroxy-11-deoxycorticosterone, **19:**191
2,3-Hydroxy-2,3-dihydrotryptophane, **15:**102
6-Hydroxy-dopamine, **55:**169
tyrosine hydroxylase induction, **70:**166
Hydroxyeicosatetrenoic acids, cholesterol ester hydrolases regulation, **69:**71
Hydroxyepilubimin, **55:**52
6-*N*-Hydroxyethyl-ATP, **35:**487
N-7-Hydroxyethyloxymethylguanine, **54:**63
Hydroxyethylthiamine diphosphate, activity of, **24:**468
Hydroxy fatty acids. *See also* Fatty acids
synthesis, **8:**335–336
5-Hydroxy ferulic acid, **55:**39
2′-Hydroxygenistein, **55:**5
structure of, **55:**6
β-Hydroxyglutamic acid, **53:**169
as substrate for glutamine synthetase, **31:** 201–203
γ-Hydroxyglutamic acid, as substrate for glutamine synthetase, **31:**201–203
N-Hydroxyglycine, **35:**156; 157; 158
γ-Hydroxyhomoarginine, **50:**152; 155
Hydroxyhydroquinone, metabolism in plants, **16:**140
L-3-(4-Hydroxy-3-hydroxymethyl-phenylalanine, **50:**166
5-Hydroxyindolacetic acid (5-HIAA), in cerebral spinal fluid, **29:**494
excretion in affective psychosis, **29:**493
metabolism of serotonin determined by excretion of, **29:**520
β-Hydroxyisobutyryl-CoA hydrolase, propinoic acid metabolism, **26:**367
β-Hydroxyisovaleryl CoA, **18:**210
in photosynthesis, **19:**44; 45
Hydroxyisovaleryl CoA carboxylase, in photosynthesis, **19:**42–45
Hydroxykynureninase, **49:**3; 36
3-Hydroxykynurenine, **17:**419–421, **19:**194
in biosynthesis of nicotinic acid, **15:**99–102
Hydroxylamine, **35:**156; 161; 170; 487; 489, **38:**277, **51:**220, **56:**22, **57:**125–126, **58:**63; 161
and acetoacetic acid metabolism, **6:**257
antagonist of asparagine and glutamine, **13:**259–260
in exchange reactions, **13:**289–295; 298; 301; 304; 308–309; 311; 312
in glutamine synthesis, **13:**276–280

role in nitrogen fixation, **13:**369–373
effect on cytochrome a_3, **20:**165; 166
effect on degradation of pectic substances, **20:**352
in identification of pectin, **20:**345
in reactivation of phosphorylated esterases, **20:**305; 307; 308
effect on Hill reaction and photosynthesis, **19:**50
effect on transamination, **19:**354–356; 367; 368; 375; 376
effect on tryptophan oxidase, **19:**122; 127
reactions with activated amino acids, **19:**236; 237; 250; 251; 253; 255
reaction with activated fatty acids, **19:**257
in transpeptidation reactions, **19:**310; 311
as enzyme inhibitor, **15:**17; 379
test for acyl mercaptides, **15:**190; 220; 226–227; 230; 231; 236–237; 239–240; 246–247; 250–251
as glutamyl acceptor, **17:**177–178; 179
in genetic studies, **17:**436–437
as inhibitor of xanthine oxidase, **17:**311
inhibition of acetylation, **6:**259; 261
of amino acid decarboxylases, **6:**18
mutagenic agent, **27:**305; 312–314; 336–342
in plant metabolism, **9:**393–396
and pyruvate metabolism, **6:**257
role in nitrogen assimilation, **16:**256
in ternary complex with DPN and ADH, **22:**351
Hydroxylamine, Hill reaction inhibited by, **30:**23
Hydroxylamine reductase, **17:**252
in electron transfer, **19:**201
Hydroxylamine reductases, PN-dependent, **26:**130–131
Hydroxylamino acids, **35:**138
reduction, **35:**138
stability, **35:**138
Hydroxylase-decarboxylase method, tyrosine hydroxylase, **70:**109–110
Hydroxylases, **47:**4
location in subcellular fraction, **24:**308; 339
6β-Hydroxylases, **20:**263; 265; 269
11α-Hydroxylases, **20:**245; 265; 269; 270; 274
11β-Hydroxylases, **20:**274
14α-Hydroxylases, **20:**263
16α-Hydroxylases, **20:**265; 270
21-Hydroxylases, **20:**265
Hydroxylation, **67:**80
of anthranilate, **41:**88
of *p*-cresol, **41:**43
of *m*-hydroxybenzoate, **41:**88
of 4-nitrocatechol, **41:**43
of proline (whole-cell systems), **41:**247
rate constant, **60:**74
of steroids by microorganism, **20:**239–251; 274–276
N-Hydroxylation, **35:**156; 158; 159; 165
inhibition by iron, **35:**167
Hydroxylation of aromatic compounds, by liver nonspecific hydroxylase, **19:**144–179
by nonenzymic model system, **19:**135–143
by peroxidase hydroxylating system, **19:**143; 174
Hydroxylation patterns, in naturally occurring compounds, **16:**141–146
Hydroxylation reactions, aromatic, **27:**527–528; 536
steroid, **27:**538
Hydroxyl group, **58:**15
ω-Hydroxyl group, and intestinal transport, **49:**78
3-Hydroxyl group of pyridoxine, relation to activity, **16:**221
Hydroxyl groups
γ-glutamyl transpeptidase catalytic sites, **72:**255–262
reactivity of in enzymic reactions, **33:** 395–399
reactivity of in organic reactions, **33:** 395–399
relation to activity of alkaline phosphatase, **10:**97–98
Hydroxyl ions, **38:**202
ε-Hydroxylnorleucine, **17:**403
Hydroxyl radical, **41:**43, **46:**166; 167; 170
orientation in aromatic compounds, **19:**136; 139; 141
Hydroxylsine, **41:**245–246
in collagen, **41:**245
derived from lysine, **41:**246
DL-*cis*-5-Hydroxylsine, **50:**125
DL-*trans*-5-Hydroxylsine, **50:**125
Hydroxyl specificity of tyrosinase, **14:**84
5-Hydroxyl-L-tryptophan, **50:**157
Hydroxylubimin, **55:**52; 73

Hydroxylysine
biosynthesis, of, **37:**289
glycosylation of, **37:**293
urinary glycosided of, **37:**312
occurrence and functions, **72:**328–334
residue structure, **72:**333–334
vertebrate proteins with, **72:**328–329; 331–334
6-Hydroxylysine, **50:**169
ε-*N*-Hydroxylysine, **35:**154; 161
Hydroxylysine β1r*O* galactosyltransferase, **52:**27
D,L-4-Hydroxymandelic acid, oxidation, **50:**326
p-Hydroxymecuribenzoate, **53:**220
p-Hydroxymercuribenzoate, **59:**123
p-(hydroxymercuri)benzoate, **71:**147
7-Hydroxymethotrexate, eukaryotic dihydrofolate reductase, ligand binding (of inhibitors), **70:**58–60
4-Hydroxymethyl-4, 5′8–8trimethylpsoralen, **60:**23
Hydroxymethyl cytosine, **17:**463
5-Hydroxymethylcytosine, **18:**164, **27:**291
in bacteriophage, **25:**398
in DNA of bacteriophage, **16:**432–433
in pyrimidine biosynthesis, **21:**290; 423
4a-Hydroxy-6-methyl-5-deazatetrahydropterin, **67:**91–93
5-Hydroxymethyldeoxycytidylic acid, biosynthesis, **21:**290; 421
5-Hydroxymethyldeoxyuridylic acid, in pyrimidine synthesis, **21:**421
2-*C*-Hydroxymethyl-D-glucose, presence of a single common carrier in translocation, **29:**53
Hydroxymethyldihydropteridine, pyrophosphate ester, **27:**424
Hydroxymethyldihydropterin, **35:**38
molecular weight of, **35:**40
purification of (table), **35:**40
pyrophosphokinase, **35:**39–40
K_m values for, **35:**40
Hydroxymethylene group, reduction of, **28:**436; 437
3-Hydroxy-3-methylglutaraldehydic acid, **23:**482
3-Hydroxy-3-methylglutaric acid
coenzyme A ester, formation, cleavage in cholesterol biosynthesis, **22:**291; 292
formation from leucine metabolites, **22:**292
hypothetical scheme for reduction, **22:**294
as precursor of mevalonic acid, **22:**284; 289
reduction to mevalonate, **22:**292; 293
stereospecificity of enzymic reduction, **22:**295
structure, **22:**289
synthesis from labeled acetate, **22:**288; 289
β-Hydroxymethylglutaryl CoA, **55:**57; 59
β-Hydroxy-β-methylglutaryl-CoA, **23:**480
in carotenogenesis, **21:**337; 338
in mevalonic acid biosynthesis, **21:**336; 337
β-Hydroxy-β-methylglutaryl CoA, in photosynthesis, **19:**44; 45
β-Hydroxy-β-methylglutaryl-CoA deacylase, **21:**338
β-Hydroxy-β-methyl-glutaryl-CoA (HMG-CoA), as precursor of rubber, **29:** 233–234
in rubber biosynthesis, **29:**230; 233
Hydroxymethylglutaryl-CoA lyase, localization of, **24:**331
Hydroxymethylglutaryl CoA reductase, **37:**140; 155–156
β-Hydroxymethylglutaryl CoA reductase, **55:**59
active *vs.* inactive forms of, **55:**61
regulation of, **55:**61
β-Hydroxy-β-methylglutaryl-CoA reductase, **45:**373
chemical mechanism of, **45:**373
kinetic mechanism of, **45:**373
Hydroxymethylglutaryl-CoA reductase, localization of, **24:**296
Hydroxymethylglutaryl-CoA synthase, localization of, **24:**331
β-Hydroxymethylglutaryl CoA synthetase, **55:**57
mechanism of action of, **55:**60
Hydroxymethyl group, reduction of, **28:**436; 437
Hydroxymethyl-methyl transfer, **21:**420–431
L-3-(3-Hydroxymethylphenyl)alanine, **50:**166
3-(3-Hydroxymethylphenyl)alanine, **50:**149; 165
cis-4-Hydroxy-methyl-L-proline, proline analog, **29:**95
Hydroxymethyltetrahydrofolic acid, **23:** 459–460. *See also* N^5,N^{10}-Methylenetetrahydrofolic acid
in choline synthesis, **21:**427
isomers and structure, **21:**386–388
in methionine synthesis, **21:**423; 424; 426
in purine biosynthesis, **21:**223
in sarcosine oxidation, **21:**429

in serine-glycine interconversion, **21:**416–418
synthesis and metabolism, **21:**224; 409; 419
in thymine synthesis, **21:**420–423
N^{10}-Hydroxymethyltetrahydrofolic acid, in pyrimidine biosynthesis, **21:**264; 289; 290
Hydroxymethyltetrahydrofolic dehydrogenase, **21:**224; 388; 401; 409; 415; 417–419
Hydroxymethyl transfer, folic acid coenzymes in, **21:**396–398; 416–429
without folic acid coenzymes, **21:**419; 420
Hydroxymethyltransferases, localization of, **24:**309
4′-Hydroxymethyl-4,5,′8-trimethylpsoralen
recovered, purity, **66:**99–100
structure, **66:**87
types of HMT-thymidine adducts, **66:**87–88
2-Hydroxynaphthyl-1-hydroxamic acid, **58:**219
2-Hydroxy-5-nitro-benzylbromide, **54:**400
2-Hydroxy-5-nitrophenyl sulfate, **43:**383
4-Hydroxynon-2-enal, **57:**388
β-Hydroxynorvaline, *See* α-Amino-β-hydroxyvaleric acid
N-Hydroxyornithine, **35:**141; 159
decomposition of, **35:**141
δ-*N*-Hydroxyornithine, **35:**159
5-Hydroxy-4-oxo-L-norvaline(HONV), **39:**205
Hydroxypantothenic acid activity, **7:**436
Hydroxyphenols, *See* Oxyphenols
p-Hydroxyphenylacetic acid in oxidation of phenylalanine and tyrosine, **14:**98; 101–102; 102–103; 111
Hydroxyphenyl compounds. *See also* individual compounds
homogentisic acid from, **14:**102–104
p-Hydroxyphenyl compounds, inhibition of melanin by, **14:**93; 95–96
(*p*-Hydroxyphenyl) hydroxyethylamine, in noradrenaline formation, **14:**117–118
p-Hydroxyphenyllactic acid, in oxidation of phenylalanine and tyrosine, **14:**98; 101–102; 108–110; 111
p-Hydroxyphenylmethyl residue, in lignin, **18:**370
Hydroxyphenylpyruvase, in tyrosine metabolism, **14:**98; 106; 109
p-Hydroxyphenylpyruvate
in amino acid metabolism, **16:**292–293
in transaminations, **16:**198; 230–231
metabolic path to citric acid cycle, **15:**184
p-Hydroxyphenylpyruvate, **19:**183–187
p-Hydroxyphenylpyruvic acid, **17:**415–416
absorption spectrum, **8:**133
in lignin study, **18:**363–366; 375
4′-Hydroxyphenylpyruvic acid, **50:**148
p-Hydroxyphenylpyruvic acid-$C^{14}OOH$, prepn. of, **18:**370
p-Hydroxyphenylpyruvic acid in oxidation of phenylalanine and tyrosine, **14:**97–98; 100; 101–102; 102–104; 105–106; 108–110; 111; 115
Hydroxyphenylserine decarboxylases, classification, **14:**256
m-Hydroxyphenyltrimethylammonium ion, effect on cholinesterase, **20:**298
Hydroxyphenyluria, occurrence of, **14:**98–102; 108–110; 111; 122–123
cis-5-Hydroxypipecolic acid, **50:**125
trans-5-Hydroxypipecolic acid, **50:**125
4-Hydroxypipecolic acid, **50:**122
trans-4-Hydroxy-L-pipecolic acid, proline analog, **29:**95
trans-5-Hydroxy-L-pipecolic acid, proline analog, **29:**95
17α-Hydroxyprogesterone, **19:**191; 192
11α-Hydroxyprogesterone, microbiological formation, **20:**237; 246; 269
11α-Hydroxy-17α-progesterone, microbiological formation, **20:**259
16α-Hydroxyprogesterone, microbiological formation, **20:**270
proof of structure, **20:**276
14α-Hydroxyprogesterone, proof of structure, **20:**276
Hydroxyproline, **37:**274, **49:**80
in actinomycin, **37:**316–317
alternate pathways of biosynthesis, **37:**293
and collagenase action, **23:**168; 170; 176; 191; 203–208; 210–213; 216
and glycine, **23:**457–458
degradation, in animals, **37:**296–300
in plants, **37:**331
degradative pathways in animals, **37:**298–300
detection of, **37:**279
determination of, **37:**279–281
inhibition of proline incorporation, **29:**103
synthesis from proline, **29:**103
metabolic relationships with other amino acids, **16:**257; 259
reaction with quinones, **16:**114; 128

Hydroxyproline (*Continued*)
metabolism in bacteria, **37:**312–317
metabolism of, **37:**282–312
natural occurrence, **37:**276–278
nomenclature of isomers, **37:**276
oxidation by tyrosinase, **14:**141
in plants, in algae, **37:**320
auxin-induced cell elongation and, **37:** 324–325
in *Avena* cytoplasm, **37:**322
biosynthesis of, **37:**325; 326; 328
in cell elongation, **37:**325
in cell wall formation, **37:**327
in cell-walls, **37:**317
chemical nature of, **37:**320; 322
in cytoplasm, **37:**320
degradation of, **37:**331
distribution of, **37:**317–331
extraction from sandal leaves, **37:**323
extraction of, **37:**321
in the free state, **37:**329
in fungi, **37:**320
glycosylation of, **37:**328
incorporation of proline analogs, **37:**330
in plasma, **37:**311
separation of isomers, **37:**281–282
urinary, chemical nature of, **37:**304
collagen pools and, **37:**304
effect of age, **37:**301
effect of body size, **37:**301
effect of diet, **37:**300; 301
effect of lathyrogens, **37:**311
in free state, **37:**308; 310
hormones and, **37:**307
in Kwashiorkor, **37:**308
in malabsorption, **37:**308
origin of, **37:**303
in pathological studies, **37:**306
ratio of collagen breakdown and, **37:**305
cis-4-Hydroxy-L-proline
bacterial degradation of, **37:**315–316
biosynthesis in sandal, **37:**329
distribution in nature, **37:**277
incorporation into plant protein, **37:**324
proline analog, **29:**95
in sandalwood, **41:**293
*cis*4-Hydroxyproline, **50:**125
trans-3-Hydroxyproline, **50:**125
trans-4-Hydroxy proline, **41:**245; 246; 277
biosynthesis of, **41:**246
formation on nascent chains, **41:**277
stabilization of triple-helix, **41:**246
trans-4-Hydroxy-L-proline
in actinomycin, **37:**316–317
biosynthesis in plants, **37:**325
biosynthesis of, **37:**282–284
degradation in bacteria, **37:**313–315
degradative pathways in animals, **37:** 298–300
distribution in nature, **37:**276
free isomer in animals, **37:**293
proline analog, **29:**94–95
4-Hydroxyproline
earthworm cuticle collagen, **72:**332
occurrence and functions, **72:**328–334
triple helix stabilization, **72:**332–333
vertebrate proteins with, **72:**328–329; 331–334
[^{14}C] Hydroxyproline, converted from, [^{14}C] proline in cell-free system, **41:**247
cis-4-Hydroxy-D-proline, distribution in nature, **37:**278
formation in bacteria, **37:**312
3-Hydroxyproline, earthworm cuticle collagen, **72:**332
cis-3-Hydroxy-L-proline, proline analog, **29:**95
trans-3-Hydroxy-L-proline, proline analog, **29:**95
Hydroxyproline catabolizing enzymes, **23:**170
Hydroxyproline-2-epimerase, absence of pyridoxal phosphate, **44:**118
active-site models, **44:**127
active-site peptides, **44:**123
deuterium effects, **44:**120
hydrogen exchange, **44:**119
as obligatory enzyme in hydroxy-L-proline utilization, **44:**117
properties, **44:**118
role of cysteines, **44:**122
specificity studies, **44:**118
Hydroxyproline ethyl ester, reaction with quinones, **16:**112
Hydroxyproline index, **37:**303
Hydroxyprolinemia, **37:**309
Hydroxyprolines, Biochemistry (Kuttan and Radhakrishnan), **37:**273
Hydroxy prolyl-tRNA, **41:**276
β-Hydroxypropionyl-CoA, in oxidation of propionic acid, **26:**367–370
15-Hydroxy-prostaglandin dehydrogenase, **46:**242
Hydroxypruvate-3-P, **51:**213; 216; 243

3-^{14}C-Hydroxypyruvate, in ribose synthesis, **28:**413
Hydroxyputrescine, **36:**207; 225
analytical procedure, **36:**206
automated ion-exchange methods, **36:**206
chromatography, **36:**206
dansy chloride, **36:**206
fluorescent, **36:**206
gas chromatography, **36:**206
ion-exchange chromatography, **36:**206
paper, **36:**206
serum amine oxidase, **36:**206
spermidine dehydrogense, **36:**206
thin-layer electrophoresis, **36:**206
concentration, **36:**206–208
bacteria, **36:**206
bacteriophages, **36:**208
fungi, **36:**206
in gram-negative bacteria, **36:**206
in gram-positive bacteria, **36:**207
virus, **36:**208
distribution, **36:**206–208
2-Hydroxyputrescine, **36:**228
2-Hydroxypyridine, **50:**5. *See also* 2-Pyridone
catalysis by, **50:**11
dissociation constant, **50:**5
3-Hydroxypyridine
as degradation product of pectins, **20:**350
dissociation constant, **50:**5
electrostatic interaction in, **50:**4
pK values of, **50:**2–5; 12
spectrum of, **50:**4
tautomerism of, **50:**1–7
5-Hydroxypyridinylalanine, as a tyrosine analog, **29:**114
β-*N*-(3-Hydroxy-4-pyridone)-alanine, *See* Mimosine
4-Hydroxy-2-pyrrolecarboxylic acid, *See* Oxyminalin
Hydroxypyruvaldehyde, **59:**130
Hydroxypyruvate, **15:**156; 172, **58:**35
in ribose synthesis, **28:**397; 410
transketolase action on, **15:**157–159; 176–177
Hydroxypyruvic acid, in photosynthesis, **19:**16; 26
β-Hydroxypyruvic acid, in transaminations, **16:**198; 204; 234
Hydroxypyruvic acid reductase, in photosynthesis, **19:**16
8-Hydroxyquinoline, **35:**282, **46:**76, **58:**221
activation of β-glucuronidase, **16:**379
effect on homogentisate oxidase, **19:**126
inhibition of chlorophyll synthesis, **46:**70
as inhibitor of metalloflavoproteins, **17:**240; 252; 258; 316
o-Hydroxyquinoline glucuronide, **17:**148
8-Hydroxyquinolineglucuronide, as substrate for β-glucuronidase, **16:**365; 382; 385
Hydroxyquinoline-*N*-oxide, **17:**69
8-Hydroxyquinoline-7-sulfonic acid, **17:**311
Hydroxyquinone, polymerization, **16:**140
6-Hydroxyskatole sulfate, in schizophrenic urine, **29:**522
20β-Hydroxysteroid dehydrogenase, PN-dependent, **26:**118; 167
3α-Hydroxysteroid dehydrogenases, **20:**260; 265; 266; 268; 269; 272
3β,17β-Hydroxysteroid dehydrogenases, **20:**260; 265; 266; 268; 269; 271–273
β-Hydroxysteroid dehydrogenases, localization of, **24:**298; 343
2-Hydroxytetradecanoic acid, CoA thioesters, **67:**401
4-Hydroxytetrahydropterin, **74:**238
structure of, **74:**237
6-Hydroxy-2,4,5-triamino-pyrimidine, inhibitory effect, **27:**424
5-Hydroxytryptamine, **44:**229, *See* Serotonin
vasoconstrictor component of serotonin, **16:**72
Hydroxytryptophan decarboxylase, localization of, **24:**330
5-Hydroxytryptophan (5-HTP), effect on depressed patients by, **29:**493
effects on reserpinized animals by, **29:**487
function of neurotransmitter, **29:**483
and LSD 25, **29:**518
Hydroxytyramine, **5:**74; 79–81
in adrenaline synthesis, **14:**118
in noradrenaline formation, **14:**117–118
formation in animal body, **5:**79–81
α-*N*-Hydroxytyrosine, **35:**155
Hydroxyurea, **65:**149
β-Hydroxyvaline, **53:**169
Hyopoxanthine, and chemiluminescence, **25:**126
Hyoscine
origin, **8:**226
from scopine, **8:**241
in *Datura,* **8:**224
Hyoscyamine. *See also* Atropine
arginine in synthesis, **8:**236–240
produced in roots, **8:**224–226
Hyoscyamus, in alkaloid studies, **8:**224

Hyperammonemia, hepatic glutamine regulation, **72:**77
Hyperbaric oxygen, **35:**315
Hypercapnia, **57:**301–302
Hyperchromic shift, from DNA degraded by acid DNase, **31:**3
Hyperglycemia, **57:**156
Hypermethioninemia, **56:**265
Hyperoxaluria, **57:**151
Hyperphenylalaninemia, **35:**314
Hyperprothrombinemia in pregnancy, **10:**19
Hypertensin, **7:**524–528; 529; 543; 544; 546–550
  action of trypsin, **7:**526
  action of tyrosinase, **7:**526; 543
  animal specificity of formation, **7:**526–528
  destruction and cytochrome c, **7:**533
  formation, **7:**520–521; 523–524; 526–528
  peroxidase action on, **14:**155–156
    tyrosinase action on, **14:**147–148; 150
  precursor of, **7:**523
  properties, **7:**524–526
  serum, inactivation by snake venoms, **8:**489
Hypertensinase, **7:**530–533
  therapeutic use, **7:**538–540
Hypertensinogen, **7:**523
Hypertensinogen and complement, **8:**103
Hypertensins, amino acid interchanges in, **24:**417; 422
Hypertension, and kidney enzymes, **7:**513–556
  diet in, **7:**544–545
  effect of cytochrome c in, **7:**544
  effect of oils, **7:**541–542
  essential, and oxidative enzymes, **7:**538
  establishment of, **7:**545–550
  human, **7:**548–550
  maintenance of, **7:**546–550
  methods for production, **7:**514–516; 545–548
  neurogenic, **7:**514
  and pyrogenic agents, **7:**540
  renal, **7:**515
  and sodium, **7:**544–545
  use of quinones in, **7:**544
  and yohimbine, **7:**547–548
Hypertension, Essential, and Kidney Enzymes (Schales), **7:**513
Hyperthermia
  and fructose 1,6-bisphosphatase, **42:**223
  malignant, **54:**163–164
Hypertrophy and hyperplasia, **4:**225; 226; 227; 242
Hyperuricemia, in gout, **41:**1
Hypochlorous acid, **58:**14
Hypochromicity in glutamate dehydrogenase-NADPH complexes, **39:**371
Hypochromism, and exciton interactions, **24:**369
Hypoferemia, **44:**217
Hypoglycemia, **57:**144; 155
Hypoglycin
  inhibition of gluconeogenesis, **52:**271
  oxidation of fatty acid, **52:**271
Hypoglycin A, **50:**119; 125; 132
Hypoglycin B, **50:**119; 125
Hypolipidemic drugs, **57:**88–89; 93
Hyponitrite, **57:**276
  role in nitrogen fixation, **13:**369–373
Hyponitrite reductase, PN-dependent, **26:**131
Hyponitrous acid in plant metabolism, **9:** 393–394
Hypophosphate analogs, **43:**21
Hypophyseal growth hormone, effect of adsorption, **28:**28
Hypophysectomy, **7:**205, **42:**73; 74; 273, **46:**406
  effect on β-glucuronidase, **16:**388
    in transaminase study, **16:**228
  effect on ribose synthesis, **28:**417
  and iodine uptake, **7:**201
Hypophysis, diabetogenic hormone of, **8:**414
Hypophysis and thyroxine uptake, **7:**209
Hypotaurine, **39:**341
Hypothalamus
  methanol-forming enzyme, activity in, **42:**274
  tyrosine hydroxylase, **70:**107–108
  uptake of dopamine in, **29:**488
Hypothiaminosis, **15:**414–415
Hypoxanthine, **15:**3, **35:**475, **51:**201, **54:**64; 68; 232, **60:**2. *See also* Xanthine oxidase
  biosynthesis, **21:**201; 202
  folate metabolism, **27:**459
  as substrate for xanthine oxidase, **17:** 238–239; 242; 298–301; 309–313; 323
    genetic studies, **17:**426–428
Hypoxanthine dehydrogenase, PN-dependent, **26:**128
Hypoxanthine-Guarine Phospho-ribosyltransferase (HGPRT), **41:**1; 2; 14–23
  activation, **41:**20
  from adenocarcinoma, **41:**14
  antibody-affinity chromatography, **41:**23

from beef liver, **41:**14
binding specificity, **41:**14
deficiency, **41:**2
from *E. coli,* **41:**15
genetic heterogeneity, **41:**19
from human erythrocyles, **41:**14
hybridization of, **41:**20
immuno reactive HGPRT protein, **41:**19
magnesium, **41:**15
metabolic consequences, **41:**22
metabolic cooperation, **41:**21
Michaelis constants, **41:**15
molecular weight of, **41:**16
mutations, **41:**18
properties of, **41:**14
from rat brain, **41:**14
from *Salmonella pombe,* **41:**14
from *Salmonella typhrimurium,* **41:**15
subcellular distribution, **41:**16
substrates for, **41:**14
subunits, **41:**16
tissue distribution, **41:**16
transformation, **41:**21
Hypoxanthine phosphoribosyltransferase (HRPT), **51:**180–181; 184
Hypoxanthine pyrophosphorylase, allosteric regulation, **28:**124
Hypoxia, **58:**85
fall in brain norepinephrine by, **29:**487
Hypo-XMP-pyrophosphorylase, inhibition by nucleotides, **28:**124
p-Hyrdoxyphenylpyruvate oxidase, **19:**183–187; 199
Hysteresis, **53:**64–66
in reactions of fructose 1,6-bisphosphatase, **53:**65–66; 75
Hystersis, **37:**556; 558; 559; 610

I

I, **35:**464
IAA, *See* Iodoacetic acid
Ice, enzyme action in, **33:**285–286
I-cell disease, **60:**124
Ices, **65:**230; 241
Ichtozyme, **23:**173
Icithiamine, **15:**400; 404
Ideal stirring, **34:**504
Iditol, **15:**145; 150
L-Iditol, **15:**118
Iditol-dehydrogenase, PN-dependent, **26:**118
Idonic acid, **60:**233
IDP, *See* Inosine diphosphate
IDPase, *See* Inosine diphosphatase
IDP-sugars, **26:**61
Iduronate2-sulfate sulfatase, **60:**253
activity
in lymphocytes, **60:**255
in serum, **60:**255
towards dermatan sulfate, **60:**255
towards heparan sulfate, **60:**255
binding sites, **60:**255
biological role, **60:**255
catalyzed reactions, **60:**254
heat stability, **60:**254
inhibition, **60:**255
isoelectric point, **60:**254
molecular weight, **60:**254
properties, **60:**254
purification, **60:**254
reaction mechanism, **60:**255
sulfate esters hydrolysis, **60:**254
Iduronic acid, **60:**225; 233
L-Iduronic acid, **60:**242
of connective tissue, **25:**440–441
α-L-Iduronidase, **60:**241
binding site, **60:**243
biological role, **60:**243
biosynthesis, **60:**242
catalyzed reactions, **60:**242
genetics, **60:**243
hydrolysis of iduronic acid, **60:**242
inhibition, **60:**243
maturation, **60:**242
molecular weight, **60:**242
reaction with alkaline phosphatase, **60:**242
receptor-mediated endocytosis, **60:**242
substrate, **60:**243
IGP, *See* Imidazole glycerol phosphate
Ile-Asp-Ala-Pro-Ser (IDAPS) sequence, fibronectin, **70:**12–13
Ileum, rabbit, transport of sodium and glucose in, **29:**59
Imidazole, **43:**341; 382, **58:**9; 28. *See also* Histidine
in acid-base catalysis, **25:**26–30
action on thiaminase, **15:**401
complex with iron and porphyrin, **15:**38

Imidazole (*Continued*)
in transacetylations, **15:**248; 261–262
affinity for hemins, role in catalysis, **14:**174; 188
in electron transfer system, **14:**40–41
biosynthesis, **21:**403
chelates, in metal ion action of enzymes, **21:**144–146
as catalyst, in acetylimidazole decomposition, **22:**81; 82
in nitrophenylacetate hydrolysis, **22:**80
catalytic properties, **20:**86–92; 94; 97; 98; 310; 315
in active centers of esterases, **20:**289; 290; 293; 300; 302; 305; 309–312; 314; 315
chemical structure and reactivity, **20:** 53–58
identification, **20:**63–96
role in biological systems, **20:**51–110
effect on fatty acid peroxidase, **19:**125
effect on LADH complexes, **27:**73–75
effect on succinic dehydrogenase, **18:**88–90
and enzyme complexes, **26:**164
and esterase activity, **24:**461–464
and light emission, **25:**142
Imidazole, Role in Biological Systems (Barnard and Stein), **20:**51
Imidazoleacetic acid, oxidation in histidine metabolism, **19:**179–181
Imidazoleacetic acid oxidase, **19:**179–181; 199
Imidazoleacetic acid riboside, labeled ribose, **28:**400; 414; 417
Imidazole acetol, **17:**406–407
Imidazole acetol (4–2′-keto-3′-hydroxypropyl)-imidazole), relation to histidine metabolism, **16:**299–300
Imidazole acetol phosphate, in transaminations, **16:**196; 236
relation to histidine metabolism, **16:**299–300
Imidazole derivatives, catalytic properties, **20:**86–92; 94; 97; 98
formation and properties of *N*-substituted, **20:**82–85
ionization constants, **20:**57; 58
Imidazole glycerol, **17:**406–407
Imidazole glycerol phosphate
in histidine biosynthesis, **21:**403–405
relation to histidine metabolism, **16:**300
Imidazole glycerol phosphate (IGP) synthase, **72:**131–136
Imidazole glycerol phosphate synthase, **66:** 249–251
Imidazole glycerol (4-(trihydroxypropyl)-imidazole), relation to histidine metabolism, **16:**299–300
Imidazole group
in electron transfer, **17:**87
hydrogen bonds in, **50:**29
Imidazole phosphate, **20:**85; 99, **51:**220
Imidazolepropionic acid, in histidine degradation, **21:**405; 406
4-Imidazolepropionic acid anilide, reaction with benzoyl chloride, **20:**83
5-Imidazolepropionic acid anilide, reaction with benzoyl chloride, **20:**83
Imidazole pyruvate, **54:**135; 136
Imidazolepyruvic acid, in transaminations, **16:**196; 235
Imidazole residue, and catalysis by RNase, **24:**446–448; 450–452; 454; 457
of chymotrypsin, **24:**460
Imidazole ring, **50:**36
as functional enzyme group, **21:**246–248
tautomerization of, **50:**2
Imidazoles
in cytochrome c, **7:**282; 284–286
link in hemoglobin, **7:**273; 274; 276; 277; 312
in peroxidase, **7:**292
genetic studies, **17:**405–407
in cytochrome *c*, **17:**349
inhibition of β-lactamase, **28:**295
2-Imidazolidone, **35:**349
L-2-Imidazolidone-4-carboxylate, **43:**538; 541
2-Imidazolidone-4-carboxylic acid, **43:**536
Imidazolium, ion, entropy for dissociation, **43:**311
Imidazolonepropionase, **74:**72
Imidazolyl lactate, conversion to histidine, **19:**350
Imidazotriazines, **17:**311
Imide, and esterase activity, **24:**462–463
Imide group, **38:**215; 216
Imide intermediates, glutamine-dependent nitrogen transfer, AS-B catalysis, **72:**188–190
Imidodiphosphate, **43:**3; 15
Imidophosphate, **43:**15
adenyl-imidophosphate (AMP-PNP), **43:**15

Imidoyl phosphates, preparation, reactions, **22:**183–186
similarity to phosphate-carbodiimide adduct, **22:**183
structure, **22:**183; 184
in uridine-5′-pyrophosphate synthesis, **22:**187
Imines
pyridoxal-P-dependent, **56:**4–5
pyridoxylidene, **56:**6
Imino acid analogs, in modifying collagen biosynthesis, **41:**267
Imino acid residues, in collagen, **41:**245; 246
Imino acids, activation and incorporation into protein, **29:**100
substrate specificity of proline sRNA synthetase, **29:**101
a-Imino acids, and *a*-amino acids, **8:**117–118
Iminodiacetate, **57:**115
α,α′-Iminodicarboxylic acids, in transamination, **16:**225–226
Imino enzyme, **33:**425; 429
N-Iminoethyl-L-orinithine, **65:**316–317
α,α′-Iminoglutaric-succinic acid, **16:**225
Iminoquinones, formation in browning of plant tissues, **16:**152
Imino-*o*-quinones character, **16:**112–113
a,b-Imino radical arrangement
hypothetical mechanism, **66:**14; 16
nonenzymatic model, **66:**14–15
Imipramine
additive effects with monoamine oxidase inhibition, **29:**485
and iminodibenzyl compounds, **29:**490
inhibition of norepinephrine by brain by, **29:**484; 492
inhibition of reserpine effects by, **29:**485
inhibitors of, **29:**483–485
mechanism of action of, **29:**483–485
anticholinergic activity of, **55:**177
structure-activity studies of, **55:**178
Immobilization, **34:**456; 513; 520. *See also* Immobilized enzymes
and allosteric mechanisms, **57:**210–212
approaches to, **57:**181–182
and conformation of enzymes, **57:**183–185; 206
definitions of, **57:**180–181
drawbacks of, **57:**182–183; 215
and folding of proteins, **57:**215
of inactive protyrosinase, **57:**210
and kinetic cooperativity, **57:**211
methods of, **57:**186–187
models
of bienzyme system, **57:**226–227
first, **57:**225–226
of mechanochemical phenomena, **57:** 229–232
of multienzyme system, **57:**227–228
and molecular mobility, **57:**211
and molecular processes, **57:**209
and oligomeric proteins, **57:**201–204
of plasminogen, **57:**210
proteases stabilized by, **57:**216
and protein-protein interactions, **57:**183; 216–217
and proteins, **57:**183–185
and refolding of proteins, **57:**212–216
and regulation of enzymatic activity, **57:**229
of trypsin, **57:**204; 209
Immobilization of Enzymes: An Approach to Fundamental Studies in Biochemistry (Martinek), **57:**179
Immobilization stress, tyrosine hydroxylase induction, **70:**166
Immobilized active coenzyme preparations, **46:**264
Immobilized B_{12} analog, **46:**258
Immobilized Coenzymes in General Ligand Affinity Chromatography and Their Use as Active Coenzymes (Mosbach), **46:**203
Immobilized enzymes, **34:**455; 494. *See also* Immobilization
application of, **57:**180
and crosslinking, bifunctional, **57:**233–234
definition of, **57:**180–183
and denaturation
protease preparation, **57:**216
protein-protein interactions, **57:**216–217
reactivation of enzymes, **57:**217–219
stabilization of enzymes, **57:**217–219
and electron microscopy, **57:**233–234
enzymatic reactions of
conformational, **57:**204–209
protein-protein interactions, **57:**209–210
sorption, biospecific, **57:**209
and kinetic cooperativity, **57:**210–212
and ligands, **57:**204–205
mechanical phenomena of
immobilization-based models of, **57:** 229–232

Immobilized enzymes (*Continued*)
regulation of enzymatic activity, **57:** 232–233
in membrane systems, **57:**219–225
in multienzyme systems, **57:**225–228
preparations of, **57:**183
properties of, conformational, **57:**183–185
and protein structure
quaternary, **57:**187–195
and solid supports, **57:**196–204
tertiary, **57:**186–187
and refolding of proteins, **57:**212–216
regulation of activity of, **57:**232–233
Immobilized Enzymes, Effect of the Microenvironment on the Mode of Action of (Katchalski, Silman, and Goldman), **34:**445
Immune globulin. *See also* Globulins; Serum proteins
separation by chromatography, **14:**326; 328; 351; 357–359
Immune hemolysis, and enzyme inhibitors, **8:**97
mechanism, **8:**96–97
Immune reactions of luminous bacteria, **7:** 255–256
Immunity
and enzyme problems, **6:**33–122
and fertilization, **9:**243; 267
Immunity, and Enzyme Problems (Sevag), **6:**33
Immunity and role of complement, **8:**97
Immunity reactions, and thiol groups, **11:** 244–245
Immunization, **41:**134
Immunoabsorption, of light vesicle preparations, adenosine receptors, **69:**114–115
Immunoadsorbents, **36:**34; 81
Immunoassay
enzyme-linked, **58:**68
of Prolyl Hydroxylase, **41:**259
Immunoblot analysis, P450c17, **68:**60
Immunoblotting, RNA polymerase II transcription, reversible phosphorylation of CTD during transcription cycle, **71:**51
Immunoblotting techniques, adenosine receptors, **69:**97; 109
Immunochemical studies, **43:**153
Immunochemical studies of papain, **19:**282
Immunochemistry, **42:**147
chemical modifications of proteins, **24:** 105–160
of yeast mannans, **40:**250
Immunochemistry, Enzymology, and Related Subjects, Chemical Modifications of Proteins and Their Significance in (Sri Ram, Bier, and Maurer), **24:**105
Immunochemistry of Yeast Mannans (Ballou), **40:**239
Immunocompetency, adenosine deaminase deficiency and, **51:**198–199
Immuno-dominant sugars, **36:**132
Immunofluorescence, **47:**366
Immunofluorescence microscopy, adenosine receptors, **69:**113
Immunoglobulin, **47:**133; 136; 138
λG-Immunoglobin, human, **41:**117
Immunogold electronmicroscopy, *Caulobacter* cell cycle, **71:**26–27
Immunological activity, of antigen films, **28:**24
Immunologically reactive polysaccharides, formation of, **11:**306–315; 327–329
Immunological polysaccharides, of fungi, **28:**336–337
Immunological properties
effect of irradiation, **9:**134
of glutamate dehydrogenase, **28:**142
of β-lactamase, **28:**253; 255–257; 270–272; 275; 304–305; 312
of tobacco mosaic virus, **15:**468–472
of mutants, **15:**473–475; 478–479; 481
Immunological properties of RNase, **24:**170; 184; 221; 222
Immunological reactions, thermodynamics, **15:**49–77
Immunological studies. *See also* Antibodies
of protein composition, **14:**277
Immunological studies, *Caulobacter* cell cycle, flagellar structure and assembly, **71:**13
Immunological test in enzymic adaptation studies, **13:**69; 72; 85–88
Immunological tests, for enzymes, **25:**330; 332–333
in glycoprotein study, **25:**447–450
steps in reactions, **25:**13–15
Immunologic character, of bacterial walls, **26:**223–226
of polysaccharides, **26:**67–68
Immunologic cross reactivity, **42:**127
Immunoprecipitation, **60:**258

Immunopressive agents, use of adenosine deaminase inhibitors as, **51:**199–200
Immunosuppressant action, **39:**224
Immunosuppressive effects, **39:**223; 224
Immunosympathectomy, with NGFAS, **31:**62
Immunosynpathectomy, effect on norepinephrine uptake, **29:**484
IMP-dehydrogenase, allosteric regulation, **28:**121–123
Impermeable membrane, **34:**513–514
IMP phosphorylase, allosteric regulation, **28:**123–124
Inactivation, **34:**455, **39:**343–345
by acid, **23:**41–44
and denaturation, **23:**31; 39–44; 46–48; 50–58
of enzyme proteins, **23:**29–82
by heat, **23:**39
by inhibitors, **23:**31; 44–48
ratio of, **23:**35; 47
and reactivation, **23:**40–41; 43–44
by urea, **23:**39–41; 43
of acid phosphatase, **10:**90
of alkaline phosphatase, by destruction or blocking of amino or thiol groups, **10:**97–98
by formation of complexes, **10:**94–97
of prothrombin, **10:**22–23
of antibiotics, by thiols, **11:**246–247
of enzymes, by lachrymating agents, **11:**239
of pectinesterase, **11:**284
of phosphorylase, **11:**304
of tobacco mosaic virus, **11:**224; 239
of wilting toxins, **11:**407–408; 412–413
of antitoxin, **7:**248
of centers of dissymmetry, **7:**80; 94–99
of coenzymes, **7:**410; 412–413; 414; 415; 427–435
of cozymase by yeast, **7:**410; 414
of enzymes, **7:**248
of luciferase, **7:**246–248; 256
of luminescence, **7:**235
of nicotinamide derivatives, and rate of growth, **7:**422–423; 434
of nicotinic acid and different subtrates, **7:**420–421
of nicotinic acid derivatives, **7:**419–420; 422–423; 434
of pantothenate, by bacteria, **7:**444–448
of pressor amines, **7:**542–545
spectra, **7:**257
of thiamine derivative, **7:**454
of virus, **7:**248
of bacteriophages, **9:**142; 155; 156
of enzymes, by heat, **9:**31–39; 57; 68; 122
by light, **9:**102; 103–109; 111–112; 128; 135; 136–137; 138–154; 159–162
of hormones, **9:**138; 142; 156
of viruses, **9:**115–116; 129; 154–161; 174; 220–228
of enzymes by radiant energy, **14:**36–37
of proteins, by adsorption, **14:**332–333
by oxidation, **14:**131–132
by peroxidase, **14:**154–158
by tyrosinase, **14:**140–142; 144–150
of penicillins, **28:**287–294
secondary, **34:**474
Inactivation of antinvasin I, **8:**485
Inactivation of enzymes. *See also* Chemical modifications of proteins and Ionizing radiations
by acylation, **24:**459–460
by ionizing radiations, effect of additives, **24:**373–375; 378
energetic requirements for, **24:**367–371
mechanisms for, **24:**360–364
molecular changes, **24:**397–399
Inactivation of prothrombin by thrombin, **16:**49
Inactivation of rennin by alkali, **15:**428
Inactive protyrosinase, **57:**210
Inborn errors of metabolism, **14:**122–123. *See also* Alcaptonuria; Hydroxyphenyluria; Phenyluria
Inclusion products, *See* Solubilization of matter
Incorporation enzyme, **24:**4
Incubation, **34:**473–474
Indican, excretion by psychotics, **29:**521
Indicator, **34:**510
Indigo carmine, as artificial electron acceptor, **30:**23
Indigotrisulfonate, chemiluminescence, **25:**150
Indium compounds, **17:**137; 143; 144
Indole, **19:**125; 128, **42:**287, **58:**35
determination, **10:**54
mechanism of formation, **10:**53–60; 62–64
precursors, **10:**52
production from tryptophan, **10:**51–82
utilization by bacteria, **10:**71–74
genetic studies, **17:**416–419; 421; 450; 458

Indole (*Continued*)
and thiaminase model, **15:**404
formation by thiaminase-producing bacteria, **15:**413
Indole-3-acetate, **57:**23–24
Indoleacetic acid
effect on plant metabolism, **9:**438
as inhibitory analogue, **6:**136
Indole-3-acetic acid, in treatment of plant diseases, **20:**399
β-Indoleacetic acid in plants, **10:**247
Indoleacetic acid oxidase system of plants, **18:**37
Indoleacteamide, in shizophrenic urine, **29:** 521–522; 528–529
Indole derivatives, *N*- and *O*-methylation of, **25:**415–416
Indole-forming system of *E. coli,* **18:**218
Indole-3-glycerol phosphate, **34:**521
Indole oxidase, **19:**125; 127; 128
Indolepyruvic acid, **58:**35
in transaminations, **16:**198; 235; 294
Indole-5,6-quinone, as intermediate in melanin formation, **16:**163–168; 171; 172a
Indole 5,6-quinone, role in tyrosine metabolism, **14:**76–77
Indole-quinone condensation, in melanin formation, **16:**165; 168
Indole ring in alkaloids, **8:**235–236
Indoles, **44:**211
formation during sclerotization, **16:**161
reaction with quinones, **16:**110–112
in synthesis of tryptophane, **16:**288; 293–294
Indole-serine ligase, **16:**293
Indolmycin, stereochemistry of biosynthesis, **50:**288
Indolybutyric acid, **19:**124
β-Indolylacetate, effect on pectic substances, **20:**355; 368; 370; 371
Indolylacetic acid, **19:**124
Indolylacetic acid oxidases, **19:**123–125; 127
3-Indolyl-1,4-benzoquinones, formation, **16:**110
Indolylbutyric acid oxidase, **19:**123
Indolylpropionic acid, **19:**124
Indolylpropionic acid oxidase, **19:**123
3-Indolylquinones, properties, **16:**111
Indomethacine, **58:**221
Indophenol dyes
as artificial electron acceptors, **30:**22
in cytochrome study, **17:**366
Indophenol oxidase, **17:**329; 336, **18:**7
and copper, **23:**304
Indophenol oxidase in sperm, **9:**370
Indophenol oxidase in tobacco leaves, **10:**421
Induced asymmetry, *See* Asymmetry
Induced biosynthesis of enzymes, **13:**67–119
Induced dissymmetry, *See* Dissymmetry
Induced enzyme synthesis, definition, **22:**101
generalized hypothesis, **22:**143
kinetics, **22:**107–117
model for, **22:**144–146
Induced fit, **35:**494, **37:**483; 606, **43:**230; 312
concept, **37:**406
definition, **37:**410, **43:**237; 241; 360
theory, **37:**400; 404; 480; 599
Induced-fit hypothesis of substrate specificity, **33:**41
Induced protein synthesis, **22:**100–156
definition, **22:**101; 102
Inducers
definition, **22:**101
gratuitous, **22:**113; 115
mechanism of action, **22:**140–145
as regulator of enzyme synthesis, **22:**144; 145
effect on synthesis of proteins, **46:**406
of β-lactamase, **28:**272–275
response to, **28:**255–257; 262–263
Inducer transport system, **22:**108–112
Inducibility, of β-lactamases, **28:**242; 248; 251–255; 268–269; 274–277; 279; 280
Inducible laminarinase, **28:**343; 348–349
Inducible permeases, **22:**109–110
Induction
of arabinose system, **30:**111–112
coordination of enzyme, **30:**89–90
enzyme, characteristics of, **30:**86–90
enzyme system for metabolizing sugars of *Escherichia coli,* **30:**81–137
of galactose system, **30:**103–105; 107–110
of α-galactosides, characteristics of, **30:**126
general discussion of, **30:**130–133
of β-glucosidase, **30:**128
internal, description of, **30:**106–107
kinetics of enzyme, **30:**88–89
of maltose system, **30:**11–120
simultaneous of lactose system, **30:**90
genetic control of, **22:**102–104
of β-lactamase, and penicillin resistance, **28:**307–308

Induction constant, definition, **28:**273
Induction of enzyme synthesis, **13:**67–119
Induction period in citric acid cycle, **10:**178–179
Induction period in enzymic oxidoreductions, **15:**3; 16–17
Induction period in tyrosin-tyrosinase mixtures, **14:**78; 87–88; 93; 95
Induction periods, photosynthesis, **1:**220–232; 246
Induction ratio, β-lactamase, **28:**264; 273
Inductive effect, **39:**355, **49:**31
Inductor in enzyme biosynthesis, **13:**68–69; 93–98; 98–100; 100–105
Industrial fermentation, **7:**557–615. *See also* Fermentation
Infants, new-born, prothrombin activation in, **16:**64
Infection of host cell by viruses, **9:**173–179
Infectivity, virus, **17:**495
Infectivity influenza virus, loss of, **9:**225–228
Influenza bacilli, effects of nicotinamide derivatives on growth of, **7:**415–416
Influenza hemagglutinin, **52:**133
Influenza virus, **15:**456; 482, **17:**495; 497, *See* Virus
  activity and particle identity, **9:**191–195
    chemical composition, **9:**229–232; 234
    effect of light on, **9:**142; 155
    homogeneity, **9:**219–220
    hydration, **9:**206; 208–211
    inactivation, **9:**221; 225–228
    physical constants, **9:**187; 188
    purification, **9:**185; 187
    sedimentation rate, **9:**208–211
    shape, **9:**206
    size, **9:**206
  neuraminidase of, **20:**136; 137; 140; 142–144
Influenza viruses, **52:**132
Information carrier in enzyme formation, **24:** 13–20
Information-storage structure, and self-replication, **27:**409
Information transmission. *See also* Coding signals
  from gene, **24:**11–20
Infrared spectroscopy, **47:**74; 76
INH, *See* Isonicotinyl hydrazide
Inhibited signal, **51:**138; 157
Inhibiting action of pantoic acid derivatives, **7:**447–453
Inhibiting concentrations of antibacterial substances, **7:**490–493
Inhibition, **1:**194; 195; 196, **34:**503–507, **39:**343. *See also* Allosteric regulation; Competitive action; Control mechanisms; Feedback regulation; Regulation; Repression
  of alkaline phosphatase, **10:**94–98
    of citric acid, **10:**215–216
    of coagulation of blood, **10:**33–38
    of enzymes, and reactivation, **10:**102–104
      by complex formation, **10:**95–97
    of fumarate oxidation by streptomycin, **10:**161
    of oxidation by nicotine, **10:**388
    of penicillin, **10:**253; 254
    of succinic dehydrogenase, **10:**174–175
    of tryptophanase, **10:**69; 70–71; 74–81
  of amylosucrase, **11:**317–318
    of coagulation, **11:**245–246
    of dextransucrase, **11:**312
    of enzyme activity by —SH reagents, **11:**232–244
    of pectinesterase, **11:**273; 280–282
    of phosphorylase, **11:**303; 317–318
    of streptogenin, **11:**408
    of thiol enzymes, by ionizing radiations, **11:**256–257
    of toxin activity, **11:**244–245
  of cell division, **9:**241; 242–243
    of egg respiration, **9:**302–304
    of enzymes, **9:**60; 68
    of proteinases, **9:**429
  cobalamin analogs, **26:**271–274
  cobinamide analogs, **26:**268–269; 274
  of collagenase, **23:**194–197
    of elastases, **23:**233–237; 242
    of lipase, **23:**148–149
  competitive, **25:**221; 222–223; 225; 227; 229; 233; 235; 241–242; 249; 258; 263; 266–267, **39:**353, **46:**391
    choice of substrate and inhibitor concentrations, **29:**18
      end product, **29:**92–93
      false end-product, **29:**117
      false feedback, **29:**92–93
      hyperbolic, **29:**10–11
      phlorizin, of translocation reaction, **29:**53
      plotting of kinetic data, **29:**18–19

Inhibition (*Continued*)
of tryptophan synthesis by analogs, **29:**121
kinetics, **14:**171
of hydrogen peroxide decomposition, **14:**176
of melanin formation, **14:**92–96
kinetics of, **25:**175; 177–178
of enzymes, **28:**44–46; 50–52; 58; 62; 65–69; 72–73; 93–94; 110–114; 124; 347–349; 393
by borate reaction, **12:**516–520
of viruses, by borate, **12:**520
in respiratory chain, **28:**159–160; 177; 178; 186; 190; 192; 194; 198–200; 202; 210–212; 214; 217; 221–222; 225
of enzymes by quinones, **16:**124; 126
of β-glucuronidase by excess substrate, **16:**369
of myosin ATPase by magnesium ions, **16:**336
of phenolase, **16:**128–129
of prothrombin activation, **16:**52; 69
sulfonamide, reversal, **16:**274
of threone activity, **16:**54–57; 76
of threonine formation, **16:**271
of threonine utilization, by methionine, **16:**269
of thrombin-fibrinogen by estrogenic compounds, **16:**85
of tyrosine decarboxylase, **16:**219; 221
in enzymic adaptation studies, **13:**77–80; 81–83; 100–105; 110
of chymotrypsin, **13:**321–343
of glutamic acid metabolism, by benzylhomocysteine sulfoxide, **13:** 256–257
by hydroxylamine, **13:**259–260
by methionine sulfoxide, **13:**255–259; 263; 278–279
of hyaluronidase action, **13:**214–219
of transferases, **13:**291; 292; 293; 299
feedback, explanation, **26:**105–107
of folate biosynthesis, **27:**417–468
of LADH, mechanism, **27:**73–75
glutathione synthesis, **73:**245–260 (*See also* Glutathione synthesis)
of invertase, **15:**275; 281; 285
of nucleosidases, **15:**107–108
of urease by silver, **15:**274; 275; 285–286
of β-lactamases, **28:**262; 274; 294–297; 309
and lipides, **8:**44
of cholinesterase, **8:**464; 470–471
competitive, **8:**473; 480–481
of dehydropeptidases, **8:**145
of hyaluronidase, **8:**473–474; 484–486; 487
of mitochondria, **18:**131
non-competitive, **25:**221; 222–223; 229; 233; 235; 249; 258; 263; 266–267
patterns of, **45:**287
by-products, **45:**288
dead-end, **45:**292; 296
double, **45:**301
induced substrate, **45:**298
mixed product, **45:**296
substrate, **45:**296; 409
and structural similarity, **7:**418
of catalase activity, **7:**298–299; 311
of microorganism growth by β-aminobutyric acid, **7:**437
by asparagine, **7:**437
by aspartic acid, **7:**437
of pantothenate assimilation, **7:**451–452
of pantothenate inactivation, **7:**450–451
substrate, **39:**354
of thymidylate synthetase, **25:**400
Inhibition, Hg^{2+}, **36:**118
Inhibition, Product, of Enzymes (Walter and Frieden), **25:**167
Inhibition, Stoichiometric, of Chymotrypsin (Balls and Jansen), **13:**321
Inhibition analysis, *See* Antimetabolites
Inhibition and combining ratio, **52:**10–11
Inhibition assay, **66:**44; 64
Inhibition by AMP, **53:**67–68
Inhibition by irradiation, **38:**416
Inhibition constant, for hapten-antibody combination, **15:**64–67
Inhibition constants, **25:**221, **46:**393
8-analinonaphthalene sulfonate, **46:**393
bilirubin, **46:**393
cephalothin, **46:**393
3,6-dibromosulfophthalein, **46:**393
indocyanine green, **46:**393
Inhibition index, **6:**138
Inhibition of methylation reactions, **51:**184
Inhibition of oxidase activity, **44:**212
Inhibition of pepsin, *See* Pepsin, inhibition of
Inhibition of ribonucleotide reductase (by dATP accumulation), **51:**190–194

Inhibitor, **39:**346
active site-directed irreversible, **50:**335
classification, **14:**242
competitive, in tyrosine oxidation, **14:**81; 84; 93–95
trypsin, separation by chromatography, **14:**342; 346
competitive, **39:**345; 346
reversible, **39:**345
Inhibitor-accelerator complexes of blood, separation by centrifuging, **16:**88–89
Inhibitor-enzyme binding, **27:**450–453
Inhibitor-enzyme interactions, steps in, **25:** 13–15
Inhibitor I (or C), respiratory, **17:**96–112; 116; 123
Inhibitor mechanism
asparagine synthetase, chemotherapeutic applications, **72:**151–153
collagen hydroxylases, **72:**357–364
Inhibitors, **7:**4; 206–207, **34:**446; 512, **48:**64. *See also* Antagonists, Anticoagulants, and Antimetabolites; Competition and Kinetics
and active center of enzyme, **24:**115
disulfides as, **24:**269
effect on RNase, **24:**235
effect on sperm, **24:**69–72
organophosphate, **24:**117
active-site-directed, **48:**46; 47; 48
adenylosuccinate synthetase (AMPSase), **73:**79–84
of *o*-aminophenylglucuronide formation, **16:**403
of blood coagulation, **16:**85–89
of clot formation, **16:**25; 29; 40
of β-glucuronidase, **16:**369; 370; 372–375; 379–382
of platelet cofactor I, **16:**32
of prothrombin activation, **16:**25; 29; 47–48
relation to hemorrhagic disease, **16:**74
of transaminase, **16:**222
use in study of phenolase complex in terminal oxidation, **16:**135–136
vitamin B_6 analogues, **16:**220–222
binding of, **48:**48
in blood, **30:**303–305
of coagulation, **30:**306
of chitin synthetase
animal, **59:**82–83
fungi, **59:**71–74
conduritol B epoxide, **48:**46
in conversion of coenzyme I to II, **7:**426–427
and Cotton effects in proteins, **27:**48
of FAH_2-reductase, structure-activity relationships, **27:**440–451
of deaminases, **9:**497
of desoxyribonuclease, **9:**519
of virus, **9:**174; 183
dihydropteridine reductase, **67:**237–238
of *E. Coli* β-galactosidase, **48:**49
and effect of nicotinic acid derivatives, **7:**422
effect on production of oxygen by illuminated chloroplasts, **12:**25–28
of acetylcholinesterase, **12:**279–290; 292–294; 299; 305–311; 318
of amylases, **12:**408–409
of carboxypeptidase, **12:**238–248
of choline acetylase, **12:**322–323
kinetics of action of, **12:**238–248; 286–290; 293–294; 296–297
effects at constant temperature and pressure, **7:**251
enzyme, and resonance structures, **10:**72–74
of acid phosphatase, **10:**90
of aconitase, **10:**192
in hydrocarbons, **10:**471
of indole formation, **10:**53–54; 58–60; 69–71; 71–74; 74–81
in enzyme complex, **26:**157; 164–165
in enzyme kinetic studies, **17:**11–12; 19–25; 32–33; 41; 45–49; 55; 58; 59
of metalloflavoproteins, **17:**237–243; 246–247; 252; 254; 256; 258; 260; 265; 275–276; 282
of respiratory chain, **17:**69; 79; 93–97
of succinoxidase system, **17:**367–371
in xanthine oxidase study, **17:**308–314; 316–318; 323
of enzymes, **48:**34; 46; 47; 48; 49; 50; 54
in enzyme studies, **25:**330; 331
products as, **25:**167–274
and role of coenzyme Q, **25:**309
by Fru-2,6-P_2, **59:**334–335
mechanisms of, **59:**335–337
glutathione S-transferases, **69:**6
of glyoxalase I, **59:**114–118
and immune hemolysis, **8:**97
of lipoxidase, **8:**447–448

Inhibitors (*Continued*)
respiratory, and phospholipide synthesis, **8:**331
inactivation by, **23:**31; 44–48
of malate synthetase, **23:**412
of photophosphorylation, **23:**386
protection from, **23:**46–48
respiratory, **23:**347; 350–352; 363–369; 375
interaction with activators, **46:**346
of Krebs cycle, **19:**359–366
of mixed function oxidases, **19:**131; 143; 176; 182; 184; 185; 189; 190; 194; 198; 199
of ornithine cycle, **19:**376–378
of oxygen transferases, **19:**81; 90–92; 97; 99; 118; 121; 122; 125–128
of papain, **19:**286; 287; 290; 306; 318–320
of photosynthesis, **19:**22; 50; 63; 64
of transamination reactions, **19:**347; 348; 354–368; 375; 376
of uricase, **19:**205
and luminescence, **7:**233–251; 254
mechanism-based, **59:**107
mixtures of two, **7:**242–244
myristoylCoA:protein, N-myristoyltransferase, **67:**399–402
natural, for protein methylase II, **42:**273
regulatory of pyruvate carboxylase, **42:**17
of oxidases in plants, **18:**37
resistance of *Arum* spadix enzymes to, **18:**46–47
respiratory, effect upon cytochrome system, **18:**6; 7; 30–35
of root tip respiration, **18:**42–43
of succinic dehydrogenase, **18:**92–94
in pantothenate study, **7:**444
phenylalanine hydroxylating system, **67:** 216–219
phoisphorylated, **48:**219
of plant toxins, **11:**407; 412
protease, **45:**393
of quinoprotein oxidoreductases, **59:**187; 200–202
radioactively labeled, **48:**46
and respiration in luminescent systems, **7:**229–231; 233; 254
respiratory, and phosphatide formation, **7:**151
and strained molecule theory of enzyme action, **15:**291–292
mechanism of action in enzymic oxidoreductions, **15:**2; 3; 17; 23; 26; 29–30; 37
of thiaminase, **15:**398–399
types, **15:**285–286; 296–297
use in kinetic studies of enzymes, **15:** 278–279; 281–282; 285–290
suicide, **59:**107
in sulfur metabolism, **7:**392
on the synthesis of bacterial cell protein, **33:**340–343
on the synthesis of gramicidin S,
actinomycin D, **33:**342–343
5-bromouridine, **33:**340
chloramphenicol, **33:**340–342; 345; 347; 348
dihydrostreptomycin, **33:**340–342
5-fluorouracil, **33:**340
mitomycin C, **33:**342–343
puromycin, **33:**340–342; 345; 347; 348
ribonuclease, **33:**348
transition-state, **59:**107
Inhibitors of amino acid transport systems, need for irreversible action, **49:**90
Inhibitory action
of *p*-aminobenzoic acid on sulfonamides, **6:**38–40; 130
analogues, **6:**129–144
drugs, **6:**40–51; 140
enzyme role, **6:**40
in enzymology, **6:**142–143
mechanism, **6:**143–144
in pharmacology, **6:**140–142
of benzedrine on amines, **6:**41–42
in chemotherapy, **6:**40–51
of chloromercuribenzoate on ATPase, **6:**167
of copper on ATPase, **6:**165
of copper sulfate on amino acid decarboxylases, **6:**17; 18
effect on growth of organisms, **6:**43–51
of ephedrine on adrenaline, **6:**40–41
of fluoride on ATPase, **6:**166
of hydrazine on amino acid decarboxylases, **6:**18
of hydrogen on butyric acid fermentation, **6:**247
of hydrogen peroxide on ATPase, **6:**167
of hydroxylamine on acetylation, **6:**259; 261
of hydroxylamine on amino acid decarboxylases, **6:**18

of iodoacetamide on ATPase, **6:**167
of iodoacetate on ATPase, **6:**166; 187
of iron sulfate on amino acid decarboxylases, **6:**17; 18
of ketoglutarate on acetylation, **6:**261
of lithium in capsule formation, **6:**98
of magnesium of ATPase, **6:**165; 185
of malonate on succinoxidase, **6:**43; 206
of mercuric chloride on amino acid decarboxylases, **6:**17; 18
molar inhibitor-antagonist ratios, **6:**44–51
of normal sera on proteinase of *Streptococcus pyogenes,* **6:**99
of oxidation on ATPase, **6:**167
of phlorizin on ATPase, **6:**166
of poisons on assimilation, **6:**287–298
of potassium cyanide on amino acid decarboxylases, **6:**17; 18
of potassium permanganate on amino acid decarboxylases, **6:**17; 18
of prostigmine on cholinesterase, **6:**209; 210; 211
on purified amino acid decarboxylases, **6:**17–21
of pyruvate on acetylation, **6:**261
of semicarbazide on amino acid decarboxylases, **6:**18
of silver nitrate on amino acid decarboxylases, **6:**17; 18
of silver on ATPase, **6:**165; 176
of sisto-amylase on amylase, **6:**425
of sulfanilamide on amino acid decarboxylases, **6:**17; 18; 19
of triphosphate on ATPase, **6:**160
of vitamins on sulfonamides, **6:**63–64
Inhibitory action of catalase derivatives, **7:**298–299
Inhibitory group, **7:**207
Inhibitory relationships, amino acids, **17:** 410–412; 416
purine-pyrimidine, **17:**430
Inhibitory side products, EPR, and, **51:**131–135
Inhibitory uncouplers, **23:**367
Initial concentration, **34:**514
Initial velocity, patterns, **45:**279–284
Initial velocity patterns, **45:**279–284
Initiation sequences, **47:**405
Inner complex, of D-amino acid oxidase, **34:**58; 64; 66
Inner-sphere complex, **59:**244
Inorganic carrier, **34:**453
Inorganic phosphate and choline phosphate, **7:**188–189
Inorganic pyrophosphate, **35:**486; 491, *See* PP_i-PFK
Inosine, **17:**426–428, **18:**222, **34:**3, **35:**475, **46:**171, **54:**64
with FU, **54:**87
phosphorolysis of, **15:**103–104
and synthesis of 2,3-diphosphoglycerate in partially carbamylated red cells, **40:**17
Inosinediphosphatase, in mitochondria, **21:**40
Inosine diphosphate, **19:**39
Inosinediphosphate, in mitochondrial carboxylations, **21:**46; 47
Inosine 5′-monophosphate, amination of, **18:**230
Inosine nucleotides, in mitochondria, **21:**10
Inosine-5′-phosphate dehydrogenase, **21:**229; 250
Inosine triphosphate, *See* ITP; ITP
Inosine triphosphate, as phosphate donor, **18:**195
Inosinetriphosphate, in mitochondrial carboxylations, **21:**46
Inosine triphosphate (ITP), phosphofructokinase inhibited by, **29:**355
Inosinic acid, biosynthesis, **16:**451, **21:**201–228; 232; 234–249; 252; 253; 399
in histidine biosynthesis, **21:**404; 405
hydrolysis, by metal ions, **21:**184
Inosinic acid and semen metabolism, **9:**353
Inosinic acid dehydrogenase, activation of, **27:**211
Inosinic acid pyrophosphorylase, **21:**209
Inosinicase, **21:**220–222; 400
Inositol, **4:**285; 304
and α-amylase, **12:**397; 398; 400
reaction with borate, **12:**495; 514
in mitochondrial lipid synthesis, **21:**52
in sperm metabolism, **24:**54; 64
Inositol epimerization, **44:**90
Inositol hexaphosphate, **51:**212; 236
Inositol monopshosphatase, human, **74:**71
m-Inositol pentaphosphate, **51:**236
Inositol phosphatide, as inhibitor of blood coagulation, **16:**87–88
Inositols, in hexose isomerization, **14:**202; 203
Insanity, manic-depressive, **29:**492
Insect cuticle, *See* Cuticle, insect
Insect cuticles, hardening, **16:**154
tanning, **16:**110

Insect digestive enzymes, *See* Digestive enzymes
Insecticide, fluoroacetate as, **18:**138
Insecticides
change in body, **24:**308
effect on sperm, **24:**71
organic phosphorus, **13:**322; 335; 336; 338
Insect proteases, *See* Proteases
Insect Proteases and Peptidases (Law, Dunn, and Kramer), **45:**389
Insects
enzyme systems in, **13:**12; 17
ATPase in muscle of, **13:**168
β-glucuronidase of, **16:**381–383
growth and development, **45:**411–413
methylated lysine and arginine derivatives in, **42:**235
Insect vectors, **9:**173–176; 181
Insemination
artificial, **9:**332; 334; 369; 381
of immotile sperm, **24:**37–38; 41
In situ activation, tyrosine hydroxylase regulation, **70:**139; 145; 147; 151; 162
Insolubilization, **34:**452
Insoluble enzymes, **33:**276
action of on soluble and insoluble substrates, **33:**269–283
Insoluble matter, solubilization, migration, and utilization, **17:**135–158
Insoluble substrates, action of insoluble enzymes on, **33:**269–283
and enzyme reactions, **33:**257
poly-β-hydroxybutyrate, **33:**264–265
of starch, **33:**260
Insulin, **3:**96–99, **15:**146; 161, **33:**422, **34:**328; 331, **36:**53; 80; 83, **46:**159, **47:**48; 79; 100–101; 124; 136, **53:**240; 268, **56:**275, **59:**365–366. *See also* Pyruvate
absorption spectrum, **4:**286
acid *vs.* alkaline, **56:**155
action of proteolytic enzymes on, **19:**302
chromatographic evaluation of purity, **19:**281
determination of amino acid sequence, **19:**298
isotope studies of biosynthesis, **19:**260
action of tyrosinase on, **14:**136; 138; 147; 149
effect on glucose diphosphate concentration, **14:**212
oxidation of tyrosine in, **14:**86
oxidized, activity of, **14:**131
adsorption studies, **14:**336
separation by chromatography, **14:**327; 328; 337; 340; 341; 342
shape of molecule, **14:**184
structure studies, **14:**277; 279; 281; 283; 284; 286; 287; 289; 292; 293; 294; 296; 301; 304; 305; 307; 309
adsorption at interfaces, **28:**12; 23–24
effect on glycogen, **28:**100
reduction of, **28:**443
amino acid ratio in, **4:**3; 4
analysis, **13:**246
beef, **33:**457; 466; 470; 483
binding to membranes, **27:**227–228
cystinedipeptide of, **27:**204
sequence analysis, **27:**229–231
biological tests of, **33:**458–459; 512
and carbohydrate metabolism, **4:**260
and coenzyme A, **57:**155–156
column chromatography, **22:**430
effect of pH on structure, **22:**426; 428
variation of amino acid composition, **22:**65
configuration of, **23:**25
elastase action on, **23:**238
crystalline, **33:**446; 483; 486
deficiency, effect on phosphotransferase activity of glucose 6-phosphatase, **34:**305
degradation of, **33:**513–514
encymatic, **33:**513–514
in diabetes, **54:**164
effect, on mammalian liver glycogen synthetase, **34:**343
on phosphorylase, **34:**399
on synthetase phosphatase, **34:**332
effect in familial periodic paralysis, **6:**215; 216; 217
effect of, **54:**157
and D-amino acid oxidase inhibitors, **57:**144
and cysteamine metabolism, **57:**155–158
negative, **57:**158–159
effect on duct-ligated dogs, **8:**189
role, **8:**413–414
effect on enzymes, **37:**159
effect on fructose diphosphatase, **52:**276
effect on glucose-6-phosphatase, **25:**235
effect on membranes, **29:**377
effect on phosphoenolpyruvate carboxykinase, **52:**298
effect on pyruvate dehydrogenase, **52:**295
effect on pyruvate kinase, **52:**304

effect on tryptophan peroxidase, **25:**377
and enzymes, **57:**158
and fat metabolism, **4:**261
in fructose-2, 6-6bisphosphate regulation, **54:**154
functions of
D-amino acid oxidase reactions, **57:** 146–147
cysteamine metabolism, **57:**153–158
glyoxylate metabolism, **57:**148–152
insulin stimulated NADPH oxidase, **57:**147–148
negative effect of, **57:**158–159
and glycogenesis, **34:**331
and glycogen metabolism, **4:**259–260
growth factors of, **57:**162–164
human, **33:**457; 458; 466; 470; 490; 492; 494
imidazole in, **20:**62; 74; 97
imidazole in polymerization of, **20:**101
ionization constants, **20:**58
zinc binding in, **20:**97
and ketogenesis, **4:**260–261
mechanism of action of, **57:**159–162
messenger system for, **57:**87; 145
modified, **33:**446; 509–517
molecular metabolism of, **57:**87
phenylalanine hydroxylase synthesis control, **67:**228–230
phosphoenolpyruvate carboxykinase (GTP), **69:**219; 240–243
and phosphorylases, **11:**304
reduction of disulfide in, **11:**228–229
role of —SH groups in synthesis of, **11:**248
and phosphorylation, **57:**159
pig, **33:**457
in regulation of glycogen synthetase, **34:**333
release from pancreatic islets, **49:**59
semisynthetic, **33:**446; 457; 486
sheep, **33:**457; 465–466; 468; 483; 490
species differences, **33:**512
structure, **15:**274
structure of, **33:**446; 447; 460
structure of molecule, **4:**4
substitution for, **33:**514–515
Insulin-A-chain, S-sulfonate, **33:**451; 452; 453; 454; 482
synthesis of, **33:**465; 484–488; 503
thiol form, **33:**454
Insulin-agarose, **36:**40
Insulin analogs, **33:**500; 517
Insulin and phosphorus turnover, **7:**119–120; 123–125
Insulin and the Stimulation of Glycogen Synthesis. The Road from Glycogen Structure to Glycogen Synthase to Cyclic Amp-Dependent Protein Kinase to Insulin Mediators (Larner), **63:**173
Insulinase, **23:**238
column chromatography, **22:**454
localization of, **24:**327
Insulin-B-chain, disulfide form, **33:**449; 451; 453; 454; 458
S-sulfonate, **33:**451; 452; 453; 454; 482; 498
synthesis of, **33:**472–473; 476; 478–479; 481; 488–491
Insulin B-chain, xoidized, **41:**181; 194
Insulin-enzyme relationship, **1:**191
Insulin (I^{131}), SH-group binding, **27:**227
Insulin-inactivating enzyme, SH-binding, **27:**219
Insulin receptor, **50:**316
Insulins, amino acid sequence, **24:**107; 416–418; 422; 425; 428–430
effect of, chemical modification of, **24:**144
effect on sperm, **24:**85
formation from polypeptides, **24:**23
helical structure, **24:**449
heterogeneity of, **24:**171
irradiation of, **24:**379–380
separation in RNase extraction, **24:**166
Insulins, Synthese des: Anfange und Fortschritte (Lubke and Klostermeyer), **33:**445
Insulin synthesis
combination of chains in, **33:**446; 448–459; 502–503
formation of disulfide bridges, **33:**454; 503; 504
partial, **33:**448–459
total, **33:**448; 460; 503–507
Intact cells, **36:**40
Integral membrane proteins, **43:**201
Integrin α4β1, fibrinogen binding inhibition, **70:**13
Integrin α5β1, fibrinogen binding inhibition, **70:**7; 10; 12
Integrin cell-surface receptors, fibronectin and, **70:**1–3. *See also* Fibronectin
Integrin αIIbβ3, fibrinogen binding inhibition, **70:**7; 11; 14
Integrin-recognition sequences, short, fibronectin, **70:**3–6

Integrin αvβ3, fibrinogen binding inhibition, **70:**7
Integument, shedding of, **45:**415
Interaction
lipid-lipid, **34:**478
protein-lipid, **34:**478
thrombin-fibrinogen mechanism for, **30:**274
Interaction energies, **51:**76; 97; 102
Interactions
α-α, α-β, β-α, β-β, **47:**72–74; 83; 87; 125
DNA-protein, **47:**79; 81
electrostatic, **47:**131
local, **47:**94
long-range, **47:**49; 94; 132
pairwise, **47:**48–49
short-range, **47:**94; 96; 132
side-chain contacts, **47:**77
clotting factors and, **30:**271–276
types of reversible macromolecular, **30:**141
Interactions, Van der Waals, **43:**337
Interactions of Dinitrogenase and Dinitrogenase Reductase (Emerich, Hageman, and Barris), **52:**1
Intercellular glutamine cycling, acinar glutamine metabolism, **72:**70–71
Interchain electron transport, **34:**139
Interchain interactions, respiratory, **28:**161; 224; 229–230
Interface, air/water, **34:**484
enzyme-substrate, **34:**491
glyceride/water, **34:**487
leucithin/water, **34:**490
lipid/water, **34:**486; 489–491
membrane-solution, **34:**504–505
oil/water, **34:**489
Interfaces
adsorption of enzymes, **28:**1–40
air-water, **28:**6–16; 21–25
oil-water, **28:**16–18; 25–27
types, **28:**6
water-solid, **28:**18–21; 28–30
between domains in proteins, **50:**24
reactions at, **11:**35–89
Interfaces, Reactions at, Relation to Biological Problems (Danielli and Davies), **11:**35
Interfacial activation, **48:**324; 340; 364; 365
Interfacial adsorption, **48:**362
of lipase, **48:**363
Interfacial denaturation, of lipase, **48:**328; 332; 334; 335; 350; 351; 352; 363; 364
Interfacial inactivation, **48:**334
of lipade, **48:**338
Interfacial tensions, values, **28:**17
Interference phenomenon in virus behavior, **9:**183
Interferon, **50:**318–320
interaction with gangliosides, **50:**318–320
Interleukin 1, **60:**226
Interlocking pathways, allosteric regulation, **28:**47–48; 64; 90–91
Intermediary metabolism, **34:**513
genetics of, **17:**394–437
Intermediate, **34:**477–478
lipid environment, **34:**477
lipid-soluble, **34:**478
Intermediate compound formation. *See also* Michaelis-Menten complex
Intermediate compound formation., in enzymic oxidations, **15:**28–29
Intermediate compound formation catalysis, **14:**163–173; 179–1980
Intermediates
accumulation, **27:**476; 487; 492; 494–495; 499–500
in enzyme action, **27:**1–36
peroxidase, **27:**120–124; 129
aspartic acid activation, β-aspartyl-AMP, **72:**158–159
definition, **16:**251–252
in biosynthesis, criteria for, **16:**250–254
enzyme reaction cycle, isotope trapping, **72:**25–26
glutamine-dependent nitrogen transfer, **72:**173–175
AS-B catalysis, **72:**188–190
glutamine synthesis, -glutamyl-phosphate intermediacy, **72:**16–18
glutamine synthetase acyl-P intermediate, borohydride trapping, **72:**26–28
Intermediates in phosphoglucomutase action, **24:**465
Intermediate steps in enzyme reactions, **25:**1–38
Intermedin, and melanin formation, **14:**91
Intermolecular cross-linking, **34:**453
Intermolecular forces. *See also* Van der Waals forces
between hapten and antibody, **15:**67
Intermolecular forces between enzyme and substrate, **12:**291–302; 305–311.

See also Chain length, Particle size, and Van der Waals' dispersion forces
Internal duplication, aminoacyl-tRNA synthetases, **40:**150
neurophysins, **40:**291; 293; 294
Internal galactosidic linkage, **36:**105
Internal methylation, **65:**259–260
Internal return, **50:**347
Internal rotation, **43:**343
freezing of, **43:**350
Internal rotation potential, and protein structure, **23:**1–27
International Rotation Potential, Possible Polypeptide Configurations of Proteins from the Viewpoint of (Mizushima and Shimanouchi), **23:**1
Interpeptide bridges, biosynthesis, ε-amino of lysine as site of initiation, **40:**123
in *A. crystallopoietes,* **40:**127
L. viridescens, **40:**127; 128; 129; 130
M. roseus, **40:**124; 125
S. aureus, **40:**121; 122; 123; 124
S. epidermis, **40:**125; 126; 127
aminoacyl-tRNA as donor substrate, **40:**130; 131
glcNAc-MurNAc(-pentapeptide)-*P*-*P*-phospholipid as acceptor substrate, **40:**133
glycyl-tRNA as donor substrate, **40:**121; 122
sequential transfer of glycine residues, **40:**123
Inter-phenylpropane unit linkage in lignin, **34:**211
Intestinal bacteria, action on hydrocarbons, **10:**476; 479
Intestinal canal, in low vertebrata, **1:**375–378
Intestinal cells, permeability, **8:**50–51
Intestinal digestion, in low vertebrata, **1:** 390–397; 407–409; 411; 412
Intestinal fluid, water exchange with blood, **13:**47–48
Intestinal HMG-CoA reductase, **38:**404
Intestinal mucosa, **2:**84; 85
glutamine tissue levels, **53:**220
and phosphatide metabolism, **7:**135–137
deaminating and decarboxylating activities, **7:**389
and plasma lecithin, **7:**156
Intestinal phosphatase, **9:**496; 497; 517
Intestinal transport, **49:**78
Intestine, Na^+-amino acid symporter in, **29:**61
Na^+-glucose symporter in, **29:**61
Intestine, water permeability of, **13:**37
Intestine and plasma phosphatides, **7:**161–163
Intracellular binding, coenzyme, **22:**373–375
Intracellular distribution of enzymes, **24:** 291–358
Intracellular hydroxylation sites, collagen hydroxylaes, **72:**338–340
Intracellular localization, of tyrosine hydroxylase, **70:**106–108
Intracellular proteinases, **1:**78–88
Intracellular transfer of DNA, **17:**459–465
Intrachain interactions, respiratory, **28:**161; 230
Intramolecular electron transfer, cytochrome *c* oxidase, electron transfer, **71:**121–124
Intramolecular reactions, **43:**271
Intramolecular transfer, **43:**496
Intravascular clotting, effect of platelet factors on, **16:**68–69
Intravenous injections, of thromboplastin, **16:**86–87
Intrinsic binding energy, **43:**222; 241; 317; 322; 323; 332; 336; 358, **51:**86; 93; 98
of substituent, **43:**359
utilization, **43:**342
Intrinsic constants of immunological reactions, **15:**51–54; 57; 61; 64–65; 71; 75
Intrinsic factor, inhibition, **38:**416
irradiation with visible light, **38:**416
Intrinsic isotope effects, *See* Isotope effects
Intrinsic pathway of blood coagulation, **48:** 279–280
Intrinsic system, *See* System, intrinsic
Inulase, **5:**118, **17:**194
cacao, **25:**466
classification, **14:**239
Inulase in tobacco, **6:**368
Inulases, in pectic enzymes, **11:**292
Inulin, **17:**192–195; 199, **49:**69
in plant wilting, **11:**417–422; 423
Inulinase
classification, **14:**239
soil, **28:**386
Inulin type polysaccharides, synthesis, **15:** 313–314
Inulobiose, **17:**203
Inulobiosyl-D-glucose, **17:**203
1-Inulobiosyl-D-glucose, **15:**314
Inulosucrase, **17:**195
Inulotriosyl-D-glucose, **17:**210

Inverse Pasteur effects, **17:**126
Inversion, in ribose synthesis, **28:**394; 398–399
Inversion of configuration, **7:**80; 84
Invertase, **5:**118, **18:**288, **34:**459; 524–525, **40:**242, **56:**155. *See also* Sucrose
action of tyrosinase on, **14:**140; 146–147
activity at low temperatures, **33:**286
on DEAE-cellulose, **33:**271
activity in frozen state, **9:**626–627
adsorption at interfaces, **28:**3
in β-glucan synthesis, **28:**340
assay of, **56:**156
cacao, **25:**466
product inhibition of, **25:**221; 224; 242
classification, **14:**246
column chromatography, **22:**437
pK value, **22:**58
DNA synthesis and, **56:**165
effect of pressure on, **7:**258
formation after enucleation, **24:**12
inactivation by radiations, **24:**371; 374–375; 377; 381; 401; 403
molecular weight from radiation studies, **24:**366
hydrolytic activity of, **56:**162
inactivation by oxidation, **14:**131
by peroxidase, **14:**154–155; 157–158
inhibition of, **56:**162; 165
isolation, **14:**400
isolation of, **56:**155
kinetics of action, **12:**195
molecular forms of, **56:**164
origin of multiple forms of, **56:**226
oxidation of tyrosine in, **14:**86
in photosynthesis, **19:**5
in plastids, **20:**211; 212; 231
properties of, **56:**159–162
purification of, **56:**157–159
regulation of, **56:**163–165
separation by chromatography, **14:**329; 342; 346; 362–363
soluble *vs.* insoluble form of, **56:**159
staining of, on gels, **56:**157
and sucrose synthesis, **10:**126–127 (*See also* Sucrase)
in *Pseudomonas saccharophila,* **10:**125
synthesizing and hydrolizing activity in living plants, **1:**336–339; 343–365
transfructosylase activity of, **56:**162
Invertase activity
action of urethan, **7:**259
effect of adsorption, **28:**30–31
Invertase in tobacco, **6:**367–368
Invertases, **15:**310; 313–314, **17:**194
change of activity with iodination, **15:**297
complex with sucrose, **17:**3
formation of oligosaccharides by, **17:** 189–190; 195–207; 219; 221; 222; 225–227
inhibition studies, **15:**275; 281; 285; 286–287; 297
mold, **17:**196–200; 202; 204; 221; 226
pH dependence studies, **15:**281–282
yeast, **17:**195–196; 198–205; 219; 221; 222; 225; 226
Invertase synthesis, effect of nucleus, **24:**20
Invertebrata, digestion, **1:**371–374
Invertebrates, laminarinases in, **28:**341; 342
in polysaccharide degradation, **28:**354
Invertebrate tissues, **59:**372
In vitro activation, tyrosine hydroxylase regulation, **70:**138; 151
In vitro studies
annexins, **71:**215
RNA polymerase II transcription, **71:**43; 48; 49–50; 69
In vivo activation, tyrosine hydroxylase regulation, **70:**127–128; 130; 166
In vivo regulation, cysteine *S*-conjugate β-lyase reactions, **72:**224–226
In vivo studies
annexins, **71:**216
RNA polymerase II transcription, **71:**51; 69; 70
Iodide
catalase action on, **19:**109
complex with carbonic anhydrase, **47:** 225–226; 228; 233
effect o fluorescence of riboflavin, **20:**32
effect on muscle proteins, **13:**154–155; 162–163
permeability of gastric mucosa, **13:**56
Iodide in blood in myotonia, **6:**214
Iodide oxidase, **9:**561–562
Iodides, **17:**143
organic, **15:**3
Iodinase, location in subcellular fraction, **24:**307
Iodinated casein, effect on xanthine oxidase, **17:**322

Iodinated compounds, effect on respiratory chain, **17:**113–114; 118
Iodinated fibrinogen, action of thrombin, **16:**37
Iodinated serum albumin, immunological reaction, **15:**74
Iodination, in identification of imidazole, **20:**68; 71; 72; 95
  of histidine and histidine derivatives, **20:**59
Iodine, **54:**400
  action on sulfhydryl enzymes, **15:**285; 289–290
    action on tyrosine in enzyme, **15:**289
    in sterilization of rennin extracts, **15:**427
  effect on pectinesterases, **20:**360
  in garlic, **11:**379
    reaction with thiols, **11:**207; 220; 225
  in labeling of active site, **22:**54; 55
Iodine cleavage, of corrinoid coenzymes, **26:**253
Iodine in thyroid metabolism, **7:**199–209
Iodine metabolism in thyroid gland, **14:**112–116
Iodine-penicilloic acid complex, in assay of β-lactamase, **28:**246–247
Iodine sensitivity, of β-lactamase, **28:**298–303
Iodine uptake, and hypophysectomy, **7:**205
  effect of thiocyanate, **7:**202–203
  effect of thiouracil, **7:**202–203; 206; 207
Iodinin, as inhibitory analogue, **6:**134; 142
Iodoacctamide, **58:**151
Iodoacetamide
  effect on ATPase, **6:**167
  reactivity, **68:**9–10
Iodoacetamide inhibition, mutant pig liver thioltransferases, **66:**174–175
Iodoacetanide, **54:**400; 402
Iodoacetate, **15:**94; 141–144; 167; 189; 225; 244; 287–290; 381–382, **54:**400
  effect on ATPase, **6:**166; 187
  effect on β-lactamases, **28:**297
  effect on mitochondrial oxidative phosphorylation, **21:**31
  effect on mold metabolism, **9:**555–556; 574–576
  effect on papain, **19:**286; 287; 290
  in identification of imidazole, **20:**75
  inhibition of *o*-aminophenylglucuronide formation, **16:**403
  inhibition of enzymes by, **11:**238–239
    reaction with thiols, **11:**213–215; 222; 226; 230
  inhibition of ribonuclease by, **29:**171
  inhibitory role in glycolysis, **5:**135; 137; 138
  reactivity, **68:**9–10
Iodoacetate and phosphorus turnover, **7:**124
Iodoacetate (C^{14}), effect on LADH, **27:**81
Iodoacetic acid, **66:**172
  effect of pH on inhibition by, **18:**34–35
  in labeling active site of ribonuclease, **22:**53
    reactions with methionine and protein constituents, **22:**251
Iodoacetic acid and yeast, **7:**189
N-iodoacetylamidoethyl-1-aminonapthalene-5-sulfonate, cytochrome *c* oxidase, **71:**92
Iodoalkanes, **54:**402
Iodoaryl group, **58:**54
Iodobenzoate in inhibitor studies, **15:**289–290
Iodobenzoic acid, **58:**53
Iodocyanine green, **46:**391; 392; 393
  dissociation constant, **46:**389
Iodoinvertase, **15:**297
Iodomethane, **46:**394; 396; 397; 406
Iodopepsin, **15:**297
p-Iodophenol, glucuronide as substrate for β-glucuronidase, **16:**382
p-Iodophenylacetic acid, **58:**53
Iodophenyl group, **58:**55
Iodosobenzene, **60:**73
Iodosobenzoic acid, reaction with thiols, **11:**207; 219–220; 222; 225
Iodothyronines, **17:**118
Iodotyrosine deiodase, classification, **14:**248
5-Iodouracil, dehalogenation of, **54:**74
Ion channel model system annexin, V, **71:** 209–239
  annexin-membrane complex, **71:**221–226
    ion channel activity, **71:**224–226
    lipids affinity, specificity, and binding mode, **71:**221–222
    lipid state, **71:**224
    structure of, **71:**222–224
  annexins, generally, **71:**214–216
  annexin V crystal structure, **71:**216–220
  background, **71:**210–214
    membrane proteins and, **71:**210–212
    structure function relationship, **71:**212–214
  mutants analysis, **71:**226–231
    central pore mutations, **71:**226–229
    charged residue mutation, **71:**230
    N-terminal mutant properties, **71:**231

Ion exchange
in nuclear uptake, **27:**259; 260
in protein chromatography, **22:**428–433
Ion exchange in fermentation, **7:**28–29
Ion exchange properties, of pectic substances, **20:**355; 357
Ion-exchange resins in blood-clotting study, **10:**15; 20
Ion exchangers
for column chromatography of enzymes, **22:**441–445
effect on pectic substances, **20:**350; 351; 370
Ionic behavior of RNase, **24:**183–184
Ionic complex, **34:**469
group, **34:**471
strength, **34:**452; 463–465; 484
type bonding, **34:**452
Ionic environment, effect on laminarinase, **28:**347
Ionic groups, of papain, **19:**285
Ionic radius, of metal ions, effect on enzyme activation, **21:**156–158; 162
Ionic strength, **39:**345. *See also* Buffer effects; Electrostatic effects; pH
effect on carboxypeptidase activity, **12:** 241–242; 252–253
effect on clotting of fibrinogen, **16:**38; 39
effect on enzyme kinetics, **17:**37–40; 59
in genetic studies, **17:**443
of papain, **19:**277
Ionic surfactants, enzyme catalysis and, **51:** 51–54
Ionization, **34:**457, **42:**109; 110
curves, **42:**95; 96; 129
at interfaces, **11:**37–47; 49–54
Ionization, ΔH of, in identification of amino acids at active site, **22:**57; 58
Ionization constant, **43:**4; 499
Ionization constants, **57:**109–110
of carboxyl groups, in aspartyl and glutamyl residues, **19:**275; 324
in papain, **19:**275; 324
for nitrogen atoms in imidazole, pyridine and pyrrole, **20:**55
in identification of imidazole, **20:**63; 65; 70; 75
of imidazole derivatives, **20:**57; 58
of pectic substances, **20:**354
of protein side chains, **20:**58; 59
of thiol group, **20:**70
of sulfhydryl groups, **19:**285; 324
Ionization potential, of metal ions, effect on enzyme activation, **21:**157–159
Ionizing groups, **34:**461
Ionizing Radiation, Effects of, on Enzymes (Augenstine), **24:**359
Ionizing radiations
effect on adsorbed enzymes, **28:**33–35
effect on enzymes, **24:**359–413
and origin of life, **27:**356; 359; 362
β-Ionone, effect on carotenogenesis, **21:**321
Ion pairing, **39:**345, **58:**52, **66:**178–179
Ion pairs, **50:**365; 366
Ion pump, nuclear membrane, **27:**259–260
Ions
accumulation in mitochondria, **21:**56; 57
distribution at interfaces, **11:**37–47; 49–54; 67–69
reactions at interfaces, **11:**69–74; 83
effect on actin polymerization, **16:**334–335
interaction with myosin and its subunits, **16:**331–333
free diffusion of, **13:**36
permeability to, **13:**48–63
metal, **44:**202
trivalent, **44:**202
Ions, K^+, **36:**95
Mn^{2+}, **36:**95
Ion transport, ATPase in, **23:**383
Ion transport, in plants, **18:**48–52
Ion transport (calcium-transporting ATPase), **51:**89–98
ability to form acyl phosphate, **51:**93; 95
chemical specificity, **51:**91
destabilization, **51:**94
equilibria and utilization of binding energies, **51:**92–98
interaction energy, **51:**96
intrinsic binding energy, **51:**93; 98
rules for vectoral coupling, **51:**90–92
I5P, *See* Inosine 5′-monophosphate
Ipomeamarone, **55:**54
biosynthesis of, **55:**76; 78
metabolism of, **55:**83
Ipomea βtus, **55:**76
IPP, *See* Isopentenyl pyrophosphate
Iproniazid, in depression, **29:**493
in schizophrenia, **29:**528
Iridin, nonhomogeneity, **16:**443

Iridophycus flaccidum, carbohydrate, oxidase in, **19:**202
Irisin, **17:**193
Iron, **15:**398; 410, **35:**136; 165; 173, **44:**188; 197, **46:**37; 70; 92. *See also* Ferric, Ferrous, and Metal ions; Ferric ion, Ferrous ion, Heme, Hemoproteins, Iron porphyrin; Ferric ions; Ferric ions, and Ferrous ions; Ferric ions and Ferrous ions; Ferrous ion; Ferrous ions
absorption, **44:**224
in aldehyde oxidase EPR study, **27:**168–172
binding, by pectic substances, **20:**356
by protein groups, **20:**97
changes in electronic configuration in, formation of porphyrin complexes, **15:**19–20
chelation of, **35:**137–139
and chloroplast activity, **12:**5–9
complex with plant toxin, **11:**413–416
in thiol oxidation, **11:**203–206; 211; 216
as component of metalloflavoproteins, **17:**237–242; 244; 245; 249; 253–264; 266–268; 273–276; 282; 284; 285
dietary, effect on xanthine oxidase, **17:**321
radioactive, intracellular distribution of cytochrome *c,* **17:**351
role in electron transfer, **17:**77; 82; 87; 110; 338–339; 354
in xanthine oxidase, **17:**294; 302–303; 305–308; 313–318; 321
concentration in melanins, **16:**171–172
effect on uptake of adenine and ribose by myosin, **16:**342
in nonenzymic transaminations, **16:**224
in cytochrome a_3, **20:**167
didates, **35:**138
in dihydroorotic dehydrogenase, **27:**172–175
and DPNH-cytochrome *c* reductase activity, **20:**169
in DPNH cytochrome reductase, **27:**179
effect on formate dehydrogenase activity, **48:**18
non-heme, **48:**17
effect on hydroxamate synthesis, **35:**165
effect on lipoxygenase synthesis, **58:**250
effect on riboflavin production, **7:**571; 572
as electron carrier, DPNH dehydrogenase, **27:**181
EPR spectrum, **27:**116; 117; 138–140
and flavine coenzymes, **22:**379; 380
in flavoproteins, **22:**392
in xanthine oxidase, **22:**382; 383; 392
in formation of ALA synthetase, **46:**87
of fumaric reductase, **18:**102
in succinic dehydrogenase, **18:**69; 74–76; 77; 79–81; 86; 94–99
in heme, **46:**38
hemin-protein, in energy transport, **14:**33; 37–42
in lipoxidase, **19:**99; 103
melanin pigmentation, **14:**94
microsomal Fe_x, **27:**190–191
in mitochondria, EPR study, **27:**187–190
in mixed function oxidases, **19:**198
mobilization, **44:**216
nonheme
in DPNH dehydrogenases, **20:**192
in respiratory chain, **20:**153; 179; 185; 188
in mitochondria, **21:**100; 112
in mitochondrial fragments, **21:**100–101; 104; 112–114
non-heme, **27:**89
in respiratory chain, **28:**157; 160; 162; 169–172; 192; 194; 209–214; 225-228
nonheme ferrous, enzymes, **74:**239
oxidation mechanism, **44:**199
radioactive, in study of mitochondrial nonheme iron, **21:**114
in respiratory chain preparations, **20:**153; 188
role in enzyme systems, **15:**3; 6–7; 14–28; 32; 33; 38–39; 42; 171; 379
role in hemin catalysis, **14:**173–176; 188
role in tetrapyrrolic compounds, **12:**13–20
in sperm, **24:**80
storage cells in liver, **44:**220
and succinic dehydrogenase activity, **20:**190
in tetrapyrroles, **46:**82
thermochemical changes in formation of porphyrin complexes, **15:**16–19
transport, **35:**136; 174
in tryptophanase, **10:**61
utilization of, **44:**214
valence in enzyme-substrate compounds, **12:**157–159; 184
and xanthine oxidase, **19:**202
in xanthine oxidase, EPR study, **27:**160–168
Iron/activity ratio, of succinic dehydrogenase, **18:**94

Iron chelatase, *See* Ferrochelatase
Iron chelates, **46:**101
Iron chelators, effect on succinate dehydrogenase, **28:**225–228
Iron complexes
 and Cotton effects, **27:**93
  EPR of, **27:**158–159
 decolorization of, **17:**142; 144; 145
Iron deficiency anemia, **46:**176
Iron enzymes, Cotton effect, **27:**87–92
Iron enzymes, in seedlings, **18:**41–42
Iron (Fe^{59}), incorporation into haemin *a,* **23:**315
Iron/flavin ratio, in succinic dehydrogenase, **18:**80
Iron flavoproteins, EPR study, **27:**119; 159; 173
Iron flavoprotein succinic dehydrogenase as, **18:**73; 80–81; 100
Iron in semen, **9:**346
Iron ions. *See also* Ferric ions and Ferrous ions
 in mitochondria, **21:**10
 model catalyses with, **21:**174; 175; 177
Iron isotopes (^{56}Fe, ^{57}Fe), in EPR study of biological system, **27:**156–160
Iron mobilization properties of Cp, **44:**206
Iron-phorphyrin, in protein enzymes, **1:**178; 179
Iron porphins (or hemes), **7:**267; 306; 307; 308–318
 complexes, **7:**306
 decomposition of, **7:**333–339
 proteins, and oxygen transport, **7:**313–317
 specialization of function, **7:**309–313
Iron-porphyrin compounds, EPR study, **27:** 138–140
Iron porphyrin proteins, **7:**267; 309; 313–317. *See also* Iron porphins
Iron porphyrins, **46:**82
 in respiratory chain, **23:**341–343; 347
Iron porphyrins and oxidation of unsaturated fatty acids, **8:**373
Iron prophyrin. *See also* Heme, Hemoproteins
 in cytochrome a_3, **20:**167
 and cytochrome oxidase, **19:**203
Iron protein
 tyrosine hydroxylase, **70:**201–206
 x-ray crystallography, **67:**318–323
Iron proteins, EPR study, **27:**153; 155–160
 non-porphyrin, **27:**155–160
Iron protoporphin, decomposition to bile pigments, **7:**333–337. *See also* Iron porphins
Iron protoporphin IX and catalases, **7:**316
Iron-protoporphyrin, **60:**36
Iron-protoporphyrin IX, in spermidine dehydrogenase, **36:**225
Iron redox system, relation to flavin, **27:**176–177
Iron requirement, for prolyl hydroxylase, **41:**252
Iron-riboflavin chelates, **17:**258
Iron salts
 effect on tryptophan metabolism, **10:**68; 71
 solubilization of, **17:**137; 142–145; 147; 148; 150–152
Iron sulfate, inhibition of amino acid decarboxylases, **6:**17; 18
Iron-sulfur clusters, lysine 2,3-aminomutase, **66:**17–26
Iron-sulfur coordination, **47:**60
Iron-sulfur protein, **47:**13
Iron-sulfur proteins
 hydrogen bonds in, **50:**29
 redox potential, **67:**310–312
Iron Superoxide Dismutase, **41:**86
Iron tetrapyrroles, **7:**308
Iron transferrin, **58:**251
Irpex lacteus, hydrolytic enzymes from, **25:**350
Irradiation. *See also* Photochemical processes, Radiations, and Ultraviolet radiations
 and effect of fluoroacetate, **18:**122
 effect of oxygen, **27:**328–330
 effect on blood coagulability, **10:**37
 effect on bound enzymes, **27:**268
 effect on pectic substances, **20:**352; 353
 effect on purine analogs, **27:**459–460
 of enzymes, effects on, **24:**359–413
  of nucleolus, effect on RNA synthesis, **24:**20
 mutations from, **6:**82–84; 92–93
 ultraviolet
  effect on skin, **14:**82–83; 88
  mutagenic, **27:**326–327; 336; 339; 341; 342
Irradiation and yeast, **7:**190
Irradiation effects, in sperm, **24:**90
Irradiation products, mutations, **27:**328–330
Irreveresible processes, entropy of, **11:**7–10
Irreversible changes, **39:**345
Irreversibly reduce chromophore, **34:**137
Irritability of cytoplasm, **8:**3; 42–47; 56
Isatin, **17:**311
 activation, **15:**29–30
Isoalloenzymes, **25:**363–365
Isoalloxazine, **59:**149

Isoalloxazine derivatives
in EPR study, **27:**134–135
in transglycosylations, **17:**219–221
Isoalloxazine moiety of FAD, as indicator in mechanism study, **27:**3; 7–16
Isoalloxazine ring, in coenzyme binding, **22:**383–387; 393
3-imino group of, **22:**383
binding site, **22:**386; 387
structure, changes in on binding, **22:**384
Isoalloxazines, as inhibitory analogues, **6:**142
Isoamylamine production, **6:**2
Isoamylase, **34:**360; 408
Isoascorbate, and *p*-hydroxyphenylpyruvate oxidase, **19:**184
in model hydroxylating system, **19:**135
Isobestic point, **42:**94; 110
Isobutanol, **17:**354
Isobutylpropenyl disulfide, isolation, **10:**259
5′-Isobutylthioadenosine, **54:**212; 236; 252
in arginine methylation inhibition, **54:**253
cleavage of, **54:**239
phosphate effect on, **54:**238
vs. methylthioadenosine, **54:**237
Isobutyrate, **58:**176
Isobutyric acid, metabolism, in mitochondria, **21:**46; 47
Isocaproate, **58:**185
Isocarboxyazid, monoamine oxidase inhibited by, **29:**528
Isochorismate, **50:**148
Isocitrase, product inhibition of, **25:**223; 243–244
Isocitratase, inhibition by C_4-acids, **23:**415–420; 422
in microorganisms, **23:**410–411*t;* 413; 415–420; 422; 425; 430; 435; 445
regulation of formation, **23:**416–420
Isocitrate
and acetate metabolism, **23:**408–420
in citric acid cycle, **15:**117; 122; 185; 191–192
and fatty acid synthesis, **26:**300–301
oxiation, **20:**172; 173
oxidative decarboxylation, **16:**254–255
D,L-Isocitrate, **51:**236
D-Isocitrate, **51:**236
$L_S(+)$-Isocitrate, **23:**410
d-Isocitrate, in photosynthesis, **19:**12; 36
D-Isocitrate, photochemical fixation of carbon dioxide in, **15:**126
threo-D_3L_3-Isocitrate, allosteric regulation, **28:**60
Isocitrate and CO_2 fixation, **12:**89–91
Isocitrate dehydrogenase, **9:**572–574, **37:**453; 497; 579, **46:**242; 477
allosteric regulation, **28:**58–60; 135–139; 144
DPN-specific, **28:**135–139; 144
and DR metabolism, **28:**443
TPN-specific, **28:**135–139
chemical mechanism of, **45:**374
kinetic mechanism of, **45:**297; 301; 308; 316; 318
fatty acid synthesis, **26:**300
PN-dependent, **26:**119; 161
in soluble space, **27:**263
Isocitrate dehydrogenase (decarboxylating), NADP-linked, location, **24:**297; 337
Isocitrate dehydrogenase (NAD-dependent), location, **24:**297
Isocitrate dehydrogenase (NADP), **37:**518
β-Isocitrate dehydrogenase (NADP), **37:**579
Isocitrate oxidation, and phosphorylation, **18:**311
Isocitrates, as solubilizing agents, **17:**419
Isocitric acid. *See also* Citric acid cycle
decarboxylation of, **15:**156
deuterium-labeled, **18:**127
equilibrium with *cis*-aconitic and citric acids, **10:**194–196
formation from citric acid, **10:**193–194
in tissues, **10:**169
oxidation, in mitochondria, **21:**13; 14; 16; 17; 28; 121; 122
in mitochondrial fragments, **21:**117
relation to oxalosuccinic acid metabolism, **10:**206
stereoisomeric forms, **10:**192–193
Isocitric acid and fatty acid catabolism, **8:** 392–401
Isocitric acid dehydrogenase, classification, **14:**254
Isocitric acid in mold metabolism, **9:**540; 550; 572–576
Isocitric dehydrogenase, **12:**89–91, **15:**97; 117; 122; 192, **17:**380; 449
and ATP, **18:**231
and citric acid cycle, **18:**244; 272–273
and fluoroacetate, **18:**118; 123–125
in plants, **18:**299; 304
in ATP synthesis, **23:**341; 346

Isocitric dehydrogenase (*Continued*)
 localization in cytoplasm, **13:**7
 pig heart, **22:**371
 in plants, **9:**412
Isocitric dehydrogenases
 diphosphopyridine nucleotide- and triphosphopyridine nucleotide-linked, **20:**172; 173
 multiple forms, **25:**347; 363
 in photosynthesis, **19:**14; 36; 53
Isocitric enzyme, *See* Isocitric dehydrogenase
d-Isocitric lyase, **23:**410
Isocitric lyase, column chromatography, **22:**481; 484
Isocitric oxidase, in mitochondria, **21:**13; 14
Isocitric r TPN transhydrogenases, **14:**254
Isocoenzyme A, synthesis by phosphoramidate method, **22:**200; 201
Isocoumarins, **55:**84
Isocyanate
 aliphatic, **54:**19–20
 alkyl, **54:**19
 analogy with carbon dioxide, **40:**13; 14; 15
  nucleophilic attack by unprotonated amino groups, **40:**13
  reactive tautomer of cyanate, **40:**4
Isocyanate derivative, **34:**449–451
Isocytochrome C's, **42:**259
Isodichroic point, **42:**108
Isoelectric focusing, **48:**33; 36; 38; 65, **57:**375, **58:**71
Isoelectric focusing studies, with hormones HCG and FSH, **41:**118
Isoelectric point, **34:**493, **42:**78; 87, **46:**385; 386–387, **58:**80
 of enzymes, **46:**386
 of papain, **19:**275
Isoelectric point of rennin, relation to activity, **15:**431
Isoenzymes
 classification of, **69:**2
 glutathione S-transferases, **69:**12; 13
 in nucleus, **27:**270–271
 of Superoxide Dismutases, **41:**67
Isoenzymes of Glutathione Transferase (Mannervik), **57:**357
Isoenzymes of GSH transferase
 characterization of
  acidic human, **57:**373
  basic human, **57:**372
  description of, **57:**363; 369
  microsomal rat, **57:**372
  near-neutral human, **57:**372–373
  rat 1-1, **57:**369
  rat 1-2, **57:**370
  rat 2-2, **57:**369–370
  rat 3-3, **57:**370
  rat 3-4, **57:**371
  rat 4-4, **57:**370–371
  rat 5-5, **57:**371
  rat 6-6, **57:**371
 nomenclature of, **57:**360–363
 occurrence of, **57:**360
 purification of, **57:**373–375
 tissue distribution of, **57:**388
Isoflavone 4′-*O*-methyl transferase, **55:**36
Isoflavonoid phytoalexins, **55:**4–49
 abiotic elicitors of, **55:**9
 biosynthesis and enzymology of, **55:**9–42
  central phenylpropanoid pathway, **55:**9–19
  de novo, **55:**9
  enzymic formation of chalcones, **55:**19–29
  isoflavone conversions, **55:**34–42
  isoflavone formation, **55:**29–34
  phenylalanine ammonia lyase in, **55:**14
 from chalcones, **55:**29; 34
 degradation of, **55:**42–49
  fungal metabolism, **55:**44–49
  host metabolism, **55:**42–44
 formation of, **55:**29–34
  aryl migration in, **55:**31
  with chalcone epoxide, **55:**31
  via spirodienone intermediates, **55:**32; 33
 historical perspectives, **55:**4
 isoprenylation of, **55:**36
 of Leguminosae, **55:**4; 6
 methylation of, **55:**36
 4′,5′-methylene-dioxy substitution of, **55:**39
 occurrence of, **55:**4–9
 prenylation of, **55:**43
 pterocarpans from, **55:**34–42
 sources of, **55:**4; 6
 structure of, **55:**4–9
 turnover of, **55:**44
Isogriseofulvin, **46:**137
Isohemagglutination, **15:**54–64; 75; 76
Isohemagglutinins, **41:**137–138
 autologons blood-cell, **41:**137
 formation, **15:**62–64
 theories, **41:**138

Isohemipinic, **34:**215
Isoionic pH, **42:**88
Isolated adipocytes, adenosine receptors, **69:**100
Isolation of enzymes, **14:**375–410
Isolation of Enzymes (Schwimmer and Pardee), **14:**375
Isoleucine, **19:**295; 296, **34:**15; 21, **35:**482; 484; 486; 489, **50:**130; 132, **51:**217, **58:**174
biosynthesis, **16:**279–284
allosteric regulation, **28:**60–61; 64; 70
isomers, in enzymic oxidation, **16:**199
in transamination reactions, **16:**190; 192; 193; 197; 207; 208; 228–230; 234
and citric acid cycle, **18:**257; 271
genetic studies, **17:**399–403; 409; 411
metabolism, in mitochondria, **21:**46; 47
in phosphorylated peptides of esterases, **20:**288
DL-Isoleucine, resolution by papain, **19:**309
L-Isoleucine, transamination reactions of, **19:**347; 348; 367
Isoleucine adenylate, synthesis by carbodiimide method, **22:**191
L-Isoleucine auxotrophs, **16:**279; 280; 284
DL-threonine as growth factor for, **16:**250
Isoleucine intermediates, bacterial lysine biosynthesis, **72:**283–285
Isoleucine transaminase, **16:**280
Isoleucine tRNA synthetase, **55:**347
Isoleucine-tyrosine reaction, **16:**208
Isoleucovorin, *See* N^5,N^{10}-Anhydroformyltetrahydrofolic acid
Isoleucyl adenylate, **43:**346
S-DL-Isoleucylglutathione, synthesis and hydrolysis, **19:**247; 248
Isoleucyl-tRNA synthetase, **37:**453; 476; 518; 579, **43:**333; 344
Isolubimin, **55:**52
Isomaltose
from amylopectin, **12:**389
from hydrolysis of starch, **11:**340–341
from maltose, **15:**312
in transglycosylations, **17:**208; 210; 212–214; 220; 227
Isomaltotriaose, **17:**211
Isomaltotriose, **15:**312
Isomerase, **5:**132
and lipolysis, **23:**157
in oxidation of hydrocarbons, **27:**499; 508
pentose-P, allosteric regulation, **28:**118
in phenylalanine metabolism, **19:**89
in plants, **18:**286
in propionic acid metabolism, **26:**309
light sensitivity, **26:**265
steroid, **20:**260; 265; 266; 268; 269; 273
Isomerase, L-arabinose, reaction catalyzed by, **30:**110
Isomerase action, in ribose biosynthesis, **28:**395–397
Isomerases, **56:**293; 401. *See also* individual enzymes
classification, **14:**221; 223; 229; 236–238; 241; 258
isotope exchange, **43:**493
isotope transfer, **43:**493
maximum velocities, **43:**493
molecular weight, **43:**493
subunit weight, **43:**493
localization of, **24:**333
phosphomannose, **14:**205
ribulose phosphate, **14:**207
Isomeric peptides, and D-amino acid oxidase, **8:**150–152
of saturated amino acids, **8:**147–152
specificity of enzyme system for, **8:**150–152
Isomerization, **46:**403, **58:**43
enzymic, **14:**193–218
of amino acids in peptide chain, **14:**265; 268–270; 293
eukaryotic dihydrofolate reductase, chemical transformation step, **70:**90–92
of glutamic acid, role of cobamide coenzyme, **26:**264
of methylmalonyl-CoA, role of cobamide coenzyme, **26:**264–265
of glyceraldehyde 3-phosphate in photosynthesis, **19:**24
of ternary inhibitor complexes, eukaryotic dihydrofolate reductase, **70:**56–58
Isomerization, Enzymic, and Related Process (Leloir), **14:**193
Isomerization of ADP, **6:**162
Isomerizations, and enzyme action, **25:**20–21
Isomeroreductase, induction of, **37:**390
Isomeroreductase acetohydroxy acid, induction of, **37:**388
Isomers
of DAP, **16:**298
positional, **43:**44
rotational, stable configurations of, **23:**7–25

Isomyristate, **58:**185
Isoniazid, **60:**39
Isonicotinic acid hydrazide
effect on carotenogenesis, **21:**319
enzyme inhibition, **25:**237–240
relation to vitamin B_6, **16:**222
Isonicotinyl hydrazide (INH), effect of transamination reactions, **19:**354; 356–358; 375
effect on vitamin B_6 elimination, **19:**356
Isonitrosoacetone, in reactivation of phosphorylated esterases, **20:**308; 312
Isooctane, **34:**469; 477
effect on DPNH oxidase and succinic oxidase, **20:**178; 181; 182
Isoparaffins, microbial oxidation, **27:**493–495
Isopencicillin *N* synthase, **74:**243
Isopenicillin *N* epimerase, **74:**134; 152
Isopenicillin *N* synthase, collagen hydroxylases, cosubstrate binding, **72:**350–357
Isopentenyl adenine, **35:**469
N^6-Isopentenyl-2-methylthioadenosine, **35:**475
Isopentenyl pyrophosphate, **55:**64
biosynthesis of rubber from, **29:**240–243 (*See also* 3-Methylbut-3-en-l-yl pyrophosphate)
conversion of mevalonate to, **29:**236–240
precursors of, **29:**240
utilization in formation of rubber, **29:**246
farnesyl pyrophosphate from, **55:**78
isomerization, **22:**305; 317
in rubber biosynthesis, **22:**329
in squalene biosynthesis, **22:**297; 304; 309–313
structure, **22:**303
evidence for, **22:**302; 303
Isopentenyl pyrophosphate-incorporating enzyme, partial purification of, **29:**242
properties of, **29:**242
Isopentenyl pyrophosphate isomerase, **22:**304; 305; 310, **55:**64; 65; 66
in squalene biosynthesis, **22:**297
Isopentenyl pyrophosphate isomerase (pig liver), **50:**274
Isopentenyl thiomethyl A, **35:**464
Isophosphorylase, **12:**412; 417–419; 422–424, **15:**309
in polysaccharide synthesis, **11:**305–306; 325–326; 366–367
Isoprenaline, **67:**148
Isoprene rule, of Ruzicka, **29:**221
Isoprenoid catabolism, **35:**388
Isoprenoid compounds, *See* Coenzyme Q
Isoprenoids
biogenesis, **22:**330; 331
synthesis in higher plants, **29:**222
Isoprenoid side chain, and Q activity, **25:**307
Isoprenoid units, in cholesterol, **22:**285
arrangement of acetate carbons in, **22:**288
in squalene, **22:**313
structure, **22:**309
Isopropanol, catalase action on, **19:**109
effect on papain, **19:**286; 287; 290
Isopropanol moiety, **46:**53
Isopropenylacetate, reaction with imidazole, **20:**84
4-Isopropyl-5,6-dihydroxyindole, melanin formation, **16:**166
7-Isopropyl-5,6-dihydroxyindole, melanin formation, **16:**166
α-Isopropylmalate synthetase, end product inhibition, **29:**146
Isopropyl methylphosphonofluoridate, *See* Sarin
Isopropyl thiogalactaside, induction of β-galactosidase by, in *Escherichia coli,* **30:**149
Isopropyl-β-thiogalactoside, in β-galactosidase induction, **22:**111
Isopropylthio-β-galactoside (IPTG), induction of lactose system with, **30:**88
reaction with lactose repressor, **30:**94–95
Isopropyltropolones, in red cedar, **16:**137
Isoproterenol, tyrosine hydroxylase induction, **70:**169
Iso-pyridoxamine, activity of, **16:**220–221
Isoquinoline compounds, oxidation, **15:**112
Isoquinoline group in alkaloids, **8:**233; 235; 236
Isoquinoline red, chemiluminescence, **25:**127
Isoquinoline structure and pyridine ring, **8:**233; 235; 236
Isorhamnetin, as sex hormone in *Chlamydomonas eugametos,* **16:**144
Isoriboflavin, **15:**89
Isosorbide dinitrate, **46:**399
Isostearate, **58:**185
Isosteres, amino acids analogs, **29:**90
Isoteres, **44:**211
Isothiocyanates, **7:**370
Isotope, tracer techniques, **34:**15

Isotope distribution, in glycine metabolism, **23:**457–462
in glycolate metabolism, **23:**435; 441–445
from labeled acetate, **23:**426–427; 432–434
in oxalate metabolism, **23:**449–450; 452
Isotope effect, **43:**498; 499
in alcohol dehydrogenase reaction, **39:**402
in glutamate dehydrogenase reaction, **39:**400
in proton transfers, **25:**22
Isotope effects. *See also* Isotope studies
absolute size of, **45:**312
absolute value of, **45:**315; 317
deuterium, **45:**312; 313
equilibrium perturbation, **45:**319
tritium, **45:**312; 313
and deuterium exchange, **59:**139
kinetic, in mechanism studies, **24:**442; 448; 468; 474
solvent, **24:**461; 472
in kinetic solvent isotope studies, **59:** 124–125; 252
primary, **59:**120–121
Isotope effects in enzymology, heavy atom effects, **46:**443–446
intrinsic effects, **46:**452–454; 460–461
calculation, **46:**452–454
measurements, **46:**449–452
noncompetitive *versus* competitive, **46:**449–452
primary hydrogen effects, **46:**419–430
equilibrium effects, **46:**427–429; 432
semiclassical theory, **46:**419–424
tunneling, **46:**424–426
secondary hydrogen effects, **46:**430–437
α effects, **46:**430–436
β effects, **46:**436
on phase transfer, **46:**437
solvent effects, **46:**437–446
in H_2O-D_2O mixtures, **46:**439–440
magnitude of kinetic effects, **46:**440–446
on pk_a *versus* k_{cat}, **46:**437–439
V_{max} *versus* V_{max}/K_m, **46:**446–448
Isotope exchange, molecular, **50:**361
positional, **50:**362–392
Isotope exchange, Zn, and binding to enzyme, **27:**79–80
blocking by substrate, **27:**72
Isotope exchange studies, **45:**301–308
microtubule dynamics, **69:**129
Isotope Exchange Studies of Enzyme Mechanisms, Positional (Rose), **50:**361
Isotope experiments
in amino acid biosynthesis, **16:**252; 256; 257; 259–261; 264–266; 269; 272; 273; 275–280; 282–286; 290–292; 296–302
in aromatic ring synthesis, **17:**414; 415
in enzyme mechanism study, **17:**2; 48; 55
Isotope incorporation into DNA, **16:**481–483
Isotope partition method, **45:**309
Isotopes, *See* Carbon 13, Carbon 14, Nitrogen 15, Oxygen 18, Phosphorus 32, and Sulfur 35; individual compounds containing isotope tracers; individual Element, eg, Carbon isotopes
distribution in acetic acid from CO_2, **12:** 110–119
Isotope studies. *See also* Isotope effects; Labeled compounds
of acetate metabolism, **14:**120–122
of active centers of esterases, **20:**284; 288; 290
of amino acid metabolism in plants, **20:**321; 322; 328
of asparagine and glutamine biosynthesis in plants, **20:**331–334
of biosynthesis of pectic substances, **20:**369; 370
of galactose metabolism, **20:**113–115; 121; 122; 125
of galactose-negative *Escherichia coli,* **20:**129; 130
of glycolic acid in plants, **20:**219
of hydrolytic action of esterases, **20:**301
of imidazole in enzymes, **20:**81
of lactose biosynthesis, **20:**123
of protein synthesis in plants, **20:**220–223
of steroid enzymes, **20:**273; 274
of aldose-ketose transformation by alkali, **14:**205
of amino acid determination, **14:**272; 273
of biosynthesis, of adenylates, **19:**251; 257; 258
of glutamine, **19:**253; 254
of glutathione, **19:**254–256
of pantothenic acid, **19:**259
of peptides, **19:**260
of proteins, **19:**259; 260
buried bases, **59:**112
of carbon dioxide, **19:**19–21; 23; 28–33; 35; 37–40; 42–45; 61; 63

Isotope studies (*Continued*)
of carotenogenesis, **21:**310; 321; 330–340; 343; 344; 346
of folic acid coenzymes and one-carbon metabolism, **21:**396–398; 416
of histidine biosynthesis and degradation, **21:**403; 407
of methionine synthesis, **21:**424–426
of mitochondrial nonheme iron, **21:**114
of mitochondrial reactions, **21:**25; 44–52; 54
of pterins, **21:**377; 393
of purine biosynthesis, **21:**201; 202; 205; 206; 212–214; 218–220; 235; 241; 243–245; 247; 249; 251; 253; 254
of pyrimidine biosynthesis, **21:**265–269; 273; 275; 279; 281–283; 285–289; 421; 423
of sarcosine oxidation, **21:**428
of synthesis of active formate, **21:**411
of citric acid cycle, **18:**248–249; 267–269; 272–274
of hexose isomerization, **14:**203
of Hill reaction, **19:**49; 53
of homogentisic acid formation, **14:**103
hydride-transfer mechanism, **59:**112
of hydrogen peroxide-ozone-water system, **19:**84; 85
of 3-hydroxyanthanilic acid-nicotinic acid relationship, **19:**94
incorporation of solvent-derived hydrogen isotope, **59:**111–112
inherently slowly exchanging bases, **59:**112
isotope effects of partitioning, **59:**113
of Krebs cycle in algae, **19:**12
of metaphosphate phosphorus, **19:**17
of methionine metabolism in vitamin B_6-deficient rats, **19:**352; 353
of model hydroxylating system, **19:**136; 137
of nonenzymic reactions between proteins and activated amino acids, **19:**261
or protein structure, **14:**281; 300; 302
of oxygen metabolism, **19:**83; 85; 96; 121; 131; 132; 176; 190–194; 205; 207
of phenylalanine acid tyrosine metabolism, **14:**104–105; 120–122
of phosphoglucomutase, reaction, **14:**209–210
of phosphoglyceric acid mutase action, **14:**213
of photosynthetic phosphorylation, **19:**55; 56; 63
of protein adsorption, **14:**345; 361; 364
of sucrose synthesis in plants, **19:**5
of transferase activity of papain, **19:**310
of tyrosinase antibodies, **14:**93; 96
Isotope study, *See* Tracer studies; individual compounds
Isotopically labeled compounds, uptake by nuclear RNA, **16:**474–475
Isotopically labeled precursors of nucleic acids, **16:**450–462
Isotopic competition
in amino acid study, **18:**256–257
in amino acid synthesis, **16:**248–249; 269; 272; 273; 278; 282; 286–287; 298
in investigation of amino acid precursors, **16:**227–228
Isotopic experiments, on carboxylation, **26:** 307–308; 311–330
Isotopic fractionation, **57:**261
Isotopic labeling, in hydrolysis of ATP, **16:**337
in study of cell nucleus, **16:**413–415; 429; 440; 446; 450–462; 471–472; 474; 475; 477; 481–485; 492
Isotopic oxygen, **19:**80
in photosynthesis, **25:**42; 58
Isotopic studies
enzyme reaction cycle intermediates, **72:** 25–26
glutamine synthetase, **72:**15–16
kinetic reactions, **72:**22–24
positional exchange experiments, **72:**28–29
Isotopic substitution, in flavin freeradical study, **27:**133–137
Isotopic tracer experiment, **59:**111
Isotopic tracers
in contractile system, **16:**342
in lignin studies, **16:**149
in melanin studies, **16:**166–167
in transamination studies, **16:**186; 191; 193; 199; 200; 216; 218; 220; 223; 226–228
and glyoxylate cycle, **23:**404–409
Isotopic tracer techniques, **39:**349
Isotropic lipid inclusions, cholesterol ester hydrolases, **69:**59; 63
Isotropic solution, **34:**487
Isovaleraldehyde, oxidation, **15:**257
Isovalerate, **17:**154, **19:**43–45
Isovaleric acid, **58:**174
bactericidal action, **4:**154; 155
and biotin enzymes, **26:**303–305

Isovaleryl chloride, reaction with imidazole, **20:**73
Isovaleryl CoA, in photosynthesis, **19:**44; 45
Isoxanthopterin, **35:**49; 71
Isozyme patterns adenosine deaminase deficiency, **51:**176–177
Isozymes, **25:**345–364, **35:**282, **48:**33
of aldolases, **31:**141
and artifacts, **25:**362
definition, **25:**330; 345–346
β-glucanases as, **28:**352
multiple forms, **25:**343
of phosphofructokinase, **48:**210
Itaconate, **15:**362
as inhibitor of succinic dehydrogenase, **18:**94
oxidation, in mitochondria, **21:**22; 24; 25
Itaconate dehydrogenase, in mitochondria, **21:**22
Itaconic acid in mold metabolism, **9:**576–577
Itaconyl-CoA, **21:**25
Itaconyltrypsin, properties of, **24:**136
Iterative structure-function studies, NMP, kinase catalytic mechanism, **73:**110–111
ITP, *See* Inosine triphosphate
ITPase activity, of actomyosin, **16:**340
ITPase activity of myosin, **13:**168–169
ITP (inosine triphosphate)
action of myosin on, **16:**336–337; 340
effect on myosin, **16:**344
effect on actin polymerization, **13:**154–155
effect on actomyosin, **13:**178
myosin activity for, **13:**167; 168–169
Ivy, antibiotic treatment of diseases, **20:**401

J

Jasmonic acid, **58:**200
Jaundice, **46:**407
obstructive
antithrombin in, **10:**36
clotting time in, **16:**33; 84
Javanicin, **7:**474; 486; 489
as wilting toxin, **11:**403; 422; 423
Jelly coat dissolving factor, **9:**349
Jelly coat of egg, **9:**245–246
chemical nature, **9:**257–258; 259–264
and cytoplasmic maturity, **9:**250–251
and fertilization, **9:**269; 274–275; 313–315
fertilizin in, **9:**258–259; 313–315
Jencks, William P., **51:**75–106
Jensen sarcoma of mice, D-leucine in, **20:**25
Jerusalem artichoke, fructose transfer by, **17:**192–195; 200; 204; 226; 229
JG22, **59:**40–44
Jimson Weed, **55:**73
Jump experiments, **23:**376–377
Juvenile hormone, absence of, **45:**415

K

K. aerogenes citrate lyase, **43:**63
Kaempferol, **55:**42
Kampferol, **17:**434
Kaolin, **48:**281
acid phosphatase adsorbed by, **33:**271
Kaolinite, **33:**267, **34:**457
7-KAP synthetase, **35:**335
Katalyse, Organishen, Chemismus der (Langenbeck), **14:**163
k_{cat} value, **45:**409
KCL-extracted axonemes, **53:**147
140 k complex, **59:**40–44
K^{111} CR, **37:**564
KDO, *See* 3-Deoxy-D-manno-octulo-sonic acid
Keilin and Hartree heart-muscle preparation, *See* Mitochondria, heart
Keilin-Hartree heart muscle preparation, **28:**159; 161–165; 169–170; 173–179; 218; 221–222. *See also* Heart muscle preparation
Keilin-Hartree preparation, oxidation of succinate by, **18:**67; 70; 71; 91–92; 105–107
Keilin's oxidase, **18:**7
Keratan sulfate, **52:**82, **60:**103; 108; 248
degradation sequence, **60:**281
structure, **60:**282
Keratin, *See* Fibrous proteins
binding of pigments, **16:**172
digestion of, **45:**397
supercontracted, elasticity, X-ray analysis, structure, **1:**7; 10
α-Keratin, configuration of, **23:**21
β-Keratin, configuration of, **23:**21–22
α-Keratin, X-ray analysis, structure, **1:**8–11
Keratinase, **23:**201; 226
from *Streptomyces fradiae,* **41:**195; 238
Keratin fibers, reduced, combination with *p*-benzoquinone, **16:**126

Keratin formation and surface SH concentration, **11:**55–56
Keratinization and tyrosinase activity, **14:**133
Keratinolytic enzyme, **23:**179
Keratins, **15:**458
solution by reduction, **11:**228
α-Keratin type fibers, spatial arrangements, **13:**158–159
β-Keratin X-ray analysis, structure, **1:**8–11
Kestose, **15:**313
Kestoses, **17:**202; 203; 204; 210
Ketene, reaction with thiols, **11:**227
Kethoxal, **35:**461, **59:**130
Ketimine, **58:**43
carboxyl group, **58:**45
conversion to aldimine, **58:**44
Ketimines, **53:**163; 170–172
from action of quinones on amino acids, **16:**120
pyridoxamine-P, **56:**4
2-Keto, 3-3desoxygalactose, 5-5lactone, formation and conversions in galactose metabolism, **20:**119; 120
α-Keto-β, β-dimethyl-γ-hydroxybutyric acid, **17:**430
l-α-Keto-β-methylvalerate, in transaminations, **16:**193; 197; 199; 201
3-Keto acid CoA-transferase, **58:**145; 182
α-Keto acid dehydrogenases, **18:**339–340
Ketoacid-lyases, localization of, **24:**331
Keto acid oxidases, **4:**210; 237–238; 248
a-Keto acid peptides and carboxypeptidase, **8:**129
Keto acids. *See also* specific acids; specific keto acids
and CO_2 fixation, **12:**46; 47
from deamination of amino acids by quinones, **16:**117–120
genetic studies, **17:**400; 411; 448–450
salts as solubilizing agents, **17:**148–149
and glutaminase activity, **13:**283–284; 284–285
role in nitrogen fixation, **13:**372
inhibition of oxidation, **9:**567–568
interaction with amino acids, **4:**77–79
in sugar nucleotides, **26:**40; 41; 51; 56
α-Keto acids. *See also* α-Ketoglutarate and Pyruvate; Transamination and individual α-keto acids
α-Keto acids, **7:**374–375; 389–390; 397; 534
catalytic decomposition, **14:**168–169; 170; 171; 181
effect on choline acetylase, **12:**322
effect on decarboxylases, **16:**219–220
in nonenzymic transamination, **16:** 223–226
role of vitamin B_6 in substitution for amino acids, **16:**211
in transaminations, **16:**218; 219–220; 226–228; 256–257; 264
formation of acyl coenzyme A, **15:**129
oxidation, **15:**121–122; 201–202; 215; 218–219
in plant metabolism, **9:**392; 394–395; 398–401; 401–403; 403–409; 411–412; 417–418; 420; 423; 432–433
relation to urea synthesis, **15:**325–327; 338–341
α-Keto acids, **35:**523, **53:**163
oxidative decarboxylation of, **18:**333–338
reductive amination of, **19:**336–338; 340; 341; 352; 354
a-Keto acids, and *a*-amino acids, **8:**117–118
and dehydropeptides, **8:**119
effect on deamidation of glutamine and asparagines, **8:**158–164
in preparation of dehydropeptides, **8:** 124–128; 128–129; 161–162
synthesis, **8:**119
use in synthesis of dehydropeptides, **8:** 128–129
α-Ketoacids, hepatic glutamine transport, glutamine transaminases, **72:**55
β-Keto acids, metabolism, **15:**185; 186; 232–244; 249; 251–252
β-Keto acids in citric acid cycle, **10:**182–186
α-Keto acids in fermenting tobacco, **10:**349; 383–386
Keto acid substrate, metal ion activation, **35:**359
β-Ketoacyl ACP reductase, **36:**298–299
β-Ketoacyl ACP synthetase, **36:**295–298
β-Ketoacyl coenzyme A, **15:**115; 213
S-β-Ketoacyl coenzyme A derivatives, **15:**232–244
α-Ketoadipamic acid, in transaminations, **16:**196
α-Ketoadipate, in lysine metabolism, **16:**296–297
in transaminations, **16:**194
β-Ketoadipate group, **50:**327
α-Ketoadipic acid, **17:**403

β-Ketoadipic acid, metabolism, **15:**184–185; 236
α-Ketoaldehyde, **59:**106; 129
Ketoaldehyde mutase dismutations, **7:**92
α-Keto-ε-aminocaproic acid, in transaminations, **16:**197; 201; 232
7-Keto-8-aminopelargonate(7-KAP), **35:**335
7-Keto-8-aminopelargonic acid, **38:**334
  condensation of alanine and pimelyl CoA, **38:**334
  7-KAP synthetase enzyme purification, **38:**336
  synthesis in cell-free extracts, **38:**334
α-Keto-δ-aminovalerianic acid in alkaloid biogenesis, **8:**238–239
α-Keto-δ-aminovaleric acid, **16:**117; 197; 201; 259
α-Keto analogues of amino acids, preparation, **16:**188–189
3-Ketoarabinitol-1,5-bisphosphate formation, **67:**14
α-Ketoburyrate apodehydrogenase, **18:**339
α-Ketobutyrate, **17:**401–402
α-Ketobutyrate, from cystathionine cleavage, **16:**273
  in amino acid metabolism, **16:**280; 282–286
  in transamination reactions, **16:**189; 194; 203; 205
α-Ketobutyrate oxidation, **18:**338
α-Ketobutyrate synthase reaction, **33:**139
α-Ketobutyric acid, oxidation, **15:**209
β-Ketocaproate, **15:**235–236; 238
α-Ketocaproic acid, **16:**197
β-Ketocaproyl coenzyme A, **15:**233; 242
α-Keto-δ-carbamidovaleric acid, in transaminations, **16:**196
α-Keto-*N*-carbobenzoxy-ε-aminocaproic acid, in transaminations, **16:**195
α-Keto-*N*-carbobenzoxy-δ-aminovaleric acid, in transaminations, **16:**195
α-Keto-*N*-chloroacetyl-ε-aminocaproic acid, in transaminations, **16:**195
α-Keto-*N*-chloroacetyl-δ-aminovaleric acid, in transaminations, **16:**195
α-Ketocyclohexylacetic acid, in transaminations, **16:**196
α-Keto-β-cyclohexylpropionic acid, in transaminations, **16:**196
Ketodecarboxylase, **58:**174
β-Ketodecarboxylases, classification, **14:**256
2-Keto-3-deoxyglucarate aldolase, catalysis of carbon-carbon cleavage/condensation reactions by, **31:**131
2-Keto-3-deoxyoctonic acid, in carbohydrate metabolism, **26:**40; 41; 51; 56
2-Keto-3-deoxy-6-phosphogluconate, in DR synthesis, **28:**422–423; 429; 437
2-keto-3-deoxy-6-phospho-Gluconate aldolase, **37:**445
2-Keto-3-deoxy-6-phosphogluconate aldolase, from *Pseudomonas fluorescens,* **31:**139
2-Keto-4-deoxy-7-phosphoheptulose, formation of, **28:**422
2-Keto-3-desoxy-6-phosphogluconate, **15:**167
α-Keto-*N*-dimethylglutaramic acid, in transaminations, **16:**196
α-Keto-γ-ethiolbutyric acid, in transaminations, **16:**196
Keto fatty acids. *See also* Fatty acids
  and fatty acid catabolism, **8:**392–401
  synthesis, **8:**335–336
  and synthesis of waxes, **8:**334–337
5-Ketofructose, **15:**171
Ketogenic activity of phenylalanine and tyrosine, **14:**74; 96–111
5-Ketogluconic acid, **15:**157
α-Ketoglutaraldehyde, in synthesis of active formate, **21:**410
α-Ketoglutaramate
  deamidation, **16:**257
    in transaminations, **16:**196; 200; 201; 227
  in transaminations, **16:**196
Ketoglutarate, **58:**178
  inhibition of choline and sulfanilamide acetylations, **6:**261
  in tyrosine transamination, **19:**183
α-Ketoglutarate
  in *Aerobacter aerogenes* mutant, **16:**252
    role in transaminations, **16:**186; 189–200; 203–209; 218; 222–223; 226; 227; 230; 233–237; 239; 262
    in synthesis of amino acids, **16:**254–255; 264; 287; 296–297
  and CO_2 fixation, **12:**46; 47; 52–53; 86; 89–91; 103–106
    oxidase, **12:**105
    oxidative decarboxylation of, **12:**92; 103–106
  and pyridine nucleotides, **15:**114–115; 117; 122; 124–126

α-Ketoglutarate (*Continued*)
oxidation, **15:**184–185; 201–202; 213; 214; 219; 221–222; 224–230; 234; 235; 253; 256; 262
relation to urea synthesis, **15:**325–327; 334–340; 342; 383
in racemization of alanine, **14:**195
in transamination, **14:**100; 105
in reduction of diphosphorpyridine nucleotide, **20:**151
α-Ketoglutarate, **46:**42; 84; 166
and acetate metabolism, **23:**408–410; 414; 425; 427; 430
and glycine metabolism, **23:**454; 455
and oxidative phosphorylation, **23:**348; 349
in synthesis, **23:**403–404
5-5C^{14}, **23:**334
cosubstrate for prolyl Hydroxylase, **41:**254
genetic studies, **17:**431; 449
oxidation, and phosphate, turnover, **17:**174–176
P:O ration from oxidation, **17:**333
as inhibitor of pyruvate carboxylase, **42:**17
in photosynthesis, **19:**13; 36; 52; 53
regulation of use, **28:**48
α-Ketoglutarate binary complex with glutamate dehydrogenase, **39:**376
effect on glutamate dehydrogenase-NADPH complexes, **39:**375
inhibition of glutamate dehydrogenase reaction, **39:**390
ternary complex with glutamate dehydrogenase and NADP, **39:**393
α-Ketoglutarate (C^{14}), in transaminations, **16:**191
α-Ketoglutarate dehydrogenase, **57:**157, **58:**146
and ATP synthesis, **23:**333–335; 388
α-Ketoglutarate dehydrogenase complex, **28:**159
α-Ketoglutarate dehydrogense, **46:**113–114
α-Ketoglutarate oxidase, **18:**118; 338
α-Ketoglutarate oxidation, and phosphorylation, **18:**311
α-Ketoglutarate synthase reaction, **33:**139–140
α-Ketoglutaric acid
in biosynthesis of glutamic acid in plants, **20:**323–327
carboxylation to oxalosuccinic acid, **10:** 208–213
from citric acid, **10:**167–168
in dismutation, **10:**190
in mold metabolism, **9:**540; 568; 572–576
α-Ketoglutaric acid, **46:**91
amination of, **18:**253; 255; 268
in lignin study, **18:**363
in mitochondrial lipid synthesis, **21:**52
oxidation, in mitochondria, **21:**13; 14; 28; 29; 33–35; 44; 46; 121
in purine biosynthesis, **21:**219
synthesis, in mitochondria, **21:**21
oxidative decarboxylation, **18:**273
in plants, **18:**297–300
reduction of disulfide, **27:**219
transamination reactions of, **19:**338; 341; 346–350; 357; 359; 360; 363; 364; 380
a-Ketoglutaric acid and fatty acid catabolism, **8:**392–401
α-Ketoglutaric acid dehydrogenase, **15:**85; 115; 122; 226–229; 235; 259
α-Ketoglutaric acid dehydrogenase complex, **26:**172
α-Ketoglutaric acid oxidase, activity of –SH group, **11:**241
α-Ketoglutaric decarboxylases, classification, **14:**256
α-Ketoglutaric dehydrogenase
effect of fluorocitric acid on, **18:**128
in mitochondria 35, **21:**35; 74; 121
Ketoglutaric oxidase, **19:**13
α-Ketoglutaric oxidase, in mitochondria, **21:**13; 14
α-Ketoglutaric oxidase, isolation, **14:**405
α-Ketoglutaric oxidases, **15:**218; 228–230
α-Ketoglutaric oxidase system, **17:**70; 114
α-Keto-δ-guanidinovaleric acid, in transaminations, **16:**195; 261
α-Ketoheptylic acid, in transaminations, **16:**194
α-Keto-β-hydroxybutyrate, **17:**401
α-Keto-β-hydroxybutyric acid, **16:**198; 234
α-Keto-ε-hydroxycaproic acid, in transaminations, **16:**196
2-Keto-4-hydroxyglutarate aldolase (KHG aldolase), inhibition by borohydride of, **31:**140
from rat or beef liver, **31:**140
2-Keto-4-hydroxy-4-methylglutarate aldolase, catalysis of carbon-carbon cleavage/ condensation reactions by, **31:**131
α-Keto-β-hydroxy-β-phenylpropionic acid, **16:**198

α-Ketoisocaproate, **19:**365
β-Ketoisocaproate, **15:**235; 238
β-Ketoisocaproic acid, **58:**181
  cleavage of, **58:**182
α-Ketoisocaproic acid, in transaminations, **16:**197; 203; 229–230
β-Ketoisocapyroyl coenzyme A, **15:**242
Ketoisoleucine, utilization, **16:**279–281; 283
α-Ketoisovaleric acid, **16:**198; 201; 203–204; 229; 285–287
α-Ketoisovaleric acid, genetic studies, **17:**411; 423; 426; 430
Ketol condensation, **15:**158
Ketol condensation in isoleucine synthesis, **16:**283
  in valine synthesis, **16:**285
Ketols, PN-dependent oxidations, **26:**118; 122
Ketomalonate, peroxidase action on, **19:**125
Ketomalonic acid oxidase, **19:**123; 125
α-Keto-γ-methiolbutyric acid, in transaminations, **16:**197
α-Keto-*N*-methyladipamic acid, in transaminations, **16:**194
α-Keto-γ-methyleneglutaramic acid, in transaminations, **16:**197
α-Keto-γ-methyleneglutaric acid, in transaminations, **16:**197; 237
α-Keto-γ-methylglutaramic acid, **16:**197; 200; 201
α-Keto-*N*-methylglutaramic acid, in transaminations, **16:**196
α-Keto-γ-methylglutaric acid, **16:**197
2-Keto-4-methylthiopbutyric acid, **54:**243; 245
d-α-Keto-β-methylvalerate, in transaminations, **16:**193; 197; 201; 229
α-Keto-β-methylvaleric acid, **17:**411
Ketone bodies, **58:**145. *See also* Acetoacetic acid and β-Hydroxybutyric acid
  and adrenaline, **8:**414
  from even-numbered fatty acids, **8:**401–409
  formation and degradation, **8:**401–409
  formation in liver, **8:**409–412
  and glycogen of liver, **8:**408
  in kidneys and muscle, **8:**409–412
  oxidation and C_6-tricarboxylic acid cycle, **8:**401–412
Ketone formation in steroids by microorganisms, **20:**251–253
Ketone metabolism, and fluoroacetate, **18:** 144–146
Ketone reductase, PN-dependent, **26:**168
Ketones and autoxidation of unsaturated fatty acids, **8:**427
Ketones from microbial oxidation of hydrocarbons, **10:**467
α-Keto-δ-nitroguanidinovaleric acid, **16:**197
Ketonuria and creatinuria, **8:**402; 407
β-Ketooctanoate, **15:**236; 238
α-Ketopantoic acid, synthesis, **21:**419
3-Ketopentose, **19:**24
Ketopentose phosphate, formation of, **28:**396
α-Ketophenylacetic acid, in transaminations, **16:**198; 201
3-Ketophosphogluconate, **15:**156; 157
2-Ketophosphogluconic acid, **15:**156; 157
11-Ketopregnanolone, **46:**158
4-Keto-L-proline, proline analog, **29:**95
β-Keto reductase, **15:**234; 243
Ketose-diphosphate-1-phosphatase, in ribose synthesis, **28:**397
Ketoses, enzymes for conversion to aldoses, **24:**333
Ketosis, **2:**269
Δ^5-3-ketosteroid isomerase, **46:**384, **67:**54
Δ^4-3-Ketosteroid reductase, PN-dependent, **26:**127–128
  specificity, **26:**60
17-ketosteroids, **46:**163
Ketosteroids, glucuronides as substrates for β-glucuronidase, **16:**368; 380; 384
Δ^5-3-ketosteroids, isomerization of, **46:**400
Δ^4-3-Ketosteroids, microbiological conversions, **20:**262
Δ^4-3-Ketosteroids, PN-reduction, **26:**127; 128
α-Ketosuccinamate, deamidation, **16:**269
  in transaminations, **16:**195; 200–202; 227
Keto sugars. *See also* individual sugars
  synthesis by transketolase, **15:**160
β-Keto thioesters, **58:**182
β-Ketothiolase, in *R. rubrum,* **23:**423
β-Ketothiolase (thiolase), **15:**233; 237; 238; 239; 240–244
β-Ketovalerate, **15:**235–236; 238
α-Ketovaleric acid, **16:**197
β-Ketovaleryl coenzyme A, **15:**241–242; 244
Ketovaline, utilization, **16:**279–281
Key spot theory of red-cell sensitization, **8:**82

Kidney
allosteric regulation of glucogenesis, **28:**87; 88; 92
and choline deficiency, **7:**140–141
and phosphatide metabolism, **7:**135–136; 142; 167
and plasma phosphatides, **7:**161–163
glucose 6-phosphatase, **34:**310–311
phosphorylase, **34:**392
trehalose in, **34:**298
and oxidation of ketone bodies, **8:**409–412
enzymes, **8:**355
renal gluconeogenesis, phosphoenolpyruvate carboxykinase (GTP), **69:**216–218; 226; 245–249; 257
Kidney, Na^+-glucose symporter in, **29:**61
rat, α-aminoisobutyric acid translocation in, **29:**61
glycine translocation in, **29:**61
lycine translocation in, **29:**61
Kidney cathepsins and renin, **7:**522; 529–530
Kidney enzymes, amino acid breakdown by, **7:**533–536
and essential hypertension, **7:**513–556
Kidney Enzymes and Essential Hypertension (Schales), **7:**513
Kidney extracts and hypertension, **7:**541–542
Kidney phosphatides and phlorizin, **7:**142
Kidney phosphofructokinase, **48:**207–208
Kidney proteinases and pepsin, **7:**529
Kidney tumor, mouse, enzymes in, **24:**322; 323
Kievitone, **55:**5
accumulation of, induction of, **55:**14; 15
chalcone isomerase and, **55:**29
prenyl group of, **55:**41
structure of, **55:**6
Kinase, **43:**11, **50:**390; 391
acetate, **50:**391
chemical properties, **50:**62–66
hexokinase, **50:**391
molecular properties, **50:**44–46
phosphoglycerate, **50:**391
phosphorylation of glycogen synthetase by, **28:**99; 101; 102; 108
production of CTP, **28:**446
physical properties, **50:**44–62
tyrosine hydroxylase activation, **70:**114
Kinase, RNA polymerase II transcription, C-terminal domain characteristics, extended structure, **71:**50
Kinase activation factor (KAF), in allosteric regulation, **28:**99; 101; 107; 108
Kinase reactions, and carbonate activation, **18:**210
in CoA synthesis, **18:**207
and group-transferring enzymes, **18:**192; 194–201
Kinases
and ADP-ATP exchange, **23:**355
for β-aspartyl phosphate, **18:**226
in ATP use, **18:**230
and carbamylation reactions, **18:**227–228
classification, **14:**226; 251
nuclear, **27:**278; 280
phosphoryl transfer by, **17:**163; 164; 168; 175; 180; 183–185
purification of, **46:**246–247
turnover number, **25:**32
Kinase systems, manganese, EPR study, **27:**140–142
Kinetically significant complexes, **39:**354
Kinetically significant transient intermediates, **39:**353
Kinetic analysis, **39:**350; 353
two-substrate, **39:**353
Kinetic behavior, **34:**495; 507; 513
first-order, **34:**520
parameters, **34:**502; 504
Kinetic behavior of inhibitors, **42:**17
Kinetic competence, **59:**245
Kinetic constants, **46:**261–263
comparison of, for soluble and insoluble enzymes (table), **33:**270
for xanthine oxidase activity, **17:**299–301
Kinetic cooperativity, **57:**210–212
resulting from ligand substitution reactions, **53:**326–327
Kinetic Data, Enzyme, Statistical Analysis of (Cleland), **29:**1
Kinetic isotope effect, using α-deuterated substrates, in D-amino acid oxidase reaction, **34:**56–58
Kinetic isotope effects
asparagine synthetase
glutamine-dependent nitrogen transfer, **72:**169; 171–172
mechanisms of, **72:**186–188
vs. papain, **72:**177–183
in enzymology, **46:**415–494
glutamine-dependent nitrogen transfer

AS-B mutants, **72:**180–183
mechanisms of, **72:**183–186
glutamine synthetase, **72:**22–24
in mechanism studies, **24:**442; 448; 468; 474
Kinetic Isotope Effects in Enzymology (Klinman), **46:**413
Kinetic mechanism, of PP ribose-P synthetase, **41:**4
Kinetic mechanisms, **46:**401
of acetate kinase, **45:**302
of aconitase, **45:**354
of alcohol dehydrogenase, **45:**316; 318
of alkaline phosphatase, **45:**357
of biotin-containing enzymes, **45:**291
of carboxypeptidase-A, **45:**359
of chymotrypsin, **45:**337
of creatine kinase, **45:**285; 289; 308
of enolase, **45:**352
equilibrium ordered, **45:**281
of fatty acid synthetase, **45:**292
of fructokinase, **45:**284; 308; 311; 324; 370
of fructose diphosphatase, **45:**285
of fumarase, **45:**318; 349–352
of galactokinase, **45:**308
of glucose-6-phoshatase, **45:**304
of glutamate dehydrogenase, **45:**320
of glutamic dehydrogenase, **45:**316; 318
of glycerokinase, **45:**284; 293; 294
of hexokinase, **45:**285; 293; 299; 305; 308; 310
of β-hydroxy-β-methylglutaryl-CoA reductase, **45:**373
of isocitrate dehydrogenase, **45:**297; 301; 308; 316; 318
of liver alcohol dehydrogenase, **45:**307
of malic dehydrogenase, **45:**316
of malic enzyme, **45:**282; 290; 294; 295; 316; 317; 319
multi-site Ping-Pong, **45:**277; 280; 282; 291; 292
of nitrogenase, **45:**282
of nucleoside diphosphate kinase, **45:**290
ordered, **45:**288; 294; 305
of phenylalanine-tRNA symthetase, **45:**295; 296
of phosphofructokinase, **45:**281; 285; 311
Ping-Pong mechanisms, **45:**280; 311
of pyruvate carboxylase, **45:**291; 304
of pyruvate dehydrogenase complex, **45:**282; 292
random, **45:**281; 284; 289
both substrates sticky, **45:**281
of saccharopine dehydrogenase, **45:**299
Theorell-Chance mechanism, **45:**283; 289; 306
of thymidylate synthetase, **45:**370
of transaminases, **45:**290
of transcarboxylase, **45:**304
of triose phosphate isomerase, **45:**326; 329; 348; 354–355
Kinetic parameters, effect of product concentration, **25:**268–269
Kinetic Problems of Transport, Some Special (Christensen), **32:**1
Kinetic properties, **51:**230–244
bisphosphoglycerate synthase, **51:**241–244
phosphoglycerate mutase, **51:**230–241
effects of anions, **51:**233–236
pattern, **51:**230–233
reactions schemes and rate equations, **51:**237–241
Kinetics. *See also* Rate constants and Rate equations
of action by, alcohol dehydrogenase, **24:**469
chymotrypsin, **24:**459–460
on RNase, **24:**234–235
of enzymic hydrolysis of cyclic phosphates, **24:**444–448; 452–455
of bacterial photosynthesis, **1:**289–301
chemical
and periodic reactions, **23:**88
of collagenase action, **23:**209–210
of photosyntheses, **1:**209–216
of DPN-cyanide formation, **15:**94
of DPNH-X formation, **15:**94
of enzymic oxidoreductions, **15:**3; 7; 14–18; 21; 28–29; 38
of flavin-DPNH reaction, **15:**89
of hydrolytic enzymes, **15:**272–290; 296–297
of thiaminase action, **15:**397–398
enzyme, **17:**1–64
of mitochondrial system, **17:**79–81; 85; 108–109
of muscle sarcosome components, **17:**83
of enzyme inhibition, **12:**238–240; 245–248; 286–290; 293–294; 296–297
of enzyme-substrate compounds in dehydrogenase systems, **12:**184–185
of hemoprotein complexes, **12:**163–171

Kinetics (*Continued*)
of invertase action, **12:**195
of peptidase action, **12:**198–200; 202–203; 209; 212; 241–242; 248–251
of enzyme reactions, **9:**25–71, **25:**1–38
of enzyme-substrate association, **9:**56–60
of photochemical denaturation, **9:**105–108; 130–131
of reversible step-wise reactions, **9:**1–23
of the steady state, **9:**49–56
of virus inactivation, **9:**174; 220–228
of phytochrome isomerization, **26:**23
of protein synthesis, **26:**107–110
of product inhibition, **25:**173–183
quasi-equilibrium, **25:**188–189; 195–196; 202–220
steady state, **25:**174; 175; 179–182; 185–188; 194–195; 199–202; 205
of proteinases, **1:**89; 90
of succinic dehydrogenase action, **18:**90–92
Kinetics, Enzyme, Effects of Temperature on (Sizer), **3:**35
Kinetics, Enzyme (Alberty), **17:**1
Kinetics and mechanism, composite active site, **49:**163; 165
fast-reaction techniques, **49:**161
steady-state kinetic studies, **49:**157–158; 162; 163
Kinetics in heterogeneous systems, **3:**40–41
Kinetics of Biological Reactions with Special Reference to Enzymic Process (Stearn), **9:**52
Kinetics of enzyme reactions
with cellulose, **33:**260–261
with coacervates, **33:**245
with elastin, elastase action of, **33:**257–258
with gelatin, **33:**257
with gels, **33:**245; 255–259; 275
with particulate enzymes, **33:**290–294
with proteins, **33:**255–259
in solutions, **33:**245
with starch, **33:**260
Kinetics of hydrolysis
of cinnamoyl-thiol subtilisin, **33:**394–395
NPA, by thiol-subtilisin, **33:**392–394
Kinetics of phosphorylase action, **11:**302
Kinetics of Substrate Reaction During Irreversible Modification of Enzyme Activity (Tsou), **61:**381
Kinetic specificity, **33:**403; 430, **53:**263
Kinetic studies, **39:**355
aldehyde oxidase, **27:**168–172
of dihydroorotic dehydrogenase, **27:** 173–175
of oxidation, EPR, **27:**125–130
of polymer propagation, **27:**404–405
of xanthine oxidase, **27:**161–166
of AMP binding to phosphorylases, **24:**274
of enzyme formation, **24:**5; 6; 13; 16; 21
of phosphorylase *a* reaction with PCMB, **24:**269
of polypeptide synthesis, **24:**3–4
of RNase, **24:**234–235; 238–239; 452–455
of mechanism of action of esterases, **20:**290–315
of degradation of pectic substances, **20:**348; 349
of liver alcohol dehydrogenase system, **20:**31–48
of papain, **19:**312–325
Kinetic Studies, Determining the Chemical Mechanisms of Enzyme-Catalyzed Reactions by (Cleland), **45:**273
Kinetoplast, Feulgen-positive reaction of, **16:**423; 425
Kininase-I, **57:**5
Kininase-II, **57:**5
Kininogen, high-molecular-weight, **48:**278; 290–291
Kinins, **48:**290
K^+Ions
effect on cellular volume, **52:**262
effect on glucogen metabolism, **52:**262
effect on glucose phosphorylation, **52:**262
effect on hormone response, **52:**262
Klason lignin, from labeled shikimic acid, **18:**367
Klebsiella aerogenes
glutamate synthase, **66:**271–275
urease from, **74:**71; 72
Klebsiella aerogenes, β-lactamase from, **28:**245; 260; 290; 293; 297
Klebsiella spp., β-lactamase of, **28:**268–297
K_m of dihydroxyacetone, **43:**494
K_m value, **34:**460; 505–507; 524
K_M values, **45:**409
$\mathrm{K_m}$ values, *See* Michaelis constants
Koji acid, **2:**183; 224
Kojibiosyl, in teichoic acids, **30:**225
Kojic acid, **7:**474; 475; 486; 489; 490; 492
biosynthesis, **15:**157; 176

Kojic acid in mold metabolism, **9:**576; 577–580
Kojitriosyl, in teichoic acids, **30:**225
15K protein, **54:**355; 356
KQAGDV sequences, fibrinogen, **70:**5; 6
Krabbe's disease, **38:**295; 298; 301; 309, **60:**111
Krebs and Johnson citric acid cycle, **8:**395–396
Krebs cycle, **43:**92, **57:**228. *See also* Citric acid cycle; Tricarboxylic acid cycle
  and dissimilation of pyruvate by liver tissue, **74:**38–44; 41
  in microorganisms, **23:**402–404
  in normal and cancer cells, **4:**206; 207; 209–211; 212–214; 222; 249
  oxidative phosphorylation in, **23:**334; 340–341; 348
Krebs cycle acids
  effect on carotenogenesis, **21:**318
    in mitochondrial lecithin synthesis, **21:**51
    oxidation, in mitochondria, **21:**12–14; 17; 20; 21
      in mitochondrial fragments, **21:**80–82; 123
  oxidation in plants, **20:**225
  and steroid 11β-hydroxylase activity, **19:**189
Krebs cycle and nitrogen metabolism, **19:**338; 339; 344; 353; 359–367; 371
  in plants, **19:**12–14
Krebs cycle enzymes
  in mitochondria, **21:**11–21
  in plants, **20:**224–227
Krebs cycle intermediates, P:O ratios from, **17:**333
Krebs-Heneseleit cycle, for urea formation, **16:**259
Krev-1, **65:**82
Kupffer cells, **46:**121; 139
Kwashiorkor, **42:**278
Kynurenate, **16:**235
Kynurenic acid, **15:**99, **17:**418; 421, **49:**28; 34
Kynureninase, **15:**102, **16:**235, **17:**416–419, **35:**111, **74:**133; 152
  cysteine *S*-conjugate β-lyase reactions
    kynurenine pyruvate aminotransferase, **72:**222–223
    rat liver metabolism, **72:**216–219
Kynureninases, **49:**1–40
  antiserum, **49:**8; 11
  apo-form, **49:**6; 20
  assay, **49:**3
    by fluorometric determination, **49:**3
    by spectrophotometric determinations, **49:**3
  bacterial kynureninase, **49:**4
  catalytic mechanisms, **49:**3; 28; 30
    a, β-elimination pathway, **49:**3; 30
    β, *r*-splitting of kynurenine, **49:**29
    β-diketone pathway, **49:**3; 29; 33
  constitutive enzyme, **49:**8
  constitutive kynureninase, **49:**36
  distribution, **49:**4
  fungal kynureninase, **49:**6
  inducible enzymes, **49:**4; 6
  inducible kynureninase, **49:**36
  kinetical analyses, **49:**33
  liver, **49:**12; 13
  mammalian kynureninase, **49:**12
  purification, **49:**4; 7
    affinity chromatography, **49:**7
  regulation mechanisms, **49:**13; 22; 36
    alanine, **49:**15; 19; 36
    catalytic center activities, **49:**26
    comparison of, **49:**23
    constitutive kynureninase, **49:**26
    half reactions, **49:**19
    hog liver kynureninase, **49:**26
    inhibition by amines, **49:**13
    inhibition by amino acids, **49:**13
    a-keto-δ-aminovalerate, **49:**19
    negative effectors, **49:**25; 36
    of *Neurospora cra*ssa, **49:**22
    ornithine, **49:**15; 17; 19; 22; 36
    overall reaction, **49:**19
    positive effectors, **49:**25; 26; 36
    of *pseudomonas kynurenin*ase, **49:**15
    Δ′-pyrroline-2-carboxylate, **49:**19; 26
    pyruvate, **49:**18; 19; 20; 36
    spectral shift, **49:**20; 22
    transmination, **49:**15; 18; 19; 20; 23; 24; 26; 36; 37
  resolution of, **49:**6
  two types of, **49:**8; 36
    immunological relation, **49:**8
    physiological functions, **49:**10
    special assay, **49:**8
  yeast kynureninase, **49:**12; 26
Kynureninases: Enzymological Properties and Regulation Mechanism (Soda and Tanizawa), **49:**1

Kynurenine, **17:**416–421, **49:**1
absorption, **49:**3
conversion of, **49:**2
hydroxylation, in mitochondria, **21:**27
in nicotinic acid, biosynthesis of, **15:**99–102; 123
metabolic pathway, **15:**184–185
in transaminations, **16:**197; 235
in tryptophane metabolism, **16:**294
L-Kynurenine, oxidation in tryptophan metabolism, **19:**194
Kynurenine formamidase, localization of, **24:**327
Kynurenine from tryptophan, **10:**56
Kynurenine 3-hydroxylase, **19:**194; 195; 199
Kynurenine-niacin pathway, **49:**2
Kynurenine pyruvate aminotransferase, cysteine *S*-conjugate β-lyase reactions, **72:** 222–223
Kynurinase, classification, **14:**249
Kyrurenine aminotransferase, **49:**29

L

L. mesenteroides, degradation of carbohydrates by, **28:**403–404
L. plantarum, degradation of carbohydrates by, **28:**403–404
pentose metabolism, **28:**421; 450
labeled, effect on lysine of *E. coli,* **16:**297
Labeled acetate, synthesis studies, **14:**120–121; 122
Labeled C, N, O, P, and S, *See* specific compounds
Labeled compounds. *See also* Isotope studies; specific compounds, eg, Acetate, labeled
Labeled phenylalanine, metabolism of, **14:**101
tyrosine from, **14:**119
use in adrenaline synthesis, **14:**116–117
Labeled phenyllactic acid, use of, **14:**99
Labeled phosphorus, and egg albumin, **7:**159–160
and egg shell, **7:**159–160
and liver, **7:**128–133
Labeled precursors, *See* Tracer studies; individual compounds
Labeled proteins, and enzyme formation, **24:**5; 6; 12; 14
Labeled tyrosine, metabolism of, **14:**101
Labeling data, interpretation, cautions, **28:** 404–405
Labial glands, **45:**405
Labile factor, *See* Factor 5; Factor V
use of term for blood clotting factor, **16:**59
Labile FAD, **34:**86
Labile sulfide, **34:**92, **39:**361
Laboratory animals, **53:**204; 205–208
Laccase, **4:**100; 135, **18:**29; 37; 310–311; 372, **19:**203
absorption bands of, **33:**195; 224; 225
broad component in EPR of, **33:**214
catalytic activity of, **33:**230–232
CD curve of, **33:**201–202
composition of copper in, **33:**196
copper atoms, roles in catalytic reaction, **33:**236–238
diamagnetic copper ions in, **33:**224–229
EPR-nondetectable copper ions in, **33:**223
extinction coefficient of, **33:**198
forms of copper in, **33:**195
kinetic studies of, **33:**223–236
nonequivalent cupric ions in, **33:**203
ORD curve of, **33:**203
oxidation-reduction potential of, **33:**207
oxygen, product of water reduction of, **33:**230–231
pH dependent reduction of, **33:**208
reaction with cyanide, **33:**218–220
reaction with fluoride, **33:**220–221
classification, **14:**254; 255
column chromatography, **22:**481; 482
comparison with tyrosinase, **4:**101
as terminal oxidase, **16:**131; 132
effect of carbon monoxide on, **16:**136
in oxidation of coniferyl alcohol, **16:**149
role in browning of plant tissues, **16:** 151–152
Laccase, EPR study, **27:**138; 147–148
Laccase-catalyzed oxidations, EPR study, **27:**125–127
Laccase-hydroquinone-oxygen system, coupling to quinone reductase, **16:**134
Lachrymating agents, inactivation of enzymes by, **11:**239
reaction with thiols, **11:**215–216; 226–227
Lac-operator, binding of lac-repressor protein, **40:**45
Lac operon, **44:**249; 253; 256

Lacquers, formation in browning of plant tissues, **16:**152
LAC repressor, **47:**77–81; 122; 136
Lacs, class of phenolic compounds, **16:**141
  role in browning of plant tissues, **16:**150–153
α-Lactalalbumin, **43:**412; 418
  amino acid sequences of, **43:**425
  effects on lactose synthetase, N-acetylactosamine synthetase activity, **43:**415
  enzymatic assay of, **43:**415
  invariant residues of, **43:**432
Lactalbumin
  amino acids in, **14:**263
  mol. weight, **1:**22; 23
α-Lactalbumin, **52:**56; 58; 59; 64; 65; 66; 67; 97
  agarose, **52:**146
Lactaldehyde, **15:**154
Lactam, formation and structure of, **29:**402–403
β-Lactamase. *See also* Penicillinase
  activity of, **28:**244–250
  amino acid analyses, **28:**270–272; 275–278; 280–284; 311
  bacillary, location, **28:**264
  biochemistry and function, **28:**237–323
  biosynthesis of, **28:**269–278
  biosynthetically modified, **28:**275–278
  broad-spectrum, **28:**251–253
  catalytic properties, **28:**287–297
  cell-bound, **28:**242; 248; 251–255; 263–269; 279; 280; 282–283; 291; 296; 305
  cellular location, **28:**263–269; 309–310
  chromatographic separation, **28:**279
  clinical significance, **28:**307–308
  conformational changes, **28:**251; 254–257; 262; 267; 272; 297–304; 306–307
  constitutive, **28:**242; 251; 273–274; 276–277; 279; 280; 282; 311
  definition, **28:**241
  detection of products of action, **28:**245–249
  dissociable, **28:**253
  distinct, of common origin, **28:**253–254
  effects of peptides on, **28:**262–263
  electrophoretic mobility, **28:**284–286
  enzyme-substrate interactions, **28:**297–304
  evolutionary perspectives, **28:**310–313
  extracellular, **28:**242; 248; 251–255; 263–265; 269; 278–280; 282–283; 288–290; 292; 296–299; 305; 309
  genetics, **28:**269–272; 280; 311–313
  immunological tests, **28:**253; 255–257; 270–272; 275; 304–305; 312
  induction of, **28:**242; 248; 251–257; 262–263; 268–269; 273–277; 279; 280; 307–308
  intracellular, **28:**242; 243; 248; 265; 309
  isolation and purification, criteria, **28:** 278–279
  kinetics of action, **28:**247–248; 251; 255; 258–261; 264; 267; 268; 287–292
  liberation of, **28:**265–266
  localization, **28:**305
  molecular forms of, **28:**254–255
  molecular properties, **28:**285–287
  nomenclature for, **28:**238–241
  occurrence of, **28:**241–244
  optical rotatory power, **28:**298–299
  peptide maps of trypsin digests, **28:**284
  physical constants, **28:**286
  physiological significance, **28:**308–309
  plurality of, **28:**250–255
  properties, related to origin, **28:**309–310
  purified, comparison, **28:**279–287
  reaction catalyzed, **28:**244–248
  response to inducer, **28:**255–257
  reversible changes, **28:**297–299
  specificity, and identity, **28:**250–255
    and structure, **28:**255–263
  staphylococcal, location, **28:**263–264
  structure and activity, **28:**257–260
  substrate for, **28:**241
  synthesis *in vivo,* **28:**274–275
  unit of activity, **28:**241
β-Lactamase activity, measurement of, **28:** 246–248
Lactamase II, **56:**293
β-*Lactamase (Penicillinase), Biochemistry and Function of* (Citri and Pollock), **28:**237
β-Lactamase preparations, purified, **28:**276–277; 279
Lactase, **5:**120, **17:**190; 219; 221
  cacao, **25:**466
  enzymic adaptation, **13:**70; 106
  in *Neurospora* mutants, **22:**25
Lactase Dehydrogenases, Structure and Function (Everse and Kaplan), **37:**61
Lactate, **5:**130, **51:**236, **53:**208; 210; 225
  bacterial fixation of carbon dioxide in, **74:** 34–36
  in complex with erythrocyte LDH-DPN, **22:**367

Lactate (*Continued*)
conversion to glucose, allosteric control, **28:**86–98
effect on myosin monolayer, **6:**183
labeled, and fluoroacetate, **18:**146
in plant respiration, **18:**303
labeled incorporation into valine, **16:**285
lactate-pyruvate-glucose relations, **5:**139
lactate-pyruvate ratio, **5:**139
oxidation in normal and cancer tissues, **4:**204; 206; 243; 244
in photosynthesis, **19:**6; 11; 12; 40; 41; 52; 53
from pyruvate, **15:**202; 207; 221
in urea synthesis, **15:**332; 338
D-Lactate, as substrate in mechanism study, **27:**2; 14–16; 32
L-Lactate, as substrate substitute, **27:**4; 14–16
Lactate and CO_2 fixation, **12:**51
Lactate dehydrogenase, **37:**412; 426; 434; 442; 453; 454; 470; 476; 482; 485; 493; 497; 518; 519; 550; 557; 580; 602; 609, **43:**354, **46:**233; 243; 270; 449, **47:**58; 60; 124; 133, **49:**230, **55:**212, **56:**292, **58:**161, **59:**193
binding of coenzymes, **37:**75–78
biological regulation of, **37:**115–124
chemical mechanism of, **45:**377
chemical mechanisms, **37:**102–114
coenzyme analogs, **37:**83–87
conformational change, **37:**81
correlation of hepatoma growth rate with, **29:**331
oxamate inhibition of, **29:**339
dissociation, **37:**67–70
distribution, **37:**63–75
D-lactate-specific, **37:**67
dogfish, **46:**255
eukaryotic dihydrofolate reductase, NADPH-binding site and NAD-binding site compared, **70:**42
evolution, **37:**124–126
hybrids of, **37:**66
immunoglobulin, **46:**244
inhibitors, **37:**87–99
isoenzymes of, **27:**270–271
of nucleus, **27:**261–262
isozymes of, **46:**237
kinetic parameters, **37:**99–102
kinetic parameters for, **25:**262–263; 267
multiple forms, **25:**334; 339–341; 345–347; 353; 364
product inhibition of, **25:**224; 259–263
location in subcellular fraction, **24:**295–296; 335
as a polymeric enzyme, **37:**65–71
primary structure, **37:**72–75
renaturation, **37:**69
sequential responsc, **37:**564; 590
structure and function, **37:**61–126
substrate inhibition, **37:**92–99
substrates, **37:**87–99
subunits, **37:**68
ternary complexes, **37:**94
tetrameric form, **37:**71
X-ray crystallography, **37:**75
D-Lactate dehydrogenase, **37:**454; 519; 551; 580, **44:**255, **46:**243
L-Lactate dehydrogenase, **46:**244
Lactate dehydrogenase isozyme-5, **37:**140
[3-^{3}H] Lactate, detritiation, **52:**255; 257
Lactate fermentation of *micrococcus lactilyticus,* **35:**406
1-^{14}C-Lactate, labeling of pentoses, **28:**426
[2-^{14}C] Lactate, measurement of Pyr/PEP cycle, **52:**257
Lactate racemase, **44:**93
absorption spectrum, **44:**94
analogy with glyoxylase, **44:**95
in *Clostridium butylicum,* **44:**94
hydroxylamine inhibition, **44:**94
internal hydride transfer model, **44:**95
intramolecular H transfer, **44:**95
lack of hydrogen exchange, **44:**94
molecular weight, **44:**94
purification, **44:**94
Lactate racemases, **59:**108
Lactates, as solubilizing agents, **17:**149
Lactate utilization, and lipoic acid, **18:**342
Lactation
phosphoenolpyruvate carboxykinase (GTP), **69:**222–223
uridine, diphosphogalacrose-4-epimerase in, **20:**119
Lactic acid
in blood
in familial periodic paralysis, **6:**215
in muscular atrophy, **6:**196; 198
in muscular dystrophy, **6:**207; 222; 223
in myasthenia gravis, **6:**212

in myotonia, **6:**214
carbon dioxide fixation in, **6:**248–250
distribution of C^{14} from glucose, **15:**176
mechanism of enzymic oxidation, **15:**8–9; 15
for thiaminase disease, **15:**415
in gluconeogenesis, **54:**167
and glycine metabolism, **23:**456
glycogen resynthesis from, **6:**270; 271
of malignant tissue, **54:**181
in plants, **18:**297–298; 300; 303
racemization, **14:**196–197
recycling of, **54:**167
Lactic acid and fat formation, **8:**315
Lactic acid bacteria
activity of lipoic acid, **18:**331
arginine biosynthesis by, **16:**259
aspartate requirement of, **16:**269; 298
replacement of B_{12} in, **16:**274
in cacao fermentation, **25:**470–475
Lactic acid dehydrogenase
disulfides, **27:**231
effect of fluoro compounds on, **18:**118; 151 152
and lipoic acid, **18:**333; 337
in plants, **18:**297; 300; 303–304
reactions of, **18:**100
nonenzymic reaction with *C*-alanylthiophenol, **19:**261
L-Lactic acid dehydrogenase, classification, **14:**253
Lactic acid in mold metabolism, **9:**576–577
Lactic acid in relation to growth of lower plants, **10:**225
Lactic acidosis, **52:**291
Lactic acid synthesis, thiol group, **27:**213
Lactic codehydrogenase, structure studies, **14:**302
Lactic-cytochrome reductase, **17:**284
Lactic dehydrogenase-elevating virus, LDH virus, **39:**226
Lactic dehydrogenases, **5:**134, **7:**425, **15:**85; 115; 120; 172; 192; 202; 204; 206–207; 210–211; 214; 216–217; 221; 228, **17:**70, **34:**448–452; 523, **46:**169
and ATP synthesis, **23:**331
and cytochrome *c*, **23:**60; 66
denaturation of, **23:**33; 48–50; 62
and glycolate metabolism, **23:**444
and chloroplast system, **12:**39
classification, **14:**255
in alanine racemization, **14:**195
isolation, **14:**398; 401
in lactic acid racemization, **14:**197
separation by chromatography, **14:**335; 350
complex with reduced diphophopyridine nucleotide, **20:**40
crustacean and mammalian, **22:**366
erythrocyte, **22:**367
heart, **22:**358–366
complexes with DPNH, acetylpyridine DPNH, **22:**358; 359
complex with DPNH, **22:**359; 362
effect of acetylation on, **22:**360; 361
effect on fluorescence polarization of DPNH, **22:**366
inactivation rate in presence and absence of coenzyme, **22:**364; 365
inhibition by various reagents, **22:**359
protection by acetylpyridine analogue of DPNH, **22:**402
rabbit skeletal muscle, **22:**366; 367
rat liver, **22:**367
reaction with pyridine nucleotide analogues, rates of, **22:**366
denaturation process, **26:**105
PN-dependent, **26:**118; 120–121; 156; 157; 161; 163
in propionic acid metabolism, **26:**287; 297; 305; 310; 316; 350–351
effect of
phosphate, **24:**81
sulfite, **24:**82
effect of polycations, **40:**41; 42
serum levels in cyanate treated sickle-cell patients, **40:**24; 25
and fat metabolism, **8:**391
in nuclei, **8:**33
inhibition by thiol reagent, **11:**237
inhibition of, **13:**332
in mitochondria, **21:**74
in photosynthesis, **19:**7; 11; 12; 52; 53
role of metal, **24:**112–113
in semen, **24:**55
Lactic oxidase, **17:**244; 245; 262; 267
in electron transfer, **19:**201
Lactic oxidative decarboxylase, **19:**202
Lactic oxidative decarboxylase, EPR study, **27:**183
Lactic racemases, **14:**258

Lactobacilli, and cobamide coenzymes, **26:**266
propionic acid metabolism, **26:**293
Lactobacilli, pentose fermentation in, **15:**176
Lactobacillus acidophilus
growth factors, **14:**49–50; 66
and mevalonic acid, **23:**472; 479; 483
Lactobacillus arabinosus, **15:**246; 345
alkali hydrolysis of membrane teichoic acid of, **30:**226
membrane teichoic acid of, **30:**225
ribitol teichoic acid of, **30:**231
asparagine metabolism in, **13:**250; 259–260
and citric acid cycle, **18:**266–267
pyrimidine metabolism, **18:**217
enzymic adaptation in, **13:**71; 79
folate use, **27:**427; 457
formation of citrulline by, **16:**265
formation of CoA, **16:**17
transamination reactions in, **16:**190–191; 193; 199
glutamine metabolism in, **13:**254–262
glutathione metabolism in, **13:**263
growth factor for, **21:**218
growth factors, **14:**49; 53; 66–69
malic enzyme in, **14:**254
racemase in, **14:**196; 197
synthesis of coenzyme A by, **14:**65
malic enzyme in, **19:**34
propionic acid metabolism, **26:**293; 326; 328; 329
transferases in, **13:**295
Lactobacillus bifidus
cell walls, **26:**198
and mevalonic acid, **23:**472
pyrimidine metabolism, **18:**217
Lactobacillus brevis, PN-dependent enzyme, **26:**118
Lactobacillus buchneri, CDP-glycerol pyrophosphorylase in the membrane fraction of, **30:**236
Lactobacillus bulgaricus
enzymic adaptation in, **13:**111–113
galactose-metabolizing enzymes of, **20:**116; 120; 121
metabolism, **20:**120; 121
pyrimidine metabolism of, **18:**217
Lactobacillus bulgaricus 09, pyrimidine biosynthesis in, **21:**266; 267; 277
Lactobacillus bulgaricus factor, *See* LBF
Lactobacillus bulgaricus Factor, Pantethine and Related Forms of (Snell and Brown), **14:**49
Lactobacillus bulgaricus factor (LBF), relation to CoA, **16:**5; 6
Lactobacillus bulgaris, β-galactosidase, **25:**336
Lactobacillus casei, **17:**319, **70:**75
asparagine metabolism in, **13:**249–250
factor, **21:**372, **27:**427–428
in flavin test, **18:**77–79
and lipoic acid, **18:**319; 331
folate inhibitor study, **27:**433; 438–439; 442–446; 455; 460
growth factors, **14:**49; 53; 66
histidine biosynthesis in, **21:**403
membrane teichoic acid of, **30:**225
quinones in, **25:**289
Lactobacillus delbrückii, **15:**249
growth factors, **14:**49–50; 66
oxidation of pyruvate by, **18:**338
phosphorolysis of pyruvate by, **18:**227
phosphorolytic cleavage in, **23:**337
pyruvic oxidase in, **19:**202
Lactobacillus delbrueckii extract, acetyl phosphate formation in, **6:**233
Lactobacillus fermenti, growth factors for, **14:**66
Lactobacillus growth factor, **9:**468; 494–495
Lactobacillus helveticus
growth factor for, **14:**50; 53; 59–60; 66; 68; 69
nucleoside synthesis in, **17:**220–221
Lactobacillus heterohiochi, **23:**473
Lactobacillus homohiochi, **23:**473
Lactobacillus lactis, growth factors for, **14:**66
Lactobacillus leichmannii, **46:**54
growth factors for, **14:**66
pentose metabolism, **28:**436; 444–447; 453
tracer studies, **28:**429–431
pyrimidine metabolism, **18:**217
reductases of, **28:**446–447
ribonucleotide reduction, vitamin B_{12}-linked, **28:**436; 444–447
vitamin B_{12}, **26:**267; 273
Lactobacillus pentosus, **18:**227
phosphoketopen-toepimerase in, **19:**28
Lactobacillus plantarum, **35:**38
alkali hydrolysis of wall teichoic acid of, **30:**229

CDP-glycerol and CDP-ribitol in, **30:** 234–235
Michaelis constant for CDP-ribitol in, **30:**240
teichoic acid biosynthesis in, **30:**239
2-deoxy-D-ribose-5-phosphate aldolase from, **31:**139
folate use, **27:**427; 456
metabolism in, **74:**34
Superoxide Dismutase from, **41:**79; 80
Lactobacillus viridescens, acceptor specificity, **40:**135
aminoacyl-tRNA specificity, **40:**132
biosynthesis of interpeptide bridges, **40:**127; 128; 129; 130
cell function, **40:**136; 137
Lacto-*N*-biose I, **52:**48–49
Lacto-*N*-difucohexaose I, **52:**48
Lactoferrin, **52:**115
human, **52:**27
Lactoflavin. *See also* Flavin, Riboflavin, and Flavoprotein
effect on luminescence, **7:**252
fluorescent spectrum, **7:**227
mode of action as coenzyme, **14:**180–183; 188
reduction, **15:**4–5
Lacto-*N*-fucopentaose I, **52:**48; 49; 62; 70
Lactogenic activity, **42:**75; 120; 127; 128; 133; 138; 140; 144; 145; 148; 149; 152; 156
Lactogenic hormone, **42:**124
Lactoglobulin, **3:**99–101
β-Lactoglobulin
adsorption studies, **14:**360
structure studies, **14:**284
binding of methyl orange, **16:**332
effect of chemical modification, **24:**145
effect of pH on structure, **22:**428
electrophoretic behavior of, **30:**162
β-Lactoglobulin A, low temperature tetramerization of, **30:**141
tetramerization of, **30:**143; 144
theoretical and experimental sedimentation and descending electrophoretic patterns of, **30:**142
Lactonase, **48:**131
Lacto-*N*-neotetraose, **52:**37; 40; 41; 43; 54
Lactones
and esterase activity, **24:**461–462
phosphorolytic cleavage, **15:**155
Lactone solutions, chemical changes in, **36:** 169–171
Lactonization at interfaces, **11:**58; 59–66
Lactonizing enzyme, **27:**511–512; 535
Lactoperoxidae, and cytochromes, **17:**375
Lactoperoxidase, **46:**62, **58:**69, **66:**53. *See also* Peroxidase
absorption spectrum, **23:**309
activity of, **14:**151
purification, **14:**381; 401
separation by chromatography, **14:**348; 352
column chromatography, **22:**480; 482
complex with hydrogen peroxide, **12:** 159–160; 164; 167; 174; 187
effect of radiations, **24:**388; 397
Lactose, **52:**36; 43
enzyme of metabolism of, **30:**84
gene map of operon, **30:**85
location of markers on the bacterial chromosome, **30:**86
enzymic synthesis, **14:**203
use by yeast, **14:**198
fermentation by thiaminase-producing bacteria, **15:**413
in polysaccharide synthesis, **15:**306; 314–315
fermentation of, **5:**103–104
formation in mammals, **20:**119; 121; 123
linkage to sialic acid, **20:**135
in transglycosylations, **17:**218–219
Lactose (C^{14}), **17:**219
Lactose phosphate, **18:**201
Lactose-1-phosphate, formation, **20:**123
Lactose synthetase, **36:**73
chemical modifications, **43:**433
enzymatic assay of, **43:**415
hormonal control, **43:**473; 476
immunochemical properties, **43:**440
mechanism of, **43:**453
physical properties of, **43:**441
subcellular organization, **43:**474
substrate specificity, **43:**453
three-dimensional structure, **43:**429
effect of α-lactalbumin, **40:**57
Lactose Synthetase (Hill and Brew), **43:**411
Lactosylceramide, **38:**297; 302, **52:**31; 32; 40; 59; 62; 72, **60:**102
Lactosylceramide-β-galactosidase, **38:**298
Lactosylceramidosis, **38:**302
Lactosylsulfatide, **60:**168

Lacto-*N*-tetraose, **52:**37; 40; 41; 43; 44; 59
S-Lactoylglutathione, **59:**106
S-(D-Lactoyl)glutathione, **59:**124
Lactoyl-gluthione lyase, **37:**454
Lactyl CoA dehydratase, **66:**35
Lactylglutathione, **23:**325
S-Lactylglutathione, glyoxalase II action on, **19:**248; 249
  synthesis, **19:**241; 249
LADH
  active center
    asymmetry of, **27:**55–56
    primary sequence, **27:**80–82
  characteristics, **27:**49–50
  complexes of, **27:**50–56; 58
    dissociation constants, **27:**63–64
    sign of Cotton effect, **27:**87
  Cotton effects in study of mechanism, **27:** 49–82
  horse, Cotton effects, **27:**49–82
  human, properties, **27:**58; 82–83
  inhibition of, mechanisms, **27:**73–75
  substrate binding, **27:**70–72
  sulfhydryl groups of, **27:**75–82
LADH-DPNH chromophore, Cotton effect, **27:**50–54; 56; 61–70
LADH-DPNH complex, association constants, **27:**69
  Cotton effect, **27:**98
  dissociation, **27:**66–68
  effect of inhibitors, **27:**73–75
  effect of sulfhydryl reagents, **27:**75–82
LADH-OP complex formation, **27:**49–50; 53–56; 58; 66; 70; 73–75
$[(LADH)^{65} Zn^{2+}{}_2]$, exchange of Zn^{2+}
  effect of coenzyme, **27:**64–65
  effect of substrate, **27:**64–65
LADH-Zn-OP chromophore, **27:**53–56; 66
LADH-Zn-OP complex, effect of inhibitors, **27:**73–75
Lamellae, **34:**483
  in chloroplasts, **30:**5 (*See also* Chloroplast)
    composition of, **30:**8–12
    lipid composition of spinach chloroplast, **30:**9
    lipid content in, **30:**8–10
Lamellar micellas, **51:**34–35
Lamellar phase, **34:**469
Lamellar structure, and chemiluminescence, **25:**127
Lamellar system, in chloroplasts, **25:**67–73
Lamellibranchs, laminarinase of, **28:**354
Lamellidens marginalis, thiaminases of, **14:**334
Laminarases, multiple forms, **25:**350
Laminaribiose, **17:**217–218
  as β-1,3 glucan, **28:**327; 329; 331; 339; 340; 344
    and laminarinase action, **28:**345; 350–353
Laminaribiose:orthophosphate glucosyltransferase, **28:**339–340
6-*O*-β-Laminaribiosyl-D-glucose, from β-1,3 glucans, **28:**331
Laminaribiosylmannitol, **28:**350
Laminaridextrins, action of β-glucanase, **28:**349
  in β-1,3 glucan synthesis, **28:**339–340
  hydrolysis, **28:**348; 350; 353
  from laminarin, **28:**349–351
  solubility and, **28:**327
Laminarin, Biochemistry of, and the Nature of Laminarinase (Bull and Chesters), **28:**325
Laminarin, in algae, seasonal variation, **28:** 330–331
  biochemistry of, **28:**325–364
  biosynthesis of, **28:**337–340
  degree of polymerization, **28:**332–334
  insoluble, **28:**327; 328; 331–333
  β-1,6 linkages, **28:**331–334
  patterns of hydrolysis, **28:**349–351
  physiological activity, **28:**356–358
  preparation of, **28:**331
  soluble, **28:**328; 331–333
  structure study, **28:**331–334
  as substrate of laminarinases, **28:**342; 344; 349
  uses of, **28:**353–358
Laminarinase, assay, **28:**344
  biochemistry of, **28:**325–364
  distribution, **28:**341–343
  ecological significance, **28:**353–356
  enzyme fractionation, **28:**349
  functions, **28:**341
  fungal, **28:**338
  kinetics, activation, **28:**347
    inhibition, **28:**347–348
    pH effect, **28:**346–347
    substrate specificity, **28:**344–346
    synthesis, **28:**348–349
    temperature, **28:**346–347
  mode of action, **28:**352–353

patterns of hydrolysis by, **28:**349–351
preparation, from higher organisms, **28:** 343–344
purification of, **28:**344
substrates for, **28:**334; 352–353
unit of activity, **28:**344
Laminarinase complex, **28:**349–353
Laminarinase components, separation of, **28:**351–352
Laminarinase fermentations, culture conditions, **28:**343
Laminarinase induction, **28:**348–349
Laminarinase system, of *Myrothecium verrucaria,* mode of action, **28:**352–353
Laminarin-enzyme digests, D-mannose in, **28:**333–334
Laminarin enzymolysis, scheme for, **28:**353
Laminarin sulfate
activation of lipoprotein lipase, **40:**33
anticoagulant action, **28:**356
Laminaritetrose, **28:**339; 352
Laminaritetrosylmannitol, **28:**350
Laminaritol, properties, **28:**333
Laminaritriose, from laminarin, **28:**331; 350; 352
in syntheses, **28:**339; 340
Laminaritriosylmannitol, **28:**350
Laminarose, properties, **28:**333
Laminin, fibronectin, ligand-binding site model, **70:**14
Langmuir adsorption equation, and protein adsorption, **28:**18
Langmuir adsorption isotherm, **33:**258
Langmuir equation, **33:**259
Langmuir isotherm, **33:**262
relation to chromatography, **14:**320–325
Langmuir's adsorption theory, application to hydrolytic enzymes, **15:**277–280
Langmuir trough, **33:**282
Lanosterol
in cholesterol biosynthesis, formation, **22:**320–324
as intermediate, **22:**286–288
removal of methyl groups from, **22:** 324–326
conversion to cholesterol, **22:**324–327
formation, **19:**192–194
formation from squalene, **22:**288; 322
location of formation, **24:**308; 339
PN-dependent reduction, **26:**127
in rubber biosynthesis, **29:**244
structure, **22:**288
synthesis from *trans*-squalene, **22:**287
Lanosterol 14α-demethylase, **68:**40
Lanthanides, model catalyses with, **21:**160
Lanthanum, fixation of nucleic acid by, **9:**486
Lanthanum compounds, **17:**137; 143–145; 147
Lanthanum ions
acetate complex of, model catalyses with, **21:**188; 189
model catalyses with, **21:**171; 175–181; 183–188; 190
and phosphate ester hydrolysis, **24:**466
physiological effects, **21:**193
Lanthanum salts
in isolation of enzymes, **14:**397
use in myosin precipitation, **13:**163
Lanthionine, **56:**39
LAO, **65:**18–19
Larvae, blowfly, (pupae and adult), rhodanese content of, **39:**337
Larval proteins of insects, **16:**156–158
Lasers, in phytochrome study, **26:**12
Latency, **34:**485
Latex, ability to metabolize sugars, **29:**225
accumulation of 3-phosphoglyceric acid in, **29:**225
acetyl-CoA in, **29:**228–229
biosynthesis of nonrubber terpenoids in, **29:**243–244
carbohydrate metabolism in, **29:**225–226
centrifugal analysis of, **29:**223–225
p-chloromercuribenzoate inhibition of isopentenyl pyrophosphate incorporation, **29:**243
hexose monophosphate shunt in, **29:**225–226
location of inorganic pyrophosphatase in, **29:**235–236
location of phosphatase in, **29:**235–236
low oxygen consumption in, **29:**228
malonate incorporation by, **29:**230–232
synthesis of *cis* and *trans* terpenoids in, **29:**248
Latex vessel, cytology of *Hevea brasiliensis,* **29:**222–225
electron micrograph of, **29:**224
Lathodoratin, **55:**84
Lathosterol, structure, **22:**327

Lathyrine, **50:**152; 155
Lathyrism
collagen defect in, **37:**310
and toxic amino acids, **29:**147
Lathyrogens, in seeds, **29:**147
Lathyrus odoratus, toxic amino acids of, **29:**147
Lauryl maltoside, cytochrome *c* oxidase, electron transfer, **71:**117
Layer, diffusion, **34:**504
unstirred, **34:**503–507
LBF, **14:**49–71, *See* Lactobacillus bulgaricus factor
LBF-1A, activity for various organisms, **14:**65–69
isolation from Ashbya gossypii, **14:**50–52
relation to coenzyme A, **14:**65
relation to pantothenic acid, **14:**53–55
LDF theory, eukaryotic dihydrofolate reductase, chemical transformation step (reactant activation), pteridine activation as hydride acceptor, **70:**80
LDH, *See* Lactic dehydrogenase
Lead
binding by protein groups, **20:**56
effect on permeability, **11:**247
effect on phosphoglucomutase activity, **14:**212
in aldehyde condensations, **14:**172
in isolation of enzymes, **14:**397
inhibition of formation of ALA dehydrase, **46:**44
inhibitor action, **15:**26
Lead ions
chelate formation with amino acids and organic phosphates, **21:**145; 147
model catalyses with, **21:**174; 175; 177
effect on, collagenase, **23:**196
elastases, **23:**234
Lead mercaptide formation, **11:**216
Lead poisoning, **18:**142, **46:**175–176, **56:**401
relation to catalysis by metal ion phosphatase models, **21:**194
Lead salts, **17:**143; 145; 152
Learning, and origin of life, **27:**408–410
Least-squares method, for statistical analysis, **29:**2–4
Leaves, cytochromes in, **18:**16; 17; 22–25; 29; 31–33; 41; 52–54
respiration in, **18:**286; 292; 294
terminal oxidases of, **18:**43–44
Leaving group, **39:**331; 355
Lecithin, **34:**473–475; 489–491. *See also* Phosphatides
association with Leparin, **16:**81
and blood clotting, **10:**27
in snake venoms, **10:**19–20
cardiolipin, **34:**484
origin of methyl groups, **25:**417–420
in plastids, **20:**204
speed of formation, **7:**137–138
in sperm metabolism, **24:**57–61
metabolism of, **24:**338
synthesis, in mitochondria, **21:**50; 51
water interface, **34:**490
Lecithinal-protein complex, **34:**474
Lecithinase, **23:**186; 367; 382
adsorbed, activity of, **28:**3; 22
column chromatography, **22:**467–469
mechanism of action, **11:**78–79
α-Lecithinase, **23:**188
Lecithinase and virulence of Clostridia, **13:**221
Lecithinase A of snake venom, **8:**460; 487; 490
Lecithinase C, in cereal seedlings, **18:**40
Lecithinases
classification, **14:**225; 244; 245; 250
effect on DPNH oxidase and succinic oxidase, **20:**178
in plastids, **20:**213
Lecithinases of snake venom, **8:**476; 486–487; 490
role in cytolysis, **8:**54
Lecithin (C^{14}), in sperm metabolism, **24:**58
Lecithinemia, **7:**156
Lecithin of egg yolk, effect on liver, **8:**191
Lecithin phosphodiesterases, classification, **14:**245
Lecithin synthesis, **18:**208
Lectins, **48:**60; 63; 64; 67; 69; 71; 72; 77; 83; 84, **50:**307
immobilized, **48:**38
Leeches, **48:**297
Leghemoglobin, **46:**88
Legume lectins, **56:**225
Leiomyosarcoma, sulfokinase in, **25:**441
Leishmania, **46:**102
Lens
regeneration from iris cells, **16:**485
structure and function, α-crystallin/small heat-shock protein, **69:**156–157

Lens S-crystallins, glutathione S-transferases, **69:**3. *See also* entries under Crystallin
Lentinus degener, ring methyl group, **25:**297
Lentinus lepideus, action on wood, **18:**354
the aromatization process in, **18:**362–366; 369
Lentinus lepideus, and lignin studies, **8:**293–295
degradation of cellulose, **8:**272–273
of glucose, **8:**272–274
Lenzites abietina, enzymic transformations of steroids by, **20:**248
Lenzites sepiaria, action on wood, **18:**352–354
Lenzites sepiaria, degradation of cellulose, **8:**272–273
of glucose, **8:**272–274
Lepidoptera, **46:**65
Leprosy bacillus, fatty acids in, **8:**335
Leptocurare, **43:**191; 193; 194; 195
Lesch-Nyhan syndrome, **41:**1; 18–20; 25–26
LET, *See* Linear energy transfer
Lethal synthesis, **18:**114; 121–122
Lettré-Ehrlich ascites cells, enzymes of, **24:**306; 307
Lettuce, antibiotic treatment of diseases, **20:**392
Leu-Asp-Val (LDV) sequence, fibronectin, **70:**2; 3
Leucaena, as a source of mimosine, **29:**108
Leucine, **34:**19, **50:**130; 132, **51:**217, **54:**48; 50; 52
bacterial decarboxylation of, **6:**2
in biosynthesis of phenylalanine in plants, **20:**329
optical antipodes in living organisms, **20:**2
in phosphorylated peptides of esterases, **20:**288
thermodynamic data on ionizations, **20:**64
catabolism (Fig. 17), **35:**385
diketopiperazine, **35:**163; 164
in cholesterol biosynthesis, **22:**283; 284; 292
control of synthesis, **29:**147
effect of analogs on synthesis, **29:**147
conversion into valine, **58:**183
effect on carotenogenesis, **21:**337; 338
enzyme for activating, **19:**258
labeled, in protein biosynthesis, **19:**261
transamination reactions of, **19:**365
from fatty acids, **58:**185
formation by leukocytes, **58:**183
genetic studies, **17:**400
glutathione S-transferases, **69:**17
labeled, in ethionine study, **25:**376; 378–379
and nitrogen balance, **58:**174
production inhibition, **58:**179
in transglutaminase reactions, **54:**37–39
in urea synthesis, **15:**341 (N^{15}); 358 (N^{15})
D-Leucine, **33:**367; 368
determination in presence of L-leucine, **20:**25
occurrence in native proteins, **20:**25
L-[^{13}N]Leucine, **57:**327–329
L-Leucine
activating effect on glutamate dehydrogenase reaction, **39:**378
binary complex with glutamate dehydrogenase, **39:**385
effect on glutamate dehydrogenase-NADPH complexes, **39:**377; 414
formation of ternary complex with glutamate dehydrogenase-ADP, **39:**387
K_A values for, **29:**45
properties of uniporter, **29:**45
oxidation in plants, **20:**320; 322
structure, **20:**2
transamination reactions of, **19:**347; 367–369
β-Leucine
interrelation with leucine, **58:**177
origin of, **58:**176
Leucine (deuterium), **16:**199
biosynthesis, **16:**286–287
in transamination reactions, **16:**190; 192; 193; 197; 199; 201; 203; 204; 207; 208; 228–230
tert-Leucine, in transaminations, **16:**193; 197
Leucine adenylate, synthesis by carbodiimide method, **22:**190
Leucine-alanine system, for transamination, **16:**204
Leucine 2,3-aminomutase, **58:**178
activity in cytosol, **58:**180
activity in mitochondria, **58:**180
in heart, **58:**180
in muscle, **58:**180
occurrence, **58:**180
Leucine aminopeptidase, **15:**292, **56:**350, **68:**153–204
activation by metal ions, **21:**150; 155
kinetics, **21:**139
bovine lens, **74:**71
catalysis, divalent metals, **68:**163–177
amastatin complex, **68:**169
amastatin interactions, **68:**173

Leucine aminopeptidase (*Continued*)
bestatin complex, **68:**168
bestatin interactions, **68:**172
differentiation and identification of binding sites, **68:**163–167
L-leucyl-L-valine, **68:**170–171; 174
roles for metals in catalysis, **68:**168–177
catalytic mechanism, **68:**186; 188
carboxypeptidase A inhibitor, **68:**198
α-chloroketones, **68:**189–190
coordinate analogue inhibitors, **68:**189
di-Zn^{2+} *versus* single-Zn^{2+} peptidases, **68:**196–203
gem-diolate group stabilization, **68:**192
gem-diolate transition state, **68:**191–192
gem-diolate-Zn^{2+} interactions, **68:**200–201
hydroxyl-Zn^{2+} interactions, **68:**194–195
phosphorus-containing inhibitors, **68:** 198–199
tetrahedral carbon *gem*-diolate, **68:**199–200
X-ray structure-based, **68:**186; 188–196
classification, **14:**247
column chromatography, **22:**455; 456
crystallization, **68:**154–155
digestion of porcine ACTH with, **29:**408
distribution of, **12:**205–206
purification of, **12:**196–198
specificity of, **12:**194; 200–205
hydrolytic activity of, **56:**350
imidazole in, **20:**98
inhibitor binding, **68:**177–187
cathecoyl dipeptide, **68:**178–179
reduced bestatin isostere, **68:**179–180
sulfur-containing, **68:**180–184
thiol inhibitors, **68:**182–183
localization of, **24:**324
partial degradation of protein by, **24:**125
metal centers of, properties of, **56:**352–355
metal content of, **56:**350–351
metalloenzyme substitutions of, **56:**304
molecular properties of, **56:**351–352
in papain and mercuripapain structure, **19:**300–301
specificity, **19:**329
regulatory role of zinc in, **56:**316
structure, **68:**154–160
active site Zn^{2+} environment, **68:**157–159
quaternary, **68:**156
treatment of γ-globulin, **41:**162
Leucine-*p*-aminophenyl-alanine copolymers, **34:**455–456
Leucine catabolism, and biotin enzymes, **26:**303–304
Leucine dehydrogenase, diaminopimelate dehydrogenase, **72:**319–320
Leucine-glutamate system for transamination, **16:**204
Leucine-isoleucine transamination, **16:**208
Leucine-α-ketoglutarate transaminase, **18:**255
Leucine metabolism, **58:**173
adenosylcobalamin, **58:**178
allosteric regulation, **28:**70
catabolic pathway, **58:**175
in cobalamin deficiency, **58:**185
and iso-fatty acid metabolism, **58:**187
β-keto pathway, **58:**176
and nerves, **58:**187
D-Leucine (N^{15}, deuterium), inactivity, **16:**218
L-Leucine (N^{15}, deuterium), in transaminations, **16:**218
Leucine (N^{15}), in transaminations, **16:**193; 218
Leucine residues, attachment to virus, **17:**487
Leucine transaminases, **16:**286
Leucinol, **35:**487
Leucinol-AMP, **35:**487
Leucoanthocyanins, from cacao, **25:**462–465
Leucocyanidins of cacao, **25:**463–465; 478–483
Leucocyte β-glucuronidase, **16:**362; 369; 387
in leukemia, **16:**397
Leucocytes, Na^+-amino acid symporter in, **29:**61
Na^+-glucose symporter in, **29:**61
Leucodopachrome, in melanin formation, **16:**164
Leucokinin, **41:**155–157; 159
fraction IV, **41:**159
properties, **41:**156
reduction after splenectomy, **41:**155
stimulation of phagocytosis, **41:**157
Leucokininase, action on Tuftsin, **41:**167
Leuconostoc citrovorum
folate requirement, **27:**421; 436–437; 441; 457
growth factor for, **21:**380; 381; 392
Leuconostoc mesenteroides, **15:**157; 176–177; 311, **19:**6–6phosphogluconic dehydrogenase of; 38
absence of aldolase in, **31:**126

action on asparagine peptides, **13**:244–245; 250–253
action on aspartic acid, **13**:250–253
action on glutamine peptides, **13**:244–245; 261–262
inhibition of growth by methionine sulfoximine, **13**:258
dextran formation in, **30**:190
dextrans from dextransucrase of, **30**:197
growth factors for, **14**:66
source of sucrose phosphorylase, **10**:123; 141
Leuconostoc species, dextran synthesis by, **15**:175; 310
Leucophaea maderae, **45**:400
Leucoplasts, formation, **20**:203
proteins in, **20**:206
Leucopterin, **35**:49
isolation, structure and synthesis, **21**:376
relation to purines, **27**:422
Leuco-riboflavin, **15**:89
Leucosin, occurrence, **28**:334
Leucovorin, *See* N^5-Formyltetrahydrofolic acid methyl synthesis, **25**:402
Leucovorin (citrovorum factor, folinic acid), **27**:426; 436–437; 439; 441–442; 454; 457; 458
Leucoytes, normal human activities of enzymes of glycogen metabolism in, **34**:347
Leucrose, **15**:311, **17**:202; 208
Leucyl, phenylalanyl-tRNA protein transferase, acceptor specificity, **40**:104; 105; 106
assay method, **40**:98
cellular function, **40**:109; 110; 111
donor substrate specificity, **40**:101; 102
inhibitors, **40**:98
molecular weight, **40**:99
mutant defective in, **40**:110; 111
peptide acceptors, **40**:105; 106
protein acceptors, **40**:104; 105
reaction requirements, **40**:98
ribosomal acceptor, **40**:109; 110
in vivo soluble protein acceptors, **40**:110
Leucyl adenylate, as activated form of leucine, **19**:258; 259
synthesis, **19**:252
Leucyl aminopeptidase, action of muscle phosphorylases, **24**:266
Leucylglycine, inactivity in transamination, **16**:239
Leucyl peptidases. *See also* Leucine amino-peptidase
reactivation of, **10**:111
Leucyl phosphate, **19**:250
Leucyl-tRNA synthetase, from *E. coli,* **40**:147
L-Leucyl-L-valine, **68**:161; 170–171
binding to leucine aminopeptidase, **68**: 186–187
interactions with leucine aminopeptidase, **68**:174
Leukemia, **42**:277
antithrombin in, **10**:37
DNA in, **16**:426
fluoropyrimidines in treatment of, **54**:62
folate metabolism, **27**:435; 436; 442; 454; 459
β-glucuronidase, **16**:397
histone in, **16**:446
lymphatic, ribose synthesis, **28**:414
rates of anaerobic glycolysis in, **29**:326–327
viral, **47**:366
Leukemia cells
friend, **46**:115
and vitamin B_{12}, **26**:267; 270
Leukemic leucocytes, histones in, **18**:175
Leukemic tissue, nucleoprotein synthesis in, **7**:187
phosphorus distribution, **7**:186–187
phosphorus turnover in, **7**:186–187
Leukocyte common antigen related protein, **68**:3
Leukocytes, lipoxygenases, **58**:258
Leukocytosis, **58**:132
Leukoderma, **14**:95. *See also* Albinism; Depigmentation; Melanin; Pigmentation
Leukophilic γ-Globulin, **41**:150; 152; 153
binding to leukocytes *in Vitro,* **41**:150
binding to leukocytes *in vivo,* **41**:152
fraction I, **41**:153
fraction IV, **41**:150
Leukotrience A_4 synthase, **58**:202
antarafacial character, **58**:285
hydrogen removal, **58**:282
Leukotriene C_4 synthase, glutathione S-transferases, **69**:4; 5; 6
Leukotrienes, **57**:377–378, **58**:193, **59**:104
cholesterol ester hydrolases regulation, **69**:71
epoxyleukotrience, **58**:208
hydrolysis product, **58**:201
leukotriene B_4, **58**:285
Leupeptin, **58**:251

Levans, **5:**116; 117; 118, **17:**193–195; 222
branching ratio of, **30:**203 (*See also* Mechanisms)
course of the subreactions in formation of, **30:**215
effect of glycerol on formation of, **30:**218
effect on dextran synthesis, **30:**191
equation for formation of, **30:**206
formation mechanism of, **30:**204
formation of, **30:**202–210
glycosidic bonds in, **30:**182
hypothesis of formation of, **30:**218
mechanism of formation of, **30:**179–221
molecular weight of, **30:**204
properties of, **30:**203
summary of formation of, **30:**208–210
enzymic synthesis, **10:**141
synthesis of, **11:**298–299; 312–315
Levansucrase, **5:**119, **17:**191; 195; 221; 222; 228
acceptor constants for, **30:**207
action of, **30:**205
on sucrose, **30:**203
from *Aerobacter* species, **30:**202
Arrhenius plot for, **30:**207
from *Bacillus* species, **30:**202
considerations on the action of, **30:**211–219
graph of action of, **30:**208
graph of temperature dependence of the reactions of, **30:**210
influence of glucose and glycerol additions on, **30:**205
initial rates of the action of, **30:**209
inversion reaction of, **30:**205
kinetics of, **30:**204–208
mechanism of, **30:**204–208
Michaelis constant for, **30:**207
properties of, from *Aerobacter levanicum,* **30:**202–203
reaction of, with sucrose, **30:**202
specificity of, **30:**203
structure of, **30:**203–204
temperature stability of, **30:**203
classification, **14:**222; 251
in polysaccharide synthesis, **11:**298–299; 312–315; 371
secondary interactions of, **33:**436
Levansucrase-F, molecular weight of, **30:**203
properties of, **30:**203
Levansucrase-S, properties of, **30:**203
Levan synthesis, **6:**300
Levan-treated rabbits, effect of penicillin, **28:**274
Levan type polysaccharides, synthesis, **15:**313; 314
Levorphanol toxicity, reversal by polyamine, **36:**239
Levulinate, **46:**115; 121
Levulinates, **17:**149
Levulinic acid, **46:**88; 131; 178
in rubber biosynthesis, **29:**236
Lewis blood groups, **52:**48
Lewis blood types, inheritance, **36:**134
Lewisite, as enzyme inhibitor, **11:**235; 238
Li^+, amino acid-dependent uptake of, **49:**81
substitution for Na^+, **49:**78
toleration by system ASC, **49:**81
Lichenase, **8:**271; 285
Lichenases, classification, **14:**246
Licheniformins, adsorption studies of, **14:**347
Lichenin, as β-1,3 glucan, **28:**329
as substrate for laminarinases, **28:**342
Lichens, **7:**485; 488
β-1,3 glucans of, **28:**329
Life, origin of, **7:**54
Lifetimes of radiation products in biological systems, **24:**384; 387; 401
Ligand affinity chromatography, general, **46:**208
Ligand binding
adenosine receptors, **69:**91–94; 108–110
eukaryotic dihydrofolate reductase, **70:**47–65
cooperativity of, **70:**52–53
generally, **70:**47–48
NADPH and NADP binding, **70:**48–50
substrate binding and tetrahydrofolate, **70:**50–52
Ligand binding (of inhibitors), eukaryotic dihydrofolate reductase, **70:**55–60
7-hydroxymethotrexate, methotrexate polyglutamates and other inhibitors, **70:**58–60
ternary inhibitor complexes, isomerization of, **70:**56–58
Ligand binding (side chain roles), eukaryotic dihydrofolate reductase
Arg^{70} role in, **70:**63–64
Glu^{30} role in, **70:**60–61
hydrophobic residues lining active site, **70:**61–63
Lys^{54} role in, **70:**64

Ligand-binding site model, fibronectin, **70:**13–15
Ligand binding sites, **48:**221
adenylosuccinate synthetase (AMPSase), **73:**64–72
chemical modification of, **48:**225–227
Ligand exclusion theory, **39:**378; 388
Ligandin, **48:**14
Ligand-induced stabilization, quantitative analysis, **37:**528
Ligandins, **46:**384; 389; 390; 407
definition, **46:**407
identity with transferase B, **46:**390
induction by phenobarbital, **46:**390
storage protein, **46:**408
Ligand interactions, glutamine synthetase, stereochemical mapping, **72:**20–22
Ligandin Y, **46:**146
Ligand preparations, general, stability of, **46:**232
Ligands
allosteric, **37:**403
isosteric, **37:**403
of cobalt atoms, in corrinoids, **26:**236–237; 252–253; 256–257; 275
general, **46:**208–221
immobilized, preparation of, **46:**222
preassembly synthesis of, **46:**221; 222
hydrophobic, **46:**389
inhibition of cytochrome *c* by, **33:**313
phenylalanine hydroxylase activation, **67:**139–146
Ligand substitution, steps in, **25:**12; 15–20
Ligand substitution chemistry
generalizations, **53:**312–316
intramolecular ligand displacements, **53:** 340–341
no stable ternary complex as intermediates, **53:**316
rate-determining step, **53:**316
Ligand Substitution Chemistry and Enzymology (Jenkins), **53:**307
Ligases, **56:**293; 401
localization of, **24:**333–335
Light, **57:**232
acceleration of autoxidation by, **15:**2
conversion into chemical energy, **18:**54–57
of CO poisoning of respiration, **18:** 30–33
effect on amino acid metabolism in plants, **20:**321
effect o lipoxidase, **20:**218
effect on carbon monoxide inhibition of cytochrome oxidase, **17:**341
effect on xanthine oxidase, **17:**311
effect on carotenogenesis, **21:**311–316
excessive, carotenoid protection against, **21:**360; 361
effect on sperm, **24:**85
reactions to in plant photosynthesis, **33:**136
Light-dependent reactions, in photosynthesis, **19:**45–65
Light effects, enhancement, **25:**62–66
on two-pigment systems in photosyntheis, **25:**62–64
Light Emission, Chemistry (McElroy and Seliger), **25:**119
Light emission, chemistry of, **25:**119–166
Light-emitting reaction of Cypridina, **25:** 157–158
Light energy. *See also* Photosynthesis and Radiation
in photosynthesis, **25:**45–46
transfer from carotenoids to chlorophylls, **21:**350–354
Light-induced absorbance changes, **51:**28–30
Light intensity, effect on protein-DNA cross-linking, **66:**125–126
Lightning bugs, alkali shift in, **25:**292
Light reaction. *See also* Bioluminescence, Light emission, and Photosynthesis
in bacteria, factors, **25:**144–156
in firefly, enzyme-catalyzed, **25:**131–134
inhibition, **25:**133–137; 142
photosynthetic, chemical separation of, **25:**65–66
Light saturation of plants, **1:**204
Light scattering, by ascites tumor cells, **29:**349
of mitochondria, **29:**67; 348
Lignification, **18:**349–378
mechanism of, **18:**359–375
observation of
by radioautographic study, **34:**209
by ultra violet microscopy, **34:**209
Lignification (Schubert and Nord), **18:**349
Lignin methyl alcohol, in air-cured tobacco, **6:**329
Lignins
biosynthesis, **16:**136; 146–150, **55:**9, **56:**204
unmethylated, role in browning of plant tissues, **16:**150–153

Lignins (*Continued*)
via hydroxycinnamic acid CoA thiol esters, **55:**17
and cellulose degradation, **8:**279–295
chemical nature, **8:**282–283
comparison of similar types, **8:**292–293
constitution, **8:**288–290
enzyme studies, **8:**290–295
linkage to cellulose, **8:**283–288
origin, **8:**280–281
changes during fermentation of tobacco, **10:**369–370; 396; 402
color reactions of, **34:**208
of dicidous wood, **34:**213
distribution of, in cell walls, **34:**209
ethanolysis of, **34:**213
hydrogenolysis of, **34:**213
occurrence of, **34:**208
oxidation with nitrobenzene in alkali, **34:**213
permanganate oxidation of, **34:**215
decomposition of, **5:**201–202
in flue-cured tobacco, **6:**364
nature of, **18:**350–359
primary building stones, **18:**370–374
secondary building stones, **18:**374–375
precursors labeled with ^{14}C, **34:**223
Lignins, Formation and Biological Degradation of (Higuchi), **34:**207
Lima beans, **39:**359
antibiotic treatment of diseases, **20:**390; 391
Lima bean trypsin inhibitor, **48:**288
Limit dextrinase, **34:**407
isolation, **14:**397; 405
Limit-dextrinase, **12:**410
Limit dextrins, *See* Dextrins, limit
Limit of rennin action on casein, **15:**431–433; 435–436
Linadryl, substitution for platelet cofactor I in formation of threone activity, **16:**54
Linaloöl, **22:**306; 307; 315
structure, **22:**306
Linalool, **50:**269
d-Linaloöl, in cacao, **25:**459
Lindane, **46:**138; 139
Linear energy transfer (LET), and enzyme inactivation, **24:**371–373; 375; 376
from different types of radiations, **24:**365
Lineweaver-Burk plots, **34:**523, **57:**117–118
of allosteric inhibition, **28:**50; 134
Linkage group, relation to chromosome, **22:**3
Linkages
in glycoproteins, **25:**442–453
split by proteinases, **1:**67
in teichoic acid, **30:**242
Linked functions, calcium ATPase, **65:**208–210
Linked genes, **13:**127–129
N-Linked oligosaccharides, **52:**94
O-Linked oligosaccharides, **52:**94
Linoleaidic acid, conformation, **58:**275
Linoleate, **58:**285
oxidation, **19:**100–102
potassium, relation to Q-enzyme, **11:**326
Linoleate peroxides, **46:**171
Linoleic acid
and ethionine metabolism, **25:**373
oxidation in plastids, **20:**217
Linolenic acid
in lamellae, **30:**9
oxidation in plastids, **20:**217
Lintner valuse, **7:**589; 593
Lipase, **5:**197, **33:**262–263, **34:**486–487
action, **34:**486
mechanism of, **48:**335–340
adsorption, on glass beads, **48:**337
on monolayers, **48:**332; 335; 349
role of essential residues in, **48:**362–363
molecular properties of, **48:**322
monoacetyl derivative of, **48:**339
pancreatic, **48:**325–326
substrates, chemical requirements for, **48:**326
action on emulsified esters, **23:**142–149
adsorption in emulsions, **23:**142–145
effect of, pilocarpine on, **23:**225
surface-active compounds on, **23:**148
gastric, **23:**130
interfacial action, **23:**142–149
intestinal, **23:**130
mode of action, **23:**152–154
pancreatic, **23:**129–161
definition of, **23:**134–135
general properties of, **23:**149–158
inhibition of, **23:**148–149
purification of, **23:**135–142
specificities, **23:**154–158
ricinus, **23:**157
serum lipoprotein, **23:**157
activity of –SH group, **11:**237; 239; 241
adsorption, **28:**3; 27

cacao, **25:**461; 466
column chromatography, **22:**447
intracellular distribution, **16:**463; 466
localization of, **24:**316; 325
low vertebrata, **1:**380; 389; 394–396; 401
in muscular atrophy, **6:**199
in muscular dystrophy, **6:**205; 207; 219; 223
oxidation-reduction balance and vitamin E, **6:**504–507
synthesis, **6:**271
in muscular dystrophy, **6:**222
in tobacco, **6:**369
pancreatic, **34:**487–488
activation by amino acids, **10:**100
relation of *p*H to activity, **10:**84
Lipase, Pancreatic (Desnuelle), **23:**129
Lipase activity, tests for measuring, **23:**136–137
Lipase-bile salt complex, **48:**349
Lipase-colipase system, **48:**340–353
Lipases
adsorption studies, **14:**334; 343
classification, **14:**244
fungal, isolation of, **14:**380
effect of freezing on, **9:**626
histochemical methods for, **9:**593
titrimetric methods, **9:**601
mechanism of action, **15:**3
in plastids, **20:**213
reaction with alkyl phosphates, **20:**90
Lipases cytochemical method for, **8:**30
pancreatic, **8:**406
serum, **8:**406
Lipase unit, **23:**137
Lipemia, alimentary, **40:**31
Lipid, **34:**456; 468; 473; 479
bilayer, **34:**483–484
biosyntheses, **34:**476
chains, **34:**468
composition, **34:**466; 483
environment, **34:**476; 478; 484; 492
intermediate, **34:**477
enzyme interaction, **34:**478
membrane, **34:**485
metabolism, **34:**478
micelles, **34:**486
microenvironment, **34:**472
mitochondria, **34:**468
molecules, **34:**446
neutral, **34:**467; 483
phase, **34:**469
protein complex, **34:**469
in respiratory chain preparations, **20:**153; 188
solution, **34:**483
Lipid A, in lipopolysaccharide, **31:**80
and polysaccharide, linkage between, **31:**83
structure of, **31:**83
Lipid bilayers, **28:**32–33
Lipid character, of phosphorylase *a,* **28:**105
Lipid complex, of cytochrome *c,* **23:**342–343
Lipid-containing particles, in reconstitution of respiratory chain, **28:**209–222
Lipid cytochrome *c,* **23:**310–311
Lipide, *See* Thromboplastic substances
in thromboplastic material, **16:**30
Lipide character, of prosthetic groups of cytochromes, **17:**344; 348
Lipide composition of cell, **16:**426
Lipide inhibitors in blood coagulation, **16:**32; 54–56; 85–89; 92
of threone activity, relation to hemophilia, **16:**56–57
Lipide-protein complex, **7:**163
Lipides. *See also* Cholesterol, Fats, Fatty acids, Lecithin, Phosphatides, Phospholipides, and Sterols
of blood, choline content, **8:**180; 197–198
effect of raw pancreas, **8:**185–186; 191
and cell structure, **8:**7–8; 9–10; 18–20; 25; 48; 50; 53–54
and complement, **8:**88
composition of nuclear and cytoplasmic types, **8:**62
distribution of liver in depancreatized dog, **8:**177–178
effect of pancreatic duct ligation, **8:**187–189
and enzyme inhibition, **8:**44
and functioning of cytoplasm, **8:**3; 35–36; 43; 44; 46–47; 50; 59
and inhibition, **8:**44
of liver, effect of choline, **8:**191–193
in duct-ligated dogs, **8:**188; 189
effect of methionine, **8:**198–200
effect of raw pancreas on, **8:**183–185
in normal and depancreatized dogs, **8:** 175–179
metabolism, effect of raw pancreas, **8:** 183–186
stimulating properties of pancreatic juice, **8:**190–191

Lipides (*Continued*)
plasma, in insulin-treated depancreatized dogs, **8:**179; 180
role in nerve conduction, **8:**46–47
synthesis, **8:**59–60; 299–341
and *Torulopsis lipofera,* **8:**301; 302; 304; 305; 306–307; 308; 309; 310; 312; 313; 315–316; 321; 325; 328
of viruses, **9:**230
rates of hemolysis, **9:**46
Lipides, Synthesis (Kleinzeller), **18:**299
Lipides and cell structure, **13:**14
Lipides and permeability, **7:**21–22
Lipide-sieve theory of cell permeability, **8:** 48–51
Lipid films, properties of, **28:**15; 19; 32–33
Lipid formation, by photosynthetic bacteria, **23:**422; 423; 429
Lipid-lipid interaction, **34:**478
of primary complexes, **28:**218–219
Lipid Metabolism, Inborn Errors of (Brady), **38:**293
Lipid monolayers, **28:**2; 14; 15
Lipid peroxidation, **46:**150; 174, **57:**377, **58:**192
ubiquinone, **65:**127
Lipid peroxides, **46:**167; 171
Lipid phosphorus and sperm metabolism, **24:**57–59
Lipid phosphorus and sperm metabolism of, **24:**338
Lipids. *See also* Lipoproteins and Phospholipids
and acetate, **23:**431–432
and cytochrome oxidase activity, **23:** 310–311
and haem a_2, **23:**313–314
and reductase activity, **23:**461
from *n*-alkanes, **27:**472; 484–493
in amino acid activation, **24:**4
metabolism of, **24:**338
in coenzyme Q separation, **25:**293
and electron transport, **25:**306–308
liver, and ethionine metabolism, **25:** 369–380
in photosynthetic structure, **25:**70–73
in electron transport, **28:**160; 162; 170; 192–194; 202; 207; 220–223
serum, effect of laminarin sulfate, **28:**356
enzyme action on, **33:**263–264
and fluoroacetate, **18:**115; 130; 139–142
in mitochondria, **21:**9; 75
synthesis, **21:**49–52; 74
in mitochondrial fragments, **21:**101; 106–109; 113
and protein synthesis, **26:**109–110
of bacterial walls, **26:**224–227
CDP derivatives in synthesis, **26:**63
Lipids affinity, annexin-membrane complex, ion channel model system annexin V, **71:**221–222
Lipid solubility, of amino acids, **49:**45–48
Lipid-soluble intermediate, **34:**478
Lipid storage diseases, **38:**293
diagnosis, **38:**308
Lipid surfaces, specific adsorption on, **28:**30–33
Lipid synthesis, by sperm, **24:**89
Lipid uncouplers, **23:**366–367; 387
Lipid/water interface, **34:**478; 486; 489–491
Lipoamide, **39:**342
Lipoamide dehydrogenase, **37:**419; 497; 519, **46:**244, **59:**146
Lipoate, **39:**333; 342; 343; 347; 360, **57:**98
oxidized, **39:**333; 342; 352; 358
persulfide, **39:**354; 362
Lipocaic, preparation and properties, **8:**193; 195; 196
Lipoextract, **34:**476
Lipoflavoprotein, mitochondrial, **22:**386
Lipogenesis, **43:**90
phosphoenolpyruvate carboxykinase (GTP), **69:**205
Lipoglycoprotein, of bacterial wall, **26:**226–227
Lipohydroperoxidase, **58:**195; 202
Lipoic acid, **17:**431, **18:**319–347, **35:**338
and acetate-replacing factor, **23:**471–472
and glycine, **23:**461
and phosphorylation, **23:**333–334; 385; 388
binding to pyruvate apooxidase, **26:**333
PN-reduction, **26:**128–129
binding with TPP, **22:**400
in oxidative decarboxylation, **22:**400
and coenzyme Q, **25:**313
and photosynthesis, **25:**99
disulfide exchanges, **27:**219
redox chain, SH-group, **27:**216
function of, **18:**333–344
α-Lipoic acid
in citric acid cycle, **15:**184; 207; 214–223; 229–230; 256
as component of respiratory chain, **20:**151

Lipoic Acid, Chemistry and Function (Reed), **18:**319
α-Lipoic acid, chemistry of, **18:**325–332
isolation of, **18:**321–324
as protogen A, **18:**320–321
β-Lipoic acid, chemistry of, **18:**325–329; 331
isolation of, **18:**322–324
as protogen-B, **18:**320
β-Lipoic acid, in citric acid cycle, **15:**214; 215
Lipoic acid-activating enzyme, **18:**343
Lipoic acid ($Lip(SH)_2$), in DR metabolism, **28:**438–441; 445; 446; 448
Lipoic Acids, Biochemistry, and Chemistry of (Schmidt, Altland, and Goedde), **32:**423
Lipoic acid (S^{35}-labeled), **18:**340; 341
Lipoic acid-splitting enzyme, **18:**339; 343
Lipoic dehydrogenase, **15:**221–223; 230
and ATP synthesis, **23:**333; 352
Lipoic dehydrogenase, EPR study, **27:**183
Lipoic transacetylase, **15:**220; 222–223
Lipoids, and origin of life, **27:**357
Lipolysis, **34:**487, **57:**144
Lipolytic activity of elastase, **23:**239; 241; 247
Lipometabolic phospholipides in muscular atrophy, **6:**199
Lipoperoxide structure, and mitochondrial swelling, **27:**229
Lipophilicity, exaggerated role, **49:**47
Lipophilic nature, **34:**446
Lipopolysaccaride, phosphomannomutase/phosphoglucomutase (PMM/PGM), **70:**231
Lipopolysaccharide, **34:**457; 474–476
biosynthesis of
in Gram-negative enteric bacteria, **31:**77–124
metabolic control of, **31:**115–117
effect of treatment with acid on, **31:**78
extract of, **31:**78
genetic control of synthesis of, **31:**100–114
genetic determination of modification enzymes of, **31:**110–111
genetics of T antigen synthesis of, **31:**114
glucose-heptose linkage of, **31:**85
heptosyl-heptose linkage of, **31:**85
interaction of phospholipid and, **31:**90
location of enzymes for synthesis of, **31:**118
and location of β-lactamase, **28:**265
mechanism of, **31:**87–88
mechanism of monosaccharide transfer in, **31:**88–92
monosaccharide composition of polysaccharide portion of, from *Salmonella,* **31:**79
O side chains of
biosynthesis of, **31:**92–99
modification of synthesized, **31:**99–100
possible methods of chain elongation in, **31:**97
role of antigen carrier lipid in, **31:**93–98
overall structure of, in salmonella, **31:**82
penetration of, into leaflet structure composed of phosphatidyl ethanolamine molecules, **31:**91
phase conversion of O antigens of, **31:**111–113
R core of
sequence of sugars in, **31:**87
structure and biosynthesis of, **31:**82–92
from *Salmonella typhimurium*
quantitative sugar composition of polysaccharide portion of, **31:**84
UDPgalactose-deficient mutant of, **31:**83
structure of antigen carrier lipid in, **31:**98–99
synthesis of nucleotide-sugar precursors control of, **31:**115
Lipopolysaccharide, Cell Wall, Biosynthesis of, in Gram-Negative Enteric Bacteria (Nikaido), **31:**77
Lipopolysaccharides, **58:**100
and murein sacculi, **26:**197; 213
of cell walls, **26:**225–226
and sugar nucleotides, **26:**47–49; 60–62; 66; 68–72
Lipopolysachharide deficiency, *Caulobacter* cell cycle, flagellar structure and assembly, **71:**9
Lipoprotein, **40:**34
acid, from nuclei, **18:**179
association with heparin, **16:**81
of cell nuclei, **16:**449
of cuticle, **16:**153; 154; 156
glucose-6-phosphatase as, **28:**98
and location of β-lactamases, **28:**265
interaction with polyanions, **40:**55; 56
Lipoprotein complex, **34:**469; 475–476; 480
electron transfer complex, **34:**479
mosaics, functional, **34:**479–481
particles, **34:**470

Lipoprotein complexes, particulate, preparation of, **28:**209–215; 218
reconstitution of respiratory chain, **28:** 209–220; 222–223; 225
Lipoprotein lipase
activation
by chondroitin sulfate, **40:**33
by dextran sulfate, **40:**33
by heparin, **40:**31; 33
by laminarin sulfate, **40:**33
effect of ionic strength, **40:**34
factors necessary for activation, **40:**33
inhibition
by heparinase, **40:**31
by protamine, **40:**31
by pyrophosphate, **40:**31
by sulfated polysaccharides, **40:**33
inhibitor binding site, **40:**34
interaction with high-density lipoproteins, **40:**34
stabilization by heparin, **40:**31; 33
Lipoprotein nature of sarcoplasm ATPase, **13:**167–168
Lipoproteins
cold lability of, **23:**381–382
and oxidative phosphorylation, **23:** 381–382
and enzyme formation, **24:**5; 24
films, **28:**16–17
isolation from plastids, **20:**207–209
in mitochondria, **21:**100
in mitochondrial fragments, **21:**100; 107–110; 115; 116; 118; 119; 124
position in mitochondrial electron transfer chain, **21:**119; 120; 122
and murein sacculi, **26:**197; 213
of cell walls, **26:**225–226
Lipoproteins and calcium, **8:**45
Liposomes, **50:**307, **58:**239
Liposome system, **34:**483–485
Lipothiamide, **17:**431, **18:**332
Lipothiamide pyrophosphate, **15:**222–223
Lipothiamide pyrophosphate (LTPP), **18:** 335–337
Lipothiamids, in energy transport, **14:**40
Lipothiamin pyrophosphate, **15:**122
Lipothiamin pyrophosphate disulfide, **15:**119
Lipotropic action, of βine, **8:**192
of choline, **8:**172
of methionine, **8:**172
of proteins, **8:**192
Lipotropic agents, and methyl groups, **12:**483
Lipotropic Substances and phosphatide turnover, **7:**138–141
Lipovitellin, **7:**163
Lipoxidase, **19:**99–104; 127
amino acids in, **14:**275
classification, **14:**253
isolation, **14:**399; 405
in plastids, **20:**217; 218; 229; 231
Lipoxidase amino acid content, **15:**296
Lipoxidase and Autoxidation of Unsaturated Fatty Acids (Bergstrm and Holman), **8:**425
Lipoxidases, **8:**373–374; 441–453
activator of, **8:**448–449
and autoxidation of unsaturated fatty acids, **8:**426–457
and carotenoid oxidase, **8:**441
and coupled reactions, **8:**451–453
distribution, **8:**442–443
and essential fatty acids, **8:**426
inhibition, **8:**447–448
properties, **8:**446–447
purification, **8:**445–446
reaction mechanism, **8:**449–451
specificity for double bond position, **8:**426
standardization, **8:**444–445
substrate specificity, **8:**443–444
and thiol compounds, **8:**373–374
Lipoxin A, **58:**206
Lipoxygenase catalysis
fatty acid activation, **58:**230
hydrogen abstraction, **58:**228
mechanism of, **58:**227
partial dioxygenase reactions, **58:**229
reversible oxygen binding, **58:**228
Lipoxygenase kinetics, **58:**222
one-site model, **58:**226
Lipoxygenase mRNA, **58:**250
biological importance, **58:**290
positional isomerism, **58:**277
stereochemistry, **58:**291
Lipoxygenases, **58:**192. *See also* Reticulocyte lipoxygenase
acetylenic substrates, **58:**215
action
on mitochondrial phospholipids, **58:**252
on plasma membranes, **58:**239
activity

in bone marrow, **58:**247
in erythrocytes, **58:**247
measurement of, **58:**195
aerobic assays, **58:**194
amino acid composition, **58:**199
in animal, **58:**193
nomenclature, **58:**286
antarafacial character, **58:**284
arachidonic acid, substrate, **58:**200
assays, **58:**193
benoxaprofen inhibitor, **58:**220
binding sites, **58:**222
chiral centers, **58:**282
complex products, **58:**291
cooxidations, **58:**210
cyanide inhibitor, **58:**221
damage to chloroplasts, **58:**244
degradation, **58:**251
dioxygenation
of arachidonic acid, **58:**299
double products, **58:**296
of methyl esters, **58:**201
enzymology, **58:**200
epoxyhydroxy derivatives of fatty acids, **58:**303
fatty acids arrangement, **58:**281
ferric form, **58:**227
ferrous form, **58:**227
fluorescence of, **58:**228
forms, **58:**227
formula for characterization, **58:**286
free radicals formation, **58:**195
hydrogen removal, **58:**278
inactivation mechanism, **58:**216
inhibition, **58:**211
inhibitors, **58:**195; 217
interaction with membranes, **58:**238
from invertebrates, **58:**262
involvement in degradation of mitochondria, **58:**247
iron binding, **58:**231
in leukocytes, **58:**258
lipohydroperoxidase activity, **58:**206
molecular properties, **58:**198
molecular weight, **58:**198
nomenclature, **58:**285
in plants, **58:**193
and platelet aggregation, **58:**258
in platelets, **58:**256
prochiral centers, **58:**285
products of, **58:**200
prostaglandin G synthase reaction, **58:**277
purple complex form, **58:**227
pyrazoline inhibitors, **58:**219
reaction
fatty acid conformation during, **58:**275
kinetics, **58:**195
mechanisms, **58:**276
with phospholipids, **58:**201
reactions catalyzed by, **58:**202
and reticulocyte maturation, **58:**247
self-inactivation, **58:**211
in soybean, **58:**194
specificity, **58:**287
stereospecificity
of hydrogen removal, **58:**287
of oxygen insertion, **58:**287
substrate, **58:**200
preparation, **58:**194
steric structure, **58:**274
suicidal character, **58:**195
suicidal substrate, **58:**215
cis-trans isomerism, **58:**285
Lipoxygenases and Their Metabolites, The Stereochemistry of the Reactions of Proposed Nomenclature of Lipoxygenases and Related Enzymes (Kuhn, Schewe, and Rapoport), **58:**273
Lipoyl chloride, **15:**222
Lipoyl dehydrogenase, copper content, **27:**146
Lipoyl dehydrogenases, absorption spectrum of, **34:**53
fluorescence of, **34:**53
Lip(SH)$_2$, *See* 6,8-Dimercaptooctanoic acid
LipsS$_2$, *See* α-Lipoic acid
Liquefaction of semen, **9:**346–348; 378
Liquid crystals, **34:**477
Liquid of crystallation, **52:**182
Liquid-trystalline phase, **34:**468
Lisinopril, **57:**9–10; 35
Lithium
effect on distribution of body sodium by, **29:**497
inhibitory action, **6:**98
Lithium aluminum hydride, effect on pectic substances, **20:**352
Lithium bromide, dissolving of fibrin, **16:**36; 39
Lithium bromide solution, and denaturation of RNase, **24:**194–196

Lithium chloride, dissolving of fibrin, **16:**36
Lithium diodosalicylate, **50:**315
Lithium hydride, in protein structure study, **14:**279; 286
Lithium in fructose-bisphosphatase activity, **54:**132
Lithium ions, **15:**246; 251; 257
  and choline acetylase activity, **12:**318
Lithium salts, **17:**149
Lithium uptake, **49:**81
Litmocidin, **7:**475; 486; 489; 490; 492
Littorina littorea, glucosulfatase of, **22:**229
Liver, **46:**42; 43, **49:**31, **59:**318–319. *See also* Metabolic regulation, and Fru-2, 6–6P_2, liver
  and acid-soluble phosphorus, **7:**120–125
    and choline deficiency, **7:**140–141
    and labeled phosphatides, **7:**128–133
    and phosphatide metabolism, **7:**135–137
    and plasma phosphatides, **7:**154; 161–163
  allosteric regulation of glucogenesis, **28:** 87–95; 98
    purine metabolism, **28:**118–121
    sugar pyrophosphorylases in, **28:**111; 112
    UDP-glucose dehydrogenase, **28:**116
  aryl sulfatase activities in different species, **22:**224
    mammalian, arylamine sulfokinase in, **22:**213
      aryl sulfatases of, **22:**216–223
      steroid alcohol sulfokinases in, **22:**213
    ox, steroid sulfatase of, **22:**227; 228
  carbon dioxide fixation in glycogen by, **74:**44–46
  dissimilation of pyruvate by, carbon dioxide fixation in, **74:**38–44; 41
  effect of rare earth metal ions on, **21:**193
  glycogen, **34:**292; 293
    glycogen embryonic, **34:**291
    glycogen synthetase, **34:**318
      mammalian, **34:**337
    homogenate, synthesis of glycogen in, **34:**296
    phosphorylase, **34:**383
    rat, activities of enzymes of glycogen metabolism in, **34:**313
  hepatic gluconeogenesis, phosphoenolpyruvate carboxykinase (GTP), **69:**214–216; 223–226; 245–249; 256; 264–265
  human, **39:**359
  and ketone body formation, **8:**409–412
    antifatty-liver factor of the pancreas, **8:**171–202
    cirrhosis, **8:**182–183
  mouse, **46:**44
  phenylalanine hydroxylase in, **74:**235
  rat, **46:**59
    proline activating enzyme, **29:**102
  ureagenesis in, **53:**228
Liver alcohol dehydrogenase, **37:**409; 445
  effect on diphosphopyridine nucleotide, **20:**43
    molecular weight, **20:**37; 38
    preparation and purification, **20:**37
  isozymes, **46:**250
  kinetic mechanism of, **45:**307
  mechanism of deduced from relaxivity data, **33:**59–62
Liver alcohol dehydrogenase system, kinetics and equilibria, **20:**31–48
Liver Alcohol Dehydrogenase System, Kinetics and Equilibria in (Theorell), **20:**32
Liver and action of vitamin K, **2:**307; 308
Liver dehydrogenase, **20:**122
Liver enzyme, effect on LBF, **14:**53–55; 64
Liver enzymes, heme biosynthetic chain, **46:**108–110
Liver esterase, and relation of binding energy to catalytic efficiency, **33:**436–437
Liver fluke, phosphofructokinase, allosteric regulation, **28:**78–79
Liver glycolysis
  in aerobiosis, **52:**282
  in anaerobiosis, **52:**282
  fatty acid synthesis, **52:**282
  lactate output, **52:**282
  maximal rate, **52:**282
Liver heme metabolism, control by heme, **46:**144
Liver hexokinase, effects of fasting and refeeding with hexoses on, **39:**313
Liver lipids, *See* Lipides of liver
Liver microsomes, **18:**48
Liver nonspecific aromatic hydroxylase, **19:**174–176; 199
Liver phosphatides, and amino acids, **7:**141
  and choline phosphate, **7:**141
  deuterium content, **7:**170
  and yolk phosphatides, **7:**157–158

Liver phosphorus turnover and dietary state of animal, **7:**120–125; 167
Liver phosphorylase, **24:**282–283; 286
  active, formation of, **28:**108
Liver phosphorylase-inactivating enzyme, **24:**280
Liver phosphorylase kinase, **24:**282
Liver phosphorylase phosphatase, **24:**282
Liver proteins, rate of renewal, **7:**176
Liver regenerating, ornithine decarboxylase in, **36:**229
Liver residue, isolation of lipoic acid from, **18:**323–324
Liver slices, in biosynthesis of cholesterol, **22:**283
Liver storage and vitamin E, **6:**488–490
Local flows, **34:**496
  pH, **34:**461
Localization of enzymes in cytoplasm, **13:**1–20
Lock-and-key model, **43:**222
Locomotion, relation to cytochromes, **17:**335
Locus
  maltose A, **30:**119
  operator, **30:**90–92; 98
  promoter, **30:**100
Locus coeruleus, tyrosine hydroxylase, **70:**107; 120
Locus effect, definition of, **33:**250
  in enzyme reactions, **33:**250–254
Locusta migratoria, **45:**395
Locusta migratoria migratorioides, **45:**413–414
Log normal distribution curves, **50:**4
Lohmann's enzyme, classification, **14:**251
Lohmann's isomerase, **14:**205
Lomofungin, **56:**107
London-van der waals forces, **34:**469
Long chain glycerides, **34:**487
Longitudinal relaxation time, correlation time for dipolar interaction, **33:**6
  correlation time for scalar interaction, **33:**7
  paramagnetic effects on, **33:**6–7
  physical description of, **33:**3–5
Long range action of enzymes, **9:**69–71
Long range forced and enzyme action, **12:**196
Long-range forces, **11:**86
Long-term glutamine regulation
  glutaminase regulation, **72:**65
  glutamine synthesis, **72:**67
Loop closure, **67:**18; 21
Lophine, chemiluminescence, **25:**121
Loroglossol, **55:**95
Low molecular weight exogenous ligands, reaction with R2, **65:**167
LSD 25, fall in brain norepinephrin by, **29:**487
  increase in brain serotonin level produced by, **29:**518–519
  inhibition of serotonin action on the uterus by, **29:**518
  in psychiatry, **29:**481
  in schizophrenia, **29:**516; 517; 534
  serotonin antagonism of, **29:**518
LTH (ovine lactogenic hormone), **42:**75; 113; 118; 119; 120; 122
LTPP, *See* Lipothiamide pyrophosphate
Lubimin, **55:**66; 73
  biosynthesis of, **55:**67; 69; 71
  metabolism of, **55:**81; 83
  structure of, **55:**52
Luciferase, **7:**23; 216–227, **15:**3; 118, **37:**520; 580, **43:**31; 43; 334
  analogs, **44:**45
    color of emitted light, **44:**44
    dioxetane mechanism, **44:**44
    effect of 3-IsoATP, **44:**46
    effect of Sigma -ATP (1,N^6-ethenodenosine triphosphate, **44:**46)
    emission spectra, mixture with dehydroluciferin, **44:**59
    essential sulfhydryl groups, **44:**52
    excited-state proton transfer, **44:**56
    hydrolase activity, **44:**40
    hydrophobic nature of active site, **44:**54
    kinetics of light production, **44:**60
    luciferin binding site, peptide from, **44:**53
    molecular weight, **44:**50
    nature of active site, **44:**52
    nature of enzyme-product complex, **44:**61
    nature of E-P complex, **44:**60
    optical rotatory dispersion studies, **44:**49
    peptide maps, **44:**50
    physical properties, **44:**48
    stoichiometry of substrate binding, **44:**54
    structure of the emitter, **44:**44
    substrate specificity, **44:**45
    sulhydryl peptide, **44:**53
    thermal inactivation, **44:**50
    tritium-hydrogen exchange studies, **44:**48
  in annelids, **25:**160
    bacterial, **25:**144–156
    Cypridina, **25:**156–158

Luciferase (*Continued*)
firefly, and luminescence, **25:**129; 131–144
Renilla, **25:**158–159
in various organisms, **25:**159–161
and immune reactions, **7:**255–256
inactivation of, **7:**246–248; 256
properties, **7:**222–223
Luciferase-L-AMP complex, reaction of ATP with, **44:**41
Luciferase-catalyzed reactions, sequence of events, postulated, **44:**63
thermodynamics, **44:**63
Luciferases, **19:**198
Luciferin, **7:**23; 216; 248; 256; 258, **44:**38
in annelids, **25:**160
bacterial, **25:**144–156
Cypridina, **25:**121; 156–158
firefly, **25:**121; 131–144
mechanism of luminescence, **25:** 124–126; 129–131
quantum yield, **25:**140–142; 144; 146
Renilla, **25:**158–159
stereospecificity of, **25:**142–143
in various organisms, **25:**159–161
dark oxidation of, **7:**217–226
effect of radiation on, **7:**219–228
energy states during oxidation, **7:**219; 222
luminescent oxidation, **7:**217–226
and naphthoquinones, **7:**257
oxidation, **9:**41, **15:**3; 113; 118
properties, **7:**217–222
quantitative estimates, **7:**217
redox potential, **7:**258
structure of molecule, **7:**218; 257
structures of, **44:**39
of monanion and dianion forms, **44:**46
Luciferin-AMP activity, **25:**129–131; 133–137; 140; 143–144
Luciferin and analogs, free energy of binding, **44:**47
Luciferin-enzyme-AMP, **25:**129; 132–136
Luciferin-enzyme complex, light emission, **25:**129; 131–137
Luciferin esters, chemiluminescence, **25:** 129–131
Luciferin-luciferase reaction, **18:**212. *See also* Luminescence
kinetics of, **7:**223
scheme, **7:**221
Luciferin-luciferase system, **7:**222; 229
effect of pressure on, **9:**68–69
Luciferin system, oxidation-reduction potential, **7:**218
D(−)-Luciferyl adenylate, **44:**38
Luciferyl adenylate, synthesis by carbodiimide method, **22:**191
Lucigenin, chemiluminescence, **25:**125
Lucilia cuprina, **59:**81–82
Lumiflavin
binding capacity, **22:**379
(^{15}N)(^{2}H), EPR study, **27:**134–135
from succinic dehydrogenase, **18:**78–79
Luminal, and flavoprotein, **7:**24
effects of oxygen consumption, **7:**24
Luminal glutamine
fate of, **53:**215
utilization of, **53:**214–218
Luminescence. *See also* Absorption spectra, Fluorescence, Luciferinluciferase reaction, Phosphorescence, Spectrophotometric determinations, and Spectroscopic studies
activation energy of, **7:**237; 242; 246
and aluminum, **7:**253
bacterial, **7:**215–264
emission spectrum, **7:**227
loss of, **7:**227
mechanism of, **7:**226–233
mechanisms controlling intensity, **7:** 233–244
relation to aerobic metabolism, **7:**231–233
respiration, **7:**23; 253–255
and calcium, **7:**253
and catalytic systems, **7:**232
chemistry of, **25:**119–166
mechanism of, **25:**137–144
stereospecificity, **25:**142–143
thermal, in photosynthetic tissues, **25:**82
use of oxygen, **25:**143; 147–149; 151–154
effect of buffers, **7:**226; 258
of *p*H on, **7:**226; 251–252
of lactoflavin, **7:**252
of pressure on, **7:**233–242; 244–251
effect of pressure, **9:**68–69
effect of temperature, **9:**41
and flavoprotein, **7:**227; 232
and mercuric chloride, **7:**25
and sodium, **7:**253
and temperature, **7:**233–242; 244–251
total, **7:**226

and urethan, **7:**237; 242–247; 258
and zinc, **7:**253
Luminescence, Bacterial (Johnson), **7:**215
Luminescence of tetrapyrrolic compounds, **12:**19–20
Luminescent oxidation, **7:**217; 218; 224–225
data from spectroscopic studies, **7:**219–221
Luminol, **58:**67
chemiluminescence, mechanism, **25:** 122–131; 143–144
structure, **25:**121
Luminol reaction, platinum as catalyst, **15:**23
Luminous bacteria, **25:**124; 125; 128; 144–156, *See* Luminescence
Luminous organisms, **25:**159–162
Lunarine, **36:**228
Lupinane structure and pyridine ring, **8:**235; 236
Lupine alkaloids, origin, **8:**226–227
Lupine, transamination reactions in, **16:**190–191
Lupinus albus, **55:**5
Lutein
in chloroplast lamellae, **30:**9; 50
in photosynthesis, **25:**71
Luteinizing hormone, effect of adsorption, **28:**28
Luteinizing hormone (LH), **41:**114–115
Luteolin, **17:**434
Luteone, **55:**5
Lutoid particles, role of, required for rubber synthesis, **29:**234–235
in rubber, **29:**223; 250
Lutropin, **50:**314
Lyases, **43:**57, **56:**293
aldolase, **56:**385–386
alliinase, **56:**10; 34–38
aminolevulinate dehydratase, **56:**397–401
carbonic anhydrase, **56:**386–397
cyanoalanine synthase, **56:**10; 28–34
cystathionase, **56:**10; 38–41
cysteine (*See* Cysteine lyase)
localization of, **24:**330–333
pyridoxal-P-dependent, **56:**7–82
serine sulfhydrase, **56:**10; 18–27
tryptophanase, **56:**11; 41; 42; 43; 45
Lyases and syntheases, classification, **14:**223; 228–229; 236–238; 256–257; 258
Lycomarasmine, as wilting toxin, **11:**402; 403–416; 422–435
complex with iron, **11:**413–416; 423
inactivation of, **11:**407–408
resistance to, **11:**429–432
specificity of, **11:**422–426
symptoms of damage by, **11:**418; 424
Lycopene, **17:**435
biosynthesis, **21:**340; 341; 344; 345
degradation products, **21:**334
structure, **21:**297
Lycopersicon esculentum, **55:**73; 76; 84
Lycopersicon esculentum in alkaloid studies, **8:**219
Lycopodium, nicotine in, **8:**223
Lymphatic leukemia, ribose synthesis, **28:**414
Lymph node cells carbohydrate metabolism, **28:**77
Lymphoblastic leukemia, **39:**222
Lymphoblasts, malignant, respiration of, **29:**338
Lymphocyte, differentiation, **58:**100
Lymphocytes, **48:**66; 75; 76; 77; 78; 80; 83–84
mouse, **48:**84
mouse spleen, **48:**72
stimulation, **48:**71
Lymphocytes with adenosylmethionine decarboxylase, **56:**272
Lymphoid organs and desoxyribonucleic acid, **7:**175–176
Lymphoma, **39:**219
6C3HED, difference spectra of the cytochromes of Ehrlich ascites tumor cell and, **29:**326; 328; 329
glucose utilization of, **29:**356
Lymphomas, mouse, **47:**364
Lymphopeptidase, specificity of, **12:**207–209
Lymphosarcoma, **39:**219; 222
enzymes in, **24:**329
mouse, **29:**338
Lymphosarcoma, Gardner, ribose synthesis, **28:**414
Lyoenzymes, isolation of, **14:**378
Lyoglycogen, **3:**133
Lyotropic series, effect on protoplasm, **7:**42
Lys^{54}, eukaryotic dihydrofolate reductase, ligand binding (side chain roles), **70:**64
lysA gene, L,L-diaminopimelate decarboxylase
bacterial lysine biosynthesis, **72:**310–314
stereochemistry, **72:**281–283
lysC-encoded aspartokinase III, bacterial amino acid biosynthesis, **72:**283–285
Lysed liver mitochondria, **39:**360
Lysin, egg membrane, **9:**349
sperm, **9:**268–271

Lysine, **7:**198; 279; 280; 287; 296; 525, **34:**33; 36, **48:**157, **50:**143; 144; 152; 168; 169; 170, **51:**217, **58:**28. *See also* Amino acids; Proteins

b-Lysine, formation mechanism, **66:**16

b-Lysine, interconversion with lysine, *See* Lysine 2,3-aminomutase

b-Lysine mutase, **66:**26–27

acetate as precursor, **16:**286; 296–298

as activator of β-glucuronidase, **16:**376

biosynthesis, **16:**295–299

in histones, **16:**445

metabolism of *Neurospora,* **16:**256; 293

in myosin, **16:**323; 327

in protamines, **16:**442

in residual protein, **16:**447

in transamination reactions, **16:**190; 192; 197; 205–206; 229; 231–232

acylation of amino group, **58:**80

analogs of, in transglutaminase reactions, **54:**36; 38; 42–45

bacterial decarboxylase, **6:**4

activation of apoenzyme by pyridoxal phosphate, **6:**27

adaptive formation in *E. coli,* **6:**8

codecarboxylase of, **6:**29

effect of age of culture on growth, **6:**12

of growth temperature, **6:**11

of inhibitors on activity, **6:**18

*p*H of growth medium, **6:**10; 16

Michaelis constant, **6:**16

purified, **6:**12–21

reversible resolution, **6:**21

specificity, **6:**15

bacterial lysine biosynthesis (*See* Bacterial lysine biosynthesis)

blocking of ε-amino groups by, **29:**434

in chloroplasts, **20:**209

ionization constants, **20:**58

in phosphorylated peptides of esterases, **20:**288

thermodynamic data on ionizations, **20:**64

use of blocking reagents with, **20:**68

and citric acid cycle, **18:**257; 271–272

and CO_2 fixation, **12:**45

in decarboxylation of oxalacetate, **12:**81

column chromatography, **22:**428; 429

decarboxylation, in *Bacterium cadaveris,* **22:**109; 110

role in ribonuclease activity, **22:**71; 72

degradation to fatty acids, role of cobamide coenzyme, **26:**266

in mureins, **26:**198–199; 202–203; 210–212; 216–219

E. coli synthesis, **25:**361

and ethionine effects, **25:**372

as equivocal model, **49:**59

genetic studies, **17:**403–405

as inhibitory analogue, **6:**133

labeled, **19:**260

in papain, **19:**284; 296; 298; 300

transamination reaction of, **19:**343

metabolism, **58:**176

oxidation of, **14:**132

pepsin activity, **14:**189

peptide derivatives of, in transglutaminase reactions, **54:**36–45

effect of leucine position on, **54:**37–39; 48

in interactions with transglutaminase intermediates, **54:**45

stereochemistry of, **54:**45–50; 52

and piperidine rings, **8:**237

and pyridine rings, **8:**237

and solanaceous alkaloids, **8:**237

and RNase activity, **24:**207; 209; 220–224; 244; 245; 247; 248; 250; 451; 455

role in enzyme activation, **34:**32

structure of, **54:**42

as transglutaminase substract, **54:**51

in UDPG analog, **14:**204

in urea synthesis, **15:**(N^{15}); 324; 338; 341 (N^{15})

β-Lysine, **38:**417

^{15}N-Lysine, **38:**434

^{15}N-β-Lysine, **38:**434

L-Lysine, oxidation in plants, **20:**320; 322

DL-Lysine, resolution by papain, **19:**309

L-Lysine, transamination reactions of, **19:**368; 381

L-Lysine-adding enzyme, **26:**50

Lysine-2,3-Aminomutase, **38:**420–425

absorption spectrum, **38:**422

bound pyridoxal phosphate, **38:**422

inhibition by carbonyl reagents, **38:**424

iron content, **38:**422

molecular weight, **38:**423

oxygen lability, **38:**422

reaction, equilibrium constant, **38:**422

S-adenosylmenthionine requirement, **38:**424

sedimentation coefficient, **38:**423

subunit size, **38:**423
sulfhydryl groups per mole, **38:**423
Lysine-2,3-aminomutase, **42:**279
Lysine 2,3-aminomutase, **66:**1–36. *See also* Adenosylcobalamin
activation, **66:**5–6
steps, **66:**19–20
cobalt, **66:**17–26
cofactor requirements, **66:**2
EPR spectrum, **66:**9–11
high-spin cobalt(II), **66:**17–18
hydrogen stereochemistry conservation, **66:**6
iron-sulfur clusters, **66:**17–26
mechanistic significance, **66:**2
metabolic significance, **66:**3
molecular properties, **66:**4–5
peptide mapping, **66:**4–5
proposed mechanism, **66:**19; 21
purification, **66:**3–4
pyridoxal phosphate and, **66:**13–17
radical intermediates, **66:**9–13
structure, **66:**13
Lysine analog, **57:**35
Lysine auxotrophs, of *E. coli,* **16:**297
of *Neurospora,* **16:**295–297
Lysine biosynthesis, **46:**83
allosteric regulation, **28:**45; 53; 60–62; 70–72
Lysine decarboxylase
adsorption studies, **14:**326
sequence of active site, **36:**213
L-Lysine decarboxylase, adsorption studies, **14:**346
L-lysine decarboxylase, from *Bacterium cadaveris,* **36:**220
Lysine fermentation, cofactor requirements, **38:**415–416
Lysine in actin, **13:**152
β-Lysine-α-ketoglutarate transaminase, **38:**438
Lysine Metabolism by Clostridia (Stadtman), **38:**413
D-Lysine mutase, **66:**26–27
D-α-Lysine mutase, **38:**430–433, **43:**7
acetylation of, **38:**431
ATP activation, **38:**432
ATP phosphonic acid analogs, **38:**432
complex of two proteins, **38:**431
diamino acid inhibition, **38:**433
fold purification, **38:**431
magnesium, ion requirement, **38:**432
monovalent cation requirement, **38:**432
NH_2OH, effect of, **38:**432
nitration of, **38:**431
relationship to -lysine mutase, **38:**433
β-Lysine mutase, **38:**426
cobamide protein moiety, **38:**429
cofactor requirements, **38:**428–430
protein components, **38:**428–430
pyridoxal phosphate requirement, **38:**428
pyruvate effects, **38:**430
spectrophotometric assay of, **38:**429
sulfhydryl protein component of, **38:**428
L-β-Lysine presence in basic antibiotics, **38:**425
roseothricin, **38:**425
streptolin, **38:**425
streptothricin, **38:**425
viomycin, **38:**425
Lysine residue
active site of enzyme, **28:**451–452
at the amino side of the splitting point, **41:**185
Lysine residues, **39:**347
Lysine-rich histone, from DNA fractionation, **18:**169; 170; 172; 175; 177
Lysines, **2:**22–24, **57:**18
Lysins, adsorption studies, **14:**364
Lysis
and calcium, **8:**94–95
effect of various ions, **8:**94–95
of red cells, formula for lytic curve, **8:**89–91
of cell walls, **28:**253; 348–349; 355–356
of fermentation organisms, **7:**564
Lysis of bacteria, agents for, **26:**213–214; 216
Lysis of cell through bacterial virus, **2:**8–9; 22–24
Lysis of platelets, by crude thrombin preparations, **16:**69
Lysolecithin, **34:**475, **55:**165
effect on mitochondrial oxidative phosphorylation, **21:**32
phenylalanine hydroxylase activation, **67:**103–106; 158–159
Lysolecithin and hemolysis, **8:**460; 487; 490
Lysolecithis, PAH and, **70:**122
Lysophosphatidate acyltransferase, **36:**302
Lysoplasmalogen inhibitory effect in sperm metabolism, **24:**59–60
Lysosomal catabolism, of asialocerulo-plasmin, **41:**109

Lysosomal enzymes
biosynthesis, **60:**219
carbohydrate composition, **60:**90
in fibroblasts, **60:**221
molecular heterogeneity, **60:**221
and nuclear purity, **27:**251–253
processing, **60:**219
synthesis, **60:**90
transport to lysosomes, **60:**221
Lysosomal esterase, **24:**317
Lysosomal fractions, effect of fructose 1,6-bisphosphatase, **42:**218
Lysosomal hydrolases, **60:**91
acid, localization of, **24:**315
localization of, **24:**323
Lysosomes, **38:**307, **42:**215, **58:**114, **60:**90
cholesterol ester hydrolases, **69:**45
enzymes in, **21:**39–41
Lysosulfatide, **60:**176
Lysozyme, **15:**274; 276, **17:**75, **18:**248, **33:**261, **34:**457; 485, **37:**412; 419; 421; 465; 471; 520; 525; 551, **41:**116, **43:**242; 250, **46:**478–479, **47:**47; 72; 124; 133, **48:**32; 43; 45; 47; 50, **50:**27; 34; 37, **52:**184; 185; 210; 214; 216; 220, **53:**262, **60:**7
action of tyrosinase on, **14:**147
classification, **14:**246
isolation, **14:**389; 403
separation by chromatography, **14:**331; 339; 341; 342; 346; 351; 353–354; 367
structure studies, **14:**274; 283; 285; 286; 287; 289; 292; 294; 296; 301; 305
use in isolation of enzymes, **14:**379
activator of glycosyl transferases, **40:**58
adsorption at interfaces, **28:**3; 20
in preparation of β-lactamase, **28:**276
in bacterial cell disruption, **33:**350
binding of substrate, **50:**34
chromatographic evaluation of purity, **19:**281
colorimetric substrates, **37:**9
column chromatography, **22:**424; 430; 475; 477
effect of pH on structure, **22:**426
effect of
chemical modification, **24:**122–123
ionizing radiations, **24:**371; 372; 376; 378; 382; 389; 395–396; 403
heat denaturation of, **23:**56
heterogeneity of, **24:**171
hydrogen bonds, **50:**27
inhibition, **13:**216
inhibition of, **36:**176
invariant residues of, **43:**432
lack of –SH groups, **11:**242
mechanism, **37:**3; 7; 51
intermediates, **37:**11
pH dependence, **37:**10
secondary isotope effect, **37:**11
model systems, **37:**12 51
multiple forms, **25:**334; 338–339; 345
proton exchange in, **50:**37
secondary interactions of, **33:**436
substrate specificity, **37:**5
T4 phage, **47:**94–97; 135
Lysozyme Action, Physical Organic Models (Dunn and Bruice), **37:**1
Lysozyme imidazole identification in, **20:**72; 94; 95
Lysozyme methyl ester, activity of, **24:**122–123
Lysozyme (muramidase), localization of, **24:**322
Lysozyme resistance, of bacteria, **26:**223–224; 227
Lysozymes
degradation of murein, **26:**200–212; 227
disulfide bridges, **26:**102
egg white, and mureinase specificity, **26:**213
of phage T4, synthesis of, **27:**317
sequence analysis, **27:**230
Lys-161 residue, dihydrodipicolinate synthase (DHDPS), **72:**289–290
Lys-99 residue, γ-glutamyl transpeptidase substrate binding, **72:**252–254
Lys-69 residue, L,L-diaminopimelate decarboxylase, **72:**312–314
ε-Lysylamine, in melanin formation, **16:**172a
ε-Lysyl amino groups, role in combination of quinones with proteins, **16:**124
role in sclerotization, **16:**160–161
Lysyl carbamate, **67:**41
Lysyl groups, modification, **53:**70
on fructose 1,6-bisphosphatase, **53:**70
Lysyl hydroxylase, **41:**270; 274; 275
catalytic reactions, **72:**326–327
cofactor requirements, **41:**275
co-substrates and reaction mechanisms, **72:**345–357
gene mutations for Ehlers-Danols syndrome, **72:**383–384
inhibition and inactivation, **72:**357–364

intracellular hydroxylation sites, **72:**338–340
molecular properties, **72:**336–338
peptide substrates, **72:**339; 343–345
regulation mechanisms, **72:**385–386
substrates, **41:**275
subunit structure, **72:**380–382
Lysyl oxidase (human placental), **59:**193
Lysyl phosphatidyl glycerol, **34:**466
Lysyl-tRNA phosphatidylglycerol transferase, acceptor substrate specificity, **40:**117
donor substrate specificity, **40:**114; 115
inhibitors, **40:**113
phosphatidyl-(2′-deoxy) glycerol as acceptor, **40:**117
phosphatidylglycerol as acceptor, **40:**117
requirements, **40:**113
site of lysine transfer to phosphatidylglycerol, **40:**117
Lysyl-tRNA synthetase, **37:**476; 494; 520; 581
inhibition by polyuridylic acid, **40:**154
Lytic activity, effect of fluoride on, **8:**95
effect of hexametaphosphate on, **8:**95
Lytic activity of thrombin on fibrin, **16:**42–43
Lytic enzymes, effect on cell walls, **26:** 213–214
α-Lytic protease, **46:**481, **53:**261

M

M. glutamicus, allosteric enzymes, **28:**73
M. *tuberculosis,* DNA components, **16:**433
M. vinacea, multimolecular forms, **36:**97
Maackiain, **55:**5
biosynthesis of, **55:**8
formation of, **55:**39; 40
in isoflavan conversion, **55:**48
structure of, **55:**7
M-AchE, **34:**526
Macrodextrin, from glycogen by α-amylase, **34:**360
Macroenvironmental parameter, **34:**446
Macroglobins, changes in schizophrenics, **29:**501–502
α_2 Macroglobulin, **48:**291
Macroglycolipids, **52:**117; 121
Macromolecular chaperones, **67:**275
Macromolecular evolution, theories of, **27:**396–397; 405–410
Macromolecular interaction, **42:**242
Macromolecular nature of catalysis, **25:**20–21
Macromolecular probes, **34:**481
structure, **34:**446
Macromolecule formation in polysaccharide synthesis, **11:**297–337
Macromolecules, **57:**158–159. *See also* Electrophoresis, moving-boundary; Interactions; Polymerization; Transport
bagshaped, and cell walls, **26:**193–232
effect of ionizing radiation on, **24:**360
interactions with each other, **30:**142–149
interactions with small uncharged molecules, **30:**149–175
reversible complex formation between, **30:**145
showing two peaks in electrophoresis, **30:**163
theory of transport of interacting systems of, **30:**139–177
Macromolecules, Affinity Chromatography of (Cuatrecasas), **36:**29
Macromolecules, Bagshaped (Wiedel and Pelzer), **26:**193
Macromolecules and energy transport in living cells, **14:**5–6; 19–42
Macrophage
human, nitric oxide synthases expression, **65:**292–293
inducible nitric oxide synthases, **65:**298–301; 311
Macrophage migration inhibitory factors (MIF), glutathione S-transferases, **69:**4
Macrophages, **46:**62, **58:**100, **59:**34, **60:**226
Magnesium, **34:**485; 490, **46:**37; 70; 96; 100
activation of alkaline phosphatase, **9:**590
effect on activation of egg, **9:**290–291
and fertilizability, **9:**248
antagonism with Ca, **25:**18–20
and bioluminescence, **25:**131–134
and product inhibition, **25:**231–232
in bacteriochlorophylls, **46:**38
and cell structure, **8:**7–8; 19–20; 45
effect on hemolytic activity of complement, **8:**93–95
and fat formation, **8:**309; 315
and ketone body formation, **8:**401
and nucleic acids, **8:**17; 19; 20; 50
and phosphatides, **8:**24; 50
chlorophyll intermediate containing, **19:**58
and glutamine synthetase activity, **19:**255
and photosynthetic enzymes, **19:**8; 21; 22; 25; 37; 41

Magnesium (*Continued*)
in transamination reactions, **19:**354; 374
in chlorophylls, **46:**38
cytochrome *c* oxidase, **71:**101; 102; 103
effect of on formation of peptide, **33:**345
ions of, **33:**369
effect on ATPase, **6:**165; 185
effect on pyruvate cleavage, **6:**239
in familial periodic paralysis, **6:**217
in muscular dystrophy, **6:**205; 222
in myasthenia gravis, **6:**212
effect on respiratory chain, **17:**116–118
in phosphate transfer from ATP, **17:**176–177; 180
formation of porphyrin chelates, **46:**82; 83
and growth of luminous cultures, **7:**253
and respiration, **7:**254
leucine aminopeptidase binding, **68:**163–164; 167
in nucleus, **27:**256
and phosphoglucomutase activity, **14:**212, **20:**213
binding by protein groups, **20:**56
in DPNH-cytochrome b_5 reductase, **20:**169; 171
effect on protein synthesis in plants, **20:**223
as inorganic complement, **14:**242
in photosynthetic phosphorylation, **20:**228
phosphomannomutase/phosphoglucomutase (PMM/PGM), **70:**229–230
requirement for hypoxanthine-guanine phospho ribosyl transferase, **41:**15
in tetrapyrroles, **46:**81; 82
Magnesium-ATP, substrate for PP-ribose-P synthetase, **41:**4
Magnesium ATP coordination isomers, **45:**285
Magnesium chelatase, **46:**90
activity, **46:**69
Magnesium chloride, effect on actomyosin dissociation, **16:**345
effect on myosin and its subunits, **16:**332; 333; 344
effect on relaxation of muscle, **16:**348–350; 352
Magnesium chlorins, **46:**67
Magnesium-2,4-divinylpheoporphyrin a_5, **46:**72
monomethyl ester, **46:**70; 71; 78; 79
Magnesium hydroxide, adsorbing agent for clotting factors, **16:**27; 28
Magnesium icons, and fluorocitrate, **18:**146–147
and formation of DNA polymer, **18:**213–215
Magnesium ion
in carboxyiation reactions, **26:**286; 289; 291; 296; 300; 305–306; 311–313; 315–316; 320–324; 326–327; 331–334; 337
and CO_2 fixation, **12:**51; 67; 80–81; 87
and choline acetylase activity, **12:**318–322
and peptidase action, **12:**195; 196; 198; 205; 220; 222; 240–241
role in tetrapyrrolic compounds, **12:**18 20
complexed with ADP, **54:**373
complexing with ATP, **16:**333
effect on actin polymerization, **16:**334–335
effect on myosin ATPase, **16:**336–337; 338; 340
in phosphorylation of pyridoxal, **16:**219
in creatine kinase activation, **54:**403–404; 405; 406
effect on actin, **13:**152–153
effect on actomyosin and ATP interaction in gels, **13:**175–176; 178
on actomyosin and ATP interaction in sols, **13:**171–173
on hyaluronidase inhibition, **13:**217
on ITPase, **13:**168
on myosin activity, **13:**164–166
on sarcoplasm ATPase, **13:**167–168
in glutamine synthesis, **13:**277; 279; 290–291
in glutathione synthesis, **13:**281
and myofibrillar structure, **13:**186–187
myosin binding of, **13:**157
in transferase system, **13:**300
for fructose-bisphosphatase activation, **54:**127; 156; 176
in fructose-bisphosphatase regulation, **54:**178
in fructose-bisphosphate activation, **54:**168
and phosphoglucomutase activity, **24:**467
effect on, ribosomes, **24:**7; 18
sperm, **24:**80
Magnesium ions
in acetate metabolism, **23:**407; 410; 412
and ATP synthesis, **23:**325–326; 332; 337–341; 359–363; 365; 368; 369; 372; 381–382; 385; 386
effect on, collagenase, **23:**195–196
elastases, **23:**234
in glycolate metabolism, **23:**436; 442–443
and oxalate metabolism, **23:**447
and swelling, **23:**385
in allosteric regulation, **28:**133

in carboxylase reaction, **28:**89
in DR metabolism, **28:**435–436; 439–440; 442; 444; 448
with glycogen synthetase, **28:**99
in isocitrate dehydrogenase action, **28:** 136–138
in phosphorylase conversion, **28:**107
in APS phosphokinase reaction, **22:**209
in arylamine sulfokinase system, **22:**213
in ATP sulfurylase system, **22:**208
in biosynthesis of *S*-adenosylmethionine, **22:**244
chelate formation with amino acids, peptides and organic phosphates, **21:**145–148; 155; 239
effect on ATPase activity, **21:**39
enzyme activation by, **21:**148; 152; 153; 155; 156; 158–160; 170
kinetics, **21:**136; 139–141
in formate activation, **21:**407; 412
in methionine synthesis, **21:**424
in mitochondria, **21:**8; 10; 100
effect of Versene on, **21:**5
in mitochondrial fragments, **21:**100
in mitochondrial reactions, **21:**24; 46
model catalyses with, **21:**175; 177
properties, **21:**157
in purine biosynthesis, **21:**204; 208; 219; 234; 241; 242; 250
in pyrimidine biosynthesis, **21:**272; 289; 421; 422
effect on alkaline phosphatase, **10:**89–93; 98–100; 102–104; 106–107; 110–111; 113
in citric acid cycle, **10:**189; 201; 208; 215
and enzyme activity, **11:**241–242
effect on membrane permeability, **11:**71
effect on myosin, **11:**230
and rennin action, **15:**433
in citric acid cycle, **15:**201; 204; 207; 213; 214; 217; 223; 225; 229; 230–231; 235; 237; 241; 249; 251; 252; 258
effect on enzymes, **15:**39–40; 154; 156; 168; 170; 172
in phosphorylations, **15:**145–147; 151–152
in pyridine nucleotide metabolism, **15:**104; 105; 111; 117; 130
and urea synthesis, **15:**327–328; 330–332; 334–335; 344–345; 365–367; 373; 379–382
role of, in glutamine synthetase, **31:**188
Magnesium mercaptide, **11:**216
Magnesium porphyrins, **46:**82
Magnesium protoporphyrin, **46:**37; 69; 70; 89
Magnesium protoporphyrin-9, **46:**100–101
Magnesium protoporphyrin methyl transferase, **46:**89
Magnesium protoporphyrin monomethyl ester, **46:**70–71
Magnesium salts, in solubility studies, **17:**138; 142; 144–149; 155
Magnesium sulfate, in biosynthesis of glutamic acid in plants, **20:**325; 326
Magnesium tetrapyrroles, **7:**307; 308; 320–321. *See also* Chlorophylls and Tetrapyrroles
Magnesium-2-vinyl-protoporphyrin 9 monomethyl ester, **46:**72
Magnetic circular dichroism (MCD) spectroscopy, cytochrome *c* oxidase, **71:**87
Magnetic measurements of hemoglobin, **7:**328
Magnetic moment of hemoproteins, **7:**269; 275–278
Magnetic properties, of catalase and peroxidase, **12:**157–159
Magnetic properties of electrons, *See* EPR spectroscopy
Magnetic susceptibility, of complex ES, **33:** 331–332
of cytochrome *c* peroxidase, **33:**319
and effective Bohr magneton numbers, **33:**320
of hemoproteins, **33:**319–325
Magnetic susceptibility studies of ferrous hemoglobin, **7:**312
Magnetometric measurements of hemoproteins, **7:**269; 275–278; 283; 291–293; 297–298
Mahler enzyme, EPR study, **27:**177; 179
Maize
genetic studies of, **13:**138
methyl transferase from, **46:**70
Maize endosperm, molecular weight of enzymes from, **46:**365
mutants of, **46:**319
Malaria, **46:**140
Malate, **50:**149, **51:**236
and acetate metabolism, **23:**405–415; 419–420; 425; 428–430
and glycolate metabolism, **23:**437–441; 446
and CO_2 fixation, **12:**46; 50–51; 53; 61–63; 63–67; 83; 85; 132

Malate (*Continued*)
and kynurenine 3-hydroxylate, **19:**194
and urea synthesis, **15:**327; 330–332; 337–339; 343–345; 359; 362; 373
in citric acid cycle, **15:**185; 194–200; 259–262
oxidation, **15:**15
L-Malate, in plants, **19:**12; 13; 33–36; 39; 52; 53
Malate-aspartate cycle, to oxidize DPNH by mitochondria, **29:**342
Malate dehydrogenase, **37:**454; 455; 494; 520; 551; 581, **46:**244; 270, **54:**172, **55:**219, **58:**239
activity of –SH group, **11:**239; 241
allosteric regulation, **28:**88
decarboxylating, **37:**582
nuclear-cytoplasmic ratio, **27:**262; 263
Malate dehydrogenase (decarboxylating), **53:**221
Malate dehydrogenase (decarboxylating), NADP-linked, location of, **24:**297; 337
Malate dehydrogenase (NAD-dependent, location of, **24:**296; 337; 341; 343
Malate dehydrogenases (MDH), multiple forms, **25:**341–343; 346–347; 354
Malate formation, enzymic mechanism, **23:** 409–413
Malate oxidation, and phosphorylation, **18:**311
Malates, **17:**144; 149
Malate synthase, **46:**464, **55:**208
Malate synthetase, and acetate metabolism, **23:**411–413; 415; 419–420; 425; 430
and glycolate metabolism, **23:**437–440; 445
Malate thiokinase, **58:**159
Malate transport, **52:**296–298
Maleate, **15:**261, **51:**236
γ-glutamyl transpeptidase activity, **72:**250
Maleate isomerase, and L-aspartase, coupling of, **74:**299
Maleate isomerization, **46:**403
Maleic acid, oxidation, in mitochondria, **21:**20; 121; 122
Maleic hydrazide
effect on carotenogenesis, **21:**319
and indolylacetic acid oxidase, **19:**127
Maleylacetoacetate, **59:**137
4-Maleylacetoacetic acid, in phenylalanine metabolism, **19:**89
Maleylacetonacetate cis-trans isomerase, **59:**136–139
Maleylacetone pseudoacid, **59:**137
Maleyl isomerases
and cis-trans isomerizations, **59:**136–139
description of, **59:**136–139
and double cis-trans isomerization, **59:**142
GS group rotation, **59:**141–143
maleylacetone moiety rotation, **59:**139
mechanisms of action for, possible, **59:**140
Maleylpyruvate, **59:**137
Maleylpyruvate *cis-trans*-isomerase, **59:**136
Malformin, **33:**368
Malic acid
changes during fermentation of tobacco, **10:**340; 364–366; 396; 402
decarboxylation, **15:**156
in plants, effect of light, **15:**162
from phenylalanine and tyrosine, **14:**104
racemization, **14:**197
in plant respiration, **18:**283; 291; 297–303; 309–310
L-Malic acid, **15:**117; 122; 124–125
Malic acid carbohydrate oxidation, **6:**359; 360
Malic acid dehydrogenase, classification, **14:**253
Malic acid in mold metabolism, **9:**540; 546; 547–550; 558–560; 567
Malic dehydrogenase, **4:**237, **9:**412; 569, **15:**85; 117; 122; 125; 172; 194–200; 331; 338, **17:**377, **18:**118; 129; 273; 298; 301, **22:**370
correlation of hepatoma growth rate with, **29:**331
and glycolate metabolism, **23:**439; 444
kinetic mechanism of, **45:**316
and malic enzyme, **12:**61; 65; 67
in mitochondria, **21:**74
and phytochrome action, **26:**27
PN-dependent, **26:**118; 156; 158
and propionic acid metabolism, **26:**294; 306; 310; 329; 350; 351
product inhibition of, **25:**224; 264
in sperm, **24:**87
Malic dehydrogenases
in *Chlorenchyma,* **20:**219
in *Escherichia coli,* **19:**40
in photosynthesis, **19:**13; 14; 39; 52; 53
Malic enzyme, **15:**97; 117; 122; 126; 172–173; 246; 345, **45:**282, **46:**244; 449; 477, **53:**222
adaptive, classification, **14:**254
classification, **14:**238; 243; 254

isolation, **14:**388
chemical mechanism of, **45:**374
and CO_2 fixation, **12:**52; 53; 54–55; 61–63; 63–67; 72; 82; 85; 90
and chloroplast system, **12:**39
correlation of hepatoma growth rate with, **29:**331
kinetic mechanism of, **45:**282; 290; 294; 295; 316; 317; 319
and propionic acid metabolism, **26:**294–295
Malic enzyme (NADP-linked dehydrogenase), location of, **24:**297
Malic enzymes, **18:**267
effect of fluorocitric acid on, **18:**128
in *Escherichia coli,* **19:**40
in photosynthesis, **19:**33–36; 53
L-Malic monomethyl ester, **15:**261
Malic r DPN transhydrogenases, classification, **14:**253
Malignancy, hyaluronidase action in, **13:** 217–218; 230
Malignant cells. *See also* Cancer, Carcinoma, Hepatoma, and Tumor
ethionine effect, **25:**373–375
Malignant nuclei, **18:**175–176
Malignant tissue, *See* Tumors
histochemical studies, **9:**593
Malignolipin, **36:**227
Malonate, **15:**3; 238; 261; 279; 327; 335–337, **51:**236
binding to aspartate transcarbamylase, **68:**82; 84; 88–90
and DPNH-cytochrome *c* reductase activity, **20:**170
effect on inactivation of succinic dehydrogenase, **20:**179
incorporation into rubber, **29:**230–232
as precursor of rubber, **29:**232
inhibition of degradation of acetic acid in yeast, **10:**174–176
inhibition of tissue respiration, **10:**168
inhibition of succinoxidase, **6:**43; 206
inhibitor of respiratory chain, **28:**190; 199–200; 225
inhibitory role in glycolysis, **5:**135
kinetics of succinic dihydrogenase action, **18:**91–92
Malonate and phosphorus turnover, **7:**124
Malonate competition, for succinic dehydrogenase, **18:**93–94
Malonates, **17:**149
Malonic acid, effect on Krebs cycle, **19:**360; 365
Malonic acid diamide, deamination, by metal ions, **21:**179
Malonic acid in mold metabolism, **9:**550–554
Malonic semialdehyde, in transaminations, **16:**194
Malony CoA, **35:**365
Malonyl-CoA, in rubber biosynthesis, **29:**230
Malonyl CoA-ACP transacylase, **36:**294–295
Malonyl CoA-CO_2 exchange reaction, **36:**275
Malonyl CoA decarboxylation, **35:**358
S-Malonyl coenzyme A, **59:**125
Malonyl dialdehyde, **58:**239
Malt, laminarinase, kinetics, **28:**346; 347
Malt and wheat amylase activity, **6:**460–463
Maltase, **5:**93; 97–103; 119; 120, **8:**271, **12:**409
cacao, **25:**466
multiple forms, **25:**349–350
intestinal, imidazole in, **20:**94
specificity, **20:**142
low vertebrata, **1:**381; 391; 392; 394; 408; 411; 412
soil, **28:**386
Maltase in tobacco, **6:**368
Maltases, **17:**190; 213–214; 221; 227
classification, **14:**245
Maltobionic acid, **17:**210
Maltodextrine-phosphorylase, properties of, **30:**118; 123
reaction catalyzed by, **30:**118
Maltodextrin phosphorylase, *E. coli,* **34:**388
Maltodextrin series, glycosyl transfers among, **17:**214–217
Maltodextrinyl transfers, oligosaccharide formation by, **17:**210–211
Maltol rearrangement of streptomycin, **10:** 156–157
Maltopentaose, in glycosyl transfers, **17:** 214–217
Maltose, **5:**97–103, **18:**290, **56:**144; 192
from amylopectin, **12:**388–389
from amylose, **12:**386–387
in synthesis of polysaccharides, **12:** 419–421
enzymes and their structural determinants, **30:**117–119
location of markers on the bacterial chromosome, **30:**86

Maltose (*Continued*)
system of utilization of, **30:**117–125
enzymic synthesis, **10:**141
mechanism of utilization by *Escherichia coli,* **10:**124
fermentation by thiaminase-producing bacteria, **15:**413
in polysaccharide synthesis, **15:**306–307; 308; 310; 311–313; 315
fermentation of, **5:**93; 114; 115
in β-glucan synthesizing system, **28:**339
in glucose transfer, **17:**206; 208–209; 211–217; 220; 223; 224; 227
in polysaccharide synthesis, **11:**298–299; 320–322; 322–324; 327–329; 330
in sperm metabolism, **24:**53
Maltose (α-1,4), **18:**222
β-Maltose, from action of β-amylase on amylopectin, **12:**393–394
from action of β-amylase on amylose, **12:**391–393
Maltose permease, **65:**32
Maltose-permease, properties of, **30:**118
Maltose phosphorylase, **17:**165; 223, **18:**222; 224
Maltose r linear dextrin transglucosidase, **14:**251
Maltose value of wheat flour, **6:**429–441
O-α-Maltosyl-(1→2)-β-D-fructofuranoside, **17:**206
Maltosyl sucrose, **15:**310
Maltotetraose, **17:**213; 214
Maltotetraose (C^{14}), from sugar nucleotides, **26:**58
Maltotriaose, in glycosyl transfers, **17:**213–217; 227
Maltotriose, **17:**209
in β-glucan synthesizing system, **28:**339
as priming agent for polysaccharide synthesis, **12:**414
from amylose, **12:**386–387
Maltotriose from maltose, **15:**307
Maltozymase
enzymic adaptation, **13:**81
induction in yeast, **22:**106; 107
Malts, **7:**590
method of evaluation, **7:**589
Maltulose, **17:**220
Malyl-coenzyme A, **23:**413
Mambin, fibrinogen binding inhibition, **70:**7
Mammalian acetylcholinesterase, **43:**109; 117; 125; 137; 138; 143; 191; 201; 203
bovine erthrocyte enzyme, **43:**149; 166; 194
Mammalian cells, P450 expression, **68:**42–50
COS cells, **68:**42–49
stable transformants, **68:**50
vaccinia virus, **68:**49–50
Mammalian dihydrofolate reductase, **70:**25. *See also* Eukaryotic dihydrofolate reductase
Mammalian folate transport, *See* Eukaryotic folate transport
Mammalian hexokinase, **39:**279–316
activity assays of, **39:**251
adenylate energy charge response and, **39:**310–311
ADP inhibition of, **39:**290–293
allosteric effects of ADP on, **39:**291–293
competitive inhibition studies of, **39:**293–294
diabetes and insulin effects on, **39:**314–316
distribution of, and, **39:**299
latency, **39:**299
synaptosomal mitochondria, **39:**299
eclectrophoresis of, **39:**281
effectors of soluble-particulate distribution of, **39:**296–303
ATP, **39:**296; 297
glucose-6-P, **39:**296; 297
Mg^2+, **39:**298; 300
NaCl, **39:**298
phosphate, **39:**297; 298; 301; 302
sucrose, **39:**298
equilibrium constant of, **39:**251
and hexokinase and glucokinase levels, **39:**313
initial rate studies of, **39:**284–285
isozyme sequence homology, **39:**284
isozymes of, differential kinetic properties, **39:**285
tissue distribution, **39:**281–282
Mg^2+-related effects on, **39:**309–312
molecular weight of, stability of, **39:**283
preparation of, **39:**280–281
product inhibition of, **39:**303–309
role in regulation, **39:**304–305
product inhibition studies of, **39:**288–293
glucose-6-*P* inhibition and, **39:**288–290
proteolytic attack of, **39:**283
soluble-particulate distribution of, **39:** 295–303
effect of detergents on, **39:**280; 299–300

in vivo state of, **39:**303
substrate specificity of, for phosphoryl donors, **39:**288
isozymic forms, **39:**285
studies of, **39:**285
sugar conformation and, **39:**286–288
Mammalian peroxisomes, **57:**87–89
Mammalian phosphofructokinase, **48:**201–214
Mammalian polynucleotide kinase, **48:**267–273
Mammalian systems, mechanism of enzyme induction in, **29:**375–379
Mammalian tissues, rhodanese activity in, **39:**336
Mammals
β-glucuronidase of, **16:**362–377
methylated lysine and arginine derivatives in, **42:**235; 236
Mammary gland, phosphoenolpyruvate carboxykinase (GTP), **69:**222–223
Mammary glands
effect of fluoroacetate on, **18:**133
and ketone body formation, **18:**146
galactose-1-phosphate uridyl transferase in, **20:**119
Mammary tumor, enzymes in, **24:**313
Man, congenital abnormalities in, **22:**3; 4
hemoglobin formation and interlocus complementation in, **22:**37; 39
hybrid haptoglobins in, **22:**36
serum cholinesterase abnormalities in, **22:** 15–17
and altered substrate affinity, **22:**19
Mandelate dehydrogenase, **50:**328
L-Mandelate dehydrogenase, **50:**333
Mandelate group, metabolic pathway, **50:**325
Mandelate metabolism, **50:**327
Mandelate racemase, **44:**91, **50:**325–358, **67:**54
absolute divalent metal ion requirement, **50:**337–340
amino acid composition, **50:**336–337
assays, **50:**332–334
biological occurrence, **50:**331
^{13}C-NMR studies, **44:**93
EPR, **44:**93
genetics, **50:**331–332
hydrogen exchange, **44:**92
kinetic scheme, **50:**341–342
Mn^{2+} binding, **44:**93
molecular weight determination, **50:**334–336
DL-α-phenylglycidate inhibition, **44:**93
properties, **44:**92
of *Pseudomonas putida,* **44:**92
rate constants, **50:**341–342
regulation, **50:**329–331
specificity, **44:**92
Mandelate Racemase (Kenyon and Hegeman), **50:**325
Mandelic acid, **15:**184–185
Mandelic acid esters, hydrolysis, by metal ions, **21:**180
D,L-Mandelic acid, oxidation, **50:**326
Mandelic acids, **17:**153
l-Mandelic nitrile, formation, **20:**5–17
d-Mandelic nitrile, kinetic studies of biosynthesis, **20:**5–17
Manduca, **59:**86–87; 91; 93
Manduca sexta, **45:**395; 418
Maneb, compatibility with antibiotics, **20:**386
Manganese, **41:**70
activation of acid phosphatase, **9:**591
in plant metabolism, **9:**396; 398
and tricarboxylic acid cycle, **9:**573
and carbon dioxide fixation, **12:**61; 71; 80–81; 87–88; 89–90; 99
effect on transmethylation, **12:**468–469
-protein complex, **12:**198–200; 205
and peptidase action, **12:**195; 196–197; 198–200; 201–203; 205; 209; 212; 216–218; 220; 222–224
and carboxylations, **26:**326
and fatty acid synthesis, **26:**296; 298; 300; 301
in *Chlorella* cells, ESR spectrum of, **30:**25
in chloroplasts, **30:**47
in photosynthesis, **30:**24
effect on ATPase, **6:**164; 165
effect on pyruvate cleavage, **6:**239
effect on peroxidase, **19:**143
and oxygen evolution in photosynthesis, **19:**57; 58
in oxygen metabolism, **19:**118; 124; 125; 127; 190; 198; 199
and photosynthetic enzymes, **19:**21; 25; 33–37; 39; 40
effect on respiratory chain, **17:**117; 118
in metalloflavoproteins, **17:**249; 252; 268
in phosphate transfer from ATP, **17:** 176–177
effect on superoxide dismutase biosynthesis, **58:**83
and esterase activity, **20:**90

Manganese (*Continued*)
binding by protein groups, **20:**97
in leucine aminopeptidase, **20:**98
of microsomes, **27:**190
in photosynthesis, **25:**69; 86; 87
in plants, **18:**302
and pyridine nucleotides, **15:**105; 117
catalysis by, **15:**41; 347; 379; 380; 398; 410
in citric acid cycle, **15:**201; 204; 213; 223; 258
as cofactor for aldolase, **15:**171
in phosphogluconic acid oxidation, **15:**156
in phosphorylations, **15:**146
in thiol oxidation, **11:**203–204; 216
Manganese-ATP complex, EPR study, **27:**140
Manganese complexes, Cotton effects, **27:**93
Manganese compounds, solubilization of, **17:**138; 142; 143; 146; 147; 150; 152
Manganese dioxide, **15:**3
Manganese (II), EPR spectrum, **27:**111; 112; 140–142
mutagen, **27:**330
in peroxidase reaction, **27:**124
Manganese ion
complexed with ADP, **54:**373
in creatine kinase activation, **54:**403–404; 408–409; 412
in fructose-bisphosphatase regulation, **54:**127; 178
in fructose-bisphosphate activation, **54:**168
Manganese ions. *See also* Manganic ions and Manganous ions
complex formation, **10:**201–202; 204–208
activation of alkaline phosphatase by, **10:**99–100; 102–104; 110–111
in carboxylation of α-ketoglutaric acid, **10:**208–209; 213
effect on tryptophan metabolism, **10:**68
in mitochondria, **21:**10
in mitochondrial reactions, **21:**38; 46
model catalyses with, **21:**174–177
Manganese protoporphyrin, **60:**70
Manganese salts, and phosphoglucomutase activity, **14:**212
in enzyme isolation, **14:**398
Manganese superoxide dismutases, **41:**84
Manganic ions, model catalyses with, **21:**177
Manganous ions
in allosteric regulation, **28:**138
effect on laminarinases, **28:**347
chelate formation with amino acids, peptides and organic phosphates, **21:**145–148; 155
enzyme activation by, **21:**142; 143; 152; 153; 155; 156; 158; 159; 224
in hydroxymethyl transfer, **21:**417; 419
in mitochondrial serine biosynthesis, **21:**429
properties, **21:**157
in synthesis of active formate, **21:**411
and denaturation of lysozyme, **23:**56
effect on, collagenase, **23:**195–196
elastases, **23:**234
and exchange reactions, **23:**360; 368
and swelling, **23:**385
effect on transferases, **13:**291; 296–297; 298; 302–303
in glutathione splitting, **13:**287
Mania, treatment with lithium salts, **29:**497
Manic depression, **74:**236
Manihot utillissima, **39:**339
Mannanase, endo-α1r6-, **40:**264
exo-α-, **40:**243; 245; 247; 251; 263
Mannan-protein complex, agglutination factors, **40:**242
cell-coating, **40:**242
enzymes, **40:**242
glycosyl-serine linkages, **40:**242
immunodominant component, **40:**246; 250
phosphodiester linkages, **40:**242; 266
Mannan-protein-glucan complex, **40:**243
Mannan-protein subunit, **40:**243
Mannheim's emulsion process in growth of yeasts, **9:**662; 665–666
Mannich reaction, **54:**72
Mannitol, **15:**167
as fructose receptor, **17:**196
genetic studies, **17:**456
as solubilizing agent, **17:**148
terminal, in laminarin, **28:**331; 333–334; 345; 350; 352; 353
Mannitol dehydrogenase, PN-dependent, **26:**118
^{14}C-Mannitol, in laminarin synthesis, **28:**338
Mannitol-6-P epoxide, **43:**512
Mannitol-1-phosphate dehydrogenase, PN-dependent, **26:**118; 161
D-Mannoheptulose, **15:**161–162
D-Mannosamine, **25:**434
Mannose, **15:**145; 146; 147; 148; 161, **45:**407; 410

fermentation by thiaminase-producing bacteria, **15:**413
in β-1,3 glucans, **28:**333–334; 336
D-Mannose. *See also* Sugar nucleotides
in lipopolysaccharide from *Salmonella,* **31:**79; 92
sugar nucleotides from, **26:**44–49
D-Mannose, in polysaccharide synthesis, **15:**306
Mannose 1-phosphate, alginate biosynthesis, **70:**224; 227
Mannose 6-phosphate, alginate biosynthesis, **70:**224
Mannose-1-phosphate, relations to other hexose monophosphates, **15:**168–170
Mannose-6-phosphate, relations to other hexose monophosphates, **15:**168
Mannose-6-phosphate-glucose-6-phosphate and fructose phosphate, **14:**195; 205–206
Mannose-phosphate isomerase, **15:**168
Mannose-6-phosphate receptors, **60:**220
Mannose phosphates, isomerization, **14:**195; 205–206; 208–209
Mannose P isomerase, **43:**492
Mannose-6-P isomerase, **43:**502
D-Mannose P isomerse, **43:**492
Mannose polysaccharides, synthesis, **28:** 113–114
α-Mannosidase, **48:**39; 53; 82, **60:**150
assay, **60:**151
biological role, **60:**154
genetics, **60:**154
inhibition, **60:**153
isoelectric point, **60:**152
kinetic data, **60:**153
lysosomal, **60:**154
molecular weight, **60:**152
properties, **60:**152
purification, **60:**151
in semen, **24:**88
localization of, **24:**322
substrates, **60:**152
β-Mannosidase, **48:**53
assay, **60:**155
biological role, **60:**157
genetics, **60:**157
inhibition of, by mannono-1:4-lactone, **36:**159
by mannone-1:4-lactone, **36:**159
isoelectric point, **60:**156
kinetic data, **60:**156
molecular weight, **60:**156
properties, **60:**156
purification, **60:**155
substrate, **60:**157
β-Mannosidase, classification, **14:**246
χ-Mannosidase, inhibition of, by mannono-1:4-lactone, **36:**159
by mannono-1:5-lactone, **36:**159
α-Mannosidases, **56:**292
assay of, **56:**206
inhibitors of, **56:**209
mechanism of action of, **56:**209
occurrence and localization of, **56:**205–206
pH effect on, **56:**208–209
properties of, **56:**208–210
purification of, **56:**206–208
regulation of, **56:**210–211
role of, **56:**210
β-Mannosidases
assay of, **56:**211–212
occurrence and localization of, **56:**211
properties of, **56:**212–213
purification of, **56:**212
role of, **56:**213–214
α-Mannosidases, classification, **14:**246
Mannoside β1r2 *N*-acetylglucosaminyltransferase, **52:**27
Mannosidosis, **60:**150; 165
Mannosidostreptomycin, structure, **10:**161–162
α1r3-Mannosyltransferase, **40:**256; 261
Mannuronate residues, exopolysaccharide alginate synthesis, **70:**234
D-Mannuronic acid, and pentose synthesis, **28:**398
Manometric methods in photosynthesis, **25:** 51–53
M-antigen and hyaluronidase action, **13:**219
MAP kinases, hepatic glutamine transport, hepatocyte hydration, **72:**52–53
Maple syrup urine disease, **35:**386, **58:**174; 187; 188
Map (phi, psi), **47:**87
MAP phosphorylation, microtubule dynamics, **69:**144
Mapping of the biotin A locus, **38:**349
complementation analysis, **38:**351
genetic map for, **38:**352
intracistronic completation, **38:**351
transduction, **38:**350
Maramus, **42:**278

Marckwald's asymmetric synthesis of (−) - valeric acid, **7:**80; 99–100
Marcus theory, cytochrome *c* oxidase, enzyme turnover control and H + /e - ratio, **71:**162–163
Margarate ester
from microbial oxidation of olefins, **27:**502
from oxidation of hydrocarbons, **27:**490
Marine pseudomonad, **49:**84
Marker enzymes, and nuclear purity, **27:** 251–253
Marker molecule, **34:**485
Maroteaux-Lamy disease, **60:**257
Marrow, **46:**142
Marrow erythroblasts, **46:**103
Marsh factor, and muscle relaxation, **16:**339; 349–352
Mass action hypothesis, of enzyme induction, **22:**117; 120
Mass action relationships, statistical, **34:**446
Mass spectrometry, **57:**259, **59:**178–179
to detect O_2-radical, **41:**37
studies of CO_2 fixation in acetate, **12:**110–119
of yeast polysaccharides, **40:**247
Mass spectroscopy, of phosphate esters, **55:** 336–337
Mast cell chymase, **57:**4
Mastocytoma cells, **58:**291
Matrix, **34:**446; 463–465; 481; 491; 521–522; 527
bound enzyme system, **34:**521
collodion, **34:**454
Mäule reaction, **34:**208
Mauve spot, in schizophrenia, **29:**522
Maximal velocity, **39:**355
limiting value of, independent nature of surrogate present, **53:**324–326
Maxwell-Boltzmann distribution, **34:**460–462
M13 DNA, **67:**448
MDR, involvement in transport, **65:**38
Mean residue molecular ellipticities, **42:**102; 140
Mechanism, **39:**209; 211. *See also* Product inhibition of enzymes
of azide inhibition, **44:**213
of biopolymer growth, **30:**179–221
of chemical reactions, **3:**37–39
of clotting of milk by rennin, **15:**440–442
concerted reaction, **36:**189
of enzyme action, and extrinsic Cotton effect, **27:**37–104
of enzyme reactions, **25:**1–38
enzymic, in citric acid cycle, **15:**183–270
of enzymic hydrolysis, **15:**271–299
of enzymic oxidoreduction, **15:**1–48
exchange reaction, **36:**189
ATP ADP, **36:**189
ATP Π, **36:**189
EP π, **36:**192
^{18}O-Π succinate carboxyl, **36:**189
succinate succinyl CoA, **36:**189
of formation of dextran and levan, **30:** 179–221
formation of enzyme-bound succinyl phosphate, **36:**193
effects of desulfo CoA on, **36:**193
high energy nonphosphorylated form of the enzyme, **36:**191
of hydrogen transfers, **26:**132–156
of inactivation of enzymes by radiations, **24:**361–364
of oxidation of aromatic diamines, **44:**211
of oxidative phosphorylation, **59:**266
phosphorylated enzyme intermediate, **36:**189–191
degree of, phosphorylation *E. coli* enzyme, **36:**190
phosphorylation porcine enzyme, **36:**191
site of phosphorylation, **36:**190
of PP-ribose-P amido-transferase, **41:**9
for product inhibition of enzyme action, **25:**183–220
of prolyl hydroxylase, **41:**273
reaction
bifunctional bonding of, **30:**183
for dextran and levan formation, **30:**183
insertion-type growth with polymer transfer, **30:**187–189
insertion type of, **30:**184
monofunctional bonding of, **30:**183
stepwise propagation of, **30:**184
of reactions concerned with carbon-nitrogen bond formation, **39:**171
arguments for concerted *versus* sequential reactions, **39:**171–177
enigmatic reactions of nucleoside triphosphates, **39:**171
related to enzyme catalysis, **24:**441–482
of RNase action, **24:**239–243; 443–458
S-phosphoryl CoA, **36:**189
study of slow reactions, **27:**1–36

substrate synergism, **36:**196
of the Succinyl CoA synthetase reaction, **36:**189–197
of synthesis, of adenosine triphosphate, **23:**323–399
of translocation catalysis, **29:**41–44
Mechanism-based inhibitors, **59:**107
Mechanism of Action of Fructose, 1,6-Bisphosphate (Benkovic and deMaine), **53:**45
Mechanism of Enzyme Action, An Approach through the Study of Slow Reactions (Yagi), **27:**1
Mechanism of Enzyme Action, Extrinstic Cotton Effects and (Ulmer and Vallee), **27:**37
Mechanisms of Biopolymer Growth: The Formation of Dextran and Levan (Ebert and Schenk), **30:**179
Mechanisms of Synthesis of Adenosine Triphosphate (Racher), **23:**323
Mechanism studies, **39:**350
Mechano-chemical coupling, **51:**104
Mechanochemical systems, **57:**229–232
Meconium mucoprotein, as substrate for neuraminidase and receptor destroying enzyme, **20:**137; 144
Mediators, in oxidation-reduction systems, **11:**203; 242–244; 249
Medicago sativa, **55:**5
isoflavonoid phytoalexin in, **55:**35; 36
Medicarpin, **55:**5; 9
detoxification of, **55:**47
formononetin in, **55:**29
fungal metabolism of, **55:**45; 47–48
hydroxylation of, **55:**48
structure of, **55:**7
vestitol interconversions with, **55:**34
Megateriopeptidase, **41:**203
Meisenheimer complex, **43:**383
Meister's theory of gutamine synthesis, **72:**9–12
Melanin
histochemical methods, **9:**598
production in *Glomerella* mutants, **22:**24
Melanine. *See also* Albinism; Depigmentation; Leukoderma; Pigmentation
definition, **14:**89–90
formation, **14:**74–96; 103; 112; 133; 137; 140; 145–146; 152
Melanins, **18:**372. *See also* Color reactions and Pigments
autoxidative degradation, **16:**168–169
chordate, **16:**163–172a
paramagnetic property, **16:**172a
relation to zinc, **16:**117
Melanoblase, use of term, **14:**75
Melanochrome, in melanin formation, **16:**164; 168
Melanocytes, **16:**169
use of term, **14:**75
Melanocyte-stimulating hormone (MHS), **14:**91; 96
variation of amino acid composition, **22:**65
Melanogenesis, difference between autoxidative and enzymic, **16:**167
Melanoids formed in aging of tobacco, **10:**408
Melanoma cells, **60:**235
Melanomas
dopa in, **14:**109
melamin formation in, **14:**94
melanin granules in, **16:**169–170
tyrosinase in, **14:**79; 83; 88; 89; 133
Melanophage, use of term, **14:**74
Melanophore hormone, *See* Hormones, pituitary
Melanophore-stimulating activity, **29:**445–456
effect of blocking the amino terminal group on, **29:**451
effect of chain length on, **29:**448
enhancement of, by alkali, **29:**451
importance of side chain function for, **29:**453
stereochemical configuration and, **29:**455
Melanoproteins, **16:**170–171
Melanosis of insect blood, **16:**154
β-Melanotropin, amino acid interchanges in, **24:**417; 430
lower peptide sequences, **24:**420–425
α-Melanotropin (α-MSH), amino acid sequence of, **29:**408–410
amino acids necessary for MSH activity in, **29:**463–464
effect of blocking the amino terminal in, **29:**449
effect of chain length on, **29:**445–449
glutamine analog of, **29:**434
importance of side-chain functions of, **29:**453
influence of side-chain functions on activity of, **29:**452
melanophore-stimulating activity of, **29:** 445–456
prolongation of activity of, **29:**452
synthesis of, **29:**434

α-Melanotropin (α-MSH), amino acid sequence of (*Continued*)
fragments of, **29:**412–433
peptides related to, **29:**392
β-Melanotropin (β-MSH), amino acid sequence of, **29:**408–410
bovine, synthesis of, **29:**436
prolongation of activity of, **29:**452
synthesis of peptides related to, **29:**392
Melanotropin peptides, general strategy of the synthesis of, **29:**410–412
Melanotropins, biological properties of, **29:** 443–468
effect of NaOH on activity of, **29:**452–453
isolation and structure of, **29:**406–410
melanophore-stimulating properties of, **29:**446
methods of peptide synthesis of, **29:**393–406
structure-activity relationships of, **29:** 391–477
synthesis of, **29:**434
α-Melanotropins, lower peptide sequences in, **24:**420–425; 430
Melanotropins (and Adenocorticotropins), Structure-Activity Relationships of, Synthetic Approach (Ramachandran and Li), **29:**391
Melanuria, occurrence, **14:**89; 94
Melatonin, source of methyl, **25:**416
Melezitose, hydrolysis, **15:**310
Melibase, action on raffinose, **10:**125
Melibiase, cacao, **25:**466
Melibiose, **56:**171, **60:**96
location of markers on the bacterial chromosome, **30:**86
Melibiose in polysaccharide synthesis, **15:**306
Melibiose zymase as adaptive enzyme, **6:**78
Melilotic acid, **18:**373
Melons, antibiotic treatment of diseases, **20:**398
Melphalan, **49:**91
Members, Natural, Translocations through (Mitchell), **29:**33
Membranal changes, effect of ethionine, **25:**377–379
Membrane, **34:**478; 499; 503; 512–516; 522–525. *See also* Permeability
alkaline phosphatase, **34:**454–456; 480–481; 504–508
animal, **34:**467
artificial, **34:**447; 485
artificial enzyme, **34:**447; 495
association of glycogen with, **34:**289
asymmetric enzyme, **34:**455
axonal, **34:**493
bacterial, **34:**467
bilayer, **34:**467–468; 483–485
biological, **34:**456–457; 467; 483; 492–494
bound enzyme, **34:**446–447; 472; 478–479
effect of detergents on, **34:**300
bound nature of glucose-6 phosphatase, **34:**299
cell, permeability, **16:**472
cellophane, diffusion of protamines through, **16:**441; 444
nuclear, diffusion through, **16:**415; 416
effect on nucleus-cytoplasm relations, **16:**476; 486–488
cell envelope, **34:**456; 467; 470
cell junction, **34:**470
chloroplast, **34:**470; 492
collodion, **34:**452–455; 508
components, **34:**479–481
conductance, **34:**485
enzyme, **34:**447; 495; 500–505; 509
enzyme reaction, consecutive, **34:**471
flow pattern across, **29:**50–51
K^+/H^+ exchange in human red cells, **29:**47
lipoprotein, translocation catalysis through, **29:**39–41
mitochondrial, dicarboxylic acids and permeability to, **29:**67–68
H^+-coupled translocation of, **29:**66–68
monocarboxylic acids and permeability to, **29:**67–68
tricarboxylic acids and permeability to, **29:**68
natural, lipoprotein, **29:**42–43
translocations through, **29:**33–87
plasma, Na^+/K^+ antiporter-ATPase, **29:**69
fraction, **34:**446
functional, **34:**481
impermeable, **34:**513–514
microsomal, **34:**478
mitochondrial, **34:**481; 492
model system, **34:**484
native, **34:**472
origin of, **38:**229
papain, **34:**505
papain-collodion, **34:**454; 495–496; 509–511
peritrophic, **45:**392

phosphatase, **34:**505
photoreceptor, **34:**470
plasma, **34:**481
preparation and structure of, **34:**454
proteins, **34:**468
structure, **34:**477; 484
synthetic, **34:**454
thickness, **34:**501
three-layer enzyme, **34:**454
two-collodion, **34:**455
unit, **34:**470–471
vesicle, **34:**480
Membrane anchors, **67:**281–282
Membrane-bound proteins, histidine permease, **65:**16–17
Membrane fraction, from protoplasts, **24:**5
Membrane fragment distribution, **43:**202
Membrane function, and ATPase, **29:**505–506
Membrane-impermeant dysprosium probes, cytochrome *c* oxidase, **71:**110
Membrane-mediated information, biotransducers of, **50:**303–320
Membrane-Mediated Information, Biotransducers of (Brady and Fishman), **50:**303
Membrane nuclear, **27:**243; 248; 259–260; 263; 265; 268–271
Membrane permeability, and cacao seed death, **25:**476; 477; 486
Membrane potential, **51:**104
Membrane proteins, ion channels and, **71:** 210–212
Membrane reconstitution, **49:**75
Membranes, **36:**40; 51, **38:**201; 203, **48:**62; 63; 64; 73; 74. *See also* Surfaces
active transport across, **25:**17
bacterial, enzymic activity, **25:**436
alteration of permeability by metal ions, **11:**70–71
animal cell, **48:**67
action of glycosidases on, **48:**51; 52
bacterial cell walls, **26:**193–232
lipoproteins, in protein synthesis, **26:** 109–110
cell, role of amide as transfer agent across, **13:**268–271
mitochondrial, **13:**9
permeability, **13:**21–65
chloroplast, subunits associated with, **30:**6–7
teichoic acid in, **30:**225
thylakoid, in chloroplasts, **30:**7
damage to, **11:**406
endogenous glycosidases in, **48:**66
erythrocytes, **48:**81
aged, **48:**77
glycolipids from, **48:**55
human, **48:**56
fluidity, **48:**71
glycosidases in investigation of, **48:**65
intracellular, **48:**72
and laminarinase synthesis, **28:**343
cell, location of β-lactamase, **28:**266
cytoplasmic, attachment of γ-penicillinase, **28:**253
mitochondrial, in respiratory chain study, **28:**164
permeability, **28:**265; 267–268
of liver, mammalian, **48:**60; 64
permeability changes in, **12:**269–270; 315–317; 332–334
plasma, **48:**80
sialic acid in, **48:**72
treatment with sialidases, **48:**68
Membranes of cells, **8:**17–20
Membrane-solution interface, **34:**504–505
Membrane surfaces, disulfide exchanges, **27:**226–229
Membrane systems
active transport in, **57:**224–225
diffusional limitations of, **57:**222–224
immobilized models in, **57:**219
kinetic properties in, **57:**221–222
microenvironmental effects on, **57:** 219–221
Membrane topology, of cytochrome P-450, **60:**63
Membrane transport, **49:**41–101
Membrane Transport (Christensen), **49:**41
Membrane-transport protein human erythrocyte, **37:**465; 552
Menadione, **41:**42, *See* Vitamin K_3
and coenzyme Q, **25:**313; 322
labeled, in coenzyme Q synthesis, **25:**301
as electron acceptor, **28:**211
Menadione (2-methyl-1,4-naphthoquinone), reduction, **16:**134–135
Menadione reductase, **20:**180
distribution, **16:**134–135
location of, **24:**304; 305
Menaphthyl sulfate, **46:**394; 396

Menaquinone, **46:**2
Menten, *See* Michaelis-Menten equation
Menthol, glucuronide as substrate for β-glucuronidase, **16:**367–369; 371–372; 377; 380
Menthyl-α-D-glucuronide, effect on β-glucuronidase, **16:**371–372
Menthyl-β-D-glucuronide, effect on β-glucuronidase activity, **16:**372; 378
MEOS, *See* Microsomal ethanol-oxidizing system
Mepacrine, effect on tryptophan metabolism, **10:**59–60; 64; 69
Meperidine esterase, classification, **14:**244
Mercaptal formation, by reaction of quinones and thiols, **16:**120–121
Mercaptans. *See also* Disulfides, Sulfhydryl groups, and Thiols
  acid constant values, **27:**202
  amperometric titration of, **11:**228
  disulfide exchanges, **27:**218; 226
  effect on acid-catalyzed disulfide exchanges, **27:**205
  and enzyme activity, **27:**212–217
  equilibrium with disulfides, **27:**199; 200; 202; 207–208
  methylation of, **25:**421–422
  reaction with RSA, **27:**210; 230–231
  reduction of proteins by, **27:**210–212
Mercaptans in plant metabolism, **10:**257
Mercaptide ions, in disulfide exchanges, **27:**201–205; 207; 223; 224; 229
Mercaptides, **11:**216–218; 220; 227–228; 230; 237–238; 239–240; 247
Mercaptoacetic acid, reaction with *o*-quinones, **16:**122
Mercaptoacylamino acids, **57:**30–31
Mercaptoalkanyl amino acids, **57:**30
Mercapto-amino acids, **56:**63–67
L-Mercaptoaminoacids, deamination of, **19:**344
4-Mercaptobenzolsulfonic acid, in disulfide exchange, **27:**203
Mercaptoethanol, **39:**342
  action of ribonuclease, **27:**219
    disulfide exchanges, **27:**214; 230
    and enzyme activity, **27:**221
β-Mercaptoethanol, **54:**402
  phosphoenolpyruvate carboxykinase (GTP), **69:**208
2-Mercaptoethanol, **18:**337, **35:**251
  cytochrome, *c* oxidase, **71:**92
β-Mercaptoethylamine
  derivative of coenzyme A, **15:**187
  relation to coenzyme A, **14:**63–65
    relation to LBF, **14:**54–55; 55–57; 67
2-Mercaptoethylamine
  relation to CoA, **16:**5–7; 10; 17
  thiol esters as acylating agents, **16:**18
Mercaptoethylamine, and protein gel formation, **27:**221–222
  and vasopressin action, **27:**227
β-Mercaptoethylamine, in synthesis of activated amino acids, **19:**239
Mercaptolysis of disulfides, **27:**230–231
Mercaptopropanoyl-L-tryptophan, **57:**31
6-Mercapto purine, as a substrate for hypoxanthine-quanine phosphoribosyl transferase, **41:**14
6-Mercaptopurine conjugates, cysteine *S*-conjugate β-lyase reactions, prodrug targeting, **72:**227–230
Mercaptopyruvate, **39:**340; 363
3-Mercaptopyruvate-cyanide-sulfur transferase, **27:**212
Mercaptopyruvate sulfurtransferase, **39:**363
Mercaptopyruvate sulfur transferase, **56:**292
Mercaptopyruvate sulfurtransferase, *E. coli*, **39:**364
β-Mercaptopyruvate sulphur-transferase, localization of, **24:**315–316
β-Mercaptopyruvic acid, in transaminations, **16:**196; 233
Mercapturate, **57:**358
Mercapturate pathway, early research, **72:** 200–203
Mercapturic acid, **7:**194, **57:**358; 376
Mercapturic Acid Biosynthesis, the Role of Glutathione S-Transferases in (Boyland and Chasseaud), **32:**173
Mercapturic acids, **46:**384
Mercurial, **39:**344
Mercurials, organic, in identification of imidazole, **20:**71; 75
Mercuri-analogs, **43:**45
Mercuribenzoate, **39:**344
p-Mercuribenzoate, **58:**152
  inhibition of aldolase by, **31:**154
Mercuribenzoate derivative, eukaryotic dihydrofolate reductase, structural

Messenger-like activity, in nuclear RNA, **27:**247; 269; 275–276; 295
Messenger RNA, *See* mRNA
Messenger RNA Species, Hormonal Modulation of, in Normal and Neoplastic Rat Liver (Feigelson and Kurtz), **47:**275
Metaarginase, **15:**376
Metabolic acidosis, **53:**213; 229–230
Metabolic alkalosis, hepatic glutamine regulation, **72:**75–77
Metabolic control, patterns of, **28:**60–71
Metabolic cycle, **34:**524
  reactions, **34:**495
Metabolic generation and utilization of phosphate bond energy, **1:**99–162
Metabolic pathway, mandelate group, **50:**325
Metabolic processes, free energy of, **11:**8–9
Metabolic quotient, definition, **3:**196
Metabolic reactions, **1:**95; 154–158
  and ATP synthesis, **23:**327–387
Metabolic regulation, and Fru-2,6-P_2
  animal tissues
    adipose tissue, **59:**371
    other, **59:**371–372
    pancreatic islets, **59:**369–370
    striated muscle, **59:**370–371
  description of, **59:**357–359
  fungal spores, **59:**377–379
  liver
    and α-adrenergic agents, **59:**366–367
    and 2,5-anhydromannitol, **59:**367–368
    and anoxia, **59:**369
    and cAMP-mediated effects, **59:**363–365
    and diabetes, **59:**365–366
    and ethanol, **59:**367
    and gluconeogenic substrates, **59:**367–368
    and glucose, effect on, **59:**360–363
    and glucose balance, control of, **59:** 359–360
    and insulin, **59:**365–366
    and nutritional status, effect on, **59:** 360–363
    and obesity, **59:**365–366
    and perinatal period, **59:**369
    and vasopressin, **59:**366–367
  plants
    metabolic activities, resumption of, **59:**380
    photosynthesis, **59:**379
  yeast
    cAMP in, **59:**373–374
    and catabolite inactivation, **59:**373–374
    and glucose in *S. cerevisae,* **59:**373–374
    and regulation of Fru-2,6-P_2 concentration, **59:**374–375
    and role of Fru-2,6-P_2, **59:**375–377
Metabolic regulator in sperm cells, **24:**50; 57; 59; 73
Metabolic spaces, in the nucleus, **27:**256–268
Metabolism, **53:**215. *See also* Carbohydrate metabolism
  aerobic, **1:**139–148
  anaerobic, **1:**132–139
  aspergillus, **2:**188–201
  bacterial, and luminescence, **7:**231–233
    of glucose by yeast, **7:**15–20
    of vitamin-like substances, **7:**409–460
  biotin, role in, **3:**289
  errors of, **58:**187
  glyoxylate, **57:**148–152
  inhibitors of, **57:**163
  methanol, **59:**128 (*See also* Metabolic regulation, and Fru-2, 6–6P_2)
  L-nitrogen-13 amino acids
    and amino-oxyacetate, **57:**326–328
    and arteriovenous differences, **57:**334–336
    characteristics of, in dogs, **57:**325–326
    and L-glutamate, **57:**331–333
    myocardial accumulation, **57:**323–325
    in pancreas, **57:**328
    in salivary glands, **57:**328–329
    and tumor uptake studies, **57:**329–331
    and L-valine, **57:**333–334
  nitrogen-13 ammonia
    and blood flow, **57:**292–294
    compartmentation of, **57:**296–299
    entry of, into brain, **57:**295–296
    and hypercapnia, **57:**301
    introduction, **57:**289–290
    results of, in brain, **57:**290–292
    routes of, **57:**299–301
  nitrogen-13 nitrate
    and cancer, **57:**279
    in intestines, **57:**281–284
    and stomach, **57:**279–281
    and urinary nitrate, **57:**282
  nitrogen-13 nitrite
    and cancer, **57:**279
    in intestines, **57:**281–284
    in stomach, **57:**279–281
    and urinary nitrate, **57:**282

of nucleic acids, **16:**450–462
nuclear, **16:**469–491
purple bacteria, **1:**234–242; 264–269; 275–281; 289; 292; 301–307
relation of rate to life span, **20:**23
sulfur bacteria, **1:**236; 265; 269; 270; 274; 278; 280; 281
Metabolism, Fermentative, Nitrogen Fixation and Photosynthesis, Chemistry and Function of Ferredoxins in (Buchanan and Arnon), **33:**119
Metabolism in Disease of Muscle (Hoagland), **6:**193
Metabolism of 2-Carbon Compounds by Microorganisms (Kornberg and Elsden), **23:**401
Metabolism of Propionic Acid (Karizo and Ochoa), **26:**283
Metabolism of Spermatozoa (Salisbury and Lodge), **24:**35
Metabolism type, and coenzyme Q level, **25:**289
Metabolites, **53:**212
control of synthesis, **25:**170–172 (*See also* Product inhibition of enzymes)
conversion into inhibitory analogues, **6:** 136–137
synthesis, respiratory enzymes in, **6:**62–63
effect on sugar pyrophosphorylases, **28:** 111–114
intermediary, effect on sperm, **24:**61–66
phosphorylated, in sperm metabolism, **24:**62–63
labeled, response of chromosomes to, **16:**485
in nuclei, **27:**264; 265
Metachromatic leukodystrophy, **38:**294; 295; 298; 302; 306; 309, **60:**168
Metahemipinic, **34:**215
Metal, **41:**59–63
Co^{++}, **41:**60; 62; 63
Cu^{++}, Cu, **41:**59–63
Fe^{++}, **41:**60
Hg^{++}, **41:**60; 62
Mg^{++}, **41:**60
Ni^{++}, **41:**60
Zn^{++}, **41:**59–63
Metal activation of enzymes, elementary steps, **25:**12; 15–20
Metal binding, **44:**213
Metal-bridge complexes, simple and cyclic, **33:**30
Metal centers, nitrogenase, **67:**312–318
Metal chelates
electron-carrier function, **27:**179
in enzyme structure, **22:**73; 74
Metal chelating agents, effect on, collagenase, **23:**195–196
elastases, **23:**234
Metal chelators, **67:**216
Metal-chelator-sensitive alkaline proteinases, **41:**184; 187; 201–207; 230–232; 237; 238
from *proteus mirabilis,* **41:**184
from *pseudomonas aeruginosa,* **41:**184; 187; 201–207
from *serratia marcescens,* **41:**184; 205
from *serratia* sp., **41:**184; 204; 230
Metal-chelator-sensitive neutral proteinases, **41:**180–191; 193; 194; 197; 200–207; 214–215; 226–230; 237; 238
from *Aeromonas proteolytica,* **41:**184
from *Aspergillus oryzae,* **41:**180; 182; 184–187; 190–191; 193; 194; 227
from *Bacillus cereus,* **41:**183
from *Bacillus megaterium,* **41:**183
from *Bacillus subtilis,* **41:**183; 187
from *Bacillus thermoproteolyticus,* **41:**183
from *Clostridium histolyticum,* **41:**181; 183
from *Micrococcus caseolyticus,* **41:**183
from *Pseudomonas aeruginosa,* **41:**184; 187; 201–207
from *Streptomyces griseus,* **41:**182; 186–189; 191; 214–215
from *Streptomyces nargensis,* **41:**184
Metal clusters
MoFeP, positioning, **67:**352–353
nitrogenase, structure and function, **67:** 309–323
D_2O exchange near 4Fe-4S cluster, **67:**315–316
electron density maps, **67:**320–323
Fe-Fe distances, **67:**313
homocitrate, **67:**317
iron-sulfur protein redox potential, **67:** 310–312
metal centers, **67:**312–318
MgATP or MgADP effect on redox potential, **67:**313–314; 316

Metal clusters (*Continued*)
MoFeS cluster, **67:**322–323
P clusters, **67:**317–318
spatial distribution model, **67:**319–320
x-ray crystallography, **67:**318–323
Metal-coenzyme bond, **27:**68–70; 72
Metal-combining agents, effect on carboxypeptidase, **12:**238–242
Metal complexes
acyl transfer in, **58:**11
chelated, and tryptophan metabolism, **10:**64
steps in formation of, **25:**12; 15–20
in thiol oxidation, **11:**204–205
with lycomarasmine, **11:**413–416; 423
Metal complex formation and myosin ATPase activity, **13:**167
Metal coordination, **58:**11
Metal electrodes, reaction with thiols in e.m.f. measurements, **11:**211–212
Metal-enzyme complex, **10:**102–104; 111; 113–117; 204–208
Metal-enzyme complexes, and mechanisms of catalysis, **24:**470
Metal enzymes, nomenclature and classification, **21:**132–136
Metal-flavin interactions, **27:**183–187
Metal flavoproteins, EPR study, **27:**160–187
Metal-histidine complexes, Cotton effects, **27:**96
Metal-H_2O_2 complexes, **15:**17
Metal ion, **39:**333; 346; 357
binding capacity, **39:**348
Metal Ion Activation of Enzymes (Malmström and Rosenberg), **21:**131
Metal-ion-promoted reactions of phosphates, **24:**466–467
Metal ion replacements, **59:**114
Metal ions, **38:**202, **48:**108, **53:**55–59. *See also* the individual ions; individual ions; individual metals
adenylosuccinate synthetase (AMPSase), **73:**77–79
alkali, model catalyses with, **21:**174
alkaline earth, model catalyses with, **21:**174
effect on protein stereochemistry, **21:**158
in enzyme activation, **21:**131–167
complex-forming properties, **21:**141–151
kinetics, **21:**136–141
mechanism of action, **21:**160–163
specificity, **21:**151–160
model catalyses with, **21:**160; 161; 169–198
properties, **21:**157
rare earth, model catalyses with, **21:** 170–188; 190
physiological effects, **21:**192–194
catalysis of indole production, **10:**68–69
in cholesterol biosynthesis, **22:**296; 297
in farnesyl pyrophosphate synthesis, **22:**319
requirement, in mevalonic kinases, **22:**299
in TPP enzymes, **22:**401
divalent
effect on phosphorylase *b,* **24:**264–265; 269; 274; 277–279; 286
and lipoic acid, **18:**340
effect on activity of pectinesterase, **11:** 280–282
activity of polygalacturonase, **11:**287–288
surface reactions, **11:**67–69; 69–74
effect on β-amylase, **12:**400–401
effect on autoxidation of glutathione, **10:** 294–295
and enzyme-inhibitor complex, **10:**96–97
in hydrocarbon oxidation, **27:**471; 485; 509; 512; 515; 522; 537
inhibition of tryptophanase action, **10:**70–71
interaction with fructose 1,6-bisphosphatase, **53:**51–52; 55–58; 72
and peptidase action, **12:**193; 194–195; 198–200; 205–206; 209; 212–216; 216–218; 220; 221–225; 238–242
and thiol groups, **11:**203–206; 216–217; 227–228; 237–238; 240; 251
toxicity of, **11:**69–74
transition, EPR, **27:**107; 118–120; 137–144
Metall-Ionen Katalysierte Vorg,nge (Bamann and Trapman), **21:**169
Metalloaldolases (class II aldolase), distribution and properties of, **31:**129–130. *See also* Aldolases
mechanism of action, **31:**130–131
related enzymes of, **31:**131
Metallobiochemistry, **56:**285
Metallobiochemistry of Zinc Enzymes, The (Vallee and Galdes), **56:**283
Metalloendopeptidases, **45:**397
Metalloenzyme, alcohol dehydrogenase as, **27:**49–50
substrate binding, **27:**72

Metalloenzymes, **57:**135–137, **58:**3
advances in study of, **56:**285
entatic state of, **56:**313; 317
inhibition of, chelating agents in, **56:**300–303
metal binding sites of, **56:**306; 310; 313
spectral parameters of, **56:**306; 307
steps in reaction, **25:**15
substrate binding, **58:**12
zinc, **56:**292–293; 304
Metallo-enzymes, **24:**112–113
Metalloenzymes, Mechanisms of Two- and Four-Electron Oxidations Catalyzed by Some (Hamilton), **32:**55
Metalloflavoproteins. *See also* Flavins, Flavin coenzyme, Flavin enzymes, Flavin nucleotides, and Flavoproteins
and PN-dependent reactions, **26:**125; 130; 131
and succinic dehydrogenase, **18:**99
Metalloflavoproteins, Nature and Function (Mahler), **17:**233
Metalloporphins, **7:**355–358
other than iron and magnesium, **7:**319–320
Metalloporphyrin coordination complexes, mechanism of enzymic oxidoreduction, **15:**16–21; 32–33; 38–39
Metalloproteases, **45:**402
Metalloproteinases, **45:**397, **53:**256
Metalloproteins. *See also* individuals
Cotton effect, **27:**49–50; 56; 58–60
EPR study, **27:**110–111; 120; 144–187
requirement of —SH group by, **11:**240–241
Metallothionein, **56:**401–407
function of, **56:**407
isolation of, **56:**401
metal centers of, properties of, **56:**403–407
metal content of, **56:**402
molecular properties of, **56:**402–403
in nucleic acid metabolism, **56:**407
pH effect on, **56:**406
spectral features of, **56:**404
Metallothionein, Cotton effect, **27:**60; 80
Metal mercaptide chromophores, at active center, **27:**78–80; 82
Metal-organic complex catalysis, **14:**173–176; 179–190
Metal oxides, in model enzyme system, **15:**21
Metal peptidases, action as acid-base catalysts, **15:**293
Metal phthalocyanines, oxidation, **15:**33; 39–40
Metal poisons, effect on photosynthesis, **12:** 25–28
Metal proteins, as enzymes, **12:**194–195
complex, bond type in, **12:**198–200; 205; 212–216
Metal requirements, pyruvate, phosphate dikinase, **45:**120–125
Metals, **49:**103; 120. *See also* individual metals, eg, Iron
for adenosylmethionine decarboxylase activity, **56:**267
binding, by imidazole, **20:**96–99
in identification of imidazole, **20:**66; 67; 71; 75
by protein groups, **20:**56; 96–99
catalysis, relation to interatomic spacing, **15:**291
catalytic activity of, **56:**303
as catalytic centers of enzymes, **56:**291; 298
colloidal, in model enzyme system, **15:**22; 26; 39
and Cotton effect in proteins, **27:**48
dose response curve to, **56:**290; 291
effect on succinic dehydrogenase, **18:**75; 88; 93–94
effect on thermostable factor, **15:**410–411
in galactosidase inhibition, **56:**175
in glucosidase inhibition, **56:**194
heavy
concentration in melanin granules, **16:** 171–172
heavy, inhibition of β-glucuronidase by, **16:**382
in nonenzymic transaminations, **16:** 223–226
relation to melanin color, **16:**172
effect on
RNase, **24:**236–237; 241; 451
sperm, **24:**71–72
effect on hydroxysteroid dehydrogenases, **20:**269
effect on pectic substance, **20:**351
role in enzyme action, **24:**112–113; 119–120; 236–237; 241; 442; 470
role in protein molecule association, **24:**119–120
in invertase inhibition, **56:**162
isolation of, **56:**291
in metalloproteins, **56:**298
in protein complexes, **56:**298

Metals (*Continued*)
role in ATP reactions, **49:**120
role in biocatalysis, **15:**19; 25; 26; 39; 41–42; 289; 296
role in dehydrogenase enzymes, **15:**2–3
role in enzyme-catalyzed substitutions at each of phosphorus atoms of ATP, **49:** 103–126
role in hydrogen transfer, **26:**171
role of carboxytransphosphoryase reaction, **45:**100–102
study of, **56:**284; 285
thiaminase inhibitors, **15:**398
Metals, Role in Enzyme-Catalyzed Substitutions of Each of the Phosphorus Atoms at ATP (Mildvan), **49:**103
Metal specificity in complex formation, **25:** 15–20
Metamorphosis, **45:**411; 413
Metanephrine
in catatonia, **29:**530
in cats, **29:**517–518
synthesis of, **25:**383
Metaphosphatase, **9:**356
*p*H optimum, **10:**84
soil, **28:**386
Metaphosphatases, classification, **14:**224; 248
Metaphosphate, **19:**17, **55:**370
effect on myosin thread extensibility, **6:**176
hydration, by metal ions, **21:**177
Metapyrocatechase (catechol-2,3-oxygenase), benzene oxidation, **27:**510
Metastable liquid-crystals, cholesterol ester hydrolases, **69:**58
Metastable species. *See also* Free radicals and Radical species
in irradiated enzymes, **24:**377–381
Metastasis, **59:**36–38
Metathrombin, **10:**38
Metazoa, Fertilization (Runnstrm), **9:**241
Meteorites, carbon compounds in, **27:**350–354; 360; 393
Methallyglycine, **50:**165
Methane, **38:**202. *See also* Hydrocarbons
anaerobic formation, form hydrocarbons, **27:**530
in cosmic bodies, **27:**349–354
dissimilative metabolism, **27:**477–480
attack by microorganism, **10:**444–445; 446–447; 449; 454–455; 458
enzymic synthesis, **26:**267
formation of, bacterial fixation of carbon dioxide and, **74:**14
production by microorganisms, **10:**478–479
Methane fermentation, **2:**146
Methane monooxygenase, **65:**164
Methane-oxidizing bacterium, **27:**477
S-Methane sulfonyl-CoA, **58:**159
Methanobacillus omelianskii, vitamin B_{12}, **26:**267
Methanobacterium omelianskii, metabolism in, **74:**14
Methanococcus vannielli, **48:**4
formate dehydrogenase of, **48:**18
as methane-producting organism, **48:**18
Methanol, **42:**269, **51:**133, **58:**6
and biological methylation, **12:**471–474
catalase action on, **19:**109
effect on papain, **19:**317; 322
catalytic decomposition, **15:**26
in pectin, **20:**342; 344; 349
Methanol blindness, **59:**128
Methanol dehydrogenase
and methane oxidation, **27:**479
and pyrrolo-quinoline quinone, **59:**171; 183–184; 197–199
and quinoprotein oxidoreductases, **59:** 195–196; 200; 204–205; 207–208
Methanol dehydrogenases, **59:**187–188
Methanol-forming enzyme, **42:**269
Methanol metabolism, **59:**128
Methanol-peroxidizing enzyme, **27:**476
Methanolysis, **58:**302
Methanomonas methanica, metabolism in, **74:**6
Methanomonas methanica, oxidation of hydrocarbons, **27:**477–479
Methanomonas methano-oxydans, oxidation of hydrocarbons, **27:**478
cis-(exo)-3,4-Methanoproline, **50:**128
cis-3,4-Methanoproline, **50:**134; 135
cis-3,4-Methano-L-proline, **50:**162; 163
Methanosarcina methanica, metabolism in, **74:**14
Methemalbumin, **46:**174
Methemerythrin, spectroscopic characteristics, **65:**160
Methemoglobin, **7:**267, **15:**18–21, **46:**142
Methemoglobinemia in abnormal tyrosine metabolism, **14:**101–111

Methemoglobin reducing factor, *See* Ferredoxin
in photosynthesis, **19:**49
Methemoglobin-reducing factor, ferredoxin as, **27:**87
Methemoglobin reductase, **19:**203
Methemoglobin reduction factor, **12:**31; 33
Methene link in heme, **7:**334–337
N^5,N^{10}-Methenyl-FAH_4, and purine metabolism, **27:**459
N^5,N^{10}-Methenyl-FAH_4 dehydrogenase, reaction, **27:**425
N^5,N^{10}-Methenyltetrahydrofolic acid, *See* N^5,N^{10}-Anhydroformyltetrahydrofolic acid
Methicillin, action of β-lactamase, **28:**239; 250–251; 253; 267; 271; 288–294; 296; 300; 301; 303–304; 306; 312; 313
Methicillinase, as β-lactamase, **28:**241
Methicillin-hydrolyzing activity, **28:**251
Methionase, **54:**248
Methionine, **5:**238, **7:**193–196, **42:**127, **46:**49; 53; 80; 87; 103, **49:**90, **50:**130; 152; 154; 168; 169, **51:**217. *See also* Ethionine and Methylation, biological; Methylation, biological, Sulfide link, Sulfur compounds; Sulfur compounds
activation of, **18:**202; 209; 211
and citric acid cycle, **18:**257
active, **26:**263
identity of, **25:**412
in methylation of nicotinamide, **15:**106; 111–112
of tobacco mosaic virus mutants, **15:**479
in *S*-adenosylmethionine synthesis, **22:**243
in spermine biosynthesis, **22:**261
analog-resistant mutants in fungi, **29:**138
analogs of, **29:**139 (*See also* Ethionine)
codons for sRNA fractions, **29:**154
control of synthesis of, **29:**139
effects on schizophrenia, **29:**527–528
increases of, in rats, **29:**528
protection during peptide synthesis, **29:**442–443
resistance to analogs in bacteria, **29:**139
assay for enzymic formation, **22:**254
biosynthesis, and methionine sulfonium compounds, **22:**263–271
formation from homocysteine, **22:**265–268
incorporation of formaldehyde in, **22:**266
metabolic functions, **22:**239
reactions with mustard gas, iodobromoacetic acid, **22:**251
and bacterial growth, **10:**274
content of plant proteins, (table), **10:**271
in plant metabolism, **10:**268–284
synthesis in living plant cells, **10:**271–274
and biological methylation, **12:**439–441; 458–469; 469–474; 478–479
biosynthesis and metabolism, **21:**223; 398; 410; 423–426; 430
chelates, in metal ion activation of enzymes, **21:**146
in pyrimidine biosynthesis, **21:**288
sparing effect on vitamin B_{12} requirement, **21:**430
conversion, to cystine, **7:**380–383
to homocysteine, **7:**375–378
to α-keto acid, **7:**374–375
conversion to cystine, **5:**238
and creatine synthesis, **6:**221
as inhibitory analogue, **6:**139
cytochrome *c* oxidase, **71:**92
demethylation, **7:**375–378
disulfide exchanges, **27:**218
and folate metabolism, **27:**426; 427; 459
and one-carbon pool, **27:**426
in DNA synthesis, **28:**423
synthesis of, **28:**445
E. coli synthesis, **25:**361
effect of vitamin B_6 on metabolism, **19:**351; 352
isopropyl ester, condensation by chymotrypsin, **19:**237
labeled, **19:**260; 352; 353
and enzymic catalysis, **24:**467
and RNase activity, **24:**203; 205; 206; 212; 214–220; 232; 243–246; 248; 249; 251; 449–450
as ethylene precursor, **54:**231; 233
and formate, **12:**48
from homocysteine, **25:**423–426
labeled, in studies of biosynthesis of pectic substances, **20:**369; 370
in studies of protein synthesis in plants, **20:**222
and liver lipids, **8:**192; 198–200
lipotropic action, **8:**172
and synthesis of choline, **8:**198
methyl group biosynthesis, **25:**396; 401–410

Methionine (*Continued*)
methylthioadenosine recycling, to, **54:** 239–244
from methylthioribose-1-p, **54:**245; 246
as noncompetitive antagonist of sulfonamide, **16:**300
biosynthesis, **16:**269–275
relation to cystine metabolism, **16:**277–279
relation to threonine and homoserine, **16:**269–271
in synthesis of lysine, **16:**298
in transamination reactions, **16:**190; 192; 197; 201; 203; 207; 239
use by *E. coli* mutants, **16:**284
oxidation of, **12:**480–481, **14:**132
oxidized derivatives of, **12:**467–469
RNA polymerase II transcription, reversible phosphorylation of CTD during transcription cycle, **71:**51
D-Methionine, **67:**170
L-Methionine, **46:**58, **67:**168–171
K_A values for, **29:**45
L-Methionine, oxidation in plants, **20:**320
D,L-Methionine, oxidation in plants, **20:**322
DL-Methionine, resolution by papain, **19:**309
Methionine activating enzyme, **15:**111–112
Methionine-activating enzyme, **25:**383–388; 412–413
Methionine acyl adenylate, formation of, **25:**382
Methionine acyl adenylate-forming enzyme, **25:**387–388
Methionine adenosyltransferase
in animal tissues, **56:**259–265
deficiency of, **56:**265
forms of, **56:**260–263
substrates and inhibitors for, **56:**264
bacterial, **56:**254–257
genetics of, **56:**254–255
kinetics of, **56:**257
metal ion requirements for, **56:**256
preparation and characterization of, **56:**255–257
localization of, **24:**312
structural formula for, **56:**265
of yeast, **56:**257–259
genetics of, **56:**258–259
purification of, **56:**257
regulation of, **56:**259
stereochemistry of, **56:**259
Methionine Adenosyltransferase (S-Adenylmethionine Synthetase) and S-Adenosylmethionine Decarboxylase (Tabor and Tabor), **56:**251
Methionine analogs, activation of, **25:**384
Methionine auxotroph, of *Neurospora,* **16:**272
Methionine bindings in corrinoids, **26:**245; 253
Methionine biosynthesis, **56:**40
Methionine (C^{14}), in coenzyme Q synthesis, **25:**298
from labeled formaldehyde, **25:**403; 404
methylation by, **25:**417; 418
Methionine (C^{14} and D-labeled), methylation by, **25:**415; 417
Methioninecarboxymethylsulfonium residues, from RNase, **24:**219
Methionine (C^{14}) metabolism, **25:**375; 379–380; 384
Methionine esters, tripeptide formation from, **13:**310
Methionine metabolism, relation to ethionine effect, **25:**369–393
Methionine r ethanolamine transmethylase, **14:**250
Methionine r glycine transmethylase, **14:**250
Methionine r guanidoacetic transmethylase, **14:**250
Methionine r nicotinamide transmethylase, **14:**250
Methionine (S^{35}), in transmethylations, **25:**412
L-Methionine-SR-sulfoximine, **43:**533, **53:**206
Methionine sulfone, **7:**373
Methionine sulfonium compounds, and methionine biosynthesis, **22:**263–271
Methionine sulfoxide, **7:**373, **22:**246, **58:**198; 213
biochemical functions, **22:**270; 271
as methyl donor, **22:**270
Methionine sulfoxide, as substitute for methionine, **22:**271
Methioninesulfoxide, reduction of, PN-dependent, **26:**129
Methionine sulfoxide (MSO), antagonist of glutamic acid, **13:**255–259; 263; 278–279
Methionine sulfoxide residues from RNase, **24:**219
Methionine sulfoximine
antagonist of glutamic acid, **13:**258

mechanism of inhibition by, on glutamine synthetase, **31:**208–212
Methionine sulfoximine (MSO), inhibition mechanism, **72:**17–20
Methionine sulfoximine phosphate, structure of, **31:**211
Methionine sulfur oxidation, **7:**372–378
mechanisms of, **7:**373–378
to sulfate, **7:**372
to taurine, **7:**372–373
to thiosulfate, **7:**372
Methionine synthesis, role of Comethylcobalamin, **26:**266–267
Methionine synthetase
in *E. coli,* **22:**122
reaction, **27:**425
Methioninyladenylate, **35:**452
Methionol, *See* γ-[Methylmercapto]-propyl alcohol
Methionyl adenylate, **35:**452
S-Methionylglutathione, synthesis, **19:**247
Methionyl-tRNA synthetase, **37:**476; 494; 521, **48:**7
Methiothepin, tyrosine hydroxylase regulation, **70:**128
4-Methlyberlliferyl glycosides, **60:**91
Methmyoglobin, adsorption at interfaces, **28:**20
Methods, iterative, fitting nonlinear equations by, **29:**5–6
least-squares, **29:**2–4
matrix, **29:**4
orcinol, in schizophrenia, **29:**505–506
thiobarbituric acid, in schizophrenia, **29:**506
Methotrexate, **37:**166–167, **43:**331, **54:**58
dihydrofolate reductase, **70:**25; 28; 84–87
with dihydrofolate reductase, **54:**82
eukaryotic dihydrofolate reductase, **70:**34–35; 55–56
with 5-fluorouracil, **54:**101
folate inhibitor, **27:**435; 437; 441; 452–456; 459; 460
FU activation with, **54:**68
ternary inhibitor complexes, isomerization of, eukaryotic dihydrofolate reductase, **70:**56–58
Methotrexate polyglutamates, eukaryotic dihydrofolate reductase, ligand binding (of inhibitors), **70:**58–60
Methoxatin, *See* Pyrrolo-quinoline quinone
Methoxy-*p*-benzoquinone, secretion by beetles, **16:**155
Methoxycarbonyl-CoA disulfide, **58:**159
Methoxycarbonylmethyl disulfide, **54:**402
Methoxyflurane, **55:**261–262
6-Methoxygossypol, **55:**80
structure of, **55:**56
10-Methoxyharmalan, formation of, **29:**520
structure of, **29:**521
6-Methoxyhemigossypol, **55:**80
structure of, **55:**56
3-Methoxy-4-hydroxymandelic acid, imipramine reduces excretion of, **29:**490. *See also* Vanillylmandelic acid (VMA)
Methoxylation of *p*-hydroxyphenylpropane unit in lignin formation, **16:**149
Methoxyl groups, ring, of coenzyme Q, synthesis, **25:**296–297; 321
Methoxyl groups and alkaloid biogenesis, **8:**234
6-Methoxymellein, **55:**84–85
Methoxyneuraminic acid, composition, properties, and structure, **20:**138; 139
2-Methoxyphaseollin isoflavan, **55:**7
8-Methoxypsoralen, structure, **66:**87
3-Methoxypyridine, dissociation constant, **50:**5
p-Methoxytyrosine, action of tyrosinase on, **14:**137
N-Methy-*N*-amidinoalanine, **54:**386; 387–388
Methyl, **57:**29–30
labile, and formate, **12:**48
N-Methylacetamide, configuration of, **23:**14–16
N-Methylacridan, model coenzyme, **15:**31–32
3-Methyladenine, **60:**2
N^6-Methyladenosine, **65:**256; 259–260
function, **65:**265–270
localization, **65:**260–261
substrate specificity, **65:**273–276
2′ and 3′-O-Methyl adenosine triphosphates
analogs containing arabinose, **43:**26
analogs containing xylose, **43:**26
N-MethylAIB uptake, **49:**61
a-Methylalanine ethyl ester, **66:**14–15
N-Methyl-amidino-*N*-methyglycine, **54:**386; 390
Methylamine, **54:**30, **58:**19
thermodynamic data on ionization, **20:**64
Methylamine dehydrogenase, **59:**191–192; 194–195; 207
Methylamine dehydrogenases, **59:**201
Methylamine oxidase, **59:**193; 208
N-Methylamino acids in transaminations, **16:**218

2-Methyl-4-amino-5-aminoethylpyrimidine, action on thiaminase, **15:**401–402
2-Methylamino-6-amino purine, source of methyl, **25:**383
2-Methylamino-6-aminopurine, synthesis, from 2,6-diaminopurine, **22:**256
a-Methyl-*b*-alanine ethyl ester, **66:**14–15
2-Methyl-*p*-aminobenzoic acid, and folate analog, **27:**429
1-Methyl-3-(3-amino-3-carboxypropyl) pseudouridine, **54:**230
Methylaminoethanols, as methyl donors, **12:**462; 464; 466
4-Methylamino group, in transaminations, **16:**220
2-Methyl-4-amino-5-hydroxymethyl-pyrimidine, action on thiaminase, **15:**401–403; 410
2-(Methylamino)-isobutyric acid, **49:**57; 71
Methyl amino-malonate, **35:**87; 88
5-Methylaminomethyl-2-thiouridine, **35:**475
N-Methylamino oxidase, **19:**203
2-Methyl-4-amino-pyrimidine, action on thiaminase, **15:**401–402; 410
Methylandrostenediol, effect on β-glucuronidase, **16:**394
Methyl anisate, from brown rot, **18:**363
Methyl-β-arabinoside, **52:**36
Methylarachidonate, and mitochondrial swelling, **27:**229
N^{ω}-Methyl-L-arginine, **65:**298
  nitric oxide synthases, inhibition, **65:**314
Methylase, **38:**80
Methylaspartase, **74:**301
β-Methylaspartase (methylaspartate ammonialyase), cobamide coenzyme-dependent, **26:**359; 363–365
β-Methylaspartate, from glutamate isomerization, **26:**343; 356; 359–365
Methylaspartate ammonialyase, cobamide coenzyme-dependent, **26:**359; 363–365
Methylaspartate mutase [L-*threo*-3-methylaspartate carboxy-amino-methylmutase], in propionic acid metabolism, **26:**344; 346; 347; 352; 358–359
  mechanism of reaction, **26:**353–354; 363–365
  properties, **26:**359–365
  stereochemistry of reaction, **26:**364–365
Methylaspartate reaction, cobalamins in, **25:**407
α-Methyl-DL-aspartic acid, effect on ornithine cycle, **19:**377; 378
Methylated amino acid, **42:**229
  in blood, **42:**276
  discovery, history of, **42:**229
  genetic analysis, **42:**239
  insensitivity to puromycins, **42:**239
  natural occurrence, **42:**229
  origin, **42:**239
  in urine, **42:**276
Methylated arginine, distribution of, **42:**263
Methylated arginine derivatives, content of various, **42:**234
Methylated bases, **35:**470–473
Methylated lysine derivatives, content of various, **42:**234
N-Methylated purines, mutagens, **27:**331
Methylated substances in plants, **8:**234
Methylathodoratin, **55:**84
Methylation, **42:**254; 261, **57:**358, **58:**15, **59:**184. *See also* Ethionine, Methionine, and Methionine metabolism; mRNA, methylation
  of amino acids, **29:**399–401
    demethylation and, in schizophrenia, **29:**526–529
  of ε-amino group of lysine residues, **42:**249
  arginine, cell cycle during, **42:**267
    in neoplasia, **42:**267
  biological, **12:**429–491, **25:**395–431
  Caulobacter cell cycle, transcription regulation, **71:**16
  de novo methyl synthesis, **25:**396–410
  of free carboxyl groups, **42:**268
  histidine residues, **42:**275
  histone, **42:**254; 255
    temporal relationship between and DNA synthesis, **42:**253; 254
  of mercaptans, **25:**421–422
  nonenzymatic, **42:**260
  in plants, **9:**439–442; 444
  protein, **42:**228; 245
    change in basicity, **42:**246
    neutralization of charge, **42:**246
    theoretical, **42:**245
  of tobacco alkaloids, **8:**240
    studies of cellulose, **8:**255–256; 280–281
  tracer studies, **25:**397–399; 401–404; 408; 409; 412; 414–415; 417–420
  transmethylation, **25:**410–427

Methylation, Biological (Challenger), **12:**429
Methylation, Biological (Greenberg), **25:**395
N-Methylation, of catechol and indole derivatives, **25:**415–416
O-Methylation, of catechol and indole derivatives, **25:**416
in coenzyme Q synthesis, **25:**296–297
Methylation of carboxybiotin-enzyme complex, **26:**327–328
Methylation of inorganic sulfur in plants, **10:**240
Methylation of lignin, **16:**150
Methylation of nicotinamide, **15:**110–112
Methylation of RNase, **24:**224–225; 246
Methylation reactions, miscellaneous, **25:** 422–423
4-Methyl-*o*-benzoquinone, reaction with salmine, **16:**126
Methyl-*p*-benzoquinone, relation to cuticle hardening, **16:**155
Methylbisglyoxalguanylhydrazone, inhibition of tumor growth by, **29:**336
3-Methylbut-3-en-l-yl pyrophosphate, in rubber biosynthesis, **29:**221
β-Methylbutyramide, **54:**13; 17
Methyl butyrate, **34:**487
with lipase, Michaelis curves, **23:**145–148
β-Methylbutyric acid, **54:**14
α-Methylbutyric acid, metabolism, in mitochondria, **21:**46; 47
Methyl carnosine, source of methyl group, **25:**383
4-Methylcatechol
formation in peroxidase-catalyzed reaction, **19:**143
phenolase action on, **16:**109–110; 158
pigment formation from, **16:**112
quinones from, **16:**122
Methyl (C^{14})-cobalamin, and methionine synthesis, **26:**267
Methylcholanthrene, **60:**39
3-Methylcholanthrene, **46:**173; 409
Methyl cinnamate, from brown rot, **18:**363
Methylcobalamin, methylation, **25:**407
N-Methylcreatine, **54:**386
conformational preferences of, **54:**379–383
in creatine kinase binding, **54:**386; 390
nmr spectroscopy of, **54:**381
3-Methylcrotonic acid, synthesis from acetate, **22:**286
β-Methylcrotonyl-CoA, **21:**214
phosphorolysis of, **23:**338
β-Methylcrotonyl CoA carboxylase
molecular characteristics, **35:**388
reaction, **35:**387
ATP-^{1} ^{4}C-ADP exchange, **35:**387
ATP-^{3} $^{2}P_i$ exchange, **35:**387
α-carboxylation, **35:**388
enzyme-^{1} $^{4}CO_2$, **35:**387
β-Methylcrotonyl-CoA carboxylase [3-Methylcrotonyl-CoA:CO_2 ligase (ADP)], biotin enzyme, **26:**297
properties of, **26:**303–305
reaction mechanism, **26:**312; 316–320; 325; 328–330
S-Methylcysteine, **22:**243, **50:**147
utilization by *Neurospora* mutants, **22:**268
S-Methyl-L-cysteine, **67:**171–172
5-Methylcytosine
in DNA, **16:**432–433
in thymine biosynthesis, **21:**423
5-Methyl cytosine, in tumor metabolism, **29:**353–354
6-Methyl-5-deazatetrahydropterin, **74:**240
binding by phenylalanine hydroxylase, **74:**240
structure of, **74:**240
5-Methyldeoxycytidine, in pyrimidine biosynthesis, **21:**288
5-Methyldeoxycytidine deaminase, **21:**288
5′-Methyl-5′-deoxy-selenoadenosine, **54:**235
5-Methyldesoxycytidylic acid, in DNA, **16:**431
N′-Methyldihydronicotinamide, **15:**93
2-Methyl-3,8-dihydroxychromone, *See* Alginetin
1-Methyl-5,6-dihydroxyindole, formation of melanins, **16:**164
2-Methyl-5,6-dihydroxyindole, formation of melanins, **16:**164
3-Methyl-5,6-dihydroxyindole, in melanin studies, **16:**168
Methyldioxolane, **55:**157
5-Methyl-1,2-dithiolane-3-valeric acid, **18:**332
Methyl donors. *See also* Methylation, biological
biological, **22:**252
requirements for, **12:**476–478
Methyl donors and choline synthesis, **8:**198
α-Methyldopa, in catatonia, **29:**530
in depression, **29:**489–490
function of neurotransmitter, **29:**483
N,N-methylenebis, **34:**452

3,3′-Methylenebis (4–4hydroxycoumarin), *See* Dicoumarol
Methylene blue
as artificial electron acceptor, **30:**22
effect on mitochondrial oxidative phosphorylation, **21:**31
as electron acceptor, **28:**172; 173; 177; 210; 222; 225
mutagenic action, **27:**327
and reduced pyridine nucleotides, **15:**89; 98
assay for pyruvic oxidase, **15:**212
Methylene blue reaction of *aspergillus* culture, **2:**224 226; 229
Methylene blue reduction in microbial attack on hydrocarbons, **10:**447; 469
β-(Methylenecyclopropyl)alanine, **50:**125
mode of action, **29:**143
α-(Methylenecyclopropyl) glycine, **50:**126
α-(Methylenecyclopropyl)glycine, inhibition of the growth of seedlings, **29:**142
Methylenedioxy groups and alkaloid biogenesis, **8:**235
Methylenediphosphonic acid, **43:**3
N^5,N^{10}-Methylene-folate-H_4, methyl synthesis, **25:**396–397; 399; 405–408
N^5,N^{10}-Methylenefolate-H_4 reductase, methyl synthesis, **25:**405; 407
γ-Methyleneglutamic, in transaminations, **16:**197; 201; 202
γ-Methyleneglutamic acid, **50:**118; 119
γ-Methyleneglutamic acid, in transaminations, **16:**197; 237
4-Methylene-DL-proline, proline analog, **29:**95
5,10-Methylenetetrahydrofolate, **54:**58
oxidation of, to dihydrofolate, **54:**70
N^5,N^{10}-Methylenetetrahydrofolic acid, interconversion with N^5,N^{10}-anhydroformyltetrahydrofolic acid, **21:**415
isolation, and properties, **21:**386; 387; 397
structure, **21:**219; 387
synthesis, **21:**386; 412
N^5,N^{10}-Methylene tetrahydropteroyl-triglutamate, methyl synthesis, **25:**405
a-Methylenic group, reactivity and autoxidation, **8:**427; 431–432; 434–435; 437–438
Methylethylacetothetin, **22:**272
Methyl excretion in muscular dystrophy, **6:**221
5-Methyl-FAH_4, activity, **27:**437
2-Methyl fatty acid racemerization, **44:**130
Methyl fluoroacetate, **18:**115
N^{10}-Methylfolate, as substrate of dihydrofolate reductase, **27:**436
N^{10}-Methylfolate-H_4, methylation, **25:**408
5-Methylfolate-H_4, methylation by, **25:**420
N^5-Methylfolate-H_4, methyl synthesis, **25:**399; 404–409
N^5-Methylfolate-H_4 (C^{14}), methylation, **25:**408
X-Methylfolic acid, antimetabolite, **27:**436
N-Methylformamide, **51:**111; 113; 126; 128; 130; 131; 140
N-Methyl formamide, effect on nucleic acid metabolism, **16:**458–459
Methyl-D-fructofuranoside 1,6-P_2, **53:**51–57; 61
2′-*O*-Methyl G, **35:**470
4-*O*-Methyl-D-galactose, presence of a single common carrier in translocation, **29:**53
Methylgalactoside-permease, description of, **30:**108–109
N-Methyl-L-glucosamine, **10:**147–149; 157–158; 160; 162
α-Methyl D-glucose, in β-glucan synthesizing system, **28:**339
6-Methylglucose, oxidation, **15:**148
3-*O*-Methyl-D-glucose, presence of a single common carrier in translocation, **29:**53
Methyl α-D-glucoside, **55:**363; 364
α-Methylglucoside, as an acceptor for dextran formation, **30:**192
effect on dextran synthesis, **30:**191
α-Methyl-D-glucoside, presence of a single common carrier in translocation, **29:**53
α-Methylglucoside in polysaccharide synthesis, **15:**306
α-Methylglucoside-6-phosphate, **44:**243
α-and β-Methyl-D-glucosides, inaction of β-glucuronidase on, **16:**371
α-Methyl glucoside-T, as acceptor in dextran formation, **30:**199
Methylglutaconic acid, **35:**139; 145
β-Methylglutaconyl-CoA, **21:**214
1,3,5-C^{14}-β-Methylglutaconyl-CoA, CO_2 transfer, **26:**312–313
Methylglutaconyl-CoA hydratase, and biotin enzymes, **26:**304
L-α-Methyl glutamate, **35:**86
γ-Methylglutamate
in transamination reactions, **16:**189; 197
in transaminations, **16:**197; 201; 202

β-Methylglutamate, in transamination reactions, **16:**189
Methyl-L-glutamate, spatial arrangements, **13:**159
threo-γ-Methyl-L-glutamic acid, as substrate for glutamine synthetase, **31:**205
α-Methylglutamic acid, as substrate for glutamine synthetase, **31:**194–195
γ-Methylglutamic acid, as substrate for glutamine synthetase, **31:**200–201
β-Methylglutamic acid, conversion of, to D-alloisoleucine, **31:**200
as substrate for glutamine synthetase, **31:** 198–200
α-Methylglutamine, **54:**19; 20
peptide derivatives of, **54:**18
γ-Methylglutamine, conversion of, to L-α-methylsuccinic acid, **31:**202
β-Methyl-γ-glutamylhydroxamate, synthesis of, **31:**199
β-Methyl-γ-glutamylhydroxamic acid, synthesis of, **31:**199
2-Methylglutarylproline, **57:**32–33; 40
N-Methylglycine, **50:**162
N-Methylglycocyamine, **54:**384
Methylglyoxal, **59:**129–130
accumulation in cells, **15:**143; 167
and glycine metabolism, **23:**456
in synthesis of *S*-lactylglutathione, **19:**241
Methylglyoxal bis(guanylhydrazone), **56:**271; 274
Methyl green, staining of kinetoplast by, **16:**423
Methyl group, **58:**15, **60:**3. *See also* 1-Carbon particles and Formate
binding energy, **43:**334; 347
and demethylation, **7:**375–378
and transmethylation, **7:**376–378
of methionine, biosynthesis, **16:**274–275
of thymine, from formate, **16:**450
ring, in coenzyme synthesis, **25:**297–298
substitution by ethyl, **25:**370–372
synthesis *de novo,* **25:**396–410
on steroid molecule, **58:**47
Methyl groups
deficiency of, **8:**172
effect on dehydrogenation of fatty acids, **8:**361; 362; 365
effect on β oxidation of fatty acids, **8:** 351–352
transfers in nornicotine synthesis, **8:** 233–235
incorporated, turnover of, **42:**244
histones in Ehrlich ascites tumor cells, **42:**244
and lipotropic properties, **12:**483
method for detecting lability, **12:**476–478
oxidation of, **12:**480–482
synthesis in body, **12:**469–474
transfer in chemical reactions, **12:** 453–455
transfer in mold metabolism, **12:**455–457
in nutrition, **22:**240
removal, in formation of cholesterol from lanosterol, **22:**324–326
Methyl group synthesis, **9:**441–442
Methyl group transfers, **50:**286–290
7-Methyl-GTP, **35:**60
Methylguanidine, **15:**361–362
2′-*O*-Methyl guanylic acid, **35:**472–473
3-Methylheptane, microbial oxidation, **27:**495
Methylheptenoe, effect on carotenogenesis, **21:**320
Methylhexanes, microbial oxidation, **27:** 493–495
Methylhioadenosine (Schlenk), **54:**195
Methylhistamine, source of methyl group, **25:**383
1-*N*-Methylhistidine, **42:**236
1-Methylhistidine, effect on fructose 1,6-bisphosphatase, **42:**202
3-Methylhistidine, effect on fructose 1,6-bisphosphatase, **42:**202
DL-α-Methylhistidine, effect on histidine synthesis, **29:**152
α-Methylhistidine, growth inhibition in *E. coli,* **29:**130
3-*N*-Methylhistidine, in actin, **42:**230
3-*N*-Methyl-L-histidine, natural occurrence, **42:**275
Methyl histidine, source of methyl group, **25:**383
2-Methyl-4-hydroxy-5-aminomethyl-pyrimidine, action on thiaminase, **15:**401
Methyl *p*-hydroxycinnamate from xylose, **8:**294–295
4-Methyl-5-β-hydroxyethyl thiazole, action on thiaminase, **15:**401–403
2-Methyl-3-hydroxy-4-hydroxymethyl-5-aminomethylpyridine, microbiological activity, **16:**220–221

2-Methyl-4-hydroxy-5-hydroxymethyl-pyrimidine, action on thiaminase, **15:**401; 410
2-Methyl-3-hydroxy-4-methyl-5-hydroxymethylpyridine, vitamin B_6 antagonist, **16:**221
4-Methyl-3-hydroxyphenylalanine, **67:**207–208
2-Methyl-4-hydroxypyrimidine, action on thiaminase, **15:**401; 410
Methyl I, **35:**469
4-Methylimidazole, reaction with benzoyl chloride, **20:**83
5-Methylimidazole, reaction with benzoyl chloride, **20:**76
4-Methylimidazole haemochrome, absorption spectrum, **23:**289–297; 306–310
Methylimidazolenicotine-hemin as catalase model, **15:**28
2-Methylindole, reaction with quinones, **16:**110
Methyl iodide, **42:**257
Methyl isooxazolone, **15:**239–240
Methyl isopropyl ketone, **58:**176; 181
O-Methylisourea, in identification of imidazole, **20:**76
β-Methyl-α-ketobutyrate, **18:**338
Methyl ketones
  from microbial oxidation of hydrocarbons, **27:**482–484
  and synthesis of waxes, **8:**334–337
    from fatty acids, **8:**350–352; 356; 357
    metabolism, **8:**353; 403
3-Methyllumiflavin, binding capacity, **22:**379
ε-*N*-Methyl-DL-lysine, **42:**229
^{14}C-Methyl-L-methionine sulfoximine, as substrate for glutamine synthetase, **31:**210
Methyl malonate, **58:**187
Methylmalonate (C^{14}), biosynthesis, **26:**286
Methylmalonate reaction, cobalamins in, **25:**407
Methylmalonyl CoA
  decarboxylase, **35:**406–410
    molecular properties, **35:**407
  mutase, **35:**379
  in photosynthesis, **19:**42; 43
  racemase, **35:**379; 402
Methylmalonyl-CoA
  decarboxylation of, **23:**338
  isomerization, role of cobamide coenzyme, **26:**264–265
Methyl malonyl CoA, and ATP, **18:**230
Methylmalonyl-CoA apomutase, absorption spectrum, **26:**351
Methylmalonyl-CoA(b), isomerization to succinyl-CoA, **26:**343–367
Methylmalonyl-CoA (C^{14}), carboxybiotin-enzyme complex formation, **26:** 320–324
  isomerization to succinyl-CoA, **26:**345; 354–357
  racemization of, **26:**334–343
Methylmalonyl-CoA (C^{13} and C^{14}), reaction mechanism, **26:**356–358
Methylmalonyl-CoA (H^3), reaction mechanism, **26:**339–340
Methylmalonyl-CoA holomutase, absorption spectrum, **26:**348–349
Methylmalonyl-CoA isomerase, role of cobamide coenzymes, **26:**264–265
Methylmalonyl-CoA mutase [2-Methylmalonyl-CoA:CoA-carbonylmutase], apoenzyme, **26:**345–346; 351
  holoenzyme, from sheep liver, **26:**345–349
  mechanism of reaction, **26:**353–359
  metabolism of propionic acid, **26:**285; 309; 334–335
  P. shermanii, **26:**349–353
  properties, **26:**345–359
Methylmalonyl CoA racemase, equilibrium, **44:**128
  from animal tissue, **44:**128
  ^{3}H incorporation, **44:**129
  from *Propioni bacterium,* **44:**128
Methylmalonyl-CoA racemase, metabolism of propionic acid, **26:**284–285; 309; 334–343; 345; 350–351
  purification, **26:**337
Methylmalonyl-CoA transcarboxylase [Methylmalonyl-CoA:pyruvate transcarboxylase], metabolism of propionic acid, **26:**285; 292; 297; 338; 350–351
  properties, **26:**307–311
  reaction mechanism, **26:**313–316; 328–331
Methyl mercaptan, formation in mushrooms, **10:**240–241
γ-[Methylmercapto]-propyl alcohol, **10:**242
6-Methylmercaptopurine riboside (MMPR), **54:**58
  Fu activation to nucleotides with, **54:**68
Methylmercury chloride, **54:**403

Methyl mercury derivative, eukaryotic dihydrofolate reductase, structural change due to derivatization with, **70:**46
Methylmethane sulfonate, mutagenic action, **27:**323–326
Methyl methanethiosulfonate, **54:**400–402, **58:**151; 159
S-Methylmethionine
  biosynthesis, **22:**245; 246
    discovery, **22:**242
    history, **22:**240
    as methyl donor, **22:**242
    in microorganisms, **22:**268; 269
    nonenzymic splitting, **22:**260; 261
    stability to alkali, acid, enzyme action, **22:**248
  and methylation, **25:**415
Methylmethionine, as methyl donor, **25:** 423–426
Methylmethionine sulfonium as methyl donor, **25:**426
1-Methyl-6-methoxy-1,2,3,4,-tetrahydro-2-carboline, identical with adrenoglomerulotropin in pineal gland, **29:**520
Methyl *p*-methoxycinnamate
  biogenesis of, **18:**362–370
  from lignin, **8:**294
    from wood, **8:**272; 293–295
Methyl *p*-methoxycinnamate from glucose, **8:**272; 273; 294–295
Methylnaphthalenes, microbial oxidation, **27:**523–525
2-Methyl-1,4-naphthoquinone, *See* Menadione
Methylnaphthoquinone, effect on hypertensin formation, **7:**524
2-Methyl-1,4-naphthoquinones, **46:**1
N′-Methylnicotinamide, **15:**87; 95
  in metabolism of nicotinamide, **15:**101 (C^{14}); 106; 109–112
N-Methyl nicotinamide, effect on Zn exchange of LADH, **27:**64–65
  and LADH complex formation, **27:**62–65
N-Methylnicotinamide, formation from nicotinamide, **22:**255
N′-Methylnicotinamide, source of methyl group, **25:**383
N′-Methylnicotinamide from nicotinic acid, **12:**457; 465; 474–478
α-Methylnorepinephrine, in catatonia, **29:**530
N-Methylnucotinamide, excretion of, **29:**529
Δ^4-9-Methyloctalindione(3,8), microbiological conversions, **20:**271; 272
Methyl orange, binding to proteins, **16:**332
Methylornithine, **56:**274
N-Methyl oxotremorine, **55:**157
Methyl parathione, **46:**395
Methylphenanthrenes, microbial oxidation, **27:**525–528
o-Methyl phenylalanine, **67:**195
p-Methylphenylalanine, **35:**286
4-Methylphenylalanine, **67:**205–206
Methyl-β-phenylpropionate, kinetics of hydrolysis, **20:**294
Methylpherase, **12:**469
Methylpherases, alkyl group transfer by, **25:**383–388
Methyl phosphate r guanidoacetic transmethylase, **14:**250
Methyl phosphate synthesis, **55:**310; 311
4-*N*-Methyl piperidyl benzylate, **55:**143
  muscarinic site binding of, **55:**150
Methyl precursors. *See also* Betaine, Choline, Methionine, and Protein
  and antifatty-liver factor of pancreas, **8:** 197–200
cis-4-Methyl-L-proline, proline analog, **29:**95
trans-4-Methyl-L-proline, proline analog, **29:**95
α-Methylpropionamide, **54:**13
α-Methylpropionic acid, **54:**14
N′-Methyl-6-pyridone-3-carboxyl-amide, **15:**109–112
N-Methylpyridoxal phosphate, in aspartate aminotransferase, **50:**23
ω-Methyl-pyridoxal 5-phosphate, preparation, **16:**215
ω-Methyl-pyridoxamine 5-phosphate, preparation, **16:**215
β-Methylpyrrolidone carboxylate, synthesis of, **31:**199
4-Methyl-*o*-quinone, reaction with indoles, **16:**111
3-Methylriboflavin, binding capacity, **22:**379
Methylsalicylic acids, oxidation of methylnaphthalenes, **27:**523–524
6-Methylsalicylic acid synthase, **55:**26
Methyl scopolamine, **55:**150
S(+)-2-Methyl serine, **35:**93
O-Methylserine, effect on *E. coli,* **29:**145

L-α-Methylsuccinic acid, conversion of γ-methylglutamine to, **31:**202
Methylsulfinyl, **57:**358
S-Methyl-sulfonium-L-methionine, **54:**223
Methylsulfonyl, **57:**358
10-Methyltetrahydrofolate, **54:**58
7-Methyltetrahydropterin, **35:**298, **67:**173
6-Methyltetrahydropterin, tyrosine hydroxylase, substrate specificity, **70:**110; 111
N^5- Methyltetrahydropteroylglutamic acid, activated methyl, **25:**396; 399; 404–409
2-Methylthiazoline, **57:**131
4-Methylthiazolium compounds as inhibitors of thiamase, **15:**399
Methylthioadenosine, **22:**241, **54:**196–265
in adenine synthesis, **54:**247
analogs and homologs of, **54:**209
biosynthesis of, **54:**209–210
chemical synthesis of, **54:**210–213
analysis of, **54:**203–208
HPLC chromatrography in, **54:**207–208
ion exchange chromatography in, **54:**206
paper chromatography in, **54:**206–207
catabolism of, **54:**232–248
abnormalities of, **54:**244
methylthioribose kinase in, **54:**239
nucleoside hydrolases in, **54:**232–235
nucleoside phsphorylases in, **54:**235–239
via recycling to methionine, **54:**239–244
cellular transport, uptake, and release of, **54:**249–250
concentration of, cellular, **54:**217
conformation of, **54:**202
as cyclic AMP phosphodiesterase inhibitor, **54:**254
deamination of, **36:**215, **54:**205
product of propylamine transferase, **36:**232
in spermidine biosynthesis, **36:**215
enzymatic decomposition of adenosylmethionine to, **54:**222–227
via azetidine-2-carboxylic acid, **54:** 225–227
via T_3-phage, **54:**224–225
enzymatic formation of, **54:**213–232
by polyamine synthesis, **54:**214–222
with putrescine, **54:**215–216
with spermidine and spermine, **54:**216
with spermidine synthase, **54:**219–220
with spermine synthase, **54:**220–221
via adenosylmethionine conversion, **54:**216; 231–232
via adenosylmethionine decarboxylase, **54:**218–219
via biotin formation, **54:**227–228
via discadenine, **54:**230–231
via uridine and pseudouridine, **54:**228–230
formation, decomposition, **22:**263
growth-promoting function of, **54:**251
history of, **54:**197–198
hydrolysis of adenosylmethionine to, **54:** 200–201
interaction of, with cells and tissues, **54:** 249–254
isolation of, from cells, **54:**199–200
in lymphocyte transformation, **54:**254
melting point of, **54:**203
in methionine biosynthesis, **22:**269
mnr spectrum of, **54:**203
nomenclature of, **54:**196–197
phosphate-dependent breakdown of, **54:**215
properties of, **54:**202–203
in rRNA morphology and metabolism, **54:**253
selenium analog of, **54:**209; 211
solubility of, **54:**203
in spermidine synthase inhibition, **54:**219
in spermine synthase inhibition, **54:**220
structure, **22:**263
structure and chemical synthesis of, **54:** 201–202
ultraviolet absorbance of, **54:**204
5′-Methylthioadenosine, adenosine deaminase action, **25:**386
in ethionine study, **25:**385–386
Methylthioadenosine nucleosidase, **54:**235; 241
role of, **54:**248
Methylthioadenosine nucleoside hydrolase, **54:**232–235
Methylthioadenosine nucleoside phosphorylase, **54:**235–239; 247; 248
Methylthioadenosine phosphorylase
hormonal effect on, **54:**253
in lymphocyte transformation, **54:**254
Methylthioadenosine sulfoxide, **54:**211; 248
8-Methylthioctic acid, **18:**332
Methyl thiocyanate, **58:**213
Methylthio derivative, **57:**358
β-Methylthiopropionic acid, methyl ester, in plants, **10:**242–243
Methylthioribose, **54:**249–250

Methylthioribose kinase, **54:**239
Methylthioribose-1-p, **54:**245; 246
Methylthiotubercidine, **54:**219; 236
O-Methylthreonine, incorporation into protein, **29:**145
as a leucine analog, **29:**145
Methyl transfer and choline, **7:**376–377
Methyl transferase, **58:**112
Methyltransferase, reactions involving *S*-adenosylmethionine, **25:**413–423
reactions not requiring *S*-adenosylmethionine, **25:**423–426
stereospecificity, **25:**413
Methyltransferases, **50:**274
location of, **24:**309
O^6-Methyl transferases, **60:**15
Methyltryptophan, **50:**169
5-Methyltryptophan, **50:**170
growth inhibition by, **29:**122
N-Methyltyramine, source of methyl group, **25:**383
α-Methyl-*m*-tyrosine, depletion of catecholamines in brain by, **29:**485
function of neurotransmitter, **29:**483
Methylumbelliferon, **60:**243
4-Methylumbelliferyl-α-D-galactopyranoside, **38:**306
5-Methyluridylic acid, in pyrimidine biosynthesis, **21:**288
γ-Methylvaleramid, **54:**13; 19
Methylxanthine r guanidoacetic transmethylase, **14:**250
metL-encoded aspartokinase, II bacterial amino acid biosynthesis, **72:**283–285
Mevaldic acid, **23:**482–483
in mevalonic acid biosynthesis, **21:**336; 337
Mevalonate, conversion to rubber by ATP stimulation, **29:**237
formation of metabolites in latex as function of time, **29:**239–240
isopentenyl pyrophosphate converted to, **29:**236–240
in rubber biosynthesis, **29:**230; 236
synthesis of, **29:**230
Mevalonate kinase, isolation from latex serum, **29:**234–235
physical properties from various sources, **29:**237; 238
separation from latex, **29:**237
Mevalonate pathway, **65:**106–107
Mevalonate pyrophosphate, decarboxylation and dehydration of, **29:**239–240
Mevalonic acid, **23:**471–483
biosynthesis, **21:**336; 337
radioactive, in study of carotenogenesis, **21:**335–337; 339; 340; 343; 346
dimethylallyl pyrophosphate from, **55:**64; 65
formation of, **55:**50; 57–59
labeled, **23:**476–479
phosphorylation of, **55:**61
precursor of steroids and terpenoids, **29:**236
in terpenoid phytoalexin synthesis, **55:**50
Mevalonic Acid, Discovery and Chemistry of (Wagner and Folkers), **23:**471
Mevalonic acid (2-C^{14}), in coenzyme Q synthesis, **25:**287; 296; 298–303
Mevalonic acid (C^{14}), sterols in flowering, **26:**28
Mevalonic acid kinase, **55:**61
activation and inhibition of, **55:**63
properties of, **55:**62
Mevalonic acid 5-phosphate, **55:**61; 63
Mevalonic dehydrogenase, **21:**337
Mevalonic dehydrogenase system, **23:**480
Mevalonic kinase, **23:**479
Mevalonic 5-pyrophosphate, **55:**63
Mevalonolactone, **50:**263
Mevalonyl pyrophosphate, **23:**324
Meyerhof oxidation quotient, Pasteur effect measured in terms of, **29:**325
M factor, and ADP-ATP exchange, **23:**385
Mg^{2+}, **53:**52; 54; 55; 57; 58; 59; 61; 63; 67; 75
effect on rate of poly (glycerol phosphate) synthesis, **30:**236
Mg^{++}, substitution by polyamines, **36:**244
MgADP, iron protein redox potential, **67:** 313–314
MgATP, **54:**404
FeP-binding, **67:**344–347
hydrolysis, **67:**357–358
iron protein redox potential, **67:**313–314
role in *nif* genes, **67:**330–331
MIA, *See* Monoiodoacetate
Mice, jimpy, **42:**266
Micelle aggragates, **34:**486
structure, **34:**468
Micelles, **43:**377; 383, **51:**33, **58:**195
activity coefficients, **43:**388; 393
binding of colipase to, **48:**355
lipase activity on, **48:**323–324; 345
destabilization, **43:**393

Micelles (*Continued*)
dissociation, **9:**6; 9–15; 17–20
entropy effects, **43:**389
free energies in, **43:**388
phospholipid, effect of antibiotics of alkali-metal permeability, **29:**47
stabilization, **43:**393
Michael addition, quinones, **65:**100
Michaelis and Menten equation, **7:**3
Michaelis complex, **51:**126–127, **57:**71; 73. *See also* Enzyme complex
activated, **27:**29–32
of D-amino and oxidase, conformational change, **27:**23–29
crystallization of, **27:**2–4; 9–10; 16–23
mechanism study, **27:**1–36
Michaelis constant, **7:**3; 392, **29:**1–2, **33:**274, **41:**4; 8; 15; 255; 263; 275, **51:**232; 233
for D-amino acid oxidase, **57:**108; 111
apparent, **34:**460–465; 502; 522–523; 525
for arylsulfatase A, **60:**175
for D-aspartate oxidase, **57:**119–120
for bradykinin analogs, **41:**263
for CDP-ribitol in *Lactobacillus plantarum* and *Staphylococcus aureus,* **30:**240
of dextran, **30:**194
for glucosylation in *Bacillus subtilis,* **30:**242
for levansucrase, **30:**207
for poly (ribitol phosphate) in *Bacillus subtilis,* **30:**242
for dephosphorylation of ITP, **16:**337
for D,L-thiazolidine-2-carboxylate, **57:**111
in exchange transport and non-exchange transport, **29:**52
for α-L-fucosidase, **60:**159
for fumarase action, **25:**180–181
of β-galactosidase, **60:**251
for β-glucuronidase, **16:**369
for L-α-hydroxy acid oxidase, **57:**127
for hypoxanthine-guanine phosphoribosyltransferase, **41:**15
for lipase, **23:**144
for lysyl hydroxylase, **41:**275
for myosin-ATPase, **16:**336
for O_2, **57:**107–108
optimal value of, **43:**329
for P-P ribose-P amidotransferase, **41:**8
for PP-ribose-P synthetase, **41:**4
for prolyl hydroxylase, **41:**255
of purified amino acid decarboxylases, **6:**16
of succinic dehydrogenase, **18:**91–92
Michaelis curves, for lipase, **23:**143–148
Michaelis-Davidson model, **34:**492
Michaelis dissociation constant, **4:**44; 55
Michaelis-Henry equation, for aspartate transcarbamylase, **28:**128
Michaelis-Menten complex, **15:**3–4; 18; 28
Michaelis-Menten constant
application to hydrolytic enzymes, **15:** 277–280
of thiaminase complexes, **15:**399
Michaelis-Menten equation, **33:**248; 262–263; 275; 403–404, **34:**462
integrated, **68:**23
kinetics, **34:**497; 502
Michaelis-Menten kinetics, **25:**169–170; 174, **33:**288; 418, **45:**429, **50:**342, **58:**234. *See also* Enzyme-substrate complex; Product inhibition of enzymes
of allosteric enzymes, **28:**49; 52; 83; 122; 131; 134; 136–137
of β-lactamase action, **28:**248
relaxation study, **25:**8
Michaelis-Menten parameters, protein tyrosine phosphatases, **68:**11–12
Michelian kinetics, enzyme action, steady-state hypothesis, **73:**26–29. *See also* V/K constant
Microbacterium audouini, collagen action, **23:**177
Microbacterium butyricum, use of acetate, **23:**411
Microbacterium canis, collagen action, **23:**177
Microbacterium gypseum, action on collagen, **23:**177
Microbacterium mentagrophytes, collagen action, **23:**177
Microbacterium rubrum, collagen action, **23:**177
Microbial antagonism, **7:**462
Microbial assimilations, **6:**269–305
acetylmethylcarbinol, **6:**300
amino acids, **6:**271; 300–305
butyric acid synthesis, **6:**271; 300
caproic acid synthesis, **6:**300
carbohydrate synthesis, **6:**271
carbon, **6:**272–287
carbon dioxide, **6:**298–299; 300
dextran synthesis, **6:**300

fat synthesis, **6:**271
glycogen resynthesis, **6:**270; 271
glycogen synthesis, **6:**299–300
levan synthesis, **6:**300
nitrogen assimilation, **6:**300–305
poisons affecting, **6:**287–298
polysaccharide synthesis, **6:**299–300
Microbial Assimilations (Clifton), **6:**269
Microbial metabolism
catabolite inactivation in, **54:**172–173
catabolite repression in, **54:**173
fructose-bisphosphatase in, **54:**168–174
Microbial oxidations, of hydrocarbons, **27:** 469–546
Microbial Oxidations of Petroleum Hydrocarbons, Mechanisms of (van der Linden and Thijsse), **27:**469
Microbial proteinases, **41:**179–239
acid proteinases, **41:**180; 185; 208–212; 234; 237
alkaline, **41:**189–195; 204–207; 216–226
amino acid sequences of, **41:**239
extracellular, **41:**180
metal proteinases, **41:**180; 183–184; 200–207; 226–232; 237; 238
neutral, **41:**200–204
primary specificity, **41:**180
secondary interaction, **41:**180
serine alkaline proteinases, **41:**180; 182; 188–190; 192; 193; 216–226; 237
serine trypsin-like proteinases, **41:**182; 188; 213–216
side-chain specificity, **41:**181
tertiary structures of, **41:**239
thiol porteinases, **41:**180; 197–200
Trypsinlike, **41:**188–189; 213–216
Microbiological activity, of soils, **28:**365–390
Microbiological degradation of cellulose, **8:**253–298
Microcalorimetry, **34:**468
Microcapsules(nylon and collodion), **34:**448
Micrococci, oxidation of hydrocarbons, **27:**474; 488; 502
Micrococcus aerogenes, N^5-formyltetrahydrofolic isomerase in, **21:**413
tetrahydrofolic formylase in, **21:**226; 407; 408
Micrococcus caseolyticus, proteinase from, **41:**183
Micrococcus cerificans, oxidation of hydrocarbons, **27:**473; 489; 490; 502
Micrococcus dehydrogenans, *See* Flavobacterium
Micrococcus denitrificans, **46:**87
use of acetate, **23:**411–413; 415; 417; 420
Micrococcus glutamics, and biosynthesis of ornithine, **33:**374
Micrococcus hyicus, wall teichoic acid of, **30:**230
Micrococcus lactilyticus
fumaric reductase in, **18:**103
succinic dehydrogenase from, **18:**73; 88
propionic acid metabolism, **26:**292
propionyl CoA carboxylase in, **19:**39
Micrococcus luteus, **37:**2
Micrococcus lysodeikticus, **34:**485
cell walls, **26:**200–204; 206–207; 210–211; 214; 217; 223
propionic acid metabolism, **26:**293
and citric acid cycle, **18:**248
nucleic acid formation in, **18:**228
CO_2 fixation by, **15:**144
β-lactamase of, **28:**242
metabolism in, **74:**23; 25; 27–28
oxalacetic carboxylase in, **19:**39
oxidoreductase from, **27:**370
separation of enzymes from, **14:**379; 405
Micrococcus pyogenes (Staphylococcus) albus, cytochromes of, **23:**266–267; 315
Micrococcus radiodurans, superoxide dismutase from, **41:**79
Micrococcus roseus, aminoacyl-tRNA specificity, **40:**130
biosynthesis on interpeptide bridges, **40:**124; 125
Micrococcus sodonensis, use of acetate, **23:**411; 412
Micrococcus sodoneusis, allosteric enzymes, **28:**69
Micrococcus sphaeroides, oxidation of hydrocarbons, **27:**472; 506; 508
Micrococcus urea, oxidation of hydrocarbons, **27:**509; 515
Microenvironment, **34:**446; 456; 494; 512; 524
effects, **34:**447
lipid, **34:**472
Microenvironmental compartment, **43:**93
Microenvironmental factors, **43:**118
Microenvironments, **43:**378

Microfibril, of residual protein, **16:**447
Microfungi, and β-lactamase, **28:**312–313
Microheterogeneous systems, **34:**446
Microinjection, microtubule dynamics, **69:**140
Micromicellar solutions of enzymes
effect of temperature on, **51:**55–58
rapid mixing of, **51:**65–70
Micromolar, definition, **7:**409
Microorganisms. *See also under* individual organism, e g, Aerobacter
action of hydrocarbons on, **10:**451–454
action on hydrocarbon derivatives, **10:** 462–464
action on hydrocarbons, **10:**443–486
criteria used in determining activity, **10:**445–448
factors favoring, **10:**448–449; 470
oxidation, **10:**454–462
products, **10:**464–470
variety attacking hydrocarbons, **10:**445; 449–451
decomposition of rubber hydrocarbons, **10:**474–476
modification of petroleum and its products, **10:**470–474
asparagine metabolism in, **13:**249–250; 250–253
enzymic adaptation in, **13:**70–73; 76
enzymic transformations of steroids by, **20:**237–282
fat synthesis by, **9:**653–704
glutamine metabolism in, **13:**253–262
glutathione metabolism in, **13:**262–263
hydrolysis of cephalosporins, **28:**242–244
laminarinases of, **28:**341–343
pentose interconversions, **28:**394
tracer studies, of DR synthesis, **28:**423–431
metabolism of 2-carbon compounds, **23:** 401–470
phenylalanine and tyrosine metabolism in, **14:**111; 112
photosynthetic, nitrogen fixation studies, **13:**357–360
relation of virulence to hyaluronidases, **13:**219–221
role in wine making, **3:**349–386
transferases in, **13:**290–296
Microorganisms, Assimilation of Hydrocarbons by (ZoBell), **10:**443
Microorganisms, Enzymic Transformation of Steroids by (Vischer and Wettstein), **20:**237
Microorganisms, Interrelations Between Growth and Metabolism of Vitamin-like Substances in (McIlwain), **7:**409
Microorganisms, Metabolism of 2-Carbon Compounds by (Kornberg and Elsden), **23:**401
Microorganisms, Na^+-dependent transport in, **49:**83
Microorganisms effect of fluoroacetate on, **18:**134–136
lipoic acid requirement, **18:**342
photosynthetic reactions, **18:**53–54; 56–57
tricarboxylic acid in, **18:**242–251
Microperoxisomes, **57:**87–89
Microscopy, Electron, of Enzymes (Haschemeyer), **33:**71
Microsomal cytochrome, **34:**482
membranes, **34:**478
phosphatidic acid phosphotase, **34:**474
phospholipid, **34:**475
Microsomal enzymes, and nuclear purity, **27:**251–253
Microsomal ethanol-oxidizing system (MEOS), **45:**437–441
induction, **45:**472
of MEOS activity by chronic ethanol ingestion, **45:**439
Microsomal forms, glutathione S-transferases, **69:**4; 6
Microsomal oxidase, liver, **46:**60
Microsomes, **46:**15, **47:**3. *See also* Cytoplasm, Particulate material and Particles; Ribosomes and Subcellular fractions
adsorptive properties, **28:**31
from *Arum* spadix, **18:**26–28
and cytochrome oxidase, **18:**15
plant, terminal oxidase of, **18:**48
catalytic system for ribonuclease, **26:** 102–103
coagulant activity of, **30:**286–287
disulfide exchanges, **27:**219
EPR study, **27:**190–191
and enzyme formation, **24:**5; 14
enzymes in, **13:**3; 6; 8–10
liver, DPNH-specifi flavoprotein of, **20:**171
liver, effect of ethionine on, **25:**377–379
methylation by, **25:**418–421

Microsomes, RNA of, **16:**438
Microsomes (egg white), peptide and protein biosynthesis in, **19:**261
Microsomes (liver), as site of peptide biosynthesis, **19:**261
Microtactoids, **9:**196–197
Microtubule
  assembly regulation, **53:**143–144
  in cellular processes, **53:**116
  nucleotide-dependent enzymes associated with, **53:**113–156
  protein, phosphorylation of, **53:**152–154
  protein components, **53:**117–119
  structure, **53:**114–117
Microtubule-associated proteins (MAPs), **53:**117–119
Microtubule dynamics, **69:**121–154
  assembly/disassembly turnover, **69:**132–136
  assembly with GTP analogues, **69:**136–139
  control options, **69:**143–145
  dynamic instability, **69:**134–136
    generally, **69:**132–133
    treadmilling, **69:**133–134
  future research directions, **69:**147–149
  generally, **69:**121–123
  GTPase reaction, pharmacologic implications of, **69:**146–147
  intracellular dynamics, **69:**139–143
  tubulin, guanine nucleotide interactions with, **69:**123–124
  tubulin-GTP/GDP cycle, bioenergetics of, **69:**124–134
Microtubulins, dynein adenosine 5′-triphosphatase, interaction with, **53:** 146–150
Midgut, **45:**401; 407
Midpiece of complement, **8:**101; 103–111
Milk
  clotting of, **15:**423–448
    coagulation by thiaminase-producing bacteria, **15:**413
  phosphatides in, **7:**166–168
Milk Clotting and Rennin (Berridge), **15:**423
Milk-clotting enzyme in cacao, **25:**466
Milk fat, precursors, **7:**167–168
Milk oligosaccharides, **25:**444–445, **52:**30
Milled wood lignin (MWL), **34:**211
Millon's reaction, for staining nuclei, **16:** 419–420
Mimosa, as a source of mimosine, **29:**108
Mimosa pudica, **20:**368
Mimosine, **50:**119; 147
  infertility in rats caused by, **29:**108
    inhibition of growth in *E. coli* by, **29:**108
    inhibition of hair growth by, **29:**108
    mechanism of toxicity, **29:**108
    naturally occurring tyrosine analog, **29:**108
    in plants, **29:**94
    reaction with pyridoxal phosphate, **29:**108
    toxic effects on animals, **29:**108
L-Mimosine, collagen hydroxylase inhibitor mechanisms, **72:**362–364
Mimusops balata, gutta obtained from, **29:**222
Mineral changes in urine, in myasthenia gravis, **6:**212
Mineral deposits, formation dates, **27:** 354–355
Mineral media, for bacteria, **27:**471–472
Minimal inhibitory concentration (MIC), of penicillins for *Staphylococci,* **28:**258
Minimal medium, definition, **16:**251
Minor loop, **35:**473
Miracil D, **40:**84
Mirror-image catalysis, **59:**110
Mirsky protein, **9:**283–285
Mirsky reagent, **18:**179
Miscellization, Solubilization and Microemulsion (ed. Mittal), **51:**31
Misorientation, **39:**354
Misrecognition, **35:**492
Mitchell copper loop model, cytochrome *c* oxidase, proton pumping, **71:**158–160
Mitochondria, **15:**33; 125–127; 129–131; 330; 334; 367, **34:**466; 470; 477; 481, **39:**359; 360, **46:**40; 59; 123. *See also* Coenzyme Q; Cytoplasm, Particulate material and Particles; Cytoplasmic particles; Granules, subcellular; Particles; Particles, cellular; Particles and Particulate material; Subcellular fractions
  accumulation of ADP in, **29:**282 (*See also* Calcium uptake, Valinomycin, Gramicidin)
    alkalinization of, during Ca^{++} uptake, **29:**280–281
    ATPase system in, **29:**73–74
    ATP hydrolysis in, **29:**271
    calcium uptake in, **29:**264–265; 271
      inhibition by oligomycin, **29:**266

Mitochondria (*Continued*)
cation and anion content of, **29:**261–263
chemical coupling hypothesis for ion transport of, **29:**298–303
citrate accumulation in, **29:**312
inhibition by 2,4-dinitrophenol and cyanide, **29:**312
decrease in respiration with oligomycin in, **29:**349
effect of antibiotics on K^+ uptake, **29:**46
efflux of Ca^{++} from, **29:**273
ejection of K^+ during calcium uptake, **29:**281
energy-dependent accumulation of divalent cations in, **29:**264–277
Sr^{++} and Mn^{++} accumulation, **29:** 273–274
energy-linked accumulation of Mg^{++}, **29:**274–275
energy-linked ion movements in systems of, **29:**259–320
folding of proton-translocating respiratory chain, **29:**306
and H^+-translocator oxidoreductases, **29:**76
hydrogen ion ejection in, **29:**278
hydrogen ion pump hypothesis, **29:** 303–311
ion translocation process in, **29:**311–312
in latex, **29:**228–229
limited loading with Ca^{++}, **29:**267–274
properties of, **29:**267–268
location of calcium phosphate in, **29:** 275–276
locus of inhibitors of, **29:**343
maintenance of electrical potential across membrane, **29:**47
malate accumulation in, **29:**312
massive loading of Ca^{++} in, **29:**264–267
movements of monovalent cations in, **29:**285–294
NAD incorporation in, **29:**311–312
and Na^+/K^+ antiporter ATPase, **29:**73
oxidation of DPNH by, **29:**342
oxidative phosphorylation and its relationship to the mechanism of ion transport of, **29:**296–314
permeability to choline, **29:**46–47
pH changes in, **29:**48
physiological aspects of ion uptake of, **29:**312–314
potassium exchange in, **29:**285
potassium ion uptake in, **29:**286–288
pyruvate oxidation *vs.* nitrogen content of, **29:**337
rat liver, action of digitonin in, **29:**294–295
action of 2,4-dinitrophenol in, **29:** 294–295
effect of parathyroid hormone on cation translocation, **29:**46
metal content of, **29:**262
rebounds and oscillations in ion movements of, **29:**294
respiration-independent K^+-H^+ exchange between the medium and, **29:**292–294
site of action in drugs, **29:**374
soluble components of, **29:**263
suspension effect in, **29:**292–294
tumor, **29:**326
failure to isolate, due to, **29:**326
ultrastructural changes after divalent cation uptake, in, **29:**275–277
uptake of anions, **29:**67; 277–278
acetone extracted, **34:**474
adrenal, oxidation of reduced diphosphopyridine nucleotide by, **20:**172
constitution of respiratory chain in, **20:**147–199
heart, analysis and enzymic activities, **20:**153; 188; 191
in study of respiratory chain, **20:**152; 155–180; 182; 184; 185; 187–189; 191–193
kidney, succinic oxidase inactivation in, **20:**176
liver, analysis and enzymic activities, **20:**153
in study of respiratory chain, **20:**152; 156–158; 160; 171–174; 183; 185
allosteric regulations, **28:**77; 78
from *Arum* spadix, **18:**25–28; 47
from barley seedlings, **18:**40
cytochrome b_7 in, **18:**25
effect of fluoroacetate on, **18:**131–133
from leaves, **18:**43–44
plant, **18:**299; 305–308; 309–310; 311
respiratory activity of, **18:**13–15; 36
and salt respiration, **18:**52
and succinic dehydrogenase, **18:**67–68; 71–74; 77; 80; 87; 104–108
in synthesis of ATP, **18:**58

and ATPase activity, **16:**342
of liver cells, **16:**366–367
melanin granules as, **16:**169–170
plant phenolases in, **16:**152
RNA of, **16:**438; 461–462
as seat of terminal oxidation, **16:**489–490
ATPase activity in, **13:**168
in ATP preparation, **23:**326; 327
contractile structure of, **23:**383–385
effect of calcium on, **23:**352; 360; 366; 368
jump experiments, **23:**376–377
oxidative phosphorylation in, **23:**334; 340–385
beef liver, **46:**58
and cell organization, **8:**24
and chromidia, **8:**4–6
and defense mechanisms, **8:**52–55
and functioning of cytoplasm, **8:**20–23
and respiratory enzymes, **8:**31–37
composition, **21:**7–10
electron micrographs, **21:**84–86
enzymic functions, **21:**11–56; 74; 81
relation to substructure, **21:**57–60; 76
identification, **21:**2; 3
ion accumulation by, **21:**56; 57
isolation and purity, **21:**4–7; 74; 75; 78; 79
metabolism in, **21:**1–72; 266; 271; 272; 428; 429
copper of, EPR study, **27:**148–155; 187–190
disulfide exchanges, **27:**219
DPN synthesis, **27:**276–277
EPR study, **27:**187–190
reflectance spectroscopy, **27:**188–190
swelling of, **27:**228–229
degradation in reticulocytes, **58:**255
electric properties, **58:**244
electron transfer system, **58:**243
in energy transport, **14:**40
enzyme action in, **33:**294–300
oscillatory states in, **33:**296; 298–300
ATP, **33:**297
conformation changes, **33:**297
damping in, **33:**299–300
proton evolution, **33:**297
sodium-ion production, **33:**297
permeability of, **33:**292
respiration of, **33:**296–297
ultrastructure of, **33:**298
enzyme adsorption, **28:**2
and enzyme formation, **24:**5
enzymes in, **13:**3; 7–15
heart, allosteric enzymes, **28:**137–138
isolated, from rat intestine, **53:**219
isolation of enzymes of, **14:**380; 398
leucine 2,3-aminomutase in, **58:**180
lipid, **34:**468
liproprotein complex from, **28:**209; 212; 214; 218
liver, effect of carcinogens on, **25:**377–379
organization of, **24:**25
oxidation and reduction of coenzymes Q, **25:**305–319
preparations, **34:**472
respiratory chain from, **28:**155–236
reticulocyte lipoxygenase action on, **58:**242
thiophosphate analogs of ADP and ATP, translocation of, **43:**13
tyrosinase in, **14:**74
yeast, oxidase of, **24:**468
Mitochondrial ATPase, **23:**360–363, **43:**29; 36; 39, **49:**223–280
catalytic properties, **49:**235–261
amino acid residues in active site, **49:** 252–254
binding sites, **49:**256–259
catalytic expressions of soluble F_1, **49:** 239–251
catalytic sites, **49:**259–261
differences in properties of soluble *versus* membrane-bound F_1, **49:**254–256
hydrolytic activity, **49:**239
inhibitors and activators, **49:**243–252
tightly bound nucleotides, **49:**236–239
inhibition by AMP-PNP, **43:**20
isolation, **49:**228–230
enzyme assay, **49:**223–230
extraction, **49:**228–229
structure, **49:**230–235
molecular weight of native enzyme, **49:**230–231
subunit stoichiometry, **49:**233–235
subunit structure, **49:**231–233
Mitochondrial ATPase (Penefsky), **49:**223
Mitochondrial cycle, **23:**345–346
Mitochondrial electron transfer, elementary particle, **28:**219
Mitochondrial electron transfer system, **34:** 474–476
membrane, **34:**481; 492
phospholipids, **34:**473

Mitochondrial enzyme, in methylation, effect of lipid solvents, **25:**418
Mitochondrial fragments. *See also* Electron transport particles and Phosphorylating electron transport particle
composition and enzymic activities, **21:**83
electron micrographs, **21:**87–93
isolation, purity and properties, **21:**75–79; 102–107
notation system for, **21:**101; 102
oxidative phosphorylation in, **21:**38; 39; 80–83; 87; 91; 102; 112; 123; 124
serine biosynthesis in, **21:**429
Mitochondrial gel, models for, **15:**27
Mitochondrial ghosts, energy-linked ion movements in, **29:**294–296
Mitochondria (liver), as site of amino acid activation, **19:**261
Mitochondrial malate dehydrogenases, multiple forms, **25:**341–343; 346–347
Mitochondrial membrane
broken, for respiratory chain study, **28:**164
inner phosphorylating apparatus of, **49:** 225–227
Mitochondrial metabolism, relation to nucleus, **16:**472–473
Mitochondrial Metabolism (Schneider), **21:**1
Mitochondrial O cycle, **65:**126–127
Mitochondrial preprations, comparison with cytochrome *c*-cytochrome oxidase complex, **28:**207–208
Mitochondrial Respiratory Chain, DPNH Dehydrogenase of (Singer and Gutman), **34:**79
Mitochondrial structure, and function, **23:** 345–346; 383–385
Mitochondrial superoxide dismutases, **41:**35; 73–78
evolutionary significance, **41:**76
intracellular distribution, **41:**73
intramitochondrial localization, **41:**75
site of synthesis, **41:**75
Mitochondrial system, ethionine effect on, **25:**380; 381
Mitochondrial Systems, Energy-Linked Ion Movements in (Lehninger, Carafoli, and Rossi), **29:**259
Mitochondrial transport
asparagine synthetase, **72:**149–151
hepatic glutamine transport
glutaminase, **72:**54
intramitochondrial glutamine concentration, **72:**47–48
plasma membrane and, **72:**45–47
Mitochondrion, **34:**479, **49:**56; 75
enzymic characterization of, **18:**14
inner membrane crossing by amino acids, **49:**56
Mitochrome, **17:**71; 84; 106
effect on mitochondrial oxidative phosphorylation, **21:**32; 37; 38
Mitomycin, mutagen, **27:**330
Mitomycin c, **60:**16
Mitomycin C, induced nuclease in the K12(λ) cell, **29:**201
Mitomycin *c,* synthesis of gramicidin, S, inhibited by, **33:**342–343
Mitosis
imidazole in, **20:**91
role of glutathione in, **11:**253–255; 257
synthesis of DNA and histone in, **18:**184–185
Mitotic rate, and pyridine nucleotides, **27:**277
MIX, *See* Molecular isotope exchange
Mixed function oxidases, classification, **19:**80
model system for, **19:**134–143
in oxygen metabolism, **19:**128–199
properties, **19:**199
Mixed function oxidation, **47:**2; 11
Mixed solvents, **45:**159
acid-base equilibria, **45:**171
catalytic activity, **45:**186
cooled solutions, preparation of, **45:**166
dielectric constant, **45:**161
enzymes, solubility of, **45:**185
experimental procedure, **45:**190
experimental value of pa_H, **45:**178
ionic strength, **45:**169
isodielectric with water, **45:**167
potentiometric determination of pa_H, **45:**173
viscosity, **45:**161
Mixed vesicles, **34:**480
MMPR, **54:**101
Mn^{2+}, **53:**51; 52; 54; 55; 56; 57; 58; 59; 61; 63; 67; 72
MnADP in creatine kinase activation, **54:**412
MN antigens, **52:**122
Mn^{2+} binding, **53:**56
MN Blood group, **52:**121–132
biochemical basis of, **52:**122–126
restoration of, **52:**126–132

Mobility, electrophoretic, **48:**59
Mobility-shift method, speed-up sequence analysis, **67:**442–443
Mobilization of plasma iron, **44:**213
Model, photomultiplier, of the clotting cascade, **30:**297–299
Model catalysts, **39:**358
Model coenzymes, of pyridine nucleotides, **15:**87; 92–93
Model complex, mechanism study, **27:**3–12; 23–28
Model compounds, haem *a*, **23:**289–312
  absorption spectra, **23:**289–297; 306–310
Model enzyme, of thiaminase, **15:**404
Model enzyme systems, **15:**2; 14–15; 21–32; 22*t*; 39
Model hydroxylating system, **19:**134–137; 139–143; 199
Model reaction, for peptide biosynthesis, **19:**263
Model reactions for RNase, **24:**251–252; 444–448; 455
Models for enzyme inactivation by ionizing radiations, **24:**361–364; 371
Models for hydrogen transfers, PN-dependent, **26:**132–156
Models of receptor-adenylate cyclase interactions
  allosteric models, **53:**28
  collision coupling model, **53:**35
  dynamic receptor affinity model, **53:**35
  floating receptor model, **53:**35
  general unifying model, **53:**33–35
  kinetic approaches, **53:**22–26
  ternary complex model, **53:**29
Model substrates, **49:**56
  amino acids as, **49:**60
  for system ASC, **49:**75–77
Model systems, **34:**446; 484; 495; 513
  of DPN action, **24:**470
    of esterases, **24:**461–463
    of phosphoglucomutase action, **24:**465–467
Model systems of oxidative phosphorylation, **23:**370; 377
Modification, chemical, **42:**126; 127
Modular mutagenesis, **69:**3
Möena anomaly, **16:**73
MoFe protein, **67:**333–339. *See also* FeP-MoFeP interaction
  metal clusters, positioning, **67:**352–353
  α-subunit, **67:**334–336; 338
  β-subunit, **67:**336–339
$\alpha_2\beta_2$ MoFe protein, **67:**316–317
  x-ray crystallography, **67:**318–323
Moffitt equation, Cotton effect, **27:**51; 52
Molar enzymic activity of ATPase, **6:**168
Molasses and yeast production, **9:**654; 668–669
Molasses in fermentation, **7:**585–587
Mold, transferring enzymes, **17:**196–200; 202; 204; 222; 226; 227
Mold bran, development, **7:**591–595
Mold collagenase, from *Trichophyton Schoenleinii,* **32:**147
Molds
  acid formation in, **9:**537–584
  collagenase activity, **23:**176–177
  wood-destroying, **8:**269–279; 290–295
    degradation of simple sugars by, **8:**271–274
Molds, Acid Formation in (Walker), **9:**537
Molecular aggregates, in DNA solutions, **18:**165–168
  of histones, **18:**170–174
Molecular association of enzymes, **24:**119–120
Molecular asymmetry, **43:**135
Molecular biology, **59:**12–16
  and origin of life, **27:**395–399
Molecular changes with inactivation of enzymes, **24:**397–399
Molecular chaperone activity, α-crystallin, **69:**163–165
Molecular docking, adenosine receptors, **69:**94
Molecular genetic analysis, *Caulobacter* cell cycle, flagellar structure and assembly, **71:**8
Molecular genetics, survey, **27:**291–295
Molecular integrity, **42:**150; 152
Molecular isotope exchange, **50:**361
Molecular Mechanisms of Coupling in Hormone Receptor-Adenylate Cyclase Systems (Stadel, De Lean, and Lefkowitz), **53:**1
Molecular orbital calculations, **47:**17
Molecular orbitals of proteins, **15:**34–35
Molecular organization, **34:**479
  oxygen, **34:**476
  topography, **34:**479
  weight, **34:**482
Molecular properties, **44:**189
Molecular Properties and Transformations of Glycogen Phosphorylase in Animal Tissues (Krebs and Fischer), **24:**263
Molecular recognition, *See* Oligosaccharides

Molecular structure, of collagen, **23:**169; 213; 217–218
of proteins, **23:**1–27
Molecular weight, **39:**193; 218; 347; 364, **41:**6; 7; 11; 16; 250; 274, **42:**78; 84; 86; 88; 147; 148
of biopolymers, **30:**189
of hypoxanthine-guanine phospho ribosyl transferase, **41:**16
of lipoxidase, **19:**99
of mercuripapain, **19:**271
of papain, **19:**269; 271; 277; 278; 283–285
of liver alcohol dehydrogenase, **20:**37; 38
of pectin, **20:**344
of sialic acid and urine mucoprotein, **20:**141
of lysyl hydroxylase, **41:**274
of PP-ribose-P amidotransferase, **41:**11
of PP-ribose-P synthetase, **41:**6; 7
of prolyl hydroxylase, **41:**250
of proteins, **1:**14; 15; 18; 19; 20
Molecular weights of enzymes, from irradiation studies, **24:**365–367
Molecular weights of proteins, **9:**123–129
Molecular weight studies, **46:**366
Molecule, concept of, in biological material, **4:**17–18
Molecule dimensions by adsorption studies, **14:**323–324; 329–331
Molecule orientations, in photosynthetic structure, **25:**71–73
Molecules, amphipathic, **34:**483
lipid, **34:**446
protein, **34:**446
Molluscs, β-glucuronidase of, **16:**381–383
Molluses, laminarinase from, **28:**342; 344; 354
mycolytic powers, **28:**355
Molting cycle, **45:**417
Molting fluid, **45:**405; 417; 418
Molting gel, **45:**417; 418
Molting insect, **45:**415–417
Molybdate
and acetyl phosphate hydrolysis, **6:**235; 236
effect on fluorocitric acid, **18:**116
Molybdenum, **48:**17
and aldehyde oxidase, **19:**202
and xanthine oxidase, **19:**202
as component of metalloflavoproteins, **17:**237–251; 253; 268–270; 276; 282; 284; 285
labeled, **17:**237; 305
in xanthine oxidase, **17:**237; 294; 302; 305–308; 313–318; 321
effect on formate dehydrogenase activity, **48:**18
in melanin, **16:**172
in plant metabolism, **9:**398
in xanthine oxidase, EPR study, **27:** 160–168
Molybdenum centers, **51:**107–165
deuterium substitution, exchange and transfer, **51:**138–143
domains and cofactors, **51:**111–113
enzymes, **51:**108–110
EPR spectra and, **51:**117–147
EXAFS method and, **51:**147–149
general features, **51:**117–121
generation of the spectra, **51:**121–122
high-pH forms, **51:**128–130
information from various techniques, **51:** 149–159
inhibitory side products, **51:**131–135
intermediates, **51:**130–131
introduction to, **51:**108–117
methods of studying, **51:**113–117
nonfunctional degradation products, **51:** 135–138
problems and interpretations, **51:**152–159
prospects, **51:**159
substitution with other isotopes, **51:** 143–147
substrate product, and anion complexes, **51:**122–128
summary and inferences, **51:**149–152
Molybdenum cofactor, **51:**111–113; 159
Molybdenum compounds, EPR study, **27:**142
Molybdenum compounds, solubilization of, **17:**138; 143; 145; 149
Molybdenum deficiency, effect on xanthine oxidase, **17:**320–321
Molybdenum domain, **51:**111–113
Molybdenum-iron flavoproteins, EPR study, **27:**119; 180; 186–187
Molybdenum (V), in aldehyde oxidase, **27:** 168–172
chemical species, EPR study, **27:**161–166
Molybdoflavoproteins, **17:**269–271; 284; 332
Momilactone, **55:**78
Momilactone A, **55:**79; 99
structure of, **55:**55

Momilactone B, **55:**79
structure of, **55:**55
Monactin, **35:**176
properties of, **29:**289
Monamine oxidase, **19:**203
Monilia, and mevalonic acid, **23:**473
Monilinia fructicola, **55:**36
plant disease caused by, **20:**399
Monkey growth hormone, **42:**75
Monoacetyl-1,4-diaminobutane, cooled, **36:**222
in *E. coli,* **36:**222
Monoacetylmonovinyldeuteroporphyrin, **23:**282
Monoacetylspermidine, cooled, **36:**222
in *E. coli,* **36:**222
Monoamine dehydrogenase, activity of –SH group, **11:**241
Monoamine oxidase, **35:**85; 86
location of, **24:**302; 330
stereochemistry of reaction, **35:**85–86
Monoamine oxidase (MAO)
mechanism of action of, **29:**483–485
mitochrondial enzyme, **29:**483
in schizophrenia, **29:**517
Monoamine oxidase (MAO) inhibitors, **29:** 483–485
additive effects of imipramine, **29:**485
effect on assay of plasma fractions in rope-climbing test, **29:**512
effect on glucose utilization, **29:**489
effect on patients with affective disorders, **29:**493–494
effect on transmitter amines, **29:**490
and LSD 25, **29:**518
Monoamine oxidases, classification, **14:**253
Monoamine r O_2 transhydrogenases, **14:**253
Monoaminooxidase, **58:**245
Mono- and dipyrroles, absorption spectra, **7:**358–362
Monocarboxylic amino acids, in transaminations, **16:**203–204
Monoclonal antibodies
cytochrome *c* oxidase, **71:**99–101
RNA polymerase II transcription, reversible phosphorylation of CTD during transcription cycle, **71:**53–55
Monoclonal antibody, fibronectin, central cell-binding domain of, **70:**10
Monodehydroascorbic acid
reduction by PN-enzymes, **26:**123–127
in respiratory chain, **20:**172
Monod-Wyman-Changeux model, of allosteric control, **28:**56–60; 131; 137
Monoesterases, soil, **28:**386
Monofluoroalanine, **55:**231
Monofluorocitric acid, **18:**116–117
Monogalactosyldiglyceride, **60:**115
in chloroplast lamellae, **30:**9
Monoglyceridase, **23:**153
Monohydric phenols, oxidation by tyrosinase
constant rate of, **4:**138–141
cresolase activity of tyrosinase measurements, **4:**116–117
high cresolase preparations, **4:**107–109
induction period, **4:**135–138
mechanism, experimental evidence, **4:**133–135
steps, **4:**135–137
structure of phenol, **4:**141–142
Monohydroperoxypolyenoic acids, **58:**277
Monohydroxyacetone phosphate, **15:**114
Monohydroxyethylmonovinyldeuteroporphyrin, **23:**282
Monoiodoacetate, action on dehydrogenases, **23:**46–48
Monoiodotyrosine
reaction with thiourea, **27:**211–212
synthesis, in mitochondria, **21:**27
Monoiodotyrosine in thyroxine formation, **14:**112–115
Monolayers. *See also* Surfaces
adsorption of ions on, **11:**67–69
complex formation in, **11:**82–85
factors affecting rate of reaction in, **11:**58–69
photochemical reactions in, **11:**67; 76; 79–82
Monomer, **57:**196–201
Monomer-dimer equilibrium, **39:**350
Monomeric forms, **42:**84; 85
Monomer units, **34:**487
N^G-Monomethyl-L-arginine, **42:**263
ε-*N*-Monomethyllysine, ratio to ε-N-dimethyllysine, **42:**256
Mononuclear cu_A models, binuclear *versus,* cytochrome *c* oxidase, structural biochemistry, **71:**93–99
Mononucleotide pyrophosphorylase, allosteric regulation, **28:**123–124

Mononucleotides, hydrolysis, by metal ions, **21:**174; 175; 184; 186
identification, with phosphatase models, **21:**192
Monooxygenase, **60:**38
Monooxygenases, **55:**248–251
cytochrome P_{450}, **55:**278
fluorinated steroids and, **55:**263
in vitamin D_3 conversion, **55:**276
Monoperphthalic acid, **35:**454
Monophenol oxidase, cacao, **25:**466
Monophenols
and indolylacetic acid oxidase, **19:**124; 127
pigments from, **16:**112
Monophosphoadenosine, **15:**83
Monosaccharide interconversions, allosteric regulation, **28:**110–113
Monosaccharides, **56:**143. *See also* Carbohydrates
analogues, asialoglycoprotein receptor specificity, **66:**44–46
formation of, **26:**37–57
hydrogen bonding, **65:**231; 233; 235–236
Monosaccharide transfer, mechanism of, **31:** 88–92
Monosialoganglioside, **48:**74
Monosialoganglioside G_{M1}, structure of, **50:**304
Monosialotrihexosylceramide (Ganglioside G_{m2}, **38:**295
Monothiols, **15:**219, **39:**360; 362
Monovalent activators, **51:**243
Monoxygenase, **46:**166
Monoxygenases, **47:**1–40
Montierella sp., action on collagen, **23:**177
Mo nucleus, **51:**115
Moracins, **55:**91
Morphine
antagonism to N-allylnormorphine, **6:**41
metabolism in plants, **16:**142
Morphine and plant metabolism, **8:**242
Morphogenesis, molecular, **26:**193–232
Morphology, and bacterial sacculi, **26:**193–232
of plant development (*See* Phytochrome)
Morquio disease
type A, **60:**263
type B, **60:**185; 253
Morris and Novikoff hepatomas, fast-growing, **42:**256
Mosaic films, **28:**9; 11–12; 23
Mosaic model, **34:**471–472
Mosquito, *See* Aedes aegypti
Mössbauer spectroscopy, **27:**107; 139, **47:**22; 26
cytochrome *c* oxidase, **71:**87; 110; 111
Motility of protoplasm, **8:**37–42
Mouse, adenylosuccinate synthetase (AMPSase) gene structure, **73:**91–93
Mouse cells, transformed, **50:**306
Mouse encephalomyelitis virus, purification, **9:**187
Mouse-spleen, PP-ribose-P amido transferase from, **41:**10
6MPH$_4$, tyrosine hydroxylase regulation, **70:**144; 204
M-protein, **37:**455
M_r enzyme, cysteine *S*-conjugate β-lyase reactions, **72:**223–224
mRNA, **24:**19, **34:**3, **38:**211, **47:**378, **57:**13; 400; 405
collagen hydroxylases
prolyl 4-hydroxylase α-subunit structure, **72:**366–368
regulation mechanisms, **72:**385–386
γ-glutamyl transpeptidase gene expression, **72:**267–269
hormone-sensitive lipase, cholesterol ester hydrolases regulation, **69:**69
levels in hepatomas, **47:**304; 306; 380
methylation, **65:**255–278
bovine prolactin mRNA, **65:**262–264
eukaryotic, **65:**256–260
prolactin precursor mRNA, **65:**265
Rous sarcoma virus, **65:**261–262
in vivo localization, **65:**260–265
relation to DNA, **47:**381
5′-terminal cap, **47:**406
tryptophan oxygenase nRNA, **47:**304; 306
tyrosine hydroxylase, **70:**172; 173; 179–182
β-MSH, amino acid sequences, **24:**435
MSH (melanocyte-stimulating hormone), **14:**91; 96
MSO, *See* Methionine sulfoxide
Mucase
and collagenase, **23:**208
and elastin, **23:**239; 241; 242
Mucases, classification, **14:**246
Mucic acid, as inhibitor of β-glucuronidase, **16:**373; 374
Mucilages, reaction with borax, **12:**502–509
Mucin, **52:**109, *See* Mucopolysaccharides
galactosyltransferase, **52:**60

oligosaccharide synthesis, **52:**112
ovine, **52:**108
ovine submaxillary, **52:**32
porcine submaxillary, **52:**27; 108
submaxillary, **52:**41; 59
submaxillary asialoovine, **52:**43
Mucinase in *Clostridium welchii,* **13:**200
Mucin core protein α1r*O*
N-acetylgalactosaminyltransferase, **52:**27
Mucins and rennin, **15:**425–426
Mucodextrinases, classification, **14:**246
Mucoidy-nonmucoidy interconversion, exopolysaccharide alginate synthesis, **70:**245–251
Mucolipidosis, **60:**136; 165
Mucolytic activity, of *C. histolyticum,* **23:**208
Mucolytic enzymes, **23:**227; 229; 232; 237; 239; 244
+*cis, cis*-Muconic acid, **19:**81–83; 85–87
Muconic acid, from microbial oxidation of catechol, **27:**508–511
cis, cis-Muconic acid, metabolic pathway, **15:**184–185
cis,cis-Muconic acid, microbial oxidation, **27:**511–512
Mucooligosaccharase in snake venoms, **8:**472
Mucopeptide, and β-lactamases, **28:**265; 309
Mucopolysaccharases. *See also* Hyaluronidases
hyaluronidase action of, **13:**209
Mucopolysaccharidase, **23:**208
Mucopolysaccharide, **16:**375
sulfated, tyrosine hydroxylase regulation, **70:**122
Mucopolysaccharides, **17:**147, **34:**457, *See* Carbohydrates and Polysaccharides
acidic, in connective tissue, **25:**434–441
action of hyaluronidases on, **13:**202; 225
Mucopolysaccharidosis, **60:**106
Mucopolysaccharidosis I H, **60:**241
Mucopolysaccharidosis I S, **60:**244
Mucopolysaccharidosis VI, **60:**257
Mucopolysaccharidosis VII, **60:**240
Mucopolysaccharse in snake venoms, **8:**472
Mucoprotein, **58:**179
Mucoproteinase, collagen, **23:**173; 208
Mucoproteins, **34:**457. *See also* specific compounds, eg, Serum Mucoprotein
as substrates for neuraminidase and receptor destroying enzyme, **20:**136; 137; 140–144
Mucor, acid formation in, **9:**538; 545–546
fat synthesis by, **9:**674
metabolism, **9:**660
Mucor corymbifer, enzymic transformations of steroids by, **20:**242; 247
Mucor griseocyanus, enzymic transformations of steroids by, **20:**247
Mucor mucedo, **59:**93
enzymic transformations of steroids by, **20:**262
Mucor parasiticus, enzymic transformations of steroids by, **20:**243; 244
Mucor rouxii:
and chitinase activity, **59:**90
and chitin synthetase, **59:**65; 68; 72
and distribution of chitin, **59:**62
Mucor species, enzymic transformations of steroids by, **20:**247
Mucosa
intestinal, translocation through the luminal border of, **29:**53
and vitamin E, **6:**510
Multicomponent systems, effect of radiations on, **24:**399
Multidrug resistance, **65:**7
Multienzyme assembly, of respiratory chain, **28:**159
Multienzyme complex, **49:**137; 139
circular dichroism spectra, **49:**139
difference absorbance spectra, **49:**139
physical studies on subunit interaction, **49:**130
resolution and dissociation, **49:**141
Multienzyme complexes, **34:**446, **36:**33, **52:**108; 112
Multienzyme systems, **57:**225–228
Multigene families, of glutathione S-transferases, **69:**3–7
Multiheaded enzymes, **23:**314–315
Multilayer cap, microtubule dynamics, **69:**127
Multilayers, **11:**86
Multimolecular forms, interconversion of, **36:**97
Multimolecular systems, and origin of life, **27:**365–371
Multiple forms of enzymes, **25:**329–368
Multiple Forms of Glycosidases in Plants, Biochemistry of (Dey and Del Campillo), **56:**141
Multiple layer films, **28:**9; 11
Multiple myelomata, clotting time in, **16:**33; 84
Multiple regression analysis, **47:**85

Multiple sclerosis, adsorption studies, **14:**361
Multiple sulfatase deficiency disorder, **60:**264
Multipoint adsorption, **57:**202
Multiprotein complexes, and oxygenation, **47:**9
Multisubstrate analogs, **43:**351; 364
Multivalency of enzymes, **17:**13
Multivalent binding, **50:**308
Multivalent feedback inhibition of enzymes, **28:**46; 65
Mumps virus, neuraminidase of, **20:**144
Mumps viruses, **52:**133
Mung bean, β-1,3 glucan of, **28:**337–339
Muramic acid (MA), formula, **26:**198
  synthesis of, **26:**49–50
Muramic acid phosphate, in Grampositive bacteria, **30:**248
Muramidase (lysozyme), localization of, **24:**322
Muramylamidase, action on muropeptides, **26:**202; 204; 212; 216
Mureinases, bacterial, **26:**202; 211–216; 222; 226–228
Mureins. *See also* Cell walls
  architecture of, **26:**200–204
  definitions, **26:**194–196
  degradation by enzymes, **26:**205; 210; 212; 219
  regular constituents of, **26:**196–199
Murein sacculus, accessories of, **26:**222–228
  definition, **26:**194–196
  growth of, **26:**215–222
Murein structure, **26:**207–211
Murexine, imidazole in, **20:**102
Murexus trunculus, **20:**102
Murine DHFR, eukaryotic dihydrofolate reductase, kinetic scheme (transient state resutls), **70:**69–70
Murine DHFR promotor, RNA polymerase II transcription, reversible phosphorylation of CTD during transcription cycle, **71:**57–58
Murine lymphoid tumors, **47:**365
Murine lymphomas, **47:**364
Murine plasmacytomas, **47:**365
Muropeptide hydrolases, bacterial, **26:**202; 211–215; 218; 220
Muropeptides, occurrence of, **26:**196
  structure, **26:**200–204; 217; 219
Musca domestica, **45:**414; 419
Muscarine, **55:**139
  structure of, **55:**140
Muscarinic agonist
  binding of, to receptor, **55:**149; 156–161
    GTP in, **55:**159
    high *vs.* low affinity sites for, **55:**156–157; 158
    thiol modifying agents, **55:**158
  structure of, **55:**140
Muscarinic antagonist
  binding of, to receptor, **55:**148
    in central nervous system, **55:**166
    concentration and affinity in, **55:**166–167
    displacement of, by drugs, **55:**150
    heterogeneity in, **55:**154–156
    isomerization of complex, **55:**150–154
    number of sites for, **55:**150
    in peripheral nervous system, **55:**167
    regional distribution of, **55:**151
    sex dimorphism in, **55:**156
  structure of, **55:**141
  x-ray irradiation affecting, **55:**170
Muscarinic receptors, **55:**137–187
  activation of, **55:**139
  in adenylate cyclase inhibition, **55:**147–148
  aging as factor in binding to, **55:**171
  agonist-antagonist binding with, **55:**161
  agonist binding to, **55:**140; 149; 156–161
  antagonist binding to (*See* Musacrinic antagonist)
  antidepressants and, **55:**177–179
  antipsychotic drugs and, **55:**179–182
  assay techniques for study of, **55:**148
  autoradiography of, **55:**168
  axonal transport and, **55:**170
  binding of ligands to, **55:**148–163
  biochemical signals coupled to, **55:**144–148
  biological architecture of, **55:**184–187
  biosynthesis of, **55:**172–175
  chronic drug treatments affecting, **55:**175
  conformational status of, **55:**156
  definition of, **55:**142
  degradation of, **55:**172–175
  desensitization of, **55:**174; 175
  diisopropylfluorophosphate effect on, **55:**175
  estradiol receptors and, **55:**161–162
  in estrous cycle, **55:**162
  in ion fluxes, **55:**145
  isomerization of antagonist complex with, **55:**150; 180
  localization of, **55:**165–172
  macromolecule interactions of, **55:**161–163

mediating cGMP increases, **55:**146–147
methodology for studying, **55:**142
molecular weight of, **55:**164
ontogenesis of, **55:**170–172
pharmacologic and surgical manipulation of, **55:**175–176
phenyclidines and, **55:**182–183
in phosphatidylinositol breakdown, **55:** 145–146
photoaffinity labeling of, **55:**163; 165
physiologic significance of, **55:**139
post *vs.* presynaptic, **55:**169–170
as protein, **55:**163–165
protein phosphorylation with, **55:**147; 186
psychoactive drug interaction with, **55:** 177–183
regional distribution of, **55:**165–169
solubilization of, **55:**164
specificity of binding of, **55:**149
tissue responses mediated via, **55:**142–144
in vivo manipulation of, **55:**172–176
Muscarinic Receptors, Biochemical Characterization of the (Sokolovsky, Gurwitz, and Kloog), **55:**137
Muscarinic response, **55:**139
Muscle, **34:**468. *See also* Actomyosin and Myosin
activation, **16:**346–347
contraction, **16:**351–354
glycerol-extracted, **16:**314; 346–354
myosin in, **16:**346–354
relaxation, **16:**347–351
carbohydrate metabolism, **28:**77–80; 83–86; 95; 99
development of, **42:**275
in glycogen synthetase, **34:**326
heart, contractility, **17:**335
striated, **59:**370–371
Muscle, Diseases of Metabolism in (Hoagland), **6:**193
Muscle adenylic acid, **12:**514–515
Muscle and oxidation of ketone bodies, **8:** 409–418
Muscle and phosphatide turnover, **7:**142
Muscle branching factor, **17:**191
Muscle contraction
binding energy in coupled vectorial processes and, **51:**79–89
destabilization energy, **51:**84–85
electrostatic destabilization, **51:**86; 87
entropy, **51:**85
intrinsic binding energy, **51:**86
model for, **51:**79
rules for the vectoral process, **51:**83–84
utilization of binding energies (steps A, B, and C), **51:**84–89
creatine kinase in, **54:**368
glycogen synthetase conversion (I to D), **28:**102
relation to phosphorylase, **24:**285–286
using ATP system, **14:**14–17
Muscle Contraction, Adenosine Triphosphate and Structural Proteins in Relation to (Needham), **12:**151
Muscle contraction, ATP and structural proteins in relation to, **13:**151–197
permeability studies, **13:**27; 36; 58–60
Muscle contraction and polypeptide chains, **8:**29
Muscle diseases and altered metabolism, **6:** 193–225
and adrenaline, **6:**215; 217
and amino acids, **6:**212; 221; 222
atrophy, **6:**195–201
classification, **6:**208
progressive muscular, **6:**218–225
succinodehydrogenase activity, **6:**196; 198; 199
and bromide, **6:**213–214
and calcium, **6:**205; 212; 217; 218; 222; 2113
and carbohydrate, **6:**212; 222
cardiac hypertrophy, **6:**201–203
and chloride, **6:**204; 205; 206; 207; 215
and cholesterol, **6:**205; 207
and collagen, **6:**223
and creatine, **6:**196; 202; 203; 204; 205; 206; 207; 212; 214; 220; 221; 222; 223
and creatine hydrate, **6:**216
and creatine phosphate, **6:**223
and dextrose, **6:**222
familial periodic paralysis, **6:**214–218
blood changes, **6:**215–218
chemical constituents of muscle, **6:**218
mechanism, **6:**215; 216
urine changes, **6:**215; 216; 217
and glucose, **6:**207; 217; 222
and glutathione, **6:**198
and glycogen, **6:**195; 196; 197; 198; 205; 223. (*See also* Muscle glycogen)
and hexose phosphate, **6:**217
and insulin, **6:**215; 216; 217

Muscle diseases and altered metabolism (*Continued*)
and iodide, **6:**214
and lactic acid, **6:**196; 198; 207; 212; 214; 215; 222; 223
and lipides, **6:**199; 205; 219; 223
and magnesium, **6:**205; 212; 217; 222
myasthenia gravis, **6:**208–212
acetylcholine metabolism, **6:**209; 210; 211
blood changes, **6:**212
cholinesterase activity, **6:**209; 210; 211
esterase activity, **6:**210
mechanism, **6:**209
urine changes, **6:**212
and myosin, **6:**224 (*See also* Myosin)
myotonia, **6:**213–214
acetylcholine concentration, **6:**213
blood changes, **6:**213
and nitrogen, **6:**207; 212; 222
and oxygen, **6:**197; 198; 206; 218; 223
and phosphate, **6:**207; 215; 217
and phosphocreatine, **6:**205; 218
and phospholipides, **6:**199; 202–203; 206
and phosphorus, **6:**203; 205; 206; 207; 211; 212; 217; 218; 222; 223
and potassium, **6:**202; 203; 205; 212; 213; 215; 216; 217; 218
and potassium chloride, **6:**215; 216
progressive muscular dystrophy, **6:**218–225
adenosine triphosphate activity, **6:**223; 224
blood changes, **6:**222–223
and diastase, **6:**222
and lipase, **6:**229
and trypsin, **6:**222
ultraviolet photomicrography, **6:**219–220
urine changes, **6:**220; 221; 222; 223
and prostigmine, **6:**209; 210; 211; 216
and protein, **6:**220
and pyridoxine, **6:**222
and quinine, **6:**214
and sodium, **6:**212; 213; 215
and sodium bicarbonate, **6:**216
and sodium chloride, **6:**202; 216
succinoxidase activity, **6:**206
and sugar, **6:**212; 215; 222
and uric acid, **6:**212
and vitamin E, **6:**203–207; 221; 222; 483–485
voluntary muscle diseases, **6:**207–225
Muscle fiber, stretching, X-ray analysis, **1:**13
Muscle fibers, **43:**17
Muscle glycogen. *See also* Muscle diseases
resynthesis from lactic acid, **6:**270; 271
synthesis, **6:**299–300
Muscle glycolysis, **5:**213
alcoholic fermentation in, **5:**213
Muscle phosphorylase, **9:**498, **17:**163
and ATP, **18:**231
electron microscopy of, **33:**102
Muscle phosphorylases, **24:**264–282
Muscle proteins, solubilization of, **17:**143; 146
Muscle rigor, delay of, **50:**230
Muscles, intersegmental, degeneration of, **45:**415
Muscle selenoprotein, **48:**15–16
Muscle tissue, phosphorus compounds, rate of renewal, **7:**113–120
Muscular activity, and phosphorylase conversion, **28:**107; 109
Muscular contraction. *See also* Myosin
and oxidative phosphorylation, **17:**129–130
and phosphate bond energy, **1:**101; 149
and phosphorylase *a*, **11:**304
thermodynamics of, **11:**23–25
Muscular contraction and phosphorylase, **12:**415–416
Muscular dystrophy, acute, and dipyrroles, **7:**338
Mushroom dehydrogenase, in oxidation of coniferyl alcohol, **16:**148
Mushroom extracts, action on coniferyl alcohol, **18:**372
Mushroom redoxase, in oxidation of coniferyl alcohol, **16:**148
Mushrooms, antibiotic treatment of diseases, **20:**402
Mustard oil glucosides, formation, **10:**265–266
Mustard oils in plant, **10:**262–268
Mustelus canis, endonuclease of, **29:**182
Musts, *See* Wine making
Mutagenesis
adenosine receptors, **69:**109; 112
fibronectin, central cell-binding domain of, **70:**8; 10
modular, **69:**3; 31–34
phosphomannomutase/phosphoglucomutase (PMM/PGM), **70:**230
phosphomannose isomerase-guanosine diphospho-D-mannose pyronphosphorylase (PMI-GMP), **70:**227–228
quantitative aspects, **27:**332–333

random, **69:**31
rubisco
random, **67:**57–58
site-directed, **67:**33–36
structure-guided, **67:**50–56
side-directed, P450 enzymes studies, **68:** 47–48
site-directed, **69:**3; 19
pig liver thioltransferase, **66:**171–172
site-specific, **69:**31; 34–36
tyrosine hydroxylase, **70:**183; 184; 186; 189–193
Mutagenesis studies
cytochrome c oxidase, **71:**91; 101
ion channels, **71:**213–214
Mutagenic arene oxides, **57:**370–371
Mutagens, reactions with nucleic acids, **27:**296
Mutant, **44:**248
of hexose phosphate transport system, **44:**247; 248
Mutant enzymes, algebraic *versus* graphic analogies, **73:**49–50
Mutant forms, **41:**8; 12
of PP-ribose-P amido transferase, **41:**12
of PP-ribose-P synthetase, **41:**8
Mutant of *Neurospora,* mi-l(poky), **42:**258
Mutants, **47:**79
asn^+, **39:**234
base analog, **27:**302; 304–305; 331; 335–342
effect of D_2O, **27:**330
C^c, in arabinose system, **30:**114
$\lambda\ B^r\ mal_B$, properties of, **30:**123
0^c, definition and properties of, **30:**90–91
deletion, in lactose system, **30:**99
i^8, in lactose operon, **30:**92
$i^8 0^c$, description of, **30:**94
i^- in lactose operon, **30:**92
i^t, properties of, **30:**96
in maltose B, **30:**121–122
MT^-, in maltose system, **30:**120–121
polar, properties of, **30:**91–92
$Rglp^n$, properties of, **30:**122
classification, **27:**335–339
defective in polymerization of repeat units, **31:**109 (*See also* individual R mutants)
defective in synthesis of transfer of O side chains, **31:**106–109
with no R core synthesis, **31:**103
phosphoglucoisomeraseless, properties of, **31:**85
synthesis of an incomplete R core in, **31:**102–106
TDPrhamnose deficient, properties of, of *Salmonella typhimurium,* **31;93**
UDPgalactose 4-epimeraseless, properties of, **31:**86
without glucosyl transferase, **31:**104
without heptosyl transferases, **31:**103
deficient, and citric acid cycle, **18:**269–270
ethionine resistant fungi, **29:**138
operator-constitutive, **29:**128
microbial, in study of amino acid biosyntheses, **16:**250–251; 255; 256; 258; 259; 261–267; 268; 269; 271–274; 275–277; 299
of *Chlamydomonas eugametos,* hormones, **16:**144
of *E. coli,* amino acid metabolism, **16:**208; 229–230
temperature-sensitive, protein tyrosine phosphatases, **68:**15
Mutants analysis (ion channel model system annexin V), **71:**226–231
central pore mutations, **71:**226–229
charged residue mutation, **71:**230
N-terminal mutant properties, **71:**231
Mutants of tobacco mosaic virus, **15:**459; 461–462; 470; 472–482
Mutant technique in genetic studies, **17:**395
Mutarotase, **15:**145; 150–151
Mutarotases, **44:**88
crystalline preparation from beef kidney, **44:**89
kinetic study, **44:**89
open-chain aldose as an intermediate, **44:**90
properties, **44:**89
sources, **44:**89
Mutarotation, and asymmetric induction, **7:**68–69
and asymmetric transformation, **7:**74
Mutarotations, of glucose, **15:**150–151
of tetramethyl glucose, catalysis of, **15:** 294–295
Mutase, ketoaldehyde, **7:**92
Mutases, in synthesis of aromatic amino acids, **28:**63
Mutation, **4:**21; 22; 23; 24; 28; 29; 32; 33, **47:**80; 81
chemical, **33:**382
definition of, **33:**381

Mutation (*Continued*)
deletion, **33:**382
point, **33:**381–382
simulated, **33:**382
at active site of biologically active proteins, **33:**381–400
racemization during, **33:**384
of soybean trypsin inhibitor, **33:**399
effect on enzyme formation, **24:**7–11
enzyme involvement in, **17:**452–455
and enzymic adaptation, **13:**91; 105–108; 112
enzyme studies of, **13:**138–145
and gene size, **13:**124; 132–134
and space relationships of genes, **13:** 134–138; 145
in genes, **3:**22–23
point, **47:**381
in viruses, **3:**31
Mutation, Chemical Basis of (Orgel), **26:**289
Mutation, Simulated at Active Site of Biologically Active Proteins (Polgar and Bender), **33:**381
Mutational mechanisms, **27:**306–332
Mutational substitutions, **27:**305–306; 311
Mutation effects, eukaryotic dihydrofolate reductase
chemical transformation step, **70:**92
kinetic scheme (transient state results), **70:** 71–73
Mutations, **38:**205–207
bacterial, and light emission, **25:**154–156
by base analogs, **27:**302; 304–305; 330; 331; 335–342
chemical basis, **27:**289–346
deletions, **27:**296; 301–305; 308; 311; 324; 325; 331; 332
insertion, **27:**296; 303–305; 308; 319; 332
lethal, **27:**311; 325; 332
nature of, **27:**295–297
repression by deoxyribonucleosides, **27:**341–342
reversion, **27:**297; 301–303; 305; 309; 317–318; 331–333; 335–342
site specificity, **27:**333–335
spontaneous, **27:**331–335
transitions, **27:**305–306; 311; 314; 316–320; 324; 327; 329; 331; 332; 335–342
transversions, **27:**305–306; 319–320; 324; 335–336; 338–340
genetic (*See* Genetic factors in enzyme synthesis)
of hypoxanthine-guanine phosphoribosyltransferase, **41:**18
from ionizing radiations, **11:**257
β-lactamase structure, **28:**270–274
lethal, appearance upon fertilization, **16:**479
light-induced, **9:**76; 156
virus, **9:**180–184
photochemically produced, **14:**35–36
protein tyrosine phosphatases, **68:**8–11
Mutations, and Enzyme Problems (Sevag), **6:**33
Mutation spots in genes, **24:**8–11
MutT pyrophosphohydrolase, **73:**183–207
active site, **73:**195–203
background, **73:**183–185
bond cleavage site, **73:**185–187
pyrophosphoryl transferring enzymes divalent cations role, **73:**187–191
pyrophosphoryl transferring enzymes power, **73:**187
solution structure, **73:**191–195
free enzyme, **73:**191–194
quaternary MutT-M^{2+}-AMPCPP-M^{2+} complex, **73:**194–195
MV_2 enzyme, cytochrome *c* oxidase, **71:**123; 125
Myasthenia gravis, **6:**208–212. *See also* Muscle diseases and altered metabolism
tyrosine metabolism in, **14:**99; 101; 108–110
Mycelianamide, **35:**155; 163
Mycetin, **7:**488
Mycobacteria, **58:**72
Mycobacteria, oxidation of DPNH, **27:**119
oxidation of hydrocarbons, **27:**471; 473–475; 477; 482–484; 498; 534
Mycobacterium
aniline and *p*-amino-benzoic acid hydroxylation by, **19:**144
glycerol phosphorylation by, **19:**17
pyrocatechase in, **19:**181
and mevalonic acid, **23:**479
Mycobacterium avium, propionic acid metabolism, **26:**298
Mycobacterium avium, folic acid biosynthesis in, **21:**394
Mycobacterium lacticola, enzymic transformations of steroids by, **20:**254; 255; 262; 271

Mycobacterium laticolum, oxidation of hydrocarbons, **27:**472
Mycobacterium paraffinicum, oxidation of hydrocarbons, **27:**475; 481
Mycobacterium phlei, **18:**163
and coenzyme Q, **25:**284; 319; 322
β-lactamase of, **28:**242
phosphorylating system from, **23:**353
Mycobacterium phlei, mitochondrial vitamin K in, **21:**113
Mycobacterium rhodochrous, oxidation of hydrocarbons, **27:**506; 507; 511; 513
Mycobacterium segmatis, enzymic transformations of steroids by, **20:**262
Mycobacterium smegmatis
β-lactamase of, **28:**242; 278
polysaccharides of, **45:**62–75
regulation of by end products, **45:**53–58
palmitoyl-CoA, control by, **45:**55–58
polysaccharide, control by, **45:**58
sites of control, **45:**54
propionic acid metabolism, **26:**291
Mycobacterium smegmatis, Control Mechanisms for Fatty Acid Synthesis in (Bloch), **45:**1
Mycobacterium smegmatis acetyl-CoA carboxylase, **45:**9
Mycobacterium smegmatis fatty acid synthetase (Type I), **45:**11–33
acetyl-CoA concentrations, **45:**13–15
coenzyme A, **45:**14
FMN, **45:**15
malonyl-CoA, **45:**14
pyridine nucleotides, **45:**15
fatty acid patterns, **45:**23
bimodel fatty acid patterns of *M. smegmatis,* **45:**24
BSA, effect on chain length, **45:**30
liposomes, **45:**31
palmitoyl-CoA thioesterase, **45:**32
polysaccharide, effects of on chain length, **45:**27
substrate concentrations, effect on fatty acid spectra, **45:**25
temperature effects, **45:**33
Kinetics of, **45:**16–23
acetyl-CoA concentrations, effect of, **45:**21–23
initial burst, **45:**16–21
multienzyme complex, isolate and purification, **45:**11–13
ionic strength, **45:**13
partial reactions catalyzed by, **45:**34–48
α, β-enoylacyl reduction, **45:**46
acetyl and malonyl transacylase sites, separate sites, **45:**36
acetyl transacylase, **45:**36
condensation reactions, **45:**44
β-hydroxyacyl dehydration, **45:**46
β-ketoacyl reduction, **45:**45
malonyl transacylase, **45:**36
partial reactions, **45:**44
transacylase, long-chain, **45:**38
transacylation and bimodal product distribution, **45:**42
Mycobacterium smegmatis fatty acid synthetase (Type II), ACP-dependent, **45:** 75–78
Mycobacterium smegmatis synthetase catalyzed palmitoyl-CoA elongation, **45:** 48–53
Mycobacterium smergmatis, **46:**355
Mycobacterium sp., superoxide dismutase from, **41:**79
Mycobacterium species, cholesterol dehydrogenase in, **20:**267–270
Mycobacterium spp.
carboxylation enzyme, **26:**297; 304; 305
β-lactamase of, **28:**242
Mycobacterium tuberculosis, **15:**193
amino acids in, **14:**266
Mycobacterium tuberculosis, β-lactamase of, **28:**242
Mycobactin, **35:**139; 154; 168; 177
Mycoderma valida, β-lactamase of, **28:**243
Mycodextran, **15:**312
Mycolysis, microbiological control, **28:**355–356
Mycophenolic acid, **7:**475; 486; 489; 490; 492
Mycoplasma laidlawii, **34:**470
Mycoribrin, inhibition of trehalose phosphate synthetase, **40:**37
Mycostatin, in treatment of plant diseases, **20:**384; 390; 392
Mycothricin complex A, in treatment of plant diseases, **20:**393
Mydriatic alkaloids, synthesis, **8:**224–225
Myelin, **34:**466, **60:**111
Myelin basic protein, tyrosine hydroxylase regulation, **70:**151
Myelogenous leukemias, absence of Tuftsin activity, **41:**174

Myeloperoxidase, **46:**62, **60:**219
absorption spectrum, **23:**309
activity of, **14:**151; 155–156
Myocardial infarction, creatine kinase in diagnosis of, **54:**369
Myofibrillar structure, biochemistry of, **13:** 184–187
Myofibrils, **43:**27
activation by salts, **13:**165–166
comparison with other fibrils, **8:**16–17
contraction, **43:**28
and enzyme formation, **24:**5
Myofibrils, ADP in, **16:**351
effect of ions on, **16:**338–339
Myogen A, isolation, **14:**401–402
Myogen carp, **47:**124; 133
Myoglobin, **7:**313, **41:**53, **43:**339, **47:**47; 49; 59; 124; 133, **54:**307, **60:**37
conduction of oxygen through solutions of, **29:**33–34
conductor of oxygen, **29:**34–35
oxygen dissociation constant, **29:**35
translocation in, **29:**35
human, electrophoresis of, **30:**168
imidazole in, **20:**98
intracellular distribution, **16:**463; 466
mechanism of action as enzyme, **15:**18–21; 33; 38
quasi lipoxygenase activity, **58:**234
and RNase structure, **24:**182
formation from polypeptide, **24:**22
lower peptide sequences in, **24:**421
sperm whole, structure, **24:**416
x-ray studies, **24:**448
structure of, **23:**2–4; 24, **26:**92; 98–100
X-ray structural determination of, **47:**47; 49; 133
Myoglobin, Cotton effect in, **27:**47; 60; 90–92
Myoglobin fluoride, EPR study, **27:**138
Myoglobin haem *a* compound, **23:**289–297; 306–310
Myoglobins, **46:**59; 65
in energy transport, **14:**33–35; 185
structure studies, **14:**284
Myohemerythrin, **47:**117–118; 120–123
Myoinositol, **60:**92
Myo-inositol, as inhibitor, **36:**119
Myoinositol-1-phosphate synthase, **46:**249
Myokinase, **5:**131, **15:**146; 232; 347–350; 380–382, **18:**196; 197
activity of –SH group in, **11:**243
classification, **14:**251
isolation, **14:**397
effect on muscle contraction, **16:**348; 350
in myosin preparations, **16:**315; 318
removal from actin, **16:**334
and light emission, **25:**137
in ribonucleotide reductase, **28:**444
Myokinase in adenylic acid formation, **6:**161; 170
actomyosin, **6:**157–158
and ADP, **6:**176
and apyrase, **6:**174
and ATP, **6:**154; 158; 160; 173; 175
and ATPase, **6:**147–189 (*See also* ATPase, in myosin)
identify with ATPase, **6:**170–174
and catalase, **6:**184
and cozymase, **6:**176
crystalline, **6:**151–155; 170–171
enzymic properties, **6:**155; 158–170
flow birefringence, **6:**154; 174–182
and lactate, **6:**183
mechanochemistry, **6:**174–184
and metaphosphate, **6:**176
monolayers, **6:**182–184
in muscular dystrophy, **6:**224
myosin B, **6:**181
preparative separation, **6:**170
and pyrophosphate, **6:**176
recrystallized, **6:**153; 155; 171
and silver, **6:**176
solubility, **6:**153–154; 155–156; 171–172
specificity, **6:**158–162
stoichiometry, **6:**182
surface viscosity, **6:**184
threads, **6:**175–178
and trypsin, **6:**184
viscosity, **6:**154; 178–182
Myosin, **4:**215; 216; 217; 218; 220; 238, **5:**134, **20:**91, **23:**341; 373; 381–382; 385, **42:**230, **43:**5; 7; 10; 11; 17; 22; 27; 28; 36; 37; 45; 355, **46:**248. *See also* Actin, Actomyosin, and Muscle; Fibrous proteins
activity as enzyme, **13:**163–167; 168–169
amino acid composition, **16:**319; 330
fractions A and B, size and shape, **16:**317

structural and functional aspects, **16:** 313–360
and ATP in muscle contraction, **13:**156–159; 161–163
biosynthesis of, **50:**223
effect of creatine on, **50:**223–234
denaturation of, **11:**223–224
reaction with alkylating agents, **11:**226
thiol groups in, **11:**221; 229–232; 244
effect of radiation on, **24:**388
volume irradiated, **24:**365–367
effect on actin polymerization, **13:**155
end groups, fluorodinitrobenzene method for, **13:**157–158
identity with adenosinetriphosphatase, **8:**31
interaction with actin and ATP, **13:** 169–180
kinetics, transient, **43:**11
in muscles, diffraction pictures, **1:**13
and product inhibition, **25:**231
solubilization of, **17:**143; 146
in special types of muscle, **13:**180–184
structure studies, **14:**279; 284
L-Myosin, **16:**314
Myosin, Adenosinetriphophatase Properties (Engelhardt), **6:**17
Myosin, Structural and Functional Aspects (Szent-Györgyi), **16:**3
Myosin and contractility of sperm tail, **9:**335; 339
Myosin ATPase, **23:**360; 381, **37:**483, **50:**377; 387–390
effect of metal ion, **25:**18–19
elementary steps in reaction, **25:**14
product inhibition, of, **25:**232–233
turnover number, **25:**32
and metal complex formation, **13:**167
in embryonic development, **13:**183
Myosin ATP-ase, **15:**351
Myosin ITPase activity, **13:**168–169
Myosins, ATPase activity, SH-group, **27:** 214–215
Myosin triphosphatase, **7:**250
Myotonia, **6:**213–214
Myristic acid, temperature sensitive auxotrophy, **67:**405–408
MyristoylCoA, **67:**389–391
acyl chain, **67:**395
analogs, unsaturated, **67:**398–399
intracellular pools, **67:**410
Nmt interaction, mutations affecting, **67:** 405–408
MyristoylCoA:protein N-myristoyltransferase, **67:**375–416
acylCoA binding domain, **67:**404
acylCoA:Nmtlp catalytic intermediate, **67:**388–391
acylCoA substrate specificities, **67:**394–399
catalytic domain mapping, **67:**402–405
function, **67:**376–381
future directions, **67:**415–416
inhibitors, **67:**399–402
mutations affecting myristoylCoA interaction, **67:**405–408
orthologous, primary structures, **67:**384
peptide binding site, **67:**404–405
peptide substrate specificity, **67:**391–394
protein n-myristoylation in *E. coli,* **67:** 382–383
protein=protein interaction mediation, **67:**381
reaction mechanism, kinetic analysis, **67:** 387–388
regulation studies in *S. cerevisiae,* **67:** 408–414
acylCoA functional segregation, **67:** 411–412
myristoylCoA intracellular pools, **67:**410
in vivo myristoylation alteration, **67:** 412–414
tetradecanoyl group, **67:**380
vegetative growth, **67:**379–380
in vitro catalytic efficiency, **67:**409
in vitro enzyme assays, **67:**381–382
yeast, initial characterization, **67:**384–387
Myronate sulfatase, classification, **14:**245
Myrosulfatase, classification, **14:**245
Myrosulfatase in plants, **10:**310
Myrothecium verrucaria, **17:**217
Mytilus edulis, succinate dehydrogenase from, **28:**180
Myxobacter AL-1, **59:**90
Myxobacter *a*-Lytic protease, **41:**179; 195; 196; 238
Myxobacter AL-1 protease I, **41:**179; 207–208; 238
Myxobacter AL-1 protease II, **41:**179; 208; 238
Myxococcus xanthus, **70:**246, **71:**3
Myxomycetes, **46:**102

N

N^{15}
use in amide metabolism studies, **13:**265; 266–267; 273–275; 280; 284
use in nitrogen fixation studies, **13:**349–360; 361–368; 371
use in transmethylation studies, **12:**463–464; 466
^{15}N in phosphatide turnover, **7:**172
Na^+, electrochemical gradient, **49:**66; 67
cotransport with, **49:**67
position taken at System A receptor site, **49:**80–81
role in transport of dicarboxylic amino acids, **49:**82–83
as surrogate for side-chain cationic group, **49:**63
Na^+, K^+-ATPase, **46:**248
Na^+ binding, position of for System ASC, **49:**77
NaCl, **34:**523
NAD, **25:**395; 397; 402; 405–408, **34:**526, **50:**311. *See also* DPN
and oxidation of hydrocarbons, **27:**479; 485; 486; 520; 522; 531; 532
NAD^+, **36:**95, **43:**296; 354
analogs of, **43:**385
NAD, NAD^+ NADH, NADP, $NADP^+$ NADPH, uses of terms, **24:**294
NAD adducts, **46:**239
NAD analogs, **46:**212–213
in enzyme study, **25:**331
malate dehydrogenase study, **25:**341–343
NAD-dependent, PQQ-containing methanol dehydrogenase, **59:**188
NAD^+-dependent alcohol dehydrogenase, **59:**201
NAD-dependent enzymes, forms of, **25:**331
Na^+-dependence of amino acid transport, in marine pseudomonad, **49:**84
Na^+-dependent inhibition
of a Na^+-dependent system, **49:**78
of a Na^+-independent system, **49:**78
Na^+-dependent transport, in microorganisms, **49:**83
NAD glucohydrolase, **37:**140
NAD-glycohydrolase, **50:**313, **58:**111
NADH, **44:**197, **49:**230. *See also* DPNH; Pyridine nucleotides
binding to regulatory site of glutamate dehydrogenase, **39:**384; 409
complexes with glutamate dehydrogenase, **39:**372
and dihydrofolate reductase, **27:**434
in respiratory chain, **28:**159–160; 164; 168; 186; 194; 215–217; 219
succinate cytochrome *c* reductase, in reconstitution of respiratory chain, **28:**215; 217
succinate oxidase, reconstitution from primary complexes, **28:**209–220; 222–223; 225
succinate reductase, in reconstitution of respiratory chain, **28:**216; 219
NADH-CoQ reductase (complex I), in reconstitution of respiratory chain, **28:**210–212; 216–220; 223
NADH cytochrome b_5 reductase
location of, **24:**304
PN-dependent, **26:**128; 166
NADH-cytochrome c-oxidoreductase, multiple forms, **25:**359
NADH-cytochrome *c* reductase
in respiratory chain, **28:**210–214; 216
NADH cytochrome *c* reductase, **34:**474
$NADH_2$ cytochrome *c* reductase, location of, **24:**303; 343
NADH dehydrogenases, in respiratory chain, **28:**159–160; 196; 208–209
NADH diaphorase, location of, **24:**305
$NADH_2$ in coacervate droplets, **27:**370
hydrocarbon oxidation, **27:**515; 532; 533; 537; 538
NADH-lipo-amide-oxidoreductase (diaphorase), multiple forms, **25:**358–359
NADH 1 mM, **49:**72
NADH oxidase, **37:**582
from heart muscle, **28:**197; 202; 207
reconstitution of, **28:**208–209
in reconstitution of respiratory chain, **28:**215–219; 230
$NADH_2$-oxidase system, in oxidation of hydrocarbons, **27:**531
NADH oxidation, mechanism, **26:**143–156
NADH pyrophosphatase, pyridine nucleotides, niacin derivatives salvage, **73:**159–161
NADH reductase, activity, **28:**209
NADH-semidehydroascorbate reductase, **66:**180
NADH-ubiquinone reductase, **34:**110
Nadi oxidase, *See* Cytochrome oxidase

Nadi reagent, insect phenolase activity for, **16:**158
NAD-kinase, hydrogen transfer, **26:**116
NAD+ kinase, quinolinate to NAD+, **73:** 155–158
NAD-linked dehydrogenases, **24:**338
NAD-nucleosidase, **37:**582
hydrogen transfers, **26:**116
localization of, **24:**323; 330
NAD or NADP as acceptor, oxidoreductases, **24:**295–299; 301; 303–304; 337; 341; 343
NADP, **25:**395–397; 401–403; 406
binary complex with glutamate dehydrogenase, **39:**393
ternary complex with glutamate dehydrogenase and glutarate, **39:**393
ternary complex with glutamate dehydrogenase and α-ketoglutarate, **39:**393
and oxidation of hydrocarbons, **27:**479; 485; 515; 522
photoreduction of, by isolated chloroplasts, **30:**3; 23
reduction by chloroplasts, role of ferredoxins in, **33:**123–125
in respiratory chain, **28:**159
$NADP^+$, **34:**520; 526
NADP analogs, **46:**213; 221
NADP binding
cooperativity, eukaryotic dihydrofolate reductase, ligand binding, **70:**52–53
eukaryotic dihydrofolate reductase, **70:**48–50
NADPH, **34:**520
and dihydrofolate reductase, **27:**434
electron transfer to thioltransferases, **66:** 151–152
eukaryotic dihydrofolate reductase chemical transformation, **70:**74
nitric oxide synthases requirement, **65:** 305–306; 319–320
oxidation, **67:**80
tyrosinase/tyrosine hydroxylation contrasted, **70:**106
NAD(P)H, glutamate synthase, **72:**112–117
$NADPH_2$, in oxidation of hydrocarbons, **27:**537
$NADPH^3$, methylation studies, **25:**401
NAD(P)H: quinone acceptor oxidoreductase, *See* DT-diaphorase
NADPH activation, by hydride donor, eukaryotic dihydrofolate reductase, chemical transformation step (reactant activation), **70:**75
NADPH analogs, difference spectra of glutamate dehydrogenase complexes with, **39:**372
NADPH binding
cooperativity, eukaryotic dihydrofolate reductase, ligand binding, **70:**52–53
eukaryotic dihydrofolate reductase, **70:**48–50
structural changes induced by, eukaryotic dihydrofolate reductase, **70:**44–45
NADPH-binding site/NAD-binding site, compared, eukaryotic dihydrofolate reductase, **70:**41–42
NADPH-cytochrome *c*, **46:**135
NADPH cytochrome *c* reductase, **37:**164; 165
cytochrome P_{450}, **46:**244
location of, **24:**304; 343
NADPH-cytochrome *P*-450, active site, **60:**59
NADPH-cytpchrome *P*-450 reductase, **60:**44
$NADPH_2$ diaphorase, location of, **24:**304–305
NADPH-L-glutamate glutamate dehydrogenase complex, **39:**388
$NADPH_2$-glutathione reductase, SH-group, **27:**215–216
$NADPH_2$-GSH (C^{14}), **27:**215
NADPH/$NADP^+$ status, cellular, **66:**155
NADPH oxidase, **57:**147–148. *See also* Respiratory burst oxidase
NADPH oxidation, **60:**72
NADPH reductase, **46:**120
NADP-specific dehydrogenases, multiple forms, **25:**363–364
NAD(P) transhydrogenase, location of, **24:**303
NAD+ pyrophosphatase, pyridine nucleotides, niacin derivatives salvage, **73:**159–161
NAD-pyrophosphorylase
hydrogen transfer, **26:**116
localization of, **24:**314
NAD-specific dehydrogenases, forms of, **25:**331; 363–364
NAD synthesis, **49:**28
NAD+ synthesis, **73:**135–182
importance of, **73:**135–136
niacin derivatives salvage, **73:**158–166
niacin utilization, **73:**164–166
pyridine nucleotides to niacins, **73:**158–164
quinolinate synthesis, **73:**136–144
aerobic pathway, **73:**140–144

NAD+ synthesis (*Continued*)
anaerobic pathway, **73:**136–140
quinolinate to NAD+, **73:**145–158
NAD+ kinase, **73:**155–158
NAD+ synthetase, **73:**153–155
NMD adenylyl transferase, **73:**149–153
phosphoribosyltransferase, **73:**145–149
NAD-synthetase, **66:**283–284, **72:**134–136
hydrogen transfer, **26:**116
quinolinate to NAD+, **73:**153–155
Nagarse, multiple forms of enzymes, **25:**338
Na^+/K^+ ATPase
erythrocyte ghost membranes, **43:**42
hepatic glutamine transport, hepatocyte hydration, **72:**49–50
Namatodes, **46:**102
L-[13N]Amino acids, **57:**328–329
NANA, *See* N-Acetylneuraminic acid
Naphthalene, **58:**40
microbial oxidation, **27:**519–522
Naphthaleneacetic acid, **18:**39
Naphthasemiquinones, in oxidation of DPNH, **27:**119
1,2-Naphthohydroquinone-phenolase-oxygen system, coupling to quinone reductase, **16:**134
1-Naphthol, quinone formation, **65:**109
1-Naphtholsulfonic acid, **18:**291
1,2-Naphthoquinone, as electron acceptor, **16:**134
Naphthoquinones, **18:**104, **65:**109. *See also* Vitamin K
and coenzyme Q, **25:**302; 305; 322
effect of luciferase on, **25:**150–151
inhibition of choline acetylase, **12:**323
as inhibitory analogue, **6:**142
and luciferin, **7:**257
reactions with indoles, **16:**111
1,4-Naphthoquinones, standard series, oxidation of reduced enzyme, **17:**277
α-Naphthoquione sulfonate, action on fibrinogen, **10:**33
α-Naphthylacetic acid, effect on pectic substances, **20:**370
α-Naphthylamine, action on thiaminase, **15:**400
2-Naphthyl-β-D-glucofururonolactone, inaction of β-glucuronidase on, **16:**371–373
2-Naphthyl-β-D-glucopyruronoside, action of β-glucuronidase on, **16:**366; 371–373
α-Naphthylthiourea, inhibition of melanin by, **14:**93; 94
2-Naphtol-3,6-disulfonic acid, as inhibitor of β-glucuronidase, **16:**374
1,4,5,8-Napthalenetetracarboxylate, **51:**236
Napthoquinones, in electron transport studies, **17:**69; 123; 368; 373
Narcosis, **9:**67–68
Narcotics, **17:**368. *See also* specific compounds, eg, Amytal
effect on cytochromes, **18:**6; 12
effect on cytochromes *a, b, c,* and c_1, **20:**150; 175; 186
effect on enzymatic processes in leaves and fruits, **1:**346; 356
effect on photosynthesis, **12:**25–28
lipoid theory, **7:**21–22
manner of action, **7:**20–25
permeation, **7:**20–25
Nargase
action on phosphorylases, **24:**271
and LADH active center sequence, **27:**81
Naringenin, **55:**19–20
formation of, **55:**22
hydroxylation of, **55:**34
Nascent protein, free energy content, **26:**96–98
relation to enzyme structure, **26:**91–98; 100
Native bromelain, **34:**465
conformation, **34:**456
enzyme, **34:**462; 504–508
membrane, **34:**472
milieu, **34:**446
particulate enzyme, **34:**447
particulate systems, **34:**456
Natural selection, **38:**201; 206; 213; 227
Navicula, fucoxanthin in, **30:**17
NCD_4, **65:**204–206
Nearest neighbor pairs, **47:**420; 421; 423
in nucleic acid strands, **47:**419
Nebularine, **54:**210
Necleophilic cleavages, **39:**330
Necrosin, action on collagen, **23:**173
Necrotic liver degeneration, **48:**2
Necrotizing factors, **23:**199
Nectria haemotococca, **55:**47
Negative stain, preparation of, **33:**79
Negative staining, analog images, **33:**97
description of, **33:**77–82
embedding of molecules in stain, **33:**79–81

interpretation of results, **33:**96–98
support films, **33:**78–79
Neisseria, carbonic anhydrase of, **47:**151
Neisseria catarrhalis, quinones in, **25:**289
Neisseria gonorrheae, glutamine metabolism in, **13:**254
Neisseria gonorrhoreae, **70:**231
Neisseria meningitidis, **17:**223, **18:**263
Neisseria perflava, amylosucrase in, **15:**309
phosphorylase in, **15:**304
Q-enzyme in, **15:**307
Nematodes, and fluoroacetate, **18:**136–137
Nembutal and hypertension, **7:**547
Neoaspergillic acid, **35:**164
Neodymium compounds, **17:**147
Neodymium ions, model catalyses with, **21:**177
Neoglycopeptides, secondary specificity with asialoglycoprotein receptor, **66:**46–47
Neoglycoproteins, **52:**114; 115; 116
Neokestose, **17:**204
Neomycin
activity, **10:**164
in schizophrenia, **29:**526
in treatment of plant diseases, **20:**392–395; 398; 402
Neoplasia, **42:**267
characterization of, **29:**322
Neoplasm. *See also* Malignancy
characteristic biochemical properties of, **29:**324–325
and their normal tissues of origin, **3:**318–334
Neoplastic cells, calcium levels in, **29:**366
regulation of enzyme activity and enzyme levels in, **29:**367–381
Neoplastic disease, and chemotherapy, **27:** 418–419
Neoplastic tissues
control mechanisms in, **29:**360–363
dehydrogenases in, **29:**360–363
isocitric, **29:**362–363
lactic, **29:**361–362
malic, **29:**361
enzyme activity in, **8:**148
EPR study, **27:**190–191
and phosphatide turnover, **7:**145–146
protein-specific methyltransferase activity, **42:**257
succinate oxidase, **28:**179
Neopterin, **35:**53
triphosphate, **35:**57
Neopyrithiamine, inhibition of thiaminase action on thiamine, **15:**399
Neostigmine, **55:**176
Neotetrazolium reductase activity, location, **24:**305
Neoxanthin, in chloroplast lamellae, **30:**9
content of, in chloroplast fractions, **30:**50
Nephrosis, and β-glucuronidase, **16:**397
Nephrotoxicity, cysteine *S*-conjugate β-lyase reactions, **72:**208–212
Nernst theory, **33:**268
Nerolidol, **22:**306; 307; 315
structure, **22:**306
Nerolidyl purophosphate, **22:**317; 318
formation from, condensation with farnesyl pyrophosphate, **22:**315–317
in squalene biosynthesis, **22:**297; 314; 315
Nerve cell, metabolic cycle of, **12:**315–317; 331–334
Nerve conduction, **7:**46
role of lipides, **8:**46–47
role of polypeptide chains, **8:**46–47
Nerve fibers
acetylcholinesterase in, **12:**270–271; 271–274
energy source in conductance by, **12:**312–315
Nerve fiber stimulation, effect of change in surface potential, **11:**76
Nerve function, use of ATP system, **14:**14–17
Nerve gases, in chemical-biological warfare, **34:**9
inhibition of enzyme activity, **34:**10
Nerve growth factor (NGF), **60:**227
biological sources of, **31:**52–56
chemical properties and metabolic effects of, **31:**51–75
in chick embyronic sensory ganglia
effect of, **31:**64; 65
incorporation of ^{14}C-acetate into total lipids with, **31:**67
on oxidation of C-1-glucose and C-6-glucose, **31:**65
time sequences of, stimulated acceleration of protein and RNA synthesis, **31:**69
effect of, on glucose metabolism, **31:**63–66
effect of puromycin and of actino-mycin on uridine-^{3}H and on L-leucine-^{14}C incorporation by sensory ganglia in medium with, **31:**70
effect of removal of salivary gland on, **31:**56

Nerve growth factor (NGF) (*Continued*)
effect on fine structure of nerve cells by, **31:**71–72
effect on protein and RNA synthesis by, **31:**67–71
ganglion cell exposed to, **31:**72
lipid metabolism affected by, **31:**66–67
metabolic effects of, on receptive nerve cells, **31:**62–72
from mouse submaxillary gland, **31:**60–62
androgen control of, **31:**54–56
disc electrophoresis on acrylamide gel of, **31:**61
molecular weight of, **31:**61–62
purification of, **31:**60
in mouse tumors (sarcoma 180 and 37), **31:**53
properties of, **31:**53
in salivary gland, **31:**56
from snake venom, **31:**57–60
disc electrophoresis on acrylamide gel of, **31:**59
molecular weight of, **31:**57
S_{20} of, **31:**57
purification of, **31:**58
tyrosine hydroxylase regulation, **70:**146; 147; 150; 151; 171–172
Nerve Growth Factor (NGF): Chemical Properties and Metabolic Effects (Angeletti, Levi-Montalcini, and Calissano), **31:**51
Nerve stimulation, tyrosine hydroxylase induction, **70:**166
Nervous tissue
choline acetylase in, **12:**323–325
citric acid level, **18:**146–150
and fat formation, **8:**318; 319
permeability studies, **13:**36; 60–63
spike potentials from, **23:**83–84; 124–125
Net rate constant, **45:**275
Network theory, of antigen-antibody reaction, **15:**68; 71–75
Neuberg's ketol, **7:**93
Neuramidinase, activity of, **25:**445–446
Neuraminic acid. *See also* Sialic acid
in glycoproteins, **25:**434; 442–446
levels in cerebral spinal fluid of schizophrenics, **29:**505–506
Neuraminidase, **38:**307, **41:**99; 100; 105; 107–109; 116; 118; 120; 126; 127, **46:**388
from *Clostridium perfringens,* **41:**108; 109; 118; 120
from influenza virus, **41:**120
substrate and mode of action, **20:**135–146
from *Vibrio cholerae,* **41:**107; 108; 120
Neuraminidase: Substrate and Mode of Action (Gottschalk), **20:**135
Neuraminidases, **36:**65, **52:**30, **60:**135
Neuraminyl lactose, specific rotation, **20:**143
Neurogenic hypertension, **7:**514
Neurohypophyseal hormone
effect on sodium flux, **13:**52; 53; 54
effect on water permeability, **13:**46
Neurolathyrogens, in seeds, **29:**147
Neuroleptics, **55:**181
muscarinic receptor interaction with, **55:**182
tyrosine hydroxylase regulation, **70:**128
Neurophysin-Hormone Systems, Molecular, Thermodynamic, and Biological Aspects of: A Model System for the Analysis of Protein-Peptide Interactions (Breslow and Burman), **63:**1
Neurophysins
amino acid composition, **40:**280; 288
amino acid sequence, **40:**288; 289
antibodies, **40:**279; 284; 295; 296
binding affinity, **40:**277; 323; 324; 329
hormonal, **40:**282; 283; 284; 287; 304; 306; 307; 309; 321
oxytocin, **40:**313
binding free energy, **40:**322; 323; 324
biological function, **40:**272
biosynthesis, **40:**285; 286
and Ca^{++}, **40:**282; 283
carboxyl groups, **40:**299; 313; 314; 315; 325; 327
circular dichroism, **40:**298; 299; 300; 301; 304; 307; 308; 309; 311; 312; 313; 315; 316; 317; 318; 319; 321; 322; 325; 326; 328
comparison amoung, **40:**228
conformation, **40:**296; 297; 298; 299; 300; 301; 302; 303; 304; 305
conformation changes, **40:**300; 301; 304
disulfide conformation, **40:**300; 301; 302; 303; 304; 305
disulfide content, **40:**273
disulfides, **40:**288; 291; 292; 298; 311; 327; 328
evolution, **40:**292; 293

evolutionary precursor, **40:**295
evolutionary significance, **40:**273
function, **40:**284; 285
α-helix, **40:**299
H^+-ion equilibria, **40:**299; 313
internal duplication, **40:**273; 291; 293; 294; 329
lipolytic activity, **40:**284; 285
as lipoprotein, **40:**287; 288
localization, **40:**276; 277; 278; 279; 280; 281; 282
molecular weight, **40:**288
multiplicity, **40:**274; 275; 276
nitrated, **40:**300; 307; 308; 309; 322; 323; 325; 328
nomenclature, **40:**274; 275; 276
nuclear magnetic resonance, **40:**299; 315; 317; 318; 319; 320; 321; 322; 324; 326; 328
number of binding sites, **40:**305; 306; 307; 308; 309; 329
pH dependence of binding, **40:**309; 310; 315; 316
polymerizing equilibria, **40:**296; 297; 311
precursor, **40:**285; 286; 304; 305; 314
purification, **40:**286; 287; 288
proteolytic degradation during, **40:**274; 275; 287
random coil structure, **40:**299
reduction of disulfides, **40:**301; 302; 303; 304
release, **40:**281; 282; 283; 284
sedimentation velocity, **40:**309; 311
species variations, **40:**274; 275; 276
structural homology with the posterior pituitary hormones, **40:**295
β-structure, **40:**299
synthesis, **40:**277; 279
tyrosine of, **40:**288; 291; 299; 300; 311; 313; 325; 326; 327; 328
Neurophysins (Breslow), **40:**271
Neurosecretion (Sach), **32:**327
Neurospora, **19:**3
amino acid metabolism in, **16:**272; 276; 283–284
mutants, amino acid metabolism in, **16:**256; 258–259; 262–264; 269; 272; 273; 278–281; 287; 292; 293; 295–297; 299; 302
cytochromes, **17:**361
diphosphopyridine nucleotidase of, **20:**116
DPN-ase in, **15:**80; 84–85; 88
enzymes and reactions of pyrimidine biosynthesis in, **21:**265; 266; 279; 281
enzyme synthesis in, **24:**9–10; 14–16; 23
enzymic adaptation in, **13:**106
genetic studies of, **13:**138; 139–145
folate use, **27:**427
formation of aromatic compounds by, **18:**362
and glycolate metabolism, **23:**435–436
hydroxyanthranilic acid metabolism in, **19:**94, **34:**468
nicotinic acid formation in, **15:**98–101, **19:**94
respiratory chain, **28:**222
Neurospora crassa, **34:**521, **46:**43; 49; 91–92, **49:**6; 37, **51:**111; 112, **58:**71, **59:**63; 68; 76; 90
adenylosuccinase mutants, **22:**12; 29; 33
and interallele complementation, **22:**34
in vitro complementation studies of, **22:**37
allosteric enzyme of, **28:**59–60; 69; 99; 131; 137
arginine-requiring mutants, **22:**27
arginino succinase formation and interallele complementation, **22:**35
argininosuccinase mutants, **22:**29
auxotropic mutants of glutamic acid dehydrogenase of, **30:**149
and biosynthesis of ornithine, **33:**374
carboxylase in mutants, **22:**25
chitin formation, **26:**59
PN-dependent enzymes, **26:**119; 130; 131
tryptophane synthetases, **26:**110–111
and citric acid cycle, **18:**251–253; 270
coenzyme Q derivative in, **25:**322
methylation in, **25:**417
multiple forms of enzymes, **25:**364
tyrosinase, **25:**359
conversion of dehydroshikimic acid in mutants, **22:**21
cytoplasmic transmission of heredity in, **22:**3; 28
effect of steroid hormones on, **20:**264
enzymic transformations of steroids by, **20:**243; 244
polygalacturonase of, **20:**363; 366
endonuclease of, **29:**197
genetic mapping of, **22:**2
genetic studies, **17:**396–414; 416–419; 422–426; 429–431; 435–443; 445–452; 454–455

Neurospora crassa (*Continued*)
inorganic nitrogen metabolism, **17:** 250–252; 283–284
glutamic dehydrogenase mutants, **22:**8–11; 29
and heat activation, **22:**19
and interallele complementation, **22:**34; 36
partial back-mutations in, **22:**8–10
isoleucine + valine-requiring mutants, **22:**33
lactase in mutants, **22:**25
loci of histidine synthesis in, **22:**37; 38
methionine mutants, utilization of sulfonium compounds, **22:**268
mutations preventing utilization of nitrate in, **22:**26
nitrogen transfer reactions in, **21:**252
purine-requiring mutants, **21:**254
oxidation of hydrocarbons, **27:**513
pantothenate-requiring mutants, **22:**27
pyrroline-5-carboxylate reductase mutants, **22:**7; 8
and alterations in activation energy, **22:**19
pyruvic acid in mutants, **22:**25
transamination in, **16:**205; 229; 230; 231; 232; 235–236
transferases in, **13:**298; 299; 302–303
tryptophan metabolism, **10:**58–60; 62
tryptophan synthetase, formation, and interallele complementation, **22:**35
inhibition, **22:**26
tryptophan synthetase mutants, **22:**11; 29
and effect of suppressor mutations, **22:**30
and production of cross-reacting material, **22:**31
tyrosinase-deficient mutants, **22:**25
tyrosinase mutants, **22:**4
tyrosinases from, variation in, **22:**5; 6; 30
Neurospora DPNase, product inhibition of, **25:**222; 237–240
Neurospora sitophila, enzymic transformations of steroids by, **20:**245
Neurosporene, synthesis and structure, **21:**342
Neurotoxins, **47:**136
in snake venoms, **8:**490–491
Neurotransmitters, amines having function as, **29:**482
and behavior, **29:**482–489
catecholamines having function as, **29:**482
mechanism of interference, **29:**483
norepinephrine interfering with, **29:**483
and schizophrenics, **29:**515–520
Neutral acid, **34:**492
lipid, **34:**467; 483
surfaces, **34:**452
Neutral CE hydrolase, cholesterol ester hydrolases, **69:**46–47
Neutral deoxyribonuclease, localization of, **24:**321
Neutral proteases, multiple forms, **25:**351
Neutral Proteinases, metal-chelator-sensitive, **41:**183–184; 200–204
Neutral salts, **39:**355
Neutron scattering studies, annexinmembrane complex, ion channel model system annexin V, **71:**222
Neutrophils, **58:**258, **60:**235, **65:**49
nitric oxide synthases, **65:**297–298
$p47^{phox}$, **65:**69–70
Newcastle disease virus, **52:**133; 137; 140
neuraminidase of, **20:**144
purification, **9:**187
NH, **57:**32–33; 40–41; 43
-NH_2 group, introduction of, PN-dependent, **26:**128
Niacin, **4:**302–303. *See also* Nicotinic acid
Niacinadenine dinucleotide, *See* NAD
Niacinadenine dinucleotide phosphate, *See* NADP
Niacin derivatives salvage, NAD+ synthesis, **73:**158–166. *See also* NAD+ synthesis
Nickel
catalysis by, **15:**25; 41
effect on pectic substances, **20:**351
in melanin, **16:**172
and RuDP carboxylase, **19:**22
Nickel, Raney, effect on tryptophan metabolism, **10:**68
Nickel compounds, **17:**138; 142; 147; 150
Nickel-copper alloys, catalysts, **15:**26
Nickel ions. *See also* Nickelous ions
activation of alkaline phosphatase by, **10:**99; 110
combination with porphyrin, **12:**18
as decarboxylating agent, **12:**72; 80
effect on elastases, **23:**234
effect on monolayer photochemical reactions, **11:**67
in thiol oxidation, **11:**203
model catalyses with, **21:**177
Nickel mercaptide, **11:**216
Nickelous ions

chelate formation with amino acids and peptides, **21:**145–147
enolase activation by, **21:**152
properties, **21:**157
Nickel oxide, **34:**456
Nicotiana, **55:**69
Nicotiana affinis in alkaloid studies, **8:**218
Nicotiana glauca in alkaloid studies, **8:**219; 223; 224; 227–231
Nicotiana glutinosa in alkaloid studies, **8:**223; 227–228; 230–231
Nicotiana rustica in alkaloid studies, **8:**219; 223
Nicotiana tobacum, alkaloids in, **8:**240–241
synthesis and distribution of nicotine, **8:** 217–223; 228–231
Nicotinamidase, hydrogen transfer, **26:**116
Nicotinamide, **5:**207–5:236, 15:80–81; 82; 91; 106; 107 (radioactive), **49:**28, **50:**311. *See also* Nicotinic acid; Pyridine nucleotides
action on thiaminase, **15:**401
effect on choline oxidation, **17:**383
reaction with ATP, kinetic studies, **17:**36
enzymatic reactions involving, **5:**207–5:232
as enzyme inhibitor, **25:**222; 237–240
formation on autoclaving nutrient media, **13:**260
in hydrogen transfer models, **26:**134–140
metabolic fate, **15:**108; 109–110 (C^{14})
metabolic interrelations of derivatives, **15:**106–108
mode of action, **14:**183
and oxidase, **19:**175; 189; 192
precursors of, **15:**103
and thiaminase model, **15:**404
treatment of schizophrenics with, **29:**527
Nicotinamide, Enzymic Reactions (Schlenk), **5:**207
Nicotinamide adenine dinucleotide (NAD+), **58:**101; 110
cysteine *S*-conjugate β-lyase reactions, rat liver with kynureninase, **72:**216–219
Nicotinamide-adenine dinucleotide phosphate (NADP), **48:**10
Nicotinamide adenine dinucleotide synthetase, *See* Desamidonicotinamide adenine dinucleotide amidotransferase
Nicotinamide and cozymase, synergism, **7:**421
Nicotinamide binding site, eukaryotic dihydrofolate reductase, **70:**40–41; 45; 46; 62–63
Nicotinamide (C^{14})
DPN synthesis, **27:**276
in DPN synthesis, **18:**206–207
as enzyme inhibitor, **25:**237
Nicotinamide coenzymes, abbreviation for, **24:**294
Nicotinamide deamidase, pyridine nucleotides, niacin derivatives salvage, **73:**161–163
Nicotinamide derivatives, effects on influenza bacilli growth, **7:**415–416
rate of inactivation, **7:**422–423
Nicotinamide methiodide, model coenzyme, **15:**31
Nicotinamide methylkinase, **15:**111
Nicotinamide methylpherase, **15:**106; 111
Nicotinamide methyltransferase, **25:**413–414
Nicotinamide mononucleotide, **15:**81; 82; 84; 89; 91; 92; 94; 97; 98; 106–108, **17:**184
biosynthesis, **15:**103–104
formation of, **18:**218
reaction with ATP, **18:**193
in synthesis of DPN, **15:**104–105
Nicotinamide nucleoside, phosphorolysis of, **18:**225
Nicotinamide nucleotide dehydrogenases, **5:**208–222
codehydrogenases, **5:**208–211
effect of yeast enzymes on, **5:**228–230
and malignant growths, **5:**218–222
methods for quantitative determination of, **5:**216–218
Nicotinamide of DPNH, and coenzyme binding, **27:**57; 58; 62; 64–65
Nicotinamide phosphoribosyltransferase, NAD+ synthesis, niacin utilization, **73:**165–166
Nicotinamide ribofuranoside, **15:**82
Nicotinamide riboside, **15:**94; 103–108, **18:**222
and PN-dependent enzymes, **26:**123
Nicotinamide riboside bond, **15:**84; 85; 90; 91; 107
Nicotinamide-sensitive nucleosidase, **15:**107
Nicotinate, **57:**147
Nicotinate phosphoribosyltransferase, NAD+ synthesis, niacin utilization, **73:**164–165
Nicotine, **55:**139. *See also* Alkaloids
and anabasine biosynthesis, **8:**229–231
biosynthesis, **8:**217–223

Nicotine (*Continued*)
and anabasine, **8:**229–231
in *Datura,* **8:**224
formation in roots of various species, **8:**223
localization of synthetic mechanism, **8:**222
methods of determination, **8:**218
and nornicotine, **8:**211; 227–231; 233–235
and origin of pyridine ring, **8:**235–236
precursors, **8:**236
rate of demethylation, **8:**234
changes during fermentation of tobacco, **10:**340; 386–394; 403
degradation in processing of tobacco, **10:** 429–433
detoxication in the animal body, **10:**429
effect of catalase on disappearance from tobacco, **10:**414
enzymic conversion, **6:**376–377
tyrosine hydroxylase regulation, **70:**150
Nicotine compounds, **17:**143
Nicotine synthesis and plant physiology, **8:** 245–246
Nicotinic acetylcholine receptor, ion channels and membrane proteins, **71:**212
Nicotinic acid, **46:**83, **57:**162. *See also* Niacin; Nicotinamide
action of *Fusaria,* **5:**169; 170; 189; 191
action on thiaminase, **15:**401
action on thermostable factor, **15:**410
biosynthesis, **15:**98–103
metabolic fate, **15:**108; 109–111 (C^{14})
and thiaminase model, **15:**404
and alkaloid synthesis, **8:**235–236
as precursor of nicotine, **8:**236
from tryptophan, **8:**235; 236
and water absorption, **8:**246
antagonism to benzamide, **10:**69
relation to thiamine in plants, **10:**246; 247
in tryptophanase, **10:**61–62
and biological methylation, **12:**457; 465; 474–478
derivatives, electrophilic carbon of, **12:**299–302
derivatives
actions of inhibitors, **7:**422
and coliform organisms, **7:**416–423
and dehydrogenation, **7:**416–417
effect on respiration, **7:**416–417
and *Hemophilus parainfluenzae,* **7:** 411–416
inactivation by *Proteus vulgaris,* **7:** 419–420
inhibition of, **7:**418–419
metabolism in microorganisms, **7:** 410–435
rates of microbial reaction with, **7:**432
metabolism, **7:**423–424
fate in animal body, **5:**225–227
formation in tryptophan metabolism, **19:** 92–96
genetic studies, **17:**418–419; 421
inhibitory analogues of, **6:**132; 139
in molds, **9:**572
synthesis, **5:**222–225
treatment of schizophrenics, **29:**527
Nicotinic acid amidase, **12:**476
Nicotinic acid derivatives, effect on cholinesterase, **20:**298
Nicotinic acid-phosphoribosyltransferase, hydrogen transfer, **26:**116
Nicotinic receptor agonist, tyrosine hydroxylase induction, **70:**169
Nicotinic receptors, **55:**139
Nicotinuric acid
and biological methylation, **12:**475–478
relation to nicotinamide, **15:**109–111
Nicotinylalanine, **49:**31
Nicotyrine, oxidation by liver, **10:**430
Niemann-Pick disease, **38:**294; 295; 298; 299
nif genes, **67:**328–333
FeMoco reactivation, **67:**330
groups, **67:**331–332
nitrogen fixation, **67:**328
products, **67:**332
Nigeran, **15:**312; 313; 314
Nigerose, **17:**214
Night clouds, **27:**352
NIH Shift, **67:**204–211
4a-hydroperoxide, **67:**209
arene oxide mechanism, **67:**205–207
aryl hydroxylases, **67:**205
aryl hydroxylation, **67:**208–209
cytochrome P-450, **67:**210
flavoprotein monooxygenase, **67:**210
4-methyl-3-hydroxyphenylalanine, **67:** 207–208
4-methylphenylalanine, **67:**205–206
NIkkomycins, **59:**72
Ninhydrin, **17:**311

action on fibrinogen, **10:**5; 33
reaction with asparagine, **20:**337; 338
Ninhydrin assay, to follow hydrolysis of peptides by proteinases, **41:**203
Nissl granules, in schizophrenics, **29:**510
Nitella clavata, antibiotic absorption by, **20:**387
Nitratase, enzymic adaptation, **13:**82; 97; 99
Nitrate assimilation, **57:**270–271
Nitrate dissimilation, **57:**271–273
Nitrate ester reductase, **46:**384; 398
Nitrate esters, **46:**398
as substrates, **46:**399
Nitrate metabolism, **17:**436–437
Nitrate nitrogen, assimilation in plants, **9:** 392–398; 413–414
Nitrate reductase, **46:**245, **51:**109; 157
classification, **14:**254
in electron transfer, **19:**201
in hydrocarbon oxidation, **27:**531
and luminescence, **25:**154
in *Neurospora* mutants, **22:**26
Nitrate reductase activity, **46:**86
Nitrate reductases, **17:**250–251; 267–269; 276; 294; 317–318; 436
PN-dependent, **26:**130–131
Nitrate reduction
metalloflavoproteins in, **17:**283
in photosynthesis, free energy of, **25:**46
in plants, mechanism, **9:**393–398
Nitrates, **51:**125; 127, **65:**287–288. *See also* Nitrogen compounds
in air-cured tobacco, **6:**343
changes during fermentation of tobacco, **10:**394–395; 402
in creatine kinase transition state complex, **54:**415
as electron acceptor, **17:**239; 240; 269; 275; 298; 316; 323
effect on pyrimidine use, **17:**429
as nitrogen source in *Neurospora* mutants, **22:**26
in plant metabolism, **10:**224–225
and plant respiration, **18:**284
reduction in microbial attack on hydrocarbons, **10:**448
Nitration, **42:**144
Nitric oxide
endothelium-derived, **65:**288; 290
inhibition of cytochrome oxidase, **17:** 339–341; 343; 345
role in shock, **65:**318
Nitric oxide donors, glutathione synthesis, **73:**259–260
Nitric oxide synthases, **65:**287–330
L-arginine binding site, **65:**307
brain
physical and kinetic parameters, **65:**296
purification, **65:**295
tissue distribution, **65:**294
chelating agents, **65:**305–306
cofactor dependence, **65:**304
constitutive, **65:**291
agonist effects, **65:**309–310
Ca^{2+} concentration and, **65:**309
purification and properties, **65:**295–298
regulation, **65:**307–310
in vivo regulation, **65:**310
endothelial, **65:**298
expression in tissue and cells, **65:**291; 293
inducible, **65:**291
H_4-biopterin, **65:**313
induction and expression, **65:**311–313
purification and properties, **65:**298–307
inhibition
by arginine analogs, **65:**314–317
by cofactor antagonists, **65:**317–318
irreversible, **65:**316
isoforms, **65:**291–295
characteristics, **65:**291–292
substrate specificity, **65:**298–299; 306–307
mechanism, **65:**319–329
arginine hydroxylation, **65:**326
constitutive nitric oxide synthases, **65:** 321–322
flavin role, **65:**328
intermediate, **65:**323
mixed-function monoxygenase activity, **65:**324
NADPH requirements, **65:**319–320
NO synthesis, **65:**320–321
oxygen source, **65:**322
stable isotope studies, **65:**322
metal ions, **65:**305–306
molecular weight, **65:**296
NADPH requirement, **65:**305–306
neutrophil, **65:**297–298
pharmacological control, **65:**318–319
pH effect, **65:**301–302
selective inhibition, **65:**316

Nitride, **19:**109
in enzymic adaptation studies, **13:**77–79; 82
Nitrile hydration, glutamine-dependent nitrogen transfer, **72:**181–183
Nitriles, use in synthesis of dehydropeptides, **8:**128–129
Nitrite reductase, **17:**252; 268; 283–284; 436, **46:**49
from *Aspergillus niger,* **46:**63
column chromatography, **22:**445; 446
in electron transfer, **19:**201
PN-dependent, **26:**130–131
Nitrites, **19:**109
formation, **46:**394; 398
in plant metabolism, **9:**393–398
p-Nitroacetanilide, **15:**211–212
Nitroacetic acid, **35:**157
p-Nitroaniline
acetylation of, **15:**211–212, **18:**129; 131
Nitroanilines, activation of thiaminase, **15:**399; 400; 410
Nitroarginine, in transaminations, **16:**197
N^{ω}–Nitro-L-arginine, nitric oxide synthases inhibition, **65:**316
Nitrobacter
metabolism in, **74:**6
Nitrobacter agilis, **71:**126
phosphorylating system from, **23:**353
Nitrobacter vulgaris, phosphoribulokinase from, **74:**100; 120; 121
Nitrobacter winogradsky, phosphoribulokinase from, regulation of, **74:**109
p-Nitrobenzoyl-L-arginine, methyl ester, **19:**306
p-Nitrobenzyl chloride, **46:**396; 397; 406
3-*o*-Nitrobenzyl-(3)-4-methylthiazolium salt, **15:**399
p-Nitrobenzylthioinosine, **54:**64
Nitroblue tetrazolium, **41:**38; 48, **58:**62
p-Nitrocarbobenzoxy group, in peptide synthesis, **29:**395–396
4-Nitrocatechol, **58:**217
Nitrocatechol sulfate, **38:**305, **60:**169; 255
p-Nitrocatechol sulfate, **38:**302
Nitro-cellulose filters, **35:**477
Nitrocresols, effects on tissues, **7:**25–27
permeability to, **7:**25–27
Nitroethane oxidase, **19:**203
P-Nitrophenyl χ-D-galactoside, as inhibitor, **36:**109
Nitroformate, **58:**81
Nitrofurfural semicarbazone, in treatment of thiaminase disease, **15:**415
Nitrogen, **46:**98
in air-cured tobacco, **6:**337–348
alkaloids, **6:**343–348
conversions, **6:**337–339
nitrates, **6:**343
protein digestion, **6:**339–343
in artificially cultured tobacco leaves, **6:** 318–321
assimilation, **6:**300–305
Caulobacter cell cycle, transcription regulation, **71:**17–18
elementary, evolution in aging of cigarette tobacco, **10:**407–408
in flue-cured tobacco, **6:**364
in green tobacco leaves, **6:**314–316
labeled, in studies of asparagine and glutamine biosyntheses in plants, **20:**331–334
in studies of glutamic acid metabolism in plants, **20:**322
in muscular dystrophy, **6:**207; 222
in myasthenia gravis, **6:**212
partition in plants, **9:**427–428
Pseudomonas putida, **70:**223
in synthetic polysaccharides, **15:**304
transfer reactions of, aspartic acid in, **21:** 252–255
glutamine in, **21:**240–252
Nitrogen-13. *See also* Nitrogen-13 labeled compounds
availability of, **57:**257–258
in blue-green algae, **57:**264–267
half-life of, **57:**252; 258; 261
and nitrogen-15, **57:**257–261
production of, **57:**252
and sensitivity of detection, **57:**258–260
sources of, **57:**262
in soybean nodules, **57:**267–268
synthesis, **57:**262; 264
in tracer work, early, **57:**252–253
Nitrogen-14, **57:**257
Nitrogen-15
in flavins, effect on EPR, **27:**134–135
and nitrate dissimilation, **57:**271–272
and nitrogen-13, **57:**257–261
and proteins in plants, **18:**284
in tracer work, early, **57:**252–253

in transamination studies, **16:**186; 193; 200; 218; 223; 226–227
Nitrogen-13 ammonia. *See also* Brain ammonia extraction; Brain ammonia metabolism
assimilation in plants and microorganisms, **57:**287–289
and blood flow, **57:**255
metabolism of, in mammals
and blood flow *vs.* brain ammonia extraction, **57:**292–294
in brain, **57:**296–302
entry of, in brain, **57:**295–296
in heart studies, **57:**255; 296–302
introduction, **57:**289–290
in liver studies, **57:**302–310
in other organ and tissue studies, **57:** 314–319
results of, in brain, **57:**290–292
in tumor studies, **57:**313–314
in murine tumor, **57:**331–332
synthesis, **57:**285–287
Nitrogen-15 ammonia, **57:**292
Nitrogen-13 as a Biochemical Tracer (Cooper, Gelbard, and Freed), **57:**251
Nitrogenase, **17:**282, **51:**109; 110; 112, **52:**1–20, **67:**299–359
complex
lifetime of, **52:**18–19
stoichiometry of, **52:**15–18
dissociable components, **67:**307–309
electron transfer, order of events associated with, **67:**355–356
$FeP_R \cdot 2MgATP$ regeneration, **67:**356–358
future prospects, **67:**359
general mechanism and partial reactions, **67:**323–328
consensus mechanism, **67:**326
one-electron transfer cycles, **67:**326–327
interaction of FeP and MoFeP during catalysis, **67:**353–356
kinetic mechanism of, **45:**282
kinetic studies, **52:**13–20
making FeP competent, **67:**349–352
metal centers, **67:**312–318
structure and function, **67:**309–323
MoFeP metal cluster positioning, **67:** 352–353
molecular genetics, **67:**328–349
FeP protein, **67:**340–349
MoFe protein, **67:**333–339
nif genes, **67:**328–333
physiological concerns, **67:**358–359
role of ferredoxins and ATP, **67:**301–307
substrates and inhibitors of, **52:**19–20
Nitrogenase activity, **46:**88
Nitrogen assimilation, **16:**254–257
Nitrogen balance, **58:**174
during fermentation of tobacco, **10:**340; 350; 377–380; 381–382; 386; 393–394
in periodic catatonia, **29:**529–530
Nitrogen-carbon bonds, formation, in purine biosynthesis, **21:**229–255
Nitrogen catabolism, in mammals, **19:**381–386
Nitrogen catabolites
allantoin transport affected by, **56:**125
in repression of allantoin enzyme production, **56:**117–120
Nitrogen compounds in tobacco, changes during fermentation, **10:**376–398; 402–403; 406–409
Nitrogen donor, ammonia as, **28:**121
Nitrogen fixation, **46:**88
comparative biochemistry of, **13:**345–375
role of hydroxylamine in, **13:**304
ferredoxins in (*See* Ferredoxins, in nitrogen fixation)
nif genes, **67:**328
Nitrogen Fixation, Comparative Biochemistry (Wilson), **13:**345
Nitrogen Fixation, Photosynthesis and Fermentative Metabolism, Chemistry and Function of Ferrodoxins in (Buchanan and Arnon), **33:**119
Nitrogen gas, effect on enzyme inactivation by radiations, **24:**376
L-Nitrogen-13 glutamate, **57:**323–326
Nitrogen-13 labeled amines, **57:**340
Nitrogen-13 labeled L-amino acids
metabolism of, in mammals
and amino acids, **57:**328–329
and AOA, **57:**326–328
and arteriovenous differences, **57:**334–336
and carbon-14 concentrations, **57:**333
and distribution measurement, **57:**333–335
and L-glutamate, **57:**331–333
and L-leucine, **57:**327–329
and tumor uptake studies, **57:**329–331
and L-valine, **57:**327–328
synthesis, **57:**319–323

Nitrogen-13 labeled compounds
  L-amino acids
    metabolism of, in mammals, **57:**323–336
    synthesis, **57:**319–323
  cautions about, **57:**262
  nitrogen-13 ammonia
    assimilation in plants and microorganisms, **57:**287–289
    and blood flow, **57:**292–294
    entry of, in brain, **57:**295–296
    in heart studies, **57:**302–310
    in liver studies, **57:**310–313
    metabolism of
      in brain, **57:**296–302
      in mammals, **57:**289–290
    in other organ and tissue studies, **57:** 314–319
    results of, in brain, **57:**290–292
    synthesis, **57:**285–287
    in tumor studies, **57:**313–314
  nitrogen-13 N_2
    in blue-green algae, **57:**264–267
    in soybean nodules, **57:**267–268
    synthesis, **57:**262; 264
  nitrogen-13 nitrate
    assimilation in plants and microorganisms, **57:**270–271
    in denitrification studies, **57:**273–279
    dissimilation reactions, **57:**271–273
    metabolism of, in mammals, **57:**279–284
    synthesis, **57:**268–270
  nitrogen-13 nitrite
    in denitrification studies, **57:**273–279
    metabolism of, in mammals, **57:**279–284
    synthesis, **57:**268–270
  nitrogen-13 NO_2, **57:**284–285
  nitrogen-13 N_2O, **57:**284–285
  N-nitroso compounds, **57:**336–340
  other, **57:**340–341
  synthesis, **57:**262–263
Nitrogen-13 labeled gases, **57:**259
Nitrogen-13 labeled glutamine, **57:**270
Nitrogen-13 labeled hydroxylamine, **57:** 340–341
Nitrogen-13 labeled *N*-nitroso compounds, **57:**336–340
Nitrogen metabolism. *See also* Asparagine; Glutamine; Glutathione
  and citric acid cycle, **18:**251–258
  and enzymic adaptation, **13:**77–80; 81
  of higher plants, **9:**391–454
  inorganic, **17:**250–252; 281–283
  link to tricarboxylic acid cycle, **28:** 143–144
    role of glutamine, allosteric regulation, **28:**48; 66–69
  in mammals, **19:**335–386
    by methods other than transamination, **19:**344–346
Nitrogen Metabolism of Higher Plants (Street), **9:**391
Nitrogen mustards. *See also* Vesicants
  effect on blood coagulability, **10:**37
  inactivation of guinea pig complement I by, **11:**245
Nitrogen-13 nitrate
  assimilation in plants and microorganisms, **57:**270–271
  in denitrification studies, **57:**273–279
  dissimilation reactions, **57:**271–273
  metabolism of, in mammals, **57:**279–284
  synthesis, **57:**268–270
Nitrogen-13 nitrite
  in denitrification studies, **57:**273–279
  metabolism of, in mammals, **57:**279–284
  synthesis, **57:**268–270
Nitrogen-13$^{13}NO_2$, **57:**284–285
Nitrogen-13$^{13}N_2O$, **57:**284–285
Nitrogenous groups, enzymes transferring, **24:**312
Nitrogen oxides
  biosynthesis, **65:**287–290
  effect on enzyme inactivation by radiations, **24:**375–376; 378
Nitrogen regulator I (NRI), glutamine synthesis expression, **72:**35–36
Nitrogen requirement, genetic studies, **17:** 448–450
Nitrogen sources, **56:**117–118
  effect on carotenogenesis, **21:**311
Nitrogen supply, effect on fat formation by microorganisms, **9:**658–661
Nitrogen transfer. *See also* Transamination
  asparagines synthetase, **72:**107–111
    catalysis mechanisms, **72:**146–147
    genetic control, **72:**148–149
  glutamine-dependent molecular mechanisms, **72:**167–190
    AS-B catalysis, **72:**188–190
    AS-B/papain comparisons, **72:**175–183

asparagine synthetase models, **72:** 167–175
heavy-atom kinetic isotope effects, **72:** 183–186
kinetic mechanism in asparagines synthetase, **72:**186–188
role of aspartate in, **16:**255–256
in synthesis of urea, **15:**320; 325; 330–331; 337–338; 340–343; 363
Nitrogen-13 urea, **57:**341
Nitroglycerin, **46:**398; 399
δ-Nitronorvaline, **35:**143
p-Nitrophenol
enzyme inhibitor, **16:**135–136
Nitrophenols
effect of pH on activity, **18:**34–35
effect on sperm respiration, **9:**369
p-Nitrophenoxide, **58:**26
p-Nitrophenyl, **60:**91
p-Nitrophenyl acetate, **43:**380, **54:**8, **58:**13. *See also* NPA
action of pancreatic enzymes on, **23:**136
p-Nitrophenylacetate
(C^{14}), in dehydrogenase study, **26:**173
p-Nitrophenyl acetate
carbonic anhydrase catalyzed hydrolysis, **47:**164; 191–194; 196–198; 206–207; 209–210; 214–215; 227–229
p-Nitrophenylacetate
hydrolysis by chymotrypsin, **20:**302; 309; 312; 314; 315
reaction with imidazole, **20:**81; 82; 86; 88; 89
Nitrophenylacetate, as quasi-substrate for chymotrypsin, **22:**52; 66; 67
p-Nitrophenyl *N*-carbobenzyloxyglycinate, hydrolysis by thiol-proteases, **33:**398
p-Nitrophenyl diphenyl phosphate, **43:**379
p-Nitrophenyl esters, **43:**382
in peptide synthesis, **29:**406
o-Nitrophenyl-α-D-fucoside, **60:**102
p-Nitrophenyl hexanoate, **43:**382
2-Nitrophenylimidazole, reaction with benzoyl chloride, **20:**83
p-Nitrophenylnitromethane, **46:**424
Nitrophenyl phosphate, **17:**165; 167; 170; 173; 174; 183
p-Nitrophenyl phosphate, **34:**505–507; 511; 524
p-Nitrophenyl propionate, carbonic anhydrase catalyzed hydrolysis, **47:**206–207; 211; 223
p-Nitrophenyl salicylate, **43:**380
p-Nitrophenyl tetrahydrophyranyl ether, **58:**28
p-Nitrophenyl trimethylacetate, **54:**8
1-Nitropropane, **46:**398
2-Nitropropane, **46:**397
Nitroprusside, **58:**82
test for –SH groups, **11:**219–220; 222; 224; 234
Nitroprusside reaction, **5:**24
p-Nitropyridine N-oxide, **46:**394
Nitroquinoline oxide, **60:**15
4-Nitroquinoline *N*-oxide, mutagen, **27:**331
Nitro reductase, PN-dependent, **26:**131
Nitrosamides, mutagenic agents, **27:**330
Nitrosoalkane dimers, **35:**139; 140; 141
Nitrosoamines, mutagenic agents, **27:**330
Nitrosocarbaryl, **57:**336–340
N-Nitroso-*N*-chloroethyl-l-chloroethyl carbamate, *See* BCNC
Nitrosodimer, **35:**141
N^{10}-Nitrosofolic acid, isolation and synthesis, **21:**373
Nitrosoguanidine, **58:**118
Nitrosomonas, metabolism in, **74:**6
Nitroso reductase, PN-dependent, **26:**131
Nitroso-tyrosine-pepsin, activity, **14:**190
Nitrosoureas, halogenated, **55:**272–274
Nitrosylferrocytochrome, cytochrome *c* oxidase, **71:**108–109
Nitrous acid, **58:**63
mutagenic agent, **27:**296; 302; 305–306; 336; 339–342
mechanism, **27:**309–312
Nitrous oxide, effect on binding of carbon dioxide to carbonic anhydrase, **47:**186
N-linked glycosylation, adenosine receptors, **69:**90
N-Linked oligosaccharides, **52:**94
NMD adenylyl transferase, quinolinate to NAD+, **73:**149–153
NMN
complex with LADH, **27:**62–65
exchange across nuclear membrane, **27:**270
NMN deamidase, pyridine nucleotides, niacin derivatives salvage, **73:**163–164
NMNH
as coenzyme for glutamate dehydrogenase, **39:**373; 381; 389; 407
NMP kinases, *See* Nucleoside monophosphate (NMP) kinases

NMR-EPR, distance measurements by, **53:** 72–74
NMR probe, **43:**40
NMR spectroscopy, **34:**468–469, **49:**104, **57:**260. *See also* NMR studies on mechanisms of enzyme action
of alkaline phosphatase, **55:**433; 436–437; 440; 441
^{13}C, **55:**415; 426–430
^{113}Cd, **55:**396; 413–426
^{35}Cl, **55:**415
of coenzyme Q, **25:**279
enzyme mechanisms and, **33:**1–70
eukaryotic dihydrofolate reductase, structural evidence from, **70:**47
^{19}F, **55:**395
and glyoxalase, I, **59:**108; 110; 116
^{17}O
of dAMP, **55:**337; 338; 339
of oxygen chiral phosphate esters, **55:** 337–340
of phosphate esters, **55:**337–340
of oxygen chiral phosphate esters, **55:** 321–335
^{31}P
of inorganic phosphate, **55:**322
of inorganic thiophosphate, **55:**328–330
of nucleoside diphosphates, **55:**325–327
of nucleotides, **55:**327–328
of phosphate diesters, **55:**322–324
of phosphate monoesters, **55:**333–335
psoralenated oligonucleotides, structure determination by NMR and X-ray crystallography, **66:**128
and pyrrolo-quinoline quinone, **59:**171–172; 178; 180
of pyruvate carboxylase, **42:**35
of RNase, **24:**181
NMR studies, **47:**74
NMR studies of mechanisms of enzyme action
and carbonyl-polarizing enzyme reactions
glyoxalase I, **59:**289–290; 292
model reactions to biotin enzymes, **59:** 297–299; 301–303
polarization process, **59:**287–288
yeast aldolase, **59:**293–297
conclusions of, **59:**303–304
description of, **59:**242–243
and nucleophilic substitutions on phosphorus
adenylate kinase, **59:**253–257; 259–263; 265–267
cyclic adenosine monophosphate-dependent protein kinase, **59:**267–269; 272–274
DNA polymerase, I, **59:**274–278; 280; 282–283
experimental criteria, **59:**245–253
mechanistic considerations of, **59:**245–247
staphylococcal nuclease, **59:**283–287
parameters of, useful, **59:**243–245
NMR Studies of the Mechanism of Enzyme Action (Mildvan and Fry), **59:**241
Nocardia
antiviral agent from, **20:**402
enzymic transformations of steroids by, **20:**262
β-lactamase of, **28:**242
oxidation of hydrocarbons, **27:**471–474; 496–499; 506; 508; 511; 516; 519; 520; 531
PN-dependent enzymes, **26:**131
Nocardia corallina, oxidation of hydrocarbons, **27:**473
Nocardia hydrocarbon-oxydans, **27:**488
Nocardia opaca, oxidation of hydro carbons, **27:**496; 518
Nocardia petroleophila, oxidation of hydrocarbons, **27:**488
Nocardia rugosa, vitamin B_{12}, **26:**260
Nocardia salmonicolor, oxidation of hydrocarbons, **27:**497
Nodoc, **34:**3
Nodule
hemoglobin in, **13:**346
in nitrogen fixation studies, **13:**352–357; 364–365; 369
transferase activity in, **13:**298; 304
Noenzymatic rate of Fe(II) oxidation, **44:**198
Noformin, in treatment of plant diseases, **20:**402
Nojirimycin, **36:**178–180
Nomenclature of enzymes, **14:**219–260
Nomenclature of Enzymes (Hoffmann-Ostenhof), **14:**219
Nonactin, properties of, **29:**289
Nonaqueous phase, **34:**476
Non-azide-sensitive ferroxidase, **44:**204
Nonblue cupric ion, *See* Cupric ion, nonblue
Noncovalent bonding, **34:**469

Noncovalent bonds, in enzyme structure, **22:**73; 74
Non-electrolytes, catalyzed translocation of, **29:**40–41
Nonexchangeable guanine nucleotide site, microtubule dynamics, **69:**123
Nonheme-iron proteins, mechanisms of deduced from relaxivity data, **33:**62–63
  outer sphere relaxation of, **33:**63
Nonheme irons, **34:**92; 132; 133; 137
  absorption spectrum characteristics of, **34:**139
  redox function of, **34:**133
  type V—SH groups and, **34:**117; 118
Nonhistone proteins, **38:**49–52; 86
Nonlinear regression, **58:**216
Non-Michaelian kinetics, **57:**378–379
Nonphosphorylating preparations, of respiratory chain, **28:**161–165
Nonpolar solvents, **51:**33
Nonproductive binding, **33:**418, **43:**172; 263; 342; 360; 381, **53:**264
Nonprotein amino acids, *See* Amino acids, nonprotein
Nonprotein Amino Acids of Plants (Fowden, Lea, and Bell), **50:**117
Nonpyridoxal amino acid racemases, comparative features, **44:**127
Nonpyridoxal racemases, **44:**117
Non-redox-active metal centers, cytochrome *c* oxidase, **71:**101–103
Nonsaturable transport, **49:**47
  representation of diffusion, **49:**47
Nonsecretors, **36:**135
Non-selenium dependent GSH peroxidase, **57:**370
Nonspecific amino acid racemase, **44:**101
Nonspecific catalysis, **39:**333
Noradrenaline, **44:**229
  from phenylalanine and tyrosine, **14:**74; 116–119
Noradrenochrome, in schizophrenia, **29:**516
Noradrenolutin, in schizophrenia, **29:**516
Norbornane amino acid, **49:**61; 71; 83; 90
19-Norcorticoides, microbiological formation, **20:**275
Norepinephrine, **67:**147–148
  biosynthesis of, **74:**236
  in catatonic reaction, **29:**530
    causes of fall in brain levels of, **29:**487
    changes in the hypothalamus by, **29:**514
    3,4-dehydroxyphenylserine decarboxylated to, in reserpinized mice, **29:**488
    increase in hypothalamus of rabbits, **29:**514
    inhibition of uptake by imipramine, **29:**484
    lack of uptake after surgical denervation, **29:**484
    location of, in brain, **29:**482
    metabolism of, in mental patients, **29:**492
    oxidation of, **29:**516
    sedation from, **29:**488
    sedatives preventing fall of, **29:**487
  effect on sperm, **24:**84
  fluorinated, **55:**239; 240
  methylation, **22:**256
  methylation of, **25:**415; 416
    synthesis of, **25:**383
  tyrosine hydroxylase, **70:**104; 145
Norite eluate factor, **21:**372
Norleucine, **17:**400
  effects of, on gramicidin S formation, **33:**340
    effects of on synthesis of gramicidin S, **33:**347
  incorporation into alkaline phosphatase, **29:**136
    inhibition of growth by, **29:**136
    as methionine antagonist, **29:**91; 135
    in synthetic ACTH, **29:**466
  in transamination reactions, **16:**190; 192; 197; 201
L-Norleucine, **67:**170
DL-Norleucine, resolution by papain, **19:**309
Normetanephrine
  imipramine-caused increase in excretion of, **29:**490
  inactive, **29:**482–483
  increased excretion in catatonia, **29:**530
Nornicotine
  changes during fermentation of tobacco, **10:**389–391
  and methyl group transfers, **8:**233–235
    biosynthesis from nicotine, **8:**227–231; 233–235
    in *Datura,* **8:**224
    similarity to anabasine, **8:**228–229
19-Norprogesterone-17α-ol, microbiological conversions, **20:**251
Norspermine, **54:**222
19-Nortestosterone, microbiological conversion, **20:**243; 244

Nortriptyline
anticholinergic activity of, **55:**177
structure-activity studies of, **55:**178
structure of, **55:**177
Norvaline, **17:**400
effects of on gramicidin S formation, **33:**340
in transamination reactions, **16:**190; 197; 201; 204
Notatin, **7:**475; 476; 489; 491; 493, **9:**560–563, **15:**148; 306, **17:**218. *See also* Glucose oxidase
classification, **14:**253
use in polysaccharide synthesis, **11:**321
Novel Aspects of the Biochemistry of the Molybdenum Cofactor (Rajagopalan), **64:**215
Novikoff ascites tumors
cytidine-diphosphate reductase of, **28:** 436–444
DR metabolism, **28:**448
Novikoff hepatoma, **42:**256
DR metabolism, **28:**436
Novikoff hepatoma ascites cells, phosphofructokinase in, **29:**355
Novobiocin
effect on murein synthesis, **26:**218
inhibition of synthesis of poly (glycerol phophate) by, **30:**244
Novo-proteinase, multiple forms, **25:**338
NPA, **33:**391
hydrolysis of, kinetics of (table), **33:**393
hydrolysis of by thiol-subtilisin, kinetics of, **33:**392–394
NPS, *See* Potassium *p*-nitrophenyl sulfate
NPS-Cl, **42:**138
NPS-HGH, **42:**138; 139; 140; 141; 143; 147
nRNA
fluoropyrimidine effect on, **54:**88
FU effect on posttranscriptional polyadenylation of, **54:**88
hypomethylation of, **54:**89
NS7, **67:**225
N,S-modified glutathiones, **59:**111
N splitting, molybdenum centers in enzymes, **51:**146
N-Terminal amino acid
for mouse GSH transferase, **57:**396–397
for rat GSH transferase, **57:**393–396
N-terminal mutant properties, ion channel model system annexin V, mutants analysis, **71:**231
N-Terminal sequence, **45:**407
N-terminus, acetylation, **66:**157
Ntn amidotransferases
asparagines synthetase, **72:**106–112
glutamine-dependent nitrogen transfer, **72:**176–183
glutamine domain residues, **72:**111–112
nitrogen transfer, **72:**107–111
future research, **72:**117–119
glucosamine 6-phosphate synthase, **72:** 104–106
glutamine and acceptor domains, **72:** 104–105
glutamine domain structure, **72:**105–106
glutamate synthase, **72:**112–117
[Fe-S] and flavin components, **72:**114–116
glutamine domain, **72:**116–117
glutamine PRPP amidotransferase, **72:**93–104
acceptor domain, **72:**99–100
allosteric transition and coordination of catalytic activities, **72:**102–104
crystal structures, **72:**96
enzyme classification, **72:**93–95
feedback inhibitor binding sites, **72:** 100–102
glutamine domain, **72:**96–99
nucleotide regulation and allosteric control, **72:**95
nitrogen transfer, **72:**139
structure-function relationship, **72:**90–93
Ntn hydrolases, **72:**117–119
Nuclear coordinate associated electron transfer (NCET), cytochrome *c* oxidase, redox linkage, **71:**168; 177
Nuclear-cytoplasmic activity ratio, of enzymes, **27:**269
Nuclear encoded subunits, cytochrome *c* oxidase, structural biochemistry, subunit III, **71:**103–108
Nuclear enzymes, **16:**413; 415–418; 462–469
control functions, **27:**275–283
solubility, **27:**250–252
specific activity, **27:**247–250
Nuclear glycolysis, **39:**302–303
Nuclear magnetic resonance (NMR), **41:**72, **57:**260
of alkaline phosphatase, **56:**344; 349

in D-amino acid oxidase reaction, **34:**55
cholesterol ester hydrolases regulation, **69:**61
histidyl ligand detection by, **56:**315
and hydride transfer, **24:**473
and thiamine action, **24:**468
of metallothionein, **56:**403
of neurophysin, **40:**299; 315; 318; 319; 320; 321; 322; 324; 326; 327; 328
of nucleotides, **27:**68
of oxytocin, **40:**322
of vasopressin, **40:**320
of yeast polysaccharides, **40:**247; 249; 256
of zinc metalloenzymes, **56:**311; 312
Nuclear-magnetic-resonance spectroscopy, *See* NMR spectroscopy; Nuclear magnetic resonance
Nuclear magnetic spectroscopy, *See* NMR spectroscopy
Nuclear membrane, permeability, **27:**263; 265; 268–271
structure, **27:**243; 248; 259–260; 269
Nuclear paramagnetic resonance, study of racemization, **26:**340
Nuclear phosphorylation, **27:**274
Nuclear poly(A)polymerase, **56:**292
Nuclear preparation, criteria of purity, **27:** 241–256
electron microscopy, **27:**244–245
Nuclear relaxation studies, **49:**117
Nuclear resonance spectroscopy, in electron-transfer study, **26:**136; 143
Nuclear Spin Relaxation Induced by Paramagnetic Probes in Study of Enzyme Mechanisms (Mildvan and Cohn), **33:**1
Nuclear spin relaxation rates, *See* Relaxation rates, nuclear
Nuclease activities, in different tissues, **27:**246
Nuclease digestion, **49:**195–197
Nuclease II
preparation of, **31:**24
spleen, chromatography of, on DEAE-Sephadex A-50, **31:**27
Nuclease P_1, **55:**375
Nucleases, **17:**141, **37:**422; 469; 525; 589, **56:**292
base specificity of, **29:**210–211
contaminants of, **29:**209
of *E. coli,* **29:**197–199
intracellular location in *E. coli,* **29:**198
ionic requirements of, **29:**209
micrococcal, **29:**191–197; 200–201; 206; 207; 209; 210
amino acid composition, **29:**192–193
ionic requirements of, **29:**196–197
liberation of mononucleotides and other products for DNA by, **29:**195
mode of action of, **29:**194–196
preparation of, **29:**191–194
specific activity of different preparations of, **29:**192
mode of action of, **29:**210–211
mung bean, **29:**169–170; 188–190; 209; 210
properties of, **29:**188
specificity of, **29:**188–189
pH optimum of, **29:**209
physical and chemical characterization of, **29:**209
of plant origin, **29:**190
site of cleavage of, **29:**210–211
streptococcal, **29:**199–200
from *Streptomyces* sp. No. 41, **29:**199
venom endonuclease, **29:**205–206
venom exonucleases, **29:**202–205
mode of action, **29:**204–205
viral, **29:**197–202
virus-induced, **29:**201–202
classification, **14:**225
in isolation of enzymes, **14:**398; 405
deficiency in inert cells, **16:**452
effect on activation of β-glucuronidase, **16:**376
effect on cell-bound β-lactamase, **28:**254–255
soil, **28:**386
effect on enzyme-binding, **27:**266
of pancreas, **23:**130–133
staphylococcal, **55:**353–356
Nuclease S_1, **55:**375
Nuclease staphylococcal, **37:**413; 434; 442; 521; 583
Nucleates, insoluble, solubilization of, **17:**137; 138; 143; 145; 147; 149
Nucleating core, **47:**132
Nuclei. *See also* Enucleation experiments
biosynthetic capacity, **27:**272–275
enzymology of, **27:**239–288
isolated, evaluation of, **27:**241–256
isolation of, **27:**240–242
permeability of, **27:**268–271
precursor molecules, **27:**269–270

Nuclei (*Continued*)
preparation from thymus, **27:**275
cell
method of preparation of, **17:**346
phosphatides in, **7:**168–169
cell isolation of, **16:**413; 414–417
interphase, isolation of chromosomes from, **16:**417; 448
isolated, incorporation of labeled amino acids, **16:**492
transfer or replacement in cell, **16:**413; 414; 470; 477; 485
and enzyme formation, **24:**5; 11–20
glycolytic activity of, **24:**336
polyamines and, **36:**250
Nucleic acid biosynthesis, effect of antimetabolites, **27:**461
Nucleic-acid-calcium-lipide compounds in nerves, **8:**47
Nucleic acid constituents, genetic studies, **17:**426–430
Nucleic acid derivatives, as solubilizing agents, **17:**136–141; 147
Nucleic acid-like macromolecules, and origin of life, **27:**396–397
Nucleic acid-like protein interaction, **27:** 395–398
Nucleic acids, **5:**19–20, **9:**455–535, **22:** 137–140, **36:**33. *See also* Chromosomes; Cytoplasm; DNA; Mutation; Nucleoproteins; Nucleus; Purine nucleotides; Pyrimidine nucleotides; Ribose; RNA
absorption spectrum, **3:**6–13; 21–23
action of phosphodiesterases on, **17:**171–173
genetic studies, **17:**455–457; 459–465
adsorption studies, **14:**344
in fractionation of enzymes, **14:**398
alkylation of, **27:**305; 320–326; 335–336; 338–342
amino acids and synthesis of, **29:**151
DNases and their use in studies of structure of, **29:**165–220
and enzyme repression, **29:**152
synthesis, effect of starvation on, **29:**150
asynchronous synthesis, **28:**431
of bacteriophage, **24:**17
in biosynthesis of proteins, **19:**264
bond with protein, **3:**17; 18
and calcium, **8:**7; 17; 19; 20; 50
chelates, in metal ion activation of enzymes, **21:**148
synthesis, in mitochondria, **21:**54
in chromosomes, **3:**13–15
cross-linking, **27:**324; 326; 330
cytochemical studies on, **9:**593
in cytoplasm, **3:**23–28
deamination of, **27:**309–311
disintegration by snake venom, **8:**460
effect of nitrous acid, **27:**309–311
effects of hydroxylamine, **27:**312–314
elimination from laminarinase preparation, **28:**343
enzymes for metabolism, **24:**341
Feulgen's reaction, **3:**4–6; 9–10
in fractionation of enzymes, **15:**155
reaction with manganous ion, **15:**204
of tobacco mosaic virus, **15:**459–461
of virus mutant, **15:**477–482
and functioning of cytoplasm, **8:**3; 4–7; 15–17; 19–23; 35; 58
hydrogen bonding, **65:**218; 224
hydrolysis by ribonuclease, mechanisms, **24:**443–458
labeling by ^{14}C-cytidine, **28:**420–421
and magnesium, **8:**17; 19; 20; 50
of nuclei, **27:**274–275
origin of, **38:**216
and origin of life, **27:**356–357; 359; 361; 363–365; 367–369; 374–375; 390; 392; 398
pentoses of, **28:**391–460
in plastids, **20:**210; 211
polyamines and, **36:**242–243
polymerization, **3:**14; 15; 20; 30; 31
and protein-bound enzymes, **27:**265
and protein synthesis, **27:**295; 396–397
radiation chemistry, **27:**330
sequence of bases, **27:**290–295; 335
in sperm, **9:**338–340
on surface of Gram-positive bacteria, **11:**56
synthesis, role of nucleosides, **28:**423; 425; 429
in thromboplastic material, **16:**30
of the cell nucleus, **16:**422–440
precursors of pyrimidines, **16:**266–268
renewal and synthesis, **16:**451–457
turnover, **7:**173–182; 188–189
in viruses, **3:**29–32
of viruses, **9:**229–234

in viruses and chromosomes, **4:**4–8
Nucleic Acids, Chemistry and Enzymology (Schlenk), **9:**455
Nucleic Acids, Primary Structure, Use of DNases in Studies of (Laskowski), **29:**165
Nucleic acid sugars, labeling by glucose-1-^{14}C, **28:**425
Nucleic acid synthesis, **27:**282–283, **38:** 198–199, **39:**236
control of, **25:**170–171
Nuclei of cells
function of enzymes in, **13:**15
isolation, **9:**597
Nuclei (P^{32}), in enzyme synthesis, **24:**15–16
Nucleoaminases, classification, **14:**246; 250
Nucleocidin, **55:**272
Nucleocytoplasmic relations, between residual protein and mass of cytoplasm, **16:**413
and distribution of enzymes, **16:**463–469
and energy-yielding reactions, **16:**468; 471–473
and genetic transmission, **16:**475–486
and nuclear membrane, **16:**476; 486–488
and nucleotide metabolism, **16:**473–474
and protein synthesis, **16:**474–475
and RNA, **16:**461–462
study by enucleation of cells, **16:**469–474
Nucleodepolymerases, classification, **14:**245
Nucleofugality, **59:**125
Nucleohistone, *See* Chromatin
Nucleoli, polyamines and, **36:**250
Nucleolus
enzymes of, **27:**268
relation to protein synthesis, **16:**475
RNA in, **16:**436; 438
Nucleolytic enzymes, use in locating nucleic acids in nucleus, **16:**417
Nucleophile-glyoxylate adducts
diversity in structure of, **57:**100–101
formation of, **57:**101–102
as substrates for D-amino acid oxidase, **57:** 99–101
as substrates for D-aspartate oxidase, **57:** 99–101
as substrates for L-α-hydroxy acid oxidase, **57:**99–100
Nucleophiles, **46:**384; 403, **57:**133–135
Nucleophilic aromatic substitution, catalytic mechanisms, **69:**21; 31–32; 35
Nucleophilic attack, **34:**486
Nucleophilic catalysis by enzymes, **24:**448; 461; 464
Nucleophilicity
of glutathione, **46:**383
of sulfhydryl group, **46:**383
S^- Nucleophilicity, **66:**154
active cite cysteines, thioltransferases, **66:**159–160
of cystine, **27:**216
and reaction rate, **27:**203
Nucleophilic strength, **39:**331
Nucleophilic substitutions on phosphorus
adenylate kinase
adenosine monophosphate-binding site of, **59:**261–262
and MgATP-binding site of enzyme and peptide fragment, **59:**253–257
and positioning of MgATP in X-ray structure, **59:**257; 259–261
and sequence homologies, with F_1 adenosine triphosphatase, Ras p21, and other nucleotide-binding proteins, **59:**262–263; 265–267
associative mechanism of, **59:**245–246
cyclic adenosine monophosphate-dependent protein kinase, **59:**267–269; 272–274
dissociative mechanism of, **59:**245
DNA polymerase I, **59:**274–278; 280; 282–283
experimental critera of
base catalysis, general, **59:**249
electron dense substrates, **59:**252–253
kinetic isotope effects, **59:**252
location of metal and other eletrophilic groups, **59:**247–248
positional isotopic exchange, **59:**249–252
reaction coordinate distance, **59:**248–249
mechanistic consideration of, **59:**245–247
monomeric metaphosphate, **59:**246
staphylococcal nuclease, **59:**283–287
trigonal bipyramidal phosphorane, **59:**246
Nucleophosphatase, **5:**22
Nucleophosphorylase, **9:**505
Nucleoprotamines in fish sperm, **9:**329; 338–339
Nucleoprotein-calcium-phosphatide of plasma membranes, **8:**49–51
Nucleoprotein coacervate, **34:**520
Nucleoprotein of liver, type of thiols in, **11:**221

Nucleoproteins, **4:**5; 11–17, **7:**187; 189
as agents of enzyme formation, **24:**5
alkali content, **4:**16
in biosynthesis of proteins, **19:**264
calcium content, **4:**15–16
in chromosomes, **3:**15–23
constitution of, **18:**176–177
in cytoplasm, **3:**6; 23–29
effect of light on, **9:**97 of viruses; 229–234
enzymes associated with, **27:**265–268
and genes, **3:**21; 28; 29–32
genetic material, **18:**161–190
imidazole in, **20:**99; 100
isolation from plastids, **20:**207
iron content, **4:**15–16
magnesium content, **4:**15–16
and myofibrillar structure, **13:**185–186
nonhistone, **18:**177–181
in nucleus, **3:**4–6; 28–29
orientation, **3:**20
and respiratory enzymes, **8:**35
solubility, **3:**19
in viruses, **3:**3; 29–32
zinc content, **4:**16
Nucleoproteins (Minsky), **3:**1
Nucleosidases, **9:**471; 472; 473; 474; 489–490; 495; 498; 499–500, **15:**81; 106–108
of *A. aerogenes,* effect on 5′-methylthioadenosine, **25:**386
Nucleoside-2′,3′-cyclic phosphates, in assay for RNase, **24:**172–173
Nucleoside diphosphatase, localization of, **24:**329
Nucleoside diphosphate glucose phosphotransferases, allosteric regulation, **28:**95; 97–98
Nucleoside diphosphate glucose transferase, in β-glucan synthesis, **28:**340
Nucleoside diphosphate kinase, **21:**280, **43:**13
localization of, **24:**314
phosphohistidine in, **36:**191
reactions of thiophosphate analogs, **43:**14
Nucleoside 5′-diphosphate kinase (NDPK), **53:**139
Nucleoside diphosphates, **55:**325–327
as phosphate donors, **18:**196–197
Nucleoside diphosphokinases, **18:**195
and biotin enzymes, **26:**312
in mitochondria, **21:**53
nuclear, **27:**278; 280
Nucleoside disphosphate kinase, **53:**139–144
exchangeable nucleotide transphosphorylation, **53:**141–143
microtubule assembly regulation, **53:** 143–144
purification and general enzyme properties, **53:**139–141
Nucleoside ligands, of corrinoid coenzymes, **26:**252–253; 275
Nucleoside monophosphate (NMP), MutT enzyme bond cleavage site, **73:**185. *See also* MutT pyrophosphohydrolase
Nucleoside monophosphate kinase, **18:**196–197, **73:**103–134
background, **73:**103–104
catalytic mechanism, **73:**110–119
active site residue roles, **73:**111–114
iterative structure-function studies, **73:** 110–111
phosphoryl transfer mechanism, **73:** 114–119
localization of, **24:**314
primary structure, **73:**104–107
adenylate kinase, **73:**104
guanylate kinase, **73:**104–107
uridylate and cytidylate kinases, **73:**107
substrate-induced changes, **73:**124–128
conformational, **73:**124–127
dynamical, **73:**127
residues in fit, **73:**127–128
substrate specificity, **73:**119–124
NMP site, **73:**120–124
NTP site, **73:**119–120
tertiary structure, **73:**108–110
backbone folding, **73:**108
substrate binding sites, **73:**108–110
Nucleoside monophosphate sugars, formation of, **26:**41
Nucleoside monophosphokinase, and biotin enzymes, **26:**312
Nucleoside oxidative enzymes, **17:**320
Nucleoside phosphoroamidates, synthesis, **22:**195–197
Nucleoside phosphorylase, **15:**104; 108, **18:**216; 224–225
classification, **14:**251
isolation, **14:**398
intracellular distribution, **16:**464; 466–468
localization of, **24:**341

Nucleoside phosphorylases, **9:**499–500, **21:**217
and RNA synthesis, **28:**419
Nucleoside-5′-polyphosphates, synthesis by carbodiimide method, **22:**181; 186; 187
Nucleosides, **3:**19, **9:**490–503, **43:**11, **47:**402. *See also* Nucleic acids; Nucleotides; Purines; Pyrimidines; individual nucleosides
abiogenic syntheses, **27:**384–385
absorption spectra, **9:**482–484; 484–485; 487–490
chemical synthesis, **9:**500–503
of cobamide coenzyme, effect on activity, **26:**352
competitive, effect on virus multiplication, **17:**479–483
of corrinoid coenzymes, **26:**250–252; 256; 262
definition, **9:**457–458
effect of hydrazine, **27:**314–315
labeled, in nucleic acid study, **16:**450–462
relation to CoA, **16:**12–16
metabolism, **9:**472–473; 474; 487–490
in nucleic acid synthesis, **28:**423; 425; 429
phosphate bonds, **9:**478–479
synthesis, with phosphatase models, **21:**192
Nucleoside triphosphatase A, chromatin-bound, **27:**267–268
fractionation, **27:**279–282
nuclear, **27:**278–282
Nucleoside triphosphatase B, chromatin-bound, **27:**267–268
nuclear, **27:**278–283
specificity, **27:**279–282
Nucleoside triphosphate (NTP), **50:**80
MutT pyrophosphohydrolase, **73:**183–184 (*See also* MutT pyrophosphohydrolase)
as phosphate donors, **18:**195
as pyrophosphate donors, **18:**198–200
role in formation, of nucleotide acid anhydrides, **18:**201–212
of polynucleotide, **18:**212–215
reactions with sugar phosphates, **28:**110–114
repression by, **27:**282–283
Nucleoside triphosphate-glucose phosphotransferases, allosteric regulation, **28:**95; 97–98
Nucleotidase, **9:**498, **56:**292
3-Nucleotidase, **15:**83
5′-Nucleotidase, **18:**216, **51:**181, **55:**369, **60:**272
action on phosphate from CoA, **16:**14
location of, in *E. coli,* **31:**116
as marker enzyme, **27:**252
in nucleus, **27:**274
5-Nucleotidase, **15:**106–108
activity, **10:**87
classification, **14:**245
isolation, **14:**401
separation by chromatography, **14:**346; 348
structure studies, **14:**309
column chromatography, **22:**435
effect on semen, **24:**86
localization of, **24:**320
in semen, **9:**353–354
b-Nucleotidase, action on CoA, **16:**16
Nucleotidases, **15:**334
Nucleotidases (phosphomonoesterases), in methyl synthesis study, **25:**397
Nucleotide acid anhydrides, synthesis of, **18:**201–212
Nucleotide bases
analytical determination, **9:**469–470
and mutations, **27:**290–295
Nucleotide binding, calcium ATPase, **65:** 196–198
Nucleotide-binding domain, of dehydrogenases, **50:**29
Nucleotide-binding site, eukaryotic dihydrofolate reductase, **70:**40–41
Nucleotide-bound sugars, modifications of, **26:**41–44
Nucleotide coenzymes, functions, structural features, **22:**158
in mevalonic kinase reaction, **22:**299; 300
synthesis, by carbodiimide method, **22:** 175–182; 186–192
by mixed anhydride method, **22:**159–174
by phosphoramidate method, **22:**192–201
Nucleotide Coenzymes, Synthesis (Baddiley and Hughes), **22:**157
Nucleotide content of erthrocytes and mononuclear cells (patients with ADA deficiency), **51:**171
Nucleotide-dependent enzymes, microtubule systems association, **53:**113–156
Nucleotide-Dependent Enzymes Associated with Microtubule Systems (Terry and Purich), **53:**113
Nucleotide doublets, and genetic code, **47:**419; 427

Nucleotide hydrolysis, microtubule dynamics, **69:**133
Nucleotide-linked sugars, **36:**137–139
Nucleotide peturbation mechanism, **68:**115–116
Nucleotide phosphatase transfer enzyme, **17:**185
Nucleotide phosphodiesters (DNA), **18:**212–215
Nucleotide phosphodiesters (RNA), **18:**228–229
Nucleotide polyphosphate activations, **17:** 176–186
Nucleotide pyrophosphatases, **15:**81; 82; 85; 94; 106–107, **18:**77, **57:**159
classification, **14:**248
effect of borate on, **12:**517
hydrogen transfers, **26:**116
localization of, **24:**323; 329
Nucleotide pyrophosphorylases, **21:**242–244
regulation, **28:**110–114; 123–124
Nucleotide regulation, glutamine PRPP amidotransferase, **72:**95
Nucleotide-*N*-ribosidase, **9:**508–509, **18:**231
Nucleotides, **9:**503–509, **35:**450, **47:**401, **48:**165–170. *See also* Deoxynucleotides; Diphosphopyridine nucleotide; Flavin nucleotides; Nucleic acids, Purines; Pyrimidines; individual nucleotides
abiogenic syntheses, **27:**384–385
absorption spectra, **9:**482–484; 487–490
accumulation in penicillin-treated bacteria, **26:**216–219
of cobalamins, **26:**240–241
and cobamide coenzyme activity, **26:**352; 360
of cobamins, **26:**240; 260–262; 275
of corrinoids, **26:**236; 239–241; 243; 244; 248; 250; 256
biosynthesis, **26:**260–262
incorporation of sugars into, **26:**39–41
murein-related, **26:**216–219
pyridine, hydrogen transfer, **26:**115–191
sugar, in carbohydrate synthesis, **26:**35–88
acid-soluble
allosteric regulation by, **28:**78–86; 88–89; 92–94; 104–106; 110–114; 124; 138–144
biosynthesis, allosteric regulation, **28:**135
form deoxyribose, **28:**419; 422
labeling from glucose-2-^{14}C, **28:**409
in nuclei, **27:**263; 264
purine, feedback inhibition by, **28:**66
reduction of, **28:**135; 436–444
ribose labeling, **28:**416–418
sugar, precursors of glycoprotein, **28:** 114–115
action of RN-ase on, **17:**171–173
flavin, metal chelates of, **17:**273–274
as inhibitor of DPNH-cytochrome reductase, **17:**258
in phosphate transfer reactions, **17:**163; 164; 167; 170–173; 175–186
purine, inhibition of metalloflavoproteins, **17:**275
complexes with LADH, **27:**50–53; 56–70
dissociation constants, **27:**64
definition, **9:**457–458
effect of enucleation on metabolism, **16:** 473–474
effect on activation of β-glucuronidase, **16:**376
labeled, in nucleic acid study, **16:**450–462
in myosin study, **16:**332–333; 340; 349
relation to CoA, **16:**12–16
triphosphorylated (*See also* ATP, ITP, and UTP)
splitting by myosin, **16:**331
effect on light emission, **25:**137
as enzyme inhibitors, **25:**222–223
FeP interaction, **67:**344–349
inhibition of enzyme, **25:**171; 226–230; 237–240
linkage in nucleic acids, **9:**515–516
metabolism, **9:**487–490
in mitochondria, effect of Versene on, **21:**5
phosphorylation, **21:**52–54
NMR of, **55:**327–328
(P^{32}), thymidine triphosphate, **25:**399
in phosphorylase *a*, **11:**304
and phosphorylase action, **24:**274–275; 279; 282
sea pansy, and light emission, **25:**158–159
synthesis, with phosphatase models, **21:**192
Nucleotides, Sugar, and Synthesis of Carbohydrates (Ginsburg), **26:**35
Nucleotide-specific phosphatases, deficiency in inert cells, **16:**452
intracellular distribution, **16:**466; 467
Nucleotide-sugar, **52:**24
Nucleotide synthesis, **27:**419
genes determining the enzymes of, **31:** 106–107

and sugar transferring enzymes, **17:**220–221; 229
Nucleotide transphosphorylation, exchangeable, **53:**141–143
Nucleotide triphosphates, allosteric regulation, **28:**81; 82; 88–89; 92–94
Nucleotidyl acylates, **18:**220
Nucleotidyl cylcase, **55:**340
Nucleotidyl hydrolases, **60:**12
Nucleotidyl polymerase, **55:**340
Nucleotidyl transferases, localization of, **24:**314
Nucleotidyl transfer reactions, **18:**193–194, **55:**340–373
enzymes catalyzing, **55:**292
with oxygen chiral phosphate esters, **55:** 341–343
in prochiral substrate conversions
to prochiral products, **55:**340
to proprochiral products, **55:**349–357
via ATP sulfurylase, **55:**347–348
via bovine spleen exonuclease, **55:**352–353
via cyclic nucleotide phosphodiesterase, **55:**349–351
via isoleucine tRNA synthetase, **55:** 347–348
via nucleotidyl cyclase, **55:**340–341
via snake venom phosphodiesterase, **55:** 351–352
via staphylococcal nuclease, **55:**353–356
Nucletropomyosin, and ATP in muscle contraction, **13:**158–159
Nucleus
absorption spectrum, **3:**6–8
isolation from cytoplasm, **3:**13–17
nucleal reaction, **3:**4–6; 10
nucleic acid content, **3:**4–6; 9–13
protein content, **3:**12–23
ultraviolet microscope investigation, **3:**6–13
Nucleus, Enzymology of (Siebert and Humphrey), **27:**239
Nucleus of cell
chemistry, **16:**411–500
interaction with cytoplasm, **8:**56–65
specific functions, **8:**55–65
Nutrients, plant, and resistance to wilt, **11:**429; 432–434; 435
Nutrition
and physiological function, **1:**187
relation to xanthine oxidase, **17:**304–306; 313; 318–322
Nutritional deficiency, **54:**250–251
Nutritional influence, in ribose synthesis, **28:**416–417
Nutritional requirements as basis for classification of organisms, **2:**143
Nutritional studies, of coenzyme Q, **25:** 301–304
of enthionine, **25:**370–372
Nutritional XO factor, **17:**237
Nutrition factors, in blood clotting, **16:**66
Nutrizeptors, antigenic activity, **6:**119
Nystatin, **34:**485, *See* Mycostatin

O

O Antigens, **34:**478
extraction of, **31:**78
phage conversion of, **31:**111–113
Obesity, **59:**365–366
Obstructive jaundice, antithrombin in, **10:**36
Ochoa reaction, **10:**199–213
Ochromonas malhamensis, vitamin B_{12}, **26:** 270–275
Ochromonas malhemensis, glycolysis in, **19:**6
Octanal (^{14}C), from octane, **27:**486
Octane (^{14}C) oxidation, enzymic, **27:**486; 537
Octane-di-2-ethylhexylsodium sulfosuccinate-water system, **51:**37
Octane-oxidizing extract, **27:**538
Octanoate, **15:**238; 244 (C^{14})
inhibition of oxidation, **18:**151
Octanoate activation, **18:**208
Octanoate-1-C^{14}, sperm metabolism of, **24:**65
Octanoic acid, oxidation, in mitochondria, **21:**21–23
Octanoic acid oxidase, localization in cytoplasm, **13:**11
Octanoic oxidase, in mitochondria, **21:**22
Octanol-water partition coefficient, **43:**335
Octylamine, precipitation of heparin, **10:**38
N-Octyloxamate, **57:**98
Octyl thiocyanate, **46:**399
Odd-number-displacement mechanism, **57:**382
Odontosyllis enopla, luminescence, **25:**160
Odoriferous products in cacao, reactions producing, **25:**460; 482
^{18}O-exchange, **49:**241, **53:**63–64
from phosphate, by fructose 1,6-bisphosphatase, **53:**63–64

O-exchange, enzyme-catalyzed, **50:**344
$^{18}O(p,n)^{18}F$ reaction, **57:**269–270
—OH group, introduction of, PN-dependent, **26:**128
O_2H radicals
EPR spectra, **27:**119
and mutations, **27:**328
OH radicals, EPR spectra, **27:**119
Oidium (Oospora) lactis, fat synthesis by, **9:**658; 660–661; 663; 666; 670–673; 686; 690–691; 698
protein synthesis by, **9:**671
Oil, crude, extent of change by microbial action, **10:**472–473
rate of oxidation by microorganisms, **10:** 471–472
release from oil-bearing sediments, **10:**473
Oils, **34:**447
depressor, substance in, **7:**542
effect on hypertension, **7:**541–542
volatile, from cacao, **25:**459
Oil/water interface, **34:**489
Oil-water interface, activity of adsorbed molecules, **28:**25–27
adsorption of enzymes, **28:**16–18
Okadaic acid, tyrosine hydroxylase regulation, **70:**150
Oken's enzyme, elastase activity, **23:**173
Okra polysaccharide, inhibition of threone activity, **16:**55
Okunuki oxidase, **23:**290
Old yellow enzyme, **18:**101
active sites of antigenicity, and coenzyme binding, **22:**390
and coenzyme binding, **22:**383; 384
groups involved in binding, **22:**386–389
fluorometric studies of kinetics, **20:**33; 34
Oleate dehydrogenase, activity of —SH group, **11:**241
Oleate in fructose-bisphosphatase regulation, **54:**142; 143–144
Olefin oxygenase, oxidation of hydrocarbons, **27:**532–533
Olefin reductase:alkane dehydrogenase, reversible, **27:**531; 534
Olefin reduction, mechanism, **24:**472–474
Olefins, microbial oxidation, **27:**499–503
Oleic acid, **46:**103, **60:**114
bactericidal action, **4:**154; 155
dissociation constant, **46:**391
oxidation in plastids, **20:**217
Oleuropein, role in browning of plant tissues, **16:**150
Oligase, **5:**90; 91; 92; 114; 119, **28:**350
2′,5′-Oligioadenylate ribonuclease L, **59:** 218–221
2′,5′-Oligioadenylate synthetase, **59:**216–217
2′,5′-Oligioadenylate system
and cell proliferation, **59:**230–335
cyclic adenosine monophosphate-dependent regulation of
behavior of, coordinated, **59:**230–235
generalization of, **59:**235–238
induction of, **59:**221–222; 224
and inhibition of 2′phosphodiesterase, **59:**224–230
description of, **59:**213–216
enzymes in, **59:**216–218
mechanisms of action, chemical, **59:**218–221
Oligodeoxynucleotide adaptors, synthetic, DNA cloning simplification, **67:**453–455
Oligodeoxynucleotide linkers, synthetic, DNA cloning simplification, **67:**452–453
Oligogalacturonic acids, in pectin, **20:**342–345; 350
Oligo-1,4 → 1,4-glucan transferase, **34:**352
Oligo-1,6-glucosidase
imidazole in, **20:**94
pK value, **22:**58; 59
Oligomeric proteins, *See* Symmetry model for oligomeric proteins
Oligomucases, classification, **14:**246
Oligomuropeptides, **26:**195; 196; 201; 202
structure, **26:**204–207; 212
Oligomycin
ATP hydrolysis inhibited by, **29:**271
in Ca^{++}-induced respiratory jump, **29:**279
decrease in mitochondrial respiration by, **29:**349
effect on ascites tumor cells, **29:**350
inhibition of calcium uptake by, **29:**266; 271
inhibition of intramitochondrial ADP phosphorylation by, **29:**349
mechanism for blocking ATPase activity by, **29:**299
mitochondria, chloroplasts, and chromatophores sensitive to, **29:**73
as replacer of ATP + Mg^{++} in Ca^{++} uptake, **29:**270

in treatment of plant diseases, **20:**390; 402; 403
Oligomycin A, inhibitor of oxidation, **23:**365; 368–369
Oligomycin-sensitivity-conferring protein, **49:**226
Oligonucleotidases, in *E. coli,* **29:**202
synergistic effect of Ca^{++} due to activation of, **29:**178
Oligonucleotide binding, **49:**214–215
Ile-tRNA synthetase-oligonucleotide binding, **49:**215
Oligonucleotide excision mechanism, **60:**25
Oligonucleotide hybridization, **49:**192–195
Oligonucleotides, **38:**223
action of RN-ase on, **17:**172
concentration effect on photocross-linking, **66:**123–125
psoralenated, structure determination by NMR and X-ray crystallography, **66:**128
single-stranded, specific cross linking to, **66:**122–123
Oligopeptide transport, **35:**200–204
α-amino group essentiality, **35:**206–208
competition phenomena, **35:**215–219
dipeptide inhibition, **35:**225
in *E. coli,* **35:**200–201
energy requirement, **35:**225–227
in lactobacilli, **35:**201–204
side chain specificity, **35:**213
size restriction, **35:**227–228
stereospecificity, **35:**222–224
terminal carboxyl group dispensability, **35:**209–213
transport deficient mutants, **35:**213–215
Oligosaccharase, **13:**210
Oligosaccharidases, classification, **14:**224; 245–246
Oligosaccharide chain, exopolysaccharide alginate synthesis, **70:**234
Oligosaccharides, **36:**102–103, **56:**143, **66:** 41–78. *See also* Carbohydrates; individual saccharides
alcoholic fermentation of, **5:**87–5:127
antenna flexibility, thermodynamics, **66:**74
asparagine-linked, **52:**97
biosynthesis, **52:**93–113
branched, **66:**42
conformation, **66:**43; 61–77
flexible linkage identification, **66:**61–62
fluorescence energy transfer measurement, **66:**62–65
solution conformation and binding affinity, **66:**76–77
steady-state end-to-end distance measurements, **66:**66–67
temperature-dependent modulation of branch flexibility, **66:**71–76
time-resolved measurement of branch flexibility, **66:**68–71
criteria differentiating mechanism of, **5:**94–97
dolichol phosphate, **52:**94; 96
equation for formation of, **30:**206
formation by transglycosylation, **17:**189–332
in β-1,3 glucan synthesis, **28:**338–340
as inhibitors of laminarinase, **28:**348
from laminarinase action, **28:**351–353
mannitol-containing, **28:**350
from glycoproteins, **25:**447–450
from hyaluronic acid, **16:**371
N-Linked, **52:**94
O-Linked, **52:**94
milk, **52:**30
of milk, **36:**139–141
N-linked, **66:**42–43
isomeric glycopeptides for defining terminal linkage specificity, **66:**50–51
potential roles for flexibility, **66:**78
solubilized asialoglycoprotein receptor, binding characteristics, **66:**52
structural requirement for asialoglycoprotein receptor binding, **66:**47–52
synthetic analogues for exploring valency, **66:**47–49
nonhydrolytic mechanism of, **5:**121
polyisoprenol-linked, **52:**76
in polysaccharide synthesis, **15:**302–303; 306; 307; 310–315
as prosthetic group of bovine submaxillary gland mucoprotein, **20:**143
recent hypotheses on nature of, **5:**107
structure-function relationships, **52:**113–144
synthesis of, **26:**63; 65, **36:**108
triantennary, structure, **66:**74–75
xylose, labeling of, **26:**58
Oligosaccharides, Alcoholic Fermentation (Liebowitz and Hestrin), **5:**87
Oligosaccharides, Formation by Enzymic Transglycosylation (Edelman), **17:**189

β-1,6-Oligosaccharides, from trehalose, **28:**338
Oligosaccharide units, in polysaccharides, **26:**58–59
Oligosulfomucase, classification, **14:**246
O-Linked oligosaccharides, **52:**94
Olivil, formation from coniferyl alcohol, **16:**148
Omadine, in treatment of plant diseases, **20:**399
Ommochrome, **17:**418
Omphalia flavida, indolylacctic acid oxidase in, **19:**124
Oncogenicity, **48:**72–74
One-acceptor mechanism, **50:**347; 351
One-carbon compounds, **15:**127
One-carbon units, folic acid coenzymes in metabolism of, **21:**396–429; 431; 432
  in carotenogenesis, **21:**339
  in purine biosynthesis, **21:**224; 225
  in pyrimidine biosynthesis, **21:**288–290
One enzyme-one linkage concept, **52:**25; 46
One enzyme-one linkage hypothesis, **52:**28
One-step mechanism, **36:**121
One-way catalysis, **43:**259
One-way enzymes, **43:**224
Oocystis apiculata, **59:**74
Ootheca of cockroach, reactions during formation, **16:**154–155; 156
Opening phenomena, in mitochondria, **21:**116; 117
Operators
  genetic properties of, **30:**99
  map position of lactose, **30:**85
Operon, **46:**86, **50:**327
  definition of, **30:**83
    galactose, genetics of, **30:**105
      map of, **30:**105
L-Ophioamino acid oxidase
  and catalase, **8:**476–477
  mechanism of action, **8:**480–481
  occurrence, **8:**482–484
  of snake venoms, **8:**476–484; 487–489
  specificity, **8:**477–479
Ophiobolus herbotrichus, steroid hydroxylation, by, **19:**191
Ophiobolus herpotrichus, enzymic transformations of steroids by, **20:**249; 251; 271; 274; 275
Ophiobolus species, enzymic transformations of steroids by, **20:**262
Ophiocholinesterase of snake venoms, **8:**471
Ophio oxidase in snake venoms, **8:**476–484; 487–489
Ophiostoma, **17:**403; 427–428
  lysine metabolism of, **16:**296
Ophiostoma, genetic studies of, **13:**139–145
Ophiostoma mutant, amino acid requirement, **16:**301
Opposing enzyme reactions, regulation, **28:** 46–47
Opsin, **42:**230; 233
Opsonic activity and calcium, **8:**111
Opsonic effects of fresh sera, relation to complement, **8:**111
Optical absorption spectra, **33:**315–319
  charge-transfer bands, of ferric hemoproteins, **33:**317
  of cytochrome c peroxidase (*See* Cytochrome c peroxidase, optical absorption spectra of)
  of high-spin ferric hemoprotein, *a* and β bands in, **33:**316
  of low-spin ferric hemoprotein, *a* and β bands in, **33:**316–317
Optical absorption spectroscopy, cytochrome c oxidase, **71:**110–111
Optical activity
  disulfide, **40:**301
  of β-1,3 glucans, **28:**330
  and mechanism of enzyme action, **27:**37–104
  in nature, **7:**104–106
  relation to aging, **20:**1–29
Optical Activity and Aging, Possible Relation Between (Kuhn), **20:**1
Optical antipodes, in living organisms, **20:**1–6; 20; 21; 26–28
Optical density
  of chlorophyll, **19:**58
  measurements in turbid solutions, **17:**75–76
Optical lability, *cis-trans* isomerism, **7:**71–72
  reduction of, **7:**76
  restricted rotation, **7:**72
  tautomeric equilibrium, **7:**71–72; 75
Optically active substances, production, **7:** 79–101
Optically labile racemates, **7:**80; 84
Optically selective activation, **7:**80; 94–99
Optically selective synthesis, **7:**80; 84–94
Optically stable racemates, resolution, **7:**79; 81–84
Optical properties

of nucleic acids and derivatives, **9:**480–490
of proteins, effects of irradiation, **9:**110
of pectic substances, **20:**353
Optical purity in living organisms, **7:**101–104
Optical rotation, specific, and denaturation, **23:**37; 73–74; 76
Optical rotatory dispersion, **34:**471, **35:**509, **39:**349, **42:**98, **47:**47; 49
of proteins, **27:**38–49; 52–53; 55–56; 61
Optical specificity of peptidases, **12:**195–196; 200; 201; 208; 211; 224; 227–228
Optical spectroscopy, **55:**407–409
to detect O_2 radical, **41:**37
EPR, **27:**117
of starch, **12:**365–367
Orange I, II, III, effect on myosin, **16:**345
Orbital steering, **43:**269
Orchinol, **55:**95
Orcinol, **18:**373
Orcinol method, in schiozphrenia, **29:**505–506
Orcylalanine, **50:**147
ORD, **39:**197
of plastocyanin, **33:**201
of stellacyanin, **33:**201
ORD curve, of azurin, **33:**201
of laccase, **33:**203
ORD spectrum, of purple intermediate, **34:**63
Organelles
enzyme adsorption, **28:**2
intracellular, enzyme distribution between, **24:**291–358
submolecular, **34:**485
Organic acids. *See also under* individual acids
changes during fermentation of tobacco, **10:**340; 364–366; 396; 402; 403
from microbial oxidation of hydrocarbons, **10:**444; 466–467
Organic compounds
evolution of, **27:**349–354
synthesis by high-energy proton bombardment, **27:**352
origin of, **38:**202
Organic hydroperoxides, **57:**370
Organic solvents, **34:**469
Organization, **34:**446
Organized structure, and origin of life, **27:** 391–393
Organomercurials, structural changes due to derivatization with, eukaryotic dihydrofolate reductase, **70:**46–47
Organo-metal complex catalysis, **14:**173–176; 179–190
Organometal compounds, disulfide reactions, **27:**207
Organophosphates
bacterial hyrdrolysis of, **74:**52
toxicity of, **74:**52
Organophosphoester, in chemical-biological warfare, **34:**9
as insecticide, **34:**9
as pesticide, **34:**9
Organ specificity, of lactate dehydrogenases, **25:**339–341
Organule, **45:**404
Orientation. *See also* Monolayers, Packing of molecules
of molecules in a monolayer, **11:**84–85
Orientation factor, in velocity of enzyme reactions, **22:**85–89
Orientation of molecules in chloroplast, **25:** 71–73
Origin of enzymes, **27:**347–380
Origin of life, **27:**347–380
experimental approaches, **27:**381–415
Origin of Life, A Theory on (Black), **38:**193
Origin of Life and the Original of Enzymes (Oparin), **27:**347
Origin of Life Problem, Experimental Approaches to (Pattee), **27:**381
Orinithine decarboxylase, **56:**273
Orinithine transaminase, **56:**134
Ornatin, fibrinogen binding inhibition, **70:**7
Ornithine, **35:**141; 159; 170, **50:**143; 145; 199, **53:**211; 222, **56:**134, **58:**178
activity in *Neurospora* mutant, **16:**256
as activator of β-glucuronidase, **16:**376
deamination, **16:**117
ω-hydroxy analogue as growth factor, **16:**296
metabolic relationship with other amino acids, **16:**257; 259–267
role of α- and δ-amino groups, **16:**264
in transamination reactions, **16:**190; 192; 197; 205; 231; 232
and alkaloid biogenesis, **8:**237–240; 242; 244
and pyridine ring, **8:**236
and pyrrolidine ring, **8:**237
and tropane rings, **8:**237
in arginine-requiring *Neurospora* mutants, **22:**27

Ornithine (*Continued*)
biosynthesis of, **33:**373–375
arginase in, **33:**374; 375
arginine desiminase in, **33:**374
ornithine transcarbamylase in, **33:**374
via *N*-acetylated derivatives, **33:**374
via glutamic-γ-semialdehyde, **33:**374
in biosynthesis of nicotinamide, **15:**103
in urea synthesis, **15:**320; 321–329; 332; 339–340; 342; 356; 363–372; 374–383
carbamylation of, **18:**227–228; 231
and citric acid cycle, **18:**253–254
genetic studies, **17:**395–399; 449
in citrulline synthesis, **17:**180–181
product of acetylornithine, deacetylase, **40:**71
transamination reactions of, **19:**343
transcarbamylase, **35:**161
D-Ornithine, **33:**367
L-Ornithine, transamination reactions of, **19:**367
Ornithine aminotransferase, **74:**133
evolution of, rate of, **74:**158
mitochondrial, species distribution of, **74:**158–159
Ornithine auxotrophs, **16:**261; 263; 264
Ornithine carbamoyltransferase, **40:**72, **53:**221
location of, **24:**309
translation of accumulated messages, **40:**85
Ornithine carbamyltransferase, **50:**141
Ornithine cycle, **9:**405, **19:**371–378; 382
Ornithine cycle enzymes, activity in liver, **39:**58
adaptation in, **39:**66; 67–69
arginase turnover, **39:**71
changes during fetal and neonatal development, in humans, **39:**75
in rats, **39:**73; 74
changes during metamorphosis of tadpoles, **39:**76
coordinated changes in, **39:**66
deficiencies, of arginase, **39:**65
of argininosuccinase, **39:**63; 64
of argininosuccinate synthetase, **39:**63; 64
of carbamyl phosphate synthetase, **39:**62
of ornithine transcarbamylase, **39:**62
disorders of amino acid metabolism, **39:**61
in earthworm, **39:**78; 79
enzyme, degradation, **39:**71
synthesis, **39:**71
turnover, **39:**71
hormonal regulation of, **39:**70
independent regulation of arginase, **39:**70
in land planarium, **39:**78
mechanism of regulation, **39:**70
protein deprivation, effect of, **39:**67–69
protein load, effect of, **39:**67–69
in purinotellic terrestrial snails, **39:**80
regulation in, **39:**66
regulation of the arginine pool, **39:**73
in ureotelic invertebrates, **39:**78
Ornithine decarboxylase, **37:**140, **55:**235
bacterial, **6:**4
activation of apoenzyme by pyridoxal phosphate, **6:**27
adaptive formation in *E. coli,* **6:**8
codecarboxylase of, **6:**29
effect of growth temperature, **6:**11
of inhibitors on acitivity, **6:**18
*p*H of growth medium, **6:**10; 16
Michaelis constant, **6:**16
purified, **6:**12–21
reversible resolution, **6:**21
specificity, **6:**15
eukaryotic, **74:**143
structural data on, **74:**134; 142
evolution of, rate of, **74:**158
prokaryotic, **74:**143; 156
structural data on, **74:**134; 142
species distribution of, **74:**158–159
L-Ornithine decarboxylases, **14:**256
Ornithine decarboxylases, biosynthetic, **36:**211
in cells *in vitro,* **36:**246
control mechanisms, **36:**216–217
feedback inhibition, **36:**216
repression, **36:**216
effect of hormones, **36:**246
embryonic tissues, **36:**246
half-life, **36:**246
inducible, **36:**211
Ornithine inhibitors, **56:**274–275
Ornithine-keto acid reaction, **16:**205
Ornithine mutase, **38:**443, **66:**26–27
reaction mechanism, **66:**28–29
Ornithine oxidation, **38:**443
Ornithine-oxo-acid aminotransferase, **53:**221
Ornithine/proline auxotrophs, in study of amino acid synthesis, **16:**262–264
Ornithine transaminase
glucose repression by, **29:**380
in microorganisms, **16:**263
Ornithine δ-transaminase, activity in arginine-requiring *Neurospora* mutants, **22:**27

Ornithine transcarbamylases
activity in arginine-requiring *Neurospora* mutants, **22:**27
synthesis, in *E. coli,* **22:**122
and biotin, **26:**295
from bovine liver, **39:**15
catalytic properties of, **39:**16
inhibitor binding, **39:**15
physical properties of, **39:**16
substrate binding, **39:**15
from rat liver, **39:**14
from *S. faecalis,* **39:**17
catalytic properties of, **39:**16
physical properties of, **39:**16
Ornithodoros moubata, **70:**4
Orosomucoid, **40:**37, **41:**110–113; 118; 119; 121–124, **52:**32
Asialo-, **41:**110–113; 119; 122–124
linkages in, **25:**444–445
removal of, **41:**121
tritiated preparations, **41:**118
Orotate, **50:**155, **55:**271
Orotate phosphoribosyl transferase, in erythrocytes, **41:**22–23
Orotic acid
(C^{14}) in study of nuclear RNA, **16:**440
conjugation with ribose-5-phosphate, **17:**184
in genetic studies, **17:**429–430
formula, **21:**263
in pyrimidine biosynthesis, **21:**264–266; 274–280; 285; 290
radioactive, in study of pyrimidine biosynthesis, **21:**264; 266; 279–281
synthesis and metabolism, in mitochondria, **21:**52; 53
PN-dependent reduction, **26:**127
precursor of pyrimidines, **15:**356 (isotopically labeled); 357
precursor of ribonucleic acid, **15:**119; 127
as precursor of pyrimidnes, **16:**266–268; 450
incorporation into RNA, **16:**461
in synthesis of pyrimidines, **18:**215–217
Orotic acid analogs, **21:**278
Orotic acid-N^{15}, incorporation into DNA, **16:**458–459
Orotidine-5′-phosphate, **18:**216
in pyrimidine biosynthesis, **21:**264; 278
Orotidine-5′-phosphate decarboxylase, **21:**278; 279
Orotidine-5′-phosphate pyrophosphorylase, **21:**278
Orotidine 5′-phosphate pyrophosphorylase, **18:**217
Orotidylate decarboxylase, **55:**271
Orotidylate pyrophosphorylase, in nucleus, **27:**274
Orotidylic decarboxylase, in erythocytes, **41:**23
Orsellinic decarboxylase, **34:**448
Orsomucoid, **56:**210
Orthoester, **37:**19
Orthoesters, A-1, **37:**27
general acid catalysis, **37:**26
intermolecular, **37:**26–28
secondary deuterium kinetic isotope effect, **37:**19
Orthogenesis, **4:**32; 33
Orthologous sequences, definition of, **74:**132
Orthomyxoviruses, *See* Influenza viruses
Orthophenanthroline, in allosteric regulation, **28:**140
Orthophosphate. *See also* Phosphate
allosteric regulation of enzymes, **28:**76–79; 81; 84–86; 95–97; 110; 112; 116–118
glutamine synthetase activity, **72:**13
as inhibitor, **46:**337
inorganic
in biosynthesis of glutamine and glutathione, **19:**253–256
and glucose 6-phosphate formation, **19:**63
labeled, **19:**254–256
and liver nonspecific hydroxylase, **19:**175
degradation, **21:**214
effect on mitochondrial coenzyme Q, **21:**112
exchange reaction with ATP, **21:**36; 39; 74; 213; 234; 235; 239; 240; 247
in formate transfer reactions, **21:**227; 228
incorporation into casein, **21:**45
incorporation into mitochondrial compounds, **21:**45; 49; 51; 54
in purine degradation, **21:**402
in muscle tissue, **6:**149; 159; 161; 187
^{18}O-Orthophosphate, **72:**15–16
Orthoquinoid dyes, cause of mutations, **6:**95; 114
Oryza sativa, **55:**79, **71:**93
Oscillograms, from nerve fibers, **23:**84

O side chains, alterations of structure of, by conversion of bacteriophages, **31:**113
biosynthesis of, **31:**106
structure of, **31:**82
Osmosis, entropy change in, **11:**15–23
Osmotic cytolysis, **7:**255
and luminous bacteria, **7:**253
Osmotic pressure, **42:**85
effect on sperm, **24:**76–78
frog oocyte, **27:**269
and luminous bacteria, **7:**253
measurements, **42:**78
Osmotic shock, **44:**240, **65:**3
O splitting, of molybdenum signals, **51:**145–146
Otala punctata, steroid sulfatase of, **22:**227; 228
Ouabain, ATPase inhibited by, **29:**69
Outer complex, of D-amino acid oxidase, **34:**64; 66
Outer phase, **34:**463
solution, **34:**461–465
substrate concentration, **34:**463
Ovalbumin, **56:**210
denaturation of, **23:**56–57
effect of chemical modification of, **24:**144
isotope studies of biosynthesis, **19:**260
optimal pH for papain action on, **19:**303
moving-boundary electrophoresis patterns of, **30:**150
surface reactions, **11:**39–41; 46–47; 67
variation in sulfur content, **15:**274
Ovalbumin A, column chromatography, **22:**477
Ovalbumin-antiovalbumin system, **15:**67–68; 71; 74
Ovarian cyst mucin, as substrate for neuraminidase and receptor destroying enzyme, **20:**137; 144
Overcrowding, fall in brain norepinephrine by, **29:**487
Overlapping α and β regions, **47:**53–55; 99; 109; 113
Oversaturated solution, **34:**487
Ovine mucin, **52:**108
Ovine submaxillary mucin, **52:**32
Ovomucin, as substrate for neuraminidase and receptor destroying enzyme, **20:**137
Ovomucinase, **23:**189
Ovomucoid, **56:**210
as antigen, effect of chemical modification, **24:**144
column chromatography, **22:**477
complexing between trypsin and, **30:**143
denaturation, and calcium, **15:**445
structure studies, **14:**283; 287; 305
Ovothiols: Biological and Chemical Perspectives (Shapiro and Hopkins), **64:**291
Oxacillin, action of β-lactamase, **28:**239; 288–289; 291–293; 300; 301; 303; 313
Oxalacetate. *See also* Aspartic acid and α-keto acids
and alternate pathways of glucose and fructose metabolism, **15:**144; 172–173
in citric acid cycle, **15:**183; 185–186; 188 (C^{14}); 191; 195 (C^{14}); 202; 203; 205; 206; 209; 210; 213; 232; 235–237; 241; 247; 249; 259; 262
and pyridine nucleotides. 117, **15:**121–122; 125
and urea synthesis, **15:**325; 330–331; 333–340; 373
in amino acid metabolism, **16:**255–256; 268; 271
in transamination reactions, **16:**186; 189; 195; 201; 202; 223; 226; 227; 233; 237
and CO_2 fixation, **12:**46; 47; 50–51; 53–54; 54–60; 61–63; 63–67; 104–105
and choline acetylation, **12:**327–329
chromatographic analysis of, **12:**76–77
energy for condensation with acetate, **12:**315; 327
role of aspartate in synthesis of, **12:**79; 81–85
role of pyruvate in synthesis of, **12:**73–80
peroxidase action on, **19:**125
in plants, **19:**38–40; 52; 53
Oxalacetate-ATP carboxylase, **15:**172
Oxalacetate carboxylase, and CO_2 fixation, **12:**53; 54–60; 63–67; 68–73; 73–74; 77–78; 80–81; 84
Oxalacetate decarboxylase, **12:**93
Oxalacetate decarboxylation, **42:**5
Oxalacetate-pyruvate condensation, effect of streptomycin on, **10:**161
Oxalacetate reaction, **45:**106
Oxalacetic acid, **17:**429–430. *See also* C_4-dicarboxylic acids; Citric acid cycle
and antiketogenesis, **8:**408
in citric acid cycle, **10:**221–222
and fatty acid catabolism, **8:**392–401

intermediate in metabolism, **8:**412–414
and ketone body formation, **8:**401–490
in mold metabolism, **9:**540; 542; 558; 564–566; 568–570; 571; 572–576
in purine biosynthesis, **21:**201
in pyrimidine biosynthesis, **21:**265
reactions in mitochondria, **21:**20; 28; 117
from pyruvic acid, **8:**401–409
role in nitrogen fixation, **13:**372
transamination reactions of, **19:**338; 359; 360
transformation into citric acid by muscles, **10:**180–182
Oxalacetic acid oxidase, **19:**123; 125
Oxalacetic carboxylase, **15:**201
isolation, **14:**379
in mitochondria, **21:**46; 47
in photosynthesis, **19:**38; 39
Oxalacetic β-decarboxylases, classification, **14:**237; 256
Oxalacetic decarboxylases, isolation of, **14:**405
Oxalacetic oxidase, in mitochondria, **21:**13; 14
Oxalates, **5:**34, **57:**98; 130–133
anticoagulant effect, **5:**34
calcium, **17:**142
labeled, **23:**449–450
metabolism by microorganisms, **23:**401; 404; 446–452
oxidation of, **23:**447–449
synthesis from, **23:**449–452
peroxidase action on, **19:**125
Oxalic acid, **57:**129
in air-cured tobacco, **6:**337
in flue-cured tobacco, **6:**364
changes during fermentation of tobacco, **10:**366; 396; 402
in dismutation, **10:**190
cleavage, by metal ions, **21:**177
formation in molds, **9:**538; 540; 542–543; 545; 547–550; 551–554; 554–560
Oxalic acid diamide, deamination, by metal ions, **21:**179
Oxalic acid oxidase, **19:**123; 125
Oxaloacetate, **50:**149, **55:**225. *See also* Carboxylation
and acetate, **23:**408–409; 414; 415; 425–426; 463
and glycolate, **23:**437–440; 445; 446; 463
in syntheses, **23:**403–404
allosteric regulation, **28:**87–92
effect on inactivation of succinic dehydrogenase, **20:**179
exchange reaction with pyruvate (C^{14}), **26:**313–314
and pantothenate, **7:**443
Oxaloacetate decarboxylase, **35:**410, **46:**477, **53:**221; 222
and biotin, **26:**295
inhibition by metal-binding agents, **35:**410
Oxaloacetic acid, **53:**169
in biosynthesis of aspartic acid in plants, **20:**327; 328
in formation of *N*-acetylneuraminic acid, **20:**140
labeled, **18:**268–269
in lignin study, **18:**363
Oxaloacetic carboxylase, **18:**118; 301
Oxaloacetic competition, for succinic dehydrogenase, **18:**93–94
Oxalopropionic acid, **53:**169
Oxalosuccinate
and acetate metabolism, **23:**409–410
in citric acid cycle, **15:**185
decarboxylation, **16:**254–255
in mitochondria, **21:**20
Oxalosuccinic acid
decarboxylation of, **12:**71–72; 86–88
enolization of, **12:**88–89
preparation of, **12:**87
influence of metals on decarboxylation, **10:**201–202
from α-ketoglutaric acid, **10:**208–213
manganese complexes, relation to carboxylase, **10:**201–202; 204–208
Ochoa reaction, **10:**199–213
quantitative determination, **10:**202
relation to citric acid cycle, **10:**169
synthesis and properties, **10:**199–208
in mold metabolism, **9:**540
oxidative decarboxylation, **15:**117
Oxalosuccinic carboxylase, **12:**87–88; 89–91, **18:**118
Oxalosuccinic β-decarboxylases, classification, **14:**256
Oxalurate
as inducer of allantoin enzyme production, **56:**108
structure of, **56:**107
transport system for, **56:**128
in urea transport, **56:**127

Oxalyl-CoA, **23:**445; 448–449; 451–452, **50:**123
N-Oxalylcysteamine, **57:**131; 139–140
S-Oxalylcysteamine, **57:**131
N-Oxalyl derivatives, **57:**135
N-Oxalylglycine, collagen hydroxylase inhibitor mechanisms, **72:**360–364
Oxalyl thioester
  and L-hydroxy acid oxidase, **57:**124–125; 127
  and nucleophiles, **57:**133–135
  structure of, **57:**124
Oxamate, lactate dehydrogenase inhibited by, **29:**339
Oxamate derivative, sepharose-linked, **46:**254
Oxamycin, **56:**61
  D-Cycloserine, murein synthesis, **26:**218
Oxatetradecanoic acids, **67:**413–414
Oxene, **46:**171
Oxene transferase, **47:**36
Oxgluciferin, structure of, **44:**39
Oxidase, L-*a*-hydroxy acid, in kidney, **8:**355
Oxidase, D-amino acid, in plants, **10:**300
  pyroracemic acid, in plants, **10:**300
Oxidase activity, **44:**194
  and copper-binding substances, **23:**305
Oxidase-dehydrogenase balance of mitochondria, **17:**93; 97; 101
Oxidase function, and lipids, **23:**313–314
Oxidase in tea-fermentation, **2:**114; 115; 120; 126
Oxidase mechanism, for respiration, **28:**156
Oxidase reaction, produce inhibition of, **25:** 184–185; 224; 249–250
Oxidases, **15:**102. *See also* Cytochrome oxidase; Verdoperoxidase; specific oxidases
  activity of hemin, **15:**27
  aerobic, **7:**295
    amine, **7:**533–538; 542–543
    amino acid, **7:**374; 379; 393; 534
    cysteine, **7:**379; 386; 388; 395
    cytochrome, **7:**24; 206; 227; 232; 316; 324; 385
    glucose, **7:**300
    sulfinic acid, **7:**387
    xanthine, **7:**300
  aerobic, in metabolism of nicotinamide, **15:**112
  a-amino acid, in kidneys, **8:**355
  classification, **14:**227–228; 230; 236; 254; 255
  classification and definition, **19:**80; 81
    in oxygen metabolism, **19:**79–233
    reactions of peroxidases, **19:**104–128
  definition, **20:**149
  enzymic adaptation, **13:**71; 74; 106
  heart muscle, **23:**290; 291; 298–302; 307
  multiple forms, **25:**359–360
  pig heart, in α-ketoglutarate oxidation system, **15:**226–230
  of plant tissues, **18:**13; 36–37
  role of iron in activity, **14:**176
  sclerotizing, phenolase complex as, **16:**158
  terminal, in plants, **16:**131–136
  in tobacco, **6:**372–375
  unsaturated-fat, in animal tissues, **8:**442–443
  yeast, effect of CO on, **23:**292; 300; 308
  of yeast mitochondria, mechanism of action, **24:**468
    role of metal, **24:**112–113
Oxidases, EPR spectroscopy, **27:**107; 119–130
  and hydrocarbon oxidation, **27:**475
  mixed function, oxidation of hydrocarbons, **27:**537
Oxidase systems, use of cytochrome *c,* **29:**218
Oxidation, **42:**127, **58:**47. *See also* Hydrogen transfers
  biological, reconstitution of chain, **28:** 155–236
  of dihydropyridines, **26:**143–156*t;* 146–153
  glutathione synthesis, **72:**240–242
  of glycolate, **23:**434–440
  by heme, **46:**103
  inhibition of ATPase activity by, **6:**167
  at interfaces, **11:**58; 59–66; 69–76
    of thiols, by oxygen, **11:**203–206; 207–210
      by ultraviolet light, **11:**210
      by various agents, **11:**206–210; 215–216; 218; 219–221; 223–226; 230–232; 235–237
  mixed-function, **47:**32–35
  one-electron, **26:**143–144
  in semen, **9:**372–374
    light-induced, **9:**83–84; 86–88; 109; 112–113; 119–120; 134
    reversible step-wise, **9:**1; 13; 15–16
  terminal, lack of enzymes in nucleus, **16:** 489–490
  two-electron, **26:**144–156
β-Oxidation, **57:**88

of fatty acids from hydrocarbons, **27:**484–490; 493
inhibitor of, **27:**500
of propionate, **26:**367–370
α-Oxidation, of *n*-alkanes, **27:**491
ω-Oxidation, of *n*-alkanes, **27:**493
Oxidation cycle, citric acid, **15:**183–270
β-Oxidation enzymes, oxidation of hydrocarbons, **27:**487; 497
α,ω-Oxidation of alkanes, **27:**486; 488; 493; 504
Oxidation of amino acids, **8:**150–152; 412–414
a Oxidation of fatty acids, **8:**354–355; 404
α,β Oxidation of fatty acids, **8:**362
α,γ Oxidation of fatty acids, **8:**354
α- and β-oxidation of fatty acids, **2:**275–279
β Oxidation of fatty acids, **8:**317; 336; 346; 347–357; 360; 364; 365; 369; 375; 376; 377; 380–387; 404; 411; 412
alternating, **8:**358–360; 404; 409–412
δ Oxidation of fatty acids, **8:**358–360
ω Oxidation of fatty acids, **8:**336; 346; 365; 372; 374–380; 380–387; 406
Oxidation of water, in photosynthesis, **25:**86–89
Oxidation-phosphorylation sequence, **23:**373–374
Oxidation quotient of *aspergillus* from respiration and aerobic fermentation and carbon monoxide, **2:**215–216
Oxidation reactions in the living cell, **1:**347
reduction reaction in anaerobic metabolism, **1:**102; 136–138
Oxidation-reduction, **39:**360
and origin of life, **27:**369–371
between thiols and quinones, **16:**120–121
in browning of plant tissues, **16:**150–153
Oxidation-reduction balance, lipide, and vitamin E, **6:**504–507
Oxidation-reduction dyes, as uncoupling agents, **23:**366
Oxidation-reduction function of hydrogenated pteridines, **25:**400–401
Oxidation-reduction potentials, **15:**5; 7–8; 15–16; 30–32; 38; 41–42
of cytochrome c, **7:**238–284
of cytochromes, **17:**333–334; 336; 347; 353; 362; 363, **18:**16; 25; 27–28
and photosynthesis, **18:**54–57
of respiratory chain, **17:**115; 120–121; 123–124
of riboflavin and its nucleotides, **17:**277
of cytochromes *a*, **23:**312–314
of DPN and TPN, **19:**51
of Hill reagents and other systems, **19:**47; 48
effect of activation energy, **9:**41–42
of glutathione and glutathione disulfide, **15:**221
of lipoic acid and reduced lipoic acid, **15:**221
for liver alcohol dehydrogenase system, **20:**44
of cytochrome c_1, **20:**173; 174
of cytochromes *a, b,* and *c,* **20:**186
for oxidized and reduced diphosphopyridine nucleotide, **20:**38; 44
of vitamin K_1, **20:**183
of pyridine nucleotides, **15:**89–90
for quinones, **16:**107–108; 113; 114
of blood, during pupation, **16:**159–160
of thiol systems, **11:**211–212; 236–237; 250–251; 252
Oxidation-reduction reactions, donor-acceptor-enzyme complex, **15:**291
enzymic, mechanism, **15:**1–48
of hexose phosphates, **15:**153–168
of hexoses, **15:**148–150
of pyridine nucleotides, **15:**87–90
Oxidation-reduction reactions in cells, **7:**46
Oxidation-reduction systems
involving thiols, **11:**203–206; 206–210; 211–212; 236–237; 241–244; 249; 250–252
kinetics, **9:**15–16
of sulfur compounds in plants, **10:**223–224
Oxidative deamination, **7:**374–375; 389; 393; 397. *See also* Transamination, Transdeamination
of glycine by *o*-quinones, **16:**117–120
Oxidative decarboxylation
in oxidation of hydrocarbons, **27:**486
in pentose synthesis, **28:**394–395
Oxidative degradation, **34:**476
Oxidative enzymes
application of EPR spectroscopy, **27:**105–198
and essential hypertension, **7:**538
intracellular distribution, **16:**468
soil, **28:**387
Oxidative phosphorylation, **24:**337, **33:**300. *See also* Photosynthetic phosphorylation
allosteric regulation, **28:**138–139
reconstitution, **28:**156; 159; 161–164; 230
ATPase of, **49:**225

Oxidative phosphorylation (*Continued*)
in ATP system for energy transport, **14:**10–14
chemical hypothesis of, **33:**295
chemical mechanism, **49:**267
chemiosmotic hypothesis for, **29:**306–307
mechanism for mitochondrial ion transport and its relationship to, **29:**296–314
chemiosmotic hypothesis of, **33:**296
effect of fluoroacetate on, **18:**133
electron transport, **23:**327; 340–385
intermediates, **23:**372–377
mechanism of, **23:**355–385
oxidation level of intermediate, **23:** 375–377
reconstitution of, **23:**372
reversal of, **23:**382–383
substrate level, **23:**327–336
theories, **23:**373–383
thermodynamic aspects, **23:**352–353
in mitochondria, **20:**156; 160; 161; 172
in plants, **20:**225–228
vitamin E in, **20:**181
vitamin K_1 in, **20:**182–185
and phosphate transfer, **17:**163–164; 174–176
metalloflavoproteins in, **17:**285–286
and the respiratory chain, **17:**65–134
in plants, **19:**16; 17; 55
and steroid 11β-hydroxylase, **19:**189
Oxidative phosphorylation system, **40:**38; 39
Oxidative processes, effect of enucleation on, **16:**471; 473
Oxidative-reductive enzymes, inhibition of, **6:**58–67
Oxidative trapping, **59:**113
N-Oxides, **60:**73
Oxidizability of hydrocarbons by microorganisms, **10:**454–462; 467
Oxidized enzyme-substrate complex, activation energy for, → purple intermediate, **34:**60
conversion of, to purple intermediate, **34:**55
interactions involved in, **34:**52
occurrence of, **34:**49; 50
Oxidizing agents
effect on pectic substances, **20:**352; 368
for thiols, **11:**203–206; 206–210; 215–216; 218; 223–226
Oxidizing power of living plants, **1:**336; 342
16α, 17α-Oxidoallopregnane-3β-ol-20–20one, microbiological conversions, **20:**255; 256
Oxidogenic substrates, **27:**121; 125
14α, 15α-Oxido-11β-hydroxy-substance S, microbiological formation, **20:**250; 251
16α,17α-Oxidopregnane-3,20-dione microbiological conversions, **20:**255; 256
16α,17α-Oxidopregnenolone, microbiological conversions, **20:**255; 256
16α,17α-Oxidoprogesterone, microbiological conversions, **20:**254–256
Oxidoreductase, NAD-H_2, in coacervate, **27:**370
Oxidoreductases, **56:**292, **58:**285
alcohol dehydrogenase, **56:**318–325
classification, **14:**227–228; 233; 235–236; 241; 242; 253–255
column chromatography, **22:**478; 485
distribution in subcellular fractions, **24:** 295–309; 333; 339
in plastids, **20:**214–219; 229
in polysaccharide synthesis, **28:**113–114
superoxide dismutase, **56:**325–329
Oxidoreductases, H^+-translocator, **29:**76–78
and chloroplasts, mitochondria, bacteria, and chromatophores, **29:**76
Oxidoreduction. *See also* Redox
in chloroplasts, **12:**1–39
in photosynthesis, **11:**117–119; 148–150; 158–160; 164; 182–189; 191–195
Oxidoreduction, Enzyme (Leach), **15:**1
Oxidoreduction, In Chloroplasts (Hill), **12:**1
Oxidoreduction loop
proton translocating, **29:**76
of the respiratory chain, **29:**77
Oxidoreductive enzyme, **34:**476
9β,11β-Oxido-substance S, microbiological formation, **20:**250; 251
14α,15α-Oxido-substance S, microbiological formation, **20:**250; 251
Oxime enzyme, coenzyme-enzyme binding, **27:**86–87
Oximes
formation role in nitrogen assimilation, **16:**256
in plant metabolism, **9:**394–395
in reactivation of phosphorylated esterases, **20:**305; 307; 308
Oxine, effect on homogentisate oxidase, **19:**126
3-Oxoadipate, from microbial oxidation of hydrocarbons, **27:**511–512; 522; 524–525
terminal respiration of, **27:**512–513

3-Oxoadipate CoA transferase, **27:**513
α-Oxoaldehydes, **59:**105
2-Oxo-aldoximes, in reactivation of phosphorylated esterases, **20:**307
Oxo-bridge exchange reaction, **65:**167
Oxodienoic acid, **58:**196
2-Oxoglutarate, **50:**132; 149; 156; 157
collagen hydroxylase
inhibitor mechanisms, **72:**358–364
reaction mechanisms, **72:**345–357
3-Oxoglutarate decarboxylase, **27:**519
4-Oxohomoarginine, **50:**152
5-Oxoprolinase
mechanism of action, **43:**540; 541
slow step in γ-glutamyl cycle, **43:**543
tissue distribution, **43:**542
5-Oxo-L-prolinase, **43:**523; 533
bacterial, **43:**549
inhibition by L-2 imidazolidone-4-carboxylate, **43:**535
purification of, **43:**534
5-Oxo-D-proline, **43:**551
formation in mammals, **43:**550
5-Oxo-L-proline, **43:**519; 536
in bacteria, **43:**547
levels, **43:**527
metabolism, **43:**522; 523
in skin, **43:**528; 529
utilization, **43:**529–532
*5-Oxo-L-*Proline (L-Pyroglutamate, L-Pyrrolidone Carboxylate), The Metabolic Formation and Utilization of (Van Der Werf and Meister), **43:**519
5-Oxoprolinuria (Pyroglutamic aciduria), **43:**543
enzyme defect in, **43:**545
Oxotremorine, **55:**176
structure of, **55:**140
Oxyanion hole, **50:**36
Oxycation radicals, charge stabilization, **66:**31
Oxycellulose from cellulose, **8:**265–266; 267
Oxygen, **47:**1
and acetyl phosphate formation, **6:**233
in familial periodic paralysis, **6:**218
and galactose fermentation, **6:**74–77
in muscular atrophy, **6:**197; 198
in muscular dystrophy, **6:**206; 223
activation energy of, **11:**97
production in photosynthesis, **11:**102; 172–173; 179; 194
tension, and thiols, **11:**252–253
uptake in photosynthesis, **11:**101
addition to double bonds, **8:**432–434; 437–439
atmospheric, reduction, **47:**8
chemistry, **47:**1
and CO_2 fixation, **18:**300
diffusion of, **29:**34
effect on carotenogenesis, **21:**314–316
radioactive, in study of mitochondrial oxidative phosphorylation, **21:**36
effect on cytochromes *a, b,* and *c,* **20:**150; 175; 186
labeled, in studies of hydrolytic action of esterases, **20:**301
in studies of steroid enzymes, **20:**274
reaction with cytochrome, **20:**150; 151; 164
effect on enzyme inactivation by radiations, **24:**375–376; 378–379
radiobiological action of, **24:**395–396
as electron acceptor with metalloflavoproteins, **17:**237; 239–241; 246; 253; 254; 257; 263; 282; 284–286; 295; 308; 315–318
reaction with xanthine dehydrogenase, **17:**297; 301
enzymic mechanisms of metabolism, **19:** 79–233
evolution in photosynthesis, **19:**57–61
as a Hill reagent, **19:**48
reduction to water, **19:**200; 201
evolution of, role of ferredoxins in, **33:** 125–127
oxidation-reduction potentials for, **33:** 231–232
reduction of, **33:**230–232
free, and origin of life, **27:**354; 355
isotopic, oxidation of hydrocarbons, **27:**479–480; 482–484; 489; 491; 493; 502–503; 509; 529; 531; 536
in hydroxamate synthesis, **35:**158; 159; 166
toxic effects of, **35:**316
labeled, in hydrolysis of ATP, **16:**337
role in formation of *o*-quinone from catechol, **16:**128–131
metabolic products of, **46:**165–175
molecular, **46:**65; 89
redox potentials of, **46:**167–168
single state, **46:**170
molecular, as hydrogen acceptor in glucose oxidase action, **15:**148 (O^{18})

Oxygen (*Continued*)
in enzymic oxidoreductions, **15:**3; 10; 11; 12 (isotopically labeled); 14–16; 18–21; 23–24; 31–32; 37; 89; 124
in pyruvic oxidase system, **15:**212; 223
production by chloroplasts, **12:**2–6; 6–9; 25–28
reduction by P450 monoxygenase, **47:**1
Oxygen and K^+ permeability, **13:**62–63
2:3-Oxygenase, splitting of catechol, **27:**522
Oxygenases, **47:**2; 6
hydrocarbon oxidation, **27:**470; 480–482; 528–530; 532–534; 536–538
location of, **24:**309
specificity, rubisco, **67:**56–60
Oxygenation, **58:**86
rubisco, **67:**10–11
Oxygen atom transfer, **51:**153
Oxygen chiral phosphate esters, **55:**291–376
configurational analysis of, **55:**321–340
in enzyme-catalyzed transfer reactions, **55:**340–369
nonenzymatic application of, **55:**369–371
spectroscopic applications of, **55:**371–372
synthesis of, **55:**294; 306–321
Oxygen Chiral Phosphate Esters (Gerlt, Coderre, and Medhi), **55:**291
Oxygen concentration and bacterial luminescene, **7:**229
Oxygen consumption
in browning of plant tissues, **16:**151–153
in plants, relation to phenolase activity, **16:**131–136
as criterion of microbial attack on hydrocarbons, **10:**446–447
in fertilized eggs, **9:**297–312
in semen, **9:**367–370; 372–374
by homogenates, dilution effect, **233**
in normal and cancer tissues, **4:**203; 204; 205
of nuclei, **27:**254
in sperm, **24:**52–68
Oxygen diffusion, influence on respiration, **7:**8
Oxygen dissociation constant, of myoglobin, **29:**35
Oxygen effect, on radiation sensitivity, **17:** 452–455
Oxygen evolution in photosynthesis, **25:**86–89
Oxygen exchange
bicarbonate-water, carbonic anhydrase catalysis of, **47:**180–184
inter-carbon dioxide, carbonic anhydrase catalysis of, **47:**181–184
with phosphate, **43:**356
water and acetate, **12:**304
water and CO_2, **17:**163; 175
water and phosphate, **17:**162–164; 175; 183
Oxygen fixation, in hydrocarbon substrates, **27:**536–538
Oxygen isotopes, **55:**293
in phosphate monoester synthesis, **55:**308
Oxygen liberation by illuminated chloroplasts, **1:**219; 220
Oxygen Metabolism (Mason), **19:**79
Oxygen (O^{18})
in enzymatic phosphate transfer, **17:**161–164; 166; 174–176; 183
in invertase study, **17:**225
labeling of ATP, **23:**327; 355; 356; 359–360; 368; 386
Oxygen ($^{34}O_2$), in leaf respiration study, **18:**33
Oxygen (${O_2}^{18}$), in photosynthesis, **25:**42
Oxygen (O_2) as acceptor, enzymes with, **24:**299–301; 305–309
Oxygen radical, **46:**174
Oxygen reduction, **17:**111–112; 364; 365, **41:**37
by photolysis, **41:**37
by radiolysis, **41:**37
by ultrasonication, **41:**37
Oxygen Reduction by the P450 Monoxygenase System (Gunsalus and Sligar), **47:**1
Oxygen tension
adaptation to, **17:**283–284
and cyanide inhibition of cytochrome oxidsae, **17:**339–340
and cytochrome *c* activity, **17:**381
effect on xanthine oxidase, **17:**311
effect on sperm, **24:**39–41; 82–83
Oxygen toxicity, of Superoxide Dismutase, **41:**84–86
Oxygen transfer, carotenoid function in, **21:**360
Oxygen transferases, classification, **19:**80
in oxygen metabolism, **19:**81–128
properties, **19:**126–128
Oxygen-transferring agents in tobacco leaves, **10:**346–349; 373–375; 382–386
Oxygen-transferring enzyme, **23:**267
in oxidation of hydrocarbons, **27:**534; 538
Oxygen-transporting enzyme, **17:**336
Oxygen uptake
of fermenting tobacco, **10:**344–349

in photosynthesis, **25:**58–60
α-Oxyheme, **46:**65; 175
Oxyhemocyanin, Cotton effect, **27:**60
Oxyhemoglobin, **15:**14; 18–21; 37
Oxyhemo globin, **41:**90
Oxyhemoglobin, **46:**169
adsorption at interfaces, **28:**20; 25; 26
rate of dissociation of, **29:**36
Oxyhemoglobin structures, **7:**275–277
Oxyhydrase, **2:**94
N-(1-Oxyl-2,2,5,5-tetramethyl-3-3pyrrolidinyl)iodoacetamide, **54:**400
Oxyluciferin, **7:**218
Oxyluciferyl adenylate, synthesis by carbodiimide method, **22:**191
Oxyminalin, in pectinesterase, **20:**360
Oxynitrilase, **7:**92–93
3-Oxy-5-oxoproline, **43:**538; 539; 540; 541
4-Oxy-5-oxoproline, **43:**538; 539; 540; 541
Oxyphenols from microbial oxidation of hydrocarbons, **10:**467–469
p-Oxyphenyl group in pepsin activity, **14:**190
Oxytetracycline. *See also* Agrimycin
compatibility with agricultural chemicals, **20:**386
effect on streptomycin-resistance of phytopathogens, **20:**387
in treatment of plant diseases, **20:**384; 392–395; 398; 400–403
Oxythiamin
inactivity as coenzyme, **14:**181
as inhibitory analogue, **6:**131
Oxytocin, **40:**272; 292; 293, **48:**10
activity, and SH-groups, **27:**228
amino acid sequence in, **24:**428; 430
lower peptide sequences in, **24:**421–425
synthesis of, **24:**107
analogs of, **29:**392
synthesis by *p*-nitrophenyl ester method of, **29:**406
use of tosyl group in synthesis of, **29:**438
8-arginine, **40:**282
binding to neurophysin, **40:**306; 307; 309
affinity, **40:**282; 284; 313; 321; 329
α-amino group, **40:**313; 314; 315; 320; 324; 327; 329
circular dichroism changes, **40:**313; 317; 319; 321; 322; 325
disulfide, **40:**321
free energy, **40:**322; 323; 324
number of sites, **40:**305; 306; 308; 309
pH dependence, **40:**309; 310; 315; 316
biosynthesis, **40:**285; 286
conformation, **40:**311
disulfide, **40:**311
evolutionary precursor, **40:**295
localization, **40:**276; 277; 278; 279; 281
nuclear magnetic resonance, **40:**321; 322
peptide analogs, **40:**305; 309; 311; 317; 318; 319; 320; 321; 322; 323; 324; 326
position 1, **40:**320; 321; 324
position 3, **40:**322; 324
precursor, **40:**286
α-amino group, **40:**314
reduction of, **28:**443
release, **40:**277; 282; 283; 284
storage, **40:**277
structural homology with neurophysin, **40:**295
structure, **40:**288; 289; 290; 291; 292; 293; 294; 295; 296
structure studies, **14:**289; 299
substitution, isoleucine, **40:**316
phenylalanine, **40:**316
tyrosine of, **40:**311; 316; 317; 318; 319; 320; 324; 326; 327
Oxyuranus scutellatus scutellatus, **46:**16
Ozonolysis, **58:**296

P

P450, **47:**24; 25; 39
catalyzed oxygenation, **47:**39
oxidation reduction potential, **47:**25
spin equilibrium, **47:**24
P45011β, **68:**44; 46
P45017α, **68:**54
expression in *Escherichia coli,* **68:**52–59
P-700, absorbancy changes due to, **30:** 35–37. *See also* Chlorophyll
as electron carrier between pigment systems, **30:**35–38
kinetics of oxidation of, **30:**35–36
oxidation of, **30:**31
transfer of electron from plastocyanin to, **30:**42
P. aeruginosa, allosteric enzymes, **28:**111
^{32}P, distribution in cell nuclei, **7:**182–183

effect on peptides of dehydrophenylalanine, **8:**134
porcine, as source of lipase, **23:**135–139
and reactions at interfaces, **11:**76–77
Pancretic fluid, column chromatography, **22:**449; 452; 462–464
Panose
from maltose, **15:**312
in transglycosylations, **17:**202; 210–213; 220
β-linked analog of, **17:**218
Pantethein, **27:**218
Pantetheine
activity for various organisms, **14:**65–69
and coenzyme A, **14:**63–65
derivatives, **14:**59–61; 61–62
relationship to LBF, **14:**55–59
activity of, **18:**207
cofactor in
gramicidin S heavy enzyme, **35:**17; 31
tyrocidine heavy enzyme, **35:**22; 31
fatty acid biosynthesis, **35:**18; 32
function in antibiotic biosynthesis, **35:**32
mechanism of formation from pantothenic acid, **16:**16–17
reaction with quinones, **16:**121
relation to CoA, **16:**4–12; 16–17
synthesis of phosphates of, **16:**10–11
thiol esters as acylating agents, **16:**18
Pantetheine 2′, 4′-hydrogen phosphate, from CoA, **16:**15–16
Pantetheine 2′-phosphate, **16:**11
Pantetheine 4′-phosphate, relation to CoA, **16:**10–12; 16–17
Pantetheine 4′-pyrophosphate, relation to CoA, **16:**15
Pantethine
activity for various organisms, **14:**65–69
growth factor for *Lactobacillus bulgaricus,* **14:**49–71
relationship to coenzyme A, **14:**62–65
to multiple naturally occurring forms of LBF, **14:**61–62
synthesis, **14:**55–59
disulfide exchanges, **27:**218
disulfide form of LBF, **16:**5
Pantethine and Related Forms of the Lactobacillus bulgaricus Factor (Snell and Brown), **14:**49
Pantoic acid, **16:**239; 281
activation by ATP, **17:**184
genetic studies, **17:**430–431; 442–445
α-amino analogue of, **16:**284
in biosynthesis of pantothenic acid, **19:**259
derivatives, inhibiting action of, **7:**447–453
and synthesis of pantothenic acid, **7:** 435–440
enzymatic formation, **53:**371–373
Pantoic acid-β-alanine condensing enzymes, **17:**442–445
Pantolactone, in synthesis of pantetheine, **16:**6
Pantothenate. *See also* Pantothenic acid
action of related substances, **7:**446–448
bacterial inactivation of, **7:**444–448
and fumarate, **7:**443
genetic studies of, **13:**141
liberation from organisms, **7:**440
from tissues, **7:**440
and oxaloacetate, **7:**443
quantity in microorganisms, **7:**440–441
biochemical activities, **7:**441–444
rates of inactivation by bacteria, **7:**439
rates of production by bacteria, **7:**437–440
requirement for, in *Neurospora* mutants, **22:**27
Pantothenate action, and type of substrate, **7:**442–444
on growth, **7:**448–453
Pantothenate deficiencies, effects on fermentation, **7:**442–444
effects on respiration in microorganisms, **7:**441–444
Pantothenate inactivation, inhibition by pantoyltaurine, **7:**450–451
Pantothenic acid, **4:**278; 279. *See also* Coenzyme A (CoA); Pantothenate; Pantothenic acid deficiency
activity for various organisms, **14:**65–69
amino acid in, **14:**266
relationship to coenzyme A, **14:**63–65; 70
to LBF, **14:**50; 53–58; 63–65
bibliography, **3:**281–287
biochemistry, **3:**259–281
biological synthesis, **3:**262; 263
biosynthesis, **19:**259
allosteric regulation, **28:**70
biosynthesis of, **53:**371–374
β-alanine, enzymatic formation of, **53:** 373–374
pantoic acid, enzymatic formation, **53:** 371–373
chemistry, **3:**254–259

Pantothenic acid (*Continued*)
and choline acetylation, **12:**320; 325–326; 329
reaction with borate, **12:**514
cleavage products, **3:**256–258
in coenzyme A, **15:**187
synthesis, **15:**253–254
distribution, **3:**263
effect on coenzyme Q, **25:**304
genetic studies, **17:**430–431; 442–445
hydrolysis, **7:**442
inhibitory analogues of, **6:**132; 137; 139; 143
isolation, **3:**254–255
metabolism and microorganism growth, **7:**435–454
microbial synthesis from β-alanine, **7:** 435–440
in mitochondria, **21:**8
optical activity, **3:**259
origin, **3:**263
pharmacology, **3:**279
physiological activity, **3:**259–263
physiological functioning, **3:**266–287
physiological importance, recognition, of, **3:**255–256
potency, testing methods, **3:**254–255
quantitative determination, **3:**264–265
resolution, **3:**259
specificity, **3:**259–263
structure, **3:**256–258
and yeast growth, **3:**255–257; 259; 261; 262; 279–280
Pantothenic acid 2′
4′-diphosphate, **16:**9
4′-hydrogen phosphate, **16:**9–10
Pantothenic Acid, Chemistry and Biochemistry (Williams), **3:**253
Pantothenic acid, CoA as derivative, **16:**3–4
formation of pantetheine from, **16:**16–17
relation to ketovaline, **16:**281
synthesis of phosphates of, **16:**8–12
Pantothenic acid and fat formation, **8:**310
Pantothenic acid conjugate, relationship to LBF, **14:**53–54
Pantothenic acid deficiency
animal experiments
dogs, **3:**275–276
fowls, **3:**268–270
hogs, **3:**276
mice, **3:**273–275
other animals, **3:**276–277
rats, **3:**270–273
in human nutrition, **3:**277–279
Pantothenic acid in citric acid cycle, **10:**221–222
Pantothenic acid in plant metabolism, **9:**407
Pantothenic acid 2′-phosphate, **16:**8
Pantothenic acid 4′-phosphate, **16:**8–9; 17
Pantothenic acid synthesis, **18:**199; 209
Pantothenic acid synthetase, in *E. coli,* **22:**4; 5
Pantothenyl azide, in synthesis of pantetheine, **16:**6
Pantothenyl cysteine
as CoA precursor, **16:**17
4′-phosphate, as growth stimulant for A. *suboxydans,* **16:**17
Pantoyl adenylate, **19:**259
Pantoyltaurine, action on growth, **7:**452–453
inhibiting action of, **7:**447–453
Papain, **1:**78–80; 85–88; 92; 93, **5:**22; 23; 46, **10:**300–303, **18:**324, **34:**448–451; 456–459; 503; 507–511, **37:**455; 552, **41:**179; 186; 198; 237, **46:**483–484, **47:**60; 124, **53:**242; 251; 254; 257; 261; 263; 266; 267; 274–278; 290. *See also* Mecuripapain
action in blood clotting, **10:**12
action on collagen, **23:**173; 176; 219
action on elastin, **23:**226–229
action on phosphorylases, **24:**271
active sites, **24:**459
effect of partial degradation of, **24:**125
inactivation by radiations, **24:**374–375
partial degradation by, **24:**129–130
water-insoluble derivative of, **24:**108
activation, **15:**292
and rennin, **15:**434
active site, location of, **22:**63; 78
column chromatography, **22:**454–456
pK value, **22:**58; 59
classification, **14:**232; 247
isolation, **14:**394
effect of DFP on, **13:**332
synthesis of amide linkage by, **13:**307–308
effect of quinones on, **11:**247
thiol and enzyme activity, **11:**234; 236; 238; 239; 241
effect on fructose 1,6-bisphosphatase, **42:**204
effect on fructose diphosphatase, **28:**93; 94
inhibitors of, **43:**302
mechanism of action, **19:**326; 327

preparation, **19:**270–274
properties, **19:**267–334
sources, **19:**268
specificity of, **12:**194
and sulfhydryl groups, **10:**227
turnover number, **25:**32
uses, **19:**268
Papain, Properties (Kimmel and Smith), **19:**267
Papain, SH-group, **27:**214
Papain and sulfhydryl groups, **9:**429
Papain-chymopapain, separation of components, **14:**383
Papain-collodion membrane, **34:**454; 495–496; 509–511
Papain mutants, glutamine-dependent nitrogen transfer, *vs.* AS-B, mechanisms, **72:** 175–183
Papaverine
metabolism in plants, **16:**142
tyrosine hydroxylase induction, **70:**169
Papaver somniferum alkaloid content, **8:**242
Papayotin, effect of radiations on, **24:**397–398
Paper chromatography, **42:**237
Papin, **33:**280
specificity of, **33:**433–434
PAPS, *See* Adenosine-3′-phosphate-5′-phosphosulfate
PAPS sulfotransferase, **74:**188; 189
Papulospora sp., β-lactamase of, **28:**243
Paracasein, **15:**436; 441–442; 443; 445
Paracentrotus lividus, azetidine-2-carboxylic acid addition to, **29:**96
Paracoccus denitrificans, **71:**84; 97; 100; 101; 102; 103; 104; 107; 117; 121; 126; 189
Paracolon bacillus, and β-lactamase, **28:**307
Paracolon group of bacteria, β-lactamase of, **28:**242
Paracrystals, **5:**7; 8
Paraffins. *See also* Hydrocarbons, aliphatic
ring-substituted, microbial oxidation, **27:** 495–498
n-Paraffins
microbial oxidation, **27:**480–493
Parafinic side chains, **34:**468
Parahematin, **7:**267
Parahematins, **15:**18
Parahemophilia, **16:**50; 65
Parallel transport systems, relation between, **49:**88–91
Paralogous sequences, definition of, **74:**132
Paralysis, familial periodic, **6:**214–218
Paramagnetic effects on T_1 and T_2, *See* Longitudinal relation time; Transverse relaxation time
Paramagnetic line-broadening NMR, **50:**352
Paramagnetic macromolecular systems, enhancement of relaxation rates in, *See* Relaxation rates, nuclear
Paramagnetic probes, effect on enchanced nuclear relaxation rates, **33:**16
manganous ion, **33:**16
NAD analog, ADP-R, **33:**59–60
Paramagnetic Probes Induce Nuclear Spin Relaxation in Study of Enzyme Mechanism (Mildvan and Cohn), **33:**1
Paramagnetic property of melanins, **16:**172a
Paramagnetic species, **34:**44; 63
Paramagnetism, and metal-flavin contact, **27:**184
Paramecia, effect of rare earth metal ions on, **21:**192
Paramecium, development of antigens in, **16:**478
Paramecium bursaria, enucleation studies, **16:**470
Parameica, genetic transformation of, **6:**89–90
Parameters, enzyme, **34:**504
kinetic, **34:**502–504
macroenvironmental, **34:**446
Paramylon, euglenid, **28:**354
as β-1,3 glucan, **28:**327–329; 334; 337; 338; 341; 342
Paramyosin, banding, **16:**320
Paramyxoviruses, **52:**132
Paraoxon, **43:**170
acetylcholinesterase reaction with, **74:**52–53; 53
bacterial hydrolysis of, **74:**52
hydrolysis catalyzed by phosphotriesterase, **74:**74–75
pH-rate profile of, **74:**54–55; 55; 80
Parapepsin I, *See* Pepsin B
Parapepsin II, *See* Pepsin C
Paraperoxidase, **7:**288–289
Parasites, plant, **11:**426–432
Parathyroid hormone, attributes of valinomycin shown by, **29:**292
effect on cations and mitochondria by, **29:**46
effect on permeability of mitochondrial membrane by, **29:**378
K^+ and Mg^{++} accumulation induced by, **29:**301
required for Mg^{++} uptake, **29:**275

Paratose, *See* 3,6-Dideoxy-D-glucose
Paricles, subcellular. *See also* Microsomes, Mitochondria, and Ribosomes
distribution of enzumes in, **24:**291–358
Parkinsonism, **74:**236
Parkinson's disease, **44:**229
Parrish-Perlin mechanism, of laminarinase action, **28:**345
Parthenium argentatum, rubber producer, **29:**222
Parthenogenesis, artificial, **9:**286–297; 319
in mammals, **9:**252
Parthenogenetic development of gametes, **9:**243
Partial concurrent solvent exchange, **50:**345
Particles. *See also* Cytoplasm, Microsomes, Mitochondria and Particulate material; Mitochondria
from Arum spadix cells, **18:**25–28
intracellular, respiratory activity of, **18:** 13–15; 18–20; 29
cellular (*See also* Mitochondria)
in arginine formation, **15:**327; 334
in citrulline synthesis, **15:**328–329; 364–365; 367
DPN-nucleosidase in, **15:**107
enzyme systems in, **15:**224
cytochrome b-c_1, in respiratory chain, **28:**190; 194–199; 204; 208–209; 223; 229
cytoplasmic, and enzymic adaptation, **13:**93; 106
devoid of succinate dehydrogenase, **28:** 173–174
elementary, in reconstitution of respiratory chain, **28:**159; 216; 219; 230
insoluble, enzymes for transaminations, **16:**208
in reconstitution of respiratory chain, lipid-containing, **28:**209–223; 225
lipid-free, **28:**167–209
submitochondrial, **28:**164–167
soil, adsorption of enzymes, **28:**367; 368
submitochondrial, electron transfer in, **27:**188
EPR spectra, **27:**117; 153–154; 188–190
α-Particles, and enzyme inactivation, **24:**373; 376
LET from, **24:**365
Particle size. *See also* Chain length, Intermolecular forces, and Van der Waals' dispersion forces
and enzyme action, **12:**381
Particle systems, and citric acid cycle, **18:** 244–246
Particulate acetylcholinesterase, **43:**105; 114; 117; 118; 139; 191; 210
Particulate aminoacyl-tRNA GlcNAc-MurNAc(-pentapeptide)-*P*-*P*-phospholipid transferases, **40:**120; 121; 122; 123; 124; 125; 126; 127; 128; 129; 130
Particulate enzymes, **33:**290–294
Particulate immobilized enzyme systems, **34:**521
Particulate material. *See also* Mitochondria
glutaminase in, **13:**283
of liver cells distribution of β-glucuronidase in, **16:**366–367
relation to tyrosine reactions, **16:**231
Particulate system, in photosynthesis, **25:**43; 67–86
Particulate units, in succinate oxidation, **18:** 104–108
(Part II) During Fermentation and Aging (Frankenburg), **10:**325
Parvalbumin, **47:**124; 133
Pasteur effect, **15:**173–174, **57:**145
allosteric regulation, **28:**71; 74–75; 78; 83–84; 86
in ascites tumor cells, **29:**348
in HeLa cells, **29:**327
neoplasms with a normal, **29:**325
and carbohydrate metabolism, **3:**180–183
inhibition, **16:**135
Pasteur effect and enzymic adaptation, **13:**77
Pasteur effect in spermatozoa, **24:**55–56
Pasteur effect with *aspergillus,* **2:**215; 216
Pasteurella pestis, **18:**273
rhamnose-utilizing mutant, **22:**33; 34
Pasteurella sp., β-lactamase of, **28:**243
Patching and capping, **50:**307
Patella vulgata, aryl sulfatase of, **22:**219; 220
steroid sulfatase of, **22:**227; 228
Pathways of enzymic processes, **24:**442
Patulin, **7:**476; 477; 487; 489; 491; 493
in treatment of plant diseases, **20:**392; 394; 395
Pauling-Corey's α-helix, **23:**2–4; 21
Pausing, RNA polymerase II transcription, potential roles, **71:**66–67
P450BM-3, **68:**41
P450c17, **68:**41–43; 48; 53
immunoblot analysis, **68:**60
P450c21, **68:**41

P450c27, **68:**41
P clusters, **67:**317–318
PCMB. *See also* p-Chloromercuribenzoate
effect on LADH-DPNH complex, **27:**75–78; 80; 82
use in study of enzyme conformation, **28:**302; 303
P4502C subfamily, **68:**47–48
P-450 Cytochromes: Structure and Function (Blake and Coon), **60:**35
P.D., in respiratory chain, **28:**159; 193–194; 196
PDIp polypeptide, protein disulfide isomerase (PDI) prolyl 4-hydroxylase β subunit structure, **72:**374–375
PE, *See* Pectinesterases
Peaches, antibiotic treatment of diseases, **20:**399; 400
Peanut plants, amides in, **13:**246
Pears, antibiotic treatment of diseases, **20:**398; 399
polygalacturonase inhibitor in, **20:**365
Peas
antibiotic treatment of diseases, **20:**392; 393
respiration in, **18:**286; 292; 294
Pectase, classification, **14:**244
Pectase in tobacco, **6:**368
Pectases, *See* Pectinesterases
Pectates, **17:**147
Pectic acid, *See* Pectic substances
Pectic acid derivatives, hydrolysis by polygalacturonases, **20:**362; 363
preparation, **20:**347
properties, **20:**353; 356; 357
Pectic enzymes, **11:**267–293, **20:**358–371
activity in frozen state, **9:**626
classification, **14:**246
Pectic Enzymes (Lineweaver and Jansen), **11:**267
Pectic substances, **20:**341–357; 367–371
nomenclature, **11:**269–270
Pectic Substances and Pectic Enzymes (Deuel and Stutz), **20:**341
Pectin, *See* Pectic substances
reactivity with borax, **12:**505
separation of enzymes by chromatography, **14:**364; 365
Pectinase
bacterial, separation by chromatography, **14:**339; 351
in cocoa fermentation, **25:**473
Pectinases, *See* Polygalacturonases
effect of adsorption, **28:**28
Pectinates, **17:**149
Pectin demethoxylases, *See* Pectinesterases
Pectin deploymerase, classification, **14:**246
Pectin depolymerases, *See* Polygalacturonases
Pectin derivatives, heparin-like action of, **20:**357
hydrolysis by polygalacturonases, **20:**362
preparation, **20:**346
properties, **20:**353; 356
Pectinesterase, **11:**277–284
assays for, **13:**323
effect of DFP on, **13:**332
cacao, **25:**466
mode of action, **11:**271–273
Pectin esterases, classification, **14:**244
separation by chromatography, **14:**342; 364–365
Pectinesterases, in hydrolysis of pectic substances, **20:**358–360; 364; 367; 368; 370
Pectin-galacturonase, adsorption studies, **14:**339
Pectin galacturonidases, classification, **14:**246
Pectin glycosidases, *See* Polygalacturonases
Pectin methoxylases, *See* Pectin esterases
Pectin-methylesterase, adsorption studies, **14:**339; 351
classification, **14:**244
Pectinol 100, polygalacturonase in, **20:**343
Pectinol K 2073, polygalacturonases of, **20:**367; 370
Pectins
in air-cured tobacco, **6:**327–330
in flue-cured tobacco, **6:**364
changes during fermentation of tobacco, **10:**340; 366–368; 396; 402
Pectolases, *See* Polygalacturonases
Pediococcus cerevisiae (Leuconostoc citrovorum) folate requirement, **27:**421; 436–437; 441; 457
Pediococcus pentosaceus, allosteric enzymes, **28:**118
PEG, **52:**200
Pelargonidin, **17:**433–434
Penatin, **9:**563, *See* Notatin
Pencillium, acid formation in, **9:**538; 547; 560–563; 565
alcohol formation in, **9:**545
fat synthesis by, **9:**663; 673–674; 691
Penicidin, **7:**488

L-Penicillamine, penicillin from, **10:**254
Penicillase, destruction of penicillin by, **10:**253
Penicillia, biological methylation in, **12:** 430–442; 446–448; 452–458
Penicillic acid, **7:**477; 486; 489; 490; 492
Penicillin, **4:**281, **33:**367. *See also* Procaine penicillin
absorption by *Nitella clavata,* **20:**387
effect of borate on production of, **12:**521–522
effect on, RNase, **24:**238
on, sperm, **24:**85
effect on bacterial wall, **26:**216–219
effect on growth zone of alga, **26:**221
effect on milk coagulation, **15:**444
effect on Staphylococci, **20:**123
inhibition of synthesis of poly (glycerol phosphate) in *Bacillus* by, **30:**244
role in penicillinase synthesis, **22:**116; 117
in treatment of plant diseases, **20:**393; 402
Penicillin A and B, *See* Notatin
Penicillin acylases, and β-lactamase, **28:**249
Penicillin amidase, and β-lactamase, **28:**238; 247–249
Penicillinase, **34:**481–482, **57:**208. *See also* β-Lactamase
adsorption studies, **14:**346
classification, **14:**247
in *B. cereus,* **22:**106; 109; 137
B. cereus mutants, **22:**21
cephalosporin C in, **22:**113
constitutive levels, **22:**104
constitutive mutants, **22:**103
and RNA transfer, **22:**140
sensitivity to RNA antagonists, **22:**138
induction in *B. cereus,* **22:**119
kinetics of, **22:**116; 117
nature of precursor in, **22:**124
stereospecificity of inducer, **22:**119; 120
ultraviolet sensitivity of, **22:**138
synthesis, **22:**144
effect of azaguanine on, **22:**139
biochemistry and function, **28:**237–323
enzymic adaptation, **13:**70; 95–98
occurrence of, **28:**241–244
specificity and identity of, **28:**250–255
steric structure, **26:**93
γ-Penicillinase, and β-lactamase, **28:**251; 253–255; 266; 288; 291; 296
Penicillinase activity, in microorganisms, **28:**243
Penicillinase-forming colonies, **28:**248
Penicillinase-loss mutants, genetic, **28:**312
Penicillin B, **15:**148
Penicillin-binding proteins (PBP), *Caulobacter* cell cycle, **71:**6–7
Penicillin-destroying agent, **28:**242
Penicillin-destroying mechanisms, in animal tissues, **28:**243
Penicillin G, in treatment of plant diseases, **20:**392
Penicillin in fructose-bisphosphatase inactivation, **54:**139
Penicillin resistance
genetic studies, **17:**455–457
and β-lactamase, **28:**307–313
bacterial resistance to, **28:**274–275
definition, **28:**241
groups of, **28:**300
hydrolytic enzymes for, **28:**238; 242; 243; 247–253
inactivation of, **28:**287–294
kinetic constants, effect of side chain on β-lactamase activity, **28:**260
β-lactamase-resistant, **28:**312
stability to β-lactamase, steric factors, **28:**258–262
structures, **28:**239
as substrate for β-lactamase, **28:**245–253; 262; 267; 287–294
effect of nucleus, **28:**260–261
effect of optical activity, **28:**260
effect of side chain, **28:**257–260; 273
Penicillins, **7:**477–479; 486; 489; 490; 492
amino acids in, **14:**266
production of chemical changes in *Staphylococcus aureus,* **14:**203–204
identification, **7:**497
in plants, **10:**227; 250–256
mechanism of synthesis *in vivo,* **10:** 255–256
relation to cysteine in plants, **10:**254; 269
types, **10:**250–251
Penicillium, **17:**396; 399
arginine metabolism in, **15:**375
glucose oxidase in, **19:**206
and mevalonic acid, **23:**473
Penicillium, aurantio-violaceum, **35:**156
resticulosum, **35:**140
Penicillium adametzi, enzymic transformations of steroids by, **20:**262

Penicillium chrysogenum, **17:**212, **18:**258
action on collagen, **23:**177
use of acetate, **23:**405; 411
penicillin amidase of, **28:**249
polygalacturonases of, **20:**365; 389
Penicillium citrinum, P1 nuclease from, **74:**71
Penicillium cyaneo fulvum, proteinase from, **41:**182
Penicillium expansum, polygalacturonases of, **20:**366; 367
Penicillium lilacinum, enzymic transformations of steroids by, **20:**262
Penicillium mutants, amino acid synthesis by, **16:**258–259; 261–262; 264
Penicillium notatum, **34:**489
enzymic transformations of steroids by, **20:**248; 263
Penicillium patulum, **55:**26
Penicillium puberulum, **7:**488
*Penicillium roquefor*ti, **49:**10
Penicillium species, enzymic transformations of steroids by, **20:**239; 245; 248; 262
pectinesterase of, **20:**360
Penicillium spiculosporium, **43:**82
Penicillium spinulosum, **17:**197
Penicillium stipitatum, tetrahydrocoenzyme Q in, **25:**322
Penicillium urticae, enzymic transformations of steroids by, **20:**248
Penicilloic acid, from β-lactamase action, **28:**244–247
Penicillopepsin, **44:**1, **53:**261; 287
charge-relay system for, **50:**13
Penicillum notatum, enzyme for glucose isomerization, **14:**193–194
Penicillum spinulosum, action on carbohydrates, **15:**314
Penillinase, **37:**466; 477; 483; 494; 497; 500; 521; 552; 553; 559–561; 583; 586; 594; 595; 601; 602
A-type ligand, **37:**561
determinants of, **37:**602
S-type ligand, **37:**516
Pentachloronitrobenzene, in treatment of plant diseases, **20:**402
Pentachlorophenol, **46:**138
effect on mitochondrial oxidative phosphorylation, **21:**31
Pentacyanoammonioferrate, **50:**136
1,4-Pentadiene, **58:**279
Pentahexosylfucoglycolipid, **38:**295
2,3,4,5,6-Penta-*O*-methyl mannitol, from laminarin, **28:**333
Pentane, **58:**207
Pentapeptides, **38:**220; 221; 222
Δ^2-Pentenylpenicillin, action of β-lactamase on, **28:**257
Pentosans
in air-cured tobacco, **6:**327
in flue-cured tobacco, **6:**369
changes during fermentation of tobacco, **10:**340; 368–369; 402
Pentose
alcoholic fermentation of, **5:**167–179; 199
colorimetric tests for, **9:**475–477
utilization by microorganisms, **9:**661; 672; 677; 679–680; 700
mechanism of breakdown in *Fusaria,* **5:** 170–171
Pentose aminoimidazole, **17:**427
Pentose cycle, non-oxidative limb, **28:** 395–397; 400; 401; 404–407; 409–413; 415–419; 428
oxidative limb of, **28:**394–395; 401; 404–405; 407–413; 415–419; 424; 428
Pentose fermentation, and phosphorolysis, **18:**227
Pentosenucleic acid, *See* RNA
Pentose 5-phosphate, as histidine precursor, **16:**300–301
Pentose phosphate cycle
allosteric regulation, **28:**116–118
in photosynthesis, **19:**14–16; 19–33
Pentose-phosphate cycle oxidative system, enzymes of, **24:**297; 309; 317; 333; 337
Pentose phosphate isomerase, **15:**154; 157; 158–160
allosteric regulation, **28:**118
Pentosephosphate pathway, **34:**219
Pentose phosphate pathway, in plants, **18:** 293–295
Pentose phosphates. *See also* individual phosphates
in photosynthesis, **25:**94–103
role in carbohydrate metabolism, **15:**121; 153–160; 163–168; 176
Pentose-5-phosphate shunt, enzymes of, thiol groups, **27:**214

Pentoses. *See also* Deoxyribose and Ribose
asynchronous synthesis, **28:**428; 431; 452–453
biosynthesis of, **28:**391–460
from ^{14}C-labeled precursors, **28:**400; 405–420; 423–435; 454
by embryonic tissues, **28:**418–419; 434–436; 448–449
exchange reactions, **28:**400–402; 407
by tumor cells, **28:**432–434
fluorinated, **55:**244
from glucose, **15:**121
of tobacco mosaic virus, **15:**459–461
interconversions among, **28:**394
origin of C atoms by various pathways, **28:**401
of *P. saccharophila,* labeling, **28:**430
purine-bound, labeling, **28:**433
pyrimidine-bound, labeling, **28:**433
of RNA, origin, **16:**451
use in condensation of acetate and CO_2, **16:**268
Pentoses and fat formation, **8:**312–314; 328
Pentoses and phosphatide formation, **7:**149
Pentose shunt, in spermatozoa, **24:**57
Pentosuria, hereditary defect, **16:**396
S-Pentosylmethionine, **54:**223
Pentosyltransferases, localization of, **24:**312
P-enzyme, **59:**352–353
P450 enzymes, **68:**37–61
cloning, **68:**38
expression
in bacteria, **68:**52–60
in baculovirus, **68:**51
eukaryotic systems, **68:**39–40
heterologous, **68:**38–39
in mammalian cells, **68:**42–50
COS cells, **68:**42–49
stable tranformants, **68:**50
vaccinia virus, **68:**49–50
systems used for, **68:**39–40
in yeast, **68:**40–42
forms, **68:**40–41
mitochondrial forms, **68:**44
signal anchor, **68:**47; 52
structure-function relationship evaluation, **68:**47–48
superfamily, **68:**37–38
Peonin, as sex hormone in *Chlamydomonas eugametos,* **16:**144
PEPA, *See* Phosphoenol pyruvic acid
PEP carboxykinase, and phytochrome action, **26:**27
PEP carboxylase, and phytochrome action, **26:**27
PEPCK-C gene transcription model, *See* Phosphoenolpyruvate carboxykinase (GTP)
Peperomia, antibiotic treatment of diseases, **20:**401
Pephydroxylation, by liver nonspecific hydroxylase, **19:**177–179
Pepper, sweet, phytoalexina of, **55:**69–73
Peppers, antibiotic treatment of diseases, **20:**391; 393; 401
PEP (phosphoenol pyruvate), **18:**301
Pepsin, **1:**39; 69–74, **5:**22, **15:**62; 272, **17:**307, **23:**130, **34:**31, **41:**179; 180; 187; 237, **43:**224, **44:**1, **53:**261; 264; 284–286; 290
absorption spectrum, **9:**143–144
photochemistry, **9:**140; 143–147
acetylation, **44:**16
acetylation of, **33:**418–419
acid catalysis of, **33:**428
action of, on protein substrates, **33:** 421–423
on pyridylalkyl ester substrates, **33:**417
on synthetic substrates, **33:**406–421
acyl enzyme intermediate of, **33:**427; 429
amino acid sequence of, **33:**406
catalysis of ^{18}O exchange, **33:**427
catalysis of transamidation reactions by, **33:**425–426
competitive inhibition of (table), **33:**426
conformational changes in, **33:**421
conformational flexibility in, **33:**421
diacylimino compound intermediate of, **33:**427
enzyme action of, deuterium-isotope effect of, **33:**426–427
equilibrium-binding, measurements of 426, studies of, **33:**419
esterase activity of, **33:**414–415
homogeneity of, **33:**405–406
inactivation of by diazo ketones (table), **33:**424
inhibition of, **33:**413; 424
and inhibition of hydrolysis, **33:**425–426
interaction with oligopeptide substrate, **33:**422

isotope exchange, **33:**425
mechanism of action of, **33:**423–429
oligopeptide substrates, **33:**420
pH dependence of, **33:**425
pH optimum of, **33:**409–410
plastein reaction of, **33:**425
properties of, **33:**405–406
secondary interaction effects on, **33:** 415–421
side-chain specificity of, **33:**407; 410–413
stereochemical specificity of, **33:**413–414
synthetic substrates of, **33:**408
acetyl derivatives, **15:**297
action on, collagen, **23:**171; 172; 215–216
elastin, **23:**225; 228; 243
action on egg albumin and insulin, **19:**302
specificity, **19:**328
action on fibrinogen, **10:**32
action on hypertensin, **7:**526
action on renin substrate, **7:**529
and kidney proteinases, **7:**529
action on ribonuclease, **24:**133; 174–176; 196; 227–229; 233; 235; 244; 246; 247; 250
activation of acetyltrypsinogen by, **24:** 126–127
effect of, acylation on, **24:**136
autolysis, **24:**125–126
ionizing radiation, **24:**378; 389; 397; 400
localization of, **24:**324
molecular weight from radiation studies, **24:**366
partial degradation by, **24:**128–131
valence of, **24:**118–119
active site, **44:**15; 27
acyl-enzyme, **44:**20; 29
adsorbed, activity of, **28:**2; 3; 16; 22–25; 29
radiosensitivity, **28:**34
amino acid sequence, **44:**2
amino-enzyme, **44:**22; 25; 29
cationic substrate, **44:**5
classification, **2:**51–61
complexing between serum albumin and, **30:**143
electrophoresis for complex formation between bovine serum albumin and, **30:**146
Michaelis-Menten complex for, **30:**147
condensation reactions, **44:**31
in denaturation study, **23:**32; 34; 37–38
effect of borate on, **12:**518
specificity of, **12:**194
effect of pH on structure, **22:**426
pK value, **22:**58
effect of phosphate esters in, **13:**332
effect on glycyldehydrophenyl-alanine, **8:**134
effect on murein sacculus, **26:**224
enzyme-substrate complex, **44:**15
esterase action, **44:**7
fluorescence measurements, **44:**9
hydrolysis of sulfite esters, **44:**21
imidazole in partial hydrolyzates of, **20:**69
imidazole reactivity in, **20:**58–61
inaction on histones, **16:**446
iodinated, **15:**289
iodination, **44:**16
in low vertebrata, **1:**372; 375; 377–379; 381; 382; 388; 389; 394; 403; 404; 406; 410
mechanism, **44:**8
model systems, **44:**18
product inhibition of, **25:**224–225
release of product, **44:**24; 26
and rennin, **15:**425; 427–428; 431; 433; 434–435
secondary interactions, **44:**16
in sequence analyses, **27:**230
solvent isotope effect, **44:**19
specificity of, **41:**180
stopped-flow measurements, **44:**12; 24
synthetic substrates, **44:**3; 27
transition state, **44:**20; 28
transpeptidation, **44:**22; 30
Pepsin action, specificity and mechanism of, **33:**401–443
Pepsin Action, Specificity and Mechanism of (Fruton), **33:**401
Pepsin action of tyrosinase on, **14:**136; 138; 140; 141; 144; 146–147
action of insulin, **14:**149
activity of amino groups, **14:**189–190
assay of activity, **14:**381–382
classification, **14:**232; 247
isolation, **14:**378; 393; 398; 402
oxidation of tyrosine in, **14:**86
peroxidase action on, **14:**155; 157–158
separation by chromatography, **14:**321–322; 328; 330; 332; 342; 345
structure studies, **14:**274; 276; 293; 294; 301

Pepsin-albumin films, action of x-rays, **28:**34
Pepsin and Related Acid Proteinases, The Mechanism of the Catalytic Action (Fruton), **44:**1
Pepsin B, **33:**405
Pepsin C, **33:**405
hydrolytic action on insulin of, **33:**422
Pepsin D, **33:**405
Pepsin-like enzymes, **45:**400
Pepsin nucleates, **17:**143
Pepsinogen, **40:**37; 38, **44:**11, **57:**210
activation of, **24:**126–127
localization of, **24:**324
Pepsitensin, **7:**529
peroxidase action on, **14:**155
Pepstatin, **44:**10; 20, **53:**284; 287
Peptic digests, **42:**152
Peptidase. *See also* individual peptidases
adsorption studies, **14:**337
classification, **14:**223–225; 230; 238; 242; 247; 250
cyanide inhibition of, **15:**287
metal ions in, **15:**272
in sperm, **24:**66
in tobacco, changes during processing, **10:**414
relation to structure of alkaline phosphatase, **10:**116–117
role of metal ions, **10:**104
Peptidase models, hydrolysis with, **21:**177; 178; 191
Peptidases, **23:**130, **35:**194–197, **45:**417; 420. *See also* specific substances
action of snake venoms on, **8:**486; 488–489
linkage attacked by, **8:**164; 165
activation by metal ions, **21:**153; 163
kinetics, **21:**136; 138; 140
bacterial, **23:**187–189; 199; 205; 210; 221; 231
classification, **2:**51
enzymatic properties, **2:**69–92
homogeneity, **2:**71–72
specificity, antipodal, **2:**83; 84
histochemical studies, **9:**605
titrimetric methods, **9:**601
in insect development, **45:**411
intracellular, **35:**195
localization in cytoplasm, **13:**8
location of, **35:**196
for muropeptides, **26:**211–215
non-obligatory coupling to peptide transport, **35:**196
presumably involved in collagen breakdown mechanisms, **34:**193–194
specificity of, **12:**191–257, **35:**196
toxic peptides, **35:**196
D-Peptidases, classification, **14:**247
Peptidases, Enzymic Properties (Johnson and Berger), **2:**69
Peptidases, Specificity (Smith), **12:**191
Peptidases and Proteases, Insect (Law, Dunn, and Kramer), **45:**389
Peptidases in tobacco, **6:**369
Peptidase systems, **2:**84–88
in animals, **2:**85; 86
of bacteria, **2:**87; 88
of fungi, **2:**86; 87
of intestinal mucosa, **2:**84; 85
of pancreas enzymes, **2:**54; 85
in plants, **2:**86
Peptide
active site, of cocoonase, **45:**411
adsorption studies, **14:**345; 347
biosynthesis, **35:**162
intermediates in
cyclization, **35:**13; 15
elongation, **35:**15
gramicidin S biosynthesis, **35:**12
initiation, **35:**14; 15
isolation, **35:**13
transport, **35:**192–194; 195
in *Bacteroides,* **35:**193
distinction from amino acid transport, **35:**193
in *E. coli,* **35:**198–199
in Lactobacilli, **35:**199–200
studies with auxotrophs, **35:**204–205
studies with homologous peptides, **35:**205–206
tyrocidine biosynthesis
cyclization, **35:**27
elongation, **35:**27–28
initiation, **35:**27–28
isolation, **35:**27
bond cleavage, chymotrypsin, **67:**221
cyclic, fibrinogen binding inhibition, **70:**6–8; 14–15
hydrolysis, **58:**6
initiation, **29:**154
Nmt substrate specificity, **67:**391–394

RGD sequence, **70:**5
Peptide aldehyde, **43:**369
Peptide amide bonds, and Cotton effect, **27:** 46–47
Peptide bond
effect of proton donors on cleavage of, next to proline residue, **29:**440
formation of, activated esters method of, **29:**405–406
azide method of, **29:**403
carbodiimides method of, **29:**405
mixed anhydride method of, **29:**405
equilibrium constant for hydrolysis, **43:**314
hydrolysis, by metal ions, **21:**160; 163; 177; 178; 182; 183; 191; 194
and internal rotation of chain, **23:**13–16
synthesis of, **38:**196; 197; 203; 212
Peptide bond cleavage, **57:**70–73
Peptide bonds, **7:**195
of amino acids, **1:**2; 3; 29
breaking of to give units of myosin, **16:** 320–326; 330
and denaturation, **26:**95
enzymatic synthesis, **1:**90–93; 94; 96
enzymes acting on, **24:**324–327
hydrolysis, **1:**67
photolysis, **9:**90–96; 118–119; 126; 130; 134; 147; 157; 160–162
in proteins, **1:**67; 69
Peptide chain
arrangement, **1:**28–35
folding, **1:**9; 29; 40
Peptide chains
order of residues in papain, **19:**297
in protein structure, **14:**278–289
order of residues in, **14:**289–302
oxidation of tyrosine in, **14:**86
stability of amino acids in, **14:**265; 268–270
Peptide chain sequence, **42:**272
Peptide chains of tobacco mosaic virus, arrangement, **15:**457; 480–482
molecular weight, **15:**462–464; 468
Peptide derivatives, synthetic. *See also* specific compounds, eg Hippurylamide
papain action on, **19:**303–306
Peptide effect, **16:**114; 126; 160
Peptide formation, and ATP, **18:**229–230
Peptide grouping, reaction with quinones, **16:**125; 126
Peptide hormones
constitution of, **24:**107
disulfide exchanges, **27:**226–228
Peptide linkage, **50:**14; 33
cis linkage, **50:**33
enolization of, **50:**14
Peptide mapping, **39:**347
tyrosine hydroxylase regulation, **70:**146
Peptide maps, **45:**410
Peptide metabolism and dehydropeptidases, **8:**151–152
Peptide peptidohydrolases, localization of, **24:**324–327
Peptides, **7:**396; 397, **49:**244–245, **50:**14; 33. *See also* Polypeptides; Proteins; Synthetic peptides, fibronectin
absorption spectra, **9:**89–91
light-induced changes in, **9:**91–96
asparagine, deamidation of, **13:**244–245
asparagine, utilization by microorganisms, **13:**244–245; 250–253
character of bond, **13:**158–159
glutamine, deamidation of, **13:**244–245; 261–262
γ-glutamyl, in transferase system, **13:** 300–301
linkage, use of term, **13:**241
synthesis, relation of asparagine, glutamine and glutathione to, **13:**237–319
synthesis of bond, **13:**280–282; 289–305; 305–310; 311–315
synthetic, in exchange reactions, **13:** 305–311
utilization by microorganisms, **13:** 244–245; 261–262
biosynthesis, in egg white microsomes, **19:**261
by intramolecular aminolysis, **19:**261; 262
model reaction for, **19:**263
role of CoA and CoA derivatives in, **19:**249
role of diacylimide bond in, **19:**260; 263
site of
chelates, in metal ion activation of enzymes, **21:**146; 147
chromatography of mixtures, **22:**419
from collagen action, **23:**207–208; 210–213
fibrous, **23:**2–4; 21–22; 25
synthetic, action of collagenase on, **23:**205–207
as complexing agents, **17:**150–151

Peptides (*Continued*)
hydrolysis by α-chymotrypsin, kinetic studies, **17:**46–49
conversion to pyrrole compounds, **8:**130
a-keto acid, **8:**129
of cystine, enzymic degradation, **8:** 145–147; 148
rate of desulfuration, **8:**135
isomeric, and D-amino acid oxidase, **8:** 150–152
of saturated amino acids, **8:**147–152
saturated, preparation, **8:**129; 133–134
effect on β-lactamase, **28:**262–263
of glutamine, assay, **16:**258
removal from fibrinogen by thrombin, **16:**36–37; 41–42
synthesis, relation to transamination, **16:**239
glycyl, hydrolysis, **15:**292
hydrodynamic volume measurement, **35:** 228–233
thyrotropin releasing hormone, **35:**204
inhibition of ATPase activity, **49:**244; 245
intermediates, countercurrent distribution coefficients of, **29:**423
irradiation of, **24:**377–379
phosphorylated, from labeled phosphorylase *a,* **24:**269–271
linkages to form proteins, **10:**286–288
mechanism of hydrolysis, **19:**311
metal binding, **50:**33
phosphorylated, hydrolysis, **68:**25
small, related to gramicidin S, **33:**351–356
synthesis, by activated amino acids, **19:**236; 241; 242; 244; 246–249
by papain, **19:**308; 309
synthesis by transmination, **4:**82; 93–94
synthetic, **19:**269; 303–306
tautomerization of, **50:**14
in transpeptidation reactions with papain, **19:**310
tryptic, **39:**347; 350
tryptophan-containing, **39:**347
in tryptic digests of papain, **19:**297–300
tyrosine-phosphorylated, **68:**21
as wilting toxins, **11:**402–403
Peptides, UDP-*N*-acetylmuramic acid, **26:**49–51
Peptide sequences. *See also* Amino acid sequences and Protein structure
in protein chains, **24:**420–425; 425–436
Peptide Structures, Identical and Analogous, in Proteins (Sorm), **24:**415
Peptide structures in proteins, identical and analogous, **24:**415–439
Peptide substrates, collagen hydroxylases, **72:**339; 341–345
cosubstrate binding, **72:**348–357
inhibition mechanisms, **72:**363–364
Peptide synthesis, **12:**192
advances in, **29:**392
aspects of methods of, **29:**393–406
tert-butyl group in, **29:**412–413
tert-butyloxycarbonyl group in, **29:** 397–398
carbobenzoxy group in, **29:**394–395; 412–413
cleavage of carbobenzoxy group by hydrogen bromide in glacial acetic acid, **29:**395
Curtis azide procedure of, **29:**411
N-ethyl-5-phenyl-isoxazolium-3′-sulfonate as a peptide-forming agent in, **29:**406
evaluation of protecting groups in, **29:** 437–443
formation of the peptide bond in, **29:** 403–406
masking of amino groups in, **29:**394–399
mixed anhydride procedure used in, **29:**401
modified carbobenzoxy groups in, **29:** 395–397
peptide-forming agents in, **29:**443
preparation, and hydrolysis of *tert*-butyl esters for, **29:**400–401
of phthalyl amino acids for, **29:**399
of *p*-toluenesulfonyl group in, **29:**398
of triphenylmethyl group for, **29:** 398–399
protection of functional groups in, **29:** 393–403
reductive cleavage by sodium in liquid ammonia in, **29:**395
removal of protecting groups with sodium in liquid ammonia in, **29:**439
removal of triphenylmethyl group in, **29:**398–399
synthesis of ACTH and MSH peptides in, **29:**406–443
total synthesis of, β-MSH, **29:**434
porcine, ACTH, **29:**433–434

analytic aspects of, **33:**462; 464; 476; 478; 502
fragment condensation, **33:**462; 473
racemization, **33:**463; 471; 476
solid phase, **33:**456; 462; 483; 498; 500; 502–503; 509
stepwise, **33:**460; 462
strategy of, **33:**459–460; 484
tactics of, **33:**460–509
Peptide Synthesis, Solid-Phase (Merrifield), **32:**173
Peptide synthetases, localization of, **24:**335
Peptide Transport (Payne and Gilvarg), **35:**187
Peptidic linkage, in glutathione, **13:**282
role of energy-rich phosphate bonds in formation of, **13:**311–315
use of term, **13:**241
Peptido-asparaginase, **39:**238
Peptidoglutaminase I, **54:**2
Peptidoglutaminase II, **54:**2
Peptidoglycan, **40:**119; 135
Peptidoglycans, **53:**258
Peptidoglycan synthesis, in teichoic acid biosynthesis, **30:**245
Peracid-supported hydroxylation, **47:**36
P/2*e* ratio, for noncyclic phosphorylation, **30:**64
Perchloroethylene, cysteine *S*-conjugate β-lyase reactions, **72:**212
Perfect enzymes, enzyme action, V/K *constant,* **73:**32–33
Perfluorocyclohexane, **60:**67
Perfluorohexane, **46:**171
Perfused livers, **44:**216
Periodate, **35:**477
oxidation, **35:**462; 474
Periodate-arsenite-basic fuchsin reagent, **48:**24
Periodic acid, effect on pectic substances, **20:**352
Periodic acid-Schiff method, cell wall study, **28:**330
Periodic enzymic reactions, **23:**83–127
Periodic Enzymic Reactions and Their Possible Applications (Christiansen), **23:**83
Periodicity. *See also* Banding and Bands
of contractile proteins, **16:**320; 325–326; 347; 352
Periodic phenomena, **23:**83–127
in living organisms, **23:**83–84; 119–122
Periodic reactions and product inhibition, **25:**172–173
Peripheral membrane proteins, **43:**201
Periplaneta americana, **45:**419
Periplasm, **34:**456
enzyme location, **28:**266
Periplasmic binding, proteins, **44:**240
Periplasmic permeases
analysis of ATP-binding pocket, **65:**28–29
energy coupling, **65:**25–28
hydrophilic components, **65:**14–16
hydrophilic membrane components, **65:**6
hydrophobic components, **65:**11–14
membrane-bound complex, **65:**19–24
membrane-bound components, **65:**17–18
membrane-bound proteins, **65:**16–17
shock-sensitive, **65:**3
structure, **65:**2–7
substrate-binding proteins, **65:**18–19
Periplasmic transport, speculative model, **65:**3–6
Peritrophic membrane, *See* Membranes, peritrophic
Permeability, **34:**475; 483; 513; 526, **59:**72. *See also* Diffusion; Membrane
active, **16:**487
of nuclear membrane, **16:**415; 416
alteration by metal ions, **11:**70–71
role of —SH proteins in, **11:**247
and amino acid metabolism, **17:**400; 404; 406; 408; 416
effect of nucleate complexes, **17:**140
barrier, **34:**472
capillary, effect of hyaluronidase on, **13:**216; 227–228
of brain tissue, **13:**62–63; 271
determination of constants from tracer experiments, **13:**25–29; 43; 44
of frog skin (*See* Frog Skin)
of synovial membranes, **13:**229–230
tracers in study of, **13:**21–65
use of constants, **13:**22–23
cell membrane, and light emission, **25:**156
changes in cacao seed fermentation, **25:**476; 477; 486
constants, **7:**14; 19
and adsorption, **7:**20–21
of capillaries, **7:**197–198
of dividing or secreting cells, **7:**174
and enzyme reactions, **7:**1–34
of kidney membranes, **7:**522
and lipides, **7:**21–22

Permeability (*Continued*)
and nucleic acid turnover, **7:**174
of plasma membranes, **7:**8; 15; 22
of protoplasm, **7:**45
erythritol, **34:**467
exaggerated, **49:**46
and functioning of cytoplasm, **8:**20; 48–51; 53; 56
of intestinal cells, **8:**50–51
glycerol, **34:**467
of membranes, and sulfonamide effect, **27:**430
of nucleus, **27:**263; 265; 268–271
of nuclear envelope, **24:**336
of sperm cells, **24:**62
and oxidative processes, **18:**243–251
urea, **34:**467
water, **34:**467
Permeability and Enzyme Reactions (Brooks), **7:**1
Permeability barrier
to large peptides, **35:**234–235
in use of mitochondria, **28:**164
Permeability barriers
of cells and cell particles, **16:**254; 255; 257; 263; 265; 266; 267; 271; 272; 276; 287; 293; 298
effect on cytochrome activity, **17:**332; 351–352
Permeability changes of cell membranes, relation to acetylcholine, **12:**269–270; 315–317; 332–334
Permeability membrane, and β-lactamases, **28:**265; 267–268
Permeability of cells, as factor in metabolism, **15:**326–327; 341–342; 357; 378; 380
Permeability of egg surface and fertilization, **9:**285–286
Permeability Studies, Applications of Tracers (Ussing), **13:**21
Permease, **34:**525
in canavanine-resistant mutants, **29:**134
for hydrophobic amino acids, **29:**91
proline, **29:**105
galactose, properties of, **30:**103
β-galactoside, function of, **30:**84
and oxidation of hydrocarbons, **27:**477
Permeases, kinetics, **22:**108–112
Permeation. *See also* Diffusion
of glucose, into *Chara ceratophylla,* **7:**18
into yeasts, **7:**13–20
of mercuric chloride, **7:**25
of narcotics, **7:**20–25
of plasma membrane, **7:**5; 10–12
through lipide layer, **7:**12–13
Pernicious anemia, metabolism of tyrosine in, **14:**99
Pernicious anemia and nucleic acids, **9:**495
Peronospora parasitica, plant disease caused by, **20:**391
Peronospora tabacina, plant disease caused by, **20:**397
Peroxidase, **5:**23, **12:**14; 16, **15:**102; 274, **18:**9; 30; 43, **34:**450, **35:**295, **46:**59–60, **52:**179, **55:**34, *See* Verdoperoxidase
action on uric acid, **19:**206
complexes with hydrogen peroxide, **19:**105; 109–114; 118–121; 123–125
hydroxylating system, **19:**143; 174; 199
kinetic constants, **19:**313
oxidase reactions of, **19:**104–128
activity, in lignin formation, **18:**373
activity with adsorbents, **33:**271
binding sites of, **24:**116
horseradish, activity of, **24:**123
location in subcellular fraction, **24:**307; 309
cacao, **25:**466; 480
intermediates in reaction, **25:**33
turnover number, **25:**32
cytochemical methods for, **8:**30
in chromidia, **8:**31
horse radish, **23:**290; 291; 295–297
horse-radish, **52:**179; 210
hydrogen transfer, **26:**116
PN-dependent, **26:**132
inhibition of, **13:**32
mechanism of action, **15:**16–18; 23; 27–29; 32–33; 37
photooxidation of Mn^{2+} stimulated by, **30:**25
in plastids, **20:**214
turnover number, **12:**171–174
Peroxidase, Cytochrome c (Yonetani), **33:**309
Peroxidase, Oxidation of Proteins by (Sizer), **14:**129
Peroxidase *a,* **23:**295
Peroxidase activity and flavor changes in vegetable tissue, **9:**615; 618–624; 628–631; 637–649
Peroxidase haem *a* compound, **23:**289–297; 306–310
Peroxidase-H_2O_2 (II), rate constant, **25:**14

Peroxidase-hydrogen peroxide compound, **7:**268; 294–295; 300
Peroxidase I, **7:**288–289; 291; 292
Peroxidase in oxidation of glutathione, **11:**249
Peroxidase in tea-fermentation, **2:**115; 119; 120
Peroxidase intermediates, **27:**120–124; 129
Peroxidase in tobacco, changes during processing, **10:**414; 415; 421
Peroxidase kinetics, **73:**3–23
  background, **73:**3–4
  discussion, **73:**20–21
  interpretation, **73:**15–20
  method, **73:**5–6
  preparation and standardization, **73:**4–5
  procedure, **73:**6–9
  results, **73:**9–15
    breakdown rate, **73:**12–15
    equilibrium, **73:**9–11
    formation rate, **73:**11–12
Peroxidase-oxidase reaction, substrate free radicals, **27:**123; 125
Peroxidase oxidation, effect on chymotrypsin, **20:**286
  in identification of imidazole, **20:**77; 78
Peroxidase protein, reactivation by hemins, **7:**287–288
Peroxidases, **7:**310; 312; 316; 317; 322–323, **47:**37. *See also* Hemoproteins, Iron porphins, Porphins, and Tetrapyrroles
  activity toward cytochromes, **17:**68; 375
    kinetic studies, **17:**4–6; 14; 55–58
    valence states, **17:**110–111
  adsorption studies, **14:**346
    classification, **14:**227–228; 236; 254
    in energy transport, **14:**39–40
    evidence of multiple nature, **14:**390
    oxidation of proteins by, **14:**129–161
    studies of catalysis by, **14:**174–176; 182–183; 190
  column chromatography, **22:**432; 470; 471; 480; 482
  effect of freezing on, **9:**625; 626
    histochemical methods for, **9:**588
    reagents for, **9:**638–649
    varieties, **9:**638–643
  and heme-linked groups, **7:**266–269; 274; 286–295
  horse-radish, composition, **7:**286–288
    spectrophotometric investigations, **7:** 288–291
    mode of action, **7:**294–295
    nature of heme-linked groups, **7:**293–294
    oxidation of phenols, **16:**127; 152
Peroxidases, Cotton effect, **27:**47; 59; 90–92
  EPR spectra, **27:**119–130; 138
  and oxidation of hydrocarbons, **27:**476–479; 491
Peroxidases in tobacco, **6:**375
Peroxidase-substrate compounds, **12:**153–190; 195
Peroxidatic oxidations, mechanisms, **19:** 104–117
Peroxidation, of lipids, **41:**43
Peroxide, cytochrome *c* oxidase, **71:**111
Peroxide compounds, **33:**325–327
  comparison of (table), **33:**330
  complex ES of, **33:**325
  Compound II of, **33:**326–327
  Compound I of, **33:**327–328
  Fe (IV), **33:**329
  Fe (IV) derivative, **33:**326–327
  Fe (V), **33:**329
  half-time of decomposition of, **33:**325
Peroxides. *See also* Hydrogen peroxide
  in disulfide exchanges, **27:**205
    from mutagenic radiations, **27:**328–329
    mutagens, **27:**331
  effect on homogentisate oxidase, **19:**126
  from enzymic oxidations, **15:**10; 12
  and luminescence, **25:**124–131; 143–144; 147–149
Peroxides and antioxidants, **8:**439–441
Peroxidic products, from autoxidation of fatty acids, **8:**426; 428–439; 449–451
  from autoxidation of unsaturated compounds, **8:**426–439; 449–451
Peroxidized β-lipoprotein, **44:**204
Peroxisomal oxidases. *See also* D-Amino acid oxidase; D-Aspartate oxidase; L-α-Hydroxy acid oxidase
  background of
    D-amino acid oxidase, **57:**89–94
    D-aspartate oxidase, **57:**94–96
    L-α-hydroxy acid oxidase, **57:**96–98
    mammalian peroxisomes, **57:**87–89
    microperoxisomes, **57:**87–89
  inhibitors
    D-amino acid oxidase, **57:**141–145
    diuretics, **57:**137–138
    L-hydroxy acid oxidase, **57:**139–141

Peroxisomal oxidases (*Continued*)
introduction to, **57:**86–87
and nucleophile glyoxylate adducts
D-amino acid oxidase, **57:**102–113
D-aspartate oxidase, **57:**113–122
L-α-hydroxy acid oxidase, **57:**122–128
suggestions for, **57:**98–102
products
and amino acid oxidase, **57:**135–137
hydrolysis of, **57:**129–133
and nucleophiles, **57:**133–135
reactions of
chemical considerations, **57:**129–132
inhibitors, **57:**137–145
Peroxisomal Oxidases and Suggestions for the Mechanism of Action of Insulin and Other Hormones (Hamilton), **57:**85
Peroxisomes, **57:**87–89. *See also* Peroxisomal oxidases
Peroxy acid, decarboxylation, **60:**74
Peroxydase, in the living cell, **1:**347
Peroxygenase, **60:**73
Peroxygenation, **60:**46
4*a*-Peroxy-H_4-biopterin, **65:**326–327; 329
Peroxy intermediate, formed by prolyl hydroxylase, **41:**272
Peroxyphenylacetic acid, decarboxylation, **60:**74
Peroxypterin-iron complex, as intermediate in pterin hydroxylases, **74:**253
Perseitol, **15:**161
Perseulose, **15:**161
Persistence, **37:**610
Persistence equation, for optical antipodes, **20:**11–16
relation to life span, **20:**21–24
Persulfides, **39:**328; 329; 342; 357; 358; 360; 364
sulfur, **39:**358
Perturbants, **42:**104; 107
solvents, **42:**106
Perturbation functions, study of fast reactions, **25:**2–9
Perturbators, action on catalase, **23:**45–46
and denaturation, **23:**44–46
Pertussis toxin, **58:**131
adenyl cyclase inhibition, **58:**133
ADP-ribosyltransferase activity, **58:**133
biological activity, **58:**132
mechanism of action, **58:**133
purification, **58:**132
structure, **58:**134
Pestalotia foedans, enzymic transformations of steroids by, **20:**245
Pestalotia funera, enzymic transformations of steroids by, **20:**248; 249
Petals, carotenogenesis in, **21:**308; 346; 347
carotenoids in, **21:**306
Petasites vulgaris, pectin in, **20:**345
PETP, *See* Phosphorylating electron transport particle
Petroleum. *See also* Hydrocarbons
action of microorganisms on, **10:**444–445; 447; 449
geomicrobiological prospecting for, **10:**449; 455
microbial modification, **10:**470–474
products, action of microorganisms on, **10:**473–474
in volcanic formations, **27:**353–354
Petroleum Hydrocarbons, Mechanism of Microbial Oxidations of (van der Linden and Thijsse), **27:**469
Petroleum hydrocarbons, microbial oxidations, **27:**469–546
Petroselinum hortense, **55:**17–18
chalcone isomerase of, **55:**27; 29
chalcone synthase of, **55:**19; 20
Peziza species, enzymic transformations of steroids by, **20:**243
PFK
affinity of, for fructose-6-p, **54:**153
AMP as effector of, **54:**179
in bumblebee flight muscle, **54:**162–163
in cancer, **54:**166
in diabetes, **54:**164; 165
fructose-2, 6–6bisphosphate effect on, **54:**153; 154; 180
glucagon effects on, **54:**150; 151
phosphorylation of, **54:**151
with starvation, **54:**158
vs. fructose-bisphosphatase activity, **54:**160
in yeast, **54:**174
PFK 1
activation factor of, **59:**323
and animal tissues, **59:**324–327
and bacteria, **59:**328
covalent modification of, **59:**320; 323
and fungi, **59:**327–328
low molecular weight stimulator of, **59:** 320–323
and plants, **59:**328

and protists, **59:**328
and stabilizing effect of Fru-2,6-P_2, on liver, **59:**328–329
PFK 2, **54:**155
and animal tissues, **59:**353–354
and fungi, **59:**354–356
and plants, **59:**356
and protists, **59:**357
and rat liver bifunctional enzyme
discovery of, **59:**341
evidence for, **59:**341–344
and histidyl-P enzyme, **59:**352–353
kinetic properties of, **59:**346–348
other reactions catalyzed by, **59:** 350–352
phosphorylation by cAMP-dependent protein kinase, **59:**344–346
PG, *See* Polygalacturonases
PGA, *See* 3-Phosphoglyceric acid
$PGAI^2$, *See* Phosphoglyceraldehyde
PGal transuridylase, *See* Galactose-1-phosphate uridyl transferase
PGal uridyl transferase, *See* Galactose-1-phosphate uridyl transferase
3-PGA-P^{4-}, *See* 3-Phosphoglyceryl phosphate
$PGF_{2\alpha}$, cholesterol ester hydrolases regulation, **69:**71
PGI_2, cholesterol ester hydrolases regulation, **69:**71
PGI_2 synthase genes, cholesterol ester hydrolases regulation, **69:**73
pH, **34:**452–461; 482–485; 492–495; 508; 523–526. *See also* Buffer effects, Electrostatic effects and Ionic strength
acid, **34:**461
activity profile, **34:**457–463; 511
adenosine receptors, **69:**92
and aggregate size of tobacco mosaic virus, **15:**450–451; 454; 469
alkalide, **34:**461
alkaline, phenylalanine hydroxylase activation, **67:**126–130
conformational change, **67:**129
fluorescence properties, **67:**127
preincubation, **67:**128
rate-limiting steps, **67:**127
substoichiometric concentration of phenylalanine, **67:**128–129
bulk (pH_b), **33:**251–252
curve, **34:**493
differential effect of on Systems A and L, **49:**88
effect of in enzyme reactions, **33:**250–251; 280
effect on carotenogenesis, **21:**318
effect on model catalyses by metal ions, **21:**172
effect on chymopapain, **19:**315
effect on papain and mercuripapain, **19:**274–281; 288; 302; 303; 305–310; 313–315; 318–324; 326
effect on transamination reactions, **19:**354
effect on degradation of pectic substances, **20:**349
effect on esterase action, **20:**291–302; 305; 308–313
effect on kinetics of liver alcohol dehydrogenase system, **20:**41; 42; 46–48
effect on pectinesterases, **20:**360
effect on polygalacturonases, **20:**365
effect on structure and viscosity of proteins, **22:**426–428
effects on aspartate transcarbamylase, **68:** 76–77
of glucose fermentation, **33:**288
local, **34:**461
mutagenic effect, **27:**305; 319–320; 335; 338
optimum, **42:**270
ph_b, **33:**251–252
relation to enzyme kinetics, **17:**26–34; 45; 46; 48–55
surface (pH_s), **33:**251–252
variation of kinetic parameters with, **29:**10
*p*H
in acetyl phosphate decomposition, **6:**235
and amino acid decarboxylases, **6:**3; 10–11; 15
in histamine production, **6:**2
at interfaces, **11:**37–47
changes in, during photosynthesis, **11:** 101–102; 108; 194
effect on fluorescence of chlorophyll, **11:**133–134
pH-activity curve, of model enzyme system, **15:**23; 28; 31
of rennin, effect of sodium chloride on, **15:**431
of thiaminases, **15:**396–397
pH-activity measurements, in identification of imidazole, **20:**63; 65; 70; 75; 77; 79; 90; 94; 95; 100; 101
in study of esterase action, **20:**292–294

Phaeohaemin *a,* potential of, **23:**312–313
Phaeohaemin *b,* potential of, **23:**312–313
Phaeophyceae, carotenoids in, **21:**298; 301–303
Phaeophyta, **46:**79
Phage, *See* Bacteriophage; Escherichia coli; λ phage
Phage CP51, **44:**176
λ phage, **38:**311
  λ *dg,* properties of, **30:**105
  and the maltose system, **30:**120–121
  properties of repressor of, **30:**101
Phage cross, **27:**300
Phage display libraries, cell-adhesive proteins, **70:**4
Phage DNA replication, **27:**299
Phage epsilon[15], conversion of O antigen by, **31:**111–112
Phage epsilon[34], conversion of O antigen by, **31:**111–112
Phage P22, effect of, on lipopolysaccharide synthesis, **31:**111
  transduction of antigens 4 and 9 in, **31:**102
Phage PBS2, **44:**181
Phage phi29, **44:**173
Phage phie, **44:**172
Phage SP01, **44:**165; 168; 169
Phage-specific enzymes, formation of, **27:**299
Phage T4, **44:**171
Phage varphi14, effect of, lipopolysaccharide synthesis, **31:**111
Phage varphi27, repressor action of, **31:**112
Phagocytes, **65:**49–50
Phagocytosis, **41:**39; 99; 112–113; 155, **58:**100, **65:**50
  of isolated γ-globulin, **41:**155
  Pseudomonas aeruginosa, **70:**223
Phagolysosomes, **46:**125; 146
Phalloidin
  amino acids in, **14:**266
  in plants, **10:**285–295–10:296
Pharmacological activity of drugs, **1:**193
Pharmacology
  Pseudomonas aeruginosa, **70:**223; 245
  tyrosine hydroxylase, **70:**128; 151; 166; 169
Phase behavior, cholesterol ester hydrolases, **69:**58
Phase boundary potentials, **9:**116–120
Phaseollidin
  prenyl group of, **55:**41
  structure of, **55:**8
Phaseollin
  hydroxylation of, **55:**48
  induction of accumulation of, **55:**106
  in isoflavin conversion, **55:**48
  metabolism of, **55:**43; 44
  monilicolon A in production of, **55:**99
  prenyl groups of, **55:**41
  structure of, **55:**8
Phaseollinisoflavan, **55:**7
Phaseolus aureus, **48:**9, **55:**27
  oxidative phosphorylation in, **20:**224
Phaseolus aureus, inhibition in, **29:**95
  proline sRNA synthetase, **29:**101
Phaseolus vulgaris, **55:**4
  chalcone synthase of, **55:**19; 20
  fungal metabolism of phytoalexins from, **55:**46
  isoflavonoid phytoalexin formation in, **55:**34; 36; 37; 38
  phaseollin of, **55:**43
  phenylalanine ammonia lyase induction in, **55:**15
Phases. *See also* Surfaces
  experimental tests for, **11:**42–47
  reactions in two surfaces, **11:**85–86
Phase transfer, **46:**435
pH dependence, of β-amylase, **15:**283
  of hydrolytic enzymes, **15:**281–285; 289; 297
pH dependence of binding of competitive inhibitor, **45:**326
pH dependence of enzyme action, **22:**85
  in identification of amino acids at active site, **22:**57–59
Phe-Arg, **57:**27
pH effects hydronium and hydroxide ions as surrogates, **53:**333
Phenanthrene, microbial oxidation, **27:**525–528
Phenanthridine, in allosteric regulation, **28:**140
Phenanthridine structure and pyridine ring, **8:**234; 236
Phenanthroline
  alcohol dehydrogenase and, **56:**320
  carbonic anhydrase and, **56:**389
  carboxypeptidase and, **56:**359
  effect on oxidase, **19:**91; 97; 126; 194
    effect on photosynthesis, **19:**50; 63
  polymerases and, **56:**338–339
o-Phenanthroline, **35:**282, **41:**183, **67:**176–177
  effect on succinic dehydrogenase, **18:**95–98

Hill reaction inhibited by, **30:**23
inhibition of aldolase by, **31:**132
as inhibitor of metalloflavoproteins, **17:**242; 258; 282
1,10-Phenanthroline, **54:**133
o-Phenanthroline complex with alcohol dehydrogenase, **26:**164–165
1,10-Phenanthroline-LADH, effect of inhibitors, **27:**73–75
1,10-Phenanthroline (OP), and LADH complex formation, **27:**49–50; 53–56; 58; 66; 70
Phenazine ethosulfate, as electron carrier, **18:**84–87
Phenazine methosulfate, **18:**68–70; 72; 75; 76; 80; 82–87; 99–100, **34:**477
Phenazinemethosulfate, **39:**333
Phenazine methosulfate, **49:**70; 71
and activity of ring A dehydrogenases, **20:**270
as electron acceptor, **17:**253; 254; 256; 285; 366, **28:**168–173; 176; 181–184; 194; 197; 210; 225; 227
Phenazine methosulfate as hydrogen carrier, **15:**85; 98
Phenazine methosulfate (PMS), as artificial electron acceptor, **30:**23
as artificial electron donor, **30:**23
Phenazine reductase, in respiratory system, **28:**168–170; 172; 186; 197; 209; 226
Phenazines, as inhibitory analogue, **6:**141; 142
Phenazines reduction, **15:**4; 42; 89
Phencyclidines, **55:**182–183
Phenethylamine, **54:**30
Pheneylephrine in fructose-2, 6–6bisphosphate regulation, **54:**154
Phenindamine, substitute for platelet cofactor I, **16:**54
Phenobarbital, **46:**117; 119; 128; 133; 134; 136; 149; 153; 156; 390, **60:**39
Phenol
effect on carotenogenesis, **21:**319; 320
microbial oxidation of hydrocarbons, **27:** 506–507
in tyrosinase action, **14:**145; 146
peroxidase action on, **14:**151–154
Phenolase, cytochemical methods for, **8:**30
Phenolase complex, **16:**105–184, **19:**131–134; 199
Phenolase Complex, Comparative Biochemistry (Mason), **16:**105
Phenolases, adsorption studies, **14:**346; 350
Phenolates, **17:**153
Phenol dehydrogenase, in oxidation of coniferyl alcohol, **16:**148
Phenol hydroxylase, **27:**508
Phenol *o*-hydroxylase, phenolase complex as, **16:**136; 145–146; 150
Phenolic acids in flue-cured tobacco, **6:**364
Phenolic components of cacao bean, **25:** 460–465; 476–480; 482–484; 487–489
Phenolic compounds. *See also* Hydroxyphenyl compounds; Tyrosine; and other specific compounds
oxidation by peroxidase and tyrosinase, **14:**132–133
Phenolic residues of proteins, role in sclerotization, **16:**162–163
Phenol o-hydroxylase, *See* Cresolase
Phenoloxidase, **45:**418
Phenol oxidase
action on coniferyl alcohol, **18:**371; 372
in apples, **18:**45–46
in leaves, **18:**43–44
in plants, **18:**29–30; 36–40; 47; 310–311
isolation, **14:**405
Phenoloxidase in tobacco, changes during processing, **10:**414
Phenol oxidase of mushroom, **16:**148
Phenol oxidases, *See* Laccase; Monohydric phenols; Polyhydric phenols; Tyrosinase
in tobacco, **6:**373–374
Phenolphthalein, **60:**238
Phenolphthalein glucuronide, as substrate for β-glucuronidase, **16:**365–372; 374; 375; 379–382; 385–386; 400
Phenolphthalein β-monoglucuronide, as assay substrate for β-glucuronidase, **16:**365
Phenolphthalein phosphate, **17:**167
Phenol production, **6:**2
Phenols, **44:**206. *See also* Diphenols, Phenolase complex, Polyphenols, and Quinones; Monohydric phenols; Polyhydric phenols; Polyphenols
in air-cured tobacco, **6:**334–337
in fire-cured tobacco, **6:**365
changes during fermentation of tobacco, **10:**340; 373–375; 417
effect on microbiological formation of hydrocortisone, **20:**246
in lignin study, **18:**370–374

Phenols (*Continued*)
o-hydroxylation, **16:**106
hydroxylation patterns in naturally occurring compounds, **16:**141–146
use in plant oxidase studies, **16:**132
reaction with peroxidase, **15:**17
Phenolsulfatase in plants, **10:**311
Phenol sulfatases, classification, **14:**245
Phenol sulfokinase, **22:**211; 212
assay, **22:**211
and assay for PAP, **22:**211
occurrence, **22:**211
reaction equilibrium, **22:**211
tyrosine, tyrosine derivatives as substrates in, **22:**212
Phenolytic enzymes, **27:**506
Phenothiazine derivatives, **44:**209
Phenothiazines, **55:**179
atropine-like action of, **55:**181
extrapyramidal side effects of, **55:**181
Phenoxymethylpenicillin, action of β-lactamase, **28:**239; 258; 288–290; 292
Phenylacetaldehyde, peroxidase action on, **19:**125
Phenylacetates, **17:**153
Phenylacetic acid, **50:**149
from ethylbenzene, **27:**517–518
oxidation of hydrocarbons, **27:**471; 473; 496; 498; 516–519
in oxidation of phenylalanine, **14:**97–101
Phenylacetoxy radical, **60:**74
Phenylalaninase in oxidation of phenylalanine, **14:**97–100; 101; 106; 108
Phenylalanine, **15:**119; 123, **42:**105; 143, **49:**92, **50:**148; 149; 164; 165; 170, **51:**217, **58:**37. *See also* Labeled phenylalanine
action of permanganate on, **14:**132
action on thiaminase, **15:**400
activity, **50:**165
analogs of, **29:**110; 115; 122
chemically synthesized analogs of, **29:**109
kinetics of incorporation into protein, **29:**112
sRNA synthetase, effectiveness of analogs on, **29:**111
substrate specificities of, **29:**113
substrate specifications of sRNA synthetase of, **29:**113
bands, **42:**108; 110
biogenesis of thiazole moiety, **53:**367
biosynthesis, **16:**287–295
allosteric regulation, **28:**62–63; 71
metabolism in plants, **16:**144–145
relation to tyrosine metabolism, **16:**301
in transamination reactions, **16:**190–192; 198; 201; 203; 207; 208; 228; 230–231; 239
and vitamin B_{12}, **26:**270
biosynthesis in plants, **20:**328; 329
catabolism of, **74:**235; 236
conversion to tyrosine, **35:**245
hydroxylase, a mixed-function oxidase, **35:**247
effect of vitamin B_6 on metabolism, **19:**351; 352
transmination reaction of, **19:**368
effect on indole production by *Escherichia coli,* **10:**77–81
in energy transport studies, **14:**34–35
entrance to citric acid cycle, **15:**184
genetic studies, **17:**412–422; 449
homeostasis, **67:**160–161
interconversion effects, **67:**141–145
labeled, in nonenzymic reactions, **19:**260
peroxidase action on, **19:**143
transamination reactions of, **19:**350; 365
metabolism, **67:**86
metabolism of, **14:**73–128
microbial degradation, **27:**518
origin of, **18:**362; 365–366
relation to lignin, **18:**374
oxidation, **21:**432
phenylalanine hydroxylase, activation, substoichiometric concentrations, **67:**128–129
racemization of, mechanism for, **33:**369
^{14}C-D-Phenylalanine, **33:**368
D-Phenylalanine, **33:**339; 367; 368; 369, **67:**168
L-Phenylalanine, **50:**166, **55:**9
incorporation of, into *Eucalyptus,* 223, into lignins, of bamboo, **34:**223
of coniferin, **34:**223
of maple plants, **34:**223
of spruce, **34:**223
by tissue cultures, **34:**223
of wheat, **34:**223
of white pine tissue culture, **34:**223
K_A values for, **29:**45
L-Phenylalanine, metabolism, **19:**87–92; 181–187

effect of vitamin B_6 on, **19:**351; 352
transmination reactions of, **19:**367–369
Phenylalanine, Metabolism (Lerner), **14:**73
DL-Phenylalanine, resolution by papain, **19:** 309
L-phenylalanine, tyrosine hydroxylase, **70:**117
Phenylalanine ammonia lyase, **55:**14
abiotic elicitors of, **55:**14
active sites of, **55:**12
biotic elicitors of, **55:**15
in control of phytoalexin synthesis, **55:**109
dose-response curves for induction of, **55:**15
in flavonoid biosynthesis, **55:**15
forms of, **55:**14
induction of, **55:**14; 15; 106–107
inhibition of, **55:**12
isoenzymes of, **55:**14
in kievitone accumulation, **55:**15
kinetics of, **55:**12
mechanism of action of, **55:**12; 14
molecular weight of, **55:**12
in phaseollin accumulation, **55:**106
properties of, **55:**13
regulation of, **55:**109–113
Phenylalanine auxotrophs, of *E. coli,* **16:** 291–292; 295
Phenylalanine (C^{14}), in coenzyme Q biosynthesis, **25:**298–299
L-Phenylalanine decarboxylases, classification, **14:**256
Phenylalanine hydroxylase, **17:**415, **19:** 181–183; 199, **55:**249, **67:**77–243
activation
binding to regulatory and catalytic sites, **67:**120–122
conformational change, **67:**116
by exposure to alkaline pH, **67:**126–130
fluorescence, **67:**117–119
lag period, **67:**110–111
by lysolecithin, **67:**103–106
model where regulatory binding sites become functional on tetramer formation, **67:**124–125
$6MPH_4$ as coenzyme, **67:**114–115
by N-ethylmaleimide, **67:**108–110
by partial proteolysis, **67:**106–108
by phenylalanine, **67:**110–126
preincubation, **67:**111–112
pterin coenzyme, **67:**113
quenching, **67:**118–119
second activating site, **67:**113
sigmoid saturation kinetics, **67:**116
stoichiometry, **67:**121
sulfhydryl groups, **67:**119
L-tryptophan effect, **67:**122–124
active site, **74:**248–249
interactions of catecholamines with, **74:**246–247; 247
iron, **74:**239–241
ligands, **74:**243–244
amino acid hydroxylation, mechanism of, **74:**254–257; 256; 257
assay, **67:**95–97
assays, **70:**110
bacterial
as nonheme iron enzyme, **74:**241
reaction catalyzed by, **74:**270
structure of, **74:**241
catalytic domain of, **74:**248–249
structure of, **74:**242–243; 243; 246–247; 247
catechol complexes formed with, **74:**239; 246–247; 247
from *Chromobacterium violaceum,* **74:**235; 240–241; 253
kinetic mechanism of, **74:**259
cupric, **74:**240–241
enzyme-bound iron roles, **67:**211–216
carbinolamine intermediate accumulation, **67:**213–214
enzymatic and nonenzymatic oxidations, **67:**215
flavin monooxygenases, **67:**211
iron-oxo species, **67:**211–212
pterin cofactor, **67:**214–215
quinoid dihydropteridine, **67:**212–213
high and low activity forms, **67:**109
histidine ligands, **74:**239–240
historical discovery of system components, **67:**79–95
4a-carbinolamine, **67:**92
4a-hydroxy-6-methyl-5-deazatetrahydropterin, **67:**91–93
catalase, **67:**94–95
cofactor, **67:**80–81
dehydratase, **67:**91
dihydrofolate reductase, **67:**83–84; 92–93
dihydropteridine reductase, **67:**87–88
dithiothreitol, **67:**94–95
dopamine β-hydroxylase, **67:**85

Phenylalanine hydroxylase (*Continued*)
glutathione peroxidase, **67:**95
hydroxylation reaction, **67:**80
NADH activity, **67:**89–90
phenylalanine hydroxylase stimulating protein, **67:**90–91
sepia, **67:**82
sepiapterin, **67:**83
tetrahydrobiopterin, **67:**84
tetrahydrofolic acid, **67:**81
tyrosine hydroxylase, **67:**85
human
active site of, catecholamines in, **74:**246–247; 247
catalytic domain of, **74:**242; 243; 248
regulatory domain of, **74:**248
structure of, **74:**241; 248
inhibitors, **67:**216–219
H_2O_2, **67:**179; 182
iron ligands, **74:**243; 244
kinetic mechanism of, **74:**258–259; 259
in liver, **74:**235
long-term regulation, **67:**226–231
cortisol, **67:**227
dexamethasone, **67:**228
fetal development, **67:**230–231
glucocorticoids, **67:**227
phenylalanine hydroxylase synthesis control, **67:**228–230
mixture of phosphorylated and nonphosphorylated species, **67:**132–133
molecular properties, **67:**98–102
amino acid composition, **67:**101–102
iron content, **67:**102
molecular weight, **67:**98–100
Stokes radius, **67:**99
$6MPH_4$-mediated reduction, **67:**176–178
NIH Shift, **67:**204–211
oxidation of tetrahydrobiopterin, in presence of tyrosine, pterin formation during, **74:**252–253
from *Pseudomonas aeruginosa,* **74:**235
structure of, **74:**242
pyrimidines as cofactors, **67:**196–204
carbonyl oxide, **67:**196
conversion of phenylalanine to tyrosine, **67:**200–203
DHPR, **67:**201
diaminodioxypyrimidine (III), **67:**201
divicine, **67:**201
NADH, **67:**199
ring-closure reaction, **67:**203–204
2,5,6-triamino-4-oxo pyrimidine, **67:**196–198
rat
activity, kinetics of, **74:**271
catalytic domain of, **74:**248–249
reaction catalyzed by, **74:**270
regulatory domain of, **74:**248–249
structure of, **74:**242; 248–249; 271
reaction catalyzed by, **74:**235; 236; 238
arene oxide intermediate in, **74:**255; 257
hydroxylating intermediate, identity of, **74:**251–254
NIH shift in, **74:**254; 255; 256
products of, **74:**238
reaction of [2,5-H2]-phenylalanine with, **74:**255–257
reduction, **67:**176–180
regulation of, **74:**270–272
model for, **74:**279–280; 280
regulatory domain of, **74:**248–249; 280–281
interaction with active site, **74:**280–281
regulatory properties, **67:**102–166
kinetic constants, **67:**165–166
lysolecithin, **67:**158–159
$6MPH_4/HB_4$, **67:**162–163
phosphorylation-dephosphorylation (*See* Phosphorylation-dephosphorylation)
rat kidney, **67:**157–162
spontaneous activation, **67:**162–166
DL-*m*-tyrosine, **67:**159
uremia, **67:**160–161
X and M removal, **67:**165
spectral properties of, **74:**239–240
steady-state mechanism, **67:**180–181
structure, **67:**219–226
$6MPH_4$-mediated reduction, **67:**224
NS7, **67:**225
peptide bond cleavage, **67:**221
tyrosine hydroxylase, **67:**221–223
structure of, **74:**241–249
substrates for, **74:**238
substrate specificity
amino acids as substrates and activators, **67:**166–172
D-methionine, **67:**170
L-methionine, **67:**168–171
L-norleucine, **67:**170
D-phenylalanine, **67:**168

phenylalaninol, **67:**169–170
S-methyl-L-cysteine, **67:**171–172
tetrahydropterin, **67:**172–175
DL-β-2-thienylalanine, **67:**172
L-tryptophan, **67:**169
L-tyrosine, **67:**169
substrate specificity of, **74:**249–251
sulfhydryl groups, **67:**106
uncoupled reactions catalyzed by, **67:**181–195
4a-carbinolamine, **67:**187; 191–192; 194–195
4a-hydroperoxide, **67:**187–188; 191; 193–194
BH_4 oxidation, **67:**191; 194–195
H_2O_2 formation, **67:**183–186
inactivation by H_2O_2, **67:**182
o-methyl phenylalanine, **67:**195
7MPH$_4$, **67:**189
P-fluorophenylalanine, **67:**181
substrates required, **67:**183–184
tetrahydropterin oxidation, **67:**184
L-tryptophan, **67:**181–182
utilization of 2,5,6-triamino-4-pyrimidinones by, **74:**254
Phenylalanine-4-hydroxylase, **37:**494
Phenylalanine hydroxylase phosphatase, **67:**133–135
Phenylalanine hydroxylase stimulating protein, **67:**90–91
Phenylalanine Hydroxylating Systems from Mammalian Liver (Kaufman), **35:**245
Phenylalanine hydroxylation
tyrosine hydroxylase, **70:**117; 118; 119
unconjugated pterins, **70:**103
L-Phenylalanine oxidase, classification, **14:**253
Phenylalanine pool, **34:**221
Phenyl alanine racemase, ATP-dependent, activation-racemization reaction, **44:**110
in synthesis of Gramicidin, S, **44:**108
thioester linkages, **44:**108
L-Phenylalanine r O_2 transhydrogenases, **14:**253
Phenylalanine-tRNA synthetase, **50:**135; 164; 165; 166
Phenylalanine-t RNA synthetase, kinetic mechanism of, **45:**295–296
Phenylalaninol, **67:**169–170
Phenylalanyl-tRNA synthetase, **43:**251; 333; 344
Phenylalkanes, microbial oxidation, **27:** 473–474; 495–498
Phenylalkylamines, **44:**211
Phenylasazone of ribose, **35:**60
Phenylbutazone, **46:**141
trans-4-Phenyl-3-buten-2-one, **46:**394; 396
Phenylbutyric acid, ad inhibitory analogue, **6:**136
Phenylcoumaran, **34:**211
p-Phenylenediamine, **15:**6
p-Phenylenediamine
in cytochrome study, **17:**356; 368; 369
inhibition of diphosphopyridine nucleotide, **6:**42
substrate free radical from, **27:**126–128
p-Phenylenediamine, effect on indolylacetic acid oxidase, **19:**127
p-Phenylene diamine oxidase activity, **44:**197
Phenylenedimaleimide, **58:**147
Phenylephrine
effect on fructose 6P cycle, **52:**286
role of Ca^{2+}, **52:**316
effect on gluconeogenesis, **52:**314
effect on glycolytic flux, **52:**306
effect on pyruvate kinase, **52:**304
β-Phenylethyl mustard oil, occurrence, **10:**265
O-Phenyl-α-D-glucopyranoside, **17:**210
Phenyl α-glucopyranoside, **48:**50
α-Phenylglucoside in polysaccharide synthesis, **15:**306
Phenylglucuronide, as substrate for β-glucuronidase, **16:**366–369; 372–374; 400
α-Phenylglyceric acid, **50:**356
α-Phenylglycidate, **50:**355
α-Phenylglycine, **17:**417
Phenylglycine, accumulation, **16:**295
in transamination reactions, **16:**190; 198
Phenylglyoxal, γ-glutamyl transpeptidase substrate binding, **72:**251–254
Phenylglyoxylic acid, decarboxylation by model enzyme, **15:**29
Phenyl groups, binding energy, **43:**347
Phenylhydrazides, of amino acids, synthesis with papain, **19:**309
Phenylhydrazine, **56:**268, **58:**66
Phenylhydrazine action on thiaminase, **15:**400
Phenyl-α-L-iduronide, **60:**243
2-Phenylimidazole, reaction with benzoyl chloride, **20:**83
Phenyliodine dichloride, **58:**52
Phenyl isocyanate, reaction with thiols, **11:**227

Phenylisothiocyanate (PITC), in identification of imidazole, **20:**95
γ-Phenyl-α-ketobutyric acid, enzymic amination, **16:**186
Phenylketonuria, **17:**415, **35:**246; 315, **67:**79; 151, **74:**235
  gene-determination of, **13:**124
Phenylketonuric infant, **49:**92
Phenyllactic acid, **17:**415
Phenyllactic acid in oxidation of phenylalanine, **14:**97–101; 108
Phenylmethane skeleton in lignin study, **18:**374
Phenylmethanesulfonlyfluoride (PMSF), **33:**386
Phenylmethanesulfonyl, **59:**123
Phenylmethane sulfonyl fluoride, **41:**208
Phenyl methanesulfonyl fluoride, γ-glutamyl transpeptidase catalytic sites, **72:** 255–262
Phenyl methylsulfonyl fluoride (PMSF), **48:**290
Phenylpantothenone, as inhibitory analogue, **6:**138–139
Phenyl penicillins, and active site of β-lactamase, **28:**260
Phenyl phosphate
  as phosphate donor, **18:**197–198
  phosphoryl donor, **17:**167; 170
Phenylphosphoric acid, in synthesis of activated amino acids, **19:**249
Phenylpropane skeletons, in lignin, **18:**358; 370–374
Phenylpropanoid class of phenolic compounds, in plant metabolism, **16:**141; 147–150
Phenyl propanoid metabolism, **56:**204
Phenylpropanoid pathway, **55:**9–19
Phenylpropionates, **17:**153
Phenylpyrimidines, folate inhibitors, **27:** 444–445
Phenylpyruvic acid, **50:**148, **58:**35
  absorption spectrum, **8:**130; 132; 137
    from dehydropeptides, **8:**134; 135; 137
    effect on deamidation of glutamine, **8:** 160–164
  and aromatic amino acids, **18:**362
  genetic studies, **17:**415–416
  metabolism, **16:**198; 230–231; 288
  peroxidase action on, **19:**125
    transamination reactions of, **19:**349; 350; 365
  transamination with, in plants, **20:**328; 329
Phenylpyruvic acid in oxidation of phenylalanine, **14:**97–101; 105–106; 108; 111
Phenylpyruvic acid oxidase, **19:**125
Phenylpyruvic oligophrenia, *See* Phenyluria
Phenylpyruvoylglycine, absorption spectra and *p*H, **8:**130
  preparation, **8:**129
Phenylserine, **53:**169
  in transaminations, **16:**193; 198
β-phenyl serine, **35:**97
6-Phenyltetrahydropterin, **67:**173
Phenylthiocarbamide, inhibition of melanin formation by, **14:**93; 94
Phenylthioisocyanate
  in protein structure studies, **14:**281–283; 284; 287
  in study of virus protein, **15:**464
Phenylthiourea, inhibitor of melanin formation, **14:**93; 94
Phenylthreonine, **53:**169
Phenyltriazine inhibitor binding, eukaryotic dihydrofolate reductase, **70:**35–37
Phenyl trimethyl ammonium ion, coupled with hapten, **15:**67
Phenylurea, Hill reaction inhibited by, **30:**23
Phenylurethane, effect on DPNH oxidase and succinic oxidase, **20:**178
Phenyluria, occurrence of, **14:**98–100; 108; 111; 122–123
Phenytoin, **60:**39
Pheophytin *a,* **46:**81
Pheophytins, **46:**81–84; 100; 102
Pherases, use of term, **14:**226
Pheynlthiourea, effect on indolylacetic acid oxidase, **19:**124
pH_g, **33:**290
Philodendron, antibiotic treatment of diseases, **20:**401
Phlorhizin
  inhibition of phosphorylase by, **12:**416
  inhibitory role in glycolysis, **5:**135
Phloridzin, **56:**219
Phlorizin, **17:**210
  and ATP, **7:**124; 125
    and kidney phosphatides, **7:**142
    and phosphatide formation, **7:**137
  effect on ATPase, **6:**166
  effects of, on sugar uniport and glucose/galactose antiport, **29:**52

inhibition of translocation reaction by, **29:**53
enhancement of electron transport inhibited by, **30:**65
and phosphorylase, **24:**273
Phloroglucinol, **18:**373
metabolism in plants, **16:**144–145; 150
Pholas dactylus, luminescence in, **25:**160–161
Pholyphosphates, effect on myosin and actomyosin, **16:**332–333
Phoma βe, plant disease caused by, **20:**394
Phoma species, nitrogen fixation studies in, **13:**347; 350–352
Phoma tuberosa, plant disease caused by, **20:**396
pH optima, cholesterol ester hydrolases regulation, **69:**50–55
Phorbol esters, tyrosine hydroxylase regulation, **70:**150
Phormidium uncinatum, cell walls, **26:**221
Phosferases, classification, **14:**220
Phoshoenolpyruvate carboxykinase (GTP), **53:**221; 222
Phosophorylation, of RNase, **24:**207; 222; 245
oxidative, and electron transport, **24:**337
Phospatatase, **34:**14; 15
Phosphagens, **1:**116–119. *See also* Phosphocreatine
relation to contraction, **16:**341
synthetic, as functional probes in intact animals, **50:**224–230
Phosphatase, **5:**120, **15:**151; 152; 153; 164; 167; 352; 394; 416
acid, **10:**85–86; 300
inactivation, **10:**90
inhibition by DFP, **13:**332
alkaline, inhibition of, **13:**321; 332
localization in cytoplasm, **13:**2; 7; 9–10; 11; 13; 14; 16
in mitochondria and lysosomes, **21:**39–41; 58
alkaline, activation by magnesium ions, kinetics, **21:**148
comparison with model catalysts, **21:**189–191
effect on hydroaromatic phosphoric acid esters, **21:**173; 190
residual activity, **21:**135
prostatic, **21:**210
action of rennin, **15:**432–434
action on glycogen synthetase, **28:**99; 101
effect on phosphorylases, **28:**109
in glucogenesis, **28:**87
in ribose synthesis, **28:**397; 411
action on pyridine nucleotides, **15:**81–82; 83; 106–108
alkaline, **10:**83–122, **15:**106
activation, by amino acids, **10:**101; 106–107
by divalent cations, **10:**98–100; 102–104; 110–111
by Mg^{++}, **10:**89–93
constitution, **10:**107–117
effect of substrates on activity, **10:**88–89
factors affecting activity, **10:**94–107
physiological and pathological chemistry, **10:**83
preparation, **10:**89–93
properties, **10:**89–93
specificity, **10:**86–89; 104–105
structure, **10:**112–117
variations due to origin, **10:**104–105
comparison of acid and alkaline, **10:**85–86
effect of borate on, **12:**517
involvement in conversion of H_2-neopterin triphosphate to folate (table), **35:**67
isodynamics, **10:**84–85
kinetics, **15:**28
in lutoid particle fraction of latex, **29:**235–236
mechanism of catalysis by, **10:**113–117
multiple *p*H optima, **10:**84–85
nonspecific, in nucleus, **27:**274
as phosphoryl-transferring enzyme, **17:** 166–168; 175
dual role, **17:**171–174
phosphate-water exchange of oxygen, **17:**162–164
transfer reactions, **17:**190–191
prostatic, **15:**440
in *Pseudomonas saccharophila,* **10:**125
renal, **7:**522
Phosphatase, Alkaline (Roche and Thoai), **10:**83
Phosphatase, RNA polymerase II transcription, C-terminal domain characteristics, extended structure, **71:**50
Phosphatase-exonuclease (exonuclease III), in *E. coli,* **29:**197
Phosphatase membrane, **34:**505
Phosphatase membrane-collodion enzymes, alkaline, **34:**495–496; 507–509

Phosphatase models, **15:**22; 23
as reagents for analysis and organic synthesis, **21:**192
hydrolysis with, **21:**170–175; 180–191
Phosphatases, **1:**336; 341; 365, **20:**90; 91. *See also* Acid phosphatase and Alkaline phosphatase; Phosphate transfer; Polyphosphatases; Pyrophosphatases; and specific enzymes; specific phosphatases
acid, **16:**367 (*See* Acid phosphatase)
alkaline, **16:**4; 7; 387; 399; 417–418; 463; 466–468 (*See* Alaline phosphatase)
action on CoA, **16:**4; 7
barley, in determination of CoA structure, **16:**14
column chromatography, **22:**432; 470; 471
organic *E. coli,* mutants deficient in, **22:**17; 18; 19
isolation from extracts of *E. coli,* **22:**18
prostatic, column chromatography, **22:**443; 469–471
in β-glucuronidase preparation, **16:**372
intestinal, action on CoA, **16:**3; 6
intracellular distribution, **16:**417–418; 463; 466–468
localization of, **24:**65; 282; 305; 316; 319–320
synthesis of, **24:**20
of snake venoms, **8:**460; 486–487
action on glycerophosphoric acid isomers, **21:**183
activation by magnesium ions, **21:**170
alkaline, **24:**9–10; 282; 318–319; 343
alkaline, adsorption studies of, **14:**323; 324; 328; 329; 332; 333–334; 343; 350; 357; 363; 365
alkaline, inactivation by oxidation, **14:**131
in glucose diphosphate synthesis, **14:**212
isolation, **14:**398; 399; 403
bacterial, disulfide bridges, **26:**102–103
catalytic activity, **24:**466
colorimetric methods, **9:**600
in *E. coli,* **14:**212
effect on, phosphorylases, **24:**282
semen, **24:**86
effect on acylation, **24:**137; 138
genetic control of, **24:**9–10
histochemical methods for, **9:**589–592
hydrolysis of glucose-4-phosphate, **14:**202
intestinal, action on coenzyme A, **14:**53–54; 61–62; 62–65; 70
as kinases, **18:**197
in polymer formation, **18:**288
in test of succinic dehydrogenase, **18:**78–79
lopster muscle, **24:**276
multiple forms of, **25:**331; 348
in photosynthesis, **25:**95
product inhibition of, **25:**178; 222; 226–236; 267
potato, test on phosphorylases, **24:**282
prostate, structure studies, **14:**302
in semen, **9:**342; 352–354, **24:**65
separation by chromatography, **14:**346; 348; 363; 365
soil, methods for, **28:**380–382; 384; 386
results, **28:**387; 388
structure studies, **14:**308
Phosphatases, P_I:H_2O exchange, **50:**377
Phosphatases in tobacco, **6:**369
Phosphatase synthesis, effect of nucleus on, **24:**20
Phosphatase transfer, **18:**197–198
Phosphate, **51:**10. *See also* Phosphorylase and Phosphorylated sugars
acid-labile, from luminescent reaction, **7:**218
inorganic, conversion to phosphatide, **7:**151–152
and egg formation, **7:**157–160
and yolk phosphatide, **7:**157–160
buffer, effect on steroid 11β-hydroxylase, **19:**190
in citric acid cycle, **10:**189–190; 211–212; 214
in creatine kinase reaction
lysyl and arginyl side chain interaction with, **54:**419
transfer of, **54:**417–418; 420–421
effect on hydrogen evolution, **6:**237
effect on pyruvate metabolism, **6:**238; 239
exchange in bacterial metabolism of acetyl phosphate, **6:**247–248
in familial periodic paralysis, **6:**215; 217
in muscular dystrophy, **6:**207
radioactive, **6:**246–247
effect on metabolism of *Torula,* **9:**661; 678–679
energy-rich, **9:**478–479
effect on RuDP carboxylase, **19:**22

effect on succinic dehydrogenase, **18**:87–90
as glucosyl acceptor, **18**:224
high energy, from particle activity, **18**: 13–15
inorganic, and energy preservation, **18**:221
and lipoic acid, **18**:340
exchange, in photosynthesis, **11**:149–150
high-energy, and CO_2 fixation, **12**:53
inorganic, **15**:103; 118; 130; 141; 170; 171; 173, **55**:474
in citric acid cycle, **15**:; 184; 185; 189; 191; 194; 196; 199–211; 217–218; 224–225; 227; 232 (radioactive P); 245 (radioactive); 246; 248; 253 (P^{32}); 258; 262
in exchange reactions, **23**:359–360; 364; 373
inhibition by, **15**:279
NMR spectroscopy of, **55**:322
in normal and cancer cells, **4**:210; 214; 215; 222; 223; 224; 226
relation to urea synthesis, **15**:330; 343–344; 346–347; 350; 352 (P^{32}) 363; 365–370; 372; 379; 380 (P^{32})
intestinal in, DR synthesis study, **28**:425
and LADH complex formation, **27**:62–63
in Mo-catalyzed reactions, **17**:239; 269–270; 276; 323
O^{18} labeled, in phospho-bond study, **17**:161–164; 166; 174–176.
muramic acid, in Grampositive bacteria, **30**:248
in myosin, **16**:332
cyclic, of pantotheinc acid, **16**:9–10; 11; 15–16
inorganic, transformation to organic, **16**:471–473
labeled, in contractile system, **16**:342
uptake by enucleated amoebae, **16**: 471–472
P^{32}, in nucleic acid study, **16**:450; 452–456; 458; 461–462
Phosphate absorbing protein, in contractile system, **16**:331
radioactive, in study of glutamine formation, **16**:257
in nitrate reductase, **22**:392
in phosphocreatinine, reservoir of, **54**:368
Pseudomonas putida, **70**:223
role in polysaccharide degradation and synthesis, **11**:366–367
surface mechanisms for uptake of, **13**:16
and synthesis of 2,3-diphosphoglycerate in partially carbamylated red cells, **40**:17
Phosphate acceptor system, **23**:350–352
Phosphate and fat formation, **8**:307–308; 328; 330
Phosphate binding, **43**:332
Phosphate-binding site, rubisco, **67**:50–51
Phosphate bond, energy-rich. *See also* Acyl phosphates and ATP
in glutamine synthesis, **13**:276–280
in glutathione synthesis, **13**:280–282
in peptide synthesis, **13**:240–241; 311–315
Phosphate bond energy, **1**:99–162
Phosphate Bond Energy, Metabolic Generation and Utilizations (Lipmann), **1**:99
Phosphate bonds
energy rich, use in glycosidic bond formation, **10**:128
generation by *S*-acyl CoA, **15**:201
energy-rich, **15**:129–131; 235
mechanism of formation, **15**:12
relation to urea synthesis, **15**:320; 327; 330–331; 334–337; 340; 343; 347–352; 363–364; 367–370; 372–373; 379; 381–382
Phosphate compounds of egg and fertilization, **9**:307–312
Phosphate concentrations, *Caulobacter* cell cycle, **71**:6
Phosphate cycle, **1**:121–131
Phosphate-dependent glutaminase, **53**:224
in small intestine, **53**:218–221
Phosphate diesters, oxygen chiral
nonenzymatic reactions of, **55**:370–371
synthesis of, **55**:312–319
via hydrolysis of cyclic, phosphoramidates, **55**:316–317
via hydrolysis of phosphorothioates, **55**:317–319
via Wittig-Staudinger chemistry, **55**: 313–316
Phosphate donors, non-nucleotides as, **18**: 197–198
nucleoside diphosphates as, **18**:196–197
nucleoside triphosphates as, **18**:195
Phosphate ester, **34**:524

Phosphate esters, **43:**332, **58:**28
cleavage study with O^{18}, **17:**161–164
effect of fluoro compounds on, **18:**149
in plant respiration, **18:**285–287; 289; 292–295; 312
group potential, **1:**103–105
oxygen chiral, **55:**291–376
configurational analysis of, **55:**321–340
development of experimental methods for, **55:**295; 296
mass spectral methods in, **55:**336–337
in enzyme-catalyzed transfer reactions, **55:**340–369
nonenzymatic application of, **55:**369–371
^{17}O NMR methods for, **55:**337–340
^{31}P NMR methods for, **55:**321–335
spectroscopic applications of, **55:**371–372
synthesis of, **55:**294; 306–321
development of experimental methods for, **55:**295; 296
synthesis, by alkaline phosphatase, **10:** 105–107
Phosphate esters as inhibitors, **13:**321–343
Phosphate group
binding energy of, **43:**348
in coenzyme binding, **22:**387–389
of FMN, **22:**393
Phosphate groups
and casein clotting, **15:**432–434; 437
in nucleoproteins, **18:**176–177
type and position in CoA, **16:**7–17
Phosphate-independent, maleate-stimulated glutaminase, **53:**223
Phosphate migration, on 3-carbon particles, **14:**195; 213–215
on sugar molecules, **14:**193–195; 199; 208–213
Phosphate monoesters, **58:**4
configurational analysis of, **55:**330–335
hydrolysis, **68:**14
hydrolysis of, **55:**382
mass spectroscopy of, **55:**336
NMR spectroscopy of, **55:**333–335
nonenzymatic reactions of, **55:**369–370
synthesis of, **55:**307–312
via cyclic hydrobenzoin triesters, **55:** 310–312
via cyclic phosphoramidates of ephedrine, **55:**308–310
via hydrogenolysis of benzyl phosphate esters, **55:**308
Phosphate (O^{18}), **18:**223
Phosphate (P^{32})
exchanges, **17:**107; 122–123; 176; 178; 185
in DPNH oxidase study, **17:**84
inorganic, exchange with nucleotide diphosphates, **18:**228–229
(inorganic)-ATP
exchange reaction, **23:**333; 348; 356–358; 365; 366; 370; 371; 374; 375; 382; 386
labeling of ATP, **23:**326
Phosphate (P^{32}), ATP exchanges, **26:**311–312; 314–316
in carboxylations, **26:**322–326
Phosphate potential, free phosphate concentration estimated by, **29:**357
influence of, on DPNH/DPN ratio, **29:** 351–354
phosphofructokinase and hexokinase influenced by, **29:**354–357
Phosphates, **46:**291
condensed
chelates, in metal ion activation of enzymes, **21:**147
organic, chelates, in metal ion activation of enzymes, **21:**147; 148
solubilization by, **17:**143–146; 147
insoluble, solubilization of, **17:**138; 142–149; 151; 152; 155
sugar, as solubilizing agents, **17:**146–147
cleavage of, **17:**225
cyclic (*See also* individual cyclic phosphates)
effect on, RNase activity, **24:**236–237
RNase denaturation, **24:**187–190; 249
sperm metabolism, **24:**80–82
radioactive, in enzyme synthesis, **24:**18
RNase activity for, **24:**164; 165; 171–173; 187–188; 235–243; 443–448; 452–455
inhibition, of leaf and algal ADP-glucose pyrophosphorylases, **46:**339
trisubstituted, as transferase substrates, **46:**394
Phosphates in energy transport in living cells, *See* ATP system
Phosphate transfer
enzymatic, **17:**159–188
in plants, **18:**311–313
Phosphate Transfer, Enzymic (Axelrod), **17:**159
Phosphate transfer reactions, **58:**12

Phosphate-transferring enzymes, activity of —SH groups in, **11:**243
Phosphate triesters, carbonic anhydrase catalyzed hydrolysis of, **47:**190
  dimethyl 2,4-dinitrophenyl phosphate, **47:**191; 193; 206–207
Phosphatidase, of pancreas, **23:**130
Phosphatidate phosphatase, localization of, **24:**320
Phosphatide metabolism, and kidney, **7:** 135–136; 142; 167
Phosphatide methylating system, **25:**417–420
Phosphatide molecules, origin, **7:**131
  rate of renewal of constituents, **7:**131–133
Phosphatides. *See also* Cephalin, Lecithin, and Sphingomyelin; Phospholipides; Phospholipids
  and calcium, **8:**7; 24; 50
  in cell nuclei, **7:**168–169
  in chick embryo, **7:**165–166
  combined, **7:**163–165
  in cytoplasm, **7:**168
  deuterium in, **7:**170
  in egg yolk, **7:**157–160; 165–166
  extraction from liver, **7:**134
  formation, and cyanide, **7:**149–150
    and adrenal glands, **7:**145
    and carbon monoxide, **7:**149–150
    and hexoses, **7:**148–149
    and pentoses, **7:**149
    and phlorizin, **7:**137
    and respiratory inhibitors, **7:**151
    and sulfide, **7:**149
    in tissue slices, **7:**147–151
  free, **7:**163–165
  and functioning of cytoplasm, **8:**3; 7; 20–24; 35
  and glycerophosphate, **7:**152
  and ingested fat, **7:**135–137
  and lipotropic substances, **7:**138–141
  and magnesium, **8:**24; 50
  in milk, **7:**166–168
  in plasma membranes, **8:**49–51
  purification, **7:**133–135
  and tumors, **7:**145–146
  turnover studies, **7:**127–173
Phosphatides (P^{32}), and ethionine metabolism, **25:**373
Phosphatide turnover, and diethylstilbestrol, **7:**160–161
  and blood, **7:**146–147
  and carcinosarcoma, **7:**138; 146; 171
  and choline analogs, **7:**172–173
  and muscle, **7:**142
  and neoplastic tissue, **7:**145–146
  and sarcoma, **7:**168–169
Phosphatidic acid, **34:**490, **54:**144
  in lipopolysaccharide synthesis, **31:**90
Phosphatidyl choline, **34:**466
Phosphatidylcholine, **46:**117, **60:**40; 70
Phosphatidyl choline
  cholesterol ester hydrolases regulation, **69:**50
  ethanolamine, **34:**466; 473–476; 484
  glycerol, **34:**466
  inositol, **34:**466; 473
  in lipopolysaccharide synthesis, **31:**90
  serine, **34:**466
Phosphatidyl choline biosynthesis, **25:**417–420
Phosphatidyl ethanolamine
  in Gram-negative bacteria, **31:**89–90
    in lipopolysaccharide synthesis, **31:**90
  synthesis, in mitochondria, **21:**51
Phosphatidylethanolamine biosynthesis, *Caulobacter* cell cycle, flagellar structure and assembly, **71:**9
Phosphatidyl-ethanolamine methyltransferase, location, **24:**309
Phosphatidyl glycerol, in lipopolysaccharide synthesis, **31:**90
Phosphatidyl inositol, **50:**317
  cholesterol ester hydrolases regulation, **69:**50
Phosphatidylinositol breakdown, **55:**145–146
Phosphatidyl serine, **54:**144
Phosphatidylserine, **67:**103
Phosphatidyl serine
  cholesterol ester hydrolases regulation, **69:**50
  reactivation of cytochrome oxidase, by, **23:**310
Phosphatidyl-L-serine, tyrosine hydroxylase, **70:**122; 187
Phosphatidyl serine decarboxylase, in mitochondria, **25:**418
Phosphatolysis, **17:**166; 169
Phosphenolypyruvate carboxykinase, **37:**556; 562
  differential inactivation, **37:**586–587
  persistence, **37:**562
Phospherases, classification, **14:**220
1-Phospho-1, 2-2(*S*)-propanediol synthesis, **55:**308; 309

Phosphoamidase action of proteolytic enzymes, **15:**433–434
Phosphoamidases, classification, **14:**224; 248
Phosphoamide bonds, in ribonucleoprotein, **18:**179–180
Phosphoamide ester, hydrolysis, by metal ions, **21:**182
Phosphoamides, hydrolysis, by metal ions, **21:**176; 194
5-Phospho-D-arabonate, **17:**146
Phosphoarginine, as phosphate donor, **9:** 478–479
  in sperm, **9:**342
Phospho-bond forming and cleavage, nature of reaction, **17:**161–170
Phosphocellulose, **47:**282
Phosphocellulose-purified tubulin, **53:**124
Phosphocholine (C^{14}), methylation study, **25:**417
Phosphocreatine, **15:**143, **17:**164; 167. *See also* Creatinephosphoric acid; Phosphagens
  conformational preferences of, **54:**378; 379–384
  in familial periodic paralysis, **6:**218
    in nutritional muscular dystrophy, **6:**205
  as phosphate donor, **18:**198
  protonation of, via histidine, **54:**395
  relation to muscle function, **16:**341; 348–351
  vs. phosphocreatinine, **54:**384
Phosphocreatine and nerve impulse propagation, **12:**313–314
Phosphocreatine-creatine kinase system, relation to reversible contraction-relaxation cycle, **16:**350
Phosphocreatine in sperm, **9:**342
Phosphocreatine transphosphorylase, **15:**143
Phosphocreatinine
  as reservoir of high-energy phosphate, **54:**368
  vs. phosphocreatine, **54:**384
5-Phosphodeoxyribosylpyrophosphate, in pyrimidine biosynthesis, **21:**285
Phosphodiester
  acyclic oxygen chiral, **55:**312
  hydrolysis of, **55:**349
  NMR spectroscopy of, **55:**322–324
  prochiral, **55:**312
  synthesis of, **55:**314
Phosphodiesterase, **9:**517–518, **56:**292
  activity of acid DNase, **31:**33–36
  from beef pancrease, **29:**182
    from germinating barley, **29:**190
    in hepatopancreas of octopus, **29:**183
  cyclic nucleotide, **55:**349–351
    bovine heart, **55:**372
    snake venom, **55:**351–352
      hydrolysis reactions of, **55:**306
  effect of borate on, **12:**518
  *p*H optimum, **10:**84
  with RNase, **24:**169; 172
2′-Phosphodiesterase, **59:**217–218
Phosphodiesterases, **17:**171–174, **43:**9
  classification, **14:**225; 245
    isolation, **14:**401
    separation by chromatography, **14:**348
    structure studies, **14:**309
  column chromatography, **22:**435; 462–469; 485–486
Phosphodiesterases Involved in DNA Repair (Weiss and Grossman), **60:**1
Phosphoenolpyruvate, **48:**160; 171
  in allosteric regulation, **28:**111–112
    synthesis, allosteric regulation, **28:**87–92
  and ATP, **23:**324–327; 332; 373; 388
    labeled, **23:**326
  carboxylation enzymes, nomenclature, **26:**294
  in fructose-bisphosphatase regulation, **54:**178; 181
    in microbial metabolism, **54:**169–172
  in fructose-bisphosphate regulation in microbial metabolism, **54:**169
  glucagon effect on, **54:**150
  in photosynthesis, **19:**11; 39; 40
  pyruvate kinase phosphorylation effect on, **54:**151
  in synthesis of *p*-AB, **27:**421
Phosphoenol pyruvate carboxykinase, **18:** 301–302
Phosphoenolpyruvate carboxykinase, **52:**298; 299
  allosteric regulation, **28:**88–89
  characteristics, **52:**298
  content, **52:**298
  correlation of hepatoma growth rate with, **29:**331
  degradation, **52:**298
  in diabetes, **54:**164
  effect of hormones, **52:**298
  ferroactivator, **52:**299
  in microbial metabolism, **54:**172
  propionic acid metabolism, **26:**294; 295; 307
  regulation, **52:**299

in starvation and diabetes, **52:**298
starvation effect on, **54:**157; 158
synthesis, **52:**298
Phosphoenolpyruvate carboxykinase (GTP), **69:**203–281
enzyme, **69:**204–227
development of, **69:**218–222
dietary and hormonal regulation of, **69:**222–227
early research on, **69:**204–205
forms of, in mammalian tissues, **69:** 206–208
kinetic properties of, **69:**208–213
metabolic role of, **69:**214–218
tissue distribution of, **69:**213–214
future research directions, **69:**260–265
hepatic gene therapy, **69:**264–265
transcriptional analysis, **69:**260–262
transgenic animals, **69:**263
gene, **69:**227–232
mRNA stability regulation, **69:**230–232
RNA processing, **69:**227–230
structure, **69:**227
gene transcription regulation, **69:**232–260
generally, in rat, **69:**232–233
hormonal control, **69:**233–244
PEPCK-C gene transcription model, **69:**255–260
PEPCK-C promotor, in transgenic mice, **69:**244–255
Phosphoenolpyruvate carboxykinase (PEPCK), hepatic glutamine regulation, **72:**65
Phosphoenolpyruvate carboxylase, propionic acid metabolism, **26:**294
Phosphoenolpyruvate carboxylating enzymes, mechanisms of deduced from relaxivity data, **33:**49
Phosphoenolpyruvate carboxytransphosphorylase, propionic acid metabolism, **26:**294; 309
Phosphoenolpyruvate carobxy-kinase, product inhibition of, **25:**224
Phosphoenol pyruvate (PEP), **18:**301
Phosphoenol pyruvic acid, **1:**111; 119–121; 128
Phosphoenolpyruvic acid
in CO_2 fixation, **15:**172
in photosynthesis, **25:**102; 103
Phosphoenolpyruvic carboxylases, in photosynthesis, **19:**38–40
Phosphoenol transphosphorylase, **15:**348
Phosphoenzyme (E-P), **51:**249
Phosphoenzyme intermediates, **55:**388–391
Phosphoenzymes, **43:**86; 91
Phosphoepoxycreatine, **54:**393
4-Phosphoerythryl-S-enzyme–SH, in photosynthesis, **25:**101
Phosphoesterases, classification, **14:**223; 224–225; 244–245
Phosphoethanolamine (C^{14}), methylation, **25:**418
Phosphofructokinase, **18:**273, **34:**325, **37:**427; 456; 466; 483, **43:**31; 39; 42; 44, **46:**246; 260, **52:**277–281, **53:**49, **55:**360
and adenosine-3′,5′-phosphate, **24:**279
localization of, **24:**313
allosteric regulation, **28:**78–86; 92–93; 122, **52:**277; 289
physiological significance, **28:**83–85
in ascites tumor cells, **29:**347
ATP inhibitor or, **29:**355
correlation of hepatoma growth rate with, **29:**331
effect of adenosine nucleotides in, **29:**359
influence of the phosphate potential on, **29:**354–357
inosine triphosphate inhibitor of, **29:**355
in Novikoff hepatoma ascites, **29:**355
reaction, cytoplasmic ATP utilization upon glucose addition by, **29:**348
steady-state ratios of, in Ehrlich ascites tumor cells, **29:**355
change in activity after glucagon, **52:**289
classification, **14:**251
in cyclic synthesis, hydrolysis of fructose 1,6-bisphosphatase, **42:**222
deficiency of, **34:**335
electron microscopy of, **33:**108
general properties, **52:**277–278
interaction with fructose 1,6-bisphosphatase, **53:**49
interconversion, **52:**279–281
CAMP, **52:**280
change in activity, **52:**279
effect of glucose, **52:**280
phosphorylation, **52:**279
kinetics of action, **28:**80–83
nuclear-cytoplasmic ratios, **27:**262
nucleic acid content, **40:**47

Phosphofructokinase (*Continued*)
regulation by fructose diphosphatase, **52:**278
stabilizing factor, **52:**278–279
stereospecificity, **52:**278
6-Phosphofructo 1-kinase, *See* PFK 1
6-Phosphofructo 2-kinase, **54:**155, *See* PFK 2
Phosphofructokinases, **48:**193–236
activity, measurement of, **48:**195–196
citrate as inhibitor of, **48:**218
allosteric properties of, **48:**217
of bacteria, **48:**216
chicken liver, **48:**208
cold treatment of, **48:**208
electron microscopy, **48:**205
erythrocyte, **48:**209–210
forms *a* and *b,* **28:**79
heart muscle, **48:**206–207
isozymes of, **48:**210
kidney, **48:**207–208
liver, **48:**207–208
mammalian, **48:**201–214
nonidentical subunits, **48:**209
phosphorylation of, **48:**232–233
procaryote, **48:**214–216
pyrophosphate-dependent, **48:**216
reaction mechanism, **48:**198–201
role in regulation of glycolysis, **48:** 227–235
skeletal muscle, **48:**201–206
specificity, anomeric, **48:**197
substrate, **48:**196–198
tumor cells, **48:**213
yeast, **48:**214
Phosphofructokinase (Uyeda), **48:**193
Phosphofructokinose, kinetic mechanism of, **45:**281; 285; 311
Phosphogalactoisomerase, isolation, **14:**405
Phosphoglucoisomerase, **59:**111
gene position of, in *Salmonella typhimurium,* **31:**102–103
Phosphoglucokinase, classification, **14:**251
Phosphoglucomulase, **43:**413
Phosphoglucomutase, **5:**131, **9:**473, **15:**147; 168–169, **18:**286, **34:**16; 17; 295; 517, **37:**423; 434; 456; 521; 600, **43:**226; 233; 235; 331, **55:**361–362, **56:**292, **59:**340
active center, **24:**131
localization of, **24:**314; 320
mechanisms of catalysis, **24:**465–467
active configuration, **22:**68
amino acid composition near active site, **22:**65
column chromatography, **22:**474; 475
kinetic analysis of activity during modification, **22:**60
labeling of active site in, **22:**50; 66; 67
proteolytic activity after cysteine treatment, **22:**70
activity of –SH group in, **11:**243
adsorption studies, **14:**362–363
classification, **14:**258
isolation, **14:**405
in isomerizations, **14:**198; 199; 208–213
correlation of hepatoma growth with, **29:**331
ethionine effects, **25:**381
inhibition, **20:**127
in plastids, **20:**211; 213; 228
serine in, **20:**88
interaction of, **51:**75–76
localization in cytoplasm, **13:**7
in photosynthesis, **19:**8
in *Pseudomonas putrefaciens,* **10:**123–124
in utilization of maltose, **10:**124
transfer of phosphate groups, **34:**16
6-Phosphogluconate
allosteric regulation, **28:**117–118
R-5-P from, **28:**393–395
in plants, **18:**294–295
Phosphogluconate dehydrogenase, **37:**466
6-Phosphogluconate dehydrogenase, **18:**294, **59:**340
ethionine effects, **25:**381
nuclear-cytoplasmic ratios, **27:**262
ribose synthesis, **28:**411
6-Phosphogluconate-dehydrogenase, correlation of hepatoma growth rate with, **29:**331. *See also* Glucose-6-phosphate dehydrogenase
Phosphogluconate dehydrogenase (decarboxylating), location of, **24:**297
6-Phosphogluconic acid, **15:**121; 153; 160; 165–167
in photosynthesis, **19:**14; 15; 37; 38
Phosphogluconic acid dehydrogenase, in photosynthesis, **19:**15; 37; 38
6-Phosphogluconic dehydrogenase, **15:**97; 117; 154; 156–157; 165; 166
Phosphogluconic fermentation shunt, **15:** 166–167

6-Phosphogluconic r TPN transhydrogenases, **14:**254
Phosphogluconolactone, **15:**155
Phosphoglucose isomerase, column, chromatography, **22:**483
Phosphoglucose isomerases, multiple forms of, **25:**333
Phosphoglyceraldehyde, **1:**125, **5:**214
3-Phosphoglyceraldehyde
oxidation, **15:**258
in photosynthesis, **25:**95; 97; 102–105
Phosphoglyceraldehyde dehydrogenase, **5:**133, **9:**574, **15:**11
in *Chlorenchyma,* **20:**219
classification, **14:**238–239; 254
effect of radiations, **24:**389; 398
inactivation by cathepsin III, **12:**205–206
intracellular distribution, **16:**468
role in anaerobic glycolysis, **5:**137
3-Phospho-D-glyceraldehyde dehydrogenase, **17:**377; 380
Phosphoglyceraldehy dedehydrogenase, activity of —SH group, **11:**241–242; 251
inhibition by ionizing radiations, **11:**256
Phosphoglycerate
aci-acid, **67:**9; 12; 36
and glycolate metabolism, **23:**441; 444
labeled, **23:**333
3-Phosphoglycerate
and oxalate metabolism, **23:**450
as solubilizing agent, **17:**146
Phosphoglycerate, ADP-, **51:**213
2-Phosphoglycerate, dehydration of, **23:**327; 339–340
D-3-Phosphoglycerate, dephosphorylation, **16:**234
Phosphoglycerate kinase
and ATP synthesis, **23:**326; 329–333
localization of, **24:**313
in photosynthesis, **19:**10; 53
3-Phosphoglycerate kinase, **43:**30; 44, **46:**246
classification, **14:**251
in ATP system for energy transport, **14:**10–12
turnover number, **24:**123
Phosphoglycerate mutase, **55:**360–361
Phosphoglycerate mutase (MPGM), **51:** 212–213; 249
active site of, **51:**227
amino acid composition, **51:**217
distribution of enzyme, **51:**216
kinetic properties, **51:**230–241
peptides, **51:**228
rapid kinetics, **51:**225
substrate specificity, **51:**215–219
Phosphoglycerate Mutases, the (Fothergill-Gilmore and Watson), **62:**227
2 - Phosphoglycerate - 3 - Phosphoglycerate transformation, **14:**195; 213–214
Phosphoglyceric acid, **1:**119; 120, **15:**129; 172
and arginine synthesis, **15:**330–331; 332–333; 335; 343–345; 347–348; 352; 368; 373
in photosynthesis, **15:**144, **25:**47; 53; 61; 94–97; 102; 103
in purine biosynthesis, **21:**203
2-Phosphoglyceric acid, **15:**172
chelates, in metal ion activation of enzymes, **21:**148
3-Phosphoglyceric acid
accumulation of, in latex, **29:**225
in photosynthesis, **19:**7; 10; 19–24; 28–32; 52; 53; 64, **25:**94–97; 102–105
2-Phosphoglyceric acid formation in photosynthesis, **12:**46; 49–50; 101; 131–132; 133–135; 140–141
Phosphoglyceric acid mutase, **14:**208; 213–215
action, **14:**213–215
classification, **14:**258
D,L-2-Phosphoglyceric acid, ternary complex with Zn ++-albumin, **21:**149
2-Phosphoglyceric dehydratases, classification, **14:**237; 256
Phosphoglyceric kinase
effect of ATP/ADP ratio, **17:**122
isolation of, **14:**398
3-Phosphoglyceric kinase, EPR study, **27:**140
3-Phosphoglyceric mutase, **15:**172
Phosphoglyceric mutase, nuclear, **27:**272
Phosphoglyceric transphosphorylase, **15:**172
3-Phosphoglycerinaldehyde dehydrogenase, PN-dependent, **26:**119; 143; 163; 164; 173
3-Phosphoglycerinaldehyde dehydrogenase, SH-groups of, **27:**212–213
Phosphoglycerolkinase, in plastids, **20:**228
Phosphoglyceromutase, **5:**133; 180, **15:**348, **19:**10, **37:**466
amino acid composition near active site, **22:**65
labeling of active site in, **22:**50
localization of, **24:**314
multiple forms, **25:**343; 363

3-Phosphoglyceryl enzyme, **23:**329
Phosphoglyceryl kinase, **18:**292
site of cleavage by, **17:**163; 164
Phosphoglyceryl phosphate, **1:**109; 111; 112; 121; 128
3-Phosphoglyceryl phosphate, in photosynthesis, **25:**102
3-Phosphoglyceryl-S-enzyme—SH, in photosynthesis, **25:**100–101; 105
3-Phosphoglyceryl-SG enzyme, **18:**220
6-Phosphoglyconate dehydrogenase, **46:**233; 245; 271
Phosphoguanidine, **54:**400
Phosphoguanidine linkages, **1:**116–119
Phosphogylcerate, biosynthesis, allosteric regulation, **28:**76; 77
Phosphohexiosomerase, in *Pseudomonas putrefaciens,* **10:**123–124
in utilization of maltose, **10:**124
Phosphohexoisomerase, **19:**8
correlation of hepatoma growth rate with, **29:**331
ethionine effects, **25:**381
in plants, **18:**286
Phosphohexokinase, **5:**132, **15:**146, **19:**8
classification, **14:**251
Phosphohexose isomerase, **5:**132, **15:**155
classification, **14:**237; 258
P′ (6-Phospho-1-hexyl)-P_2-(6-amino-1-hexyl)-pyrophosphate, **46:**366
3-Phosphohistidine, **51:**220
Phospho homoserine isomerase, **35:**97
Phosphohydrolases, **34:**474
Phospho-hydrolytic enzyme, **17:**171
Phosphohydroxypyruvate, in transamination, **16:**236
Phosphoinositide, **54:**144
Phospho-2-keto-3-deoxygluconate aldolase, **37:**553
Phospho-2-keto-3-deoxyheptonate aldolase, **37:**457; 495; 521; 553
Phosphoketolase
in ATP synthesis, **23:**337; 339
in photosynthesis, **25:**99
and TPP, **22:**401
Phosphoketopentoepimerase, **19:**15; 28
Phosphoketopentose epimerase, in photosynthesis, **25:**95
Phosphokinase, **15:**105
Phosphokinase enzymes, **23:**360
Phosphokinases
activation by metal ions, **21:**150; 155
classification, **14:**220; 251
enzymic adaptation, **13:**70
L-Phospholactate, **43:**297
Phospholipase, **34:**472–474; 486, **46:**173
column chromatography, **22:**467–469; 476
Phospholipase, A_2, **51:**41; 49
Phospholipase A_2, **58:**202; 252
Phospholipase A, action, **34:**475
isoenzymes, **34:**131
Phospholipase C, **34:**471; 475, **41:**120, **55:**146, **57:**161
from *Bacillus cereus,* **74:**71
localization of, **24:**321
Phospholipase cyclic nucleotide, **56:**292
Phospholipase D, **34:**490, **55:**348
Phospholipase films, activity of, **28:**22
Phospholipases, **17:**367, **33:**263, *See* Lecithinases
classification, **14:**225; 244; 245; 250
Phospholipid, **34:**457; 467–477; 483–486
in the conversion of Factor II, **30:**280–281
counter-ions, **34:**490
micelles, **34:**469
model, **34:**483
Phospholipid choline, ethyl derivatives of, **25:**383
Phospholipide biosynthesis, **18:**207–208
Phospholipide oxidation in sperm, **9:**341; 372–373
Phospholipides. *See also* Lipides and Phosphatides; Phosphatides
in cardiac muscle hypertrophy, **6:**202–203
in muscular atrophy, **6:**199
in muscular dystrophy, **6:**206
and choline content of blood, **8:**180; 197–198
in livers of depancreatized dogs, **8:**177–178
synthesis, **8:**330–331
Phospholipids, **1:**4; 5, **46:**4; 5, **54:**143; 181, **58:**113; 211. *See also* Depot fats, transport; Phosphatides
in amino acid transport, **29:**367
annexins, **71:**214–215
and coenzyme Q, **25:**306
cytochrome *c* oxidase, electron transfer, **71:**118
in electron transport, **28:**160
ion channel model system annexin V, **71:**221
and methyl groups, pathways, **25:**420

in mitochondria, **21:**8; 9; 109
and sperm metabolism, **24:**57–61
metabolism of, **24:**338
structures from, **27:**391
in succinic oxidase system, **20:**178
synthesis in plastids, **20:**224
tyrosine hydroxylase regulation, **70:**122–123
Phospholipids (P^{32}) and sperm metabolism, **24:**58
Phospholipids (phosphatidyl glycerols), in chloroplast lamellae, **30:**9
Phospholipsae A_2, annexin crystal structure, **71:**217
Phosphomannoisomerase, map position for gene of, **31:**108
reaction catalyzed by, **31:**107
Phosphomannomutase, reaction catalyzed by, **31:**107
Phosphomannomutase/phosphoglucomutase (PMM/PGM), exopolysaccharide alginate synthesis, biosynthetic enzymes, **70:**229–232
Phosphomannose isomerase, **14:**205, **56:**293
column chromatography, **22:**483
Phosphomannose isomerase-guanosine diphospho-D-mannose pyrophosphorylase (PMI-GMP), exopolysaccharide alginate synthesis, biosynthetic enzymes, **70:**224–229
5-Phosphomevalonate, phosphorylation, **22:**301
in squalene biosynthesis, **22:**297–300
identification, **22:**298; 299
structure, **22:**303
5-Phosphomevalonate kinase, **55:**63
Phosphomevalonate kinase, isolation of, from latex serum, **29:**234–235
properties of, **29:**239
Phosphomevalonic kinase, function, assay, **22:**301; 302
in squalene biosynthesis, **22:**297
Phosphomonoesterase, **9:**517–518. *See also* Alkaline phosphatase
action on CoA, **16:**16
alkaline, specificity, **10:**86–89
classification, **10:**86–88
Phosphomonoesterases
classification, **14:**225; 232; 245
column chromatography, **22:**469–472
in methyl synthesis study, **25:**397
phosphoryl transfers, **17:**167; 170–171; 174
Phosphomutase, **15:**165
Phosphomutases, classification, **14:**226; 229; 241
N-Phosphonacetyl-L-aspartate
binding to active site, **68:**82–83
relationship with aspartate transcarbamylase substrates, **68:**74–75
Phosphonate analogs, **43:**3; 6
6-Phosphoniosinic acid, in purine biosynthesis, **21:**253; 254
Phosphonoacetamide, binding to aspartate transcarbamylase, **68:**82; 84
Phosphonoacetic acid, **42:**29
Phospho-organic compounds, **1:**103–110; 113
Phosphopantetheine, **43:**63; 66
Phosphopanthetheine, **57:**128
Phosphopantotheine, **18:**202
Phosphopantothenate, **43:**63
Phosphopentoisomerase, in photosynthesis, **25:**95
Phosphopentokinase
in photosynthesis, **25:**95
in plastids, **20:**228
Phosphopeptides from labeled phosphorylase *a*, **24:**269–271
Phosphopeptone, action of rennin on, **15:**433
Phosphopherases, **6:**149; 170
Phosphoprotein kinase, **38:**77
Phosphoprotein phosphatases, action on phosphorylases, **24:**282
localization of, **24:**320–321; 343
Phosphoproteins, **38:**76
hydrolysis, by metal ions, **21:**193
inorganic phosphate incorporation into, in mitochondria, **21:**45
Phosphoproteins and phosphatides, **7:**165
Phosphopyridoxamine, **43:**346
Phosphopyruvate, **5:**133, **15:**172–173, **17:**167
as phosphate donor, **18:**197
Phosphopyruvate carboxylase, **37:**457; 554; 584
Phosphopyruvate carboxylase [GTP-oxaloacetate carboxylyase (transphosphorylating)], *See* Phosphoenolpyruvate carboxykinase
Phosphopyruvate carboxylase [Orthophosphate:oxaloacetate carboxylase (phosphorylating)], *See* Phosphoenolpyruvate carboxylase
Phosphopyruvate hydratase, **37:**467
localization of, **24:**332; 335

Phosphopyruvate phosphokinase, activity of —SH group in, **11:**237; 239; 243
Phosphopyruvic acid, chelates, in metal ion activation of enzymes, **21:**148
Phosphopyruvic r ADP transphosphatases, **14:**251
Phosphopyruvic transphosphorylase, **15:** 172–173
Phosphoramidate method, for adenosine-5′-polyphosphates synthesis, **22:**199
for adenosine-5′-pyrophosphate synthesis, **22:**198
for adenosine-5′-triphosphate synthesis, **22:**198; 199
for coenzyme A, 3′-dephospho-coenzyme A synthesis, **22:**200; 201
for FAD synthesis, **22:**199; 200
for nucleotide coenzymes synthesis, **22:** 192–201
advantages of, **22:**194; 195
benzyl phosphoramidate in, **22:**194
salts, solvents for, **22:**197
water in, **22:**197
for uridine diphosphate glucose synthesis, **22:**199
Phosphoramidates, **56:**341
cyclic, **55:**308–310
in phosphate diester synthesis, **55:**316–317
Phosphoramidates, N-substituted, preparation, **22:**197
Phosphoramide mustard, **55:**278
Phosphoramidic acid, structure, **22:**193
Phosphoramidon, **57:**67; 70
Phosphorescence, *See* Luminescence
Phosphorescence spectra of porphins, **7:** 350–351
Phosphoriboisomerase, **19:**15; 21; 27, **23:**450
in plants, **18:**293–294
Phosphoribomutase, isolation, **14:**398
Phosphoribose binding, in cobalamins, **26:** 271–273
5-Phosphoribose pyrophosphokinase, **21:**241
Phosphoribose pyrophosphorylating enzyme, **17:**185
Phosphoribosyladenosine triphosphate synthetase, **37:**427
5-Phosphoribosylamine, **18:**217
in purine biosynthesis, **21:**204; 208; 210; 211; 234; 235; 238–247
N-(5′-phosphoribosyl)-anthranilate, **34:**521
Phosphoribosyl-ATP-pyrophosphorylase, allosteric regulation, **28:**124–126
histidine inhibition, **28:**44
Phosphoribosyl-ATP pyrophosphorylase, and histidine biosynthesis, **29:**129
Phosphoribosyl-ATP synthetase, interaction with RNA, **40:**45
5-Phosphoribosyl *N*-glycosides, of non-purine compounds, **18:**217
5-Phospho-ribosyl-l-pyrophosphate, precursor of pteridines, **35:**52
Phosphoribosyl pyrophosphate, **23:**324
Phosphoribosylpyrophosphate, **41:**3, **49:**281; 297. *See also* PRPP synthetase
determinants, **49:**299–303
generation of, **49:**299
intracellular concentration, **49:**298
regulatory role, **49:**283
in purine synthesis, **49:**283
in pyrimidine synthesis, **49:**284
synthesis of, **49:**297; 303
in mammalian cells, **49:**281–306
utilization of, **49:**282; 302–303
in other phosphoribosyltransferase reactions, **49:**282
in purine nucleotide synthesis, **49:**282
in pyrimidine nucleotide synthesis, **49:**283
Phosphoribosyl pyrophosphate, in allosteric regulation, **28:**120
nucleotides from, **28:**422
5-Phosphoribosylpyrophosphate, in amino acid and nucleotide syntheses, **21:**208
in histidine biosynthesis, **21:**404; 405
in purine biosynthesis, **21:**204; 207–210; 217; 218; 229; 240; 247
in pyrimidine biosynthesis, **21:**264; 277; 278; 281
5-Phosphoribosyl-1-pyrophosphate, in folate synthesis, **27:**424
Phosphoribosyl pyrophosphate, in nucleotide formation, **18:**215–219
synthesis of, **18:**199–200
Phosphoribosylpyrophosphate amidotransferase, **53:**223
inhibition of, **25:**171
5-Phosphoribosylpyrophosphate amidotransferase, **21:**208; 244; 245
5-Phosphoribosyl-pyrophosphate amidotransferase, **39:**94; 102
alternative steps in purine biosynthesis, **39:**100

analogs of glutamine, reaction with, **39:**97
cations, role of, **39:**96
conformational states of, **39:**98
properties of the enzyme, **39:**95
ribose-5-phosphate aminotransferase, relationship to, **39:**100
subunit structure, **39:**95
Phosphoribosyl pyrophosphate amido-transferase, (PP-ribose-P amido-transferase; EC 2.4.2.14), **41:**1; 2; 8–14
from *aerobacter aerogenes,* **41:**10
from bakers yeast, **41:**10
from chicken liver, **41:**10
defective, **41:**2
Hill coefficients, **41:**100
from human lymphoblasts, **41:**10
from human placenta, **41:**10
inhibition of, **41:**10
mechanism of, **41:**9
Michaelis constant of, **41:**8
molecular heterogeneity, **41:**11
molecular weight of, **41:**11
from mouse-spleen, **41:**10
mutant forms, **41:**12
from pigeon liver, **41:**100
rate of synthesis or degradation, **41:**11
from rat liver, **41:**10
regulation, **41:**9
from *Salmonella pombe,* 10, **41:**10
subunits, **41:**11
Phosphoribosyl pyrophosphate (PRPP), **54:**58; 64; 68
concentration of, in tissues, **54:**68; 69
Phosphoribosylpyrophosphate Synthesis in Mammalian Cells (Becker, Raivio, and Seegmiller), **49:**281
Phosphoribosyl pyrophosphate Synthetase, (PP-ribose-P synthetase; EC 2.7.6.1), **41:**1–8
association and dissociation, **41:**6
cation requirement, **41:**9
from Ehrlich ascites tumor cells, **41:**4
from human erthocytes, **41:**4
increased, **41:**2
Kinetic mechanism, **41:**4
Michaelis constants for, **41:**4
molecular weight of, **41:**6; 7
mutant form, **41:**8
from pigeon liver, **41:**9
regulation of, **41:**5
regulation of synthesis and degradation, **41:**6
from *Salmonella typhimurium,* **41:**4
specific genetic alterations of, **41:**7
subunits, **41:**7
Phosphoribosyltransfease, quinolinate to NAD+, **73:**145–149
Phosphoribosyl transferase, binding of RNA, **40:**48
Phosphoribosyltransferase (PRTase), acceptor domain, glutamine PRPP amidotransferase active sites, **72:**99–100
5-Phosphoriboxylpyrophosphate, **17:**418
Phosphoribulokinase, **19:**21; 32; 64, **23:**450
active site, **74:**113
integrity of, evaluation of, **74:**115–116
structure of, **74:**117–120; 118
from *Alcaligenes eutrophus (Ralstonii eutrophus),* **74:**96; 97; 98; 99; 100; 117; 120; 121
regulation of, **74:**110
amino acid sequences of, **74:**99–100
from *Arabidopsis,* **74:**100; 120; 121
catalytic mechanism of, **74:**113
characterization of, **74:**97–98
from *Chlamydomonas,* **74:**100; 114; 117; 120; 121
in multienzyme complex, **74:**104–105
from *Chlorogloeopsis fritschii,* **74:**98
from *Chromatium,* **74:**97
eukaryotic, **74:**97
regulation of, **74:**102–105
structural basis for, **74:**105–109
structure of, **74:**105–109
genes, **74:**99
and glyceraldehyde 3-phosphate, in multienzyme complex, **74:**103–105
from *Heterosigma carterae,* **74:**99–100
from *Hydrogenomonas eutrophia,* regulation of, **74:**110
from *Hydrogenomonas facilis,* regulation of, **74:**109
ice plant, **74:**100; 120; 121
invariant acidic residues of, **74:**120
isolation of, **74:**97
metabolic role of, **74:**96
molecular characteristics of, **74:**98–100; 99
in multienxyme complex, **74:**103–105
mutagenesis
results, interpretation of, **74:**116–122
targets for, selection of, **74:**116–117

Phosphoribulokinase (*Continued*)
from *Nitroacter vulgaris,* **74:**100; 120; 121
from *Nitrobacter winogradsky,* regulation of, **74:**109
pea, **74:**100; 103; 120; 121
physiologic role of, **74:**96; 103
prokaryotic, **74:**97–98
and eukaryotic, comparison of, **74:**98–100; 100
regulation of, **74:**109–110; 110–113; 111; 112
structure of, **74:**100–102; 101; 110–113; 111; 112
from *Pseudomonas facilis,* regulation of, **74:**109
reaction catalyzed by, **74:**96
regulatory mechanisms of, **74:**102–113
from *Rhodobacter sphaeroides,* **74:**98–99; 100; 117; 120; 121
overexpression of, **74:**115
regulation of, **74:**109–110; 111; 112
structure of, **74:**100–102; 101; 109; 111; 113–114
from *Rhodopseudomonas capsulata,* **74:**98; 100; 120; 121
regulation of, **74:**110
from *Scenedesmus,* in multienzyme complex, **74:**103–105
spinach, **74:**97–98; 100; 103–105; 116; 120; 121
structure/function correlations for, **74:** 114–115
analogies in homologous enzyme reactions, **74:**122–124; 123
structure of, **74:**98–100
and regulation of, **74:**105–109
substrate specificity of, **74:**96
subunit structure, **74:**100–102; 101; 102
from *Synechococcus,* **74:**114
from *Synechocystis,* **74:**100; 120; 121
from *Thiobacillus neapolitanus,* regulation of, **74:**109
Walker B motif, **74:**123; 124
identification of, **74:**121
wheat, **74:**100; 120; 121
from *Xanthobacter flavus,* **74:**100; 120; 121
Phosphoric acid, **1:**356–361
in synthesis of activated amino acids, **19:**250
6-Phosphoric acid, **58:**28
Phosphoric acid anhydrides, and pyrophosphorolysis, **18:**206–208
Phosphoric acid esters
action of RN-ase on, **24:**443–448
hydrolysis, by metal ions, **21:**170–175; 180–182; 184–188; 190; 191
by organic complexes of metal ions, **21:**188; 189
Phosphoric acid in nucleic acid, **9:**477–480
Phosphoric compounds, and origin of life, **27:**360–362
Phosphoric diester hydrolases, localization of, **24:**321
Phosphoric esters, in solubility studies, **17:** 146–149; 151; 152
Phosphoric monoester hydrolases, localization of, **24:**318–321
Phosphorochloridate-phosphate condensation, in nucleotide synthesis, **22:**162–164
Phosphorochloridates
nucleotide coenzymes synthesis, **22:**166–171; 173; 174
general structure, **22:**160
preparation, **22:**161
Phosphoroclastic reaction in bacterial metabolism of acetyl phosphate, **6:**236–239
reversibility, **6:**244
Phosphoroclastic reactions in CO_2 fixation, **12:**92; 93–101
Phosphorolysis, **5:**110–114. *See also* Phosphate transfer
of C—C, C—N, C—S bonds, **23:**327; 336–339
use of labeled (O^{18}) phosphate, **17:** 162–164
Phosphorolysis in synthesis of sucrose, **10:** 126–129
Phosphoroselenate, chiral, **55:**312
Phosphorothioates, **55:**296–306, **56:**341
analogs of adenine nucleotides, **55:**298–306
hydrolysis of, **55:**317–319
via alkaline phosphatase, **55:**387; 390
preparation of, **55:**294
ribonuclease A, **55:**296–298
separation of diastereomers of, **55:**298
Phosphorous NMR, **50:**377; 378
O-chemical shift, **50:**377
^{18}O-chemical shift, **50:**378
Phosphorus, **51:**147
acid-soluble, turnover in liver, **7:**120–125; 167

in cardiac muscle hypertrophy, **6:**203
in familial periodic paralysis, **6:**217; 218
in muscular dystrophy, **6:**205; 206; 207; 222; 223
in myasthenia gravis, **6:**211; 212
labeled, in identification of active centers of esterases, **20:**284; 288
in studies of phospholipid synthesis in plastids, **20:**224
in studies of uridine diphosphogalactose formation, **20:**113
in studies of uridine diphosphoglucose pyrophosphorylase, **20:**121
in plant metabolism, **9:**397; 427; 436–437
radioactive, **5:**135; 136; 137; 142
Phosphorus 32. Entries will be found under specific compounds.
Phosphorus compounds
in chick embryo, **7:**166
in egg yolk, **7:**157–160; 165–166
rate of renewal, **7:**113–120
turnover studies, **7:**113–192
nomenclature, **16:**8
Phosphorus-containing groups, enzymes for transferring, **24:**312–315
Phosphorus dehalogenases, classification, **14:**248
Phosphorus metabolism in brain, **7:**126–127; 143–145
Phosphorus oxychloride, in synthesis of activated amino acids, **19:**249
Phosphorus (P^{32})
in DNA duplication study, **18:**183
effect on DNA, **24:**13
in kinetic studies, **17:**48
in virus multiplication study, **17:**494
in study of nuclear RNA, **16:**440
in transamination studies, **16:**216; 218
uptake in dividing cells, **16:**429
Phosphorus (^{32}P), in DNA, **27:**241; 456
folate inhibitors study, **27:**456
inorganic, uptake in nucleus, **27:**274
and soluble space, **27:**265
test for DPN synthesis, **27:**276
Phosphorus (P^{32}) exchange, in ADP, **17:**185
in ATP, **17:**176
in glutamine synthesis, **17:**178
in oxidative phosphorylations, **17:**107; 122–123
Phosphorus (P^{32}) labeled
ATP
in phosphorylase *b* to *a* reaction, **24:** 277–279; 286
in preparation of phosphorylase *a*, **24:** 269–270
diisopropyl fluorophosphate, **24:**131
nucleus, in enzyme synthesis, **24:**15–16; 18
phospholipids, **24:**58
phosphorylase *a*, **24:**269–271; 280
Phosphorus turnover, and fluoride, **7:**123; 124; 151
in cell nuclei, **7:**182–186
dietary state of animal and liver, **7:**120–125
glucose effect on, **7:**118–125
and insulin, **7:**119–120; 123–125
and iodoacetate, **7:**124
in leukemic tissue, **7:**186–187
and lipotropic substances, **7:**138–141
in yeast, **7:**189–190
Phosphorycholine transfer, **18:**208
Phosphorylase, **5:**110; 119, **34:**14; 15; 363, **35:**127
a, **37:**559
acetylation, **34:**366
action of, **12:**411; 412–419
purity of, **12:**401
active site, **34:**366
activities of tetrameric form, **34:**379
adsorption studies, **14:**346; 365
amino acids in, **14:**275
classification, **14:**221; 222; 226; 251; 252
aggregation sites, **34:**378
alkaline, **34:**304
allosteric nature and binding sites, **34:**366
AMP site, **34:**370
bacterial and fungal, **34:**388
blocking of SH groups, **34:**375
bovine muscle, **34:**392
b to *a* interconversion, involvement of adrenal corticosteroids, **34:**399
in carbohydrate changes, **26:**36–37
conformational changes induced by AMP, **34:**371
crustacean and insect, **34:**389
crystallization of, **34:**391
de nova synthesis by, **34:**323
dimeric-tetrameric interconversion, **34:**364
dimer-tetramer interconversion, **34:**378; 379
dinitrophenylation of, **34:**368
effect, of AMP analogs on, **34:**374
of EDTA in, **34:**392

Phosphorylase (*Continued*)
of insulin on, **34:**399
of SH reagents on, **34:**367
in embryonic and neural tissue, **34:**391
frog and fish, **34:**391
gel electrophoresis of, **34:**388
heart, effect of hormones on, **34:**394
in heart during ischemia, **34:**395
heart muscle, isozymes forms, **34:**394
hepatic, ethionine effect on, **25:**380–382
product inhibition of, **25:**246
histochemical localization, **34:**400
human muscle, tetramer to dimer interconversion, **34:**394
inhibition by dinitrophenol, **34:**375
inhibition by metabolites, **34:**371
kidney, **34:**392
kinetic mechanism of, **34:**370
in liver, **34:**383
active and inactive forms, **34:**384
subcellular fractionation, **34:**384
lobster muscle, dimer → tetramer interconversion, **34:**389
localization in cytoplasm, **13:**7
in maize, **34:**388
maltodextrin, **34:**388
in muscle during exercise, **34:**393
Neurospora crassa, **34:**389
nonenzyme *b* → *a* interconversion, **34:**382
in nucleoside formation, **17:**428
nomenclature, **17:**169
in oligosaccharide formation, **17:**191; 195; 209; 211; 215–217; 221; 223; 228
pig muscle, **34:**393
plant and algae, **34:**387
in plants, **19:**4; 5
in plastids, **20:**211; 212; 228
muscle inhibition, **20:**70
in poikilothermic species, **34:**378
potato, acceptor specificity, **34:**388
action mechanism, **34:**388
degradation role of, **34:**387
pyridoxal site, **34:**376
rabbit muscle, **34:**363
interconversion of *a* and *b* forms, **34:**380
molecular weight and substructure, **34:**363
rat diaphram, effect of glucose 1-phosphate and glucose 6-phosphate on, **34:**395–396
rat muscle, **34:**393
relation to glucuronic acid transferring action, **16:**403
seryl site, **34:**376
sex organ, **34:**392
starch, effect of arsenate on, **10:**127
sucrose, as transglucosidase, **10:**138–141
use in synthesis of disaccharides, **10:**123–143
in synthesis of starch type polysaccharides, **15:**303–306; 307–309
term, **18:**192
in thyroid gland, **34:**396
yeast, **34:**389
Phosphorylase, Glycogen, Molecular Properties and Transformations of, in Animal Tissues (Krebs and Fischer), **24:**263
α-1,4-Phosphorylase, in production of amylopectin, **12:**357
Phosphorylase *a*
allosteric regulation, **28:**54; 55; 80; 101; 104–106
interconversion with *b,* **28:**106–109
nature of, **11:**303–304
properties of, **12:**415–416
solubility of, **23:**72
stimulation of production by adrenaline and cyclic AMP, **34:**398
Phosphorylase activity, regulation of, physiological significance, **28:**109–110
Phosphorylase *a* (P^{32}), conversion to phosphorylase *b,* **24:**280
phosphopeptides from, **24:**269–271
Phosphorylase *a* phosphatase, **34:**382
dog liver, activation and inactivation of, **34:**385
effect of glucose on, **34:**385
Phosphorylase *a* to *b* reaction, **24:**279–280
Phosphorylase b, **37:**483; 486, **43:**40, **52:**192; 239
change to a, **26:**302
effects of polycations, **40:**44
Phosphorylase *b,* **46:**481
allosteric regulation, **28:**54; 55; 80; 101; 104–106
interconversion with *a,* **28:**106–109
conversion to phosphorylase *a,* **24:**264–265; 277–279
nature of, **11:**304
properties of, **12:**415–416
as a protein phosphatase, **34:**413

Phosphorylase *b* kinase, **24:**270; 277–279, **34:**328–329; 380
activation by analogs of cyclic AMP, **34:**381
activity, **24:**277–279
in allosteric regulation, **28:**107–109
effect on glycogen synthetase, **28:**99; 101; 104
deficiency, **34:**335
inactivation by dephosphorylation, **34:**382
in I-strain mice, **34:**383
and muscle contraction, **24:**285–286
phosphorylation during activation, **34:**382
specificity of, **24:**280–282
Phosphorylase *b* kinase kinase, **28:**107
Phosphorylase *b* kinase-kinase, **34:**381
activation of fat-cell lipase by, **34:**381
Phosphorylase *b* kinase phosphatase, **34:**382
Phosphorylase *b*-$Metal_2^{++}$-AMP_2 crystals, **24:**265
Phosphorylase *b* phosphatase, allosteric regulation, **28:**104
Phosphorylase b/phosphorylase a, **43:**15
Phosphorylase *c,* **11:**304
of lobster muscle, **24:**276
Phosphorylase (glucosyltransferase), in coacervates, **27:**368
Phosphorylase kinase, **50:**41–108
activation of, **34:**380–381
chemical properties, **50:**41; 65
amino acid composition, **50:**65
sulfhydryl groups, **50:**66
dissociation of, **50:**57–59
influence of ATP, **50:**58
influence of salts, **50:**57
effectors of, **50:**86–95
electron microscopy, **50:**53
enzymatic covalent modification of, **50:** 96–108
activation by proteolysis, **50:**96–97
autophosphorylation, **50:**101–106
dephosphorylation, **50:**107–108
phosphorylation by cAMP-dependent protein kinase, **50:**97–101
and glycogenolysis, **24:**285
in preparation of phosphorylase *a,* **24:**265; 275; 276
glycogen phosphorylase *b,* influence of on quaternary structure of, **50:**58
heart type, **50:**56
isoenzymes, **50:**55
kinetics, **50:**77–81
with ATP, **50:**77
effect of metal ions, **50:**78
effect of pH, **50:**77
mechanism of, **50:**79
phosphorylase *b,* **50:**77
reverse reaction, **50:**80
tetradecapeptide, use as substrate, **50:**79
liver type, **50:**56
phosphorylase phosphatase, **34:**380
physical properties, **50:**50; 51
enzyme molecular weight, **50:**50
sedimentation coefficient, **50:**50
subunit molecular weight, **50:**51
purification of, **50:**49–50
from red and white skeletal muscle, **50:**55
antibody, **50:**55
α-subunit in, **50:**55
regulation of activity, **50:**86
Ca^{2+}, **50:**86–90
glucose-6-P, **50:**94
glycogen, **50:**92
Mg^{2+}, **50:**90–92
phosphate, **50:**92; 93; 94
salts, **50:**92; 93
regulatory and chemical properties, **50:** 41–108
roles of subunits of, **50:**59–62
ATP site, **50:**60
Ca^{2+} binding, **50:**61
catalytic subunit, **50:**59
specificity, **50:**72; 73; 75; 76
comparison with cAMP-dependent protein kinase, **50:**75; 76
use of synthetic peptides as substrates, **50:**72; 73
subunit stoichiometry, **50:**53
α-subunit, **50:**53
β-subunit, **50:**53
δ-subunit, **50:**53
γ-subunit, **50:**53
thiophosphorylation, **43:**15
trypsin-activated, **50:**58
Phosphorylase Kinase, Chemical and Regulatory Properties of, and Cyclic AMP-Dependent Protein Kinase (Carlson, Bechtel, and Graves), **50:**41

Phosphorylase phosphatase, action of hormones on, **34:**398
in blood cells, **34:**397
blood platelets, **34:**398
effect of ions on, **34:**390
mechanism of inactivation of, in pigeon muscle, **34:**397
Phosphorylase phosphatase (PR enzyme), **24:**279–282
action of phosphorylase *a,* **24:**270
and muscle contraction, **24:**285–286
specificity of, **24:**280–282
Phosphorylases
activity of —SH group in, **11:**243
degradation of polysaccharides by, **11:**301; 316–317; 359–361
effect on amylose, **11:**359–361
preparation of, **11:**364
in synthesis of polysaccharides, **11:** 298–299; 299–304; 304–306; 309; 316; 317–318; 321; 324–326; 330; 348–358; 366–367; 369–370
in synthesis of sucrose, **11:**370–373
and altered pyridoxal phosphate, **22:**396; 397
conversion of b to a, **22:**397; 398
function of phosphate group in, **22:**397
pyridoxal phosphate in, **22:**396–398
spectrum, as function of pH, **22:**397
in biosynthetic reactions, **18:**191–240
interconversion of *a* and *b,* **28:**106–109
in nucleotide synthesis, **27:**419
and nutritional state, **24:**284
catalytic activity, **24:**272; 467
changes in forms, **24:**284–286
formation after enucleation, **24:**12
glycogen, molecular properties and transformations, **24:**263–290
liver, **24:**282–283; 286
lobster muscle, **24:**276
muscle, amino acid composition, **24:** 266–268
end groups, **24:**266–268
role in glycogenolysis, **24:**283–286
valence of, **24:**118–119
Phosphorylases *a* and *b,* **24:**24
molecular constants, **24:**265
skeletal muscle, interconversion of, **24:** 277–282
human, **24:**265–267; 275–276
rabbit, preparation and properties, **24:** 264–275
Phosphorylases and Pyrophosphorylases in Biosynthetic Reactions (Kornberg), **18:**191
Phosphorylase synthesis, effect of nucleus, **24:**20
Phosphorylase systems, in β-1,3 glucan synthesis, **28:**338–340
Phosphorylated carbohydrates, in metabolism of aromatic amino acids, **16:**291
Phosphorylated compounds, **23:**323–399
and glycolate metabolism, **23:**437; 439
high energy, Table, **23:**324
Phosphorylated neopterin, **35:**56
Phosphorylated sugars in polysaccharide synthesis, **11:**298–299; 299–306; 309; 316; 317–318; 321; 324–326; 329; 330; 366–367
Phosphorylating apparatus, of inner mitochondrial membrane, **49:**225–227
Phosphorylating electron transport particle, properties, **21:**59; 60; 79; 80
Phosphorylating enzyme, **15:**225–232; 235; 253; 262
Phosphorylating enzymes, and semen metabolism, **9:**337; 354; 355–364
Phosphorylating preparations, of respiratory chain, **28:**156; 159; 161–164; 230
Phosphorylating process, reversal of, **23:** 382–383
Phosphorylating sites, in respiratory chain, **23:**347–354
Phosphorylating triose phosphate r DPN transhydrogenases, **14:**238–239; 254
Phosphorylation, **1:**112; 120; 124–126, **5:**111, **34:**492, **58:**4, **59:**337–338. *See also* Oxidative phosphorylation and Photosynthetic phosphorylation; Transphosphorylation, ATP
of acetate, **6:**256–257
of acetic acid, in keto acid synthesis, **6:**246
and pyruvate oxidation, **6:**254–256
and active transfer of glucose, **8:**51
of glucose and diabetes, **8:**403
adenosine receptors, **69:**101; 103; 104; 112
of ADP in nucleus, **27:**272
in coacervates, **27:**368–371
and origin of life, **27:**361
and adrenaline oxidation, **17:**350–351
and ascorbic acid oxidation, **17:**350–351

oxidative, and the respiratory chain, **17:** 65–134
metalloflavoproteins in, **17:**285–286
photosynthetic, **17:**282–283
role of cytochromes in, **17:**333–334; 363
aerobic, **3:**170–172
Caulobacter cell cycle, transcription regulation, **71:**18; 23
by cell nucleus, **16:**471–473
enzymic, pyridoxal, **16:**211; 219
of pyridoxamine, **16:**219
of imidazole derivatives, **16:**299–300
of sugar derivatives, relation to transamination, **16:**236
coupled with oxidative decarboxylation, **15:**224–230; 230–232
of hexose monophosphates, **15:**151–153
of hexoses, **15:**142–143; 145–148
oxidative, production of ATP, **15:**129–131; 177–178
and crotonyl-S. CoA, **23:**432
C-terminal domain, **71:**42 (*See also* RNA polymerase II transcription)
cyclic, **23:**386; 424
in chloroplasts, **30:**63–64
and glycine metabolism, **23:**461
oxidative, and electron transport, **23:**327; 340–385
substrate level, **23:**327–336; 374; 376
photosynthetic, **23:**327; 385–387
site of ATP formation in, **30:**67–70
effects of desaspidin on, **30:**68–69
of esterases, **20:**304–312
and fermentation, **7:**27–30
and glucose absorption, **7:**119–120
and phosphorus turnover, **7:**126
of fructose-6-P, control, **28:**78–86
of glucose, allosteric regulation, **28:**75–76
of glycogen synthetase, **28:**99; 101; 102; 108
oxidative, allosteric regulation, **28:**138–139
in *Fusaria,* **5:**179–181
of glucose by glucose 6-phosphatase, **34:**301
MAP, microtubule dynamics, **69:**144
mechanisms, **5:**136; 137
and mold metabolism, **9:**571–576; 580
and protein synthesis, **9:**437
non-, mechanisms, **5:**136; 137
noncyclic
Hill reaction accompanied by, **30:**64
number and location of sites for ATP formation in, **30:**64–66
oxidative
in microsomes, **21:**29; 30
in mitochondria, **21:**28–38; 113; 119; 124
in mitochondrial fragments, **21:**38; 39; 80–83; 87; 91; 102; 112; 123; 124
in nuclei, **21:**29
in photosynthesis, **25:**58–62
role of coenzyme Q, **25:**286; 319–322
and passage of sugar into yeast cell, **13:**16
phenylalanine hydroxylase, *in vivo,* **67:** 135–138
in photosynthesis, **30:**3; 63–70
photosynthetic, **21:**357, **25:**43; 53–62
pseudocyclic, ATP formation coupled to NADP+ reduction with, **30:**66–67
reversible, of CTD during transcription cycle, RNA polymerase II transcription, **71:** 51–59 (*See also* RNA polymerase II transcription)
types of, **30:**63–64
tyrosine hydroxylase, **70:**114; 127–165; 187
Phosphorylation, Oxidative, Electron Transport and (Green), **21:**73
Phosphorylation, Oxidative, Respiratory Chain and (Chance and Williams), **17:**65
Phosphorylation coupled to proton transport, **51:**99–105
ADP-ATP exchange, **51:**104
binding and equilibria, **51:**101–105
chemiosmotic system, **51:**99; 104–105
exchange of ^{18}O, **51:**101
exchanges of $H_2\,^{18}O$ into phosphate, **51:** 103–104
H_2O.ATP exchange, **51:**104
H_2O.P exchange, **51:**104
intrinsic binding energy, **51:**100; 101
mechano-chemical coupling, **51:**104
membrane potential, **51:**104
P_I-ATP exchange, **51:**104
proton gradient, **51:**102; 103; 104
proton-motive force, **51:**104
rules for, **51:**100–101
simple hypothesis for, **51:**99–100
uncouplers, **51:**104
Phosphorylation-dephosphorylation, phenylalanine hydroxylase regulation, **67:**130–157
active conformation stabilization, **67:**141–142

Phosphorylation-dephosphorylation, phenylalanine hydroxylase regulation (*Continued*)
blood glucagon levels, **67:**151–152
Ca^{2+}-mediated phosphorylation, **67:**149–150
cAMP-dependent kinase, **67:**131–132
cAMP-independent protein kinase, **67:**148
E and E^1, **67:**142–144
enzyme E-P, **67:**142–144
epinephrine, **67:**146–147
glucagon addition, **67:**135–137
glucagon and *in vivo* activity, **67:**138–139
glucagon-mediated phosphorylation, **67:** 152–153
ligand effect, **67:**139–146
norepinephrine, **67:**147–148
other hormone effects, **67:**146–150
phenylalanine and BH_4, **67:**141–145
phenylketonuria, **67:**151
phosphatase, **67:**133–135
physiological significance, **67:**150–154
question of regulation, **67:**154–157
streptozotocin, **67:**137
tetrahydropterins, **67:**140
in vivo phosphorylation, **67:**135–138
Phosphorylation-dimerization, of phosphorylase *b,* **28:**104
Phosphorylation efficiency, **23:**348–349; 353–354; 363; 367–368; 370–371
Phosphorylation of glucose
glucose 6-phosphate, a necessary step, **52:**260
role of glucose 6-phosphatose, **52:**260
Phosphorylation of Tb*E*GDP, **53:**141–143
Phosphorylation-oxidation sequence, **23:** 373–374
Phosphorylation potential, definition of, **29:**351
Phosphorylative and non-phosphorylative carbohydrate fermentation, **3:**156–160
Phosphorylative enzyme system, **4:**202; 206; 208; 213
Phosphorylcholine, **17:**436
in mitochondrial lecithin synthesis, **21:**50; 51
in semen, **24:**64–65
and semen metabolism, **9:**342; 352–353
S-Phosphoryl-CoA, in succinate activation, **21:**409
Phosphoryl coenzyme A, **15:**252
Phosphoryl-containing anhydrides, enzymes acting on, **24:**328–330
Phosphoryl-enzyme, **35:**346
Phosphoryl-enzyme derivative, allosteric regulation, **28:**97–98
Phosphoryl ethanolamine and phosphatides, **7:**152
Phosphoryl group, **60:**3
Phosphoryl group acceptors, **28:**95–98
Phosphoryl groups, transfer of, **18:**200–201
N-Phosphoryl imidazoles, preparation and properties, **20:**85; 86; 89; 91; 92
N^{10}-Phosphoryltetrahydrofolic acid, in formate activation, **21:**226; 408; 409
Phosphoryl transfer, **17:**161; 166–168; 170. *See also* Phosphate transfer
with imidazole derivatives, **20:**85; 86; 91; 311
Phosphoryl transfer mechanism, NMP kinase catalytic mechanism, **73:**114–119
Phosphoryl transfer reactions
enzymes catalyzing, **55:**292
glucokinase in, **55:**360
monomeric metaphosphate, in, **55:**370
with oxygen chiral phosphate esters, **55:** 341–343
phosphofructokinases, **55:**360
in proprochiral substrate conversions
to proprochiral products, **55:**356–364
to proproprochiral products, **55:**364–369
via acetate kinase, **55:**359–360
via acid phosphatase, **55:**364–366
via alkaline phosphatase, **55:**364–366; 390
via ATPase, **55:**367–368
via creatine kinase, **55:**359–360
via glucose 6-phosphatase, **55:**367–368
via glucose phosphotransferase system, **55:**362–364
via glycerol kinase, **55:**357–358
via GTPase, **55:**367–368
via hexokinase, **55:**358–359
via 5′-nucleotidase, **55:**368–369
via phosphoglucomutase, **55:**361–362
via phosphoglycerate mutases, **55:**360–361
via pyruvate kinase, **55:**358–359
via T_4 polynucleotide kinase, **55:**359
Phosphosaccharose, **1:**360
Phosphoserine, **16:**234, **53:**169, **58:**156
D-Phosphoserine, **16:**236
o-Phosphoserine, chelates, in metal ion activation of enzymes, **21:**148
Phosphoserine aminotransferase, **74:**133
evolution of, rate of, **74:**158
phylogenetic distribution of, **74:**157

O-Phosphoserine phosphatase, product inhibition of, **25:**197; 222; 236; 237; 239
Phosphoserine phospholase, **56:**19–20
5-Phosphoshikimic acid, **17:**414, **18:**363–366
in metabolism of aromatic amino acids, **16:**288; 290
Phospho terminology, **17:**160
Phosphotransacetylase, **23:**411, **43:**67. *See also* Transacetylase
in citric acid cycle, **15:**184; 188–189; 191–195; 198–200; 203–211; 217–220; 245–251; 254–259
classification, **14:**251
and fluoroacetyl-CoA, **18:**129
and lipoic acid, **18:**333; 336
as phosphorylase, **18:**192–226
product inhibition of, **25:**224; 242
test for CoA, **16:**16
Phosphotransferase, **17:**175
activity of glucose 6-phosphatase, **34:**305
carbamyl phosphate as a phosphate donor for, **34:**308
effect of Mg^2 + on, **34:**308
extracted from carrots, **29:**190
Phosphotransferases, localization of, **24:** 312–314
Phosphotransferase system, **16:**404, **44:**238
Phosphotriesterase
active sire residues, **74:**54–56
managenesis of, **74:**61–62
activity of, heavy atom isotope effects on, **74:**82–84; 84
amino acid sequence of, **74:**53–54; 54
apoenzyme
reconstitution of, **74:**56–58; 57
structure of, **74:**62–64; 63; 64
binuclear metal center in, **74:**60–62; 65; 66; 71; 72
in hydrolysis of organophosphates, **74:** 81–82
cadmium-113-NMR spectroscopy of, **74:** 58–60; 59
carbamylated lysine residue of, **74:**71
modifications to, **74:**73–74
electron paramagnetic resonance, spectroscopy of, **74:**60–61; 61
gene, **74:**53
genes for, **74:**53
inactivation
by diethylpyrocarbonate, **74:**55–56
by mechanism-based inhibitors, **74:**84–86
by *o*-phenanthroline, **74:**55–56
metabolic function of, **74:**52–53
metal-substituted forms of
activity of, **74:**57
Cd^{2+}/Cd^{2+}-containing, structure of, **74:** 64–68; 66; 67; 68
Cd^{2+}/Zn^{2+}-containing hybrid, **74:**59; 60
kinetic constants for, **74:**57–58; 58
metal ions used for, **74:**57–58; 58
Mn^{2+}/Mn^{2+}-containing, electron paramagnetic resonance spectrum of, **74:**60; 61
Zn^{2+}/Zn^{2+}-containing, structure of, **74:** 68–71; 69; 70
organophosphate hydrolysis catalyzed by, **74:**52–53
binucler metal center in, **74:**81–82
heavy atom isotope effects on, **74:**82–84; 84
kinetic analysis of, **74:**76–77; 78
mechanism of, **74:**55–56; 56; 74–88
model for, **74:**56
pathway of, **74:**74–75
solvent viscosity and, **74:**78–80; 79
stereochemistry at phosphorus center and, **74:**74–75
substrate specificity of, **74:**88–91
and other enzymes, comparison of, **74:**71–73
paraoxon hydrolysis catalyzed by, **74:**74–75
pathway of, **74:**74–75
pH-rate profile of, **74:**54–55; 55; 80
stereochemistry at phosphorus center and, **74:**74–75
from *Pseudomonas diminuta*
amino acid sequence for, **74:**53; 54
gene for, **74:**53
purification of, **74:**53–54
reaction catalyzed by, **74:**52
reaction mechanism of, **74:**74–88
inhibitors based on, **74:**84–86
kinetic analysis of, **74:**76–77; 78
model for, **74:**87–88
pH-rate profiles of, **74:**54–55; 55; 80
rate-limiting steps, determination of, **74:**76–77; 78
solvent viscosity and, **74:**78–80; 79
stereochemistry at phosphorus center and, **74:**74–75
structure of, **74:**53–74

Phosphotriesterase (*Continued*)
substrate specificity of, **74:**88–91
alterations to, **74:**90–91
and urease, comparison of, **74:**71–73; 72; 74
Phosphotriesterase homology protein, of *Escherichia coli,* **74:**71–72
Phosphotrioseisomerase, **18:**292
Phosphotriose isomerase
in photosynthesis, **25:**95
stability, **14:**390
Phosphotungstate, **17:**143
Phosphotyrosine, positioning in peptide sequence, **68:**27–28
Photinus pyrales, **44:**38
Photoactivation, microtubule dynamics, **69:**141
Photoadducts, psoralen-DNA, preparation, **66:**91–92
Photoaffinity cross-linking experiments, adenosine receptors, **69:**89–90
Photoaffinity labeling, RNA polymerase II transcription, reversible phosphorylation of CTD during transcription cycle, **71:**52–53
Photoaffinity-labeling techniques, adenosine receptors, **69:**90
Photoaffinity labelling
determining asialoglycoprotein receptor binding site topography, **66:**54–57
quantitative aspects, **66:**57–58
rat asialoglycoprotein receptor stoichiometry, **66:**58–60
Photoaffinity labels, **43:**42
Photobacterium leiognathi, **58:**73
Photobacterium phosphoreum, emission spectrum, **7:**257
Photobiological sensitization, by drugs, **11:**67
by plant toxins, **11:**410
Photobleaching recovery, **50:**308
Photochemical action spectrum, of heart muscle and yeast, **20:**165
Photochemical action spectrum of CO compound of cytochrome oxidase, **17:**341
Photochemical efficiency of photosynthetic process, **25:**48–51
Photochemical inactivation spectrum
of *Proteus phosphoreum,* **7:**257
of *Proteus splendidum,* **7:**257
Photochemical oxidation-reduction of thiols, **11:**210; 226
Photochemical processes. *See also* Irradiation, Radiations, and Ultraviolet radiations
and origin of life, **27:**355; 358; 370–371
Photochemical reactions, **12:**1–39
and heterocyclic ring structures, **8:**207
in alkaloid biogenesis, **8:**207
in monolayers, **11:**67; 76; 79–82
Photochemical reactions in plants, **9:**395–397; 414–417
Photochemical reactions of dihydropyridines, **24:**471–472
Photochemical studies of energy transport in living cells, **14:**31–37
Photochemical Systems of Photosynthesis (Boardman), **30:**1
Photochemistry
of amino acids, **9:**77–89
of peptides, **9:**89–96
of proteins, **9:**96–170
mutations, **27:**326–327
of thymine, **27:**318
of phytochrome action, **26:**11–15
psoralens, **66:**87–90
Photochemistry
of Enzymes, Proteins, and Viruses (McLaren), **9:**75
of Sulfur *(Bersin),* **10:**223
Photochlorophyllid, **46:**74
Photoconductivity, in photosynthetic tissues, **25:**82
Photocontrol, of plant processes, **26:**1–33
Photocrosslinking, **49:**197–200
Photodissociation, of CO-cytochrome a_3, **28:**228–229
Photodynamic action, of methylene blue, **27:**327
Photodynamic sensitization, carotenoid protection against, **21:**357; 358
Photoelectron transport reactions, **25:**53–66
Photoelectron transport scheme, **25:**89–94
Photoelectron transport system, **25:**43–45; 53–65
location of, **25:**67–69
physical and chemical properties, **25:**67–86
Photoinduced changes in absorption spectra, **25:**73–79
Photoisomerization of phytochrome, *See* Phytochrome
Photokinesis, carotenoid function in, **21:**356
Photoluminescence in photosynthetic tissues, **25:**82

Photolyases, **60:**16
Photolysis
  of amino acids, **9:**81–89
    of peptides, **9:**91–96
    of proteins, **9:**110; 119; 126; 147; 157; 160–162
    of riboflavin, **9:**474–475
  of disulfides, **27:**206–207
Photometabolism
  of acetate, **23:**420–431
  effect of fluoroacetate, **18:**135
Photomorphogenesis, **26:**1–33
Photooxidation
  carotenoid function in, **21:**354
  in conformation study, **28:**302
  effect on chymotrypsin, **20:**286; 289
    in identification of imidazole, **20:**68; 72; 76–78; 80; 81; 90; 94; 95
  of phosphoglucomutase, **24:**467
    of RNase, **24:**203; 204; 217; 218; 224; 245
  and superoxide dismutase, **41:**67
Photo-oxidation, of cytochrome b_{559}, in chloroplasts, **33:**136
Photo-oxidation in plants, **1:**232–234
Photoperiodism, effect of phytochrome, **26:**1–33
Photophosphorylation, **18:**59, **23:**385–387, **25:**43; 53–66; 90, **34:**492, **49:**269, **50:**377; 387–390, **58:**245
  cyclic, **23:**424, **25:**53–58; 60–62; 64–66; 80; 90; 92; 93
  ferredoxins in (*See* Ferredoxins in photophosphorylation)
  oxidative, **25:**58–62; 69; 90; 93
  and quantasomes, **25:**69–70
Photoreceptor membrane, **34:**470
Photoreduction in algae, **1:**243–248
Photorespiration, **57:**88
Photoreversibility of CO poisoning of respiration, **18:**5; 30–33; 37
Photosensitization, and chemiluminescence, **25:**126–128
Photosensitization of proteins, **9:**134–136; 160–161
Photosynthesis, **12:**31–37; 46, **15:**121; 126–127, **54:**175, **59:**379. *See also* Photochemical reactions
  and acetyl phosphate anabolism, **6:**248–250
  action spectra of, **30:**61–63 (*See also* Chloroplasts, Lamellae, Chlorophyll, Photosystem, and Photochemical systems)
  apparatus in, **30:**4–16
  Calvin cycle in, **30:**3
  carbon reduction cycle of, **30:**3
  chromatic transient in, **30:**17
  CO_2 fixation with soluble stroma in, **30:**6
  coupling of photochemical systems and rate-limiting reactions in, **30:**42–43
  electron carriers between the pigment assemblies in, **30:**26–40
  electron flow from Z^- to $NADP^+$ in, **30:** 40–41
  electron transport in, **30:**16–43
  Emerson enhancement effect in, **30:**3; 17
  energy conversion in, **30:**16–43
  energy transfer in, **30:**18–20
  fractionation of the photochemical systems of, **30:**43–56
  general formulation for, **30:**2
  Hill reaction in, **30:**3
  Krasnovsky reaction in, **30:**20
  light reactions of, **30:**6
  in Mn-deficient algae, **30:**25–26
  ^{18}O exchange in, **30:**2–3
  overall equation for, **30:**2
  oxygen evolution sequence in electron transport in, **30:**24–26
  photochemical systems of, **30:**1–79
  photosynthetic unit in, **30:**18–20
  quantum efficiency of, **30:**3; 56–61
  resonance transfer in, **30:**20
  role of manganese in, **30:**24
  two light reactions required in, **30:**16–18
anaerobic, **1:**242–255
bacterial, **45:**146, **74:**7
Blackman period, **1:**206–209; 213; 215
carbon dioxide in, **74:**6
carotenoid function in, **21:**349–354
  pterins in, **21:**377
chemical kinetics, **1:**209–216; 247; 248
CO_2 fixation in, **12:**49–50; 101; 129–145
and cytochrome, **18:**52–59
  lipoic acid in, **18:**340–342
  and respiration, **18:**285; 290–291; 303
effect of two lights, **25:**62–66
  energetics, **25:**39–117
  maximum efficiency, **25:**48–51
  oxygen test, **25:**152

Photosynthesis (*Continued*)
effect on amino acid metabolism in plants, **20:**321
enzymic aspects, **19:**1–77
and fluorescence, **1:**216; 223
and fluorescence of chlorophyll, **11:**91–99
in bacteria, **11:**133–139; 148–149; 181
carbon-dioxide uptake in, **11:**153; 155–158; 189–195
in diatoms, **11:**142–147
energy transfer in, **11:**117–119; 127–130; 133–134; 137; 145; 148–149; 150; 154–156; 166–167; 182–189; 191–195
gas exchange in, **11:**101–102; 120–121; 123; 140; 153; 155–158; 172–173; 179; 189–195
mechanism of, **11:**2; 4; 8–9; 95–96; 100; 103–105; 107; 108–109; 122–123; 127–129; 133–134; 137; 145; 147–148; 150; 154–156; 158–160; 167–170; 176–178; 182–189; 191–197
oxygen production in, **11:**102; 172–173; 179; 194
oxygen uptake in, **11:**101
phosphate exchange in, **11:**150
redox phenomena in, **11:**117–119; 148–150; 157–160; 164; 176–177; 180; 182–189; 191–195
and starch, **11:**341–343
glyceric acids in, **15:**172
heptulose formation in, **15:**160–163
invertase, **1:**350; 352; 353
light and dark reactions, **1:**205; 206; 287
metabolism of the purple bacteria, **1:**234–242; 264
phosphoglyceric acid in, **15:**144; 156
in plants, **1:**232–234
relation to Ochoa reaction, **10:**212–213
saturation phenomena, **1:**204–220
Photosynthesis, and Chlorophyll Fluorescence (Wassink), **11:**91
Photosynthesis, Bacterial (van Neil), **1:**263
Photosynthesis: Energetics and Related Topics (Bassham), **25:**39
Photosynthesis, Enzymic Aspects of (Vishniac, Horecker, and Ochoa), **19:**1
Photosynthesis, Facts and Interpretations (Frank and Gaffron), **1:**199
Photosynthesis, Nitrogen Fixation, and Fermentative Metabolism, Chemistry and Function of Ferrodoxins in (Buchanan and Arnon), **33:**119
Photosynthesis and pressure, **7:**249
Photosynthesis in heterotrophic assimilation of CO_2, **2:**138–140; 143
Photosynthetic bacteria, **1:**263–328
anaerobic oxidations of aromatic compounds, **27:**535–536
metabolism of acetate, **23:**420–431
nitrogen fixation studies in, **13:**357–360
Photosynthetic electron transport, **58:**244; 245
Photosynthetic enzymes, in plastids, **20:**227; 228
Photosynthetic phosphorylation, **17:**282–283, **19:**54–56; 63–65. *See also* Oxidative phosphorylation
and ATP synthesis, **23:**327; 385–387
in plastids, **20:**228
vitamin K_3, in, **20:**183
Photosynthetic pyridine nucleotide reductase (PPNR), *See* Ferredoxin
ferredoxin as, **27:**87
Photosynthetic reaction center, **46:**83
Photosynthetic reduction of cytochrome *c,* **18:**44
Photosynthetic unit, **1:**209–212; 221
Photosynthin, **1:**312, **21:**299
Photosystem 1, action spectrum for, **30:**62–63
structural organization of, **30:**55–56
Photosystem 2, action spectrum for, **30:**62–63
coupling between, and photosystem 1, **30:**42
E_0' as the primary reductant of, **30:**29
fluorescence as an indicator of election flow in, **30:**27–29
function of plastoquinone in, **30:**29–31
primary reductant of, **30:**26–27
structural organization of, **30:**55–56
Photosystem reaction centers, **46:**81
Phototaxis, carotenoid function in, **21:**354–357
Phototropism, carotenoid function in, **21:**358; 359
Photoxidase, in photosynthetic bacteria, **18:** 56–57
Photoxidation, of amino acids, **9:**83–84; 86–88
of proteins, **9:**109; 112–113; 119–120; 134
pH profiles, **59:**124
activity lost at low and high pH, **45:**343
of PK's for competitive inhibitor, **45:**326
of V, **45:**329; 330; 336; 340
of V/K, **45:**329; 330; 331; 340
pH-rate profiles, glutathione S-transferases, **69:**20

pH_s, **33:**251–252
pH-stat assay, **43:**117
pH studies, tyrosine hydroxylase, **70:**113–114; 127; 130; 133; 135; 136; 137; 157; 183
Phthalate, **51:**236
Phthalate-utilizing organisms, **18:**151
Phthalic acid, affinity for β-glucuronidase, **16:**367
as inhibitor of β-glucuronidase, **16:**374
3-β-Phthalimidoethyl-4-methylthiazolium salt, **15:**399
Phthalocyanines, metal, oxidation of, **15:**33; 39–40
tetrasulfonated, metal complexes, **15:**33
Phthium species, plant disease caused by, **20:**394
Phtic acid, **51:**212
pH values, micromicellar system, **51:**39–40
pH variation studies, **45:**320–348
Phycobilins, **46:**38; 39; 64–68; 93
pigments, **46:**65
quantum conversion by, **25:**63–64
Phycobiliproteins, **46:**64; 93
Phycochromes, **46:**93
Phycocyan (blue), separation by chromatography, **14:**349
Phycocyanin, **46:**92; 93
Phycocyanins, chromophores of, **26:**20–21
Phycocyanobilin, **46:**67; 92; 93
Phycoerthyrin, in *Porphyrium,* **30:**17
Phycoerythrin, **46:**68; 93; 102
synthesis, **46:**92
Phycoerythrin (red), separation by chromatography, **14:**326; 328; 349
Phycoerythrins, chromophores of, **26:**21
assay and separation, **26:**15–18; 29
Phycoerythrobilin, **46:**67; 68; 92
Phycomyces, **7:**454
Phycomyces blakesleeanus, **59:**62; 65; 377–379
enzymic transformations of steroids by, **20:**243; 248
Phygon, compatibility with antibiotics, **20:**396
Phylloquinone, **46:**1
Phylloquinone reductase, **20:**184
Phylogeny, and related protein structures, **24:**419; 423
Phymatotrichum omnivorum, pectin in, **20:**344
Physarum polycephalum, **46:**102; 154
Physical-chemical properties, ADA deficiency, **51:**174; 179
Physicochemical parameters of bovine RNase, **24:**177–181
Physiological efficiency of enzyme, definition, **28:**287
Physiological functions, **39:**361; 362
Phytase
of cacao, **25:**465–466
classification, **14:**234; 239; 245
soil, **28:**386
Phytases, specificity, **10:**87
Phytates, solubility studies, **17:**146; 149
Phythium de Baryanum, polygalacturonases of, **20:**365; 366
Phytinase, classification, **14:**239
Phytin phosphoester hydrolase, classification, **14:**234
Phytoalexins
accumulation of, induction of, **55:**98; 99–104
abiotic elicitors in, **55:**101–102
biogenic amines in, **55:**103; 104
biotic elicitors, **55:**103
ethylene in, **55:**103–104
factors in, **55:**97–98
glucan elicitors in, **55:**99–100
glycoproteins in, **55:**99
levels in, **55:**97
localization of, **55:**98
monilicolon A in, **55:**99
specificity of, **55:**98
suppressors of, **55:**100
viral, **55:**102
acetate-polymalonate, **55:**84
via fatty acids, **55:**84; 90
via polyketide intermediates, **55:**84
biosynthesis of
control sites for, **55:**108; 109
enzymes in, regulation of, **55:**109–114
induction of enzymes in, **55:**105–109
definition of, **55:**2; 97
dihydrophenanthrene, **55:**9
biosynthesis of, **55:**95; 96
diterpenoid, **55:**49; 78–80
historical perspectives on, **55:**2
induction process of, **55:**3
isoflavonoid (*See* Isoflavonoid phytoalexins)
isolation and identification of, **55:**2
metabolism of, regulation of, **55:**97–114
for plant disease resistance, **55:**2–3
polyacetylenic, **55:**84
in resistance to infection, **55:**97

Phytoalexins (*Continued*)
sesquiterpenoid, **55:**49; 66–78
precursor to, **55:**66
shikimate-polymalonate derivatives, **55:**84; 91–95
stilbene, **55:**84
terpenoid, **55:**49–83
biosynthesis and enzymology of, **55:** 50–81
classification of, **55:**49
degradation of, **55:**81–83
of eggplant, **55:**73–76
of Jimson weed, **55:**73–76
of potato, **55:**66–69
sources of, **55:**50; 51–56
structure of, **55:**50; 51–56
of sweet potato, **55:**69–73; 76–78
of tobacco, **55:**69–73
of tomato, **55:**73–76
Phytoalexins: Enzymology and Molecular Biology (Dixon, Dey, and Lamb), **55:**1
Phytochemical reduction, asymmetric, **7:**91–92
Phytochemistry of sulfur, **10:**223–323
Phytochrome, **46:**38; 65; 68; 101, **56:**164
action spectra of, **26:**3–6
biochemically active form, **26:**23–24
molecular rearrangement, **26:**10–11
and plant growth, **26:**1–33
reversibility of photoreaction, **26:**9–15; 18–23
Phytochrome, Control of Plant Growth and Development (Siegelman and Hendricks), **26:**1
Phytochrome chromophore, photoreversibility of, **26:**18–21
Phytoene, analysis, composition and structure, **21:**342; 343
Phytoene series, in carotenogenesis, **21:**340; 341; 343–346
Phytoflavin
photoreduction of $NADP^+$ catalyzed by, **30:**16
as substitute for ferredoxin in nitrogen fixation, **33:**145
Phytofluene, structure, **21:**343
Phytol, **46:**74; 75; 76; 77; 79; 94; 100
Phytol pyrophosphate, **46:**75
Phytomycin, composition, **20:**384
in treatment of plant diseases, **20:**390
Phytophora infestans, **55:**2; 66
Phytophora megasperma, **55:**99
Phytophthora infestans, plant disease caused by, **20:**391
Phytophthora phaseoli, plant disease caused by, **20:**390
Phytoproteases, **10:**300
Phytuberin, **55:**66; 69
biosynthesis of, **55:**72
structure of, **55:**51
Phytuberol, **55:**69
biosynthesis of, **55:**72
structure of, **55:**51
Phytylation, **46:**76
P_i, *See* Phosphate, inorganic
as inhibitor, **46:**335
Picolinhydroxamic acid, in reactivation of phosphorylated esterases, **20:**307
Picolinic acid, formation in tryptophan metabolism, **19:**93; 94; 96; 97
Piericidin, **34:**102; 103; 111; 123; 134; 137; 142; 145; 146
binding of, **34:**103
combined effects of with mercurials, **34:** 120–121
inhibition by, **34:**104–107; 109
interrelation of the effects of mercurials, **34:**120
lipid involvement in binding, **34:**107
reversal of inhibition by, **34:**107
role of protein in specific binding sites of, **34:**108
specific binding sites for, **34:**117
specific sites of, **34:**104
unspecific sites of, **34:**104
^{14}C-Piericidin, **34:**103; 104; 108; 120
Piericidin A, inhibition site of, **34:**101–111
Piericidin binding, SH groups in, **34:**108
and type V—SH, **34:**120
Pieris brassica, **45:**395; 398
Pieris brassicae, **59:**82
Pieris, origin of pteridines in, **27:**422
Pigeon liver, PP-ribose-P amido transferase from, **41:**10
PP-ribose-P synthetase from, **41:**9
Pigeon liver extract, formation of CoA in, **16:**6; 11
Pigment
changes during photosynthesis, **25:** 73–81
and semiconduction, in photosynthesis, **25:**81–86

in genetic studies, **17:**420; 421; 427; 428; 432–435
Pigmentation. *See also* Albinism; Leukoderma; Melanin; Pigmentation
control of formation in man, **14:**91–96
hormone control of, **14:**90–91
in normal skin, **14:**82–83
Pigmentation, in *Drosophila,* **17:**447–448
Pigment formation in *Pseudomonas,* **27:** 521–522
Pigment granules, and tyrosinase action, **14:**133; 137; 145–146
Pigment molecules. *See also* Chlorophyll, Chloroplasts, Coenzyme Q, and Photosynthesis
orientation, in chloroplasts, **25:**70–73
and photosynthesis, **25:**44–45; 50–51; 62–66
Pigment P-700, in photosynthesis, **25:**64–66; 75; 78; 79; 84; 85; 89–92
Pigments. *See also* Melanins, Quinones, Polyphenols, and Phenolic substances; Phytochrome; individual pigments
changes during fermentation of tobacco, **10:**372–373
of galls, **16:**137
formation, by action of phenolase, **16:**112; 114–117
from peroxidases acting on phenols, **16:**127; 152
from *o*-quinones, **16:**117–120; 122
during sclerotization of cuticle, **16:**154; 156; 160; 162
green sulfur bacteria, **1:**312; 313
from microbial action on hydrocarbons, **10:**469–470; 479
protein complexes, **1:**311; 312
purple bacteria, **1:**235; 292; 293; 307–314
and vitamin E balance, **6:**509
plant, in air-cured tobacco, **6:**333–334
Pigments in *Fusarium* fats, **9:**686–689
Pigment systems, irradiation of, **24:**380
Pigment Y, structure, **21:**348
Pigs, copper deficient, **44:**217
Pilin protein, *Caulobacter* cell cycle, **71:** 29–30
Pilocarpate, ionization constant, **20:**58
Pilocarpine, **55:**139; 157
imidazole in, **20:**102
ionization constant, **20:**58
structure of, **55:**140
Pilocarpine injections, effect on pancrease, **16:**421–422
Pimelic acid, **35:**334
Pimelic acid biosynthesis, from lower and higher homologues, **38:**331
isotopic studies, **38:**329
Pinacol rearrangement, in synthesis of isoleucine and valine, **16:**283–286
Ping-pong, **35:**361–363
acetyl CoA carboxylase, **35:**362
hybrid mechanism, **35:**362
molecular basis, **35:**363
transcarboxylase, **35:**362
Ping-pong mechanism, **57:**379–382
of aminoacyl-tRNA synthetase, **40:**172; 173
Ping-pong mechanisms, **45:**277; 280; 282; 290; 291; 292; 301; 311
dihydrodipicolinate synthase (DHDPS), **72:**287–290
γ-glutamyl transpeptidase, **72:**247–250
catalytic sites, **72:**254–262
multisite, **45:**277; 292
pyruvate, phosphate dikinase, **45:**118–119
Pinguianain, action on collagen, **23:**176
Pink spot, incidence of, **29:**525
relation to schizophrenia, **29:**524–526
Pinoresinol, formation from coniferyl alcohol, **16:**148
Pinoresinol-type structure, **34:**211
Pinosylvin, **55:**92
Pipecolic acid, **17:**403, **50:**122; 124; 125
form lysine, **16:**232
2,4-*cis*-4,5-diOH-Pipecolic acid, **50:**124
4-OH-Pipecolic acid, **50:**124
5-*OH*-Pipecolic acid, **50:**124
L-Pipecolic acid, proline analog, **29:**95
Δ'-Piperdine-2-carboxylate, **50:**124
Δ'-Piperdine-6-carboxylate, **50:**125
Piperidine, mechanism of catalysis of aldehyde condensation by, **14:**168
Piperidine ring and lysine, **8:**237
2-Piperidone-6-carboxylate, **43:**538; 539; 540; 541
Pipsyl technique, protein structure studeis in, **14:**272; 273; 279; 284; 300; 301–302
Pirenzepine, **55:**155
Piritrexim, eukaryotic dihydrofolate reductase, ligand binding (side chain roles), **70:**62

PIR (Pepsin-inactivated ribonuclease), **24:** 227–229; 232; 247; 249
Pisatin, **55:**2
  accumulation of, **55:**14
  biogenic amines affecting, **55:**104
  biosynthesis of, **55:**18
  formation of, **55:**39
  metabolism of, **55:**47
  pterocarpan from, **55:**47
  structure of, **55:**7
Pisum sativum, **55:**4; 5, **71:**84; 93
  chalcone synthase of, **55:**19
PITC, *See* Phenylisothiocyanate
Pithomyces chartasum, **55:**39
Pitressin, destruction in tissues, **14:**131
Pituitary factor, effect on hexokinases, **15:**146
Pituitary gland. *See also* Hormones, pituitary
  effect on thyroid, **14:**113–116
Pituitary gland, methanol-forming enzyme activity in, **42:**274
Pituitary hormone, activity of degradation products, **16:**48
Pituitary hormones
  activity of films, **28:**23; 28
  amino acid sequence, **24:**418–419; 422–425; 430
PIVKA, **46:**9
PIX, *See* Positional isotope exchange
pK_a, **57:**55–56
PKA, tyrosine hydroxylase regulation, **70:** 130–131; 135–136; 137; 138; 139; 141; 147; 152; 154; 155; 157; 161; 163; 168; 177; 181
PKC, tyrosine hydroxylase regulation, **70:**154
PK's, **45:**322
  of catalytic groups on enzymes, **45:**322
  temperature dependence of, **45:**322
pK's of amino acid groups, in identification of amino acids at active site, **22:**57–59
pK value, **34:**461–462; 492
pK value, of 3-hydroxypyridine, **50:**2–15. *See also* Dissociation constants
Placental microvilli, **49:**69
Placental toxin, as thromboplastin extract, **16:**87
Plakalbumin, structure studies, **14:**299; 300–301
Planetary landings, **27:**394–395
Planets, atmospheres, hydrocarbons in, **27:**350
Plant cells, amino acid transport systems, **49:**57
Plant Diseases, Antibiotics and (Tanner and Beesch), **20:**341
Plant enzymes, and soil activity, **28:**368
Planteose, **17:**210
Plant Growth, Control by Phytochrome (Siegelman and Hendricks), **26:**1
Plant growth, effect of phytochrome, **26:**1–33
Plant leaf ADP-Glucose pyrophosphorylases, **46:**335
  3-P-glycerate activation, **46:**339
  phosphate inhibition, **46:**339
Plant metabolism
  asparagines synthetase, **72:**149–151
  and phenolase activity, **16:**131–153
Plant pathogens, control of, **28:**355–356
Plant photosynthesis, light reactions in, **33:**136
Plants, **46:**42, **50:**117, **59:**328; 356; 379–380
  accumulation of selenium in, **29:**141
    growth regulator from *Aspergillus,* **29:**144
    protective mechanisms against toxic amino acids, in, **29:**94
    toxicity of *O*-methylthreonine, **29:**145
  allosteric enzymes, **28:**111; 112
    phosphofructokinases, allosteric control, **28:**78–79
    synthesis of UDP-xylose, **28:**116
    uronic acid pathway to pentoses, **28:** 398–399
  amide metabolism in, **13:**263–266; 303
    glutamotransferase in, **13:**296–298
    mitochondria in, **13:**13
    permeability of cells, **13:**43
  antibiotic treatment of diseases, **20:**383–406
    biosynthesis and metabolism of amides and amino acids in, **20:**319–340
    transformations of pectic substances in, **20:**367–371
  carbonic anhydrase of, **47:**151; 215
  cells, **46:**88
  chloroplasts in, **30:**4–5
  effect of fluoroacetate on, **18:**138
    higher, cytochrome in, **18:**1–64
    respiration in, **18:**281–318
  enzymic reactions in, **19:**1–77
  eukaryotic, **46:**84
  ferrochelatase fractions, **46:**59
  β-glucuronidase of, **16:**381
  glutamate in, **46:**83
  growth, effect of rare earth metal ions on, **21:**192
    higher, carotenogenesis in, **21:**310–314; 316; 317–320; 322; 323; 340

carotenoid functions in, **21**:352
carotenoids in, **21**:298–300
higher, **46**:94
ADP-glucose pyrophosphorylase of, **46**:335
iron-containing superoxide dismutase, **58**:73
leguminous, **46**:87
root nodules of, **46**:87
methylated lysine and arginine derivatives in, **42**:234; 235
nitrogen metabolism, **9**:391–454
non-protein amino acids of, **50**:117–171
biosynthesis, **50**:129–133
chemotaxonomy, **50**:133–135
as indices of plant relationships, **50**:135
sulfur metabolism, **10**:223–323
tetrapyrrole pathway, **46**:100–101
thiaminase in, **15**:390; 391; 394; 397; 398; 400; 401; 403; 406–411; 415
Plants, Green , Antibacterial Substances from (Kavanagh), **7**:461
Plants, Higher, Cytochrome in (Hartree), **18**:1
Plants, Higher, Nitrogen Metabolism (Street), **9**:391
Plants, Higher, Reaction Paths in the Respiration of (James), **18**:281
Plants, Living, Enzymic Process (Kurssanov), **1**:329
Plants, Some Problems in Pathological Wilting of (Gäumann), **11**:401
Plant viruses, **9**:173–177. *See also* individual viruses
purification, **9**:184–187
Plasma, **51**:170; 174
choline in, **8**:180; 197–198
fractionation and complement, **8**:102–103
lipides, in insulin-treated depancreatized dogs, **8**:179; 180
membranes, and cell functioning, **8**:17–20; 49–51
phosphatides in, **8**:49–51
proteins, synthesis, **8**:54–55
coagulation of, **5**:33–65
cofactor, **10**:37–38
decalcification by ion-exchange resins, **10**:15
by phosphates, **10**:15
proteins, binding to Ca^{++}, **10**:14
thromboplastin and autocatalysis, **10**:7–13
thromboplastinogen in, **10**:7–13; 16–17; 27–28
globulin fractions of, **5**:35; 42; 44
seminal, **9**:342–344; 355–357
stored, prothrombin in, **16**:58–59; 70
and vitamin E synergies, **6**:499–503
Plasma Ac-globulin, *See* Ac-globulin
Plasma amine oxidase (bovine), **59**:186; 193
Plasma factor, *See* Platelet cofactor I
Plasmagenes from minced cells, **8**:6–7
Plasma glycoproteins, **25**:446–450, **41**:99–127
Plasma iron levels, regulation, **44**:217
Plasmakinin, identity, **10**:10
Plasma kinins, analogs of, **29**:392
Plasma lecithin and intestinal mucosa, **7**:156
Plasmalogen
in mitochondrial lipoproteins, **21**:108; 109
in sperm metabolism, **24**:58–61
Plasma membrane, **34**:481, **59**:74–75
energization oxidation-reduction system, **49**:70
of hepatocyte, **49**:69
penicillinase, **34**:482
permeability, **7**:8; 15; 22
permeation of, **7**:5; 10–12
structure of, **7**:12–13
rat liver, **34**:474
Plasma membrane receptor, **57**:146–147
Plasma membranes
of animal cells oxidation-reduction activity in, **49**:72
binding by, **41**:99; 119–121
Plasma membrane transport, hepatic glutamine transport, **72**:45–47
Plasma phosphatides, **7**:167
and choline, **7**:139–140
deuterium content, **7**:170
and diethylstilbestrol, **7**:161
and gastrointestinal tract, **7**:154
interchange with tissue, **7**:152–154
and intestine, **7**:161–163
and kidney, **7**:161–163
and liver, **7**:154; 161–163
origin, **7**:161–163
and yolk phosphatides, **7**:157–158
Plasma proteins, survival curves of asialo derivatives, **41**:110–112
Plasma protein turnover, **7**:196–198
Plasma-serum difference, in PTC patients, **16**:73–75
Plasma thromboplastin antecedent, sensitivity to glass, **30**:261

Plasma thromboplastin antecedent (PTA), **48:**278
hemophilia-like disease, **16:**73
hemorrhagic disease, **16:**74–75
Plasma thromboplastin component (PTC), *See* Platelet cofactor II
Plasma trace proteins, equilibrium conditions, **16:**75–76
Plasmid, **58:**87
β-lactamase synthesis, **28:**269–270
Plasmin and blood clotting, **10:**12; 22; 32
Plasminogen, **48:**292, **57:**210
activation of, L-aspartase in, **74:**333–334
Plasminogen proactivator, **48:**288
Plasmodia, pyrimethamine resistance, **27:**455
sulfonamide resistant, **27:**431
Plasmodium berghei, folate metabolism, **27:**444; 445
Plasmodium gallinaceum, folate metabolism, **27:**444–446; 455; 456
Plasmolysis of fertilized egg, **9:**282–283
Plastein, **53:**241; 272; 285
Plastein formation by proteinases, **13:** 305–306
Plastein reaction of pepsin, **33:**425
Plasteins, **9:**428–429
Plasticizers, relation to contractility, **16:** 339–340; 347
Plastic surfaces, proteolytic fragments, **70:**8
Plastids, **46:**59. *See also* Chloroplasts; Phytochrome
enzymic properties, **20:**201–236
isolation, **20:**204; 205
origin and structure, **20:**202–204
from etiolated leaves, **18:**24–25
separation from leaves, **18:**43–44
starch formation in, **18:**291
Plastids, Enzymology of (Sissakian), **20:**201
Plastocyanin, **47:**84; 88; 136, **56:**314
composition of copper in, **33:**196
extinction coefficient of, **33:**198
forms of copper in, **33:**200
ORD curves of, **33:**201
oxidation-reduction potential of, **33:**207
studies of, **33:**208
pH dependent reduction of, **33:**208
from photosystem 1 particles derived by Triton X-100 treatment, **30:**56
P-700 accepting electron from, **30:**22
properties of, **30:**14–15
role of, **30:**38–39
Plastocyanin, EPR study, **27:**150
Plastoquinone
in algae oxidation and reduction of, **30:**31
in *Anacystis nidulans,* **30:**31
in chloroplast lamellae, **30:**9
extraction of, **30:**29
function of, **30:**29–31
requirement for Hill reaction activity for, **30:**29
in chloroplasts, **25:**66; 69
photoinduced changes, **25:**78; 80
relation to coenzyme Q, **25:**286; 291–292; 300; 301; 307; 322
Platelet-aggregating activity, **48:**300
Platelet aggregation assay, fibronectin, central cell-binding domain of, **70:**11
Platelet cofactor I, in blood clotting, **16:**26; 30–32; 68–69; 86; 89
relation to hemophilia, **16:**56–57
role in prothrombin activation, **16:**25; 49; 52–55; 70; 76; 91–92
Platelet cofactor II, inhibitor of, **16:**86
purification, **16:**32
role in prothrombin activation, **16:**25; 49; 72–75; 91–92
Platelet deficiency, bleeding tendency in, **16:** 67–68
Platelet enzyme, **10:**11–12
Platelet factor 4, antiheparin activity, **16:**70–71
Platelet factor 3, equilibrium conditions among trace proteins, **16:**76
role in activation of prothrombin, **16:**52–55; 70–71; 91
Platelet factor 2, role in blood clotting, **16:**70; 92
Platelet factor 1, role in prothrombin activation, **16:**70; 91
Platelet factors, in blood clotting, **16:**70–72; 91–92
Platelets, **41:**170
aggregation of, **58:**258
binding to, **59:**25–26
blood, and thromboplastinogen, **10:**9–12
protection by silicone film, **10:**16–17
role in clotting, **10:**16–17
in clotting reaction, **30:**259
and fibronectin receptor, **59:**48–49
and fibronectin synthetic peptides, **59:**31–34
lipoxygenase in, **58:**256
stereochemistry, **58:**288

preparation of concentrates of, **16:**69–70
role in blood clotting, **16:**30; 49; 52–57; 64–72
selective epoxidation, **58:**57
Platinous ions, chelate formation with amino acids, **21:**145
Platinum, colloidal, **15:**3; 23; 26
Platinum electrodes, reaction with thiols, **11:**212
Pleated sheets, **50:**35
Pleurotin, **7:**479; 480; 487; 489; 491; 493
Ployglutamate, D-configuration, formation, **16:**206; 237
Ploynucleotides, DNA as, **16:**432; 435
RNA as, **16:**437–438
PLP, *See* Pyridoxal-5′-phosphate
Plums, antibiotic treatment of diseases, **20:**399; 400
Plus-and-minus method, coupled to improved gel system, **67:**443–446
Plutonium (Pu^{239}), fission fragments from, **24:**365
PMB, **35:**452; 487
PMS-ascorbate, **49:**72
PNA, **48:**72; 77; 84. *See also* p-Nitrophenylacetate
PN-enzymes, purified (Table), **26:**118–119
reactions catalyzed by, **26:**117–132
Pneumococcal transformations, DNA as effective agent, **16:**412
Pneumococci, **60:**284
polysaccharides, synthesis of, **26:**63; 66; 72–73
transglycosylases of, **26:**60–61
Pneumococci and polysaccharides, **11:**307–308; 327
Pneumococci, enzymic adaptation in, **13:**70
hyaluronidase in, **13:**222–223
Pneumococcus
genetic transformation of, **6:**86–89
ribitol phosphate in, **30:**233
teichoic acid of, **30:**232
Pneumococcus, sulfonamide resistant, **27:**432
Pneumococcus-transforming principle, **18:**168
Pneumococcus-transforming principles, genetic studies of, **13:**145–146
Pneumococcus type I, hydrolysis by polygalacturonases, **20:**362
uridine diphosphogalacturonic acid in, **20:**369
Pneumococcus type II, neuraminidase of, **20:**144
Pneumococcus type III, DNA components, **16:**433
Pneumococcus type III polysaccharide, data on complement studies, **8:**99–100
Pneumococcus type XIV, uridine diphosphogalactose-4-epimerase of, **20:**119; 120
Pneumocystis carinii, **70:**25; 48; 50; 59; 70–71
Pneumonia virus, *See* Virus
PNH-cytochrome reductases, **26:**125; 130–131
PNH-oxidoreductases, **26:**125
P1 nuclease, from *Penicillium citrinum,* **74:**71
POF, *See* Pyruvate oxidation factor
Point mutation, *See* Mutation, point
Poisoning, effect on nuclei, **27:**268
Poisons, metabolic, effect on sperm, **24:**60; 63; 68; 69–72
Poisson distribution function, for distribution of galactosyl residues, **41:**107
Poky system in *Neurospora,* **17:**437–440; 454
Polar activation, **7:**76–79
Polar bonds, of proteins, **23:**5
Polar groups, **34:**468–471
and enzyme activity, **12:**290–311
and enzyme-substrate complex, **12:** 193–194; 199–200; 213
and specificity of carboxypeptidase action, **12:**226–227; 237–238
surfaces, **34:**484
Polar groups in muscle proteins, **13:**188
Polarimetric measurements of protein asymmetry, **27:**40
Polarity
asymmetry regulation and, Caulobacter cell cycle (*See also* Caulobacter cell cycle)
asymmetry regulation and, *Caulobacter* cell cycle, **71:**1–39
microtubule dynamics, **69:**124
Polarity changes, and agglutination, **15:**51
and immunological precipitation, **15:**68
Polarity of molecule and inhibitor activity, **10:**72–74
Polarizability, photoinduced, **25:**82
Polarized fluorescence in chloroplasts, **25:**72–73
Polarized light
and origin of optical activity, **7:**104–106
use in asymmetric decomposition, **7:**94
use in asymmetric synthesis, **7:**84–85
in study of starch, **12:**365–367

Polarizing light microscopy, cholesterol ester hydrolases, **69:**58
Polarographic procedures, **39:**333
Polarography in determination of vitamin K_1, **2:**305
Polarography in histochemical studies, **9:** 604–605
Poliomyelitis virus, effect of light on, **9:**142
purification, **9:**187
Pollen, terminal oxidases of, **18:**16; 46
Pollen cells, chromatin transfer in, **16:**423
Pollock effect, **13:**93–98; 99; 114–115
Pollock effects, **22:**116; 117
Polyacetylenic phytoalexins, **55:**84
Polyacrylamide, **34:**521
gel, **34:**448
Polyacrylamide electrophoresis, **39:**333
Polyacrylamide gel electrophoresis, **57:**14
Polyacrylamide gels, **42:**79; 80
assay for superoxide dismutase, **41:**49
Polyacrylanide-gel electrophorests, with superoxide dismutase, **41:**58; 64
Polyadenine, and origin of life, **27:**361; 367
Polyadenylate-synthesizing enzyme, chromatin space, **27:**268
Polyadenylic acid, nuclear enzyme for, **27:**275
Polyaffinity theory of enzyme-substrate complex, **12:**195–196; 200
Polyalanine, effect on RNase, **24:**209; 223; 245; 248; 250
Polyalcohol derivative, **34:**453
Polyalcohols, PN-dependent oxidations, **26:**118; 121; 122
Polyamines, effect on aminoacyl-tRNA synthetase, **40:**176; 177
Polyamine synthesis
from adenosylmethionine, **54:**216
from methylthioadenosine, **54:**217–218
as methylthioadenosine sources, **54:** 213–222
from putrescine, **54:**214
from spermidine and spermine, **54:**216
via adenosylmethionine decarboxylase, **54:**218
Polyamino acids, **47:**48–49
structures from, **27:**391
synthetic, Cotton effect, **27:**45; 48; 95
Poly-α-amino acids, effect on blood coagulation, **16:**88
Poly-*p*-aminostyrene, **34:**449; 453
Polyanionic derivative, **34:**458–461
Polyanions
activations
effect on ATPase, **40:**39; 40
effect on cells, **40:**51; 52; 53; 54; 55
effect on oxidative phosphorylation system, **40:**38; 39
effect on rabbit-muscle aldolase, **40:**37
effect on trehalose phosphate synthetase, **40:**36
effect on ulcer formation, **40:**56
interaction with lipoproteins, **40:**55; 56
effect on, denaturation of RNase, **24:**187–190; 192–193
PR-enzyme, **24:**280
RNase, **24:**219; 238
ganglioside sialidase inhibition, **60:**149
tyrosine hydroxylase regulation, **70:**122–127
Poly(A)RNA, **57:**399
Polyaromatic hydrocarbons, microbial degradation, **27:**507; 508; 528
Poly-γ-benzyl-L-glutamate, spatial arrangements, **13:**159
Polybrominated biphenyls, **60:**39
Polycarboxylic acids, effect on malonyl-CoA formation, **26:**301
Polycationic derivative, **34:**453; 460–461
Polycations
activations
effect on acid phosphatase, **40:**42
effect on aldolase, **40:**41; 42
effect on α- and β-amylase, **40:**41; 42
effect on hyaluronidase, **40:**40; 41
effect on lactic dehydrogenase, **40:**41; 42
effect on phosphorylase b, **40:**44
effect on polynucleotide phosphorylase, **40:**42; 43; 44
effect on sialyl transferase c, **40:**44
effect on gelation of pectic substances, **20:**357
Polychlorinated biphenyls, **46:**138
Polycyclic aromatic hydrocarbons, **46:**390
Polydenate ligands, complex formation with, *See* Complex formation with polydenate ligands
Polydeoxynucleotides, **60:**15
Polydeoxyribonucleotides, **18:**212–215
Polydiazonium salt, **34:**449–453
Polyelectrolyte, **34:**459
carriers, artificial, **34:**456
effect, **34:**459

environment, **34:**456; 491
field, **34:**486
phase, **34:**460–461
Polyelectrolytes
activated enzymes, **40:**32
inhibition of enzymes, **40:**50
polyanion activations, **40:**31; 32; 33; 34; 35; 36; 37; 38; 39; 40
lipoprotein lipase, **40:**31; 32; 33; 34; 35
pepsinogen, **40:**37; 38
polycation activations, **40:**40; 41; 42; 43; 44
β-glucuronidase, **40:**41
polynucleotide interaction with protein, **40:**29
requirements as enzyme effectors, **40:**30
interactions of, **18:**165–166
Polyelectrolytes, Interactions with Enzymes and Proteins (Elbein), **40:**29
Polyene antibiotics, **34:**483–485
Polyethene sulfonic acid, inhibitor for DNases, **29:**172
Polyethylene glycol, **52:**198
Poly (ethylene glycol), **47:**83
Polyethylene glycol dehydrogenase, **59:**194
Polyfluoroalanines, **55:**227–234
Polyfructosides, synthesis of, **11:**313
Polyfunctional catalysis, **50:**11
Polyfunctional catalysis by hydrolytic enzymes, **15:**294–295
Poly (galactosylglycerol phosphate), synthesis in *Bacillus licheniformis,* **30:**237
Polygalacturonase, **11:**285–288
adsorption on substrate, **14:**401
separation by chromatography, **14:**343; 351; 364–365
mode of action, **11:**273–276
Polygalacturonases, in hydrolysis of pectic substances, **20:**343; 360–368; 370
Polygalacturonic acid, *See* Pectic substances
Polygalacturonidase, cacao, **25:**466
Polygalitol, **17:**210
Poly (glucosylglycerol phosphate), synthesis in *Bacillus licheniformis,* **30:**237
Polyglutamate of *B. anthracis,* linkages in, **14:**278
Polyglutamic acid, Cotton effect in, **27:**47–48
Poly (glycerol phosphate), biosynthesis of, in *Bacillus licheniformis or Bacillus subtilis,* **30:**235–236
formulation of synthesis of, **30:**236
inhibition of synthesis of, **30:**244
Polyglycine, structure of, **23:**2
Polyglycines, irradiation of, **24:**377
Polygonatum multiflorum, imino acid in, **29:**95; 100; 101
proline sRNA synthetase, **29:**101
Polyhexametaphosphate (Calgon), effect on myosin, **16:**333
Polyhydric phenols. *See also* Catechol oxidation
oxidation by tyrosinase, **4:**103–110; 130; 133–135
phenol structure and tyrosinase, **4:**131–132; 142
Polyhydroxy alcohols, effect on β-glucuronidase, **16:**376
Poly-β-hydroxybutyric acid, and acetate metabolism, **23:**422–424
Polyhydroxy compounds, as coenzyme or substrate, reaction with borate, **12:** 516–518
simple reaction with borate, **11:**494–12:502
Polyisoprene, formation of, **29:**249
methyl groups of acetate derived from, **29:**226
ozonolysis of, **29:**236
Polyisoprenoid alcohol, **34:**478
Polyisoprenoid synthesis, and mevalonic acid, **23:**476
Polyisoprenol-linked oligosaccharides, **52:**76
Poly(l-proline), prolyl 4-hydroxylase inhibition and inactivation, **72:**358–364
Polylysine, **18:**169
complexes with DNA, **40:**45
effect on cells and tissues, **40:**57
effect on hyaluronidase, **40:**41
effect on polynucleotide phosphorylase, **40:**42; 43; 58
effect on sialyl transferase c, **40:**44
interaction with chondroitin sulfate, **40:**57
Polymannuronic acid-5′-epimerase, **44:**91
Ca^{2+} dependence, **44:**91
tritium incorporation, **44:**91
Polymer adsorption, **28:**18–19
β-Polymerase, coded for by phage genome, **31:**112
α-Polymerase, repressor of, **31:**112
Polymerase chain reaction, **67:**449
Polymerase chain-reaction techniques
adenosine receptors, **69:**85; 105–107
cholesterol ester hydrolases regulation, **69:**69

Polymerases, **56:**292
characteristics of, **56:**335
classes of, **56:**334
DNA, **56:**292; 334
inhibition of, **56:**338
metal centers of, properties of, **56:**337–341
metal content of, **56:**335–336
molecular properties of, **56:**336–337
nuclear poly(A), **56:**292; 334
origin of, **38:**227
RNA, **56:**292; 332
spectral features of, **56:**338
Polymerase-type catalyst, and origin of life, **27:**400
Polymer chemistry, and origin of life, **27:** 401–405
Polymeric esters, enzyme action, **33:**264–265
Polymeric matrics, cross-linking, **34:**452
networks, cross-linked, **34:**447
Polymer interactions, and molecular evolution, **27:**409–410
Polymerization, **53:**255; 256; 277. *See also* Depolymerase action
of assembled microtubules, **53:**130–139
degree of, **30:**143
in dextrans, the number average degree of, **30:**198
equilibrium, constant for, **30:**143
and dissociation of phosphorylases *a* and *b*, **24:**267; 269; 271; 275
of proteins, **24:**23; 119–120
and disulfides, **27:**220
of olefins, **27:**206
and origin of life, **27:**358; 360–361; 363–371; 374–375; 392; 395–399
guanine nucleotides measured by turbidity, **53:**128
at interfaces, **11:**63–64
in synthesis of polysaccharides, **11:** 297–337
mechanism of, **47:**352–353
microtubule protein, **53:**152–154
and multiple enzyme forms, **25:**362
reversible, of methyltransferase, **25:** 425–426
during sclerotization process, **16:**161–162
of actin, **16:**333–335; 342
of activated fibrinogen, **16:**37–38; 41–42
of indole-5, 6–6quinone in melanin formation, **16:**163–168
of quinone derivatives, **16:**110; 140; 148; 163–168
of tubulin GTP, **53:**142
of virus soluble antigens, **17:**488–489; 492
Polymerization, in Enzymic Synthesis of Polysaccharides (Hehre), **11:**297
Polymerization and autoxidation of unsaturated fatty acids, **8:**428; 438–439
Polymerization intermediates, in action of thrombin on fibrinogen, **16:**39–40; 41–42; 44
Polymerization of action, **13:**152–156; 182
Polymerization of proteins, imidazole in, **20:**101
Polymerization of vinyl compounds, **15:**7
Polymerization process, exopolysaccharide alginate synthesis, **70:**234
Polymerized enzyme gel, **34:**452
Polymer molecules, synthetic, configuration and internal rotation, **23:**11–13
Polymer propagation, steric control, **27:**402
Polymers
of bacterial cell walls, **26:**193–232
linear, sequential control, **27:**402
N-containing, abiogenic syntheses, **27:**386
three-dimensional folding, **27:**387; 401–405
properties of enzymes adsorbed on, **33:** 271–272
simple, biosynthesis of, by plants, **16:** 136–140; 151–152
water-insoluble, in purification of proteins, **24:**108
Polymer types of melanins, **16:**172a
Polymines, effect on ribosomes, **24:**7; 22
Polymixines, amino acids in, **14:**266
Polymorphonuclear leukocycle, **41:**153; 154; 178
basophil, **41:**153
bovine, **41:**153
canine, **41:**153
cosmophil, **41:**153
human, **41:**153
stimulation of the phagocutic activity, **41:** 154; 178
Polymorphonuclear leukocytes, **46:**62
Polymyxin, in treatment of plant diseases, **20:**392; 402
Polymyxin B, effect on K^+ by, **29:**46
Polymyxin D, **33:**367–368

Polynucleotide deligase, **38:**75
Polynucleotide dephosphohydrolase, **34:**520
Polynucleotide kinase, **48:**245–275, **60:**8
applicable to long polymers, **29:**208
calf thymus, **48:**271
mammalian, **48:**267–273
rat liver, **48:**264–271
reaction, **48:**246
in vivo function, **48:**272
Polynucleotide Kinase (Kleppe and Lillehaug), **48:**245
Polynucleotide nuclease, effect of radiations on, **24:**397
Polynucleotide phosphorylase, **17:**463, **18:** 228–229, **37:**498, **43:**32, **50:**377
addition to coacervate, **27:**369
of nucleus, **27:**275
effect of ionic strength, **40:**43
effect of polycations, **40:**42; 43; 44; 58
secondary interactions of, **33:**436
Polynucleotides, **9:**509–523, **38:**199
activators of enzymes, **40:**48; 49; 50
complexes of enzymes, **40:**48; 49; 50
specific interactions with enzymes, **40:**44; 45; 46; 47; 48; 49; 50
formation of, **17:**172; 185
and origin of life, **27:**357; 360–361; 364–365; 369; 374–375; 385; 392; 397; 401
radiation chemistry, **27:**330
precursors of, **18:**196
Polynucleotides, Interactions with Enzymes and Proteins (Elbein), **40:**29
Polyoldehydrogenase, PN-dependent, **26:**121
Polyols
oxidation, in mitochondria, **21:**25
in sugar nucleotides, **26:**54–55
synthesis of, **26:**63; 66; 73–76
Polyornithine, complexes with DNA, **40:**45
inhibition of polyanion activation, **40:**36
Polyoxins, **59:**71–73
Polypeptidase, **12:**206–210
yeast, **12:**210–211
Polypeptidases, **2:**72–81. *See also* Proteolytic enzymes
aminopeptidases, **2:**76–81
carboxypolypeptidases, **2:**72–76
classification, **2:**51
Polypeptide, **47:**49
Polypeptide antibiotics, **34:**483
derivative, **34:**461
Polypeptide catalyst, and origin of life, **27:**398
Polypeptide chain
action of tyrosinase on tyrosyl groups in, **14:**139; 142; 150
as apoenzyme structure, **14:**184–185
reactive grouping, **14:**130–131
amino acid sequence, of protamines, **16:** 442–443
angles of internal rotation in, **23:**15
axes of, **23:**13–18
possible configurations of, **23:**18–23
configuration of, **13:**158–159; 188
shortening and lengthening, **67:**271–272
and sugar nucleotides, **26:**49–50
formation of disulfide bridges, **26:**100–103
enzymes from, **26:**89–114
Polypeptide chains
and functioning of cytoplasm, **8:**9–14; 14–17; 24; 29; 37–39; 41–42; 49–51
role in nerve conduction, **8:**46–47
reinitiation of, **47:**417
termination, **47:**409–411
Polypeptide Configurations, Possible, of Proteins from the Viewpoint of Internal Rotation Potential (Mizushima and Shimanouchi), **23:**1
Polypeptide configurations of proteins, **23:**1–27
Polypeptide formation, in enzyme synthesis, **24:**21–22
Polypeptides, **38:**199. *See also* Gramicidin; Peptides; Proteins; Tyrocidine
adsorption studies, **14:**343–344; 347; 350
cyclol pattern, **4:**5; 6
cytochrome *c* oxidase, **71:**83–85; 128–129
irradiation of, **24:**377–379
α-keratin pattern, **4:**6
natural, D-amino acids in, **16:**250
and origin of life, **27:**369; 374–375; 383–384; 392; 401
synthetic, spectropolarimetric studies, **27:**45–46
thermal synthesis, **27:**358
protein structure from, **24:**22–25
release from template, **24:**22–25
spontaneous formation of, **38:**204
sulfur-containing, **10:**288–297
synthesis of, **38:**206
Polypeptides of cacao, **25:**465–468

Polypeptide structure, in energy transport, **14**:22–25; 27–29; 35; 41
Polypeptide synthesis, **24**:2–4; 11–20
Polyphenol oxidase, **17**:453
of cacao, **25**:465–468; 480; 482–483; 485; 487; 488
product inhibition of, **25**:250
in oxidation of amino acids in plants, **20**:321; 322
in plastids, **20**:215; 217
in plant respiration, **10**:374–375; 417, **18**:37; 41; 310–311; 372
Polyphenoloxidase, soil, **28**:387
Polyphenol oxidase in plants, **9**:403; 441
Polyphenol oxidases, classification, **14**:254; 255
Polyphenol oxidases in tobacco, **6**:373–374
Polyphenol-oxidizing enzymes, classification, **14**:255
Polyphenols. *See also* Diphenols, Phenolase complex, Phenols, and Quinones
activity during fermentation of tobacco, **10**:346–349; 373–375; 381; 396; 417
from biological oxidation of benzene, **10**:469
condensation with amino acids, **10**:381; 382–386; 408; 429
in air-cured tobacco, **6**:334–337
in flue-cured tobacco, **6**:364
extractable from plants, **16**:151–153
in artropods, **16**:153; 155
formation of quinones from, **16**:136
in plants, **16**:133
reaction with borate, **12**:495
Polyphenols as inhibitors of hyaluronidase activity, **13**:214
Polyphenols of cacao, **25**:460–465; 476–480; 482–484; 487–489
Polyphosphatase
chromatographic properties of, **31**:35
inhibitors of, **31**:35
p-nitrophenyl esters as substrates for, **31**:35
nucleoside, bis-(*p*-nitrophenyl) phosphate as substrate for, **31**:35
pH optimum of, **31**:35
sedimentation coefficient of, **31**:35
substrates for, **31**:35
thermal inactivation of, **31**:35
Polyphosphatases, **18**:193. *See also* Pyrophosphatase
classification, **14**:224; 248
specificity, **10**:86
Polyphosphate analogs Ap_4, Ap_5, A, **43**:23
Polyphosphate compounds. *See also* ATP
and muscle contraction, **13**:192–193
Polyphosphates. *See also* ATP, ADP, ITP, Pyrophosphate, Triphosphate and UTP; Phosphates, condensed
alkaline earth salts of, **17**:155–156
and origin of life, **27**:360–362
Polyploid cells, DNA in, **16**:228; 417; 422; 425–426; 429; 456; 483
Polyporus, quinones in, **25**:288
Polyreaction, definition of, **30**:183
Poly (ribitol phosphate), *in vitro* synthesis of, **30**:239
Poly (ribitol phosphate) synthetase, pH optimum of, **30**:240
site of, **30**:239–240
in *Staphylococcus aureus,* properties of, **30**:240
Polyribonucleotides, synthesis of, **18**:192; 212; 228–229
Polysaccharase, in animal utilization of laminarin, **28**:355
in barley endosperm, **28**:342
of soil organisms, **28**:354
Polysaccharase inducers, **28**:348–349
Polysaccharidases
classification, **4**:70
enzymic adaptation, **13**:70
Polysaccharide
and acetate metabolism, **23**:404–405; 423
ribose phosphate, RN-ase action on, **17**: 172
Polysaccharide epimerase, **44**:91
Polysaccharide phosphorylase, **18**:222
physiological role, **28**:419
Polysaccharide phosphorylases, **24**:264
Polysaccharides, **3**:109–135, **5**:88; 90, **52**:25, **56**:143. *See also* Amylopectin, Amylose, Dextrins, Glycogen, Starch; Carbohydrates; Glucans; Glycogen; Laminarin
abiogenic syntheses, **27**:386
antigenic properties, **26**:67–78
as antigens, **24**:115
bacterial
amino sugars in, **25**:433
from connective tissue, **25**:434–441

of glycoproteins, **25:**442–453
heptoses in, **15:**162
enzymic synthesis, **15:**301–318
native, weak bonds in, **15:**304
reaction with borate, **12:**513–514
in bacterial walls, **26:**223; 225–226
blood group reaction with borate, **12:**509–513
borate complex, nature of, **12:**507–509
capsular, distribution of sugars in, **26:**47
synthesis of, **26:**72–74
classification, **14:**224; 246
from ^{14}C-mannitol, **28:**338
complex, allosteric regulation, **28:**113–114
in cuticle of anthropod, **16:**153
inhibition of threone activity by, **16:**55
in prothrombin molecule, **16:**29
degradation patterns, **28:**348–351; 353
distribution of sugars in, **26:**47–49
effect on microbiological formation of hydrocortisone, **20:**246
enzymic synthesis of, **11:**297–337; 339–375
in plant wilting, **11:**416–422
of fungi, immunological, **28:**336–337
β-1,3 glucans, **28:**326–330; 335–337
laminarinase action, **28:**344–346
hydrogen bonding in, **28:**327; 337
laminarin-type, uses, **28:**353–358
lipid-containing, **28:**265
pentoses of, **28:**398
primers in synthesis of, **26:**58–59
reaction with borate, **12:**502–514
as receptor sites for viruses, **12:**520–521
reserve, related to laminarin, **28:**334
structural specificity of, **26:**67–78
synthesis of, **12:**419–421, **18:**221–222, **26:**41; 63–78, **28:**113–114; 337–340
by transglycosylation, **17:**191; 192; 209; 214–217; 221; 228–229
mixed, **17:**147
Polysaccharides, Enzymic Synthesis of (Hehre), **11:**297
Polysaccharides, Enzymic Synthesis of (Stacey), **15:**301
Polysaccharide sulfuric acid esters, **10:**23–24 235
Polysaccharide synthesis, **6:**299–300
comparison with sucrose synthesis, **10:** 140–141
energy source, **10:**128–129; 141
Polysomes, **60:**219
Polyspermy, prevention, **9:**244–245; 271; 317; 319
Polysulfates, **18:**169
Polysulfides, **39:**328; 330; 361; 362
inorganic, **39:**340
Polysulfide sulfur, **39:**340
Polysulfur radicals, **27:**206
Polytene chromosomes, banding, **16:**460–461
Polyteny, relation to DNA content of cell, **16:**430
Polythionates, **39:**330; 361
Polythionic acids in plant metabolism, **10:** 229–232
Polytoma uvella, use of acetate, **23:**411; 412
Polytomella coeca, use of labeled acetate, **23:**404
Polytomella coecae, phosphorylase in, **15:**304
Q-enzyme in, **15:**307
Polytripeptides, as inhibitors of prolyl hydroxylase, **41:**267
Polyuridylic acid, inhibitor of lysyl-tRNA synthetase, **40:**154
Polyvalent acids, reversible step-wise dissociation, **9:**1; 4–6; 13–14
titration curves, **9:**8–9
Polyvinyl alcohol, reaction with borax, **12:** 506–507
Polyvinyl chloride, configurations of, **23:**13
Polyvinylpyrrolidone, in isolation of mitochondria, **21:**5
Polyvinyl pyrrolidone (PVP), in subcellular fractionation, **24:**307
Pons medulla, tyrosine hydroxylase, **70:**107
P/O ratios, in Krebs cycle, **23:**348–349; 353–354; 363; 367–368; 370–371
Porcine Elastase, **41:**196
Porcine submaxillary gland mucoprotein, as substrate for neuraminidase and receptor destroying enzyme, **20:**144
Porcine submaxillary mucin, **52:**27; 108
Pore radii, **34:**454
Porgesterone, inhibitory uncoupler, **23:**367
Poria vaillantii, action on wood, **18:**352–354
Poria vaillantii, degradation, of cellulose, **8:**272–273; 278–279
of glucose, **8:**272–274
of wood, **8:**271–272
Porphine-like substances, abiogenic syntheses, **27:**386

Porphins, **7**:267–270; 272; 286; 288; 296–297. *See also* Metalloporphins and Tetrapyrroles
absorption spectra, **7**:340–350
decomposition, **7**:333–339
destruction, **7**:337–338
distribution, **7**:318–319
fluorescence, **7**:318–319
fluorescence spectra, **7**:348–350
phosphorescence spectra, **7**:350–351
ring structure, **7**:340–341
separation, **7**:351–355
by chromatography, **7**:355
side chains, **7**:322–328
solubilities and *p*H, **7**:351–355
synthesis, **7**:328–333
vinyl groups of, **7**:325; 333
Porphobilin, **46**:163
Porphobilinogen, **46**:48; 49; 55; 102; 107; 162; 163
formation of, **46**:39; 42–44
[^{13}C] Porphobilinogen, **46**:47
Porphobilinogen deaminase, localization of, **24**:332
Porphobilinogen synthase, localization of, **24**:332
Porphyra tenera, cytochromes of, **25**: 75–76
Porphyria, **7**:319; 330; 338; 355, **46**:107; 160–165
acute intermittent, **46**:115; 140; 160; 161–163; 164; 178
coproporphria, **46**:161; 162–163
cutanea tarda, **46**:140; 164
diseases, of erythrocytes, **46**:165
variegate or congential cutaneous hepatic, **46**:160; 163
Porphyria yezoensis, **46**:49
Porphyridium, action spectrum for cytochrome oxidation in, **30**:61
cytochrome oxidation in, **30**:38
Porphyridium cruentum
cytochrome *c* in, **30**:35
lamellae of, **30**:7
photosynthetic reactions of, **18**:53–54
Porphyrin, absorption spectrum, **4**:276; 305; 306
Porphyrin *a,* **23**:270–284; 310, **46**:60
Porphyrin *a* α, **23**:281–282; 284
Porphyrin *a* β, **23**:281–282; 284
Porphyrin biosynthesis, **23**:304; 455
Porphyrin complexes, roles in oxidoreduction, **15**:16–21; 32–33; 38–39
Porphyrindin reaction with thiols, **11**:207; 219–220; 222–223; 224; 230–232
Porphyrin peptides, from cytochrome *c,* **18**:79
Porphyrin ring, in corrinoids, biogenesis, **26**:258–260
relation to vitamin B_{12}, **26**:234–235; 256
Porphyrins, **46**:37; 82; 121; 163, **59**:117, *See* Hemins; Porphins; Tetrapyrrolic compounds
chromatography, **46**:179
of cytochrome, **17**:338–339; 343–345; 357
of cytochromes, **23**:269
inhibitory analogues of, **6**:134; 143
iron, **8**:373
manganese complexes of, **33**:315; 319
precursors of, **18**:258; 259
X porphyrin, **46**:163; 165
Porphyrins, EPR, **27**:138–140
and origin of life, **27**:363; 370–371
Porphyrins in thiol oxidation, **11**:206; 226
Porphyrinuria, **7**:320
Porphyrium, phycoerythrin in, **30**:17
Portal circulation, **39**:359
Positional isotope exchange, **50**:361–392
glutamine synthesis, **72**:28–29
Positional Isotope Exchange Studies on Enzyme Mechanisms (Rose), **50**:361
Positive effectors, **48**:219–225
Posterior pituitary gland, **42**:269
Postmitochondrial supernatant, **46**:23
Posttranslateral regulation, glutathione synthesis, **73**:236
Post-translational modification of proteins, **67**:265–284
biological functions
covalently attached cofactors, **67**:280–281
cross-links, **67**:278–280
membrane anchors, **67**:281–282
protein turnover, **67**:283
regulatory on-off switches, **67**:276–277
reversible reactions, **67**:276–278
spontaneous reactions, **67**:278
ubiquitin pathway of degradation, **67**:283
unidirectional reactions, **67**:278–283
polypeptide chain shortening and lengthening, **67**:271–272
protein sequencing, **67**:271
reversible, **67**:276–278

secondary amino acids, **67:**267–271
specificity, **67:**272–276
co- and post-translational reactions, **67:**274–276
pre-translational reactions, **67:**274
Posttranslational modification process, **42:**241
Post-translational modifications, adenosine receptors, **69:**89–91
Posttranslational modifications, RNA polymerase II transcription, C-terminal domain characteristics, **71:**49–50
Potassium, **34:**483–485
in affective disorders, **29:**495
in cardiac muscle hypertrophy, **6:**202; 203
in familial periodic paralysis, **6:**215; 216; 217; 218
inhibition of amino acid decarboxylases, **6:**17; 18
in muscular dystrophy, **6:**205
in myasthenia gravis, **6:**212
in myotonia, **6:**213
chloride, effect on muscle proteins, **13:** 169–170; 173–177
effect on sperm, **24:**40; 44; 49; 73; 78–80
in nucleus, **27:**256–259
Potassium chloride
effect on gel-sol transformation, **16:**331
effect on myosin and its subunits, **16:**320; 321; 329; 332; 348–350
in isolation of mitochondria, **21:**5
Potassium cyanide. *See also* Cyanide, Hydrogen cyanide
effect on steriod 11β-hydroxylase, **19:**199
Potassium cyanide as inhibitor for diamine-oxidase, **2:**101
Potassium dehydroepiandrosterone sulfate, formation in steroid alcohol sulfokinase reaction, **22:**212; 213
in steroid sulfatase assay, **22:**227
Potassium depolarization, tyrosine hydroxylase regulation, **70:**128–129
Potassium ethyl xanthate. *See also* Ethyl xanthate
effect on indolylacetic acid oxidase, **19:**124
Potassium exchange, **51:**99
Potassium in plant metabolism, **9:**397; 427; 437; 438
Potassium iodide
effect on actin polymerization, **13:**154–155
use in muscle extraction, **13:**162–163
effect on polymerization of actin, **16:**333
Potassium ion
for fructose-bisphosphatase activity, **54:**127; 129; 132
in fructose-bisphosphatase regulation, **54:**178
Potassium ions, **44:**257, **55:**145
chelate formation with organic phosphates, **21:**147
enzyme activation by, **21:**143; 153; 154; 158
in mitochondria, **21:**10
properties, **21:**157
in purine biosynthesis, **21:**204; 213; 221–223; 229; 400
concentration changes in tryptophanase activity, **10:**78
effect on actin polymerization, **16:**335
effect on enzymes, **15:**116; 173; 246; 251; 257; 258; 335
effect on myosin, **11:**230–231
and fermentation, **7:**27–30
and glycogen synthesis, **7:**27–28
and luminescence, **7:**253
and myofibrillar structure, **13:**186–187
effect on actin, **13:**152–153; 155
on actomyosin, **13:**160
on myosin activity, **13:**157; 164–166; 168–169
permeability studies, **13:**27; 28; 55–56; 58–63
Potassium ions and choline acetylase activity, **12:**316; 318–322
Potassium (K +) efflux, hepatic glutamine transport, hepatocyte hydration, **72:** 49–50
Potassium *p*-nitrophenyl sulfate, in phenol sulfokinase reaction, **22:**211
Potassium phosphate, **58:**152
Potassium salts, **17:**149
Potato
Phytoalexins of, **55:**66–69
sweet, phytoalexins of, **55:**76–78
Potato aldehyde oxidase, reaction with borate, **12:**516–517
Potatoes, ALA synthetase in, **46:**41
Potatoes, antibiotic treatment of diseases, **20:**392; 395; 396
Potato inhibitor, **41:**182; 204
Potato latent mosaic virus, mutations, **9:**181

Potato tubers, cytochromes and oxidases in, **18:**19; 29
respiration in, **18:**286; 292
terminal oxidases of, **18:**38–39
Potato viruses, **15:**449–450; 453
Potential, **34:**490–493
electric, maintenance across mitochondrial membrane, **29:**47
gradient, **34:**494
ζ-Potential, **33:**251; 263; 267
Poteriochromonas, **59:**74
P/O value, calculation, **17:**123–124
Power-compensating automatic calorimeter, **51:**7
PP, *See* Pyrophosphate, inorganic
PP_7, **57:**391
$p22^{phox}$, **65:**78–80
hydropathy and glycosylation sites, **65:**77
$p32^{phox}$, **65:**74–75
$p45^{phox}$, **65:**79–81
$p47^{phox}$, **65:**67–72
cloning, **65:**70–73
$p67^{phox}$, **65:**72–74
PP_i, role in metabolism of, bacterial photosynthesis, **45:**146–147
E. histolytica, **45:**131; 132; 140–146
P. shermanii, **45:**131; 132; 134–140
PP_i acetate kinase, **45:**132–133
PP_i and metabolism, **45:**149
PP_i-PFK
and effect of Fru-2,6-P_2, **59:**330–331
and interconversion with ATP-PFK, **59:**332
occurrence of, **59:**329–330
properties of, **59:**329–330
role of, **59:**332–334
PP_i -phosphofructokinase, **45:**130–132
chemical characteristics, **45:**131–132
physical characteristics, **45:**131–132
physiological role, **45:**130
PPNR. *See also* Ferredoxin
in photosynthesis, **25:**55; 66; 91–94
PP-ribose-P aminotransferase, **41:**14
PQQ, *See* Pyrrolo-quinoline quionone
PQQH
and glutathione reductase, **59:**149
and pyrrolo-quinoline quinone, **59:**173–174; 178–179; 182
and quinoprotein oxidoreductases, **59:** 194–195; 197
$PQQH_2$
and pyrrolo-quinoline quinone, **59:**173–174; 178–179; 182; 185–186; 297
and quinoprotein oxidoreductases, **59:**194; 203
$PQQH_4$, **59:**180; 182–184
Praseodymium compounds, **17:**147
Praseodymium ions, model catalyses with, **21:**177
Prealbumin, **46:**150
Preangiotonin, **7:**523
Prebiotic soup, **38:**203
Precanavalin, **56:**225
Precipitation, effect on pectic substances, **20:**353
Precipitin reaction, with yeast mannans, **40:**250
Precipitin reactions, quantitative aspects, **8:**72; 84–85
Preconditioning, **37:**482; 610
Precursor, enzyme, and enzymic adaptation, **13:**81–88
Predictive accuracy, **47:**47; 48; 50; 55–58; 92–97; 114; 121–123; 125; 131
correlation coefficient, **47:**57; 92–94; 96
overpredictions, **47:**47–49; 56; 86; 87; 130
quality of predictions, **47:**56; 88; 92–94; 122–124
random guessing, **47:**48; 57–58
Predictive methods, protein conformation, **47:**48; 50. *See also* Protein predicators
anti-helical and helical pairs, **47:**47; 88
Bayesian pattern recognition, **47:**86
comparison of, **47:**85–97
computerized models, **47:**49–50; 97; 102–111; 124–125; 132
conformational parameters, **47:**50–58
energy minimization, **47:**47; 87; 132; 135
helical wheels, **47:**48; 68; 89
helix terminal initiation, **47:**86; 131
histograms, **47:**94
informational theory, **47:**48
Ising matrix (sigma, *s),* **47:**48; 85
joint histograms, **47:**90–91; 94–95; 132
matching helical fragments, **47:**47
multiple regression, **47:**85
multi-state model, **47:**87
nine-state model, **47:**87
nonapeptide, **47:**86
statistical mechanical, **47:**98
stereochemical theory, **47:**88
tripeptide breakers, **47:**48; 50; 89; 96
Prednisolone, microbiological formation, **20:**260; 275

Prednisone, formula, **20:**241
microbiological formation, **20:**260; 261; 265; 275
Pre-enzyme, in enzyme synthesis, **22:**136; 137; 145; 146
properties, **22:**136
Preferential synthesis, of enzyme protein in starved cells, **22:**125; 126
Prefolic A, identity of, **25:**408–409
$\Delta^{4,6}$-Pregnadiene-17α, 21–21diol-3, 11, 20–20trione, microbiological conversions, **20:**254
$\Delta^{1,4}$-Pregnadiene-17α,21-diol-3,11,20-trione, *See* Prednisone
5β-Pregnan-3α, 17α-diol-11,20 dione, **46:** 156
Pregnancies, monitoring, **38:**308
Pregnancy
β-glucuronidase in, **16:**387–388; 391; 394; 397–399
sensitivity to thromboplastin injection, **16:**87
and vitamin E, **6:**507–509
Pregnandiol, **46:**158
5β-Pregnane-11, 20–20dione, **46:**158
Pregnanediol, glucuronide as substrate for β-glucuronidase, **16:**368; 380; 384–385
Pregnane-3,20-dione, formula, **20:**240
microbiological conversions, **20:**273
Pregnane-11α-ol-3,20-dione, microbiological conversions, **20:**273
Pregnane-16α-ol-3,20-dione, microbiological formation, **20:**257
5β-Pregnane-3β-hydroxy-20-one, **46:**159
Pregnane-3,11,20-trione, microbiological conversions, **20:**273
Δ^4-Pregnene-11α,21-diol-3,20-dione, *See* 11-Epicorticosterone
Δ^4-Pregnene-11β,21-diol-3,20-dione, *See* Corticosterone
Δ^4-Pregnene-17α,21-diol-3,20-dione, *See* Substance S
Δ^4-Pregnene-17α,21-diol-3,11,20-trione, *See* Cortisone
Δ^4-Pregnene-3,20-dione, *See* Progesterone
Δ^4-Pregnene-21-ol-3, 20–20dione, *See* Cortexone
Δ^4-Pregnene-20β-ol-3-one, microbiological formation, **20:**253
Δ^4-Pregnene-11β,17α-21-triol-3,20-dione, *See* Hydrocortisone
Pregnenolone, **46:**156
effect on β-glucuronidase, **16:**388
microbiological conversions, **20:**243; 259
Pregnenolone-16α-carbonitrile, **60:**39
Pregnenolone 16α-carbonitrile, **46:**156
Prehypertensin, **7:**523
Preincubation, **34:**473
Preinduction effects, in induced enzyme synthesis, **22:**111; 112; 116
Preinitiation complex
differential interaction with, RNA polymerase II transcription, reversible phosphorylation of CTD during transcription cycle, **71:**58–59
RNA polymerase II transcription, reversible phosphorylation of CTD during transcription cycle, **71:**55–58
Prekallikrein, **48:**278; 290–291
Premature infants, tyrosine metabolism in, **14:**99; 101; 108–110
Prenatal diagnosis, ADA deficiency, **51:**170
Prenyl transferase, **55:**265
Substrate specificity of, **55:**41
PR enzyme, *See* Phosphorylase phosphatase
Preparation and structure of enzyme membranes, **34:**454
Prephenate, **50:**148
Prephenate dehydratase, associated with chorismate mutase, **29:**119
effect of phenylalanine and tyrosine analogs on, **29:**120
Prephenic acid, **18:**362
genetic studies, **17:**414; 416; 417; 421
Prephenic acid in metabolism of aromatic amino acids, **16:**230; 288; 292; 294
Prephenoloxidase, activating enzyme, **45:**419
activation of, **45:**419
activator, **45:**418–419
from *B. mori,* **45:**418
Prephenylalanine, *See* Prephenic acid
Preprotonation, eukaryotic dihydrofolate reductase, chemical transformation step (proton donation), **70:**81–83
Preservatives, hydrocarbon, attack by microorganisms, **10:**451–452; 452–453; 453–454; 474
Pressor agents, therapeutic measures, **7:** 538–545

Pressor amines, **7:**533–538; 541; 542–545
Pressor substances
  dopa, **5:**81
  hydroxytyramine, **5:**74; 79
  tryptamine, **5:**71
  tyramine, **5:**72; 79
Pressure
  and activation of enzymes, **7:**250
    and denaturation of proteins, **7:**248–251
    effects on bacteria, **7:**249
      on enzyme, systems, **7:**249
      on invertase, **7:**258
      on muscle contraction, **7:**249
      on nerve, action potentials, **7:**249
      on oxygen consumption, **7:**249
    and fermentation, **7:**249
    and luminescence, **7:**233–242; 244–251
    and photosynthesis, **7:**249
  effect in enzyme action, **9:**66–69
    effect in enzyme activation, **9:**30–31
  effect on ATPase action of myosin and actomyosin, **16:**336
    use in developing tension in actomyosin, **16:**346
  effect on enzyme action, **12:**382
  film, effect on rate of interfacial reaction, **11:**58–64
Pressure jump, relaxation methods, **25:**2–9
Pre-translational reactions, **67:**274
Primary *n*-alcohols, **34:**488
Primary amino acid racemases, binding specificity, **44:**101
  pyridoxal phosphate binding, **44:**100
  pyridoxal stereochemistry, two-base mechanism, **44:**115
  stereochemical aspects, **44:**113
  sulfhydryl groups, **44:**107
Primary complexes, lipoprotein, preparation of, **28:**209–215
  reconstitution of NADH, succinate oxidase from, **28:**209–220; 222–223; 225
Primary phosphorylation, **1:**124–126
Primary specificity, **53:**259; 260
Primary structure, **39:**200; 346, **42:**123; 124
  HCS, **42:**114; 118
  HGH, **42:**97; 98; 114; 118
  LTH, **42:**114; 118
  SGH, **42:**114; 118
Primary-valence catalysis, **14:**163–173; 179–190
Primer
  in dextran formation, **30:**190
  for glycogen biosynthesis, **34:**351
    for glycogen synthesis, formation of, **34:**351
    requirement, of glycogen synthetase, **34:**321
      of potato phosphorylase, **34:**387
  in murein synthesis, **26:**227–228
    in polysaccharide synthesis, **26:**58–59
Primer-enzyme complex, **25:**246
Primer molecules, *See* Receptor molecules
Primers, *See* Chain initiators
  in polysaccharide synthesis, **28:**338; 340
Primers in polysaccharide synthesis, **11:** 300–302; 309–310; 314; 318; 321–323; 353–358; 359–361
Primeval broth, formation of, **27:**354–365
Priming, respiratory burst oxidase, **65:**58–59
Priming agents for phospholytic synthesis of glucosidic bonds, **12:**413–414; 416; 417
Principles and Procedures in the Isolation of Enzymes (Schwinner and Pardee), **14:**375
Pro, **47:**52; 53; 67; 72; 101; 104
Proaccelerin, **48:**278
  blood clotting factor, **16:**60
Proaccelerin and blood clotting, **10:**28
Proactinomyces erythropolis, enzymic transformations of steroids by, **20:**258
Proactinomyces opacus, oxidation of hydrocarbons, **27:**496; 531
Proactinomyces roseus, enzymic transformations of steroids by, **20:**242; 243
Probability profiles, **47:**88; 97; 126
Procaine penicillin. *See also* Penicillin
  in treatment of plant diseases, **20:**393
Procaine penicillin G, in treatment of plant diseases, **20:**401
Procain esterase, classification, **14:**244
Procarboxylase, **34:**32
  enzymic activation, **34:**35
Procarboxypeptidase, activation of, **24:** 126–127
Procarboxypeptidase A, **56:**292
  column chromatography, **22:**451
Procarboxypeptidase B, **56:**293
  column chromatography, **22:**450–452
Procarboxypeptidases, **23:**131–133; 136
Procarboxypeptidases A and B, localization of, **24:**324

Procaryote phosphofructokinase, **48:**214–216
Procollagen, **37:**332, **41:**280, **42:**167
biological implications, **42:**184
Ehlers-Danols syndrome, **42:**186
heritable defects in collagen, **42:**185
hydroxylysine-deficient collagen, **42:**186
macromolecular assembly, **42:**184
transport, fibril formation, **42:**185
biosynthesis, **42:**179
chemistry, **42:**178
conversion to collagen, **42:**180
historical background, **42:**169
peptidase, **37:**332
registration peptides, **42:**171
size, **42:**174
Procollagenases, **23:**167
Procollagen (Martin, Byers, and Piez), **42:**167
Procollagen peptidase, **41:**280
Procollagens, **23:**167–168
Proconvertin, **48:**278
properties of, **30:**262 (*See also* Factor VII)
use of term, **16:**66; 67
Proconvertin deficiency disease, **16:**65
Procoonase, activation, **45:**405; 409
chemistry, **45:**407
NH_2-terminal sequences of, **45:**408
production in galeae, **45:**404; 407
synthesis and secretion, **45:**403–405
Prodrug targeting, cysteine *S*-conjugate β-lyase reactions, **72:**226–230
Product, **34:**495–503; 513; 517
concentration of, **34:**502
local, **34:**498
Product complexes, glutathione S-transferases, **69:**23
Product dissociation, **43:**349
Product inhibition, **39:**204
biological significance of, **25:**170–173
of enzymes, **25:**167–274
kinetic consequences of, **25:**173–183
mechanisms for, **25:**183–220
prevalence of, **25:**220–267
types of reactions, **25:**183–220
Proelastase, **23:**223–226; 236, **34:**32
enzymic activation, **34:**35
Proenzyme, activation, **17:**447–448
Profibrin, **10:**29–30; 31
Profiles, concentration, **34:**495
Proflavin, **43:**235
Progesterone, **2:**331; 336; 347; 349, **46:**159
absorption spectrum, **4:**299
bound to receptor, **46:**249
effect on β-glucuronidase, **16:**387–388; 390; 394; 399–400
effect on mitochondrial oxidative phosphorylation, **21:**31
effect on *Neurospora crassa,* **20:**264
formula, **20:**240
microbiological formation and conversions, **20:**237; 242–249; 251; 253; 254; 257; 259; 261; 263; 264; 269; 270; 273; 275
hydroxylation, **19:**191
Progesterone derivatives, microbiological conversions, **20:**251
Progesterone-11α-ol, microbiological conversions, **20:**249
Progesterone-11β-ol, microbiological conversions, **20:**251
Progesterone-16α-ol, microbiological conversions, **20:**242; 254
Progesterone-17α-ol, microbiological conversions, **20:**242; 247; 251
17α-Progesterone-11α-ol, microbiological formation, **20:**257
Progesterone-11-one, microbiological conversions, **20:**251
Programmed cell death, **45:**414
Programs, FORTRAN, for fitting enzyme kinetic data, **29:**23–31
Proinhibitors, collagen hydroxylase inhibitor mechanisms, **72:**362–364
Proinsulin, **33:**449; 460; 484, **47:**84; 125–126; 136
C-chain, **33:**460; 509
synthesis of, **33:**507; 509
three-dimensional structure of, **47:**84; 125
β-turn regions in homologs of, **47:**131
Proinvasin I in snake venoms, **8:**484–486; 487
Prokaryotic folate transport, *See* Bacterial folate transport
Prolactin precursor mRNA, methylation, **65:**265
Prolidase, **23:**205
classification, **14:**247
metal-protein complex, **12:**215–216
specificity of, **12:**219–221; 224
Prolinase, **12:**221, **23:**205
classification, **14:**247
linkage attacked by, **8:**164; 165

Proline, **17:**395–399, **35:**482, **37:**247, **42:**118, **47:**47; 49; 51–53; 63; 67; 72; 101; 104; 113, **50:**125; 144; 145; 152; 154; 161; 162; 164, **51:**217, **53:**208; 209; 210; 222, **56:**118
activating enzyme from rat liver, **29:**102
analogs of, **29:**94–108
abnormal development produced by, **29:**96
activation and incorporation into protein, **29:**100
inhibition of growth and development by, **29:**95–100
mechanism of resistance to, **29:**106
molecular properties, **29:**101
chemical structures of, **29:**94–95
control of biosynthesis of, **29:**106
effect of analogs on protein structure, **29:**98
effect on analogs on biosynthesis of, **29:**106
evolution of sRNA synthetase, **29:**102
mutants resistant to analogs of, **29:**106
overproduction by *E. coli* strains, **29:**107
as precursor to hydroxyproline, **29:**103
uptake of analogs, **29:**105
amide, **35:**488
analog, **57:**36
analogs and collagen synthesis, **37:**282
analogs of, in collagen extrusion, **37:**289
biosynthesis, **16:**205; 257–259; 262–264
and citric acid cycle, **18:**253–254; 256–257
control patterns for transport and catabolism of, **56:**132; 133
and glycine, **23:**457; 458
as helix breaker, **47:**52
and inhibition of C-terminal, **57:**35–37
metabolism, **16:**231; 255
and nicotine biosynthesis, **8:**245
and pyridine ring, **8:**236
nitrogen catabolite repression of, **56:**124
as N-terminal amino acid, of protamines, **16:**444
of salmine, **16:**126
oxidation by tyrosinase, **14:**141
oxidation of, **24:**340
in papain, **19:**284
in phosphorylated peptides of esterases, **20:**288
reaction with quinones, **16:**114; 118; 128
residue, **57:**31
substituents, **57:**29
in synthesis, of nicotinamide, **15:**103
metabolic pathway, **15:**185
in tobacco mosaic virus, **15:**462–464
of urea, **15:**324
transport of, **56:**124
D-Proline, **57:**91
L-Proline, **50:**163
L-Proline, oxidation in plants, **20:**320; 322
Proline analogs, as inhibitors of prolyl hydroxylase, **41:**266
Proline auxotrophs, of *E. coli,* **16:**258; 259
Proline/glutamate auxotrophs, **16:**259; 264
Proline hydroxylation, in cell-free systems, **37:**284–293
hydroxyprolyl-sRNA, **37:**286
mechanism of, **37:**285
in plants, **37:**326
on polyribosomes, **37:**283–284
and secretion of collagen, **37:**288
size of the precursor, **37:**285–289
stage of hydroxylation, **37:**285–289
in vivo, in animals, **37:**282
Proline hydroxyproline pyrrolidine, reactions with quinones, **16:**112
Proline/ornithine auxotrophs, **16:**262–264
Proline-ornithine-glutamate, transaminations, **16:**231
Proline-ornithine interconversion, **16:**205; 231
L-Proline oxidase, classification, **14:**253
Proline oxidase, of a cyclophorase preparation, **16:**259
Proline racemase, **44:**124
active-site models, **44:**127
binding of pyrrole-2-carboxylate, **44:**126
hydrogen exchange, **44:**125
inhibition, **44:**125
kinetic anomally, **44:**120
Δ'-pyrroline-2-carboxylate as inhibitor, **44:**125
sarcosine as substrate, **44:**126
sulfhydryl requirement, **44:**125
Proline-reductase system, **18:**342
Proline residues, and collagenase action, **23:**191; 205–208; 211–212; 218
L-Proline r O_2 transhydrogenases, **14:**253
Prolines
cytochrome *c* oxidase, **71:**91
RNA polymerase II transcription, C-terminal domain characteristics, extended structure, **71:**50

Prolipase, **23:**131
Prolyl hydroxylase, **41:**245; 293
activation by ascorbate, **41:**284–286
activation by lactate, **41:**283; 284
amino acid composition, **41:**248
antibodies, **41:**251
from chick-embryo, **41:**248
crossreacting protein, **41:**285
elevated levels, **41:**290
enzyme immuno assay, **41:**259
inactivation by dithiothreitol, **41:**251; 286
inhibitors, **41:**266
a-keto glutarate as cosubstrate, **41:**254
localization, **41:**278
molecular weight, **41:**250
occurrence, **41:**247
other names for, **41:**246
peroxy intermediate, **41:**272
purification and properties, **41:**248
from rat-skin, **41:**248
requirement for *a*-ketoglutarate, **41:**253
requirement for Fe^{+2}, **41:**252
requirement for molecular oxygen, **41:**247; 252
requirement for reducing agent, **41:**253
stimulation by bovine serum albumin, **41:**255
stimulation by catalse, **41:**255
stimulation by dithiothreitol, **41:**256
structure, **41:**251
substrates, **41:**260
subunits, **41:**250; 288
and superoxide ion, **41:**273
Prolyl Hydroxylase (Cardinale and Undenfriend), **41:**245
Prolyl 3-hydroxylases
catalytic reactions, **72:**326–327
co-substrates and reaction mechanisms, **72:**345–357
inhibition and inactivation, **72:**357–364
intracellular hydroxylation sites, **72:**338–340
molecular properties, **72:**338
peptide substrates, **72:**339; 341–345
regulation mechanisms, **72:**384–386
Prolyl 4-hydroxylases
catalytic reactions, **72:**326–327
co-substrates and reaction mechanisms, **72:**345–357
inhibition and inactivation, **72:**357–364
intracellular hydroxylation sites, **72:**338–340
molecular properties, **72:**334–336
peptide substrates, **72:**339; 341–345
regulation mechanisms, **72:**384–386
α-subunit structure, **72:**364–368
Prolyl-tRNA, **41:**276
Prolyl-tRNA synthetase, **50:**161; 162; 163; 164
Promazine, **55:**182
Promitochondria, **46:**91
Promoters, γ-glutamyl transpeptidase gene expression, **72:**265–267
Promotor clearance, RNA polymerase II transcription, potential roles, **71:**62–64
Promotor interactions, RNA polymerase II transcription, potential roles, **71:**61–62
Promotor specificity, C-terminal domain characteristics, RNA polymerase II transcription, **71:**46–49
Promyelocytes, **60:**219
Pronase, **34:**485, **35:**338, **41:**112; 124
action on biotin complex, **26:**328; 331
Prontosil, **27:**418; 420
Propanediol phosphate, **15:**114
Propanediol phosphate and liver phosphorus, **7:**123
Propanol, catalase action on, **19:**109
Properdin system, effect of glucans, **28:**357–358
Properties, enzyme, **42:**265
Prophyrin, of aldehyde oxidase, **17:**244–245; 268
role in electron transfer, **17:**110; 338–339
Propimate metabolism, **35:**379
Propionaldehyde, DR-aldolase substrate, **28:**450
Propionaldehyde, oxidation, **15:**256; 257; 258
Propionate
in citric acid cycle, **15:**249; 251
mechanism of oxidation, **15:**7
formation in photosynthesis, **19:**40–42
labeled, **19:**42
Propionate (C^{14}), and carboxylation, **26:**308
formation of odd-numbered fatty acids, **26:**370
β-oxidation pathway, **26:**367–368
Propionbacterium, and citric acid cycle, **18:**271
Propionbacterium shermanii, **59:**330
Propionibacteria, **45:**94
control mechanisms in, **45:**94
electron transport system, **45:**137
fermentation phosphorylation scheme, **45:**135
Propionibacterium arabinosum
propionic acid metabolism, **26:**308
and succinate fermentation, **74:**19

Propionibacterium pentosaceum
metabolism in, **74:**21
propionic acid metabolism, **26:**292
propionyl CoA, carboxylase in, **19:**40–42
Propionibacterium shermanii, **46:**51; 54; 87
metabolism of, and P-P_1, **45:**134–140
pyruvate, phosphate dikinase, in, **45:**108
as source of vitamin B_{12}, **46:**87
and succinate fermentation, **74:**18–19
Propionic acid
from amino acids, **26:**344
catabolic pathways (summary), **26:**369
formation in *P. shermanii,* **26:**309
metabolism, **26:**283–378
in man, **26:**344
β-oxidation, **26:**367–370
role of cobamide coenzymes, **26:**265; 285; 335; 343–367
from succinic, by bacteria, **26:**292
bacterial fixation of carbon dioxide in, mechanism of, **74:**29–34
bacterium, **2:**135–137; 144; 147; 151
location of fixed carbon, **2:**160–161
mechanism of CO_2 fixation, **2:**161–166
carboxylation, in mitochondria, **21:**46; 47
metal ion complexes of, model catalyses with, **21:**189
fermentation, mechanism of, **74:**14–17
location of fixed carbon in, **74:**28–29
from oxidation of hydrocarbons, **27:**487; 490; 494
Propionic Acid, Metabolizm (Kaziro and Ochoa), **26:**283
Propionic acid bacteria, *See* Propionibacteria
isotope studies on, **15:**177
Propionic acid fermentation, **35:**402
fixation of formaldehyde in, **12:**127–128
Propionic acid groups of protoporphin, **7:** 325–326
Propionic acid oxidase, **19:**203
Propionyl CoA, **18:**230, **43:**76; 80
biotin content, **35:**383
carboxylase, **35:**379
holocarboxylase-synthesizing system, **35:**338
in photosynthesis, **19:**41–43
physical-chemical characteristics, **35:**383
reaction
arsenate-simulated ATPaSe, **35:**381
ATP-$^{3\ 2}$-P-and ATP-$^{3\ 2}$-labeled ADP exchange, **35:**380
bicarbonate-carboxylating species, **35:**382
Propionyl-CoA
carboxylation of, **26:**284; 286–334
phosphorolysis of, **23:**338
Propionyl-CoA apocarboxylase, activation of, **26:**331–334
Propionyl-CoA (C^{14}), exchange reactions, **26:**313–316
Propionyl-CoA carboxylase, **37:**554
Propionyl CoA carboxylase, in photosynthesis, **19:**40–43; 59; 60
Propionyl-CoA carboxylase [propionyl-CoA:CO_2 ligase (ADP)], amino acid composition, **26:**288
carboxybiotin complex, **26:**320–330
eaction mechanism, **26:**311–316; 320–330
formation of holocarboxylase, **26:**331–334
metabolism of propionic acid, **26:**284–285; 297; 334–343; 345
properties of, **26:**286–292; 300; 305; 310–311
catalytic constants, **26:**291; 315
Propionyl-CoA (C^{14})-methylmalonyl-CoA exchange rate, **26:**313–316
Propionyl-CoA oxidation, of hydrocarbons, **27:**472
Propionyl coenzyme A, **15:**197; 209; 236; 237; 244 (S^{35})
Propionyl-L-glutamate, **15:**371
Propionyl phosphate, **15:**209
Propiothetin, *See* Dimethyl-βpropiothetin
Pro-prochiral center, properties of, **50:**245–247
Pro-prothrombinase, **10:**28
Propylamino groups, decarboxylation and transfer, **22:**261; 262
Propylbenzilyl choline mustard, **55:**141; 150
Propyl benzyloylcholine mustard, **55:**163
S-(*n*-Propyl)glutathione, **59:**116
γ-Propylideneglutamic acid, **50:**118
Propyl isopropylacetamide, **46:**128; 137
Prorennin, **15:**425
Pro-81 residue, triad amidotransferase, **72:**139
Proserozyme, **10:**13; 27–28
Prostaglandin A_1, **46:**396
Prostaglandin cyclooxygenase, **55:**199
Prostaglandin endoperoxide, **46:**173
Prostaglandins, **57:**377–378, **59:**133
Prostaglandin synthase, **58:**215
Prostaglandin synthesis, **58:**233
Prostate phosphatase, **9:**519
Prostatic phosphatase, **15:**440

Prosthestic groups, hydrogen bonds to, **50:**28
Prosthetic group, **39:**347
  use in enzyme terminology, **14:**240–243
Prosthetic group removing enzyme, **12:**415–417
Prosthetic groups
  attachment to protein, **24:**23–24
    effects on proteins, **24:**110–111
  biological function, **1:**4
  catalase, **1:**170
  of cytochromes *a,* **23:**265–321
  enzymes, **1:**177; 178; 186; 187; 190; 192; 193
  protein, **1:**4
  of succinic dehydrogenase, **18:**76–81; 93–94
  visual purple, **1:**190
Prosthetic groups of enzymes, mechanisms of action, **17:**273–281
Prostigmine
  effect on cholinesterase, **20:**295; 298
  effect on esterase activity, **6:**210
    on muscle cholinesterase, **6:**209; 210; 211
  effect on hydrolysis of acetylcholine, **12:**284; 286–290; 292–293; 299; 305–306
  in familial periodic paralysis, **6:**216
Protacatechuic oxidase, enzymic adaptation, **13:**71
Protactinomycin, **7:**480; 487; 489; 491; 493
Protamine
  action of tyrosinase on, **14:**150
    use in isolation of enzymes, **14:**398
  as activator of β-glucuronidase, **16:**376
    of cell nucleus, **16:**413; 441–444; 446
    effect on heparin action, **16:**88
  effect on blood coagulability, **10:**37
    precipitation of fibrinogen, **10:**35
  effect on β-glucuronidase, **40:**40
    effect on hyaluronidase, **40:**41
  reaction with phosphorylase *b,* **24:**265; 274
Protamine-ATP compound, **17:**144
Protamine-heparin compound, **10:**34
Protamine nucleates, **17:**137; 143; 145
Protamines, associated with DNA, **18:**162; 170–172; 176–177
  in purification of succinic dehydrogenase, **18:**72
Protamine salmine, activation of phosphorylase b, **40:**44
Protaminobacter species, enzymic transformations of steroids by, **20:**262
Protease, **5:**23
  in amino acid assessment of cytochrome, c_1, **54:**306
  chromatin, **38:**81–82
  in fructose-bisphosphatase regulation, **54:**172
  neutral, **56:**293; 304
Protease action, on protamines, **16:**442
Protease activity
  effect of radiations, **24:**388; 392–394
  of imidazole, **20:**74; 76–82
Protease from Russell's viper venom (RVV-X), **48:**296
Protease inhibitor, **43:**124; 129
Protease inhibitors, **55:**3
Proteases, **1:**63, **15:**272, **33:**271, **46:**481–484, **47:**281, **50:**12; 13. *See also* Proteinases; Proteinases and Proteolytic enzymes; Proteolytic enzymes
  acid, **45:**414
    antibodies, **45:**410
    cathepsin-like, **45:**413
    in digestion of endocuticle, **45:**417
    digestive, **45:**390–402
    fecal, **45:**401; 402
    inhibitors, **45:**393
    insect, and peptidases, **45:**389–425
    in insect development, **45:**411
    in insect molting, **45:**415–420
    lysosomal, **45:**413–415
    secretion, **45:**400–401
      control of, **45:**400
      stimulation, **45:**401
  active sites of, **24:**459
    localization of, **24:**315; 325
    mechanisms of catalysis, **24:**458–464
  activity in frozen state, **9:**626–627
    titrimetric methods, **9:**601
  adsorbed, activity of, **28:**28
    soil, methods for, **28:**384–386
      results, **28:**387
  amino acid composition near active site, **22:**65
    labeling of active site by diisopropylfluorophosphate, **22:**52
  brain-specific protective, **29:**503
  in casein preparations, **15:**431
  endogenous, adenosine receptors, **69:**96
  linkage attacked by, **8:**164; 165
    of snake venom, **8:**460; 476; 486–490
  in living plants, **1:**336; 339; 340; 352–355; 361; 364; 365
  mechanism of action, **15:**3

Proteases (*Continued*)
multiple forms, **25:**351; 352
product inhibition of, **25:**221–225; 229
sulfonyl derivatives of, **33:**385–386
thromboplastic activity, **10:**12
transfer reactions, **17:**190–191
Proteases and Peptidases, Insect (Law, Dunn, and Kramer), **45:**389
Proteases and proteolysis, **43:**107; 109; 114; 129; 134; 136; 137; 151; 203
polypeptide degradation, **43:**149
trypsin, **43:**153
Proteases in tobacco, **6:**369
Protease zymogens of pancreas, **24:**166
Protecting groups, **33:**459; 462; 463; 466; 479; 481–482; 484; 488; 491–493; 500
Protection against denaturatuon, **43:**155; 193
Protection of ascorbic acid, by glutathione, **11:**250
of –SH groups in enzymes, **11:**234; 242–243; 250–251; 256–257
Protective agents, and ionizing radiations, **24:**362; 374–375; 376; 380–381; 386; 391–392; 394–397
Protective factor X, **35:**323
Protein, **34:**451–455; 467–470; 476–477; 484–486
acid hydrolyzate, analysis of, **42:**237
aggregated, **42:**77; 80; 127
aggregation, **47:**81
biological synthesis, **47:**380–381
chain reversal, **47:**60–64
conformational changes, **47:**74
DNA interactions, **47:**79; 81
folding, **47:**46; 88; 133–135
N- and C-terminals, **47:**94; 96
predicted structures, **47:**45; 97; 137–139
secondary structure, **47:**52; 88; 92–93
sequence determinations, **47:**46
synthesis, origin of, **47:**425
X-ray atomic coordinates, **47:**58; 70
boundaries, **34:**454
density, **43:**337
fraction of occupied space, **43:**248
packing density, **43:**248
dietary, effect on xanthine oxidase, **17:** 319–321
role in electron transfer, **17:**110; 337–338
effect of analogs on structure of, **29:**153
microtubule, **53:**117–119
molecules, **34:**446
phosphorylation of, **38:**76–78
synthesis, **38:**198–199; 200; 201
turnover, **37:**138–143
Protein, Vitamin K dependent, **48:**277
Protein A, **48:**20
Protein adsorption, specificity of, **28:**20
γ-Proteinaes, of *Cl. histolyticum,* **23:**198; 202
Protein amino acids, **50:**140; 143
intermediates in, **50:**143–145
modification of, **50:**145
syntheses, **50:**143
Protein analysis, and disulfides, **27:**229–232
Protein apoR2, **65:**166
Proteinase
B. subtillis neutral, **53:**283
cacao, **25:**466; 480; 486
multiple forms, **25:**337–338; 350–351
of *Streptococcus pyogenes,* inhibitory action of sera on, **6:**99
from the venom of *Gotalus atrox,* **41:**201
λ-Proteinase, **23:**186–188; 191–197; 198
δ-Proteinase, **23:**187–189; 195; 198; 200
Proteinase activity, **45:**417
Proteinase C, from *Saccharomyces Gerevisiae,* **41:**197
Proteinase-Catalyzed Synthesis of Peptide Bonds (Fruton), **53:**239
Proteinase I, localization of, **24:**326
Proteinase II, localization of, **24:**327
Proteinase inhibitors, **47:**125–130; 136
amino acid sequences, **47:**127
B-turn occurrence in, **47:**127
sequence homology of, **47:**125
Proteinase K, **58:**248
Proteinases, **2:**69; 70, **56:**2, **58:**251, **60:**227. *See also* Microbial proteinases; Proteases; Proteolytic enzymes; specific enzymes
action on
denatured proteins, **23:**31–36; 48; 50–51; 55; 64
native proteins, **23:**32–36
adsorption on substrate, **14:**401; 405
in *B. subtilis,* **14:**300–301
bacterial, action on
collagen, **23:**171; 177–181; 183–189; 194; 196–200; 202; 220–221
cytochrome *c,* **23:**62; 64–66; 67
denatured proteins, **23:**31–45; 70
elastin, **23:**227; 229–231
reduced, **23:**62–64
Taka-α-amylase, **23:**75–76

in carious lesions, **23:**203
Ca-stabilized, **23:**188
catalysis of exchange reactions, **13:**305–310
inhibition, **13:**321–343
localization in cytoplasm, **13:**7; 10; 12
classification, **1:**64, **2:**51; 53
clostridial, **23:**178–179; 184–189; 191; 194; 197; 198; 200; 202
complex with DFP, **23:**72–74
ferrous-activated, **23:**188
gastro-intestinal, **1:**69–78
hydrolysis, **1:**65; 66
intracellular, **1:**78–85; 88
kinetics, **1:**89; 90
mode of action, **14:**183; 190
pancreatic, **12:**192
in proton structure study, **14:**267; 285
separation by chromatography, **14:**329; 337; 343; 345; 349; 363
specificity, **1:**63–98
specificity of, **33:**430
stereochemical specificity, **1:**75; 82; 84; 85; 87; 93–94
substrate specificity, **23:**76–78
and synthesis of proteins, **9:**428–430
inhibition, **9:**429
Proteinases, Microbial, Comparative Specificity of (Morihara), **41:**179
Proteinases, Specificity (Bergmann and Fruton), **1:**63
Protein association in the glutamate dehydrogenase reaction, **39:**411
Protein bonds, **1:**39; 40
Protein-bound iodine (PBI), in catatonia, **29:**530
Protein breathing, **59:**286
Protein-Ca^{++} binding, **10:**14
Protein-carbohydrate interactions, in cells, **20:**93; 94
Protein-carbohydrate linkage in glycoproteins, **25:**450–453
Protein chemical modification, **53:**69–72
Protein chemistry, glutathione synthesis, **73:**216–220
Protein-chitin complex, **16:**162
Protein-coenzyme interaction, **15:**37–42
Protein cofactors, **60:**177–186
purification, **60:**178
Protein combination with porphyrins, **12:**14
Protein complex, **43:**201; 202
Protein composition, and enzymic function, **27:**39–41
Protein compounds, solubilization of, **17:**137; 143–146
Protein configuration and thiol groups, **11:**219
Protein conformation
and coenzyme binding, **22:**345; 346
and enzymic activity, **22:**361; 362
as function of energy, **22:**74–76
and Cotton effect, **27:**46–48
Protein conformation and enzyme action, **25:** 20–21
Protein coupling, adenosine receptors, **69:**97
Protein denaturation, **3:**43–45
Protein disulfide, formation, dehydroascorbate and protein disulfide isomerase, **66:** 181–182
Protein disulfide isomerase
DHA reductase activity, **66:**181–182
protein disulfide formation, **66:**181–182
Protein disulfide isomerase (PDI), prolyl 4-hydroxylase β subunit, **72:**335–336
cosubstrate values, **72:**348–357
function, **72:**375–380
intracellular hydroxylation sites, **72:**338–340
regulation mechanisms, **72:**385–386
structure, **72:**368–375
Protein-enzyme competition, in sulfonamide action, **6:**65–67
Protein (enzyme) surface, diagram of, **22:**47; 48
Protein error, with indicators, **11:**47; 55
Protéines, Structure des (Desnuelle), **14:**261
Protein fibrils. *See also* Fibrils
structural periodicity, **8:**14–17
Protein films. *See also* Film formation
relaxation times, **28:**16
spatial arrangements, models, **28:**9–12
Protein-forming system, irreversibility, **22:** 127–129
Protein gels, and SH-groups, **27:**221–222
Protein inhibitor, of blood coagulation, **16:**87
Protein inhibitors, **49:**259
Protein kinase, **34:**329; 382, **46:**150, **50:**315
calmodulin-dependent, aromatic amino acid hydroxylases as substrates for, **74:** 272–273; 276–277; 281
cAMP dependent, **50:**41–108
autophosphorylation of, **50:**95–96
chemical properties, **50:**62–66
amino acid composition, **50:**62
chemical modification, **50:**65
covalently bound phosphate, **50:**63
sulfhydryl groups, **50:**64

Protein kinase (*Continued*)
isoenzymes, **50:**54–57
antibody, **50:**54
type I, **50:**54
type II, **50:**54
kinetics, **50:**80; 81
cAMP as activator of native enzyme, **50:**80
effect of nucleoside triphosphates, **50:**80
effects of divalent cations, **50:**80
mechanism of, **50:**80
reversibility of reaction, **50:**81
physical properties, **50:**47; 48
molecular weight, **50:**47
SDS gel electrophoresis, **50:**49
sedimentation constant, **50:**47
purification, **50:**44
of catalytic subunit, **50:**45
of holoenzyme, **50:**44
of regulatory subunit, **50:**46
regulation of activity, **50:**82–108
cyclic AMP binding, **50:**82
effect of MgATP, **50:**84–85
heat-stable protein inhibitor, **50:**85
specificity, **50:**66; 75
comparison with phosphorylase kinase, **50:**74; 76
protein substrates, **50:**67; 68; 69; 70; 71
use of synthetic peptides as substrates, **50:**71
subunit stoichiometry, **50:**50–53
cAMP-dependent, aromatic amino acid hydroxylases as substrates for, **74:**270; 272; 274; 278; 281
cholesterol ester hydrolases, **69:**46
Protein Kinase, Cyclic AMP-Dependent; Chemical and Regulatory Properties of Phosphorylase Kinase and (Carlson, Bechtel, and Graves), **50:**41
Protein kinase C, cholesterol ester hydrolases regulation, **69:**74
Protein kinase C (PKC), tyrosine hydroxylase regulation, **70:**143; 145; 146; 148
Protein-ligand interactions, alloplastic, **37:**404
autoplastic, **37:**404
autosteric, **37:**403
homosteric, **37:**403
plasticity, **37:**403
Protein ligase problem, glutamine synthesis, **72:**38–40
Protein-lipid interaction, **34:**478, **58:**239
Protein localization (*Caulobacter* cell cycle), **71:**25–30
chemotaxis proteins, targeting of, to swarmer pole, **71:**26–28
flagellar assmebly and, **71:**28–29
generally, **71:**25–26
Protein metabolism
alterations in neoplastic cells in, **29:**363–367
and carbohydrate metabolism, **3:**245–246
Protein methylase I, **42:**257
assay for, **42:**247
distribution of, **42:**249; 264
Protein methylase II, **42:**257; 268
activity, **42:**256; 273
animal age, **42:**273
assay for, **42:**248
catalytic properties, **42:**270
distribution of, **42:**249
molecular properties, **42:**271
properties of, **42:**270
purification of, **42:**269
Protein methylase III, **42:**250; 257
assay for, **42:**248
distribution of, **42:**249
K_m value of, for S-adenosyl-L-methionine, **42:**252
properties, **42:**251
purification, **42:**251
Protein Methylation: Chemical, Enzymological, and Biological Significance (Paik and Kim), **42:**227
Protein-methyl ester, **42:**272
Protein moiety of enzyme, function, **15:**11; 18; 27; 30; 32–42
Protein molecules
association of, **24:**119–120
folding of polypeptide chain, **24:**22–25
enzyme, conformational changes, **27:**23–29
evolution of, **47:**417
Protein-nucleic acid interactions, **49:**189
synthetase-tRNA interaction, **49:**189
Protein phosphatase-1, mammalian, **74:**71
Protein phosphokinase, action on phosphorylases, **24:**282
Protein polysulfide, **39:**340
Protein predictors, **47:**47; 48; 50; 86; 87; 88. *See also* Predictive methods

Argos-Schwarz-Schwarz, **47:**132
Barry-Friedman, **47:**88; 90–95; 132
Blagdon-Goodman, **47:**86; 131
Burgess-Ponnuswamy-Scheraga, **47:**86; 90–97; 132
Chou-Fasman (computer method), **47:**97; 102–111
Chou-Fasman (empirical rules), **47:**50–58; 88–102; 132; 134; 137–139
Finkelstein-Ptitsyn, **47:**48; 88–97
Froimowitz-Fasman, **47:**85–86
Guzzo, **47:**47; 93; 95
Haggis, **47:**48
Kabat-Wu, **47:**50; 89–96; 102; 132
Kotelchuck-Scheraga, **47:**47; 90–92
Krigbaum-Knutton, **47:**85
Leberman, **47:**48; 93; 95
Lewis-Bradbury, **47:**131
Lewis-Go-Go-Kotelchuck-Scheraga, **47:**48; 90–92
Lewis-Scheraga, **47:**48; 131
Lim, **47:**88–97
Low-Lovell-Rudko, **47:**47; 75; 134
Nagano, **47:**49; 87–97; 132
Periti, **47:**86
Periti-Quagliarotti-Liquori, **47:**47
Prothero, **47:**47; 93; 95
Ptitsyn-Finkelstein, **47:**49; 93; 95; 96
Robson-Pain, **47:**48; 86–87; 90–92
Schellman, **47:**93; 95–96
Schiffer-Edmundson, **47:**48; 75
Tanaka-Scheraga, **47:**98
Wu-Kabat, **47:**48; 50
Zimm-Bragg, **47:**48
Protein-protein interaction, **34:**478
Protein-protein interactions, **40:**57; 58; 59
Protein radicals, effect of irradiation, **24:**385
Protein-ribonucleate complex, in autolysis, **6:**82
Proteins, **46:**384–389, **50:**21; 23; 24. *See also* Amino acids; Carotenoproteins, Lipoproteins, Phosphoproteins, Pteroproteins and specific compounds; Corpuscular proteins; Enzymes; Fibrous proteins; Immobilization; Isolation of enzymes, and specific proteins; individual proteins; specific proteins
absorption in depancreatized dogs, **8:**180–181
antigenic action, **8:**54–55
and fatty acid manufacture, **8:**316
foreign, as cell poisons, **8:**54–55
and functioning of cytoplasm, **8:**9–10; 18–23; 46–47; 57–58
lipotropic effect, **8:**192
mechanism of synthesis, **8:**58–59
metabolism and alkaloid biogenesis, **8:**206; 207; 212; 217; 228; 234; 238; 239; 241–245
and purine metabolism, **8:**242
absorption spectra, **4:**279; 307, **9:**96–97
accommodation of, by water pools, **51:**40–47
acetylation of, **38:**78–80
conformation in chromatin, **38:**59–61
distribution on DNA in chromatin, **38:** 67–70
DNA binding sites, **38:**61–64
methylation of, **38:**80–81
origin of, **38:**216
role in chromatin structure, **38:**64–67
acylation of, **20:**73–75
biosynthesis in plants, **20:**335; 336
effect on pectin substances, **20:**356
imidazole reactivity in, **20:**58–61
native, D-glutamic acid and D-leucine in, **20:**25
optical purity in living organism, **20:**2
in plastids, **20:**204; 206–209; 229
synthesis in plants, **20:**219–223
adsorption studies, **14:**319–374
aggregation of, **27:**220–226
asymmetry, **27:**38–41; 46–49; 52–53; 55–56; 61
biosynthesis in nucleus, **27:**275
coagulation, and —SH groups, **27:**223–225
Cotton effects, **27:**41; 46–49; 51–52
denaturation, **27:**45–47; 52; 77–78; 82; 84–85; 88; 149–150; 159; 222–226; 230
disulfide bridges, **27:**206
disulfide exchanges, **27:**220–229
with lower molecular weight substances, **27:**210–217
in disulfide exchanges, **27:**203
disulfides of, enzymic reduction, **27:** 217–219
gel formation, **27:**221–222
globular, effect of urea, **27:**221
metal, EPR, **27:**144–187
Mössbauer spectroscopy, **27:**107
non-enzymic, Cotton effects in complexes of, **27:**60

Proteins (*Continued*)
optical rotatory dispersion of, **27:**38–49; 52–53; 55–56; 61
and origin of life, **27:**357; 358; 361; 363; 366–371; 374–375; 390; 392; 404
reactions with disulfides, **27:**212–217
reactions with mercaptans, **27:**210–212; 230–231
sequence analysis, enzymic hydrolysis, **27:**229–230
SH-groups, and denaturation, **27:**222–226; 230
in vitro synthesizing systems, **27:**290; 297
allostery in, **50:**30–33
amino acid composition, **46:**387–389
amino acid incorporation into, in mitochondria, **21:**44
amino acid residue in
chemical method for replacement of, **33:**383–399
enzymic method for replacement of, **33:**399
amino acids, **1:**2; 64
in, **58:**174
p-aminobenzoic acid in, **6:**39–40
analysis of hydrolyzates, **6:**15
denaturation by irradiation, **6:**82–84; 92–93
digestion in air-cured tobacco, **6:**339–343
effect in progressive muscular dystrophy, **6:**220
reactivated enzyme, **6:**83
analysis, **1:**2
antibody to transferases, **46:**389
associated with DNA, basic, **18:**169–177
neutral, **18:**175–176
nonhistone, **18:**177–181
association-dissociation, **1:**17; 18
autosynthesis, **4:**8–11
bacterial cell (*See* Bacterial cell protein)
binding, **46:**390, **49:**59
non-heme-iron, **49:**73
biologically active, simulated mutation at active site of, **33:**381–400
biosynthesis, **9:**654; 655–658; 658–661; 665; 669; 671; 675–676; 679–680
by activated amino acids, **19:**260; 264
in egg white chromosomes, **19:**261
biosynthesis of, **57:**155
of cacao, **25:**465–468; 480; 484
rate constants for reactions, **25:**9–15
and cell preparation, insulin-sensitive, **57:**158
chelating properties, **21:**143
chemical and x-ray evidence on structure, **3:**63–108
chromatography, **22:**419–437
chromosomal, **16:**412–413
contractile, **16:**316–317; 319 (*See also* Actomyosin and Myosin)
cytoplasmic, comparison with residual, **16:**447
fibrous (*See also* Actomyosin and Myosin)
fibrous α-structure of myosin, **16:**318–319; 325–326
formation of sclerotin from, **16:**153–163
localization in nuclei of cells, **16:**417–419
of melanin granules, **16:**169
nuclear, **16:**413; 415; 416; 418; 420; 426; 436; 440–450; 458
reactions with quinones, **16:**122–127; 128
separation by chromatography, **16:**238
water-soluble, quinone tanning of, **16:**156
and citric acid cycle, **18:**267
classification, **1:**15; 16; 20; 23
collagen group, **3:**91–96
concentration effect on photocross-linking, **66:**123–125
conformational changes in, **50:**30–33
conformation change, and solubility, **23:** 70–72
conjugated, effect of oxidation state on, **23:**64; 67–68
digestion of, method of estimation, **23:**35
enzymes, denaturation of, **23:**29–82
inactivation of, **23:**29–82
rotatory properties of, **23:**37; 72–76
solubility, and oxidation state, **23:**72
globular, denaturation of, **23:**30–32; 38; 48; 50; 55
molecular structure, and denaturation, **23:**50; 55; 70–72
and inactivation, **23:**31; 40–41; 46–48
and oxidation state, **23:**58–70
secondary, **23:**50
and solubility, **23:**70–72
polypeptide configurations, **23:**1–27
constituents of chromatin, **4:**11–17
crystallized
from aqueous-organic mixtures, **52:**197
from salt solutions, **52:**197; 199
cyclol-cage structure, **1:**29–33
cyclol pattern, **4:**4; 5; 6

denaturation, **1:**35–40, **7:**235; 248–251, **9:**31–39; 66–69; 98–99; 101–108; 112; 113; 124–127; 128–131; 133–137; 141; 143; 146–148; 159–162
dissociation, **7:**39
egg white, **7:**196
hydrolysis, **7:**56
protoplasm, **7:**54
shape, **7:**61
synthesis, **7:**56
denaturation of relation to thiols, **11:**220–223; 224; 226; 255
enzymic activity and thiol groups, **11:** 232–244
mercaptide formation, **11:**227–228
nonenzymic activities of –SH groups, **11:**244–248
oxidation of thiol groups in, **11:**207; 210; 219–221; 223–226
reactions with alkylating agents, **11:** 226–227
reactions with reducing agents, **11:**228–229
sulfur distribution in, **11:**229
surface reactions, **11:**39–41; 46–47; 47–56; 67; 71; 76; 81–82; 83
types of –SH groups in, **11:**219–221; 233–235
destruction by radiation, **14:**35–37
determination of amino acid sequence, **19:**298
distinction between enzymic and non-enzymic, **15:**36–37; 296–297
role in energy transfer, **15:**295–296
of tobacco mosaic virus, **15:**461–464
of virus (TMV) mutant, **15:**479–482
disulfide bridges in, **10:**278; 286–288
DNA cross-linking, via psoralens, **66:** 119–126
domains of, **50:**24
effect of metal ions on stereochemistry, **21:**158; 161; 162
of egg, changes following fertilization, **9:** 283–285; 315–317
electrophoretic behavior of, in carboxylic acid-containing media, **30:**162–163
electrostatic forces in, **50:**24
in energy transport, **14:**6; 19–42
enzyme action on, **33:**255–259
enzymes, **1:**177; 178
enzymic, isolation, **14:**375–410
fibrous, **1:**5–13, **4:**2
β-Galactosid binding, **52:**116
globular, **1:**13–40, **4:**2
growing at subzero temperatures, **52:**220
hydration, **1:**19
hydrogen-bonded network of, **50:**24
hydrogen bonding, **65:**218; 224
hydrogen bonds in, **50:**3; 23
immobilized, **57:**183–185
interactions of, with other proteins, **57:** 209–210; 216–217
interfaces between domains in, **50:**24
hydrogen bonds extending across, **50:**24
isoelectric point, **46:**386
keratin-myosin group, **3:**76–91
α-keratin pattern, **4:**6
light-induced changes in, **9:**75–76; 108–162
long-range hydrolysis, **9:**69–71
metabolic activity of amide nitrogen of, **13:**266–267
of muscle, **13:**151–197
synthesis, genetic studies, **13:**146–147
relation of asparagine, glutamine and glutathione to, **13:**237–319
specific protein synthesis in biosynthesis of enzymes, **13:**69–73; 77; 80; 83–88
metabolism, **1:**95
microtubule-associated, microtubule dynamics, **69:**144
mitochondrial, effect on mitochondrial oxidative phosphorylation, **21:**32
mol.-weight, **1:**19; 20
molecular structure, **4:**2–4
molecular weight, **4:**2; 3
native, **57:**185
new techniques for study of structure, **14:**261–318
2^n3^m rule, **3:**66–71; 88–91; 99; 100; 104; 105
nonenzymic reactions with activated amino acids, **19:**261
nonfibrous, **4:**2; 14
non-heme-iron, **39:**361
and nucleic acids, **4:**4–8
nucleo- (*See* Nucleoproteins)
genic significance, **18:**184–186
oligomeric
assembly of, **57:**201–204
enzyme activity of, **57:**196–201
structure of, **57:**187–193
oxidation by tyrosinase and peroxidase, **14:**129–161
papain action on, **19:**302; 303
peptide linkage, **1:**2; 67; 69

Proteins (*Continued*)
phenylalanine or tyrosine metabolism in, **14:**74; 119–120
with phospholipids, **1:**4; 5
in plant metabolism, **9:**391–439
in plant respiration, **18:**283–284
and plant viruses, **17:**483–489
as complexing agents, **17:**149–151
serologically related to plant viruses, **17:**483–496
solubilization of, **17:**143; 146
titration curves, **17:**38
properties, **5:**37
proportion of amino acid residues, **3:**64–71; 74; 77–82; 94–97; 100
prosthetic groups, **1:**4
purification, **46:**248–249; 384–386
rates of degradation, **37:**139
refolding of, **57:**212–216
of respiratory chain, **28:**160–161
ribonucleo-, **4:**211; 229; 230; 231
secondary amino acids, **67:**267–271
self-organizing, **57:**213–214
in seminal plasma, **9:**342–343
serum, column chromatography, **22:**432
shape of the molecules, **1:**23–28
side chains, **1:**3; 4
silk fibroin, **3:**73–76
sizing, **52:**118
in solution cf. crystalline, **26:**92–94
as solvents, **50:**21
in sperm, **9:**338–339
storage, **46:**407
structure, **1:**1–42, **50:**23
effect on chromatography, **22:**425–428
structure of
and cross-linking, **57:**233–234
and electro microscopy, **57:**234
and immobilization, **57:**183–185
quaternary, **57:**187–195
solid supports, **57:**196–204
tertiary, **57:**126–127
and x-ray analysis, **57:**233
subunit structure, **46:**387
sugar-specific binding, **52:**113
synthesis, **1:**17; 154, **9:**421–426; 428–439; 654–661; 665; 669; 671; 675–676; 679–680
vitamin B_{12} in, **21:**431
thromboplastic, **5:**35–39; 46; 56; 57
turnover, **67:**283
rates of in liver, **46:**120
twofold axis in, **50:**24
viruses, **1:**50
vitamin K-dependent, in blood coagulation, **46:**3–7
vitamin K-dependent carboxylation, DT-diaphorase and, **65:**115–116
in vitro synthesis, **9:**428–430
x-ray classification of fibrous proteins, **3:** 71–73 (*See also* Fibrous proteins)
x-ray evidence and stoichiometry of, **3:**63–108
zig-zag peptide chains, **1:**32–34
zinc
chelating agent effect on, **56:**303
hydrolases, **56:**341–384
isolation of, **56:**298–300
isomerases and ligases, **56:**401
lyases, **56:**383–401
mechanism of action of, **56:**408–411
metallothionein, **56:**401–407
oxidoreductase, **56:**318–327
transferases, **56:**329–341
of unknown function, **56:**401–407
Proteins, Adsorption Studies (Zittle), **14:**319
Proteins, biologically active, chemical modification of, **24:**114–125
chemical modifications and biological activity, **24:**105–160
effect of ionizing radiations on, **24:**359–413
electrolyte properties and biological function, **24:**109
enzyme activity, effect of chemical modifications, **24:**105–160
functional groups of, **24:**105–160
identical and analogous peptide structures, **24:**415–439
labeled, and enzyme formation, **24:**5; 6; 12; 14
metabolism of, **24:**340
proteolytic modification of, **24:**125–126
stabilization of, **24:**111
Proteins, Chemical Modifications of, and Their significance in Enzymology, Immunochemistry, and Related Subjects (Sri Ram, Bier, and Maures), **24:**105
Proteins, Enzymes, Denaturation and Inactivation of (Okunuki), **23:**29
Proteins, Identical and Analogous Peptide Structures in (Sorm), **24:**415

Proteins, Induces Synthesis (Halvorson), **22:**99
Proteins, Interactions with Polyelectrolytes and Polynucleotides (Elbein), **40:**29
Proteins, Oxidation by Tyrosinase and Peroxidase (Sizer), **14:**129
Proteins, Phtochemistry (McLaren), **9:**75
Proteins, Possible Polypeptide Configurations of, from the Viewpoint of Internal Rotation Potential (Mizushima and Shimanouchi), **23:**1
Proteins, Prediction of the Secondary Structure of, from their Amino Acid Sequence (Chou and Fasman), **47:**45
Proteins, Stoichiometry, X-Rays and (Astbury), **3:**63
Proteins, Structural, and Adenosine Triphosphate in Relation to Muscle Contraction (Needham), **13:**151
Protein solutions, viscosity, and SH-groups, **27:**222
Protein starvation, effect on nuclei, **27:**268
Protein structure, **43:**247. *See also* Amino acid sequences and Peptide sequences
  and catalytic activity, **24:**105–261; 448–452
  catalytic diversity evolution, **69:**37
  covalent, **24:**173–177; 248–251
  evolutionary concept of, **24:**415–439
  gene structure and, **69:**38–40
  molecular association, **24:**119–120
  molecule conformation, **24:**200–202; 246–248
  from polypeptide, **24:**22–25
  primary, **24:**415–439
  and prosthetic groups, **24:**23–24; 110–111
  regularities in, **24:**436–437
  of RNase, **24:**23; 173–177; 200–202; 211; 215–216; 221; 226–227; 246–251
    relation to function, **24:**243–252; 448–452
  terminology, **26:**91–98
    of enzymes, **26:**89–114
  tertiary, effect of radiations, **24:**394
Protein Structure (Bull), **1:**1
Protein structures, in films, **28:**4–5; 7; 8; 11–12; 17–21; 23–25; 31–32
Protein-substrate interaction, **15:**37–42
Protein sulfur, rate of replacement, **7:**194–198
Protein synthesis, **8:**54–55, **39:**222; 236; 237, **43:**25
  as basis for memory, **8:**64
  in cancerous tissues, **8:**61
  current model of, **22:**129–137
    template mechanism of, **22:**132–134
  effect of amino acid analogs on, **29:**150
    initiation, **29:**134
    polypeptide chain initiation, **29:**154
    termination, **29:**154
  enzymes from different sources, **25:**333
    and ethionine effects, **25:**372–373; 376–382; 388
    and photosynthesis, **25:**47; 55; 62; 71
  and fat formation, **8:**310–312
  inhibitors of, effect on β-lactamase, **28:**274
  microbial, and oxidation of alkanes, **27:**472; 492
    role of nucleic acids, **27:**295; 396–397
  and nucleotides, **17:**464
    and RNA, **17:**478
  rate of, **26:**107–110
    and vitamin B_{12}, **26:**254; 267–270
  relation to histone, **16:**446
    relation to transamination, **16:**228; 239
    role of nucleus in, **16:**440; 474–475; 491–492
  by sperm, **24:**88
    coding signals, **24:**427; 437
    in organisms, **24:**1–33
Protein Templates, Polypeptide Synthesis on: The Enzymatic Synthesis of Gramicidin S and Tyrocidine (Lipmann, Gevers, Kleinkauf, and Roskoski), **35:**1
Protein tyrosine kinase, substrate specificity, **68:**24
Protein tyrosine phosphatases, **68:**1–31
  catalyst for hydrolysis, **68:**12–13; 23
  catalytic domain, **68:**4–5
  chemical modification, **68:**8–11
  cytoplasmic, **68:**3
  domain alignment, **68:**17–18
  general acid-general base, **68:**17
  histidine residue, **68:**10
  hydrolysis of phosphate monoesters, **68:**14
  kinetic studies, **68:**11–14
  Michaelis-Menten parameters, **68:**11–12
  mutant, temperature-sensitive, **68:**15; 17
  phosphoenzyme intermediate, **68:**6–8
  phosphorylcysteine intermediate, **68:**6
  positional information, **68:**29–30
  receptor-like, **68:**2
    CD45, **68:**2–3
    leukocyte common antigen related protein, **68:**3

Protein tyrosine phosphatases (*Continued*)
tandem domains, **68:**29–30
residues important in structure and/or catalysis, **68:**14–19
sequence alignment, **68:**14–16
SH2-containing proteins, **68:**29–30
Protein X, of tobacco mosaic virus, **17:**487–489
Proteodermatan sulfate, **60:**224
internalized, **60:**225
Proteoglycans, **52:**25; 81
aggregates, **60:**271
binding kinetics, **60:**224
from cartilage, **60:**226
degradation
by nerve growth factor, **60:**227
partial, **60:**226
mobilization from tissue, **60:**226
Proteoheparan sulfate, **60:**225; 234
internalized, **60:**225
Proteokeratan sulfate, **60:**249
Proteolysis, **1:**344; 345, **33:**266, **46:**4, **57:**161, **58:**117
ATP-dependent, **58:**254
autolytic, activation energy, **15:**274
partial, phenylalanine hydroxylase activation, **67:**106–108
of pepsin on egg albumin, **33:**249
rate in LDH, **22:**364; 365
and reticulocyte maturation, **58:**253; 254
of tryptophan synthase, **49:**166
Proteolysis studies, guanosine diphospho-D-mannose dehydrogenase, **70:**233
Proteolytic action of thrombin, **16:**42
Proteolytic attack, **57:**208
Proteolytic cleavage, **67:**271
Proteolytic coefficient, **33:**404, *See* Specific activity
Proteolytic degradation, **44:**191
Proteolytic degradation of RNase, **24:**227–235; 244
Proteolytic enzyme, **34:**455
Proteolytic enzymes, **1:**63; 78, **2:**49–67, **42:**261. *See also* Collagenase and Elastase; Proteases; Proteinases
activation, **2:**61–64
classification, **2:**49–67
coupled reactions, **2:**64–67
differential modification and inactivation, **37:**473–477
elastolytic activity, **23:**226; 227; 229; 239; 242; 243
heterospecificity, **2:**52–61
homospecificity, **2:**52–61
larval, collagenase activity, **23:**174–175
plant, action on, collagen, **23:**176
elastin, **23:**226
meat, **23:**245
Proteolytic enzymes and milk clotting, **15:**424; 431–432; 434–435; 444
Proteolytic fragments, fibronectin, central cell-binding domain of, **70:**8
Proteolytic modification of proteins, **24:** 125–131
Protesinases, in plastids, **20:**220
Protesterone, tyrosine hydroxylase induction, **70:**172
Proteus
and amino acid decarboxylases, **6:**7
elastase in, **23:**228
Proteus mirabilis, **41:**184
β-lactamase, **28:**268; 290
Proteus morgani, β-lactamase, **28:**268; 290; 293; 297
Proteus phosphoreum, **7:**257
Proteus rettgeri, β-lactamase, **28:**268; 290
Proteus sp., β-lactamase of, **28:**243; 268
Proteus splendidum, emission spectrum, **7:**257
Proteus vulgaris
amino acid metabolism in, **16:**233
aryl sulfatase of, **22:**219; 220
chondrosulfatase of, **22:**229
chondrosulfatase from, **25:**439
quinones in, **25:**289
collagenase activity, **23:**180
cytochromes of, **23:**267; 270
β-lactamase, **28:**268; 290; 297
metabolism in, **74:**21; 34–36
oxidation of pyruvate by, **18:**338
succinic dehydrogenase of, **18:**92
pyruvic oxidase in, **19:**202
Proteus vulgaris, 83–85, **15:**118; 119; 125–126; 193; 223–224; 416
Proteus vulgaris, antimetabolite effect of hydroxylamine on, **13:**259–260
enzymic adaptation in, **13:**86
glutamine synthesis in extracts of, **13:**279
transferases in, **13:**290–291; 293; 294; 295–296; 298
ultrasonic disintegration, **13:**296

Proteus vulgaris, folate metabolism, **27:**435; 445–447; 451–452; 454; 460
Proteus vulgaris, inactivation of niacin derivatives, **7:**419–420
Prothrombin, **5:**33; 34; 41–42; 46, **46:**1; 3, **48:**278, **52:**34; 158. *See also* Antithrombin and Thrombin; Blood, coagulation of; Thrombin; Vitamin K
abnormal, **46:**2; 6; 7–14
action of tyrosinase on, **14:**147–148
separation by chromatography, **14:**326; 342; 344; 346
activation, **16:**25; 30–31; 44–75; 82; 91–92
chemistry, **16:**32
derivatives, **16:**49–50
determination of concentration in plasma, **16:**59; 64–67; 70; 85; 90
distribution in centrifugal field, **16:**89
equilibrium conditions among trace proteins, **16:**75–76
inactivation by thrombin, **16:**48–49
inhibitors of change to thrombin, **16:**25; 29; 76–86; 88
purification, **16:**27–29
stability in stored plasma, **16:**59
activation by various enzymes, **10:**12–13; 21–22
activation of, **30:**276–283 (*See also* Factor II)
complex of, and Factor II, **30:**277–278
Factor V of, **30:**278–279
Factor X and activated Factor X (X_a) of, **30:**279–280
mechanism of Factor II conversion of, **30:**281–283
as parent molecule of Factors VII, IX, and X, **30:**265
phospholipid of, **30:**280–281
of plasma, **30:**257
properties of, **30:**262
adsorption on barium sulfate, **22:**456
aglycoprothrombin, **46:**13
asialoprothrombin, **46:**13
calcium-binding peptide, **46:**12
calcium ion binding, **46:**5; 10
conversion to thrombin, **10:**16–17; 21
role of Ca^{++}, **10:**14–15
determination, **2:**293; 301–311, **10:**17–20
dicoumarol-induced, **46:**9; 11; 13
purification, **46:**10
formation of, **5:**55
fragment, **46:**1; 5; 11
and globulin, **5:**42
in hemophilic plasma, **10:**10–11
inhibition of action, **5:**47; 48
level and liver, **2:**308; 314; 315
liver precursor, **46:**14–17
mechanism of activation, **48:**303–305
nature, **10:**3; 23–28
in newborn, **2:**312–314
normal, **46:**21
phospholipid binding, **46:**5
plasma concentration, **46:**6
preparation, **10:**20–23
removal of sialic acid to reduce circulation time, **41:**126
in schizophrenia, **29:**507
and thromboplastic protein, **5:**35
and vitamin K, **2:**287; 288; 307–309, **5:**41; 55
Prothrombin, PIVKA, **46:**9
precursor, **46:**2; 9; 15
prethrombin, **46:**1; 5; 11
purification, **46:**6; 10
intrahepatic precursor, **46:**15; 16
tryptic peptides, **46:**12
vitamin K-dependent structures, **46:**12
Prothrombin-A, *See* Factor 5
Prothrombin activator capacity of ether extracted plasma or serum, **16:**86
Prothrombinase, **10:**28
Prothrombin-B, **10:**24
Prothrombin derivative I, formation, **16:**47–48
Prothrombinogenase, **10:**28
Prothrombin time
effect of glass on, **16:**66–67
and Schütz law, **10:**5
relation to prothrombin concentration, **10:**17–20
Prothrombin time test, description of, **30:**286
Prothrombokinase, *See* Thromboplastinogen
Prothrombokinin, *See* Thromboplastinogen
Protista, **7:**52
Protists, **59:**328; 357
Protoanemonin, **7:**480; 487; 489; 491; 493
Protobionts, evolution of, **27:**371–375
Protocatechuate, anaerobic decarboxylation, **27:**535
from toluene, **27:**515
Protocatechuic acid, **17:**414, **18:**374
action of phenolase on, **16:**110
role in browning of plant tissues, **16:**150

Protocatechuic acid (*Continued*)
role in sclerotization, **16:**160–161
use in plant oxidase studies, **16:**132
oxidation, **19:**98; 99
Protocatechuic acid -4-O-β-glucoside, **45:**419
Protocatechuic acid oxidase, **19:**97–99; 126
formation in *Neurospora* mutants, **22:**21
Protochlorophyll, **46:**74; 75; 76
Protochlorophyllide, **46:**70; 72; 76; 78; 79; 94; 95; 96; 100; 101. *See also* Magnesium-2-vinylphenoporphyrin a5 monomethyl ester
Protochlorophyllide-holochrome, **46:**73
Protochlorophyll protein complex, **19:**61; 62
Protocollagen, **41:**278; 281; 282
Protocollagen hydroxylase, **37:**289–293
assay of, **37:**290
cofactor requirements, **37:**291
distribution, **37:**291
purification of, **37:**291
substrate specificity, **37:**292
Protodeum, **45:**391
Protofibril, during contraction, **16:**352
Protogen, **15:**214–215, **18:**320. *See also* Lipoic acid
Protogen-A (α-lipoic acid), **18:**320–321; 324
Protogen-B (β-lipoic acid), **18:**320; 324
Protohaematin compounds, **23:**294; 295
Protohaem compounds, **23:**304; 312–314
Protohaemin, **23:**271; 277; 283; 285–288; 315
Protohematin, in respiratory chain preparations, **20:**153
Protoheme, **46:**59; 101
Protohemin, **15:**37
Protolytic reactions, mechanism, **25:**21–30
Protomer-polymer equilibrium, **35:**373
Protomers, **57:**196–201
Protomyosins, properties, **16:**317; 328–330; 332; 354
Proton, shift, during transport, **49:**65
Proton abstraction, **43:**511
Proton acceptor, **51:**154
Protonation, reverse, **45:**327; 347
Protonation-deprotonation rates, **25:**22
Proton donation, eukaryotic dihydrofolate reductase, chemical transformation step, **70:**81–90
Proton equilibria, **47:**27
α-Proton exchange, **53:**163; 164; 173–178; 196
β-Proton exchange, **53:**163; 164; 189–191
Proton exchange with the solvent, **38:**434
Proton flow, medium for, **49:**72
Proton gradient, **51:**102; 103
Proton-motive force, **51:**104
Proton occupation numbers, **34:**492
Proton pump, **60:**91
Proton pumping, cytochrome *c* oxidase, **71:** 125–162
chemical models, **71:**135–138
Babcock model, **71:**143–146
Chan CU_A model, **71:**146–149
Chan CU_A model (alternate), **71:**150–152
Chan CU_B model, **71:**152–154
generally, **71:**135
Mitchell copper loop model, **71:**158–160
Rousseau models, **71:**138–143
summarized, **71:**160–162
Wilkström model, **71:**154–158
Woodruff model, **71:**135–138
problem definition, **71:**125–127
restrictions on potential models, **71:**127–134
Protons
high-energy, syntheses by, **27:**352
transport of, **50:**30
Protons (7 m.e.v.), LET from, **24:**365
Proton splittings, **51:**119; 124
Proton tautomerism, asymmetric induction in, **35:**124
conservation of proton, **35:**104–105; 114; 125
possible mechanisms, **35:**102; 122
Proton transfer in homogeneous catalysis, **25:**12; 21–32
to imidazole, rates, **25:**26–30
intramolecular, **25:**29–30
Proton transport, phosphorylation coupled to, **51:**99–105
ADP-ATP exchange, **51:**104
binding and equilibria, **51:**101–105
chemiosmotic system, **51:**99; 104–105
exchange of, ^{18}O, **51:**101
exchanges of $H_2\,^{18}O$ into phosphate, **51:** 103–104
H_2O.ATP exchange, **51:**104
H_2O.P exchange, **51:**104
intrinsic binding energy, **51:**100; 101
mechano-chemical coupling, **51:**104
membrane potential, **51:**104
P_I-ATP exchange, **51:**104
proton gradient, **51:**102; 103; 104
proton-motive force, **51:**104

rules for vectorial coupling, **51:**100–101
simple hypothesis, **51:**99–100
uncouplers, **51:**104
Protopectin, *See* Pectic substances
Protopectinase, **11:**289
classification, **14:**246
Protopectinases, **20:**364
Protoplasm. *See also* Cytoplasm
contractility and hydration, **8:**37–42
dynamic state of structure, **8:**26–30
energetic state, **1:**333; 334
Protoplasm, Properties, with Special Reference to Influence of Enzymic Reaction (Seifriz), **7:**35
Protoplasm and enzymic reactions, **7:**35–36
anesthesia, **7:**41; 52–54
contractility, **7:**42–43
double refraction, **7:**60
elasticity, **7:**40–43; 60
energy sources, **7:**54
extensibility, **7:**41
fluidity, **7:**60
gelation and coagulation, **7:**40
permeability, **7:**45
rhythm, **7:**50–52
rigidity, **7:**60
spirality, **7:**44
thixotropy, **7:**40
torsion, **7:**44
viscosity (*See* Viscosity of protoplasm)
Protoplasmic colloids and aging, **8:**26–27
Protoplasmic structures, changes in, **8:**62–65
Protoplasts, **20:**203
Protoporphin, and vitamin deficiencies, **7:**318
propionic acid groups, **7:**325–326
Protoporphin IX, distribution, **7:**318
properties, **7:**306–307
Protoporphyrin, **17:**436; 442, **23:**285; 286; 288, **46:**37; 90; 94; 102; 107; 115; 142; 165
monoester of, **46:**69
Protoporphyrin-9, **46:**54–58; 62; 69; 80; 86; 87; 90; 92; 93; 100; 101; 102
metabolism, products of, **46:**59–69
Protoporphyrin IX, formation of, **24:**300; 314–342
Protoporphyrinogen, **46:**57
Protoporphyrinogen-9
oxidase, **46:**91
oxidation, **46:**58
Protoporphyrinogen oxidase, **46:**54
Protoporphyrins acetate in biosynthesis of, **12:**312
Prototheca zopfi, glycolysis in, **19:**6
Protozoa, **46:**102
enzymic transformations of steroids by, **20:**260
Protransglutaminase, coagulation factor XIII, **38:**111
Protransglutaminases, activation, **38:**175–177
Protriptyline, **55:**177
Protyrosinase, activation, **14:**133
Proximity effect, **43:**285
transferases, **46:**404
in velocity of enzyme reaction, **22:**83–89
PRPP, **54:**58, *See* 5-Phosphoribosylpyrophosphate
increasing levels of, methods of, **54:**101
PRPP synthetase, **49:**284–297
control of amount of enzyme, **49:**291
effectors of activity, **49:**285–288
activators, **49:**287–288
inhibitors, **49:**286–287
substrates, **49:**285
genetic control, **49:**295
quaternary structure and activity, **49:**290
requirements and characteristics of enzyme reaction, **49:**284–285
structure of the enzyme, **49:**288–289
PR transferase, **38:**19; 21
bifunctional protein, **38:**20
genetic analysis, **38:**21
latency, **38:**29
proteolytic digestion, **38:**20; 21
purification, **38:**34
role in aggregate formation, **38:**22
Trp B region, **38:**1; 22
tryptophan inhibition, **38:**22; 23
Prulourasin, **18:**71
Prunasin, **56:**199
Ps. *fluorescens,* transaminations in, **16:**190–191; 192
Psalliota compestris, quinones in, **25:**288
P450scc, **68:**44; 46; 58
P-S difference, *See* Plasma-serum difference
Psedomonas fluorescens, L-aspartase from, **74:**308–309
characteristics of, **74:**297; 298
Psedomonas testosteroni, enzymes in, **20:** 265–269; 271; 272; 276
enzymic transformations of steroids by, **20:**262

Pseudoacid, **59:**137
Pseudo-allelism, **13:**135
Pseudoamylose, **11:**324–325
Pseudo-ATP, **43:**22
Pseudocholinesterase
active center of, **24:**434
amino acid composition near active site, **22:**65
Pseudo-cholinesterases, classification, **14:**244
Pseudocollagenase, **34:**162
Pseudocyclic photophosphorylation, *See* Ferredoxins, in photophosphorylation, oxygen-linked noncyclic
Pseudoferroxidase activity, **44:**203; 205
γ-Pseudoglobulin, bovine, moving-boundary electrophoresis patterns of, **30:**152
Pseudohemoglobin, **18:**57
Pseudolipohydroperoxidase, **58:**260
Pseudomonads, use of glycine, **23:**453–454
Pseudomonads, oxidation of hydrocarbons, **27:**474; 508; 510; 515; 517; 520
Pseudomonas, **17:**418, **46:**173
and citric acid cycle, **18:**247; 250; 258
column chromatography of enzyme of, **22:**484
pyrocatechase in, **19:**81
Pseudomonas, 175, **15:**177; 184–186
Pseudomonas aeruginosa, **15:**416, **39:**338, **71:**92; 188
acetate metabolism, **23:**410; 411
cytochrome of, **23:**267; 312; 314–315
proteinases from, **23:**180; 227; 228
phenylalanine hydroxylase from, **74:**235
structure of, **74:**242
Pseudomonas aeruginosa, exopolysaccharide alginate synthesis in. *See also* Exopolysaccharide alginate synthesis
Pseudomonas aeruginosa, PN-dependent enzymes, **26:**130
Pseudomonas aeruginosa, Alkaline proteinase from, **41:**184; 187; 201; 207
Pseudomonas aeruginosa, exopolysaccharide alginate synthesis in, **70:**221–254
Pseudomonas aeruginosa, oxidation of hydrocarbons, **27:**471; 472; 476; 484–487; 493; 500; 503; 506; 513; 515; 519; 520; 523; 525; 531; 533
Pseudomonas apii, plant disease caused by, **20:**392
Pseudomonas, 2-carbon metabolism in, **23:** 435–437; 440–446; 450
elastases in, **23:**227–229
proteinases in, **23:**180; 183
use of acetate, **23:**413–416; 418–420
Pseudomonas citronellolis, pyruvat carboxylase of, **26:**303; 306
Pseudomonas convexa, oxidation of hydrocarbons, **27:**515
Pseudomonas desmolyticum, oxidation of hydrocarbons, **27:**523
Pseudomonas diminuta
hydrolysis of organophosphates, **74:**52
phosphotriesterase from
amino acid sequence for, **74:**53; 54
gene for, **74:**53
Pseudomonase exotoxin A, **58:**115
active sites, **58:**117
amino acid composition, **58:**116
and denaturing agents, **58:**117
immunization against, **58:**116
lethal dose, **58:**116
mechanism of action, **58:**116
protein synthesis inhibition, **58:**116
purification, **58:**116
Pseudomonas, enzymic adaptation, **13:**74
Pseudomonas facilis, phosphoribulokinase from, regulation of, **74:**109
Pseudomonas fluorescens, **17:**284, **23:**411; 412, **70:**222; 223; 245
allosteric enzymes, **28:**69
imidazoleacetic acid oxidase in, **19:**179
protocatechuic acid oxidase in, **19:**97; 98
plant disease caused by, **20:**395
Pseudomonas fluorescens, 102, **15:**117; 157; 177; 193
*Pseudomonas fluoresc*ens, **49:**4; 29; 31
Pseudomonas fluorescens, enzymic adaptation in, **13:**71
Pseudomonas fluorescens, histidine degradation in, **21:**407; 414
Pseudomonas fluorescens, 2-keto-3-deoxy-6-phosphogluconate aldolase from, **31:**139
Pseudomonas fluorescens, oxidation of hydrocarbons, **27:**512; 514; 515; 518; 536
Pseudomonas fluorescens, propionic acid metabolism, **26:**368–370
Pseudomonas infection, **58:**115
Pseudomonas KB 1, use of labeled acetate, **23:**405; 406
Pseudomonas lachrymans, plant disease caused by, **20:**392

Pseudomonas marginalis, **49:**5; 31; 37
Pseudomonas marginalis, plant disease caused by, **20:**401
Pseudomonas medicaginis, var. *phaseolicola,* plant disease caused by, **20:**389
Pseudomonas mendocina, **70:**222; 223
Pseudomonas methanica, oxidation of hydrocarbons, **27:**472; 476–482
Pseudomonas morsprunorum, plant disease caused by, **20:**400
Pseudomonas myxogenes, **23:**180
Pseudomonas ovalis Chester, use of acetate, **23:**408–409; 411; 412; 416–420
Pseudomonas ovalis, PN-dependent enzymes, **26:**118
Pseudomonas oxalaticus, metabolism of oxalate, **23:**448–452
Pseudomonas oxidase, **23:**312; 314–315
Pseudomonas, oxidation of hydrocarbons, **27:** 471–473; 476; 479; 484–488; 492–494; 498; 499; 502; 508–511; 514–516; 518; 522; 524; 527; 531–533; 537; 538
Pseudomonas phaseolicola, plant disease caused by, **20:**388
Pseudomonas, prephenate dehydratase, effect of analogs on, **29:**120
Pseudomonas propanicum, oxidation of hydrocarbons, **27:**471
Pseudomonas pseudomallei, **23:**180; 227
Pseudomonas putida, **50:**327; 329; 330, **51:**50; 64, **60:**38, **70:**223; 238–239; 245
Pseudomonas putrefaciens, enzymic adaptation in, **13:**70
utilization of sucrose by, **13:**253
Pseudomonas putrefaciens, source of sucrose phosphorylase, **10:**123–124
Pseudomonas putrifaciens, **23:**411
Pseudomonas putrifaciens, glucokinase in, **15:**147
Pseudomonas pyocyanea
β-lactamase from, **28:**245; 251; 259–261; 268; 290; 293; 297
quinones in, **25:**289
Pseudomonas pyocyanea, oxidation of hydrocarbons, **27:**489
Pseudomonas saccharophila, **18:**286
pentose metabolism, tracer studies, **28:** 409–410; 413; 422–423; 429; 430
Pseudomonas saccharophila, carbohydrate metabolism in, **15:**160; 166
Pseudomonas saccharophila, enzymic adaptation in, **13:**70
utilization of sucrose by, **13:**253
Pseudomonas saccharophila, source of sucrose phosphorylase, **10:**123–143
Pseudomonas solanacerum, **71:**11
Pseudomonas sp., β-lactamase of, **28:**243
Pseudomonas sp., Superoxide Dismutse from, **41:**79
Pseudomonas species, enzymic transformations of steroids by, **20:**252; 258
pectinesterase of, **20:**360
Pseudomonas species, blue proteins, EPR study, **27:**138
Pseudomonas stutzeri, **71:**93
Pseudomonas stutzeri, oxidation of hydrocarbons, **27:**475; 530
Pseudomonas stutzeri, PN-dependent enzymes, **26:**131
Pseudomonas syringae, **70:**223; 246
Pseudomonas syringae, plant disease caused by, **20:**399
Pseudomonas tabacci, plant disease caused by, **20:**397
Pseudomonas tabaci, methionine mutants, utilization of thetins by, **22:**268
*Pseudomonas-taetrol*ens, **49:**24
Pseudoperonospor cubensis, plant disease caused by, **20:**392; 398
Pseudopyridoxine, nature, **6:**26
Pseudorotation, **50:**388
Pseudosubstrates, **44:**195; 196
Pseudouridine, **35:**470; 474–475, **54:**87–88; 228–230
of tRNA, fluoropyrimidine effect on, **54:**89
Pseudovitamin B_{12} coenzymes, and β-methylaspartate mutase, **26:**359–360; 362
Psi, **35:**464; 474; 475
Psilocin, increase in brain serotonin level produced by, **29:**518–519
Psilocybin, in schizophrenia, **29:**534
Psoralenated oligonucleotides, structure determination by NMR and X-ray crystallography, **66:**128
Psoralen-DNA adduct, spectra, **66:**92–93
Psoralen-DNA cross-links
HPLC, **66:**97–98
large-scale synthesis, **66:**95–100

Psoralens, **66:**85–141
base-catalyzed reversal, **66:**89–90
biological process applications, **66:**90–91
diadduct, large-scale synthesis, **66:**95–100
DNA
kinetic scheme, **66:**88–89
photoadducts, **66:**91–92
furan-side monoadducts
HPLC, **66:**94–95
large-scale synthesis, **66:**92–95
future directions, **66:**126–127
photochemistry, **66:**87–90
preparation, rationale behind, **66:**100–103
protein-DNA cross-linking, **66:**119–126
DNA furan-side monoadduct, **66:**120–122
light intensity effect, **66:**125–126
oligo and protein concentration effect on photocross-linking, **66:**123–125
specific cross-linking to single-stranded oligo, **66:**122–123
usefulness, **66:**126
spin-labeled
application to DNA dynamics study, **66:**128–130; 139–141
EPR spectra, **66:**137–139
forms of, **66:**130–131
photochemical characterization, **66:** 135–137
photoreactivity with dsDNA, **66:**134–135
potential future applications, **66:**139–141
synthesis, **66:**128–133
structure, **66:**86–87
vitiligo and psoriasis treatment, **66:**86
Psychoses, biochemistry of catatonia, **29:** 529–531
biochemistry of functional, **29:**479–553
functional, **29:**480
involutional, **29:**492
manic-depressive, **29:**515
mineral metabolism in, **29:**494
organic, **29:**480
schizo-affective, **29:**492
schizophrenia, **29:**499–536
Psychoses, Functional, Biochemistry of (Weil-Malherbe), **29:**479
Psychosine, **60:**115
formation of, **24:**311
PTA, *See* Plasma thromboplastin antecedent properties of (*See also* Factor XI)
P.T.A., properties of, **30:**262
PTC autoprothrombin II, *See* Factor IX
PTC (plasma thromboplastin component), *See* Platelet cofactor II
Pteridine activation, as hydride acceptor, eukaryotic dihydrofolate reductase, chemical transformation step (reactant activation), **70:**78–81
Pteridine portion of folic acid
biosynthesis of, **53:**355–359
dihydroneopterin triphosphate, as precursor, **53:**357
Pteridine ring, eukaryotic dihydrofolate reductase, **70:**58; 63
Pteridines, **35:**35
antagonists to, **27:**460
biogenesis of, **27:**422–424
condensation with *p*-AB, **27:**422–424
and folate system, **27:**446; 447
relation to purines, **27:**422; 424
hydrogenated, oxidation-reduction function, **25:**400–401
Pteridines, Biosynthesis of (Brown), **35:**35
Pteridine substrates, tetrahydrofolate and, eukaryotic dihydrofolate reductase, **70:**50–52
Pteridine trihydroxypropyl-phosphate, in folate biogenesis, **27:**423
Pteridophyta, carotenoids in, **21:**300
Pteridylthiazolium compound, folic acid from, **15:**404
Pterin, **35:**35
enzymic synthesis of, **35:**50–69
Pterin-CH_2 OP, **35:**39
Pterin-CH_2 OPP, **35:**39
Pterin coenzyme, tyrosine hydroxylase regulation, **70:**136–137
Pterin cofactor, **67:**214–215
tyrosine hydroxylase regulation, **70:**122–123
Pterin-dependent hydroxylases, *See* Phenylalanine hydroxylase; Tryptophan hydroxylase; Tyrosine hydroxylase
Pterine substrates, tyrosine hydroxylase, substrate specificity, **70:**110; 117
Pterin ring, eukaryotic dihydrofolate reductase, ligand binding (side chain roles), **70:**61
Pterins
absorption spectra, **21:**388
isolation, synthesis and properties, **21:** 376–378

interconversion of, **35:**69–74
unconjugated, phenylalanine hydroxylation, **70:**103
as xanthine oxidase inhibitors, **17:**310
oxidation by xanthine oxidase, **17:**238; 240; 298–299
Pteroate, as substrate of dihydrofolate reductase, **27:**436
Pterobilin, **46:**65
Pterocarpans
fungal hydroxylation of, **55:**48
isoflavone conversion to, **55:**34
Pteroic acid, and folate, **27:**430
Pteroproteins, definition, **21:**371
reactions catalyzed by, **21:**396–432
Pterostichus melanis, **45:**394
Pterostilbene, **55:**92
Pteroylglutamate cofactors for methyl synthesis, *See* Folic acid coenzymes
Pteroylglutamates, occurrence, **27:**421
Pteroylglutamic acid. *See also* Folate and Folic acid; Folic acid
effect on xanthine oxidase, **17:**310; 321
sulfanilamide analogs, **27:**430–431
Pteroylpolyglutamates, enzymatic synthesis of, **53:**359–363
^{32}P transfer, from chromatin, **7:**187–189
to hepatic tissue, **7:**187–189
Ptyalin, adsorption studies, **14:**329
Puberulic acid, **7:**481; 486; 489; 490; 492
tropolone derivative in fungi, **16:**137
Puberulonic acid, **7:**481; 486; 489; 490; 492
tropolone derivative in fungi, **16:**137
Puccinia graminis, plant disease caused by, **20:**401
Puccinia species, plant disease caused by, **20:**401
Puccinia striiformis, **46:**340
Pulcherrimin, **35:**167
Pulcherriminic acid, **35:**155; 163
biosynthesis, **35:**163
Pullulan, **34:**408
Pullulanase, **34:**406
Pulse radiolysis, **41:**37; 44; 58
Pulse rate, temperature dependence, **23:**122–124
Pump, potassium, activation of, by valinomycin, **29:**46
Pump-and-leak hypothesis, **49:**46
Purcell measurement of relaxation rates, *See* Relaxation rates, nuclear, Carr-Purcell pulsed method of measurement of
Purification
by affinity chromatography, **48:**33
of biological materials, **48:**33; 37
glycosidases, **48:**36
enzyme, **42:**265
glutathione synthesis, **73:**216–220
respiratory burst oxidase, **65:**60–62
Purification methods, **42:**75
Purine analogs
and effect of x-irradiation, **27:**459–460
FAH_2-reductase inhibitors, structure-activity relation, **27:**444–445
effect on enzyme synthesis, **24:**16–18
effect on virus multiplication, **17:**479–483; 496
Purine bases, introduction of -OH or -NH_2, **26:**128
Purine biosynthesis
allosteric regulation, **28:**65–66; 118
regulation of, **25:**170–171
Purine-bound pentoses, **28:**407–409; 414; 418–419; 425; 433
Purine cobamide coenzyme, absorption spectrum, **26:**248–249
Purine cobamide enzyme, and methylmalonyl-CoA mutase, **26:**347; 352
Purine-deoxyribose bond, lability of, **28:**433
Purine derivatives
fluorinated, **55:**268–272
as xanthine oxidase inhibitors, **17:**309
Purine desoxyribosides, **9:**493
Purine disulfide, **43:**39
analog, **43:**37
Purine hydroxylase, **19:**203
Purine metabolism, **41:**1; 2
aberration in, **41:**2
allosteric regulation, **28:**118–126
disorders of, **41:**1
ultimate product of, **41:**1
Purine-modified analogs, **43:**28
Purine moieties of DPNH, and Cotton effect, **27:**61
Purine mononucleotides, an LADH complex formation, **27:**68–70
Purine nucleosidase, **9:**496, **15:**103
Purine nucleoside phosphorylase, **18:**216; 222; 225
localization of, **24:**312
Purine nucleotide binding site, **28:**141
Purine nucleotide diesters, **17:**172

Purine nucleotide phosphorylase (PNP), **51:**168; 180–181
T-cell immunodeficiency associated with, **51:**200–201
Purine nucleotide pyrophosphorylase, **18:**216
Purine nucleotides. *See also* Allosteric regulation; Deoxyribose; Ribose; individual nucleotides
6-amino, allosteric regulation, **28:**66; 120–121
and Cotton effect in LADH complexes, **27:**59; 61–70
formation of, **18:**217–218
in hepatoma cells, **28:**135
6-hydroxy, allosteric regulation, **28:**66; 120–121
inhibition of metalloflavoproteins, **17:**275
interconversions, **28:**120–123
metabolism, allosteric regulation, **28:**66; 119; 121–124; 132; 142
Purine pyrophosphorylase, allosteric regulation, **28:**123–124
Purine ribonucleoside triphosphate, **58:**142
Purine ribonucleotides, synthesis, **28:**419
Purine riboside phosphorylase, effect on 5′-methylthiodenosine, **25:**386
Purine ribosides, **9:**490–493
Purines, **4:**305; 307–309, **38:**214. *See also* Nucleic acids, Nucleosides, and Nucleotides; individual purines, Nucleosides, Nucleotide bases, Nucleotides
abiogenic syntheses, **27:**384–385
absorption spectra, **9:**480–484
acetic acid from, **12:**107–108; 121–127
biosynthesis of, **12:**43; 45; 48; 84–85
breakdown of, **12:**451–453
and CO_2 fixation, **12:**107–108; 121–127
and amino acid sequence, **24:**9–11; 15
specificity of RNase action, **24:**165; 171
biosynthesis, **9:**460–463
enzymic reactions in, **21:**199–261; 396; 398–401; 410
degradation, **21:**223; 227; 401–403
cleavage by microorganisms, **23:**327
and glycine, **23:**458
and folate metabolism, **27:**426; 427; 444–445; 455; 458–460
glycine as precursor, **18:**258
ureides of, **18:**259
from hydrolysis of CoA, **16:**4
of DNA, **16:**430–436
incorporation of formate, **16:**274
labeled, in nucleic acid study, **16:**450–462
noncompetitive antagonists of sulfonamide, **16:**300
of nuclear RNA, **16:**437–440
in transaminations, **16:**237
imidazole in, **20:**102; 103
as inhibitors of uric acid formation, **17:**310
genetic studies, **17:**426–428; 431; 459–465
as substrates of xanthine oxidase, **17:** 238–241; 298–301; 309; 317–318; 322
in transglycosylations, **17:**220–221
inhibitory analogues of, **6:**134
metabolism, **9:**441–443
methylation of, **25:**414–415
in nucleic acids, **9:**522
occurrence, **9:**458–460
in plants, **8:**234
relation to protein metabolism, **8:**242
precursors of pteridines, **35:**48
relation to pteridines, **27:**422–424
tautomerism of, **50:**36
and vitamin B_{12}, **26:**267
Purines, H-8 exchange reactions of, **49:**202
Purines, Synthesis, Enzymic Reactions in (Buchanan and Hartman), **21:**199
Purine salvage pathway
immunodeficiency syndromes associated with, **51:**200–202
role of adenosine deaminase in, **51:**179–182
Purine skeleton, source of atoms, **16:**450–451
Purity of enzymes, evidence, **14:**388–390; 401–404
Purity tests for enzymes, **12:**401–404
Puromycin, **42:**240; 268
effect of, on uridine-^{3}H and on L-leucine-^{14}C incorporation by sensory ganglia in medium with nerve growth factor, **31:**70
effect on corticosterone synthesis by, **29:**470
effects on enzyme synthesis by, **29:**376
inhibition of ACTH by, **29:**470
as inhibitor, **29:**201
synthesis of gramicidin S, inhibited by, **33:**340–342; 345; 347; 348
tyrocidine synthesis, effect on, **33:**350
Purple bacteria
fluorescence, **1:**313; 314

metabolism, **1:**234–242; 264–269; 275; 280; 281; 289; 290; 291; 301–307; 320
pigments, **1:**235; 292; 293; 307–314
Purple complec, conversion into paramagnetic species, storage in the dark, **34:**44
crystallization, **34:**43
Purple intermediate, absorption spectrum of, using α-aminobutyric acid as substrate, **34:**43
activation energy, for conversion of, to fully reduced enzyme, **34:**61
for $E_{ox} \cdot S \rightarrow$, **34:**60
CD spectrum of, **34:**63
conversion of, oxidized enzyme-substrate complex to, **34:**55
to fully reduced enzyme, **34:**60
effect of pH, on rate of conversion of, to fully reduced enzyme, **34:**61
on rate of formation of, **34:**58
electronic interaction involved in, **34:**63; 66
formation of, with D-leucine or D-valine, **34:**50
ORD spectrum of, **34:**63
reaction of, with oxygen, **34:**67
reversible reaction between, and fully reduced enzyme, **34:**63
Purple sulfur bacteria, metabolism, **1:**236; 269; 270; 274; 278; 280; 281; 284; 285; 289; 292; 302
Purpurogallin, in plant metabolism, **16:**136–138
Putidaredoxin, **47:**13; 14, **48:**10, **60:**46
P450 electron transfer, **47:**32–33
spectral properties of, **47:**14
Putrescin and diamine-oxidase, **2:**93; 95; 96; 98; 99; 103–107
Putrescine, **55:**104, **56:**266; 267, **57:**103; 112
occurrence of, **54:**216
polyamine synthesis from, **54:**214
in spermine biosynthesis, **22:**262
Putrescine and alkaloid biosynthesis, **8:** 237–240; 242
Putrescine production, **6:**4
PYCA, *See* Pyrrol-2-carboxylic acid
Pycnodothis species, enzymic transformations of steroids by, **20:**262
Pycnosporium species, enzymic transformations of steroids by, **20:**247
Pyocyanin, membranes of chloroplasts permeability barrier for, **30:**6
Pyocyanine, **15:**4
Pyocyanin system, cozymase inactivation, **7:**427–435
Pyranose-furanose, equilibrium mixture of D-fructose, **10:**132
Pyranose ring, configurations, **23:**23
Pyranose ring structure, relation to β-glucuronidase action, **16:**371–373
Pyranoside acceptors, **18:**224–225
Pyrazolines, **58:**219
Pyrazolylalanine, **50:**147
Pyrichrome, **15:**402
Pyricularia oryzae, **55:**79
Pyridien nucleotide transhydrogenase, **20:**172; 173
Pyridine. *See also* N′-Methylnicotinamide, Nicotinic acid, Nicotinuric acid, Trigonelline
action on thermostable factor, **15:**410
action with thiaminase, **15:**401–403; 407; 410–412
activator of thiaminase model, **15:**404
in air-cured tobacco, **6:**348
effect on degradation of pectin substances, **20:**351
pie-electron density of, **20:**54
in reactivation of phosphorylated esterases, **20:**308
interrelationship of carbohydrate metabolism and nucleotides of, **29:**341
methylated derivatives of, **12:**474–478
Pyridine ACA, action of β-lactamase, **28:**259
3-Pyridine aldehydedeamino-DPN, lactate dehydrogenase coenzyme, **27:**271
Pyridine 3-aldehyde DPNH, enzymebound, Cotton effect, **27:**57; 58
Pyridine-2-aldoxime methiodide, in reactivation of phosphorylated esterases, **20:**306; 307
Pyridine alkaloids, and dehydrogenases, **8:** 246–247
relation to tropane alkaloids, **8:**224
and tryptophan, **8:**239
Pyridine-2-carboxaldehyde, **58:**12
Pyridine carboxaldehydes, carbonic anhydrase catalyzed hydration, **47:**187; 206–207
Pyridine coenzymes
binding of, **22:**340–375
binding to lactate dehydrogenase, **37:**75–87
Pyridine compounds, reoxidation, **26:**154
Pyridine 2,4-dicarboxylate, collagen hydroxylase inhibitor mechanisms, **72:**360–364

Pyridine 2,5-dicarboxylate, collagen hydroxylase inhibitor mechanisms, **72:**361–364
2,6-Pyridine dicarboxylate, dihydrodipicolinate reductase interaction, **72:**297
Pyridine haematohaemochrome, **23:**313
Pyridine haemochrome a_2, **23:**312
Pyridine haemochrome *a,* absorption spectrum, **23:**289–297; 306–311
oxidation potential, **23:**281; 313
Pyridine-hemin, as catalase model, **15:**28
Pyridine hemochromogens, **54:**278; 280
from cytochromes, **17:**343–344; 357
Pyridine nitrogen, in imidazole, **20:**53; 55–57; 59; 84–86; 91; 96
Pyridine nucleotide and luminescent system, **7:**232
Pyridine nucleotide binding site, **28:**140–141
Pyridine nucleotide coenzymes, **15:**79–139
mechanism of action, **15:**7–16
of nuclei, **27:**276–278
Pyridine Nucleotide Coenzymes (Singer and Kearney), **15:**79
Pyridine-nucleotide-cytochrome reductase, as metalloflavoprotein, **17:**262; 264; 283
Pyridine nucleotide reactions, with no other cofactor, **26:**118–120; 122–123
with other cofactors, **26:**120–132
Pyridine nucleotide reductase, photosynthetic, **27:**87
Pyridine nucleotides. *See also* Diphosphopyridine nucleotide; Respiratory chain; Triphosphopyridine nucleotides
action of DPNH-Q reductase on, **25:**313–315
allosteric regulation, **28:**84; 94; 106; 140–144
classification in enzyme system, **14:**241
and dehydrogenase inhibition, **25:**251–266
and enzyme catalysis mechanisms, **24:**442; 469–475
as enzyme inhibitors, **25:**223
as hydrogen acceptors, **18:**12–13
hydrogen transfer, **26:**115–191
with isocitrate dehydrogenase, **28:**135–139
and light emission, **25:**149–151; 155; 160–161
and methyl group synthesis, **25:**400–401
nuclear and cytoplasmic, **27:**272–275
as oxidants in Hill reaction, **18:**58–59
and oxidative phosphorylation, **17:**334
in kinetic studies, **17:**36; 42–45
and phosphate transfers, **17:**175; 184–186
in respiratory chain, **17:**70–71
spectra, **17:**72–77
as substrates for xanthine oxidase, **17:** 298–299
oxidized, molecular structure, **22:**339
reduced, molecular structure, **22:**339
in squalene biosynthesis, **22:**308
state of, in mitochondria, **22:**374
in spermatids, **22:**375
in pentose metabolism, **28:**453
in photophosphorylation, **23:**386–387
in respiratory chain, **23:**341; 344–347
in plants, **18:**292–296; 299; 303–304; 306–310; 312
reduction by illuminated chloroplast, **18:**44
reduction of cytochrome b_3, **18:**48
role of catalysis by, **15:**112–131
Pyridine nucleotides, NAD+ synthesis, niacin dervatives salvage, **73:**158–164
Pyridine nucleotide transhydrogenase, **15:**117; 131
Pyridine nucleotide transhydrogenases, classification, **14:**227
Pyridine protohaemochrome, **23:**311
Pyridine ring, hydroxylation, **27:**536
Pyridine ring in alkaloids, **8:**235–236; 237
Pyridine-3-sulfonic acid, as inhibitory analogue, **6:**132
Pyridinium glycoside, α configuration, **18:**206
Pyridinium ion, and ADH binding, **22:**356
Pyridinium ring, enzyme binding, Cotton effect, **27:**55; 56; 58
Pyridinium salts, reduction of, **26:**133–143
Pyridinnucleotiden, di Wasserstoffubertragung mit (Sund, Diekmann, und Wallenfels), **26:**115
Pyridinoprotein dehydrogenases, **15:**89; 96–98; 112–131
use in estimation of DPN and TPN, **15:**96–98
Pyridinoprotein enzymes, **15:**112–131
mechanism of action, **15:**5; 8–16; 17
test for, **15:**89; 96–98
Pyrido (2,3-d) pyrimidines, folate antagonists, **27:**446–447; 449–452
2-Pyridone, **50:**5; 11. *See also* 2-Hydroxypyridine
catalysis by, **50:**11
dissociation constant, **50:**5

Pyridones, relation to nicotinamide, **15:**88; 109–112
Pyridoxal, **6:**26; 30, **46:**39, **53:**163. *See also* Pyridoxal phosphate, Pyridoxamine, Pyridoxamine phosphate, Pyridoxine and Vitamin B_6
activation of tyrosine decarboxylation, **6:** 26–29; 30
aldehyde oxidase substrate, **17:**246; 322; 382
enzyme complex, Cotton effect, **27:**59
and glycine metabolism, **23:**454; 457; 460
mode of action as coenzyme, **14:**174; 180–183; 190
phosphate, genetic studies, **17:**410; 418; 450
and phosphorylase activity, **24:**272
reaction with glutamic acid, **22:**393; 394
relation to glycine and serine metabolism, **16:**277
relation to transaminations, **16:**210; 211; 219; 223–226
Pyridoxal amino acid racemase, *See* Primary amino acid racemases
Pyridoxal catalysis, **53:**163–198
Pyridoxal-catalyzed dealdolation, **53:**163; 185–189
Pyridoxal-catalyzed dephosphonylation, **53:**192–195
Pyridoxal enzyme, Cotton effect, **27:**86–87
Pyridoxal kinase, **46:**114
localization of, **24:**313
Pyridoxal kinase activity, substrates for, **16:**219
Pyridoxal-5-p, **54:**139
Pyridoxal-P-dependent aldimines, **56:**4; 7
Pyridoxal-P-dependent enzymes
classification of, **56:**3–7
function of, **56:**3
lyases (*See* Pyridoxal-P-dependent lyases)
reactions catalyzed by, **56:**5
in β replacement reactions
carbonyl reagent interactions of, **56:**57–59
cation effects on, **56:**43
coenzyme analog interactions of, **56:**54–55
cycloserine enantiomer interactions of, **56:**61–63
holo-apoenzyme equilibria of, **56:**42–43
isotopic exchange of hydrogenatoms and B substituents of, **56:**50–54
mercapto-amino acid and aminothiol interactions of, **56:**63–67
Michael addition reactions of, **56:**67–70
oligomer-subunit equilibria of, **56:**42
quaternary structure of, **56:**42
relative reaction rates and substrate affinities, **56:**46–49
size and stability parameters of, **56:**42; 44–45
sodium borohydride reactions of, **56:**59–60
spectral features of, **56:**49–50
steady state kinetics of, **56:**70–76
sulfhydryl reagent reactions of, **56:**60–61
Pyridoxal-P-dependent lyases, **56:**7–82
alliinase, **56:**10; 34–38
mechanisms for, **56:**7–13
physicochemical and catalytic features of, **56:**42–76
subtypes of, **56:**10–11
vs. other replacement reactions, **56:**34–41
β-cyanoalanine synthase, **56:**10; 28–34
cycloserine inactivation of, **56:**79
cystathionase, **56:**10; 38–41
cysteine lyase, **56:**10; 14–18
in elimination reactions, **56:**7–13
in β-replacement reactions, **56:**13–34
serine sulfhydrase, **56:**10; 18–27
tryptophanase, **56:**11; 14; 42; 43; 45
Pyridoxal phosphate, **6:**27–29, **7:**535, **46:**114; 344; 371; 373, **53:**165, **58:**34; 178. *See also* Pyridoxal, pyridoxamine, Pyridoxamine phosphate, Pyridoxine and Vitamin B_6; Pyridoxine, Vitamin B_6
and acetylornithine aminotransferase, **40:**71
in amino acid metabolism, **9:**401; 407
aminomutase reactions, **66:**31
analogs of, **50:**22
tautomerism in, **50:**1–37
analogues, **35:**116–117
binding to protein, **35:**127
optical properties of bound cofactor (table), **35:**125
Schiff base metal complexes, **35:**119
b-imino radical rearrangement, **66:**31–32
binding, **22:**393–398
4-formyl derivatives, **22:**394; 395
function, in phosphorylases, **22:**396–398
in homoserine deaminase-cystathionase, **22:**396
linkage to cystathionase, glutamicaspartic transaminase, **22:**395
molecular structure, **22:**339
products of reaction with albumins, **22:**396

Pyridoxal phosphate (*Continued*)
role of formyl group in, **22:**394; 395; 398
as catalyst for decarboxylations of amino acids, **16:**119
in amino acid metabolism, **16:**271; 273
as coenzyme in transaminations, **16:**191; 192; 205; 211; 213; 215–219; 219–220; 224–225; 236; 238
as cofactor for tryptophane desmolase, **16:**293–294
preparation, **16:**214–215
classification as coenzyme, **14:**241
in enzymic racemization of amino acids, **14:**193; 195–196
in oxidation of phenylalanine and tyrosine, **14:**97; 98; 100; 102; 105
and cobamide coenzyme action, **26:**365
codecarboxylase activity, **6:**27
enzyme bound, Cotton effect, **27:**86–87
and glutamine decarboxylase, **23:**382
in glycine biosynthesis, **21:**417; 418
in serine biosynthesis and metabolism, **21:**224
lysine 2,3-aminomutase and, **66:**13–17
methyl synthesis, **25:**402; 403; 406
with multiple forms of enzymes, **25:**364
tumor content of, **25:**378
reaction with mimosine, **29:**108
relationship to transamination reactions, **19:**349
Schiff base, **53:**167
stability, **6:**28
and urea formation, **15:**343; 379
requirement for kynureninase, **15:**102
Pyridoxal-5′-phosphate, **53:**84
localization, **27:**276
Pyridoxal 3-phosphate, **16:**214
Pyridoxal 5′-phosphate, **49:**5; 6; 13; 17; 18; 19; 20; 22; 24; 26; 28; 29; 37
aldimine linkage, **49:**6
half of sites′ reactivity, **49:**6
hydration of, **50:**7
spectrum of, **50:**7
tautomerism in, **50:**7
interconversion reactions, **49:**21
modification of, **49:**18
in phosphorylase, **34:**376
Pyridoxal 5-phosphate, **16:**214
Pyridoxal Phosphate, Tautomerism in, and in Enzymatic Catalysis (Metzler), **50:**1
Pyridoxal phosphate and CO_2 fixation, **12:**55
Pyridoxal Phosphate Catalysis, Stereochemical Aspects of (Dunathan), **35:**79
Pyridoxal phosphate-dependent α, β-elimination, **50:**283–286
Pyridoxal-5′-phosphate-dependent catalytic antibodies, **74:**170–173; 172; 173; 174
Pyridoxal-5′-phosphate-dependent enzymes
acting on amino acids
aldimine intermediates, **74:**130; 144–145; 145–146; 146–147; 147
mechanistic features of, **74:**130
reaction pathways of, **74:**130; 131
active-site mutations, engineering of, **74:** 167–170
D-alanine aminotransferase family, **74:** 141–142
denotations of, used by various authors, **74:**139; 140
evolutionary pedigree of, **74:**154; 155
structural data on, **74:**134
alanine racemase family, **74:**142
denotations of, used by various authors, **74:**139; 140
evolutionary pedigree of, **74:**154–155
structural data on, **74:**134; 141
aldimine intermediates
cisoid conformation of, **74:**144; 146–147; 147
external, **74:**130; 143; 144; 192–194; 193; 195; 196
internal, **74:**130; 144–145; 145–146; 191; 192
amino acid sequences
data on, **74:**132–135
profile analysis of, **74:**135–138
catalytic mechanisms, evolutionary divergence in, **74:**155–156
common properties, evolutionary constraints due to pyridoxal-5′-phosphate and, **74:**143–149
emergence
hypothetical scenario for, **74:**139
time of, **74:**156–157; 157
engineering, of, **74:**163–170
evolutionary convergence among, examples of, **74:**142–143; 143
evolutionary lineages of, **74:**139–142
evolution of
experimental simulation of, **74:**163–173

rates if, **74:**158–159; 159–160; 161–163; 162
families
denotations of, used by various authors, **74:**139; 140
evolutionary pedigrees, and functional specialization, **74:**146–155
evolutionary trees of, construction of, **74:**136–138
evolution of, **74:**131–132
profile analysis of, **74:**135–138; 137
sequence profile analysis of, **74:**135–138; 137
structural data on, **74:**132–135; 133–134
α family, **74:**137–141
denotations of, used by various authors, **74:**139; 140
evolutionary pedigree of, **74:**149–152; 150–151
structural data on, **74:**133–134
β family, **74:**141
denotations of, used by various authors, **74:**139; 140
evolutionary pedigree of, **74:**152; 154–155
structural data on, **74:**134
functional specialization
evolutionary course of, **74:**152–155
during phylogenesis, **74:**157–159
geminal diamine intermediates, **74:**192–194; 193; 221–222
with homologous sequences in all three biological kingdoms, **74:**156–157; 157
isoenzymes, homologous heterotopic, evolution of, **74:**160–163; 162
molecular evolution of, **74:**129–132; 176–179; 178
not acting on amino acid, substrates, **74:** 174–175
physiological roles of, **74:**130
protonation at C4′, stereochemistry of, **74:**147–149
pyridoxal-5′-phospate as cofactor for, alternatives to, **74:**175–176
quinonoid intermediate, **74:**195; 196; 224
reaction intermediates, **74:**191; 192; 193; 195; 196
reactions catalyzed by, **74:**130; 131; 143; 144
non-B6 enzymes catalyzing, **74:**175–176; 176
β-replacement reactions, **74:**191–197; 194; 195
sequences
data on, **74:**132–135; 133–134
family profile analysis of, **74:**135–138; 137
homologous, **74:**132
orthologous, **74:**132
paralogous, **74:**132
profile analysis of, **74:**135
structural data on, **74:**132–135; 133–134
γ subfamily of, **74:**140; 151
substrate specificity of
change in, engineering of, **74:**163–167
specialization for, **74:**153–155
Pyridoxal phosphate dependent reaction, **49:**169
inhibition by pyridoxal phosphate analogs, **49:**169
of tryptophan synthase, **49:**145–158
Pyridoxal phosphate-enzyme complex, **16:**216; 217
Pyridoxal phosphate in tryptophan metabolism, **10:**59; 61–62; 62–64; 65; 70
Pyridoxal phosphate (P^{32}), in transamination studies, **16:**216; 218
Pyridoxal phosphate (PLP)
bacterial lysine biosynthesis, **72:**281–283
L,L-diaminopimelate decarboxylase, **72:**311–314
L,L-diaminopimelate epimerase, **72:**303–307
cysteins *S*-conjugate β-lyase reactions
bacteria and fungi, **72:**215
glutamine transaminase K in rat kidney, **72:**219–222
high M_r enzyme, **72:**223–224
human liver, **72:**215–216
kynurenine pyruvate aminotransferase, **72:**222–223
rat liver with kynureninase, **72:**216–219
in vivo regulation, **72:**224–226
glutamine-dependent nitrogen transfer, **72:**191
Pyridoxal-5′-phosphate (PLP), binding to phoshorylase *a*, **24:**266
in phosphorylase, **24:**271–274
Pyridoxal-pyridoxamine conversion, function, **10:**62–63
Pyridoxamine, **6:**26; 30, **53:**163. *See also* Pyridoxal, Pyridoxal phosphate, Pyridoxamine phosphate, Pyridoxine and Vitamin B_6
attachment to β-cyclodextrin, **58:**35
conversion to cyclodextrin derivatives, **58:**36

Pyridoxamine (*Continued*)
derivatives, **58:**42
in transamination reactions, **16:**190; 210; 219; 223–226; 237
substrate for pyridoxal kinase, **16:**219
$C_4{}'$-t_1-Pyridoxamine, **35:**101
Pyridoxamine derivatives, **43:**346
Pyridoxamine phosphate, **50:**17, **58:**35. *See also* Pyridoxal, Pyridoxal phosphate, Pyridoxamine, Pyridoxine and Vitamin B_6
coenzyme in transaminations, **16:**211; 215–219; 219–220; 224–225; 238
preparation and characterization, **16:** 214–215
enzymebound, Cotton effect, **27:**86
tautomerism of, **50:**4
Pyridoxamine 5′-phosphate, **35:**510, **49:**20; 22; 24; 26; 36
Pyridoxamine phosphate-enzyme complex, **16:**216; 217
Pyridoxamine phosphate (P^{32}), in transaminations, **16:**216; 218
Pyridoxamine 5′-phosphate (PMP), cystein *S*-conjugate β-lyase reactions, **72:** 219–222
Pyridoxamine-P-ketimines, **56:**4
Pyridoxamine pyruvate aminotransferase, **74:**148
Pyridoxamine-pyruvate transaminase, **43:**353
Pyridoxamine pyruvate transamination, **35:**103; 104; 105
mechanism of action, **35:**103
stereochemistry of transamination, **35:**99
Pyridoxine, **4:**303–304, **17:**436. *See also* Pyridoxal, Pyridoxal phosphate, Pyridoxamine, Pyridoxamine phosphate and Vitamin B_6; Pyridoxal phosphate, Vitamin B_6; Vitamin B_6
action on thiaminase, **15:**401
as codecarboxylase factor in growth medium, **6:**8; 25; 30
in muscular dystrophy, **6:**222
streptococci deficient in, tyrosine decarboxylase activity of, **6:**25–26
dissociation constant, **50:**5
tautomerism of, **50:**4
and metabolism of D- and L-amino acids, **19:**352
in plants, with relation to thiamine, **10:**246; 247
reaction with borate, **12:**515
substrate for pyridoxal kinase, **16:**219
and transamination reactions, **19:**349; 351
and urea formation, **19:**345
and urea synthesis, **15:**342–343
Pyridoxine and decarboxylation, **7:**454
Pyridoxine and fat formation, **8:**310; 316; 330
Pyridoxine deficiency, **49:**13
Pyridoxine phosphate-enzyme complex, **16:**217
Pyridoxyl alanine, **43:**353
Pyridoxylamines, inactivity, **16:**221
Pyridoxylamino acids, inactivity, **16:**221
Pyridoxylglutamic acid, **16:**221; 225
Pyridoxylideneglutamic acid, **16:**221; 225
ε-*N*-Pyridoxyl-lysyl-phenylalanine, from phosphorylase, **24:**273
Pyridyl carbonates, carbonic anhydrase catalyzed hydrolysis, **47:**206–207; 209
Pyridylpyridinium ion, **39:**346
Pyrimethamine, inhibitor of folate system, **27:**431; 439–443; 447; 448; 452; 453; 455; 456; 458; 460
potentiator for sulfonamides, **27:**455
Pyrimidine, fluorinated
anabolism of, **54:**64–69
biochemical mechanisms of action of, **54:**61
biochemical parameters predicting tumor responsiveness to, **54:**102–109
anabolic enzymes, activity of, **54:**106–108
dUMP accumulation, **54:**105
FdUMP levels, **54:**103–105
reduced folates availability, **54:**105–106
thymidylate synthetase, active, **54:**106
in vitro tests, **54:**108
chemotherapeutically active, **54:**62–63
cytotoxic and chemotherapeutic effects of, **54:**92–102
RNA incorporation of, biochemical effects of, **54:**86–92
structures of, **54:**60
synthesis of, **54:**58
uptake and activation of, **54:**63–69
Pyrimidine amination, and ATP, **18:**231
Pyrimidine analogs
effect on enzyme synthesis, **24:**16–18
effect on virus multiplication, **17:**479–483; 496
as folate antagonists, **27:**437–451; 455; 458
Pyrimidine bases, in meteorites, **27:**350
Pyrimidine biosynthesis, **56:**329

Pyrimidine-bound pentose, **28:**407–409; 418–419; 433
Pyrimidine deoxyribonucleoside 2′-hydroxylase, **41:**270
Pyrimidine derivatives, and thiaminase, **15:** 399–405; 410–414; 416
Pyrimidine desoxyribosides, **9:**494–495
Pyrimidine-2-hydroxamin acid, in reactivation of phosphorylated esterases, **20:**307
Pyrimidine moiety, **53:**364–366
Pyrimidine mononucleotides, and LADH complex formation, **27:**68
Pyrimidine nucleoside-2′,3′-cyclic phosphate, **34:**501
Pyrimidine nucleoside kinase, **21:**281
Pyrimidine nucleoside kinases, allosteric regulation, **28:**135
Pyrimidine nucleoside phosphorylase, **21:**281
Pyrimidine nucleoside phosphorylases, **18:**216; 222
Pyrimidine nucleosides, in nucleic acids, **9:**522
Pyrimidine nucleotides. *See also* Allosteric regulation; Deoxyribose; Ribose; individual nucleotides
  and Cotton effect in LADH complexes, **27:**61–64
  of DNA, source, **28:**420–421; 423
  effect on purine pyrophosphorylases, **28:**124
  in hepatoma cells, **28:**135
  inhibition of pyrimidine synthesis, **28:**45
  metabolism, allosteric regulation, **28:**126–135
  in nucleic acids, **9:**522
Pyrimidine nucleotide synthesis, **18:**215–219
Pyrimidine oxidase, properties, **19:**203
Pyrimidine-3′-phosphate diester, RN-ase as enzyme for, **17:**171–172
Pyrimidine phosphoribosyl transferase, **54:**64; 65
  in assessing tumor responsiveness to fluoropyrimidines, **54:**107–108
  in FU activation, **54:**66–68
Pyrimidine ribosides, **9:**493–494
Pyrimidines, **4:**300; 307–309. *See also* Nucleic acids, Nucleosides, and Nucleotides; individual pyrimidines, Nucleosides, Nucleotide bases, Nucleotides
  abiogenic syntheses, **27:**384–385
  absorption spectra, **9:**480–484
  and amino acid sequence, **24:**9–11; 15
    specificity of RNase action, **24:**165; 241; 242
  and argininosuccinic acid, **15:**354; 356–357
    of tobacco mosaic virus, **15:**460–461
  biosynthesis, **9:**467–468, **16:**266–268
    allosteric regulation, **28:**45; 126–135
    of DNA, **16:**430–436
    enzymic reactions in, **21:**263–294; 420–423
    labeled, in nucleic acid study, **16:**450–462
    of nuclear RNA, **16:**437–440
    in transaminations, **16:**237
  biosynthesis of, **12:**45
  cofactor activity with phenylalanine hydroxylase, **67:**175
  degradation, **9:**468–469
  and dietary purines, **9:**463
  fluorinated, **55:**268–271
  in genetic studies, **17:**399; 429–431; 459–465
    in transglycosylations, **17:**220–221
  methylation of, **25:**414–415
  occurrence, **9:**466–467
  as phenylalanine hydroxylase cofactors, **67:**196–1204
  in plants, **9:**442–443
  synthesis of, **18:**257; 261
  tautomerism of, **50:**36
Pyrimidines, Enzymic Synthesis (Reichard), **21:**263
Pyrimidines, H-5 exchange of, **49:**208; 212; 213
Pyrimidines and vitamin B_{12}, **26:**267
Pyrimidine starvation, **51:**184
Pyrimidylaniline, **15:**413
Pyrimidylmethanol, **15:**400; 411–413
Pyrithiamin, as inhibitory analogue, **6:**131; 137; 143; 144
Pyrocatechase
  effect on phenolase complex, **19:**131; 199
    mechanism of action, **19:**83; 85–87
    properties, **19:**81; 126
  enzymic adaptation, **13:**74
Pyrocatechase (catechol-1,2-oxygenase), benzene oxidation, **27:**509
Pyrocatechin, in oxidation of amino acids in plants, **20:**321; 322
Pyrodinium bahamence, luminescence in, **25:**161
Pyrogallol, **18:**373
  catalase action on, **19:**109
    effect on indolylacetic acid oxidase, **19:**127
    peroxidase action on, **19:**106

Pyrogallol (*Continued*)
chemiluminescence, **25:**121
inhibition of β-lactamase, **28:**295
in plant metabolism, **16:**136–137; 152
Pyrogenic agents and hypertension, **7:**540
L-Pyroglutamate, **43:**519
Pyronin-methyl green staining of cell nuclei, **16:**417
Pyronin staining, of chromosomes, **16:**460
of RNA, **16:**436
Pyrophosphatase, **6:**149, **15:**250–251; 253; 351, **17:**171, **55:**386, **56:**293. *See also* Polyphosphatase
action on UDPG, **14:**199
adsorption studies, **14:**346
amino acids in, **14:**275
classification, **14:**224; 248
effect on light emission, **25:**133; 135–137
and transmethylation, **25:**413
inorganic, **18:**205; 219–221, **20:**122, **23:**360
in muscle, **16:**350
potato, action on CoA, **16:**12; 13
snake venom, action on CoA, **16:**12; 14
in sugar nucleotide synthesis, **26:**57
isolation, **14:**402; 405
latex, **29:**235–236
in mitochondria, **21:**40
yeast, inhibition by DFP, **13:**332
Pyrophosphatases
alkaline, role of metal ions, **10:**104
specificity, **10:**86
in ribonucleotide reductase, **28:**444
soil, **28:**386
Pyrophosphate, **15:**94; 106–107; 351, **35:**488; 489, **43:**346, **60:**174. *See also* Adenylpyrophosphate
and ADH binding, **22:**356
binding by myosin and actomyosin, **16:**333
inorganic, effect on myosin, **16:**344
in citric acid cycle, **15:**184; 202–203; 223; 231–232; 234; 249; 252 (P^{32})
effect on LADH complexes, **27:**73–75
effect on LADH-DPNH complex, **27:**61–64
effect on myosin thread extensibility, **6:**176
formation in galactose metabolism, **20:**126
inorganic, uridine diphosphoglucose pyrophosphorylase action on, **20:**121
formation of, **43:**357
as inhibitor of metalloflavoproteins, **17:**239; 258; 260; 269; 275
inorganic
allosteric regulation, **28:**95–98
in ATP system, **14:**17
labeled, **19:**251; 257; 258
and papain activity, **19:**286
metabolism of, **18:**193; 219–221, **45:**85 (*See also* PP_i)
as succinic dehydrogenase inhibitor, **18:**94
reaction with diphosphopyridine nucleotide, **15:**81–82; 104–105 ($P^{32}P^{32}$)
turnover studies, **7:**114–126
Pyrophosphate: oxaloacetate carboxylyase, *See* Phosphoenolpyruvate carboxytransphosphorylase
Pyrophosphate-Acetate Kinase, Properties of, and Role in the Metabolism of Inorganic Pyrophosphate (Wood, O′Brien, and Michaels), **45:**85
Pyrophosphate and lysis, **8:**94–95
Pyrophosphate-ATP exchange, **17:**181–184
Pyrophosphate binding motif, AS-B form, asparagines synthetase, **72:**165–167
Pyrophosphate binding region, eukaryotic dihydrofolate reductase, NADPH binding, structural changes induced by, **70:**45
Pyrophosphate donors, nucleoside triphosphates as, **18:**198–200
Pyrophosphate:fructose 6-phosphate phosphotransferase, *See* PP_i-PPK
Pyrophosphate fructose-6-P_2 phosphotransferase, *See* PP_i-PFK
Pyrophosphate-glucose phosphotransferase reaction, **28:**97
Pyrophosphate hydrolase, localization of, **24:**328–329
Pyrophosphate (K+), **51:**236
Pyrophosphate linkage in CoA, **16:**12–14
Pyrophosphate moiety of DPNH, and enzyme binding, **27:**56
Pyrophosphate (P^{32}) carboxylations, **26:**333
Pyrophosphate-phosphorfructokinase, Properties of, and Role in the Metabolism of Inorganic Pyrophosphate (Wood, O′Brien, and Michales), **45:**85
Pyrophosphate phosphotransferase, acceptor specificity, **34:**302
identity with glucose 6-phosphatase, **34:**302

Pyrophosphate ($P^{32}P^{32}$), **18:**202–203
Pyrophosphate (PP^{32}), and light emission, **25:**135
Pyrophosphate (P^{32}-P^{32}) exchange with ATP, **17:**181–182
Pyrophosphates. *See also* ADP and ATP
and coenzyme Q, **25:**320–322
and light emission, **25:**131–136
Pyrophosphomevalonate decarboxylase, **55:**108
Pyrophosphoric acid, hydrolysis, by metal ions, **21:**175; 182
Pyrophosphoric acid esters, hydrolysis, by metal ions, **21:**175
Pyrophosphorolysis of pyridine nucleotides, **15:**91; 104–105; 106–107
Pyrophosphorolyzing enzymes, **18:**199
Pyrophosphorylase, CDP-glycerol, in the cells of *Bacillus licheni-formis or Bacillus subtilis,* **30:**236
in the membrane fraction of *Staphylococcus lactis* and *Lactobacillus buchneri,* **30:**236
Pyrophosphorylase, term, **18:**193
Pyrophosphorylases
mononucleotide, regulation, **28:**123–124
sugar, **28:**110–114
in nucleotide synthesis, **27:**419
regulation of, by metabolites, **34:**315
in synthesis of sugar nucleotides, **26:**39–57; 67
Pyrophosphorylases and Phosphorylases in Biosynthetic Reactions (Kornberg), **18:**191
Pyrophosphorylases in biosynthetic reactions, **18:**191–240
Pyrophosphoryl coenzyme A, **15:**252
1-Pyrophosphoryl-ribose-5-phosphate synthetase, mechanism of, **49:**113
Pyrophosphoryl transfer, **49:**113–116
Pyrophosphoryl transferring enzymes
catalytic power, MutT pyrophosphohydrolase, **73:**187
divalent cations role, MutT pyrophosphohydrolase, **73:**187–191
Pyrophyllite, in treatment of plant diseases, **20:**390; 399
Pyroracemic acid, in plants, **10:**300
relation to fatty acid metabolism, **10:**186
Pyrrol compounds, melanins as, **16:**167
Pyrrole, **46:**102
activation of enzyme, **7:**537
formation, **7:**329
and glycine, **7:**329
pie-electron density of, **20:**54
reactions with quinones, **16:**112
Pyrrole-2-carboxylic acid, formation and conversions, **20:**139; 140
Pyrrole compounds from peptides, **8:**130
Pyrrole nitrogen, in imidazole, **20:**53; 55–57; 83; 84; 86
Pyrrole tricarboxylic acid, from *Sepia* melanin, **16:**172
Pyrrolidine, **50:**164
pigment from, **16:**112
Pyrrolidine ring and ornithine, **8:**237
Pyrrolidone carboxylate, glutamine synthetase reaction, **72:**16–17
Pyrrolidone carboxylates, **43:**315; 319
Pyrrolidone carboxylic acid, γ-glutamyl derivatives from, **31:**186–187
L-Pyrrolidone carboxylpeptidase (5-Oxo-L-prolyl carboxylyl peptidase, **43:**523; 525
Pyrrolidone carboxyl-ribonuclease, **24:**171
3-Pyrroline, **50:**164
Pyrrolineacetic acid, **38:**438
O-aminobenzaldehyde adduct, **38:**438
Pyrroline-5-carboxylate reductase, **53:**221
localization of, **24:**302
Neurospora mutants, **22:**7; 8; 19
Pyrroline carboxylate reductase, feedback effects of proline inhibition of, **29:**367
Δ'-Pyrroline-5-carboxylic acid, **40:**70
Δ^1-Pyrroline-5-carboxylic acid, in proline biosynthesis, **16:**258
Pyrrolo-quinoline quinol, *See* $PQQH_2$
Pyrrolo-quinoline quinone
absorption coefficients of, **59:**177
absorption spectra of, **59:**176–178
in active site, **59:**194–196
adduct extraction, **59:**184–185; 197; 200; 202–204
and adduct formation, **59:**179–181
and alkaline degradation, **59:**183
analogs of, **59:**174–176; 200
analysis of, **59:**184–186
and apoenzyme, **59:**185–186
binding of, **59:**194–195
biological assay of, **59:**185–186
and borate monester, **59:**180
covalently-bound, **59:**192–194
and 4-desoxy-PQQ, **59:**206

Pyrrolo-quinoline quinone (*Continued*)
and 5-desoxy-$PQQH_2$, **59:**176
and 7,9-didecarboxy-3-alkyl derivative, **59:**175
distribution of, **59:**186
and ENDOR spectroscopy, **59:**179
and ESR spectroscopy, **59:**171; 178–179; 183; 195
and esters, **59:**175; 178–179; 200; 206
3-nitro, **59:**183
excretion of, **59:**174
fluorescence spectra of, **59:**176–178
fluroescing derivative of, **59:**183–185
and HPLC, **59:**185
and hydrate, **59:**177
hydrazone formation, **59:**182; 185; 193
interaction of, with other cofactors, **59:** 195–196
and mass spectrometry, **59:**172–173; 178–179
methylation, **59:**184
and NMR spectroscopy, **59:**171–172; 178; 180
and nucleophilic agents, **59:**179–180
occurrence of, **59:**186
open ring structures, **59:**175; 184; 202; 206
perspectives on, **59:**186
production of, **59:**174
properties of, physical, **59:**176–179
reactivity of, chemical, **59:**179–180; 182–184
redox potentials of, **59:**179
reduction of, **59:**179; 180; 182
ring opening of, **59:**182–184
structure elucidation of, **59:**171–174
synthesis of, **59:**174–176
and vitamins, **59:**186; 209
Pyrrolo-quinone semiquinone, *See* PQQH
Pyruvate, **15:**166–167 (C^{14}); 171–173, **43:**11; 44, **46:**323, **50:**144; 147
and acetate, **23:**433; 462–464
and glycine, **23:**457; 459; 462–464
and glycolate, **23:**437–441; 444–446; 462–464
phosphorolytic cleavage, **23:**336–339
acetate fixation in, **6:**245
bacterial fermentation to acetyl phosphate and formate, **6:**238; 239
carbon dioxide fixation in carboxyl, **6:**246
cleavage in *Clostridium butylicum,* **6:**237
reversibility, **6:**246–248
effect on sulfanilamide acetylation, **6:**258
and formate equilibrium, **6:**245
formation with acetyl phosphate plus formate, **6:**245
inhibition of choline acetylation, **6:**261
metabolism, and bacterial acetyl phosphate formation, **6:**237–238
effect of phosphate, **6:**239
and hydroxylamine reactivity, **6:**257
and thiaminpyrophosphate enzyme, **6:**238
oxidation, **6:**233
and phosphorylation, **6:**254–256
acetylation of, **5:**149–150
as activator, **46:**328–333
in alanine racemization, **14:**195
in lactic acid racemization, **14:**197
in allosteric regulation, **28:**91; 111–112
conversion to glucose, allosteric control, **28:**86–98
anaerobic decomposition, **3:**161–163
in blood, **5:**139–143
carboxylase, **52:**295–296
characteristics and regulation, **52:**295
effect of glucagon, **52:**296
inhibition in hypoglycin poisioning, **52:**271
phosphorylation, **52:**296
in citric acid cycle, **15:**183; 185; 186; 201; 224; 258
classical and nonclassical, active, **42:**7
and CO_2 fixation, **12:**46; 47; 51; 53; 61–63; 94–95
anaerobic dismutation of, **12:**101–103
fixation of formate in, **12:**95–101
role in oxalacetate synthesis, **12:**73–80
decarboxylation of, **5:**145–147
anaerobic and aerobic, **5:**148
dehydrogenase
characteristics, **52:**293
effect on glucagon, **52:**294
effects of ligand on state of activity, **52:**293; 294
phosphorylation, **52:**293
in starvation and diabetes, **52:**295
dismutation of, **5:**144–145
effect on deamidation of asparagine and glutamine, **8:**152
formation, **67:**15
formation from acetate and CO_2, **16:**268
in amino acid metabolism, **16:**117; 275; 276; 284–286; 291; 294; 302

(2-C^{14}), in transamination, **16:**223
labeled, incorporation into valine, **16:**285
in transamination reactions, **16:**189; 194; 199; 203–205; 219; 220; 222–223; 226; 233; 234; 236; 237
formation in carbohydrate fermentation, **3:**151–160
from glucose, **5:**141
insulin action and diabetes, **5:**141; 142
kinase, **52:**300–304
change in kinase properties after phosphorylation, **52:**301
effect of hormones on activity, **52:**303–304
effect of ligands, **52:**300
effect of ligands on interconversion, **52:**302–303
general properties, **52:**300–301
isoenzymes, **52:**300
phosphorylation dephosphorylation, **52:**301
in starvation and diabetes, **52:**304
kinase flux
effect of glucagon on, **52:**305
effect of starvation on, **52:**305
effect of substrates on, **52:**305
measurement of, **52:**305; 306
mechanism of formation from lactic acid, **15:**8–9
metabolism, **5:**130–5:159
anabolic, **5:**130–143
catabolic, **5:**143
metabolized, **67:**305–306
oxidation, **15:**121–122; 201; 208 (C^{14}); 213 (C^{14}); 245; 254; 256; 262
oxidation and reversible red. ox. systems, **3:**167–169
oxidation in normal and cancer tissues, **4:**204; 205; 206; 209–211; 212; 237; 238; 243; 244
oxidation of, **5:**147–149
in tumor mitochondria, **29:**344
PEP cycle, demonstration, **52:**305
phosphate dikinase, **45:**106; 108–129
crystals of, **45:**111
enzyme, function of, **45:**127–219
EPR spectra, **45:**127; 128
mechanism of reaction, **45:**113
metal requirements, **45:**120–125
molecular weight, **45:**122
purification, **45:**110
pyrophosphoryl from, isolation, **45:**115
stability, **45:**110–112
stereospecificity of reaction, **45:**106
subunit properties, **45:**112
sulfhydryl groups, **45:**112
transition-state analog complex, **45:** 126–127
tri(Uni Uni) Ping-Pong mechanism, **45:**118–119
phosphoroclastic splitting of, **5:**145
and pyridine nucleotides, **15:**117; 124–126
in reduction of diphosphophyridine nucleotide, **20:**151
relation to urea synthesis, **15:**325; 327; 332–333; 335; 336; 337; 342; 343–344; 368
in sperm metabolism, **24:**61–64; 68
and synthesis of 2,3-diphosphoglycerate in partially carbamylated red cells, **40:**17
transport, **52:**292–293
characteristics, **52:**292
effect of hormones, **52:**292
Pyruvate, Phosphate Dikinase, Properties of, and Role in the Metabolism of Inorganic Pyrophosphate (Wood, O'Brien and Michaels), **45:**85
Pyruvate-acetaldehyde condensation, **17:**402
Pyruvate-amino acid reactions, in transaminations, **16:**203–204
Pyruvate and action of pantothenate, **7:**442–444
Pyruvate apodehydrogenase, **18:**339
Pyruvate apooxidase, binding of lipoic acid, **26:**333
Pyruvate binding, dihydrodipicolinate synthase (DHDPS), **72:**287–290
Pyruvate (C^{14}), exchange reaction with oxaloacetate, **26:**313–314
Pyruvate carboxykinase, allosteric regulation, **28:**122
Pyruvate carboxylase, **17:**425, **35:**390–400, **37:**467; 554, **46:**465, **56:**293
activated by acetyl CoA, **35:**391
activation of, **42:**40
by monovalent cations, **42:**32
allosteric regulation, **28:**53–56; 88–92; 122
association of chicken liver, **42:**57
baker's yeast, **35:**397
aspartate inhibition, **35:**398
long chain fatty acyl CoA activation, **35:**397
catalytic properties, **42:**4
cold lability, **35:**396

Pyruvate carboxylase (*Continued*)
conformational changes, **42:**63
in diabetes, **54:**164; 165
dissociation, effect of, **42:**49
effector control, **35:**396–398
acetyl CoA activation, **35:**396
β-hydroxybutyryl CoA/acetoacetyl CoA ratio, **35:**397
effects of fluctuation in nutritional and physiological state, **35:**398
electron microscopy of, **33:**99–100
mechanism of deduced from relaxivity data, **33:**52–55
glucagon effect on, **54:**150
gluconeogenic role, **35:**398–400
kinetic mechanism of, **45:**291; 304
modification of sulfhydryl groups, effect of, **42:**58
molecular, characteristics, **35:**395–396
weight of, **35:**395
molecular weight, **42:**20
polypeptide composition of, **42:**23
rate equation, **42:**11
reaction, **35:**392
ATP-^{1} ^{4}C-ADP exchange, **35:**393
ATP-^{3} $^{2}P_i$ exchange, **35:**392
enzyme-CO_2, **35:**392
metal ion in second partial reaction, **35:**393
oxaloacetate-^{1} ^{4}C-pyruvate exchange, **35:**393
retention of configuration, **35:**361
reaction mechanism, **42:**5
regulatory role in gluconeogenesis, **35:**398
control by metabolite effectors, **35:**399
sedimentation of, **42:**20
starvation effect on, **54:**157
substrate for, **42:**30
subunits, number and arrangement, **42:**18
transcarboxylation site of, **42:**33
Pyruvate Carboxylase: An Evaluation of the Relationships Between Structure and Mechanism and Between Structure and Catalytic Activity (Utter, Barden, and Taylor), **42:**1
Pyruvate carboxylase [Pyruvate:CO_2 ligase (ADP)], properties, **26:**305–307
propionic acid metabolism, **26:**293; 297; 303; 311
reaction mechanism, **26:**312–316
Pyruvate decarboxylase, **43:**250, **46:**477, **55:**213; 214
Pyruvate dehydrogenase
effect of fluoroacetate, **18:**118; 146
reversibility, **23:**426
Pyruvate dehydrogenase complex, **57:**151–152
kinetic mechanism of, **45:**282; 292
Pyruvate dismutation, activity of –SH group, **11:**241
Pyruvate esters, carbonic anhydrase catalyzed hydration and hydrolysis, **47:**188
Pyruvate formate, *S*-adenosylmethionine, **66:**33–34
Pyruvate formate lyase, **66:**34
1-^{14}C-Pyruvate, in pentose synthesis, **28:**431
2-^{14}C-Pyruvate, in pentose synthesis, **28:**429
Pyruvate kinase, **34:**450; 523, **37:**421; 427; 435; 554; 584, **43:**34; 297, **46:**247; 460–462, **49:**119, **50:**364; 365; 372; 395, **53:**221; 222, **55:**212, **58:**161, **59:**338–339; 364
allosteric regulation, **28:**86; 88–89; 122
and ATP synthesis, **23:**326; 332; 339; 361
classification, **14:**251
correlation of hepatoma growth rate with, **29:**331
in diabetes, **54:**164
glucagon effects on, **54:**150; 151
intracellular distribution, **16:**468
levels in cyanate treated red cells, **40:**17
localization of, **24:**313
turnover number, **24:**123
mechanism, **49:**116
muscle, mechanism of deduced from relaxivity data, **33:**45–47
yeast, mechanism of deduced from relaxivity data, **33:**48–49
phosphorylation of, **54:**151
in phosphoryl transfer reactions, **55:**358–359
in propionic acid metabolism, **26:**287; 294; 297; 305; 307; 316
starvation effect on, **54:**157
^{14}C-Pyruvate, labeling of pentoses, **28:**419; 434–435
Pyruvate Metabolism (Stotz), **5:**129
Pyruvate-oxalacetate condensation, effect of streptomycin on, **10:**161
Pyruvate oxidase, **55:**213
activity of –SH group, **11:**236; 238; 241
and fluoropyruvic acid, **18:**151
and lipoic acid, **18:**334–335
Pyruvate oxidase complex, and glycolate metabolism, **23:**439

Pyruvate oxidation, **1:**141
and lipoic acid, **18:**333–338; 339
Pyruvate oxidation factor, **15:**214–215, **18:**319; 320; 322–324; 331. *See also* Lipoic acid
Pyruvate oxidation system, **18:**333–336
Pyruvate phospherase, **17:**163
Pyruvates, complexes of, **17:**149
Pyruvate synthase reaction, **33:**139
Pyruvate-valine reaction, **16:**203
Pyruvic acid, **5:**69; 129; 168–179; 181; 183; 186; 187; 198; 241, **49:**28, **53:**169, **58:**35. *See also* Citrtic acid cycle; Fatty acids, Keto acids, Keto fatty acids, and Pyruvate
absorption spectrum, **8:**131; 136
biosynthesis, **21:**431
oxidation, in mitochondria, **21:**34; 36; 121
in carbohydrate oxidation, **6:**359–360
decarboxylation, thiamin stimulation of, **6:**55–56
carbonic anhydrase catalyzed hydration of, **47:**187
carboxylation, **8:**401–409
catabolism, **8:**388–389
cosubstrate in deamidation, **8:**158–164
from cystine peptides, **8:**146–147
decarboxylation, **15:**29; 158
degradation in animal tissues, **10:**180–182
from dehydropeptides, **8:**135; 136; 137; 142–143; 146–147; 152; 159
from diacylaminopropionic acids, **8:**153–158
from diglycylaminopropionic acid, **8:**153–158
enzyme production of, **5:**241
and fat formation, **8:**315; 327–330
and fatty acid catabolism, **8:**392–401
formation and conversions in galactose metabolism, **20:**119; 120
in formation of pyrrole-2-carboxylic acid, **20:**139; 140
in plants, in biosynthesis of alanine, **20:**300
transamination with, **20:**329; 330
formation from 2-phosphoglyceric acid, **5:**241
formation in *Escherichia coli,* **19:**40
in photosynthesis, **19:**7; 10–13; 17; 33–35; 52; 53
transamination reactions of, **19:**338; 346; 349; 350; 355–364; 370
in genetic studies, **17:**423–426
in histidine decarboxylase action, **5:**69; 70
labeled, in plants, **18:**299; 300; 302–303
in lignin study, **18:**363; 365
phosphoroclastic cleavage of, **18:**227
in plants, **18:**291–293; 295–304; 312
in *Neurospora* mutants, **22:**25
and pantothenic acid in carbo-hydrate metabolism, **3:**266–267
from peptides of cystine, **8:**159
in tryptophan metabolism, **10:**61–62; 63–64; 68
Pyruvic acid-2-C^{14}, **18:**303
Pyruvic acid in CO_2 fixation, **2:**154–158; 162–167; 170; 174
Pyruvic acid in mold metabolism, **9:**540; 558–559; 566; 567–570; 571–576
Pyruvic-3-C^{14}, **18:**299; 303
Pyruvic carboxylase, **19:**11; 13
in plants, **18:**297–298; 313
Pyruvic decarboxylase, in plants, **18:**295–296
Pyruvic decarboxylases, classification, **14:**237; 239; 241; 256
Pyruvic decarboxylation, in plants, **18:**304; 313
Pyruvic dehydrogenase, **9:**571, **15:**115; 119; 121; 129; 259
formation of, **24:**24
in mitochondria, **21:**74; 121
Pyruvic kinase, **19:**10
residual activity, **21:**135
Pyruvic kinase, EPR study, **27:**140
nuclear-cytoplasmic ratio, **27:**262
Pyruvic oxidase, **15:**212–213; 218; 223; 229
in dihydrofolic acid biosynthesis, **21:**395
Pyruvic oxidases, **19:**202
Pyruvic phosphokinase, **15:**246
Pyruvic phosphopherase, **17:**169
Pyruvic r CoA transacetylase, classification, **14:**251
Pyruvoylalanine, conversion to alanylalanine, **16:**239
Pyruvoylgylcine, preparation and properties, **8:**129–130
Pyruvoylpeptides, and dehydropeptides, **8:**119; 129–130
conversion to saturated peptides, **8:**129
preparation and properties, **8:**129–130

Q

Q-base, **47:**388
QC, *See* Ubichromenol
Q-Enzyme, **18:**288
in transglycosylation study, **17:**191; 215

Q-enzyme, **15:**304; 307–309; 309–310, **34:**354
action of, **12:**411; 421–423; 424
effect of borate on, **12:**518
branching of amylopectin, **34:**356
classification, **14:**251
isolation, **14:**394; 405
in plants, **19:**5
in polysaccharide synthesis, **11:**298–299; 318; 323–324; 324–326; 362–365; 369–370
phosphate requirement, **11:**366–367
reversibility of action, **11:**367–369
QH_2-cytochrome *c* reductase (complex III), in reconstitution of respiratory chain, **28:**210–214; 216–220; 223
Quantasomes
arrangement of subunits in, **30:**56
in chloroplasts, **30:**7
chemical nature, **25:**69–70
ESR production in, **25:**83; 84
in photosynthesis, **25:**43; 67–69
Quantum, efficiencies and action spectra, **30:** 56–63
Quantum conversion, in photosynthesis, **25:** 84–86
apparatus, **25:**67–73
efficiency, **25:**48–51
mechanisms, **25:**84–86
solid-state phenomena, **25:**81–86
Quantum-conversion systems, and plant pigments, **25:**62–64
Quantum mechanics, application to metal protein catalysis, **17:**281
Quantum requirement of photosynthesis, **25:** 48–54
Quantum requirements
of Hill reaction and photosynthesis, **19:**50
of Hill reaction with DCIP, **30:**59
of Hill reaction with ferricyanide, **30:**59
of photoreduction of cytochrome *c,* **30:** 58–59
of photoreduction of $NADP^+$, **30:**58–59
in photosystem 2, **30:**27
for photosystem 1 particles, **30:**59–61
Quantum yield, of delayed photophosphorylation, **25:**57–58
of photosynthesis, **25:**51–53
Quantum yields
in denaturation, **9:**105–108
of enzyme inactivation, **9:**113; 141; 143–154
of peptides, **9:**92–93
of proteins, **9:**157–161
in surface films, **9:**121–124
of fluorescence of chloroplast fractions, **30:**51
Quartz, **34:**452
Quartz in model enzyme system, **15:**22
Quasi-stationary state, **34:**513
Quasi-substrate, definition, **22:**52
Quatenary ammonium compounds, effect on cholinesterase, **20:**295–298; 306; 307
Quaternary ammonium salt hydrates, **65:** 243–244
polyhedra, **65:**246; 248
Quaternary ω-N atom, effect of, **49:**65
Quaternary structure, **39:**350
of allosteric enzymes, **28:**53–56
glutathione S-transferases, **69:**8–11
Quercetin, **17:**434. *See also* Polyphenols
metabolism in plants, **16:**141; 144; 145; 151
oxidation by peroxidases, **14:**151
reaction with borate, **12:**500; 522
in tobacco leaves, **10:**347–349; 373–375
Quinacillin, action of β-lactamase, **28:**269
Quinacrine
inhibition of β-lactamase, **28:**295
as inhibitor of metalloflavoproteins, **17:**237; 243; 246; 259; 312–313; 384
Quinate dehydrogenase, **34:**220
Quinazolines, effect on folate systems, **27:**440; 446; 451–452
Quinic acid, **34:**219
in aromatic ring synthesis, **17:**413; 421
in metabolism of aromatic amino acids, **16:**287–289
Quinic dehydrogenase, **16:**289, **17:**413
Quinidine
origin, **8:**233
in synthesis of *d*-mandelic nitrile, **20:**7
Quinine, **17:**384
effect on actin polymerization, **13:**155
effect on myotonia, **6:**214
inhibition of β-lactamase, **28:**295
origin, **8:**231–233
in synthesis of *d*-mandelic nitrile, **20:**7
Quinine nucleates, **17:**147
Quinine-oxidizing enzyme, **17:**382
classification, **14:**253
Quinine oxidizing enzyme, and *N′*-methylnicotinamide, **15:**87–88; 106

in metabolism of nicotinamide, **15:**112
Quinine r O_2 transhydrogenase, **14:**253
Quinochrome, **15:**403
Quinohemoprotein ethanol dehydrogenases, **59:**189; 195–196; 205–206
Quinoid compounds, toxicity, **15:**6
o-Quinoid dyes, cause of mutations, **6:**95; 114
Quinoid H_2-biopterin, **65:**303; 328
Quinol, **18:**373
Quinolinate synthesis, **73:**136–144. *See also* NAD+ synthesis
aerobic pathway, **73:**140–144
anaerobic pathway, **73:**136–140
NAD+ synthesis, **73:**145–158
Quinolinc acid, **49:**12
Quinoline
action on thermostable factor, **15:**410
action with thiaminase, **15:**400–403; 410–412
and thiaminase model, **15:**404
methylated derivatives of, **12:**474–478
Quinoline compounds, oxidation of, **15:**112
Quinoline pathway, **49:**4
Quinoline structure and pyridine ring, **8:**235; 236
Quinolinic acid, **17:**419; 421
in biosynthesis of nicotinic acid, **15:**99–103
formation in tryptophan metabolism, **19:** 93–96
3-, 5-5, 6-6, 7-7, and 8-8Quinolinols, identification of glucuronides, **16:**383
Quinol phosphate, in oxidative phosphorylation, **23:**377
Quinol phosphates, and oxidative phosphorylation, **25:**320–322
Quinone, as hydrogen acceptor in photosynthesis, **15:**126
as oxidizing agents, **15:**3; 89; 128
Quinone as acceptor with oxidoreductases, **24:**304
Quinone formation, in cacao drying, **25:**483–484
Quinone-producing enzymes, as terminal oxidases, **16:**131–136
Quinone reductase, PN-dependent, **26:**119; 126
Quinone reductases, **65:**97–129. *See also* DT-diaphorase
role in browning of plant tissues, **16:**152
specificity toward electron acceptors, **16:**134
Quinones, **18:**93, **57:**376, **58:**69, **65:**97–129. *See also* o- and *p*-Benzoquinone, Diphenols; Coenzyme Q; Phenolase complex; Polyphenols; Semiquinones; specific quinones
as artificial electron acceptors, **30:**22–23
in photochemical systems, **30:**50
autoxidative formation of melanin, **16:**167
biosynthesis, **16:**136–140
chemistry, **16:**107–131
displacements, on, **16:**127–128
enzymic formation, **16:**128–131; 134; 136
from oxidation of catechins, **16:**151
in pigment formation (*See* Melanins and Pigments)
reactions with amino acids, **16:**112–114; 117
reactions with thiols, **16:**121–122; 126
in biological systems, **14:**132; 141; 145; 149; 152; 153
biosynthesis
naturally occurring, **65:**106–108
from xenobiotics, **65:**108–110
chromatographc separation, **25:**293–294
classification, **65:**98
coenzyme Q relation, **25:**277
condensation with amino acids, **10:**381; 382–386; 429
conjugation, **65:**119–120
as coupling agent, **23:**369–370; 380–381
in photophosphorylation, **23:**386
in respiratory chain, **23:**346–347; 373; 376–378
effect of mitochondria, **25:**305–306
effect on homogentisate oxidase, **19:**92
in photosynthesis, **19:**51
effect on superoxide dismutase biosynthesis, **58:**83
in electron transport studies, **17:**69; 118–119; 123; 239; 246; 257–259; 264; 266; 277–279; 285; 298
EPR spectra, **27:**119; 123; 125–126
and luminescence, **25:**149–151
metabolism, DT-diaphorase and, **65:**116–119
Michael addition, **65:**100
from microbial oxidation of hydrocarbons, **27:**520–522; 526–527
and oxidative phosphorylation, **25:**319–322
in photosynthesis, **25:**66; 69–70
redox chemistry, **65:**104–106; 128
redox transitions, **65:**101–104; 110–119; 128–129

Quinones (*Continued*)
reduction, one- and two-electron transfer processes, **65:**99–101
reduction by PN-enzymes, **26:**123; 126
role in tyrosine metabolism, **14:**75–96; 97–98; 105; 110–111
thiolate anion, **65:**104–106
use in hypertension, **7:**544
as wilting toxins, **11:**403
reactions with thiols, **11:**218; 247
o-Quinones
from catechols, **16:**128–131
reaction with indoles, **16:**111
reaction with salmine, **16:**126
reactivity, **16:**106; 110
reduction of, coupling to oxidations, **16:**134
sclerotizing, **16:**154–156; 160–162
from substituted catechols, **16:**138–140
as oxidation catalysts, **14:**166–168; 178
Quinones as hydrogen acceptors in chloroplast system, **12:**8–9; 26–28
Quinone tanning, **14:**133; 140; 145
Quinone thioether derivatives, **65:**103–104
Quinonimines, formation, **16:**108–110
role in pigment formation, **16:**119
p-Quinonimmonium phenolate zwitterion, **16:**113
Quinonoid dihydropteridine, nonenzymatic rearrangement, **67:**212–213
Quinonoid dihydropterin, **74:**238
Quinonoid dyes in oxidoreduction system, **15:**14; 16
Quinoprotein oxidoreductases
absorption spectra of, **59:**189–191; 194; 200; 206
active site of, **59:**194
and apoenzyme, **59:**189–190; 194; 200
binding of, **59:**194–195
bioenergetical role of, in bacteria, **59:** 207–208
characteristics of
alcohol dehydrogenases, **59:**187–189
aldose dehydrogenases, **59:**189–191
amine dehydrogenases, **59:**191–192
amine oxidases, **59:**192–193
other, **59:**193–194
cofactor interaction of, **59:**195–196
description of, **59:**186–187
and ESR spectroscopy, **59:**171–172; 196; 198; 202
and hydrazone formation, **59:**192–193; 201–202
inhibitors of, **59:**187; 200–202
mechanisms proposed, **59:**203–206
mechanistic studies of
activity of analogs, **59:**200
inhibitors, **59:**200–202
mechanisms proposed, **59:**203–206
model systems, **59:**292–293
redox forms, **59:**197–199
suicidal substrates, **59:**200–202
model systems of, **59:**202–203
perspectives on, **59:**208–209
and pyrrolo-quinoline quinone in active site, **59:**194–196
suicide substrates of, **59:**187; 199; 200–202
Quinoproteins
description of, **59:**170–171
pyrrolo-quinoline quinone
in active site, **59:**194–196
analysis of, **59:**184–186
occurrence of, **59:**186
perspectives on, **59:**186
production of, **59:**174
properties of, physical, **59:**176–179
reactivity of, chemical, **59:**179–180; 182–184
structure elucidation of, **59:**171–174
synthesis of, **59:**174–176
quinoprotein oxidoreductases
bioenergetic role of, in bacteria, **59:** 207–208
characteristics of, **59:**187–194
description of, **59:**186–187
mechanistic studies of, **59:**197–206
perspectives on, **59:**208–209
Quinothiamine, **15:**403
Quinuclidine structure and pyridine ring, **8:**235; 236
Quinuclidinyl benzilate, **55:**143
muscarinic site binding of, **55:**150
structure of, **55:**141
Quisqualic acid, **50:**147

R

RAA system, **57:**3–4; 6
Rabbit muscle, extraction of fiber proteins from, **13:**161–162

Rabbit-muscle aldolase, **40:**37
Rabbit muscle enzymes, in degradation of DR-5-P, **28:**403
Rabbit papilloma virus, activity and particle identity, **9:**193–194
hydration, **9:**204–206; 210; 211
physical constants, **9:**187; 188
purification, **9:**187
shape, **9:**203–206
size, **9:**204–206
Racemases, **35:**106–107, **55:**227; 228–229
for amino acids, **16:**206–207; 250; 271
classification, **14:**221; 223; 229; 236–238; 258, **44:**70
for alanine, **14:**182; 196
definition, **44:**70
for glutamic acid, **14:**196
for lactic acid, **14:**197
summary of hydrogen exchange features, **44:**130
localization of, **24:**333
Racemization, **7:**74–75, **53:**163; 164; 173–178; 196
of α-amino acids, **14:**193; 195–196
of amino acids during hydrolysis, **14:**263; 265; 269–270
of α-hydroxy acids, **14:**193; 196–197
carbobenzoxylation and, **29:**394
during formation of the peptide bond, **29:**410–411
tests for, **29:**410–411
triethylamine hydrochloride and, **29:**395
kinetic studies of, **20:**8–16
thermodynamics of, **20:**4; 5
of phenylalanine, mechanism for, **33:**369
Racemization of keto acids, **16:**199
Racemization of phenylalanine, in gramicidin S biosynthesis, **35:**5
in tyrocidine biosynthesis, **35:**22
Rachitic rats and phosphatide turnover, **7:**137
Radiant energy (electromagnetic) transport in living cells, **14:**3; 19–42
Radiation
effect of phytochrome (*See* Phytochrome)
effects on bacterial luminescence, **7:**227
effects on cell nucleus, **7:**177
on luciferin, **7:**219; 228
on luminous bacteria, **7:**255
from protoplasm, **7:**49
on protoplasm, **7:**48–50
enzyme involvement, **17:**452–455
ionizing, bleeding from exposure to, **16:**67
Radiation, Ionizing, the Effects of, on Enzymes (Augenstine), **24:**359
Radiation and enzymic adaptation, **13:**80
Radiation damage, **52:**184
Radiations. *See also* Irradiation, Photochemical processes, and Ultraviolet radiations
dosage for inactivation of enzymes by, **24:**367–371
ionizing, effect on enzymes, **24:**359–413
electromagnetic, and photosynthesis, **25:** 39–117
high-energy, mutagenic effect, **27:**328–330
ionizing, effect on thiols, **11:**255–257
short-wave, sulfur radical formation, **27:**206
ultraviolet, mutagenic effect, **27:**314; 326–327; 336; 339; 341; 342
Radiation sensitivity, effect of adsorption, **28:**5; 33–35
Radical mechanisms, in NADH oxidation, **26:**143–156
Radical species. *See also* Free radical formation and Free radical reactions
inactivation of enzymes by, **24:**394–397
Radioactive carbon, in study of methionine synthesis, **16:**272
Radioactive compounds, in lipase specificity test, **23:**156
Radioactive D-coniferin, incorporation into lignin, **16:**149
Radioactive cytochrome *c*, intracellular distribution, **17:**351
Radioactive DFP, in elastase study, **23:**233
Radioactive DNA, in duplication study, **18:** 182–183
Radioactive elements, effects on protoplasm, **7:**48
Radioactive Indicators, Applications in Turnover Studies (Hevsey), **7:**111
Radioactive indicators in turnover studies, **7:**111–214
Radioactive iodine and thyroid function, **14:**96
Radioactive isotopes, *See* individual labeled compounds
Radioactively labeled tyrosinase antibodies, inhibitor of melanin formation, **14:**93; 96
Radioactive phosphate, **6:**246; 247
in study of glutamine formation, **16:**257

Radioactive phosphorus, in transamination study, **16:**216; 218
Radioactive pyruvate, in transaminations, **16:**220
Radioactive radiations, ^{60}Co, and origin of life, **27:**360
^{40}K, and origin of life, **27:**359
Radioactive sulfur, in amino acid synthesis study, **16:**272
Radioactive tracers. *See also* Isotopic distribution and individual elements; Isotopic tracers
in lipase study, **23:**152
protein biosynthesis, rate, **23:**134
Radioactivity, **57:**252
Radioautograph, in carbon dioxide (C^{14}) fixation, **25:**61
Radioautography of protein synthesis, **24:**14; 15
Radiobiological reactions, temperature dependence, **24:**363; 364; 368; 371–373; 392
Radiochromatograms, protein in adsorption studies, **14:**345; 361; 364
Radioimmunoassay, **58:**67
Radiolabeling studies, *Caulobacter* cell cycle, flagellar structure and assembly, **71:**10
Radiolysis, **58:**66
Radionuclides
positron-emitting, **57:**254–257
production of, **57:**252
Radiophosphate and coenzymes, **7:**190
Radiophosphorus turnover in bone tissue, **7:**113
Radiosensitive protein, effect of β-glucan, **28:**357
Radiosensitivity, effect of acetylation of enzymes, **24:**391
surface effect, **24:**399–401
Radishes, antibiotic treatment of disease, **20:**392; 393
Raffinase, cacao, **25:**466
Raffinose, **56:**159
enzymic hydrolysis, **15:**283; 284
fermentation by thiaminase-producing bacteria, **15:**413
fermentation of, **5:**105; 106
in formation of fructose polymer, **10:**141
metabolism in *Pseudomonas saccharophila,* **10:**124–125
in isolation of mitochondria, **21:**5
in polysaccharide synthesis, **11:**298–299; 313; 316
in transglycosylations, **17:**194; 196–198; 200; 205; 220; 223
Raffinose utilization, in *E. coli,* mutants, **22:** 22–24
Raman spectroscopy, respiratory burst oxidase, **65:**55
Ra mutants, definition of, **31:**103
Rana, origin of pteridines in, **27:**422
Rana pipiens, melanophore-stimulating activity in, **29:**445
Random coil confirmation, **34:**471–472
Random coil regions, **47:**60; 64; 83
Randomization, of ^{14}C
cori cycle, **52:**257
enzymes involved, **52:**256
in glucose from galactose, **52:**255
glucose 6-p phosphatase specificity, **52:**256
measurement of fructose 6P cycle, **52:**255; 283
in metabolism of fructose, **52:**255
reactions involved, **52:**255
Random sequential mechanism, **46:**401
Random sequential reaction, **59:**129
Raney nickel, effect on tryptophan metabolism, **10:**68
Ranol sulfate, synthesis, **22:**213
Raphanin, occurrence, properties, **10:**267
Raphanus sativus, pectin biosynthesis by, **20:**369
pectin in, **20:**344
Rapidly labeled RNA, **27:**274
Rapid quench experiments, *D*-ribulose-1,5-bisphosphate, **67:**5; 7
Rapid spectra, **51:**125; 127; 128; 130; 136
Rare earth metals, activating effect, **15:**40
Ras p21 system, **59:**262–263; 265–267
Rat
autoperfused jejunal segments, **53:**209; 215; 225
autoperfused jejunal segments *in situ,* **53:**206
citrulline relased by isolated, **53:**229
citrulline uptake in kidneys of, **53:**229
gastrointestinal tract, glutaminase localization, **53:**219
gut preparation, isolated, perfused, **53:**208
intestinal preparation, isolated, perfused, **53:**205
intestine, isolated mitochondria from, **53:**219
Rat brain

developing, **42:**273
fetal, **42:**256
hypoxanthine-guanine phosphoribosyltransferase from, **41:**14
Rat-brain citrate synthase, **43:**79
Rate constant, **34:**517
for superoxide dismutase, **41:**45
Rate constants. *See also* Kinetics
of enzyme-substrate reactions, **25:**13–15
of glutamic-aspartic transaminase, **25:**33
magnitude of, **25:**9–12
of metal-enzyme interaction, **25:**12; 15–20
Rate-controlling reactions of bacterial luminescence, **7:**233–234
Rate enhancement, transferases, **46:**403; 404
Rate equation, **57:**379–380
Rate equations. *See also* Elementary steps in enzyme reactions and Product inhibition; Kinetics
derivation of, **45:**275–277
in linear form, **29:**7–11
nonlinear, fitting directly, **29:**11–14
steady state for enzymic reactions, **29:**21
product inhibition
integrated, corrected, **25:**177–178
ordered sequence, **25:**189–194; 196–199
quasi-equilibrium, **25:**188–189; 195–196; 202–220
sequence, **25:**189–194; 196–199
steady state, **25:**174; 175; 179–182; 185–188; 194–195; 199–202; 205
Rate-limiting step, **39:**355
Rate-limiting steps, bisphosphoglycerate synthase, **51:**223–226
Rate of anomerization, **44:**141
Rate studies, of enzyme reactions, **25:**1–38
Rate studies of product inhibition, near equilibrium, **25:**179–182
Rat kidney, cysteine *S*-congjugate β-lyase reactions
glutamine transaminase K, **72:**219–222
high M_r enzyme, **72:**223–224
kynurenine pyruvate aminotransferase, **72:**222–223
prodrug targeting, **72:**226–230
Rat liver, PP-ribose-P amidotransferase from, **41:**10
Rat liver, regenerating, in hypophysectomized animals, **29:**377
Rat liver F_1, **49:**231
Rat liver microsomal glucose-6-phosphotase, **34:**474
plasma membrane, **34:**474
Rat liver polynucleotide kinase, **48:**264–271
inhibitors, **48:**270
physicochemical properties, **48:**268; 269
purification, **48:**267; 268
reaction conditions, **48:**269
specificity, **48:**270
Rat mammary gland trisaccharide, **20:**143
Rats, DR synthesis, tracer studies, **28:**431–433
ribose synthesis, **28:**414–416
Rats, wealing, **42:**229
Rat-skin enzyme, amino acid composition of prolyl hydroxylase from, **41:**248
Ravenelin, **5:**193
γ-Rays. *See also* Cobalt (Co^{60}) γ-rays
inactivation of enzymes by, **24:**365; 368; 372; 374–376; 378; 379–380; 388–389; 400–401
α-Rays, irradiation of enzymes by, **24:**394
R-2-Benzylsuccinic acid, **57:**28; 50
Rb mutants, definition of, **31:**103
RCAM, **42:**85; 128; 132
RCAM derivative, **42:**128; 137; 146; 147
RCAM-HGH, **42:**133
Rc mutants, definition of, **31:**103
RCOM, **42:**85; 132; 133
RCOM derivative, **42:**128; 146; 147
RCOM-HGH, **42:**129
R core, structure of, **31:**88
RDE, *See* Receptor destroying enzyme
Rd mutants, definition of, **31:**103
Reactant activation, eukaryotic dihydrofolate reductase, chemical transformation step, **70:**74–81
Reactant ground state
reciprocal stabilization, enzyme action, **73:**47–49
stabilization, enzyme action, **73:**43–46
Reaction, acceptor, in polymer synthesis, **30:**188
hydrolysis, in polymer synthesis, **30:**188
inversion, equation for, **30:**206
propagation, in polymer synthesis, **30:**189
Reaction chains, concentrations, **7:**6–8
Reaction mechanism, of tryptophan synthase, **49:**154–157
Reaction order, entropy of activation, **43:**309
Reaction Pathways and Mechanisms of Pyridoxal Catalysis (Martell), **53:**163

Reaction rate, **34:**487; 500; 503–507; 510; 515–524
time-dependent values of, **23:**109–112
velocity, **34:**463
Reaction rates, *See* Kinetics
Reaction rates in suspensions of cells, **7:**5–9
Reactions
anticoupled solute translocation, **29:**38–39; 49–53 (*See also* Translocation)
antiport, **29:**38–39; 49–53
contransport, **29:**38
coupling of artificial, **29:**51–52
countertransport, **29:**38–39
enzyme-linked solute translocation, **29:**38
facilitated diffusion, **29:**38
group translocation, **29:**38
metabolic control by energy-utilizing, **29:**346–357
noncoupled solute translocation, **29:**38; 44–49
primary translocation, **29:**38; 69–78
proton-coupled solute translocation, **29:** 65–68
secondary translocation, **29:**38; 44–68
sym-coupled solute translocation, **29:**38; 53–65
symport, **29:**38; 53–65
uniport, **29:**38; 44–49
enzymic, elementary steps in, **25:**1–38
simultaneous, **9:**53–56
successive, **9:**1–23; 49–53
Reactions and the Structures of Molybdenum Centers in Enzymes, The (Bray), **51:**107
Reactions at interfaces, **11:**35–89
Reactivation, **34:**475, **39:**345
of alkaline phosphatase, **10:**110–111; 116–117
of enzymes, **10:**102–104; 110–111; 116–117
of cozymase, **7:**430–432; 434
of peroxidase protein, **7:**287–288
reconstitution as, **28:**189
Reactivation inactivation, **43:**68
Reactivation of –SH enzymes by glutathione, **11:**250–251
Recapture synthesis, disposing of oxidation products of fatty acids, **2:**274; 275; 279–281
Reception cone and fertilization, **9:**244–245; 268
Receptor, **42:**149; 158; 159; 160
Receptor binding, **47:**76–77; 125; 136
Receptor dephosphorylation, adenosine receptors, **69:**114
Receptor destroying enzyme. *See also* Neuraminidase
effect on erythrocytes, **20:**136
Receptor molecules, in polysaccharide synthesis, **15:**302–306; 308; 310; 315 (C^{14})
Receptor phosphorylation, adenosine receptors, **69:**104
Receptors, **48:**59; 60; 64, **53:**5–8
α-adrenergic, **53:**31
β-adrenergic, **53:**7
on cell membranes, **27:**226–228; 432; 445–450
gluczgon receptor, **53:**23
for hormones, **48:**78
insulin, **48:**79
ligand binding studies, **53:**26–32
Na^+ effects, **53:**31
prostaglandin, **53:**31
tautomerism in, **50:**32
Receptor sequestration, adenosine receptors, **69:**112–115
Receptor site, amino acid sequence, **27:**80–82
Recombinant DNA procedures, **67:**448
Recombinant DNA technology, dihydrofolate reductase studies, **70:**25
Reconstituted activity, **34:**480
Reconstitution, **34:**473–474
of cysteine oxidation system, **28:**167–168
or respiratory chain, **28:**155–236
definition, **28:**167
from its parts, **28:**167–209
from primary lipoprotein complexes, **28:**209–220; 222–223; 225
of succinate-cytochrome *c* reductase, **28:**190–203; 222–223
of succinate oxidase, **28:**158; 167–190; 208–209; 220–223; 225; 229
reversible, **28:**186–190; 225
Reconstitution of adenylate cyclase, **53:**16–22
cell fusion, **53:**16–17
genetic approach, **53:**20–22
solubilized components, **53:**17–20
Reconstitution work, respiratory chain, comparison, **28:**222–225
Red alga *Porphyra yezoensis,* **46:**49
Red Blood Cell, **34:**534

Red blood cells, **34**:479, **46**:39, *See* Erythrocytes
Red cells. *See also* Erythrocytes
capacity for carbon dioxide of partially carbamylated, **40**:15
cyanate treated, effect on metabolites and enzymes, **40**:15
effect on metabolism and function, **40**:17
oxygen affinity, **40**:17
in vitro survival time in patients, **40**:22
oxygen affinity of sickle-cell patients, **40**:17
partially carbamylated, and Bohr effect, **40**:17
effect on pyruvate kinase, **40**:17
and synthesis of 2,3-diphosphoglycerate, **40**:17
survival in sickle-cell-anemia patients, **40**:24; 25
permeability of, **11**:247
role of glutathione in production of, **11**:253
sickling, **11**:253
relation to complement and hemolysin, **8**: 79–82
sensitization for hemolysis, **8**:81–82
Red cell surface, relation of hemolysin to, **8**:81
Red corpuscles, labeling by radio phosphorus, **7**:113
Red Enzyme, *See* Ferredoxin
Redistribution reactions among carbohydrates, **11**:321; 324
Redogenic substrates, **27**:121; 125
Redoxases, definition, **10**:306
sulfur specific, **10**:306–310
Redox catalysis, **47**:6
Redox catalysts chain, SH-groups, **27**:216
Redox chemistry, quinines, **65**:104–106; 128
Redox dyes, **34**:484
Redox indicators, surface reactions, **11**:55
Redox linkage, extended model of, cytochrome *c* oxidase, **71**:167–180
Redox-linked proton pump, redox loop compared, **71**:126–127
Redox phenomena. *See also* Oxidation-reduction systems
in photosynthesis, **11**:117–119; 148–150; 157–160; 164; 176–177; 180; 182–189; 191–195
Redox potential. *See also* Electrode potential
changes during photosynthesis, **11**:148–149; 157; 185; 193
iron-sulfur proteins, **67**:310–312
Redox potentials, **59**:179, *See* Oxidation-reduction; Oxidation-reduction potentials
of cytochromes *a*, **23**:312–314
for disulfide exchanges, **27**:208
of DPN-ADH complexes, **22**:353
of enzyme-PN complex, **26**:157–164
of hydrogen transfer models, **26**:140
of hemoproteins, **7**:269; 272–273; 283–284
of iron porphin compounds, **7**:311
of luciferin, **7**:258
and sulfur metabolism in plants, **10**:228–229; 289–295
of the cysteine-cystine system, **10**:281–282
Redox systems
kinetics, **9**:15–16
with PN-dependent enzymes, **26**:121–125
Redox transitions
quinones, **65**:101–104; 110–119; 128–129
semi-and hydroquinones, **65**:101–104
Red particle fraction, of mitochondria, evaluation of purity, **21**:76
properties, **21**:103; 104; 114
separation of two forms, **21**:76; 77; 104
Red pigments of purple bacteria, **1**:292; 293; 308; 311; 312
Red radial, **34**:64
Red shift in enzyme coenzyme complexes, **39**:371
Reducase in plants, **9**:393; 398
Reduced dishosphopyridine nucleotide, *See* Diphosphopyridine nucleotide, reduced
Reduced pyridine nucleotide r nitrate transhydrogenases, **14**:254
Reducing agents
effect on pectic substances, **20**:352
and oxidase activity, **19**:128; 199
and papain activity, **19**:277; 286–290
Reductase, **5**:174; 175; 181, **39**:362, **53**:349
action on disulfides, **27**:228
ferredoxin-NADP
NADPH-cytochrome *f* reductase activity exhibited by, **30**:41
properties of, **30**:16
reduction of $NADP^+$ mediated by, **30**:22
photosynthetic pyridine nucleotide (PPNR) (*See* Ferredoxin)
Reductase activity, and lipid, **23**:461
Reductase inhibitors, in folic acid system, **27**:419; 436–453

Reductase particles, reconstituted, **28:**218
Reductases, **7:**91
  of *L. leichmannii,* **28:**446–447
  in plants, **18:**309–310
Reductases, NADPH-dependent, **24:**304; 338; 343
Reductase system, in pyrimidine metabolism, **28:**135
Reductic acid, as degradation product of pectins, **20:**349; 350
  effect on pectic substances, **20:**352
Reduction, **42:**128. *See also* Hydrogen transfers
  asymmetric phytochemical, **7:**91–92
  one-electron transfer, **26:**133–135
  of pyridinium salts, model systems, **26:** 133–143
  reactions of PN-dependent enzymes, **26:** 121–133
  two-electron transfer, **26:**135–140
Reduction, R2 iron center, **65:**167–170
Reduction potentials, in photosynthesis, **25:** 55–56; 58; 84–86; 89–94
  relation to absorption spectra, **25:**79
Reductive amination, *See* Amination, reductive; Transamination, Transreamination
Reductive carboxylic acid cycle, **33:**140–143
  activities of constituent enzymes in (table), **33:**143
  relation to reductive pentose phosphate cycle, **33:**143
  short cycle, **33:**140–143
Reductive cleavage, **39:**363
Reductive mechanisms, for DR-5-P synthesis, enzymes for, **28:**393
Reductive pentose phosphate cycle, relation to reductive carboxylic acid cycle, **33:**143
Reductones, effect on pectic substances, **20:**352; 368
Refinement, **52:**229; 230
Reflectance spectra, **27:**117
Reflectance spectroscopy, copper in cytochromes, **27:**153
  study of dihydroorotic dehydrogenase, **27:**173–175
  of submitochondrial particles, **27:**188–190
Regeneration, of enzyme activity, **9:**628–629
Regional distribution, of tyrosine hydroxylase, **70:**106–108
Regneration of cell from fragments, **16:**469–485
Regulation. *See also* Allosteric regulation; Control mechanisms; Feedback regulation; Inhibition; Repression
  allosteric of phosphorylase by AMP, **34:**370
    of enzymes of glycogen metabolism, **34:**409
    of glycogen metabolism in *Neurospora crassa,* **34:**389
    of glycogen structure by catabolic enzymes, **34:**352
    of glycogen synthesis in bacteria, **34:**314
    of glycogen synthetase activity, **34:**325
    of the interconversion of glycogen synthetase by metabolites, **34:**337
    to phosphorylase regulation, **34:**410
    of pyrophosphorylases by metabolites, **34:**315
  of blood sugar level, **28:**95–97
  enzyme, end-product, **28:**43; 70–71; 100; 128; 134; 135
  of enzyme activity, **28:**41–154
  enzymes susceptible to, **28:**44–48
  of α-galactosides, **30:**126–127
  genetics of, **30:**90–93
  of β-glucosidase system, **30:**129
  of hexokinase, in erythrocytes, **28:**76–77; 85
  of maltose, **30:**124
  positive
    in arabinose system, **30:**114–116
    in maltose system, **30:**119–120
  of PP-ribose-P amidotransferase, **41:**9
  of rhamnose system, **30:**117
Regulation of activity, threonine deaminases, **37:**363–378
Regulation of Enzymes by Enzyme-Catalyzed Chemical Modification (Holzer), **32:**297
Regulation of synthesis and degradation, of PP-ribose-P synthetase, **41:**6
Regulations of Adenosine Diphosphate Glucose Pyrophosphorylase (Preiss), **46:**315
Regulations of the 2′,5′-Oligoadenylate System by Cyclic Adenosine Monophosphate-Dependent Phosphorylation (Itkes and Severin), **59:**213
Regulators of enzyme reactions, products as, **25:**167–274
Regulatory enzymes, unique properties, **28:** 48–56

Regulatory influences, in synthesis of DR-5-P, **28**:393
Regulatory mechanisms. *See also* Control mechanisms
and acetyl-CoA carboxylase, **26**:302–303
in ribonucleotide reduction, **28**:447–449
Regulatory protein
definition of, **37**:363
models for
allosteric, **37**:371
Koshland, Nemethy and Filmer, **37**:371
Monod, Wyman and Changeux, **37**:371
steady-state data-vs-presteady-state kinetics, **37**:371
guanine nucleotide, **53**:11–15
Regulatory sites. *See also* Active sites; Allosteric sites; Binding sites
of enzymes, **28**:48–51; 53–58; 60
Regulon, **50**:329
Rejuvenation of aged amylose, **12**:352–353; 387–388; 393
Relaxation
of muscle, **16**:347–351
periodic, in the study of photosynthesis, **30**:24
Relaxation rates, nuclear, applications of to macromolecular systems, **33**:26–38. *See also* Longitudinal relaxation time; Transverse relaxation time
binding constants and number of binding sites, for binary complexes, **33**:26
for higher complexes, **33**:27
for ternary complexes, **33**:26–27
Carr-Purcell pulsed method of measurement of, **33**:18–20
chemical exchange effect on, **33**:7
combination methods (pulsed and continuous waves) of measurement of, **33**:24–25
continuous wave methods of measurement of, **33**:20–24
coordination schemes in ternary complexes, **33**:30–32
determination of molecular parameters from, **33**:10–16
distances in binary and ternary complexes, **33**:33–38
effect of paramagnetic probes on manganous ion, **33**:16
enhancement of in paramagnetic macromolecular systems, **33**:8–16
frequency dependence of, **33**:11–13
induced by paramagnetic probes in enzyme mechanisms, **33**:1–70
kinetic parameters of protein-ligand interactions (table), **33**:28–29
ligand exchange kinetics, **33**:27–28
ligand-exchanges rates, **33**:3
manganese-ligand distances (table), **33**:34
mechanisms deduced from, **33**:38–64
pulsed methods of measurement, **33**:17–20
ratio of T_{1p}/T_{2p}, **33**:14–16
structural and conformational properties, **33**:28–38
temperature dependence of, **33**:13–14
Relaxation spectrometry. *See also* Absorption spectra
mechanisms of enzyme reactions, **25**:1–38
Relaxation time, **43**:297
steady-state and product inhibition, **25**: 179–182
Relaxing agents, effect on ATPase, **16**:339
Release enzyme, **24**:22
Release factors, **47**:410
Renal antipressor agent, **7**:541
Renal carcinomas
cysteine *S*-conjugate, β-lyase reactions, **72**:211–212
prodrug targeting, **72**:226–230
Renal clearance, **42**:277; 278
Renal extracts, therapeutic use, **7**:538–540
Renal gluconeogenesis, phosphoenolpyruvate carboxykinase (GTP), **69**:216–218; 226; 245–249; 257
Renal hypertension, *See* Hypertension
Renal phosphatase, **7**:522
Renal transport, **18**:134
Renal tubular resorption, of norbornane amino acid, **49**:83
Renaturation of RNase, **24**:198
Renilla reniformis (sea pansy), light emission in, **25**:158–159
Renin, **7**:514; 516–524; 539–542; 546–550, **33**:405, **53**:287
action of tyrosinase on, **14**:147–149
animal specificity of, **7**:517; 520; 526–528
assay, **7**:516–517
enzymic nature, **7**:520–523
hydrolytic action on insulin of, **33**:423
from ischemic kidneys, **7**:522
and kidney cathepsins, **7**:522; 529–530

Renin (*Continued*)
preparation, **7:**518–519
source and properties, **7:**516
tolerance to, **7:**547
Renin-activator, **7:**523
Renin-angiotensin-aldosterone system, **57:**3–4; 6
Renin-hypertension-hypertensinase system, **7:**516–524
Renin substrate, **7:**520–521; 523–524; 526–528; 547
action of pepsin, **7:**529
Rennet, commercial, **15:**425; 431
curd, variations in, **15:**443–444
Rennin, **15:**274, **34:**450
and the clotting of milk, **15:**423–448
separation by chromatography, **14:**349; 361
separation from pepsin, **14:**398
structure studies, **14:**303
Rennin, Clotting of Milk and (Berridge), **15:**423
Rennin like enzymes, **41:**238
R-Enzyme, **18:**289
R-enzyme, **15:**309
debranching action, **11:**368
R - Enzyme, **34:**406
REOX, **42:**85
Reoxidation, **42:**134; 136
protein, **42:**137
Reporter group, **39:**346
Repression. *See also* Allosteric regulation; Control mechanisms; Feedback regulation; Inhibition; Regulation
of arginine enzymes, **29:**133
of aspartokinase, **29:**144
coordinate, of enzymes of tryptophan synthesis, **29:**121
of cystathionine enzymes, **29:**140
enzyme, **29:**92; 370–371
of enzymes of methionine synthesis, **29:**139
of enzyme synthesis by amino acid analogs, **29:**93
in histidine biosynthesis, **29:**126
nucleic acids and, **29:**152
of tryptophan biosynthetic enzymes, **29:**121
of tyrosine sensitive DAHP synthetase, **29:**119
of enzyme synthesis, allosteric regulation, **28:**61; 70–73
and nucleoside triphosphates, **27:**282–283
repression-derepression range, **40:**83
transcriptional, **40:**79; 80; 81
translational, **40:**81; 82; 83; 84; 85; 86; 87
in synthesis of DR-5-P, **28:**393
Repression and endporduct inhibition, relationship between, **37:**386
Repression-induction, in enzyme synthesis, **50:**327
Repressive complex, **40:**78
Repressive Control of Arginine Biosynthesis Enzymes (Vogel and Vogel), **40:**65
Repressor
definition of, **30:**83
descriptions of, **30:**82
estimated molecular weight of, **30:**97
interaction with inducer, **30:**94–98
interaction with operator, **30:**98–101
isolation of, **30:**95
of maltose system, **30:**123
properties of, **30:**92–93
reaction with DNA, **30:**101
stereospecificity of, **30:**93–94
structure of, **30:**97
of α-polymerase, **31:**112
of synthesis of transacetylase, **31:**112
Repressor hypothesis, of enzyme synthesis, **22:**143
Repressors. *See also* Arginine repressor; Enzyme repressor
of enzyme production, **22:**105
in β-galactoside formation in *E. coli,* **22:**20
recognition sites, **40:**76; 77; 78; 79
Reproduction, carotenoid function in, **21:**359; 360
Reproduction, DNA duplication, **18:**181–183
Reptiles, digestion, **1:**410–412
food, **1:**410
intestinal canal, **1:**378
Resazurin in dehydrogenation studies, **8:** 274–278
Reserpine
in catonia, **29:**530
in depression, **29:**489–490
inhibitors of, **29:**483–485
and LSD 25, **29:**518
mechanism of action of, **29:**483–485
relation to catecholamines in the brain, **29:**483
tyrosine hydroxylase induction, **70:**166
tyrosine hydroxylase regulation, **70:**128
Reservoir function, of amino acids, in liver, **49:**44

Respiration in spermatozoa, **24:**43; 46; 52–86
Respiration of Aspergillus (Tamiya), **2:**187
Respiration of Higher Plants, Reaction Paths in (James), **18:**283
Respiratory activity, of intracellular particles, **18:**13–15; 18–20
Respiratory adaptation, metalloflavoproteins in, **17:**283–284
Respiratory burst oxidase, **65:**49–84
activation, deactivation, and inactivation, **65:**57–59
butyl isonitrile, **65:**53; 55
cell-free, **65:**62–67
CGD classification, **65:**66–67
kinetics, **65:**63–66
CO binding, **65:**53; 55
cofactors, **65:**54–55
complexes, **65:**82–84
cytosolic components, **65:**67–75
$p32^{phox}$, **65:**74–75
$p47^{phox}$, **65:**67–72
$p67^{phox}$, **65:**72–74
enzymologic properties, **65:**52–57
H_2O_2 production, **65:**53–54
inactivation by diphenyleneiodonium, **65:**80
inhibition, **65:**53
membrane-associated components, **65:**75–81
$gp91^{phox}$, **65:**75–78
$p22^{phox}$, **65:**78–79
$p45^{phox}$, **65:**79–81
NADPH-binding component, **65:**74–75
plasma membrane, **65:**52–53
priming, **65:**58–59
purification, **65:**60–62
Raman spectroscopy, **65:**55
respiratory burst oxidase and, **65:**53
reversibility of activation, **65:**59
Respiratory chain
and ATP synthesis, **23:**341–354
crossover points in, **23:**350–352
inhibitor action sites, **23:**347
phosphorylating sites in, **23:**347–354
constitution, **28:**159–161
effect of cyanide, **28:**174; 205–206; 208; 221; 225–228
fragmentation, by bile salts, **28:**190–192; 196; 202; 204–210; 214–215
sequential, **28:**190–197
reconstitution of, **28:**155–236
structural protein of, **28:**160–161
constitution in animal tissues, **20:**147–199
in higher plants, **18:**11–13
and oxidative phosphorylation, **17:**65–134
molar ratios of components, **17:**77
Respiratory Chain
in Animal Tissues, Constitution (Slater), **20:**141
and Oxidative Phosphorylation *(Chance and Williams),* **17:**65
Respiratory Chain, Reconstitution of the (King), **28:**155
Respiratory chain components, criteria for, **20:**151; 152
Respiratory chain preparations, beef heart, composition of, **28:**161–165
Respiratory components, linkages between, **28:**225–228
Respiratory control, loss of, **17:**117–118
Respiratory control ratio (RCR), cytochrome *c* oxidase
enzyme turnover control and H + /e - ratio, **71:**162
redox linkage, **71:**178
Respiratory enzyme of Warburg, **7:**295
Respiratory enzymes
and cancer, **4:**202; 208; 222; 223; 233
in energy transport studies, **14:**32–33; 37–42
study of catalysis by, **14:**173–176; 182; 188
enzymic adaptation, **13:**75–77; 84–85; 93; 106
inhibition by carcinogenic chemicals, **4:** 248–250
and mutation, **17:**452–455
Respiratory enzymes and cell structure, **8:**7; 20; 31–37
and chromidia, **8:**7; 31–37
and mitochondria, **8:**31–37
Respiratory ferment, **23:**266
Respiratory fuel, in plants, **18:**282–284
Respiratory inhibitors, for plant tissues, **18:** 30–35; 41–42
Respiratory inhibitors and phosphatide formation, **7:**151
Respiratory particles, extraction of coenzyme Q, **25:**306–308
Respiratory pigments, in nuclei, **27:**254
Respiratory quotient
and fertilization of egg, **9:**297
of sperm, **9:**372–373

of neoplasm, **29:**358
Respiratory quotient and fat synthesis, **8:**300; 302–305; 307; 310
Respiratory quotient of *aspergilli,* **2:**188–195
Respiratory quotients, in microbial action on hydrocarbons, **10:**446–447; 465–466
of tobacco during fermentation, **10:**342–349
Respiratory quotients in sperm, **24:**66–68
Respiratory systems, association with melanin granules, **16:**169
Respiratory systems of egg, **9:**302–312
Respirometry in histochemical studies, **9:** 602–603
Restoration of activity, **34:**473
Restriant of exodus, **49:**89
in regulation of nutrient and metabolite distribution, **49:**89
Results, evaluation of kinetic, **29:**22–23
Resveratrol, **55:**92
in viniferin synthesis, **55:**94
Retardation factor, **43:**126; 129
Retarders, of clot formation, **16:**40
Reticulinase, **23:**201
Reticulocyte extracts, release enzyme in, **24:**22
Reticulocyte lipoxygenase, **58:**191–272
action on liver mitochondria, **58:**239
activity *in vivo,* **58:**251
amino acid composition, **58:**199
arachidonic acid conversion, **58:**209
arachidonic acid substrate, **58:**194
carboxypeptidase activity, **58:**200
chromatographic purification, **58:**197
cooxidizing properties, **58:**211
hydrogen peroxide inhibition, **58:**221
hydroxamic acids inhibitors, **58:**218
immunological reactivity, **58:**214
inactivation, **58:**198; 212
inactivation rate, **58:**216
inhibitors, **58:**195
isoelectric point, **58:**199
kinetic model, **58:**223; 224
leukotriene A_4 synthase activity, **58:**208
linoleic acid substrate, **58:**194
liposomes effect on, **58:**245
membranes interaction, **58:**238
methionine of, **58:**212
metochondrial respiration inhibition, **58:**196
molecular weight, **58:**198
multiple dioxygenation products, **58:**205
4-nitrocatechol inhibitors, **58:**218
physiology of, **58:**246
positional specificity, **58:**203
purification, **58:**196
from rabbit reticulocytes, **58:**196
stereospecificity, **58:**203
substrate preparation, **58:**194
synthesis regulation of, **58:**247
turnover number, **58:**197
Reticulocyte Lipoxygenase: Enzymology and Physiology of: Comparison with Other Lipoxygenases (Schewe, Rapoport, and Kühn), **58:**191
Reticulocyte lysates, **58:**116
Reticulocytes, **46:**103; 150
cell biology, **58:**246
hemoglobin formation in, **24:**12; 14; 19–20; 21
inactivity, **58:**246
iron uptake, **58:**254
maturation, **58:**246
mitochondria degradation, **58:**255
mitochondria in, **58:**241
protein synthesis, **58:**246
ribonucleases, in, **58:**250
ubiquitin-ATP-proteinase, **58:**254
uptake of labeled amino acids, **16:**491–492
Reticuloendothelial cells, **41:**112–113
Reticuloendothelial stimulating effects, of β-glucans, **28:**357
Reticuloendothelial system, RES, **39:**228
Retina
effect of change in surface potential, **11:**76
permeability of, **13:**62–63
Retinal cells, neural regeneration from pigmented retinal cells, **16:**485
Retrograded starch and aging of amylose, **12:**351–353
Retrolental fibroplasia, **41:**90
Reversal, of catalase action, **12:**175–176
of peroxidase action, **12:**175–176
Reversal of enzymic reactions, **25:**175–177; 179–182
Reversal of reactions, control, **28:**46–47; 86–88
Reverse cholesterol transport, cholesterol ester hydrolases, **69:**45
Reverse micelles, **51:**30–31
catalysis in, **51:**47–55
defined, **51:**36
Reverse protonation, *See* Protonation

Reverse reactions. *See also* Kinetics, enzyme kinetics of, **17:**10–11; 34–37; 50–53
Reverse transcriptase, **56:**292
Reverse transcription/polymerase chain-reaction technique, adenosine receptors, **69:**85; 105–107. *See also* Polymerase chain-reaction techniques
Reversibility
of action of starch-splitting enzymes, **12:** 410–424
of enzyme-substrate complexes, **12:**165
of amylomaltase action, **11:**321
of amylosucrase action, **11:**319
of *B. macerans* amylase action, **11:**322–324
of enzyme inhibition, **11:**239–240
of levansucrase action, **11:**314–315
of mercaptide formation, **11:**216–218; 227; 237
of Q-enzyme action, **11:**367–369
of thiol oxidation, **11:**211–212; 228–229; 236–237; 249
of cleavage of tobacco mosaic virus, **15:** 468–470
of immunologic precipitations, **15:**67–68
of inhibitor action, **15:**286
of isohemagglutination, **15:**54–56
of proteinase action, **13:**305–310
of transferase action, **13:**293–294; 301
of succinic dehydrogenase action, **18:** 81–84
Reversibility of phosphorylation process, **17:**122–123
Reversibility of product inhibition, **25:**175–177; 183
Reversibility of thrombin-fibrinogen interaction, **16:**40–41
Reversible reaction, between purple intermediate and fully reduced enzyme, **34:**63
Reversible Step Reactions (Michaelis), **9:**1
Reversible transitions, of β-lactamase, **28:** 297–299
rfa mutants, antigenic properties of, **31:**106
definition of, **31:**104
rfc locus, definition of, **31:**109
Rhammnulose-1-phosphate aldolase, **56:**293
Rhamnetin, metabolism in plants, **16:**141
Rhamnose, **17:**196
as carbon source for *Pasteurella* mutant, **22:**33; 34
resistance to, *Salmonella* mutants, **22:**34
in pectin, **20:**344
L-Rhamnose
enzymes and their structural determinants, **30:**116–117
gene map of system, **30:**116
genetics of regulation, **30:**117
location of markers on the bacterial chromosome, **30:**86
system of utilization of, **30:**116–117
in lipopolysaccharide from *Salmonella,* **31:**79; 92
L-Rhamnose-isomerase, reaction catalyzed by, **30:**116
Rhamnosidases, **56:**221
L-Rhamnulokinase, reaction catalyzed by, **30:**116
L-Rhamnulose-1-phosphate-aldolase, reaction catalyzed by, **30:**116
Rhamulose, **18:**287
Rhapontium, **55:**92; 94
Rhein, **34:**121–122; 134
Rheumatic fever, tyrosine metabolism in, **14:**99; 108–110
Rheumatoid disease, hyaluronidase inhibition in, **13:**218; 220–221
Rhizobium bacteroids, **46:**54; 87–88
Rhizobium japonicum, in nitrogen fixation studies, **13:**354–355
Rhizobium japonicum, superoxide dismutase from, **41:**79
Rhizobium nodM, glucosamine 6-phosphate synthase, **66:**267–268
Rhizobium trifolii, growth and nitrogen fixation, **13:**347; 356
Rhizoctonia ferrugina, enzymic transformations of steroids by, **20:**239
Rhizoctonia solani, plant disease caused by, **20:**392; 394
Rhizoctonia species, enzymic transformations of steroids by, **20:**239
Rhizopterin, isolation, structure, synthesis and properties, **21:**374
Rhizopterin (N^{10}-formyl folic acid), **27:**422
Rhizopus arrhizus
enzymes in, **20:**269
enzymic transformations of steroids by, **20:**242; 243; 245; 273; 274
steroid hydroxylation by, **19:**191
Rhizopus cambodjae, enzymic transformations of steroids by, **20:**242
Rhizopus nigricans, **23:**411

steroid hydroxylation by, **19:**191
Rhizopus nigricans, enzymic transformations of steroids by, **20:**243–246; 257; 259; 262; 264; 271; 274
Rhizopus-pepsin, **44:**1; 6; 8
Rhizopus pepsin, **53:**261; 287
Rhizopus species, enzymic transformations of steroids by, **20:**237; 243–245
ergosterol in, **20:**264
Rhizopus stolonifer, **55:**99
*Rhizopus stoloni*fer, **49:**10
Rhizopus tritici, polygalacturonase in, **20:**343; 367
Rhodamine B, chemiluminescence, **25:**123; 127
Rhodanase, **7:**400
in mitochondria, **21:**40
Rhodanese, **39:**328; 332; 335; 346; 350; 362
action, mechanism of, **39:**344
active site of, **39:**345
activity, **39:**341; 348; 349
animal, **39:**336
assay, **39:**334
B. subtilis, **39:**338
bovine liver, **39:**347
catalysis, **39:**357; 358
catalyzed cleavage, **39:**362
catalyzed reaction, **39:**360
catalyzed reduction, **39:**353
classification, **14:**227; 252
crystalline, **39:**333; 345
crystals, **39:**349
discovery of, **39:**335
of *ferrobacillus ferrooxidans(Thiobacillus ferrooxidans),* **39:**338
from human liver, **39:**335
inactivation of, **39:**344
intracellular localization of, **39:**337
liver, **39:**336
localization of, **24:**315
metal ion content of, **39:**348
microbial, **39:**337
in a mitochondrial enzyme, **39:**337
plant, **39:**337
S, **39:**339; 340; 358
substrates, **39:**342
sulfhydryl groups, **39:**345
sulfur-substituted, **39:**341; 343; 345; 353; 357
of *thiobacillus denitrificans,* **39:**338
zinc-free, **39:**348
Rhodanese-like, catalysts, **39:**334
reactions, **39:**358
Rhodanese (thiosulfate-cyanide-sulfur transferase), **27:**212
Rhodanese (Westley), **39:**327
Rhodanide, inhibition of certain bacteria, **10:**231
Rhodobacter capsulatus, **71:**87; 101; 117; 156
Rhodobacter sphaeroides, **71:**87; 101; 117; 156; 188
Rhodobacter sphaeroides, phosphoribulokinase from, **74:**98–99; 100; 117; 120; 121
overexpression of, **74:**115
regulation of, **74:**109–110; 111; 112
structure of, **74:**100–102; 101; 109; 111; 113–114
Rhodococcus bronchialis, **58:**73
Rhodomicrobium vannielli, **46:**332
Rhodophyta, **46:**79
Rhodopseudomonads, oxidation of hydrocarbons, **27:**535
Rhodopseudomonas, ADP-glucose pyrophosphorylases, **46:**332
activation by fructose-P_2, **46:**332
activation by fructose-6-P, **46:**332
activation by pyruvate, **46:**332
Rhodopseudomonas, carotenoid-deficient mutant of, **19:**62
Rhodopseudomonas capsulata, **46:**328; 330; 331
ADP-glucose pyrophosphorylase, inhibitors of, **46:**330
phosphoribulokinase from, **74:**98; 100; 120; 121
regulation of, **74:**110
Rhodopseudomonas capsulatus
allosteric enzyme of, **28:**65; 70; 72; 73
use of acetate, **23:**411; 425; 430
Rhodopseudomonas palustric, use of acetate, **23:**411; 425; 430
Rhodopseudomonas palustris, **46:**76; 332, **54:**169
Rhodopseudomonas spheroides, **46:**39; 42; 44; 69; 70; 76; 84; 332–333; 367
enzymes, **46:**332
photosynthesis in, **18:**56–57
regulation of tetrapyrrole biosynthesis in, **46:**90
use of acetate, **23:**422; 425
Rhodopseudomonas spp., **23:**412
Rhodopseudomonas viridus, **46:**76

Rhodopsin, **58:**128
adenosine receptors, **69:**104
photoreversibility of, **26:**23
and retinal ATPase activity, **6:**185
Rhodopsin cycle, function of DPN in, **15:**128
Rhodoseptoria species, enzymic transformations of steroids by, **20:**247
Rhodospirillum, **34:**474
Rhodospirillum ADP-glucose pyrophosphorylases, **46:**323–328
activators of, **46:**332
activator specificity, **46:**323–325
kinetic properties of, **46:**325–328
molecular weight of enzymes from, **46:**367
Rhodospirillum haemoprotein, **23:**267
Rhodospirillum molischianum, **46:**324
ADP glucose pyrophosphorylases, activation by pyruvate, **46:**324
molecular weight of enzymes from, **46:**367
Rhodospirillum rubrum, **15:**251, **46:**76; 84, **54:**169, **59:**330
activation spectra, **22:**373; 374
ADP-glucose pyrophosphorylases, **46:**323
activation by pyruvate, **46:**323
kinetic studies, **46:**369
reaction mechanism, **46:**369
allosteric enzymes of, **28:**54; 70–73; 111; 112
cytochrome *c* in, **19:**57
cytochrome c_2 in, **19:**49
Hill reaction in, **19:**57
molecular weight of homogeneous enzymes form, **46:**367
photosynthetic phosphorylation by, **17:** 282–283
photosynthetic reactions of, **18:**53–54; 56–57; 135
propionic acid metabolism, **26:**291
use of acetate, **23:**421–430
Rhodospirillum rubrum, ESR measurements in, **25:**83; 84
photosynthesis in, **25:**76–77
pigment system, **25:**65
Rhodospirillum tenue
ADP-glucose pyrophosphorylase activation, **46:**325
inhibition by AMP, **46:**327
molecular weight of enzymes from, **46:**367
Rhodotorula glutinis, fat synthesis by, **9:** 674–681; 690
Rhodotorulic acid, **35:**163
Rhodoviolascin, **1:**311
Rhodoxanthin, structure, **21:**299
Rhus vernicifera, blue proteins, EPR study, **27:**138
Rhynchosciara angelae, chromosome banding in, **16:**460–461
Rhyodophyceae, carotenoids in, **21:**298; 301; 302
Rhythm
biological, and product inhibition, **25:** 172–173
in protoplasm, **7:**50–52
Rhythmic changes in nerve conduction, **8:**46–47
Ribitol, **9:**474–475
cytidine diphosphate, structure of, **30:**224
Ribitol dehydrogenase, product inhibition of, **25:**224; 253–259; 261; 266
Ribitol phosphate, in pneumococci, **30:**233
Ribityl group, in coenzyme binding, **22:**387
Ribityl group of riboflavin, origin of, **53:** 348–349
Riboflavin, **58:**64. *See also* Flavin, Flavoprotein, and Lactoflavin; Flavins, Flavo-proteins, and Leuco-riboflavin; Flavins and Flavoproteins; Metalloflavoproteins
absorption spectrum, **4:**301–302; 305
analogues, binding capacity, **22:**397
binding capacity, **22:**378; 379
fluorescence, changes with binding, **22:**376
quenching of, **22:**377
isoalloxazine ring in, **22:**383
protein binding of, **22:**388; 390
ribityl group of, **22:**387
semiquinoid form, **22:**377
antagonism to mepacrine, **10:**69
in tryptophanase, **10:**60–62
biosynthesis of, **53:**346–353
5-amino-2,4-dioxy-6-ribityl-aminopyrimidine, **53:**349–351
regulation of, **53:**352–353
change upon coupling with protein, **15:**38
chemiluminescence, **25:**123; 125; 149; 150
effect on coenzyme Q synthesis, **25:**304
deaminase, **53:**349
dietary, and choline oxidation, **17:**384
glucosides of, **17:**219–221; 229
oxidation-reduction potential, **17:**271; 277
relation to xanthine oxidase, **17:**303–304; 307; 313; 319; 321

effect of halogen salts on fluorescence, **20:**32
effect on carotenogenesis, **21:**321
function in phototropism, **21:**358; 359
in mitochondria, **21:**8
EPR study, **27:**131; 135–136; 183
Eremothecium ashbyii, **7:**572
and fat formation, **8:**310
in snake venoms, **8:**484
from fermentation by-products, **7:**570–571
microbial synthesis and breakdown, **7:**454
in molds, **9:**572
photolysis, **9:**474–475
in normal and cancer tissues, **4:**246; 247
in oxidation of hydrocarbons, **27:**485
produced by yeast, **7:**572–573
reaction with borate, **12:**500; 501; 514; 515–516
reductase, **53:**349
reduction of, **15:**4–5; 89
sensitivity to iron, **7:**571
Riboflavin coenzymes, **9:**474
Riboflavin deficiency, effect on succinic dehydrogenase, **18:**76
Riboflavin derivatives, and succinic dehydrogenase, **18:**76–79
Riboflavin glucoside transferase, **17:**229
Riboflavin-metal complexes, **17:**273
Riboflavin 5′-monosulfate (FMS), in ternary complex formation, **27:**5–7
Riboflavin nucleotides, states of oxidation, **15:**4–6; 89
Riboflavin operon, **53:**353
Riboflavin phosphate, **7:**375, **18:**202, *See* Flavin mononucleotide
Riboflavin phosphate in amino acid metabolism, **9:**402
Riboflavin synthase, **53:**351
Riboflavin tetrabutyrate, absorption spectrum of, **34:**54
α-D-Ribofuranose 1-phosphate, from ribonucleoside, **18:**224
β-D-Ribofuranose 1-phosphate, inertness of, **18:**224
Ribokinase, enzymic adaptation, **13:**70
Ribonuclease, **5:**22; 23, **9:**514; 518–519, **15:**274; 276; 461, **17:**171–172, **33:**430, **34:**450–451; 501, **37:**419; 422; 457; 478, **43:**341; 360, **47:**47; 49; 55; 60; 84; 124; 133; 136; 138–139, **52:**27, **58:**28; 250
A, **52:**185; 186; 192; 198; 201; 202; 203; 208; 209; 212; 214; 223; 226; 235–237; 238–239
denaturation, **52:**205; 207
absorption spectrum and inactivation, **24:** 369–370
amino acid sequence, **24:**173–176; 200–202; 416; 418; 421–424; 426–427; 430; 432–435; 448
chemical modification of, **24:**113–114; 131–134; 145; 199–227; 244; 248
denaturation of, **24:**184–197
depolymerase action of, **24:**173; 214
derivatives of, **24:**204–213
effect of, ionic strength on activity, **24:**236
ionizing radiations, **24:**365–367; 369–370; 374–375; 382–383; 389; 395–398; 404
irradiation, **24:**203; 204; 217; 218; 224; 227
methylation, **24:**224–225; 246
partial degradation, **24:**125; 227–235
various substances on activity, **24:** 235–243
entropy of activation, **24:**455–458
and enzyme synthesis, **24:**15–17
heterogeneity of, **24:**167–172
isolation and purification, **24:**165–172
kinetic studies, **24:**234–235; 238–239; 452–455
localization of, **24:**314; 315; 325; 343
mechanism of action, **24:**239–243; 251–252; 443–458; 465; 467
methods of assay, **24:**172–173
models of, **24:**251–252; 444–448; 455
physicochemical characterization, **24:** 177–184
regularities in, **24:**436–437
species difference, **24:**141
specificity of action, **24:**165
structure, secondary and tertiary, **24:**23
structure-activity relationships, **24:**116; 243–252; 448–452
structure and function, **24:**161–261
volume irradiated, **24:**365–367
action of subtilisin on, **22:**70; 76
active site, labeling of, **22:**53
location of, **22:**78
column chromatography, **22:**424; 430; 431; 440; 448; 451; 462–467

Ribonuclease (*Continued*)
effect of pH on viscosity, **22:**426–427
noncovalent forces in, **22:**74
pH dependence of methionine reaction, **22:**60
reversible denaturation, **22:**77
structure, **22:**73
covalent, **22:**70; 71
relation to activity, **22:**70–72
variation of amino acid composition, **22:**65
activity, relation to SH-group, **27:**221
from different tissues, **27:**246
disulfides, **27:**219; 230; 231
in nucleolus, **27:**268
in solubilization of enzymes, **27:**266
adsorption at interfaces, **28:**3; 28–30
pancreatic, effect of antisera, **28:**306
adsorption to ion-exchange resins, **33:**272
chemical mechanism of, **45:**359
chromatographic evaluation of purity, **19:**280
determination of amino acid sequence, **19:**297; 298
isotope studies of biosynthesis, **19:**259
in chromidia, **8:**31
in nuclei, **8:**33
and cyclic phosphate formation, **18:**198
and polymer formation, **18:**229
in prepn. of DNA, **18:**164
cytochemical studies, **9:**605
effect of penicillin on activity, **10:**254
effect on β-glucuronidase, **40:**40
effect on peptide synthesis, **33:**345–346
effect on protein synthesis, **17:**478
effect on virus RNA, **17:**493; 494; 496
genetic effects, **17:**464
imidazole in, **20:**69; 76; 77
in plastids, **20:**210; 211; 214
in mitochondria and lysosomes, **21:**39–41; 58
of pancreas, **23:**130; 132
rate of biosynthesis, **23:**134
uncoupling action, **23:**367; 369
preparation of, **33:**277
in protein syntheis study, **16:**491
in RNA study, **16:**436; 438
purity of, **12:**401
reaction mechanism, **58:**30
reversible denaturation of, **26:**91; 93; 100–103
steric structure, **26:**93; 95–96; 107
S, **52:**185; 192; 201; 204
synthesis of gramicidin S, inhibited by, **33:**348
tertiary structure of, **47:**47
tyrocidine synthesis effect on, **33:**350
Ribonuclease, Structure and Function of (Scheraga and Rupley), **24:**161
Ribonuclease A, **37:**467; 477, **53:**247, **55:** 296–298
active aggregates, **40:**59
chemical modification of, **24:**204; 214–217; 221; 222; 225; 227
covalent structure of, **24:**173–177
disulfide bridges of, **24:**197
exchange experiments, **24:**192–193
immunological properties, **24:**184
limited proteolysis of, **24:**227–235
preparation of, **24:**172
reoxidation of, **24:**200
column chromatography, **22:**464; 465
effects of ethionine on, **25:**376
Ribonuclease B
denaturation spectral shifts, **24:**196
effect of radiations on, **24:**227
Ribonuclease C, denaturation blue shifts, **24:**196
Ribonuclease H, **42:**279
Ribonuclease P, Enzyme with Catalytic RNA Subunit (Altman), **62:**1
Ribonuclease S, **37:**413; 478; 593
by action of subtilisin, **24:**230–232
immunological properties, **24:**184
reactivity, **24:**202; 204; 216; 247; 449
column chromatography, **22:**467
Ribonucleases
classification, **14:**245
isolation, **14:**389; 396; 397
separation by chromatography, **14:**322; 331; 333; 337; 339; 347–353; 367
structure studies, **14:**274; 283; 287; 293; 305; 306–307
multiple forms, **25:**337; 349; 362
product inhibition of, **25:**224; 226–229
separation of multiple forms, **25:**332
turnover number, **25:**32
Ribonuclease S′, **24:**231–232
Ribonucleases A and B
difference between, **24:**171
separation and properties, **24:**167–172
Ribonuclease T, **35:**478
Ribonuclease T_1, **37:**435; 467; 477
Ribonucleic acid, *See* RNA
Ribonucleic acid pyrimidines, biosynthesis, **21:**277–281

Ribonucleic Acids and Virus Multiplication (Jeéner), **17:**477
Ribonucleoprotein, in cell nuclei, **18:**179–181; 185
Ribonucleoprotein particles
and protein synthesis, **24:**5–7
as site of peptide biosynthesis in liver cytoplasm, **19:**260
Ribonucleoproteins, **4:**211; 229; 230; 231, **58:**248
Ribonucleoside diphosphate reductase, **48:**21, **65:**147–180
catalytic reaction, **65:**148–150
conserved cysteines, **65:**153–154
disulfide/dithiol interchanges in electron transfer, **65:**149
DR metabolism, **28:**449
model, **65:**151–152
redox-active dithiols, **65:**153–155
structure, **65:**150–153
terminology, **65:**151
Ribonucleoside diphosphates, in DR metabolism, **28:**438–449; 453
Ribonucleoside phosphates, reduction of, **28:**453
Ribonucleosides, labeled
reduction of, **28:**420–421; 434; 453
synthesis of, **28:**419
Ribonucleotide, into DNA, terminal enzyme, **27:**268
metabolism in nuclei, **27:**274–275
Ribonucleotide phosphorylase, **18:**223–224
Ribonucleotide reductase, **43:**26, **59:**146, **66:**34–35
B_{12} requirement, **28:**423; 429–431; 436; 440; 446–447; 453
E. coli, inhibition by cisplatin, **66:**183
of *L. leichmannii,* **28:**444–447
linked to thioredoxin, **28:**436–444
mammalian, thioltransferase and glutaredoxin in, **66:**167–168
regulation of, **28:**447–449
types, **28:**445
Ribonucleotide reductase, T-4, induced, **46:**249
Ribonucleotide Reductase (Stubbe), **63:**349
Ribonucleotides
coenzyme action, **18:**207–208
reduction of, **18:**219
feedback inhibition by, **28:**66
labeling of, **28:**408–409
reduction of, **28:**420–421; 426; 429; 431–436; 443–449
PN-dependent reduction, **26:**128
D-Ribopyranose 1-phosphate, **18:**225
Ribose, **17:**196; 418. *See also* Carbohydrates and Pentose; Nucleic acids
biosynthesis of, **28:**391–460
contribution of various pathways, **28:** 399–405
exchange reactions, **28:**400–402; 407
from labeled precursors, **28:**406–408; 414–416; 424–431
pathways, methods of assessment, **28:** 399–400
bound
DR synthesis from, **28:**420–421
synthesis, **28:**407–409; 414; 418–419; 425; 433
from 1-^{14}C-glucose, **28:**424–426
from ^{14}C-glucose, **28:**406–408; 426–431
degradation of, **28:**403–404
entropy for ionization, **43:**311
fate of C-1, **18:**295
in lignin study, **18:**363–364
from glucose, **16:**249; 474
precursors of, **16:**451
uptake by myosin and actomyosin, **16:**342
from lactate-1-^{14}C, **28:**426
and LADH complex formation, **27:**62–63
metabolism, **9:**471–474, **15:**82; 149; 164; 175–176
in myosin, **13:**166
and phosphoric acid esters, **9:**470–471; 472–474
purine-bound, synthesis, **28:**407–409; 414; 418–419; 425; 433
pyrimidine-bound, synthesis, **28:**407–409; 418–419
reduction scheme, **28:**436; 437
in RNA, labeled, **28:**411; 414–416
L-Ribose, as degradation product of pectins, **20:**351
D-Ribose, in histidine synthesis, **16:**300
of RNA, **16:**437
Ribose and Deoxyribose, Biosynthesis of (Sable), **28:**391
Ribose-1–^{1} ^{4}C, incorporation into pteridines, **35:**49
Ribose 1,5-diphosphate, **18:**199; 217

^{14}C-Ribose, DR synthesis, by tumorbearing rats, **28:**431–433
labeling of pentoses, **28:**434–435
Ribose-modified ATP analogs, **43:**24
Ribose nucleoside diphosphate, **23:**324
Ribose nucleoside triphosphate, **23:**324
Ribose nucleotides, **9:**503–507. *See also* Nucleotides
elementary composition, **9:**521
D-Ribose-5-P, **43:**493
Ribose phosphate, **15:**172
Ribose-1-phosphate
in imidazole biosynthesis, **21:**403
in pyrimidine biosynthesis, **21:**282
in synthesis of nicotinamide mononucleotide, **15:**103–104; 106
phosphoglucomutatse action on, **15:** 168–169
Ribose-5-phosphate, **17:**184; 406, **46:**53. *See also* Ribose
allosteric regulation, **28:**111–112; 118
anaerobic catabolism, **28:**395–397
in glucose metabolism, **15:**153–154; 157–160; 163–166; 167
in histidine biosynthesis, **21:**404; 405
in purine biosynthesis, **21:**202–205; 207; 209; 211; 229; 232; 240; 241; 265
in pyrimidine biosynthesis, **21:**265; 278; 285
and LADH complex formation, **27:**62–63
from 6-PGA, **28:**393–395
in photosynthesis, **19:**15; 21; 26; 27; 29–31
in plant respiration, **18:**286; 294; 312
pyrophosphorylation of, **18:**199–200
Ribose phosphate isomerase, **14:**205
localization of, **24:**333
in ribose synthesis, **28:**396–397
Ribose phosphate polysaccharide, **17:**172
Ribose-1-phosphate r purine transribosidase, **14:**251
Ribose-5-phosphate r ribulose-5-phosphate, **14:**195; 205; 207–208
Ribose-5-phosphate r TPN transhydrogenases, **14:**254
Ribose phosphates, isomerization, **14:**209
Ribose ring, flexibility of, **50:**33
Riboses, **46:**293
Ribose-S-phosphate, substrate for PP-ribose-P synthetase, **41:**3–4
Ribose-S-phosphate aminotransferase, **41:**14
Ribosomal proteins, **42:**230, **47:**46; 84; 139
Ribosomal RNA, *See* rRNA
Ribosomes, **35:**472, **38:**222, **47:**380. *See also* Particles, subcellular
base pairing, **50:**36
characteristics, **22:**133
role in protein synthesis, **22:**133; 144
as templates for protein synthesis, **22:**135; 136; 145; 146
and enzyme formation, **24:**4–7; 14–20
glycerol teichoic acid in preparations of, **30:**245
70S, in chloroplasts, **30:**26
interconversion of cobamides, **28:**447
γ-penicillinase in, **28:**253
and polypeptide synthesis, **24:**3
and protein synthesis, **26:**109–110
sedimentation pattern, **24:**18
Ribosome template, in polypeptide formation, **24:**21–22
Ribosyl-linked AMP, **46:**209
Ribosylthymine synthetase, tRNA, **54:**89
inhibition of, **54:**90
Ribothymidine, **54:**89; 90
Ribothymidylic acid, **35:**471
Ribotide, definition, **21:**200
L-Ribulokinase, reaction catalyzed by, **30:**110
Ribulose, **43:**493
D-Ribulose, in histidine synthesis, **16:**300
Ribulose 1,5-biphosphate, production of, **74:** 95–96
D-Ribulose-1,5-bisphosphate, **67:**2–3
active-site orientation, **67:**5
C3 deprotonation, **67:**5; 7; 23; 45; 51; 53
enediol(ate)
detection, **67:**5; 7
geometry, **67:**9–10
intermediate, **67:**14
reactivity, **67:**45–46
Ribulose-1,5-bisphosphate carboxylase oxygenase, *See* Rubisco
Ribulose diphosphate, **15:**154
in photosynthesis, **25:**94–96; 99; 105
Ribulose-1,5-diphosphate, **48:**171; 172
in plant respiration, **18:**285
in plants, **18:**294; 312
Ribulose 1,5-diphosphate (RuDP)
in photosynthesis, **19:**20–23; 28–32; 64; 65

Ribulose-1,5-diphosphate carboxylase, **37:**477; 498; 522; 584. *See also* Carboxydismutase
in photosynthesis, **19:**22; 32; 33; 62; 64; 65
Ribulose diphosphate cycle, **23:**450
Ribulose-5-P, **43:**493
D-Ribulose-5-P-3-epimerase, **26:**42
hydrogen exchange, **44:**87
two-site mechanism, **44:**88
L-Ribulose-5-P-3-epimerase, **44:**88
L-Ribulose-5-P-4-epimerase, **26:**42, **44:**86
aldol-like cleavage-condensation model, **44:**87
bacterial sources, **44:**86
intramolecular hydrogen transfer, **44:**86
isotope effect, **44:**86
metal-ion requirement, **44:**86
Ribulose-5-phosphate (Ru-5-P)
in glucose metabolism, **15:**153–154; 156–160; 165–166; 167
in pentose cycle, **28:**395
in photosynthesis, **25:**94; 95; 105
D-Ribulose-5-phosphate, **15:**117; 121
D-Ribulose-5-phosphate, **17:**406
Ribulose phosphate epimerase, localization of, **24:**333
L-Ribulose-5-phosphate-4-epimerase, reaction catalyzed by, **30:**110
Ribulose-5-phosphate r ribose-5-phosphate, **14:**195; 205; 207–208
Ribulose-5-phosphate (Ru-5-P)
in photosynthesis, **19:**15; 21; 28; 30; 37; 38
Ribulose 1-phosphorothioate 5-phosphate, **67:**13
Rice dwarf disease virus, transmission, **9:** 174–175
Rice phytoalexins, **55:**79–80
Ricinas communis, **54:**175
Ricinus communis, **48:**71, **55:**78–79
Rickettsia, **23:**267
Rieske protein, **54:**290
Rifampicin, **44:**166; 168; 170
Rifampsin, **40:**79; 84
Rifamycin, **58:**67
Rigidity of protoplasm, **7:**60
Rigor, effect on transverse pattern, of muscle fiber, **16:**347
Rigor mortis, muscle protein in, **13:**162; 179–180; 190
Riley agent, **39:**226
Rimocidin, in treatment of plant disease, **20:**390; 392; 396; 401–403
Ring A-dehydrogenases, **20:**265; 266; 270
Ring cleavage, in hydrocarbons, **27:**508–511; 515; 519; 522; 524; 535; 536
Ring formation, rate dependence on ring size, **22:**83; 84
Ring strain, in lipoic acid, **18:**329–330
Ring strain in cyclic phosphates, **24:**445
Rishitin, **55:**66
biosynthesis of, **55:**67; 68; 70
metabolism of, **55:**81
norsesquiterpenoid, **55:**50
in potato tubers, **55:**50
structure of, **55:**51
Rishitinol, **55:**51
Rishitione, **55:**51
R-Isopropanolamine, **46:**53
Ristocetin, effect on synthesis of poly (glycerol phosphate) in *Bacillus,* **30:**244
R-layer, *See* Murein sacculus
R mutants, definition of, **31:**100
RNA, **15:**127, **38:**226, **44:**168, **58:**28, **59:**215. *See also* DNA, Mutation and Nucleic acids; Nucleic acids; Nucleic acids Ribose; Ribonucleic acid
activation of pepsinogen, **40:**38
bound to thiol-disulfide exchange protein, **40:**47
in the oxidative phosphorylation system, **40:**38; 39
transfer, acylation by mutant synthetases, **40:**207; 208
8-azaguanine in, **28:**278
base pairing, **50:**36
and carbohydrate synthesis, **26:**67
in chromatin, **38:**52–54
and composition of template for protein synthesis, **22:**132
soluble amino acid transfer to, **22:**131; 132
synthesis during enzyme induction, **22:**137–140
transfer experiments, **22:**139; 140
containing incorporated F, **54:**58; 61
content of sperm, **24:**90
as coupling agents, **23:**369; 372
effect of amino acid analogs on synthesis, **29:**148
ethyl derivative of, **25:**383–389
methylation of, **25:**414–415

RNA (*Continued*)
fluoropyrimidine effect on, **54:**88; 91–92
fluoropyrimidine incorporation into, **54:**86–92
and functioning of cytoplasm, **8:**5–7; 9–11; 14; 25–26; 31–33; 46–47; 55
as genetic material, **24:**9–11
in assay of RNase, **24:**172–173; 235–238
of bacteriophage, **24:**17
messenger, **24:**19
steps in hydrolysis of, **24:**163–164
transfer (soluble), **24:**3
with histone, coacervate, **27:**367; 369
hydrolysis, **58:**30
hydrolysis, by metal ions, **21:**173; 174; 186
labeled, **28:**411; 414–417; 419
messenger, **27:**247; 269; 275–276; 295
microsomal, **27:**219
in microsomes, **13:**8; 9
nuclear, **27:**247; 260; 269–270; 274–275
and origin of enzymes, **27:**374
in plastids, **20:**210; 211
polymerase, **38:**71; 84
predecessor of, **38:**226
regradation of, **58:**250
ribose biosynthesis, **28:**392–419
ribosomal, **27:**247
role in molecular genetics, **27:**291; 295; 310–312
separation from DNA, **18:**163
synthesis of, **18:**228–229
in sperm, **9:**338
in tobacco mosaic virus, **15:**459–461
transfer, **46:**279
aminoacylation of, **46:**300–301
conformation variability in, **46:**296–298
life cycle of, **46:**280–282
modification of, **46:**307–308
processing of, **46:**305–307
secondary structure of, **46:**283–286
classification of, **46:**286
sequence and cloverleaf, **46:**283–285
symmetry in structure, **46:**293
tertiary structure of yeast tRNA, **46:**286–295
three-dimensional structure of, **46:**279–315
ultraviolet irradiation, **27:**327
of virus, **7:**191
rate of renewal of, **7:**179–181
and virus multiplication, **17:**477–498
genetic studies, **17:**459–465
solubilizing power, **17:**137; 147
in vitro replication, **27:**290; 295
RNA degradation, **60:**12
RNA:DNA ratio, in evaluation of nuclear purity, **27:**242; 245–247
RNA-DNA relations in chromosomes of *Rhynchosciara angelae,* **16:**460–461
^{14}C-RNA, from ^{14}C-uracil, **28:**435–436
RNA messenger, **50:**160
RNA methylase, in nucleolus, **27:**268
RNA nucleotidyl transferase, **43:**25
RNA polymerase, **43:**34, **44:**166; 169; 170, **56:**292
effect of ultraviolet, **27:**327
effect of various conditions, **27:**268
of nucleus, **27:**274
metal centers of, properties of, **56:**337
metal content of, **56:**335–336
metalloenzyme substitutions of, **56:**304
molecular properties of, **56:**336–337
sigma subunit, **44:**166; 168; 170
spectral features of, **56:**338
α subunit, **44:**166; 167
β subunit, **44:**166; 167; 168
β′ subunit, **44:**166; 167
δ subunit, **44:**167; 169
omega subunit, **44:**166; 167
RNA polymerase I, **47:**288; 289
effect of α amanitin, **47:**289
effect of cycloheximide, **47:**289
increase of rRNA synthesis by glucocorticoids, **47:**289
RNA polymerase II transcription, **71:**41–77
background, **71:**42
C-terminal domain
characteristics of, **71:**44–51
generally, **71:**44–46
promotor specificity, **71:**46–49
unique and essential qualifites, **71:**46
DNA binding, **71:**50–51
C-terminal domain characteristics
DNA binding, **71:**50–51
extended structure, **71:**50
posttranslational modifications, **71:**49–50
differential interaction with preinitiation complex
CTD kinase, **71:**58–59
RNA polymerases IIO and IIA, **71:**58–59
early stages, involvement in, **71:**53–55
generally, **71:**51

photoaffinity labeling, identification of transcriptionally active RNA polymerase by, **71:**52–53
preinitiation complexes, **71:**55–58
multicomponent process of, **71:**43–44
phosphorylation and dephosphorylation enzymology, **71:**69–72
potential roles, **71:**60–69
elongation, **71:**64–69
generally, **71:**60
initiation, **71:**60–64
reversible phosphorylation of CTD during transcription cycle, **71:**51–59
RNA-protein code, **27:**297
RNA pyrimidines, precursors, **28:**423
RNA (ribonucleic acid), **59:**215
adenylic acids from, **16:**14
cytoplasmic, **16:**423; 438; 439; 461–462; 473
role in protein synthesis, **16:**474–475; 491–492
in isolated chromosome, **16:**447
in isolated nuclei, **16:**415; 426; 436–437
in melanin granules, **16:**169
metabolic heterogeneity, **16:**461–462
methods for locating in cell, **16:**417; 422; 436
in nondividing cells, incorporation of P^{32}, **16:**455–456
nuclear, composition, **16:**437–440
isotope uptake, **16:**472; 483
origin of pentose for, **16:**451
role in physiology of nucleus, **16:**471; 474; 475
species and tissue specificity of, **16:** 438–440
yeast, as activator of β-glucuronidase, **16:**376
RNase and DNase, **46:**249
RNase L, *See* 2′,5′-oligioadenylate ribonuclease L
RNase-s-peptide, **36:**77
RNase-s-protein, **36:**77
RNA synthesis, **24:**14–20, **27:**282–283
RNA transcripts, γ-glutamyl transpeptidase gene expression, **72:**265–267
RNA transfer, **50:**160; 161; 164; 166
Roach, *See* Leucophaea moderae
Roboflavin, inhibitory analogues of, **6:**131–132; 141; 142
Roe, *See* Hartley-Roe equation
Roentgen rays, effect on desoxyribonucleic acid turnover, **7:**177–179
effect on phosphatide turnover, **7:**168–169
Role of Futile Cycles in the Regulation of Carbohydrate Metabolism in the Liver, The (Hue), **52:**247
Root nodules, leguminous plants, **46:**87
soybean, **46:**88
Roots
carotenogenesis in, **21:**308; 328; 333; 335
in cell-free extracts, **21:**346
carotenoids in, **21:**306
cereal, salt respiration in, **18:**49–52
cytochromes and oxidases in, **18:**16; 19–22; 29; 31; 33; 42; 44–45; 49–52
toxic effect of thiosulfate on, **10:**231
Rope-climbing test, in schizophrenia, **29:**523
Rosaniline, **39:**333
Rose, Zelda B., **51:**211–253
Rose bengal, **46:**392
Roses, antibiotic treatment of diseases, **20:**400
Rosinduline, **15:**4
Rosindulin 2G, chemiluminescence, **25:**150
Rotational isomers, stable configurations of, **23:**7–25
Rotation dispersion, of enzyme-coenzyme complex, **26:**164–165
Rotatory dispersion titration, of enzyme by coenzyme, **27:**52–53; 61–68
Rotatory properties, of enzyme proteins, **23:**37; 72–76
Rotenone, **34:**87; 89; 91; 96; 102; 103; 111; 123; 134; 137; 139; 145; 146
binding of, **34:**103
cells treated with, **29:**344
inhibition by, **34:**104–107; 109
by leak of, **34:**139
inhibition site of, **34:**101–111
lipid involvement in binding, **34:**107
reversal of inhibition by, **34:**107
role of protein in specific binding sites of, **34:**108
specific binding sites for, **34:**117
specific sites of, **34:**104
unspecific sites, **34:**104
^{14}C-Rotenone, **34:**102; 103; 104; 110
Rous sarcoma, *See* Virus
Rous sarcoma virus, mRNA methylation, **65:**261–262

Rousseau models, cytochrome *c* oxidase, proton pumping, **71:**138–143
Roux sarcoma virus, effect of light, **9:**142; 156
R-5-P, *See* Ribose-5-phosphate
11R-Phyton, **46:**72
rRNA
  5-fluoroorotic acid effects on, **54:**86
  5-fluorouracil incorporation into, **54:**86–88
  maturation of, **54:**86–88
  methylation, **54:**87
RSA, *See* Bovine serum albumin
Rubber. *See also* Hydrocarbons
  acetyl-CoA as precursor of, **29:**229
    ATP stimulation of mavelonate to, **29:**237
    biosynthesis of, **29:**221–257
    cis configuration of, **29:**222
    differences from other terpenoids, **29:**222
    effect of enzyme concentration on incorporation of C^{14}-isopentenyl pyrophosphate in, **29:**242
    β-hydroxy-β-methylglutaryl-CoA as precursor of, **29:**233–234
    mechanisms of biogenesis in, **29:**248–250
    molecular weight distributions of, **29:**222
    role of lutoid particle for synthesis of, **29:**234–235
    role of pyruvate in synthesis of, **29:**233
    trees producing, **29:**222
    utilization of dimethylallyl pyrophosphate and isopentenyl pyrophosphate in formation of, **29:**246
  action of microorganisms on, **10:**444–445; 447; 470; 474–476
Rubber, Biosynthesis of (Archer and Audley), **29:**221
Rubber biosynthesis, on existing rubber particles, **29:**241
  initiation of new molecules and particles in, **29:**244–247
  intermediates in, **29:**231
  from isopentenyl pyrophosphate, **29:**240–243
  in latex *in vitro,* **29:**226–227
  metabolic regulation of, **29:**250–252
  site of formation of precursors of, **29:**234–236
  stereochemistry of, **29:**247–248
  utilization of acetate and pyruvate in, **29:** 226–229
  utilization of malonate in, **29:**230–233
Rubber hydrocarbon, structure, **22:**330
  synthesis from mevalonic acid, **22:**329
Rubber transferase, inhibitors of, **29:**243. *See also* Isopentenyl pyrophosphate-incorporating enzyme
  stereospecificity of, **29:**247
Rubidium ions, **15:**116; 251; 257
  and choline acetylase activity, **12:**318
  enzyme activation by, **21:**153; 158
    properties, **21:**157
    in purine biosynthesis, **21:**213
Rubisco, **67:**1–62
  activation, **67:**12; 26; 42
  active-site
    characterization with affinity labels, **67:** 27–30
    characterization with general reagents, **67:**25–27
    independent *versus* shared, **67:**47
    location and intersubunit interactions, **67:**47–50
    lysines, chemical properties, **67:**30–33
    pK_a values, **67:**30–31
    residue function, **67:**44–47
  alternate substrates, **67:**11–13
  amino acid sequences, **67:**24–25
  carboxy intermediate
    trapping, **67:**7–8
    turnover, **67:**9; 36
  carboxylase/oxygenase specificity, **67:**56–60
  carboxylation, **67:**5–10
  catalysis minimal requirements, **67:**15–16
  chemical modification, site-directed mutants, **67:**39–40
  chemical rescue, **67:**37–40; 59–60
  cross-linking, **67:**32
  crystal structures, **67:**16–25
    active site, **67:**19–25
    L_2 and L_8S_8 enzymes, **67:**16–18
    L_8S_8 enzymes, **67:**18–19
  fallover, **67:**13
  functions, **67:**3–4
  hybrid, subunit expression, **67:**47–49
  inactivation, **67:**26–27
  kinetic isotope effects, **67:**12
  kinetic mechanism, **67:**11–13
  L_8 core, **67:**19; 59
  loop 6, **67:**18; 46; 52; 55
    carboxylase/oxygenase specificity role, **67:**59
  loop closure, **67:**35; 46

L_8S_8 hybrids
carboxylase/oxygenase specificity role, **67:**59
crystal structure, **67:**18–19
Mg^+ ligands, **67:**20
role, **67:**11; 15
mutant properties, **67:**48; 54
oxygenase activity, **67:**2–3
oxygenation, **67:**10–11
partial reactions, **67:**15–16; 36
pH effects, **67:**32
phosphate binding site, **67:**22–23
random mutagenesis, **67:**57–58
side reactions, **67:**13–15
site-directed mutagenesis, **67:**33–36; 40–50
active-site location and intersubunit interactions, **67:**47–50
active-site residue function, **67:**44–47
carboxylase/oxygenase specificity, **67:** 59–60
permanently activated carboxylase, **67:**41
validation of residue importance, **67:** 40–44
S subunit, **67:**19
structure-guided mutagenesis, **67:**50–56
intersubunit and electrostatic interactions, **67:**55–56
invisible base, **67:**51–55
phosphate-binding site, **67:**50–51
temperature-sensitive, photosynthetic-deficient mutants, **67:**58
Rubner coefficient, **2:**198–199
Rubredoxin, **47:**58; 60; 124
Rubreserine, spectrum, **16:**113
Rubrofusarin, **5:**189; 191; 192–196, **11:**403
RuDP, *See* Ribulose diphosphate; Ribulose 1,5-diphosphate
Rumen β-glucuronidase, **16:**377; 380; 381
Ruminants, propionic acid metabolism, **26:** 283–284; 344
Ru-5-P, *See* Ribulose-5-phosphate
Rutabages, antibiotic treatment of diseases, **20:**396
Ruta gravelones, **55:**41
Ruthenocene, **58:**25
Rutin. *See also* Polyphenols
inhibition of hyaluronidase by, **13:**216
reaction with borate, **12:**500
synthesis of, **26:**61
from tea leaves, **16:**151
in tobacco leaves, **10:**347–349; 373–376

S

^{35}S, **39:**352
S^{35}, in glutathione stydy, **13:**275
S1, nuclease, progressive deletion to facilitate DNA sequence analysis, **67:**451
SA. *See also* Coenzyme Q
relation to coenzyme Q, **25:**277–278
Saccharase, **23:**135, *See* Sucrase
adsorption at interfaces, **28:**3; 25
determintion in soil, colorimetric methods of, **28:**376–379; 381–383; 386
results of methods compared, **28:**387
volumetric methods of, **28:**373–376; 386
pH dependence of, **15:**284
reactivation, **10:**103 (*See also* Invertase, Saccharase)
Saccharase-sucrose complex, active form of, **17:**55
Saccharate, affinity for β-glucuronidase, **16:**367
effect on formation of *o*-aminophenylglucuronide, **16:**403
inhibition of β-glucuronidase by, **16:**372; 373; 381; 382; 385; 402
Saccharic acid, salts, **17:**149
Saccharic acid in mold metabolism, **9:**543
Saccharides
cell surface, **48:**83
complex, synthesis of, **26:**48–49
synthesis of, **26:**35–88
Saccharifying amylase, *See* β-Amylase
Saccharo-1
effect on formation of *o*-aminophenyl-glucuronide, **16:**403
4-4lactone, as inhibitor of β-glucuronidase, **16:**373; 374
Saccharomyces
cerevisiae, growth factors for, **14:**66
use of acetate, **23:**406; 411; 412; 420
Saccharomyces carlsbergensis, **54:**174
activity of LBF for, **14:**53; 66
allosteric enzymes, **28:**84
Saccharomyces cereviciae, **46:**49; 91

Saccharomyces cerevisiae, **54:**173, **56:**91–134, **70:**25, **71:**84
adenosylmethionine analongs from, **54:**209
adenosylmethionine decarboxylase of, **56:**266; 269–270
allantoin degradation by, **56:**92–98
allosteric enzymes, **28:**70; 72; 73
and citric acid cycle, **18:**244–246; 271
enzyme production control in, **56:**104–120
genetic studies, **17:**428; 441–442
in metabolic homeostasis, **56:**129–132
methionine adenosyltransferase of, **56:** 257–259
nitrogen sources for, **56:**129–130
use of biotin, **26:**333
Saccharomyces cerevisiae:
and chitinase activity, **59:**90; 93–94
and chitinases, **59:**86; 90
and chitin synthetase
and cloning, **59:**80
and distribution, intracellular, **59:**74–76
and inhibitors, **59:**73–74
particulate preparations of, **59:**63; 65
purified preparations of, **59:**68–69; 71
vectorial synthesis of, with membranes, **59:**79
zymogen form of, **59:**66
and chitosan, **59:**84
and distribution of chitin, **59:**61–62
and FBPase 1, **59:**338
functional segregation of acylCoAs, **67:** 411–412
and glucose, **59:**373–374
Nmtlp regulation, *in vivo* study, **67:** 408–414
and structure of chitin, **59:**60
temperature sensitive myristic acid auxotrophy, **67:**405–408
*Saccharomyces cerevis*iae, **49:**12; 28
Saccharomyces cerevisiae, in *S*-adenosylmethionine synthesis, **22:**247
β-glucosidase inducers and complexants in, **22:**118; 119
α-glucosidase induction in, **22:**113; 117
Saccharomyces cerevisiae, action of amitrole on, **29:**125
ethionine resistant mutants, **29:**139
α-thioethyl-D-glucophyranoside in, **29:**45
Saccharomyces cerevisiae, enzymic adaptation in, **13:**70; 71
Saccharomyces cerevisiae, galactose-1-phosphate uridyl transferase of, **20:**114
galactose metabolism in galactokinase-deficient mutant, **20:**127; 128
*Saccharomyces cerevisia*l, coenzyme Q in, **25:**288
Saccharomyces cervisiae, serine proteinase from, **41:**182; 197
superoxide dismutase from, **41:**79; 82
Saccharomyces chevalieri, enzymic adaptation in, **13:**91
Saccharomyces drosophilarum, **23:**411; 412
Saccharomyces, fat synthesis by, **9:**674–681
Saccharomyces fragilis, **17:**218–219
galactose-1-phosphate-glucose-1-phosphate transformation in, **14:**198; 199
growth factors for, **14:**66
galactose-1-phosphate uridyl transferase of, **20:**114
polygalacturonase of, **20:**343; 363; 365; 366
uridine diphosphogalactose-4-epimerase of, **20:**115
β-galactosidase, **25:**336
Saccharomyces fragilis, enzymic adaptation in, **13:**70; 111–113
Saccharomyces fragilis, in galactan synthesis, **15:**314–315
Saccharomyces marxianis, galactose-1-phosphate-glucose-1-phosphate transformation in, **14:**198
Saccharopine dehydrogenase, kinetic mechanism of, **45:**299
Saccharose
as acceptor in dextran formation, **30:**199
enzymatic synthesis, **1:**336; 338; 343; 345; 350; 356–361
Sacchromyces enzymes, **59:**87
Sacculus, and bacterial cell walls, **26:**193–232
definition, **26:**193–196
Sacroplasma reticulum, **34:**470; 474
vesicles, **34:**474–475
Safflower, **51:**5
antibiotic treatment of disease, **20:**394
Safranine T, chemiluminescence, **25:**150
Safynol, **55:**90
Sakaguchi reaction, inhibition, **10:**288
Salicin, **18:**283, **56:**199
fermentation by thiaminase-producing bacteria, **15:**413

in β-glucan synthesizing system, **28:**339
Salicinase, separation by chromatography, **14:**349
Salicylaldehyde, oxidation, **15:**257
Salicylaldehyde, Schiff bases of, **50:**9
Salicylaldoxime, **17:**252
electron flow from water to $NADP^+$ inhibited by, **30:**38
as an inhibitor of plastocyanin, **30:**38
Salicylate, **17:**154, **54:**139
Salicylate hydroxylase, **37:**435; 495; 523; 584
requirements, **27:**515
Salicylhydroxamic acid, **58:**218
Salicylic acid
inhibition of β-lactamase, **28:**295
from microbial oxidation of hydrocarbons, **27:**514–515; 525–527
from microbial oxidation of naphthalene, **27:**519–522
peroxidase action on, **19:**143
ω-*N*-Salicylidene derivatives of amino acids, **16:**232
Salivary gland, **45:**401; 414
regression of, **45:**414
Salivary glands in low vertebrata, **1:**374; 375
Salmine, **3:**71
amino acid composition, **16:**442–444
reaction with quinones, **16:**114; 126
in enzyme isolation, **14:**398
structure studies, **14:**281; 284
reaction with phosphorylase *b,* **24:**274
Salmonella, **58:**118
cell wall, **26:**228
lipopolysaccharide synthesis, **26:**47–49; 66; 68–72
roughness, carbohydrate synthesis, **26:**69
monosaccharide composition of polysaccharide portion of lipopolysaccharide from, **31:**79
R mutants of, **31:**80
S form of, **31:**80
Salmonella anatum, wild-type lipopolysaccharide of, **31:**81
Salmonella champaign, allosteric enzymes, **28:**111; 114
Salmonella enteritidis, mutants of, which cannot synthesize UDP-galactose, **31:**80
Salmonella gallinarum, cell walls, **26:**202
Salmonella hvittingfoss, allosteric enzymes, **28:**111; 114
Salmonella interitidis, enzymic adaptation in, **13:**86
Salmonella newington, use of, in lipopolysaccharide synthesis, **31:**93–94
Salmonella paratyphi, **15:**416
Salmonella paratyphis, allosteric enzymes, **28:**111
Salmonella pombe, **41:**10; 14
hypoxanthine-guanine phosphoribosyl transferase from, **41:**14
PP-ribose-P amidotransferase from, **41:**10
Salmonella schottmuelleri, **15:**416
Salmonella serpens, **42:**253
Salmonella sp., β-lactamase of, **28:**243
Salmonella typhimurium, **34:**475
Salmonella typhimurium, **42:**253, **46:**47; 341; 376, **71:**8; 11
O-acetylserine sulfhydrylase from, **74:** 189–191
isozymes, sequence alignment between, **74:**197; 198
allosteric enzymes, **28:**69; 70; 125
pentose metabolism, **28:**435
APS kinase gene in, **74:**189
ATP-sulfurylase gene in, **74:**189
carbohydrate synthesis, **26:**68
L-cysteine biosynthesis in, **74:**188–191
pathways of, **74:**186; 188–189
regulation of, **74:**189–191
cysteine regulon and related genes in, **74:**189
genetic studies, **17:**457–459
serine acetyltransferase from, **74:**189; 190
sulfate permease from, **74:**186
gene for, **74:**189
sulfite reductase gene in, **74:**189
Salmonella typhimurium, action of amitrole on, **29:**125
control of histidine biosynthesis by analogs in, **29:**126–127
control of leucine synthesis in, **29:**147
mutants resistant to histidine analogs, **29:**128
Salmonella typhimurium, biosynthesis of O side chains in, **31:**95
chromosome map of, **31:**101
2-deoxy-D-ribose-5-phosphate aldolase from, **31:**139
mutants of, which cannot synthesize UDPgalactose, **31:**80
partial map of *rfb* region of chromosome of, **31:**108

Salmonella typhimurium, biosynthesis of O side chains in (*Continued*)
pathways of biosynthesis of nucleotide sugars in, **31:**107
quantitative sugar composition of polysaccharide portion of lipopolysaccharide from, **31:**84
Salmonella typhimurium, histidine biosynthesis in, **21:**404
purine-requiring mutant, **21:**254
pyrimidine biosynthesis in, **21:**286
Salmonella typhimurium, mutations, **27:**309
Salmonella typhinurium, **41:**4; 5; 14; 79
and genetic mapping, **22:**2; 3
histidine mutants, and interallele complementation, **22:**35
and position effect, **22:**38
mapping of loci concerned with histidine, tryptophan synthesis, **22:**38
rhamnose-resistant mutants, and mapping of loci, **22:**38
hypoxanthine-guanine phosphoribosyl transferase from, **41:**14
PP-ribose-P synthetase from, **41:**4; 5
superoxide dismutase from, **41:**79
Salmonella typhosa, mutant unable to utilize xylose, **22:**33
rhamnose-resistant mutants, **22:**34
Salmonella typimurim, **70:**231
Salmonella urbana, allosteric enzymes, **28:**111
Salt bonds, between protein and DNA, **18:** 176–177; 179
Salt concentration, **35:**490
Salt effect and clotting of blood, **10:**30–32
Salt extraction, **34:**484
Salting out, effect on adsorption and elution of proteins, **14:**326–329; 331–332; 337; 343; 355–357; 363–364
Salt linkages, in enzyme structure, **22:**73; 74
Salt respiration, in plants, **18:**45; 48–52
Salts
effect on DNA, **18:**165; 166; 171
hydrotropic, **17:**152–156
Salyrgan
effect on actin polymerization, **13:**155
effect on actomyosin, **13:**161
effect on muscle volume, **13:**174
in study of contraction, **16:**339
Salyrganic acid, **57:**137–138
Samarium ions, model catalyses with, **21:**177
Sandhoff's disease, **60:**126; 165; 248
Sanfilippo disease
type A, **60:**268
type B, **60:**244
type C, **60:**270
type D, **60:**264
Saponification, of pectic substances, **20:**348; 349; 358; 359; 370
Saponins, **17:**147
Sap stream, and anion transport, **18:**52
Sarcina lutea
cytochromes of, **23:**266; 270; 309
quinones in, **25:**289
Sarcoma. *See also* Tumors
and desoxyribonucleic acid, **7:**175–179
phosphatide turnover, **7:**168–169
uptake of ^{32}P by nuclei, **7:**184; 185
mouse, glutamohydroxamic acid formation in, **13:**298–299
rat, histone of, **18:**174
treatment with alliin, **11:**397–398
Sarcoma 180
effect of β-glucan, **28:**357
folate metabolism, **27:**459
Sarcoma, DNA content, **16:**426
Sarcophaga bullata, **45:**419–420
Sarcoplasm, ATPase activity of, **13:**167–168
Sarcoplasmic reticulum, calcium ATPase, **65:**186–196
Sarcoplasmic reticulum calciummagnesium ATPase of, **51:**89–90
Sarcosine
and formate, **12:**48
as detoxicating agent, **12:**451
as methyl donor, **12:**464; 477
metabolism, formyl derivatives, of THFA in, **21:**223; 428; 429
oxidation, in mitochondria, **21:**22; 25; 26; 428
oxidation of, **12:**481
reaction in phenolase complex system, **16:**118
Sarcosine oxidase
in mitochondria, **21:**22; 25; 429
properties, **19:**203
Sarcosine oxidation system, **23:**350
Sarcosomes. *See also* Cell particles, Mitochondria, Mitochrome and Ultramicrosomes
association with respiratory chain, **17:**66; 74; 83; 114; 130

oxidation in, **17:**333
Sarcosomes and mitochondria, **13:**12; 17
Sarcosyldehydroalanine and dehydropeptidases, **8:**140
Sarin, **34:**9
effect on esterases, **20:**288–290; 305
Sativan, **55:**29
structure of, **55:**7
from vestitol methylation, **55:**35
Saturase for fatty acids, **8:**317
Saturation, **34:**454
Saturation phenomena in photosynthesis, **1:**204–220
Sauerstoffübertragende Ferment, *See* Cytochrome oxidase
Saulenchromatographie von Enzymen (Turba), **22:**417
SC, *See* Ubichromenol
Scalding. *See also* Blanching conditions
and flavor retention in vegetables, **9:**627–631; 631–637; 637–649
Scenedesmus
carbon dioxide fixation by, **19:**20; 21
Hill reaction in, **19:**50
decrease in quantum efficiency of Hill reaction in, **30:**25
effect of cycloheximide on Hill reaction in, **30:**25–26
removal of ^{54}Mn from, **30:**26
phosphoribulokinase from, in multienzyme complex, **74:**103–105
Scenedesmus cells, lipoic acid effect on, **18:**341
Scenedesmus obliquus, **46:**337
Scenedesmus obliquus, phosphorylation in, **27:**361
SC factor, *See* Cytochrome-linking factor and Succinic dehydrogenasecytochrome c combining factor
in respiratory chain, **20:**175; 177
preparation and composition, **20:**188–191
Schardinger dextrinogenase, isolation of, **14:**386–387; 405
Schardinger dextrins, **12:**365, **17:**210; 213; 215
in starch synthesis, **11:**354
synthesis of, **11:**300; 322–324; 328
Schardinger enzyme, *See* Xanthine oxidase
Schardinger reaction, **15:**14; 23
Schardinger's enzyme, classification, **14:**253
Scheie disease, **60:**244
Schiff base, **34:**453, **49:**13; 29, **53:**163; 164
in aldolase action, **28:**450–452
formation of, in aldolase reaction, **31:**128
pyridoxal phosphate, **53:**167
Schiff base formation, **35:**532
Schiff bases, **50:**8; 9
by action of quinones on amino acids, **16:**118; 120
formation in transmainations, **16:**209–210; 218; 224–226
from pyridoxal phosphate and serine, **16:**294
conformations of, **50:**15
with pyridoxal phosphate, **50:**8
of salicylaldehyde, **50:**9
tautomerism in, **50:**9–11
Schiff's base, as intermediary in transminations, **4:**79–81; 83; 84
Schiff's reagent, use in identification of nucleosides and nucleotides, **16:**13
Schistosoma enzymes, **23:**175
Schistosoma mansoni, hyaluronidases in, **13:**203
Schizokinen, **35:**169
Schizophrenia
albumin-globulin patterns in, **29:**501–502
aromaturia of, **29:**520–526
arousal theory of, **29:**531–536
fusion threshold of paired light flashes, **29:**532–533
palmar skin conductance in, **29:**532–533
biochemistry of, **29:**499–536
blood enzymes in, **29:**505–507
blood proteins in, **29:**501–515
cerebral spinal fluid proteins in, **29:** 501–515
ceruloplasmin levels in, **29:**506–507
changes in serum proteins in, **29:**501
changes in the hypothalmus in, **29:**514
EEG records of, **29:**533
effects of methoinine on, **29:**527–528
effects on carbohydrate metabolism by, **29:**512
excretion of 3,4-dimethoxyphenyl-ethylamine in urine in, **29:**500
genetic factor in, **29:**499–501
immunological reactions in, **29:**502–505
increased melanogenesis in, **29:**520
lactate/pyruvate ratio in, **29:**512–513
level of red cell ADP in, **29:**500
methylation and demethylation in, **29:** 526–529

Schizophrenia (*Continued*)
and neurotransmitters, **29:**515–520
plasma factors in, **29:**510–512; 514–515
relation of arousal level to performance in, **29:**531
as social and psychological problem, **29:**480
symptoms of incipient, **29:**533
toxic factors in plasma of, **29:**507–515
chronic, **42:**277
Schizophyllum commune, **59:**62
Schizosaccharomyces pombe, x-ray mutations in, **27:**328
Schlerotization, **45:**417
of procuticle, **45:**419
Schradan, change in body, **24:**308
Schütz law and clotting time of blood, **10:**4–5
Schütz's law, application to hydrolytic enzymes, **15:**272–273
Sciara coprophilia, **42:**245; 254
Scleroproteins, properties of, **30:**180
Sclerotan (kinkaku glucan), structural study, **28:**337
substrate for laminarinases, **28:**342; 344
Sclerotina frutigena, polugalacturonases of, **20:**365
Sclerotina homoeocarpa, plant disease caused by, **20:**401
Sclerotina laxa, polygalacturonases of, **20:**365
Sclerotina libertiana, polygalacturonases of, **20:**366
Sclerotization of arthropod cuticle, **16:**153–163
Sclerotizing enzyme, phenolase complex as, **16:**158
SCN, **39:**332
Scopine esterification, **8:**241
Scopolamine, **55:**139; 148
binding of, to muscarinic receptors, **55:** 153–154
in muscarinic receptor manipulation, **55:**176
in schizophrenia, **29:**534
structure of, **55:**141
Scopolia in alkaloid studies, **8:**224
Scutellarin, *See* 4′,5,6,7-Tetrahydroxy-flavone glucuronide
Scylla serrata, **59:**60
SDC, *See* Succinic dehydrogenase complex; Succinic dehydrogenase complex and Succinic dehydrogenasecytochrome c-combining factor
SDS, **54:**286
cytochrome c_1 changes from, **54:**281–293
S_E2-Reactions, in disulfide exchanges, **27:**205
Sea urchin, *See* Paracentrotus lividus
Seaweeds, decomposition, uses, **28:**353–355
extracts, types of hydrolysis by, **28:**350
laminarinase preparation from, **28:**343–344
Sebacic acid, effect on enzyme complex, **26:**164
Secobarbital, **46:**128
Secondary and tertiary structures, **39:**349
Secondary binding sites, **41:**213
Secondary hydrogen isotope effects, *See* Isotope effects
Secondary inactivation, **34:**474
Secondary interactions of pepsin catalysis, **33:**436–438
Secondary specificity, **53:**262; 267
Secondary structure, **42:**83; 98; 102; 110; 143; 159
Secondary Structure of Proteins, Prediction of, from Their Amino Acid Sequence (Chou and Fasman), **47:**45
Secondary structures, in proteins, **47:**52; 65; 74–75; 78; 85; 92–94; 116; 128–129; 134–137
use of multiple regression analysis, **47:**85
Second-coordination sphere, **59:**116; 122; 244
Secretin, **3:**68, **47:**136
Secretion
of digestive enzymes, **1:**373; 374
of gastric and pancreatic juice, **1:**384–387; 402; 403
Secretion process, use of ATP system, **14:**14–17
Secretors, **36:**135
Secretor status, **52:**47
Sedation, correlation with lowering brain serotonin, **29:**486
due to activation of trophotropic system, **29:**489
in schizophrenia, **29:**488
Sedation threshold, in schizophrenics, **29:**535
Sedimentaion velocity, effect of ionizing radiations on, **24:**360
Sedimentation
computation of theoretical, **30:**145
Gilbert theory of, **30:**144
theory of, of reversibly polymerizing macromolecular systems, **30:**141
velocity, of a monomer in instantaneous equilibrium with its polymer, **30:**143

of HGH, **42:**77
of papain and mercuripapain, **19:**276–279
Sedimentation coefficient, **39:**350
Sedimentation equilibrium, **42:**78; 84, **43:**70
Sedimentation pattern, and denaturation, **24:**228
of ribosomes, **24:**18
of RNase, **24:**168–169; 178
Sedimentation rates, of influenza virus, **9:** 208–211
of irradiated proteins, **9:**124–126; 129
of vaccinia virus, **9:**208
Sedimentation study of Michaelis complex, **27:**18–19; 28–29
Sedimentation velocity, **42:**77; 84; 148
Sedimentation velocity-diffuse methods, **39:**350
Sedoheptitol, **15:**161
Sedoheptulose, **15:**159; 161; 162
fate of C-1, **18:**295
in lignin study, **18:**363–365
Sedoheptulose diphosphatase, in ribose synthesis, **28:**411
Sedoheptulose diphosphate, **15:**163; 164
Sedoheptulose-1,7-diphosphate, **17:**413
Sedoheptulose 1,7-diphosphate
in photosynthesis, **19:**26; 27; 29–31
substrate of fructose diphosphatase, **28:**94
Sedoheptulose-1,7-diphosphate in synthesis of aromatic amino acids, **16:**291
Sedoheptulose diphosphate-1-phosphatase, **25:**96
Sedoheptulose 7-P, synthesis of, by aldolase, **31:**126
Sedoheptulose phosphate
origin, **14:**208
in plant respiration, **18:**286; 294; 312
Sedoheptulose-1-phosphate, **15:**154; 163
Sedoheptulose-7-phosphate, **15:**159
allosteric regulation, **28:**118
Sedoheptulose 7-phosphate, in photosynthesis, **19:**26–28; 30
Sedoheptulose phosphates, in ribose synthesis, **28:**396; 397
Sedormid, **46:**155
Sedum prealtum, heptulose metabolism in, **15:**162
Seed. *See also* Phytochrome
cytochromes and oxidases of, **18:**9; 17; 19; 29
germination, biochemical action of phytochrome, **26:**26–27
germination control, **26:**1; 7–9
Seedlings
glutamotransferase in, **13:**296–298; 303
respiration in, **18:**286
terminal oxidases in, **18:**39–43
Seed maturation, **36:**122
Seegmiller, J. Edwin, **51:**167–210
Seitz filters, use in isolation of clotting factors, **16:**26; 27; 63; 68; 73; 84
Selenate, **48:**4
Selenic acid, **48:**15
Seleninic acid, **59:**145
Selenite
effect on 16-α-hydroxylase, **20:**270
utilization of, by *Clostridium sticklandii* and *Methanococcus vannielli,* **48:**4
Selenium, **48:**17, **59:**105; 143
accumulation in *Astragalus,* **29:**140
accumulation in plants, **29:**141
analogs containing, **29:**140
as an essential nutrient, **29:**140
toxicity of, **29:**140
analogs, synthetic, of natural products, **48:**10
biochemical relation to α-tocopherol, **48:**4
compounds, and biological methylation, **12:**433–434; 435; 438; 444; 449; 453–457; 458
effect on formate dehydrogenase activity, **48:**18
inhibition of enzymes by, **11:**238
minimal fatal dose, **48:**5–6
nutrition, **48:**2–4
poisoning, **48:**5
acute, **48:**5
ameliorative effects of arsenite, **48:**5
protective effect against mercury toxicity, **48:**5
^{75}Se, **48:**1
^{77}Se, **48:**1
toxicity, **48:**4–6
Selenium-accumulator plants, **48:**5
Selenium adenosylselenomethionine, **48:**9
Selenium aminoethyl-selenocysteine, **48:**23
Selenium carboxyethyl-selenocysteine, **48:**23
Selenium carboxymethyl-selenocysteine, **48:**20; 23
Selenium-dependent biochemical processes, **48:**1–24
Selenium Dependent Biochemical Processes, Some (Stadtman), **48:**1
Selenium-dependent enzymes, **48:**3; 4
Selenium dioxide, **48:**1
reaction with water to form H_2SeO_3, **48:**1

Selenium enzymes, **59:**143; 156
^{75}Selenium-labeled protein, inhibition of, by antibiotics, **48:**22
synthesis, **48:**22
Selenium mercaptide formation, **11:**217; 218
Selenium methylselenocysteine, accumulation of, by *Astragalus* and *Stanleya,* **48:**5
γ-glutamyl peptides of, **48:**5
Selenoalanine, **56:**27
Selenocystahione, accumulation of, by *Astragalus* and *Stanleya,* **48:**5
Selenocysteine, **56:**26–27
Selenocysteine-charged tRNA^{cys}, **48:**9
Selenocysteine lyase, **56:**27
Selenocysteine synthase, **74:**152
Selenoether, derivatives, comparison of, **48:**23
Selenolate, **59:**145
Selenomethionine, **56:**253
action of enzymes on, **25:**384
in *S*-aldenosylmethionine biosynthesis, **22:**244
effects of, **29:**140
prevention of cell division in *Chlorella,* **29:**142
Selenoproteins, **48:**20; 22; 24
antibodies prepared to, **48:**22
12,000-dalton (protein A), **48:**20
glycoprotein structure, **48:**24
muscle selenoprotein, **48:**15–16
Selenosulfide, **59:**145
4-Selenouracil, **48:**9
Self-assembly, **38:**200; 213
of proteins and nucleic acids, **38:**201
Self-irradiation of cells, to secure mutants, **49:**91
Semen. *See also* Spermatozoa
anions in, **24:**80–82
cations in, **24:**78–80
enzymes in, **24:**65; 86–88
hydrogen-ion concentration, **24:**75–76
metabolism, **9:**329–390
Semen, Metabolism (Mann), **9:**329
Semiacetals, PN-dependent oxidation, **26:**119; 122
Semialdehyde, **50:**152
succinic, oxidation, **15:**258
Semicarbazide, **15:**379
Semicarbazide inhibition of amino acid decarboxylases, **6:**18
Semiclathrates, hydrogen bonding, **65:** 240–245
Semiconduction, in protein material, **24:**369; 381
Semiconduction in chlorophyll, **25:**81–82
Semiconductor properties of proteins, relation to energy transport in living cells, **14:**25–29
Semiconductors
in oxidoreduction mechanism, **15:**25; 26; 33–37; 40
proteins as, **17:**281; 338
Semimercaptal formation, **11:**218–219
Semimercaptals, **15:**256–257
Seminal fluid, and nutrition of sperm, **9:**255
Seminal plasma, **57:**12
effect on sperm, **24:**43–45
hyaluronidase in, **13:**227
Seminolipid, **60:**177
Semiquinoid intermediates, from DPNH dehydrogenase, **27:**178
Semiquinone, *See* PQQH
in dihydroorotic dehydrogenase action, **27:**173–175
of flavin, electron transfer, **27:**166; 172; 183–187
succinic dehydrogenase, **27:**176
Semiquinone formation, EPR spectra, **27:** 119–120; 122–125; 128
in flavin compounds, **27:**130–134
Semiquinones
autoxidation, **65:**121
in enzymic oxidoreductions, **15:**4–6; 11–12; 18; 42; 87
formation in oxidation of catechol, **16:**129
in melanin formation, **16:**172a
from oxidation of coniferyl alcohol, **16:**149
from quinone-hydroquinone reactions, **16:**108
in oxidation-reduction mechanisms, **17:**272; 278–279
redox transitions, **65:**101–104
unstable, relaxation methods, **25:**9; 31
Semliki Forest virus, **52:**141
Sendai virus, **52:**133; 137; 142, **58:**114
Sensitivity, **34:**455
Sensitization, photobiological, **11:**67
Sensitized fluorescence, **22:**361
Sensitized photooxidation, **35:**466
Sephadex, **34:**453; 521
Sephadex G-200, **57:**8
Sepharose, **34:**449; 455; 520–521

Sepharose 6B, **57:**10
Sepharose CL-4B, **57:**9–10
Sepharose-linked oxamate derivative, **46:**254
Sepia, **67:**82
Sepia, melanin of, **16:**172
Sepiapterin, **35:**49; 261; 285, **67:**83
biosynthesis of, **35:**69
reductase, **35:**262; 285
Sepiomelanic acid, **16:**172
Septomyxa affinis, enzymic transformations of steroids by, **20:**262
Septoria nodorum, **55:**46
Septum formation, in bacterial growth, **26:** 221–222
Sequence alignments, glutathione S-transferases, **69:**3–4
Sequence analysis, **39:**347
Sequence analysis of proteins, and disulfides, **27:**229–231
Sequence homologies
and adenylate kinase, **59:**262–263
and fibronectin, **59:**9; 14
Sequence homology, **42:**114; 117, **48:**284, **50:**314
Sequence studies of tryptophan synthase, **49:**179–181
Sequency analysis, tyrosine hydroxylase, **70:**179
Sequential fragmentation, of respiratory chain, **28:**190–197
Sequential induction, microbial oxidation of hydrocarbons, **27:**485; 506; 514; 516
Sequential mechanism, **57:**379; 381–383
Sequential reactions, **34:**445
reassociation, **34:**476
Sericin, water-soluble component of spun silk, **16:**157
Sericinase, **45:**403
Serime alkaline proteinases, **41:**180; 182–197; 214; 215; 220–223; 227; 237; 238
from *acremonium kiliense,* **41:**182
from *alkalophilic bacillus,* **41:**190
from *arthrobacter,* **41:**182
from *aspergillus melleus,* **41:**221; 223; 227
from *aspergillus oryzae,* **41:**180; 182; 184–187; 190; 191; 193; 194; 227
from *aspergillus sojae,* **41:**190; 221; 223; 227
from *bacillus subtilis,* **41:**182; 183; 187
from *a*-chymotrypsin, **41:**193; 221; 227
from pancreatic elastase, **41:**195; 196; 221; 237; 238
from *Penicillim cyaneofulvum,* **41:**182
from *saccharomyces cerevisiae,* **41:**182; 197
from *streptomyces fradiae,* **41:**182; 192; 193; 220; 222; 223; 227
from *streptomyces griseus,* **41:**182; 186–189; 191; 214–215
from *streptomyces rectus,* **41:**182
from subtilisin BPN, **41:**192; 222; 227
from subtilisin Carlsberg, **41:**180; 222; 227
Serime proteinases, **41:**188–197; 213–226
Serine, **34:**2–29; 453, **50:**130; 145; 155, **51:**217, **53:**169, **60:**252
ability to form hydrogen bonds, **34:**5; 7
and acetate, **23:**432–433
and glycine, **23:**457
and glycolate, **23:**435–436
in active centers of esterases, **20:**284; 288–290; 309; 311; 313; 315
in biosynthesis of glutamic acid in plants, **20:**325
in biosynthesis of pectic substances, **20:**369
in chloroplasts, **20:**209
in enzymes containing imidazole, **20:**81; 82; 88–90; 209; 284; 288-290; 309; 311; 313; 315
metal binding by, **20:**67
in phosphorylated peptides of esterases, **20:**188
use of blocking reagents with, **20:**68
at active site of enzyme action, **22:**65–69; 80; 81; 88
role in enzyme action, **22:**66; 67; 69
and biological methylation 439, **12:**449–450; 469–474; 478–479
biosynthesis of, **12:**43; 48–49; 95; 116
CO_2 fixation in, **12:**132; 142
deamination of, **12:**82–85
biosynthesis, **16:**274–277
condensation with indole, **16:**293–294
with deuterium, condensation with indole, **16:**294
inactivity in *Neurospora* mutant, **16:**256
labeled, conversion to glycine, **16:**276
in residual protein, **16:**447
in synthesis of methionine, **16:**272–274
in synthesis of thymine, **16:**450
in transamination reactions, **16:**191; 198; 201; 204; 207; 234

Serine (*Continued*)
biosynthesis and metabolism, **21**:223–225; 396; 398; 401; 410; 416–418
in biosynthesis of folic acid coenzymes, **21**:394
deamination, by metal ions, **21**:178
in methionine biosynthesis, **21**:423; 424; 430
in purine biosynthesis, **21**:223
in pyrimidine biosynthesis, **21**:288; 289 420–422
synthesis, in mitochondria and mitochondrial fragments, **21**:25; 26; 428; 429
in brain phosphatides, **7**:165
in transsulfuration, **7**:380–383
and chymotrypsin activity, **24**:459
and enzyme activity, **24**:465; 467
and esterase activity, **24**:461; 464
concentration in proteins, **34**:3
conversion of to L-cysteine, **33**:383–385
conversion to hydroxypyruvate, **58**:35
cysteine lyase and, **56**:18
deaminase for, **14**:346
oxidation of, **14**:132
dehydrase, **35**:107
hydroxymethylase, **35**:93
sulfhydrase, **35**:107
transhydroxymethylase, **35**:94; 95; 97
effect on indole production from tryptophan, **10**:58–60; 62; 77
relation to sulfur metabolism, **10**:280–281
in tryptophan synthesis, **10**:71–74
genetic studies, **17**:409–410; 418; 421; 422
synthesis of, **17**:184
glycine as precursor, **18**:258
hydrogen bonds generated by, **34**:25
hydrolases, **34**:5
linkage of phosphopantetheine, **34**:17–20
in methylation, **25**:397; 402; 404; 406; 420
and one-carbon pool, **27**:426; 459
from methanol assimilation, **27**:479
in papain, **19**:296
phosphorylation of, **58**:4
source of active aldehyde, **15**:158
structure, **34**:7
sulfhydrase and, **56**:18; 19; 22
L-Serine, deamination of, **19**:344
l-Serine, formation, **20**:14
DL-Serine, resolution by papain, **19**:309
d-Serine, synthesis, **20**:14
(3S,2S)-Serine-3-t_1, **35**:111
3-t_1-Serine, **35**:97
Serine acetyltransferase
reaction catalyzed by, **74**:188
regulation on, **74**:190
from *Salmonella typhimurium*, **74**:189; 190
L-Serine acetyltransferase, **50**:145
Serine A esterase, **20**:90
Serine-alanine transamination, **16**:204; 207; 234
Serine aldolase, *See* Serine hydroxymethylase; Serine hydroxymethyl transferase
reaction, **27**:425
Serine amide, ionization constant, **20**:58
Serine-amide-ketal hypothesis for RNase action, **24**:243
Serine aminotransferase, **74**:133
Serine auxotrophs of *Neurospora*, **16**:273
Serine (C^{14}), methylation by, **25**:402; 418
Serine-carbohydrate complex from glycoprotein, **25**:452
Serine deaminase, **37**:350–354
L-Serine deaminase, induction in *E. coli*, **22**:119
Serine dehydrase, **23**:459
Serine dehydratase, **49**:29
apoenzyme-holoenzyme equilibria of, **56**:43
cation effects on, **56**:43
characteristics of, **56**:41
classification, **14**:256
molecular parameters of, **56**:45
purification of, **56**:41
replacing agents for, **56**:11
size and stability of, **56**:42
substrates for, primary, **56**:11
D-Serine dehydratase, **74**:134; 152
L-Serine dehydratase, **37**:585, **74**:134; 152
iron-sulfur cluster-dependent, **74**:176
Serine dehydrogenase, oscillations in activity of, **29**:371
Serine/glycine auxotrops, **16**:275–277
Serine hydrolase, **56**:19–20
Serine hydrolases, **43**:164; 367
Serine hydroxymethylase, **21**:224; 225; 393; 401; 417–419; 424; 429, **55**:229
Serine hydroxymethyltransferase, **53**:83–110, **74**:148
active-site structure, **53**:105–107
apoenzyme, properties of, **53**:85–86
assay procedures, **53**:86–87
control mechanism, **53**:107–110

pyridoxal phosphate, role of, **53:**88–93
stereochemical studies, **53:**99–104
tetrahydrofolate, role of, **53:**93–99
Serinehydroxymethyltransferase, location of, **24:**309
Serine Hydroxymethyltransferase (Schirch), **53:**83
Serine hydroxymethyl transferase (serine aldolase), methyl synthesis, **25:**402; 406
Serine-indole ligase, **16:**293
Serine oxidase, **21:**418
Serine phosphate, shape of, **34:**27
structure of, **34:**23–25; 28–29
Serine phosphoric acid, from active site of enzyme, **15:**288
Serine proteases, **48:**281, **50:**2; 13; 34
binding of substrates, **50:**27
charge-relay system of, **50:**2; 12; 34
hydrogen-bonded network of, **50:**34
oxyanion hole of, **50:**13
role in blood coagulation cascade, **48:** 277–311
Serine Proteases in Blood Coagulation Cascade, The Role of (Davie, Fujikawa, Kurachi, and Kisiel), **48:**277
Serine proteinases, **53:**256; 268–274
Serine pyruvate aminotransferase
evolution of, rate of, **74:**158; 159
species distribution of, **74:**158
Serine residue, as part of active site of enzymes, **33:**385
Serine residues
γ-glutamyl transpeptidase catalytic sites, **72:**255–262
Ser-234 residue, pyrophosphate binding motif, **72:**167–168
Serine residues in RNase, chemical modification of, **24:**211; 226
Serine RNA polymerase, II transcription, C-terminal domain characteristics, posttranslational modifications, **71:** 49–50
Serine sulfhydrase
amino acid analysis of, **56:**22
apoenzyme-holoenzyme equilibria of, **56:**43
from baker's yeast, **56:**22–23
cation effect on, **56:**43
from chicken liver, **56:**22
coenzyme analogs and, **56:**55
cosubstrate specificities of, **56:**47
cystathionine β-synthase and, **56:**25–26
in elimination of amino acid substitutents, **56:**77
historical perspectives on, **56:**18–19
inhibitors of, **56:**22
isotopic exchange of hydrogen atoms with substitutents of, **56:**51
Michaelis and affinity constants for, **56:**48
molecular parameters of, **56:**44
purification of, **56:**20
replacing agents for, **56:**10
spectral features of, **56:**49
steady state kinetics of, **56:**74–75
substrates for, primary, **56:**10
relative activities with, **56:**21
Serine sulfhydrolyases, **56:**23
Serine transhydroxymethylase, **23:**458
Serine trypsin like proteinases, **41:**182; 186; 188; 189; 191; 213–216
from *streptomyces erythreus,* **41:**182; 186; 188–189
from *streptomyces fradiae,* **41:**182; 186; 188–189
from *streptomyces griseus,* **41:**182; 186–188; 191
Serological activity, of antigen films, **28:**24
Serologically reactive polysaccharides, formation of, **11:**306–315; 327–329
Serologically specific polysaccharides, synthesis, **26:**67–68
Serological properties of blood group substances, **12:**513
Serotonin
association with sedation, **29:**488
in the brain, **29:**482–483
depletion by reserpine, **29:**485–486
effects of decrease in the brain, **29:**487
in schizophrenia, **29:**518
similarity to hallucinogenic drugs, **29:**516
biosynthesis of, **74:**236
and phosphorylase, **24:**286
possible occurrence in platelets, **16:**72
Serotonin and Peptide Immunoneuromodulators: Recent Discoveries and New Ideas (Silverman and Karnovsky), **62:**203
Serozyme, **10:**13. *See also* Prothrombin
Ser-Pro-Phe, **57:**16
Serratia enzyme, **59:**86–87
Serratia marcescens, **18:**263, **23:**411
phosphatases, **26:**102

Serratia marcescens, ADP-glucose pyrophosphorylase of, **46:**352
molecular weight of enzymes from, **46:**367
Serratia marcescens, proteinase from, **41:**184; 205
Serratia sp., **41:**184; 204; 230
β-lactamase of, **28:**242
Serum
anticoagulant in, **16:**86–87
correction of prothrombin conversion in dicoumarol plasma, **16:**60; 63–64
cholinesterase in, **8:**462–467
collagenase inhibitors in, **23:**197; 198
milk (*See also* Whey)
and rennin action, **15:**439–440
nonspecific inhibition of hyaluronidase by, **13:**216–219
Serum Ac-globulin, *See* Ac-globulin
Serum albumin, **41:**116, **46:**140, *See* Albumin; Albumin, serum; Albumin, serum, and Serum proteins; Albumins; Bovine serum albumin (RSA)
as activator of β-glucuronidase, **16:**376–377
copper complex, **27:**93
lower peptide sequences in, **24:**420
reactions with disulfides, **27:**212
reactions with mercaptans, **27:**210–212
in restoration of mitochondrial oxidative phosphorylation, **21:**38
SH-groups, **27:**221
Serum albumin cholinesterase, **34:**448–450
Serum albumin fraction (Armour fraction V), association of serum factor with, **16:**44
Serum albumin haemochrome *a,* absorption spectrum, **23:**289–297; 306–310
Serum albumin system, bovine, antigens and antibodies, **24:**141–152
Serum cholinesterase, **43:**106; 130; 132; 141, *See* Cholinesterase
Serum complement, effect of diisopropyl fluorophosphate on, **20:**89
Serum copper, **44:**214
Serum elastase inhibitor, **23:**236
Serum esterase, effect of curare compounds on, **12:**308
effect of eserine on, **12:**286
effect on choline esters, **12:**265–270
Serum factor, *See* Fibrin-stabilizing factor
Serum factor SH-compound, **27:**220
Serumfestigkeit, **6:**119
Serum globulin, *See* Globulin, serum, Immune globulin, and Serum proteins
denaturation, **7:**248
Serum glycoproteins, method for labeling, **41:**105
Serum iron, **44:**218
Serum lactate dehydrogenase, human, **46:**255
Serum mucoprotein, a substrate for neuraminidase and receptor destroying enzymes, **20:**137; 144
Serum protein, reactions with disulfides, **27:**212
Serum proteins. *See also* individual proteins
adsorption at interfaces, **28:**13; 15; 16; 20
adsorption studies, **14:**328; 331; 360–361; 362; 363–364
interaction with polyanions, **40:**55; 56
in vivo carbamylation with cyanate, **40:**12
Serum proteins and complement, **8:**101–103
Serum-resistance, **6:**119
Serum survival time, of plasma glycoproteins, **41:**99
Seryl group, effect on imidazole, **20:**59; 60
Seryl-tRNA synthetase, **35:**463
Sesquiterpenoid phytoalexins, **55:**49
of eggplant, **55:**73–76
of potato, **55:**66–69
precursors to, **55:**66
of sweet pepper, **55:**69
of sweet potato, **55:**76
Sesquiterpenoids, condensation to squalene, **22:**313–319
Sevag gel, **18:**171–172
Sevag process, for deoxyribonucleic acid prepn., **18:**163
Sewage, relation to microbial metabolism of hydrocarbons, **10:**449; 451; 463; 465–466
Sex differences, in ethionine effects, **25:** 372–380; 381–382
Sex hormones, control of hepatic α 2u-globulin nRNA, **47:**297
Sex hormones and cholinesterase, **8:**468
SGH (ovine pituitary growth hormone), **42:**113; 118; 120; 122
S_H2-Reactions, in disulfide exchanges, **27:**206
Shadow casting, description of, **33:**75–77
mica-substrate technique, **33:**76
potential applications of, **33:**76–77

potential solutions to the resolution problem of, **33:**77
shadow-transfer method, **33:**76
SH agents, and phosphorylases, **18:**226–227
Shape, **39:**195
Shape of protein molecules, **1:**23–28
Shardinger dextrins, in polysaccharide synthesis, **15:**305; 307
Sheep erythrocytes, **49:**81
transport in, **49:**79
Sheep submaxillary gland mucin, composition of *N*-acetylneuraminic acid in, **20:**138
Sheep submaxillary gland mucoprotein, as substrate for neuraminidase and receptor destroying enzymes, **20:**137
Sheet (β-), *See* β-sheet
Shellfish, thiaminase in, **15:**389; 391; 393*t;* 394–398; 400–405; 415
SH group. *See also* Thiol
and fluoro compounds, **18:**152
of succinic dehydrogenase, **18:**75–76; 80; 82; 93
SH groups and enzyme activity, **26:**100–103
Shield base, **59:**125
Shigella dysenteriae, **15:**416, **46:**341
cytochrome a_2 in, **23:**266
Shigella flexneri, **71:**11
cell walls, **26:**202
Shigella flexneri, enzymic adaptation in, **13:**86
Shigella sonnei, β-galactosidase, **25:**336
Shigella sonnei, enzymic adaptation in, **13:**86; 111–112
Shigella sp., β-lactamase of, **28:**243
Shikimate, **23:**325, **34:**521
in metabolism of aromatic amino acids, **16:**287–291; 293–294
labeled, as source of tyrosine, **16:**287
Shikimate dehydrogenase, **34:**220
Shikimate pathway, **34:**217–221
Shikimic acid, **34:**218, **50:**148
in *p*-AB biosynthesis, **27:**421
in aromatic ring synthesis, **17:**412–414; 417; 421
conversion of, into lignin, **34:**219
distribution of, **34:**219
incorporation of, to L-phenylalanine and L-tyrosine by *Salvia Splendens,* **34:**221
labeled, in lignin study, **18:**366–370
in lignin study, **18:**362; 365–366; 375
Shikimic acid in synthesis of aromatic nucleus, **14:**120–122
Shikimic acid pathway, **55:**16
Shock
hemorrhagic, **7:**197
traumatic, **7:**196; 198
role of enzymes in, **8:**461; 490
Short chain glycerides, **34:**487
Short integrin-recognition sequences, fibronectin, **70:**3–6
Short-term glutamine regulation
glutaminase regulation, **72:**64–65
glutamine synthesis, **72:**66–67
Showdomycin-5′-phosphate, **54:**80
Shrinkage, of mitochondria, **23:**383–385
Sialic acid, **20:**135; 136; 138; 140–142, **41:**99; 100, **48:**42; 53; 56; 59; 60; 61; 64; 65; 66; 67; 68; 69; 70; 71; 73–80
derivatives of neuraminic acid, **25:**442
and immunogenicity, **48:**74–76
stimulation of aldolase, **40:**37
in transformed cells, **48:**72
Sialic acid linkage in glycoproteins, **25:**443–446
Sialic acid residues, **41:**99; 105; 106; 117; 118
enzymatic replacement of, **41:**99; 105–106
tritiation of, **41:**117–118
Sialidase, *See* Neuraminidase
Sialidases, **48:**52; 53; 60; 65; 66–71; 73; 75; 76; 78–80; 82, **60:**103; 135
activity in transformed cells, **48:**74
assay, **60:**139
biological role, **60:**142
catalytic mechanism, **60:**140
ganglioside, **60:**143
assay, **60:**145
biological role, **60:**149
deficiency, **60:**149
inhibition, **60:**148
kinetic data, **60:**147
molecular weight, **60:**146
optimal activity, **60:**146
properties, **60:**146
purification, **60:**144
substrate specificity, **60:**148
genetic, **60:**142
inhibition, **60:**140
isoenzymes, **60:**140
kinetic data, **60:**141
lysosomal, **60:**136
molecular weight, **60:**140

Sialidases (*Continued*)
properties, **60:**139
purification, **60:**137
substrate specificity, **60:**141
Sialidosis, **60:**135; 165
Sialoproteins, methods of labeling, **41:**117–118
N-Sialyl-*N*-acetylgalactosamine, linkage in, **25:**446
Sialyllactose, **52:**30; 32
3′- Sialyllactose, **52:**31; 38
Sialyltransferase, from rat liver, **41:**105; 126
Sialyl transferase c, effects of polycations, **40:**44
Sialyltransferases, **52:**25; 28–44
number and type, **52:**28–34
Sickle-cell anemia
absence of tuftsin activity, **41:**174
genetics of, **13:**141–142
protein synthesis in, **16:**475
Sickle-cell-anemia, clinical manifestations, **40:**1; 2
oxygen affinity, **40:**17; 19
patients treated with cyanate, hemoglobin levels, **40:**24; 25
hematocrit, **40:**24; 25
red cell survival, **40:**24; 25
serum bilirubin, **40:**24; 25
serum lactic dehydrogenase, **40:**24; 25
survival of cyanate treated red cells, **40:**22; 25
urea treatment, **40:**2
Sickle-cell hemoglobin, chemical basis, **24:**419
Sickle cells, carbamylation and preservation of normal cell form, **40:**3
Sickling of red cells, effect of thiols on, **11:**253
Side-chain contacts, **47:**77
Side chains, amino acids, **1:**3; 4; 40
Side-chain specificity, and microbial proteinases, **41:**181
Sideroblastic anemia, **46:**177
Siderochrome, **35:**140; 170
Siderophilin, **17:**273
Sigma constants, **46:**400
Sigma factor, exopolysaccharide alginate synthesis, **70:**246–251
Sigma minus values, **46:**400
Sigmoidal response, of enzyme to substrate, **28:**49–50; 52; 57–60; 73; 81–83; 85; 103; 105; 112; 115; 122; 128–130; 132; 134; 136–137
Signal transducers, cholesterol ester hydrolases regulation, **69:**69
Silastic, **34:**513
Silica gel, **34:**452; 521
Silicate in Mo-catalyzed reactions, **17:**269; 276
Silicic acid, model catalyses with, **21:**175
physiological effects, **21:**194
Silicomolybdate, as active carrier, **17:**239; 269; 270
Silicon compounds
solubilization of, **17:**138; 142; 144; 147; 149
unsaturated, chemiluminescence of, **25:** 126–128
Silicone oils, **51:**5–6; 33; 50
advantages of, **51:**16
Silicones, use in study of blood clotting, **10:** 15–17
Silicone-treated glassware, blood clotting in, **16:**67; 79
Silicosis, relation to catalyses by metal ion phosphatase models, **21:**194
Silk, water-soluble protein of, **16:**157
Silk fibroin, **3:**72; 73–76, **34:**3; 7
chymotryptic digests of, **24:**128–131
structure of, **23:**2; 22; 25
X-ray analysis, structure, **1:**7
Silkmoth, *See* Antheraea pernyi
Siloxen, chemiluminescence, **25:**121; 126–128
Silver
effect on amino acid decarboxylases, **6:**17; 18
effect on ATPase, **6:**165; 176
effect on indole oxidase, **19:**127; 128
effect on pyrocatechase, **19:**81; 126
in identification of imidazole, **20:**71
and pyridine nucleotide coenzymes, **15:**85; 96
enzyme inhibition by, **15:**274; 275; 285–286; 288
Silver and amylase activity, **12:**400; 401
Silver compounds, solubilization of, **17:**138; 145
Silver ions
effect on, collagenase, **23:**196
elastases, **23:**234
effect on β-glucuronidase, **16:**379
effect on mitochondrial oxidative phosphorylation, **21:**31
effect on muscle extensibility with ATP, **13:**173
in enzyme kinetic studies, **17:**58

Silver ions and tyrosinase action, **14:**86; 93; 94; 144; 146
Silver mercaptide, **11:**216; 228
Silver nitrate, effect on steroid isomerase, **20:**269
Simazin, Hill reaction inhibited by, **30:**23; 40
 photooxidation of Mn^{2+} inhibited by, **30:**25
Simulated mutation, *See* Mutation, simulated
Simultaneous reactions, kinetics, **9:**53–56
Sinalbin, occurrence, **10:**266
Sinapic acids
 esters of, **55:**9
 formation of, **55:**39
Sinapis alba, myrosulfatase of, **22:**230
Sinapyl alcohol, **18:**371; 375
Single chain human Cp, **44:**192
 two forms, **44:**193
Single crystal optical spectra, **47:**18; 19
Single-displacement mechanism, **57:**382
Single electron transfers and degradation of tryptophan, **10:**62–64; 69
Single oxygen, **41:**44
Singlet oxygen, **59:**126
Singlet state, excited, effect of ionizing radiations, **24:**364
Sinigrin, occurrence, **10:**260
Sinistrin, in garlic, **11:**379
Siroheme, **46:**49; 50; 60; 103
Sirohydrochlorin, **46:**49; 51; 83
Siroporphyrin, **46:**103
Sisto-amylase, inhibition of amylase, **6:**425
Site-directed mutagenesis
 cholesterol ester hydrolases regulation, **69:**68
 glutathione S-transferases, **69:**3; 19
Site L, availability of space at, **49:**58
Size, **39:**195
Size exclusion chromatography, RNA polymerase II transcription, reversible phosphorylation of CTD during transcription cycle, **71:**55–58
Size of protein molecules, *See* Mol weight
Sizing protein crystals, **52:**188
Skeletal muscle, **53:**214
Skeletal muscle phosphofructokinase, **48:**201–206
Skin tumors, **42:**274
Slab gel electrophoresis, **57:**11
Slaphylococcus cells, **18:**184
Slater factor, in plants, **18:**12; 306
Slater's factor, in respiratory chain, **17:**69; 79; 81; 82; 98; 106; 364; 368–370; 375; 384
Slime mold, rhythms, **7:**50
Slime mold cells, **59:**34–35
Slowly exchanging base, **59:**125
Slow reactions, in mechanism study, **27:**1–36
Slow signal, desulfo enzyme, **51:**136–137
SLR factor, **21:**372
Sluggishness, in industrial fermentation, **7:**263–264
Small heat-shock proteins (sHSPs)
 α-crystallin and (*See also* entries under Crystallin)
 α-crystallin and, **69:**157–165
Small intestine, **34:**524
S-MDA papain, **34:**522
S-MDA resins, **34:**453–456
Smith degradation procedure, applied to β-1,3 glucans, **28:**332–333; 335
Smoking properties, effect of fermentation, **10:**433–434
Smoluchowski relationship, **33:**251
Smooth endoplasmic reticulum, **46:**116; 117; 118; 119; 132; 135; 144; 149
S_N2-Reactions, in disulfide exchanges, **27:**203; 204
Snake toxin, **17:**367
Snake venom, **5:**46; 79; 228, **34:**491, **57:**25–26
 action in blood clotting, **10:**12; 27
  thromboplastin in, **10:**6; 19–20
 action on surface films, **11:**77–79
 esterase in, **12:**270
 and heparin, **8:**476
  biological significance of enzymes in, **8:**486–491
  disintegration of nucleic acids by, **8:**460
  enzymes, **8:**459–495
  and hylauronic acid, **8:**472; 487
  spreading factor, **8:**471–476; 487
 isolation of enzymes from, **14:**377; 401; 405
  separation by chromatography, **14:**332; 348; 361
 pyrophosphatase action on CoA, **16:**12; 14
Snake venom *(Crotalus atrox)* neutral proteinase from, **41:**187; 201
Snake venom films, activity of, **28:**22
Snake venom 5′-nucleotidase, **43:**33
Snake venom phosphodiesterase, **43:**43, **55:**351–352
 hydrolysis reaction of, **55:**306

Snake Venoms, Enzymes (Zeller), **8:**459
SO_3^{2-}, **39:**341; 343; 363
$^{35}SO_3^{2-}$, **39:**341; 344
Soaps, insoluble, solubilization of, **17:**138; 142; 149; 155
Sodium, **34:**485. *See also* Na
in affective disorders, **29:**495
distribution in the body, **29:**494
effect of lithium on body, **29:**497
metabolism, **29:**495
residual, in depression, **29:**494–495
space, in manic and depressed patients, **29:**494
deoxycholate, **34:**481
in familial periodic paralysis, **6:**215; 216
in myasthenia gravis, **6:**212
in myotonia, **6:**213
in hypertension, **7:**544–545
and luminescence, **7:**253
ratio to potassium in sperm, **24:**78–79
reduction with, **33:**455; 466; 481; 491–492; 502
Sodium-22, in nuclei, **27:**258–259
Sodium adenylates, in synthesis of valyl adenylate, **19:**251
Sodium amytal, in schizophrenics, **29:**535
Sodium ascorbate, as reducing agent in Hill reaction, **30:**23
Sodium borohydride, **46:**344; 371, **60:**233
Sodium boron hydride, effect on pectic substances, **20:**352
Sodium bromide, dissolving of fibrin, **16:**36; 39
Sodium butyrate, tyrosine hydroxylase induction, **70:**169
Sodium chloride
in cardiac muscle hypertrophy, **6:**202
E. coli resistance to, **6:**97
as inhibitor of β-glucuronidase, **16:**380
effect on gel-sol transformation, **16:**331
and muscle contraction, **13:**189
effect on hyaluronidase activity, **13:**213; 217
Sodium cholate, **60:**96
cholesterol ester hydrolases regulation, **69:**48
Sodium citrate solution, activation of prothrombin in, **16:**44–48; 51
Sodium concentration, and nuclear purity, **27:**247–248
Sodium dodecylsulfate, **47:**74, **49:**231; 232; 233, **58:**106
effect on RNase activity, **24:**196; 197; 238; 249
gel electrophoresis, **49:**232
Sodium dodecylsulfate-polyacrylamide gel, for determination of subunits of PP-ribose-P synthetase, **41:**7
Sodium hydrosulfide, action on thiaminase, **15:**401
Sodium iodide, dissolving of fibrin, **16:**36
Sodium ions, **15:**246; 251; 257, **55:**145
active transport of, **13:**48–49; 51–57
effect on actin, **13:**152–153
permeability studies, **13:**27; 28; 58–63
chelate formation with organic phosphates, **21:**147
effect on potassium ion-activated enzymes, **21:**153; 156
β-galactosidase activation by, **21:**154
in mitochondria, **21:**10
properties, **21:**157
and choline acetylase activity, **12:**316; 318–322
effect on actin polymerization, **16:**335
for fructose-bisphosphatase activity, **54:**129; 132
in fructose-bisphosphatase regulation, **54:**178
imidazole in facilitated diffusion of, **20:**96
Sodium (Na^+) transport, hepatic glutamine transport
hepatocyte hydration, **72:**48–53
sodium dependence and independence, **72:**45–47
Sodium pectate, structure, **20:**343; 344
Sodium periodate, **60:**73
Sodium-potassium ion transporting ATPase, **51:**98–99
Sodium-potassium transport ATPase, **43:**356
Sodium salts, **17:**149
Sodium space, in nucleus, **27:**256–261
Sodium taurocholate, cholesterol ester hydrolases regulation, **69:**48
Sodium thionate, as inhibitor of clot formation, **16:**40
Sodium thiosulfate, effect on succinic dehydrogenase, **20:**155
Soil, activities of enzymes, **28:**371–385
biological activity, enzyme methods for, **28:**365–390
enzyme analysis, comparison of results, **28:**387–388
polysaccharide degradation, **28:**354

Soil-borne plant infections, **11:**431–432
Soil enzymes, detection of, **28:**371–387
Soil processes, effect of solubilization on, **17:**138; 149
Soil sampling, **28:**368–369
Solanaceous alkaloids. *See also* Tropane alkaloids
arginine in synthesis, **8:**236–240
Solanachromene, relation to coenzyme Q, **25:**286
Solanesol, relation to coenzyme Q, **25:**287
Solanine in potato and tomato, **8:**220
Solanum lycopersicum, polygalacturonases of, **20:**366
Solanum melongena, **55:**73
Solanum nigrum in alkaloid studies, **8:**219
Solanum tuberosum, **55:**66
sesquiterpenoid levels in, **55:**81
Solavetivone, **55:**53; 68; 69
rishitin from, **55:**70
Solid enzymes, reactions of, **33:**284–285
Solid phase synthesis, **36:**278
Solid-state assemblages, and enzyme activity, **28:**2
Solid-state phenomena in photosynthesis, **25:**81–86
Solid supports, **34:**452
Solomon-Bloembergen equations, in enhancement theory, **33:**9
Solubility, of pectic substances, **20:**353
Solubility methods, in enzymes isolation, **14:**392–396
Solubility of proteins
and denaturation, **23:**70–72
light-induced changes in, **9:**129–131
Solubility test for purity of enzymes, **14:**389
Solubilization
of cytochromes, **17:**343; 346–348; 351–352; 353–355; 357; 367
of matter and its migration and utilization, **17:**135–158
of succinic dehydrogenase, **18:**66–67; 70–73
Solubilization, Migration, and Utilization of Insoluble Matter in Nature (Mandl and Neuberg), **17:**135
Soluble blood group substances, **36:**132
Soluble components, **34:**495
enzyme, **34:**447
Soluble enzymes, nuclear-cytoplasmic ratio, **27:**262
and nuclear purity, **27:**247–248; 250–252
of nucleus, **27:**261–265
Soluble factors, as coupling agents, **23:**371–372
in oxidative phosphorylation, **23:**377–378
Soluble-particular distribution of mammalian hexokinase, **39:**295–303
Soluble proteins, of nucleus, distribution of types, **27:**265
Soluble space, of nucleus, **27:**246; 261–265
Soluble substrates, action of insoluble enzymes on, **33:**269–283
Solute, hydrophilic, translocation of, **29:**39–40
Solvation, **34:**492
Solvation effects, **43:**290
glutathione S-transferases, **69:**21–22
Solvent effect on denaturation of RNase, **24:**194–197; 241
Solvent effects, **43:**284; 287; 288; 291
Solvent exchange, partial concurrent, **50:**345
Solvent isotope effect, in enzyme catalysis, **24:**461; 472
Solvent isotope effects, *See* Isotope effects
Solvent perturbation, **45:**323
Solvents, **34:**477, **58:**66
mixed (*See* Mixed solvents)
role in enzymic catalysis, **15:**35–36
role in enzyme kinetics, **15:**276; 284–285; 292; 295
Somatic cells, histone in, **16:**446
quantity of DNA in nucleus, **16:**412; 424–429
Somatostatin, **47:**84
Sonication, **34:**473
Sonic disintegration of cells, **25:**398
Sonic disruption, **34:**481
Sonic vibrations, effect on mitochondrial enzymic activities, **21:**31; 57; 58; 80–82
Sophora, **55:**5
Sophorose, **17:**217–218
Sorangium proteinase, specificity of, **33:** 432–433
Sorbitan tristearate (Span 65), **51:**5–6
Sorbitol, **15:**118; 145; 150 (C^{14}), **17:**148; 196, **18:**283
metabolism by sperm, **24:**54; 61
Sorbitol dehydrogenase, **15:**118; 145; 150
in sperm, **24:**54
Sorbitol Dehydrogenase (Jeffery and Jörnvall), **61:**47
Sorbitol-P, **43:**494

Sorbitol-6-P, **43:**502
Sorbitol-6-phosphate dehydrogenase,
PN-dependent, **26:**118
Sorbitol-T, as acceptor in dextran formation,
30:199
Sorbose, **15:**118; 145; 147; 150; 312, **18:**287
Sorbose-1-phosphate, **15:**154
Soret band, origin, **17:**338
Sorghum, antibiotic treatment of diseases,
20:394
Sound, **57:**233
Sound absorption, relaxation methods, **25:**2–9
Southern bean mosaic virus, activity and particle
identity, **9:**191; 194–195
crystallization, **9:**199–201
growth curve, **9:**179–180
homogeneity, **9:**220
hydration, **9:**204–206; 209
physical constants, **9:**187; 188
purification, **9:**185; 186
shape, **9:**203–206
size, **9:**204–206
Southern blotting, tyrosine hydroxylase, **70:**179
Soyasapogenol A, structure, **22:**328
synthesis from acetate, mevalonate, **22:**329
Soybean agglutinin, **48:**57; 72; 84
Soybean callus, ALA synthetase in, **46:**41
Soybean leaves, inorganic nitrogen metabolism,
17:250–252; 268
Soybean phosphatides, **34:**473
Soybean root nodules, **46:**88
Soybean trypsin inhibitor, **23:**76, **41:**182; 204
effect on thrombin formation, **16:**47–48; 54
simulated mutation of, **33:**399
Space group, **39:**349
Spatial distribution of functional groups of
proteins, **24:**109
SPCA, *See* Factor VII
Species differentiation of DNAs, **16:**432; 435
Species specificity, **42:**75
in fertilization, **9:**267; 268–275
of fibrinogen, **10:**31
lack of, for succinate oxidase fractions,
28:179–180
and N-terminal residues of fibrinogen and
fibrin, **16:**37
of RNAs, **16:**438–440
and stability of serum Ac-globulin, **16:**62
and thrombin requirement for clotting,
16:84
and thromboplastin, **16:**65–66
of thromboplastin action, **10:**6
Species-specific proteins, role of cell nucleus in
synthesis, **8:**56
Specific acid catalysis, **37:**14–21
Specific activity
definition, **13:**25
of enzymes, **14:**384–387; 388; 404
of enzymes, and nuclear purity, **27:**247–250
of papain and mercuripapain, **19:**270–272;
287–289; 292; 305–307
relationship to sulfhydryl-protein ratio,
19:327; 328
Specific inhibition, of constitutive enzyme
formation, **22:**121; 122
Specificity, **37:**602; 603, **48:**54; 61; 63; 71; 79,
53:259; 263
of action of hydrocarbons on microorganisms,
10:452–453
of alkaline phosphatase, **10:**86–89;
104–105
of microbial attack on hydrocarbons,
10:454–456
of sucrose phosphorylase, **10:**129–131
of adsorbent, **14:**337–338; 341–344; 364; 367
of enzyme action, classification, **14:**
234–239
of tyrosinase for hydroxyls, **14:**84
and affinity of enzyme for substrate, **12:**
406–407
of amylases, **12:**406–407
of an enzyme, **37:**602
of antibodies, **37:**604
antigen-antibody, **11:**66
of alliinase, **11:**393–397
of lycomarasmine, **11:**422–426
of pectinesterase, **11:**271–272; 279–280
of plant toxins, **11:**422–426
of polygalacturonase, **11:**273
binding, **37:**592; 602; 603; 604
blood-type, **48:**56
of catalases and peroxidases, **12:**174–175
and catalysis, **43:**305
catalytic theory, **37:**605
chemical, **51:**78; 91
of choline acetylase, **12:**325–327
of cholinesterases, **12:**264–270
classic theory, **37:**605
of dehydrogenases, **7:**413–414
of renin, **7:**517; 520; 526–528

of dehydropeptidases, **8:**118–119; 134–143
dependence on anomeric configuration of glycosyl bond, **48:**38
of enzymes, **18:**223–225, **48:**31; 32; 39; 40; 41
of tissues, DNA studies, **18:**174–176
evolution of, **37:**607; 608
of free radical action in oxidoreductions, **15:**12–13
substrate, for thiaminase, **15:**405
functional, **37:**482; 603; 604; 605; 606; 608
of β-glucuronidase, **16:**371–373; 380; 383
of glycosidases, **48:**46
of induction of enzyme synthesis, **13:**100–105
structure, in enzymic adaptation, **13:** 108–115
of transferases, **13:**294
of maltase and neuraminidase, **20:**142
of pectinesterases, **20:**358–360
of polygalacturonases, **20:**361–365
of steroid enzymes, **20:**271–273
mechanism, **9:**60
of oxidases, **19:**89; 98; 131; 181; 183; 184; 190; 191; 194
of papain, **19:**302–312
of peptidases, **12:**191–257
primary, **53:**259; 260
prosthetic group, **1:**177; 178
proteinases, **1:**63–98
receptor, **37:**603; 604; 605; 608
secondary, **37:**567, **53:**262; 267
serological, **37:**608
template theory, **37:**400
Specificity factor, **67:**56; 58
Specificity function, of amino acids in enzymes, **22:**69; 70; 90
Specificity of enzyme systems, and genes, **8:**205
cis-trans, **8:**363–364; 444
for double-bond position, **8:**366; 373–374; 426; 443–444
for hydrogen isotopes, **8:**364–365
for isomeric peptides, **8:**150–152
Specificity of L-ophioamino acid oxidase, **8:**477–479
Specificity patterns, of β-lactamase, **28:**255–257
Spectra, *See* Optical absorption spectra
difference, **17:**73–75
of cytochromes, **17:**338–345; 347; 354–363; 375
of enzyme-substrate complexes, **17:**42–43; 55–56
of metalloflavoproteins, **17:**236–238; 241–245; 254–255; 261–265; 276–277; 301–307; 315
or respiratory components, **17:**70–77; 93–97
of xanthine oxidase, **17:**301–307; 315
ultraviolet, of corrinoids, **26:**247–249
Spectral changes, in pyridine nucleotides on binding, **22:**341–343
Spectral difference, ultraviolet, **39:**346
Spectral shifts, in RNase denaturation, **24:** 195–197
of PIR, **24:**228; 232
Spectral studies, of phytochrome, **26:**3–9; 15–19
Spectrometry. *See also* Absorption spectra
relaxation, mechanisms of enzyme reactions, **25:**1–38
Spectrometry, Relaxation, Study of Elementary Steps in Enzyme Reactions by (Eigen and Hammes), **25:**1
Spectrophorometric assay, coupled, tyrosine hydroxylase, **70:**110
Spectrophotometric determinations. *See also* Absorption spectra
of hemoproteins, **7:**269; 274–275; 280–282; 288–291
Spectrophotometric investigations and cytochrome c, **7:**280–282
Spectrophotometric method, to determine esterase activity, **41:**199
Spectrophotometric method for RNase assay, **24:**172–173
Spectrophotometric studies, of cells, **16:** 419–422; 426–429
Spectrophotometric titration, **42:**93; 96; 129; 144; 145
Spectrophotometry
comparison with fluorimetry, **20:**33; 34
disulfide exchanges, **27:**208
flash, photosynthetic electron transport by, **30:**24
steady-state difference, photosynthetic electron transport by, **30:**24
Spectropolarimetry, **27:**39–41; 49. *See also* Cotton effect
Spectroscopic studies. *See also* Absorption spectra
of electron transfers, **26:**134–140; 143; 156–157
of luminescent oxidation, **7:**219–221

Spectroscopy
  atomic, **56:**286–287
  bindings in ternary complex, **27:**4–16
    EPR, oxidative enzymes, **27:**105–198
  and thermodynamic functions, **11:**13
Spectroscopy, EPR, Contributions to Our Knowledge of Oxidative Enzymes (Beinert and Palmer), **27:**105
Spectrum, action, for cytochrome oxidation in *Porphyridium,* **30:**61
  for the Emerson enhancement effect in *Chlorella,* **30:**61
  for ferricyanide reduction in a normal Hill reaction, **30:**63
  for photosystems 1 and 2, **30:**62–63
S-Peptide, and S-protein, **24:**230–232; 249
S-peptide from RNase, **24:**449–450; 454; 457
  reactivity, **24:**205; 206; 209; 210; 216–217; 219; 222; 224–225
Sperimine, **55:**104
Sperm
  adenosine triphosphate in, **9:**340–342
    characteristics, **9:**334–338
    density, and fertilization, **9:**350–351
      and metabolic studies, **9:**336–337; 359–360; 367–370
      and semen quality, **9:**330–331
    dilution factor, **9:**334; 368–370
    effect of arsenicals on, **9:**345
      factors affecting, **9:**346; 355; 367–372; 375–382
    effect of atmospheric pressure, **9:**334
      of bacterial contamination, **9:**334; 374
      of egg water on, **9:**252; 255–268; 271; 314; 369
      of *p*H changes, **9:**333–334
      of irradiation, **9:**334
    electrical changes in, **9:**330
    fertility and semen quality, **9:**330–334; 344; 378–382
    fertilizing power, **9:**252–255
    incorporation of, **9:**268–275
    motility, **9:**255; 257–258
      and adenosine triphosphate, **9:**341
      and semen quality, **9:**330–332
      and seminal plasma, **9:**342
    nucleic acid, **9:**338–340
    nutritive requirements, **9:**332–334
    permeability, **9:**333
    ripening process, **9:**336; 368
    separation of parts, **9:**598
    survival, **9:**378–382
    temperature shock, **9:**333
    viability, **9:**330–334; 368–369
  cytochromes in, **17:**335
  imidazole in motility of, **20:**103
  regulation of respiration by —SH groups, **11:**252
Spermatogenesis, protein change during, **18:**185
Spermatozoa
  adenosine receptors, **69:**108
  aerobic glycolysis, species variations, **24:** 52–55
    ATPase in, **24:**329
    dilution effect, **24:**72–74
    endogenous substract, **24:**57–61
    enzymes in, **24:**86–88; 300; 306
    epididymal, collection of, **24:**44–45
    and fertility, **24:**36–38
      and respiration, **24:**43; 46; 52–68
    fertility, and motility, **24:**36–38
    mammalian, types, **24:**43–45
    metabolism of, **24:**35–104
    motility, **24:**39–40; 46–51
      effect of pH, **24:**75–76
      effect to temperature, **24:**74
      energy supplies for, **24:**69
      in female tract, **24:**41
      in male reproductive tract, **24:**39–41
    oxidizable substrates, **24:**61–66
    respiration and motility, **24:**43; 46; 52–68
    syntheses by, **24:**88–89
Spermatozoa, Metabolism of (Salisbury and Lodge), **24:**35
Sperm cells, changes on fertilization, **16:** 479–480
  DNA content of nucleus, **16:**412; 424–425; 427; 430
  nuclear proteins of, **16:**441–444; 446; 450
  RNA content of nucleus, **16:**437
Sperm cells, DNA in, **18:**162; 167; 177–178; 180
Spermicidal agents, **24:**69–72
Spermidine, **36:**203–251, **54:**216, **55:**104, **60:**10
  acetylation, **36:**222–223; 225
  as activator of β-glucuronidase, **16:**376
  analytical procedures, **36:**206
    automated ion-exchange methods, **36:**206

chromatography, **36:**206
dansyl chloride, **36:**206
fluorescent, **36:**206
gas chromatography, **36:**206
ion-exchange chromatography, **36:**206
paper, **36:**206
serum amine oxidase, **36:**206
spermidine dehydrogenase, **36:**206
thin-layer electrophoresis, **36:**206
assay of *O*-phthaladehyde, **36:**206
binding to nucleic acids, **36:**242–243
biosynthesis, **22:**261; 262, **36:**214; 215–218; 220–221
in animals, **36:**231–232
propylamine transferase, **36:**232
S-adenosylmethionine decarboxylase, **36:**231
control mechanisms, **36:**215; 217–218
in *E. coli,* **36:**208
formula, **22:**261
propylamine transferase, **36:**215; 221
S-adenosylmethionine decarboxylase, **36:**214; 220
cell division and inhibition of microcyst formation, **36:**240–241
concentration, **36:**206–208; 227–228
in animals, **36:**227
bacteriophages, **36:**208
becteria, **36:**206
fungi, **36:**206
in gram-negative bacteria, **36:**206
in gram-positive bacteria, **36:**207
in plants, **36:**228
virus, **36:**208
conversion to toxic product by amine oxidase, **36:**236
dehydrogenase- *serratia marcescens,* **36:**225
distribution, **36:**206–208
effect of, on bacteriophages T4rII, **36:**237
on DNA-RNA enzymes, **36:**243; 245
on nucleic acid degradation, **36:**245–246
on nucleic acid synthesis, **36:**245–246
on protein synthesis, **36:**244
on ribosome structure, **36:**244
on stringent *E. coli,* **36:**245–246
effect on β-glucuronidase, **40:**40
formation of, **54:**214; 220
glutathionylspermidine, **36:**221–222
growth effects, **36:**238
growth requirements of mutant strains of *E. coli,* **36:**218
inhibitory effect on bacteriophages, **36:**237
localization, **36:**207
metabolism, **36:**224–226
amine oxidase (beef serum), **36:**235
amine oxidases, **36:**232–236
in animals and plants, **36:**232–237
diamine oxidase, **36:**235
in *E. coli,* **36:**221–223
in vivo in animals, **36:**237
mutants lacking, **36:**218–219
product of propyl-amine transferase, **36:**215
protection of bacteriophage T5, **36:**237
redistribution, **36:**207
S-adenosylmethionine, **36:**214
decarboxylation, **36:**214
synthesis, **36:**214
stabilization of nucleic acids, **36:**242–243
substrate of propyl-amine transferase, **36:**215
toxicity of, **36:**236
uptake by *E. coli,* **36:**226
Spermidine and spermine, **56:**265; 269; 274; 275
Spermidine dehydrogen, use of, in assay of spermidine, **36:**206
Spermidine dehydrogenase, **36:**225–226
Spermidine synthase, **54:**219–220
Spermine, **5:**83, **36:**203–251, **54:**216
acetylation, **36:**222–223
as activator of β-glucuronidase, **16:**376
analytical procedures, **36:**206
automated ion-exchange methods, **36:**206
chromatography, **36:**206
dansyl chloride, **36:**206
fluorescent, **36:**206
gas chromatography, **36:**206
ion-exchange chromatography, **36:**206
paper, **36:**206
serum amine oxidase, **36:**206
spermidine dehydrogenase, **36:**206
thin-layer electrophoresis, **36:**206
binding to, nuclecic acids, **36:**242–243
biosynthesis, **22:**261; 262
formula, **22:**262
biosynthesis of, in animals, **36:**231–232
propylamine transferase, **36:**232
S-adenosylmethionine decarboxylase, **36:**231
concentration, **36:**206–208; 227–228
in animals, **36:**227

Spermine (*Continued*)
bacteria, **36:**206
bacteriophages, **36:**208
fungi, **36:**206
in gram-negative bacteria, **36:**206
in gram-positive bacteria, **36:**207
in plants, **36:**228
virus, **36:**208
conversion to toxic product by amine oxidase, **36:**236
distribution, **36:**206–208
effect of, on bacteriophage T4rII, **36:**237
on DNA-RNA enzymes, **36:**243; 245
on protein synthesis, **36:**244
on ribosome structure, **36:**244
effects on aminoacyl-tRNA synthetases, **40:**176; 177
effects on β-glucuronidase, **40:**49
formation of, **54:**214
growth effect, **36:**238
growth requirements for, **36:**218
growth requirements of mutant strains of *E. coli,* **36:**218
inhibitory effect on bacteriophages, **36:**237
localization, **36:**207
metabolism of, in animals and plants, **36:** 232–237
amine oxidase (beef serum), **36:**235
amine oxidases, **36:**232–236
in vivo in animals, **36:**237
mutants lacking, **36:**218–219
phospholipids, **36:**241
product of propylamine transferase, **36:**215
protection of bacteriophage T5, **36:**237
redistribution, **36:**207
stabilization, of halophilic organism, **36:**240
of membrane-ATPase complex, **36:**241
of nucleic acids, **36:**242–243
by polyamines, **36:**240; 241
of spherophasts, **36:**241
toxicity, **36:**236; 238
antimutagenic effect, **36:**238
bacteria, **36:**238
cell culture, **36:**238
kidney, **36:**238
uptake by *E. coli.,* **36:**226
Spermine oxidase, use of in assay of spermidine and spermine, **36:**206
Spermine synthase, **54:**220–221
Sperm number, relation to hyaluronidase activity, **13:**227
Spermosin and ATPase, **6:**185
Sphacelotheca sorghi, plant disease caused by, **20:**394
Spheroidenone, biosynthesis, **21:**347; 348
structure, **21:**348
Spheroplast, definition, **26:**227
Spheroplasts, **36:**241
Spherulites, **34:**483
Sphigomyelin, **67:**103
Sphingomyelin, **34:**466, **38:**295; 298; 299. *See also* Phosphatides
rate of renewal, **7:**138
synthesis, in mitochondria, **21:**51
Sphingomyelinase, **38:**299; 308
Sphingosine, **60:**112
effect on blood coagulation, **16:**88
Spicaria species, enzymic transformations of steroids by, **20:**248
Spike potentials from nerves, **23:**83–84; 124–125
Spikes, in periodic reactions, **23:**112–117
Spinach
absorbancy changes of chlorophyll *b* in chloroplasts of, **30:**31–32
analyses for Mn in chloroplast of, **30:**25
difference spectrum for chloroplasts of, **30:**32
effect of manganese deficiency in, **30:** 24–25
fast light-induced reactions of cytochrome *f* in chloroplasts of, **30:**37
light minus dark difference spectra of chloroplasts of, **30:**36
treatment of chloroplasts of, with Triton X-100, **30:**54
carbonic anhydrase, **47:**151; 215
sulfide reductase of, **46:**63
Spinach chloroplast, **54:**175–177
Spinach leaf, molecular weight of homogeneous enzymes from, **46:**367
Spinacia oleracea, **54:**176, **55:**19
chloroplast structure, **25:**67–69
Spinacia oleracea, PN-dependent enzyme, **26:**119
Spin equilibrium, **47:**24; 28
Spin label, bovine serum albumin, **34:**468
fatty acids, **34:**468
mitochondria, **34:**468
studies, **34:**468

Spin-lattice relaxation time, *See* Longitudinal relaxation time
Spin state of cytochrome *c* peroxidase, in electron paramagnetic resonance spectra, **33:**322; 325
  high-and-low spin isomers, **33:**320; 322
  high-spin, **33:**320; 322
  low-spin, **33:**320; 322
Spinulosin, **7:**482; 486; 489; 490; 492
Spirality of protoplasm, **7:**44
Spirea aruncus, **7:**482; 487; 493
Spirilloxanthin, **1:**310; 311
  biosynthesis, **21:**344; 345; 348; 349
    effect on photodynamic sensitization, **21:**357
    structure, **21:**297
Spirillum itersonii, **46:**59; 62; 88
Spirillum sp. cell walls, **26:**202
Spirochaeta cytophaya action on cellulose, **8:**267
Spirodienone intermediates, **55:**32; 33
Spirographis haem, **23:**269
Spirographis heme, *See* Heme s
Spleen, **47:**368
Splenectomy, **41:**147–150
  post splenectomy γ-globulin, **41:**148
  presplenectomy γ-globulin, **41:**148
Splenectromized humans, lack of tuftsin activity, **41:**174
Splitting enzyme, in arginine synthesis, **16:**265, **17:**399
Splitting enzymes, classification, **14:**221
Splitting reaction, **15:**344–345; 347; 352–353; 356; 358–363; 373–374
Spongin, **23:**204–205
Spongonucleosides, and DR metabolism, **28:**436
SpooOA transcription, *Caulobacter* cell cycle, transcription regulation, **71:**17
Spore outgrowth, **44:**175; 176
Spores, mechanism of release, **26:**213
Sporulation, **44:**165; 169; 170
  genetic link to antibiotic production, **35:**3
Spreading factor of snake venoms, **8:**471–476; 487
Spreading reaction of hyaluronidases, **13:** 224–226
Spread monolayers, and adsorbed films, **28:** 13–16
S-Protein, from RNase, **24:**196–197; 247; 249; 449–450; 457
  reactivity, **24:**202; 204; 216–217; 219; 221; 222; 224–225; 227; 231–232; 250
Squalene
  biosynthesis, general mechanism of, **22:** 309–320
    inhibition by iodoacetamide, **22:**319
    from mevalonate, **22:**295–309
  in carotenogenesis, **21:**344; 346
  in cholesterol biosynthesis, **22:**285–288
  conversion to lanosterol, **22:**322
    location, **24:**308; 339
  cyclization, **22:**320–324
    mechanism, **22:**323; 324
  in eburicoic acid synthesis, **22:**329
  labeled efficiency of conversion to cholesterol, **22:**287
    formation from labeled mevalonate, **22:**290; 318
    synthesis from labeled acetate, **22:**287
  oxidation, **19:**192–194
trans-Squalene, conversion to lanosterol, **22:**287
Squalene oxidocyclase, **22:**322
Squalene oxidocyclase I, **19:**192–194; 199
Squalene synthetase system, in squalene biosynthesis, **22:**297
Squash crystal mosaic virus, physical constants, **9:**187; 188
SRB proteins, RNA polymerase II transcription, potential roles, **71:**60–61
sRNA, and amino acid analogs, **29:**150
  effect on histidine repression, **29:**152
sRNA synthetase, phenylalanine, substrate specificities, **29:**113
SSO_3^{2-}, **39:**339; 341; 355
  oxidative metabolism of, **39:**363
$^{35}SSO^{2-}{}_3$, **39:**352
$^{35}SSO^2{}_{3-}$, **39:**361
Stability, **42:**83; 100; 108; 110; 130; 131
  of papain, **19:**279–280
Stability of enzymes in isolation procedures, **14:**390–391
Stabilizing factor, **48:**234
Stable conversion factor, *See* Factor VII
Stable factor, serum, activation of prothrombin by, **16:**63–64
Stachydrine in plants, **8:**234
Stachyose, **17:**197, **56:**159
Staining techniques, **17:**137
  for enzymic histo-and cytochemistry, **9:** 586–593

Stalk formation, *Caulobacter* cell cycle, **71:**5–7
Standard state, **43:**283; 319; 330
Stanleya, **48:**5
Staphylocoagulase, **46:**10
action on fibrinogen, **10:**33; 35
Staphylococcal enterotoxin B, **47:**85
Staphylococcal nuclease, **36:**40; 56–58; 75; 77, **47:**59; 62; 124, **53:**269, **55:**353–356
and catalysis by approximation, **59:**286
and criteria of mechanism, **59:**284
and D21E, **59:**286
and D40E, **59:**286
and D21Y, **59:**286
and E43D, **59:**285
and E43N, **59:**285
and E43Q, **59:**285
and general base catalysis, **59:**285
and metal catalysis by CA^{2+}, **59:**286
overall rate acceleration by, **59:**287
and protein breathing, **59:**286
and reaction coordinate distance, **59:**284
and reaction mechanism, **59:**283–284
and R35G, **59:**287
and R87G, **59:**287
and role of Ca^{2+}, **59:**284
secondary interactions of, **33:**436
and site-specific mutants, **59:**284–285
and symmetry and rigidity of ligands of Mn^{2+}, **59:**285
and transition state stabilization, **59:**287
and X-ray structure, **59:**283
Staphylococcal proteinase, **41:**179; 195–196; 237
Staphylococci, **60:**284
uridine nucleotide accumulation in, **20:**123
Staphylococci, protein synthesis in, **24:**23–24
Staphylococci, hyaluronidases in, **13:**221
Staphylococci producing arylamines, **6:**103–107
Staphylococcus, **58:**118
Staphylococcus, synthesis of cell walls in, **22:**128
Staphylococcus albus
cytochromes of, **23:**266–267; 270; 287; 315
β-lactamase of, **28:**242; 264; 292; 296
quinones in, **25:**289
Staphylococcus aureus, **46:**86; 102
acceptor specificity, **40:**132; 133; 134
aminoacyl-tRNA specificity, **40:**130
biosynthesis of interpeptide bridges, **40:**121; 122; 123; 124
accumulation of sugar nucleotides, **26:**36
cell walls, **26:**198; 210–211
mureinases, **26:**212
murein sacculi, growth, **26:**216–218
sugar nucleotide-forming enzymes, **26:**50; 73–76
allosteric enzymes, **28:**79
β-lactamase of, **28:**242; 252; 260; 262–264; 269; 270; 277; 280–289; 292; 294; 296; 303–305; 311; 312
bacteriostatic effect of gramicidin, **4:**189
exchange-incorporation reaction in, **22:**128
glycine metabolism in, **23:**453; 456
glycosyl transfer to teichoic acid in, **30:**240
Michaelis constant for CDP-ribitol in, **30:**240
transferase enzyme pattern in, **30:**242
wall ribitol teichoic acid of, **30:**231
heme biosynthesis in, **46:**86
hexose isomerization in, **14:**203–204
metabolism with and without gramicidin, **4:**186; 187
nature of gramicidin poisoning, **4:**191
nature of tyrocidine poisoning, **4:**193; 194
oxygen and phosphate uptake, **4:**188; 189
Staphylococcus aureus, 378, **15:**416
Staphylococcus aureus, enzyme synthesis in, **24:**12
Staphylococcus aureus, folate metabolism, **27:**435; 445–447; 451–452; 457
Staphylococcus aureus, glutamine synthesis in extracts of, **13:**278–279
Staphylococcus aureus, phosphate catalyzed against arsenate, **29:**49
Staphylococcus aureus V8 proteinase, **53:**261
Staphylococcus candidus, metabolism in, **74:**34–36
Staphylococcus epidermis, aminoscyl-tRNA specificity, **40:**131
biosynthesis of interpeptide bridges, **40:**125; 126; 127
cell function, **40:**135
Staphylococcus hemolyticus, **15:**378
Staphylococcus lactis, CDP-glycerol pyrophosphorylase in the membrane of, **30:**236
elucidation of the linkage between peptidoglycan and teichoic acid in, **30:**249
isolation of teichoic acid from, **30:**238

membrane teichoic acid of, **30:**225
teichoic acid of, **30:**248
wall teichoic acid of, **30:**229
Staphylococcus, mutant of, **46:**60
Staphylococcus spp., β-lactamase of, **28:**242; 258; 263; 267; 271; 273; 274; 278; 279; 301; 312–313
Starch. *See also* Amylopectin, Amylose, Polysaccharides; Amylose, Amylopectin, and Polysaccharides
biological transformations in, **11:**339–375
biosynthesis, ADPG as glucosyl donor for, **34:**314
synthesis, phosphorylase in, **34:**387–388
biosynthesis in algae, **46:**318
biosynthesis in plants, **46:**318; 340
synthesis in *Chlorella pyrenoidosa,* **46:**340
in coacervate, **27:**368–369
for column chromatography of α-amylase, **22:**447
effect on dextran synthesis, **30:**191
enzyme action on, **33:**260
hydrolysis of, **33:**281
insoluble substrates of, **33:**260
formation of crystalline dextrins from, **11:**322
glycogen compared, **3:**119; 122–126
hydrolysis by thiaminase-producing bacteria, **15:**413
from labeled acetate, **23:**404–405
mechanism of synthesis, **11:**353–358
oxidized, hydrolysis by polygalacturonases, **20:**362
phosphorolysis in plants, **20:**212; 213
in plant respiration, **18:**286–291
primers for, **26:**58
sugar nucleotides in synthesis, **26:**61–62; 66
synthetic, properties of, **11:**352–353
in transglycosylations, **17:**191; 214; 229
waxy maize, **17:**215
Starch
chemistry of, **12:**341–377
enzymes of degradation and synthesis, **12:**379–428
hydrolysis, **12:**381–410
Starch, Biological Transformations (Peat), **11:**339
Starch *(α)* (amylose), synthesis of, **18:**222
Starch Chemistry (Meyer and Gibbons), **12:**341
Starch Degradation and Synthesis, Enzymes (Bernfeld), **12:**379
Starch formation, inhibitor of, **18:**288
Starch gel, **34:**448
adhesive properties, **12:**369; 370
Starch gel electrophoresis, adenosine deaminase deficiency, **51:**176–177
Starch-glycogen class of polysaccharides. *See also* Amylopectin-type polysaccharides and Amylopolysaccharides, branched, Amylopectin, Amylose, Polysaccharides, and Starch
formation of, **11:**299–306; 315–324; 330
Starch hydrolases, **18:**290
Starch phosphorylase, in plants, **18:**286; 287; 289
Starch synthetase, **36:**71
Starch type polysaccharides, enzymic synthesis, **15:**303–309
Starvation
effect on pentose synthesis, **28:**415; 432
phosphorylase *a* increase in, **11:**304
Starvation phase, metabolism of mold in, **9:**552–554; 559
Stationary state, **34:**496; 502–504; 511–514
Statistical Analysis of Enzyme Kinetic Data (Cleland), **29:**1
Statistical mass action relationships, **34:**446
Statistical mechanics, and entropy, **11:**5–7; 14–15; 16–19; 24–25; 30–31
Stawberry-yellow virus, **20:**402
Steady state, in nitrogen metabolism of plants, **9:**423–428
kinetics, biological, **9:**49–56
Steady-state behavior, in oxidative phosphorylations, **17:**88–112
Steady-state conditions, **52:**190
Steady-state end-to-end distance measurements, oligosaccharides, **66:**66–67
Steady state in reaction chains, **7:**6–9
Steady-state kinetics, *See* Kinetics
cytochrome *c* oxidase, electron transfer, **71:**116–118
glutathione synthesis, catalytic mechanism, **73:**241–243
hypothesis, enzyme action, **73:**26–29
Steady-state values of respiratory components, **17:**120–122
Stearate, calcium, solubilization of, **17:**142; 149
Stearic acid, **58:**51
Stearic anilide, photochemistry, **9:**93; 118
Stearic dehydrogenase, activity of —SH group, **11:**241

Stearyl coenzyme A, **15:**233
Stellacyanin, absorption bands of, **33:**195
composition of copper in, **33:**196
extinction coefficient of, **33:**198
forms of copper in, **33:**195
ORD curves of, **33:**201
oxidation-reduction potential of, **33:**207
Stem, **35:**468
Stemphylium botryosum, **55:**46
Step reactions, kinetics, **9:**1–23; 49–53
Steptomyces lavendulae, enzymic transformations of steroids by, **20:**253; 254; 261–263; 265; 270
Stereochemical configuration, effect on flexibility of hydrocarbon chain at interfaces, **11:**63; 64–66
Stereochemical disturbances, **42:**246
Stereochemical mapping, glutamine synthetase, ligand interactions, **72:**20–22
Stereochemical specificity, **15:**15; 22; 27; 40–42
Stereochemistry, **43:**500; 501
of hydrogen transfers with PN-dehydrogenases, **26:**156–171
of methylaspartate mutase reaction, **26:**364–365
of tryptophan synthase, **49:**154–157
Stereochemistry of, decarboxylation, **35:**84–92
reaction at the amino acid β carbon, **35:**97
Stereo-configuration of the amino acids of proteins, **1:**2
Stereoisomeric configuration of carotenes, **17:**435
Stereoisomers, of *S*-adenosylmethionine, **22:**246
Stereopopulation control, **43:**269
Stereoselectivity, xenobiotic substrate specificity and mechanism, **69:**32
Stereospecificity
of DR-aldolase, **28:**450–451
of enzyme action, **27:**38–41; 48–49; 55–56
in enzyme induction, **22:**117; 119; 120; 146
in HMG reduction, **22:**295
Stereospecific oxidation, of branched hydrocarbons, **27:**495
Stereospecificty of reactions, pyruvate phosphate dikinase, **45:**106
Steric effects
of inhibitor action, **15:**285; 288; 290; 297
in penicillins, in stability to β-lactamase, **28:**258–262
of substrate analog, on conformational change, **28:**304
Steric factor
and catalytic activity, **15:**25
in enzyme activity, **15:**290–292; 294
in digestion of esters, **11:**76–77
Steric factor in enzyme reactions, **9:**30; 32; 33; 43
Steric hindrance, **60:**97
Steric probes, **43:**46
Steric structure of enzymes, **26:**89–114
Sterility syndrome, and vitamin E, **6:**482–483
Steriod metabolism, pyridine nucleotide function in, **15:**127
Steroid, thymus involution, **47:**361
Steroid alcohol sulfokinases, **22:**212; 213
in formation of DHAS, **22:**212; 213
in ranol sulfate synthesis, **22:**213
specificity, **22:**213
Steroidal diamines, **36:**228; 238
Steroid assay, use of β-glucuronidase, **16:**363; 377; 383–385
Steroid biosynthesis, **47:**5
Steroid enzymes, **20:**264–276
Steroid epimerzation, **44:**97
Steroid glucuronides, hydrolysis, **16:**377; 382–385; 402
Steroid β-glucuronides, hydrolysis, **16:**363; 378
Steroid hormones, *See* Hormones
Steroid Δ^4-5α-H reductase activity, **46:**162
Steroid hydroxylase, **19:**189–192; 199
Steroid isomerase, **43:**514, **46:**384
Steroid metabolism. *See also* Androgens and Estrogens
relation to β-glucuronidase, **16:**363; 386–394; 398–402
Steroidogenesis, **50:**315
by ACTH, **29:**459
Steroidogenic potency, effect of modification of amino terminal region of ACTH on activity, **29:**464
of various chain length ACTH peptides, **29:**460–461
Steroids, **46:**116; 135; 139; 156; 159; 388, **57:**376
in allosteric regulation, **28:**140
benzophenone attachment to, **58:**49
5B-H steroids, **46:**157; 158; 159
chlorination of, **58:**53
double bond insertion, **58:**49

enzymic transformation by microorganism, **20:**237–282
estimation in urine, **20:**276
resolution of racemic mixtures, **20:**271; 272
fluorinated, **55:**263–264
hydroxylation, **19:**187–194
metabolism, **19:**188
in mitochondria, **21:**27
hydroxylations, **27:**538
unsaturated, epoxide formation, **27:**499
metabolism of, **24:**338–339
11-β-hydroxylation of, location, **24:**308
methyl groups, **58:**47
Steroids, Enzymic and Transformations by Microorganisms (Vischer and Wettstein), **20:**237
Steroids, PN-dependent dehydrogenation, **26:**127
Steroid sulfatases, **22:**227; 228
assay with DHS, **22:**227
and cortisone-21-sulfate, **22:**228
of mollusks, **22:**227
of ox liver, **22:**227
properties, **22:**227
specificity, **22:**227; 228
Steroid sulfates, **46:**393
occurrence, **22:**206
Sterol, in chloroplast lamellae, **30:**9
Sterols
and control of flowering, **26:**28–29
oxidation at an interface, **11:**74–76
synthesis, **8:**331–333
Sterospecificity, of hydrogen transfer from reduced DPN (deuterium transfer), **15:**88
Stilbene, **7:**488
Stilbene phytoalexins, **55:**9; 84
from shikimate-polymalonate pathway, **55:**91–92
Stilbene synthase, **55:**92
activity of, **55:**94
isolation of, **55:**94
reaction sequence of, **55:**93
Stilbestrol
effect on β-glucuronidase, **16:**388; 389; 394
growth rate *vs.* inhibition of glycolysis by, **29:**334–337
inhibition of glycolysis by, **29:**335
Stipatatic acid, tropolone derivative in plants, **16:**137
Stirring, **34:**525
ideal, **34:**504
Stock food supplements, seawed meal as, **28:**354–355
Stoichiomertric inhibitor of Cp, **44:**213
Stoichiometric product inhibition, **25:**177
Stoichiometry, in evaluation of mitochondrial purity, **21:**76
Stoichiometry of proteins and x-rays, **3:**63–108. *See also* Proteins
Stoichiometry of Proteins (Astbury), **3:**63
Stomodeum, **45:**391
Stomoxys calcitrans, **59:**81–82
Stopped-flow, **43:**11
Straight-chain fatty acids, as substrate-substitute, **34:**52
Strain, **43:**239, **59:**148
Strained molecule theory of enzyme action, **15:**290–292
Strain energy, of 1,2-dithiolane, **18:**329–330
Stratification of cell constituents *in vivo,* **13:**4–5
Strawberries, antibiotic treatment of diseases, **20:**402
Strepogenin, **35:**201
inhibition of, **11:**408
Streptamine, from streptomycin, **10:**149–150
synthesis, **10:**158–160
Streptidine, **10:**147–149; 149–150; 151; 160
linkage to streptobiosamine in streptomycin, **10:**155–156
synthesis, **10:**158–160
Streptobacterium plantarum, folate use, **27:**427
Streptobiosamine, **10:**148–149; 151; 154–155; 160; 161–162
linkage to streptidine in streptomycin, **10:**155–156
Streptococcal glycolysis, **7:**410; 445–448; 450–453
Streptococcal proteinase, **41:**179; 181; 199–200; 237, **53:**261; 279
specificity of, **33:**433
Streptococci
group A, hyaluronic acid synthesis, **26:**59
polysaccharide synthesis, **26:**66
propionic acid metabolism, **26:**293
protein in cell walls, **26:**223–224
isolation of protease from, **14:**405
Streptococci, and amino acid decarboxylases, **6:**6; 12
purification of tyrosine in, **6:**13
pyridoxine-deficient, tyrosine decarboxylase activity of, **6:**25–26

Streptococci, group A, hyaluronic acid synthesis, **25:**436
Streptococci, hyaluronidases in, **13:**219–221; 223; 226
Streptococcus allantoicus, **23:**338
Streptococcus bovis, superoxide dismutase from, **41:**79
Streptococcus cremoris, **18:**342
Streptococcus faecalis
alanine racemization by, **14:**195
culture of, **14:**49; 66
biotinidase from, **26:**325
effect of lipoic acid on, **18:**319; 320; 332–335; 338; 339
pentose metabolism, **28:**410–411; 413
teichoic acids in, **30:**246
thymidylate synthetase, **25:**398
Streptococcus faecalis, D-alanine requirement of, **16:**250
transaminations in, **16:**193; 211–213; 215; 219; 221
Streptococcus faecalis, action on arginine, **15:**376–379
citric acid cycle, **15:**201–203; 208–209; 214–216; 219; 223; 246
Streptococcus faecalis, enzymic adaptation in, **13:**71
Streptococcus faecalis, folate metabolism, **27:**427–429; 435; 438–441; 454–455; 457; 460
Streptococcus faecalis, growth factor for, **21:**374; 381; 391
Streptococcus faecalis R, carbamyl phosphate synthesis in, **21:**268
Streptococcus faecalis, superoxide dismutase from, **41:**79; 82; 83; 84
Streptococcus fecalis, **34:**485
Streptococcus fradiae, source of CoA, **16:**2–3
Streptococcus hemolyticus, 246, **15:**250
Streptococcus hemolyticus, utilization of glutamine by, **13:**253–254; 258; 260–262
Streptococcus lactis
and lipoic acid, **18:**321; 332
metabolism in, **74:**34
Streptococcus lactis, 214, **15:**378; 380
Streptococcus lactis, asparagine metabolism in, **13:**249–250
Streptococcus lactis R, growth factor for, **21:**372
Streptococcus lactis, superoxide from, **41:**79
Streptococcus mitis, α-1,6-glucosidase and transferase in, **34:**406
glycogen reserve, **34:**291
pullulanase, **34:**408; 409
Streptococcus mitis, superoxide dismutase from, **41:**79
Streptococcus mutans, superoxide dismutase from, **41:**71; 79
Streptococcus paracitrovorus, metabolism in, **74:**21; 34–36
Streptococcus pyogenes
cell walls, **26:**221
glucuronidase in, **16:**378
inhibitory action of normal sera on proteinase of, **6:**99
Streptococcus thermophilus, **15:**444
Streptodornase, commercial nuclease produced by *Streptococci,* **29:**199
Streptogenin, effect on *L. casei,* **13:**250
Streptokinase, **34:**450
Streptoline, amino acids in, **14:**266
Streptomyces, action on collagen, **23:**177
cytochromes of, **23:**267
Streptomyces albus, cell walls, **26:**202
Streptomyces antibiotics, **10:**145–166
Streptomyces aureus, enzymic transformations of steroids by, **20:**248
Streptomyces auriofaciens, enzymic transformations of steroids by, **20:** 242–244
Streptomyces cacaoi, **59:**72
Streptomyces chitinase, **59:**90
Streptomyces coelicolor, **70:**246
enzymic transformations of steroids by, **20:**254; 271
Streptomyces enzyme, **59:**86
Streptomyces erythreus, proteinase from, **41:**182; 186; 188–189
Streptomyces fradiae, **41:**186; 188–189; 192–195
enzymic transformations of steroids by, **20:**247
Streptomyces griseus, **20:**389, **53:**261
pronase from, **26:**328
sugar nucleotides in, **26:**45; 46
and vitamin B_{12}, **26:**260
wall teichoic acid of, **30:**232
Streptomyces griseus, proteinase from, **41:**182; 186–189; 191; 214–215

Streptomyces hydrogenans, enzymic transformations of steroids by, **20:**254; 255
Streptomyces hydrogenans, PN-dependent enzymes, **26:**118; 167
Streptomyces lavendulae, **38:**425
Streptomyces naraensis, proteinase from, **41:**184
Streptomyces rectus, proteinase from, **41:**182
Streptomyces roseocheomogenus, enzymic transformations of steroids by, **20:**249
Streptomyces sp., β-lactamase of, **28:**242
Streptomyces species, enzymes in, **20:**265; 270
  enzymic transformations of steroids by, **20:**239; 248; 249; 254; 257; 259; 270; 274
Streptomyces tendae, **59:**72
Streptomycete preparation, laminarin hydrolysis by, **28:**350
Streptomycin, **7:**482; 483; 487; 489; 490; 493, **10:**145–166, **25:**433. *See also* Accostrep, Agrimycin, Agristrep, AS-15, Phytomycin, and Streptospray
  colorimetric determination, **10:**157
  effect on carotenogenesis, **21:**318; 319; 355
  effect on RNase, **24:**238
  effect on sperm, **24:**85
  effect on synthesis of poly (glycerol phosphate) in *Bacillus,* **30:**244
  evidence of linkage, **10:**155–156
  hydrolysis products, **10:**149–155
  in preparation of β-lactamase, **28:**277
  relation of structure to antibiotic activity, **10:**160–161
  site of action, **10:**161
  structure, **10:**147–149
  synthesis of gramicidin S, effect on, **33:**345
  and translational repression by arginine, **40:**81
  and virus precipitation, **17:**485; 493
Streptomycin
  compatibility with agricultural chemicals, **20:**386; 396
  phytopathogens resistant to, **20:**387; 394
  translocation in plants, **20:**387–390; 394
  in treatment of plant diseases, **20:**384; 386; 388–403
Streptomycin and Other Streptomyces Antibiotics (Brink and Folkers), **10:**145
Streptomycin nucleate, solubilization of, **17:**137; 145
Streptomycin resistance, **17:**445; 455–457
Streptonigrin, **41:**42; 86
Streptose, **10:**147–149; 152–154
Streptospray, composition, **20:**384
Streptothricin, **7:**483; 484; 487; 489; 491; 493, **10:**146
  amino acids in, **14:**266
  in treatment of plant diseases, **20:**394
Streptozotocin, **55:**272
Strepturea from streptomycin, **10:**149–150
Stress
  associated with brain norepinephrin and serotonin levels, **29:**487
  tyrosine hydroxylase induction, **70:**166
Stretching of hair, **1:**8
Striated muscle, **59:**370–371
Strigomonas fasciculata, **46:**102
Stroma, of the chloroplasts, **30:**5
Strong charge-transfer complex, of D-amino acid oxidase, **34:**58; 64
Strontium, substitution for calcium in calcium-fibrin, **16:**43
Strontium and luminescence, **7:**253
Strontium compounds, **17:**151
Strontium ions
  chelate formation with adenosine phosphates, **21:**147
  promotion of citric acid formation, **10:**215
Strophantine-G, effect on heart muscle protein, **13:**182
Structural arrangements of cell constituents and cell metabolism, **13:**1–20
Structural Basis for Catalysis by Tryptophan Synthase (Miles), **64:**93
Structural domains, **47:**130; 133–134; 136
  and tertiary folding, **47:**133
Structural inhibitors of peptidase, **12:**242–248
Structural proteins and ATP in muscle contraction, **13:**151–197
Structural similarity principle, **27:**391
Structural specificity, of polysaccharides, **26:** 67–78
Structural unit, **34:**471
Structure, **34:**445–446
  of bacterial cell walls, **26:**193–232
  of enzymes, three-dimensional, **26:**89–114
  macromolecular, **34:**446
  solving and refining, **52:**227–230
  subcellular, **34:**446
Structure-activity relationships, of FAH_2-reductase inhibitors, **27:**440–451

Structure-activity relationships of RNase, **24:**116; 243–252; 448–452
Structure-Activity Relationships of the Adrenocorticotropins and Melanotropins: The Synthetic Approach (Ramachandran and Li), **29:**391
Structure and activity relationships of enzymes, **25:**330
Structure and Function of Ribonuclease (Scheraga and Rupley), **24:**161
Structure and Mechanism of F_0-F_1-Type ATP Synthase and ATPases (Penefsky and Cross), **64:**173
Structure-breaking effects, **43:**293
Structure determination, three-dimensional, **39:**349
Structure-function relationship, amidotransferases, **72:**89–93
Structure-function relationships, **34:**483
Structure-guided mutagenesis, rubisco, **67:** 50–56
Structure-linked latency, **34:**485
Structure of Enzymes, Secondary and Tertiary, Formation of (Staub), **26:**89
Structure of succinyl CoA synthetase, **36:** 196–200
  E. coli enzyme, **36:**196
  incubation of the enzyme with urea and iodoacetamide, **36:**198
  isozymes, **36:**198
  mammalian enzyme, **36:**196
    molecular weight, **36:**197
  merthiolate dissociation of, **36:**197
  molecular weight, **36:**197
  pHMB and sulfhydryl groups, **36:**197
  pHMB dissociation of, **36:**197
  subunits, **36:**197
    molecular weights of, **36:**198
  treatment of the enzyme with succinic anhydride, **36:**198
Structure of Yeast Mannans (Ballou), **40:**141
Structure oxidizability, of hydrocarbons, **27:**473–474
Structures, secondary and tertiary, **39:**349
Structure specificity and enzymic adaptation, **13:**108–115
Strychnine, **18:**283
  effect on actin polymerization, **13:**155
    effect on muscle, **13:**162
Strychnine compounds, **17:**143; 147
Stuart factor, **48:**278
Stuart power factor, properties of, **30:**262. *See also* Factor X
Styryl quinoline, inhibition of tumor growth by, **29:**336
Subcellular enzymes, **33:**288–294
Subcellular fractions. *See also* Microsomes, Mitochondria, and Ribosomes
  distribution of enzymes between, **24:**291–358
Subcellular ordinals, **34:**456
  structure, **34:**446
Subendothelial extracellular matrix, **60:**235
Sublingual glycoprotein, **52:**34
Submaxillary asialoovine mucin, **52:**43
Submaxillary gland, mouse
  after testosterone injection, **31:**54
  androgen control of nerve growth factor content in, **31:**54–56
  cross sections of, **31:**55
  nerve growth factor in, **31:**60–62
Submaxillary mucin, **52:**41; 59
Submerged culture of microorganisms, **9:**658; 662; 664–666; 681–684
Submitochondrial particles, **34:**456; 481–484
  for respiratory chain, **28:**164–167
Submitochondrial systems, energy-linked ion movements in, **29:**294–296
Subsites, **43:**298
  definition, **39:**380
Substance E, microbiological formation, **20:**255
Substance P, **57:**16
Substance S, formula, **20:**241
  microbiological conversions, **20:**239; 243–249; 251; 252; 254; 263; 264; 270; 271; 273; 275
Substances and enzymes, **1:**177–198
Substance U, microbiological formation, **20:**255
Substantia nigra, tyrosine hydroxylase, **70:**107; 120
Substilisin, **47:**124
N′-Substituted nicotinamide compounds, **15:**92; 94–95; 97
Substrate, **34:**446; 452; 462–465; 486–487; 495–504; 510–515; 521–525, **44:**195
  activation in glutamate dehydrogenase reaction, **39:**410
    cyanide, **39:**348
    specificity, **39:**354
    sulfur, **39:**345
      atoms, **39:**346; 349

as adsorbent for enzymes, **14:**341–344; 367
in location of chromatographed enzyme, **14:**345
role in enzyme isolation, **14:**380; 401
choice of concentrations, for complex cases, **29:**16–19
for kinetic data, **29:**14–16
concentration, **34:**487
effect in kinetic studies, **17:**8–13; 50; 58
ionization, effect in kinetic studies, **17:**33–34
mechanisms of interactions, **17:**272–281
effect on model catalyses by metal ions, **21:**172; 173
hydrophobic, **34:**478
local, **34:**498
uncharged, **34:**460
Substrate activation, **45:**409, **46:**402
enzymes exhibiting ping-pong kinetics, **53:**327
predicted from ligand substitution chemistry, **53:**315
and surrogate activation, maximum velocity result, **53:**326
Substrate-activation mechanism, **44:**198
Substrate analog, complex with protein, Cotton effect, **27:**87
Substrate analogs, **43:**340, **53:**52; 53; 54; 62
in allosteric regulation, **28:**130–131; 137
and enzyme conformation, **28:**300–304
of isocitrate dehydrogenase, **28:**59–60
for β-lactamase, **28:**241
Substrate as inductor, **13:**68–69; 73; 100–105; 107
Substrate binding
γ-glutamyl transpeptidase, **72:**247–250
active sites, **72:**251–254
to LADH, **27:**70–72
structural changes induced by, eukaryotic dihydrofolate reductase, **70:**45–46
tetrahydrofolate and, eukaryotic dihydrofolate reductase, **70:**50–52
Substrate binding monitoring, **52:**223; 224
Substrate-binding proteins, periplasmic permeases, **65:**18–19
Substrate binding sites, NMP kinase tertiary structure, **73:**108–110
Substrate-bridge complexes, **33:**30
Substrate concentration and rate of reaction, **7:**4
Substrate-enzyme combination, mechanism, **10:**72–74
Substrate-enzyme complex in energy transport, **14:**38–40
Substrate free radicals, EPR spectra, **27:** 119–130
Substrate homologs, effect on LADH complex, **27:**64–65; 71–72; 78
Substrate-independent rate determining step, **44:**201
rate limiting step, **44:**202
slowest step, **44:**201
Substrate-induced changes, NMP kinases, **73:**124–128
Substrate-induced fit, theory of, **31:**151
Substrate-induced irreversible modification by L-2-amino-4-methoxy-*trans*-3-butenoic acid, **49:**173
by *a*-cyanoglycine, **49:**173
of tryptophan synthase, **49:**172
Substrate inhibition, **15:**23; 279; 287; 291–292, **43:**105; 155
with xanthine oxidase, **17:**317–318
Substrate-metal complexes, in hydrolysis by alkaline phosphatases, **21:**189
in metal ion activation of enzymes, **21:** 139–141; 144–148; 155; 156
in model catalyses by metal ions, **21:** 180–183; 189–191
Substrate orienting, effect of, **28:**299–302
Substrate product, anion complexes and, **51:**122–128
Substrate profile, of β-lactamase, **28:**255; 267; 269; 272; 279; 282; 287; 306; 309
Substrate protection of enzyme, **15:**287–288
Substrate protein, **42:**272
Substrate-rate saturation curve, **57:**379–380
Substrates, **48:**34; 38–51
adenylosuccinate synthetase (AMPSase), **73:**79–84
as alkylating agents, **46:**408
electrophilicity as a criterion for, **46:**398
binding of, **48:**51
binding pocket, **48:**610
for catalase and peroxidase, **19:**104–109
effect on papain specificity, **19:**312
for transketolase, **19:**26
chromophoric, **48:**34
complex, **48:**54–57
and Cotton effect in proteins, **27:**48

Substrates (*Continued*)
of dihydrofolate reductases, **27:**434–436
for *E. coli* β-glucuronidase, **16:**380; 384
for mammalian β-glucuronidase, **16:** 365–373
effect on enzymes denaturation, **23:**48–50
molecular weight, higher, **48:**53
molecular weight, low, **48:**51–54
particulate, **33:**259 (*See also* Insoluble substrates; Soluble substrates)
partitioning, enzyme catalysis and, **51:**49–51
simple, **48:**38
specificity, **48:**38–43
phosphofructokinase, **48:**196–198
sticky, **45:**281; 309; 311; 331; 338; 346
Substrates in semen metabolism, **9:**372–374
Substrate specificity, **45:**409, **46:**395
of collagenase, **23:**201–209
of elastase, **23:**237–239
glutathione synthesis, catalytic mechanism, **73:**236–240
Substrate stereospecificity, for hydrogen transfer, **26:**166–169
Substrate studies, alternate, **45:**284–287
Substrate-substitute, **34:**42; 52
Substrate substitute, in ternary complex, **27:** 3–12
Substrate synergism, **35:**358, **43:**266; 307
Subtiline, structure studies, **14:**266; 299
Subtilisin, **23:**179, **34:**10; 13, **43:**298, **53:**261; 265, **54:**145; 146
action on RNase, **24:**114; 133; 202; 216; 230–232; 235; 248
effect on protein antigens, **24:**145
in enzyme formation, **24:**23
modification of, **24:**449
adsorbed on kaolinite, **33:**272
O-cinnamoyl, spectra of, **33:**389
comparison with thiol-subtilisin (table), **33:**388
produced by *B. subtilis,* **385**
specificity of, **33:**432–433
transformation into thiol-subtilisin, **33:**383
amino acid composition, **34:**13
amino acid sequence in, **34:**22
crystallographic studies, **34:**21
effect on fructose 1,6-bisphosphatase, **42:**205
effect on ribonuclease, **22:**70; 76
effect on ribonuclease A, **22:**466; 467
in fructose-bisphosphatase regulation, **54:**181
hydrogen bonding to amide hydrogen, **43:**310
multiple forms, **25:**338
specificity, **19:**328
in yeast mannan isolation, **40:**242
Subtilisin A, **25:**338
Subtilisin B, **25:**338
Subtilisin BPN′, **53:**264; 273–274; 290
Subtilisin BPN, amino acid sequences, **41:**190
hydrolysis by, **41:**217; 218
Inactivation of, **41:**225
from serine proteinase, **41:**220
X-ray crystallography, **41:**190
X-ray studies, **41:**238
Subtilisin Carlsberg, **41:**180; 186; 190; 220; 222; 227
Subtilisin Novo, **41:**190
Subtilisins, **41:**180; 186; 190; 191; 217; 218; 220; 222; 225; 227; 238
Subtilisin-treated ribonuclease, **26:**102
Subtilism cleavage, **53:**50; 51
Subtilopeptidase A, (Subtilisin Carlsberg), **34:**450; 459; 462
Subtilopeptidase B (Subtilisin Novo), **34:**451
Subunit, **34:**471
interactions, **35:**510–512
structure of aspartate-β-decarboxylase, **35:**523–537
model, **34:**472
β-subunit, MoFeP protein, **67:**337–339
amino acid sequences, **67:**337–338
α-subunit, MoFe protein, **67:**334–336; 338
amino acid sequences, **67:**334–335
Subunit cooperativety, **43:**171
Subunit dissociation, **43:**155
Subunit interaction, tryptophan synthase, **49:**160–166
composite active site of tryptophan synthase, **49:**161
Subunit properties, pyruvate, phosphate dikinase, **45:**112
Subunits, **39:**193–196; 198; 199; 218; 350, **41:**7; 11; 16; 250; 288, **42:**271
of hypoxanthine-guanine phosphoribosyl transferase, **41:**16
identical, **39:**347
identity of, **39:**200
of PP-ribose-P amidotransferase, **41:**11
of PP-ribose-P synthetase, **41:**7
of prolyl hydroxylase, **41:**250; 288

Subunit stoichiometry, counting procedures for determining number of subunits, **33:** 94–95
molecular weight methods, **33:**91–94
sedimentation-equilibrium technique, errors in, **33:**91–93
Subunit structure
carboxytransphosphorylase, **45:**90
γ-glutamyl transpeptidase light subunit, **72:**250–251; 254–262
lysyl hydroxylase, **72:**337; 380–382
prolyl 4-hydroxylases
α subunits, **72:**334–336; 364–368
β subunits, **72:**335–336
protein disulfide isomerase/β subunit polypeptide
function, **72:**375–380
structure, **72:**368–375
Subunit structure of allosteric enzymes, **28:**51; 53–58; 68; 90; 104–106; 125; 129–130
Subunit studies, **46:**366
Subzero temperature
aqueous media withstanding, **51:**2–5
enzyme-catalyzed reactions at, **51:**58–65
BAEE, **51:**58–61
substrate intermediates, **51:**61–65
fluorescence intensity, **51:**8–9
Subzero temperatures, Arrhenium plots, **45:**201; 231; 233
dielectric constant, **45:**164
electron transport chain, of liver microsomes, **45:**210; 253
of mitochondria, **45:**248
electron transport chain of photosystems in chloroplasts, **45:**205
fast techniques, **45:**357
flash photolysis, **45:**252
intermediates, of bacterial cytochrome P_{450}, **45:**228; 253
of bioluminescent reactions, **45:**216
of chymotrypsin catalyzed reactions, **45:**235
of peroxidases, **45:**214; 261
measurement, **45:**195
methodology, **45:**191
nitrogen, gaseous, as coolant, **45:**192
liquid, as coolant, **45:**193
pa_H, variations in, **45:**180
rapid freezing, **45:**245
rapid mixing, **45:**261
reduction of reaction, **45:**200
slow kinetics, **45:**250
temperature jump, **45:**264
temperature recordings, **45:**195
temporal resolution of reactions, **45:**212; 216; 228; 235
trapping of productive intermediaries, **45:**236
trapping techniques, **45:**248
x-ray crystallography of proteins, **45:**242
Subzero Temperatures, Enzymology at (Douzou), **45:**157
Successive reactions, *See* Step reactions
Succinae-phenazine reductase, in respiratory system, **28:**168–170
Succinate, **34:**476–477, **46:**114, **50:**149, **51:**236
and acetate, **23:**405–406; 408–420; 425; 428; 430
DPN reduction by, **23:**382–383
and glycine, **23:**455–456
and glycolate, **23:**441; 445
and oxalate, **23:**448–449; 452
and oxidative phosphorylation, **23:**348; 349; 354; 357; 362; 367; 370; 383; 385
in citric acid cycle, **15:**185; 201; 224–232; 235–236; 238; 240 (C^{14}); 252 (C^{14}); 253; 261; 262; 336; 337; 339
dehydrogenase, **34:**481
dideuteris, **18:**100–101
labeled, in citric acid cycle, **18:**249
in plants respiration, **18:**299; 300; 303
effect of malonate on dehydrogenation, **19:**365
formation from aromatic compounds, **19:**82
and kynurenine 3-hydroxylase, **19:**194
in photosynthesis, **19:**13; 40–43; 52; 53
effect on DPNH-cytochrome *c* reductase, **20:**170
effect on succinic dehydrogenase, **20:**155
oxidation in respiratory chain, **20:**154–167; 173–187
fixed carbon in, location of, **74:**20–21
fluorinated, **55:**216–217
formation of, in bacterial fixation of carbon dioxide, **74:**14–20
oxidation, **34:**473; 480
oxidation in normal and cancer tissues, **4:**204; 205; 209
in preparation of succinate dehydrogenase, **28:**184–186
in respiratory chain, **28:**159–160; 164; 173; 186

Succinate and CO_2 fixation, **12:**49; 51; 52–53; 103–106; 132
Succinate (C^{14}), inhibition studies, **25:**244
Succinate-coenzyme Q reductase, EPR study, **27:**177; 183
Succinate-CoQ reductase (complex II), preparation, **28:**209–211; 216–220; 223
Succinate-cytochrome c reductase, **37:**261
Succinate-cytochrome *c* reductase, absorption spectra, **28:**191
composition and activity, **28:**192
effect on cytochrome *c*-cytochrome oxidase activity, **28:**208
intact, kinetic constants, **28:**200–201
turnover numbers, **28:**201–203
reconstituted, kinetic constants, **28:**200–201
turnover numbers, **28:**201–203
reconstitution of, **28:**190–203; 222–223
in reconstitution of respiratory chain, **28:** 216–218; 220–222
variable adsorption of cytochrome *c,* **28:**204
Succinate-cytochrome *c* reductase activity, **17:**363; 367, **34:**473–474
Succinate dehydrogenase, **37:**189–226; 435, **54:**274, **55:**216, **74:**303
activation, **37:**240
by ATP, ITP and IDP, **37:**224
energy, **37:**235; 239; 243
in higher plants, **37:**243
by pH and anions, **37:**229–235
by reduced coenzyme Q_{10}, **37:**219–223
by substrates and competitive inhibitors, **37:**217
activation of, **37:**209
activity measurements, factors affecting, **37:**244–247
with alkalitreated heart muscle preparation, **28:**174–179
characteristics, **28:**170–172
with cyanide particle, **28:**180–181
effect of iron chelators, **28:**225–228
flavin component of, **28:**172; 178; 201
incorporation of ^{14}C from cyanide, **28:**225; 227
method for reconstitutively active enzyme, **28:**184–186
particles devoid of, **28:**173–174
preparation, need for succinate preincubation, **28:**184–186
reactions with artificial electron acceptors, **28:**176–178; 181–184
in reconstitution of succinate oxidase, **28:**186–190; 218–220; 222–225; 229
with cyanide particle, **28:**180–181
removal from cyanide particle, **28:**173–174
in respiratory chain, **28:**160; 166–190; 194; 196–199; 202; 209; 225–228; 230
in reversible dissociation of succinic oxidase, **28:**186–190
species specificity, **28:**179–180
stability, **28:**176–177; 208; 225–226
effect of air, **28:**182–184
effect of temperature, **28:**183
with respect to artificial activities, **28:**181–184
with respect to reconstitutive activity, **28:**181–184
succinate requirement, **28:**184–186
association-dissociation, **37:**199
catalytic properties, **37:**244–253
composition, **37:**193
conformation changes, **37:**215; 239; 240; 242
deactivation, **37:**240
deactivation of, **37:**218; 222; 230; 235–238
from *E. coli,* **37:**262
effect of phospholipids, **37:**246
energy of activation, **37:**242; 244
EPR signals, **37:**210; 251; 252
EPR spectra, **37:**210–211
evolution, **37:**261–263
inhibitors of, **37:**250
K_I, **37:**250
K_D, **37:**212; 249
kinetic constants, **37:**247–250
K_m, **37:**250
for succinate, **37:**249
labile sulfide in, **37:**193
mechanism of action, **37:**250
mechanisms for regulation, **37:**241
molecular basis of regulation, **37:**242–243
molecular parameters, **37:**208–217
molecular weight, **37:**193–208
from *mytilus californianus,* **37:**263
non-heme iron in, **37:**193
optical spectrum, **37:**208–210
particulate preparations, **37:**190–191
peptide sequence of flavin site, **37:**205–208
physiological control, **37:**239–242

reaction mechanism, **37**:250–253
reactions with SH reagents, **37**:211–213
reaction with cyanide, **37**:213–217
reconstitutive properties, **37**:253
regulation of, **37**:242
regulatory properties, **37**:217
regulatory significance of activation, **37**:219
relation of, activation to spectral changes, **37**:210
oxalacetate to the activation process, **37**:238–239
from *Saccharomyces cerevisiae,* **37**:262
soluble preparations, **37**:191
specificity, **37**:247–250
structure, **37**:193–208
structure of the covalently bound flavin, **37**:200–205
subunit composition, **37**:194
thermo-dynamic parameters for the activation and deactivation of, **37**:242
turnover number, **37**:191; 247; 249
variation of sedimentation coefficient with protein concentration, **37**:199
Succinate dehydrogenase flavin, ENDOR studies, **37**:204
Succinate dehydrogenase particle, in reconstitution of succinatecytochrome *c* reductase, **28**:222
Succinate Dehydrogenase (Singer, Kearney, and Kenney), **37**:189
Succinate-dehydrogenating enzymes, reconstitutive activity of, **28**:168–170; 181–184
Succinate formation, enzymic mechanism, **23**:409–413
Succinate-fumarate cycle, **17**:376
Succinate-glycine cycle, **23**:455–456
Succinate half esters, **43**:271
Succinate-methylene blue reductase, removal from particle, **28**:173–174
Succinate oxidase
cross-reconstitution, **28**:179–180
effect of cytochrome *c*-cytochrome oxidase complex, **28**:207
fragmentation, **28**:190; 194
intact and reconstituted compared, **28**: 176–179; 224
reconstitution of, **28**:158; 167–190; 208–209; 218–225; 229
reversible, **28**:186–190; 225
systematic scheme, **28**:197
in reconstitution of respiratory chain, **28**:216; 218–220; 227; 230
and cytochrome *c,* **23**:343
Succinate oxidation, and phosphorylation, **18**:311
Succinate-oxidizing enzyme, in respiratory chain, **28**:214
Succinate requiring *Neurospora,* **17**:426
Succinates, complexes of, **17**:149; 152
oxidation, and phosphate turnover, **17**:175
Succinate semi-aldehyde dehydrogenase, **37**:477; 501
Succindialdehyde and alkaloid biogenesis, **8**:238
Succinic acid. *See also* Citric acid cycle
activation, **21**:409
effect on carotenogenesis, **21**:318
in incorporation of valine into cytochorome *c,* in mitochondria, **21**:45
oxidation, in mitochondria, **21**:12; 13; 34; 35
in mitochondrial fragments, **21**:38; 81–83; 101; 103–107; 111; 112; 115; 116; 118–122
synthesis, in mitochondria, **21**:46; 47
degradation in yeast, **10**:173–176
formation in molds, **9**:538; 540; 545–546; 547–550; 555; 558–559; 564–566
mechanism of oxidation, **15**:7; 14
metabolism, **16**:255
thio ester with CoA, **16**:18
Succinic acid, PN-dependent oxidation, **26**:127
Succinic acid derivatives, **18**:258–259
Succinic acid diamide, deamination, by metal ions, **21**:179
Succinic acid from phenol, in *Vibrio,* **14**:111; 112
Succinic anhydride, **43**:271
Succinic-bc_1 dehydrogenase complex, isolation and properties, **21**:107; 114
Succinic-coenzyme Q reductase, **25**:311–312; 314; 317–319
Succinic-cytochrome *c* reductase, definition, **20**:149
inactivation, **20**:177
preparation and composition, **20**:187–189
Succinic-cytochrome c reductase and coenzyme Q, **25**:305; 308–311; 318
Succinic-cytochrome-oxidase system, **12**:13
Succinic-cytochrome reductase, **17**:284

Succinic dehydrogenase, **7:**281, **15:**6; 14, **17:**69; 70; 82; 350; 355; 363–374; 438–442, **18:**65–111, **43:**363. *See also* Succinic oxidase; Succinoxidase system
absorption spectra, **20:**190
in *Chlorenchyma,* **20:**219
inactivation, **20:**161; 175; 176; 178
isolation and composition, **20:**155; 188–190
in respiratory chain, **20:**154–156; 159; 160; 185–187
activity of —SH group, **11:**235; 236; 238–239; 240–241; 243
inhibition by ionizing radiations, **11:**256
in cereal roots, **18:**22
and citric acid cycle, **18:**244; 272–273
and coenzyme Q, **25:**311–312
cyanide inhibition of, **22:**382
flavines in, **22:**384
identification of amino acids at active site, **22:**63
and cytochrome *c,* **23:**65–66
in respiratory chain, **23:**343–344
effect of fluorocitric acid on, **18:**128
in electron transfer, **19:**201
kinetic constants, **19:**313
in photosynthesis, **19:**13; 143
general properties of, **18:**73–101
inhibition, **13:**332
localization in cytoplasm, **13:**12; 13
and isotope specificity, **8:**364
in chromidia, **8:**31
mechanism of action, **8:**383–384
kinetic studies, **15:**279
in melanin granules, **16:**169; 172
intracellular distribution, **16:**468
as metalloflavoprotein, **17:**253–254; 270–271
in mitochondria, **21:**12; 13; 20; 74; 98
in mitochondrial fragments, **21:**98; 99; 104
nonheme iron in, **21:**114
in mitochondrial oxidations, **21:**116; 118
molecular weight, **21:**98
position in mitochondrial electron transfer chain, **21:**119; 120; 122
in plants, **18:**298–299; 306–308; 310
preparative methods, **18:**67–73
protection by substrate, **15:**287
in respiratory chain, **18:**12–13; 22
in sperm, **9:**370, **24:**86–87
location of, **24:**300; 303; 337
terminology, **18:**66–67
Succinic dehydrogenase, EPR study, **27:** 176–177; 183
Succinic dehydrogenase complex, *See* Succinic-DPNH bc* dehydrogenase complex
Succinic dehydrogenase complex (SDC), **18:**67; 68–70; 71; 73; 77; 80; 86; 88; 105–107
inactivation, **20:**177
preparation and composition, **20:**187–191
as metalloflavoprotein, **17:**254–257; 262; 270–271; 275; 281; 284–285
Succinic dehydrogenase-cytochrome *c* combining factor, **17:**372–374
Succinic dehydrogenase flavoprotein, and coenzyme Q, **25:**277; 311–312; 315; 319
Succinic dehydrogenases, classification, **14:**255
separation by chromatography, **14:**350
Succinic Dehydrogenase (Singer, Kearney, and Massey), **18:**65
Succinic-DPNH *bc** dehydrogenase complex, isolation and properties, **21:**103–106; 114; 115
Succinic oxidase
histochemical methods for, **9:**588
in plants, **9:**408–409; 412
kinetic studies, **15:**280
protection by substrate, **15:**11
localization in cytoplasm, **13:**11; 12
Succinic oxidase activity, and coenzyme Q, **25:**306–308; 310; 317–319
Succinic oxidase system, **17:**69–71; 81–82; 437–440
activity in respiratory chain preparations, **20:**153
components, **20:**150–151; 154–167; 173–187
preparation and composition, **20:** 187–193
definition, **20:**149
and cytochromes, **17:**347; 359–375; 380
steady-state oxidation-reduction levels, **17:**88–89; 91
Succinic-phenazine methosulfate reductase activity, **25:**312
Succinic semialdehyde, in transaminations, **16:**194; 205
Succinic semialdehyde-DPT, **15:**230
Succinic thiokinase, **36:**183
Succinodehydrogenase activity in muscular atrophy, **6:**196; 198; 199

Succinoxidase, **7:**394, **37:**253; 254; 260; 261, *See* Succinoxidase system
activity of, **37:**255
activity of –SH groups, **11:**234; 237–238
adsorption studies, **14:**356
of cereal embryos and seedlings, **18:**39–40
and colloidal structure, **18:**104–108
effect of Fe-chelating agents, **18:**94
effect of fluoroacetate on, **18:**118
in plants, **18:**22; 25; 31; 46; 58; 298
and succinic dehydrogenase, **18:**66–73; 90
location of, **24:**302; 306
malonate inhibition of, **6:**43; 206
in muscular dystrophy, **6:**206
in mitochondria, **21:**13; 58; 59
effect of lecithinase on, **21:**9
effect of sonic vibrations on, **21:**57
effect of Versene on, **21:**5
Succinoxidase activity
effect of cytochrome *c,* **17:**332–333
in particles, **18:**13–15; 27
Succinoxidase chain, electron carriers, **18:**86
Succinoxidase system
classification, **14:**232–233; 255
cytochrome c, **4:**205; 210; 212; 215; 233; 234; 236; 237; 239; 240; 241; 242; 243
cytochrome oxidase, **4:**230; 234; 235–236; 237; 240; 241
and cytochromes, **18:**13; 17
in plant homogenates, **18:**18–20
and salt respiration, **18:**52
in normal and cancer tissues, **4:**233; 240–243; 247
succinic dehydrogenase, **4:**204; 205; 212; 230; 231; 235; 236; 237; 241; 242; 248; 249
succinoxidase, **4:**205–249
Succinylation, **34:**469–470, **58:**80
Succinyl CoA, **46:**39; 40; 91; 103; 111; 113; 114; 123; 177
in lysine synthesis, **16:**297
in photosynthesis, **19:**41–43
Succinyl-CoA, **17:**114
from methylmalonyl-CoA(b), **26:**343–367
in propionic acid metabolism, **26:**284–286; 292; 307–309; 334–343
in mitochondrial oxidation of α-ketoglutarate, **21:**33; 35
metabolism, **21:**227; 266
in synthesis of active formate, **21:**410; 411
Succinyl CoA: acetoacetate CoA transferase, **43:**245
Succinyl CoA, oxidation of 3-oxoadipate, **27:**512–513
Succinyl-CoA (C^{14}), from methylmalonyl-CoA, **26:**354–357
from propionyl-CoA, **26:**342; 345
Succinyl-CoA deacylase, **21:**74; 409
Succinyl CoA deacylase, classification, **14:**249
Succinyl-CoA (H^3), reaction mechanism, **26:**339–340
Succinyl-CoA phosphorylation system, site of cleavage by, **17:**163
Succinyl-CoA synthesis, **58:**159
Succinyl CoA synthetase, **36:**183–200
antibody to, **36:**197
ATPase activity of the enzyme, **36:**192
equilibrium constant, **36:**184
function of the enzyme, **36:**184–185
Michaelis constants of substrates, **36:**189
occurrence of, **36:**185
purification of, **36:**185
reaction catalyzed by this enzyme, **36:**183
specific enzymatic activities of, **36:**186
substrate specificity, arsenate, **36:**187
azaGDP, **36:**186
azaGTP, **36:**186
Ca^2+, **36:**187
Co^2+, **36:**187
coenzyme A, **36:**187
dephospho CoA, **36:**187
divalent metal ion, **36:**187
inorganic phosphate, **36:**187
Mg^2+, **36:**187
Mn^2+, **36:**187
nucleoside diphosphate, **36:**186–187
nucleoside triphosphate, **36:**186–187
succinate, **36:**185–186
6-thioguanosine triphosphate, **36:**187
Zn^2+, **36:**187
Succinyl-CoA synthetase
adenosine triphosphate binding, **58:**161
aggregated form, **58:**162
alternating sites cooperativity model, **58:**160
binding sites, **58:**155
CoA binding site, **58:**158
in control of citric acid cycle, **58:**146
in *E. coli,* **58:**147
amino acid analysis, **58:**147
binding sites, **58:**148

Succinyl-CoA synthetase (*Continued*)
chemical modification, **58:**151
crystallization, **58:**152
cysteine residues, **58:**151
dephosphorylation, **58:**150
disulfide bonds, **58:**151
extinction coefficient, **58:**147
fluorescence, **58:**151
half-sites reactivity, **58:**148
inhibitors, **58:**152
manganese ion binding, **58:**148
methionine residue, **58:**147
molecular weight, **58:**147
phosphorylation stoichiometry, **58:**148
refolding, **58:**150
structure, **58:**147
substrate specificity, **58:**148
subunit autophosphorylation, **58:**150
subunits, **58:**147
sulfhydryl group, **58:**151
tryptophane modification, **58:**151
tryptophan residue, **58:**147
unfolding, **58:**150
from gram-negative bacteria, **58:**144
GTP specific, **58:**145
histidien residue, **58:**155
histidine phosphorylation site, **58:**159
history, **58:**142
inhibitors, **58:**159
interaction with α-ketoglutarate dehydrogenase, **58:**146
large forms, **58:**144
in mitochondrial membranes, **58:**146
molecular mechanism, **58:**155
nicotinamide triphosphate binding site, **58:**157
NMR studies, **58:**156
nucleoside triphosphate specificity, **58:**144
phosphate binding site, **58:**158
phosphorylation, **58:**145
phosphorylation site, **58:**155
in pig heart, **58:**152
active sites, **58:**154
amino acid composition, **58:**153
disulfide bonding, **58:**154
extinction coefficient, **58:**153
fluorescence studies, **58:**153
interconvertible forms, **58:**154
intermolecular associations, **58:**146
molecular weight, **58:**153
phosphorylation stoichiometry, **58:**153
structure, **58:**153
substrate specificity, **58:**153
subunit refolding, **58:**155
sulfhydruyl groups, **58:**154
tryptophan residues, **58:**153
from plants, **58:**144
properties, **58:**147
reaction equilibrium constant, **58:**145
from *Saccharomyces,* **58:**144
size of, **58:**143
small forms, **58:**144
substrates, **58:**144
succinate binding site, **58:**158
succinyl binding sites, **58:**158
sulfhydryl group, **58:**158
tissue distribution, **58:**145
Succinyl-CoA Synthetase Structure-Function Relationships and Other Considerations (Nishimura), **58:**141
S-Succinyl coenzyme A, **15:**115; 122; 129; 184; 185; 199; 201; 224–242; 252–253; 256; 262
Succinyl coenzyme A, and ATP, **18:**229–230
and lipoic acid, **18:**338
Succinyl-coenzyme A, in ATP synthesis, **23:**334
and glycine, **23:**455–456
and oxalate, **23:**448; 451–452
Succinyl coenzyme A deacylase, **15:**225–229; 231; 236
Succinyl Coenzyme A Synthetase, Mechanism of Action and Other Properties (Nishimura and Grinnell), **36:**183
Succinyl-β-ketoacyl coenzyme A transferase, *See* Coenzyme A transferase
N-Succinyl-L,L-diaminopimelate aminotransferase
bacterial lysine biosynthesis, **72:**302–303
chemical mechanism, **72:**302–304
N-Succinyl-L,L-diaminopimelate descruccinylase
amino acid sequences, **72:**305–307
bacterial lysine biosynthesis, **72:**303; 305–307
chemical mechanism, **72:**305–306
substrates, **72:**303; 305
Succinyl-*S*-lipoic acid, **15:**230
Succinyl phosphate, **15:**232, **18:**231, **58:**143
cyclic form of, **36:**195
enzymatic synthesis of succinyl CoA form, **36:**195

phosphorylated enzyme intermediate, **36:**195
dephosphorylation by organic acids, **36:**195
synthesis of ATP from, and effect of, desulfo CoA, **36:**194
Succinyl thiokinase, **46:**114; 233; 247
reaction mechanism of, **31:**189
in study with ^{18}O phosphate, **31:**189
Succinyltrypsin, denaturation of, **24:**137; 138
Sucrase, **34:**481. *See also* Invertase; Invertase, Saccharase
classification, **14:**246
in tobacco, changes during processing, **10:**414; 423
Sucrase adsorption studies, **14:**325; 344
Sucrase in tobacco, **6:**367–368
Sucrease, pH dependence, **15:**284
Sucrose, **5:**120, **34:**480, **56:**159
action of dextransucrase on, **30:**190
action of levansucrase on, **30:**203
bond energy, **17:**190
release of cytochrome *c* from mitochondria by, **17:**351–352
in transglycosylations, **17:**192–208; 210; 220–226; 228
fermentation by thiaminase-producing bacteria, **15:**413
in polysaccharide synthesis, **15:**306; 309–310; 311; 313–314
fermentation of, **5:**104
formation in plants, **19:**5
gradient, **34:**480
hydrolysis, **34:**459
impermeability of nuclear membrane to, **16:**487
from photosynthesis, **11:**343
in polysaccharide synthesis, **11:**298–299; 306–315; 316–320; 327; 329; 330–331; 370–373
in plant respiration, **18:**285–291
from transglycosidation, **18:**201
polymeric degradation of, **5:**116–119
synthesis, **10:**123–132; 138–141
energy for, **10:**128–129
mechanism, **10:**126–129
synthetic, structure, **10:**131–132
Sucrose, CO_2 fixation in, **12:**132
Sucrose *(α)*, synthesis of, **18:**222
Sucrose-density-gradreit centrifugation, for studying binding activity, **41:**119
Sucrose α-glucosidase, classification, **14:**245
Sucrose phosphate, **18:**201; 285–287; 289
synthesis, **20:**122; 123
Sucrose phosphorylase, **15:**245, **17:**163; 165; 169; 228, **18:**201; 205; 222; 223–224; 286, **54:**10, *See* Phosphorylase
classification, **14:**251; 252
enzymic adaptation, **13:**70
labeling of active site in, **22:**51
Sucrose r dextran transglucosidase, **14:**251
Sucrose r glycogen transglucosidase, **14:**251
Sucrose r levan transfructosidase, **14:**251
Sucrose r levan transglucosidase, **14:**236
Sucrose r phosphate transglucosidase, **14:**251
Sucrose solutions, effect on thymus chromosomes, **16:**447
use in isolation of cell nuclei, **16:**415
Sucrose synthase, **56:**155
Sucrose synthesis, adsorption of enzymes, **28:**31
Sugar
effect on gelation of pectic substances, **20:** 357
in familial periodic paralysis, **6:**215
in muscular dystrophy, **6:**222
in myasthenia gravis, **6:**212
Sugar acids, as inhibitors of β-glucuronidase, **16:**373
Sugar alcohols, hydrogen bonding, **65:**231; 233
Sugar amines. *See also* Amino sugars and Hexosamines
nucleotides of, **26:**37; 40; 44; 46; 49–50; 55–56; 73; 76–78
Sugar beet curly top virus, mutations, **9:**181
transmission, **9:**175
Sugar beet leaf enzyme, in fructose transfers, **17:**200; 205
Sugar cane, lignin studies, **18:**354–357; 366–370
Sugar cane mosaic virus, mutations, **9:**181
Sugar chemistry, **58:**38
Sugar components of CoA, **16:**4
Sugar dinucleotide prophosphorylases, **28:**111
Sugar nucleotide pyrophosphorylase, transfer action, **26:**39
Sugar nucleotide pyrophosphorylases, allosteric regulation, **28:**110–114

Sugar nucleotides
accumulations in bacterial mutants, **26:**69
from *N*-acetyl-D-glucosamine, **26:**40; 49–50; 59; 73–75
and carbohydrate synthesis, **26:**35–88
enzymic synthesis (Table), **26:**52–56
formation and conversion of, **26:**39–57
from D-glucosamine, **26:**49–50
from D-glucose and D-mannose, **26:**44–49
glycosyl transfer from, **26:**57–78
occurrence of, **26:**50–57
specificity of transglycosylases, **26:**59–62
precursors of glycoprotein, **28:**114–115
Sugar Nucleotides, and Synthesis of Carbohydrates (Ginsberg), **26:**35
Sugar nucleotides (C^{14}), in synthesis of starch, **26:**58
Sugar phosphate derivatives, relation to transamination, **16:**236–237
Sugar phosphates
in carbon reduction pathway, **25:**94–106
in photosynthesis, **25:**42; 47
hydrolysis, by metal ion, **21:**184; 186; 193
reaction with nucleoside triphosphates, **28:**110–114
as solubilizing agents, **17:**146–147; 149
Sugar residues, **34:**475
Sugars, **66:**42. *See also* Oligosaccharides and individual sugars; specific sugars
actively transported, **29:**53
amino, recent developments, **25:**433–456
free energy of formation, **25:**46–48; 102; 104
aromatic compounds from, **18:**362–370
complexes with tetraborate, **12:**496–499
alcohols, complexes with tetraborate, **12:**496–499
fixation of CO_2 in, **12:**130–131
labeling by C^{14} in photosynthesis, **12:** 138–140
enzyme system for the metabolism of
of Escherichia coli (*See also* the individual sugars)
of *Escherichia coli,* **30:**81–137
genetic map of inducible enzymes for the metabolism of, **30:**86
fermented by thiaminase-producing bacteria, **15:**413
interconversion in nature, **14:**206–208
isomerization, **14:**193–195; 197–213
nucleic acid, biosynthesis, **16:**451
nucleotide, pathways of biosynthesis of, in *Salmonella typhimurium,* **31:**107
nucleotide-bound, modifications of, **26:**41–44
reaction with amino acids in aging of tobacco, **10:**408
in solubility studies, **17:**138; 148
labeled, in inversion study, **17:**191; 196
in transglycosylation study, **17:**196; 198–200; 204; 209; 214; 219; 224
Sugars (C^{14}), in plants, **18:**290–291
Sugar-specific binding proteins, **52:**113
Sugar transphosphorylase activity, of glucose 6-phosphatase, **28:**95–98
Suicidality, **74:**236
Suicide inhibitors, **59:**107
Sulfadiazine, folate inhibitor, **27:**458
synergistic action with trimethoprim, **27:**454; 456; 460
Sulfaguanidine, **27:**428–430
Sulfamatase (*N*-Sulfatase), and *Flavobacterium,* **22:**230
Sulfamates (*N*-Sulfates), types, occurrence, **22:**206; 207
Sulfane, **39:**330
atom, transferable, **39:**344
donor, **39:**352; 364
groups, **39:**328
reductase, **39:**342
sulfur, **39:**341; 346; 349; 358; 360; 361; 364
transferable, **39:**346
transfer, **39:**362
transferase, **39:**362–364
Sulfanes, **39:**329; 363
Sulfanilamide, **27:**420; 422; 428; 430; 432; 457, **35:**45. *See also* Sulfonamides
acetylation, **6:**257–260; 261–262, **10:** 221–222, **15:**186; 189–190; 200; 213
action on thiaminase, **15:**400
inhibition of purified amino acid decarboxylases, **6:**17; 18; 19
acetylation of, **18:**131
effect on luminous bacteria, **7:**252
effect on sperm, **24:**85
inhibition of carbonic anhydrase by, **47:**213; 224
and luminescence, **7:**237–240; 242–244; 246; 247
theory of action, **1:**165
use in determination of CoA, **16:**2

Sulfanilamide antagonists and growth, **27:**459
Sulfanilamide genetic studies, **17:**411; 455
Sulfanilic acid, **35:**45
  and folate metabolism, **27:**428–430
  inhibition of β-lactamase, **28:**295
Sulfanil-ovalbumin, as antigen, **15:**70
2-Sulfatase, **60:**290
6-Sulfatase, **60:**290
Sulfatase C, **60:**256
Sulfatase deficiency, **60:**255
Sulfatases, classification, **14:**245
Sulfatases A, B, and C, localization of, **24:** 321–322
Sulfatases in plants, **10:**310–311
Sulfate
  in cystine synthesis, **16:**277–279
    labeled, effect on heparin production, **16:**471
    in methionine synthesis, **16:**272
  transfer from PAPS, **22:**210–214
Sulfate-activating system, **18:**210
Sulfate activation, **17:**183
Sulfated mucopolysaccharide, tyrosine hydroxylase regulation, **70:**122
Sulfate esterification, **18:**209–210
Sulfate esters, **57:**377
  hydrolysis, **22:**215–231
    synthesis, **22:**207–215
      by routes not directly involving PAPS, **22:**214; 215
      sulfate-activating enzyme in, **22:**207
      sulfate-transferring enzyme in, **22:**207
Sulfate Esters, Synthesis and Hydrolysis (Roy), **22:**205
Sulfate formation, **46:**394
Sulfate ions, in ADP sulfurylase reaction, **22:**209
  in ATP sulfurylase reaction, **22:**208
Sulfate permease
  from *Escherichia coli,* **74:**186
  from *Salmonella typhimurium,* **74:**186
    gene for, **74:**189
Sulfate reduction in photosynthesis, free energy of, **25:**46
Sulfates, solubilization of, **17:**138; 142; 145; 146; 148; 149; 152
Sulfates in plant metabolism, **10:**223–224; 225; 229–232; 308
  reduction in microbial attack on hydrocarbons, **10:**448; 449; 470; 479
Sulfate transfer from PAPS, **25:**440
Sulfathiazole, **35:**45
  action on thiaminase, **15:**400
  and folate metabolism, **27:**427–428; 457
Sulfatidase, **38:**298; 302; 305; 307
  hog-kidney tissue, **38:**307
Sulfatide, **38:**295; 309, **60:**167
Sulfatide-G_{M1} activator, **60:**178
  assay, **60:**182
  binding studies, **60:**183
  deficiency, **60:**185
  genetics, **60:**184
  heat stability, **60:**183
  kinetics, **60:**184
  molecular weight, **60:**183
  physiological role, **60:**185
  properties, **60:**183
  purification, **60:**182
^{35}S-Sulfatide, **38:**302
Sulfenic acids, **7:**386–388; 395
Sulfenium cations, in disulfide exchanges, **27:**205–206
Sulfenyl, atoms, **39:**330
  sulfur, **39:**329; 330; 362
  thiosulfate, **39:**351
Sulfenylchlorides, in disulfide exchanges, **27:**205–206
Sulfhydrase, serine, *See* Serine sulfhydrase
Sulfhydryl, **36:**33
  enzymes, **39:**344
    group, **39:**346; 347; 351
      active site, **39:**347; 360
      essential, **39:**344–346; 357
    loss, **39:**345
    reagents, **39:**345
    regain, **39:**345
  proteins, **36:**74
Sulfhydryl agents
  effect on cathepsin III, **12:**205–206
  effect on RNase, **24:**238
    effect on sperm, **24:**69–72
  and phosphorylases, **18:**226–227
Sulfhydryl compounds, **15:**187; 189; 196; 213; 220–221; 228; 230–231; 239; 241; 244; 246; 248; 258–259; 261. *See also* Coenzyme A, Cysteine, Glutathione, Lipoic acid, and Thiol
  action on thiaminase, **15:**400–402; 404
  foreign, methylation of, **25:**421–422
  maintenance of reduced state, **15:**126
  oxidoreduction at interface, **15:**40

Sulfhydryl-directed reagents, **39:**344
Sulfhydryl enzyme, β-glucuronidase as, **16:**379
Sulfhydryl enzymes, **23:**148; 176; 226
effect of radiations on, **24:**381; 396
RNase as, **24:**199
inhibition, **15:**285–290
mechanism of action, **15:**10–11
Sulfhydryl enzymes and carcinogenic agents, **4:**238; 249–250
Sulfhydryl group, **46:**383, **58:**111. *See also* Thiol entries
and activity of hyaluronidase, **13:**214–215
and actomyosin formation, **13:**160–161
and enzyme inhibition, **13:**332
in glutamine synthesis, **13:**280
and muscle extensibility, **13:**173–175
and myosin deaminase activity, **13:**169
effect of fluoropyruvate on, **18:**152
of succinic dehydrogenase, **18:**75–76; 80; 82; 93
in Succinyl-CoA synthetase, **58:**151
Sulfhydryl groups, **7:**39, **27:**88; 212–219; 221–228; 230. *See also* Coenzyme A (CoA); Cysteine; Cysteine, Cystine; Disulfides, Mercaptans, and Thiols; Glutathione; Thiol groups and Thiols; individual compounds
and active acetate, **12:**329–330
activity in coenzyme A, **14:**182
and activity of hydroxysteroid dehydrogenases, **20:**269
in liver alcohol dehydrogenases, **20:**27
albsence in some arthropodins, **16:**162
activity in CoA, **16:**18
binding of quinonoid pigments, **16:**172
in melanin formation, **16:**172a
reactions with quinones, **16:**120–122; 126
relation to action polymerization, **16:**335
relation to thrombin production, **16:**29
and amylase activity, **12:**399; 400–401
of aspartate transcarbamylase, **28:**129
of β-lactamase, **28:**297
of laminarinase, **28:**347
and biological activity of proteins, **14:**131; 140; 149
and biological methylation, **12:**436–437; 441–442
and blood clotting, **9:**278–279, **10:**33
effect of light on, **9:**91; 129; 133; 134; 135
and fertilization membrane, **9:**278–279
in plant metabolism, **10:**268–284
and protein synthesis, **9:**428–429
in protein synthesis, **10:**286–288
relation to activity of alkaline phosphatase, **10:**97–98; 101
and semen metabolism, **9:**345
and catalase action, **14:**153
and choline acetylation, **12:**318–319
of coenzyme A, **14:**63–65
of cytochrome reductase, **22:**372
in flavine coenzymes, **22:**380–382
in LDH from various sources, **22:**359; 360
role in GAD, **22:**367; 368
of TPD, **22:**343–346
in yeast ADH, **22:**354–356
and denaturation, **23:**46–48; 76
titration of, **23:**37
and effect of radiations, **24:**391; 395–398
effect on sperm glycolysis, **24:**50
of phosphorylases, **24:**268–269
effect on some enzyme, **14:**390
and enzyme activity, **26:**100–103
of enzymes, **7:**394; 399
as inhibitors of plant toxins, **11:**407; 412
ionization constants, **19:**285; 324
of LADH, **27:**75–82
in metalloflavoproteins, **17:**238; 240; 245; 246; 251; 254; 256; 258; 259; 265; 271; 285
in xanthine oxidase, **17:**311; 315
in mixed function oxidases, **19:**176; 184; 185; 199
modification, on fructose 1,6-bisphosphtase, **53:**71–72
in oxygen transferases, **19:**83; 91; 97; 126
of pantetheine, **14:**59–61
and papain, **19:**269; 286–294; 324; 326; 328
partition at interfaces, **11:**47–56
and peroxidase action, **14:**158
and phosphoribulokinase, **19:**64
in proteinase activity, **14:**190
in protein structure, **14:**288–289
reactivity of in enzymic reactions, **33:**395–399
reactivity of in organic reactions, **33:** 395–399
and RuDP carboxylase, **19:**64
tyrosinase action on, **14:**133; 140; 144
and tyrosinase activity, **14:**88; 90; 91; 92; 93; 94
and urease, **19:**287

Sulfhydryl groups and luminescence, **25:**147
in photosynthesis, **25:**98–101
Sulfhydryl groups and myosin acitivity, **6:** 166–168
Sulfhydryl groups in chromidia, **8:**7; 34
Sulfhydryl modification, **49:**168; 170
by bromoacetyl pyridoximine phosphate, **49:**170
of the *a* subunit of tryptophan synthase, **49:**168
Sulfhydryl-protein ratio, and specific activity of papain, **19:**327; 328
Sulfhydryl reagents, **48:**225
and DPNH-MDH complex, **22:**370
and inhibition of heart LDH, **22:**359
reactions with LDH-DPN, -DPNH complexes, **22:**366; 367
effect on myosin, **16:**344
inactivation of ATPase by, **16:**335
inhibition of lipase by, **23:**148
Sulfhydyl group modification, **49:**172
by 5,5′-dithiobis(2-dinitrobenzoic acid), **49:**172
of holo β_2 subunit, **49:**172
by N-ethylmaleimide, **49:**172
by nitrothiocyanobenzoic acid, **49:**172
Sulfide, **7:**206; 207, **39:**329; 333; 352; 358–360; 362. *See also* Hydrogen sulfide
complex with carbonic anhydrase, **47:** 225–226; 229–230; 232
effect on tryptophan oxidase, **19:**122; 127
and formation of phosphatides, **7:**149
inhibition of metallic catalysts, **15:**26 (*See also* Disulfides)
inorganic, **39:**342
ion activity electrode, **39:**334
labile, **39:**339
in respiratory chain, **28:**170–172; 194; 226
in microbial metabolism, **16:**278
utilization, **7:**193
Sulfide-disulfide system, and hydrogen transfer, **26:**121; 129
Sulfide link. *See also* Disulfide link, Methionine, Sulfhydryl group
fission of, **12:**438–441
Sulfides, solubility, and metal ion toxicity, **17:**60
solubilization of, **17:**138; 142; 144; 145; 150
Sulfides in amino acid and peptide complexes, **10:**232
Sulfinates, **39:**333
alkyl and aryl, **39:**341; 362
organic, **39:**328
Sulfinic acid oxidase, **7:**387
Sulfinic acids, **7:**386–389; 401
β-Sulfinylpyruvic acid, **15:**118; 119; 124–125
in transaminations, **16:**196; 233
β-Sulfinylpyruvic dehydrogenase, **15:**118
Sulfite, **39:**328; 333; 355; 360; 362
in cystine metabolism, **16:**278
effect on metabolism of molds, **9:**548–549; 567
liquor, use in fat biosynthesis, **9:**654; 655; 661; 665–666; 668–669; 672; 674; 677–678
in hydrogen transfer model, **26:**138–140
oxidase, **39:**361
and papain, **19:**286
Sulfite-cytochrome *c* reductase system, localization of, **24:**306
Sulfite formation in the intermediate metabolism of sulfur in plants, **10:**287
Sulfite oxidase, **51:**109; 111; 112; 113; 122; 125; 127; 129; 157
in electron transfer, **19:**201
location in subcellular fraction, **24:**306; 341
xanthine oxidase as, **17:**239; 240; 323
Sulfite reductase, **46:**49
of spinach, **46:**63
Sulfite reductase, PN-dependent, **26:**129–131
Sulfite reductase gene, in *Salmonella typhimurium,* **74:**189
Sulfites, **15:**2; 3
solubilization of, **17:**142
Sulfitolysis, **33:**451; 493; 498, **42:**127
6-Sulfo-*N*-acetylglucosaminides, **60:**132
Sulfocholine, biological activity, **12:**483–484
N-Sulfoglucosamine sulfohydrolase, **60:**268
biological role, **60:**270
catalyzed reactions, **60:**269
optimum pH, **60:**269
Sulfoglutathione, reduction, PN-dependent, **26:**129
γ-Sulfo-α-ketobutyric acid, in transaminations, **16:**196
4-Sulfokinase, and amino sugars, **25:**440
6-Sulfokinase, and amino sugars, **25:**440
Sulfokinases, and dermatan sulfate, **25:**441
and PAP, **25:**158–159
Sulfolipid, in chloroplast lamellae, **30:**9
Sulfolobux acidocaldarius, **71:**188–189; 190

Sulfomucases, classification, **14:**246
Sulfomucodextrinase, classification, **14:**246
Sulfonamide, noncompetitive antagonists of, **16:**300
  reveral of inhibition, **16:**274
Sulfonamide-*p*AB relationship, **27:**418; 426–431; 457
Sulfonamide acetylating activity, localization of, **24:**310; 334
Sulfonamide analogs, of folic acid, **27:**428–431
Sulfonamide compounds, effect on sperm metabolism, **24:**49
Sulfonamides, **7:**207, **57:**138
  and action of nicotinic acid derivatives, **7:**418–419
  antagonism to *p*-aminobenzoic acid, **6:**38–40
    antagonism to vitamins, **6:**63–64
    as antibacterial agents, **6:**35–44; 58–67
      mode of action, **6:**37–38
      protein-enzyme competition, **6:**65–67
    inhibition of bacterial respiratory enzymes, **6:**58–67
    as inhibitory analogues, **6:**130–131; 132–133; 137; 139; 140; 144
    mutations caused by, **6:**94
    resistance of cells to, **6:**102–112
    selective action on bacteria, **6:**64
  assay, **7:**252
  binding to carbonic anhydrase, **47:**174–175; 213–224
    inhibition of carbonicanhydrase by, **47:**150; 202–204; 208; 212–215; 223–224
  effects on synthesis of H_2-pteroate and H_2-folate, **35:**44–46
  and folate synthesis, **27:**434; 457–460
    mechanism of action, **27:**420; 422; 426–433
    resistance to, **27:**431–432; 453
    synergism with reductase inhibitors, **27:**455–456; 460
  growth inhibition by and *S*-methylmethionine reversal, **22:**242
    resistance to, in *Diplococcus pneumoniae* mutants, **22:**17
  as inhibitors of H_2pteroate synthetase, **53:**355
Sulfonamides (S^{35}), **27:**428; 429; 432
Sulfonatability of oils, relation to toxicity, **10:**454
Sulfonate esters, carbonic anhydrase catalyzed hydrolysis, **47:**189; 205–207
Sulfonate group, **58:**25
Sulfonates, **17:**138; 149
Sulfonic acid group, as substitute for carboxyl group, **49:**76
Sulfonic acids, as inhibitory analogues, **6:**140
Sulfonic acids in plant metabolism, **10:**236
Sulfonium compounds
  biochemical functions, **22:**251–263
    as biological methyl donors, **22:**252
    biosynthesis, **22:**243–246
    general structure, **22:**238
    natural, discovery, **22:**241–243
    preparation, chemical properties, **22:** 246–251
    utilization by microorganisms, **22:**268–270
  derivatives of ethionine, **25:**382–388
Sulfonium Compounds, Biochemistry (Shapiro and Schlenk), **22:**237
Sulfonyl derivates of proteases, **33:**385–386
N-Sulfonyl imidazoles, **20:**92
β-Sulfonylpyruvic acid, **15:**119
Sulfoproteins, **7:**196–197
β-Sulfopyruvic acid, in transaminations, **16:**196; 233
Sulfoxide lyases, classification, **14:**229; 257
Sulfoxides
  diastereoisomers of, **22:**270
  enzymic cleavage of, **11:**393–395
  nonnitrogenous alicyclic, in plant metabolism, **10:**257–259
  reduction, PN-dependent, **26:**129
Sulfoxycysteine, **10:**282
Sulfur, **39:**328, **51:**150, **59:**105; 107; 301
  acceptor, **39:**340
    substrate, **39:**362
  chemistry, **39:**329
  colloidal, **7:**385–386
    in cytochrome, **7:**280
    elementary, **7:**193
    organic, oxidation in animals, **7:**369–407
    rate of replacement in proteins, **7:**194–198
  compounds in higher plants, **10:**227
    deficiency, effect on living cell, **10:**224; 225
    elementary, and plant metabolism, **10:** 229–232
    in the formation of chlorophyll in plants, **10:**224
    fungicidal action, **10:**238
    metabolism in green plants, effect of *p*H on influence of redox potential, **10:**228–229

of plants, enzymes involved, **10:**306–311
phytochemical cycle, **10:**228
phytochemistry, **10:**223–323
of proteins, effect of microorganisms, **10:**230
donor, **39:**341
substrate, **39:**340; 353; 357; 358; 362; 363
elemental, **39:**336; 352; 358; 361
elementary, in microbial metabolism, **16:**278
in melanin-protein binding, **16:**170–171
radioactive, use in study of methionine synthesis, **16:**272
(S^{35}), in *E. coli* metabolism, **16:**460
sources of cystine, **16:**277–279
transfer of, **16:**271–273; 278
enzyme, **39:**358; 364
intermediate, **39:**352
hydrogen acceptor, **5:**176–178; 198
metabolism, **39:**359
oxygenase, **39:**361
in papain, **19:**284
in plant metabolism, **9:**397–398
soluble organic, and protein synthesis, **9:**425–426
reduced, **39:**360
donor of, **39:**360
relation to tyrosinase activity, **17:**445
rhodanese, **39:**354; 357; 364
role in dehydrogenations by *Fusaria,* **5:** 189–192
transfer, **39:**341
transferase, thiosulfate-cyanide, **39:**359
Sulfur 35. Entries will be found under-specific compounds.
Sulfur, Organic, Oxidation in Animals (Fromageot), **7:**369
Sulfur, Phytochemistry (Bersin), **10:**223
Sulfur amino acids, genetic studies, **17:**402; 407–412
Sulfur anions, effects on sperm, **24:**81–82
Sulfurase, occurrence, **10:**307–308
Sulfur bacteria, **10:**232–234
metabolism, **1:**236; 265; 269; 270; 278; 280; 281
oxidation of thiosulfate, **10:**231
Sulfur-cobalt binding, in corrinoids, **26:** 246–250; 252–255; 258
Sulfur compounds
in body, diagram of changes, **7:**401
miscellaneous, **7:**402–404
turnover, **7:**192–199
in urine, **7:**370–371
enzyme reactions on, **5:**237–5:246
mammalian, **5:**238–246
sulfur bacteria, action on, **5:**237
Sulfur Compounds, Enzyme Reactions (Smythe), **5:**237
Sulfur compounds and biological methylation, **12:**436–437; 438–441; 441–442; 443–446; 446–447; 454–457; 458–474; 478–479; 482–486
Sulfur-containing amino acids, **11:**390–397
and growth, **7:**376; 383
Sulfur-containing groups, enzymes for transfer, **24:**315–316
Sulfur cycle, **74:**186–188
Sulfuric acid anhydrides, and pyrophosphorolysis, **18:**202; 203; 209–212
Sulfuric acid esters, in plant metabolism, **10:**234–236
in the organs of higher plants, **10:**226–227
Sulfuric esterases, classification, **14:**224; 245
Sulfuric ester hydrolases, localization of, **24:**321–322
Sulfur metabolism, enzymes for, **24:**341
Sulfur metabolism and pyridine nucleotides, **15:**124–126
Sulfurous acid in plant metabolism, **10:**236–238
Sulfur radicals, in disulfide exchanges, **27:** 206–207
Sulfur-substituted, enzymes, **39:**346; 349; 362–364
intermediate, **39:**347; 349
enzymic, **39:**352
Sulfur-sulfur bond, radical scission, **66:**153
Sulfur-sulfur bonds, **39:**329; 330; 364
scission, **39:**355; 356
Sulfurtransferases, localization of, **24:**315
Sulfurylase, ATP, **55:**347–348
Sulindac sulfide, **55:**198–199
Sulphydryl reagents, effect of, on activity, **36:**114
Supercooling, **51:**6–9; 28
calorimetric determinations, **51:**6–8
fluorescence probing, **51:**8–9
Superior cervical ganglia, tyrosine hydroxylase induction, **70:**165–166
Superoxide, **60:**75
in protoplasm, **46:**169
reaction with H_2O_2, **46:**170

Superoxide anion, **46:**165
Superoxide dismutase, **41:**35–90, **44:**189, **46:**169, **47:**117–123; 133, **56:**292, **58:**61–97, **67:**179
  active sites, **58:**75
  activity and mechanism, **41:**58–63
  activity stain, **41:**51
  amino acid sequences, **58:**73
  apoenzyme, **41:**60
  arginine position, **58:**76
  aspartate residue, **58:**74
  assay on polyacrylamide gels, **41:**49
  assays, **58:**61
  assay sensitivity, **58:**65
  assays for, **41:**35; 44–51
  atomic absorption spectrophotometry, **58:**76
  autoxidation assay, **58:**63
  autoxidation inhibition, **58:**62
  azide inhibitors, **58:**82
  bacterial, **41:**35; 70–73
    from *E. coli,* **41:**70
    the ferrienzyme, **41:**71
    intracellular localization, **41:**73
    the manganese enzyme, **41:**70
    from *Streptococcus mutans,* **41:**71
  in *Bacterioides fragilis,* **58:**77
  biological importance of, **41:**35; 78–88
  biosynthesis, **58:**83
    control of, **58:**84
    copper deficiency, **58:**84
    and 2,4-dinitrophenol, **58:**84
    effect of chelating agents, **58:**85
    magnesium deficiency, **58:**84
    and sulfur dioxide, **58:**84
    vitamin E deficiency, **58:**84
  from bovine blood, **41:**58
  from chicken liver, **41:**58
  circular dichroism spectra, **41:**60
  concerted antioxidant system, **65:**120–124
  Copper-Zinc, **41:**35; 51–70
  copper-zinc containing, **58:**70; 71
    active sites, **58:**78
    in bovine, **58:**78
    copper removal, **58:**77
    lysine residues, **58:**80
    phosphate inhibitors, **58:**82
    quantum chemical studies, **58:**82
    redox titration, **58:**78
  cyanide inhibition, **41:**45
  in desulfovibrio, **58:**75
  dianisidine autoxidation inhibition, **58:**64
  effects of metals, **41:**61
  electromorphs, **58:**71
  epinephrine autoxidation inhibition, **58:**62
  in eukaryotes, **58:**73
  evolutionary aspects, **58:**73
  ferrienzyme, **41:**86
  hematoxylin autoxidation inhibition, **58:**63
  histidine residues, **58:**74
  historical background, **41:**51–56
  historical perspectives on, **56:**325
  history, **58:**70
  homologies, **58:**74
  from human blood, **41:**58
  hybridization, **58:**71
  induction of, **41:**82–84
  inhibition by cyanide, **41:**67
  inhibitors, **58:**82
  intracellular production, **58:**83
  iron containing, **58:**72
    inhibitors specificity, **58:**82
    metal remaining, **58:**76
    x-ray crystallography, **58:**75
  Isoenzymes, **41:**67
  isolation procedures, **41:**56–58
  in liver mitochondria, **58:**75
  luminol assay, **58:**67
  lysine residues, **58:**80
  manganese containing, **58:**72
    x-ray crystallography, **58:**76
  mechanism of action of, **56:**329
  metal binding sites, **58:**76
    manganese-iron competition for, **58:**85
  metal centers of, properties of, **56:**327–329
  metal content of, **56:**326
  metalloenzyme substitutions of, **56:**304
  metal removal, **58:**76
  Mitochondrial, **41:**35; 73–78
    evolutionary significance, **41:**76
    intracellular distribution, **41:**73
    intramitochondrial localization, **41:**75
    site of synthesis, **41:**75
  mitroblue tetrazolium assay, **58:**62
  molecular properties of, **56:**326–327
  nuclear magnetic resonance of, **56:**311
  O_2-the substrate, **41:**35; 37–44
  oxygen toxicity, **41:**84–86
  as a Probe for O_2-, **41:**89–90
  pro-oxidant activity, **65:**124
  pyrogallol autoxidation inhibition, **58:**62

radioimmunoassay, **58:**67
rate constant for, **41:**45
reaction mechanisms, **58:**76
copper reduction, **58:**79
eqilibrium, **58:**81
imidazolate role in, **58:**79
pH dependence, **58:**80
reaction with oxygen, **58:**79
rifamycin assay, **58:**67
roles of Cu^{++} and Zn^{++}, **41:**60
stability, **41:**63
structure, **41:**64, **58:**73
tetrazolium oxidase, **41:**69–70
tryptophan position, **58:**76
types, **58:**70
types of, **56:**326
ultraviolet absorbance, **41:**60
from wheat germ, **41:**58
xanthine oxidase/cytochrome C assay, **58:**65
X-ray-crystallography, **41:**67
x-ray diffraction study, **58:**75
Superoxide Dismutases (Fridovich), **41:**35, **58:**61
Superoxide ion, and proly hydroxylase, **41:**273
Superoxide ion, O_2-, **46:**166; 167; 168
Superoxide radical, **41:**37; 39; 41; 46, **58:**61
biological sources of O_2-, **41:**39
the dangers of O_2-, **41:**41
decay of, **41:**46
toxicity of oxygen, **41:**39; 41
Superoxoferriheme, **41:**90
Superprecipitation of actomyosin, **16:**346
Suramin, affinity for β-glucuronidase, **16:**367
as inhibitor of β-glucuronidase, **16:**374
Surface active agents, effect on gelation of pectic substances, **20:**357
effect on pectinesterases, **20:**360
Surface-active agents, effect on sperm, **24:**70
Surface array biogenesis, *Caulobacter* cell cycle, **71:**6
Surface asymmetries, in distribution of transport systems, **49:**89
Surface effects, **34:**446
catalytic, **28:**32
in soils, **28:**367
tension, **34:**467
Surface effects in radiosensitivity, **24:**399–401
Surface energy, definition, **28:**6
Surface potential, and photoxidation, **11:**76
in monolayers, **11:**82–85
and speed of hemolysis, **11:**78–79
Surface pressure, **48:**329; 332–335; 349; 351; 352; 363
in monolayers, **11:**58–64; 82–85
Surfaces
acetylcholinesterase in, **12:**271–274; 281
reactions at, **11:**35–89
Surface structures, detachment of enzyme, **28:**266
Surface tension, definition, **28:**6
Surface tension in enzyme reactions, **33:**263
Surface tension of proteins, **9:**116–118
Surfactants, in cocoonase preparations, **45:**409
Surrogate cooperativity, resulting from ligand substitution reactions, **53:**326–327
Surrogate ligands
activating Cat ligands distinguishing from, **53:**338
characteristic kinetic properties, **53:** 322–330
definition, **53:**317
direct and allosteric surrogates, **53:**331
effects on double reciprocal plots, **53:** 318–319; 322
other properties of, **53:**330–331
Svedberg number series, **4:**2; 3; 6
Swainsonine, **56:**210, **60:**154
Swarmer cell, *Caulobacter* cell cycle, **71:**3; 4
Sweat. *See also* Fermentation of tobacco
of cigar leaf tobaccos, **10:**358–361; 361–399
Sweat processes for tobacco, **10:**327–333
microbial activity, **10:**409–412; 415–425
Swelling, of mitochondria, **23:**383–385, **29:** 67–68
Swelling of starch granule, **12:**367–369
Swine pepsin, specificity, **1:**68–71
Sylvic acid, **17:**154
Symmetrical structures, in glucagon, **24:**431
in RNase, **24:**427
Symmetry, rotational, **50:**367
torsional, **50:**363–366
Symmetry model, **33:**95. *See also* Dissociation of proteins
Symmetry model for oligomeric proteins, quasi-equivalence, **33:**90–91
spatial-equivalence theory, **33:**89–90
steric hindrance in, **33:**90
Sympatholytic agents, **7:**548

Symport, alternative mechanisms of coupling in, **29:**54–59
definition of, **29:**38
diagram of, **29:**55
electrogenic property of Na^+-glucose, **29:** 59–61
Na^+-amino acid, **29:**61–65
in Ehrlich ascites cells, brain cortex, thymus nuclei, intestine, and leucocytes, **29:**61
Na^+-glucose, **29:**53–54
in intestine, kidney, diaphragm, leucocytes, and Ehrlich ascites tumor, **29:**61
Symporter, Na^+-amino acid, **29:**63
Synapses, **43:**111; 112; 113; 200; 203
Synaptosomes, **50:**318
Syncatalytic, changes in conformation, **37:**503
Synechococcus, phosphoribulokinase from, **74:**114
Synechococcus 6301, ADP-Glucose pyrophosphorylase from, **46:**338
Synechocystis, phosphoribulokinase from, **74:**100; 120; 121
Syneresis of milk clot, **15:**438; 442
Synergism, **7:**244
between cozymase and nicotinamide, **7:**421
sulfonamides and reductase inhibitors, **27:**455–456; 460
Synergism among factors affecting alkaline phosphatase, **10:**102–104
Synergistic effect, **7:**593; 595
Synergistic inhibitors, **43:**178
Synergy site, fibronectin, ligand-binding site model, **70:**15
Synovial fluids, hyaluronidase action on, **13:**200; 218; 229–230
Synthase, *See* Tryptophane desmolase
Synthease, **5:**107
Syntheases and lyases, classification, **14:**223; 228–229; 236–238; 256–257; 258
Syntheses, and 4-C-dicarboxylic acids, **23:** 403–404; 413–415
from glycolate, **23:**440–446
from oxalate, **23:**449–452
of polyisoprenoid compounds, **23:**476
Synthesis
of disaccharides with bacterial enzymes, **10:**123–143
of polysaccharides by enzymes, **10:** 140–141
peptide, aspects of methods of, **29:**393–406
Synthesis of polysaccharides, **11:**297–337; 348–373
Synthesis of starch, **12:**410–427
Synthetase
mammalian liver, methylation, **25:**408
poly (ribitol phosphate)
pH optimum of, **30:**240
properties of, in *Staphylococcus aureus,* **30:**240
site of, **30:**239–240
tRNA, **56:**293
tryptophan, chromatography of, **30:**149
Synthetase D phosphatase, **34:**330; 332
Synthetase I kinase, **34:**329; 332
activation by cyclic AMP, **34:**329
Synthetase kinase, effect of cyclic AMP on, **34:**336
Synthetase phosphatase, **34:**344
Synthetases, Aminoacyl-tRNA (Kisselev and Favorova), **40:**141
Synthetases, in aromatic amino acid synthesis, **28:**63
Synthetic enzyme, **34:**494
membrane, **34:**454; 494
micromolecules, **34:**454
redox dyes, **34:**477
Synthetic fragment, **42:**154; 157
Synthetic peptides, fibronectin
development of, **59:**35–36
fibroblastic cells, **59:**26–31
macrophages, **59:**34
metastasis of, **59:**36–38
platelets, **59:**31–34
slime mold cells, **59:**34–35
Syringaldehyde, **34:**213
relation to lignin, **18:**355–358; 360–361
Syringic acid, **18:**374
Syringin, **56:**201
labeled, preparation of, **18:**372–373
Syringyl residues, in lignin, **18:**366; 370–374
Syringyl type lignin, **16:**147
System
antiport, diagram of, **29:**51
arabinose
characteristics of induction of, **30:**111–112
genetics and regulation of, **30:**112–116
extrinsic
activation of Factor X by, **30:**284–287
components of, **30:**269

definition of, **30:**268
intrinsic
activation of Factor X by, **30:**287–296
components of, **30:**269
definition of, **30:**268
discovery of, **30:**259
maltose, induction of, **30:**119–120
melibiose, determinants of structure of, **30:**127
photochemical, activities of, **30:**46–47
carotenoids in, **30:**50
chlorophyll-protein complexes of, **30:** 54–55
cytochromes in, **30:**48–49
fluorescence properties of, **30:**50–53
fractionation of, **30:**43–56
particles obtained by the action of Triton X-100 in, **30:**54
particles obtained by the digitonin procedure in, **30:**44–53
of photosynthesis, **30:**1–79
preparation and chlorophyll composition of, **30:**44–45
quinones in, **30:**50
rate-limiting reactions and coupling of, **30:**42–43
separation by density gradient centrifugation of, **30:**53
trace metal and P-700 contents of, **30:** 47–48
rhamnose, enzymes of, **30:**116
System ASC, for AIB, **49:**84
Systemes enzymatiques inducibles du metabolisme des oses chez Escherichia coli (Buttin), **30:**81
Systemic effect, and growth rate of tumor, **3:**338–339
reversibility of, **3:**339–340
System L, **49:**86; 87; 88; 91
and blood-brain barrier, **49:**91
fallacious views on, **49:**86; 87; 88
Systems, microheterogeneous, **34:**446
model, **34:**446
Systems A and L, cooperative relation between, **49:**89
pumping of amino acids in cells, **49:**90
System *N* transporter
glutamine synthetase and hepatocyte heterogeneity, **72:**61–63
hepatic glutamine transport, **72:**45–47
hepatocyte hydration, **72:**48–53
System *n* transporter
glutamine synthetase and hepatocyte heterogeneity, **72:**61–63
hepatic glutamine transport, **72:**45–47
Szent-Györgyi cycle, **3:**236–238

T

T. thermophilus, **71:**186
T. tonsurans, collagen action, **23:**177
TAA, *See* Taka-α-amylase
Tabakmosaikvirus und seiner Multanten, Structure (Schramm), **15:**449
Tabun, **34:**9
Tachyphylaxis, **7:**547
Tactoids, **5:**10
Tadpole liver, **42:**256
Taka amylase, **56:**210
Taka-α-amylase, denaturation of, **23:** 32–45
oxidized and reduced, **23:**74–76
Taka-amylase, linkages in, **25:**452
Taka amylase A, **37:**495
Takadiastase
action on penicillin, **28:**243
column chromatography, **22:**345
deaminase in, **15:**80
phosphatase in, **15:**416
in thiaminase determination, **15:**395
inhibition of penicillin, **10:**253
in plants, **10:**311
Taka-diastase, **17:**167; 197
fractionation, **14:**308
isolation, **14:**378
separation by chromatography, **14:**348; 363–364
Taka-diastase deaminase, action on diphosphate from CoA, **16:**14
Takamine-HT, **23:**179; 227
Taka-proteases, multiple forms, **25:**351
L-Talose in polysaccharide synthesis, **15:**306
Tamm, *See* Eigen-Tamm mechanism, formation of E-Mn-S complexes for
Tannase, classification, **14:**244
separation by chromatography, **14:**349
Tannin, reaction with borax, **12:**495; 506–507
Tannines, metabolism, **16:**133; 136; 141–143; 150–153

Tanning, of cuticle, **16:**110; 153–163
of proteins by quinones, **16:**123–125; 153–163
Tanning reaction in cacao, **25:**479–480
Tannins
in air-cured tobacco, **6:**334–337
in flue-cured tobacco, **6:**364
effect on polygalacturonases, **20:**365; 370
in tea, **2:**121–125 (*See also* Tea-fermentation)
condensation, **2:**124; 125
constitution, **2:**121
oxidation, **2:**121; 122; 128–131
polymerization, **2:**122; 123
tannin titer, **2:**122
Tannins in tobacco, **10:**347–349; 373–375
Tannins of cacao, **25:**463; 478
T1 Antigens, genetics of, **31:**114
Tapioca, **39:**339; 359
Taraxacum kok-saghyz, rubber-producing flower, **29:**222
Taraxein, as a brain-specific antibody, **29:**509
effect on cell cultures of, **29:**510
in schizophrenics, **29:**508–510
Target size analysis, **53:**32
Target volumes, effect of radiations on enzymes, **24:**365–367; 371
for trypsin, effect of concentration, **24:** 386–387; 390
Tartaric acid, **18:**283
metal ion complexes of, model catalyses with, **21:**188
in mold metabolism, **9:**549; 576
Tartrate, **15:**261
oxidation, in mitochondria, **21:**22; 24; 25
D-Tartrate, **51:**236
L-Tartrate, **51:**266
meso-Tartrate, **51:**236, **57:**116; 121
effect on Krebs cycle, **19:**360
effect on transamination, **19:**363; 364; 368
Tartrate dehydrogenase, in mitochondria, **21:** 22
Tartrates, in solubility studies, solubilization by, **17:**144; 149; 152
Tartronic semialdehyde, and glycolate metabolism, **23:**442–444; 446
and oxalate metabolism, **23:**450
Tartronic semialdehyde reductase, **23:**445; 454
Tartronsemialdehydreductase, *See* Glyceric acid dehydrogenase
Tatigkeit in Boden, die Bestimmung der Biologischen, mit Enzymmethoden (Hofmann and Hoffmann), **28:**365
Taurine, **5:**81–82; 239, **49:**57; 59, **56:**18
from cysteic acid, **7:**389
from cysteine, **7:**383; 385; 387
from cystine, **7:**393; 394; 396
form homocysteine, **7:**378; 379
from methionine, **7:**372–373
and retina, **49:**93
and thiaminase, **15:**402; 404
Taurine acyltransferase, localization of, **24:**311
Taurocholate
effect on lipolysis, **23:**150–152
formation of, **24:**311
Taurocholic acid formation, and CoA, **18:**208
Taurocyamine, in creatine kinase reaction, **54:**384
Taurodeoxycholate, **48:**348
Tautomeric catalysis, **50:**11–14
Tautomeric ratio, **50:**3
Tautomerism, and resonance, **50:**1
Tautomerism in Pyridoxal Phosphate and in Enzymatic Catalysis (Metzler), **50:**1
Tautomerization, **49:**34; 35, **50:**1–37
in enzymatic catalysis, **50:**1–37
of peptides, **50:**14
of purines, **50:**36
in pyridoxal phosphate, **50:**1–37
of pyrimidines, **50:**36
Taxifolin, **55:**48
metabolism in plants, **16:**141
Taxiphyllin, **56:**201
Taxol, GTPase reaction, microtubule dynamics, **69:**146
Tay-Sachs
disease, **38:**294; 295; 298; 300; 301; 308; 310
B-variant, **38:**301; 307
O-variant, **38:**301; 307
variant, **38:**294; 295; 298
Tay-Sachs disease, **60:**126
TCA, *See* Trichloroacetic acid
T-cell immunodeficiency, deficiency of purine nucleoside phosphorylase and, **51:** 200–201
T-Coliphages, pentose metabolism, **28:**423–428
TDP, *See* Thymidine-5′-pyrophosphate
TDP-D-glucose-2-epimerase, **26:**53
TDPglucose oxidoreductase, reaction catalyzed by, **31:**107

TDP-D-glucose pyrophosphorylase, allosteric regulation, **28:**111; 113
TDPglucose pyrophosphorylase, reaction catalyzed by, **31:**109
TDPrhamnose, accumulation of, in mutants, **31:**116
  role of, in lipopolysaccharide synthesis, **31:**92
TDPrhamnose synthetase, reaction catalyzed by, **31:**107
TDP-sugars, **26:**64; 66
TdR, **54:**58
  with 5-fluorouracil, **54:**100–101
  in reversing cytotoxic effects of fluoropyrimidines, **54:**98; 99
Tea-fermentation, **2:**113–133
  chemical changes during, **2:**120–126
  mechanism, **2:**126–132
  optimum *p*H range, **2:**117
  oxidizing enzymes, properties and nature, **2:**115–120
  and respiration, **2:**126–132
Tea Fermentation, Chemistry (Roberts), **2:**113
Tea leaf oxidase, role in browning, **16:**151
Tea leaves, components. *See also* Tea-fermentation
  carbohydrates, **2:**125
  ether soluble matter, **2:**125; 129
  of fresh leaves, **2:**120
  nitrogenous compounds, **2:**125
  tannins, **2:**121–125
Teeth
  effect of fluorovalerate on, **18:**138–139
  and vitamin E, **6:**509
Teichoic acid, **30:**223–253
  of *Actinomyces streptomycini,* **30:**231
  of *Actinomyces violaceus,* **30:**231
  of *Bacillus licheniformis,* **30:**228
  of *Bacillus stearothermophilus,* **30:**227
  of *Bacillus subtilis,* **30:**225; 230
    the effect of phosphate deficiency on synthesis of, **30:**250
  biosynthesis of, **30:**234–245
    in *Lactobacillus plantarum,* **30:**239
  cellular location of, **30:**245–250
  complete extraction of, in dilute trichloroacetic acid, **30:**247–248
  extraction with trichloroacetic acid, **30:**246
  in Gram-positive bacteria, **30:**243
  of group D streptococci, **30:**227
  isolation of, **30:**224
  of *Lactobacillus arabinosus,* **30:**225; 231
  of *Lactobacillus casei,* **30:**225
  of *Lactobacillus plantarum,* **30:**229
  location of, in *Bacillus megaterium,* **30:**245
  mechanism of biosynthesis of, **30:**237–238
  in membranes, **30:**245–246
  of *Micrococcus hyicus,* **30:**230
  phenol extraction of, **30:**246
  of *Pneumococcus,* **30:**232
  stability to alkali, **30:**226
  of *Staphylococcus aureus,* **30:**231
    glycosyl transfer to, **30:**240
  of *Staphylococcus lactis,* **30:**225; 229
    the elucidation of the linkage between peptidoglycan and, **30:**249
    isolation of, **30:**238
  stereochemistry of, **30:**229
  of *Streptomyces griseus,* **30:**232
  structure of, **30:**224–234
  in the wall, **30:**247–250
  in wall polysaccharide, **30:**250
  from walls of *Actinomyces antibioticus,* **30:**227
  in walls of *Staphylococcus lactis,* **30:**248
  in walls oxidized with sodium metaperiodate, **30:**248
Teichoic acids, in bacterial walls, **26:**223
  of *S. aureus,* **26:**73–76
Teichoic Acids (Archibald, Baddiley, and Blumson), **30:**223
Tellurium, inhibition of enzymes by, **11:**238
  mercaptide formation, **11:**217; 218
Tellurium compounds and biological methylation, **12:**433–434; 435; 438; 442; 444; 449; 453–457; 458
Temperature, **34:**452; 471. *See also* Freezing, Heat denaturation, Thermal coefficients, Thermoluminescence
  and denaturation of proteins, **7:**248–251
    and enzyme activity, **7:**259
    and luminescence, **7:**223–242; 244–251
  effect on, denaturation of RNase, **24:**188–193; 196; 249–251
    sperm, **24:**74–75
  effect on carotenogenesis, **21:**316–318
    effect on model catalyses by metal ions, **21:**172; 173
  effect on degradation of pectic substances, **20:**349; 350

Temperature (*Continued*)
effect on pectinesterases, **20:**360
effect on polygalacturonases, **20:**365
effect on luminescence, **9:**41
effect on rates of reaction, **9:**31–49
effect on papain, **19:**280; 315–317; 320–324
effect on rate of interfacial reaction, **11:** 58–64
Temperature coefficient, of fluorescence decay, **11:**96; 101
Temperature dependence
of chemical kinetics
of enzyme kinetics, **3:**41–62
general, **3:**36–41
of hydrolytic enzymes, **15:**279–280
of mutation rate, **15:**474
of red cell agglutination, **15:**55–64
of thiaminase, **15:**397
Temperature dependence in radiation effect on enzymes, **24:**363; 364; 368; 371–373; 392
Temperature-dependent modulation, branch flexibility, oligosaccharides, **66:**71–76
Temperature effects on physiological phenomena, **3:**59–60
Temperature jump, relaxation methods, **25:**2–9; 20; 31; 33
Temperature-jump relaxation of carbondioxide-bicarbonate system, effect of carbonic anhydrase on, **47:**184–186
Template
for enzyme synthesis, **22:**135; 145; 146
information of polypeptides, **24:**21–25
Template-directed assembly, microtubule dynamics, **69:**132
Template mechanism, of enzyme synthesis, **22:**132–134; 139; 140
Temporal regulation, *Caulobacter* cell cycle, transcription regulation, **71:**16
Tenascin, fibronectin, central cell-binding domain of, **70:**11
Tenebrio larvae, **59:**81
Tenebrio molitor, **45:**393
Tenotomy and muscular atrophy, **6:**196; 200; 201
Tension, development in actomyosin, **16:** 346–347
of actin-free myosin, **16:**345
of myofibrils, relations to ATPase activity, **16:**338–339
TEPP, *See* Tetraethylpyrophosphate
TEPP (tetraethyl pyrophosphate), effect on hydrolysis of acetylcholine, **12:**285; 288–290; 308–311
Teprotide, **57:**26
Terephthalanide, inhibition of tumor growth of, **29:**336
N-Terminal amines, in melanin formation, **16:**172a
reactivity with quinones, **16:**126
N-Terminal amino acids, of arthropodin, **16:**157; 160; 161
of fibrinogen and fibrin, **16:**37; 40–42
of myosin and its subunits, **16:**319; 323; 327
N-Terminal aspartic acid, metal binding by, **20:**66
Terminal deoxynucleotidyl transferase, *See* Deoxynucleotidyl transferase, terminal
Terminal Deoxynucleotidyl Transferase: Biological Studies (Bollum), **47:**374
5′ Terminal dinucleotide, **35:**467
Terminal dNT transferase, **56:**292
Terminal enzymes, of seedlings, sequence of, **18:**41–42
N-Terminal histidine, acetyl and phosphoryl derivatives of, **20:**84
metal binding by, **20:**56; 66; 71
occurrence, **20:**71; 94–97
C-Terminal histidine, *N*-substituted derivatives of, **20:**84
Terminal oxidase, **17:**66; 67; 114
and coenzyme Q, **25:**290
and pure nuclei, **27:**254
steady-state level of copper, **27:**188; 190
Terminal oxidases, **23:**266–268; 312. *See also* Cytochromes, and Cytochrome oxidase
phenol oxidases as, **18:**310–311
in plants, **18:**25; 35–48
Terminal oxidation, by phenolase complex catalysis, **16:**131–136; 146
N-Terminal proline, in protamines, **16:**444
reaction with quinones, **16:**114
Terminal reducing group, reduction of, **36:**104
Terminal residues, in papain and mercuripapain, **19:**294–301; 324
C-Terminal residues, in protein structure studies, **14:**280; 285–288; 289–302; 309–310
N-Terminal residues, in protein structure studies, **14:**280–285; 289–302; 309–310

Termination, of polymer chains, **30:**187
Ternary complex, **34:**52; 476. *See also* Enzyme complex
conformational change in enzyme, **27:**23–29
crystallization of, **27:**2–4; 9–10; 16–23
formation, in slow reaction, **27:**4–13
with LADH, **27:**65; 72; 78
Ternary complexes, **22:**351
with cyanide, **22:**352
of ethyl mercaptan, **22:**352
Ternary inhibitor complexes, isomerization of, eukaryotic dihydrofolate reductase, **70:**56–58
Teropterin
folate activity, **27:**430
in histidine degradation, **21:**406
isolation, **21:**371
Teropterin (C^{14}), from adenine-2-C^{14}, **27:**423
Terpenoid acids, in squalene biosynthesis, **22:**319; 320
Terpenoid chain biosynthesis, of coenzyme Q, **25:**299–304
Terpenoid phytoalexin, **55:**49–83
biosynthesis of, **55:**50–81
formation of precursors, **55:**50; 57–66
general outline of, **55:**57
sites of, **55:**63
classification of, **55:**49
degradation of, **55:**81–83
via fungal metabolism, **55:**83
via host metabolism, **55:**81–83
of eggplant, **55:**73–76
enzymology of, **55:**50–81
of Jimson weed, **55:**73–76
of potato, **55:**66–69
sources of, **55:**50; 51–56
structure of, **55:**50; 51–56
of sweet pepper, **55:**69–73
of sweet potato, **55:**76–78
of tobacco, **55:**69–73
of tomato, **55:**73–76
Terpenoids, *See* Coenzyme Q
nonrubber, **29:**243–244
Terramycin, effect on carotenogenesis, **21:**319
Terregens factor, **35:**168
Terthienyl in plants, **10:**256–257
Tertiary folding, of proteins, **47:**88
Tertiary structure, **42:**83; 105; 111; 113; 130; 137; 139; 140; 143; 159, **47:**47; 76; 87–88; 114; 135–137
of allosteric enzymes, **28:**53–56
of β-lactamase, **28:**299; 305
glutathione S-transferases, **69:**8–11
Testane-3α, 11β-diol-17–17one, microbiological conversions, **20:**272
Testane-3, 17–17dione, microbiological formation, **20:**257
Testane-3α-ol-17-one(5-cis), microbiological conversions, **20:**272
Testololactone, microbiological formation, **20:**261; 263
Testosterone, **15:**127, **46:**118; 119, **56:**275
effect on β-glucuronidase, **16:**388; 394; 400
effect on *Neurospora crassa,* **20:**264
microbiological formation and conversions, **20:**249; 257; 259; 263; 273; 276
effect on sperm, **24:**84
tyrosine hydroxylase induction, **70:**172
Testosterone propionate, effect on β-glucuronidase levels, **16:**389–390
Testosteronpropionate, **46:**158
Tetanolysin, activation energy, **9:**46
Tetanus toxin, bound to gangliosides, **50:**318
Tetrabenzine, and LSD 25, **29:**518
Tetrachlorobenzoquinone, *See* Chloranil
Tetrachlorodibenzo-*p*-dioxin, **60:**39
2,3,7,8-Tetrachlorodibenzo-*p*-dioxin, **46:**139; 406
Tetrachloroethylene, **51:**33
Tetracycline
and translational repression by arginine, **40:**81; 82
in treatment of plant diseases, **20:**392
Tetradecanal-oxidizing extract, **27:**486
Tetradecane ($^{14}C_1$-labeled), oxidation of, **27:**490–491
Tetradecanoly phorbol acetate (TPA), tyrosine hydroxylase transcriptional regulation, **70:**174
Tetradecapeptide, **50:**79
N,N′-Tetraethylcystamine, disulfide exchanges, **27:**209; 213; 219
Tetraethyldiuramdisulfide, exchanges, **27:**203
Tetraethylpyrophosphate, effect on cholinesterase, **20:**289; 305–307; 312
Tetraethyl pyrophosphonate, as inhibitor of serum cholinesterase, **22:**15
Tetraethyltin, change in cells, **24:**308
Tetraglycine, methyl ester, **19:**237

Tetrahedral addition intermediate, **43:**261; 370; 371
Tetrahedral intermediate, **43:**165; 171; 172; 173; 199
Tetraheme protein, **17:**345
Tetrahydroallicin, **11:**385
Tetrahydrobiopterin, **35:**285, **46:**166, **67:**84, **74:**238
  bioperin binding, eukaryotic dihydrofolate reductase, **70:**32
  structure of, **74:**237
Tetrahydrobiopterin, NO synthesis enhancement, **65:**302–304; 319
Tetrahydrocoenzyme Q_{10}, occurrence, **25:**322
Tetrahydrocortisone, glucuronide as substrate for β-glucuronidase, **16:**380; 383
Tetrahydrodipicolinate *N*-succinyltransferase
  bacterial lysine biosynthesis, **72:**298–302
  chemical mechanism, **72:**299–301
  substrate identification, **72:**298–299
  three-dimensional structure, **72:**301–302
Tetrahydrofolate, **35:**249, **53:**84; 93–99, **67:**175
  amino acid analogs, **38:**278
    carbon-6 hydrogen, **38:**242
    depletion of, **38:**270
    hydrogen transfer, **38:**238; 240
    iminium derivative, **38:**256
    lysine analogs, **38:**279
    N^5, N^{10} -methylene derivative, **38:**238
    polyglutamate derivatives, **38:**280
    synthesis from dihydrofolate, **38:**240
    synthesis with $NaBH_4$, **38:**240
    thyminyl derivatives, **38:**254
    tritium labeled, **38:**240; 242; 243
  eukaryotic dihydrofolate reductase, substrate binding, **70:**50–52
  occurrence, **27:**421
    synthesis in *E. coli,* **27:**424; 426
Tetrahydrofolate-containing coenzymes, **27:**419; 425; 426
Tetrahydrofolate dehydrogenase, location of, **24:**302
Tetrahydrofolic acid, **17:**184, **67:**81
  amination of, **18:**230
  biosynthesis, **21:**379; 380; 394; 395
    formylation, **21:**407–409
    in histidine biosynthesis and degradation, **21:**405–407
    in interconversion of active formate and active formaldehyde, **21:**415
    in mitochondrial serine biosynthesis, **21:**429
    in purine biosynthesis degradation, **21:**203; 204; 206; 207; 222; 223; 400; 402
    in pyrimidine biosynthesis, **21:**289; 290; 421–423
    in pyruvic acid biosynthesis, **21:**431
    structure, **21:**219; 379
    in tyrosine biosynthesis, **21:**432
  and glycine, **23:**457; 459–460
Tetrahydrofolic acid formylase, binding to FAH_4, **27:**421
Tetrahydrofolic formylase, **21:**225; 226; 407; 417
  and ATP synthesis, **23:**333; 338
Tetrahydrofuran, cytochrome *c* oxidase, **71:**138
N-1-Tetrahydrofuranosyl-FU, **54:**58; 62
2,3,4,5-Tetrahydrofurantetracarboxylate (diastereomers), **51:**236
4-(α-Tetrahydrofuryl) butyric acid, **18:**328; 330
Tetrahydroneopterin, **35:**285
Tetrahydroporphyrin, **46:**49
Tetrahydroprotoporphyrin-9, **46:**58; 86; 91
Tetrahydropteridine derivative, and hydrogen transfer, **26:**125
Tetrahydropterin, **67:**140
  enzymatic and nonenzymatic oxidations, **67:**215
  enzyme-catalyzed oxidation of, **74:**251–252
  oxidation, **67:**184
  phenylalanine hydroxylase specificity, **67:**172–175
  reaction of oxygen with, mechanism of, **74:**251
  tyrosinase/tyrosine hydroxylation and, **70:**106; 108
  tyrosine hydroxylase regulation, **70:**131
5,6,7,8-Tetrahydropterin, **67:**81–82
Tetrahydropterin 4a-carbinolamine, **67:**92; 187; 191–192; 194–195
Tetrahydropterin 4a-hydroperoxide, **67:** 187–188; 191; 193–194; 209
Tetrahydropterin coenzyme, tyrosine hydroxylase, **70:**200–201
  substrate specificity, **70:**110–116
Tetrahydropterin hydroperoxide, **35:**299
Tetrahydropteroic acid, conversion to FAH_4, **27:**424
5,6,7,8-Tetrahydropteroylglutamic acid, *See* Folate-H_4 don

Tetrahydropteroyltriglutamate
binding to formylase, **27:**421
methyl synthesis, **25:**405
Tetrahydropyridines, formation of, **26:**140–143
4′,5,6,7-Tetrahydroxyflavone glucuronide (scutellarin), hydrolysis, **16:**381
Tetrahymena, ethionine in proteins of, **22:**271
Tetrahymena geleii
effects of ethionine, **25:**376
and lipoic acid, **18:**320; 324; 331–332; 339; 342
Tetrahymena geleii, 214, **15:**375
Tetrahymena piriformis, use of acetate, **23:**411; 412
Tetralin hydroperoxide, chemiluminescence, **25:**121
Tetram, **55:**176
Tetrameric structure L_2 H_2, **44:**192
Tetramethylammonium hydroxide, **65:**246
Tetramethylglucose, mutarotation of, **15:** 294–295
Tetramethylglucose mutarotation, **43:**383
N,N,N′,N′-Tetramethyl-*p*-phenylenediamine (TMPD), as artificial electron donor, **30:**23
Tetranitromethane, **39:**203, **41:**38; 48, **58:**81, **59:**115
Tetranitromethane (TNM), effect on blood clotting, **10:**23
Tetranucleotide hypothesis, **9:**509–511
Tetrapeptides, **38:**204, **47:**61–62; 105
breakers, **47:**52–53; 101; 102
Tetrapeptide sequences, *See* Peptide sequences
Tetrapyrroles, **46:**37; 50; 66. *See also* Bile pigments, Catalases, Cytochromes, Hematin compounds, Hemes, Hemochromes, Hemochromogens, Hemoglobin, Hemoproteins, Iron porphins, Metalloporphins, Peroxidases, and Porphins
absorption spectra and structural characteristics, **7:**340–344
activity and side chain composition, **7:** 314–315; 322–324
biosynthesis, **46:**90
regulation in greening plant plastids, **46:**97
regulation in *Rhodopseudomonas spheroides,* **46:**90
di-, and trypyrroles, interrelationships, **7:** 330–332
distribution, structure, and properties, **7:** 305–368
linear, **46:**66
magnesium, **7:**307; 308; 320–321
naturally occurring, distribution, structure, and function, **7:**307–322
open-chain, **7:**321; 333; 358–363
pathways, **46:**101
pants, **46:**100–101
regulation, **46:**84
physical properties, **7:**339–363
Tetrapyrroles, Distribution, Structure, and Properties (Granick and Gilder), **7:**305
Tetrapyrrolic compounds. *See also* Catalase, Chlorophyll, Hemoglobin, Hemoproteins, Peroxidase
iron and magnesium in, **12:**13–20
Tetrasaccharide, **60:**239
Tetrasaccharides
hydrogen bonding, **65:**234; 238
from transglycosylations, **17:**200–202; 205; 212–213; 221
Tetrasodium imidodiphosphate, structure, **43:**15
Tetrathionase, enzymic adaptation, **13:**82
Tetrathionate
oxidation of thiols by, **10:**230–231
reaction with thiols, **11:**222; 226
Tetrathionate ions, S exchange in, **27:**203; 213
Tetrazolium dyes, test for coenzyme Q, **25:**292
Tetrazolium oxidase, **58:**72
superoxide dismutase, **41:**69–70
genetic variants of, **41:**69
Tetrazolium reduction inhibitor, **41:**69
Tetrazolium salts, reduction of, **15:**10; 31
Tetronic acid, effect on carotenogenesis, **21:**319
Tetrose
from glucose, **15:**121
in ribose synthesis, **28:**401; 402
Tetrose phosphate
in formation of sedoheptulose-1,7-diphosphate, **16:**291
in glucose-6-phosphate oxidation shunt, **15:**154; 164–166
T-even bacteriophage, effect of acridines, **27:**309
nucleotides of, **27:**291
structures, **27:**298–299
Texture, **34:**521
TF antigen, **52:**122
T-fibrin, *See* Thrombin fibrin

T-forms, definition of, **31:**114
Thalidomide, acylation of diamines and polyamines, **36:**240
Thalium in fructose-bisphosphatase activity, **54:**132
Thallophyta, carotenoids in, **21:**300
Thaw rigor of muscle, **13:**179–180; 190
Theiler's virus, *See* Virus
Thenoyltrifluoroacetone, inhibitor, **28:**160; 177; 178; 190; 199; 202; 212
Theobroma cacao, curing of seed, **25:**457–492
Theobromine, cacao content, **25:**464; 469; 484
Theone activity, components, **16:**70–71; 76
  relation, to hemophilia, **16:**56–57
    to thrombin formation, **16:**53–56; 91
Theonine, as artificial electron acceptor, **30:**22
Theophylline, **60:**47
Theorell-chance mechanism, **45:**282; 289; 306
  irreversible, **39:**351
Theories, of coagulation, basis for, **30:**263–270
Therapeutic applications of lipoic acid, **18:** 343–344
Therapy, ADA deficiency and, **51:**196–198
Thermal coefficients of radiation inactivation of enzymes, **24:**371–373
Thermal conductivity, in photosynthetic tissues, **25:**82
Thermal data, and thermodynamic functions, **11:**12–13
Thermal denaturation, **35:**478
Thermal energy, and origin of life, **27:**358; 382–386; 392; 405
Thermal energy transport in living cells, **14:**3–6; 14
Thermal luminescence, in photosynthetic tissues, **25:**82
Thermal stability, **34:**456
  of adsorbed enzymes, **28:**6; 35
Thermobacterium acidophilus, and mevalonic acid, **23:**472
Thermochemical data, for cytochromes, **17:** 333–334; 347
Thermodynamic, and activation parameters, **39:**349
  parameters, **39:**356
Thermodynamic aspects, of denaturation of α-amylase, **23:**50–58
Thermodynamic aspects of photosynthesis, **25:**39–117
Thermodynamic constants, in simultaneous reactions, **9:**54–56
  of enzyme activation, **9:**28–31
  of enzyme-catalyzed reactions, **9:**62–65
  of enzyme inactivation, **9:**32; 33; 36
  of enzyme reactions, effect of pressure on, **9:**66–69
  of enzyme-substrate association, **9:**58–60
  of protein denaturation, **9:**32; 33
  in successive reactions, **9:**51–53
Thermodynamic cycle, congruent, enzyme action, **73:**39–41
Thermodynamic data
  on protein denaturation, **23:**52; 55
  relating to hydrolytic enzymes, **15:** 279–280
Thermodynamic functions, and biological reactions, **11:**27–30
  definitions, **11:**3–4
  experimental determination of, **11:**10–13
Thermodynamic hypothesis of protein folding, **57:**212–214
Thermodynamic measurements, in identification of imidazole, **20:**64; 65
Thermodynamic parameters, **43:**287
Thermodynamics
  cytochrome *c* oxidase, catalytic cycle, **71:**114–116
  irreversible, application to photosynthesis, **25:**48–51
  and natural selection, **38:**207–208
    2nd law of, **38:**196–197
    synthesis, **38:**259
    Thymidine, carbon-11 label, **38:**283
Thermodynamics and living matter, **7:**59
Thermodynamics of acetyl phosphate, **6:**241; 246; 249
Thermodynamics of enzyme reactions, **33:** 247–254
Thermodynamics of immunological reactions, **15:**49–77
Thermodynamics of phosphorylation reactions, **17:**119–124
Thermodynamic studies, eukaryotic dihydrofolate reductase, NADPH and NADP binding, **70:**48; 49
Thermodynamique des reactions immunologiques (Wurmser), **15:**49
Thermogenesis, role of fructose 1,6-bisphosphatase, **42:**223

Thermogenesis as measure of fermentative activities in tobacco, **10:**415–416
Thermoluminescence of dried enzymes, irradiation effect, **24:**379–380
Thermolysin, **41:**180; 187; 203, **47:**57–60; 124, **50:**35, **53:**261; 265; 282; 283
carboxypeptidase homology with, **56:**381
and catalytic process of converting enzyme, **57:**15
catalytic role of zinc in, **56:**316
and Glu-143, **57:**23
inhibition of, **56:**380–381
inhibitors of, **57:**67–70
metal centers of, properties of, **56:**379–382
metal content of, **56:**376–377
metalloenzyme substitutions of, **56:**304
molecular properties of, **56:**377–379
peptidase and esterase activity of, **56:**381
peptide cleavage by, **57:**70–73
and zinc metalloproteases, **57:**20
Thermophiles, glucosamine 6-phosphate synthase, **72:**104
Thermophilic bacteria action on cellulose, **8:**267–268; 280
Thermophills phosphofructokinase, **48:**215–216
Thermostability, differential inactivation, effect of ligands on, by heat, **37:**490–495
Thermostable factor of antithiamine factor, **15:**406–411
Thermotropic mesomorphism, **34:**468
Thetin-homocysteine methyl-transferase, **25:**424–426
Thetin-homocysteinemethyl transferase, sulfur acids of, **27:**220; 231
Thetins. *See also* Dimethylacetothetin
alkyl group transfer from, **25:**383–384
and methylation, **25:**415; 424–426
as methyl donors, **25:**423–426
biological importance of, **12:**482–486
and biological methylation, **12:**443; 445–446; 446–447; 454–455; 466
biological oxidation of, **12:**481
in microorganisms, **22:**268
in plants, **10:**241
THFA, *See* Tetrahydrofolic acid
Thiabutane, structure studies, **23:**24
Thiaisoleucine, **37:**389
Thialysine, **49:**63; 65
Thiamin, **5:**141–143; 148; 168; 169; 177; 203
action on *Fusaria,* **5:**169; 177; 203
active group for coenzyme functioning, **14:**181
derivative inactivated by yeast phosphatase, **7:**454
metabolism, **7:**198–199
synthesis and breakdown by *Phycomyces,* **7:**454
dietary effect on coenzyme Q, **25:**304
as growth factor, **6:**39
inhibitory analogues of, **6:**131; 143
stimulation of pyruvic acid decarboxylation, **6:**55–56; 81
Thiaminase, **15:**389–421
Thiaminase disease, **15:**414–415
Thiaminase (Fujita), **15:**389
Thiaminase I, and base exchange reaction, **15:**414
Thiaminase II, **15:**411–414
effect of amines on, **15:**410–411
Thiaminase model, **15:**404
Thiaminases, adsorption studies, **14:**334
classification, **14:**246; 250
Thiamine. *See also* Diphosphothiamine
2, 3 attachment to cyclodextrim, **58:**46
absorption spectrum, **4:**298–300
action of sulfite on, **15:**404
action of thiaminase on, **15:**399–404; 406–407; 411–414; 415–416
biological activity with negative tests, **15:**416–418
deficiency, and thiaminase, **15:**409–411; 414–415
disappearance from colon, **15:**390
and free radical formation, **15:**4
in pyruvate oxidation, **15:**214–215
and urea synthesis, **15:**342
biosynthesis of, **53:**363–371
4-aminoimidozole ribonucleotide, precursor of, **53:**364
biogenesis of thiazole moiety, **53:** 366–371
phenylalanine, precursor of, **53:**367
pyrimidine moiety, **53:**364–366
formate, precursor of, **53:**364
carbohydrate resynthesis, **1:**147
charge-relay system, **50:**36
cleavage products, absorption spectrum, **4:**298–300
complex with nucleates, **17:**138
genetic studies, **17:**411; 431

Thiamine (*Continued*)
deficiency of, relation of transamination, **16:**228
effect on carotenogenesis, **21:**321; 322
and fat formation, **8:**310; 322; 330
effect on acetylation of amines, **8:**388–389
free, reactions, **22:**399
gasometric measurement, **9:**603
and lipoic acid, **18:**332
in normal and cancer tissues, **4:**240; 247; 248
in plants, **19:**18
Thiamine deficiency, effect on pentose synthesis, **28:**432
Thiamine derivative, and pyruvic dehydrogenase action, **15:**122
Thiamine disulfide, and thiaminase, **15:**405
biological activity, **15:**416–417
Thiamine in plants, **10:**227; 243–247
relation to cysteine in plants, **10:**269
Thiamine methyl disulfide, from action of allinase, **15:**417–418
Thiamine phosphorylase, **15:**416
and urea synthesis, **15:**342
Thiamine propyl disulfide, and thiaminase, **15:**405
from action of allinase, **15:**417–418
Thiamine pyrophosphatase, localization of, **24:**330
Thiamine pyrophosphate, **15:**141; 154; 416, **58:**2; 46. *See also* Active glycolaldehyde; Cocarboxylase and Diphosphothiamine
in ATP synthesis, **23:**334; 337
and glycolate metabolism, **23:**436; 443
and oxalate metabolism, **23:**447–448
binding, **22:**398–403
and trypsin action, **22:**402
in α-carboxylase, **22:**401
in decarboxylation, **27:**519
localization, **27:**276
effect of diet on level of, **22:**402
enzymes containing, metal requirements of, **22:**401
functional groups of, **22:**399; 400
hydrolysis, by metal ions, **21:**175; 176; 193
synthesis, in mitochondria, **21:**54
and lipoic acid, **22:**400
and mechanisms of catalysis, **24:**467–469
model studies, **58:**3
molecular structure, **22:**339
in pentose synthesis, **28:**396; 416–417
and phosphorylation, **22:**400
in photosynthesis, **19:**7; 11; 13; 18; 25, **25:**98–101
primary amino group of, **22:**398
pyrophosphate group of, and binding, **22:**400
and thiamine antagonists, **22:**402
Thiamine pyrophosphate pyruvate adduct, **43:**250
Thiamine pyrophosphate (TPP), and lipoic acid, **18:**333–338
synthesis of, **18:**199
Thiamine-thiaminase complex, **15:**399
Thiamine triphosphate, **22:**403
Thiamin pyrophosphate, classification as coenzyme, **14:**241
Thiaminpyrophosphate enzyme and pyruvate cleavage, **6:**238
Thiazole, genetic studies, **17:**411–412; 431
Thiazolealanine, histidine analog, **28:**126
Thiazole derivatives, and thiaminase, **15:** 399–404; 411; 416
Thiazole moiety, biogenesis of, **53:**366–371
Thiazolidine-2-carboxylate
chemical synthesis of, **57:**104–106
and cysteamine-glyoxylate adducts, **57:**106
ionization constants of, **57:**109–110
racemization of, **57:**106
structure of, **57:**104
as substrate for D-amino acid oxidase, **57:**108–111
and sulfhydryl reagents, **57:**106–107
in tissues, **57:**112–113
in vivo, **57:**112
D-Thiazolidine-4-carboxylate, **57:**109
Thiazolidine 4-carboxylic acid, **50:**162
L-Thiazolidine carboxylic acid, oxidation, in mitochondria, **21:**26
L-Thiazolidine-4-carboxylic acid (thioproline), proline analog, **29:**95
Thiazolidine-2,4-dicarboxylates, **57:**113–114
Thiazoline-2-carboxylate
and catalase, **57:**116
hydrolysis of, **57:**131–132
and inhibition of dopamine β-hydroxylase, **57:**136
as inhibitor of L-amino acid oxidase, **57:** 108–111
as inhibitor of L-α-hydroxy acid oxidase, **57:**140
and metal ion complexing agents, **57:**135–136

and nucleophiles, **57:**134
products of, **57:**129; 135
Δ^2-Thiazoline-2-carboxylate, **57:**104–105
Thiazolines, **57:**133
Thiazolium, **58:**3
Thickness, **34:**467; 506
DL-β-2-Thienylalanine, **67:**172
β-2-Thienylalanine, **35:**286
β-Thienylalanine, effect on phenylalanine sRNA synthetase, **29:**112
inhibition of DAHP synthetase, **29:**117
resistance to, **29:**122
β-2-Thienylalanine, in transaminations, **16:**198; 237
β-2-Thienylpyruvic acid, **16:**198
DL-β-2-thienylserine, effects on synthesis of gramicidin S, **33:**347
Thioaldehydes, synthesis in plants, **10:**259–260
Thio AMP, **35:**463
Thioarsenites, **11:**217–218
Thiobacilli, transaldolase in, **19:**27
Thiobacillus, **39:**337; 361
Thiobacillus neapolitanus, phosphoribulokinase from, regulation of, **74:**109
Thiobacillus novellus, **39:**338; 361
Thiobacillus thio-oxidans, metabolism in, **74:**6
Thiobacteriales, metabolism in, **74:**7
Thiobarbituric acid method, in schizophrenia, **29:**506
Thiobases, **35:**475
Thiocapsa pfennigii, **46:**77
Thiocarbamate, effect on phenolase complex, **19:**199
Thiocarboxyl group, **49:**76
2-Thio-5-carboxymethyl-uridine methyl ester, **35:**475
Thiochrome method, **15:**405; 410; 417
Thiochrome reaction, for fluorometric assay of cocarboxylast and related compounds, **20:**33
Thioctamide-requiring dehydrogenase, position in mitochondrial electron transfer chain, **21:**121; 122
Thioctic acid, *See* α-Lipoic acid
4-Thioctic acid, **18:**328
5-Thioctic acid, **18:**328
6-Thioctic acid, **18:**320; 328
Thiocyanate, **39:**328; 329
and disulfides of proteins, **27:**211
structures from, **27:**391
effect on indole oxidase, **19:**127; 128
effect on iodine uptake, **7:**202–203
in urine, **7:**400
effect on thyroxine synthesis, **14:**115–116; 123
organic, **39:**351
sensitive electrode, **39:**334
Thiocyanates, organic, **46:**400
as substrates, **46:**399
Thiocyanatopurine-ribose triphosphate, **43:**39
Thiocyanoalanine, **56:**30; 31
Thiocyantae, **39:**336
Thiocystine, **39:**330; 341
2-Thiocytidine, **35:**475
Thio-disulfide regulation, enzymes, **66:**184–187
Thio-disulfide status, intracellular, **66:**184
6-Thioenosine triphosphate, **43:**29; 37
Thioenzymes of plant origin, **10:**300–306
Thioester, amino acid, **35:**10
characteristics of bond, **35:**10
peptide, **35:**12
Thioesterases. *See also* Glyoxalases I and II
action on *S*-alanyl CoA, **19:**249
action on *S*-aminoacyl derivatives of glutathione, **19:**247; 248
classification, **14:**225; 249
Thioester formation, and ATP, **18:**229–230
and phosphorylases, **18:**226–227
Thioester phosphorylase, **18:**192; 228
Thioesters, **15:**189–190; 220; 232; 239–240; 240–244; 248
Thioethanol, and oxidation of hydrocarbons, **27:**485
Thioethanolamine, in coenzyme A, **15:**187
Thioether, **58:**35
Thioetherases, classification, **14:**225; 249
Thioethers, aliphatic, formation in plant cells, **10:**239–240
in the formation of sulfoxides and sulfones, **10:**239
role in metabolism of plants, **10:**238–257
Thio ethers, autoxidation, **8:**440
Thioethers, of, N-acetylcysteine, **46:**384
α-Thioethyl-D-glucopyranoside, translocation catalyst of *Saccharomyces cerevisiae,* **29:**45
β-Thiogalactoside, accumulation by *E. coli,* **22:**109
Thiogalactoside, in β-galactosidase system, **22:**113

Thio-β-galactosides, in induction of lactose system, **30:**86
Thio-β-galactoside transacetylase, reaction catalyzed by, **30:**85
Thiogel, reaction of with trypsin, **33:**257
Thioglucosidase in plants, **10:**310
β-Thioglucosides, as inducers and complexants of β-glucosidase, **22:**118; 119
Thioglucuronide, **57:**358
Thioglutamine, glutamine-dependent nitrogen transfer, AS-B *vs.* papain, **72:**178–183
Thioglycol, acid-base catalysis, **25:**26–27
Thioglycol, PN-dependent oxidation, **26:**118; 122
Thioglycolate, **39:**342
  and papain, **19:**286
Thioglycolic acid, **15:**219; 248
  effect on growth of pneumococcus, **10:**225
  and heat denaturation, **27:**226
    from insulin, **27:**229
  reaction with quinones, **16:**121
Thioglycolic acid-cystine reaction, **27:**201
Thioglycosidases, classification, **14:**224; 225; 249
Thio groups, *See* Sulfhydryl groups
6-Thioguanine
  and folate metabolism, **27:**460
  as a substrate for Hypoxanthine-guanien Phosphoribo-syltransferase, **41:**14
Thioindigo test, **18:**113
Thioketenes, cysteine *S*-conjugate β-lyase reactions, **72:**206–207
  toxicity studies, **72:**210–212
Thiokinase, column chromatography, **22:**483
Thiol, *See* Sulfhydryl group
Thiolacetate, **33:**384
  and papain, **19:**286
Thiolacetate ion, reaction of phenylmethane-sulfonyl proteases with, **33:**386–387
  reaction of *O*-tosyl-serine peptides with, **33:**386
Thiolacetic acid as acetylating agent, **12:** 329–330
Thiol-alcalase, chromatography of, **33:**391
  preparation of, **33:**387
Thiolase, **15:**233; 237; 238; 239; 240–244, **58:**182
  in oxidation of 3-oxoadipate, **27:**513
β-Thiolase, and propionic acid metabolism, **26:**296–297
Thiolate anion, **65:**104–106
  asparagines synthetase, glutamine-dependent nitrogen transfer, **72:**169–170
  catalytic mechanisms, **69:**21
  glutathione S-transferases, **69:**18; 19
Thiol-carbonyl adducts, **57:**123
Thiol compound and lipoxidases, **8:**373–374
Thiol compounds, in addition complexes with DPN, **22:**352
Thiol-directed reagents, **39:**344
Thiol-disulfide exchange, **27:**200–204; 207–209, **66:**150–151; 153
  dehydroascorbate reductase, **66:**187
Thiol-Disulfide Exchange, Molecular and Cellular Aspects of (Gilbert), **63:**69
Thiol-disulfide exchange protein, RNA content, **40:**47
Thiol-disulfide exchange proteins, **59:**106
Thiol-enzymes, activity of with synthetically formed active site, **33:**391–399
  chromatography of, **33:**390–391
  preparation of, **33:**385–391
Thiolesterase, **57:**159
Thiolester hydrolases, localization of, **24:**318
Thiolester rearrangements, **59:**119
Thiol esters, **23:**325; 327; 329; 335–339
Thiolesters, **59:**125; 126–127
Thiol esters. *See also* Coenzyme A
  formation by papain-substrate complex, **19:**326–329
    papain action on, **19:**307; 308
  properties, **16:**18–19
Thiol esters of dicarboxylic acids, hydrolysis, **26:**350
Thiol form of thiamine, **15:**418
Thiol-glyoxylate adducts, **57:**122; 124; 128
Thiol group, *See* Sulfhydryl groups
  and activity of esterases, **20:**285
    differentiation from imidazaole, **20:**70–72; 75
    iodination of, **20:**72
    ionization constant of, **20:**70
    metal binding by, **20:**56
    reactivation of, **20:**66
    roles in biological systems, **20:**52
  blocking, **27:**212–213
    and enzyme activity, **27:**212–217
    and microbial oxidation of hydrocarbons, **27:**485; 509; 513; 515

Thiol groups, **11:**201–266. *See also* Cysteine, Glutathione, and Sulfhydryl groups; Sulfhydryl groups
relation to activity of alkaline phosphatase, **10:**97–98; 101
Thiol Groups of Biological Importance (Barron), **11:**201
Thiol groups of reduced wool, **15:**288
Thiolhistidine, **7:**370
8-Thioloctanoic acids, **18:**332
Thiol proteases, hydrolysis of *p*-nitrophenyl *N*-carbobenzyloxyglycinate by, **33:**398
Thiolprotection, **33:**460; 466; 481; 493–494; 498; 503; 507
Thiol proteinases, **41:**180; 197–200
Thiol reagent, phosphoenolpyruvate carboxykinase (GTP), **69:**208
Thiol reagents, *See* Sulfhydryl agents
Thiols, **39:**328; 333; 342; 358, **58:**37; 106. *See also* Disulfides, Mercaptans, and Sulfhydryl groups; Sulfhydryl groups
and D-amino acid oxidase inhibitors, **57:** 141–142
autoxidation, **15:**41
cysteine *S*-conjugate β-lyase reactions
stability, **72:**201–203
toxic effects, **72:**208–212
unstable reactions, **72:**204–207
cysteins *S*-conjugate β-lyase reactions
stability, **72:**201–203
toxic effects, **72:**208–212
unstable reactions, **72:**204–207
and L-α hydroxy acid oxidase, **57:**126
measurement of, **27:**231–232
and oxalyl thioesters, **57:**134–135
oxidation in plant metabolism, **10:**226
partition at interfaces, **11:**47–56
role of in carboxytransphosphorylase reaction, **45:**100–102
substrate for, **57:**383
Thiol-subtilisin, catalysis of NPA hydrolysis by, kinetics of, **33:**392–394
chromatography of, **33:**390–391
cinnamoyl, deacylation of, **33:**390
kinetics of hydrolysis of, **33:**394–395
S-cinnamoyl, spectra of, **33:**389
comparison with subtilisin (table), **33:**388
transformation of subtilisin to, **33:**38
Thiol transacetylase, and fluoroacetyl-CoA, **18:**129
and lipoic acid, **18:**337
Thioltransferases, **66:**149–188
active cite cysteines, S^- nucleophilicity, **66:**159–160
apparent pK_a of Cys 22, **66:**160; 178–179
electron transfer from NADPH, **66:**151–152
functions, **66:**182–183
Hill coefficient, **66:**177
history, **66:**152–155
human, **66:**161
erythrocyte, kinetic analysis, **66:**175–177
inhibitors, **66:**174–175; 183
intrinsic dehydroascorbate reductase activity, **66:**179–181
ion pairing, **66:**178–179
in mammalian ribonucleotide reductase, **66:**167–168
non-Michaelis behavior, **66:**177
pig liver
active center, **66:**171–172
alkylation, **66:**160
amino acid sequence, **66:**170
catalytic mechanism, **66:**172–174
gene, **66:**169
high-level expression in *E. coli,* **66:** 169–171
immunological characterization, **66:** 160–161
inhibitors, **66:**183
iodoacetamide inhibition of mutants, **66:**174–175
mutants, **66:**172
pI values, **66:**178
purification and sequence analysis, **66:** 157–161
proposed reaction scheme, **66:**175–176
rat, purification, characterization, and properties, **66:**155–156
references, **66:**151
reversible oxidation, **66:**150
reversible reaction between cytoplasmic proteins and glutathione, **66:**154–155
S^- nucleophilicity, **66:**154
yeast, **66:**161–163
Thiolutin, in treatment of plant diseases, **20:**401
Thiomethyladenine, **17:**409
5-Thiomethyladenosine hydrolase, **55:**271
Thiomethylation, **42:**269
Thiomethylgalactoside (TMG) permease I, **36:**94

Thiomethylgalactoside (TMG) permease II, **36:**94
2-Thiomethyl-6-isopentenyl adenine, **35:**469
Thiomethylpentose, **9:**475
Thiomethyl products
cystein *S*-conjugate β-lyase reactions, **72:**201–203
cysteine *S*-conjugate β-lyase reactions, **72:**206–207
toxicity studies, **72:**207–212; 210–212
Thionase, action of, **12:**440–441
Thionein, **56:**403
function of, **56:**407
pH effects and, **56:**406
Thionicotinamide-DPNH, enzymebound, Cotton effect, **27:**57; 58
in EPR study, **27:**181
Thiopanic acid, as inhibitory analogue, **6:**132; 137; 140; 143
Thiophenol and derivatives, in syntheses of activated amino acids, **19:**239; 240; 247; 249; 251
Thiophile, **39:**343; 355
Thiophiles, **39:**364
Thiophilic, anions, **39:**328; 341; 342; 362
group, **39:**356
series, **39:**362
Thiophilicity, **39:**331–332; 342
series, **39:**331
Thiophosphate
configurational analysis of, **55:**328–330
hydrolysis of, **55:**390
synthesis of, **55:**319–320
Thiophosphate analogs, **43:**8
polynucleotides, **43:**8
Thiophosphate ester, hydrolysis of, **58:**4
Thiophosphorylation, **58:**163
Thiophosphoryl-CoA, **58:**142
Thioproline, effect on *E. coli,* **29:**96
isolation of sRNA for, **29:**102
β-Thiopropionic acid, reaction with quinones, **16:**121
Thiopurine ATP analogs, chromophoric substrates, **43:**29
Thio reagents, differential modification and inactivation, **37:**446–458
Thioredoxin, **47:**111–117; 118; 120; 122–123, **54:**175; 176; 177, **59:**146, **65:**148, **66:**185–186
amino acid sequence of, **47:**112
conformational profiles of, **47:**114; 115; 116
in DR metabolism, **28:**436–444; 446; 454
E. coli, **66:**163–164
glutathione S-transferases, **69:**15
mutants, **66:**171
protein disulfide isomerase (PDI) prolyl 4-hydroxylase β subunit, **72:**369–375
secondary structures in, **47:**116
tertiary structures, **66:**164
β turns of, **47:**120
Thioredoxin reductase, **46:**245, **59:**146, **66:** 185–186
in DR metabolism, **28:**440–443; 446; 454
Thiorhodaceae, **1:**236; 280; 281; 285; 290; 302; 308, **39:**338
carotenoids in, **21:**303; 304
Thiorhodaceae
metabolism in, **74:**7
use of acetate, **23:**421; 430–431
5-Thioribulose-1,5-bisphosphate, **67:**13
Thioridazine
affinity of, for muscarinic receptors, **55:**82
side effects of, **55:**181
structure of, **55:**180
Thiosemiacetals, PN-dependent oxidation, **26:**119; 122
Thiosinamine, germination promoting properties, **10:**260–261
Thiosine, **35:**482
Thiosulfate, **39:**328; 343; 345; 353; 360; 361
action on thiaminase, **15:**401
binding, **39:**355; 356
from cysteine, **7:**383; 384; 387–388
metabolism, **7:**399–400
from methionine, **7:**372
in cystine metabolism, **16:**278; 279
effect on higher plants, **10:**231
oxidation of, **39:**361
reductase, **39:**339; 352; 360
activity of rhodanese, **39:**342
gluthathione-dependent, **39:**343; 360; 363
reduction, **39:**363
arsenite inhibition of, **39:**342
reductive cleavage of, **39:**342
substrate, **39:**345
Thiosulfate-cyanide, reaction, **39:**333; 358
nonenzymic catalysis of, **39:**338
nonspecific catalysts for, **39:**335
rhodanese-catalyzed, **39:**353; 356

Thiosulfate-cyanide-sulfur transferase (Rhodanese), **27:**212
Thiosulfate-lipoate reaction, rhodanesecatalyzed, **39:**343; 350
Thiosulfate r cyanide transsulfurase, **14:**252
Thiosulfates, reduction, PN-dependent, **26:**129
Thiosulfinic acid esters, disulfide exchanges, **27:**213
Thiosulfonate, aryl, **39:**341
cyanide reaction, **39:**353
substrates, **39:**355
Thiosulfonates, **39:**328; 330; 340; 353; 354; 361
aromatic, **39:**354
aryl, **39:**333
organic, **39:**332
Thiosulphate sulphurtransferase (rhodanese), localization of, **24:**315
Thiotaurine, **39:**341
Thiothiamine, and thiaminase, **15:**405
Thiouracil, **35:**476
effect on iodine uptake, **7:**202–203; 206; 207
effect on virus multiplication, **17:**480–483
inhibitor of melanin formation, **14:**93; 94; 123
inhibition of thyroxine synthesis by derivatives of, **14:**116
methylation, of, **25:**415
2-Thiouracil, effect on enzyme synthesis, **24:**16
Thiouracil-S^{35}, **17:**481
Thiouracils and biological methylation, **12:** 436–441–12:442
Thiourea
and disulfides of proteins, **27:**211–212
inhibition of urease, **17:**59
solubilization by, **17:**151
reaction with *o*-quinones, **16:**122
Thiourea and biological methylation, **12:**436; 441–442
Thiourea desulphhydrase, localization of, **24:**333
Thiourea in plants, **10:**260–262
Thioureas, inhibition of melanin formation by, **14:**93–94
peroxidase action on, **14:**152
Thiouridine, **35:**474
in anticodons, **47:**384
2-Thiouridine, **35:**475
4-Thiouridine, **35:**465; 466; 475; 476, **54:**89
Thioxotropy of protoplasm, **8:**27–28
Third esters, **34:**526
Thixotropy of protoplasm, **7:**40
Thiyl radical, **27:**200
Thiyl radical in enzymic oxidoreductions, **15:**9–11
Thompson, Linda F., **51:**167–210
Thorium compounds, **17:**145; 147
Thorium ions, model catalyses with, **21:**174; 175; 177; 178
Thpp, *See* Thiamine pyrophosphate
*thr*A-encoded aspartokinase I, bacterial amino acid biosynthesis, **72:**283–285
Three-dimensional structure, cytochrome *c* oxidase, **71:**108–110
Three-layer enzyme membrane, **34:**454
Threonine, **35:**487; 488, **46:**53, **50:**144; 152, **51:**217, **53:**169, **60:**252
and acetate metabolism, **23:**432–433
and glycine metabolism, **23:**456
aldolase (*See* Serine hydroxymethylase)
as amino-terminal in tuftsin, **41:**163
biosynthesis, **16:**268–271
allosteric regulation, **28:**45; 53; 60–62; 70–72
inactivity in *Neurospora* mutant, **16:**256
product inhibition of, **25:**171–172
E. coli synthesis, **25:**361
in glycopeptide, **25:**452
recemase for, **16:**250; 271
relation to metabolism of other amino acids, **16:**269–271; 277; 282–285; 298
in transamination reactions, **16:**191; 192; 198; 228; 234–235
in chloroplasts, **20:**209
in phosphorylated peptides of esterases, **20:**288
and citric acid cycle, **18:**257
deaminase for adsorption studies, **14:**346
oxidation of, **14:**132
deamination of, **12:**82–84
dehydrase, **35:**107
effect of streptomycin on oxidation, **10:**161
isopropyl ester, **19:**237
in papain, **19:**296
in pyrimidine biosynthesis, **21:**281
synthetase, **35:**112
of tobacco mosaic virus, **15:**462–464
of virus (TMV) mutants, **15:**474
DL-Threonine
as growth factor, **16:**250
resolution by papain, **19:**309

Threonine, C-Terminal, detachment from virus, **17:**478
genetic studies, **17:**401–403; 405; 407–412; 429
labeled, **17:**401
Threonine, RNA polymerase II transcription, C-terminal domain characteristics, posttranslational modifications, **71:** 49–50
D-Threonine, use by *E. coli* mutants, **16:**284; 285
Threonine adenylate, synthesis by carbodiimide method, **22:**191
Threonine aldolase, **74:**133; 152
in pentose metabolism, **28:**422; 431
Threonine auxotroph, of *E. coli,* **16:**271
Threonine catobolism, pathway, **37:**365
Threonine deaminase, **37:**483
allosteric regulation, **28:**44; 52–54; 64
binding of t-RNA, **40:**49
interaction with RNA, **40:**45
biodegradative, **37:**379–381
ADP as activator, **37:**364
effect on aggregation, **37:**367
mechanism of, **37:**366–369
AMP as activator, **37:**364
effect on aggregation state, **37:**365
effect on CD, **37:**366
effect on SH groups, **37:**365
effect on substrate affinity, **37:**366
mechanism of, **37:**365–366
specificity of the AMP requirement, **37:**366
apomonomer, **37:**380
reconstituted holomonomer, **37:**380
behavior, during centrifugation, **37:**381
during gel filtration, **37:**381
Cl. tatanamorphum, **37:**380
Cl. tetanomorphum, **37:**366
cooperative substrate binding, **37:**367
E. coli, **37:**381
effect of energy charge, **37:**365
molecular weight, **37:**380
effect of ADP, **37:**380
effect of storage, **37:**380
in plants, **37:**369
reaction mechanism, **37:**354–363
regulation of formation, **37:**384–386
catabolite repression of, **37:**385
conditions for induction, **37:**384
effect of cyclic AMP, **37:**385
resolution of, **37:**367; 380
role of, **37:**365
serine deaminases, distribution of, **37:** 350–354
structural features of, **37:**379–381
effect of AMP, **37:**379
suicide reaction, **37:**361
biosynthetic, **37:**386–390
apodimers, **37:**382
disulfide bonds in, **37:**382
of B. *licheniforms,* **37:**375
effect of monovalent cations, **37:**375
of B. *stearothermophilus,* **37:**375
B. subtilis, **37:**373; 377; 378; 382
substrate dependent activation, **37:**373
cold sensitivity, in *B. subtilis,* **37:**383
in *R. rubrum,* **37:**383
reversal by brushite, **37:**383
tetramer-dimer dissociation in, **37:**383
cooperative substrate binding, **37:**370
desensitization, **37:**375
distribution of, **37:**350–354
E. coli, **37:**372; 376; 378; 382
effect of *ilv,* **37:**388
accumulation of enzyme forming potential, **37:**388
genetic control, **37:**388
effect of isoleucine, **37:**369
cooperativity of threonine binding, **37:**372
effect of $S_{0.5}$, **37:**371
effect of monovalent cations, **37:**375
effect of valine, **37:**369
genetic specification, **37:**388
hyperbolic substrate saturation, **37:**370
immature enzyme
binding of leucyl-*t*RNA, **37:**387
E. coli, **37:**387
effect of isoleucine and valine, **37:**387
maturation of, role in repression, **37:**387
molecular weight, **37:**382
effect of effectors, **37:**382
pleiotropic mutation affecting, **37:**388
R. rubrum, **37:**375; 384
molecular weight, **37:**383
resolution of, **37:**384
reaction mechanism, **37:**354–363
regulation of formation, **37:**385–386
E. coli, **37:**385
genetic control, **37:**385–386

multivalent repression, **37:**385
S. typhimurium, **37:**385
role in repression, **37:**386–390
E. coli, **37:**388
S. typhimurium, **37:**386
yeast, **37:**389; 390
S. typhimurium, **37:**372; 378; 382
binding sites, **37:**382
resolution of, **37:**382
stability, **37:**376
effect of isoleucine, **37:**376
structural features, **37:**381–384
of yeast, **37:**376
effect of isoleucine and valine, **37:**376
in *E. coli,* suppression by isoleucine, **22:**33
multiple forms, **25:**364
product inhibition of, **25:**172
L-Threonine deaminase, desensitization of, **26:**106–107
Threonine Deaminases (Umbarger), **37:**349
Threonine dehydrase
degradative, allosteric regulation, **28:**54
dietary induction of, **29:**372
glucose repression by, **29:**370; 380
effect of ethionine on, **25:**380
Threonine dehydratase, **37:**495; 554; 585, **56:**43
Threonine dehydrogenase, **46:**245
Threonine intermediates, bacterial lysine biosynthesis, **72:**283–285
L-Threonine (N^{15}), in corrin ring biogenesis, **26:**260
Threonine residues
dihydrodipicolinate synthase (DHDPS), **72:**290
γ-glutamyl transpeptidase catalytic sites, **72:**256–262
of RNase, chemical modification of, **24:**211; 226
Threonine synthase, **74:**134; 141
phylogenetic distribution of, **74:**157
L-Threotrihydroxypropyl-H_2-pterin, **35:**53
Threshold value, of wilting toxins, **11:**410–413
Thrombin, **5:**32; 33; 34; 42–43; 55–59, **34:**10; 13; 21; 450, **46:**1; 3, **53:**265. *See also* Antithrombin and Prothrombin; Prothrombin
action of, **5:**57–59
action of tyrosinase on, **14:**147–148
oxidation of tyrosine in, **14:**86
structure studies, **14:**285
action on blood platelets, **10:**11–12
action on synthetic substrates, **16:**33–34
adsorption on fibrin, **16:**78
adsorption on glass, **16:**78–79
effect activity of platelets, **16:**69
effect on Ac-globulins, **16:**61–62; 91–92
effect on concentration on clotting time, **16:**33; 38
enzyme activity, **16:**24–25; 33–34; 36–44
equilibrium conditions among trace proteins, **16:**76
formation from prothrombin, **16:**44–57
inactivation, **16:**25; 84
interaction with fibrinogen, **16:**35–42
lytic effect on fibrin polymer, **16:**42–44
neutralization by antithrombin-III, **16:** 79–80
amino acid composition, **34:**13
amino acid composition near active site, **22:**65
column chromatography, **22:**456
pK value, **22:**58
amino acid sequence in, **34:**22
autocatalytic formation, **10:**8
changing concentration of, following glass contact activation at zero time, **30:**260
chromatography of, **30:**272
esterase activity of, **30:**258
properties and interactions of, **30:**272–274
from prothrombin, **30:**256; 265
resemblance to trypsin, **30:**272
from clotting globulin, **10:**10
coinhibitor, **10:**35–36
formation of, **5:**55–57
and globulin, **5:**35
imidazole in, **20:**76; 77
inhibitory action in blood coagulation, **5:**48
mechanism of formation, **10:**25
preparation, **10:**20–23
reactions in formation, **10:**17
reaction with fibrinogen, **10:**32
standardization with ninhydrin, **10:**5
Thrombin and complement, **8:**103
Thrombin-antithrombin complex, **16:**85
Thrombin fibrin (T-fibrin), relation to Ca-fibrin, **16:**43–44; 92
Thrombinogenase, **10:**28
Thrombocytes, **41:**170
effect of pectic on, **20:**357
Thrombocytopenia, relation to platelet deficiency, **16:**67–68

Thrombogene, as clotting factor, **16:**60
Thrombokinase, *See* Thromboplastin
from damaged cells and platelets, **30:**256
origin of term, **30:**258
effect on plasma Acglobulin, **16:**61
use of term, **16:**30
and elastase, **23:**239
Thrombophilic γ-Globulin, **41:**171; 172
binding to thrombocytes *in vitro,* **41:**171
binding to thrombocytes *in vivo,* **41:**171
fraction I, **41:**172
fraction IV, **41:**172
Thromboplastic substances, **5:**32; 34–41; 45–49. *See also* Papain; Snake venom; Thromboplastin; Trypsin
lipide, **5:**39–41; 56
papain, **5:**46
protein, **5:**35–39; 46; 56; 57
snake venom, **5:**46
trypsin, **5:**45
Thromboplastin, **5:**32; 56, **10:**2; 3–7; 7–13; 16–17; 19–20. *See also* Antithromboplastin, Prothrombin, Thrombin; Blood, coagulation of
equilibrium conditions in trace protein interactions, **16:**75–76
plasmatic cofactor of, **16:**58; 59
purification, **16:**30
relation to platelet factors, **16:**70
role in prothrombin activation, **16:**25; 49–52; 62; 91; 92
toxic reaction to intravenous injection, **16:**86–87
origin of term, **30:**258
Thromboplastinogen, **10:**2; 7–13; 16–17. *See also* Thromboplastin
relation to factor *5,* **10:**27–28
Thrombosis, antithrombin in, **10:**37
Thromboxane, **58:**258
synthesis, **58:**291
Thromboxanes, **46:**167; 174
Thrombozyme in blood, **10:**13
Thylakoid membranes, **58:**245
Thylakoids, in chloroplasts, **30:**5
Thymectomized mice, effects of normal mouse serum on, **41:**173
Thymidine, **18:**222, **51:**198, **54:**58
of DNA, biosynthesis of, **28:**423
with 5-fluorouracil, **54:**100–101
growth factor, **9:**468; 494–495
phosphorylase of, **17:**482
radioactive, in DNA duplication study, **18:**183
in reversal of toxicity of fluoropyrimidines, **54:**94–95
sparing effect on vitamin B_{12} requirement, **21:**430
biosynthesis, **21:**420–423
Thymidine cyclic 3′, 5′-monophosphorothioate, **55:**345
Thymidine deoxyribokinase, feedback effects of TTP on, **29:**367
Thymidine derivatives, in allosteric regulation, **28:**135
effect on asparatate trancarbamylase *in vivo,* **28:**132
^{14}C-Thymidine, incorporation into DNA, **28:**429
Thymidine kinase, **54:**64; 65
in assessing tumor responsiveness to fluoropyrimidines, **54:**101; 107
Thymidine-*p*-nitrophenyl ester, as substrate for acid DNase, **31:**33
Thymidine 3′-(4-nitrophenyl phosphate), **55:**352
Thymidine 5′-(4-nitrophenyl phosphate), **55:**315
Thymidine 3′-(4-nitrophenyl phosphorothioate), **55:**352–353
Thymidine phosphorylase, **54:**65
Thymidine-5′-pyrophosphate, synthesis by mixed anhydride method, **22:**170
Thymidine triphosphate, biosynthesis, **21:**421
Thymidine-5′-triphosphate, synthesis by mixed anhydride method, **22:**171
Michelson modification of, **22:**171
Thymidine triphosphate (TPPP) (C^{14}, P^{32}), incorporation into DNA, **18:**213–215
Thymidylate, **54:**58
Thymidylate biosynthesis, **25:**396–401
Thymidylate kinase, **21:**421
Thymidylate synthase, **55:**269–271
Thymidylate synthetase, **38:**235, **45:**369, **46:**485
B. subtilis, **38:**258
chemical mechanism of, **45:**370
in Chinese hamster cells, **38:**266
immunologic studies, **38:**262
isotopic assay, **38:**243
kinetic mechanism of, **45:**370
L. casei, **38:**263
lymphocytic leukemia, **38:**266
in methylation, **25:**397–401
model reactions, **38:**253–256
mouse fibroblasts, **38:**266

in neuroblastoma cells, **38:**266
phage, **38:**261
in rat hepatomas, **38:**267
regenerating liver, **38:**265
regulatory aspects, **38:**261
relation to folates, **27:**425; 434; 437; 458
reversible inactivation, **38:**251
SH groups, number of, **38:**249
spectrophotometric assay, **38:**245
subunits, **38:**250
target in cancer chemotherapy, **38:**268
temperature-sensitive mutants, **38:**260
thiols, effect of, **38:**248
Thymidylate Synthetase (Friedkin), **38:**235
Thymidylate synthetase (TS), **54:**58
activation of C-5 position of, **54:**71–75
chemotherapeutic effect of, **54:**92
in conversion of dUMP to dTMP, **54:**70
in dUMP binding, **54:**93
FdUMP interaction with, **54:**77–78; 79–80; 81
inhibition of, **54:**59; 61; 62; 63
chemotherapeutic effect of, **54:**92
chtostatic effect of, **54:**70
metabolic limitation of, **54:**93–97
kinetic studies on activity of, **54:**82
measurement of, after drug treatment, **54:**106
Thymidylic acid
biosynthesis, **21:**264; 265; 288–290; 420–423
in DNA, **16:**431
Thymidylic acid kinase, methylation study, **25:**399
Thymine
biosynthesis, **21:**287–290; 398; 410; 416; 420–423
formula, **21:**263
in pyrimidine biosynthesis, **21:**290
and DNA structure, **18:**164
in DR metabolism, **28:**448–449
and folate metabolism, **27:**426; 427; 437; 455; 458–460
photochemistry of, **27:**318
genetic studies, **17:**429; 463
replacement by bromouracil, **17:**483
incorporation of C^{14}-formate, **16:**457–458
of DNA, **16:**431–433
origin of methyl group, **16:**450
isolation, **38:**236
mutant requiring, **38:**258
prebiotic synthesis, **38:**236
methyl group biosynthesis, **25:**396–401
role of cobamide coenzymes, **26:**264
in tumor metabolism, **29:**353–354
Thymine analogs, as folate antagonists, **27:**437; 460
Thymine (C^{14}), methylation study, **25:**399
Thyminedeoxyribose-5′-phosphate, *See* dTMP
Thymine glycol, **60:**15
Thymine glycol-DNA glycosylase, **60:**6; 13
Thymine 7-hydroxylase, **41:**270
Thymineless death, **38:**270
5-Thyminyltetrahydrofolate, **54:**77
Thymocytes, **47:**359
Thymol, **18:**373
Thymonucleic acid. *See also* Nucleic acid and Nucleoproteins
in chromosomes, **8:**5; 6
Thymonucleodepolymerase, adsorption studies, **14:**346
Thymus, **47:**357
cortex, **47:**368
involution, **47:**361
terminal deoxynucleotidyl transferase, **47:**357
Thymus nuclei, Na^{+}-amino acid symporter in, **29:**61
Thyroblobulin, **36:**74
Thyroglobulin, **4:**286, **52:**59
amino acids in, **14:**263; 266
separation by chromatography, **14:**350
in vivo acceptor in arginine transfer reaction, **40:**108
site of arginylation, **40:**96
Thyroid, effect on familial periodic paralysis, **6:**214
on myotonia, **6:**214
Thyroid activity, effect of thiourea, **27:**212
Thyroidectomized animals, and diiodotyrosine, **7:**203–205
and thyroxine, **7:**203–205
Thyroidectomy, **46:**406
effect on nuclei, **27:**268
Thyroid gland, and tyrosine metabolism, **14:**94; 96
iodine-containing compounds in extracts, **14:**113
Thyroid glands, **50:**314
Thyroid hormone
effect on oxidative phosphorylation, **17:**124
phosphoenolpyruvate carboxykinase (GTP), **69:**240

Thyroid hormones, **46:**390, **47:**294–297
control of hepatic α 2u-globulin mRNA, **47:**294–297
nuclear receptors for, **47:**295
Thyroid metabolism, **7:**199–209
Thyroid-releasing factor, **55:**237
Thyroid-stimulating hormone, *See* TSH
Thyroid stimulation and cold, **7:**199
Thyronine, dissociation constant, **46:**391
Thyrotropic hormone, **7:**199–202
effect on iodine turnover, **7:**199–202
effects of carbon monoxide, **7:**206
Thyrotropin, **50:**314
Thyroxin
in allosteric regulation, **28:**140; 141
effect on ribose synthesis, **28:**417
and mitochondrial swelling, **27:**228
Thyroxine, **49:**28
absorption spectrum, **4:**277–286
activity of α-keto analogue, **16:**186
in transamination, **16:**237
adsorption studies, **14:**361
in oxidative phosphorylation, **14:**13
and peroxidase action, **14:**151
from phenylalanine and tyrosine, **14:**74; 111–116; 123
effect on mitochondrial oxidative phosphorylation, **21:**31; 32
metabolism, in mitochondria, **21:**27
effect on respiratory chain, **17:**113–114
effect on sperm, **24:**83–84
effect on sperm respiration, **9:**369
effect on synthesis of coenzyme Q, **25:**304
enzymatic induction by, **29:**374; 378
use in catatonia, **29:**529–530
prosthetic group, **1:**191; 192
and thyroidectomized animals, **7:**203–205
injected, **7:**208–209
turnover, **7:**199–209
as uncoupler, **23:**366; 378; 383
Thyroxine-binding globulin, **36:**76
Thyroxine-magnesium antagonism, **17:**118
Tibia test, **42:**74
Tight-binding complexes, **52:**11–12
Tiglic acid, **50:**130; 132
Tillandsia usenoides, **7:**488
Time dependence, **34:**517
Time-resolved measurement, branch flexibility, oligosaccharides, **66:**68–71
Time resolved resonance Raman (TR3) spectroscopy, cytochrome *c* oxidase, **71:**122–123
Tin, effect on tryptophan metabolism, **10:**68
Tineola bisselliella, **45:**395; 396; 399
enzymes from, **45:**397
Tin mercaptide, **11:**216
Tin salts, **17:**137; 145
Tiron, **41:**53
Tissue culture, **46:**179
use of collagenase, **23:**218
Tissue factor, **48:**278; 281; 307
Tissue fractionation, enzyme distribution, **24:**291–358
Tissue homogenates and phosphatide formation, **7:**149–150
Tissue proteolytic enzymes, specificity, **1:**81
Tissues
adipose, **59:**372
animal
and metabolic regulation, **59:**369–372
and PFK 1, **59:**324–327
invertebrate, **59:**372
larval, histolysis of, **45:**414
Tissues, Ethionine Effects on (Stekol), **25:**369
Tissue slices, diiodotyrosine formation in, **7:**205–208
phosphatide formation in, **7:**147–151
thyroxine formation in, **7:**205–208
Tissue-specific gene expression
γ-glutamyl transpeptidase, **72:**265–267
lysyl hydroxylase, **72:**337–338
Tissue specificity, DNA studies, **18:**174–176
Titanium, in melanin, **16:**172
Titanium compounds, **17:**137
Titanium ions, model catalyses with, **21:**177
Titration curves, of dibasic acids, **9:**8–9
of enzyme substrate, **9:**17–20
of micellar complexes, **9:**10–15
of redox systems, **9:**16
Titration curves of enzyme by coenzyme, **27:**52–53
Titration of nucleic acids, **9:**512–514
Titrimetric determinations of hemoproteins, **7:**269; 273–274
Titrium-release assay, tyrosine hydroxylase, **70:**109
Titys toxin, **55:**174
TK, *See* Transketolase
TMP, *See* Thymidylic acid

TMPD, use as acceptor in respiratory chain, **28:**159; 172; 207
Tn antigen, **52:**122
TNM, *See* Tetranitromethane
Tobacco
air curing, **6:**324–362
effect on Dynamic Group, **6:**349–362
carbohydrate oxidation, **6:**358–362
conversions, **6:**349
unidentified compounds, **6:**353–356
effect on Nitrogen Group, **6:**337–348
conversions, **6:**337–339
protein digestion, **6:**339–343
effect on Static Group, **6:**327–337
alkaloids, **6:**343–348
amylase, **6:**367
antibiotic treatment of diseases, **20:**396–398; 402
artificially cultures leaves, **6:**317–323
Dynamic Group, **6:**321–323
Nitrogen Group, **6:**318–321
Static Group, **6:**318
carbohydrases, **6:**366–368
carbohydrates, **6:**327–330; 349–353
catalase, **6:**375
chlorophyllase, **6:**369
curing, **6:**323–377
chemical effects, **6:**323–365
enzymic processes, **6:**365–377
general characteristics, **6:**323–324
dehydrogenases, **6:**371–372
desmolases, **6:**369
emulsin, **6:**368
enzymic conversions in leaves, **6:**365–366
esterases, **6:**368–369
ether-soluble organic acids, **6:**356–358
ether-soluble substances, **6:**330–333
fire curing, **6:**365
flavoproteins, **6:**372
flue curing, **6:**362–365
green leaf components, **6:**312–317
Dynamic Group, **6:**316–317
Nitrogen Group, **6:**314–316
Static Group, **6:**313–314
hydrolases, **6:**366–369
inulase, **6:**368
leaf enzymes in curing, **6:**366
lipase, **6:**369
maltase, **6:**368
nicotine conversion, **6:**376
nitrates, **6:**343
oxalic acid, **6:**337
oxidases, **6:**372–375
pectase, **6:**368
pectins, **6:**327–330
peptidases, **6:**369
peroxidases, **6:**375
phenols, **6:**334–337; 373
phosphatase, **6:**369
pigments, **6:**333–334
polyphenols, **6:**334–337; 373
proteases, **6:**369
resins, **6:**332–333
sucrase, **6:**367
tannins, **6:**334–337
types, **6:**311–312
volatile oils, **6:**331–333
zymase complex, **6:**370
Tobacco alkaloids, **8:**211; 217–224; 227–231; 234; 235–236; 240–241; 245–246
Tobacco hornworm, *See* Manduca sexta
Tobacco leaf, chemical changes during fermentation and aging, **10:**325–441
Tobacco Leaf Chemistry: (Part I) During Curing (Frankenburg), **6:**309
Tobacco mosaic virus, **3:**103–104, **4:**4; 5; 6; 7, *5:*1–5:26, 7:190–191, **20:**402, **41:**201, **47:**84; 135, *See* Virus
activity, **1:**45
aggregation, **9:**212–215; 217–218
chemical composition, **9:**229–234
crystallization, **9:**195–197; 199
cytidylic acid from, **9:**520
disintegration, **9:**221–225
hydration, **9:**203–215
identity as nucleoprotein, **9:**190–195
inactivation, **9:**221–225
mutations, **9:**180–182
photochemistry, **9:**97; 115; 129; 142; 155–156
physical constants, **9:**187; 188
purification, **9:**184–187
shape, **9:**203–215
size, **9:**203–215
transmission, **9:**176
amino acid content, **4:**7
anisotropy of flow, **5:**8; 9; 11
centrifugation, **1:**47
chemical constitution and multiplication, **17:**478–483

Tobacco mosaic virus (*Continued*)
proteins of, **17:**484–495
chemical modification of, **24:**108–109
chemical properties, **5:**18–26
composition, **5:**18–20
disintegrating agents, **5:**22
effects of urea, **5:**21–22
formation of complexes, **5:**22–23
modifications of, **5:**23–26
stability, **5:**20–22
denaturation, **7:**248
effect of nitrous acid, **27:**310–312
inactivation by thiol reagent, **11:**224; 239
nucleic acid content, **4:**5
paracrystals, **5:**7; 8
physical properties, **5:**7–12
anisotropy of flow, **5:**8; 9; 11
effect of *p*H on birefringence, **5:**9
measurements, **5:**12
ultracentrifugation, **5:**11; 12
radiation, **1:**48
size and shape, **1:**56
size of virus particle, **5:**12–17
breadth measurements, **5:**12–14
bulk measurements, **5:**17
length measurements, **5:**14–17
Tobacco Mosaic Virus, Physical and Chemical Properties (Pirie), **5:**1
Tobacco mosaic virus (TMV), structure, **15:**449–484
double refraction, **15:**455–456; 459
Tobacco necrosis virus, activity and particle identity, **9:**194–195
crystallization, **9:**200–202
physical constants, **9:**187; 188
purification, **9:**186
Tobacco phytoalexins, **55:**69–73
α-Tocopherol, *See* Vitamin E
α-Tocopherol, *See* Vitamin E
α-Tocopherol, **46:**171, **48:**4
in chloroplast lamellae, **30:**9
and coenzyme Q, **25:**285; 286; 292; 293; 303–304; 319–320
in mitochondrial fragments, **21:**113
β-Tocopherol, *See* Vitamin E
γ-Tocopherol, *See* Vitamin E
D-α-Tocopherol, and glycine, **23:**461
α-Tocopherolhydroquinone, *See* Vitamin E
α-Tocopherol hydroquinone, **6:**477; 478
function, **6:**479
γ-Tocopherol hydroquinone, **6:**477; 478
function, **6:**479
Tocopherol in energy transport, **14:**40
Tocopherol Interrelationships (Hickman and Harris), **6:**469
Tocopherol phosphate, **67:**104
α-Tocopherolquinone, *See* Vitamin E
α-Tocopherol quinone, **6:**477; 478; 480
function, **6:**479
as inhibitory analogue, **6:**135; 141
Δ-Tocopherolquinone, **20:**182
γ-Tocopherol quinone, **6:**477; 478; 480
function, **6:**479
α-Tocopherolquinone, in chloroplast lamellae, **30:**9
Tocopherols
absorption spectra, **4:**291–293
interrelationship, **6:**469–524 (*See also* Vitamin E)
labeled, in coenzyme Q synthesis, **25:**301
separation from coenzyme Q, **25:** 293–294
Tocopherols as antioxidants, **8:**441; 447–448
α-Tocopherylquinone, and DPNH-Q reductase, **25:**313
relation to coenzyme Q, **25:**287
Tollens-Hiller-Pacsu concept of cellulose structure, **8:**259–262
Toluene
2, 4-4diiso cyanate, as a coupling agent, **41:**116
microbial oxidation, **27:**513–516
Toluene immersion, **43:**114; 116
p-Toluene sulfonyl arginine methyl ester (TAME), as substrate for thrombin, **30:**258
p-Toluenesulfonyl group, preparation of, **29:**398
p-Toluenethiosulfonate, **39:**342
Toluidine, inhibition of invertase by, **15:**287
Toluidine blue
effect on blood coagulability, **10:**37
effect on sperm nucleus, **16:**479
p-Toluquinone, as electron acceptor, **16:**134
Toluylene blue, as artificial electron acceptor, **30:**22
Tolypothrix tenuis, **46:**92
Tomatin, **11:**427
Tomatin (lycopersicin), **7:**488
Tomato bushy stunt virus, **5:**1–5:6
activity and particle identity, **9:**194–195

crystallization, **9:**198–199
effect of light on, **9:**142
homogeneity, **9:**219–220
hydration, **9:**204–206
physical constants, **9:**187; 188
purification, **9:**186
shape, **9:**203–206
size, **9:**204–206
chemical composition, **5:**4–5
physical properties, **5:**2–4
stability, **5:**5–6
Tomato Bushy Stunt Virus, Physical and Chemical Properties (Pirie), **5:**1
Tomatoes
antibiotic treatment of diseases, **20:**391–393; 402
carotenogenesis in, **21:**333–335; 341; 344
genetic study of carotenoids in, **21:**325–328
Tomato phytoalexins, **55:**73
Tomato wilt, **11:**403–437
Tonin, **57:**4
Topography, **34:**481
active site, **46:**256
binding site, **46:**256
Torpedo marmorata, **43:**106; 133
Torsional strain, **59:**148
Torsion angles, **59:**148
Torsion in protoplasm, **7:**44
Torula, coenzyme Q in, **21:**111
Torula, transaldolase in, **19:**27
Torula cremoris, **18:**332
growth factors for, **14:**66
Torula histolytica, hyaluronidase action on, **13:**230
Torula lactosa, growth factors for, **14:**66
Torularhodin, structure, **21:**306
Torula utilis
amino acid metabolism in, **16:**231
ribose synthesis by, **28:**408
Torula utilis, coenzyme Q in, **25:**288; 293–295
Torula utilis No. 3, production for food, **7:**580
Torula utilis (Torulopsis utilis), fat synthesis by, **9:**656–658; 661; 662; 664–665; 674–681; 690–691; 698–699
protein synthesis by, **9:**654; 655–658; 661; 665; 675–676; 679–680
Torula yeast, fumarase from, **25:**347
Torulopsis, tyrosine from labeled acetate, **14:**120–121
Torulopsis lipofera and synthesis of lipides, **8:**301; 302; 304; 305; 306–307; 308; 309; 310; 312; 313; 315–316; 321; 325; 328
Torulopsis magnolia, oxidation of hydrocarbons, **27:**493
Torulopsis utilis, **17:**396–398, **18:**31
5′-*O*-Tosyl-2,′ 3′-*O*-isoproplyidene adenosine, **54:**211
Tosyl-L-arginine, methyl ester, **19:**306; 307
Tosyl arginine methyl ester, **48:**288
Tosylarginine methyl ester, hydrolysis by thrombin, **16:**33–34; 37; 41–42
Tosyl group. *See also p*-Toluenesulfonyl group
Tosyl-L-lysine chloromethyl ketone, **41:**182
Tosyl-L-phenylalanine chloromethyl ketone, **41:**190
Tosyl-L-phenylalanyl diazomethane, **44:**11; 17
Toxic Amino Acids, Action as Antimetabolites (Fowden, Lewis, and Tristram), **29:**89
Toxic effects, **39:**222
Toxicity
of amino acid analogs, **29:**92
of collagenase, **23:**199–201
effect of detergents, **11:**85
of metal ions, **11:**69–74
of hydrocarbons, **10:**451–452; 452–453; 453–454
or organic halogen compounds, **10:** 277–278
selective, as basis for chemotherapy, **27:**419
Toxicity of active constituent of *Dichapetalum cymosum,* **18:**113–150
Toxicity of borate, **12:**521–523
Toxicity of Dichapetalum cymosum (Peters), **18:**131
Toxic peptides, alanyllysine, **35:**199
glycylleucine, **35:**199
lysylalanine, **35:**199
lysylleucine, **35:**199
triornithine, **35:**197
amide, **35:**211
trivaline, **35:**197
D-toxin, ion channels, **71:**214
Toxin-antitoxin reaction, **15:**69; 74–75
Toxinase, **23:**199
Toxins
action of peroxidase on, **12:**187
bacterial

Toxins (*Continued*)
effect on mitochondrial oxidative phosphorylation, **21:**31
peroxidase action on, **14:**153; 155–156
diphtheria, adsorption studies, **14:**332; 338
peroxidase action on, **14:**154–156
tetanus, peroxidase action on, **14:**155
clostridial, **23:**184–189
role of –SH groups in, **11:**244–245
wilting, **11:**401–437
κ-Toxins, **23:**185; 186
Toxoplasmosis, and folate inhibitors, **27:**458
Toyocamycin, **54:**88
Toyocamycin phosphate, **43:**35
TPD, *See* Triosephosphate dehydrogenase
T2-phage, mutations, **27:**311
T4 phage lysozyme, **47:**93–97; 135
T4 phage thioredoxin, **59:**148
TPN. *See also* Pyridine nucleotides; Triphosphopyridine nucleotide; Triphosphorpyridine nucleotide
in energy transport, **14:**13
in oxidation on phenylalanine, **14:**106
in neoplastic tissue, **29:**363
TPN-cytochrome c reductase, classification, **14:**227; 254
isolation, **14:**398; 405
TPN-cytochrome reductase, localization in cytoplasm, **13:**11
TPN-dehydrogenases, in pre-cytochrome system, **17:**363–365
TPNH, **35:**246, *See* Triphosphopyridine nucleotide, reduced
in deoxyribonucleotide formation, **29:**354
high levels of ATP elevated by, **29:**353
in DR metabolism, **28:**435–444; 446; 454
and light emission, **25:**151; 154; 161
and photosynthesis, **25:**42–44; 53–62; 65–66; 68; 71; 88–95; 102; 103; 105; 106
reduction of cytochrome b_5, **17:**362
TPN(H), Cotton effects, **27:**94
TPNH-cytochorme *c* reductase, in mitochondria, **21:**14; 15; 17; 18; 21; 58
TPNH-cytochrome *c* reductase, **20:**170
TPNH cytochrome c reductase, EPR study, **27:**131; 183
TPNH-cytochrome reductase, **17:**264; 350; 364; 365
in electron transfer, **19:**201
properties, **19:**203
TPNH dehydrogenase, EPR study, **27:**182
TPNH-dependent system, aspartate reduction, **17:**410
TPNH oxidase, **19:**203. *See also* Yellow enzymes
in steroid 11β-hydroxylase system, **19:**192
TPNH-oxidase system, **17:**380
$TPNH_2$ r cytochrome c transelectronases, **14:** 254
TPNH-reductases, in inorganic nitrogen metabolism, **17:**250–252
TPN-isocitric dehydrogenase, in mitochondria, **21:**13; 14; 17; 21; 58; 74
TPN-linked reductase, in aromatic ring synthesis, **17:**413
TPN-reducing factor, ferredoxin as, **27:**87
TPN-specific cytochrome *c* reductase, **15:**97–98
TPN-specific isocitrate dehydrogenase, **28:** 135–139
TPN-specific isocitric dehydrogenase, **15:**97
TPN-synthesizing enzyme, **15:**105
TPN (TPNH). *See also* NADP (NADPH, $NADPH_2$)
of nucleus, **27:**272
TPN (triphosphopyridine nucleotide). *See also* Coenzyme II
and CO_2 fixation, **12:**46; 51–52; 54–55; 58; 61–63; 63–67; 85; 89–91
as hydrogen carrier for glutamic dehydrogenase, **16:**255
comparison with CoA, **16:**14
in glucose oxidation in nuclei, **16:**474
in plant oxidations, **16:**133–135
in synthesis of amino acids, **16:**270; 271; 289
T_2 Polynucleotide kinase, **48:**246; 266
T_4 Polynucleotide kinase, **48:**247–266; 272; 273
activators, **48:**262; 263
application, **48:**264; 265
assay, **48:**247
endonuclease activity, **48:**265
enzymatic activities, other, **48:**265; 266
inhibitors, **48:**262; 263
kinetics, **48:**252–256
mechanism, **48:**252–256
molecular weight, **48:**249
pH optimum, **48:**252; 254
3′-phosphatase activity, **48:**265; 266

physicochemical properties, **48:**248–251
polyamines, **48:**263
purification, **48:**247; 248
reaction products, **48:**252
reversibility, **48:**252; 253
salt, **48:**262
SH- groups, **48:**249; 251
specificity, **48:**256; 257–261
spectrum, **48:**250
subunits, **48:**249; 251
T^4 polynucleotide kinase, **55:**359
TPP, *See* Thiamine pyrophosphate
TΨC loop, **35:**467; 468; 473
Tr, *See* Trypsin
Trace concentration, definition, **1:**188
Trace metals, **1:**189
Tracer, biochemical
introduction, **57:**252–255
and nitrogen-13, **57:**257–261
and nitrogen-15, **57:**257–261
and nitrogen-13 labeled compounds
nitrogen-13 ammonia, **57:**285–319
nitrogen-13 labeled L-amino acids, **57:** 319–336
nitrogen-13 labeled *N*-nitroso compounds, **57:**336–340
nitrogen-13 N_2, **57:**262–268
nitrogen-13 nitrate, **57:**271–284
nitrogen-13 nitrite, **57:**268–271; 273–284
nitrogen-13 N_2O, **57:**284–285
nitrogen-13 $^{13}NO_2$, **57:**284–285
other, **57:**340–341
and radionuclides, **57:**255–257
Tracer distribution, methods of analysis, **28:**402–404
Tracer elements, *See* Carbon 13, Carbon 14, Nitrogen 15, Oxygen 18, Phosphorus 32, and Sulfur 35
Tracers. *See also* Isotope studies; Isotope studies and Labeled compounds; Isotopic tracers; N^{15} and S^{35}
in permeability studies, **13:**21–65
Tracers, Application in Permeability Studies (Ussing), **13:**21
Tracer studies. *See also* label compounds
of algae, **28:**338
DR synthesis, **28:**393
in animals, **28:**431–435
in microorganisms, **28:**420–431; 435–436; 453
immunochemical study, **28:**305
of precursor of exo-β-lactamase, **28:**254
ribose synthesis, in animals, **28:**413–419; 431–435
methods, **28:**399–405
in microorganisms, **28:**405–413
biotin enzymes, **26:**307–308; 311–330
of ethionine effects, **25:**375–380
penicillin effect on cell walls, **26:**216
of phosphate to ATP, **17:**116
Tracer techniques
and CO_2 fixation, **2:**137; 146–148; 152–154; 160–164; 166; 168; 170; 172; 176
oxidation of hydrocarbons, **27:**485
Trace Substances, Enzymes and (Green), **1:**177
Tracheal cytotoxin, **58:**131
Tradescantia poludosa, **17:**453
Traffic ATPases, **65:**1–41
a-carbon backbones, **65:**20
consensus amino acids, **65:**22–23; 26
essentiality of binding protein, **65:**31–33
eukaryotic hydrophobic moieties, **65:**36–37
evolution of gene organization, **65:**10
formation of ATP-binding site, **65:**29–30
future prospects, **65:**40–41
interaction of conserved component with binding protein, **65:**30–31
MDR involvement in transport, **65:**38
multiple sequence alignment, ATP-binding domains, **65:**22; 24–25
prokaryotic, **65:**39
relevance to eukaryotic carriers, **65:**34; 36–40
reversibility, **65:**33–35
tertiary structure, **65:**18–24
three-dimensional model, **65:**20
universality of structure, **65:**7–11
Transacetylase, **12:**327, **15:**114; 129; 231; 236.. *See also* Phosphotransacetylase
and ATP synthesis, **23:**335; 339
modification, **15:**394–395
thio-β-galactoside, reaction catalyzed by, **30:**85
Transacetylases, classification, **14:**226
Transacetylation
of amino acids, **6:**262
coenzyme for, **14:**180–183
as phosphoroylsis, **18:**226
Transaconitate, **15:**362
Trans-Aconitate, **51:**236
Transacylases, classification, **14:**251; 252

Transacylation
and coenzyme A, **14:**241
Transacylation, in activated amino acids, **19:**238–259; 261–263
of glycine by CoA derivatives, **19:**249
Transadenylases, classification, **14:**226; 252
Transaldimination, **49:**34
Transaldolase, **15:**154; 160; 163–164; 166; 170–171; 176, **37:**442; 523; 585
from *Candida utilis,* **31:**138
role of essential histidine residue in, **31:**173
in carbohydrate metabolism, **26:**37
location of, **24:**309
in photosynthesis, **19:**15; 16; 27; 30; 31, **25:**96
in ribose synthesis, **28:**396–397; 402; 410–411; 415–416
Transalkylation, **25:**381–388. *See also* Ethionine, Methionine, and Methylation
Transamidation
catalyzed by papain, **33:**434
by chymotrypsin, **13:**309
Transamidinase, **14:**253
product inhibition of, **25:**223; 242–243
Transamidination, **1:**157; 158, **15:**374
Transaminase, **5:**154; 155
activity of —SH groups, **11:**239; 241
allosteric regulation, **28:**114–115
artificial, **58:**37
and mimosine, **29:**109
and pyridoxal phosphate, **10:**62; 70
glutamic-aspartic, and pyridoxamine phosphate, **10:**63; 65
in plants, **10:**300
and succinate oxidation, **23:**382
D-Transaminase, **16:**237–239
Transaminases, **4:**82–91, **15:**123; 330–331; 334; 337–341; 373, **50:**156. *See also* Aspartate aminotransferase; Transamination; Transaminations and individual transaminases
and citric acid cycle, **18:**254–255
effect of fluoroacetate on, **18:**118
classification, **14:**221; 226; 241; 252; 253
in alanine racemization, **14:**195
in oxidation of phenylalanine and tyrosine, **14:**97–100; 105–106
column chromatography, **22:**459; 460; 485
site of cyanide inhibition of, **22:**394
genetic studies, **17:**398; 400–402; 406; 449
and gluconeogenesis, **4:**212
inactivation of, **55:**227
by fluorinated substrate analogs, **55:** 229–230; 234
instability, **16:**187
role in amino acid synthesis, **16:**255; 257; 280–281
terminology, **16:**187–188
isolation, **4:**86–87
kinetic mechanism of, **45:**290
kinetics of reactions, **4:**87–88
mechanism for, **25:**243; 249
in mitochondria, **21:**41; 42
multiple forms, **25:**360
turnover number, **25:**32
and plant metabolism, **9:**398–401
specificity, **4:**87–88; 94
in sperm, **24:**66; 89
localization of, **24:**312; 340
Transaminases A, B, and C, of *E. coli,* **16:** 280–281
Transamination, **1:**154; 155; 157; 158, **3:** 215–216, **16:**185–246, **35:**507, **49:**29, **53:**163; 164; 173–178; 196, **57:**148–149, **58:**35. *See also* Transdeamination, Transreamination
allostreric regulation, **28:**66–69
in biosynthesis of amino acids, **16:**254–257; 261–264; 268; 275–276; 292; 294; 296; 299
in cancer tissue, **4:**91–92
enzyme catalyzed, **4:**82–91
analytical methods, **4:**85–86
coefficient for various tissues, **4:**88–91
comparison with model transmination, **4:**94–95
mechanism, **4:**82–83
or presence of dicarboxylic acid, **4:**85
in tissue preparations, **4:**82–83
formation of aminoacyl mercaptans during, **19:**240; 241
in genetic studies, **17:**397
and glycine metabolism, **23:**454
kinetic studies, **16:**222–223
mechanism, **16:**209–226
mechanism of, **35:**120
in mitochondria, **21:**41; 42
model reactions, **35:**119–125
in model systems, **4:**75–82

and acidity, **4:**78; 84
carbon dioxide evolution, rate, **4:**77; 79
and decarboxylation, **4:**76–82
mechanism, **4:**79–81
in nitrogen metabolism, **19:**335–386
nonenzymic, **16:**119; 188; 204; 223–226
in oxidation of phenylalanine and tyrosine, **14:**97–100; 102; 105–106
and oxidative deamination, **4:**92
in plants, **9:**398–401; 401–403; 405–406; 407; 419–420; 433, **20:**324–330; 334; 335
in protein synthesis, **4:**93–99
and pyridoxine, **7:**454
in muscle, **7:**390
reaction intermediates, **58:**45
relationship with decarboxylation, **16:** 219–220
role of amides in, **13:**265; 272; 284–285
schematic mechanism for, **25:**34
in seeds and plants, **4:**92–93
stereochemistry of, **35:**97–106
transition state for, **58:**41
Transamination, Enzymic, Dynamic Three-Dimensional Model (Ivanov and Karpeisky), **32:**21
Transamination (Meister), **16:**185
Transamination Reaction (Herbst), **4:**75
Transcarbamylation, **15:**364
Transcarbohydrases, classification, **14:**226
Transcarboxylase, **35:**332; 401–406, **45:**304; 379, **47:**127, **56:**292. *See also* specific enzymes
biotin content, **35:**406
carboxyl-carrier polypeptide, **35:**406
chemical mechanism, **45:**379; 380
enzyme-metal-substrate bridge complex, **35:**405
in formation of propionic acid, **26:**292; 307–311; 328; 329
kinetic mechanism of, **45:**304
metal content, **35:**404
molecular characteristics, **35:**405–406
ping-pong mechanism, **35:**405
reaction, **35:**402–405
enzyme-$^{14}CO_2$, **35:**404
equilibrium constant, **35:**403
metal ion role in half-reaction, **35:**404
substrate specificity, **35:**403
subunits, **35:**406
Transcarboxylase (C^{14}), and biotin (C^{14}), **26:**331
Transcarboxylations, tracer experiments, **26:**311–316
Transcarboxylation site of pyruvate carboxylase, **42:**33
Transcriptase, reverse, **56:**292
Transcription, **44:**165; 166; 168; 169; 170
Transcription, RNA polymerase II transcription, **71:**41–77. *See also* RNA polymerase II transcription
Transcriptional regulation
α-crystallin/small heat-shock protein, **69:** 165–188
tyrosine hydroxylase, **70:**172–179
Transcriptional regulation (*Caulobacter* cell cycle), **71:**15–25
asymmetric expression of non-flagellar genes, **71:**24
cell-cycle regulation of flagellar gene expression, **71:**15–21
polarized expression of flagellar genes, **71:**21–23
polarized transcription basis, **71:**24–25
Transcriptional repression, **40:**79; 80; 81
Transcrotonate, **15:**362
Transdeamination
definition, **19:**341
in transamination reactions, **19:**342; 343; 346; 347; 357; 367–371; 373; 376
Transducin, **58:**128
Transduction
chemical genetics, **17:**457–459
of β-lactamase production, **28:**269–272; 280
Transelectronases, classification, **14:**227–228; 254; 255
Transesterification by choline acetylase, **12:**326–327; 327–331
Transethylation, **22:**272
Transferase, **34:**476
activity, **34:**476
enzyme, **34:**475
protein, **34:**476
Transferase activity
of cathepsin, C, chymotrypsin, and trypsin, **19:**310
of papain, **19:**309–311
regulation of bacterial form glutamine synthetase, **72:**33–35

Transferase B, **46:**390; 391; 406
CD spectrum, **46:**391
concentration in liver, **46:**405
half-life, **46:**406
identity with ligandin, **46:**390
mRNA for, **46:**406
Transferase enzymes, and oxalate metabolism, **23:**448–449
Transferase reactions, **18:**198
Transferases, **46:**391; 406; 408. *See also* Transferring enzymes and Transglycosylation
action, **26:**59–60
from *P. shermanii,* **26:**263
alkylation of, **46:**408
aminoacyl-tRNA N-acetylmuramyl pentapeptide, **40:**118
alanyl-tRNA UDP-N-acetylmuramyl-pentapeptide, **40:**127
aminoacyl-tRNA phosphatidylglycerol, **40:**111
aminoacyl-tRNA protein, **40:**94
particulate aminoacyl-tRNA GlcNAc-MurNAc(-pentapeptide)-*P*-*P*-phospholipid, **40:**120
classification, **14:**223; 225–227; 235–236; 241; 242; 250–253
covalent binding to, **46:**408
dissociation constants, **46:**391
as distinguished from hydrolases, **33:**430
fluorescence, **46:**391
hydrolytic enzymes as, **17:**221–229
specificity of, **17:**169; 170–171; 179
inducation with phenobarbital, **46:**406
location of, **24:**309–316
and peptide synthesis, **13:**289–305; 311–315
in plants, **18:**287; 288
sex differences, **46:**406
Transferases, Aminoacyl-tRNA (Soffer), **40:**91
Transferase system (GTF)(GSH), **13:**300–301
Transfer enzyme, in polypeptide synthesis, **24:**3
Transfer reactions, of corrinoids, **26:**253–255
Transferrin, **52:**32; 107
asialo-, **41:**121–123
Cotton effect, **27:**60; 93
EPR study, **27:**138; 140
Transferring activity, of phosphatases, **18:** 197–198
Transferring enzyme *(Bacillus macerans),* oligosaccharides, from, **17:**210–211; 226
Transferring enzymes
classification, **14:**221; 223
and kinases, **18:**195–201
from molds, **17:**211–213
in oligosaccharide formation, **17:** 190–232
product inhibition of, **25:**222–223; 242–249
requiring —SH groups, **11:**243–244
Transferrin receptors, adenosine receptors, **69:**113–114
Transferrin saturation, **44:**223
Transfer RNA, *See* tRNA
Transfer RNA and Its Functional Implications, Three-Dimensional Structure of (Kim), **46:**277
Transfer RNAs, Understanding the Recognition of, by Aminoacyl Transfer RNA Synthetases (Schimmel), **49:**187
Transfer site, in mitochondria, **23:**379–380
Transformations
chemical genetics, **17:**455–457
enzymic, relaxation methods, **25:**32–35
Transforming enzyme, **12:**98
Transforming principle of bacteria, DNA as, **16:**412; 429
Transformylations, **18:**211
Transfructosidase, **15:**314, **17:**197
Transfructosylase, **17:**200; 205–207; 222; 226, **18:**288. *See also* Fructose transfer
Transfusions, of hemophiliacs, **16:**57
platelet, in bleeding from exposure to ionizing radiations, **16:**67
Transgalactosylases, **17:**218–219; 227
Transgalactosylation, **15:**314–315
Transgenic mice
α-crystallin/small heat-shock protein, **69:**165; 173; 183
phosphoenolpyruvate carboxykinase (GTP), **69:**244–255
Transglucosidase, **11:**372–373
action of sucrose phosphorylase, **10:**140–141
Transglucosidases, **12:**410–424
definition, **12:**380; 410
enzymic adaptation, **13:**70
nonphosphorolytic, **12:**419–423
in polysaccharide synthesis, **15:**312
Transglucosidation, **13:**312; 314, **15:**374
Transglucosylase activity
in invertase preparations, **17:**206–207
in polysaccharide synthesis, **28:**340

Trans-β-glucosylase activity, in laminarinase, **28:**350–351
Transglucosylases, **17:**226–227. *See also* Glucose transfer
Trans-β-glucosylytic synthesis of laminarin, **28:**338
Transglucuronosylases, microsomal, **26:**59–60
Transglutaminase, **36:**239
activity of, toward aliphatic amides, **54:**12
amide site of, **54:**10–25
conformational aspects of, **54:**12–17
with peptide derivatives of glutamine, **54:**17–19
and stereospecificity between L- and D-glutamine residues, **54:**21–25
substrate conformation at, **54:**19
in amine reactions, **54:**30
amine site of, **54:**30–50
in aliphatic amine reactions, **54:**33
conformational aspects of, **54:**33
in α,ω-diaminomonocarboxylic acid derivative reactions, **54:**33–36
in lysine peptide derivative and analog reactions, **54:**36–45
stereochemical aspects, **54:**45–50
substrate conformation at, **54:**33
in bacteria, **38:**187
distribution of, **38:**110–112
distribution of, in animal tissues, **54:**2
forms of, catalytically active, **54:**4–6
general mechanism of, **38:**112
glutamine substrates binding site of, **54:**25–30
guinea pig hair follicle, **38:**112; 180–183
catalytic properties of, **38:**181
kinetic constants for, **38:**181–183
mechanism of, **38:**181–183
molecular properties of, **38:**181
purification of, **38:**181
guinea pig liver, **38:**111
active site titrations of, **38:**160
acyl-enzymes, **38:**142
aliphatic amides as substrates for, **38:**156
amine substrates for, **38:**124
antiserum for, **38:**114
assays for, **38:**113
catalytic properties of, **38:**118–132
chemical mapping of the active site, **38:**155
chemical modification of, **38:**150–159
by 5,5′-dithiobis(2-nitrobenzoic acid), **38:**152
by *p*-mercuribenzoic acid, **38:**153
by tetrathionate, **38:**151
chemical properties of, **38:**116
difference spectrum, calcium induced, **38:**135
effect of pH, **38:**125–127
ester substrates for, **38:**122–123
glutamine substrates for, **38:**118–122
hydrolysis reaction of, **38:**120
inactivation, by copper salts, **38:**150
by iodacetamide, **38:**150
initial velocity pattern for, **38:**128
isotope exchange by, **38:**131–132
kinetic properties of, **38:**125–132
liberation of ammonia by, **38:**118
liver esterase in, **38:**114
metal ions, activation constants, **38:**137
binding to, **38:**133
dissociation constants, **38:**134
mechanism of activation by, **38:**137
role of, in catalytic activity, **38:**133
methylglutamine peptide derivatives as substrates for, **38:**156
monomeric structure for, **38:**118
physical properties of, **38:**114
product inhibition of, **38:**130
products of action on glutamine substrates, **38:**120
protein substrates for, **38:**122
purification of, **38:**113
reporter group studies of, **38:**147
reversibility of transfer reaction of, **38:**120
specificity of, **38:**121
transfer reaction of, **38:**120
as guinea pig liver enzyme, **54:**4; 5; 51
in activation of, by isocyanates, **54:**19–20
in catalysis of *p*-nitrophenyl acetate, **54:**7; 8
in catalysis of *p*-nitrophenyl trimethylacetate, **54:**8
glutamine peptide derivatives as substrates for, **54:**17
in glutamine residue modification, **54:**11
specificity of lysine derivatives and analogs for, **54:**42
substrate specificity of, **54:**11; 25–29
human blood plasma, activation by calcium ions, **38:**178
catalytic properties of, **38:**177–178
kinetic constants for, **38:**178
mechanism of, **38:**177–178

Transglutaminase (*Continued*)
physical and chemical properties of, **38:**167–170
purification of, **38:**166
role in insoluble fibrin formation, **38:** 160–166
subunit structures of, **38:**171–175
human blood platelet, catalyic properties of, **38:**177–178
mechanism of, **38:**177–178
physical of chemical properties of, **38:** 167–170
purification of, **38:**166
subunit structures of, **38:**171–175
human placenta, physical and chemical properties of, **38:**167–170
purification of, **38:**166
subunit structures of, **38:**171–175
human uterus, subunit structures of, **38:** 171–175
intermediates in reactions of
acylamido interaction in, **54:**45
in amine substrate efficacy, **54:**31–32
with lysine containing extended sites, **54:**36; 38–39
with lysine peptide derivatives, **54:**45–50; 52
in invertebrates, **38:**187
isotope exchange by, **38:**113
in mammals, **38:**187
modification of proteins and peptides by, **38:**183–186
molecular features of, **54:**4–6
nomenclature of, **54:**4
plasma factor XIIIa, human, **54:**4; 5–6; 51
reaction catalyzed by, **38:**110–112
role of epsilon(γ-glutamyl)lysine crosslink formation, **38:**187
specificity of, **54:**2; 3
at glutamine carboxamide side chains, **54:**24
for glutamine substrates, **54:**45
as liver enzyme *vs.* blood coagulation factor, **54:**51
for lysine peptide derivatives, **54:**45–47
for optionally active amino acid residues, **54:**47
specificity of lysine derivatives for, **54:**42
substrate specificity of, **54:**25–30
structural analysis of proteins and polypeptides by, **38:**183–186
Transglutaminase-Catalyzed ε-(γ Glutamyl) Lysine Bond Formation: Mechanism and Basis for (Falk), **54:**1
Transglutaminases, Molecular and Catalytic Properties of (Folk and Chung), **38:**109
Transglutamine
intermediates in reactions of, **54:**7–8; 9
mechanism of action of, **54:**7–10
kinetic, **54:**9–10
with stable enzyme-substrate intermediate, **54:**7–9
reactions catalyzed by, **54:**6–7
Transglycosidases, **11:**330; 365, **60:**273
classification, **14:**226
product inhibition, **25:**190
in synthesis of di-and polysaccharides, **10:**138–141
Transglycosidation, *See* Transglycosylation
Transglycosylase, **11:**330
Caulobacter cell cycle, **71:**6
β-Transglycosylase, in carbohydrate synthesis, **26:**75–76
Transglycosylases, **17:**189–232, **18:**200–201; 223–225
in carbohydrate synthesis, **26:**68; 70; 72–76; 78; 79
specificity of, **26:**59–62
classification, **14:**225; 226; 251; 252
Trans-*O*-glycosylases, **17:**221
Trans-α-glycosylases, **17:**218; 219
Transglycosylation, **60:**229. *See also* Glycosyl donors and Sugar nucleotides
in carbohydrate synthesis, **26:**36–37
and DPN synthesis, **18:**206–207
intramolecular, in murein, **26:**208–209
in murein synthesis, **26:**220
oligosaccharide formation by, **17:**189–232
in polysaccharide synthesis, **15:**303; 307; 311–312
rate constant, **60:**230
Transglycosylation reactions, **18:**200–201
Transhydrogenase, **23:**345, **46:**245; 255; 256
and active ribonuclease, **26:**103
Transhydrogenase activity, **34:**87
Transhydrogenase reaction, **26:**167
Transhydrogenases
classification, **14:**227–228; 236; 253–254; 255
disulfides, **27:**218
in photosynthesis, **25:**97
Transhydrogenases, TPNH-DPN, **23:**451–452

Transient absorption studies, cytochrome *c* oxidase, **71:**123–124
Transient kinetics, cytochrome *c* oxidase, electron transfer, **71:**118–124
Transimination, **15:**325–327; 331–332
Transimination process, **50:**20; 21
Transimination reaction, **50:**16
Transinhibition, **56:**123–124
in allantoate transport, **56:**126
in allantoin transport, **56:**125–126
Transition metals, EPR spectroscopy, **27:**107; 118–120; 137–144
Transition state, **43:**371, **58:**11
activity coefficients, **43:**377
binding, enzyme action, algebraic *versus* graphic analogies, **73:**50–51
complementarity to, **43:**259; 351
exchange, **43:**366
reciprocal stabilization, enzyme action, **73:**47–49
stabilization, enzyme action, **73:**46–47
Transition-state analog, **43:**239; 241; 362; 494, **59:**156
lactone tetrasaccharide, **43:**243
Transition-state analog complex, **45:**125–127
Transition-state inhibitors, **59:**107
Transition temperature, **34:**468
Transketolase, **15:**154; 157–160; 163; 165; 166; 170–171; 176–177. *See also* Active glycolaldehyde
in ATP synthesis, **23:**337
in carbohydrate metabolism, **26:**37
in formation of 2-keto-4-deoxy-7-phosphoheptulose, **28:**422
in ribose synthesis, **28:**396–397; 402; 410–411; 416; 418
location of, **24:**309
mechanisms of catalysis, **24:**468
in photosynthesis, **19:**15; 25–27; 30; 31, **25:**95; 98; 99
in plants, **18:**293–294
Transketolase-aldolase-phosphatase pathway, in ribose synthesis, **28:**397
Transketolases, and TPP, **22:**401
Transketolase-transaldolase pathway, ribose synthesis, **28:**410–411; 414–416; 418
Translatable messages, **40:**85
Translational repression, **40:**81; 82; 83; 84; 85; 86; 87
Translation of phiX174 genes, reinitiation during, **47:**415–419
Translocation
amino acid, in mouse brain, **29:**63 (*See also* Myoglobin, reactions)
for amino acids in Ehrlich ascites cells, **29:**62
anti-coupled solute, **29:**49–53
chemically specific, diagram of, **29:**42
difference between primary and secondary, **29:**78–79
D-glucose, **29:**44
of glucose and nonfermentable sugars through intestinal mucosa, **29:**53
of glycine by pigeon red cells, **29:**63–64
L-leucine, **29:**44
multiple systems, **29:**64–65
noncoupled solute, **29:**44–49
of oxygen by myoglobin, **29:**35
potassium, effect of valinomycin and gramicidin on, **29:**48
primary, **29:**69–78
proton-coupled solute, **29:**65–68
of protons during ATP hydrolysis, **29:**74
secondary, **29:**44–68; 78
sym-coupled solute, **29:**53–65
through natural membranes, **29:**33–87
of antibiotics in plants, **20:**386–390; 393; 394
Translocation model, T7 RNA polymerase, elongation complexes, **66:**117–119
Translocation of insoluble matter, **17:** 135–158
Translocation of materials in mammalian tissues, effect of amides, **13:**268–271
Translocations through Natural Membranes (Mitchell), **29:**33
Translocators, ADP/ATP, **29:**66
anion, **29:**67
definition of, **29:**78
β-galactoside of *E. coli,* **29:**65–66
L-proline in mouse pancreas, **29:**62–63
Transmembrane potential, **49:**66; 67
Transmembrane region arrangement, adenosine receptors, **69:**92
Transmembrane signalling, **50:**304
Transmethylases, **12:**485, **22:**252–254; 258
classification, **14:**226; 250; 252
Transmethylating coenzyme, **25:**410

Transmethylation, **1:**155–157, **7:**376–378, **11:**390, **12:**455; 458–464; 464–469; 469–474; 482–486, **15:**112; 400, **25:**396; 410–427, **29:**137
with *S*-adenosylmethionine, **22:**254–257
biochemical criteria for, **22:**252
with dimethylacetothetin, dimethyl-β-propiothetin, **22:**252; 253
mechanism of, **22:**258
with *S*-methylmethionine, **22:**253; 259
in biosynthesis of pectic substances, **20:**369
compounds methylated by (table), **25:**411
Transmethylation deficiency, genetic studies, **17:**408–409
Transmethylation in plants, **8:**234
Trans-β-methylglutaconic acids, **35:**139; 145
Transmidation, **53:**250; 251; 253; 255; 257; 277
Transmination, **15:**125–126; 158; 325; 330–331; 333–343; 363
Transmission of information from gene, **24:** 11–20
Trans-N-glycosylase, **17:**221; 229
Transpeptidase, *Caulobacter* cell cycle, **71:**6
Transpeptidation, **13:**309, **19:**310–311, **53:**250; 251; 252; 253; 258
in plants, **20:**223
Transpeptidation during hydrolysis of proteins, **14:**293–294
Transphosphatases, classification, **14:**220; 221; 226; 251; 252; 253
Transphosphorase, in acetylation, **12:**330
Transphosphorylases
classification, **14:**220
and light emission, **25:**137
product inhibition of, **25:**223; 244–245
Transphosphorylating enzymes, variety of turnover number, **24:**123
Transphosphorylation, **1:**126–131, **13:**314
in esterases, **20:**289; 290
in muscle, **16:**340
nucleotide, **53:**141–143
Transphosphorylations, **9:**506
Transphosphorylation systems in citric-acid cycle, **10:**189–190
Transport, **34:**512; 526, **38:**365–368
active, across cell membranes, **25:**17
coupling of Na^+ and sugar, **29:**53–54
electron, stoichiometric relationship between Ca^{++} uptake and, **29:**266–267
electron and hydrogen, **29:**43
ion, chemical coupling hypothesis for, **29:**298–303
L-leucine, of human and rabbit erythrocytes, **29:**45
metabolic control and membrane, **29:** 346–347
of nucleotides, **29:**66
of proline in *E. coli,* **29:**105
relation to control of enzyme synthesis, **29:**379–381
sugar, in erythrocytes, **29:**45
electron (*See also* Cytochromes; Ferredoxin; Macromolecules; Moving-boundary electrophoresis; Plastocyanin)
extension of theory of mass, **30:**171–175
formulation of theory of electrophoresis, **30:**152–154
involvement of plastoquinone in, **30:**29–30
moving-boundary, experiments of, **30:** 142–143
protein components of the chain of, **30:** 12–16
theory of, of interacting systems of biological macromolecules, **30:**139–177
irreversible inactivation by *p*-nitro-phenyl ester of biotin, **38:**367
nuclear, **27:**259–260
regulation by biotin, **38:**366
renal, and fluoroacetate, **18:**134
with resting cells of *L. arabinosus,* **38:**365
in *S. cerevisiae,* **38:**366
Transport, Some Special Kinetic Problems of (Christensen), **32:**1
Transport mechanisms, and folic acid, **27:**430; 437; 455–458
Transport of circulating glycoproteins, **41:** 99–127
Transport of Folate Compounds in Bacterial and Mammalian Cells (Huennekens, Vitols, and Henderson), **47:**313
Transport of Interaction Systems of Biological Macromolecules, Theory of (Canna and Goad), **30:**139
Transport proteins, **36:**31
Transport systems
amino acid translocation and, **53:**208
incompletely identified, **49:**83
confusion of unidentified systems with ASC, **49:**83

Transport systems A and L, differentiation, **49:**54–55
Transreamination, definition, **19:**341
in transamination reactions, **19:**342; 346; 366; 373; 376; 381
trans-Stilbene oxide, **57:**404
Transsulfhydrase, **14:**253
Transsulfurases, **14:**227; 252
Transsulfuration, **7:**380–383, **22:**264
ethionine interference with, **22:**272
in *Neurospora* mutants, **22:**265
Trans-synaptic induction, tyrosine hydroxylase induction, **70:**166–168
trans Terpenoids, **29:**243; 246
Transuridylase, in formation of uridine diphosphogalactose, **20:**114
Transverse relaxation time, correlation time for dipolar interaction, **33:**6
correlation time for scalar interaction, **33:**7
paramagnetic effects on, **33:**6–7
physical description of, **33:**3–5
Trasferase
aspartate transcarbamylase, **56:**329–334
mercaptopyruvate sulfur, **56:**292
nucleotidyl polymerase, **56:**334–341
terminal dNT, **56:**292
Traumatized tissue and electric potentials, **7:**48
Trauton's rule, **43:**283
Tray process for growth of microorganisms, **9:**657; 662–663; 668–670; 671
Treadmilling, microtubule dynamics, **69:** 133–134
Trehalase, **5:**120
cacao, **25:**466
classification, **14:**245
Trehalases
assay of, **56:**215–216
inhibitors of, **56:**219
mechanism of action of, **56:**218
occurrence and localization of, **56:**214–215
origin of multiple forms of, **56:**227
properties of, **56:**217–218
purification of, **56:**216
role of, **56:**218–219
in sugar transport, **56:**218–219
Trehalose, **18:**272
fermentation of, **5:**107; 113; 178
in kidney, **34:**298
α α′-Trehalose, **17:**214; 220
β β′-Trehalose, **17:**217
β β-Trehalose, as glucosyl donor, in laminarin synthesis, **28:**338
Trehalose-6-phosphatase, in insects, **28:**103
Trehalose phosphate, **18:**201, **20:**122; 123
Trehalose-6-phosphate, hydrolysis, by metal ions, **21:**184; 185
Trehalose phosphate synthetase, **40:**35; 36; 37
active aggregates, **40:**59
effect on polyanions, **40:**58
Trehalose phosphorylase, **59:**340
Trehalose synthetase, allosteric regulation, **28:**52; 102–103
Treponema pallidum, hyaluronidases in, **13:**203
Triacetic acid, **15:**241
Triacetic acid hydrolyzing enzyme, **14:**107
Triacetic acid lactonase, localization of, **24:**318
Triacetin, lipase action on, Michaelis curves, **23:**145–148
Triactin, properties of, **29:**289
Triad amidotransferases
aminodeoxychorismate synthase, **72:** 136–138
ammonia transfer theories, **72:**138–139
anthranilate synthase, **72:**136–138
carbamyl phosphate synthetase, **72:**124–131
synthetase III form, **72:**131
type II enzymes, **72:**125–131
GMP synthetase, **72:**119–124
human strains of, **72:**123–124
structure and mechanism, **72:**120–123
imidazole glycerol phosphate (IGP) synthase, **72:**131–134
NAD synthetase, **72:**134–136
structural characteristics, **72:**90–93
3,4,4′-Triaminodiphenyl sulfone, inhibitor of thrombin formation, **16:**47–48; 54
2,4,5-Triamino 6-hydroxypyrimidine, **67:**175
2,5,6-Triamino-4-oxo pyrimidine, **67:**196–198
Triangular reaction, kinetic treatment of, **23:** 92–95
mathematical treatment of, **23:**89–122
Triaryltetrazolium salts, **15:**8
Triatoma infestans, **46:**102
Triazine, eukaryotic dihydrofolate reductase, ligand binding (of inhibitors), **70:**60
Triazines, folate inhibitors, **27:**438; 440; 441; 444–446; 450–452; 455–456
Triazolopyrimidines, as inhibitory analogues, **6:**134
Tribolium larvae, **59:**81

Tri-branched photoaffinity labelling reagents, determining asialoglycoprotein receptor binding site topography, **66:**54–57
Tribrityrin emulsion, lipase action, Michaelis curve, **23:**143–144
Tributylphosphine, **58:**152
Tributyrin, lipolysis of, **23:**152
Tricarballylate, **51:**236
Tricarballylic acid in mold metabolism, **9:**549
2,7,9-Tricarboxyl-aH-pyrrolo(2,3-f)quinoline-4,5-dione, *See* Pyrolo-quinoline quinone
Tricarboxylic acid. *See under* individual acids
Tricarboxylic acid activators, **35:**365
Tricarboxylic acid cycle, **3:**235–241, **8:** 392–401; 401–409; 409–412. *See also* Carbon dioxide fixation and under individual members of cycle; Citric acid cycle; compounds
  allosteric control of, **28:**48; 65; 74; 91–92; 94; 143–144
    coordination with glycolysis, **28:**78; 86
    and ribose synthesis, **28:**412
  in animal tissues, **3:**238–241
  and biosyntheses, **18:**241–280
  and CO_2 conversions, **12:**45; 46; 47; 49–50; 105
  effect of fluoro compounds on, **18:**118–137; 140–141; 144–146; 152
  enzymes of, **24:**297; 337
  and fat synthesis, **8:**328
  and fatty acid synthesis, **26:**302–303
  fluoroacetate as indicator of, **18:**132–137
  genetic studies, **17:**425–426
    salts of components as solubilizing agents, **17:**149
  localization of enzymes, **13:**12
  in microorganisms, **3:**240–241, **23:**402–420; 424–431; 434; 435; 439–446; 462–464
  in mold metabolism, **9:**540–541; 566; 572–576
    in sperm, **9:**378
  in molds, **3:**241
  oxidation of 3-oxoadipate, **27:**506; 512–513
  and oxidation of proteins, **3:**246
  and photosynthesis, **18:**340
  in plant material, **3:**241
  in plants, **18:**298–303; 308
  and propionic acid metabolism, **26:**368–369
  and pyridine nucleotides, **15:**121–122; 124–125
  relation to amino acid synthesis, **16:**226; 249; 252; 254–256; 260; 265; 268; 285; 286; 290; 297
  relation to tobacco curing, **10:**326–327
  and Szent-Györgyi cycle, **3:**236–238
Tricarboxylic acid cycle (TCA)
  algR2 gene, **70:**238–245
  Pseudomonas putida, **70:**223–224
Tricarboxylic acids
  in ATP synthesis, **23:**334; 340–341; 348
  β-carboxylation in, **12:**86–92
2,4,5-Tricarboxypyrrole, from *Sepia* melanin, **16:**172
Trichloroacetic acid
  in denaturation study, **23:**32–36; 38
    effect on cytochrome *c,* **23:**58–69
  effect on acetyl phosphatase, **6:**253
  solubility of intermediates in prothrombin-thrombin change, **16:**46–47; 51
Trichloroethylene
  cystein *S*-conjugate β-lyase reactions, **72:**204–207
  toxic effects, **72:**210–212
2,3′,6-Trichlorophenolindophenol, **17:**251
2,3,6-Trichlorophenolindophenol (TCIP), photoreduction of, **30:**23
Trichoderma viride, enzymic transformations of steroids by, **20:**248; 249
Trichodermin, **56:**107–108
Trichomonas foetus, enzymic transformations of steroids by, **20:**255
Trichomonas gallinae, enzymic transformations of steroids by, **20:**255
Trichophyton, and staphylococcal infections, **28:**312–313
Trichothecium roseum, enzymic transformations of steroids by, **20:**242; 247–249; 252
Triethylcholine, metabolism of, **25:**375
Trifluoperazine, **55:**181
  tyrosine hydroxylase regulation, **70:**151
Trifluopromazine, **55:**182
Trifluoroacetic acid, in mixed anhydride of synthesis, **22:**165; 166; 172; 173
Trifluoroacetyl thioester, in identification of imidazole, **20:**76
Trifluoroalanine, **55:**231; 234
Trifluorodeoxyuridylate, **38:**253
5′,5,′5′-Trifluoro-DL-leucine, incorporation into protein, **29:**146
2-Trifluoromethyl adenosine, **55:**272

5-Trifluoromethyl-2'-deoxyuridine, **54:**58
for herpes keratitis, **54:**59
5-Trifluoromethyluracil, mutagen, **27:**331
Trifluoropromazine, use in schizophrenia, **29:**501
Trifluorothymidine, **54:**58
Trifluorothymidine-5'-triphosphate, **54:**58
Trifolirhizin, **55:**5
Trifolium pratense, **55:**5
maackiain formation in, **55:**40
Trifolium repens, **55:**34
Triglochinin, **56:**201
Triglyceridase, **23:**153
Triglycerides
lipase action on
mode of action, **23:**152–154
specificity of, **23:**154–158
natural, structure of, **23:**157–158
synthesis, in mitochondria, **21:**50; 52
Triglyceride transfer, protein disulfide isomerase (PDI) prolyl 4-hydroxylase β subunit function, **72:**378–380
Triglycerites, **34:**478
Triglycine, ionization constant, **20:**58
Trigonellin, **15:**109–111
Trigonelline, formation of, **12:**457; 474–478
Trigonelline in plants, **8:**234
1,2,4-Trihydroxybenzene, oxidation, **16:**140
2′,4,4′-Trihydroxychalcone, **55:**33
Trihydroxycinnamyl alcohol, **18:**372
transformation to lignin-like material, **16:**148
Trihydroxyeicosanoids, **58:**304
5,6,7-Trihydroxyflavone glucuronide (baicalin), hydrolysis, **16:**381
Trihydroxypropylpterin, **35:**53
Triiodothyronine, **46:**160
dissociation constant, **46:**391
metabolism, in mitochondria, **21:**27
3,5,3′ (I^{131})-Triiodothyronine, identification of its glucuronide, **16:**383
Trimethoprim, **38:**270
dihydrofolate reductase, **70:**28
eukaryotic dihydrofolate reductase, ligand binding (of inhibitors), **70:**55–56; 58
inhibitor, **27:**440
species differences, **27:**450–451; 454; 456; 460
synergism with sulfonamides, **27:**455–456
Trimethoprim binding, eukaryotic dihydrofolate reductase, **70:**37–40
3,4,5-Trimethoxyphenylacetic acid, oxidation of mescalin to, in schizophrenics, **29:**526–527
Trimethylacetic acid, **54:**14
4-*N*-Trimethylaminobutyrate, **42:**259
Trimethylgallic acid, **18:**374
3,4,5-Trimethylgallic acid, **34:**215
2,3,4-Tri-*O*-methyl-D-glucose, from laminarin, **28:**332
ε-*N*-Trimethyl-L-5-hydroxylysine, **42:**260
in the diatom, **42:**233
2, 4, 5–5Trimethyl imidazole, reaction with benzoyl chloride, **20:**83
Trimethyl lock, **43:**274
ε-*N*-Trimethyllysine, **42:**258; 259
biological significance, **42:**258
Trimethylpyruvic acid, in transaminations, **16:**197; 201
Trimetrexate, eukaryotic dihydrofolate reductase, ligand binding (side chain roles), **70:**62
2,4,6-Trinitrobenzenesulfonate, **59:**124; 149, **67:**30–32
Trinitroglycerin, **46:**397
Trinitrophenylation, **65:**197–198
Trinucleotides, **17:**464
Triokinase mutation, **18:**255
Triolein, effect of taurocholate on, **23:** 150–152
Triolein emulsion, lipase action, Michaelis curve, **23:**143–144
Triose
fluorinated, **55:**240–244
2,3-labeled, from acetate, **28:**412
Triosehosphate isomerase, **37:**436; 555
Triose isomerase, **55:**243
Triose phosphate, **15:**169–170; 171–172
in formation of sedoheptulose-1,7-diphosphate, **16:**291
in glucose-6-phosphate oxidation shunt, **15:**154; 156–157; 159–160; 163; 164 (C^{14})
oxidation, **15:**15
in pentose metabolism, **28:**395–397
Triosephosphate, isomerase, in photosynthesis, **19:**9; 24; 53
Triose phosphate dehydrogenase, **15:**143; 146; 171; 258–259; 338
Triosephosphate dehydrogenase, **22:**340–348
amino acid determination in, **14:**267; 275

Triosephosphate dehydrogenase (*Continued*)
in ATP system for energy transport, **14:** 10–12
classification, **14:**238
complex with DPN, binding sites, **22:**340; 341
dissociation constant, **22:**340; 341
stoichiometry, **22:**340
complex with DPN, DPN analogue, DPNH, and spectral changes, **22:**341–343
complex with DPNH, dissociation constant, **22:**341
effect of radiations on, **24:**399
location of, **24:**298
esterolytic activity of, **22:**347; 348
functional groups of, **22:**343–345
glutathione in, **22:**343; 344
hydrolytic activity of, **22:**347
inactivation by ionizing, **22:**344
labeling of active site in, **22:**51; 52
nuclear-cytoplasmic ratios, **27:**262
Triose phosphate dehydrogenase
as phosphorylase, **18:**226
in plants, **18:**292–293; 304
in photosynthesis, **25:**71; 95
Triosephosphate dehydrogenase
sulfhydryl group in, **22:**343–345
tryptic digestion, **22:**345; 346
tyrosine, tryptophan content of, **22:**366
yeast, complex with DPN, effect of pH on, **22:**344; 345
Triose phosphate dehydrogenase, activation of, **29:**353
Triose phosphate dehydrogenase, effect on diphosphopyridine nucleotide, **20:**43
Triose phosphate dehydrogenase, inhibition by CO_2O, **11:**236
Triosephosphate dehydrogenases, isotope studies of biosynthesis, **19:**260
in photosynthesis, **19:**9; 10; 32; 53
Triose phosphate in photosynthesis, **12:** 134–135
Triose phosphate isomerase, **5:**132, **14:**205, **15:**171, **45:**354, **46:**465–468, **47:**85; 117–123, **55:**242
Triosephosphate isomerase, **67:**15; 17; 23; 46; 52–53
Triose phosphate isomerase
chemical mechanism of, **45:**356
classification, **14:**258
kinetic mechanism of, **45:**326; 329; 348; 354–355
Triosephosphate isomerase, localization of, **24:**333; 335
Triose phosphate lyases, classification, **14:**228; 256
Triose phosphate mutases, and phosphate transfers, **18:**198
Triose phosphate oxidation, mechanism of, **18:**226
Triose phosphates
in photosynthesis, **25:**94–96
in plant respiration, **18:**292–295; 309; 312
Triose reductone, EPR study, **27:**124
Tripeptide, **38:**220; 221
formula of, **38:**216
Tripeptides, **47:**48; 50; 102–105; 130
enzyme action on, **12:**206–210
enzymic formation from methionine esters, **13:**310
Tripeptide sequences. *See also* Peptide sequences
in protein chains, **24:**420–425
repetition in chain, **24:**434–436
Triphenylcarbonium icons, **43:**386
Triphenylmethyl group, preparation of, **29:** 398–399
Triphenylphosphine oxide, **58:**66
Triphosphatase, classification, **14:**248
Triphosphatases, *See* Nucleoside triphosphatases
Triphosphate, *See* Phosphates, condensed
effect on myosin, **16:**344
inorganic
binding by myosin and actomyosin, **16:**333
splitting by myosin, **16:**337; 340
hydrolysis and myosin activity, **6:**147–189
inhibition of ATPase, **6:**160
Triphosphate subsite, MutT pyrophosphohydrolase, active site, **73:**199–203
Triphosphopyridine nucleotide, **15:**80–81; 84; 87–98; 103; 105–108; 112–131; 141; 145; 148–150; 154–156; 171; 173; 175; 177–178. *See also* Coenzyme II; Nucleotides and Pyridine nucleotides; TPN; TPN-cytochrome reductase; TPNH
and ATP synthesis, **23:**335; 345; 346
and glycine, **23:**457; 460
and oxalate, **23:**450–452

reduction in chloroplasts, **23**:385–387
biosynthesis, **15**:105
in carotenogenesis, **21**:346
in methionine synthesis, **21**:424
in mitochondria, **21**:15
in mitochondrial reactions, **21**:25; 27
reduced, in fatty acid synthesis, **21**:49
in mevalonic acid biosynthesis, **21**:337
in mitochondria, **21**:15
in pyrimidine biosynthesis, **21**:274; 275; 287
in serine biosynthesis, **21**:401
in THFA biosynthesis, **21**:395
in synthesis and metabolism of active formate, **21**:224; 409; 415
in citric acid cycle, **15**:184; 205; 256–258
enzymes using, **15**:114–118*t;* 121
evidence for structure, **15**:82–83
and GAD, **22**:368; 369
3′-isomer, **22**:371
reduced, in cholesterol biosynthesis, **22**:292; 293; 295–297
fluorescence shift with isocitric dehydrogenase, **22**:371
and GAD, **22**:368; 369
in squalene biosynthesis, **22**:307; 313; 314
synthesis by carbodiimide method, **22**:188
and pyrophosphorolysis, **18**:207
reduced (TPNH), and activity of hydroxylases, **20**:274
oxidation in animal tissues, **20**:172; 173
and tricarboxylic acid cycle, **9**:573–574
Triphosphopyridine nucleotide and mechanisms of enzyme catalysis, **24**:472
Triphosphopyridine nucleotide-cytochrome reductase, **18**:12; 44; 106
Triphosphopyridine nucleotide enzymes, in metabolism of glutathione, **11**:249
Triphosphopyridine nucleotide (TPN)
in oxygen metabolism, **19**:175; 189; 194; 202
in photosynthesis, **19**:3; 10; 11; 14; 15; 18; 23; 32–38; 48; 51–54; 65
reduced (TPNH), in oxygen metabolism, **19**:175–177; 189–192; 194; 195; 198; 199
in photosynthesis, **19**:2; 23; 24; 29; 30; 33–38; 51–54; 56; 65
reduction in plastids, **20**:219
Triphosphopyridinoprotein enzymes, **1**:181
Triple-helix configuration, of collagen, **41**:246; 264
of protocollagen, **41**:279; 280; 281
Triplet code, for sequence of amino acids in proteins, **22**:135
Triplet states
and effects of ionizing radiations on enzymes, **24**:362; 364; 380
of flavins, EPR study, **27**:191
Tripolyphosphate, **56**:264
Tripolyphosphoric acid, hydrolysis, by metal ions, **21**:175; 182
Tripyrroles, di-, and tetrapyrroles, **7**:330–332
TRIS, *See* Tris (hydroxymethyl) aminomethane
effect on succinic dehydrogenase, **18**:88–90
Trisaccharide
β-(2→1) linked, in Jerusalem artichoke, **17**:193
neuraminidase action on, **20**:143
Trisaccharides
from fructose transfers, **17**:194–197; 199–205; 221; 224
from glucose transfer, **17**:207; 211–218
hydrogen bonding, **65**:234; 238
Tris (hydroxymethyl) aminomethane, **15**:184; 208; 231; 344; 347
protection of pyridine nucleotides, **15**:90
Tris-(hydroxymethyl)-aminomethane, in buffers for kinetic studies, **17**:31; 49; 51–52; 59
Trisulfide, **39**:364
enzymic intermediate, **39**:351
Trisulfides, organic, **46**:40
Triterpenoid phytoalexins, **55**:49; 80–81
Tritiated water, in pentose metabolism, **28**:453
Tritiation, of asialoglycorproteins, **41**:100–103
Tritium
released during hydroxylation of proline, **41**:257
in studies of steroid enzymes, **20**:273
Tritium exchange, **42**:104, **50**:345
Tritium exchange with solvent, **24**:472
Tritium labeling, **49**:200–205
cobamide coenzyme reaction, **26**:358–359; 365
in corrinoid study, **26**:257
in isomerase reaction, **26**:265
methylmalonyl-CoA racemization, **26:** 339–340; 343
in PN-dependent reactions, **26**:127–128
for coenzyme Q synthesis, **25**:300–301

Tritium labeling (*Continued*)
in methyl synthesis, **25:**401
microbial oxidation of benzene, **27:**510
Triton S-100, **34:**477
Triton X-100, **34:**485, **47:**323; 324
Triton-X100, **57:**8
Triton X-100
effect on inhibitor of β-glucuronidase, **16:**375
particles in chloroplasts obtained by the action of, **30:**54
treatment of chloroplasts with, **30:**55
Trityl group, *See* Triphenylmethyl group
Trityl mercaptan, **33:**384
tRNA, **34:**3, **35:**453–480; 487, **38:**211; 219; 222, **47:**381–382, **48:**10, **49:**187; 188; 189; 191
alanine tRNA, **35:**464; 465; 491
aminoacyl, **49:**189
anticodon, loop, **35:**474
region, **35:**463; 464
stem, **35:**474
arginine tRNA, **35:**462; 464
base sequence, **24:**3
bromination of, **35:**464
clover-leaf formula for, **47:**382
cloverleaf structure, **35:**453
cyanoethylation, **35:**464
cytosine, **35:**461
diagrammatic representation of sequences, **49:**188
dihydrouracil loop, **35:**474; 478
diol, **35:**462
E. coli, **35:**491
during evolution, **47:**401
fluoropyrimidine effects on, **54:**88–91
formylmethionine, **35:**465
fragments from, **35:**466
FU incorporation into, **54:**88
hybrids, **35:**469
hydroxylamine, **35:**464
isoleucine, **35:**486
leucine, **35:**472
lysine, **35:**465
methyl-deficient, **35:**470–472; 490
$tRNA^{phe}$, **35:**472
modifications, **65:**277
periodate-treated, **35:**487
phenylalanine, **35:**462; 471
quarters, **35:**468
recognition, **49:**189
serine, **35:**490
sulfur-deficient, **35:**476
3′-terminus, **35:**479
5′-terminus, **35:**478
thiouridine, **35:**461
three-dimensional folding of tRNA, **49:**206
three-dimensional model, **35:**462
tyrosine, **35:**463; 464; 469
valine, **35:**461; 464; 465; 467; 476; 491
yeast alanine, **35:**468
$tRNA^{Arg}$, **35:**478
$tRNA^{E.\ Coli}{}_{Tyr}$, **35:**473
$tRNA^{fMet}$, **35:**465; 466; 468
$tRNA^{Ile}$, **35:**471
$tRNA^{Leu}$, **35:**472
$tRNA^{Met}$, **35:**465
tRNA Methyltransferases (Kerr and Borek), **36:**1
tRNA of mycooplasma, **36:**6
$tRNA_{Phe}$, **35:**492
$tRNA^{Phe}$, **35:**468; 469; 471; 472; 474
$tRNA^{Phe}{}_{Yeast}$, **35:**463
T7 RNA polymerase
cross-linked template
presence of two molecules on, **66:**108–110
sequence, **66:**106
DNaseI footprint, **66:**105–107
elongation complexes, **66:**103–119
chromatographic separation, **66:**114–117
contact with DNA, **66:**105–108
dynamics in DNaseI footprint, **66:**110–112
gel shift, **66:**112–114
stability, **66:**112–114
translocation model, **66:**117–119
tRNAs, understanding recognition of by aminoacyl transfer RNA synthetases, **49:**187–222
tRNA sulfur transferase, **48:**7
tRNA synthetases, **49:**189
aminoacyl (*See* Aminoacyl tRNA synthetases)
$tRNA^{Tyr}$, **35:**467
$tRNA^{Tyr}{}_{E.\ coli}$, **35:**476
$tRNA_{Val}$, **35:**492
$tRNA^{Val}$, **35:**465; 467; 478; 492
$tRNA^{Val}{}_{yeast}$, **35:**463
Trojan-horse inhibitors, **43:**340
Tromexan, effect on blood clotting, **16:**61; 64
Tropacocaine esterase, classification, **14:**244
Tropane alkaloids. *See also* Atropine, Hyoscine, and Hyoscyamine
esterification, **8:**240–241

relation to pyridine alkaloids, **8:**224
synthesis, **8:**224–226
Tropane ring and ornithine, **8:**237
Tropane structure and pyridine ring, **8:**235; 236
Trophic hormones, **50:**304; 314–318
mechanism of action, **50:**314–318
Trophotropic system, activation of, **29:**489
decreased activity of, **29:**487
serotonin effect on, **29:**486
Tropic acid, esterification with tropine, **8:** 240–241
Tropine esterification, **8:**240–241
Tropocollagen units, **23:**213–217
Tropolone, **57:**142–143
Tropolones, biosynthesis, **16:**136–140; 152
Tropomyosin, **57:**207
amino acid composition, **16:**319; 330
relation to myosin, **16:**330–331
and ATP in muscle contraction, **13:**158–159; 182
polar groups in, **13:**188
structure studies, **14:**279; 284
α-Tropomyosin, **47:**84
Troponin, **47:**133, **57:**207
Troponin C, annexin crystal structure, **71:**217
TrpG, amphibolic, aminodeoxy-chorismate synthase, **66:**227–228
True L-asparaginase, phosphate dependent, **39:**186
True asparaginases, **39:**186
Trypan blue, effect on myosin, **16:**345
Trypanosoma, **46:**102
Trypanosoma brucei, **59:**328, **71:**49
Trypanosoma lewisi, carbon dioxide fixation by, **74:**46–47
Trypansomes, genetic transformation of, **6:** 90–91; 95; 113; 118
Trypsin, **5:**22; 45, **23:**130; 136, **33:**272; 277, **34:**10; 13; 21; 32; 448–452; 458; 482, **36:**41; 62, **37:**419; 436; 443; 458; 468; 504; 523; 587, **41:**120; 167; 179; 180; 186; 198; 214; 215; 237, **43:**164; 169; 186, **44:**29, **45:**394, **46:**482, **48:**302, **53:**251; 252; 257; 261; 264; 265; 268; 277, **57:**184; 204; 209; 216, **60:**7
absorption spectrum, **4:**306
acetylation of, **13:**328
action with synthetic peptides, **13:**305
effect of phosphate esters on, **13:**332–333; 336
use in histological techniques, **13:**225
action of on gels, **33:**255–257
action of tyrosinase on, **14:**138; 141; 144; 146–147
adsorption studies, **14:**322–323; 326; 334; 338; 339; 340; 365
classification, **14:**232; 247
effect on tyrosinase action, **14:**150
isolation, **14:**398
peroxidase action on, **14:**155; 157–158
role in enzyme isolation, **14:**380
structure studies, **14:**285; 301
action on
alkaline phosphatase, **24:**9–10
inactivation of, **24:**117
localization of, **24:**325
lower peptide sequences, **24:**420; 430; 434
molecular weight from radiation studies, **24:**366
partial degradation by, **24:**128–131
polymer derivative of, **24:**108
properties, **24:**127
valence of, **24:**118–119
collagen, **23:**164; 167–170; 172–174; 180; 182; 202; 204; 205; 209; 212
collagenase, **23:**194
cytochrome *c,* **23:**62; 66; 67; 70; 71
denatured proteins, **23:**31–36
elastin, **23:**222; 225; 228; 238
proelastase, **23:**225; 236
spongin, **23:**205
phosphorylase *a,* **24:**270; 271
ribonucleases, **24:**133; 174–176; 198; 231; 232–235; 247; 251
action on bacterial wall, **26:**224
action on denatured protein, **26:**101; 105
and biotin-enzyme complex, **26:**325; 329; 330
disulfide bridges, **26:**102
action on benzoylarginine esters, **15:**292
and rennin, **15:**432; 433; 434
action on egg albumin, **19:**302
esterase activity of, **19:**306
in papain structure, **19:**297; 298
in transpeptidation, **19:**310; 311
action on hypertensin, **7:**526
and proteinases from kidney, **7:**529
action on tuftsin, **41:**167
active center, **23:**233

Trypsin (*Continued*)
acylation of, **24:**116; 124
adsorbed
activity of, **28:**22; 23; 25; 28; 30
radiosensitivity, **28:**34–35
adsorption at interfaces, **28:**2; 3; 8; 10; 13; 20–23; 25; 28; 30
amino acid composition, **34:**13
amino acid composition near active site, **22:**65; 68
pK value, **22:**58
reversible denaturation, **22:**76
stabilization by calcium ion, **22:**74
amino acid sequence, **24:**465
amino acid sequence in, **34:**22
classification, **2:**51–61
clotting of plasma by, **10:**12; 21–22
complexing between ovomucoid and, **30:**143
control of pancreas enzymes, **34:**33
denaturation of, **23:**34; 37–38
digestion of hexokinase by, **23:**48
digestion of myosin by, **16:**320–321; 323–324; 330
digestion of protomyosins, by, **16:**329
effect on β-glucuronidase, **16:**381
effect on rate of thrombin formation, **16:**45
inaction on deaminase from myosin, **16:**337
removal of nuclear residual protein, **16:**447
dilatometric methods, **9:**605
photochemistry of, **9:**141; 147–150
reversible inactivation, **9:**33–34
effect of, autolysis, **24:**125
chemical modification, **24:**134–141
on binding sites, **24:**116–117
ionizing radiation, **24:**367; 368; 370; 371; 372; 374–377; 379–380; 382–383; 391–393; 395–396; 398–401; 404
concentration effect, **24:**384–390
effect on cell-bound β-lactamase, **28:**254–255
effect on copper-protein, **23:**305
effect on glycyldehydrophenylalanine, **8:**134
in chromidia, **8:**31
effect on myosin monolayer, **6:**184
effect on protein-mercaptan reaction, **27:**210
LADH study, **27:**81–82
sulfur amino acids of, **27:**231
hydrogen-bonding network, **50:**35
hydrolysis of cytochrome *c* by, **17:**348
imidazole in, **20:**74–79; 86
mechanism of hydrolysis by, **20:**283–318
phosphorylated, amino acids in, **20:**288; 290
stability, **20:**285
immobilization of in acrylamide gels, **33:**278
inactivation of catalase, **1:**165
inhibitor of, **45:**393
soybean, **45:**409
inhibitors, **43:**314
insect, **45:**395
in low vertebrata, **1:**380; 383; 391–395; 408; 412
multiple forms, **25:**337–338; 350–351
product inhibition of, **25:**221; 222; 224; 225
in Rnase study, **25:**337
turnover number, **25:**32
in preparation of β-lactamase, **28:**266; 276; 282–284
in preparation of succinic dehydrogenase, **18:**70–71; 77
reaction of with Thiogel, **33:**257
resistance of tissues to, **23:**175; 180; 182; 218; 219
secretion in muscular dystrophy, **6:**222
self-activation, **34:**33
soybean inhibitor, simulated mutation of, **33:**399
specificity, **1:**75; 76
specificity of, **12:**194; 206; 228–229, **33:**432
synthetic substrates of, **33:**432
test for conformational change, **28:**99; 107; 108; 125
with Thiogel, **33:**262–263
tyrosine hydroxylase, **70:**182; 186
tyrosine hydroxylase regulation, **70:**50; 145; 148
uncoupling action, **23:**367
Trypsin-activatable proteolytic activity, localization of, **24:**316; 325
Trypsin activity, requirements for, **23:**77
Trypsin-agar complexes, **28:**35
Trypsin-chymotrypsin, separation of components, **14:**383
Trypsin inhibitor, **47:**58–59; 72; 78; 87; 98; 102; 104–111; 124; 126–130
column chromatography, **22:**453
separation by chromatography, **14:**342; 346
structure studies, **14:**283
soybean, **16:**47–48; 54
α Trypsin inhibitor, **48:**293

Trypsin inhibitors, **53:**247; 249
action on, collagenase, **23:**194
elastase, **23:**235
Trypsinlike proteinases, **41:**188–189; 237
Trypsinogen, **23:**131–134; 148, **34:**32, **57:**214
activation of, **24:**24; 126–127
amino acid sequence, **24:**416; 421; 422; 432–435
localization of, **24:**325
column chromatography, **22:**451
structure studies, **14:**283; 285
Trypsinogen-activating ability, **41:**211
Trypsinolysis, **58:**106
Trypsin-ovomucoid system, **24:**124
Trypsin substrate, **23:**240
Trypsin substrates, effect on PR-enzyme, **24:**280
Trypsin-trypsinogen reaction, **17:**448
Tryptamine, **5:**71
excretion in schizophrenics, **29:**528–529
inhibition of tryptophan transfer to sRNA, **29:**155
Tryptase
plasma, **10:**13
in tyrosinase preparation, **14:**150
Tryptic cytochrome *c,* **23:**68–69
Tryptic digestion, **42:**135; 140; 144
of DPNH oxidase complex, **17:**261–263
of succinic dehydrogenase complex, **17:**256
of TPD, **22:**345
of TPD-DPN complex, **22:**345; 346
Tryptic digests, **42:**152
of bovine pituitary growth hormone, **42:**153
Tryptic hydrolysates, **39:**348
Tryptic hydrolysis, **42:**131
Tryptic peptides, **46:**12
Tryptophan, **35:**286, **39:**347, **42:**88; 105; 118; 127, **49:**67; 90, **50:**149; 156; 157; 168; 169, **51:**217, **58:**35
in affective disorders, **29:**493
analogs of, **29:**121
incorporated into protein, **29:**112
excretion of, in schizophrenics, **29:**528
excretion of metabolites by depressed patients, **29:**493
resistance to analogs of, **29:**123
in chymotrypsin, **20:**78
use of blocking reagents with, **20:**68; 72; 77; 81
and color of riboflavin, **22:**377; 378
synthesis in *E. coli,* **22:**122
control of level in brain, **49:**92
critical, **39:**346
enzyme for activating, **19:**258
labeled, **19:**94
metabolism, **19:**92–97; 121–123; 194
peroxidase action on, **19:**123
in formation of arylamines, **6:**103–104
glutathione S-transferases, **69:**17
hydroxylation of, to 5-hydroxytryptophan, **74:**236
as inhibitory analogue, **6:**139
oxidation of, **24:**340
and pyridine alkaloids, **8:**239
relation to nicotinic acid, **8:**235; 236
synthesis in *Neurospora,* **8:**235
and various heterocycles in the plant cell, **8:**236
spectral perturbation in glutamate dehydrogenase, **39:**372
synthesis, **21:**208
Tryptophan-86, **42:**111; 112
chemical reactivity, **42:**138
D-Tryptophan, **33:**368
effect of vitamin B_6 on metabolism, **19:**351; 352
DL-Tryptophan
resolution by papain, **19:**309
L-Tryptophan, **67:**169; 181–182
effect of vitamin B_6 on metabolism, **19:**351; 352
transamination reactions of, **19:**367
phenylalanine binding effect, **67:**122–124
L-Tryptophan, oxidation in plants, **20:**320
L-tryptophan, tyrosine hydroxylase, **70:**116
Tryptophan analog, **57:**36
Tryptophanase, **35:**107, **49:**29, *See* Tryptophan indolelyase (tryptophanase)
activation
by analogues of pyridoxal-P, **42:**321
by cations, **42:**321
adduct with NH_3^+, **42:**327
and pyruvate, **42:**327
spectrum of, **42:**327
amino acid composition, **42:**294
amino terminus of, **42:**307
analogue coenzymes, **42:**319
apoenzyme, **42:**292
apoenzyme-holoenzyme equilibria of, **56:**42
cation effects on, **56:**43
cation specificity, **42:**320

Tryptophanase (*Continued*)
characteristics of, **56:**41
coenzyme specificity, **42:**318; 319
degradation of various substrates, **42:**312
distribution in bacteria, **42:**290
electron microscopy of, **33:**104
α-β elimination, **42:**312
historical background, **42:**287
immunological differences in, **42:**292
inhibition, **16:**222
products of action, **16:**293–294
inhibitors, **42:**315
isotopic exchange of hydrogen atoms with substituents of, **56:**53
kinetic parameters, **42:**314
mechanism
catalytic of, **42:**323; 325
α-aminocrotonate and, **42:**325
labilization of α-H, **42:**325
labilization of β-substrate, **42:**325
tetrahedral intermediates in, **42:**323
intermediates in, **42:**323
molecular parameters of, **56:**45
nature of the enzyme, **10:**60–62
preparation, **10:**65
properties, **10:**65–68
P-pyridoxyl binding site, **42:**306
comparison with other enzymes, **42:**308
product inhibition of, **25:**224
properties, comparative from various sources, **42:**293
of purified, **42:**291
Pyridoxal-P analogues as coenzymes of, **42:**319
Pyridoxal-P and, **42:**295
Pyridoxal-P binding, **42:**296; 297; 298; 299; 306
monovalent cations, effect on heat stability, **42:**299
SH compound and, **42:**298
SH group reactivity, effect of, **42:**306
stoichiometry rate, **42:**296
quasisubstrate, **42:**315
reactions, catalyzed, α-β elimination, **42:**310
β-replacement, **42:**310
synthesis of tryptophane, **42:**310
reduction, **42:**297
with sodium borohydride, **42:**297
replacing agents for, **56:**11
reversibility of reaction, **42:**327
sequence studies, **42:**306; 307
size and stability of, **56:**42
spectrum of, **42:**297; 298
alanine, effect of, **42:**317
monovalent cations, effect of, **42:**298
pH, effects of, **42:**297
pyridoxal-P analogues, **42:**321
substrate, effect on, **42:**317
stability of, to heat, **42:**300
to SDS, **42:**300
substrates, **42:**315
substrates for, primary, **56:**11
substrate specificity, **42:**311; 314
subunit interactions, **42:**295
dissociation into, **42:**295
sulfhydryl groups, **42:**305; 306
activity, **42:**306
reduction with DTT, **42:**306
titration with DTNB, **42:**306
synthesis from indole, **42:**313
Tryptophanase: Structure, Catalytic Activities, and Mechanism of Action (Snell), **42:**287
Tryptophanase-tryptophan reaction, **10:**51–82
Tryptophanase-Tryptophan Reaction (Happold), **10:**51
Tryptophan bands, **42:**108
Tryptophan biosynthesis, **56:**41
Tryptophan catabolism, **49:**4
Tryptophan decarboxylase, **5:**71
Tryptophan dioxygenase, **41:**40; 41
inhibition by superoxide dismutase, **41:**40
from intestine, **41:**40
from *Pseudomonas acidivorans,* **41:**41
from rat liver, **41:**41
Tryptophane
absorption spectrum, **4:**277; 307
action on thiaminase, **15:**401
metabolic pathways, **15:**184–185
as nicotinic acid precursor, **15:**98; 99 (N^{15}); 101 (C^{14})
biosynthesis, **16:**277; 287–295
allosteric regulation, **28:**62–63; 71
role of gultamine, **28:**66–69
in chromosomins, **16:**450
reaction with quinones, **16:**110
in residual protein, **16:**447
in transamination reactions, **16:**191; 192; 198; 201; 207; 208; 222; 235; 239
in energy transport studies, **14:**34–35
oxidation by permanganate, **14:**132; 133

peroxidase action on, **14:**151–154
genetic studies, **17:**412–422; 447; 450–452; 458
genetic studies of, **13:**141
oxidase, enzymic adaptation, **13:**71; 73–74
peroxidase, enzymic adaptation, **13:**71; 74
of a nuclear protein, **18:**176; 179; 181
origin of, **18:**366
Tryptophane, Cotton effects, **27:**95
Tryptophane auxotrophs, **16:**293
L-Tryptophane decarboxylases, classification, **14:**256
Tryptophane-deficiency, relation to xanthine oxidase, **17:**319
Tryptophane desmolase, **16:**293
genetic studies, **17:**424; 450–452
Tryptophane hydrolase, classification, **14:**249
L-Tryptophane oxidase, classification, **14:**253
Tryptophane peroxidase, **17:**417
Tryptophane peroxidase activity, effect of ethionine on, **25:**376–378; 380
Tryptophane rings, arrangement in virus (TMV) molecules, **15:**458–459
L-Tryptophane r O_2 transhydrogenases, **14:**253
Tryptophane synthetase, in mutant *Neurospora crassa,* **26:**110–111
Tryptophan hydroxylase
amino acid hydroxylation, mechanism of, **74:**254–257; 255; 256
in brain, **74:**236
from *Chromobacterium violaceum,* **74:**236
hydroxylating intermediate, **74:**262
instability of, **74:**236
kinetic mechanism of, **74:**257–258; 258–259; 259
mechanism of action, recent studies of, **74:**257–259; 262
rabbit
kinetic mechanism of, **74:**259
structure of, **74:**241; 243
reaction catalyzed by, **74:**236
NIH shift in, **74:**254; 255; 256
regulation of, **74:**281
structure of, **74:**241–242; 242–243
substrate specificity of, **74:**249–250
Tryptophan hydroxylation, **70:**103
Tryptophan indole-lyase (tryptophanase), **74:**133; 152; 156
Tryptophan movement, and albumin binding, **49:**92
Tryptophan-NAD pathway, **49:**9; 28; 36
Tryptophan oxidase, **19:**121–123; 127
Tryptophan oxygenase, **37:**140; 147; 420; 427, **47:**277; 284
action of glucocorticoids, **47:**277
estimation of level of functional mRNA, **47:**287
minimal deviation hepatomas, **47:**304
Tryptophan oxygenase mRNA, **47:**284; 304
control by glucocorticoids, **47:**284
level of activity in minimal deviation hepatomas, **47:**304
Tryptophan peroxidase, location of, **24:**307; 309
Tryptophan pyrollase, induction by cortisone of, **29:**375–376
inhibition of, **29:**529
in manic and depressed patients, **29:**493
in neoplastic cells, **29:**375
Tryptophan pyrrolase, **46:**60; 147; 150; 166. *See also* Dioxygenase
location of, **24:**307; 309
Tryptophan residue, **42:**138; 140; 144
modification of, **42:**144
Tryptophan residues, calcium ATPase, **65:**203
Tryptophan side-chain oxidase, **59:**193
Tryptophan synthase, **37:**420; 458; 520; 555; 585, **49:**127–185, **56:**43
catalytic properties of *E.* coli, **49:**145
chemical modification, **49:**168
kinetics and mechanism, **49:**158–160
limited proteolysis, **49:**166
purification of, **49:**131
$a_2\beta_2$ complex, **49:**131
E. coli subunits and complex, **49:**130–145
other sources, **49:**135–136
β_2 subunit, **49:**131
a subunit, **49:**131
pyridoxal phosphate-dependent reactions of β_2 subunit, **49:**145–166
spectral intermediates, **49:**149–154
structure of *E.* coli subunits and complex, **49:**136–145
$a_2\beta_2$ complex, **49:**137
β_2 subunit, **49:**136
a subunit, **49:**136
β-subunit, **74:**134; 148; 156; 197
kinetic mechanism of, **74:**206; 208–209
K87T mutant of, **74:**202–203; 224
phylogenetic distribution of, **74:**157
β-replacement reactions, **74:**197; 224

Tryptophan synthase (*Continued*)
structure of, **74:**200; 202–203
subunit interaction, **49:**160–166
Tryptophan Synthase: Structure, Function and Subunit Interaction (Miles), **49:**127
Tryptophan synthetase, **35:**110, **58:**47
chromatography of, **30:**149
formation of, **24:**14; 24
genetic control of, **24:**9
genetic loci of, **29:**121
Tryptophan synthetase, A and B components of, in *E. coli,* **22:**12–15
E. coli mutants, **22:**12–15
and interlocus complementation, **22:**37
mapping of mutations sites in, **22:**14
nonrevertible, **22:**15
properties, **22:**13
suppressor mutations in, **22:**30; 32
ultraviolet-induced reversion in, **22:**14
formation in *Neurospora,* and interallele complementation, **22:**35
measurement of properties, **22:**30; 31
mutants, cross-reacting material in, **22:**11
Neurospora mutants, **22:**11; 29
and production of cross-reacting material, **22:**31
variation of activity and inhibitor formation in *Neurospora,* **22:**26
zinc sensitivity, **22:**11; 31
Tryptophanyl fluorescence, **42:**139
Tryptophanyl-*t*RNA synthetase, **37:**443
Tryptophyl residues, **39:**346
Trytic hydrolysis of benzoyl arginine ethyl ester (BAEE), **51:**51; 52; 58–61
Trytophan, biosynthesis, **10:**71–74
in fibrin, **10:**30
metabolism, **10:**52–82
Trytophan in fructose-bisphosphatase, **54:** 149–150
TS, *See* Thymidylate synthetase
Tsetse fly, *See* Glossina morsitans
TSH
desensitization, **50:**317
in tyrosine metabolism, **14:**113–116
Tsuchihashi procedure, **41:**74
TTP (^{32}P), incorporation into DNA, **27:**241
Tubercidin, **43:**35
Tubercle bacillus, **9:**134
Tuberculin, separation by chromatography, **14:**349
Tuberculosis, relation to β-glucuronidase, **16:**397
d-Tubocurarine, **55:**139
Tubulin
enzymatic modification of, microtubule dynamics, **69:**144
GTPase activity, **53:**119–139
guanine nucleotide interactions with, microtubule dynamics, **69:**123–124
phosphocellulose-purified, **53:**124
self-assembly, **53:**124–130
Tubulin-GTP/GDP cycle, bioenergetics of, microtubule dynamics, **69:**124–132
Tuftsin, **41:**158–170
amino-terminal threonine, **41:**163
analogs, **41:**164
as a Biological Entity, **41:**164
biological properties, **41:**159
carboxy-terminal Arginine, **41:**162
carboxy-terminal Fc-fragment, **41:**160
chemical properties, **41:**161
chemical synthesis, **41:**168
chemotaxis of, **41:**177
deficiency following removal or curtailment of splenic function, **41:**170
effect on chemotactic migration, **41:**177
effect on migration of leukocytes, **41:**164
effect on viability of leukocytes, **41:**164
enzymatic destruction and resistance, **41:**159
ester linkage with H-chain, **41:**163
γG, myeloma H-chain, **41:**160
hydroxylamine cleavage, **41:**163
Inhibition of, **41:**177
Isolation of Thr-Lys-Pro-Arg, **41:**158
properties of, **41:**159
radiochemotactatic assay, **41:**177
release by leucokininase, **41:**167
release by trypsin, **41:**167
sequence of amino acids, **41:**158
Tuftsin-deficiency syndrome, **41:**159; 169
Tumor-bearing rats, pentose synthesis by, **28:**431–433
Tumor cells, regulation of hexokinase, **28:**76; 77; 85
uridine kinases in, **28:**135
Tumor enzymology
bibliography, **3:**345–348
enzymatic activity of normal tissues of normal and cancer-bearing animals, **3:**335–340
enzymatic activity of tumor-bearing and normal tissues, **3:**315–334

neoplasms and their normal tissues of origin, **3**:318–334
systemic effect, reversibility of, **3**:339–340
systemic effect and growth rate of tumor, **3**:338–339
Tumor Enzymology, Progress in (Wenner), **29**:321
Tumor Enzymology (Greenstein), **3**:315
Tumor extracts, specificity, **1**:83; 85
Tumor inhibition, effects of β-glucan, **28**:357
Tumor necrosis factor, **60**:227, **65**:318
Tumor preparations, reduction of ribonucleotides, **28**:436–444
Tumors, **39**:219. *See also* Cancer, Carcinoma, Hepatoma, and Malignant cells; Cancer and Carcinomas; Sarcoma
adenosylmethionine decarboxylase of, **56**:272–273
aerobic glycolysis in, **29**:324–330 (*See also* Hepatomas, Neoplasms)
agents inhibiting growth of, **29**:336
energy metabolism of, **29**:324–367
glycolysis in minimal deviation, **29**: 330–337
incorporation of amino acids into proteins of, **29**:365
multisite control of glycolysis of, **29**: 357–358
plasma threonine levels in, **29**:364
progress in enzymology of, **29**:321–390
protein control of, **29**:364
threonine dehydrase in, **29**:364
uptake of amino acids inhibited by cyanide and dinitrophenol in, **29**:365
condensing enzyme in, **15**:193
phosphorylating enzymes in, **15**:146
role of free radicals in, **15**:6
dehydropeptidase activity, **8**:164
effect of fluoro compounds on, **18**:155
nucleoprotein of, **18**:164; 174–176; 185–186
effect of pectin on, **20**:257
enzyme cofactors in, **25**:377–378
and enzymes, **7**:397
and phosphatides, **7**:145–146
enzymes of, **24**:296; 297; 299–301; 303; 305–307; 312–316; 318–324; 326–329; 335
ethionine-induced, **25**:373–375
hepatic, **46**:119
methionine adenosyltransferase of, **56**:263
and thiol groups, **11**:255
treatment with alliin, **11**:397
and xanthine oxidase, **17**:323
Tumors, DNA in, **16**:426; 433
histones in, **16**:446
transaminase systems in, **16**:209
Tumor tissue, DR-5-P aldolase in, **28**:449
synthesis of pentoses, **28**:414; 417; 419; 433–434
Tumor tissue, thymidylate synthetase, **25**:398
Tungstate, as molybdenum substitute, **17**:270
ingested, effect on xanthine oxidase, **17**:314; 321
Tungsten, **51**:112; 115
effect on formate dehydrogenase activity, **48**:18
incorporation in formate dehydrogenases, **48**:19
^{185}W, **48**:18
Tunicamycin, **56**:150
Tunneling, **46**:424–426
Turacin, **20**:167
Turanose, **17**:210; 220
Turbidity, hydrotropic clarification of, **17**:155
in optical density measurements, **17**:75–76
Turn (β-), *See* β-turn
Turnip yellow mosaic virus, **17**:484–485; 488–491; 495; 496
crystallization, **9**:202–203
physical constants, **9**:187; 188
purification, **9**:187
Turnover, *See* Methyl groups incorporated, turnover of
Turnover number, **43**:139, **50**:346
Turnover number for enzyme efficiency, **1**:178
Turnover number of catalases and peroxidases, **12**:171–174
Turnover numbers of enzymes, **25**:32
Turnover of the prosthetic group, **36**:287
Turnover Studies, Applications of Radioactive Indicators (Hevsey), **7**:111
Turnover studies, by radioactive indicators, **7**:111–214. *See also under* specific compound
of phosphorus-containing carbohydrates, **7**:113–127
of sulfur compounds, **7**:192–199

Tween, **34:**479
Tween 20, exopolysaccharide alginate synthesis, **70:**240–245
Two-acceptor mechanism, **50:**347; 351
Two- and Four-Electron Oxidations Catalyzed by Some Metalloenzymes, Mechanism of (Hamilton), **32:**55
Two-carbon compounds
  activation and oxidation, **15:**245–259
    in condensing reaction, **15:**186–201
  in ribose synthesis, **28:**395–396; 401; 402
Two-Carbon Compounds, The Metabolism of, by Microorganisms (Kornberg and Elsden), **23:**401
Two-collodion membrane, **34:**455
Two-enzyme system, **34:**515–520
Twofold axis, in proteins, **50:**24
Two-site hypothesis, for allosteric regulation, **28:**58
Two-step mechanism, **36:**120
Two-substrate studies, **39:**354
Type III connecting sequence (IIICS) region, fibronectin, **70:**12–13
Typhus bacillus, absorption of the agglutinin on, **15:**54
Tyr-37, **57:**23
Tyr-157, **57:**23
Tyr-248, **57:**21–24
Tyramine, **5:**72; 79
  action of peroxidase on, **14:**152
    in noradrenaline formation, **14:**117–118
    in tyrosinase action, **14:**145; 146
  production, **6:**2; 4
Tyramine-1-d_1, **35:**85
Tyramine oxidase, in mitochondria, **21:**22
Tyrocidine, **36:**304. *See also* Tyrothricin
  action on bacteria, **4:**154; 169–173; 185; 193
  amino acid sequence of, **33:**337–338
    synthesis of, effect of chloramphenicol on, **33:**350
      effect of puromycin on, **33:**350
      effect of ribonuclease on, **33:**350
  biosynthesis, **35:**20–28
    activation of amino acids, **35:**27
    ATP-PP_i exchange, **35:**8
    cyclization, **35:**27
    elongation, **35:**27–28; 31
    enzyme fractions complementary, **35:**25
  chemical nature, **4:**154; 166–169; 193
  constituents, **4:**166; 167; 168; 169
  crystalline, **4:**161
  detergent action, **4:**193
  effect on bacterial infections, **4:**175; 176
  effect on cellular metabolism, **4:**184–185
  inhibition, **4:**173; 174; 175
  isolation, **4:**154; 158
  mode of action, **4:**192–193
  molecular weight, **4:**166
  physical properties, **4:**159
  purification, **4:**158
  resistance of enzymatic hydrolysis, **4:**169
  respiration of bacteria and cells in presence of, **4:**184–185
  solubility, **4:**158; 159; 160
  structure, **35:**2
  structure and biological activity, **4:**182–184
  structure studies, **14:**266; 279; 299
  toxicity for animal cell and tissues, **4:** 178–182
Tyrocidine (Hotchkiss), **4:**153
Tyrosinase, **7:**529, **17:**150; 445–448, **20:**285, **36:**72, **56:**41. *See also* Catecholase activity; Catechol oxidation; Monohydric phenols; Polyhydric phenols
  action on fibrinogen, **16:**37
    Gastrophilus, **16:**155
    inactivation of thrombin, **16:**84
    in melanin granules, **16:**172
    role in sclerotization process, **16:**159
  action on glycine, **12:**449
  action on hypertensin, **7:**526; 543
  action on RNase, **24:**211; 226
    effect of radiations on, **24:**391
  activity measurements, **4:**115–116
  activity ratio (cresolase:catecholase), **4:** 105–109; 127; 146–148
  affinities of different phenols for, **4:**141–142
  classification, **14:**254; 255
    in melanin formation, **14:**74–96; 123
    oxidation of proteins by, **14:**129–161
    source, **14:**377
  column chromatography, **22:**479; 480
    in melanin-deficient *Glomerella* mutants, **22:**24
    in *Neurospora,* **22:**5–7; 30
    in *Neurospora* mutants, **22:**25
  copper content, **4:**105–109; 127; 147; 148
  discovery, **4:**100; 101
  dual action (*See* Catecholase activity; Monohydric phenols)

effect of freezing on, **9:**625
gasometric methods, **9:**601
enzymic adaptation, **13:**106
inhibition, **13:**216
EPR-nondetectable copper ions in, **33:**223
EPR study, **27:**138
homogeneity, **4:**127; 146–149
inactivation, **4:**101; 102; 110; 117; 123; 125–132; 142; 143
isolation and purification, **4:**104–109; 126; 146–148
lack of Cotton effect, **27:**93
mode of action, **4:**102–104
multiple forms, **25:**359
oxidation of anthocyanins, **25:**483
mushroom, action on catechols, **10:**383
nature, **4:**145
occurrence, **4:**101; 102; 104
in common mushroom, **4:**100; 102; 104; 106; 107; 108; 109; 115; 117; 120; 124; 146
in potatoes, **4:**102; 104; 106; 117; 120; 142–145
in wild mushroom, **4:**102; 109–110
oxidation of catechol, **27:**128
and peroxide formation, **4:**102; 110–112
physical properties, **4:**146–149
in plants, **18:**37; 310–311
preparations of, **4:**104–110
as respiratory enzyme in plants, **4:**142–145
stability, **4:**104; 126; 127
in tobacco, **6:**373–374
tyrosine hydroxylation contrasted, **70:**106
Tyrosinase, Oxidation of Proteins by (Sizer), **14:**129
Tyrosinase (Nelson and Dawson), **4:**99
Tyrosine, **5:**71–72; 80, **34:**21; 453, **35:**487, **42:**129, **48:**158, **50:**148; 149, **51:**217, **57:**18
absorption spectrum, **4:**277; 307
action of tyrosinase on, **14:**133–135; 137; 138
derivatives, action of tyrosinase on, **14:**134–135; 137–139; 139–151
in energy transport studies, **14:**34–35
labeled, **14:**101
metabolism, **14:**73–128
oxidation by permanganate, **14:**132
and pepsin activity, **14:**189–190
peroxidase action on, **14:**151–154
action on thiaminase, **15:**400
entrance to citric acid cycle, **15:**184
iodo derivative from iodinated pepsin, **15:**289
metabolism, **15:**119; 123
role in myoglobin, **15:**33
activation of apoenzyme by pyridoxal phosphate, **6:**27
active site, glutathione S-transferases, **69:**16
activity in pyridoxine-deficient streptococci, **6:**25–26
and activity of chymotrypsin, **20:**285
diazonium coupling with, **20:**59
metal binding by, **20:**67
use of blocking reagents with, **20:**68; 71–73; 75; 80; 94
adaptive formation in *E. coli,* **6:**8
and affinity of amylases for substrate, **12:**407
analogs of, **29:**114; 115; 118; 122
chemically synthesized analogs of, **29:**109
bacterial decarboxylase, **6:**4
biogenesis of thiazole moiety, **53:**367
biosynthesis, **16:**287–295, **21:**432
allosteric regulation, **28:**62–63; 71
metabolism in plants, **16:**132; 133; 144–145; 150
N-terminal residues of fibrinogen and fibrin, **16:**37
relations to phenylalanine metabolism, **16:**301
release from prothrombin during activation, **16:**51
role in hardening of cuticle, **16:**154–156
role in sclerotization process, **16:**158–160; 163
in transamination reactions, **16:**191; 192; 198; 201; 203; 207; 208; 230–231
biosynthesis, and vitamin B_{12}, **26:**270
(carboxy-C-14), in melanin studies, **16:**166–167
catalytic mechanisms, **69:**21
codecarboxylase of, **6:**29
Cotton effects, **27:**95
cytochrome *c* oxidase, **71:**138–139; 141; 147; 149
decarboxylase, **35:**85
stereochemistry of reaction, **35:**85
effect of age of culture on growth, **6:**12
effect on indole production by *Escherichia coli,* **10:**77–81
effect on tryptophan peroxidase, **25:**377

Tyrosine (*Continued*)
labeled, in coenzyme Q synthesis, **25:** 298–299
formation
measurement, **67:**96–97
from phenylalanine, **67:**78
genetic studies, **17:**412–422
*p*H of growth medium, **6:**10; 16
hydroxylase, **35:**288
hydroxylation of, to dihydroxyphenylalanine, **74:**236
inhibitors affecting activity, **6:**18
Michaelis constant, **6:**16
mobilization, **45:**419
$6MPH_4$ oxidation, **67:**189–190
origin of, **18:**362; 365–366
relation to lignin, **18:**374
in papain, **19:**284
peroxidase action on, **19:**108; 143
transamination, **19:**183
phenolic hydroxyl group of, **22:**422
phenylalanine conversion, **67:**200–203
in potato, substrate for tyrosinase, **4:**100; 115; 144–145
purified, **6:**12–21
pyridoxal activation of, **6:**26–29; 30
residues, **34:**453
reversible resolution, **6:**21
RNA polymerase II transcription, C-terminal domain characteristics, posttranslational modifications, **71:**49–50
role of conserved, in glutathione S-transferases catalysis, **69:**18–20
specificity, **6:**15
vitamin B factors in growth medium, **6:**9
L-Tyrosine
oxidation in plants, **20:**320; 322
D-Tyrosine, **35:**481
DL-*m*-Tyrosine, **67:**159
L-Tyrosine, **45:**419, **50:**166, **67:**169
β-alanyl, **45:**419; 420
formation in phenylalanine metabolism, **19:**88; 181–183
phenolase complex specificity for, **19:**131
transamination reactions of, **19:**367
L-tyrosine
conversion of L-dopa, **70:**105
tyrosine hydroxylase, **70:**116–117
m-Tyrosine
effect on reserpinized mice, **29:**488
peroxidase action on, **19:**143
tyrosine hydroxylase, **70:**117
meta-Tyrosine, conversion to DOPA, **67:**110
Tyrosine, Metabolism (Lerner), **14:**73
Tyrosine adenylate, synthesis by carbodiimide method, **22:**191
Tyrosine aminotransferase, **36:**67; 83, **37:**140, **42:**254, **43:**346, **46:**150, **47:**281
induction of, **37:**160
Tyrosine apodecarboxylase, assay for pyridoxal phosphate, **16:**219
Tyrosine auxotroph, of *Neurospora,* **16:**292–293
Tyrosine band, **42:**110
Tyrosine decarboxylase, **5:**71–72; 76; 77
activity with pyridoxal phosphate, **16:** 219–220
inhibition, **16:**219; 221
relation to pyridoxine, **16:**211; 215; 216
L-Tyrosine decarboxylases, classification, **14:**256
Tyrosine ethyl ester
action of tyrosinase on, **14:**134–135; 137
formation of melanin, **16:**165
Tyrosine hydroxyl, **58:**5
Tyrosine hydroxylase, **36:**74, **67:**85, **70:**103–220
active site
amino acid residues in, **74:**261–262
interactions of dihydropterin with, **74:** 244–245; 245
iron, **74:**239–240
structure of, **74:**244–245
activity of, phosphorylation and, **74:**275–277
amino acid hydroxylation mechanism of, **74:**254–257
assay methods, **70:**108–110
background, **70:**103–106
catalytic domain of, structure of, **74:**242–243; 243; 244; 245
catecholamine binding to different redox states, phosphorylation and, **74:**278–279
catechol complexes formed with, **74:**239
formation of tyrosine and *m*-hydroxyphenylalanine from phenylalanine, arene oxide intermediate, **74:**263; 264
genetics and, **70:**179–182
hydroxylating intermediate, partitioning of, between amino acid hydroxylation and unproductive decay, **74:**266
hydroxylation mechanism of, **74:**263–268; 265; 266; 268

imidazole ligands, **74:**239
induction of, **70:**165–172
intracellular localization and regional distribution, **70:**106–108
iron ligands, **74:**243
kinetic mechanism of, **74:**257–258; 258; 259
kinetics, phosphorylation and, **74:**275–278
mechanism of action, **70:**194–206
recent studies of, **74:**257–270; 264; 265
neurochemical properties of, **74:**237
pH optimum for, **74:**275
phosphorylation of, **74:**272–278
and activity, **74:**275
and interconversion of ferrous and ferric forms, **74:**278–279
and kinetics, **74:**275
pH and, **74:**275–276
phosphorylation sites, **74:**272–275; 274–275
properties of, **70:**120–121
amino acid composition, **70:**120–121
size, **70:**120
proteolysis, **67:**221–222
rat
catalytic domain of, **74:**242–243
kinetic mechanism of, **74:**259
structure of, **74:**241
reaction catalyzed by, **74:**236
reaction of oxygen with, mechanism of, **74:**260–261
regulation of, **74:**272–281
model for, **74:**279–280; 280
regulation of activity, **70:**122–165
heparin and other polyanions, **70:**122–127
phosphorylation, **70:**127–165
substrate and coenzyme availability, **70:**122
regulatory domain, **67:**223
regulatory domain of, **74:**280–281
interaction with active site, **74:**280–281
spectroscopic properties of, **74:**239
structural studies, **70:**182–194
structure of, **74:**241–249
substrates of, products from, **74:**265; 266
substrate specificity, **70:**110–119
amino acid substrate, **70:**116–119
tetrahydropterin coenzyme, **70:**110–116
substrate specificity of, **74:**250
tissue distribution of, **74:**236
transcriptional regulation mechanism, **70:**172–179
Tyrosine iodination, **60:**24
Tyrosine ionization, **42:**95
Tyrosine α-ketoglutaric transaminase, product inhibition of, **25:**224; 241–242
Tyrosine kinase activity, cholesterol ester hydrolases regulation, **69:**74
L-Tyrosine ligase (ADP-forming), **53:**154–156
Tyrosine methyl ester, in transaminations, **16:**193
Tyrosine-OH in phosphatases, **15:**28
Tyrosine oxidases, classification, **14:**254; 255
Tyrosine phenol-lyase, **56:**43, **74:**133; 149; 152; 156
active site, structure of, **74:**164; 165
as alanine racemase, **44:**111
conversion to dicarboxylic amino acid β-lyase, **74:**163–167
β-elimination reactions catalyzed by, kinetic parameters or, **74:**166
reaction catalyzed by, **74:**163
Tyrosine-*o*-phosphate, **45:**419
Tyrosine pool, **34:**221
Tyrosine residues, **42:**93; 94; 105; 112; 143; 145
ionized, **42:**108
in RNase denaturation, **24:**185–197
in RNase structure, **24:**211; 215–216; 221; 226–229; 231; 232; 248–251
Tyrosine sRNA synthetase, 3–3fluorotyrosine as a substrate, **29:**114
Tyrosinosis, **5:**80, **17:**415
use of term, **14:**109
Tyrosyl groups, modification, 1, 6-6bisphosphatase, **53:**70–71
L-Tyrosyl-O-acetyl-dopamine, **45:**419
Tyrosyl radical, **65:**177–180
scavengers, **65:**178–179
Tyrosyl residues modification, **53:**70–71
Tyrosyl-*t*RNA synthetase, **37:**420
Tyrosyl-tRNA synthetase, **50:**165
Tyrothricin. *See also* Gramicidin; Tyrocidine
bacterial infections, effect on, **4:**175; 176
composition, **4:**159
concentration and effect on bacteria, **4:**169; 170; 171; 173
from different cultures of *B. brevis,* **4:**160
inhibition, **4:**174; 175
isolation, **4:**157
resistance to enzymatic hydrolysis, **4:**169
structure studies, **14:**299
toxicity for animal cells and tissues, **4:** 178–182
in treatment of plant diseases, **20:**402

Tyrothricin (Hotchkiss), **4:**513
Tyr-106 residue, dihydrodipicolinate synthase (DHDPS), **72:**290
Tyr-133 residue, dihydrodipicolinate synthase (DHDPS), **72:**289–290
Tyvelose, *See* 3,6-Dideoxy-D-mannose

U

Ubichromenol
  and coenzyme Q, **25:**277; 284–285; 289; 290; 292; 293; 304
Ubiquinol, **65:**125–126
  cytochrome *c* oxidase, **71:**95; 190
Ubiquinone, **20:**182, **22:**330, **25:**275–328, **58:**242. *See also* Coenzyme Q
  in aldehyde oxidase, **27:**172
    in respiratory chain, EPR study, **27:** 177–178; 180
  antioxidant function, **65:**124–128
  biosynthesis, **65:**106–108
  lipid peroxidation, **65:**127
  methylation of, **25:**423
  reduction, **65:**98
  as respiratory catalyst, **23:**346–347
  synthetic homologs of, **23:**370
Ubiquinone (Hatefi), **25:**275
Ubiquinone oxidase, location of, **24:**307
Ubiquinone protein C, **54:**290
Ubiquitin-ATP-proteinase, **58:**254
Ubiquitin pathway of degradation, **67:**283
Ubisemiquinone, aldehyde oxidase, EPR study, **27:**172
UDP
  Glucose, **52:**24
  glucuronic acid-agarose, **52:**92; 148–149
  hexanolamine-agarose, **52:**60; 65; 74
  reduction of, **28:**443; 445; 447; 449
UDP-*N*-acetyglucosamine, in carbohydrate synthesis, **26:**37; 39–40; 46; 49–50; 59
UDP-*N*-Acetylglucosamine, **59:**62
UDP-*N*-acetyl-D-glucosamine, synthesis of, **28:**114–115
UDP-*N*-acetyl-D-glucosamine epimerase, **44:**81
  2-acetoamidoglucal as intermediate, **44:**83
  H transfer, **44:**83
  kinetic study, **44:**83
  mechanism, **44:**82
  regulation, **44:**82
  UDP-*N*-acetylmannosamine as intermediate, **44:**82
  UDP inhibition, **44:**84
UDP-*N*-acetyl-D-glucosamine-4-epimerase, **26:**49–50; 55
UDP-*N*-acetyl-D-glucosamine pyrophosphorylase, **26:**55
UDP-acetylmuramic acid, *in vitro* synthesis, **26:**198
  in penicillin-treated bacteria, **26:**216
UDP-*N*-acetylmuramic acid, synthesis of, **26:**36; 49–50
UDP-*N*-acetylmuramic acid peptides, synthesis of, **26:**36; 49–51
UDP-*N*-acteylglucosamine-2-epimerase, allosteric regulation, **28:**115
UDP-L-arabinose pyrophosphorylase, **26:**52
UDPG, *See* Uridine diphosphate glucose; Uridine diphosphoglucose
UDPGal, *See* Uridine diphosphogalactose
UDP-galactose
  4–4epimerase, gene position of, in *Salmonella typhimurium,* **31:**102–103
  mutants without, **31:**102
  role of, in lipopolysaccharide synthesis, **31:**92
  in teichoic acid biosynthesis, **30:**237
UDP-D-galactose, *See* Sugar nucleotides
UDP-galactose: glucose-β 4-galactosyl transferase, **43:**411
UDP-galactose: N-acetyl glucosamine-β 4-galactosyl transferase, **43:**414
UDP-galactose-4-epimerase, **37:**420; 437; 585, *See* UDP glucose-4-epimerase
  rate-limiting step in the galactose pathway represented by, **29:**367
UDP-D-galactose-4-epimerase, **26:**41–42; 69
UDP-D-galactose pyrophosphorylase, **26:**52
UDP-galactosyltransferase, localization of, **24:**311
UDP-D-galacturonic acid pyrophosphorylase, **26:**54
UDPG analog, in hexose phosphate isomerization, **14:**203–204
UDPG dehydrogenase, location of, **24:**295
UDPG-fructose 6-phosphate transglycosylase, in plants, **18:**286
UDPG-fructose transglycolase, in plants, **18:**286
UDPG-glycogen glucosyltransferase (glycogen synthetase), allosteric regulation, **28:** 98–102

UDPGlcNAcpyrophosphorylase, in the membrane fraction of *Staphylococcus lactis,* **30:**236
UDP-D-glucosamine-4-epimerase, **26:**55
UDP-D-glucosamine pyrophosphorylase, **26:**50; 55
UDP-glucose
  allosteric regulation, **28:**97; 99–100; 102–103; 106; 110–112
  substrate for trehalose phosphate synthetase, **40:**35
  in teichoic acid biosynthesis, **30:**238
UDP-D-glucose. *See also* Sugar nucleotides
  enzymes for reduction, **26:**54
  metabolism scheme, **26:**45
UDP-glucose dehydrogenase, **46:**134; 245
  allosteric regulation, **28:**116
UDP-D-glucose dehydrogenase, carbohydrate metabolism, **26:**42–43; 53; 72–73
UDP-glucose epimerase, **43:**413
UDP-D-glucose-4-epimerase, **26:**53
  catalytic model, **44:**80
    concerted reduction, **44:**75
    hydrogen transfer, **44:**74
    inactivation by reduction, **44:**75
    inhibition by 5-phosphorylated sugars, **44:**76
    isotope effect, **44:**78
    kinetic constants, **44:**73
    from mammalian sources, **44:**80
    NAD, **44:**71; 73
    oxidized intermediate, **44:**77
    reactivation by 4-keto sugars, **44:**77
UDP glucose-α-glucan glucosyltransferase, localization of, **24:**311
UDP-glucose β-1,3:glucan β-3-glucosyltransferase systems, **28:**338
UDP-glucose pyrophosphorylase, **43:**413
  allosteric regulation, **28:**110–113
  genetics of, **31:**103
  in nucleus, **27:**274
UDP-D-glucose pyrophosphorylase, in carbohydrate metabolism, **26:**39; 44–45; 52; 62
UDP-glucose transferase, role of glucose, **24:**284
UDP-glucose transferase systems, **28:**338–340
UDP-glucose-UDP-galactose epimerase, hydrogen transfer, **26:**173
UDP glucoside dehydrogenase, location of, **24:**295
UDP-glucuronate, **46:**407
UDP-D-glucuronate-5′-epimerase, **44:**81
UDP-D-glucuronic acid, in carbohydrate synthesis, **26:**39
UDP-D-glucuronic acid decarboxylase, **26:**52
UDP-D-glucuronic acid-5-epimerase, **26:**54
UDP-D-glucuronic acid-4-epimerase, carbohydrate synthesis, **26:**54; 73
UDP-D-glucuronic acid pyrophosphorylase, **26:**53
UDP-glucuronyl transferase, **46:**118; 119
  localization of, **24:**311
UDPG phosphorylase, **59:**339–340
UDPG pyrophosphorylase, **34:**311–315
  inhibition of, **34:**314
  localization of, **24:**314
UDPG (uridine diphosphate-glucose)
  classification as coenzyme, **14:**241
  in hexose phosphate isomerization, **14:**193; 198–204; 214
UDP-linked transferase systems, in β-1,3 glucan synthesis, **28:**338–340
UDP-MurNAc-pentapeptide, acceptor substrate for alanyl-tRNA UDP-MuNAc-pentapeptide transferase, **40:**128
  site of acylation, **40:**128
UDP-P-glucose, allosteric regulation, **28:**111
UDP sugars, *See* Sugar nucleotides
UDP-xylose, in allosteric regulation, **28:**116
UDP-D-xylose-4-epimerase, in carbohydrate synthesis, **26:**52
UDP-D-xylose pyrophosphorylase, in synthesis of sugar nucleotides, **26:**52
Ulcer formation, effect of polyanions, **40:**56
Ultracentrifugation, **58:**122–146
  for mol. weight of globular proteins, **1:**14; 15
Ultramicrosomes, **17:**346; 352; 358; 361–362; 378. *See also* Cell particles, Mitochondria, Mitochrome and Sarcosomes
Ultrasonication, of oxygenated water, **41:**89
Ultrasonic disintegration of *Proteus vulgaris,* **13:**296
Ultrasonic treatment, **43:**136
Ultrasonic vibrations, effect on hyaluronic acid, **13:**202
  use in study of cell structures, **13:**11

Ultrasonic waves, relaxation methods, **25:** 2–9; 20
Ultrasound, **57:**233
Ultraviolet absorbance, of superoxide dismutase, **41:**60
Ultraviolet absorption of vitamins K_1 and K_2, **2:**304; 305
Ultraviolet absorption spectra, HGH, **42:**88; 90
concentration of protein, **42:**90
molar extinction coefficient, **42:**90
Ultraviolet-induced mutations, **27:**314; 326–327; 336; 339; 341; 342
Ultraviolet irradiation, *See* Irradiation
Ultraviolet light
effect on folic acid, **21:**373
effect on mitochondrial oxidative phosphorylation, **21:**31; 38; 113
identification of folic acid compounds, **21:**390
effect on thiols, **11:**210
sensitivity of proteins to, **10:**288
Ultraviolet radiation
mutations caused by, **6:**82–84; 92–93
photomicrography in muscular dystrophy, **6:**219–220
Ultraviolet radiations. *See also* Irradiation, Photochemical processes, and Radiations
effect on adsorbed enzymes, **28:**33–35
effects on enzymes, **24:**360; 363; 364; 367; 370; 372; 380; 383; 393; 397; 400–401
and origin of life, **27:**355–360; 362; 369–371; 382–386
Ulva lobata, small light-induced absorbancy changes in, **30:**38
UM, *See* Urine mucoprotein
Umbelliferone, **55:**95
Umbelliferose, **56:**162
UMP, *See* Uridylic acid
reduction of, **28:**443; 445
UMP-2–^{14}C, conversion to deoxyribotide, **28:**434
UMP (uridine monophosphate), from UDPG, **14:**200–201
Uncharged substrate, **34:**460
Uncompetitive inhibition component, **43:**160
Unconjugated, pterins, **35:**36
Unconjugated pterins, phenylalanine hydroxylation, **70:**103
Uncoupler binding, **49:**258
Uncoupling, of oxidation and phosphorylation, **23:**330–332; 334; 349; 360; 363–369; 374; 383–385
of photophosphorylation, **23:**386
Uncoupling agents, **23:**349; 360; 363–369
Uncoupling reactions, phosphorylation and oxidation, **17:**99; 108–109; 112–119
Undecane (^{14}C), in microbial oxidation, **27:**483; 488
Undecaprenol phosphate, **52:**33
Under hydroxylation, of collagen, **41:**265; 276; 281
Undirectionality of enzyme reactions, **28:**46–47
Unesterified fatty acids, **53:**225
Unfreezable water, **51:**2–5
Uniport, cation conductors in, **29:**46
circulating carrier type, **29:**44–46
definition of, **29:**38
D-glucose, **29:**52
single channel, **29:**46–49
sugar, effects of phlorizin on, **29:**52
α-thioethyl-D-glucophyranoside translocation catalyst of *Sacchromyces cerevisiae,* **29:**45
Unitary hypothesis of enzyme formation, **22:**104; 105
Unit cell dimensions, **39:**349
Unit-membrane, **34:**470–471
Unlinked biotin cistrons, **38:**353
bio H locus, **38:**353
bir mutants, **38:**353
repressor mutants, (R^-), **38:**353
Unsaturated groupings and biochemical asymmetric synthesis, **7:**91–94
α, β-Unsaturated ketones
hydride transfer to, **24:**472–474
xenobiotic substrate specificity and mechanism, **69:**32
α, β-Unsaturation, during transamination, **16:**199
Unsaturation, PN reduction, **26:**127–128
Unstirred layer, **34:**503–507
U5P, *See* Uridine 5′-monophosphate
Uphill transport, **49:**65
Uptake, glutamine, **53:**204; 205
Uptake of glutamine by mammalian tissues, **13:**268–271
Uracil, **38:**213; 215–218; 221, **50:**155
5-benzyluracil, **38:**220; 221
biosynthesis of, **54:**74

catabolism of, **24:**328; 341
derivatives, **38:**218; 228
exclusion from DNA, **38:**269
formula, **21:**263
in pyrimidine biosynthesis, **21:**281–285; 290; 420; 421
synthesis, in liver cells, **21:**52; 53
Fu substitution of, in rRNA, **54:**86
in genetic studies, **17:**429–430; 463
in virus multiplication, **17:**480–483
5-isobutyluracil, **38:**220
5-isopropyluracil, **38:**220
mispairing frequency of, in RNA, **54:**91
of nuclear RNA, **16:**437; 439
5-phenyluracil, **38:**218; 219; 220; 221
in retarding cytotoxic effects of fluoropyrimidines, **54:**99
substituted, **38:**222
^{14}C-Uracil, in DR metabolism, **28:**435–436
Uracil-2-C^{14}, **17:**429
Uracil (C^{14}-labeled), in enzyme synthesis, **24:**18
Uracildeoxyribo-5′-phosphate, *See* dUMP
Uracil derivatives, PN-dependent reduction, **26:**127
Uracil-5-methyltransferase, tRNA, **54:**89
Uracil nucleotides, **55:**271
Uracil of tobacco mosaic virus, **15:**460
Uracil permease, **56:**125
Uranium compounds, solubilization of, **17:**137; 138; 142; 143; 145; 147
Uranium ions, model catalyses with, **21:**176
Uranyl acetate, staining of myosin by, **16:**320
Uranyl complex, as molybdenum substitute, **17:**270
Uranyl nitrate, effect on enzymes of *Cunninghamella blakesleeana,* **20:**270
Urate oxidase, **57:**87
location of, **24:**302; 305; 307; 340; 341; 343
Urdine-2′,3′-cyclic phosphate. *See also* Cyclic phosphates
in assay of RNase, **24:**172–173; 237–242
Urea, **43:**332, **50:**141
actomyosin dissociation by, **13:**161
breakdown of myosin by, **13:**157–158
effect on actin polymerization, **13:**154–155
relaxing agent for muscle, **13:**192
from allantoin degradation, **56:**92
allophanate hydrolase activity with, **56:**104; 105
from arginine catabolism, **56:**134
changes in urine, in myasthenia gravis, **6:**212
as compared with cyanate
effects on deoxyhemoglobin S gelling, **40:**3; 4
treatment for sickle-cell disease, **40:**2; 3
concentration in kidney, **20:**17; 18
degradation of, **56:**94; 95
denaturation by, **23:**35; 37–41; 43–58
denaturation of RNase by, **24:**185–188; 196; 199–200
effect on activity of RNase, **24:**215; 221; 237; 247; 249–250
effect on ATPase, **6:**166
effect on clot formation, **16:**35; 39; 40; 43
treatment of myosin by, **16:**326–330; 344
effect on DNA, **18:**167
histone aggregates, **18:**172
effect on hydrolytic enzymes, **15:**291–292
synthesis, **15:**319–387
effect on papain, **19:**280; 293
formation, **19:**243; 253; 371–381
effect on protein-mercaptan reaction, **27:**210
effect on —SH groups of proteins, **11:**221; 222
effects on tobacco mosaic virus, **5:**21–22
formation of, **74:**37–38
in prevention of sickling, **40:**3
and purine metabolism, **9:**463
and pyrimidine metabolism, **9:**469
solubilization of proteins by, **17:**151
transport system for, **56:**126–128
Urea amidolyase, **56:**94; 95
immunoprecipitation of, **56:**95; 96
purification of, **56:**95
Urea carboxylase, **56:**94
immunoprecipitation of, **56:**97
kinetic constant for, **56:**98
structural genes for, **56:**100
Urea cycle
carbamyl phosphate synthetase I relation to *E. coli* carbamyl phosphate synthetase, **66:**236–238
enzymes for, **24:**340–341
Urea cycle enzymes
carbamyl phosphate synthetase, **72:**131
hepatocyte heterogeneity and enzyme distribution, **72:**56–61
Urea formation, **16:**227; 229; 259
alternate pathways of, **39:**81; 82
in vivo investigation of, **39:**81; 82
Ureagenesis, in liver, **53:**228

Urea-*N*-glycoside, **60:**13
Urea-hydrocarbon complexes, **65:**241
Urea permeability, **34:**467
Urea precursors, products of intestinal glutamine metabolism as, **53:**228
Ureas
peroxidase-inhibiting and antithyroid action, **10:**261
substituted, effects on urease, **17:**59
Urease, **5:**23, **10:**227; 300; 303–306, **15:**62; 375, **19:**287, **34:**448; 451, **35:**400, **56:**94
action of on urea, **33:**270–271
effects of low humidity on, **33:**284
activation by amino acids, **10:**100
active groups, **15:**275–277; 289–290; 291–292
activity of —SH group, **11:**235; 236; 239; 241
effect of quinones on, **11:**247
reactivation by glutathione, **11:**233
reduction of disulfide in, **11:**228–229
types of thiols in, **11:**221; 234
adsorption at interfaces, **28:**3
estimation of soil enzymes, **28:**372; 375–376
in soil, determination of, colorimetric method, **28:**383–384; 386
methods for, **28:**386
results, **28:**387
adsorption studies, **14:**360
classification, **14:**234; 247
inactivation by radiations, **14:**36–37
isolation, **14:**398–399
mode of action, **14:**183
structure studies, **14:**302; 303
crystallization, **24:**107
effect of chemical modification, **24:**145
inactivation by radiations, **24:**374–375
molecular weight from radiation studies, **24:**36
valence of, **24:**118
cytochemical method for, **8:**30
in belladonna, **8:**239; 244
linkage attacked by, **8:**164–165
toxic effect, **8:**486
effect of borate on, **12:**518
effect of DFP on, **13:**332
elementary steps in reaction of, **25:**14
multiple forms of enzymes, **25:**362
product inhibition of, **25:**222; 240–241
turnover number, **25:**32
histochemical methods for, **9:**589
photochemistry of, **9:**141; 151–154
in plants, **9:**405; 442
titrimetric methods, **9:**601
inhibition, **20:**370
kinetic studies, **17:**38; 55; 58–60
kinetic studies on, **15:**273–277; 282–287
and phosphotriesterase, comparison of, **74:**71–73; 72; 74
pK value, **22:**58
substrate inhibition of, kinetics, **21:**137
Urease films, activity of, **28:**24–25; 29
Urea solutions, formation of cyanate, **40:**3
Urea synthesis
acinar glutamine metabolism
acid-base homeostasis, **72:**71–75
glutaminase ammonia amplification system, **72:**69–70
glutamine synthase ammonia scavenging, **72:**68–69
hepatic pathophysiology, **72:**77
metabolites of, **39:**46
Urea Synthesis and Metabolism of Arginine and Citrulline (Ratner), **15:**319
Ureides, of purines, **18:**295
of pyrimidines, **18:**261
synthesis of, **18:**277–278
α-Ureidoglutaric acid, *See* Carbamylglutamic acid
Ureidoglycollate, **56:**92
Ureidoglycollate hydrolase, **56:**92
kinetic constant for, **56:**98
structural genes for, **56:**98; 99
O-Ureidohomoserine, **50:**141
O-Ureidohomoserine synthetase, **50:**141
3-Ureidopropionase, localization of, **24:**327
Ureido ring, **35:**332
Ureidosuccinate, **17:**399
synthesis and metabolism, in mitochondria, **21:**52
synthesis of, **24:**309; 313
Ureidosuccinate (carbamylaspartate), as precursor of pyrimidines, **16:**266–268
Ureidosuccinic acid, **15:**357
and nucleic acid synthesis, **25:**170–171
Uremia, **67:**160–161
Urethane, in electron transport studies, **17:**67; 69; 361
solubilization by, **17:**138; 151

Urethan (ethyl carbamate), action on invertase activity, **7:**259
effects on luminescent bacteria, **7:**23–24
effects on yeast, **7:**22–23
and flavoprotein, **7:**22
and luminescence, **7:**237; 242–247; 258
Urethans, **19:**50
Uric acid, **41:**1, **51:**126. *See also* Purines; Purines and Xanthine oxidase
biological degradation, **9:**463–466
in breakdown of purines, **12:**451–453
synthesis of, and CO_2 fixation, **12:**121–127
catalase action on, **19:**109
oxidation, **19:**204–206
peroxidase action on, **19:**108; 206
changes in urine, in myasthenia gravis, **6:**212
excretion of, **17:**320–321
and glycine metabolism, **23:**455
oxidation by O_2 (isotopically labeled), **15:**12
Uricase, **9:**442; 463–465; 489; 491, **15:**12, **17:**280, **18:**118, **19:**202; 205; 206
activity in man and monkey, **41:**1
adsorption studies, **14:**346
intracellular distribution, **16:**463; 467
localization in cytoplasm, **13:**11
location of, **24:**307
in mitochondria and lysosomes, **21:**39–41; 58
occurrence of, **12:**451–452
Uricase in nuclei, **8:**33
Uridenediphospho Galactose: Metabolism, Enzymology, and Biology (Kalckar), **20:**111
Uridine, **35:**465, **54:**58
conversion of, to pseudouridine, **54:**87–88
C-5 proton of, exchanging, **54:**80
in DR metabolism, **28:**435
in nucleic acid synthesis, **28:**423
FU effect on nRNA incorporation of, **54:**88
in genetic studies, **17:**429–430
genetic studies of, **13:**142
in reversal of toxicity of fluoropyrimidines, **54:**94–95; 98
synthesis of, **18:**222
^{15}N-Uridine, incorporation into nucleic acids, **28:**420–421
Uridine 2′,3′-cyclic phosphorothioate, **55:**294; 297
Uridine diphophoglucose dehydrogenase, in determination of uridine diphosphoglucose, **20:**177
immunization studies with, **20:**120
Uridine diphosphate
conversion to uridine triphosphate, **20:**122
in biosynthesis of pectic substances, **20:**369
phosphorylation, in mitochondria, **21:**53
in polynucleotide formation, **18:**229
Uridine diphosphate *N*-acetylgalactosamine sulfate, from connective tissue, **25:**439
Uridine diphosphate *N*-acetylglucosamine, **18:**202
synthesis of, **25:**436–437
Uridine diphosphate *N*-acetylglucosamine-6-fucose, isolation of, **25:**449–450
Uridine diphosphate-L-arabionose, formation of, **28:**398
Uridine diphosphate choline, synthesis by carbodiimide method, **22:**188
Uridine diphosphate derivatives of amino sugars, in carbohydrate transfers, **25:**449–450
Uridine diphosphate esters, and transglycosidation, **18:**201
Uridine diphosphate galactose, **18:**202; 207–208
pyrophosphorylase, **18:**207
synthesis by mixed anhydride method, **22:**173
Uridine diphosphate-glucose, *See* UDPG
Uridine-diphosphate-glucose, comparison with CoA, **16:**15
Uridinediphosphate glucose, in plants, **19:**5
Uridine diphosphate glucose, structure, **22:**173
synthesis, by carbodiimide method, **22:**177; 188
by mixed anhydride method, **22:**173
by phosphoramidate method, **22:**197–199
Uridine diphosphate glucose pyrophosphorylase, **18:**204; 207
Uridine diphosphate glucose (UDPG), in plant respiration, **18:**287; 289; 290; 312
and pyrophosphorylase action, **18:**202; 207–208
Uridine diphosphate-D-glucuronate, pentose from, **28:**398–399
Uridine diphosphate glucuronic acid, **18:**207
in synthesis of hyaluronic acid, **25:**436–437
Uridine diphosphate L-iduronic acid, in dermatan sulfate, **25:**441
Uridine diphosphate-D-xylose, epimerization of, **28:**398
Uridine diphosphogalactose, **15:**168–169
in biosynthesis of pectic substances, **20:**370
metabolism, **20:**111–134

Uridine diphosphogalactose-4-epimerase, induction in galactokinase-deficient *Saccharomyces cerevisiae,* **20:**128
in formation of uridine diphosphogalactose, **20:**114–118
levels in normal and galactosemic subjects, **20:**125; 126
physiological function, **20:**118–120
Uridinediphosphogalactose-4-epimerase, properties of, **30:**102
reaction catalyzed by, **30:**102
Uridinediphosphogalactose-pyrophosphorylase, reaction catalyzed by, **30:**102
Uridine diphosphogalacturonic acid, in biosynthesis of pectic substances, **20:**369–371
Uridine diphosphoglucose, **15:**168–169
biosynthesis, **20:**122
in biosynthesis of pectic substances, **20:**370
formation and conversions in galactose metabolism, **20:**112; 113; 115; 117; 118; 122
pentose from, **28:**398–399
Uridinediphosphoglucose phosphorylase, column chromatography, **22:**471; 472
Uridinediphosphoglucose pyrophosphorylase, **17:**185
Uridine diphosphoglucose pyrophosphorylase, functions and mechanism of action, **20:**121–123
in formation of uridine diphosphoglucose, **20:**112; 122
in galactokinase-deficient *Saccharomyces cerevisiae,* **20:**128
levels in normal and galactosemic subjects, **20:**125; 126
Uridine diphosphoglucuronic acid, formation, **20:**122; 123
in biosynthesis of pectic substances, **20:**371
Uridine disphosphate glucose pyrophosphorylase, mechanism of deduced from relaxivity data, **33:**44–45
^{14}C-Uridine, DNA synthesis, **28:**420–421; 423
Uridine in cofactor of glucuronic acid transferring systems, **16:**404
Uridine in pyrimidine biosynthesis, **21:**282; 285; 286
Uridine kinase, **21:**282; 284, **54:**64; 65
in assessing tumor responsiveness to fluoropyrimidines, **54:**101; 107
in nucleus, **27:**274
Uridine kinases, in ascites tumor cells, **28:**135
Uridine monophosphate, *See* UMP
Uridine 5′-monophosphate, phosphorylation of, **18:**196–197
Uridine-5′-monophosphate (UMP), **54:**58
Uridine nucleotides, in mitochondria, **21:**10
Uridine-2′,3′-phosphate-5′-diphosphate, structure, **22:**170
synthesis by mixed anhydride method, **22:**170
Uridine 5′-phosphate pyrophosphorylase, **18:**216–217
Uridine phosphorylase, **54:**65
Uridine-5′-pyrophosphate, structure, **22:**169
synthesis, by carbodiimide method, **22:**186
by imidoyl phosphate method, **22:**187
by mixed anhydride method, **22:**168; 169
Uridine - 5 - (sialyl - galactosy - *N*-acetylglucosaminyl pyrophosphate), in carbohydrate transfer, **25:**445
Uridine triphosphate, *See* UTP
formation and conversions in galactose metabolism, **20:**112; 114; 122; 126
in purine biosynthesis, **21:**216
in pyrimidine biosynthesis, **21:**264; 280
Uridine-5′-triphosphate, synthesis by mixed anhydride method, **22:**170
Uridine-5′-triphosphate amidotransferase, GTP, special role in glutamine reaction, **39:**130; 135
half-of-the-site reactivity, **39:**133
kinetic analysis of the reaction, **39:**132
liver, comparison of liver and bacterial enzymes, **39:**137
enzyme from mammalian source, **39:**137
negative cooperativity, **39:**133
properties of the reaction, **39:**132
ATP-ADP exchange, **39:**132
glutamine and ammonia as amino donors, **39:**133
transfer of carbonyl oxygen of UTP to ATP, **39:**132
sources, **39:**130; 137
stoichiometry of the reaction, **39:**130
subunits, association and dissociation, **39:**136
Uridine triphosphate (UTP), amination of, **18:**230–231
and phosphate transfer, **18:**195; 196–197

in plant respiration, **18:**287; 289; 312
and pyrophosphorolysis, **18:**204–206
Uridylate kinase, NMP kinase primary structure, **73:**107
Uridylic acid, **15:**357
biosynthesis, **21:**264–266; 277–282; 284; 285
hydrolysis, by metal ions, **21:**184; 185
phosphorylation, **21:**264; 280
in pyrimidine biosynthesis, **21:**288; 290
conversion of orotic acid to, **16:**266
RNA, P^{32} uptake, **16:**462
^{14}C-Uridylic acid, DNA synthesis, **28:** 420–421
Uridyl transferase, **18:**204–206, **26:**39
Uridyltransferase, galactose-1-phosphate, properties of, **30:**101–102
reaction catalyzed by, **30:**101–102
Urinary cytosine, **51:**174
Urine, **51:**170; 181
of ADA-deficient patients, excretion of adenine compounds in, **51:**172–173
adsorption studies, **14:**365
amino acids in, **14:**266
estimation of steroids in, **20:**276
Urine mucoprotein (UM), alkaline degradation, **20:**139
composition of prosthetic group, **20:**138; 140; 141
molecular weight, **20:**141
as substrate for neuraminidase and receptor destroying enzyme, **20:**137; 144
Urine triphosphate, *See* UTP
Urobilin pigment, **46:**67
Urocamase, **55:**238
Urocanase
classification, **14:**247
ethionine effect on, **25:**380
Urocanic acid, **15:**185–186
in histidine degradation, **21:**405
Uromyces phaseoli, var. *typica,* plant disease caused by, **20:**390
Uronic acid
of β-glucan, **28:**336; 337
in wall polysaccharide, **30:**250
Uronic acid derivatives, as solubilizing agents, **17:**147–148
Uronic acid pathways, in ribose synthesis, **28:**398–401; 405; 414; 418
Uronic acids. *See also* Carbohydrates
in air-cured tobacco, **6:**351; 355
in bacterial cell walls, **26:**223
in carbohydrate metabolism, **26:**39; 41–47; 53–54; 59; 60; 66; 72–73
origin, **14:**207
Uronolactonase, localization of, **24:**317
β-Uronoside linkage, identification of, **16:**383
Uroporphin I, **7:**319
Uroporphyrin, **46:**83; 102; 107
Uroporphyrin I, **46:**164
Uroporphyrin III, **46:**138
and vitamin B_{12}, **26:**234–235
Uroporphyrinogen, **46:**102; 103; 138
formation of, **24:**332
Uroporphyrinogen decarboxylase, **46:**54; 86; 107; 139; 151; 164
Uroporphyrinogen I, **46:**46–47; 165
Uroporphyrinogen III, **46:**47; 50; 54; 55; 56; 81; 83
conversion to cobyrinic acid, **46:**51
cosynthetase reaction, **46:**81
decarboxylation to coproporphyrinogen III, **46:**54–56
formation of, **46:**39; 46–49
metabolism products of, **46:**49–58
Uroporphyrinogen-III coproporphyrinogen-III, **50:**273
Uroporphyrinogen III cosynthetase, **46:**46; 47; 48; 164
Uroporphyrinogen III methylase, **74:**188
Uroporphyrinogen I synthetase, **46:**46; 47; 48; 81; 161; 162
Uroporphyrinogen synthetase, **46:**103; 107
Uroporpyrinogen decarboxylase, localization of, **24:**330
Uroxanic acid and purine metabolism, **9:**465–466
Urushiol, role in browning of plant tissues, **16:**150
Urushiol oxidases, classification, **14:**254; 255
Uscharin in plant metabolism, **10:**256
Usnic acid, effect on sperm, **24:**85
Ustilago, **35:**176
sphaerogena, **35:**136; 140; 145; 159; 170; 174
Ustilago maydis, plant disease caused by, **20:**393
Ustilago maydis, genetic studies of, **13:**139
Ustilago zea, coenzymes Q in, **25:**289
Ustin, **7:**484; 486; 489
Uteroferrin, **65:**164–165
Uterus, actin content during pregnancy, **16:**343
actomyosin of, **16:**331
Uterus muscle proteins, **13:**180–182

Utilization of Binding Energy in Coupled Vectorial Processes (Jencks), **51:**75
UTP, *See* Uridine triphosphate
UTP dephosphorylating enzymes, **27:**274
UTP (uridine triphosphate), action of myosin and its subunits on, **16:**337; 340
effect on myosin, **16:**344
UTP (uridine triphosphate) synthesis, **14:**203
UV difference spectra, **39:**197

V

Vaccinia virus
activity and particle identity, **9:**193–195
chemical composition, **9:**229
dilution curves, **9:**176–177
effect of light on, **9:**142
homogeneity, **9:**219–220
physical constants, **9:**178; 188
purification, **9:**186–187
sedimentation rate, **9:**208
glutaredoxin, **66:**165
Vacuole, and plant respiration, **18:**299
Vagus nerve, **34:**468
Val-38, **57:**23
Val-Arg-Gly-Gin, tyrosine hydroxylase, **70:**182
Valence, of proteins, effect on chromatography, **22:**421–425
Valence of biologically active proteins, **24:** 118–120
Valence state of metals in metalloflavoproteins, **17:**268–270; 276
Valence types in nucleoprotein (TMV), **15:**461
Valeric acid, **17:**154
metabolism, in mitochondria, **21:**46; 47
δ Valerolactam, **43:**316
δ-Valerolactam, configuration of, **23:**14–15
Valine, **35:**481; 482; 484; 486, **50:**130, **51:**217, **53:**169, **58:**174
biosynthesis, **16:**279–281; 285–286
allosteric regulation, **28:**64; 70
in transamination reactions, **16:**191; 192; 198; 203; 207–209; 228–230; 239
and vitamin B_{12}, **26:**270
in biosynthesis of phenylalanine, in plants, **20:**329
in phosphoylated peptides of esterases, **20:**288
thermodynamic data in ionizations, **20:**67
conversion into leucine, **58:**183
and ethionine effects, **25:**372
genetic studies, **17:**399–403; 409; 411; 430–431; 449
labeled, in protein biosynthesis, **19:**261
in mercuripapain, **19:**301
metabolism, in mitochondria, **21:**44–47
D-Valine, **33:**367
L-Valine
K_A values for, **29:**45
transamination reactions of, **19:**347
L-[^{13}N]Valine, **57:**327–328
DL-Valine, resolution by papain, **19:**309
Valine adenylate, synthesis by carbodiimide method, **22:**190
Valine-alanine, α-aminobutyrate system, for transamination, **16:**203–204; 208; 229
Valine auxotrophs, **16:**279; 280
Valine transaminase, **16:**280–281
Valinomycin, **33:**296, **34:**483, **35:**178
action of, in rat liver mitochondria, **29:**294
action on K^+ translocation, **29:**48
as cation conductor, **29:**46
discrimination between Na^+ and K^+ by, **29:**48–49
effect on K^+, **29:**46
effect on permeability of mitochondria to choline, **29:**46–47
H^+ ejection by rat liver mitochondria induced by, **29:**290–291
inducing action of, **29:**285
induction of monovalent cation uptake by, in mitochondria, **29:**287–292
K^+ and Mg^{++} accumulation induced by, **29:**301
K^+ efflux from phospholipid micellar systems stimulated by, **29:**291
K^+ uptake by rat liver mitochondria induced by, **29:**288; 290–291
mitochondria treated by, **29:**75
structure of, **29:**48
Val-Trp, **57:**36
S-DL-Valyglutathione, synthesis and hydrolysis, **19:**247; 248
S-Valyl-*N*-acetylcystine, effect of carbon dioxide on, **19:**235
Valyl adenylate, activated form of valine, **19:**258
synthesis, **19:**251
Valyl adenylate complex, **35:**447
molecular weight of, **35:**447

S-Valyl-*N*-alanylglycylcystine, **19:**262
Valylcysteine, **19:**242
Valylglycine, in transamination, **16:**239
S-Valyl-*N*-glycylcystine, **19:**244
Valyl-tRNA synthetase, **43:**251; 334
binding energy, **43:**334
Valyl-t-RNA synthetase, **46:**249
Vanadate, **57:**161
Vanadium complexes in thiol oxidation, **11:**204
Vanadium compounds, **17:**138; 142
Vanadium compounds, EPR study, **27:**143–144
Vanadium mercaptide, **11:**216
Vanadyl ions, **57:**161
Vancomycin, inhibition of synthesis of poly (glycerol phosphate) in *Bacillus* by, **30:**244
Van der Waals' dispersion forces. *See also* Chain length, Intermolecular forces, and Particle size
effect on enzyme action, **12:**201–203; 247–248; 251; 252; 295; 306
van der Waals force, **38:**200; 220
Van der Waals forces. *See also* Intermolecular forces
in catalysis by enzymes, **14:**188–189
and enzyme activity, **15:**292
in nucleoproteins, **15:**461
between protein and coenzyme, **15:**30
van der Waals forces, effects in solubilization of matter, **17:**140
Vanillic acid, **18:**374
Vanillin, **34:**213
from lignin, **18:**355–358
in lignin formation, **18:**360–361; 367–369; 370; 372; 374
peroxidase action on, **19:**106
Vanillin (C^{14}), as lignin precursor, **18:**374
Vanilloyl methyl ketone, **34:**213
Vanillymandelic acid (VMA), in depression, **29:**492. *See also* 3-Methoxy-4-hydroxymandelic acid
excretion in catatonics, **29:**530
Van Niel's theory of photosynthesis, **1:**234–242
Vapor tension, **34:**512
Variance, contours of equal, **29:**22
square root of, **29:**6
Variations in system, among tissues, **49:**86
Vascular permeability, **58:**120
Vasoconstrictor substance in platelets, **16:**72
Vasoexciter material (VEM), **7:**522–523
Vasointestinal peptide receptor, **55:**162
Vasopressin, **40:**272; 292; 293; 296, **59:** 366–367
amino acid interchanges in, **24:**417; 422–425
lower peptide sequences in, **24:**421–425
synthesis of, **24:**107
analogs of, **29:**392
lysine, use of tosyl group in synthesis of, **29:**438
synthesis by *p*-nitrophenyl ester method of, **29:**406
binding to neurophysin, affinity, **40:**282; 284; 306; 307; 309; 321; 329
α-amino group in, **40:**314; 315; 320; 324; 327; 329
circular dichroism changes, **40:**313; 317; 319; 321; 322; 325
disulfide, **40:**321
free energy, **40:**322; 323; 324
number of sites, **40:**305; 306; 307; 308; 309
pH dependence, **40:**309; 310; 315; 316
biosynthesis, **40:**285; 286
conformation, **40:**311
disulfide, **40:**311
effect on fructose 6P cycle role of Ca^{2+}, **52:**316
effect on gluconeogenesis, **52:**314
effect on glycolytic flux, **52:**306; 315
effect on pyruvate dehydrogenase activity, **52:**291
effect on pyruvate kinase, **52:**304; 315
effect on pyruvate metabolism in in mitochondria, **52:**315
evolutionary precursor, **40:**295
localization, **40:**276; 277; 278; 279; 281
nuclear magnetic resonance, **40:**320
peptide analogs, **40:**305; 309; 311; 317; 318; 319; 320; 321; 322; 323; 324; 326
phenylalanine of, **40:**320
position 1, **40:**320; 321; 324
position 3, **40:**322; 324
precursor, **40:**286
α-amino group, **40:**314
release, **40:**277; 281; 282; 283; 284
storage, **40:**277
structural homology, with neurophysin, **40:**295
tyrosine of, **40:**311; 316; 317; 318; 319; 320; 324; 326; 327

Vasopressin (H^3), SH-group binding, **27:** 227–228
Vasopressin in fructose-2, 6–6bisphosphatase regulation, **54:**154
Vasopressins
dimers of, **27:**220
structure studies, **14:**277; 292; 299
Vasotocin, **40:**282
Vassopressin, **57:**164
Vectorial flow, **34:**479
Vectorial reactions, **33:**279–280; 291
Vegetable Tissue, Frozen Activity in (Joslyn), **9:**613
Veillonella alcalescens, superoxide dismutase, from, **41:**80
Veillonella gazogenes, propionyl CoA carboxylase in, **19:**40; 42
Velocity, initial, of enzyme reactions, **29:**1–2
maximum, of an enzyme reaction, **29:**1–2
method of measurement of, **29:**14–15
Velocity constants, *See* Kinetics
Venom, Russell's viper, in activation of Factor X, **30:**280
to define the coagulant action with a serum factor, **30:**267
Venom, snake, distribution of nerve-growth-promoting factor in, **31:**57
nerve growth factor in, **31:**53; 57–60
Venom phosphodiesterase, tryptophan and throsine residues required by, **29:**204
Venoms, collagenolytic activity, **23:**173
uncoupling action, **23:**367
Veratric, **34:**215
Veratridine, **55:**162
tyrosine hydroxylase regulation, **70:**140; 150
Veratrine, effect on action polymerization, **13:**155
Verbascose, **56:**159
Verdauung bei den neiderer Vertbraten (Vonk), **1:**371
Verdins, spectral properties of, **26:**21
Verdoglobin, **7:**313
Verdoperoxidase, **3:**137–148, **17:**68; 375, *See* Myeloperoxidase
activity, **3:**143–145
analysis, **3:**142–143
band spectrum, **3:**141–142; 143; 146
bibliography, **3:**147–148
inactivation, **3:**145
oxidatic and perioxidatic reactions, **3:** 137–139; 143–145
preparation, **3:**138–141; 146
relation to cytochrome, **10:**61
spectroscopic investigation, **3:**141–142
Verdoperoxidase (Agner), **3:**137
Versene, *See* Ethylenediaminetetra-acetic acid; Ethylenediamine tetracetic acid
Versene EDG, **17:**240
Versene (ethylenediaminetetracetic acid), effect on L-histidine, **19:**378
effect on mercuripapain, **19:**279
effect on oxidase, **19:**90; 97; 126; 127; 190; 199
in model hydroxylating system, **19:**135; 141–143
and papain, **19:**270; 274; 276; 279; 288–291; 293; 304; 305
in transamination reactions, **19:**354; 374
Versene (sodium ethylene tetraacetate), activation of β-glucuronidase, **16:**379
effect on myosin ATPase, **16:**335–336
Vertebrata, low
anatomy of alimentary canal, **1:**374–378
digestion, **1:**371–417
Vertebrata, Lower, Digestion in (Vonk), **1:**371
Vertebrata and invertebrata, digestion, **1:**371–374
Vertebrate dihydrofolate reductase, bacterial DHFR structures compared, **70:**42–44
Vertebrate heptocyte, **44:**225
Vertebrate proteins, collagen hydroxylase occurrence, **72:**328–334
cosubstrate values, **72:**345–348
Verticillium albo-atrum, plant disease caused by, **20:**392
Verticillium dehliae, polygalacturonases of, **20:**366; 369
Very-low-density lipoprotein, **40:**34
Very Rapid signal, **51:**133; 143; 152; 155; 158
O splitting, **51:**145–146
xanthine oxidase, **51:**130–131
Vesicants, inactivation of enzymes by, **11:**239
reaction with thiols, **11:**215–216; 226–227
Vesicle, **34:**484
fraction enzyme, **34:**482
fraction penicillinase, **34:**482
membrane, **34:**480
mixed, **34:**480
Vesicle-associated membrane proteins (VAMPs), adenosine receptors, **69:**114

Vesiculase and semen metabolism, **9:**347–348
Vespa crabo, **45:**396
Vespa orientatis, **45:**395; 396; 398; 402
Vestitol, **55:**29
  medicarpin, interconversions with, **55:**34
  methylation of, **55:**35
  structure of, **55:**6
V factor, **7:**414–415
  and coenzymes I and II, **7:**411
Vibration limit, **59:**122
Vibrio
  and glycolate metabolism, **23:**439
    use of acetate, **23:**411; 412
  phenylalanine and tyrosine metabolism in, **14:**111; 112
  protocatechuic acid oxidase in, **19:**97; 98
    pyrocatechase in, **19:**81
Vibrio cholera, **23:**180; 183; 228
Vibrio cholerae, **70:**231
  enterotoxins, **58:**118
  receptor destroying enzymes of, **20:**136; 140–144
Vibrio cholerae, neuraminidase activity from, **41:**107; 108; 120
Vibrio metchnikovi, **23:**183
Vibrio, oxidation of hydrocarbons, **27:**508; 512; 514; 518
Vibrio phosphorescens, luminescence in, **25:**152
Vicia faba, **55:**9; 84; 85
Vicia sativa, neurotoxins in, **29:**147
Vignafuran, **55:**84; 91
Vigna sinesis, **55:**14
Vigna unguiculata, **71:**83
Villus columnar cells (enterocytes), **53:**205
Vinblastine, **55:**173
Vinblastine, GTPase reaction, microtubule dynamics, **69:**146
Vinca alkaloids, inhibition of tumor growth by, **29:**336
Viniferins, **55:**92
  synthesis of, **55:**94–95
Vinyl compounds, polymerization, **15:**16–17
Vinylglycine, in isoleucine synthesis, **16:**285
Vinylglyoxylate, cysteine *S*-conjugate β-lyase reactions, prodrug targeting, **72:** 227–230
Vinyl groups of porphins, **7:**325; 333
2-Vinylporphyrinogen III, **46:**56
4-Vinylporphyrinogen III, **46:**56–57
Vinyl sulfide in plants, **10:**242
Viobin, elastase from, **23:**229
Viokase, elastase from, **23:**173
Violaxanthin
  in chloroplast lamellae, **30:**9
    content of, in chloroplast fractions, **30:**50
  in photosynthesis, **25:**71
Violence, **74:**236
Viologen, benzyl, as artificial electron acceptors, **30:**23
Viomycin, amino acids in, **14:**266
Viperidae, nerve growth factor in venom of, **31:**57
Viral deoxyribose, *See* Bacteriophage
Viral DNA, study of, **16:**431–433
Viral hemagglutinins, **52:**134
Viral leukemias, **47:**366
Viral neuraminidases, **52:**138–141
Viral RNA, **16:**438
Viridin, **7:**484; 485; 487; 489
Viroptic, **54:**59; 61
Virulence factors, *Caulobacter* cell cycle, flagellar structure and assembly, **71:**11
Virulence of microorganisms, relation to hyaluronidases, **13:**219–221
Virulent principle of virus, identification with characteristic particle, **9:**187–195
Virus
  inactivation, **7:**248
    reproduction, **7:**190–191
    tobacco mosaic, **7:**190–191
  tobacco mosaic, **15:**449–484, **23:**382 (*See also* Bacteriophage)
    inactivation by thiol reagent, **11:**224; 239
Virus Activity, Physicohemical Aspects (Holzapfel), **1:**43
Virus cultures, use of collagenase, **23:**218
Viruses, **9:**171–240. *See also* Bacteriophage; Phage
  activation, **1:**49
  activity and electrical charges, **1:**45
  activity and *p*H, **1:**44
  autosynthesis, **4:**4–8
  bacteriophage T4, mutation study, **27:** 297–301
  chemical change and self-reproduction, **13:**123–124; 145–146
  chemical modification, **24:**109
  and chromosomes, **4:**5
  crystal form, **1:**59

Viruses (*Continued*)
cucumber, amino acid interchanges, **24:**422; 431–432
effect of pectic on, **20:**357
effects of light on, **9:**97; 101; 105; 115; 129; 134; 142; 154–161
effects of nitrous acid, **27:**309–312
electronic microscope, **1:**57
enzymic adaptation of bacteriophages, **13:** 79–80
Equine influenza, **52:**137
and fluoroacetate, **18:**142
freezing, **1:**47
and functioning of cytoplasm, **8:**2; 3
and genes, **3:**3; 4
genetic information of, **27:**290; 293; 295
genetic material of, **24:**9–11
influenza, **4:**8, **52:**132
adsorption studies, **14:**356–357
isolation, **14:**398; 399–400
inhibition by *p*-aminobenzoic acid, **6:**39; 53
intracellular dynamic state of, **13:**145
molecular weight, **4:**6; 7
mouse encephalomyelitis, **4:**8
mumps, **52:**133
Newcastle disease, **52:**140
nucleic acid content, **3:**29–32
nucleic acids of, **9:**479
physico-chemical problems, **1:**43–62
pneumonia, adsorption studies, **14:**356–357
and radiation, **1:**48
reaction with borate, **12:**520–521
receptors, **52:**141
Rous sarcoma, adsorption studies, **14:**356
Semliki Forest, **52:**141
Sendai, **52:**133; 137; 142
separation by chromatography, **14:**342; 347; 355–357
shape and size of particles, **1:**55
single-stranded DNA, **27:**340–342
structure, **4:**5; 6; 7
Theiler's, adsorption studies, **14:**357
tobacco mosaic, **4:**4–8
action of tyrosinase on, **14:**150
adsorption studies, **14:**357
structure studies, **14:**287
chemical modification, **24:**108–109
structure, **24:**416; 418; 422; 431–432
vaccinia, hyaluronidase action with, **13:**221
Viruses, Nature of (Lauffer, Price, and Petre), **9:**171
Viruses, Photochemistry (McLaren), **9:**75
Virus infections, and xanthine oxidase, **17:**324
Virus multiplication, and ribonucleic acids, **17:**477–498
Virus Multiplication, Ribonucleic Acids and (Jeener), **17:**477
Virus proteins
aggregations, **1:**50
electrical charges, **1:**46
nucleic acids, **9:**512
in plants, **10:**297–299
Virus RNA, as information carrier, **24:**15
Visceral RNA, labeled ribose, **28:**414; 417
Viscosity, **42:**128, **47:**76
anomalous, **7:**37–39
of aqueous microphases, **51:**38–39
of cytoplasm, **8:**28
of pectic substances, **20:**354
of proteins, **9:**116; 122
of protein solutions
effect of ionizing radiations on, **24:**360
and SH-groups, **27:**222
of protoplasm, **7:**36–40
of starch suspensions, **12:**369–370; 381; 391
structural, **7:**38; 39
Viscotoxin in plants, **10:**285; 296–297
Vision, **69:**156
Visual Pigment Research, Aspects of (Morton and Pitt), **32:**97
Visual purple
surface photoxidation of, **11:**76
Vitamin A as prosthetic group, **1:**190
Vitamin, panththenic acid, **34:**19
Vitamin A, **15:**128, **46:**139
absorption spectrum, **4:**281–285
antimony trichloride reaction, **4:**282–284
bioassay, **6:**490–494
in blood plasma, **6:**501
and coenzyme Q, **25:**284; 289; 300; 302–304
and fat formation, **8:**310
destruction during autoxidation, **8:**440; 441; 443
as prosthetic group of visual purple, **1:**190
and vitamin A_2, **4:**284; 285
and vitamin E, **6:**486–510
Vitamin A esterase, localization of, **24:**317
Vitamin B. *See also* *p*-Aminobenzoic acid; Folic acid; Inositol

factors in growth medium, effect on amino acid decarboxylase activity, **6:**9; 25
Vitamin B_1, *See* Thiamin; Thiamine
and inhibitor for diamine-oxidase, **2:**100
Vitamin B_2, *See* Riboflavin
Vitamin B_6. *See also* Pyridoxal; Pyridoxamine
deficiency and phosphorylase amount, **24:**284
effect on aspartic acid β-decarboxylase, **16:**220
relation to transamination, **16:**187; 210–213; 215; 220; 228–229; 238
effect on metabolism of D- and L- amino acids, **19:**352
effect on nitrogen metabolism, **19:**353
effect on synthesis of essential amino acids, **19:**350; 351
effect on urea formation, **19:**353
enzyme activity, general scheme for control of, **35:**525
5′-phosphate derivatives, **35:**512
in mitochondria, **21:**8
and transamination reactions, **19:**348–350
Vitamin B_{12}, **9:**495, **46:**38; 51–54; 82; 87; 114, **51:**37; 40–41. *See also* Cobalamin, Cobamide, Cobinamide, Coenzymes, and Corrinoids
advances, **26:**233–281
analog, immobilized, **46:**258
biosynthesis, **46:**39
in chick diet, **14:**69
cobamide coenzyme biosynthesis, **26:** 366–367
deficiency, effect on xanthine oxidase, **17:**320
in deoxyribose and protein synthesis, **21:**431
isolation, with phosphatase models, **21:**192
in mitochondria, **21:**8
in one-carbon metabolism, **21:**430
in DNA synthesis, **28:**423; 429–431
effect on methylmalonyl-CoA mutase, **26:**346
effect on RNase, **24:**238
and fluoroacetate, **18:**135
α-glycosidic linkage in, **18:**225
genetic studies, **17:**408–409; 427
and *p*-hydroxyphenylpyruvate oxidase, **19:**184
and indole oxidase, **19:**127; 128
imidazole in, **20:**102
in methioine synthesis, **25:**403; 404–408
in methionine biosynthesis, **22:**266
and β-methylaspartate mutase action, **26:** 359–365
and propionic acid metabolism, **26:**344
and protein synthesis, **26:**254; 267–270
reaction with borate, **12:**514
role in biological methylation, **12:**474
role in metabolic processes, **26:**264–270
structure, **26:**234–237
synthesis, **46:**89
in synthesis of methionine, **16:**270; 274
and urea synthesis, **15:**343
Vitamin B_{13}, **23:**471–472
Vitamin B_6 analogues, in transamination study, **16:**220–222
*Vitamin B*12 (Bernhauer, Müller, and Wagner), **26:**233
Vitamin B_6-catalyzed reactions, general mechanism for, **16:**224–226
Vitamin B_c conjugate, **21:**371; 372
Vitamin B_{12} (Co^{60}), in assay, **25:**409
Vitamin B_{12} coenzyme
in DR metabolism, **28:**440; 453
and propionic acid metabolism, **26:**284; 343; 351–352
structure, **26:**234–238
Vitamin B complex, *See* individual vitamins, e g, Aneurin, Nicotinic acid
Vitamin B_{12} deficiency, **58:**187
Vitamin B_6 derivative in phosphorylase, **24:** 271–273
Vitamin B_{12} enzyme
and methionine synthesis, **26:**266–267
in methyl synthesis, **25:**396; 400; 405–410
Vitamin B_6 enzymes, inhibition of, **16:**222
Vitamin B_{12}-linked ribonucleotide reductase, in DR metabolism, **28:**436; 444–447
Vitamin B_{12r}
absorption spectrum, **26:**247–249
photolysis, **26:**251
synthesis, **26:**242–243
Vitamin C, *See* Ascorbic acid
absorption spectrum, **4:**296–298
and glutathione, **11:**250
synthesis of, **11:**291
and vitamin E, **6:**504
Vitamin D
absorption spectra, **4:**289–292
in antimony trichloride reaction, **4:**292
of D_2, **4:**291; 292
of D_3, **4:**292

Vitamin D (*Continued*)
of ergosterol, **4:**290; 291
and vitamin E, **6:**503
Vitamin D_3, fluorinated, **55:**276–277
Vitamin deficiencies, and protoporphin, **7:**318
Vitamin E, **6:**477–524, **65:**126–127
absorption spectrum, **4:**292–293
action on stomach, **6:**494–495
physiological phenomena, **6:**488–498
and anoxia, **6:**506
in blood plasma, **6:**498–503
and cancer, **6:**504; 507
in chick, **6:**485–486
deficiency, and muscular dystrophy, **6:** 203–207
deficiency of, relation to transamination, **16:**228
in DPNH oxidase system, **17:**82
dystrophy syndrome in human, **6:**483–485
effect on basal metabolism, **6:**481–482
effect on gaseous exchange of dystrophic muscle, **6:**206
on succinoxidase activity of dystrophic muscle, **6:**206
functions, **6:**480–510
in hay fever, **6:**510
and hormones, **6:**503–504
inhibitory analogues, **6:**135; 141
in lipide oxidation-reduction balance, **6:** 504–507
and liver storage, **6:**488–490
and mucosa, **6:**510
and muscular atrophy, **6:**485
and muscular dystrophy, **6:**203–207; 221; 222
natural, **6:**478
pigment effects, **6:**509
in pregnancy, **6:**507–509
relation to coenzyme Q, **25:**286; 289; 295; 301; 303–304; 320
requirements, **6:**511–514
in respiratory chain, **20:**153; 180–182; 185
sparing action on fatty acids, **6:**494
and starvation, **6:**507
in sterility syndrome, **6:**482–483
structural analogue of, **6:**135; 141
and teeth, **6:**509
tissue slice experiments, **6:**481
in various diets, **6:**515–518
and vitamin A, **6:**486–510
and vitamin C, **6:**504
and vitamin D, **6:**503
vitamin interrelationships, **6:**503
Vitamin E deficiency, effect on xanthine oxidase, **17:**322
Vitamin E-deficient diet, **42:**276
Vitamin-enzyme relationship, **1:**187
Vitamin H, **35:**323. *See also* Biotin
Vitamin K, **46:**1; 2. *See also* Prothrombin
absorption spectrum, **4:**293–296
as artificial electron acceptors, **30:**23
in chloroplast lamellae, **30:**9
in Factor VII synthesis, **30:**286
biosynthesis, **65:**108
and blood clotting, **10:**3; 20; 24; 27
chemistry and physiology, **2:**286–318
and coenzyme Q, **25:**284; 286–287; 289–293; 295; 300–302; 304; 313; 319–322
and photophosphorylation, **25:**53; 58; 69
color reactions, **2:**305; 306
determination by chemical and physical means, **2:**304–306
effect on clotting factors, **16:**59; 64; 66
lack of reactivity with thiols, **16:**121
2,3-epoxide of, **46:**25
and glycine, **23:**461
and oxidative phosphorylation, **23:**370; 377–378
inhibitory analogues of, **6:**134–135; 138; 139; 142
isolation of extraction, **2:**297; 298; 306
in mitochondria, **21:**8
in mitochondrial fragments, **21:**113
in photosynthesis, **21:**352
in restoration of mitochondrial oxidative phosphorylation, **21:**38
occurrence in nature, **2:**296–297
in photophosphorylation, **18:**58–59
in photosynthesis, **19:**55; 56
in photosynthetic phosphorylation, **20:**228
in respiratory chain, **20:**180
redox potential, **2:**304
reduction, **65:**98
structure, synthesis, **2:**298–300
titration as hydroquinone, **2:**306
Vitamin K_1
in electron-transport system, **17:**81–82
in reactivation of DPHN-cytochrome *c* reductase, **20:**181
in respiratory chain, **20:**182–185

Vitamin K_3
in phostosynthetic phosphorylation, **20:**183
in respiratory chain, **20:**180; 184
in photosynthesis, **19:**55; 64; 65
Vitamin K, Prothrombin, and γ-Carboxyglutamic Acid (Stenflo), **46:**1
Vitamin K-active compounds, activity, **2:**301
chemistry, **2:**301–303
redox potentials, **2:**304
Vitamin K activity
curative test, **2:**291–293
its determination in animal experiments, **2:**291–295
prophylactic test, **2:**291
units, **2:**295; 296
Vitamin K antagonists, effect on respiratory chain, **17:**118
Vitamin K (Dam), **2:**285
Vitamin K-deficiency in humans, **2:**309–314
alimentary avitaminosis, **2:**309
cholemic bleeding tendency, **2:**309–311
and coagulation anomaly, **2:**287
hemorrhagic diathesis, **2:**311
hypoprothrombinemia of newborn, **2:**311–314
and liver function, **2:**314
and other hemorrhagic diseases, **2:**314
Vitamin K-deficient rats, **46:**15
Vitamin K-dependent carboxylase, **46:**22–26
peptide substrate, **46:**24
requirements of, **46:**24–25
Vitamin K dependent proteins, **48:**277
Vitamin K in animal organism, **2:**286–309
chicks as experimental animals, **2:**287–289; 291–293
diet for artificial development of deficiency, **2:**288–291
early investigations, **2:**287
mode of action, **2:**306–309
rats and rabbits as experimental animals, **2:**290
role of putrefaction in deficiency diet, **2:**290
Vitamin K in bacteria, **2:**317; 318
Vitamin K_1 in energy transport, **14:**40
Vitamin K in green plants, **2:**315–317
Vitamin-like substances
and growth of microorganisms, **7:**409–460
metabolic changes in, **7:**454
Vitamin-like Substances, Metabolism of, Relation to Growth in Microorganisms (McIlwain), **7:**409
Vitamin M, **21:**372
Vitamin P, effect on hyaluronidase, **13:**216
Vitamin P-P, *See* Niacin
Vitamin-rich by-products, from fermentation, **7:**569–573
Vitamins. *See also* Choline; Lipoic acid, *p*-Aminobenzoic acid; Pantothenic acid; specific vitamins
absorption spectra, **4:**277; 278–285; 289–304
acceptance, **6:**474
activity, **6:**469–477
antagonism to sulfonamides, **6:**63–64
chick antidermatitis vitamin (*See* Pantothenic acid)
commercial handling, **6:**473
complexes with nucleates, **17:**138
genetic studies, **17:**430–431
conveyance, **6:**473
cooking, **6:**473
definition, **1:**190
destruction, **6:**474
diffusion, **6:**473
digestion, **6:**473
and enzyme formation, **6:**474–476
excretion, **6:**474
and fat formation, **8:**310; 316; 322; 330
destruction by unsaturated fats, **8:**440; 441; 443
and functioning of cytoplasm, **8:**20; 34
and genes, **8:**205
in garlic, **11:**379
genesis, **6:**472
harvest, **6:**472
ingestion, **6:**473
liberation, **6:**473
necessary in diet, **1:**185; 187
prosthetic groups of enzymes, **1:**186; 187; 190
rejection, **6:**473
and semen metabolism, **9:**346
storage, **6:**473
synthesis by yeast, acquired ability, **6:**90
transport, **6:**473
and urea synthesis, **15:**342–343
utilization, **6:**472–473
Vitamins, Absorption Spectra (Brode), **4:**269
Vitellin
association with a phospholipid, **1:**5
egg yolk, **7:**163–165
turnover, **7:**163–165

Vitelline membrane, and cross fertilization, **9:**271–273
changes following fertilization, **9:**275–286
dissolution, **9:**249; 268–271; 294
of egg, **9:**245; 246; 294, **271**
changes during maturation, **9:**247–248
and fertilization, **9:**314–315
Vitis vinifera, **55:**92
Vitreous humor, hyaluronidase action on, **13:**229
Vitridine, tyrosine hydroxylase regulation, **70:**139
Vitronectin, RGD sequence in, **70:**2
V/K constant
algebraic definition, enzyme action, **73:**29–31
significance of, enzyme action, **73:**31–33
V/K constant, significance of, enzyme action. *See also* Enzyme action
Vmax valves, *See* Michaelis constant
Volcanoes, and origin of life, **27:**353–354; 356; 358; 366; 389
Volemitol, **15:**161
Volemose, **15:**161
Voltage-gated ion channels, **71:**213
Volumetric methods, for soil enzymes, **28:** 373–376
Von Gierke's Type I glycogen storage disease, **54:**166
Von Willebrand's disease, **48:**300
VPO, *See* Verdoperoxidase

W

Waldenase reaction, in glucose-6-phosphate oxidation shunt, **15:**164
Walden inversion in cellulose synthesis, **15:**315
Walden inversions, in ribose synthesis, **28:**398
Waldhof-I.G. Process for yeast growth, **9:** 665–666; 675–676
Walker carcinoma 256, nitrogen in, **29:**363–364
Walnuts, antibiotic treatment of diseases, **20:**309
Warburg-Christian enzyme-coenzyme system in plants, **10:**309
Warburg's respiratory enzyme, **7:**295
Warburg's yellow enzyme. *See also* Yellow enzyme
classification, **14:**253; 255
Warburg theory, of carcinogenesis, **29:**324
opposition to, **29:**326–327
Warfarin, **46:**15; 17; 25
Warfarin-resistant rats, **46:**25
Wasserstoffubertragung mit Pyridinnucleotiden (Sund, Diekmann, and Wallenfels), **26:**115
Water, **34:**513. *See also* Aqueous solutions
absorption by pectic substances, **20:**354
as degradation product of pectins, **20:**349
labeled in studies of steroid enzymes, **20:**274
in studies of uridine diphosphogalactose-4-epimerase action, **20:**115
solubility of pectic substances in, **20:**353
activated products, by radiations, **24:**383–384; 394–397
bound and unbound, **65:**247–249
deuterated (*See* Deuterium oxide)
dielectric constant of, as function of temperature, **51:**12
distribution in affective disorders, **29:**494
effect on enzyme radiosensitivity, **24:**401–403
entropy, **43:**307
entropy of vaporization, **43:**285
flux ratio for, **13:**33–39
permeability to, **13:**42–48 (*See also* Frog skin)
formation by oxygen reduction, **19:**201; 202
labeled, in model hydroxylating system, **19:**137
in oxygen metabolism, **19:**83; 96; 132; 190; 192–194; 205; 207
in photosynthesis, **19:**23; 49
in transpeptidation, **19:**311
hydrogen bonding patterns, **65:**246
insoluble carrier, **34:**452
insoluble enzyme derivatives, **34:**447–448; 453
linear chains, **65:**229
movement, **65:**226
oxidation in photosynthesis, **25:**86–89
oxygen exchange with CO_2, **17:**163; 175
oxygen exchange with phosphate, **17:** 162–164; 175; 183
permeability, **34:**467
radioactive, in study of pyrimidine biosynthesis, **21:**268; 269; 273
role in enzymic hydrolysis, **15:**276; 284–285; 291–292; 295; 296
role in oxidoreductions, **15:**13
Water balance, in mechanical wilting, **11:** 417–422

in pathological wilting, **11:**403–406; 408–409
Water concentration
in cardiac muscle hypertrophy, **6:**202; 203
in muscular dystrophy, **6:**204; 223
Water dimmer, energy curve, **65:**220–223
Water droplets, **51:**6–12
calorimetric determinations, **51:**6–8
fluorescence probing, **51:**8–9
physical-chemical properties of, **51:**9–12
supercooling and freezing, **51:**6–9
Water fungus, *Blastocladiella emersonii,* **42:**230
Water-in-oil emulsions, **51:**5–30
applications, **51:**13–30
bacterial cytochrome P_{450}, **51:**14–16
bacterial luciferase, **51:**16–19
flash photolysis, **51:**20–28
high spin-low spin conversion, **51:**13–14
light-induced processes, **51:**28–30
emulsion procedure, **51:**5–6
physical-chemical properties, **51:**6–12
solutions of proteins, **51:**12
Water-in-oil microemulsions, **51:**30–90
catalysis in reverse micelles, **51:**47–55
enzyme, **51:**48–55
micellar enzymatic activity, **51:**47–48
micella formation (nonaqueous media), **51:**30–37
amphiphilic compounds, **51:**32–33
micelles, **51:**33–37
nonpolar solvents, **51:**33
rapid mixing and related problems, **51:**65–70
temperature and, **51:**55–65
bacterial cytochrome P_{450}, **51:**64–65
bacterial luciferase, **51:**63–64
effects (on micromicellar solutions), **51:**55–58
enzyme-substrate intermediates, **51:**61–65
horse radish peroxidase, **51:**61–63
hydrolytic enzymes, **51:**58–61
subzero, enzyme-catalyzed reactions, **51:**58–65
water pools and, **51:**37–47
accommodation of proteins, **51:**40–47
dielectric constants, **51:**37–38
pH values, **51:**39–40
physical-chemical properties, **51:**37–40
viscosities, **51:**38–39
^{3}H-Water, in pentose metabolism, **28:**453
Watermelon, antibiotic treatment of diseases, **20:**393
Water molecules, as surrogates, **53:**333–336
Water (O^{18}), in exchange reactions, **23:**327; 355; 356; 359–360; 364; 368; 373; 378; 386
Water (O^{16}), in photosynthesis, **25:**42
Water pools, **51:**37–47
accommodation of proteins by, **51:**40–47
dielectric constants, **51:**37–38
pH values, **51:**39–40
physical-chemical properties, **51:**37–40
viscosities, **51:**38–39
Water retention in tobacco during fermentation, **10:**354–355
Water-solid interface, **28:**18–21
activity of adsorbed molecules, **28:**28–30
Water-to-surfactant ratio, enzyme activity and, **51:**48–49
Water transport, in plants, **18:**48–52
Water uptake, by mitochondria, **23:**383–385
Water Vibrio, **23:**183
Waxes
changes during fermentation of tobacco, **10:**370–372; 396
constituents, **8:**332; 333–337
in epicuticle of arthropod, **16:**153
Wax moth, **46:**396
Weak seam theory, **37:**489
Weerman test, **16:**371
Weichherz's equations, **7:**5–6; 20
Weidenhagen theory, **4:**68–69
Wheat, Amylases, and Their Significance in Milling and Baking Technology (Geddes), **6:**415
Wheat, antibiotic treatment of diseases, **20:**393; 394
Wheat amylases, **6:**415–463
in breadmaking, **6:**443–463
during fermentation, **6:**452–456
gas production, **6:**444–447
malt supplements, **6:**460–463
during oven baking, **6:**452–456
process of, **6:**443–444
significance, **6:**456–458
and flour gassing power, **6:**441–443
in germinated wheat, **6:**424–428
and maltose value, **6:**429–441
effect of milling treatment, **6:**437–441
effect of wheat environment, **6:**436–437
effect on wheat variety, **6:**436–437
sound wheat, **6:**424–428
variations in activity, **6:**428

Wheat germ aggluthin, **36:**76
Wheat germ agglutinin, **48:**71
Wheat germ cytochrome *c* oxidase complex, **71:**144
Wheat-germ translational system, **47:**294
Wheat in fermentation, **7:**587–591
Wheat leaves, **46:**340
  starch accumulation, **46:**340
Wheat lipase, pK value, **22:**58
Whey. *See also* Serum, milk
  from rennin clotting, **15:**432; 443
White muscle disease, **48:**2
White rot, **18:**351
Whooping cough, **58:**131
Wien effect, **34:**494
Wighteone
  prenyl groups of, **55:**41
  structure of, **55:**6
Wilkström model, cytochrome *c* oxidase, proton pumping, **71:**154–158
Willardine, **50:**147
Willia, and mevalonic acid, **23:**473
Williams, R.J.P., hypothesis, **49:**74
  relation to P. Mitchell hypothesis, **49:**74
Wills factor, **21:**372
Wilson's diseae, **44:**225; 230
Wilson's disease, appearance of, **29:**480
Wilting, mechanical, **11:**416–422
  pathological, **11:**401–437
  physiological, **11:**406–407
Wilting, Pathological, in Plants (G,umann), **11:**401
Winemaking. *See also* Alcoholic fermentation; Wine spoilage; Yeasts
  bibliography, **3:**383–386
  enzymes in, **3:**381–384
  malic acid destruction by bacteria, **3:**380–381
  microbial fermentation, **3:**349–361
  microbial spoilage, **3:**362–380
  microorganisms, role of, in, **3:**349–381
Wine Making, Role of Microorganisms and Enzymes in (Cruess), **3:**349
Wine spoilage
  bitter ferment of wine, **3:**377–378
  hair bacillus of fortified wine, **3:**379–380
  by lactic bacteria, **3:**367–374; 379
  lactic souring, **3:**367–374
  mannitic disease of wine, **3:**374–377
  Pousse disease of wine, **3:**367–374
  slimy, **3:**378–379
  Tourne disease of wine, **3:**367–374
  by vinegar bacteria, **3:**364–367
  by yeast films, **3:**362–364
Winter effect, **15:**35
Wittig-Staudinger reaction, **55:**312; 313–316
Wobble hypothesis, **47:**383
Wobble rule, **47:**387
Wojnowicia graminis, enzymic transformations of steroids by, **20:**249; 251
Wood, lignin formation of, **18:**349–378
Wood-destroying molds, **8:**269–279
  metabolism, **8:**273–279
Wood in fermentation, **7:**574–585
Wood molds and ethyl alcohol, **8:**273–279
Woodruff model, cytochrome *c* oxidase, proton pumping, **71:**135–138
Woodward's reagent, **34:**449–453
Woodward's reagent K, *See* N-Ethyl-5-phenyl-isoxazolium-3′-sulfonate
Wood-Werkman reaction, **35:**323; 390
  oxaloacetate-H^1 $^3CO_3$-exchange reaction, **35:**323
Wool
  amino acids in, **14:**266
  enzymic digestion, **15:**292
    reactivity of thiol groups, **15:**288
  spatial arrangements of polypeptide chain, **13:**159
Wound healing in plants, **16:**153
Wruster's blue, electron acceptor, **28:**159; 172; 177; 178; 194
Wyerol, **55:**85; 90
Wyerone, **55:**9; 85
  biosynthesis of, **55:**90
Wyerone acid, **55:**85; 90
Wyerone epoxide, **55:**85; 90

X

Xanthine, **7:**208, **51:**126
  folate metabolism, **27:**459
  in genetic studies, **17:**427–428
  in purine biosynthesis and degradation, **21:**229; 401
  as a substrate for hypoxanthine-guanine phosphoribosyl transferase, **41:**14
  to uric acid, **41:**24
Xanthine antagonists, adenosine receptors, **69:**94

Xanthine dehydrogenase, **17:**241–242; 297; 308; 311–313; 317–318, **51:**109; 111; 131; 135
absorption spectra, **17:**302–303
deficiency in *Drosophila* mutants, **22:**29
summary and inferences, **51:**149–151
Xanthine oxidase, **9:**442; 489–490; 491, **15:**3; 16–17, **18:**106; 118, **37:**458, **41:**1; 2; 24; 38; 39; 53; 55; 57, **46:**170, **51:**109; 110; 112; 113; 114; 122; 125; 127; 129; 140; 152; 154–157; 159, **58:**63, **65:**54
absorption spectrum, **22:**377
active site, **22:**382
borate inhibition of, **22:**387
cyanide inhibition of, **22:**382
effect of sulfhydryl reagents on, **22:**380
iron in, **22:**383
and binding, **22:**380
pK value, **22:**58
red color, **22:**377; 379
sulfhydryl groups in, **22:**380
activity, effect of ethionine on, **25:**378; 380; 381
and luminescence, **25:**126; 148
product inhibition of, **25:**221; 224; 249–250
activity of –SH group, **11:**241
adsorption studies, **14:**326; 346
classification, **14:**253
isolation, **14:**401
contamination, **58:**69
C splittings, **51:**143–145
desulfo enzyme, **51:**136
effect of borate on, **12:**517
inactivation by radiation, **24:**377; 394
location of, **24:**299; 341
molecular weight from radiation studies, **24:**366
Inhibited signals from, **51:**131–135
relation between Rapid type 1 and Rapid type 2 signals, **51:**154
substrate protection of, **15:**287
substrates of, **51:**134–135
Very Rapid intermediate in, **51:**130–131
Xanthine oxidase, EPR study, **27:**141; 142; 160–169; 171; 172; 183
nonheme iron, **27:**89
stoichiometry of components, **27:**164–165
Xanthine Oxidase Chemistry and Biochemistry (De Renzo), **17:**293
Xanthine oxidase factor, **17:**304–305
Xanthine oxidases, **7:**300
in electron transfer, **19:**201
properties, **19:**202
as metalloflavoproteins, **17:**235–242; 249; 268–270; 276; 277; 281–282; 284
chemistry and biochemistry, **17:**293–328
in pre-cytochrome system, **17:**365; 380–381
redox potential, **17:**121
Xanthine oxidases, PN-dependent, **26:**125; 128; 131
Xanthine r O_2 transhydrogenases, **14:**253
Xanthinuria, **41:**1; 25–26
Xanthobacter flavus, phosphoribulokinase from, **74:**100; 120; 121
Xanthogenates, **17:**149
Xanthomonas campestris, **70:**231
plant disease caused by, **20:**396
Xanthomonas phaseoli
plant disease caused by, **20:**389
use of acetate, **23:**411; 412
var. *sojense,* plant disease caused by, **20:**389
*Xanthomonas pr*uni, **49:**5
Xanthomonas pruni, plant disease caused by, **20:**399
Xanthomonas species, streptomycin-resistant strains, **20:**387
Xanthomonas translucens, plant disease caused by, **20:**394
Xanthomonas vesticatoria, plant disease caused by, **20:**391
Xanthones
mode of action in *Fusaria,* **5:**195–196
role in dehydrogenations by *Fusaria,* **5:** 189–192
Xanthophyceae, carotenoids in, **21:**298–300
Xanthophyll, in air-cured tobacco, **6:**334
Xanthophyll in tobacco, **10:**372
Xanthophylls
biosynthesis, **21:**346–349
in algae, **21:**301
definition, **21:**297
degradation, **46:**100
Xanthopterin, **17:**310, **35:**36
biogenesis of, **27:**423–424
in folic acid biogenesis, **27:**422; 436
in biosynthesis of folic acid and folic acid derivatives, **21:**394
isolation, structure, synthesis and properties, **21:**376; 377

Xanthosine 5′-monophosphate, amination of, **18:**230
Xanthosine-5′-phosphate aminase, **21:**251, *See* Xanthylate amidotransferase
Xanthosine 5′-phosphate aminase, allosteric regulation, **28:**121–123
Xanthotoxin, **55:**95
Xanthurenic acid, **15:**99
  excretion of, **16:**222
Xanthylate amidotransferase, adenyl xanthylate as an intermediate, **39:**122
  properties of the reaction, **39:**120
  psicofuranine, inhibition by, **39:**123
  sources, **39:**119–122
Xanthylic acid, in purine biosynthesis, **21:**204; 229; 241; 250–252
X chromosome, **48:**100
Xenobiotics, **57:**358–359; 379
  detoxification, **69:**6
  glutathione synthesis, **72:**241–242
Xenobiotic substrate binding site, glutathione S-transferases, **69:**12; 25
Xenobiotic substrate specificity and mechanism, **69:**23–36
  binding site location, **69:**23–24
  binding site structure and specificity, **69:** 24–30
  role of individual residues in, **69:**30–36
Xenon, **43:**339
  action as photocatalyst, **12:**20
Xenopus laevis, **71:**100
Xenopus leavi, melanophore-stimulating activity in, **29:**445
Xenopus oocytes, **70:**180
Xenopus, origin of pteridines in, **27:**422
X-irradiation
  effect on xanthine oxidase, **17:**322
  role of purines, **27:**459–460
XMP adenylate, GMP synthetase, **72:**121–124
XMP-pyrophosphatase, inhibition by nucleotides, **28:**124
XOF, *See* Xanthine oxidase factor
X-protein, of tobacco mosaic virus, **15:**470
X-radiation, effect on yeast cells, **24:**12
X-ray absorption fine structure (EXAFS) spectroscopy, cytochrome *c* oxidase, **71:**84; 87; 92; 111
X-ray analysis, **57:**186; 204; 233
  protein aggregation, **27:**223
X-Ray Cryoenzymology (Fink and Petsko), **52:**177
X-ray crystallography, **39:**349
  cholesterol ester hydrolases regulation, **69:**68
  MoFeP and FeP, **67:**318–323
  psoralenated oligonucleotides, structure determination by NMR and X-ray crystallography, **66:**128
  of superoxide dismutase, **41:**67
  three-dimensional structure, **39:**364
X-ray diffraction, **52:**182
  of alkaline phosphatase, **56:**343
  analysis of protein structure, **33:**403
  of aspartate carbamylase, **56:**334
  of carbonic anhydrase, **56:**391
  of carboxypeptidase, **56:**359–360
  of hemoglobin, **40:**19; 20
  of metallothionein, **56:**403
  of methionine adenosyltransferase, **56:**255
  of thermolysin, **56:**381
  of zinc enzymes, **56:**312; 313
X-ray mutations, **27:**328–330
X-ray patterns, of β-glucans, **28:**327
X-ray photoelectron spectroscopy, **59:**145
X-rays, **47:**58; 62, *See* Roentgen rays
  action on pepsin-albumin films, **28:**34
  analysis, **47:**59; 77; 83; 102; 114; 119; 130; 133
  atomic coordinates, **47:**58; 61–62; 70
  crystallographic data for proteins, **24:**448
  crystallography, **47:**56; 74; 81; 85; 94; 121; 136; 153
  diffraction, **47:**46–47; 56; 81; 83; 135
  effect on
    bovine serum albumin, **24:**382–383
    enzymes, **24:**368; 370; 372; 376; 393–397
    relation to concentration, **24:**386–389
  effect on mitochondrial oxidative phosphorylation, **21:**31
  LET from, **24:**365
  mutations caused by, **6:**82–84; 92–93
  and origin of life, **27:**382; 383; 386
  results, **47:**85; 114; 117–121
  structural determinations, **47:**47; 84; 102; 114; 117; 118; 134; 135
  and structure of proteins, **3:**63–108
  studies of starch, **12:**365–367
  and viruses, **1:**48
X-Rays and Stoichiometry of Proteins (Astbury), **3:**63
X-ray structure determination, **43:**221
Xylanase, **56:**223

soil, method for, **28:**386
β-1,3-Xylanases, multiple forms, **25:**350
β-1,4-Xylanases, multiple forms, **25:**350
Xylitol, oxidation, in mitochondria, **21:**22; 25
Xylitol dehydrogenase, in mitochondria, **21:**22
Xyloglucans, **56:**222
3-(Xylopyranosyl)-α- or -β-D-glucopyranoside, in action of β-glucuronidase, **16:**371
p-Xyloquinone, as electron acceptor, **16:**134
Xylose, **5:**170; 171; 172, **52:**25
degradation by fungi, **8:**294–295
fate of C-1, **18:**295
UDP ester of, **18:**201
in genetic studies, **17:**423–425; 457
in tranglycosylations, **17:**197; 209; 215
oxidation, **15:**148; 149
D-Xylose
origin, **14:**206–208
in polysaccharide synthesis, **15:**306; 312
D-Xylose, **43:**493
Xylose derivative of coenzyme I, **15:**31
Xylose isomerase, **43:**496
D-Xylose isomerase, mechanism of deduced from relaxivity data, **33:**51–52
β-Xylosidase, **60:**276
β-D-xylosidase, **36:**74
Xylosidases, **56:**221–223
β-Xyloside β1r4 galactosyltrasnferase, **52:**27; 87
Xylosylserine, **60:**277
Xylosyltransferase, **52:**83; 85; 86; 87; 146
β-Xylosyltransferase, **52:**84–87
Xylulose, **43:**493
in plant respiration, **18:**287
synthesis, in mitochondria, **21:**22
D-Xylulose
from glucose, **28:**394; 398–399
transketolase action on, **19:**26
D-Xylulose-1,5-bisphosphate, **67:**12
formation, **67:**14; 36; 46
as rubisco substitute, **67:**52–53
Xylulose-5-P
phosphorolysis of, **23:**324; 327; 337
synthesis of, **28:**410
Xylulose-1-phosphate, **15:**154
origin, **14:**208
Xylulose-5-phosphate, **15:**159
Xylulose 5-phosphate
phosphorolysis of, **18:**227
in photosynthesis, **19:**15; 25–31
Xylulose-5-phosphate-3-epimerase, in ribose synthesis, **28:**396–397

Y

YADH-DPNH, Cotton effect, **27:**58; 83–85
YADH (yeast alcohol dehydrogenase), complexes of, Cotton effect, **27:**83–85
Yeast, **46:**38; 39; 55; 62; 91; 103, **59:**373–377. *See also* Fermentation; Saccharomyces; Torulopsis lipofera, composition of fats, 8:320–8:322,8:325,8:328; Yeast aldolase; specific Yeasts; specific yeasts, eg, Torula
action on nicotine, **10:**432
adenylosuccinate synthetase (AMPSase) gene structure, **73:**90–91
ADP sulfurylase in, **22:**29
and alanine racemization, **14:**195
alcohol dehydrogenase system in, **20:**40; 43; 47
cytochrome *b* reduction in, **20:**156
cytochrome *c* of, **20:**193
enzymic transformations of steroids by, **20:**252–256; 258; 273
galactose-1-phosphate uridyl transferase in, **20:**113; 114
photochemical action spectrum of, **20:**165
TPNH-cytochrome *c* reductase of, **20:**170
uridine diphosphogalactose-4-epimerase of, **20:**116; 117; 119
uridine diphosphoglucose pyrophosphorylase in, **20:**121
alcoholic fermentation, **5:**213
allantoin degradative system of, **56:**91–134
allosteric control, **28:**77–83; 89–92; 99
cell walls, β-1,3 glucan of, **28:**327; 330; 335
induction of β-glucanases, **28:**348–349
β-1,3 glucans of, **28:**328; 335–336
isocitrate dehydrogenase, **28:**135–139
laminarinase of, **28:**342; 345
p-aminobenzoic acid in, **6:**40
α- and β-glucosidase in, **22:**106
APS phosphokinase in, **22:**209
aspartate-carbamyl transferase in, **21:**272
folic acid and folic acid conjugates in, **21:**371; 372
nitrogen transfer reactions in, **21:**252

Yeast (*Continued*)
ATPase in, **6:**186
citric acid cycle in, **10:**169–177; 178–180; 189–190
constitutive β-glucosidase mutants, **22:**103
constitutive β-glucosidase synthesis in, **22:**103
and copper porphins, **7:**320
and coproporphin I, **7:**337
and cozymase inactivation, **7:**410
and cyanide, **7:**189
cytochromes in, **18:**24; 26; 33
effect of fluoroacetate on, **18:**134–135
cytoplasmic transmission of heredity in, **22:**3; 28
dipeptidase, **2:**81
polypeptidase, **2:**80
effect of urethan, **7:**22–23
enzyme for synthesis of glucose diphosphate, **14:**211–212
enzyme levels during synchronous growth of, **22:**127
enzymes of, adsorption studies, **14:**362–363
enzymic phosphorylation of sugar, **13:**16
genetic studies of, **13:**139; 143
fat synthesis by, **9:**653–704
fermentations in breadmaking, **6:**447–452
fodder, **7:**579–580
and folate, **27:**427
mutations, **27:**308
oxidation of hydrocarbons, **27:**472; 474
free amino acid pools in, **22:**125
fructose-bisphosphatase of, **54:**173
galacton permease-deficient mutant, **22:**21; 22
γ-GCS genes, glutathione synthesis, cloning of, **73:**229–230
genetic mapping in, **22:**2
β-glucosidase, ribosomal-bound, **22:**146
α-glucosidase induction in, **22:**127; 138
β-glucosidase induction in, **22:**114–116
β-glucosidase synthesis in, inhibition of, **22:**115; 121
glucoside hydrolytic ability in, **22:**22
glucoside permeability and strain difference in, **22:**22
α-glucoside permease in, **22:**109
glycogen deficient mutants, **34:**326
glycogen synthetase, **34:**324
growth stimulation of, **16:**1
mitochondrial metabolism of, **16:**472; 490
RNA of, **16:**438
source of CoA, **16:**2–3
in study of amino acid metabolism, **16:**269; 271; 276; 282; 283; 286; 290; 294; 296–300
tetraploid, DNA in, **16:**425
inhibition of gluconeogenesis in, **54:**174
invertase, **17:**195–196; 198–205; 219; 222; 225; 226
metabolic control by ADP level, **17:** 125–126
respiratory adaptation, **17:**284
and iodoacetic acid, **7:**189
and irradiation, **7:**190
isocitratase in, **23:**411
isolation of hexokinase from, **14:**376; 402
lactic dehydrogenases in, **14:**255
malate synthetase in, **23:**412
maltozymase induction in, **22:**106; 107
metabolism of glucose, **7:**15–20
metaphosphate phosphorus in, **19:**17
methionine adenosyltransferase of, **56:** 257–259
and nicotinic acid derivatives, **7:**423–432
nitrogen turnover in, **22:**125; 126
PAPS-synthesizing enzymes in, **22:**207
permeability, **7:**14–15
peroxidase in, **14:**151
phosphatase, **7:**454
phosphorus turnover, **7:**189–190
phosphorylating system from, **23:**353
photosyntheic enzymes in, **19:**9; 25; 27; 36; 37
production for food, **7:**579–580; 587
and RNA, tertiary structure of, **46:**286–295
role in cacao curing, **25:**470–475
as source of ATP sulfurylase, **22:**208
as source of enzymes, **14:**376–377; 402; 405
sulfur metabolism, **10:**227; 230; 238; 308–309
superoxide dismutase, **58:**71
thromboplastin in, **10:**6
UDPG analog in, **14:**204
use of labeled acetate by, **14:**120–121
utilization of galactose by, **14:**197–198
Yeast alcohol dehydrogenase, **18:**118, **46:** 468–471

Yeast aldolase. *See also* Yeast
and catalysis, **59:**293–294
and coordination geometry, **59:**294
and intervening imidazole ligand, **59:**295
mechanism of deduced from relaxivity data, **33:**55–56
metal-ligands donated by, **59:**294
second-sphere substrate, **59:**295
Yeast carboxylase, **18:**118
Yeast cells, **71:**46; 107
binding of alcian-blue, **40:**261
Yeast cell wall, mannan-protein-glucan complex, **40:**243
Yeast dihydrofolate reductase, **70:**25. *See also* Eukaryotic dihydrofolate reductase
Yeast enolase, multiple enzyme forms, **25:**352
Yeast enzymes, fermentation of R-5-P, **28:**395
Yeast extract
acetate-replacing factors in, **18:**321–322
lipoic acid in, **18:**321–323
growth factor for *L. bulgaricus,* **14:**49–50; 61
Yeast-β-glucan, effect on tumors, **28:**357
Yeast growth, and pantothenic acid, **3:**255–257; 259; 261; 262; 279–280
Yeast hexokinase, **37:**562, **39:**252–279, **43:**27
abortive complex formation and, **39:**268
abortive ternary complex, and, **39:**268
formation of, **39:**259
activity assays of, **39:**251
alternative substrates, studies of, **39:**262
ATPase activity of, **39:**276
ATP:phosphohydrolase activity of, glucose-6-*P* effects on, **39:**278
michaelis constants for, **39:**276
pentose activation of, **39:**272; 277; 278
relation to hexokinase kinetic mechanism, **39:**276
binding studies of, by fluorescence titrations, **39:**273
equilibrium dialysis, **39:**273
phosphate effects on, **39:**273
comparisons of rapid-equilibrium and steady state approximations by, **39:**266–267
competitive inhibition studies of, **39:**259
competitive inhibitor effects, **39:**267; 268
computer simulation studies of, **39:**266
crystal structure of, **39:**255
equilibrium constant of, **39:**251
equilibrium exchange kinetic studies of, **39:**263
comparison of pathway exchange rates, **39:**263; 266
detection of rate limiting steps, **39:**263
determination of apparent dissociation constants by, **39:**264–265
substrate synergism and, **39:**270; 271
equilibrium exchange reactions catalyzed by, substrate synergism and, **39:**270–271
glucosyl-enzyme concept of, **39:**257
initial rate studies of, **39:**257–258
isozymes of, **39:**253; 254; 275
isozymic forms of, **39:**253
limited proteolysis of, **39:**253
mechanism of deduced from relaxivity data, **33:**44
metal ion requirement of, **39:**274–275
Michaelis constant of, **39:**274; 275
molecular weight of, **39:**254–255
the phosphoryl-enzyme concept and, **39:**269–272
phosphoryl-enzyme concept of, **39:**256
preparation of, **39:**252
product inhibition studies of, **39:**258–259
regulatory properties of, **39:**279
burst phenomena and, **39:**279
PH effects on, **39:**279
polyanion activation of, **39:**279
relative phosphorylation rates of substrates with, **39:**275
specific activity of, **39:**254
substrate specificity studies of, for hexoses, **39:**274
of isozyme forms, **39:**275
for phosphoryl donors, **39:**274
subunits of, **39:**255
sulfhydryl reactivity of, **39:**255–256
Yeast hexokinases, multiple forms, **25:** 351–352
Yeast lactic dehydrogenase, **17:**372
Yeast laminarinase, β-glucosidases of, **28:**345
Yeast-lysing enzymes, of *Bacillus circulans,* **28:**355
Yeast mannan
biosynthesis, **40:**264
definition, **40:**240
enzymes, **40:**242
amino acid compositions, **40:**250
degradative, **40:**262
haptenic groups, **40:**251
immunochemistry, **40:**250

Yeast mannan (*Continued*)
isolation, **40:**242; 243
linkages, immunogenic characteristics, **40:**251
mutant hints, **40:**265; 266; 267
mutants
binding of alcian-blue, **40:**262
isolation, **40:**257
and mannan biosynthesis, **40:**265; 266; 267; 268
mannanless, **40:**259
selection procedures, **40:**257
oligosaccharides of bakers′ yeast, **40:**243
polysaccharides, linkages, **40:**243; 244; 246; 255; 258; 259; 261; 263; 265
variability, **40:**246; 247
protein component, **40:**249
structural analysis, **40:**247
structure
gene mapping, **40:**255
genetic control, **40:**255
macromolecular, **40:**240
polymorphism of, **40:**245
of *S. cerevisiae* X2180 wild-type, **40:**259
Yeast Mannans, Structure, Immunochemistry, and Genetic Control of (Ballou), **40:**239
Yeast mating cells, wall-softening enzyme, **28:**348–349
Yeast nucleotidyl transferase, complexes with tRNA, **40:**49
Yeast oxidase, **23:**269; 292; 300; 308
Yeast phosphatases, multiple forms, **25:**348
Yeast phosphofructokinase, **48:**214
Yeast polypeptidase, **12:**210–211
Yeasts in wine making, **3:**349–361. *See also* Alcoholic fermentation
acetic acid, toxicity to, **3:**358–359
alcohol contents obtained, **3:**359–360
clouding of wine, caused by, **3:**364
in content of grapes, **3:**352–356
film-forming yeasts, **3:**360–364
pure cultures, importance of, **3:**356–357
sedimentation caused by, **3:**364
Yeast zymosan, effect on properdin system, **28:**357–358
Yellow enzyme, **3:**70, **17:**121, **19:**201. *See also* Fatty acyl CoA dehydrogenase, TPNH oxidase; Old yellow enzyme; Warburg's yellow enzyme
adsorption studies, **14:**332–333
copper content, **27:**146
steps in reaction, **25:**14
structure studies, **14:**284
Yellow enzyme protein, stability, **6:**162
Yellow mealworm, *See* Tenebrio molitor
Yellow respiratory enzyme, **15:**4; 38. *See also* Flavin adenine dinucleotide and Zwischenferment
Yersinia pseudotuberculosis, **70:**3
Yersinia spp., **71:**11
Yohimbine
and hypertension, **7:**547–548
increase in brain serotonin level produced by, **29:**518–519
Yolk, egg
phosphorus compounds in, **7:**157–160; 165–166
vitellin of, **7:**163–165
Yoshida ascites tumor
pentose labeling, **28:**431–434
ribose synthesis, **28:**417
Y-Protein, **46:**389; 390
Yttrium compounds, **17:**147
Yttrium ions, model catalyses with, **21:**175; 177

Z

Zea mays, **54:**177, **71:**93
coenzyme Q in, **25:**288
Zein
amino acids in, **14:**263
dicarboxylic amino acids in, **20:**319
effect of irradiation, **24:**378
Zellen, Probleme des Energietransports innerhalf lebender (Bucher), **14:**1
Zellweger disease, **57:**89
Z-enzyme, **18:**289
separation by chromatography, **14:**348
soya bean, exohydrolysis by, **28:**350
from soybeans, **11:**347
Zero transmembrane potential cytochrome *c* oxidase, enzyme turnover control and H + /e - ratio, **71:**162
Zervamicin, ion channels, **71:**214
Zimm-Bragg parameters (σ, *s),* **47:**48–49; 85
Zinc, **15:**171; 241; 251; 379; 380
absorption of, **56:**407
and activity of 6β-hydroxylase, **20:**270
binding, by pectic substances, **20:**356
by protein groups, **20:**97

effect on fluorescence of diphosphopyridine nucleotide-linked enzymes, **20:**40
effect on intestinal maltase, and oligo-1, 6-6glucosidase, **20:**94
effect on pectic substances, **20:**351
in alcohol dehydrogenase, **27:**49–50; 53–56; 58
binding to enzyme, **27:**78–80
and structure of YADH, **27:**83–85
in alcohol dehydrogenase systems, **17:**45
in carbonic anhydrase, **17:**150
in genetic studies, **17:**424
analytical techniques for detection of, **56:**286–288
atomic spectroscopy in examination of, **56:**286
binding properties of, **56:**303–306
binding sites in enzymes, **58:**9
biochemistry of, **56:**291–317
biological roles of, **56:**290–291
in carbonic anhydrase, **58:**9
catalysis of dopachrome decarboxylative rearrangement, **16:**117
concentration in melanins, **16:**171–172
catalytic role of, **56:**316
chelating agents affecting, **56:**300–303
chemistry of, analytical, **56:**286–288
contamination control in study of, **56:**288–289
cytochrome *c* oxidase, **71:**101; 103
deficiency of, **56:**291
dose/response curve to, **56:**290; 291
effect on permeability, **11:**247
entatic state of, **56:**313–315
enzymatic role of, **56:**285–286; 290
and enzyme catalysis, **24:**442; 470
effect on RNase activity, **24:**236–237; 241
role in enzymes, **24:**112–113; 119–120
in sperm, **24:**80
enzymes, **58:**5
and fat formation, **8:**309
formation of porphyrin chelates, **46:**82
in genetic expression, **56:**286; 291
glycocorticoid effect on, **56:**407
ligands binding, **56:**315–316
and light emission, **25:**142
in liver ADH, **22:**352
in liver GAD, **22:**368
in MDH, **22:**370
sensitivity, in tryptophan synthetase mutants, **22:**11; 31
in TPD, **22:**345
in yeast ADH, **22:**355–357
and luminescence, **7:**253
metabolism of, abnormal, **56:**290
metalloenzymes with, **56:**316–317
in metalloproteins, **56:**298; 300
in metal protein complexes, **56:**300
electrophilic catalysis of, **56:**408
inhibition of, chelating agents in, **56:**300
metal substitutions in, **56:**304
nuclear magnetic resonance of, **56:**311–312
role of, **56:**316–317
x-ray diffraction studies of, **56:**312
noncatalytic role of, **56:**316–317
nutritional role of, **56:**290
in photosynthesis, **19:**50
regulatory role of, **56:**316
salts, effect on activity of dehydropeptidase preparation, **8:**143
in semen, **9:**346
spectroscopic properties of, **56:**303–313
structural role of, **56:**316
as trace element, **56:**284–286
in uricase, **9:**465
Zinc acetate, **58:**116
Zinc cations, collagen hydroxylase inhibitor mechanisms, **72:**358–364
Zinc-coenzyme association constants, **27:**69
Zinc compounds, solubilization of, **17:**138; 142; 143
Zinc-deficient *Neurospora crassa,* **15:**107
Zinc(II)
and angiotensin-converting enzyme, **57:**12
dissociation of, **57:**19–20
and inhibition of carboxypeptidase-A, **57:**24–25
and peptide cleavage by thermolysin, **57:**71; 73
Zinc ion, **39:**348, **47:**154; 155; 159
binding site for, **39:**348
in carbonic anhydrase, **47:**150; 154–155; 159; 162
in creatine kinase activation, **54:**404; 405
in fructose-bisphosphatase activity, **54:**133–135
kinetic and binding constants of, **54:**134
physiologic significance of, **54:**135–136

Zinc ion (*Continued*)
in fructose-bisphosphatase regulation, **54:**178; 181
and peptidase action, **12:**195; 210; 216–218; 220; 222–223
in chloroplasts, **12:**27
as decarboxylating agent, **12:**80; 87
Zinc ion-albumin complex, binding with DL-2-phosphoglyceric acid, **21:**149
Zinc ions
in allosteric regulation, **28:**65; 140–142
chelate formation with amino acids, peptides and fatty acids, **21:**145–148
coordination number, **21:**158
enzyme activation by, **21:**142; 143; 152; 153; 155; 156; 158; 159
in mitochondria, **21:**10
model catalyses with, **21:**174–177
properties, **21:**157
in citric acid cycle, **10:**201
effect on, collagenase, **23:**195–196
elastasis, **23:**234
effect on alkaline phosphatase, **10:**96; 99; 102–104; 106–107; 110
effect on hyaluronidase activity, **13:**214
effect on myosin ATPase, **13:**167
effect on transferases, **13:**297
Zinc ions and isolation of enzymes, **14:**396
Zinc mercaptide, **11:**216
Zinc peptide, from LADH, **27:**81
Zinc phthalocyanine, oxidation, **15:**33; 39–40
Zinc proteins
chelating agents removing metal from, **56:**303
hydrolases, **56:**341–384
alkaline phosphatase, **56:**341–350
angiotensin converting enzyme, **56:** 355–358
carboxypeptidase A, **56:**358–374
collagenases, **56:**382–384
leucine aminopeptidase, **56:**350–355
thermolysin, **56:**374–382
isolation of, **56:**298–300
isomerases and ligases, **56:**401
lyases, **56:**383–401
aldolase, **56:**385–388
aminolevulinate dehydratase, **56:**397–401
carbonic anhydrase, **56:**388–397
mechanism of enzymatic action of, **56:** 408–411
oxidoreductases, **56:**318–327
alcohol dehydrogenase, **56:**318–325
superoxide dismutase, **56:**325–329
transferases, **56:**329–341
aspartate transcarbamylase, **56:**329–334
nucleotidyl polymerases, **56:**334–341
of unknown function, **56:**401–407
metallothionein, **56:**401–407
Zinc protoporphyrin, **46:**59; 165; 175
Zinc pyridinecarboxaldoxime, **58:**12
Zinc tetraphenyl porphine, chemiluminescence, **25:**121; 123
Zinc (^{65}Zn), coenzyme complexes with, **27:** 69–70
Zinc (^{65}Zn) hydroxide, effect of DPNH, **27:**69
Zinc (^{65}Zu) exchange, in LADH complexes, **27:**64–65; 72; 79–81
Zirconium compounds, **17:**147
Zirconium ions, model catalyses with, **21:**174; 175; 177; 178
Zn^{2+}, **53:**51; 52; 55; 56; 57; 58; 61; 75
Z-Oligopeptides, as synthetic substrates for proteinases, **41:**231–234
Zoxazolamine hydroxylase, **46:**134
ZPCK, **43:**341
Z-Tripeptides, **41:**205; 206; 209–212; 214–234
kinetics of hydrolysis of, **41:**206; 209–212; 214–234
Zwischenferment, **15:**116, **48:**98. *See also* Yellow respiratory enzyme
action on G-6-P, **28:**394
classification, **14:**235; 254
Zwitterion, **34:**492
Zwitterionic lipids, ion channel model system annexin V, **71:**221
Zygosaccharomyces marxianus, growth factors for, **14:**66
Zymase, **5:**88–91; 93; 94; 114
inadequacy of hydrolase-zymase mechanism, **5:**97–307
Zymase complex, in tobacco, **6:**370
Zymases. *See also under* individual enzymes
effect of freezing on, **9:**625
enzymatic adaptation, **13:**105
Zymase system, classification, **14:**233
Zymasis, and acid metabolism in molds, **9:** 545–550; 564–566; 569–570
Zymobacterium oroticum, iron enzyme from, **27:**89
Zymobacterium oroticum, PN-dependent enzyme, **26:**119

Zymobacterium oroticum, enzymes and reactions of pyrimidine synthesis in, **21:**274–276
Zymobacterium oroticum, superoxide dismutase from, **41:**79
Zymogen, **57:**210
Zymogen activation, change in optical rotation, **27:**47
Zymogen granules, of pancreas, **23:**131
Zymogen phenomenon, in identification of imidazole, **20:**78; 79
Zymogens, **45:**407; 410. *See also* Apoenzymes and individuals
  activation, **45:**404
  activation of, **24:**24; 126–127
  cells, **45:**407
Zymohexase, **18:**292
  histochemical methods for, **9:**592–593
  in plants, **10:**300
  role of metal ions, **10:**104
Zymosan, effect on properdin system, **28:** 357–358
  RE stimulating effect, **28:**357
Zymosan-treated complement, **8:**104–111
Zymosterol, in cholesterol biosynthesis, **22:**326
  structure, **22:**325